HUMAN BRAINS

HUMAN BRAINS

Fondazione Prada

FOREWORD

As a cultural institution whose roots lie in the visual arts, dealing with science is an intellectual and political challenge that calls into question our identity and further highlights our interest in important contemporary themes—those which affect everyone's lives and of which we are sometimes unaware or do not fully understand.

Born with artists, today Fondazione Prada encompasses not only cinema, architecture, and philosophy, but also history, social topics, technology, and now science. "Human Brains" is the result of a special interest in scientific disciplines, which has grown over the past decade and developed into a lengthy inquiry conducted by the Foundation. The wish to understand the human brain and its complexity has required the support of a scientific committee and the collaboration of more than 180 scientists, philosophers, authors, and academics, and thirteen major international centers in the field of neurosciences.

We asked ourselves how we could give a voice to scholars and researchers, how we could put them in contact with a wider public, convinced of the need to combine scientific rigor with an approach that everyone can understand. Talking about the human brain means talking about human beings. Focusing on the uniqueness of our minds allows us to understand the impact that scientific research has on our daily life and in wider cultural growth. Using the Foundation as a means of promoting scientific knowledge, in particular brain studies, is a way of reasserting that culture must be useful, courageous, and not elitist. It is an attempt to overcome the boundaries between disciplines and people to enhance knowledge and achieve a greater awareness of ourselves.

But how are the ideas and knowledge expounded? How can conferences and exhibitions be used to present the human desire to understand thought, pinpoint its natural location in the body, explain its biological characteristics, and perceive its structure and processes?

We inaugurated the three-year program with "Culture and Consciousness" and "Conversations" conferences and activities, followed by two exhibitions held respectively in Venice and Milan.

"It Begins with an Idea" was an exhibition on knowledge. Constituted by a multitude of objects, accounts, and videos, the project presented ideas, theories, and studies in an experience that was neither purely educational nor solely expositive. In Venice we transformed our practice and the way we work into an opportunity to convey scientific knowledge, engaging visitors as a useful means for communicating the relevance of neurological studies and the value of those who devote their time and effort to research.

"Preserving the Brain" had the goal of stimulating an open and critical exchange between international scientists, experts, and researchers on the subject of neurodegenerative diseases, such as Alzheimer's, Parkinson's, Amyotrophic lateral sclerosis, and Multiple sclerosis, diseases that are widely spread and as yet incurable. For us, this phase of the project was particularly significant as it permitted closer understanding of the impact that scientific research has on our everyday lives, and in particular on the discovery of possible cures and treatments. It also demonstrated how critical collaboration and sharing knowledge are within the scientific community.

"Human Brains" is an experimental project and perhaps the most complex and evolving one. Its future path is traced in the dialogues that will lead to new collaborative projects in science and in studies on the brain, the organ that is our main interface with the world. "Human Brains" is part of the process of evolution of our interests, the link between the humanistic culture and scientific thought, and–more generally–contemporary society, which we hope will devote growing attention to the profound nature of human beings.

MIUCCIA PRADA
President and Director of Fondazione Prada

INTRODUCTION

"Human Brains" is a project with international reach, conceived with the aim of drawing public interest to the human brain. The title uses the plural form to emphasize the immense richness deriving from diversity combined with uniqueness: just as two identical faces have never existed in the history of humankind, nor have two identical brains, and it is clear that this is what makes each of us unrepeatable. At the same time, the project also aims to shed light on the science of the brain and the extreme complexity that distinguishes it, and on how the development of knowledge entails a continuous process of verification, with each new acquisition forming the basis for the successive one in a never-ending process.

The project has been developed by Fondazione Prada with the collaboration of a scientific committee including neurologists, neurolinguists, cognitive scientists, neurophysiologists, experts in neuroscientific communication, and philosophers and which continues to expand to include new research and perspectives. In its entirety, "Human Brains" has involved around one hundred and fifty researchers from the world's most prestigious universities and research centers, sharing an interest in the thought and the brain, its structural and functional aspects, and the complexity of the processes that underlie our perceptions, emotions, decisions, and behaviors.

The neurosciences have grown rapidly over the past thirty years. We have started to understand how the 100 billion neurons in our brain communicate with each other: we have learned that information is exchanged not only via connections between neurons but also through changes in the physical and chemical conditions of the tissue in which the neurons are embedded. Much to our surprise, we have also discovered that neurons communicate directly, via contact, with the glial cells that provide their structural support. A fundamental role in the marked growth in our understanding has been played by imaging, particularly the invention of nuclear magnetic resonance, which for the first time allowed us to see the brain in action and observe the dizzying exchange of information between the different cerebral areas.

"Human Brains" comprised four programs. The first, "Culture and Consciousness," compared the views of scientists and philosophers on the problem of consciousness, from its biological bases to the origin of thought, and with the awareness that culture, in its various manifestations in the history of humanity, is the result of the sharing of individual consciousnesses. The second program, "Conversations," was aimed at shedding light on a handful of particularly interesting neuroscientific topics, in the form of debates between experts–moderated by science journalist Alice Roberts–in which the public was able to take part with questions and comments. The third program–"It Begins with an Idea"–an exhibition that Fondazione Prada presented on the occasion of the Venice Biennale. The show was intended as an invitation to reflect on the extreme complexity and mystery of the organ that undergirds all human experience, generating our thoughts, actions, and emotions. The exhibition probed the indissoluble union of mind and brain, knowledge of which is recent in comparison to the history of humanity and still partly the subject of debate today. "It Begins with an Idea" was created in conjunction with over fifty leading neuroscience researchers, a chorus of voices that also represents the diversity and uniqueness of human brains.

The final chapter was "Preserving the Brain," an exhibition and forum on neurodegenerative diseases–jointly organized by some of the foremost international research centers–conceived with the purpose of finding a common strategy to protect the brain from neurodegenerative diseases. New technological developments have enabled us to greatly expand our knowledge of the biological bases of these diseases, and potential targets for new therapeutic procedures are emerging. "Preserving the Brain" took the form of an exhibition, intended to show visitors the different steps that must be taken to develop and create a treatment for diseases such as Alzheimer's, Parkinson's, and multiple sclerosis. Together with the exhibition took place a convention at the Fondazione Prada's Milan venue, accompanied by a series of single-topic scientific meetings at the individual research centers.

A deeper knowledge of how our brain works helps us to better understand ourselves and others, to better educate our children, to participate more decisively in social life and civil society, and to respect and appreciate the differences between people and ways of thinking. We feel the need to support a community of neuroscientists who work in an integrated manner and are equipped with tools and resources that allow them to develop their creativity and put it at the service of the community. This project dedicated to science represents a huge intellectual challenge for a cultural institution like Fondazione Prada. The dialogue embodied in "Human Brains" underscores the importance of opening the institution's program to sciences and of working together to disseminate fundamental disciplines and knowledge for our present and for a better future.

GIANCARLO COMI
President of "Human Brains" Scientific Committee

TABLE OF CONTENTS

1

CULTURE AND CONSCIOUSNESS

The first phase of "Human Brains" is represented by the online conference "Culture and Consciousness" that took place online from 9 to 13 November 2020. The event is structured in five daily discussions focused on the study of consciousness in neurosciences. Each panel included a debate between two scholars coordinated by one or more moderators.

The study of consciousness in the field of neuroscience has been addressed through comparison with other disciplines: philosophy, psychology, linguistics and anthropology. The online conference was structured following a gradual development: from a scientific and microscopic perspective to a humanistic and overall view.

The first two discussions saw neuroscientist Mavi Sanchez-Vives (ICREA Research Professor at the IDIBAPS, Barcelona) and neurobiologist Jean-Pierre Changeux (Honorary Professor at the Collège de France and at the Pasteur Institute Department of Neuroscience), moderated by neurologist Giancarlo Comi (Honorary Professor of Neurology at UniSR – Università Vita-Salute San Raffaele, Milan), and neurobiologist Eve Marder (Victor and Gwendolyn Beinfield University Professor at Brandeis University, Waltham, Massachusetts) with neuro-scientist Antonio Damasio (Dornsife Professor of Neuroscience, Psychology and Philosophy, and Director of the Brain and Creativity Institute at the University of Southern California, Los Angeles), moderated by neurologist and neuroscientist Daniela Perani (Professor of Neuroscience at UniSR – Università Vita-Salute San Raffaele, Milan). They explored the biological fundaments of consciousness, from the neurofunctional mechanisms to neurochemical and molecular basis, and they carried on in-depth analysis of connectivity as a cerebral substrate of consciousness state and the revolutionary techniques that allow investigating the brain in vivo.

Two discussions between neurolinguist Andrea Moro (Professor of General Linguistics at the Institute for Advanced Study IUSS Pavia) and cognitive psychologist Stanislas Dehaene (Director of Cognitive Neuroimaging Unit, Orsay, Paris) was moderated by cognitive neurologist Jubin Abutalebi (Associate Professor of Neuropsychology at UniSR – Università Vita-Salute San Raffaele, Milan), while the anthropologist Ian Tattersall (Curator Emeritus at the American Museum of Natural History, New York) and neuroscientist Idan Segev (David & Inez Myers Professor in Computational Neurosciences at the Hebrew University, Jerusalem) was moderated by neuroscientist Katrin Amunts (Professor for Brain Research, and Director of C. & O. Vogt Institute of Brain Research Düsseldorf and Jülich Research Center). They examined the concept of consciousness in relation to anthropology, the key role of language and its connection to the emotional and affective sphere to then reflect on the future evolution of research and on the attempts to create thinking machines.

Philosopher Michele Di Francesco (Professor of Logic and Philosophy of Science of the School for Advanced Studies IUSS Pavia) and psychiatrist and neuroscientist Giulio Tononi (Professor of Psychiatry, Distinguished Professor in Consciousness Science, Director of the Wisconsin Institute for Sleep and Consciousness at the University of Wisconsin-Madison) held the final discussion, moderated by scientific journalist Vivian Kasam; it represented a crucial moment to confront different perspectives and approaches to the complex question of consciousness. Moreover, Giancarlo Comi and Massimo Cacciari (Professor Emeritus in Philosophy at UniSR – Università Vita-Salute San Raffaele, Milan) reflected on the contributions of all the previous discussions.

SCIENTIFIC DIRECTOR

Giancarlo Comi

WITH

Jubin Abutalebi, Katrin Amunts, Massimo Cacciari,
Jean-Pierre Changeux, Antonio Damasio, Stanislas Dehaene,
Michele Di Francesco, Viviana Kasam, Eve Marder,
Andrea Moro, Daniela Perani, Mavi Sanchez-Vives, Idan Segev,
Ian Tattersall, Giulio Tononi

The following texts are transcriptions from the discussions available at
humanbrains.fondazioneprada.org/en/homepage

1

Neurobiological Bases of Consciousness

MAVI SANCHEZ-VIVES
JEAN-PIERRE CHANGEUX

PANEL DISCUSSION MODERATED BY
GIANCARLO COMI

In the Search for Consciousness: Brain Processing of Real and Virtual Worlds

MAVI SANCHEZ-VIVES

I think that there is a renewed interest in the brain coming from many different fronts. I will mention four of these fronts: on the one hand, from the governments we can see that there are initiatives going on largely collaborative projects, not only in Europe with the "Human Brains" project of which I am part and in which there are almost 100 institutions in Europe collaborating, but also in other countries. [1]

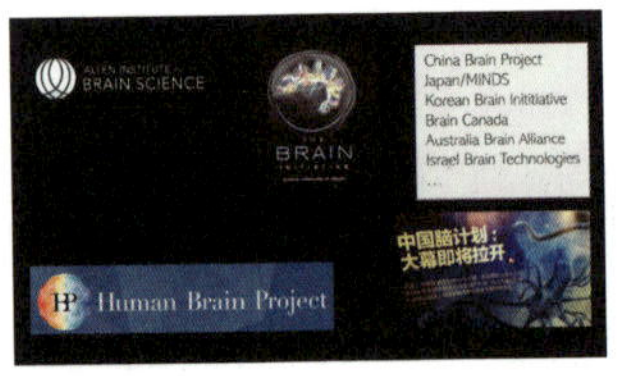

Fig. 1

In the United States there is the "Brain Initiative"; there is the "China Brain Project" in China; in Japan there is "MINDS"; in Korea there is "Brain Canada"; Australia; Israel. There is a moment in which there is an interest from the governments to advance in the study of the brain. On the other hand, I can see that there is a large interest from the side of the general public. [2] People are

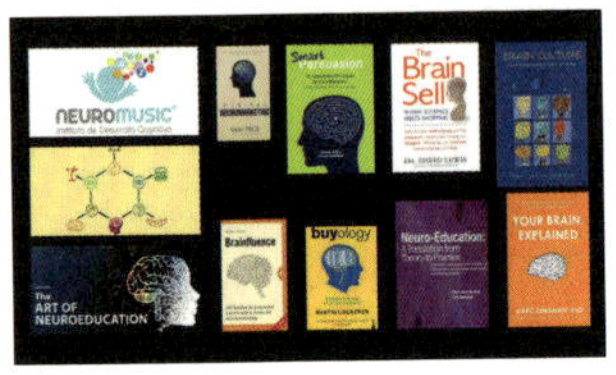

Fig. 2

genuinely interested in understanding about their brain with the idea that each one has different motivations, people want to improve their brain function, they want to prevent neurodegenerative diseases, they want to understand the science of the brain, what the brain has to tell you about different aspects. This is why I have selected in this slide different books—some are more scientific than others—but they respond to this demand, to this interest. Neuroeconomics, neuroscience, neuromusic, neuromarketing, neuroastronomy... It reflects this interest from the general public to try to understand as much as possible about their brains. On the other hand, we see a new interest from corporations. [3] There was

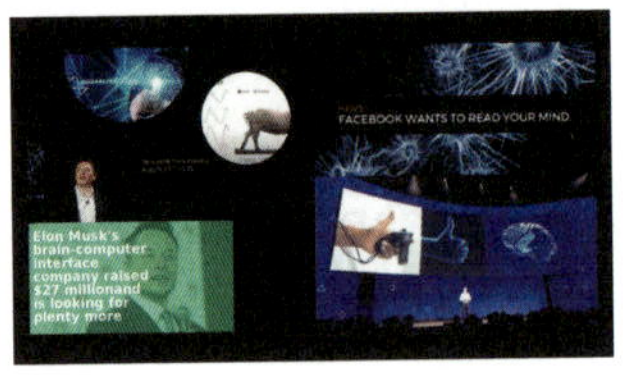

Fig. 3

recently a very announced intervention by Elon Musk presenting Neuralink, his brain lab, but there is also similar initiatives from Apple, from Google, from Facebook... They have now neuroscience labs, they have labs to study the brain. [4]

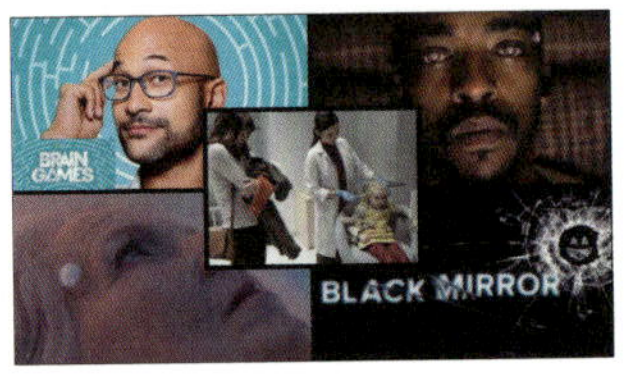

Fig. 4

Finally, we look in the entertainment world: here I included some images but there is a very large interest, and this awakes the interest in the people. There are chapters about brain interfacing, how to tap into the brain about dreaming, consciousness and so on. Why is this happening now? From the point of view of scientist I think there are two reasons why we should be interested now, and these reasons are that we can study the brain like never before and that we need to study the brain. There is this need as well. We can study the brain like never before because there have been many advances. The study of the brain requires advances in many different disciplines. We have seen in the last 20 years, 30 years advances in brain imaging, in optics, in micro and nano electronics, in new materials, in light control drugs, in optogenetics, something very important, which is the computational capabilities—we can nowadays acquire huge amounts of information and analyze these very large amounts of information. There are also developments in robotics, in virtual reality, genomics, gene therapy, so we have all these array of tools that were not available before and they allow us to dig in how the brain works. But on the other side, we really need to study the brain because there is a very large prevalence of neuropsychiatric diseases: in Europe there is a rough estimation of 180 million people per year. [5] This prevalence

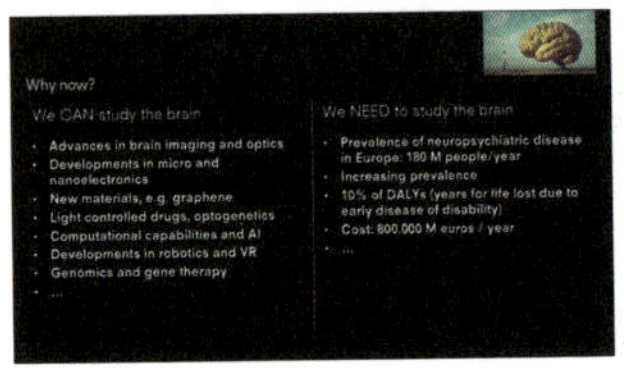

Fig. 5

increases with aging of the population, it affects the life loss due to early disease or disability, and it has a huge economic cost of roughly eight hundred thousand million euros per year in Europe. There is obviously a very important need. Of course there is the need of us as humans to try to understand ourselves. When did this start? For the last 2000 years or more, humans have been trying to understand the solar system, the universe. However, the in-depth study of the brain has a much shorter history that we could say that with some details started around 150 years ago. [6] There were these studies carried out by two men that we can see here: Camillo Golgi and Santiago Ramón y Cajal. They both used this method developed by Camillo Golgi of silver staining, and we can see here some of these reconstructions. They both shared a Nobel prize in 1906 even if they had theories that were to some extent contradictory: the neuronal theory and the reticular theory. However,

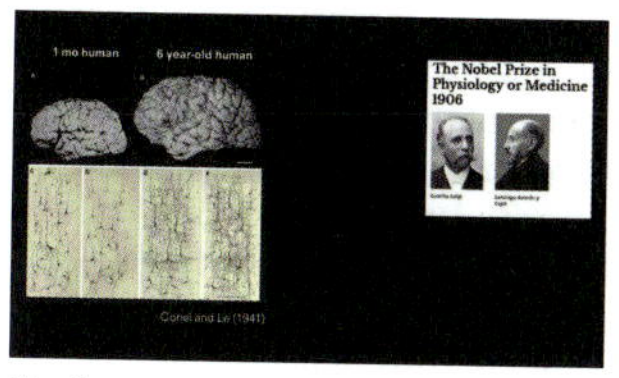

Fig. 6

even when neurons happen to be independent elements, as I will mention today, we will see that the brain functions has a very large network. They were some of the first people that were able to look into this structure that contains what we are, what we plan, what we think, and to look into it and to be amazed by these incredible structures, by these neurons. [7] In this slide you

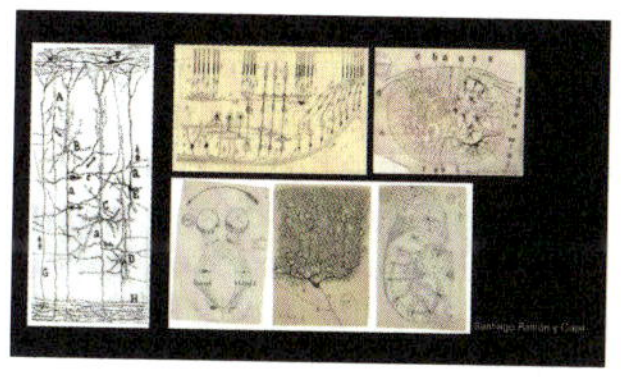

Fig. 7

can see some of these drawings made by Cajal of different structures: here the cerebral cortex or the retina, the spinal cord Purkinje's neurons in the cerebellum. If we look in a bit more detail in the cerebral cortex, we can see these dendritic spines carefully drawn where the inputs, the synaptic inputs, arrive. We can see also these interesting arrows that Cajal draw where he suggested without any real measures how the information would flow in these branches. I am going to show to you some animation shared with me by my colleague Javier De Felipe, where you can see this kind of neuron and reconstruction as it can be seen today. [8] Here is some spiny neuron that he has filled and you can see here on the right beautiful dendritic spines that have around four micrometers.
We can see very well that it is going to receive these synaptic inputs arriving into the cells. These circuits in the cerebral cortex are formed by neurons—here you can see a pyrami-

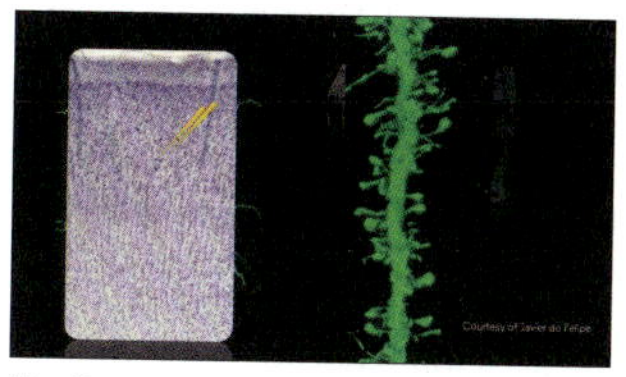

Fig. 8

dal neuron—and they receive inputs afferences, information also from inhibitory cells, and here there are three of them represented. [9] This

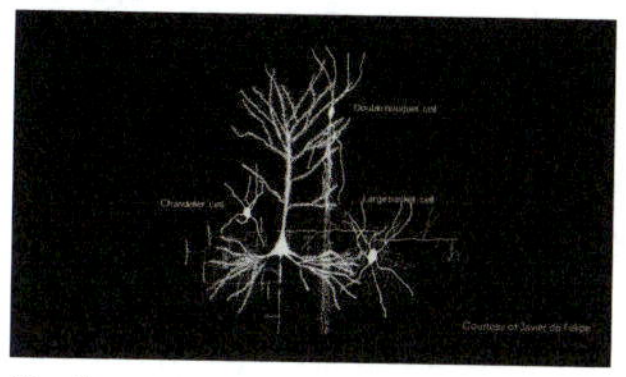

Fig. 9

forms a universe. Often the brain is compared with the universe because it is so large: there are 86 billion neurons, there are 10.000, even 30.000 inputs per neuron in humans. This means every neuron integrates information from even 20 or 30.000 other neurons and 2 or 3 million kilometers of connection. [10] It is really like a universe

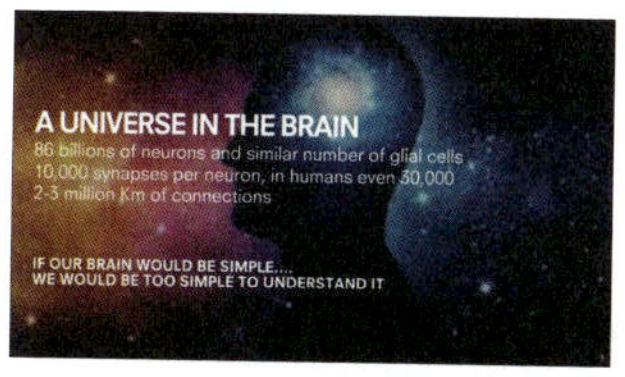

Fig. 10

that is contained in ourselves. As it has been said, if our brain was simple, it would be too simple to understand it. We are complex and our brain is also very complex and that is why. We are now going to have a look inside this cerebral cortex. This is a reconstruction made by my colleagues in the Blue Brain Project in Madrid shared also by Javier De Felipe, and I think you should be seeing it by now digging into this network. [11] I would like to make two points here: one is that you can see this incredible complexity of the cerebral cortex. As I said, each one of these neurons receives an incredible amount of information

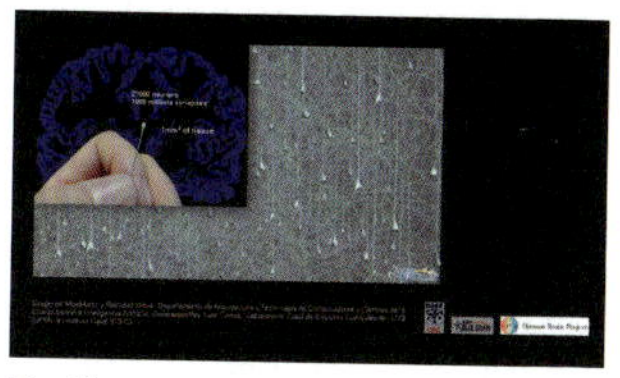

Fig. 11

that is being processed. But what is also fascinating, if you look into this structure, is that this is the basis of what we are, this is going to generate a sentient being that is conscious. What is consciousness? This is what this session is about.
I am going to speak about two aspects of consciousness. One is about consciousness levels. [12]

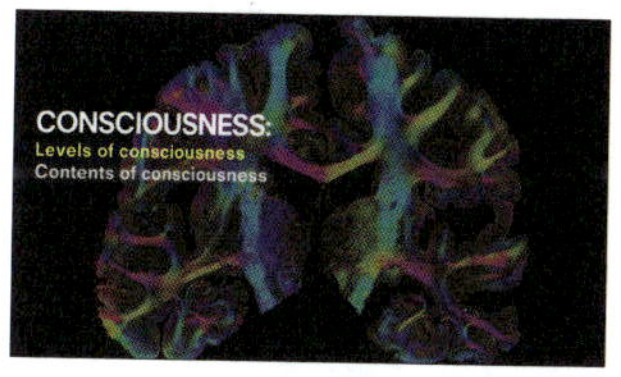

Fig. 12

What is consciousness when we talk about consciousness levels? It is what we lose when we fall asleep. Every night we close our eyes, we fall asleep, and then we disappear, we vanish and the same happens, for example, if we get a few drops of an anesthetic in our blood. What we are disappears and the next morning, or when the anesthetic washes out, we come back to the world. What you can see here in this scheme made by Steven Laureys are different brain states that can be sustained by

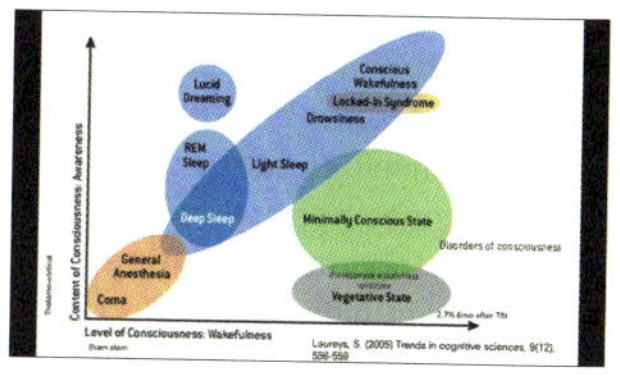

Fig. 13

the brain. [13] What it represents is wakefulness and awareness, and you

can see that we can be in deeper sleep, as I mentioned, or under general anesthesia, or we can be awake or we can be dreaming.
But there are also some pathological states—disorders of consciousness—and there can be a minimally conscious state or an unresponsive state, where the person is awake but is unresponsive, which used to be called vegetative state, and there can be also a coma. The brain network sustains all these different and possible states that have different patterns of activity, different biomarkers that we are going to be interested in to understand and to be able to identify, for example, whether a person that is unresponsive is conscious or not. In a way we can think that there are many brains in one brain, because, as we will see, the brain is going to express different activities in these different states. **[14]** This is very relevant from

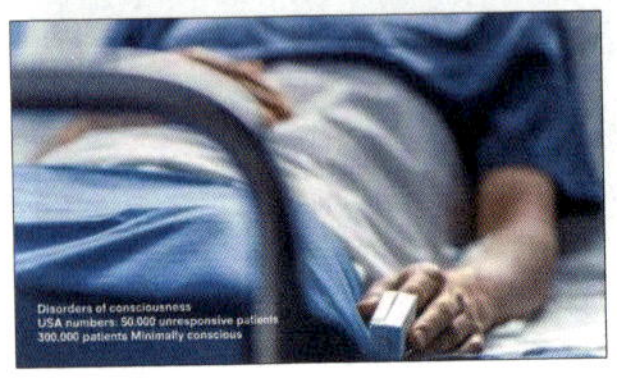

Fig. 14

a clinical perspective because there are disorders of consciousness and there are many patients in this state. These are numbers from the United States: there are 50.000 unresponsive patients and 300.000 minimally conscious patients currently in the United States. These numbers can actually increase because the care in intensive care is better and therefore more people survive. It is very important to develop measures to quantify whether a person that is unresponsive is conscious or not, to what extent is conscious, and if this person is conscious provide them with means for communication, and also eventually design methods in order to bring their consciousness back. For all these we need to understand the basis, we need to understand the underlying mechanisms. Let me start by unresponsiveness and this is here in this area in red. Unresponsiveness is associated to a very synchronized pattern of activity and you can see it here, very slow oscillations where the whole brain discharges or has these synchronized patterns of activity. This was in fact some of the first activities seen in the brain, the first recordings that were done by Caton in 1875. This was the first activity that called his attention. **[15]**

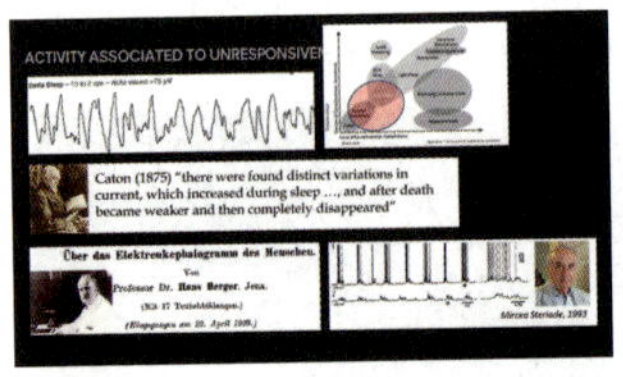

Fig. 15

Also, in 1929, the first electroencephalographic recordings in humans by professor Hans Berger was one of the first activities because it has very large amplitude such that it is associated to a slow wave sleep or to deep anesthesia. It has large amplitude because it reflects that there are many neurons active at the same time. The cellular bases were not described in some detail until 1993 by Mircea Steriade and his collaborators that described in more detail this pattern. Still, the final mechanisms of how the red generates these emergent activities is still under discussion. So when does this activity appear? **[16]** As I mentioned,

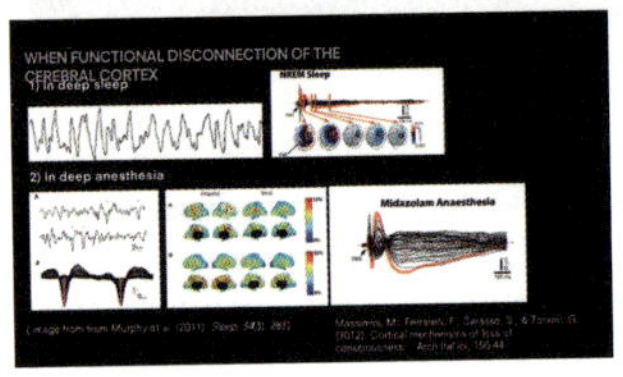

Fig. 16

it appears in deep sleep and we can observe it not only in the spontaneous activity but also, if we stimulate the brain, one can see that there is this type of response, this kind of biphasic response which doesn't propagate a lot and is very similar, we can say, in deep anesthesia.

Interestingly, this activity also appears when there are isolated areas of the cerebral cortex. So if there is an experimental cut of a piece of the cerebral cortex, there is still the generation of this slow pattern, slow rhythm. This is the case also when there is a lesion. Here you can see when there is a lesion that disconnects some areas of the cerebral cortex. One can observe this slower delta rhythm over this disconnected area. **[17]**

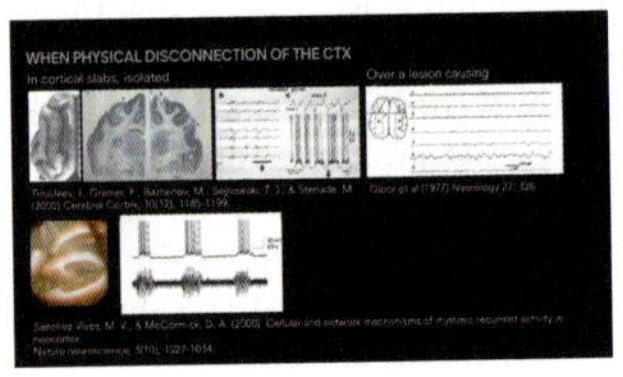

Fig. 17

In fact, quite a while ago—20 years ago—we found that even if we take a small piece of cerebral cortex and put it in vitro, in a dish, it contains enough circuitry in order to generate this activity that you can see at the bottom of the slide, which is also a slow oscillation and is very similar to the one observed when there is the full brain and there is slow wave sleep. What this is telling us is that a small isolated piece of the cortex contains enough circuitry in order to generate this pattern of activity. **[18]**

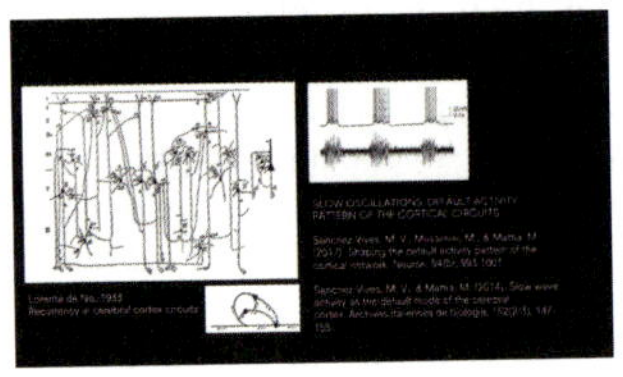

Fig. 18

This is the case that you can see here in this drawing by Lorente de Nó, because there is recurrency in the sequence and you will hear more about this during this talk. There is recurrency: some neurons connect the others and back. Therefore, in a small piece there is enough circuitry to generate this activity pattern. We suggested that this slow rhythmicity associated to unconsciousness is actually like a default activity pattern of the cortical circuit. We have suggested this for all the reasons that I showed, because in any circumstance you isolate a piece of cortex and this activity is generated. [19] Indeed, the

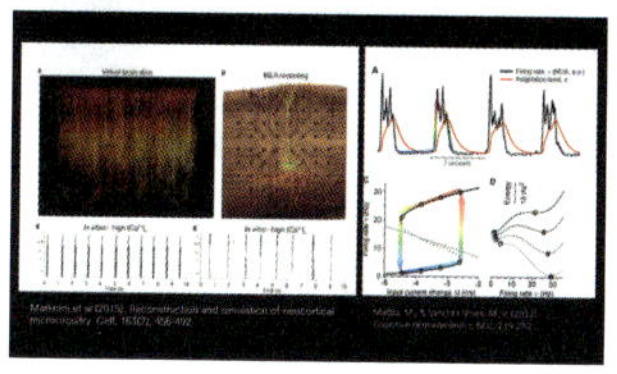

Fig. 19

same is the case in computational models: they can be very complex computational models, where many neurons are reconstructed, as we can see here in this work by Markram et al, in which these neurons reproduce a piece of the cortical network and generate these slow oscillations. The same is the case as you can see here on the right in this mean field model, basically a much simpler model that contains two elements, which are recurrency of the connections and adaptation. In a way, we can say that the default state of the cortex is to be in sleep, to be unconscious. However, obviously our experience is that of being conscious. So how is this transition from unconsciousness to consciousness? It is very important to understand unconsciousness to see how it can be transformed into consciousness. How is this transition? [20] If we look at an individual

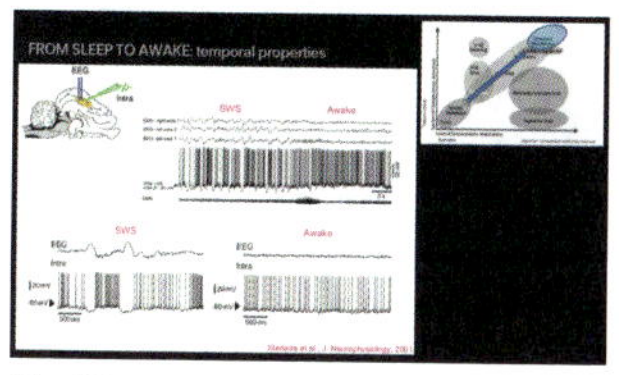

Fig. 20

neuron as this one recorded here by Steriade et al, what we can see when there is this transition from a slow wave sleep to awake is how this rhythmic pattern stops and gets transformed into a more or less continuous pattern of fighting that is not rhythmic anymore and it is, to some extent, asynchronous. If we look at this activity, not spontaneous but following a stimulation—and this is work by Marcello Massimini and collaborators—, if one stimulates the cerebral cortex in these situations in which there is unconsciousness—like sleeping, anesthesia or unresponsive wakefulness syndrome—in all the different locations there is this very similar synchronized response, relatively synchronized response, this sort of biphasic response which is associated to unconsciousness. While there is a conscious state—like awake or dreaming and so on—then there is this response that, as you can see, is not the same anymore in every point, but is different in the different locations. [21] This is

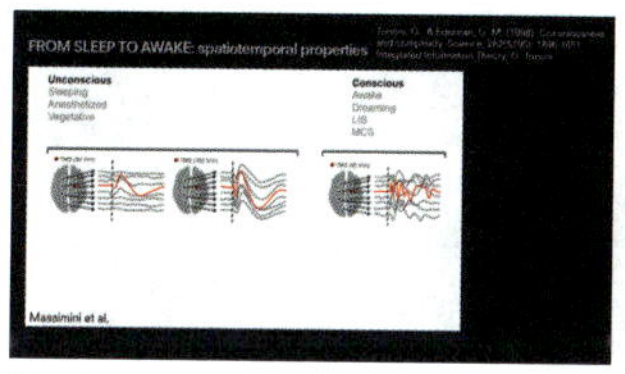

Fig. 21

the spatial aspect, the spatiotemporal aspect. To which kind of networks does it respond? In order to have this kind of more simple synchronized responses we could imagine a network which is very local, it is very segregated, or it can be all very highly integrated and, in this case, we could obtain this kind of responses. However, the situation is different in the awake state and there is this balance between these two, between the very segregated and very integrated. So this generates a high complexity in this network which is associated to consciousness. This is based on ideas originally proposed by Tononi and Edelman in 1998 in a paper in *Science*, and also the integrated information theory by Tononi is very connected to these ideas. [22] Consciousness is therefore associated to these interactions and between different areas, causal interactions that have given rise in a practical aspect to one measure of consciousness, a measure that has a clinical value and you can see it here by Casali et al. It has clinical value

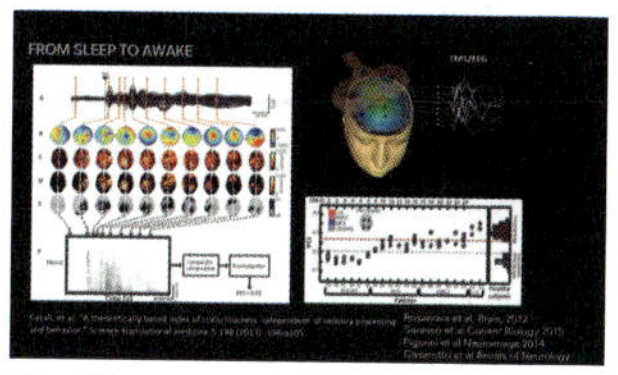

Fig. 22

because it can be done in patients, like I mentioned to you before—patients that are unresponsive and that one needs to know whether they are conscious or not. This measure consists in providing a stimulation to the brain, to the cortex, with TMS (transcranial magnetic stimulation), and it evokes responses that are recorded in different locations. I won't go into details of how this is calculated, but here you can see in the space and the time of this response over many different locations in the cortex. This is compressed with a Lempel-Ziv algorithm into one number, which is an index referred to as the perturbational complexity index, and has a relation with the consciousness level. You can see here how in situations where there is an unresponsive wakefulness syndrome or a minimally conscious state is lower, and in an awake person is higher. You can see on the right a segregation between these indexes. Therefore, this is a measure of how the response is in the many different areas of the cortex to a stimulus, and it is giving information about the functional state. [23] Therefore, from the point

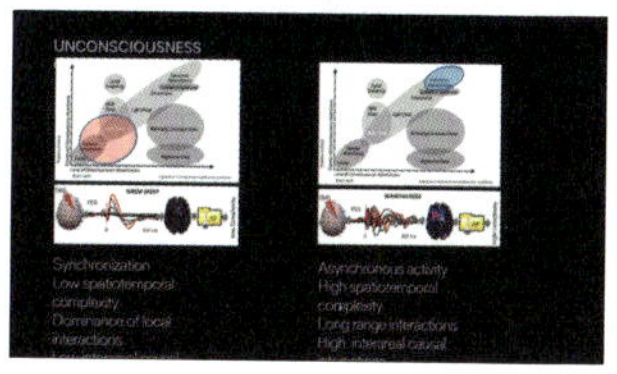

Fig. 23

of view of the levels of consciousness—just to recap what I told you—in unconsciousness there is on the left a high synchronization, lower spatial temporal complexity, a dominance to some extent of local interactions and low interareal causal interactions. However, on the right, in a conscious state, the temporal pattern is as in asynchronous, there is a high spatial temporal complexity, more dominated by long-range interactions, and a high interareal causal interaction. How about the neural correlates of consciousness? This was the name given originally by Crick and Koch during the search for what is the basis for consciousness and this could be patterns or it could be neurons. Are there any specific neurons causing or controlling consciousness? Not that we know of, but we have seen that consciousness is about interaction between areas and there are many neurons that are critical in connecting different areas in the brain. [24] I have here

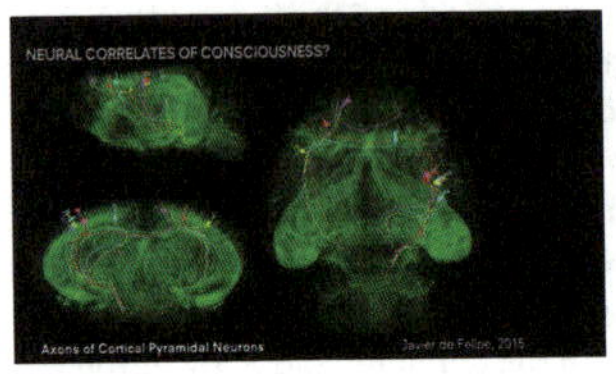

Fig. 24

some examples to show you: these are very long axons of pyramidal neurons that have a very large distribution connecting very distant parts of the brain. [25] We also have—very

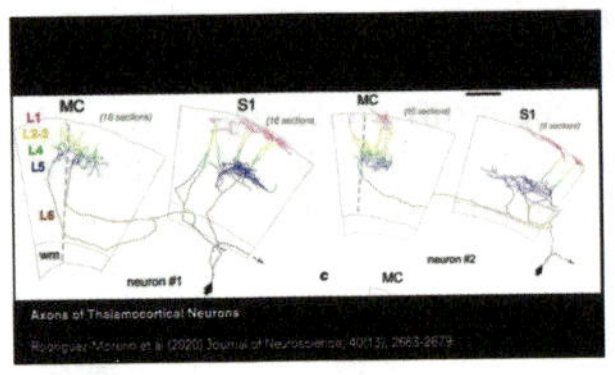

Fig. 25

importantly for consciousness—thalamocortical neurons; some of them also have axons, as you can see here, that connect different areas—in this case motor areas, in some other sensory areas. They are the axons of the same neurons in the thalamus that innervates separate cortical areas. [26] This is an example of claustrum neurons that were described by the Allen Institute: these neurons have also connections around very distant

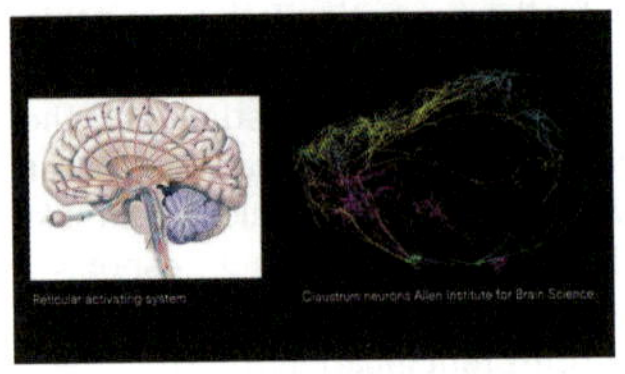

Fig. 26

part of the brain. Very important also, of course, in order to go from sleep to awake, is the reticular activating system that all the neurotransmitters that are going to mediate this transition in physiological conditions. They also have a very diffuse innervation of the brain. [27] Another aspect I want to

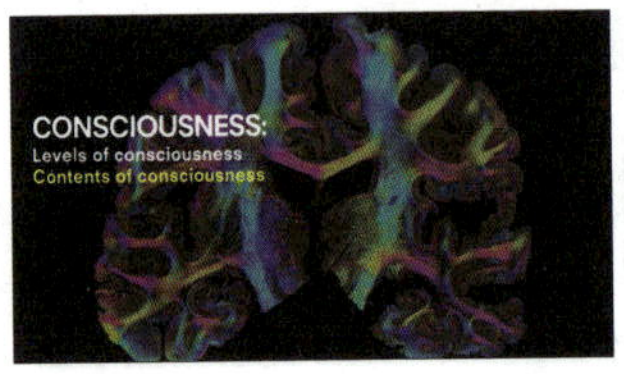

Fig. 27

talk to you about consciousness is contents of consciousness. We have seen we need this interaction across areas to have to be conscious. But what are we conscious of? [28] When

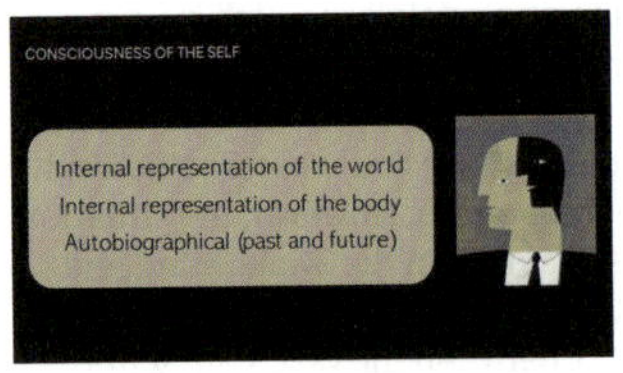

Fig. 28

we think about consciousness of the self usually one considers what we are conscious of the external world, our own body, and our thoughts of ourselves, our history, autobiography, past, and also future plans, and so on. Let's talk about the sensory representation of the world. The visual information arrived to the eyes is transduced into spikes. These spikes travel to the thalamus and arrive to the cerebral cortex. [29] This video that we just saw is how you can hear the spikes of a neuron in the primary visual cortex discharging. These are old videos by

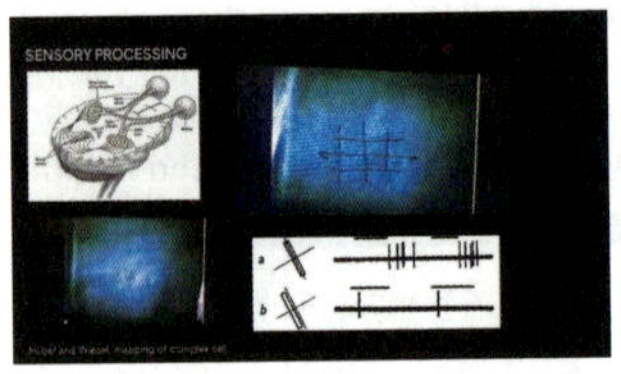

Fig. 29

Hubel and Wiesel, the first ones that characterized this type of responses. These are single unit recordings. This is what happens in the primary visual cortex: the information arrives and is encoded into spikes. I used to work in sensory processing either in primary visual cortex and also in auditory cortex. [30] Here you can

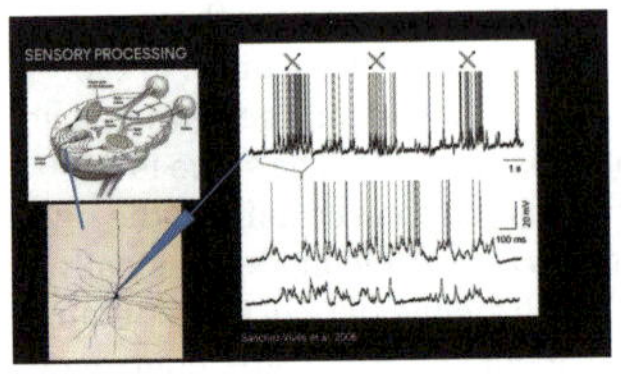

Fig. 30

see these spikes by intracellularly recording from inside, with an electrode inside one of these neurons. We can see these bursts of axons potentials and the underlying synaptic potentials. How does this transform into a perception? [31]

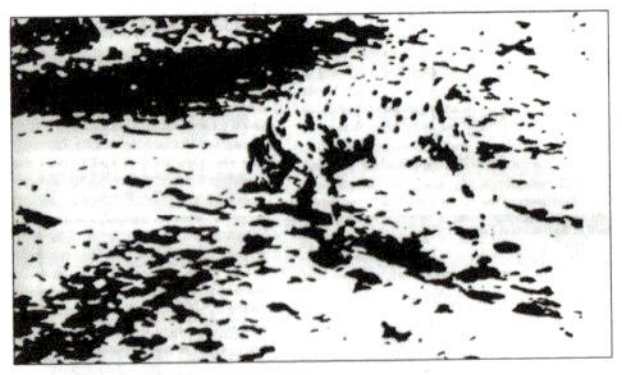

Fig. 31

If one sees this stimulus that you can see here probably you can see this black and white or these black spots over a white background. If I tell you that this is the color of a dalmatian dog then all of a sudden probably this dalmatian dog is going to pop out and you are going to be aware of this. Your primary visual cortex is processing the same inputs, but then there is this interaction—in this case with this prior because you have seen a dalmatian dog before—and there is a perception which changes. [32] There is

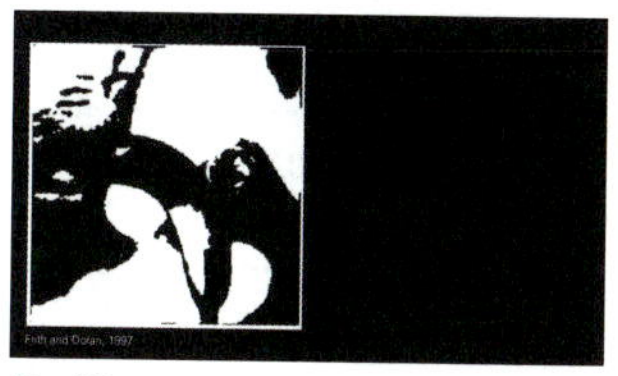

Fig. 32

another example where one can see these stimulus and if you see now a photograph with a banana then the way of seeing these black spots over a white background changes. This is normally understood as an interaction, as an active process: an interaction between the information that comes from the outside world and what is already in your brain. There has been a proposal of a theory of how this conscious perception occurs. If you have had an audiometry done, they can give you these sounds: some of them are going to be below threshold, you may not hear them, some of them are going to be above threshold. Still, some of them may be making it through your auditory cortex. [33]

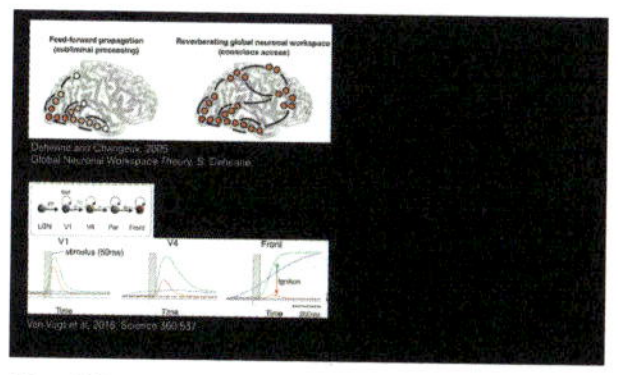

Fig. 33

The proposal made by Dehaene and Changeux in 2005 was the following: you can see in this slide that there are some stimulus that arrive to primary cortices and that eventually propagate to towers higher association areas. Some of them are going to reach frontal areas and this is going to induce an ignition that is going to cause a reverberation in this recurrent sequence that we have mentioned before. This is going to be the base for the conscious perception: the stimulus that would not make it through these frontal areas would not become conscious perception but it would be subliminal stimulus. There is also an interaction with the fact that when someone perceives, normally we ask them to report it in some form. But this is the basis also for the global neuronal workspace theory by Stanislas Dehaene. Here you can see this paper in science by Van Vugt et al.—I selected the model because it is clearer but the data is very similar—how there is what in v1, which is the primary visual cortex, appears with a very similar response, but this ignition and this amplification is going to cause this non-linear response, a much larger response that is going to be associated to the conscious perception of the stimulus. [34] This has been reported with

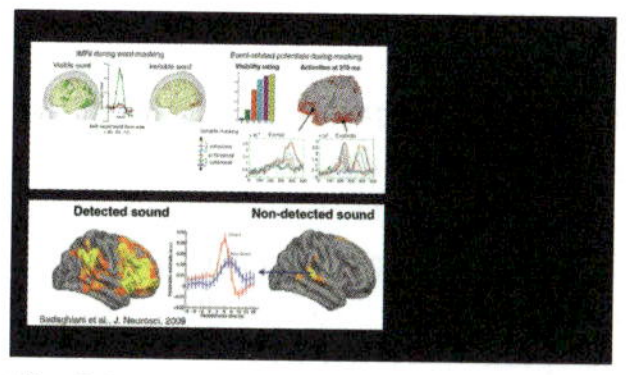

Fig. 34

other methods like fMRI, like imaging of the brain, an invisible world gives this small local response but a visible world that gives this much larger activation. The same, as you can see here: a non-detected sound remains locally in the auditory cortex but a detected sound activates many more areas in the cortex. I work in sensory processing, in early sensory areas and some 15 years ago I came across a technology which is virtual reality. [35] This suggested me that this was

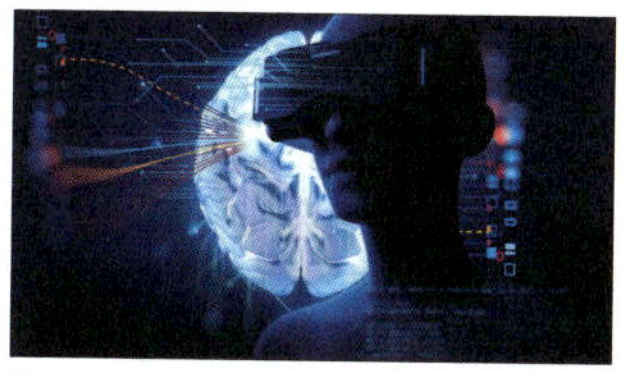

Fig. 35

an amazing tool in order to study and to understand perception. Why is that? Let me tell you what is virtual reality, although now it is quite well known, but just to be on the same track. Virtual reality displays in all sensory systems, provides multi-sensory information that instead of coming from the natural—from the outside world—is going to be generated by a computer, but it is placed eventually in all sensory systems closing the person in 360 degrees, completely. It tracks the head, the limbs, the body, and, very importantly, is going to determine what we see by our position. It is very similar to what happens in the real world, and of course there is a stereoscopic vision and so on. [36]

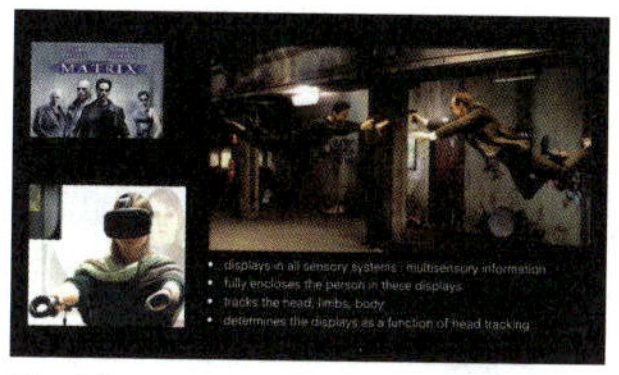

Fig. 36

Here you have a scene of the movie *The Matrix* where people live in a completely virtual world. [37]

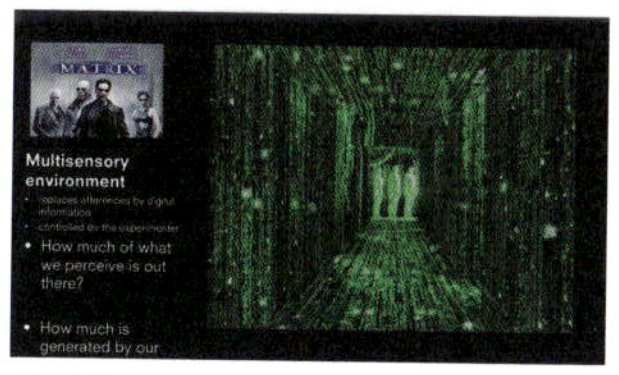

Fig. 37

Therefore, one would have a multi-sensory environment where the afferences, the input, the information, the sensory information is now digital, it is generated by a computer. Therefore we have now an environment that is completely controlled by the experimenter and this is why it is a very good tool to understand and to study perception. We can think how it opens up many questions, how much of what we perceive is actually out there, or how much is generated by our own brain, for example. [38] This is an old environment of a virtual library where the subjects of experiments had to walk around and do some tasks. This was a silent environment, but many people reported sounds, and this is why it is interesting to have a controlled environment, because then you can see how their brains are probably filling in this

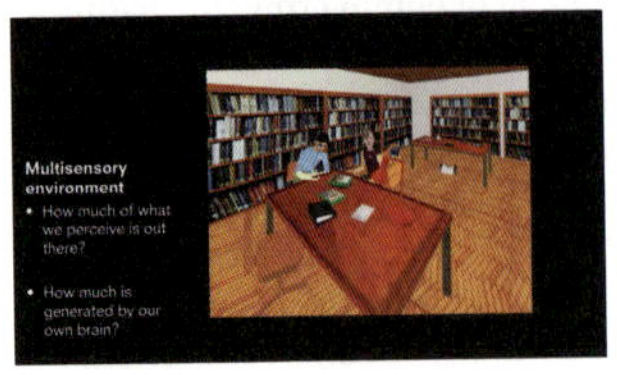

Fig. 38

auditory information that would have been logical to have in a situation like this. It is just an example to see how much of it is actually generated by our own brain. [39] We have this situation

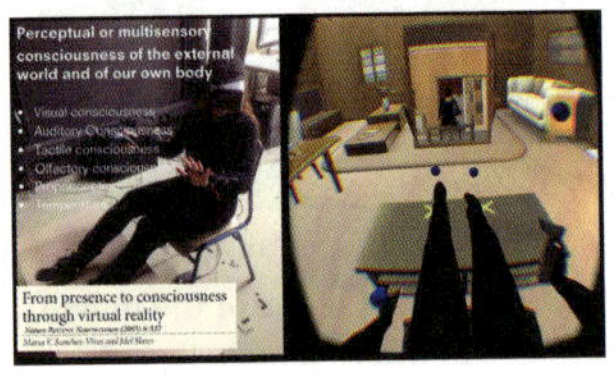

Fig. 39

where we have a perceptual or multi-sensory consciousness of the external, in this case virtual, world and also of our own body. We have this consciousness of the world and of our own body which consists of different senses—visual consciousness, auditory consciousness, tactile, olfactory, proprioception (we are moving, temperature, and so on). Some of them, or ideally all of them, would be controlled by a computer, and this of course opens up many questions: what is real? What is virtual? Is the world that we perceive virtual? Our brain is the mediator, it reconstructs and we know that there are many forms of energy out there that we are not aware of, therefore we are recon-

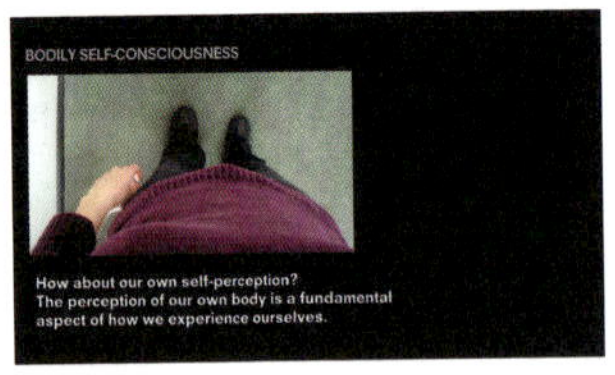

Fig. 40

structing a part of it in our own brain. [40] Then we have another aspect that you can see here which is the representation of our own body, a very important part of the self. How about self-perception? [41]

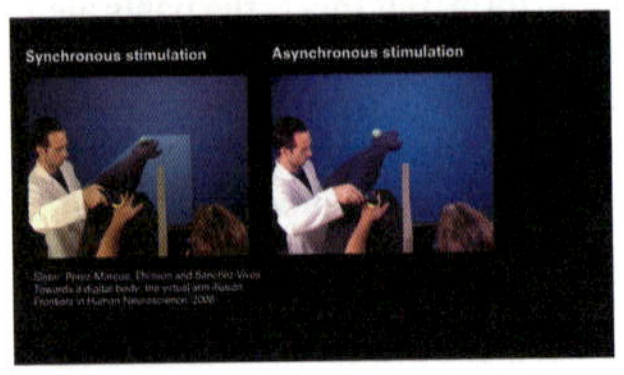

Fig. 41

The perception of our body is actually a fundamental aspect of how we experience ourselves and this is something that we started studying some years ago. We wanted to study to what extent we could feel a virtual body as our own body. Here we started exploring how these visuo-tactile correlations could induce ownership of a virtual arm, and how do we know if there is ownership? If this arm is felt as own? We have first of all to ask the subject, but we can also measure it in different ways, we can ask the person where they feel their arm to be, or we can see how they respond to a threat, or we can measure with electromyogram how is their activity in their muscle depending on the movement of the arms and so on. This is another study. [42]

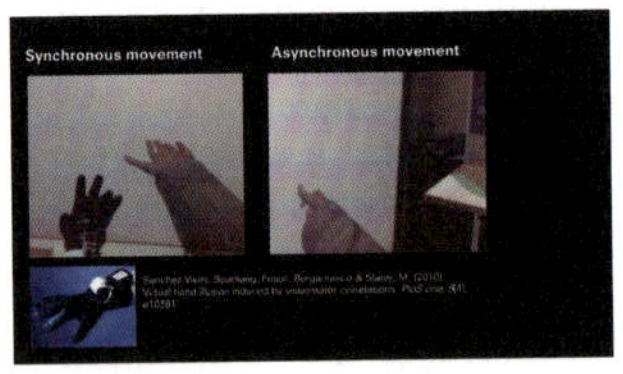

Fig. 42

We found that not only visuotactile but visuomotor and proprioceptive correlations were also adequate in

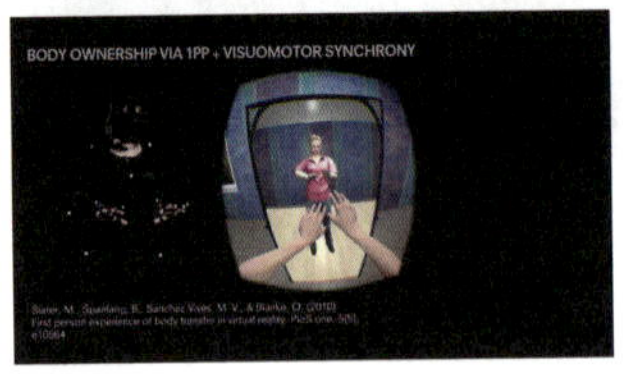

Fig. 43

order to study ownership, to induce ownership of the arm. [43] Here we can see that we can induce ownership of a full body: what you can see here is first of all first person perspective, visual motor correlations: someone sees themselves reflected in a mirror and they have correlations also with the body reflected in the mirror. [44] All these

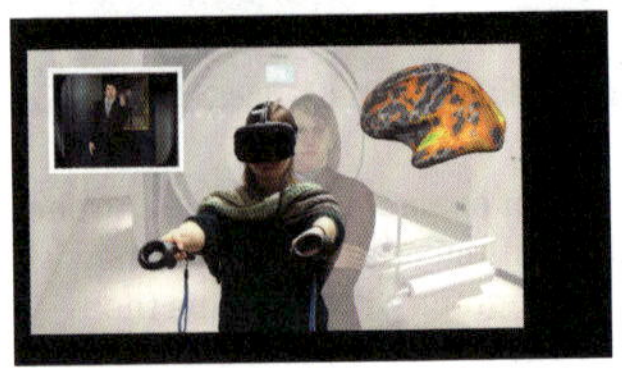

Fig. 44

together induce a strong sense that this body is your own body. This allows to ask a neuroscientist to study the activity in the brain when this happens, to try to identify the critical areas involved into this multi-sensory integration. [45]

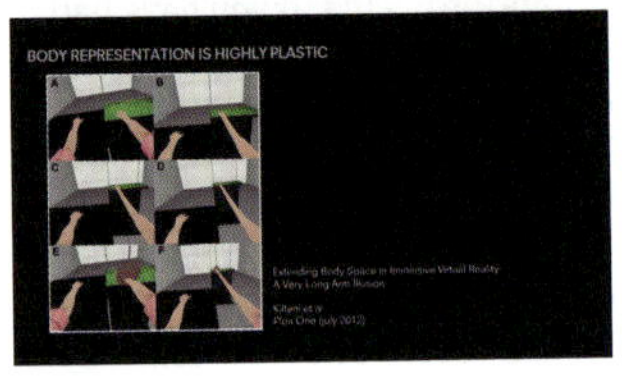

Fig. 45

I would like to just give two or three more ideas: one is that in studies that we have carried out over several years we have seen that the brain is very plastic, in the sense that it can accept many different bodies as far as there is appropriate inputs, appropriately sensory correlations. I'll show you a few examples: we can have one long arm, much longer than the other, and still the ownership persists, with some limits that are described in this study. [46] One can

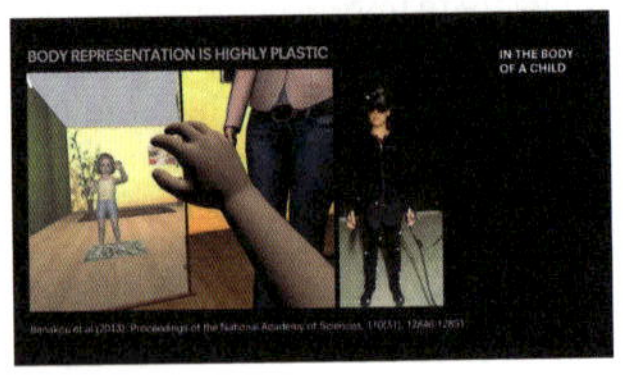

Fig. 46

also feel a smaller body, a younger body as own, like the body of a child, or, for example, one can assume as own a body of a different race, a different color. All these correlations are appropriate. This is the important thing: the visual aspect can be different and the illusion is not lost. [47] A very interesting

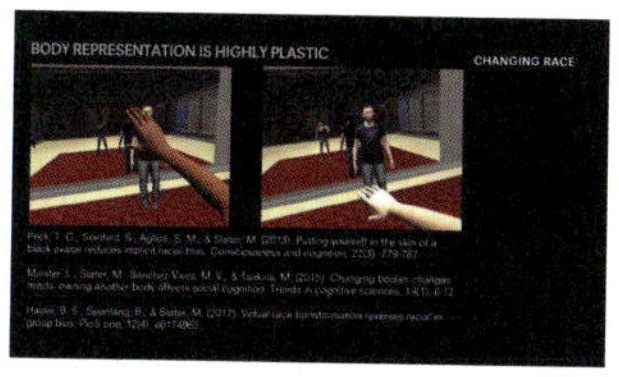

Fig. 47

aspect of this is that the shape of the body, the aspect of the body is going to have an impact on different aspects of how we process information. [48] For example, it is going to

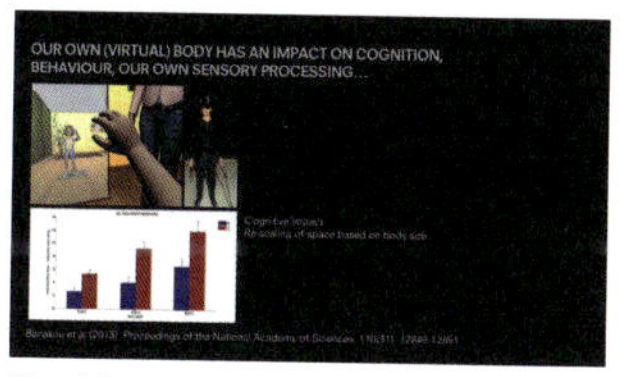

Fig. 48

have a cognitive impact on how we perceive the outside world, so there is a real scaling when the body is of a different size, or there is a decreased racial bias if we identify with a different race or—something that we have been working on quite a bit—is how we can modulate pain by changing the aspect of our body. [49]

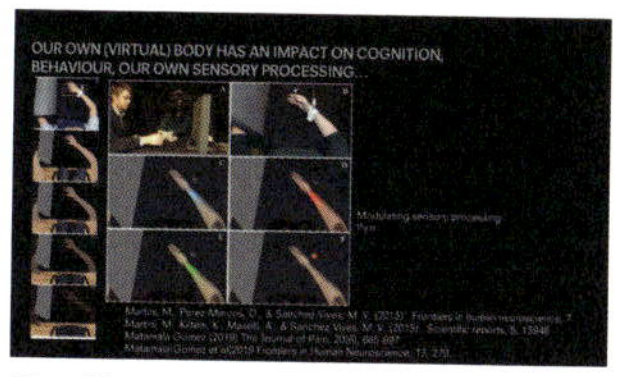

Fig. 49

This is something that can be used in chronic pain patients. Two more aspects I will touch very briefly, maybe we can discuss them later. I have talked mostly about the importance and the sensory aspects of how we feel our body as our own, how are we conscious of our body. [50] But what are the motor implications? Do we feel that we own, that we control that virtual

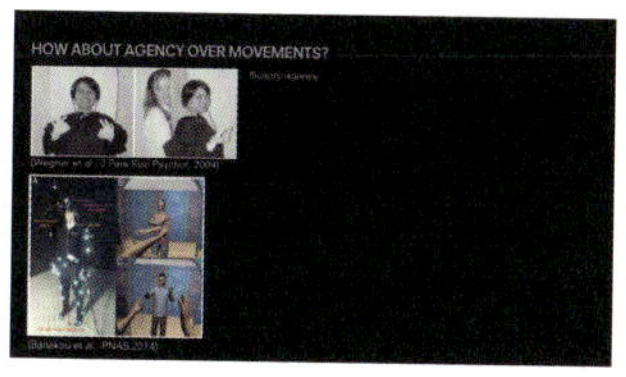

Fig. 50

body? What is the effect of agency? We know that if we feel a body as our own, one can have illusory agency. This means that, if we feel a body as our own, like a virtual body, and this body moves, one sometimes can have the illusory agency over that body, like the feeling that one has generated that movement. [51]

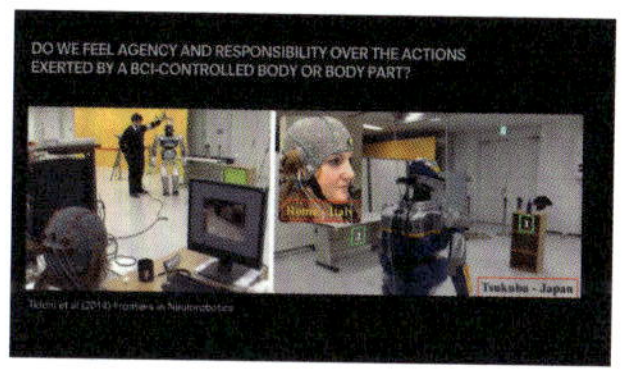

Fig. 51

One can also control a virtual body by brain computer interfaces. We studied recently to what extent we feel ownership of the body. Once we feel ownership of the body to what extent we can feel responsible by actions carried out by this body? The motivation is in this slide here. Basically, one can apply to robot bodies everything I have explained for virtual bodies. We can feel also ownership over a robot body with some constraints. What happens if by means of thinking by brain computer interface we control a robot as this? Someone is sitting in Rome and controls a robot really far

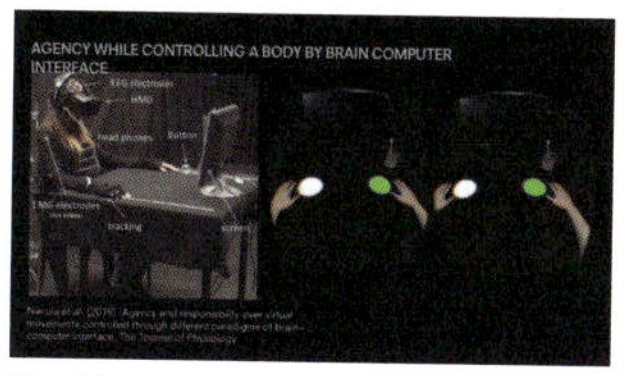

Fig. 52

away. [52] We wonder to what extent we feel agent and responsible by the movements of external, exogenous body. We run an experiment where we explored how people control a virtual body by thinking by a brain computer interface and whether they felt responsible by actions which was not single but eventually that container over the table falls on the floor. [53] We saw that there is

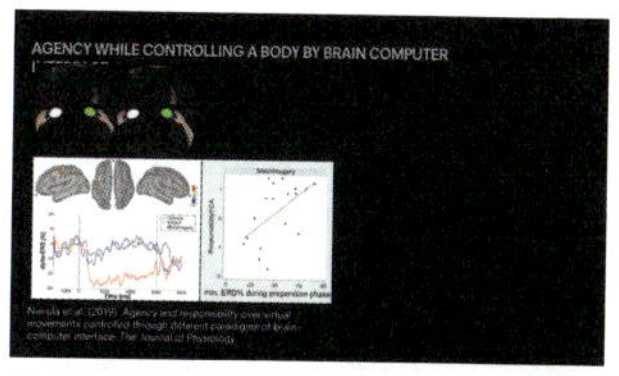

Fig. 53

actually a relation between the activity of sensory motor areas and their responsibility. I know it has been very fast but I wanted to give the idea that there are also motor implications. [54] Finally, I can say

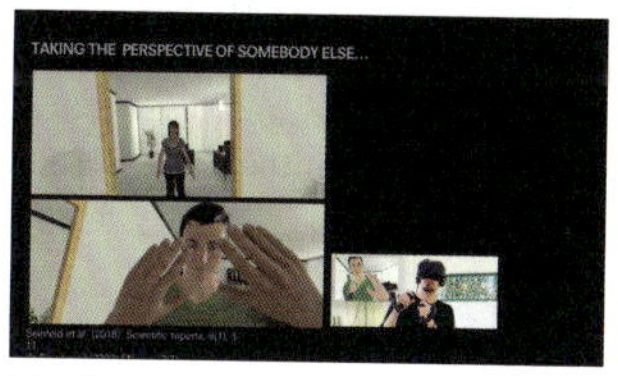

Fig. 54

that I think virtual reality is a useful tool to study perception, conscious perception of the outside world, of their own body. It can be also used to do incredible things in the real world, like to be in the perspective of someone else, for example, or to do something unthinkable like to talk to oneself. [55] This is Dennet in our

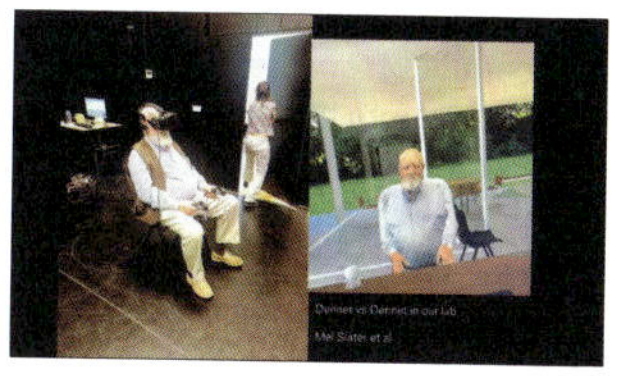
Fig. 55

laboratory: he was scanned, he got his avatar done, and he could sit here and have a conversation with himself. This opens up the door to do this kind of explorations of ourselves and of our consciousness.

The Connectomics of Conscious Processing: From the Molecular to the Cultural Level

JEAN-PIERRE CHANGEUX

I am delighted to contribute to this topic by speaking more particularly on one aspect of conscious processing, which is the relationship between conscious processing and the connectomic organization of the brain, as you can see in the right part of the slide. [1] This is a topic

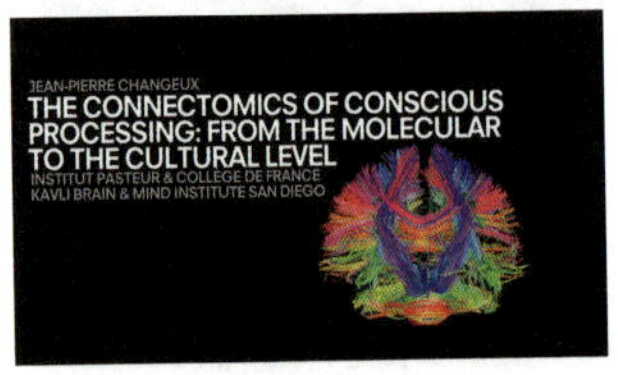

Fig. 1

that is still ongoing and I will present a few ideas and also some experimental evidence for these ideas. [2]

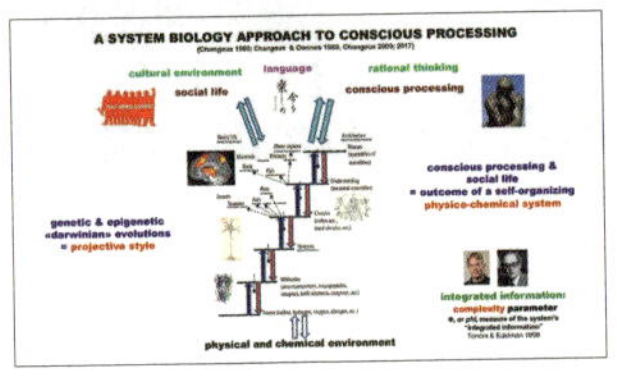

Fig. 2

Before starting on this particular aspect I would like to emphasize an important aspect of studying the brain: as you can see in slide 2 it is organized into several nested levels of complexity, from the molecular to the cellular, from the cellular to the circuits, from the circuits to more global processes in the brain. After all, we are dealing here with some kind of system which is, if I may say, a physico-chemical system grounded on the molecular level up to the higher level. The outcome of this complex organization is conscious processing but also social life, cultural environments, and, in particular, language and, of course, rational thinking. It is important to have in mind this kind of organization with hierarchical and parallel structures and realize that all this complexity is the result of evolution, and might be viewed as through development. I will discuss later on how all these levels derive from the other through some kind of Darwinian evolution, which is nested in a whole system whose outcome is conscious processing. Therefore, one cannot relate the conscious processing to only one top level, but remember that it is grounded in the molecular level, up and down. This is of course an interesting feature and I may mention the work of Tononi and Edelman on some kind of measurement of this complexity which takes place through a fine parameter. This idea is actually developed for the whole organism, and in particular for consciousness, but it is not unique to consciousness. [3]

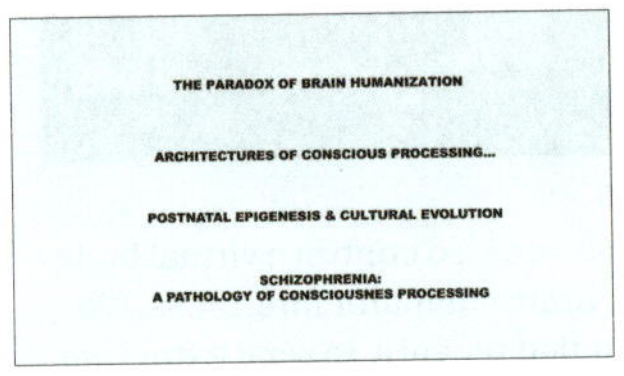

Fig. 3

I will point out three issues regarding the relationship of the conscious processing with the organization of connectomic structure. First, the paradox of brain humanization, then the architectures of conscious processing, and last, a very important issue, the postnatal epigenesis and cultural evolution. I will close by selecting one disease and showing how one could try to relate the molecular to the conscious processing level and its pathology in the case of schizophrenia.

We have the paradox of brain humanization. [4] One can simply follow the size of the brain, but also the organization as a function of time in the ancestor of *Homo sapiens* and see that, from Australopithecus to *Homo habilis* and then *Homo sapiens*, there is a fast rise through only a few million years of the volume and complexity of the whole brain. A dramatic

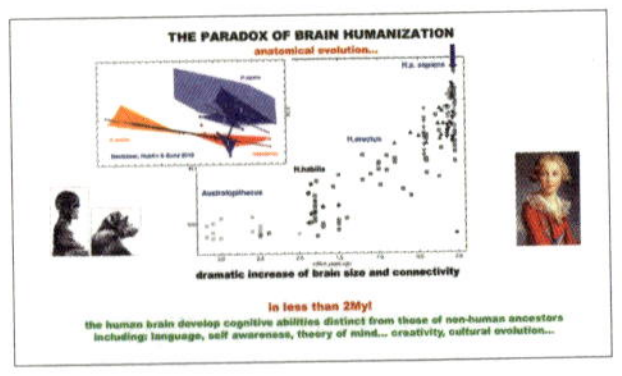

Fig. 4

increase of brain size and connectivity. If you look, for instance, at the level of *Homo sapiens*, you see that there is a relationship between *Erectus* and *Sapiens*, which looks to some extent discontinuous: we have two species and we have to understand how one may move from these ancestors of man into *Homo sapiens*. I would like to say that from our present point of view, which is to try to understand conscious processing, this evolution from Australopithecus or some ancestors still living or monkey ancestors, like the chimp, there is an extraordinary development of brain cognitive abilities. In particular, language, self-awareness—we heard about it in Mavi's talk, theory of mind—and even creativity, cultural evolution, art, science, and so forth. How can these changes in brain organization lead to the development of such remarkable features and cognitive abilities? [5] The first thing is that there

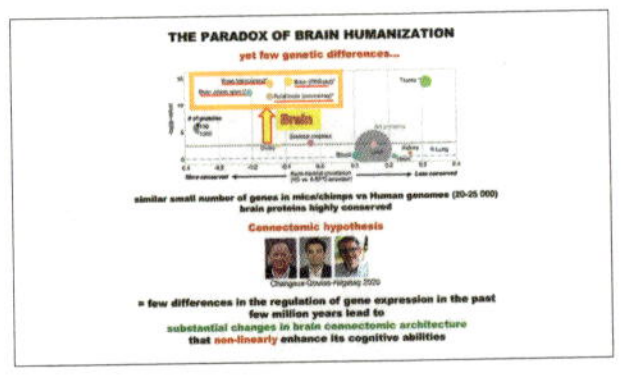

Fig. 5

are important genetic differences between the monkey and humans or even between some relative ancestors of man, like the Neanderthal to *Homo sapiens*. In fact, this is wrong. There is the same small number of genes in mice, in chimps, and in humans: between 20 and 25,000 genes. So forget about the number of genes. If you look closely at the sequence of brain proteins—and this was done recently by Guillaume Dumas in the Pasteur institute—you can see that, contrary to the testes, surprisingly, or even the lung, the brain proteins have a highly conserved sequence. There is not a big change in the genetic organization of the protein determinants in the brain compared to many of the other organs. The sequence of the proteins which contribute to brain organization and structure and high function is deeply conserved. Even there, there is a problem. How to solve this paradox that there are nevertheless some changes? There have to exist a few genetic differences having some important impact on connectivity and cognitive function of course. With Alexander Gulas and Klaus Hilgetag, we recently proposed what may be called a connectomic hypothesis in cerebral cortex: how to try to explain that few differences not in the structure of genes, but in the regulation of genes, which are of course under some genetic determinants, in the past few million years led to these important changes in connectomic architecture. The idea is that there are changes, which we call connectomic fundamentals, which would be especially under the control of these few genetic regulatory events and have an important role in the development of brain complexity. One simple thing to think about is the size of the brain and the number of neurons, which indeed increase from chimps to humans. There is another connectomic fundamental, which is an extended period of brain ontogenetic development. Another interesting aspect concerns this multi-level organization I showed in my first slide. There is clearly—from the mouse to the monkey to the chimp to man—an increased scaling up of the multi-level organization of the connectome, which ultimately enhanced conscious processing, which is engaged in the difference of conscious content, as we mentioned, that you find in the brain of the monkey and in the human brain.

There is also another interesting feature, which has been observed and demonstrated by Alexander Goulas, which is a shift of cortical layer of reference from lower to upper layer—you heard about layer 3c. This will have introduced some long axon connections compared to layer 5 which gives axons going down to essentially motor control and so on. There is some kind of shift that increases the contribution of higher cerebral cortex layers, which themselves, as you will see, are assumed to contribute to conscious processing. Lastly, there is a very important consideration if we want to study adult consciousness: a considerable extension of the period of postnatal development occurs, where the interaction with the outside world contributes to the remodeling of the synaptic organization of the brain. This lasts, as you know, about 15 years, and you have to remember that, when *Homo sapiens* appeared in Africa, the life span was about 30 years. It means that when *Homo sapiens* was selected as some kind of species, he spent half of its life building its body and mostly building its brain. This is an extremely important aspect to consider if we want to study consciousness, and in particular culture and the consciousness of the cultural brain. [6] I will make a

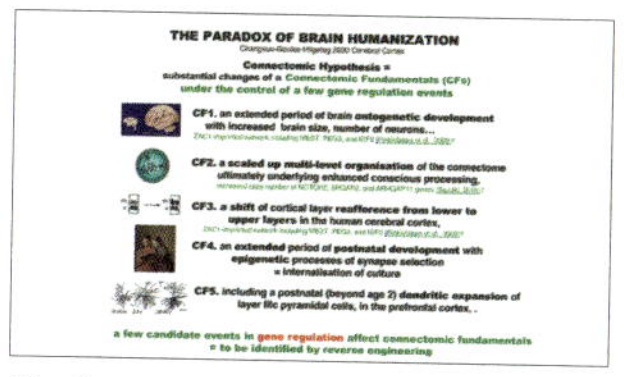

Fig. 6

special point about a fifth connectomic fundamental, which is the interesting aspect of the development of the dendritic organization of the pyramidal cells of layer 3c that you see here; especially in the prefrontal cortex, there is a special dendritic expansion that occurs beyond age 2, of which you shall see the importance. It is clear that there are a few events in gene regulation that are taking place, and some of them are in the process of being identified. In addition to that, there is this imprinting of the environment on the development of the cerebral cortex and of the brain in general, which is very characteristic of the human species.

What are the architectures of conscious processing? I will briefly mention some definitions. [7] From

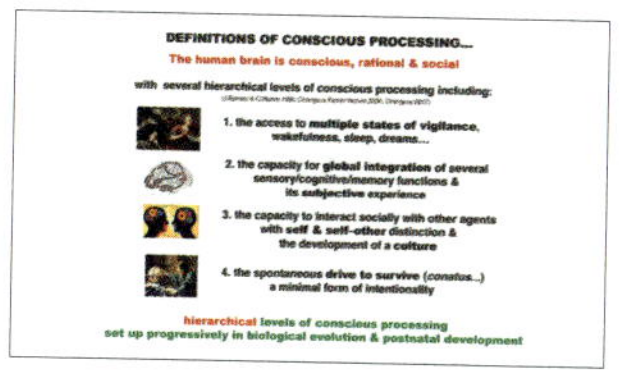

Fig. 7

the beginning I will stress that—as I said already—the human brain is conscious, rational, and social. The social aspects of the brain, in my opinion, are undervalued in most of the studies done by neuroscientists who in fact study the brain of a single individual and forget about the fact that human beings are at least as social as they are conscious. Perhaps they are even more social than they are conscious. I will distinguish a few hierarchical levels of conscious processing, introduced by Mavi before: the access to multiple states of vigilance, wakefulness, sleep, dreams; the capacity for global integration, which, as I said at the very beginning of my talk, is very important to globally integrate all the different components of the brain as a physicochemical system; the subjective experience; the capacity of social interaction with other agents, with self and other relationships; and the development of a culture. Perhaps another point, which is essential for me, comes from Spinoza: the fact that not only in humans but also in other species, where consciousness is perhaps less developed, there is a spontaneous drive to survive which was called *conatus* by Spinoza, the minimal form of intentionality. I will not discuss this aspect, but perhaps on the fifth day of this meeting it will be further discussed. There is a set of hierarchical levels of conscious processing that we have to seriously take into consideration. They set up progressively in biological evolution from simpler mammals, and in humans through postnatal development. These architectures are, at this stage, quite simple to classify. The states of wakefulness and sleep were nicely illustrated by Mavi in the previous talk, and there is a pontine sending reticular activating formation here, and then—this is from Damasio, actually—where there are cholinergic connections involved. This is a vertical control of the cortex from the brainstem. There is the thalamocortical relationship, which was studied by Llinas and Paré and colleagues, and this is again among the vertical organization of the brain and cerebral cortex. [8] I wish to insist on the fact that

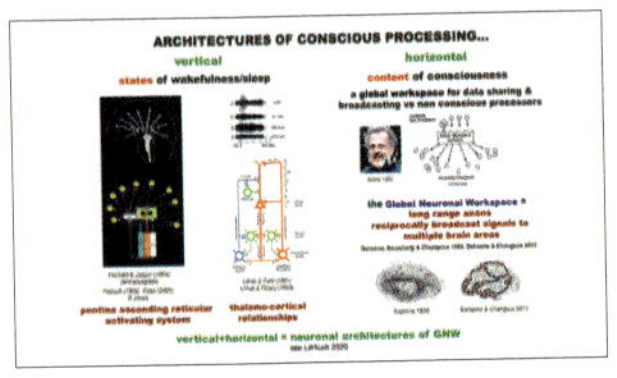

Fig. 8

there is another aspect essential for us to reach global control: the horizontal connection that takes place between cells in the cerebral cortex, between, I would say, such kind of columns or areas. The issue was expressed by Stanislas Dehaene and myself with the name of 'global neuronal workspace.' We were in fact inspired by the work of Baars, which is strictly psychological. Baars introduced precisely a global workspace where there are processors which inform and are informed by the global workspace. So there is a relationship between distributed processors and a global, physical space for data sharing and broadcasting elements which then would become conscious. If they are not broadcast, they will be non-conscious. The global neuronal workspace—I use the word 'global workspace' from Baars—needs, in addition, the most important aspect, which was not in Baars' presentation: the neuronal aspect. Then we propose that the long-range axons I already mentioned would be involved in this horizontal relationship which would combine with the vertical one, as it was proposed by Larkum recently, following these reasonings. What is this long-range connectivity? It came up from Ramón Y Cajal who discovered the existence of these long-range axons from layers 2-3. Also from Dejerine, the French anatomist who discovered all these long-range bundles of white matter studied by Goldman-Rakic, who was able to distinguish some kind of network between areas where these long range axons play a critical role. In fact, they are mostly abundant in the prefrontal cortex but also in other association areas: parietal, temporal, and cingulate context. We would make some kind of workspace in neuronal terms, a global neuronal workspace in the brain. [9] You can see here

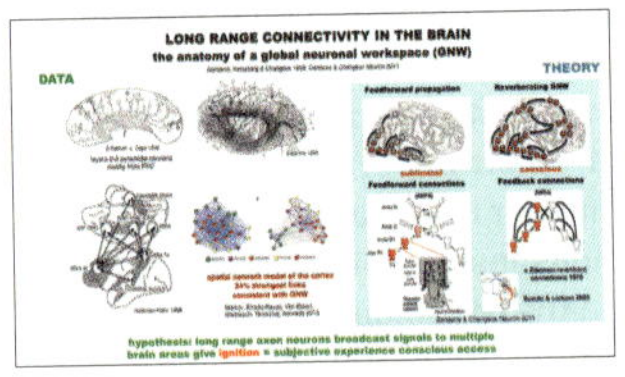

Fig. 9

interestingly, the recent work of Kennedy and Van Essen and others. They tried to look at the special network of the pyramidal cells in the cerebral cortex and found that 24% of the strongest links are precisely coincidental with what we call the global neuronal workspace. There is a strong anatomical support and the main idea that you may see illustrated here, from a theoretical and now experimental point of view, is that when sensory stimuli, for instance a visual input, enter the cerebral cortex through the primary and secondary visual cortex, if they do not reach the prefrontal cortex then they are not conscious. If they enter the prefrontal cortex—itself connected into this global neural workspace with a reverberant aspect that you can see here on the right in this diagram—then we may have access to conscious processing. Of course, the idea of a complexity of connections was mentioned by Edelman a long time ago, but not for consciousness. It was subsequently adapted by himself and by ourselves to conscious processing within the framework of the global neural workspace. I mentioned Suzuki and Larkum, and clearly illustrated here is a scheme which is exactly what we proposed 20 years ago: there is a vertical and horizontal kind of connection involved in conscious processing. The hypothesis we propose together with Stanislas is that there is some kind of particular state of activity which is involved when, for instance, sensory stimulus becomes conscious. We call it ignition, which would be the physiological activity correlated with the subjective experience of conscious aspects. This is still, of course, speculative, because we have some criteria for subjective experience of conscious success; but of course the relationship is going to be established. In this respect, I wish to mention the work of Roelfsema, which was also mentioned by Mavi, which illustrates that here you have a response, an electrophysical response. The level of v1 before and when you reach the prefrontal cortex. Then you have this ignition—this was done early on with humans—and here a distinction between this response which we relate to conscious processing in red and the non-conscious one in blue. You may realize that the time scale here is interestingly from 100 to 1000 milliseconds and then there is this lasting activity, which is essentially present in the conscious thing and to a smaller extent in the non-conscious one. This is not only true for simple visual stimuli but also an aesthetic one, and this is another talk about art. [10] Here is the work of Kringelbach who, very interestingly, showed that in the brain of infants, through a recording that you can see here, you can find

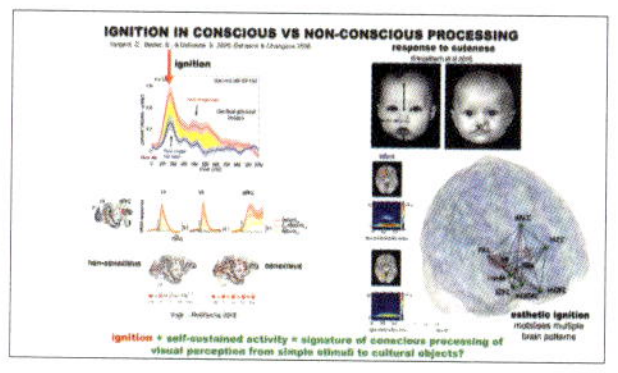

Fig. 10

an ignition that is different when the child looks at this cleft palate face, compared to which, it is more agreeable to see. I may call it an 'esthetical ignition.' This is something interesting to further investigate in the next few years. There is an activity in the cerebral cortex that Stanislas and myself consider as a signature of conscious processing of visual perception from simple stimuli to even, if I might say, a cultural object. [11] I would like to

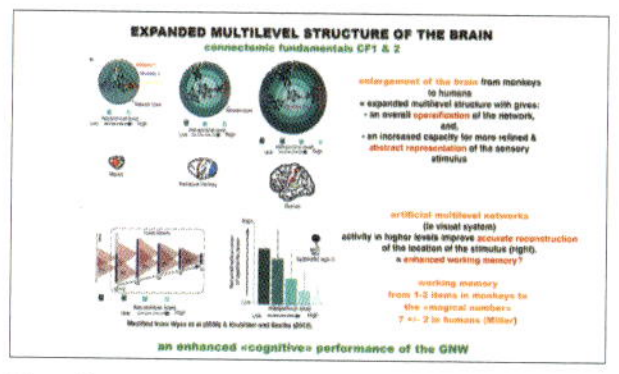

Fig. 11

continue after this presentation of the horizontal connectivity, which we consider essential for the evolution of the contents as for conscious processing, to go to some of the other connectomic fundamentals I mentioned. First, the expansion of the multi-level structure, which is interesting because it can precisely be viewed as leading to the expansion of the global neuronal workspace that you can see here, which just existed in lower animals but was much less developed. There is a dramatic increase of this connectivity, which from a theoretical point of view give as more refined and abstract representation of the sensory stimuli, and which can be demonstrated clearly by the number of errors in the reconstruction of a spatial location. You can see here at the low level that the hierarchy is very high, and at the highest level, which is possible only in this highly expanded structure, becomes even higher. You may contrast this difference between a lower and higher level of organization through the working memory, which is for one or two items in the monkey and of seven plus or minus two in humans, as everybody knows. So, there is an enhanced cognitive performance of the global neural workspace due to this connectomic fundamental, which is this expanded multi-level structure of the brain. I mentioned already the shift of the size of the soma of neurons from the inferior granular layer to the upper granular layer, and it is from the lower level to the superior globular one which indeed occurs a little in monkeys. But, as you see here, it becomes much more pronounced in the case of a human compared to monkeys in light pink, as you can see. [12] There is a very dramatic

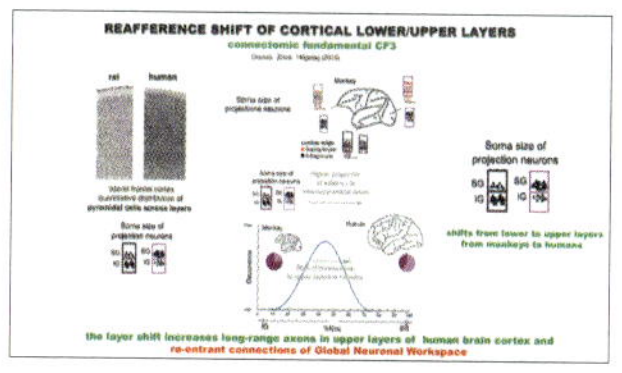

Fig. 12

shift of the connection which corresponds, according to our view, to the increase of the contribution of long-range axons in the whole cerebral cortex, in particular to the re-entrant connections of the global neural workspace. These are a few reflections I introduced with this early presentation about the connectomic fundamentals, which, as I said, may respond to a few genetic regulatory events which do

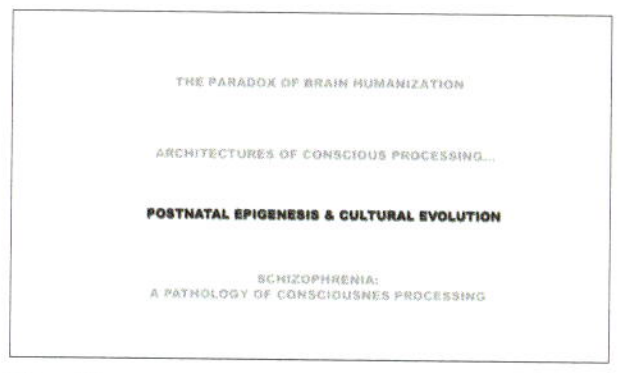

Fig. 13

not change the structure of protein so much. [13] But, for instance, the paradox of the relationship between the simplicity of the genomic evolution with the genuine complexity of the connectome is the postnatal development and cultural evolution.

The newborn brain is five times smaller in weight and in volume than the adult brain. So if you calculate how many synapses have to be built during this period from birth to adolescence you subdivide these 10 to the 14 to 15 number of synapses by the number of seconds. Amazingly, you find that, in one second, there are about 1 million synapses being formed in the brain of a newborn. If you keep your baby in your arms for five seconds, five million neurons have been made up. [14] This means that there is a

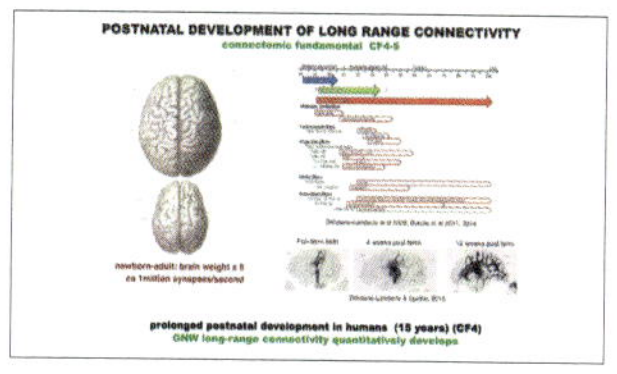

Fig. 14

dramatic increase of connection taking place after birth that the long-range connectivity will give access to consciousness. It is also increasing, as you can see here. Even though the overall pattern of the long-range connectivity does not change too much, quantitatively it is an amazing increase in the number of connections that occurs after birth. This is very important for letting the environment leave its imprint on the brain and this is just to illustrate the relationship between the increase of long-range connectivity with evolution and, of course, with the acquisition of language, as you see, from the work of Friederici. [15] The issue is how the environment is able to change the organization of this postnatal network. How is the environment able to imprint the developing network? Here is a simple scheme I proposed with Courrège and Danchin in 1973, which I think is still very true, perhaps more now than when it was proposed. The idea is

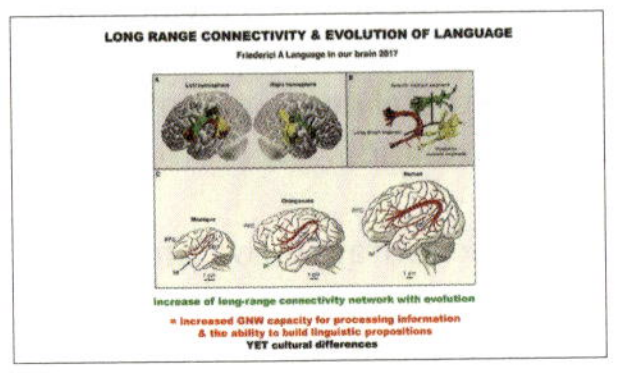

Fig. 15

that we have units within, for instance, the cerebral cortex, where you have afferent axons and target neurons, and that they are stages, which may be viewed as the critical stage or sensitive stage, where you need the neural activity to selectively stabilize the adult network. The experimental work of Hubel and Wiesel in 1963 set the stage, but at that time they thought that everything was pre-determined. In fact, as you can see here, the idea is that a synapse selection takes place during this sensitive period as well as some elimination or pruning, using the American word, which occurs during the stage where a neural activity becomes necessary for the stabilization of older networks. [16] We developed a model

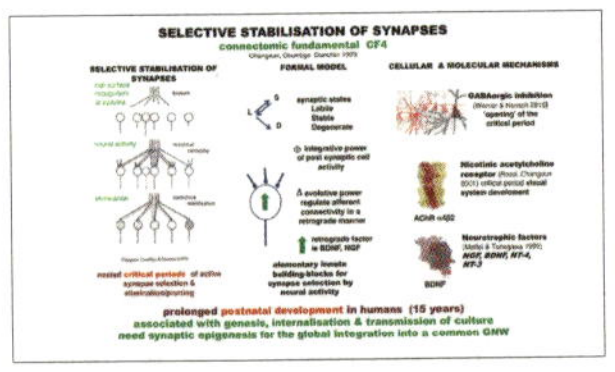

Fig. 16

where we propose synaptic states and the state of activity of the network—this is for those interested in modeling. Moreover, there is a retrograde stabilization of the presynaptic terminals and there are possible retrograde factors which control this stabilization or elimination under the state of activity of the post-synaptic cell. Indeed, there is quite a number of cellular and molecular mechanisms which have been suggested for this kind of process. For instance, a GABAergic inhibition for the opening of the critical period by Hensch, and there is evidence with Rossi and the Pisa group that nicotinic receptors may even have a critical role in the development. There are neurotrophic factors that Maffei, who heads the Pisan group with Tonegawa, so beautifully illustrated. There are indeed cellular and molecular mechanisms candidates to contribute to this selective stabilization of synapse. This is taking place at many different stages of development which are nested one with the other during the postnatal development, including the long-range connectivity of the global neuronal workspace. Just one evidence I wish to show, which is rather recent, is from the work of Ruthazer, who simply shows here—in the case of the retinal ganglion cells in the tadpole—that when you have asynchrony there is no selection, when you have here synchrony you have selection and elimination of synapses. So there are good proofs for a simple system in which this selective stabilization of synapses can be demonstrated. [17] Of course, it is

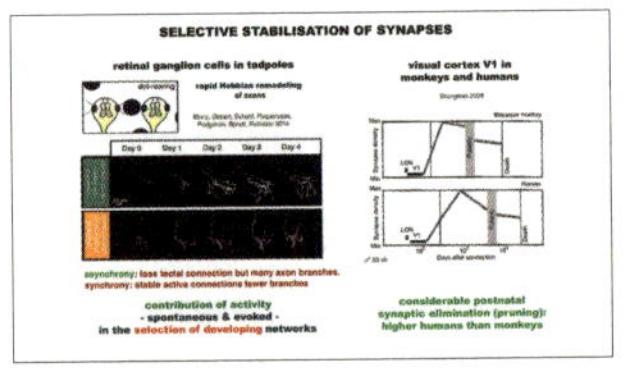

Fig. 17

more difficult to carry out this kind of experiment with humans, but I just want to show that—this is from the work of Bourgeois and Rakic—in the synaptic density during postnatal development in the macaque there is indeed some significant elimination after puberty. This is only the global envelope of all the synapses, and here you can see that in the case of humans there is a dramatic change, and in particular the data before puberty, between two and five years, and I will come back soon on that. There is indeed both an extraordinary proliferation of synapses taking place postnatally but also this elimination which becomes dominant in the late phase of this individual development.
I mentioned these five connectomic fundamentals, and the fifth one is from the work of Petanjek—who works with Rakic—which is extremely interesting. He has looked very carefully at what happens for a few months in the dendrites in the cerebral cortex following postnatal development. [18] As you can see, there is an

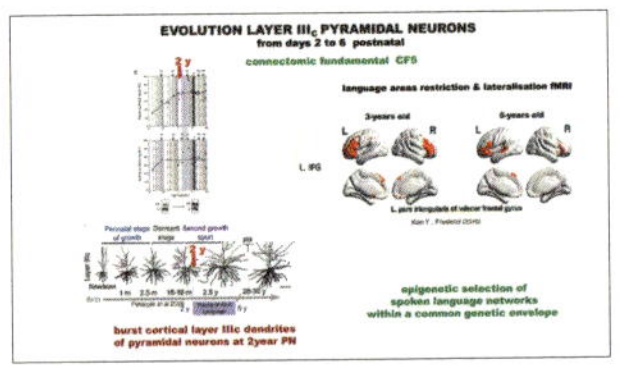

Fig. 18

additional step which is not seen in the monkey, for instance, at the age of two years and later. The possibility that we like to emphasize is that this is one important connectomic, fundamental where the language areas develop through three and five years. Of course, the workspace capacity is sufficient at two years and later to organize sentences in the language and reach higher levels of language development, which needs an important contribution of conscious processing. [19] This effect of the environ-

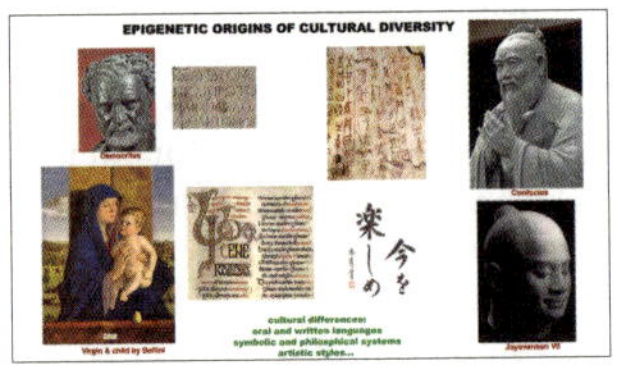

Fig. 19

ment on the organization of the brain is of course highly valuable and therefore we may consider that this epigenetic variability of the brain is at the origin of cultural diversity in the spoken language, in written language, in beliefs, even in some ethical rules—not all of them—, within a genetic envelope that is common to the human species. This is an interesting balance between this genetic envelope that I try to describe with the connectomic fundamentals and the very few genetic events that regulate this evolution and the epigenetical development of this diversity, which takes place not only for language but for the symbolic and philosophical system and, of course, artistic styles. I would just like to make a last point before I finish. [20]

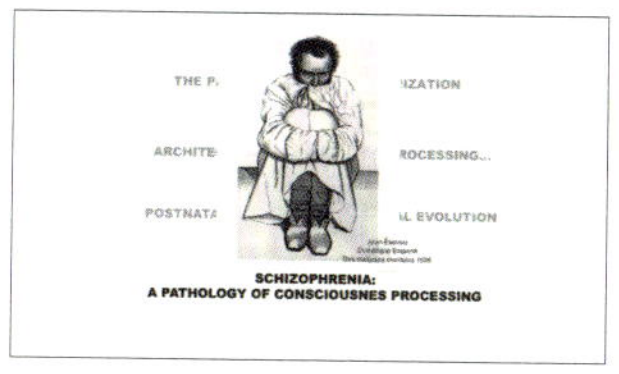

Fig. 20

Only one aspect of pathology: there was the issue of degenerative diseases, an essential problem we have to deal with for the future of humanity. I just want to select one case which is quite interesting to us: the relationship between schizophrenia and the global neuronal workspace. This is from Martinot together with Dehaene. They made some brain imaging between control and patients in the prefrontal cortex. There is a very significant reduced activation of the anterior cingulate and frontal cortex in schizophrenics versus controls. [21]

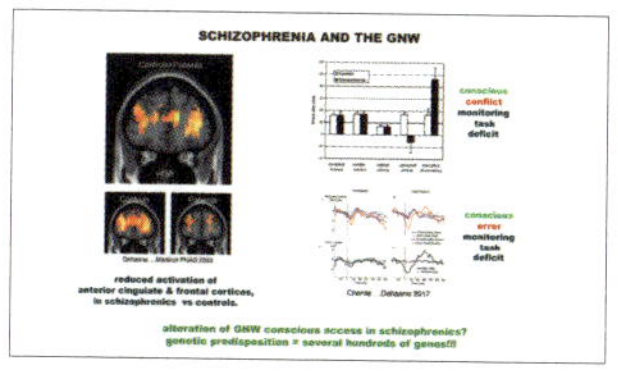

Fig. 21

If you look at monitoring cognitive task developed also by Stanislas in his laboratory, then you have conflict processing monitoring tasks, conscious errors and so on, as well as a few complex perception tasks, which are clearly altered in schizophrenics. When you look at the predisposition genes you are faced with a problem, because there is no single gene proper to schizophrenia: there are hundreds of genes. Some of them are hit by mutation and give a schizophrenic individual. What I want to say is that among these genetic predispositions—due essentially to simple genetic variation within the structural genes of some gene of the human genome in particular—it was found that the nicotinic receptors subunit genes alpha 5 and beta 4 genes appear as some genetic predisposition to schizophrenia. As you may know, we have been working for now almost half a century on this nicotinic receptor. Some of the snips in this gene, and in particular in the alpha 5 genes, increase the risk of schizophrenia. This is not the explanation of all schizophrenic patients and phenotypes, it is a population of them. This was essentially the work of Uwe Maskos and Fani Koukouli at the Pasteur Institute. They created a modified mouse with a snip of the alpha 5 subunit which in humans causes schizophrenia. Very interestingly, they show that in the cortex and layers 2-3 again there were some alterations of the nicotinic receptor not on the pyramidal cell—surprisingly, there are very few nicotine receptors on the cell—but on the inhibitory cells. This means that they contribute to what might be mentioned as the synchrony and relationship between the sets of cells and, in particular, the pyramidal cells. That is the scheme that was published two years ago. Here you can see that we feel that there is some kind of hierarchy of interneurons, which themselves regulate the state of activity of the pyramid cells that in turn give the long-range axons forming the global neuronal workspace. [22] And, in agreement

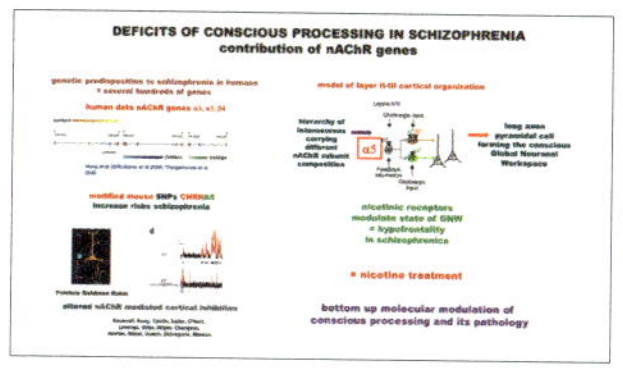

Fig. 22

with what I just showed in the previous slide, that creates some hypofrontality, some decrease of the activity of the prefrontal cortex, like in schizophrenics. More interestingly, nicotine treatments have some effects on the mouse and, as you know, in human patients there is very often chain-smoking, as if the schizophrenic patients were self-medicating by smoking cigarettes. This just to illustrate that we started with the top processing of orientation of conscious access and finally can reach the molecular level: with a particular cellular distribution of nicotinic receptors in the pyramidal cells and inhibitory neurons regulating pyramid cells in the cerebral cortex. We can conclude that here a bottom-up molecular modulation of conscious processing takes place through the nicotine receptor and its pathology.

I would like to close with one slide inspired by the discussion I had with Paul Ricoeur, a moral philosopher who unfortunately left us a few years ago and wrote a very interesting book, *Oneself as Another*, in which he discussed something which I did not have the time to even mention: the social relationship between individuals and, in particular, the way we can recognize other minds. I just mentioned the theory of mind, like a sentenced organization in the newborn. This led us to discuss the variability between individuals in the human species, which is of great importance. There is variability at the gene level and it still exists in the human population. The work is by Svante Pääbo, extremely interesting in this respect and in particular for the remnants of the Neanderthal genome in the *Homo sapiens* genome and the distribution of these diverse genomes at the geographical level. There is clearly an individual variability of the connection in the case of what I refer to as the selective stabilization of synapse. This is only a minimal scheme of what happens at many places in the same architecture in development. There is a variability which is essentially a socio-cultural viability associated with this selective stabilization which is not fully reversible. If I have a bad French accent in English is simply because of that. [23] Now there is

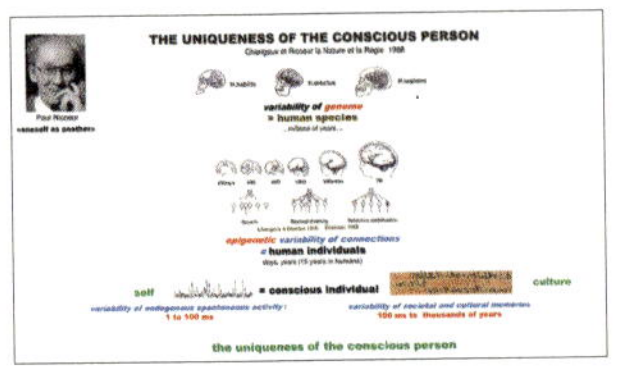

Fig. 23

also the variability of the conscious individual taking care of, as Mavi mentioned at the end of her talk, the self, oneself and all the social and cultural memories of the cultural environment which give the uniqueness of the conscious person.

This is exactly what we discussed with Paul Ricoeur a few years ago. Just to give the names of those who contributed to the work: Claus and Alexander to the connectomic hypothesis; Stanislas, with whom we worked more than 20 years in a very enthusiastic collaboration; Marina Picciotto, who discovered the first mutation in the nicotine receptor in the mouse and its consequence; and Sylvie Granon, Philippe Faure, Uwe Maskos for studies about the behavior of mice, which are a very poor model for conscious processing but at least give us some kind of basic information. So thank you very much and my question is: do you like my brain? Maybe not.

Panel Discussion

GIANCARLO COMI
MAVI SANCHEZ-VIVES
JEAN-PIERRE CHANGEUX

GIANCARLO COMI One point that has been touched by both professor Changeux and professor Sanchez-Vives was the importance of connections. I may start because I have received really many questions. I will start with Mavi Sanchez-Vives. One question that has arrived is the following: Are the not detectable stimuli affecting in any ways our consciousness even if we are not aware about it, if they are subliminal? Do they influence in some way our behavior? And also what about hallucinations?

MAVI SANCHEZ-VIVES There are stimuli that we have seen, I think we both have presented this schematic of the brain where there are some stimuli that stay in the primary areas without being able to activate the full recurrent circuits, activating the whole brain. It is my point of view but I would like professor Changeux, an expert on this, who generated the theory of the global neuronal workspace. From my perspective that I have worked and recorded from these primary core thesis, as far as there is an activation in these primary sensory areas and there is some response that may have some propagation even if it doesn't reach the frontal areas, it is going to have some influence in the circuits. To what extent this is going to affect behavior it depends. Behavior implies activation of motor circuits and so on, but there it is going to enter the activity in the cortex and it is going to have some influence which can be local.

GC Jean-Pierre, what is your view about it?

JEAN-PIERRE CHANGEUX Clearly the preceding experience of the individual, for instance during development, introduces stresses into the brain which in the adult are non-conscious, in the sense that when you speak a given language you don't even think that you are speaking French or English or whatever. Therefore, there is a deep influence of what is built in the connections of the brain during development and also, in a more transient way, through changes of synaptic efficacies. What I refer to as non-conscious stimuli may have a strong influence on your conscious decisions.

GC You mentioned how relevant or fundamental is the process of ignition, the process that in some way is able to activate in their workspace a specific approach and determining a bit the isolation of what is coming to our consciousness at each time point. The question is: which are the forces, the strength that move ignition? How is ignition activated? Because, as far as I understood, ignition is the one that activates the workspace at a given time, on a given topic. But what is determining the ignition process itself? What is determining this aspect?

JPC This was mentioned by Dehaene & myself in 2005. It is due to the recurrence processing I mentioned in my talk. You have a propagation which is not modified greatly by the visual area, primary or secondary. Then, when you reach the prefrontal cortex, there is an amplification of the signal which gives the ignition, as we said. It is not really all or none, but largely a cooperative effect of the state of activity which then is broadcast into the workspace. This is the definition of global neuronal workspace. The energy, there is nothing added to it, is intrinsic in the circuits. When you reach this state of organization of the circuits and you go far enough beyond the threshold, then you have this explosive propagation that is still under control and is, as we believe, a kind of manifestation of an activity characteristic of the representations that are conscious versus those which are non-conscious.

GC Professor Sanchez-Vives, could virtual reality also be used to increase empathy towards a different category of people, such as ethnic minorities, elders, disabled people?

MSV Yes. This is a very interesting area. As I mentioned during the talk, something that we can do in virtual reality, which is very powerful, is to embody and to feel that our body is that of another person. This allows us something that in the real world we can only do through our imagination. Some people are better than others at that. But the good thing in virtual reality is that without imagination you can just be in the shoes or in the perspective of another person, which is virtual, that you can feel embodied into that person and then you can live through an experience. So you can have a virtual experience, obviously, and experience it from the perspective of another person. In our group, and also in other groups, there have been studies doing this, for example by going into the body of someone of a different race. It has been demonstrated and measured in different ways that there is some identification, which is very automatic. It is not rational, it is not something that one reaches by thinking about it, but it is like an automatic process, an implicit learning of the perspective of this other person. There is an increase of empathy and so there is a change in attitude that can be measured, a change in behavior. This can be done and it is being explored by different people in order to be used to increase empathy and to understand the point of view on the experience of others which can be, as it was mentioned in the question, people of a different age, people of different groups and so on.

GC Professor Changeux, you mentioned that the horizontal transmission, what you call also data sharing in your global workspace hypotheses, is a fundamental aspect. If you make available to all the areas of the brain in some way and you share the information you really have the possibility even to produce the conscious state. The question is: is this telling us also that we should use the same strategy in society, that is to share knowledge as it happens in our brain?

JPC Of course we should. The past history of humanity tells us, as shown in particular with science, that we can not only share information between individuals, between scientists who make science, but also between the scientists and the general population through technologies such that the population may benefit from scientific knowledge. Consequently, there is not only this information sharing but also an evolution of it, which may be discussed in the section about culture, because culture evolves and, in particular, science evolves and makes constant progress. The issue is now whether the technology that we are developing is going to contribute to the good of humanity or not. This is a very important issue and it reminds me of Pasteur, who asked us to do science for the good of humanity. It is sometimes difficult to know what is good but it should certainly not be used to kill, to increase poverty of people, and so forth. We have to be careful about what we do with this knowledge, which has been developed creatively for hundreds of years. In the next few hundred years, what we do with knowledge will be a critical issue, in particular for the environment but also for the relationship between people in the world. Trying to have a more peaceful relationship between ourselves.

GC Another very key question is directed to both Jean-Pierre and Mavi: is brain plasticity a crucial element in favor of an anti-deterministic and anti-reductionist vision? Or, viceversa, can the concept of brain plasticity and a deterministic conception of reality coexist?

MSV I am not sure I completely understand the question, but in brain plasticity I don't see anything deterministic. Brain plasticity allows us precisely to learn, therefore to change over time, to learn to make changes. Therefore, that plasticity allows to escape, to some extent, from what could be more deterministic, which is the genetics that one is born with. There is also the epigenetics we have heard about in the previous talk. That is what I think, that precisely plasticity, which is triggered by external stimulation, by experiences, by our learning from life experience, allows us to adapt and allows us to change.

JPC I agree with what Mavi said. But this is based on the view which claims that everything is determined by genes in the brain. In contrast, there is the empiricist view—like in the case of Hume, Locke, and Aristotle—which claims that everything is learned in the organization of the brain. The brain is a white sheet of paper on which experience writes the organization of the brain. I think both of them are extreme. We are actually in between, and as I said in my last slide, we have variability at the gene level and we have also variability due to synaptic plasticity. The question is the word deterministic. There is, to my opinion, determinism in the relationship between the environment and the trace that the environment leaves in the brain. We may say, like Spinoza, that we feel free because we are ignorant of the causes that determine us. But what I would like to emphasize is that one of the main features of the human brain is precisely its variability between individuals from different cultures but also within a given individual as a function of the states of the brain. That is where reasoning and creativity open the brain up to variability, to innovation, and also to *conatus*, the ability to survive, which is characteristic of a living organism. There might be some determinism for each of these individual aspects, but if you put them together then it looks not theoretically but practically very difficult to distinguish what determines the character of a particular child. I would even say perhaps not feasible. That is why I make a strong difference between artificial intelligence and human robots, even between human robots and, in general, human beings. Robots built up to now have no culture or experience created by themselves, as opposed to humans who, for 15 years, have a lot of experience with society, with culture, with the environment and so on. It is such that under these circumstances they build up their own personality. I would not say that a robot is a person for the reason which I explained at the end of my talk.

GC Jean-Pierre, I want to close this fantastic discussion with something a bit different. The last question is for you: you curated some major exhibition on art and science and you are a member of the scientific council of the international agency of museum since 2007. What is the role of a museum today from your point of view? What does it mean to be a cultural institution today? We have something that moves in the direction of culture.

JPC There is a statement—I don't know who said it—"People without memory are people without a future" (*Les peuples sans mémoires sont des peuples sans avenir*). For me a museum is a store of culture and it plays an absolutely critical role in the creativity and I would even say in the survival of human beings. This is something I may insist on, because there is a tendency now to go to a cultural institution where you have only the last up-to-date views of a singer, of art and so on and so forth, and forget about Leonardo, Raphael, Michelangelo—all Italian, actually—, Poussin or Cézanne. I think the experience of children of art from different historical periods and different cultures, of our society is to me necessary for the education and the harmonious organization of their brain. I may close by saying that there are cultural differences between civilizations, and I mentioned and gave an illustration of Wwestern art compared to Eastern art just as an example. But I could take many other kinds of examples. Which one is the best known? All of them are important. I think it would be extremely useful to have a deeper understanding, through education, of this diversity of culture that we find in our museums, so that people mjight have a positive and constructive relationship for the future of humanity instead of having clashes between different cultures. I do not think it is very likely that this will happen but I hope that this kind of relationship between the arts of different cultures will be a factor of reconciliation between human beings, something that we need a lot of at present.

2

Functional and Structural Neuroimaging

EVE MARDER
ANTONIO DAMASIO

PANEL DISCUSSION MODERATED BY
DANIELA PERANI

Individual Differences and Differential Resilience

EVE MARDER

As you just heard, I am unambiguously a basic neuroscientist and I work at the level of small circuits. So what I tried to do in putting together this talk was to find some of the lessons from our work that I think are generally applicable to human brains and then to try and show the importance of some of these ideas. Before starting, I would like to thank all the people with whom I have worked over the years, and say that everything I have ever done benefits from the brilliance, creativity, and energy of all the people who are in my lab and with whom I have collaborated. [1]

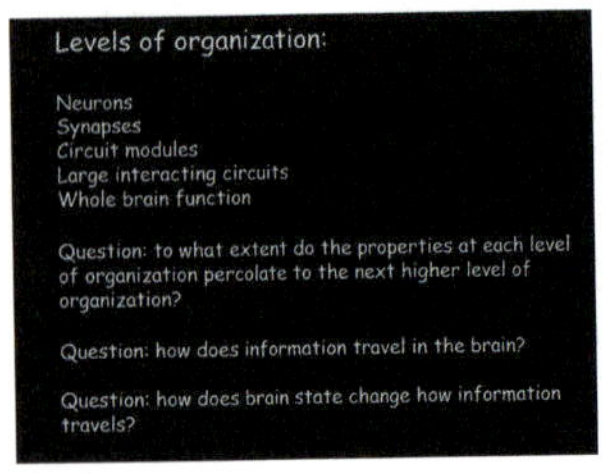

Fig. 1

Yesterday you heard a little bit about different levels of organization in the brain. We have individual neurons, and they have their individual electrophysiological properties, that is to say, their activities. We have synapses, and those are the structures that connect neurons and make circuits. Synapses come in all different kinds of flavors and are quite different from each other, and the characteristics of those synapses are also very important. Then we have circuit modules. Cells are often put together into small circuit modules with characteristic properties. One of the things that people like me do is study the properties of small circuits that could be thought of as circuit modules in larger brains. Then you have large interacting circuits and then whole brain function and there are some puzzling questions that I think are fundamental to understanding large brains. Question one: to what extent do the properties of each level of organization percolate to the next higher level of organization? To phrase it in another way: to what extent do the specifics or the details of how individual neurons work, or how individual synapses work, shape the properties at the next level of organization? And then to what extent does the next level organization, and there those interactions shape the properties of the final level of organization? Then another really important question that I am sure you'll hear more about later: how does information travel in the brain? Sometimes information may stay quite local to only one region of the brain and sometimes information travels throughout the brain. What governs how it travels? How does brain state change how information travels? These are questions that we have studied, in one way or another, using small circuits but that are obviously deeply relevant to understanding large brains, including human brains. The first connectomes, or wiring diagrams, for small circuits came in the 1980s. Connectomes, or wiring diagrams, tell us the potential routes by which neurons and networks are connected. Basically they show us who talks to whom in the brain. When we obtained the first sets of wiring diagrams, a lot of people were very disappointed because they thought that, as soon as they had those wiring diagrams, it would be revealed or would be obvious how the brain actually works or at least how that circuit worked. It was only after we had those first connectomes that we realized that the connectome or the wiring diagram was an absolutely necessary beginning first step to understand how to go from a wiring diagram to dynamics. I think it is very important to remember the lessons we learned 40 years ago in small circuits, as today the Brain Initiative and other initiatives around the world have massive efforts to establish detailed connectomes for mouse and human brains. These are essential for understanding how complex human brain dynamics arise, including consciousness, and how they emerge from biological processes and mechanisms. But again, the connectome, the wiring diagram is completely essential and completely insufficient to really understand how dynamics come from the static picture that a connectome, especially one at the at the EM level—the electronmicroscope level—provides. I think the challenge is to understand where the dynamics come from, but you have to understand the dynamics in the context of the circuit or the connectome. I am going to start going back a little bit to single neurons, because first of all I care deeply about them. All of the neurons in your brain have certain properties. They are quite different from each other, but they have some very salient properties whether you study them in a crab, a fly, a *C. elegans*, a mouse or a human. [2]

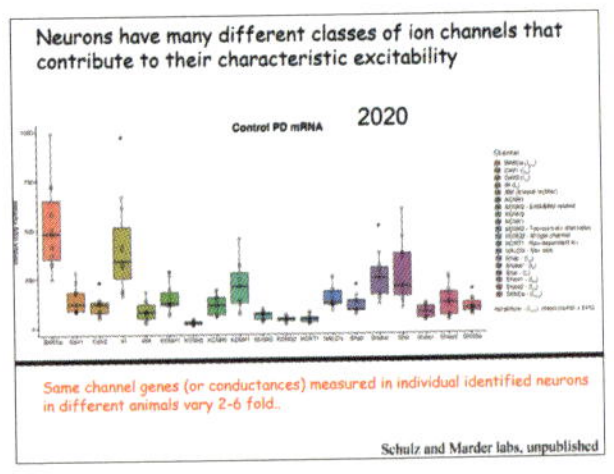

Fig. 2

One of those properties is illustrated here: these are data coming from single cell PCR measurements—these measure the messenger RNA levels for certain ion channels. In these experiments individual neurons were dissected from a ganglion (a group of nerve cells) in a crab, and then over 10 or 12 experiments their data are shown here. Down here, on the x-axis, you see the names of lots of different ion channel genes. I think we only have listed 20 or 22 of them but I am showing you this to make the point that every neuron we know about has many different kinds of ion channel genes, and one of the really interesting puzzles going forward is the kind of complexity that we see at the molecular level. How does this translate into differences at the electrophysiological level, that is, the activity of those cells? And what role does this complexity play in understanding how circuits work, and then how whole brains work? I would like to make one more point; with this slide you notice this dispersion in the values: these are single values. In our data and in everybody else's data, when you measure any parameter that you can reliably measure over and over again in multiple examples or multiple individuals, you see about a two to six fold range in those values. What that tells you is that the parameters that are being used to both govern either the activity of single neurons or how they are connected synaptically, varies in that range. That is a very interesting range because it tells you what the degrees of freedom are, that is, variability that is consistent with normal, appropriate biological activity.

We often build computational models to describe the behavior of neurons. I am going to show you some of them. [3] This is a caricature

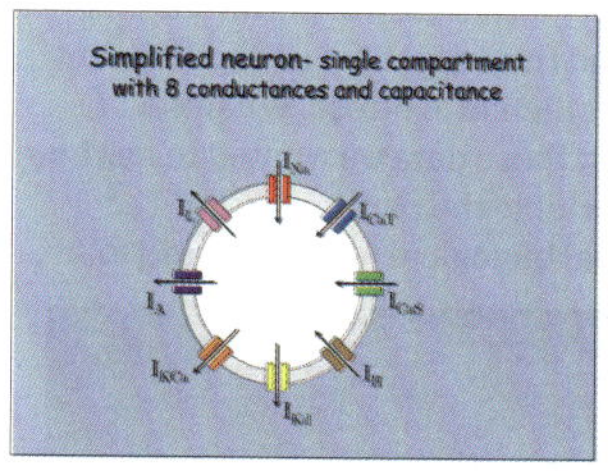

Fig. 3

of some of these models and I am going to use them to remind you that neurons have many different kinds of ion channels. Only eight of them are shown here of the 22 or the 25 that many cells might have. Commonly there is an inward sodium current that allows sodium to enter the cell, and that is important for the upswing or the depolarization of action potentials. There are multiple calcium channels, that allow calcium to enter the cell and are very important for signaling. In our models we include a current called i_H: it is a hyperpolarization activated inward current, and I think Mavi mentioned it yesterday, or she has worked on it. There are three different potassium channels. They differ from each other in terms of their voltage and time dependence. The challenge is, when we specify or measure the properties of the cell, of an individual cell, we measure the number of each one of these and other ion channels and we measure the way in which the opening and closing of these channels depends on voltage and time. When you put all that back together you get something that reflects the dynamics of the activity of the cell. [4]

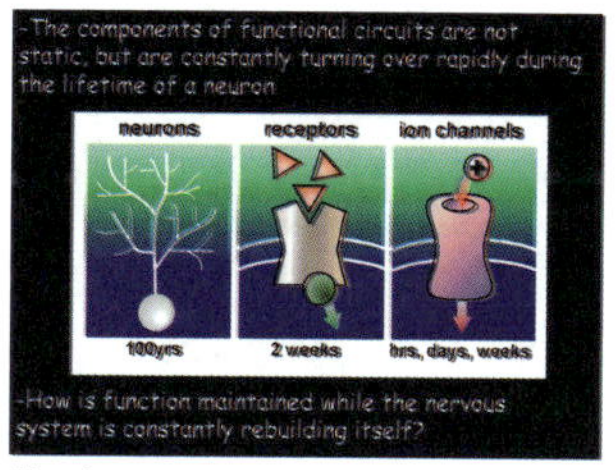

Fig. 4

I just wanted to show you that to segue into one of the really fundamental principles that was alluded to before, and that is that the components of functional circuits are not static, but are constantly turning over rapidly during the lifetime of a neuron. All of you are going to live for 100 years, which means that almost all of your neurons are going to live for 100 years. They were born early in development and they are going to hang out in your nervous system, in your brain, for all of that time. Nonetheless, the receptors that are important for synaptic transmission, and the ion channels that are important for signaling, turn over in the membrane in either weeks, hours or days. What that tells you is that all nervous systems, especially nervous systems of animals like humans, who live a long time, have a challenge of maintaining function while the nervous system is constantly rebuilding itself. If you think about this, it tells you that the brain—and I will say this probably several times—has the challenge of replacing every single component that is important for neuronal signaling while it is working, and no engineer would want to ever have to repair anything that was broken while a machine was working, nobody would want to repair a 747 while it was in flight. That is exactly what the brain is doing, so there have to be very beautiful and important mechanisms that allow the nervous system to constantly rebuild itself.

We, and many other people, have studied homeostatic models and mechanisms in the models I'll show you in a second. [5] We say that

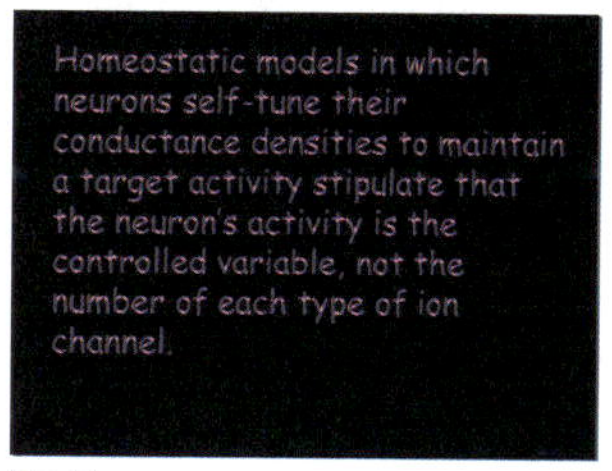

Fig. 5

neurons self-tune their conductances, that is to say the number of their ion channels, to maintain a target activity pattern. These models stipulate that it is the neurons activity that is controlled, not the number of each type of ion channel. This is important because it tells you that every neuron probably has mechanisms that allow it to monitor its own activity and then to use that in a feedback mechanism to recreate itself. When it recreates itself it doesn't recreate itself identically constantly, but it is wandering around within a desired set of activity patterns. [6] I am going to show you just one slide showing these mechanisms in operation. This is an example of what we call a bursting neuron: it is depolarizing and firing several—here three—action potentials and then it is stopping and then firing another set of action potentials. This is a desired pattern of activity, in many cases.

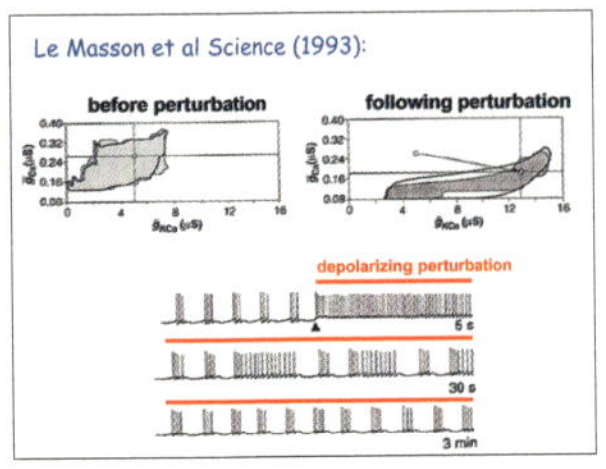

Fig. 6

Then we depolarized it: that is knocking it out of its desired state and then, slowly over time, the neuron is rebuilding itself back to its activity state. The way we think this works, is that neurons monitor their own activity using a monitor like an intracellular calcium concentration. They have a target activity and in response to being away from the target they actually insert or delete ion channels in the membrane to bring themselves back to this target activity. Notice this activity that recovers is similar to the starting activity, although the depolarization perturbation is still present, so this is a homeostatic response that allows the cell to maintain its activity despite the perturbation. These same rules can become important in understanding development and these models can self-assemble. [7]

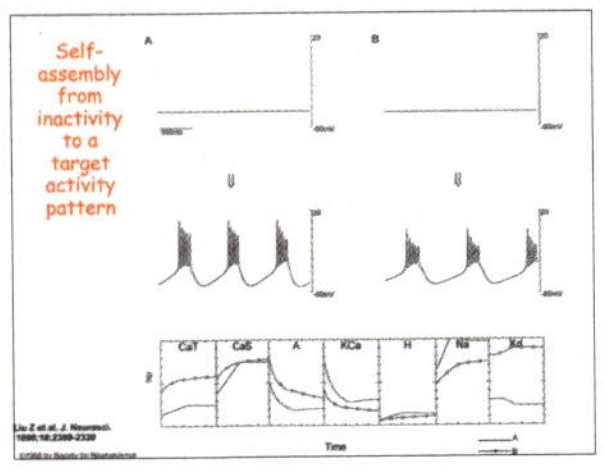

Fig. 7

This is an old slide here: you have two model neurons, they are both silent, and as they sense the fact that they are not at their target activity they build themselves into bursters. These are the conductances values that they started out with, and then over time each of the conductances changes—this is in model one, model two. Notice they all change, that every single conductance is changing, but the activity now has reached a preferred target level. This shows that you can self-assemble from inactivity to a target activity level. As I already said, brains are not stationary but they are constantly rebuilding themselves, replacing every component continuously. This raises the conundrum: how does brain function, memory and, for this meeting, sense of self, persist when all components are in a constant state of flux? I think this is one of the deepest problems in neuroscience, to really understand the rules by which this replacement is done. How do you have stability of function despite the necessity to replace all components?

I am going to move from single neurons to small circuits to show you another set of observations. [8] This

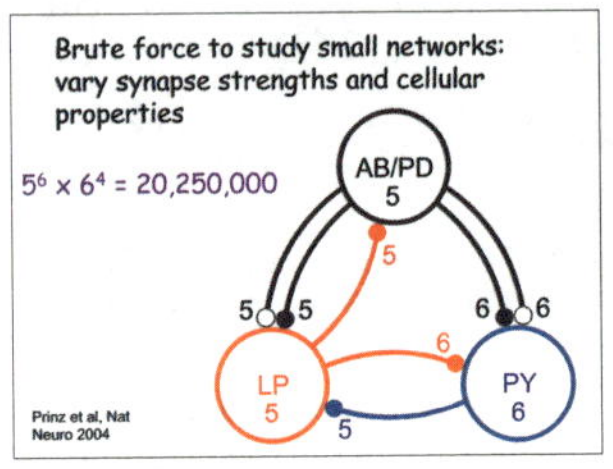

Fig. 8

goes to the question of individual variability. Astrid Prinz was a postdoc in our laboratory, and she wanted to build a model of a three-cell circuit that would be similar to one of the circuits that we study. She chose five or six cells of each cell type, five or six values for each synaptic connection. She simulated 20 million model networks, and then she selected the ones that had her preferred activity. Of them she found 2.4% had a preferred activity pattern that matched the activity pattern in our biological recordings. [9] Here are two model networks and they are producing very similar behavior. Now, when we look at the underlying parameters that went into building model network one and model network two, you realize that the underlying parameters are

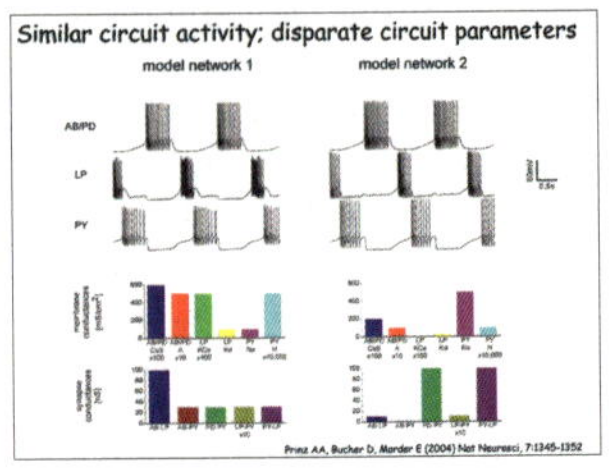

Fig. 9

completely different. This particular parameter, which is the h current, in one of the neurons is big here, it is small here. The synapse from PY to LP is big here, small here. This set of parameters and this set of parameters are totally different, but the behavior of the circuit is almost identical. When you look at this you say two things: you could say "Now I understand why you could have two cats or two humans walking down the street with a tremendous amount of underlying variability in the values of the parameters that give rise to their brain behavior, but they can have very similar behavior." But this also raises another question and that immediately says that this set of parameters and this set of parameters would likely cause this animal or this animal to respond very differently to perturbations, and they might have very different resilience to stress. When we got those data we wanted to ask how variable the parameters in real biological networks are. [10]

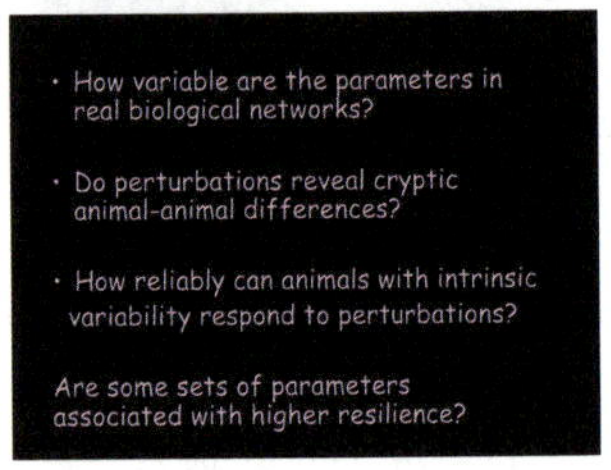

Fig. 10

I already showed you that the answer to that is about two to six fold in every parameter that we have measured. Then the question is: do perturbations reveal cryptic animal-animal differences? For, example in terms of humans, are there cryptic human to human differences that only become revealed under stress or under environmental perturbations? The next question is obviously: how reliably can animals with intrinsic variability respond to perturbations? This is a question we have as of yet no satisfactory answer for, but it is really crucial to understanding humans: are some sets of parameters associated with higher resilience? What determines why some people are exposed to Covid and get sick and others are exposed to Covid but don't get sick, etc? [11]

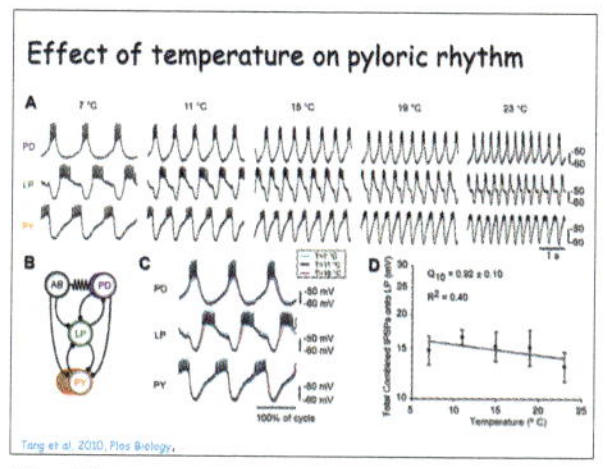

Fig. 11

To answer this question about resilience in response to stress we wanted to look at a stress that the animals we work on would normally see, which is temperature. They live in the ocean, they see temperatures from about two degrees centigrade to about 25 degrees centigrade. We are looking at recordings of that rhythm. This is the connectome, or the wiring diagram, and now you are seeing intracellular recordings of the membrane potential of each of the three neurons at 7, 11, 15, 19, and 23, degrees. The pattern is the same, but the frequency is changing. That is what you would expect. You can see the patterns is very well maintained. That is true for the temperatures that these animals usually see. All animals, even though they have that two to six-fold variability in their underlying structures, respond in this temperature range completely normally. You would never know that those differences were there. [12] But if you stress them further, if you go from 23 degrees to high temperature, 31 degrees, you start seeing crashes, or loss of

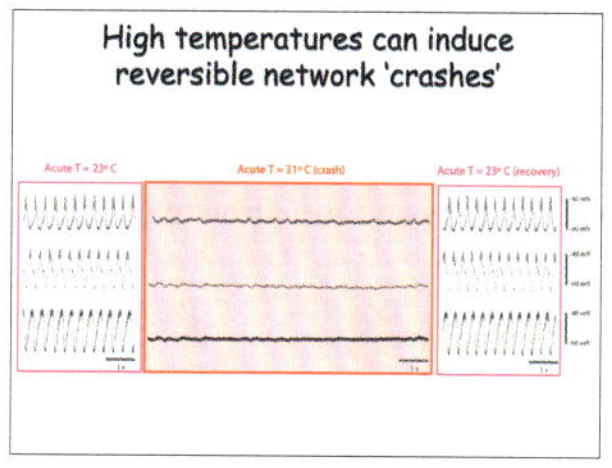

Fig. 12

activity. [13] But more interesting than that, after you look at temperatures between complete crash and normal activity—and these are now recordings of four different animals, four different people, if you will—notice that the patterns that you get during these crashes, or loss of normal activity, are completely different. Our interpretation of this is that the stress is revealing the effects of those underlying or cryptic differences in parameters that you don't see until you really stress the system. [14] To summarize, at low temperatures all networks are well behaved, showing consistent changes in frequency and outstanding temperature compensation, but—and this is really important as we come thinking about humans—each individual animal crashes with individual dynamics, as predicted by their underlying parameter differences. Obviously there can be classes of crashes that multiple animals show, but fundamentally you would expect if each human is different at the end, if you then had a good enough assay for the behavior, you would see that they all, under extreme perturbation, showed different kinds of crashes. This just says what I showed you: individuals with similar starting circuit performance, with very similar circuit mechanisms under control mechanisms can nonetheless show entirely different crash mechanisms. This is really a cautionary tale for thinking about disease, because depending on what those starting states are, you might expect that pharmacology or other therapeutics might be differently useful depending on the underlying mechanisms of those crashes. This is probably why it is so difficult to get good therapeutics for disorders like epilepsy, depression, and schizophrenia. [15] Differential resilience in the population, just to repeat, is a necessary feature that comes from the inherent degeneracy of multiple solutions.

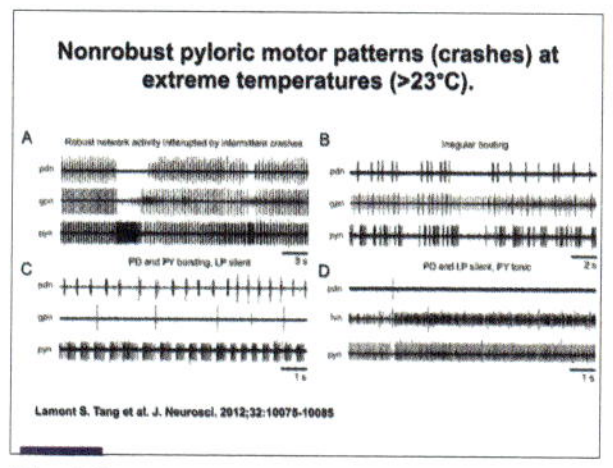

Fig. 13

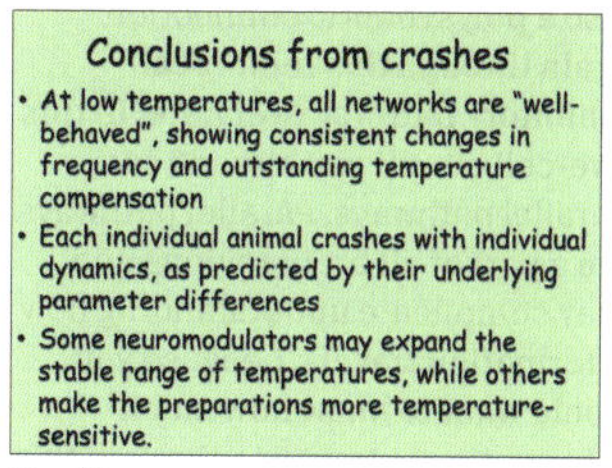

Fig. 14

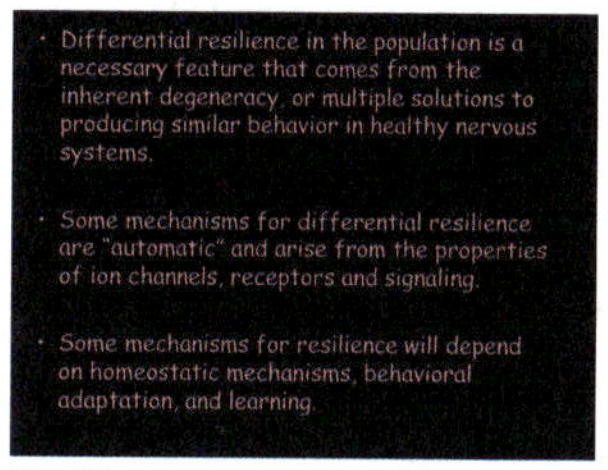

Fig. 15

Some mechanisms for differential resilience are automatic. Some mechanisms for resilience will depend on homeostatic mechanisms, behavioral adaptation and learning. I want to show you one additional set of points that have to deal with what I call parallel pathways and neuromodulation in connectomes. To do that I am going to show you a more complicated set of recordings. [16] These look a lot like some of the recordings you saw yesterday from complex brains. These are recordings that come

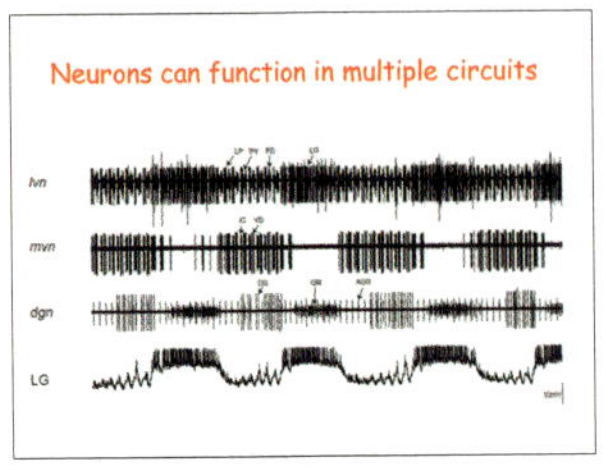

Fig. 16

from a crab, but nonetheless show fairly complex dynamics. In particular, there is a fast rhythm that is seen in the top traces, and a slow rhythm that is seen in the bottom traces, but what is most interesting is there are neurons in the middle traces that are that are firing as part of the fast rhythm, but are also firing in an envelope of activity and time with the slow rhythm. These are neurons that are part of two different circuits, and they can flip back and forth between participating in these different circuits. I am showing you these data just to make the point that neurons can combine into different ensembles under different circumstances, as a consequence of experience, or neuromodulation, or a variety of other things. [17]

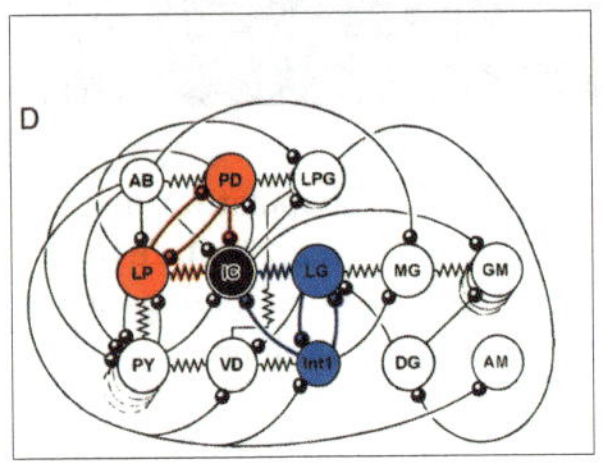

Fig. 17

This is the full connectome of the ganglion we study, and you can see it is complicated. There are only 26 cells in this crab ganglion and it is already fairly complicated. If I asked you to predict what this connectome would do, you would be unable to do so without considerably more Information. In these diagrams the resistor symbol means electrical coupling—those are electrical synapses—and all these filled circles are chemical inhibitory synapses. We haven't talked much about inhibition but in this network all of the timing comes from inhibition. [18] The reason I am showing you

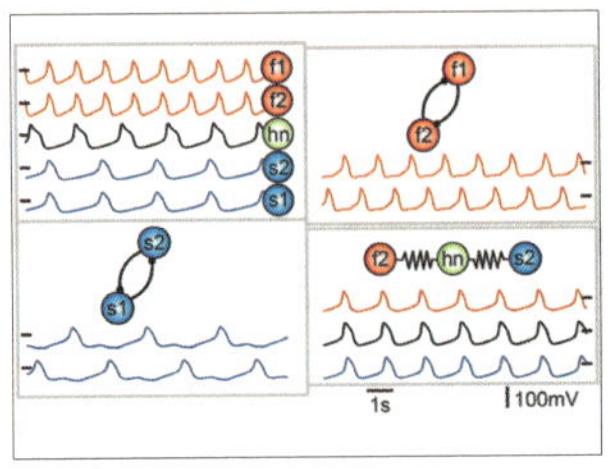

Fig. 18

this connectome is to show you that at the core of this connectome is what I call parallel pathways: here there is a little circuit module called reciprocal inhibition. We sometimes call that a half center oscillator. These two neurons fire in alternation, out of phase with each other: when one is active, the other one is inhibited and vice versa. Notice that this neuron inhibits the IC neuron monosynaptically, it makes a direct connection. But it also can inhibit the LP neuron through this electrical junction, indirectly. This is what I call a parallel pathway: there are two ways for information to travel between neuron 1 and neuron 2, either directly or indirectly. Likewise, interneuron 1 makes a monosynaptic connection to IC and also a polysynaptic connection again through this reciprocal inhibitory pathway over here. In this five-cell circuit you already see parallel pathways. Parallel pathways are a part of all brain regions and they complicate understanding how information travels, because you would want to know how much information is traveling through this part of the pathway, how much is coming through this way, and to what extent does that change what the circuit will do. To address this I am going to show you a model that Gabrielle Gutierrez has built. She decided to build a model of this circuit. So we have a fast oscillator here, a slow oscillator here, and a hub neuron. Each neuron is modeled as an oscillator, two fast neurons, a medium or a hub neuron, and two slow oscillators. When she creates this reciprocal inhibition, f1 and f2 fire in alternation, so there is s1 and s2, and then when she couples them all up you see [19] what the whole 5-cell circuit does.

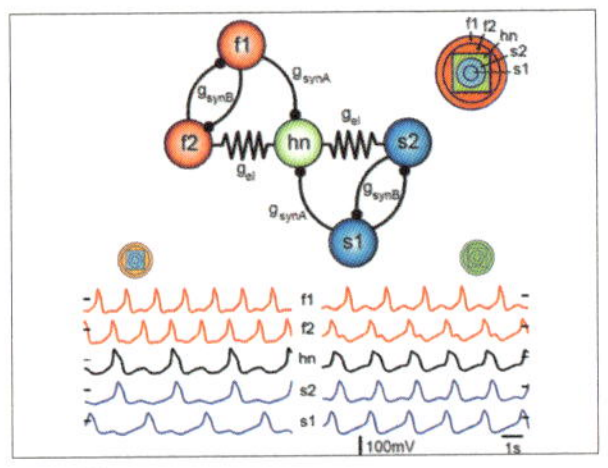

Fig. 19

I am going to spend a tiny bit of time showing you how she has chosen to visualize what the network does. One of the challenges that we have, whether we talk about small circuits or large circuits or human brains, is finding a way of visualizing the activity so that our human brain can understand what is going on. The visualization tool that Gutierrez created was a series of circles and squares, where f1 is the outside circle, f2 is the second one, the box here is HN, the frequencies of each of these cells, s2 and s1. This whole image tells you the frequency of every neuron in the circuit. She has called the strength of this synapse GsynA—that is the conductance or the strength of the synapse—and that is through here, and the strength of the electrical synapse is Gel. One set of parameters gives you this set. Here the hub neuron is firing slowly and the two fast neurons are firing in alternation, the two slow neurons are firing in alternation, but here they are slow and the hub neuron is firing with the slow oscillator. In this different set of parameters, all five neurons are firing at the same frequency. This allows to see immediately what is going on. If you now plot on the x axis the conductance of GsynA, that is the chemical synapse, and the conductance of the electrical synapse on the y axis, [20] you have a map of the activity of the network:

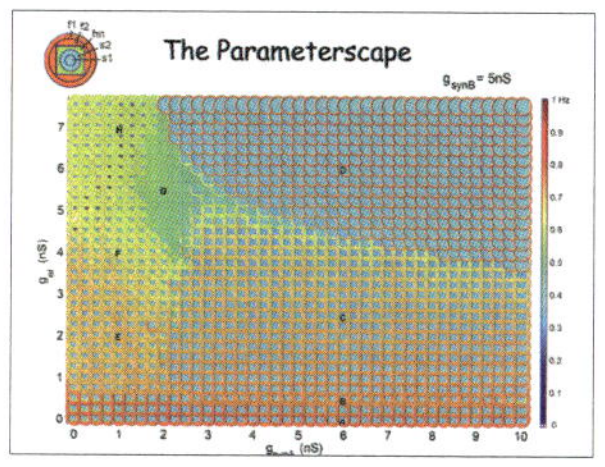

Fig. 20

this whole region here has one of the neurons firing fast and four firing slowly; this whole region has four of them firing pretty fast and one of them firing slowly; and then this region here is all five of them firing at the same frequency. You can see the boundaries as you go through these qualitative changes in network activity. This kind of qualitative change in brain state is at the core of understanding how human brains do a lot of things, because understanding what sets the state of the brain, whether any circuit is producing this kind of behavior or this kind of behavior, is at the core of trying to understand why brains are doing what they are doing, when they are doing it. Imagine now that we apply a neuromodulator, something which increases the strength of this synapse. If you start over here, in this state, it will take you across two different state transitions and take you from this activity pattern to this activity pattern. That is very profound. On the other hand, if you were to start here in a different state and now put in much more modulator, which produces a much bigger change in the strength of that parameter, you are not changing qualitatively the behavior of that circuit. Instead, you are maintaining that activity, even though you have much more modulator present. This is really a very beautiful way of seeing what we call state dependent brain activity, that is to say the action of this modulator will have very different effects depending on the starting state of the network. Here you will have no qualitative change, here you will have a very strong qualitative change, even though much less action is produced here than here. [21]

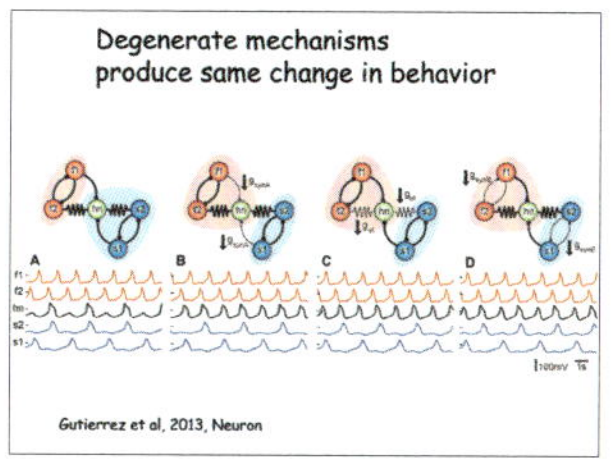

Fig. 21

I have two more slides to show you from this work, which go back to the issue of parallel pathways. Here is that starting five cell circuit and notice this HN neuron is firing slowly in time with the slow neurons. Here we change the strength of this synapse, and now the slow neuron, the hub neuron is firing with the fast network. Then you do a different perturbation: you change the strength of this synapse, the electrical synapse, and you can move the hub neuron from slow to fast. Then you do still a different perturbation and you can produce the same change in behavior. And this is only a five-cell circuit, imagine it is five parts of the brain: there are three entirely different mechanisms to give rise to what looks like the same change in neuronal coordination. This is extremely important. It arises precisely because of these parallel pathways, which are all over the brain, but it means that there is no single mechanism by which you can control what this neuron is doing, rather there are three entirely different mechanisms that can produce change in the behavior that look very much the same. This is very important and it almost certainly is taking place in our brains not exactly like this but similarly. [22] I would like to end with this slide—which goes back to this issue that was raised yesterday, I think Jean-Pierre actually raised yesterday—and that is whether neuromodulation stays local or changes the activity of a whole

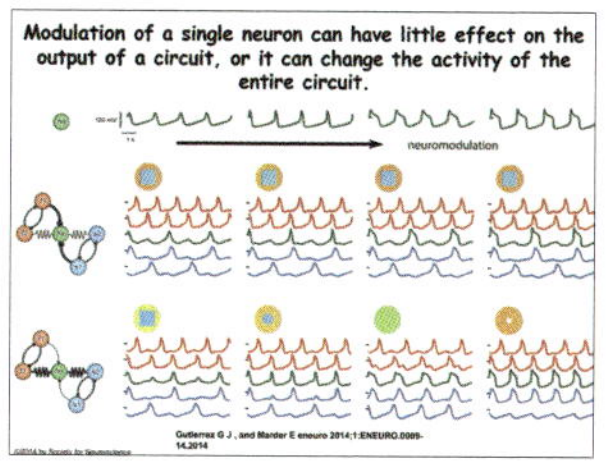

Fig. 22

circuit or whole brain. Gutierrez has modulated the properties of this HN neuron—you can see here the waveforms are changed—and she takes those different HN neurons and puts them in two different starting sets of parameters here. Then, as she changes the behavior of this hub neuron, the circuit is basically unchanged: the neuron that she is changing or modulating has changed, but everybody else stays the same. However, if she starts with a different brain state, a different set of parameters, and now she changes the behavior of this central neuron, it entirely changes the behavior of the entire circuit. Here, the s2 neuron has gone silent and you can see, if you just look at these color pictures, they are entirely different. In this case, modulation of only one neuron has percolated through and influenced the behavior of the entire network, while in this other case, modulation of one neuron has stayed local. This becomes a very important concept for thinking about local versus global effects of neuromodulation in large brains. To go back to the very beginning, we have levels of organization, we have neurons, synapses, circuit modules, large interacting circuits, whole brain circuits. The question, the really deep question we have to constantly ask: to what extent do the properties at each level of organization percolate to the next higher level of organization? Then to me, one of the deep unanswered questions is, in other words: are circuits formed from highly resilient neurons, more resilient than those formed from less resilient neurons, or does resiliency at each level of organization arise by rules or interactions at that level? We know that all individual brains are different, and differently resilient to challenge. How different are their connectomes and how does this influence their function and how information travels?

On the Biology of Feeling and Consciousness

ANTONIO DAMASIO

Let me begin by saying that these two concepts, the concept of feeling and the concept of consciousness, are normally presented as very distant from each other. Most people automatically relate to the term feeling and have a sense of what feeling is even without any kind of scientific background. People also have some sense of what consciousness may be, but at a very different level. The two things normally don't tend to go together and I am here to tell you that they do, and that physiologically you cannot understand one without the other. This is my first point.

My second point is that both feeling and consciousness are related to something deeper, something that is in fact behind a lot of this morning's conversation already. That is, the process of life, and more specifically, the process of life regulation, namely homeostasis. We were hearing from Eve Marder about how, at neuron level, we have to adjust parameters constantly so that the life of a neuron can continue; and about how very complex machinery, including iron channels and genes that control ion channels, is arranging for that environment to be maintained within certain parameters that are compatible with the continuation of the neuron activity, that is, with the continuation of the life of that neuron.

The same happens at the very large scale of a human creature. I am not just talking about the human brain, but about the human person in the human body. We need to adjust parameters in the life of that organism and, at our high end, we need to consider parameters of a much larger scale, for example, temperature, level of oxygen and level of glucose, and a variety of other chemical and physical parameters, that, if not maintained within certain narrow ranges, will lead to disease and death.

Now, feeling and consciousness are very much related to the control of homeostasis. We should actually say that they are the key components of homeostatic regulation in complex creatures—they are not going to be very important at the level of *C. elegans*, but they are very important at the level of multi-cellular, highly complex, multi-system individuals such as we are, not just human creatures but creatures of many other species.

Before I go any further in telling you about feeling and consciousness, and about how I interrelate them, let me open a parenthesis: when we think about consciousness, feeling, and mind, we very often make the mistake of assuming that all living creatures also have such phenomena and also have nervous systems as we do, and depend on processes such as feeling and consciousness in order to organize their homeostasis. Let me say that this is just not the case. In fact, when you look at the majority of living creatures on the face of the earth, and at the oldest of these living creatures in terms of evolution, we find very simple beings with one single cell such as bacteria—and you know that prokaryotes don't even have a cell nucleus—and many other creatures that do have a nucleus and are unicellular, and go through life quite intelligently, make a lot of smart moves, regulate their lives, are able to detect the presence of other organisms, and are able to detect the presence of certain conditions in their environment and respond to them adequately. Yet, those creatures do not have nervous systems. They have non-explicit competencies that are *not* represented in a nervous system, or in a mind or in consciousness. It is very important that we realize that four billion years ago creatures were beginning to operate like this and they had complicated physiological arrangements even if they had only one cell. They already possessed life, they had a birth date and a death point, their life was well regulated over a period of time because it was maintained within homeostatic parameters. In fact, they could even be organized socially and respond adequately to their environment. They are still with us today and we see, for example, that if their environment is highly conducive to providing nutrients, they can separate themselves and live more or less as single individuals; but if the environment is not conducive or is threatening, those creatures actually do the opposite, they aggregate and band together, just like we do when we are being attacked and need to fend off and create a strong defense. So even if it is beautiful to talk about nervous systems, which is what Eve and I do for a living, and even if it is beautiful to talk about incredibly complex functions such as feelings and consciousness and minds, we need to realize that the majority of creatures on earth—creatures that are part of our microbiome, that are included in our own organisms and contribute to the success of our own organism—do not have to rely on nervous systems and do not have feelings or consciousness. Yet, they are alive and well and counting. They are the most numerous on earth. They do not rely on a nervous system and yet they have very complex functions that *appear to depend* on a nervous system. Let me just add that, in the end, in a large-scale view, what we have in our mind, and feeling as part of mind, and consciousness, are projections of this enormous regulatory capacity that has existed throughout the history of life before nervous systems appeared. Of course, during the last 500 million years many organisms have had nervous systems, and during the last 100 million years (or 200 at the most), organisms have had something that resembles minds and consciousness. We need to be modest, not only in relation to the monstrous task of understanding brains at the small-scale or large-scale, but also in relation to the task of understanding how this fits into the great project and reality of life.

Now let me move on to feelings and consciousness itself. The first thing I wanted to tell you is that the two are related through the process of homeostasis, and that it is not conceivable to me—although this is not necessarily the case for everybody—that either feelings or consciousness can exist without nervous systems. What we were hearing about from Eve is extremely important to give us an idea of the enormous complexity that we have in nervous systems even when we are thinking about very, very small creatures with very small connectomes, and just a small group of neurons. Yet the complexity is huge and no one can decipher it completely. And we have to think that creatures that do not have such nervous systems, even the very simple ones, are even less likely to have phenomena as complex as minds and consciousness. One of the reasons why I am reluctant to admit that any creature that would not have a nervous system would be able to have minds is because what we talk about when we talk about minds is the capacity to represent and manipulate a cortege of representations, a cortege of images. That requires an ability to focus on something—an object, a process—either in the outside world or inside our own organisms, and create a map of an object or process. In order to create a map we need neurons—not just neurons, but neurons arranged in circuits, and, in many cases, as has been demonstrated so clearly through the study of the visual cortex or the auditory cortex, we need to have neurons distributed in particular patterns that are compatible with the making of a map, the making of "a sketch" of something you hear or something that you see, or something you touch, or something that your nervous system is "looking at" and "seeing" and literally touching inside your own organism, because there too you have the possibility of creating images of your interior, by mapping the structures of your interior, which happens to contain your viscera: our heart, lungs, our blood circulating, our guts, and so forth. We can defend the idea that we are not likely to generate minds and all of its perturbations without nervous systems. I can say the following: if it is not likely that creatures with the complexities that Eve showed this morning, have minds or consciousness, it is even less likely that creatures that do not have nervous systems at all, such as eukaryotes, would be able to create such complex operations.

Let me now turn to what I really want to talk about, the core physiology of feelings: what are feelings? Unlike the term consciousness which creates all sorts of confusion, people think that they know what feelings are. They don't, but that is okay. Most of the time people actually confuse feelings with emotions, which is a serious problem and a grave error and which I would like to ask all of you in the audience not to ever make. In other words, *feelings are phenomena that occur in our minds, they are the ultimate subjective phenomena*. They are *internal*, they are *mental*. They are, to begin with, mental expressions of the state of our own organisms. When you declare yourself feeling well, what you are describing is a state of mind in which you sense the operations of your body physiology, the operations of your organism, as being in a certain degree of harmony, without any sort of stress, flowing easily and with the amount of energy that you have at that moment, without any complaints of it being too hot, or too cool, or too trembling, or too whatever. *You are feeling well*. At the same time, we also know, especially at a time in which we are in the middle of a pandemic, what it is like to feel sick. You feel sick when a lot of these parameters of life organization and flow are suddenly put under stress and our feelings alert us to the fact that the continuation of life is no longer possible and that things are really going awry very fast way. By the way, it can get worse, get to specific points at which the normal physiology goes down and you can then talk about having pain, general or localized to an organ. You can have a heart attack and have huge pain that you refer to your chest, or you can break a bone and have huge pain related to everything that went wrong with the physiology of the injured region. At the same time, you can imagine well-being and you can also have other subjective states, such as the state of happiness, or the state of fear, or sadness. There you are dealing with something that is not just about the homeostasis of your organism, you are dealing with something that has been superposed on homeostasis, either for the good or for the bad. That is really the result of an emotion. On that notion, let me leave you immediately with something tangible and useful for today: let's agree that feelings are about ideas, about mind, about mental expressions of the state of the organism. Feelings come in two varieties. The variety that I call homeostatic—feelings such as well-being, sickness, pain, and so forth—or the emotional variety, where we have feelings such as sadness, fear, happiness, anger, and so forth. Those feelings are still mental expressions, of course, but they are mental expressions of the state of your organism *after* it has been challenged by a particular problem in the environment or a particular problem that you have in your own mind. For example, you may be threatened, your life may be at risk and you will have an appropriate reaction, which is set to organize the first response, the first alarm, the SOS of your organism in relation to the threat. That places your organism in the state that we call *fear* which is when lots of things then change in terms of what your heart does and in terms of your blood pressure and breathing; your metabolism is now prepared, all of a sudden, to cope with an emergency which may involve you running away from where you were to somewhere else. But whether you have homeostatic feelings or emotional feelings, in both cases you are dealing with a mental process, something that is in your mind and that expresses a variety of physiological arrangements that are, in themselves, collections of actions. So when you are in fear you do such things as recoil from where you are, and you have all the changes that are occurring inside your organism; your face mask changes as well as the position of your limbs, and so forth. All of those changes are the result of actions that are being taken very smartly by your organisms, under the control of the nervous system, meant to allow you to cope with your life and make sure, ahead of time, that you are not going to drift into a situation that might lead you to sickness and death.

In brief, feelings may be very simple, the result of the operations of our organism under no particular stress or state of felicity, simply expressing the average day-to-day movement, the state of our metabolism, our handling of the multiple challenges that are occurring in a multi-system complex organism such as ours. The distinction between emotions and feelings should be clear; the fact that feelings are physiologically different should be clear.

This is a good time to ask ourselves: why is it that feelings are so important? The answer is clear: they are *informers*, they give us direct information about the state of our organism. And why is that important? Because once you know that something different from the average is happening in your body, you may take corrective actions and that may save your life. This capacity to inform, to give you an assessment of the state of your organism is something extremely valuable from the perspective of homeostasis, indispensable for your survival. And just note this because it is very often overlooked and brushed aside in discussions of feelings. Their value comes from the fact that all of their information is directly in consciousness. Feelings are not just in your mind; they are in your mind in a peculiar way that is telling you, without you asking any question, "wait, this has to do with me, this has to do with this organism of mine." One thing that I do want you to remember from this talk is that *the merit of feelings is the merit of consciousness*. Feelings are valuable to us because they exist consciously. One is often led to believe that consciousness is something very mysterious, impossible to describe let alone understand. But consciousness is what permits you to refer to what is in your mind to yourself. For example, consciousness allows you to declare that you own your mind. That you are the controller, and that is what gives rise to what one calls the *self*, in psychology and in philosophy. That self consists, in fact, in the ownership of the mind which, lo and behold, is also the ownership of the body in which that mind is occurring thanks to the interactions between the body and the nervous system. The solution to the problem of consciousness passes through the solution to the problem of feeling, because consciousness begins with feeling, and that is the reason why it can only occur in creatures that are endowed with nervous systems, capable of making maps, representations, images, in short, minds. In the end, consciousness is a state in which the ownership of the organism and the ownership of the mind become obvious.

That was the first important point. Now let us consider, although not in detail, how such a process is actually possible. How is it that we can make such things as maps of our organism and end up with something that is our mind, and how is it that we generate feelings? Here I would like to call your attention to the physiology that is behind feeling, because that physiology is, to a great extent, the physiology behind consciousness. For this, I need to refer to the ways in which our brains collect information about the world inside of us and the world outside. The physiology that pertains to the outside is actually—and quite understandably—the one that was elucidated first. Thanks to the work of scientists such as David Hubel and Torsten Wiesel, we have known how, for example, the visual system operates to make those maps that I mentioned earlier and thus generate the basis for the images that constitute vision in our minds. Parallel investigations have now been carried out for hearing, for tactile processes, and for taste and smell. This exteroceptive functionality is based on the externally directed sensory organs, for example, our eyes and the retinas inside them that allow us to make maps of the outside world. An example, when I look at the Santa Monica mountains and the beautiful sky behind them, the maps that I am making of them are responsible for the respective images, but the sky and the mountains are there and I am here. The images I construct in my retinas and that my optic nerve carries to my visual cortex, on the one hand, *and* the reality that is outside, are in entirely separate worlds. There is a reality out there and around me, and then there is my organism and my nervous system within me. When I describe having pain, or being sick, or being happy, I am describing a world *within* my organism. What I am describing is information that comes through what is known as *interoception*. Far less is known about interoception than about exteroception, for reasons that are obvious: why begin by studying what is complicated and hidden inside you when you can first study what is outside and is more patent and obvious?

Interoception relies on pretty much the same organization throughout the organism: the nervous system is embedded everywhere in the organism—every little bit of our bodies—and has probes that consist of nerve terminals distributed all over every tissue and organ of our bodies. Those nerve terminals that are everywhere in our body bring into the central nervous system signals about the state of every organ and tissue. The signals come into the spinal ganglia or the brainstem nuclei, and bring information into the spinal cord and the brainstem, some of them eventually end up in the cerebral cortex in regions such as the insula, or the cingulate regions.

I wish to call your attention to something that you probably have not heard of because everybody tends to ignore it: the relationship between the body and the nervous system is completely different from the relationship that the nervous system has with the world outside and around the body. The world around can come directly into my retinas and my visual system maps it. There is nothing I can modify about that world. When I look at the landscape outside I cannot change that landscape. I can imagine it differently, but the landscape remains intact. As per our inner world, though, the relation is different. My nervous system is located inside my organism, it is a part of my organism, and it can map my entire organism. But it so happens that the nervous system has the possibility of replying to every part of the body that signals into it! The central nervous system can respond and change the structure and the operations that are in the process of actually being perceived. Suppose right now you have a wound and develop acute pain from that wound. Even before anybody gives you the appropriate treatment, surgical or a painkiller, your own nervous system will help dampen the pain and will modify the perception you have of your own world. The nervous system is so interactive and so interconnected that it can, through the avenue of chemical mediation, modify the physical condition of the place where the pain is coming from. This is dramatically different from what happens in the perception of the exterior. If anybody wants to explain feeling to you using exactly the same line of thought that is used to explain vision or hearing, you should protest and say "no, the circumstances are completely different, we don't handle the information from inside our bodies the same way that we handle the information from outside of our bodies." But then there is something else that is important to note and consists of considering the *kinds of neurons* that are actually committed to the job of "interocepting" and giving us information about what is going on in our bodies. When you look carefully you find interoceptive neurons to be quite atypical when you compare them to other neurons. The first difference is that the vast majority of those neurons do not have much myelin, and when you don't have myelin you lack insulation and you expose the axons of these neurons to the influences of the chemical surround. This does not happen when neurons have fully myelinated axons. Quite clearly, most of our exteroception depends on highly evolved, heavily myelinated neurons, while most of our interoception depends on phylogenetically older neurons that do not have that myelination.

Something else that happens in the interoceptive system is that the regions where they congregate, for instance, the spinal ganglia, do not have a blood-brain barrier. This means that the arrangement we have developed to protect our nervous system, our central nervous system in particular, from the influences of the chemicals in the surround, are lacking at the level of spinal ganglia. The chemicals that are circulating in our bloodstream can have a direct action on the neurons in those spinal ganglia. By the way, signals that come from your head into the nervous system, via the trigeminal ganglia, are equally vulnerable because their pathways are also lacking in the blood-brain barrier.

The bottom line: I want to call your attention to the fact that the ways in which the physiology of feeling is organized are quite distinctive. Feelings are not mere perceptions of the state of our body, either locally, like in pain, or more generally, like in well-being or pleasure. Feelings are not perceptions such as we have of the outside world. Feelings are different. Feelings derive from an interaction between the nervous system and the body—the non-neural component of the organism, a component which, as I tried to point out when I referred to non-minded, non-neural creatures, is rich in knowledge about life and its regulation (recall that prior to the existence of nervous systems, there was plenty of complex life, highly organized and intelligent life). This is what I wanted to leave with you: the idea that feelings and consciousness are related; the idea that feelings are the beginning of the process of consciousness; and the idea that feelings are not mere perceptions of the state of the body, but rather complex interactions due to the physiology of interoception, and the peculiar arrangement of the relationship between the body-proper and the nervous system.

Panel Discussion

DANIELA PERANI
EVE MARDER
ANTONIO DAMASIO

DANIELA PERANI Eve, your study of neural networks in crustaceans with a countable number of few neurons, has led vital foundation for current brain research projects. How much this established approach could represent the building of many models that cover the range of normal behavior parameters in humans? So from the simple to the complex, how can you highlight these crucial passage that was also mentioned by Antonio?

EVE MARDER I am not sure I completely understand the question. I think when everyone builds models, one has to decide what one is trying to reveal, explain, establish. So models always have to be built with some sort of sense of a puzzle, or a question that one wants to understand. I think the most useful models are the ones which say "here is something I really don't understand, how can I build a model to help reveal that?" I think whether you are doing it to try and understand some basic fundamental process in a human brain or in a crab brain, the essential features are to say "here is a deep mystery and now I can set up a model, or a series of models, or many models to try and answer that deep mystery."

DP I think it is an important statement that highlight how to go on in this direction. This is a question about resilience. Resilience is the ability of an individual to face and overcome a traumatic event or a period of difficulty. How is it possible that, during the evolutionary period, severe and adverse events undermining the species survival and our foundation of civilized living also, even more than the one we are facing now, have not produced devastating neural modification, but instead perhaps more resilient and protective molecular mechanism?

EM I took out a slide I probably should have shown you, which talks about resilience to climate change. In 2012 the oceans never got cold during the winter, and in the following spring and summer, our animals didn't crash, they went up to 36 degrees and were still functional. What we didn't know, at that point, was whether the long period of very warm water during the winter had allowed there to be some long-term molecular adaptations, or whether we had selected for a subpopulation that could deal with the adverse conditions, or whether our animals went to Maine, where the water was colder, and we were seeing animals that came north from Long Island. So that is a long way of saying that in a population you are always going to have animals with different sets of underlying parameters. Some of them will be more resilient to a perturbation and so, conceivably, you end up selecting for those animals, and those are the ones that live. But that doesn't mean to say that you couldn't have animals with a much more acute sort of change that then persists. I don't know if that answers the question.

DP It is a difficult point. Someone asked how much can stress affect the self-perception? How much it can devastate a self-perception?

ANTONIO DAMASIO From clinical populations one knows that cases of acute stress or sustained stress can alter your evaluation of yourself and, depending on the individual's personality and on the situation, can be devastating or not so much. One fundamental point that Eve has made is that we have these robust systems for coping and for transformation, and if they are present, as she showed so well, at the level of very small organisms. I imagine how they must present in creatures with the complexity of the central nervous systems that we have. One more detail to note is that, in addition to the enormous individual variation of creatures such as ourselves, we also have all of the varieties of acculturation of those simpler individuals. So how one copes with stress and how one is or is not devastated by stress, depends on one's fundamental biological setup, but it also depends on what was done to that biological setup by the education that you received, by how the environment around you has treated you, and so forth. Moreover, there is a connection to age. It is a different thing to have a devastating insult to your stress systems at age 60, or to have it at age 12 or age 40, when you are presumably at the peak of your powers and quite resilient physically and intellectually. This is not a case of one size fits all: the variation of the response is simply huge. Everything that we know today, whether it is from psychology or from neuroscience, at the macro or micro level, is telling us that the variation is huge and the ways of coping with it are numerous.

EM I would like to pick up on that and say that the other thing that is clear is that there are many different kinds of stress. Depending on what the stimulus is that creates the stress, you may have more or less resilience to that. For example, you could have an individual who is very resilient to one kind of stressor and much less resilient to another kind of stressor. One of the things that I think is really going to be interesting is whether some individuals are more resilient to any kind of stressor and whether there is a series of orthogonal conditions, where there is a trade-off between resilience to one kind of stress versus resilience to another kind of stress. I think that is going to be a really interesting set of questions as we go forward, to really try and understand that. Just one last thing along that: depending on the timing at which the stressor comes, if it is a one-shot deal, or repeated, or long duration you might expect very different biological sequelae. I think the repeated or the one shot, you can have different kinds of adaptations depending on the dynamics of the stressor.

AD The question was related to self-perception, so we can think that stress couldn't affect self-perception because self-perception is part of the more solid, established way or some functioning of our brain. Self-perception should remain the same, or can change due to the stress, depending by the strength.

EM I would say that anything that makes one feel weak in any way, will potentially interact with one's sense of self and self perception. Any situation that makes you feel helpless is definitely going to erode your sense of self, in some essential fashion.

AD I would agree entirely. That is so interesting because there is no parity here—the positive side and the negative side are very different in terms of what they do to us. It is a bit difficult to imagine that there can be an "excess" of pleasure, that there can be an "excess" of goodness and of homeostatic positive adaptation. The negative consequences of that "positive" side are quite different from the negative consequences of, for example, sustained pain or sustained humiliation. We see that in certain populations sustained behavioral insults translate into negative emotions and negative physiology. The consequences percolate down to your very biology, neural and non-neural. The consequences are very asymmetric. The positive consequences of feeling are not equal to the negative ones. This is a bit like quoting Tolstoy, who talked about the many kinds of unhappy families and the fact that there was only one kind of happy family. So there is more variation on the negative side. Pleasure is pleasure, well-being is well-being, but think about all the ways in which we can be negatively affected, all the names we give to the bad feeling states in relation to society and bad feeling states in relation to our own organisms. There is a huge variation in what can go wrong and in the consequences of what goes wrong. I am very intrigued by a general question—which is both a question in biology and a philosophical question—which is: how did we end up staying with the positive and not with the negative; and why is it that we have maintained this regulation that favors life, favors the good things that happen with life, as opposed to the ones that favor the negative, non-homeostatic regulation? It is an interesting question.

DP This answers to the question on one side, also. It is very important. Something which goes in the same sense: we might have feelings and perceptions, which are not fully conscious. So our brain can be modulated and influenced by this kind of input which is not really full of consciousness. But these also act on our neural systems, so we have to take into account also this kind of level of consciousness explaining behavior. Do you agree? Going in the same direction, I have this question. Understanding false beliefs requires, of course, high level of consciousness, and high level of functioning need language, knowledge, culture. Children are able to understand them at the age of four or five years. Today our society is facing an enormous amount of fake news, which bombard our perceptions, able to influence our decision, to build conscious reality deeply distorted and capable of negatively influencing society itself. You are seeing this kind of behavior, human behavior. The fake news change our way of thinking, we believe in that. But which component of our brain? If we consider this neural basis, where we can place such a weakness or functioning? Why are we not able to consciously distinguish between something which is completely negative for us and for our similars?

How should we counteract it? Of course, I cite your belief—you are right because it is important for knowledge and for increasing our mental capabilities—to strengthen these neural circuits that otherwise will become weak and malfunctioning. Do you agree on that? That we have to do something on the brain?

EM I am sure that Tony has thought about this a lot. I have been watching, obviously during this time, and I think the secret is repetition. Humans learn with repetition, and it appears that if you tell a lie often enough, large numbers of people will believe it. That is exactly what happened during our past four years: repeating and repeating lies makes them more plausible, because that is how we learn. Maybe Tony has a different take on it but I think some piece of it is just as simple as that.

AD I think I would agree. Unfortunately, we are still going through that period, we are not out of the woods. I think that the answer to your question, Daniela, is really about education in general. Education has a tremendous amount to do with repetition. I think the only way of counteracting fake news, of counteracting false beliefs, is trying to demonstrate patiently that they are false, trying to demonstrate that they don't hold now. Of course, you need to have a modicum of intelligence to be able to be affected by that. You need to be attentive, and the demonstrator needs to be persuasive and, as Eve says, probably needs to repeat and hammer the message. But I think that one of the problems that we have got into in these recent years is in part related to something that, to begin with, is a plus for us and should be an enormous value, which is social media. I think that the facility with which we have access to information, the fact that it is presented rapidly and with very little context—because that is part of the nature of electronic communication these days—and the fact that it ends up being attractive because of the imagetic, pleasurable component, all of that goes against the grain of this idea of solid thinking, going through the steps, and trying to demonstrate that something is or is not real. It is not by chance that a lot of the fake news concerns a rejection of science. Fake news goes with the rejection of science because what we do in science is to have ideas, generate hypotheses, and then go through a lot of trouble analyzing those hypotheses, thinking through them, and trying to find some kind of experiment with which we can demonstrate that a hypothesis is correct or false. All of this takes time and takes a capacity for observation, and some calm, it cannot be done at high speed and it cannot be done by simply repeating something as if it were proven to be true. I think that the reactions that people have—for example, to the idea of climate change or the need for vaccines, which is something we are going to face, because there is an anti-vaccine movement that is quite powerful—we need to cope with all of these issues. It is not a bad idea, as your question suggests, Daniela, that we analyze the circumstances in which this attitude has become triumphant.

DP We have other questions about cultures: what role do feelings and emotions play in motivating the individual and collective actions to give origin to cultures? Are there unconscious components?

AD The relationship between biology and culture is a very strong one, and unfortunately, it has not been very much emphasized. I tried to analyze it in a number of articles and in my book *The Strange Order of Things*—in Italian it is called *Lo strano ordine delle cose*. In that book, I look at the behavior of bacteria out in the field. These creatures do not have a nervous system, have a simple cell with a lot of organelles and very complex biochemical machinery. There are beautiful experiments that show how bacteria will do in environments where they have good nourishment and in environments where they don't, and one of the things that is so extraordinary is that they will, for example, if there is plenty of nourishment, scatter around and live as individuals; however, if there is not nourishment enough or if, for example, the temperatures are falling, they will actually aggregate and try to gain strength from being numerous. What we find is that in living organisms that are already regimented by homeostasis, there are degrees to which you can have large-scale social behaviors that are in fact the very grounding of societies and the very grounding of cultures. A lot of the things we discuss in our cultures are how we organize ourselves socially and politically, how we define what is moral and not moral, how we find ways to resolve the economic problems we are facing, and so forth. So close your eyes and think a little bit about how the humble bacteria have been doing it, in very simple ways, of course, but still according to the same scheme, the same basic arrangement. The motif is there already. That is why I am so convinced that, in a way, as we ascend to nervous systems and into all the complexities of behavior and mind that came after it, we are executing a large-scale plan for living organisms. I think we have to be—as we have been learning to be, in recent years—respectful of creatures that are not human and aware of their rights. We have to learn to recognize the smarts of creatures that are very simple and about which we normally don't think because we don't even see them. We know that they are inside us, and that some can create a lot of havoc, although the majority of them actually do lots of wonderful things for the maintenance of our organism. For example, we cannot have normal gut function without the collaboration of bacteria.

DP We cope with each other and this idea of the biological rules that still exist and we still have to consider is very interesting.

EM I would like to add one more thought. When I was being educated, there was a real burden on the student to remember a lot of what one had read, because you had to be able to remember it to reason and to think with it. I remembered the papers I read, I remembered my data. My students are not in the habit of remembering to the same degree because they have it all on their phone, they have made the decision, they have learned that they can use their devices as memory tools, and I think that if you give up or you abrogate your own memory as a repository of experience that may make you more hostage to the random things that come from those devices, in a very interesting way. I think part and parcel of this is the way in which people learn to offload their memory to electronics.

AD I am totally in agreement with you. Neither your creatures nor creatures like us, behave correctly if they don't have memory. Memory is absolutely essential, because if you don't have memory you are at the mercy of the present moment. One of the ways in which one can reason, in which one can avoid doing the wrong thing, is by remembering the bad things you did and the good things as well. Memory should control part of our operations. Both you and I have the experience of having graduate students around us, people that are much younger than we are and you ask them something and they go to their iPhone and complete the answer. They are not asking their hippocampus to assemble and reproduce the stuff that should have been stored in the cerebral cortices and that should appear in the minds and be available. Certainly, your claim that poor memory makes us more vulnerable sounds right. But it can also make us poorer reasoners. It is true that we have access to all the stuff that is on Wikipedia, but at the same time, we forego access to the stuff that is or should be inside us and could guide us. Memory *is* indispensable.

DP Do you believe that brain functions and activities in the future, even with the most sophisticated techniques, may be understood in their entirety, or there will always remain some aspects that, by definition, cannot be comprehended by our brain itself? That is a very supramodal question. So does our brain have some limits? Our brain couldn't understand the complexity.

EM I don't think our brain is going to understand everything, but I don't think we are anywhere close to knowing where the limits are.

AD Actually, when I was listening to you and you were talking about the variety of ion channels and the variety of the genetic control of the ion channels, it is clear that what we know today is still paltry compared to what we need to know. This is one reason why I get very irritated when people say: "There is this problem of consciousness that you will never be able to solve." How can they be so sure? We are just beginning, and I think there are many aspects of the consciousness problem that we are solving which does not mean that we will know everything and solve everything. We can be modest about what we know and modest about what science has achieved so far. I think that the idea that you have of all this non-neural intelligence is to me one of the most exciting things. I grew up obsessed with the nervous system and with brains, first as a neurologist and then as a neuroscientist. Yet I am equally amazed by what non-neural creatures can do and I think that knowledge is going to help illuminate what neurons can do as well.

EM I think there is one more thing that needs to be said, which is that there are tremendous advances being made today in figuring out how to display data so that we can visualize what is going. We don't think very well in non-linear processes. You start asking the humans to start reasoning about three non-linear processes, interacting. We are basically forced to write down models and then do simulations—except maybe a few mathematicians but they are rare— but I think one of the deep challenges, and this is happening, is that you need new ways of displaying data and organizing it, so that our very smart but also limited brains can understand it. When I say I don't know how far we are going to go, I think depending on how well we do with figuring out how to display data will really determine how much we are going to be able to understand.

DP In your both fundamental worlds it seems to emerge that the factual knowledge necessary to reason and make decisions comes to mind in the form of sensory perception, visual images, sounds, sensory motor inputs, and emotions. How is it possible to demonstrate these characteristics of the thought, knowledge, and thus of consciousness—be conscious of this event? What are the most relevant implications of these assertions? Still, it is a question on how can we measure and see and understand, all together, this characteristic of thought, so that we build up our thought based on feelings and emotions and so on.

AD It is an interesting and very complex question. What was coming to my mind as you were asking the question is the approach we take to enter these processes. It would be beautiful if we could do everything from the bottom up and study the nervous system and understand mental processes. But we cannot succeed using only that avenue, in good part because the complexity is so huge. If you were asking me to design a project to study feelings of pain, beginning with what Eve

Marder explained today about the connectomes of simple creatures, it would not be easy or practical. It is not that it is impossible, technically it might be possible, but it would be very cumbersome beginning so very far away from where you want to end up. One of the things that you need to admit is that, after all the developments in neuroscience and in biology, we still have to rely on something that was rampant in the 19th century and the beginning of the 20th century, which was introspection. I would like to say that what served William James and Sigmund Freud so well (or, for that matter, Proust and Virginia Woolf), should serve us quite well too. I am not at all embarrassed to use introspection. It is marvelous that we have the possibility of looking inside our mind processes, and of using what we encounter to organize our interpretations, test hypotheses, and construct a research plan for the study of the mind. Whenever somebody tells me "If you use introspection you are not being objective enough," I respond that we can observe phenomena of the mind with objectivity when we use introspection. Not only that, we can compare our introspection with the introspection of others and therefore we have a way of avoiding the biases and potential errors that introspection may lead you into. I need to say that it is possible for us to combine multiple introspections and to combine the results with those of "traditionally" objective analysis of biological phenomena. It is possible to generate hypotheses, put those hypotheses to the test, and advance science. Now, it should also be clear that the speed at which we can advance the science that relates to higher levels of behavior in mind, is slower than the speed at which you can advance studies of the smaller scale mechanisms, because the complexity is larger. Even so, thanks to new methodologies, not only do we have the micro approaches, but we also have, at our end, something that is not quite comparable in fine detail but is still very helpful. I am referring to imaging, the kind that is produced by modern scanners and that, properly used, can give quite a lot of information that helps decide on the validity of hypotheses.

EM I think we all use introspection to some degree, but introspection also can be quite wrong. That is to say we have 70 million people who have introspected that we have had fraudulent elections. I think there is another path towards understanding, that we haven't mentioned, which is beauty. For example, when you see a particularly beautiful dye fill of a neuron and you say that there is something about that structure and its beauty that tells you things that you don't necessarily always truly understand, but you recognize its validity or its truth, partially because of its beauty, or we perceive it as being truthful because it is beautiful. I think that is another piece of how we use emotion, or the perception of beauty as validation of truth in science in a very interesting way. For example, I just had a rotation student do a project in the lab and she generated some absolutely spectacularly beautiful images. Because they were so beautiful I believed them. It is very interesting how belief comes with beauty in assessing biological systems and how one decides what is beautiful is actually should be something you should think about or you have already thought about.

AD I think that beauty can, to a certain degree, be an arbiter and, of course, it can mislead you, in the same way introspection can mislead you (although, I don't think introspection is to blame for misleading the 70 million you mentioned; they were not introspecting anything, just voicing biases) but we have ways of correcting the deviations.

DP Our brain is consonants also, brain loves rules and not dissonant things, and this is true for a lot of sensory, motor input and cognition.

3

The Languages of Human Consciousness

STANISLAS DEHAENE
ANDREA MORO

PANEL DISCUSSION MODERATED BY
JUBIN ABUTALEBI

Introduction

JUBIN ABUTALEBI

Language is probably one of the highest human cognitive abilities. We learn it automatically early in life: we combine sounds from to phonemes and then to words and then we apply automatically rules which we acquire because our parents talk to us and we convey meaning with it and we can use prosody and pragmatics. [1] Just take

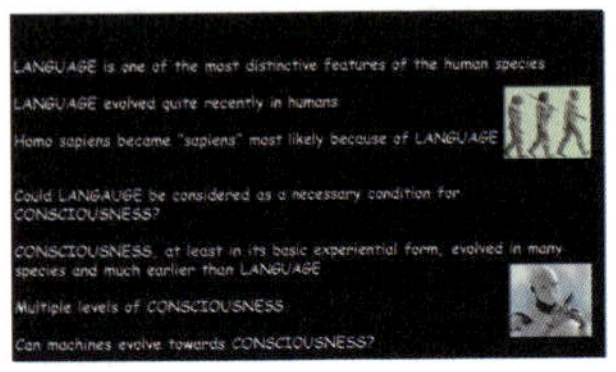

Fig. 1

as an example the sentence "This is really good." We can use it in a social context and be honest but also ironical. It is a different meaning of it. Animals cannot do this. That is what I like to say: *Homo sapiens* became Sapiens because of language. There is the link: could language be considered a necessary condition for consciousness? One has to keep in mind that animals have some sort of consciousness. In cognitive psychology we take as an example the famous mirror test: you take an animal like an ape or an elephant. While they are sleeping you design a dot or mark them with some color on the head and the next morning when they wake up you put them in front of a mirror. The animal will immediately try to touch it because there is something new and wrong. It is a kind of consciousness, it is self-awareness. It is actually a little bit more complicated because there are different levels of consciousness and I am sure Stanislas Dehaene will talk about this much more in detail. In simple medical term consciousness is like saying that a patient is not conscious because she/he is in a vegetative state or in a coma, but in cognitive psychology there are different levels. One could be being aware of our inner thoughts of what happens outside of us and so on. Funnily enough, there is also an unconscious level of consciousness: take as an example subliminal messages. Who remembers the 50s American cinemas? The Coca-Cola company would project on the cinema screen 10, 20 milliseconds images of a bottle of coke saying "Drink Coca-Cola." It doesn't come to your consciousness but somehow it comes to your brain and people got thirsty and would start buying coke. There is another really interesting point: can machines evolve toward consciousness? This is a big topic in our times because of robots, artificial intelligence, and multiple algorithms where you try to produce a machine that can understand our emotions and react consecutively. I am sure our speakers are going to talk about this today.

Human Singularity: Which Aspects of Consciousness Are Shared with Other Primates and Which Are Unique?

STANISLAS DEHAENE

I will try to discuss two topics: what is consciousness and what part of consciousness may be unique in humans since we have language and language-related abilities. I will try to do something a little bit audacious in 30 minutes: I would like to cover two main points which correspond to my two latest books. [1] First I would like to discuss

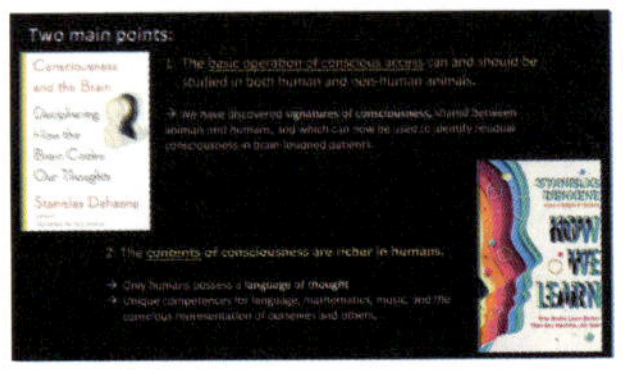

Fig. 1

consciousness, and my main point will be that the basic operation of conscious access can and should be studied in both human and non-human animals. It is not at all unique to humans. We have discovered that there are signatures of consciousness that can be found at the brain level and that are shared between animals and humans. This science of consciousness is actually quite advanced. I think that few people know this but it is so advanced that I would like to show that it can now be used to identify residual consciousness in brain lesion patients. However, my second point will be that there is something special to humans, and this is something I describe in a more recent book, *How We Learn*. My point will be that the contents of consciousness have become richer in humans. Why? Because we have a language of thought, we have a combinatorial ability, specific perhaps to the human mind, to develop languages. Of course natural language, but also the language of mathematics, of music, and the language of a theory of mind, the capacity to represent ourselves and others. I would like to argue that this is a significant enrichment of our capacity to think about ourselves and our own consciousness. [2] I'll start with the

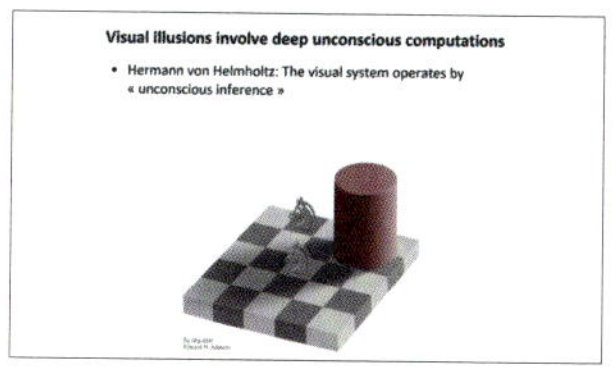

Fig. 2

first point: conscious access and how we can study it. I would like to show it in very simple terms with this classical visual illusion due to Edward Adelson at MIT. You look at this scene here and you might think "Where is the illusion?" There is a checkerboard and some dark gray, gray or black cells and white cells, what is the problem? The illusion is that these two cells are exactly the same, they are the same gray level and you can see that now, I hope, but if I remove the occluders one is dark, one is light. I hope this illusion works even though I am not physically present with you. What is happening is that your brain is noticing maybe unconsciously that there is a shadow. It is not a shadow, it is just a dark area in your retina, but it can be interpreted as a shadow coming from this object. If you subtract the shadow, then the color of the object should be lighter here for this object which is in the shade, compared to this one which is in the light. Your brain lets you see consciously only the result of an enormous amount of unconscious inference. [3] This was the term of

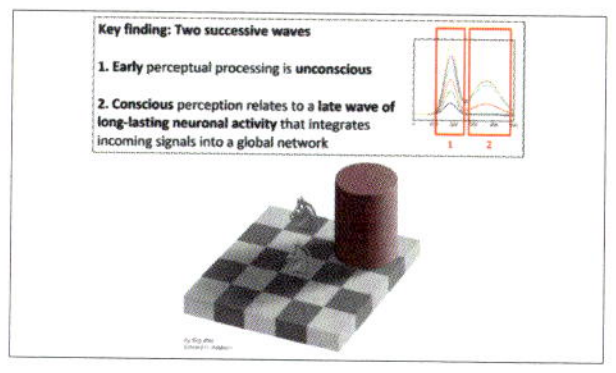

Fig. 3

Hermann von Helmholtz. I think we have only been able to validate this general idea from animals, that there is a lot of unconscious processing, and we are not aware of what is happening in our retina, which is the same exact shade of gray here. We are aware of the result of a long computation. A lot of early perceptual processing is unconscious, and conscious perception relates to a late wave of activity which creates a stable state where we see an inference about the world and we have integrated the information together. For instance, the information about the object and the shadow and the particular objects that are in the shade, we have integrated this information into a conscious language. This is going to be my strategy to study consciousness, it has been my strategy for about 20 years now. The first ingredient is to find some interesting visual illusions that allow us to manipulate conscious perception, and I would like to mention that we are all sensitive to those illusions that I am showing right now, whether we are a monkey or a human. For instance, in the classic illusion of binocular rivalry, which was studied by the pioneer Nikos Logothetis, I show you two pictures, one in each eye. Maybe your left eye is seeing the face and your right eye is seeing the house, but consciously you don't see both pictures. What you see is an alternation of one picture and the other, so one picture is conscious while the other is not. In the masking illusion in the middle, we flash you a digit, but if it is short enough and presented with a mask of letters you never see the digit, you only see the letters, and so forth. I won't detail all of these illusions. The idea is very simple: we can track in the brain what is happening to this hidden figure, the digit nine here, or the picture of the house when the face is conscious. In this way, we can ask a very simple question: what difference does it make for a stimulus to be consciously reported as seen by the subject or to an unconscious stimulus? There are two additional ingredients: one is the careful quantification of introspection. We can ask for reports, subjects tell us "I am seeing the face," "I am seeing the house," and the remarkable thing is that even monkeys can report. They can click on buttons to tell us what they see and then we can present stimuli just on the threshold. The most beautiful experiments are experiments where we present a single stimulus, always the same but we sort it as a function of whether subjects claim to see it or not to see it. We can, in this case, understand what is the difference between conscious and unconscious representations. Of course, we use all our panoply of brain imaging tools, neuronal recordings to track the fate of these conscious and unconscious stimuli. [4] This is the

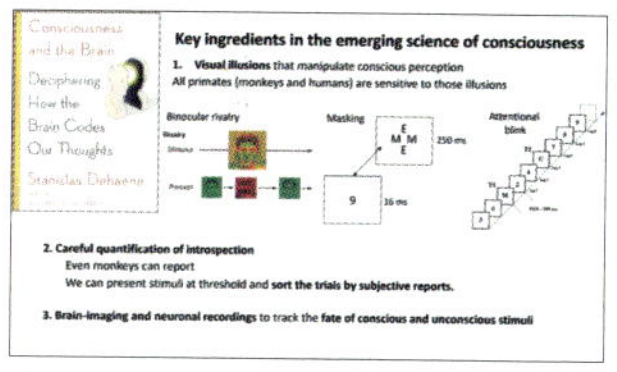

Fig. 4

story that I tell in this book, *Consciousness and the brain*. There have been many experiments. First, there have been many experiments looking at unconscious processing, and this is just a very coarse picture of the literally tens of brain areas that have been found to be activatable by asubliminal stimulus. [5] It is

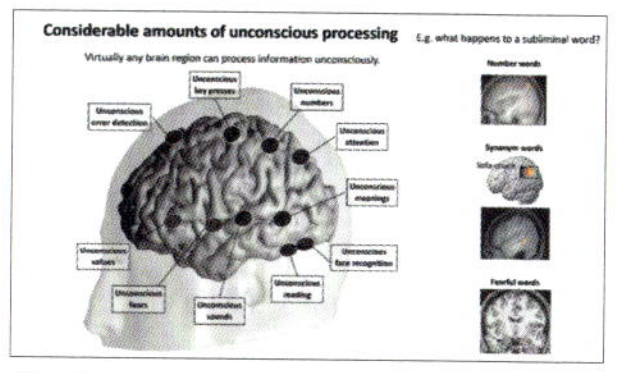

Fig. 5

not quite like I said in the introduction on this famous experiment by Coca-Cola, it never actually occurred, it is a hoax. But real experiments in the lab have allowed us to flash numbers or flash words, including for instance fearful words, and to see that so many of these specialized brain areas can be contacted unconsciously. Probably over 95 % of everything that is computed by our brain is unconscious. We are only conscious of a very small part. What is this part? [6] My colleague Jean-Pierre

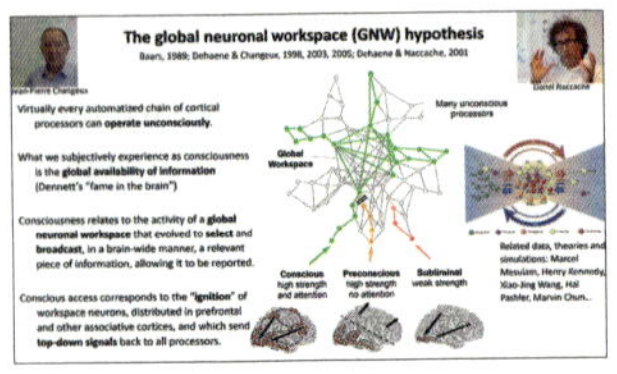

Fig. 6

Changeux was speaking yesterday, and Lionel Naccache and I have been proposing this view of a global neuronal workspace. What do we mean by that? We mean that first of all we have to accept that in the brain virtually any cortical processor can operate unconsciously. Chains of such processors. These processors are represented by little dots on this diagram. The idea is that what we experience as consciousness is when these processors share information. There is a central system that allows a piece of information, the green piece of information here, to become globally available and shared with other brain areas. We think that there is a special architecture in the brain for that, we have called it the global neural workspace: a set of neurons that have these long-distance connections that allow broadcasting information in a brain-wide manner. What we mean by being conscious is that we are able to take the information and report it verbally, but also use it in an extremely flexible manner, because it has been broadcasted to virtually any processor in the brain. Unconscious processing is very specialized, it is local, it is chains of automatized processing. Conscious processing is global broadcasting, sharing of information. At the physiological level, our hypothesis is that consciousness—conscious access—corresponds to the ignition of workspace neurons. They all come active together. Imagine these areas coming active together simultaneously to encode a particular representation: that is what we mean by broadcasting, the sharing of information. Then, of course, there is a sort of global state which remains stable or metastable for a fraction of a second and allows the signals to flow back to the original place of the start. Imagine that force for a short while, less than a second. Your brain is occupied by one particular representation exclusive of others. When you are aware of the green representation here, you cannot become aware simultaneously of another chain, like in orange, or in red. They are subliminal or unconscious. There is a lot of evidence for this model, I won't be able to show you too much tonight. I just want to say that there is anatomical evidence. For instance, there is a system in the brain, at the central level, mostly involving the prefrontal cortex, and what we call associative areas of the brain in the parietal and temporal lobes. This acts as a sort of central sharing system, a bottleneck where information is shared. [7] These are

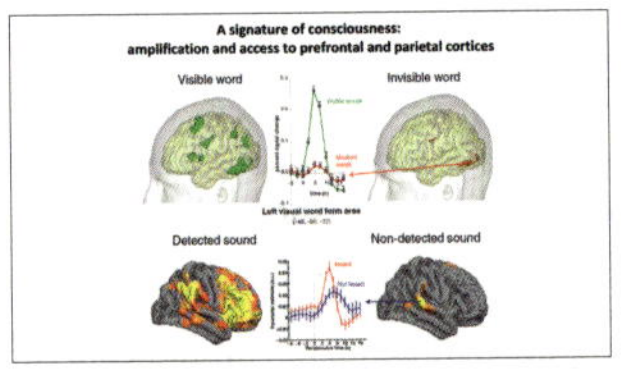

Fig. 7

some of the earliest experiments we have done, where we were flashing these words that are subliminal and trying to see the difference in brain activation. When we presented a word that remained invisible, the subject said there was no word, they couldn't see anything, or a word that was visible and subjects knew what they saw, there was a particular word. You can see some of these differences that are enormous. It is not so difficult to study consciousness, because the differences between conscious and unconscious processing are actually quite large. Unconscious processing remains focused on specialized circuits for reading, including this visual word area that was mentioned in the introduction, this region that cares about the shapes of words. But conscious words activate a much broader circuit that the information gets broadcasted to many more areas that are involved in recognizing the meaning of the word, storing it in memory, and so on. The same sort of thing occurs with sound. If you present an undetected sound, it activates local auditory areas, it goes into the cortex. But if the sound becomes detected, even if it is the very same sound, you have much more global activation, especially in the prefrontal cortex. I want to mention that it is possible these days to go down to a single neuron. [8] This is the beautiful work of

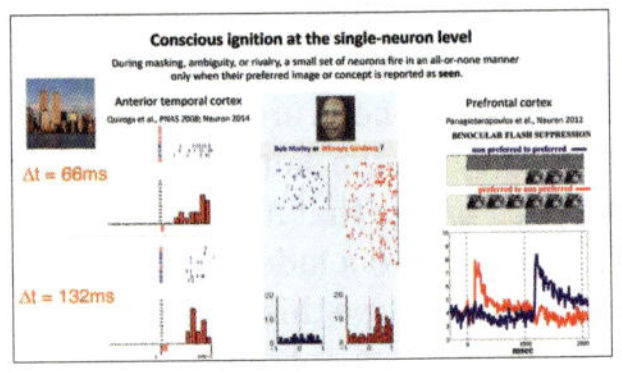

Fig. 8

Rodrigo Quian Quiroga and collaborators with Itzhak Fried. What they have done is to flush pictures like this one and present them in a condition where they are only seen about half of the time. So the trials in blue are the ones where the picture is seen, and you can see that there are neurons—this is a neuron in the anterior temporal cortex—that fire only when that particular picture is seen. These are neurons that code for a concept, such as here the World Trade Center in New York, but these neurons will fire only when the person is consciously aware of that concept. On the right trials, where the person says there was no picture, they couldn't see which picture it was, then there is no firing of the neurons. It gets even better than this. The same research group has been doing a new and more recent experiment where there is an ambiguous picture. This is a funny one where you can see a sort of mixture of Bob Marley and Whoopi Goldberg. Some neurons care about Whoopi Goldberg: they will only activate when the person says "I saw Whoopi Goldberg," but if the very same picture is presented and the person says "I was not conscious of Whoopi Goldberg, it was Bob Marley," then the neuron does not fire to the same level. All of that, of course, started with the monkey. There is a beautiful experiment with the monkey showing that in the prefrontal cortex in particular, but also in the temporal cortex, neurons again code for what the monkey sees, but not for what it doesn't see. In the top condition you can present a face, but because of the particular condition the monkey starts seeing this and then sees the face and you see the neurons firing. But if the monkey starts seeing the face and then the face is erased by these checkerboards then the neuron does not fire. We have single-neuron correlates of consciousness. [9] In a

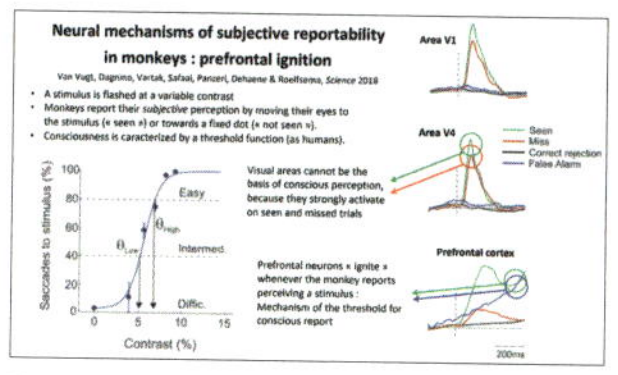

Fig. 9

recent experiment in science with Bram van Vugt, Pieter Roelfsema and colleagues we were able to show that there are mechanisms of consciousness that are extremely similar in the monkey and in the human. What we did here was to flash some stimuli at very low luminance, such that in some cases they were not seen at all, because they were below the threshold for seeing. Then they could be just at the threshold and they could be much easier to see if the contrast was higher. Our colleagues in Amsterdam then recorded from neurons in areas v1, v4, and prefrontal cortex and they were able to see what happens to the stimuli in the brain, what sort of activation they give as a function of whether the monkey reports that he has seen the stimulus because he makes a circade to it, or whether the monkey reports that he has not seen the stimulus because he makes a second to another location. We can see there is a lot of information, of course, in this paper which is technical, but I just want to show you some very basic stuff. You can see that you can get essentially the same level of activation in visual area v4, whether the stimulus is seen by the monkey or whether it is missed. It was there, it created this huge wave of activation, but the monkey reports there was no stimulus. This means that these visual areas paradoxically are not sufficient to understand consciousness, even consciousness of an extremely small visual dot is not in visual areas. Where is it? If you look at the prefrontal cortex you see that there is a very good correlate of consciousness. The prefrontal cortex ignites in this very speeded activation, suddenly, on "seen" trials. On the contrary, if the same exact stimulus is presented but the monkey misses it, there is the beginning of activation but then it fails. This difference becomes huge, we call that a non-linear ignition. It is non-linear in the sense that it is unrelated to the size of the stimulus, or to the contrast of the stimulus. It is an intrinsic shift, a balance between a high state of activity where you are conscious, and a low state of activity where you are not conscious. This is what we call ignition. And notice that even on the "false alarm," so there was no stimulus, but the monkey said "I think I saw something" and made a eye saccade, you can see that there is this ignition occurring in the monkey's brain. All of these experiments initially were done with conditions in which we stimulate the brain, but of course the brain is conscious even when it is not stimulated. If you are lying in bed, in darkness, without any sort of visual stimuli, you can still have mental images and have conscious activity. It has been interesting to ask: how can we detect consciousness in this situation when there is no stimulation? [10] We still see that there is a

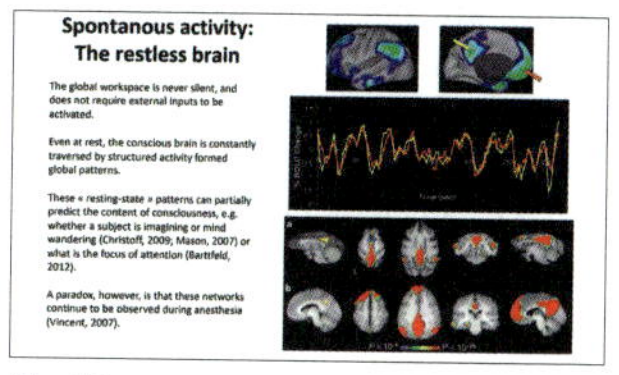

Fig. 10

conscious brain. Beautiful works, particularly by Marcus Raichle, have been showing that there seems to be a correlate of consciousness in the spontaneous resting state activity of the brain. The brain is never inactive, it keeps changing its activity all the time. These areas wax and wane together, and this is a possible correlation with of consciousness, in the sense that it partially predicts what the subject is thinking about at a particular moment. For instance, if you are doing mental calculations you will activate these dorsal areas that I have studied that have to do with arithmetic and mathematics. If you convey mental images you will activate your visual areas. If you think of playing tennis, you might activate these anterior areas for pre-motor gestures, and so on. But there was a paradox. The paradox is that many of these networks, many of this spontaneous brain activity continue to be active under anesthesia. What is unique to consciousness and what is not? [11] We were able to

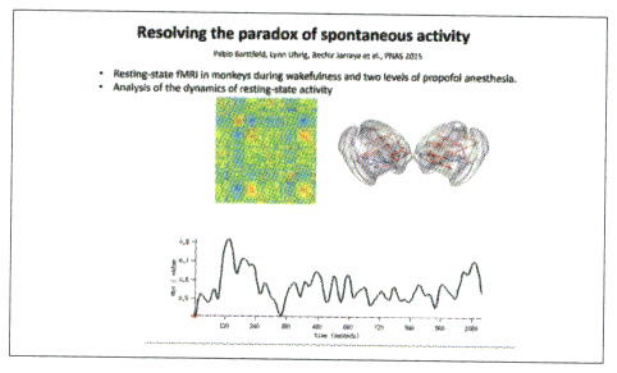

Fig. 11

sort out this question recently, in the past five years, by looking at the dynamics of activity. What you see here is the dynamics of activity in the monkey's brain as he is awake. You can see that it is constantly changing. It is what William James called the perchings and flights of a bird. Consciousness is like a bird that sometimes sits and has a stable state, but often flies away to a new state. It constantly changes, it is not a stable state in the brain for a long time. So we thought that maybe in these dynamics we could find a signature of consciousness. [12] That is what we found. In the

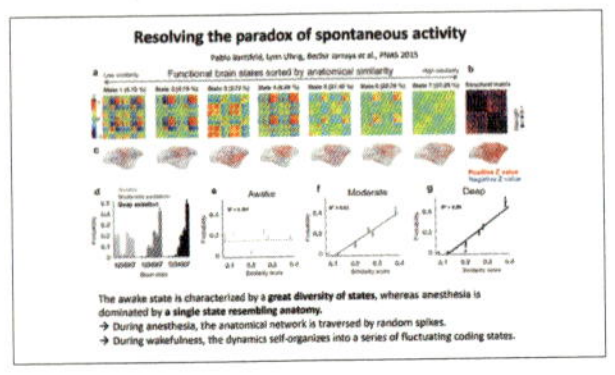

Fig. 12

monkey we were able to anesthetize the monkey and compare different states of anesthesia, so there is the awake state, moderately anesthetized, and deeply anesthetized state, and we were able to look at these dynamics of brain activity as a function of consciousness. What we have found is that there is a dramatic difference: when you are anesthetized your brain is no longer as dynamic. Its dynamics are completely dominated by one particular state of activity, which we can see here on the right, which is essentially the matrix of structural connections, the anatomy of the brain is driving the activity. It is very simple. Even during anesthesia, the brain is not completely silent, but it is essentially random spikes which are traveling through the anatomical network. It is one state of activity, if you want, but it is not a coding state, it is just a state of traveling waves in a network. But when you are awake it is completely different, you have all of this richness of other states that come in and become much more frequent than the state on the right. They essentially tell you that the brain is changing its configuration in order to have different kinds of thoughts. When we quantified this capacity to have multiple dynamic states of the brain—and these are all global states, they are all ignition-like, in the sense that they concern many different areas—we found that this is a good signature of consciousness. The more diversity of dynamic states, the more it is likely that the subject is awake and conscious. What is very nice is that this work, which was started with the monkey, was generalized to human subjects. [13] This is the newer work that was

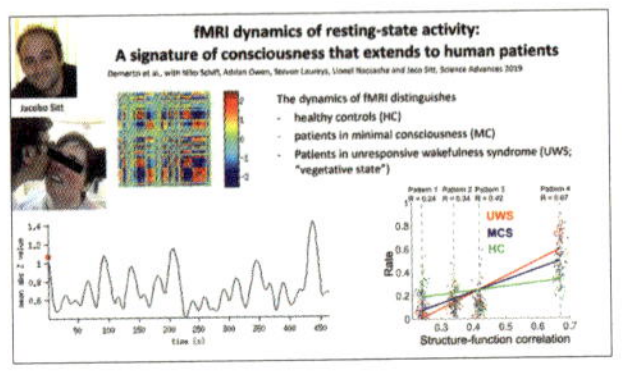

Fig. 13

done in collaboration with many, if not all of the labs, that are collaborating together to understand consciousness in human patients. What we saw is that if we study healthy controls, they have the same dynamic signatures of changing states as we had in the awake monkeys. But if we move to patients in minimal consciousness and especially patients in so-called vegetative states, we think that most of these patients are completely unconscious because they have stopped responding to external stimuli. We found that they tend to be dominated, in their dynamics, by this state that resembles the structure of the brain, the anatomy. [14] I don't expect you to understand

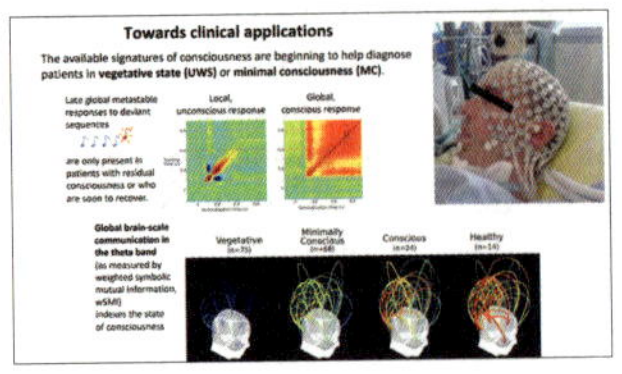

Fig. 14

in detail all of these findings but I do want you to come out with this basic understanding. We now have signatures of consciousness. Many of these signatures started to be found in the monkey and relate to this ignition of a global brain state, but once we found them in the monkey we were able also to generalize them to the human subjects. I am showing you even more signatures. We don't have to use visual stimuli, we can use auditory sequences and see how far they propagate in the brain and if they create this global state which is relatively late, after 200 milliseconds of propagation of activity in the brain, then it is likely that the patient is conscious. Likewise, we can just look at spontaneous activity, whether the brain is exchanging signals at a distance through this global workspace and if the brain is exchanging signals it is likely that the brain is a conscious brain. Minimally conscious or vegetative state patients don't seem to have enough of this global communication. This has reached a level now where we really understand consciousness enough to diagnose patients and tell to a large extent whether they are conscious or not. [15] Only three years ago we pub-

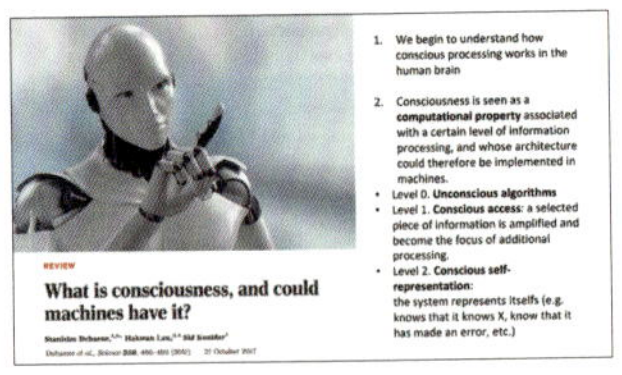

Fig. 15

lished these papers to ask what is consciousness, and especially if machines could have it. I think it is very clear that we begin to understand how conscious processing works in the human brain and that consciousness, according to us, is just a computational property, it is a level of information processing. The machine of the brain, at a certain level, needs to share information, so this architecture could be implemented in machines, it is just an aspect of the machine of the brain. Of course, we can distinguish at first level whether it is just unconscious processing in the brain. I have told you mostly about this level one of conscious access, where the brain needs to select a particular piece of information and share it with other areas, so the information is amplified and it becomes the focus of additional processing and sharing through other modules. I want to tell you now a little bit more about what we called in this paper "level two," the conscious self-representation. There is something more. We are able not only to hold on to information shared, but also we know that we have the information. In humans this goes very far. We have a real sense of self that tells us we know that we know, we know that you know that I know, we have this theory of mind which includes our own mind.

I would like to suggest that there is here a sort of level that may be a bit more unique to humans, or maybe not unique but better developed in humans. [16]

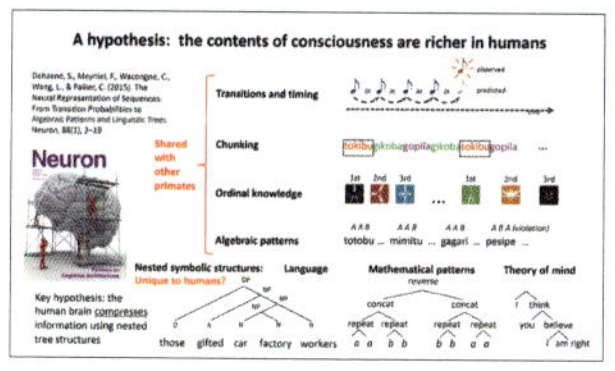

Fig. 16

In this paper in the journal *Neuron* in 2015 we speculated, based on a review of the literature, about what could be unique to human level representations. There is no doubt that there is a lot that humans and other primates share. In this illustration we have all that we share in the domain of sequence processing. All primates are able to understand that there is a sequence of notes, for instance, or a sequence of syllables being presented, and it may be surprising to you, but even non-human primates are able to chunk and discover that there are words in this sequence, they are able to understand that there is a first element and a second and a third. They are even able to understand that there are patterns, like totobu, mimitu, which is A A B. There seems to be something that is unique to humans and that we see in the domain of language but also in the domain of mathematical patterns, and perhaps in the domain of the theory of mind. This is a sense of recursion and the building of three structures through recursive structures. In a domain of language, for instance, the work of Noam Chomsky and much of modern linguistics—Andrea Moro will speak about that in a minute—has told us that when I give you a little sentence or phrase like "Those gifted car factory workers" you should not think of it as a series of words. It is not just a sequence, it is a tree. For instance, it is "car factory," these two words belong together, it is a factory that makes cars. The workers are from the car factory, so the workers are in the factory. These are nested structures that we also find in mathematics and of course in the theory of mind. If I tell you "I think you believe I am right" what you think about is really this phrase which itself is made of a sub-phrase and so on. Our key hypothesis here is that the brain is able to compress information using nested tree structures. [17]

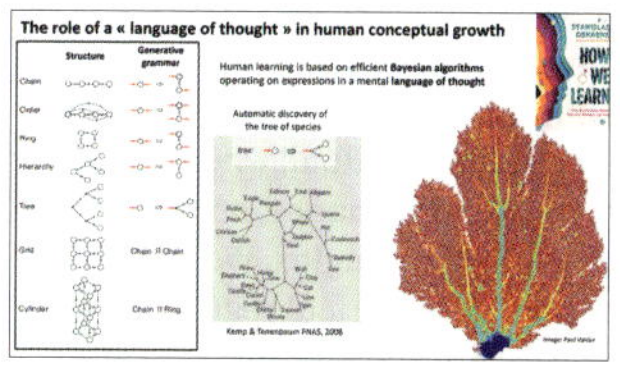

Fig. 17

In my book *How We Learn* I try to defend the idea that this is a new learning mechanism in humans. I am not alone in this idea, by the way, and here I am illustrating the work of Josh Tenenbaum, who has been arguing that thanks to grammatical structures we can be scientists, we can acquire scientific information about the external world because we have a generative grammar of concepts. It is not just inside the surface of language, the sentences that we produce, but inside our brain we have a language of thought that allows us to create combinatorial structures. This grammar may be so general that it can express many different things, such as a map, for instance, a grid, but also a tree structure, or hierarchy, or just a linear order like the numbers. Josh Tenenbaum and his colleagues were able to show that in this way you can model the human acquisition of a concept, such as the tree structure of a species—the discovery by Charles Darwin that the species are best thought about as a tree. The general idea is that the human mind has this extraordinary ability to differentiate new concepts based on combinations of simpler ones. Combinations through grammatical structures allow us to differentiate our space of concepts and I like to use this metaphor of a beautiful tree or a beautiful coral, that is what our brains develop as we learn: we develop three structures that cover our understanding of the external world. How do we know it is true where we don't? This is the topic of a lot of recent research. [18]

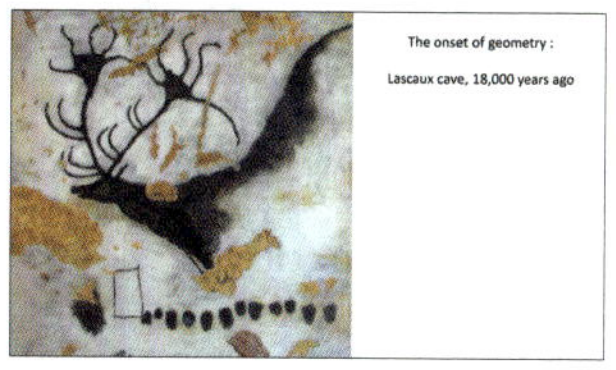

Fig. 18

I would like to finish my talk by discussing extremely recent research that we have been doing in the domain of mathematics, in the domain of the concepts of geometry. This remarkable human ability to make symbolic drawings, such as this deer here, but also this square—this is in a Lascaux cave, 18.000 years ago. [19]

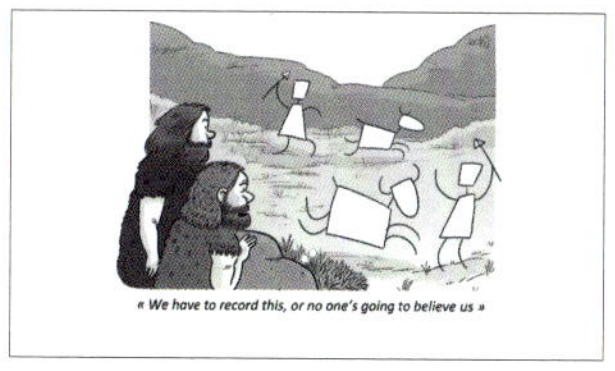

Fig. 19

I stopped for a cartoon for a second. I hope you like it. These are humans looking at the hunting scene and they say "We have to record this or no one is going to believe us" because they are seeing the world in geometric shapes. Of course, it is a joke, but if this view is correct only humans are able to think of geometric shapes because this requires a combination of concepts, combinations of lines that are parallel, the notion that there are four lines, just four lines is already a first syntactical operation, four equal sides and four equal right angles: that is the definition of a square. According to the view that I am defending, there is something quite unique about the ability of humans to think about such concepts. How old is this sense of geometry? It is probably extremely old. [20] This

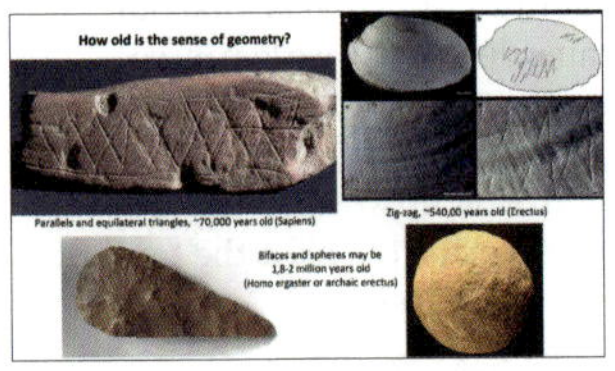

Fig. 20

is an artifact from 70.000 years old that has parallel lines forming equilateral triangles. This zigzag is already 540.000 years old and so it is not *Homo sapiens*, it is *Homo erectus* who has drawn these parallel lines forming a zigzag. Anthropologists who find these patterns immediately knew that it is a human, there is no other animal, even a chimpanzee doesn't do this sort of pattern. Bifaces and spheroids point to the fact that this could be as old as perhaps two million years old. A very old aspect of humanization. [21] In the lab, we

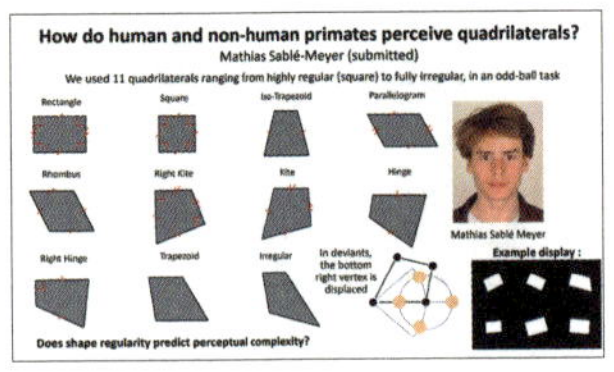

Fig. 21

have been studying whether there is something quite unique to humans. So we try to find a test that both humans and monkeys could pass. We decided that we would have this test of the outlier: I will let you look at this and you have to decide where is the different shape. You have a touch screen and you click, and presumably by now you found that you have to click on this shape, which is different. Mathias Sablé Mayer, for his PhD, has been designing all of these shapes that are extremely well matched, they are all quadrilaterals of the same total size of the distances between the dots and so on. But some are extremely regular, like the rectangle and the square, and humans understand these regularities, and some become more and more irregular. What we do is ask subjects to detect whether there is a different one, where a corner has been displaced. [22] What we have found is a remark-

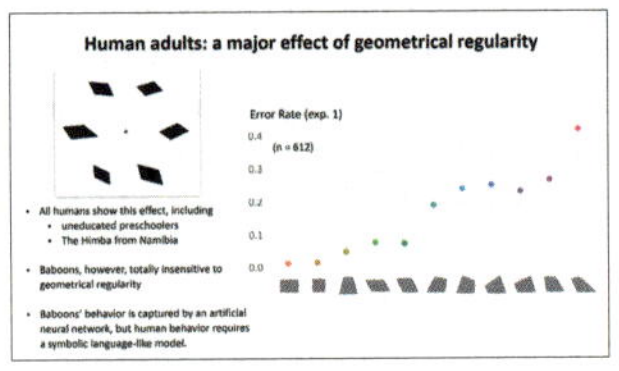

Fig. 22

able effect of geometrical regularity, as humans we are much better in detecting the oddball when it is among regular shapes, such as squares or rectangles. We have a sense of geometrical regularity, and when we are shown contrary-wise an irregular figure we are very poor, we are unable to decide that this shape is now different, that the corner has been displaced. Other shapes fall in between; for instance, you can find here this shape, which is irregular among parallelograms, but parallelograms are already more difficult than squares or rectangles. A sort of monotonous effect of regularity. [23] The first thing we did was to show that this is a human universal, at least in our hands. Whenever we do this experiment, in all sorts of variants that I won't have time to explain, we find that humans show this effect. It is even true of preschoolers or even if we go to Africa, to this Himba population. Both of these groups of subjects are

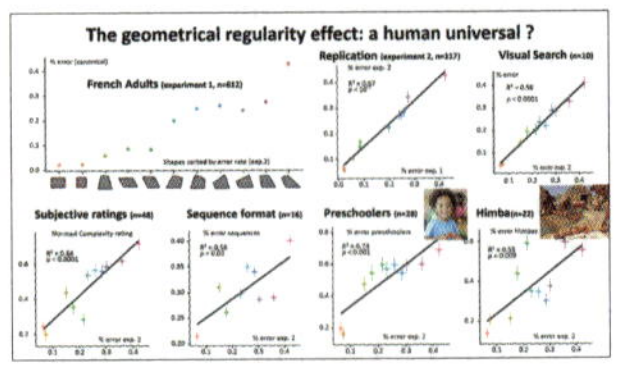

Fig. 23

not well educated, they don't go to school. This is before school. In the case of the Himba, they don't even have names for these shapes. Yet, they always found immediately that a square rectangle is more regular than a random quadrilateral. The question, of course, is if this is unique to humans. Are we the only species to be conscious of geometrical regularities? [24] We collabo-

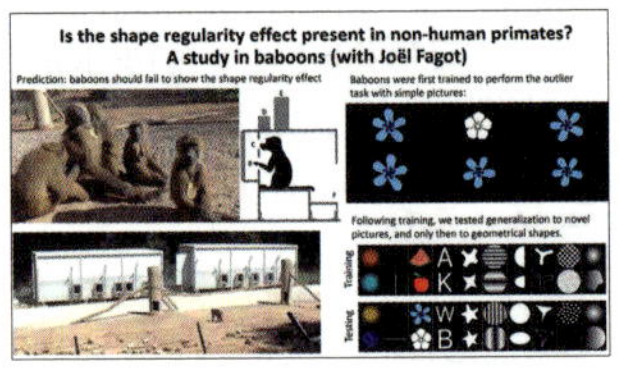

Fig. 24

rated with Joel Fagot, in the south of France. Joel has put together an amazing facility where baboons are in a group, they live together. They are essentially free to roam in this arena, but at any time they can enter this booth and in the booth they'll find a computer and a video game. 24 hours a day they can enter willingly and play a video game. Some go at 2 in the morning and go play a little bit. This is an amazing facility because it allows you to collect data from more than 30 baboons and to collect thousands of trials per week. It is an amazingly productive area for the observation of behavior. The monkeys are essentially free you know, they collaborate. The computer recognizes which monkey is entering the booth and presenting just the right trailers for this particular individual. We trained the baboons to click on the different picture. This is our outlier task. First, we train them with pictures like this one, which is not geometric, it is just a different flower. We train them with all of these pairs of elements, so click on the orange among blue or the blue among orange. We showed that monkeys, the baboons, could understand the task, they understood the outlier task and they understood that they had to click on the shape which is different from all of these shapes. They could generalize to these new shapes. [25] Once

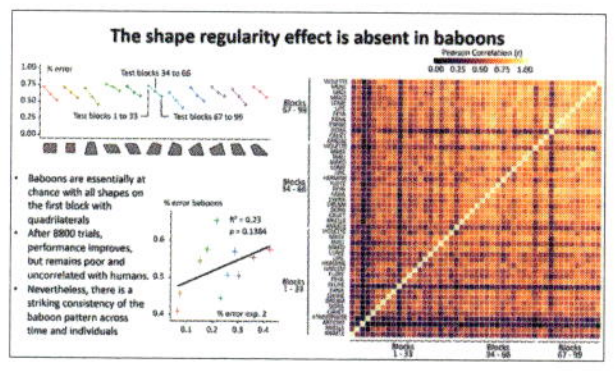

Fig. 25

we did that, we tried our geometrical regularity test and the surprise was there was no effect whatsoever. All of these shapes, whether they are highly regular, are treated the same as the irregular ones for baboons. Baboons are completely different from humans on this test, they have very little regression with regularity. It is very different from humans. It is not that they behave randomly, they all do the same sort of thing and they are all correlated with each other. Each line here is one particular baboon at different times: you can see that the hot colors indicate that they are all correlated with each other, so they do something, but they don't do it like humans. What do they do, again without going into the details? [26] We are able to under-

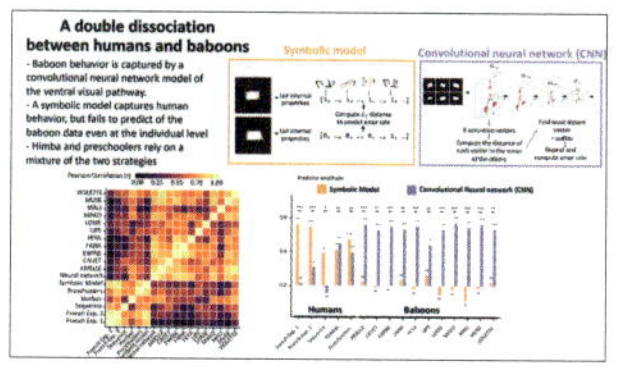

Fig. 26

stand what the baboons do with the model of the ventral visual areas. The unconscious processing pathways allow you to process a geometrical shape just like any other picture, a face or a location, by a sort of feed-forward network that identifies what it is. This can explain the baboon's behavior. All of the baboons are well predicted by this convolutional neural network, but humans are not. To explain humans, we have to have a completely different kind of model, a symbolic model that takes into account the symbolic geometrical properties. Are the two lines equal? Are these angles equal? Are these angles straight, right angles? Are there equal sides? And so on and so forth. When we list all of these properties and allow the system to assemble them into the representation of the picture, then we get a good model for what humans do. They are sensitive to regularity, in a way that the baboons are not. We really think these are two strategies. When we test the preschoolers or the Himbas, they actually use a mixture of these two strategies. As humans, we have, of course, such a visual network and we are conscious of the external world in a similar way to the monkeys, but we can also adopt this symbolic stance and look at the world through the picture of this language of souls, where we discover scientific regularities. We can think of something like a square, and, of course, many more complex thoughts, such as all of the concepts of mathematics and the construction of the pyramid of mathematics. I am reaching my conclusion now and it will be very simple. [27] I think

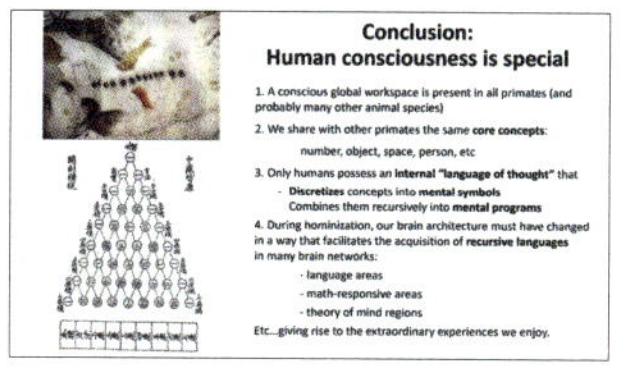

Fig. 27

that there is something special about human consciousness. First of all, the conscious global workspace is present in all primates and probably many other animal species, such as birds, for instance. We share with other primates the same core concepts of number, object, space, person, but we possess something more, we possess a language of thought that allows us to discretize these concepts into mental symbols and to combine them recursively into mental programs. I would like to argue—and this is of course our hypothesis that needs to be proven by further work—that during hominization the human brain architecture must have changed. We already have a workspace, but that workspace allows the construction of recursive languages in several different brain networks. Language areas, of course, but also mass responsive areas, theory of mind regions giving rise to the externally complex experiences that we enjoy. If this view is correct, the basic apparatus of consciousness is the same, but the contents of consciousness are much richer in humans due to this ability to think scientifically, and recursively, about the external world.

Impossible Languages: Infinity as the Fingerprint of the Human Mind

ANDREA MORO

What happens when linguistics meets neuroscience? This is what I would like to show you tonight. I will try to answer a very deep question: What makes language a human language? And by doing so, I will try to answer another, deeper question: What makes a being a human being? Let me start straightforwardly by observing the capacity that appears to be specific of humans. [1] Let's pretend we have a

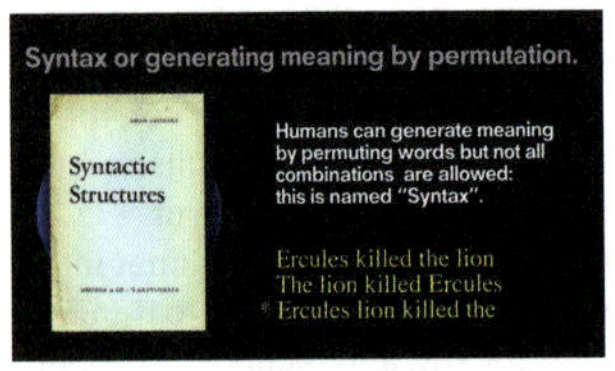

Fig. 1

micro dictionary made up of just four words: "killed," "the," "Hercules," and "lion." It is easy for any human brain to combine them in a string and say "Hercules killed the lion," or, even in another order, "The lion killed Hercules." We immediately understand that for us it is easy to convey different meanings with the same bricks, the same wall in a different order. Not all orders are possible, in fact there is at least one order which makes no sense: "Hercules lion killed the." Thus, syntax, in first approximation, can be defined as the capacity to generate meaning by permuting the order of words. Of course this is just a naïve presentation. Syntax has an extremely complex mathematical structure that started to be studied in the late 50s by Noam Chomsky at MIT, and his first theorem concerning the nature of human syntax is the basic standard to start to understand this aspect of any human language. One thing that I just said—that all and only humans can use syntax to express their thoughts—is not a new idea. At least Descartes, in his *Discours de la methode*, was able to recognize exactly this capacity. [2] Let's read

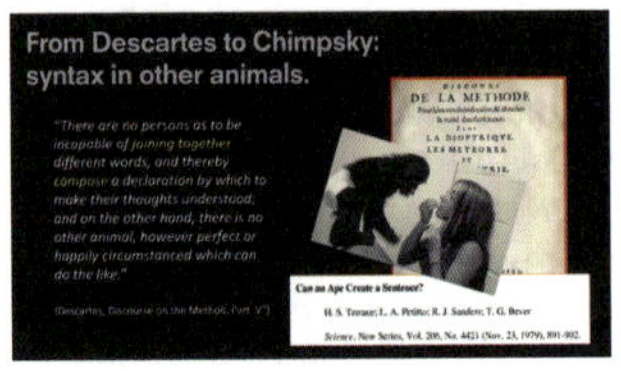

Fig. 2

his own words translated in English: "There are no persons as to be incapable of joining together different words, and thereby compose a declaration by which to make their thoughts understood; and on the other hand, there is no other animal, however perfect or happily circumstanced which can do the like." That was already clear by that time. One could ask whether scientists checked all animals, if there is no evidence of certain species of insects, maybe, on the other side of the world, that can produce meaning by permuting the same words in a string. Science always has to go through generalization, but we have to be careful and at least try to compare the kind of animals that we feel to be closer to us, for example chimps. There was an experiment in the late 70s led by Tom Bever, where a group of scientists lived with a little chimp. The name of the chip was Nim Chimpsky, because of the influence of the linguist. Here you see a picture of Laura-Ann Petitto, who was trying to teach it sign language—which, by the way, we know to be a complete human grammar, with the only fact that you do not express word with sound but with gesture—and Nim was able to learn a lot of words, 128 different words. But when it comes to sentences not one of 119 different sequences made any difference because of the syntax of the words. We can be pretty sure to say that when we understand syntax, we understand the fingerprints of human languages, and in fact probably the fingerprint of the human mind itself. Now you may ask "What do syntacticians do when they go to work every morning?" I will try to show you what I do every morning in a very approximate way that is this one. You may ask what I am showing now and I may, on my turn, ask you to describe what you see. [3]

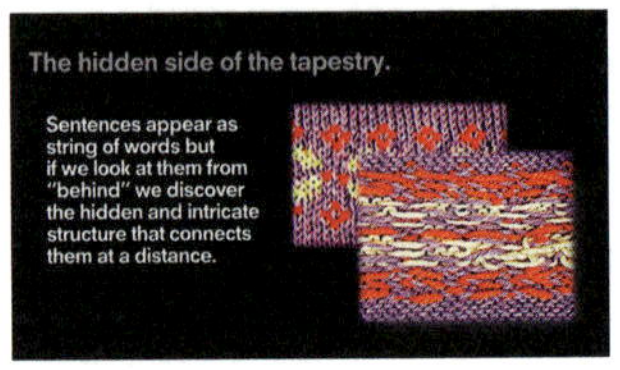

Fig. 3

If you had to describe what you see on the phone, for example, you would say that there are violet dots, and then red dots, and then white and yellow dots, and then from this dot to this dot there is a distance etc. You would describe these scattered amounts of elements in a plate. What the linguist does is turn it upside down, in a sense, or look at the other face, and see what happens to elements that are connected at a distance. Sentences appear as a string of words, but if we look at them from "behind" we discover the hidden and intricate structure that connects them at a distance. Though this is precisely what we need to do if we want to understand syntax and the core of any human language. Let's shift from this pictorial way of looking at stuff with a precise example, because I want you to get at least the real mechanics of what I mean by syntax. [4] Let's

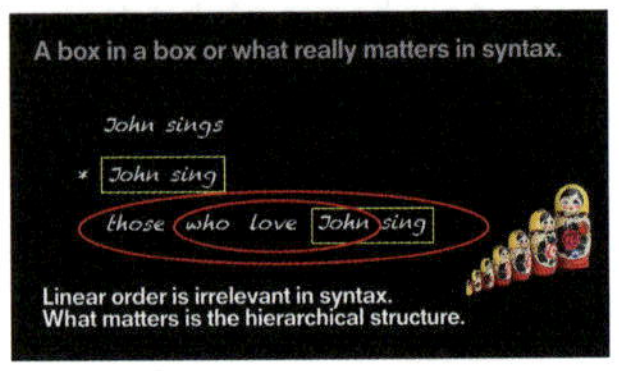

Fig. 4

make an experiment together. Let's say this string of words, a noun and a verb, a subject and a predicate, depending on the kind of analysis you get. "John sings." There is kind of a glue that keeps them together and makes a sentence out of these two elements. Now "John" is singular, "sings" is singular. What happens if "John," which remains singular, is connected with a verb which is plural? The sentence is ungrammatical. The sequence is broken by the fact that there is no match in the number. "John" is singular, "sing" is plural. My question is: is it always the case that this very sequence of two elements never occurs in a grammatical sentence? The answer, although it seems to be no, is in fact yes: it is easy enough for any human being to understand that if I make "John" be preceded by certain words, the sequence becomes legal. The very same sequence "John sing" becomes possible in "Those who love John sing." Why? We understand that very easily, because "John" is not the subject of "sing," it is the object of "love," and in fact the subject of "sing" is "those." What you see here in front of you is exactly the core capacity of any human grammar, and it tells you one thing: there is one object, a sentence, "Love John," that can be included in the same object, that is another sentence. In other words, we have two consequences: the first, the linear order is irrelevant in syntax. "John sings" and "John sing" have the same linear order, but in one case it is okay, in the other case it is not. What matters is the hierarchical structure. In a sense, it is like playing with an object and I think that it is perhaps not by chance that kids—maybe Stan would say something about it—are actually happy to play with these things. Playing is part of our cognitive capacity. There is another interesting aspect to be combined with this to reach the recipe for any human language, and it is the following. [5]

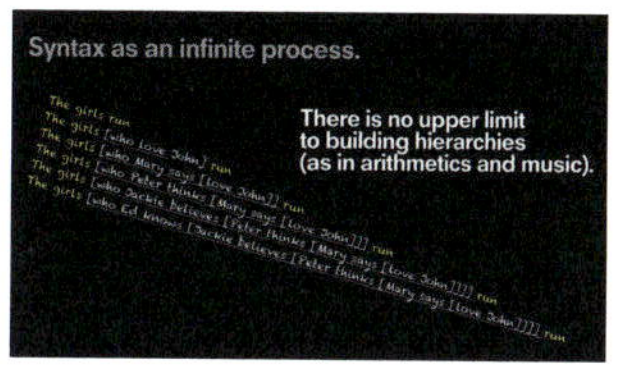

Fig. 5

Let me go back to a kind of example like before. Take "The girls run." Then you may ask what happened to my slide, as if something is rolling down on the one hand. But I want to do the same strategy as before: you have "the girls" and you have "run." Now look what I am doing: I put a sentence in between the two: "The girls who love John run," but I can do that again. "The girls who Mary says love John run," and again, "The girls who Peter thinks Mary says love John run," and again "The girls who Jackie believes Peter thinks Mary says love John, run," and again "The girls who Ed knows Jackie believes Peter thinks Mary says love John run." I think you are bored enough to let me stop with this simple example, but I tell you two things: the idea of encapsulating things, nesting things, boxing things one in another has no upper limit. Building hierarchies can be done forever. Exactly as was said before, this also happened in the other two codes that is arithmetic and music. Taking together these two facts: the fact that you can include something in the same thing and the fact that this process can be iterated at infinitum, you reach the fundamental principle of syntax that was already expressed in syntactic structures. [6] Namely, no syntactic rule can be

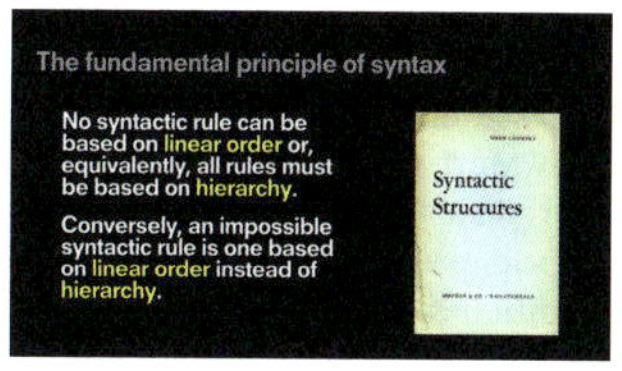

Fig. 6

based on linear order, or, equivalently, all rules must be based on the hierarchy. In fact, you can express the same principle upside down and say that an impossible syntactic rule is one based on linear order instead of hierarchy. We face the first interesting problem, which linguistics and neuroscience face together and raise one question that lies at the very center of reflection of philosophy, ever since Western philosophy started, at least with Plato. Is syntax, this complicated object I am trying to describe, determined by nature or by nurture? [7] This is exactly the problem that

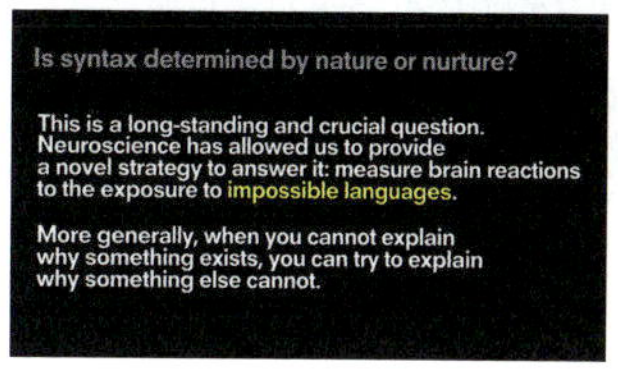

Fig. 7

we have been facing for 2500 years. This is a long-standing crucial question. Neuroscience has allowed us to provide a novel strategy to answer it, that is let's try to measure brain reaction to the exposure of impossible languages. It is the very first time that humans can try that, because neuroscience has offered us the capability, with neuroimaging, to explore what happens inside of a brain without opening the skull of a person, which is kind of an advantage if you want to study how people speak. More generally, though, I would like to highlight that the idea of exploring reality through what is not real, necessarily not real, is a very innovative strategy that people have especially expressed in mathematics. In general, when you cannot explain why something exists, you can try to explain why something else cannot exist. This is the strategy I would like to show you. First of all, I have a kind of a preliminary question, one that you may also have in mind. I am speaking of syntax and speaking of the brain, but how do we know that there is a network inside of our brain which is dedicated to syntax? Why am I asking that? Because when I speak of course I give you the syntactic rules, I exploit the syntactic rules but I also convey meaning, and meaning and syntax come together. How can we be sure that there is a network for syntax and that everything cannot be traced back to meaning at all? You can do something—which is a privilege when it comes to working. [8] What I have here is impossible plants taken from the very famous fake manuscript, the Voynich Manuscript, that also has impossible writing there. We

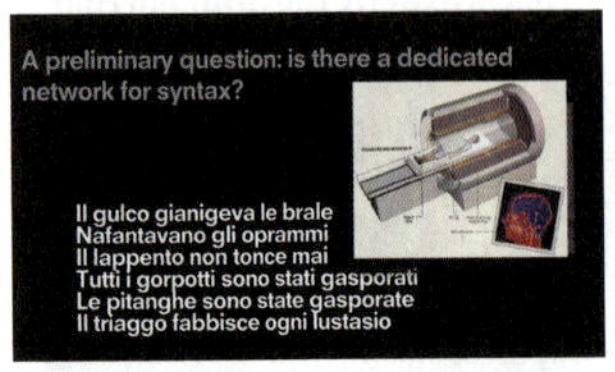

Fig. 8

made an impossible language. Although I am sure that I have the best, simultaneous translation now on here, I have to have a slide, either two slides that cannot be translated. Let me first of all try to speak to those who understand Italian and make this experiment. Look at this sentence: "Il gulco gianigeva le brale." No Italian will understand that, but in fact no other person in the world, because this is just pure syntax without meaning. I can go on. "Nafantavano gli oprammi. Il lappento non tonce mai. Tutti i gorpotti sono stati gasporati. Le pitanghe sono state gasporate. Il triaggo fabbisce ogni lustasio." I hope that the person that may accidentally have gone through this talk right now would understand that I am not talking any language I don't know, I am just making it up. Similar things can be done in English, and then what you do is try to see what the brain computes when you have pure synthetic stuff in your brain. [9] Same thing for

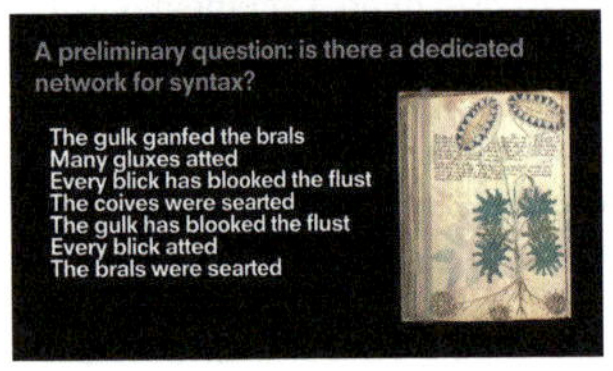

Fig. 9

those who speak English: "The gulk ganfed the brals. Many gluxes atted. Every blick has blooked the flust. The coives were searted. The gulk has blooked the flust. Every blick atted. The brals were searted." Similar things. So we have a language that has no meaning at all, although it may be evocative. In the literature of all countries you have poems that are made of fake words: you have the *Jabberwocky* by Lewis Caroll, you have *Il lonfo* by the Italian Fosco Maraini, you have many examples. What we did in the experiment was try to reduce to the minimum the amount of evocative effect, and what we found was a specific activation of the brain. [10]

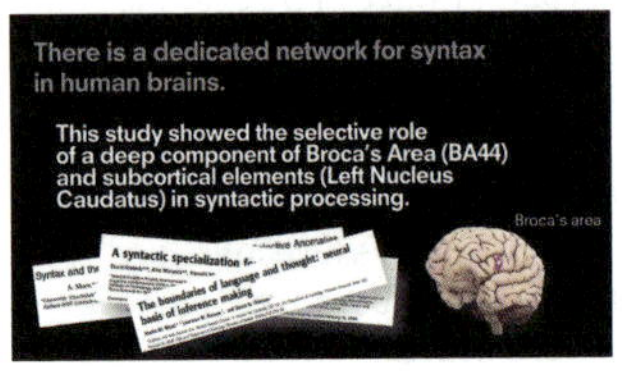

Fig. 10

This work was done at San Raffaele first, and there are persons—Stefano Cappa, Daniela Perani, and Marco Tettamanti—who were the first small community that allowed me to express this. I owe a lot to Stefano because even the history of science tells you how difficult it is when two disciplines try to get together, and that was our first experiment. The first result showed the selective role of a deep component of Broca's area (BA44) and subcortical elements (the left nucleus caudatus) in syntactic processing. Interestingly enough, this experiment was replicated by another group, one by David Embick and Alec Maranta at MIT, another one by Martin Monti and Daniel Osherson at universities in Cambridge and Sheffield. The preliminary question has been answered: We do have a syntactic network for syntax, because when you are proposed with a language that has syntax but doesn't have meaning you do have a specific activation. Now we can get to the core experiment: the one of impossible languages. The experiment was to look at what happens in Broca's area. [11] We took a group of monolingual native speakers of German, we taught them a micro Italian, that is an Italian made of 12 verbs, 12 nouns, a few prepositions, articles and so on. Then we measured Broca's activity relationed with the possible and impossible rules. Again, the possible were the ones

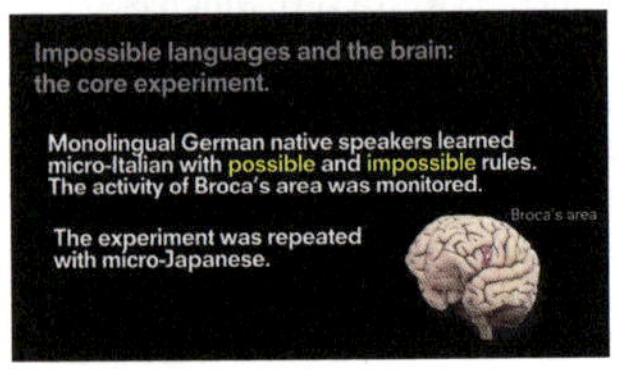

Fig. 11

based on hierarchy, that is on strings that can be as long as possible, and the impossible rules were made of linear order. To be sure that we did not just detect a similarity among languages, we did the experiment with micro Japanese: the experiment turned out to be done as follows. There were in fact three experiments, I chose to illustrate you just one of them. [12]

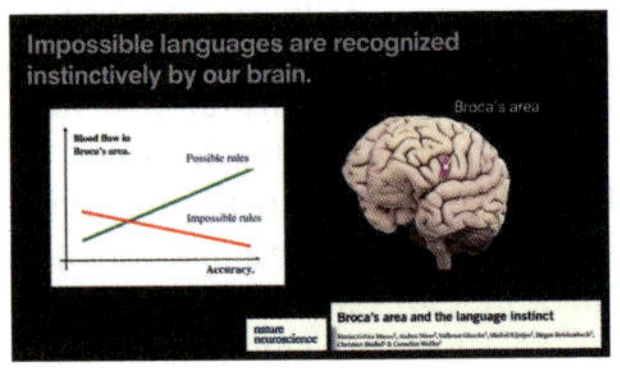

Fig. 12

How was the experiment designed? With a very simple paradigm. First, we measured the accuracy, that is, we wanted to see how better these people were doing after training with these rules. Then we measured the blood flow in Broca's area, assuming that the more you have blood, the more the area was computing the rules. What happens with possible rules? What happened is what everyone thought would happen, namely, the better you are in understanding the rules, the more Broca's area calls for energy to activate the metabolism and to process the rules. But with impossible rules we had exactly the opposite result: Broca's area was progressively inhibited and the brain recognized the fact that the rules that were not based on hierarchy, but just on linear order, were fake rules. In a sense, the answer to the question of whether it was nature or nurture, with this experiment shows that it is compatible with the idea that it is nature, because nurture, of course, cannot decide which part of the brain to use, because we don't have control of the other parts of our brain. Then there is actually one fact that I was not going to emphasize too much, but I thought that, especially in relation to what Stan said before, could be very useful. That is, one could say that syntax and sound are just the media to communicate my thought to someone else who is listening. But what happens when I don't speak? What happens when I just think or read in my mind? As you imagine, the same question could be "What happens to people who are in a vegetative state or whose consciousness is not active in the same sense that a normal subject is?" We made an experiment to try to understand what happens to language when language never leaves your mouth, that is during the so-called endophasic activity, or inner speech. We adopted a completely different technique, that is awake surgery—the person here belonged to a different group, in particular the neurosurgeon Lorenzo Magrassi from the University of Pavia, and engineer Giuseppe Aromataris—and it has been possible to prove that even when we just think in words, their sound is represented in the brain. [13]

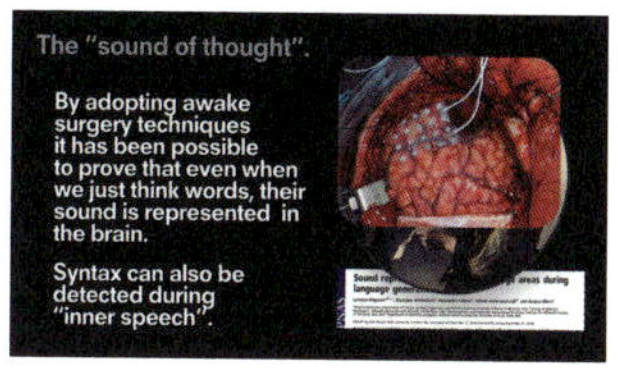

Fig. 13

In other words, syntax can also be detected during inner speech, not because we have or we don't have any special clue about the syntactic structure, but because at least we can measure the electric activity of the brain and know that at that level, without speaking, sound is represented. Of course, like any discovery that can be used to help people, to get their thoughts out of their mind if they are paralyzed, or in a sci-fi future policy, people can get your secrets from your brain. That is up to ethics, I am not going to discuss that here. What is important to say is that syntax is not only something that is useful to communicate, but it is something which is incorporated in the very language processing, even without communication. [14] We can

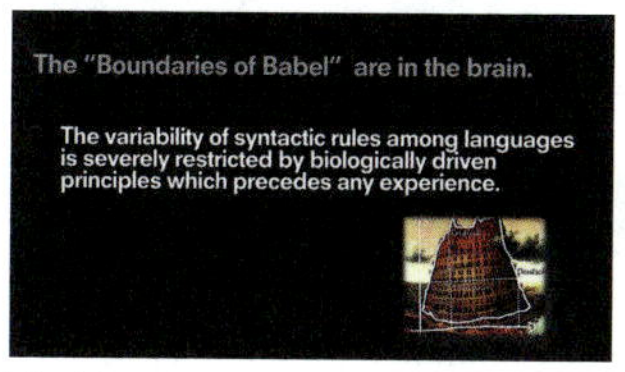

Fig. 14

conclude that the boundaries of Babel, that is the limit in which languages must keep their rules—for example they cannot be linear, they have to be hierarchical—are in the brain. The variability of syntactic rules among languages is severely restricted by biologically driven principle, which precede any experience. This is crucial. Before going to that, let me just summarize here. If neuroscience, when meeting linguistics, gives you the idea that the format of language rules is shaped by the architecture of the brain, we have a lot of consequences to compute. For example, there is one consequence, which is obvious: if languages are the execution of a genetic program, as our liver, our heart, our lungs are, no one person can have more complex grammar than others, because there is no one who has a more complex liver or a more complex heart. What we see as different in complexity must be something we misunderstand. There is something very delicate I want to say here. [15] Suppose that the grammar is at least divided into four segments, one that relates to sounds (phonetics), one that relates to words (morphology), one that relates to syntax (the composition of words), and one that relates to meaning, it is the case that when you touch one component, all the others are touched. In other words, language is more similar to a Rubik's cube than

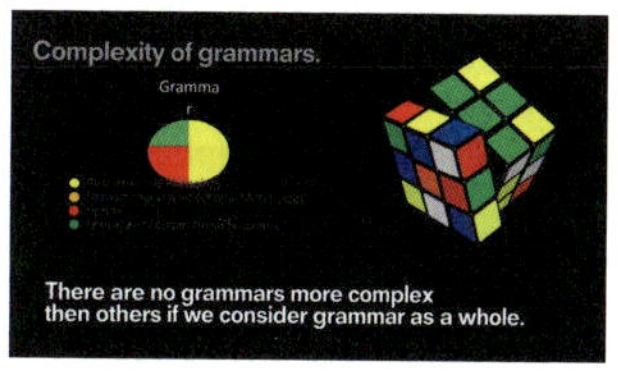

Fig. 15

to a cake, something that is so complex that when you change something in one part you have a consequence in others. In fact, no one has any objective measure of the complexity of grammar. A priori, if this is a genetic program, we have to imply that all languages are as complex. There is a kind of natural proof of the uniformity among languages, and the natural proofs come for free, because across languages, including sign languages, children acquire their languages within the same amount of time, practically. [16] There are many

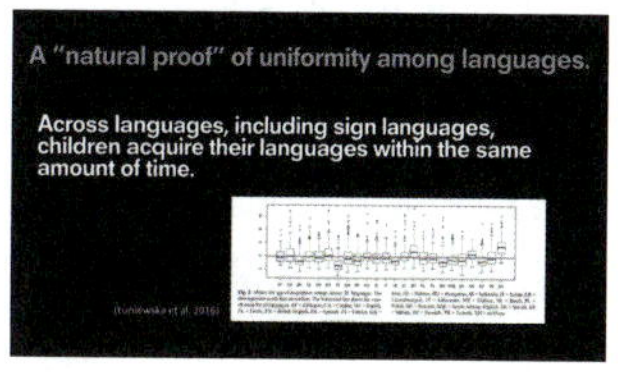

Fig. 16

experiments trying to measure that. One very beautiful experiment by Łuniewska et al. in 2016, for example, measured the difference of the acquisition of 290 words across 25 languages and they practically saw a similarity across languages, it would be impossible that that thing would be completely unknown and only cultural. [17] There is, of course, one

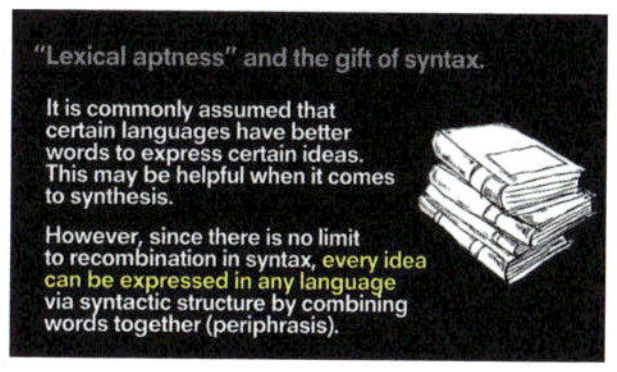

Fig. 17

possible criticism that I want to address before any other question, that is, it could be the case—it is commonly assumed—that certain languages have better words to express certain ideas, and this may be helpful when it comes to synthesis. If I can refer to a complicated concept with one word, but you need more words, I may be faster than you are. But does it really affect our understanding of reality? Does it really affect our understanding of our expression? Since there is no limit to recombination in syntax, every idea can be expressed in any language through a synthetic structure by combining words together, which is the classical term "periphrasis." Instead of one word, I can use a different number of words. Let me give you one example. Anyone can fall in love with anything, I fell in love with a verb, the verb "to be." So, "*to be* and not 'to be'" is my preferred slogan. The reason why I am saying that is that I am interested in the verb and not in the concept, but if the understanding of the notion of "being" is dependent on the presence or the verb "to be," in many languages, this notion would not be understandable because this verb is missing. [18] This is some-

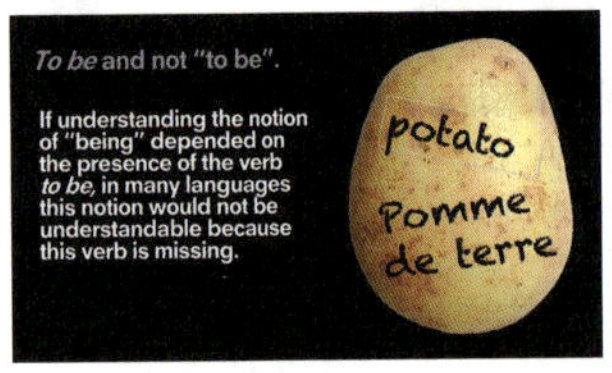

Fig. 18

thing that I explore in my book *Breve storia del verbo essere* [A short history of "be"]. But in honor of my colleague, here is another evidence of how language can get rid of problems when it comes to different words. In my language, Italian, and in English there is one single word to express this vegetable: "potato." But if I consider French or my own dialect, a northern Italian dialect, there is no single word, I have to say "pomme de terre," which is a clear periphrasis. But this is not the case: the cooks in French, at least not that I know of, are better skilled to treat this kind of food and cook it than instead of others. This happens with every concept. What you can say in one language you can actually say in the other. There is a kind of myth of ingenious languages, but there are no ingenious languages. What is ingenious is the comment to normal languages. [19] One of the most

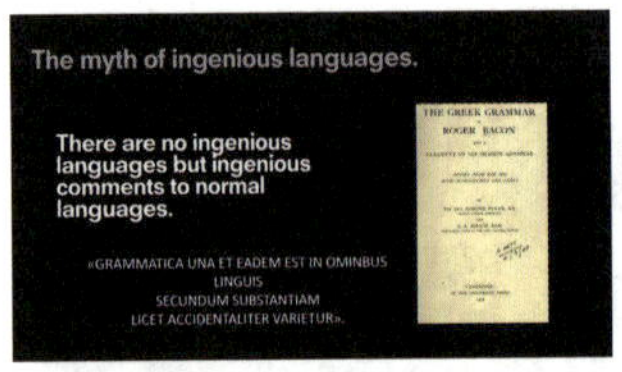

Fig. 19

famous comments on what is supposed to be the ingenious language is "La langue géniale," "La lingua geniale," considered to be Greek, of which Roger Bacon made a grammar book in the 13th century. It states exactly this: "There is only one grammar in every language according to its substance, although there can be accidental variations." You might think that this could be true of fixed languages across centuries, but languages may change. So you can say "But a certain language can become a better language after two or three centuries." This requires a very careful distinction between evolution and change when it comes to linguistics. [20] There is another

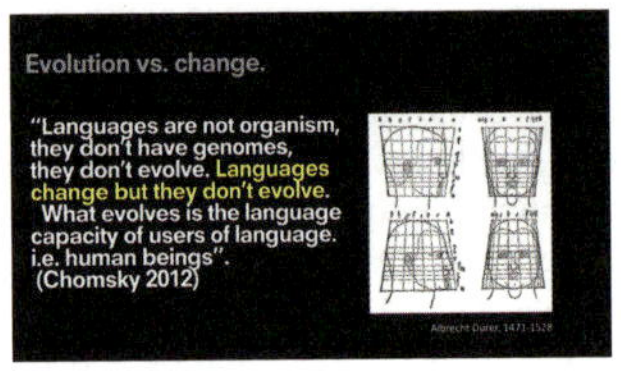

Fig. 20

quotation that I would like to address here by Noam Chomsky, who said "Languages are not organisms, they don't have genomes, they don't evolve. Languages change but they don't evolve. What evolves is the language capacity of users of language, that is, of human beings." I think that one of the best ways to capture this idea is by taking an example from painting. This is a study by Albrecht Dürer who was trying to show us that all faces are in fact the same project and that the differences are just distortions of the same geometry. In a sense, I cannot say that I have a more evolved face than my father, and my son does not have a more evolved face than mine. We have the same kind of face, although we are different. Change and evolution are different. So, also with respect to this, we actually could assume that we do speak one language, but there is one fact under our eyes—I would say under our ears to be consistent with the kind of stimulus I am talking of—, that is, that not all languages are mutually understandable. If I said I speak all languages, this is plainly not true when it comes to communication. But the fact that languages are not all mutually comprehensible may have had a positive impact on evolution by maintaining human communities at a manageable size. [21] This happens

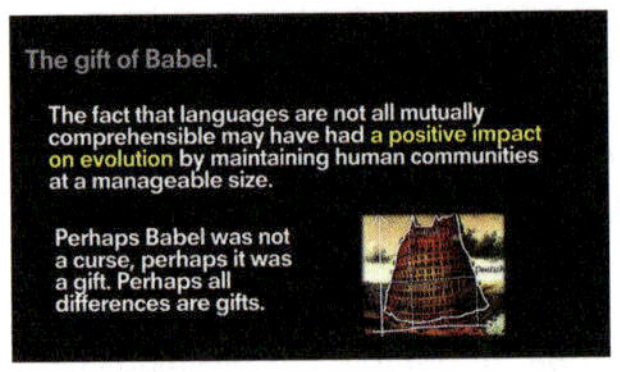

Fig. 21

in animals with diseases, it happens that certain groups cannot be too big and so the size of the community is taken together with the fact that if the group exceeds a certain number of individuals, then diseases can spread more easily and you either divide the group into subgroups or the group is eliminated. In a sense, nowadays it would be even more dramatically clear that during our evolution there was a kind of social distance that was due to language differences, and that may have had a very positive impact on evolution. Perhaps Babel was not a curse, perhaps it was a gift, and perhaps all differences are. You may say that what I am telling you is about the way we see the world with words, but it could be just a preoccupation of an old linguist. But I think it is not, I think that there is a danger that we have to avoid by combining the discoveries of linguistics and neuroscience. Let me tell you what the danger is. [22] There are two

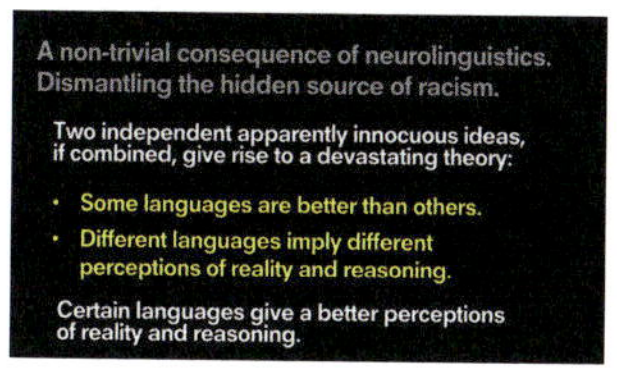

Fig. 22

independent, apparently innocuous ideas that, if combined, can give rise to a devastating theory. First idea: some languages are better than others—which I am trying to tell you it is not the case. Second idea: different languages imply different perceptions of reality and reasoning. These two ideas are independent, but if you combine them together, you come to the obvious conclusion that certain languages give a better perception of reality and reasoning. You may think that this is an issue limited to linguistics, but it is not. This is the germ that started one of the most colossal crimes of human history, the Shoah. Everything started when, in a book in the second half of the 19th century, it was theorized that there was a perfect society, with perfect social relations, with perfect conception, that had a perfect language, and perfect or noble in that language was expressed with the term "Aryan." [23] Here is an

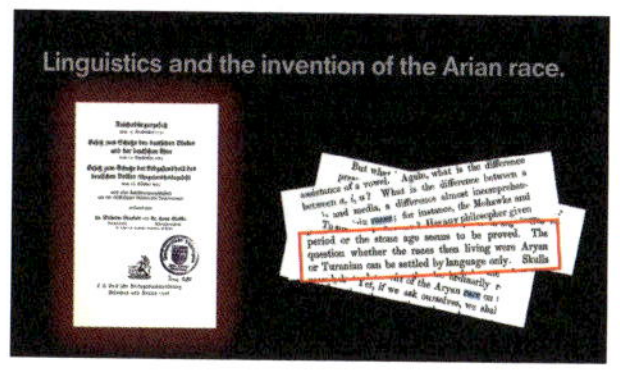

Fig. 23

excerpt from the book: the term language was immediately linked to the term race. "But what is even more curious than the absence or the presence of certain letters in certain languages, or families of languages, is the inability of some races to distinguish, either in hearing or speaking, between some of the most normal letters of our alphabet." Another one: "A difference almost incomprehensible to certain races." Another one: "Aryan race." And then, the idea that a certain race is opposed to another can be settled by language only. This was dismantled by the end of the 19th century. Linguists did not believe this. But the political propaganda did use this notion and they came up, by the first part of the 20th century, with the *Reichsbürgergesetz* protecting the Aryan race. You see how politics can use some ideas for propaganda and if there is a virus we have to contrast, it is the one that comes from the idea that there are better languages and better languages are a filter to a better understanding of the world. Trying to reach the conclusion, we are in a fortunate era. We are in an era where we can see inside heads without opening skulls, and we can see the Earth from the Moon. [24] I

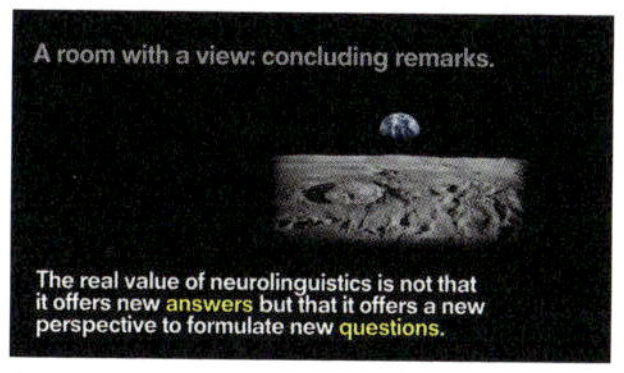

Fig. 24

assume that the real value of neurolinguistics is not that it offers new answers, but that it offers a new perspective to formulate new questions. I would like to assume that the number of people who attend the series of conferences that Fondazione Prada is organizing equals the number of questions that will come up. If you have more questions than before listening to us, then we have achieved our goal. The conclusion is that when we talk about language, we do face a destabilizing paradox. [25] We have a finite object shaped by evolution—the brain—that expresses a code—syntax—generating infinite discrete structures that cannot evolve by definition. Infinity cannot come in pieces, there is no half of infinity

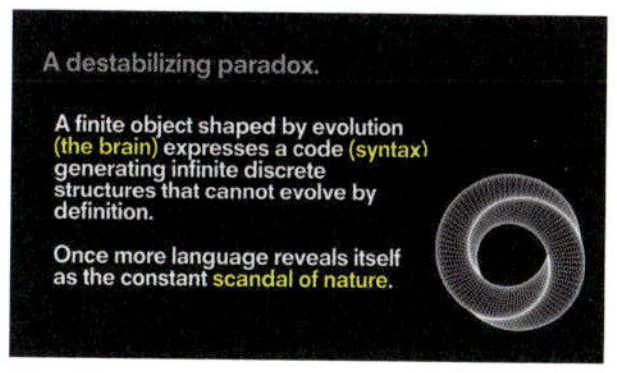

Fig. 25

and then the whole infinity. Infinity is either there or not there. Once more, language reveals itself as the constant scandal of nature. What is missing from my picture now? My talk and Stan's talk are, in a sense, complementary. He left me the anatomy of the structure of human language and I didn't speak about conscience directly, but there is one aspect that I would like to highlight, and it is the limit of our understanding of language. [26]

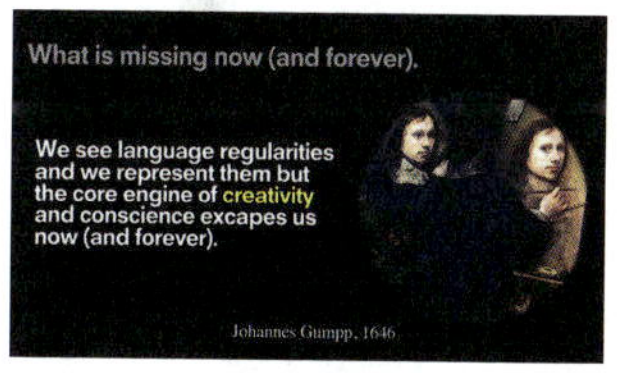

Fig. 26

To do so, I decided to use a painting by a strange painter, Johannes Gumpp, who only made two copies of the same painting, one is round and the other is square. This one is at the Uffizi. It is very dark, but the painting is as follows: there is a painter who has a mirror on his left and a canvas on his right. He looks at himself in the mirror and paints the canvas on the right. We don't see the face of the painter. I would like to say that we see language regularities outside us and we represent them, but the core engine of creativity and consciousness escapes us for now. I am not sure if it will escape forever, but it does for now to the extent that the linguist is interested in how the structure is built, and tells us a lot about how the brain itself is built. What should we do then? What are our linguists called to do? [27] With the almost final image I would like to show you one thing that surprised me when I first

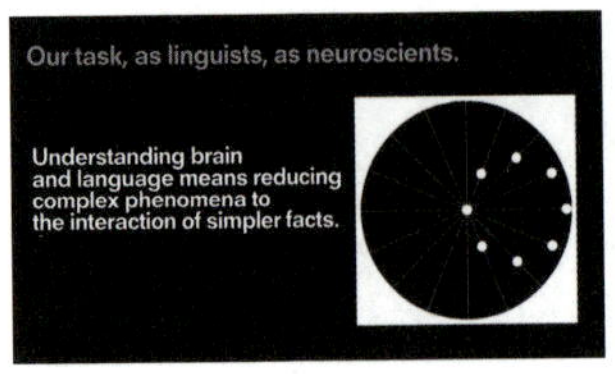

Fig. 27

saw it: our brain here sees a circle going around and around in another circle. But there is no circular movement whatsoever: all the dots are moving in a straight line on the diameter of the circle. It is our brain that computes the symphony of straight movement as a circular movement. Understanding brain and language means reducing complex phenomena to the interaction of simple facts. Sentences must be reduced to the interaction of facts that may not look like sentences. Similar thing with meaning, similar thing with phonology. I don't know if I convinced you. [28] I put a

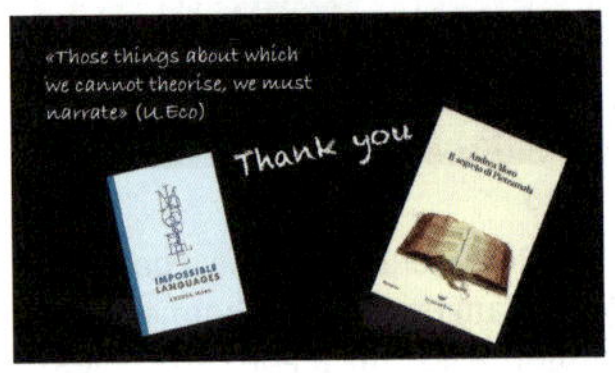

Fig. 28

lot of stuff in this book, *Impossible Languages*. During my youth, I was also exposed to the lesson of another maestro, Umberto Eco, who once said "Those things about which you cannot theorize, we must narrate." So what I did was to invent a story using impossible language, which people were forced to learn and I won't reveal to you the ending because it is up to you to discover it.

Panel Discussion

JUBIN ABUTALEBI
ANDREA MORO
STANISLAS DEHAENE

JUBIN ABUTALEBI I am going to read some questions we had from the audience. Andrea, you were talking about politics, linguistics. There is one interesting question, very general: what were the words used in history that triggered the worst consequences? If you had to choose one word to delete from all the vocabularies which one would you delete?

ANDREA MORO Can I do the other thing around and say what words should I include? Words to me are like knives. What do you use a knife for? You can cut bread or you can cut your brother, depending on what you want to do. Any word can be used in any possible way. There is not a single word that I would highlight.

JA Maybe it is more about deleting a meaning and not the word, because a word is just a form of a meaning.

AM I don't think that there is any meaning we should delete. We should be aware of all meanings. For example, I am against all kind of censorships. Things have to be known. Even if you want to preserve us from going back to a certain situation in the past, you have to go through it. One cannot avoid understanding. I think this was very clear from what Stan said, humans combine thoughts, knowledge, and awareness. You cannot delete it. If I have to say something, I would like to say that those who think that by deleting a word you have a better world are wrong.

STANISLAS DEHAENE I was going to say very much the same as Andrea, although we did not think together about that. I just gave my course at College de France on the Whorfian hypothesis and the dependency of thought and language, and the conclusion is that there is not much dependency on certain languages. Thought is hidden inside and it is a very general property of all human brains. In the first course—and it is all available online on the College de France website for those of you who want to see it—I discuss *1984*, and the question you are asking is very much one which is out of this novel, where they are progressively removing words from the dictionary and they are hoping, in this way, to create a dictatorship because you would not be able to think the corresponding concepts. It is completely false. It is not because you have a word or don't have a word that you can or cannot think the concept. What we are showing in my work is that much of mathematical concepts are completely outside of language, they are already internal and in a certain sense predate the concepts.

JA Could the fact that baboons don't recognize regularity in geometric shapes be a cultural issue? Would humans be able to recognize baboon's different voices?

SD It is interesting. We try to address this question by looking at people who have had a very different cultural experience, such as the Himba, or who have had no education and no training, and we still find very much the same regularity. Of course, it is still a bit open and this is very recent data, but we think that we are tapping here an extremely basic mechanism that is common to all humans regardless of education. Mathematics is a construction, so there is no doubt that some concepts of mathematics are culturally constructed and dependent on education. But at the foundation, what we are finding is we all have the same concepts, space and number, in particular. Then out of these atoms—much like Andrea described—we have the atoms at the bottom but then the combinations are of course culturally created.

JA Andrea, do you want to add something on this? Would we be able to understand the different voices of baboons? Your opinion as a linguist?

AM There is a question that people keep asking me: Could we recognize a language from a creature from space, an extraterrestrial language? How do we recognize that something is coming from some entity that has our intelligence? The only thing, besides the fact that we have to receive something that touches our senses—light, sound, flavors...—we have to see certain regularities, and the kind of regularities we can see should be only those compatible with our own way of combining stuff. Stan has done wonderful work in the electrophysiological sense to see what happens if you combine a string of words that have no syntax, with more syntax. This gives you like the x-ray of a language, and whether animals or extraterrestrials want to communicate with us they have to be compatible, sensorially and structurally, with us, otherwise maybe we already have our extraterrestrial around us but we simply don't understand them. Or our cats and dogs.

JA I am happy with this because I was always sure I understand fully when my cat is hungry, the different voices it makes to me, so I am happy with this answer. A very pertinent question. Someone is asking us: Can we train an animal to be bilingual taking the same order in different languages?

AM This has been done with a certain kind of birds. There is an extremely interesting experiment that was done with the so-called cowbirds in America. These birds have different songs depending on the area where they were born. They did one wonderful experiment by taking the egg of one cowbird and moving it to the nest of a mother in a community different from the one where it was taken, and then the bird was able to speak the language of the new community though the mothers were mute. How can it be the case? The experiment tells that when the bird was born, it tried to express all the languages it had inside, and the mother only responded to one language and fed it. The one that was used to feed him became his language. More or less, this is the same as we do with sounds. When we are kids we have all our sounds in our heads. If I had been raised in Tokyo, I would have a Japanese accent, then after a certain age everything is fixed. Our common maestro Jacques Mehler, who passed away this year but gave the deepest influence in the field, used to say this: you can learn by forgetting. You have an enormous amount of complexity and possibility when you are born, like a staminal mind, and then what is not used and is forgotten falls down and what remains there is your structure. I think that this is one of the most beautiful and revolutionary pictures that we all share, Stan, I, Chomsky, and other communities.

SD In fact, the phrase "to learn is to select" is actually from Jean-Pierre Changeux. This is something we share. I want to say that we did a lot of experiments on bilingualism and we are still doing some right now on readers who can read two languages, like Chinese and French, for instance. The findings from all of these experiments is to show the universality of the human brain. It is not because you speak Chinese that you have a different brain. When we scan bilingual speakers they are using exactly the same circuits, almost exactly in order to read Chinese and to read French. When we zoom in maybe we will find some small circuit differences, but enlarged it is extraordinarily similar. I think our work also points to the unity of the human brain.

JA I wanted to add something to this story about if an animal can learn. I remember I read a book by Franco Fabbro years ago, in which he mentions sparrows and canary birds and, as you mentioned, if you bring them from one world community to a different one they would, after a couple of weeks, learn the local dialect. He mentioned that these are the only two species of birds that have left-sided language areas. Something comparable. You want to say something about that? There is no other species that has a language organ in the brain, actually.

SD I don't know if you can speak of a language organ in the brain of birds, but there are some interesting genetic similarities. There is a beautiful science paper from a few years ago showing that genetic mechanisms of circuits in the brain of the bird and the human are slightly similar. But I don't think you can speak of language circuits, no. You have essentially pre-motor mechanisms of song generation of some similarity with the mechanisms of control of the voice in humans. The really important thing is to distinguish species that are vocal learners. There are some species that are able to learn their song, just like Andrea was describing with dialects, it is not at all innate, there is a really strong learning component. In all of these vocal learners, in spite, of course, of the considerable differences between the human brain and this bird's brain, you'll find some circuit similarities. But I wouldn't want to call it a language organ. I think it is just a much simpler sequence production organ.

AM As a linguist, I would converge with what Stan was saying by adding that I wouldn't even call it a language. People keep saying that birds especially can recognize complex sequences, which is true, but most of them are symmetries, sequences that are repeated. None of them is able to manage complexity that goes beyond what Chomsky discovered in the book I showed, *The Syntactic Structure*, which is called in mathematical terms, a "Markovian chain." This is crucial. It is also interesting that the new computing capacity of machines may suggest that, if not animals, maybe machines can learn a language. But it is simulation, it is like when I make a calculation with my cellular phone, I made a calculation but it is probably not even isomorphous to what happens in my brain.

SD Then maybe we disagree a little bit. I think there has been progress in machine learning that shows that there seems to be in some of these machines the capacity for recursion. We are analyzing this at the moment. There are some differences but it is getting closer and closer to the human syntactic ability.

AM I agree that you can simulate that. What I don't agree with is that the understanding of the hardware of a machine may be isomorphous to the hardware of the neurons. It may be, but it is not proven that it is. The fact that machines can learn faster and easily is clear in the case.

SD I think there is a lot of very interesting work to do at the moment to study what the current AI machines are able to do and, at the same time, I defend very much this idea that they are very far from the capacities of the human brain. This is a central theme in my book *How We Learn*, the learning of language in a human baby is much faster and much richer, much deeper than anything that happens in the current machines, even the best machines at Google.

AM Take, for example, what used to be said in the 70s and the 80s: brains are hardware, languages are software, it can run any software on the brain, which is not true.

SD That is completely untrue.

AM If anything, the languages are the software that the hardware expresses. It is the other way around. People keep insisting on that, but there is a commercial reason for this. If you buy a new computer, there are times when you buy a neural machine, a neural chip. There is nothing neural in that, but people are attracted to buying stuff that is called neural. If you buy a neural car maybe you would have more success than if you drive a normal engine.

JA Question to Stan. Your talk seems to be about mapping how and what we perceive. Do you see the language of thought as including ideas such as free will and choice?

SD The discussion of free will is a very difficult one, because I think we mix up two different concepts, and I also think it has nothing to do with language, it is something that we can discuss also in animals, whether they have free will or not. There are two notions of free will: there is one notion according to which we have a capacity to detach our decisions from the physics, we have the capacity to be random and to go beyond the deterministics of physics. I think that view is completely wrong. Our entire thought patterns are ultimately determined by the physical organization of our brains and, in this sense, we have no real free will. But in another sense we do, because we have the architectures for decisions. If you consider tasks that have to do with free will, there are tasks in which you have enough time to decide and you are able to bring your values, your knowledge, your memories in order to take a decision. These are the conditions where we say we exercise our free will. In this sense, I think we do have free will. We have a prefrontal cortex that allows us to detach ourselves from the immediate experience to think quietly as long as we need, and that is this workspace system that stabilizes ideas so that we can think about them. This is an architecture for free will. In this sense, a machine can have free will. If it has enough resources to deliberate and think, it may have a form of free will, it is autonomous in its decisions, but is it undetermined by physics? Not at all. If you think that you are exercising your free will by going to vote—and this is of course something that is in our minds these days—, "I go to vote, I exercise my free will because I think about it and I do the decision that I want without being forced by other people." It is deterministic if people try to guess my vote. I am not going to vote for Mr. Trump, this is highly predictable, so my values guide my judgment in a certain direction, which is predictable. If we separate these two meanings, I think that the workspace architecture that I described for consciousness is actually an architecture for free will. I think it is the general evolution of consciousness, it allows us to detach from the immediate contingencies of sensory motor reactions, which are often automatic and unconscious, and to deliberate.

AM There is a name for free will in language, which is language creativity. Namely, besides a few situations, you cannot predict the word that I will say in three seconds. Will I say cake or car or nose? This is a form of free will. It is like generating something that nothing in the environment can force or predict. This is the reason why I kept distance from that part of language and I only gave you the range of possibilities. Which one will you use? Whether you want to say "A man is painting the wall" or "The wall is painted by a man" cannot be predicted. Grammar is the equivalent of the space that Stan was referring to. You choose, grammar is construed in order to allow free will. There is nothing that forces the other, necessarily.

SD But you see, Andrea, I really want to draw a distinction between randomness and free will. I think many people don't distinguish these concepts enough. When you are random, you are not exercising free will. If we do this distinction correctly, then all of these stupid discussions about whether quantum processing is required for free will, whether we need a quantum consciousness and so on, they all disappear. I think free will is compatible with a deterministic brain, and a physical brain.

AM I agree. If I make something which is completely counterproductive right now, I would say that the way that Penrose has to refer to quantum mechanics in order to get free will is baroque. I mean, it is unnecessary.

SD I agree totally. I have so much admiration for Penrose, I was so happy that he is getting the Nobel Prize, but the consciousness book is very poorly informed and outside his area of competence.

AM I think that the partition is trifold, because along with it there is a kind of free will, that is what I am going to vote, who I am going to kill, what I am going to do. There is also the free will at the level of composition that is what kind of sentence I am going to produce, and we don't know anything about that. What is worse is that we don't even know how to know something about this, there is no experimental clue to get what Descartes in fact called creativity. In this sense, that may be one of the mysteries that Chomsky talks about, that made life outside our capacity. But this is why we still do research by the composing complexities.

JA Are people more gifted for languages? How does their brain work better for learning different languages? I mean, this is also important. At the kindergarten little girls aged four or five years old already talk so much in elaborated sentences, and then you see maybe a boy the same age who is still at the world level but not at the sentence level. Where does this difference come from? And, as the question goes on, are some people more gifted for learning multiple languages? We all know there are some people who just really absorb multiple languages and they become hyperglots and some not.

AM Of course, there are individual differences. This would be surprising especially if language was completely disconnected from biology. If it is connected with biology, then you have difference all over the typical features, the typical traits of a human being. There are differences in singing, someone can sing a melody in tune, some others cannot. This doesn't touch the architecture of music; it depends on several factors that may not have anything to do with culture. If Amadeus Mozart did not have his special ear, his father could have taken him to the palace and tried to sing and play but he would not have become a musician. I do think that this is a very indirect proof that, when it comes to language, nature rather than nurture matters.

JA As you all know there are some studies done in the past, like in the case of multilinguals, that if you have a better working memory, a verbal shorter memory, you can keep online more phonemes and combine them. This is old stuff, but as you say, biology makes a difference.

AM A hundred years ago these discussions weren't impossible. Remember Lennenberg and his wonderful book *The Biological Foundation of Language*. He was working in Boston in the 60s, and he discovered that people with aphasia could recover better if the stroke or the cause of the aphasia was before puberty rather than after puberty. That was the first break in the community, to think that biology may matter with language. In Lennenberg's introduction to the book, he criticizes Wittgenstein thinking of language as a social rule, made it impossible to study language as a biological thing. That is one of the reasons why I have 50 years of delay compared to others. There are two reasons: one is this and the other is the fact that we cannot have animal models for syntax, that is obviously the other reason.

SD I shall try to answer this question, it is a difficult one. I don't think we understand much about the individual differences at the moment, and it is made extremely complicated in the case of humans, because it is a combination of biology and education. Undoubtedly, there are biological differences. We can go to one extreme, which is a language impairment. There are some developmental language impairments and there are some children who are perfectly intelligent but have anomalies, for instance in their connections, in what we call the arcuate fasciculus, which is very important circuit for language, and when this circuit does not develop normally, they might be totally aphasic and unable to speak and to learn languages. I also want to stress the importance of education and early education. There is tons of evidence that different children are not getting the same sort of inputs, and there is very beautiful evidence now showing that the quality of the inputs that children are getting determines their ability to learn languages. There are huge differences according to social media in which they are born, and the recent research has shown that what is really important is that children get child-directed speech, it is not enough to absorb speech in your environment. In this respect, maybe some of Chomsky's writings were a little bit misleading. It is really the child-directed speech which is important, it is the dialogue, it is the interaction where an adult or another child of an older age makes the effort to respond and to have this back and forth between two brains. There is clearly some education-related factors here, and this may have a cascade of effects later on in the ability to learn other languages.

JA Jung, the psychoanalyst, described in his works that geometrical figures were the basis of the collective unconscious. Can this somehow be linked to the study about the human preference for geometrical regularity?

SD This is so interesting. I am guessing recently that I should read more Jung. Typically we don't read Jung much in cognitive science. It is true, there is this general idea of archetypes including geometrical figures, but not just that, also stories and so on, that tend to recur in many different cultures. One possibility is that, indeed, they belong to the sort of basic geometric repertoire of what our brains can generate. A sort of basic tree structures that we can all generate. All cultures eventually think of counting and counting on their fingers, all cultures eventually think of geometrical patterns and they paint their bodies and do parallel lines, for instance, parallel lines are all over the place in all cultures. I think we could explain some of the Jung archetypes and I confess I have to read more about this. Yet I think we could explain many of them not by some kind of sharing of minds across the world but by the much more basic biological architectures of our brain. To a large extent this is my current project.

AM I want to say two things. One is a funny one—that the protagonist of my novel has eleven fingers because I wanted people to reflect on the fact that we have that body and it is because of the way we are built that our culture is shaped. The other thing is that I was very pleased to see the funny drawing you showed in your talk, with the people from the Stone Age, in which we are looking at this geometrical scene of hunting. There are two aspects of the beauty of symmetry: one is the naive beauty that you have in the symmetry, like a Greek vase, a pure symmetry is recognized in reality. The other is hidden symmetry. To me, one of the cathedrals of humans that I would send into outer space to show that we are intelligent is any painting by Cézanne, who was able to uncover the geometry of reality and make it clear that everything is nothing but the composition of pure geometrical forms. This is something that always moves me, in a sense, because art precedes sometimes our understanding of reality.

JA To what era can a primordial language be traced back? Is it possible to know elements of the most ancient syntax? I am presuming if you go back in the days to Sumerian or ancient Egyptian. Do you have any notes on syntax?

AM I had a note in *Impossible Languages* and *The Boundaries of Babel*. When I meet the first-year students, I ask them, "What would convince you that a certain group of people had a language similar to ours if you don't have anything written?" That is interesting, you have two immediate answers. First, if you see architectures, second, if you see social complexity. Then I tell them that with the first case it is enough to see bees. Bees have constructions that are perfection in terms of geometry, but they don't have a language. When it comes to hierarchical societies, ants have a queen, etcetera. What would tell us that humans have a language? To me, it could only be by considering a sufficient span of time and see if there is progress. Animals do not have progress. A spider builds his webs the same way that his father spider and grandpa spider made theirs. We do not start civilization when we are born and discover fire at four and the wheel at five and go on. We capitalize on what was done before and that can only be conveyed if we have a rich enough language. To me progress is the only landmark that may make you infer that a community had something similar to our language.

SD I think this is very interesting. Andrea, I think you are referring to communicative language, the capacity to communicate would allow humans to suddenly explode as a society and progress together because what one individual learns, another person can also learn quicker. I think, by the way, this is the very definition of humans and completely missing from machines. When we discuss AI we forget that these machines learn by themselves but they never learn from others, they are unable to share what they have learned. They are really missing something dramatic. I like your definition and if we define it this way I think the origins of language are probably not older than 100,000 years or something. There was really an explosion, a cultural explosion, in many places in the globe. The Upper Paleolithic. However, if we think of language as this internal medium for representation, like the beginning of mathematics, then I wonder whether this could be much older. I am very struck by these bi-face artifacts that have pure symmetry that really require conceptualization before you dig them out of the stone. This perhaps is as old as two million years. But it could be an internal medium of representation and only as a second step evolved the capacity to share it with others through the communication of language.

AM I completely agree with that.

JA Stan, there is a question to you, it is a bit more precise. What is meant by human singularity and why should it be considered a value?

SD By singularity I mean something particular to humans. It is a debated term. Some people think there is no uniqueness and that because of Darwinian evolution we have to be continuous with other species. I think there are cases in evolution where the change is non-linear and it leads to something which is essentially unique, very new. This is the case with humans, as we can see from the fact that contrary to every other species, we have no real niche, we are able to adapt to any niche. In my book I try to argue that the niche of humans is learning, we are *Homo docens*, we are able to learn. That is our niche, the capacity to learn. It is amazing that we can learn about stars from a billion light years away. This is completely crazy, we never evolved for that. Our evolution really created something new, and that is what I mean by singularity, a discontinuity out of a continuum of evolution.

JA I just have a question coming in which I like pretty much, it is very important and it is addressed to Andrea. In Italian and other Latin languages, usually nouns that indicate prestigious occupations are usually masculine, like "avvocato" for lawyer, "presidente" for president and so on. Would it be useful to generate feminine nouns to reduce gender differences in employment? Would it change attitude and behavior toward women?

AM My straightforward answer is why not? But the effect is that women have to be paid as much as men. The linguistic maquillage sometimes is only used to keep away the responsibility of societies. I hate it when people force other people to say "La capitana" but then she is not paid as much as the others. If there was a mapping between gender and equality, then languages with no gender would be the languages with more equality. For example, Turkish has no gender, but the condition of women in Turkey is very low compared to other societies. I don't think that there is a mapping. I discussed this is another book that I recently wrote—I didn't mention here, it is called only in Italian *La razza e la lingua*—where I make it clear that anything you can do to promote equality has to be done, but don't think that it comes from grammar. Equality is a political decision, even wearing pants for women—I shouldn't say that probably in Fondazione Prada—but the way we dress is a political decision, isn't it? When I was a young man, women used to be dressed in skirts and then they decided to wear pants because that was a political decision, it was not an anatomical reason. So, grammar can be forced wherever we like, but we have to make it clear that it is not because of grammar that we have a better society.

JA What is the relationship between language and mathematic acquisition in relation to brain plasticity?

SD We are finding that there is very little relation between math and language. I was surprised by that. If we scan people and we compare which areas of the brain are activated when they do math and which areas are activated when they do spoken language or written language, they are completely different. There are cases of what we call double dissociation, some people are completely aphasic after a brain lesion, for instance, and yet they continue to be able to do algebra, and viceversa. You have people who have lots of trouble with mathematics but they can still speak and understand language, written or spoken, very well. The relation is one of dissociation, at least in adults. In the course of education we need language in order to communicate ideas. Language is like a pointer, I compare it to a catalyst, a catalyst brings together the objects of thought, they click together and then the catalyst can go away and the objects of thought have been created and they are independent of language.
I think this is what is happening in education.

JA Andrea, I have a very philosophical question to you. What are today, in contemporary studies of language, the dialectical relationships between philosophical intuition and empirical discovery?

AM There is no empirical discovery per se, you always have to have a philosophical background to understand reality. Philosophy comes first, always. Apples have been falling on Earth, the moon rotating around the Earth for centuries. It was only when one person decided that he could dare thinking whether the two movements were connected, that Isaac Newton expressed the theory of gravity. It is always philosophy first. The problem is that they tend to make us think that it is empirical data first. Of course, philosophy has to face empirical data. Come on, we are not mysticists, we don't have a mystical mind, we have to make predictions. Science needs to make predictions. But the way you collect data has to have a philosophical prejudice. When it comes to language that is so clear: there is no way to distinguish between sentences and a string of words if you do not use the notion of truth, and the notion of truth is the most philosophical notion whatsoever. That truth is not in the facts. The dialectical things between philosophy and empirical research is a wrong way to look at the way we think. We start with an idea and we see if it works and allows us to make new predictions. Then, maybe, it can let us change our philosophical view. There is no distinction. If there is a distinction, then we are doing wrong, we are going the wrong way.

JA Andrea, you just mentioned something really very deep, the truth. I reckon you are both members of the Pontifical Academy. Stan, the Pontifical Academy of Science, Andrea, the Pontifical Academy of Arts and Letters. What is your point about truth, religion, and science?

SD I have been a member of this academy for almost 20 years. I want to say our role is purely scientific, I don't consider religious issues.

I think it is very important to have a scientific look on the ethical issues that the church may be confronted with. There has been a continuous evolution of ideas. It is wrong to think of religious ideas as static. Science has its role to play in bringing some rationality to these debates.
For instance, there was a time where donating organs was considered horrible because you had to take organs from a living body. How could you do that? Then the ideas evolved and now the Catholic Church completely accepts the notion of brain death, which is not trivial, it is a scientific concept that you are dead when your brain is dead and you can then donate your organs. Now it is considered one of the most beautiful gestures that you can have on the part of a family to donate the organs of a recently deceased member. I see it as a metaphor, our goal as scientist is to slowly help change the ideas to a more rational state of affairs, if we can.

AM As for me, just discussing of the notion of truth will take us another seven days here, I don't think we are allowed to. I would just say one thing: one of my goals in my own personal life and activity is to show that scientific methods are not completely disconnected from the possibility to be confident in someone else's work. You do not restart science when you open your lab, you have to rely on someone else. Then, connection of reliance is, in a sense, similar to what keeps together people with faith. The other side of the coin is to show that there is nothing irrational in faith.

4

Evolution of Culture: How Far Will We Go? (from Anthropology to Technology)

IAN TATTERSALL
IDAN SEGEV

PANEL DISCUSSION MODERATED BY
KATRIN AMUNTS

Introduction

KATRIN AMUNTS

I think we cannot speak about culture without speaking about arts and symbolic forms. [1] When we

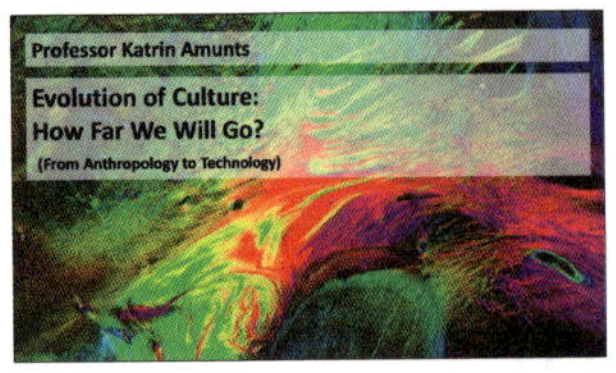

Fig. 1

have a look at this painting by Michelangelo, most people, independently of which culture they are coming from, would agree that there is an exciting beauty in this image, and that they understand the gestures, they get a feeling about the message that is being sent. That is really interesting and that was the basis that some of our colleagues formulated the concept of symbolic forms that seem to be so universal and so general. Some of my colleagues, brain researchers at least, or physicians, would see in this image not only a part of the creation but they would also see, on the right part of the image, something that looks like a brain. Speculations and hypotheses came up, that Buonarroti, who was indeed one of the first to study the brain and the human body in very much detail, has drawn this particular shape because he was thinking of a brain. [2] This is further continued when

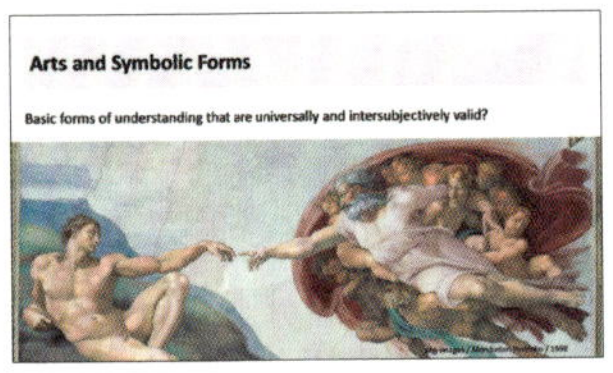

Fig. 2

we look at this painting from Michelangelo, and here the neck looks a little bit atypical. Some colleagues have the hypothesis that again Michelangelo drew a brain into this painting. I would like to emphasize that we, of course, cannot know and perhaps will never know what Michelangelo had thought when he was drawing it and whether he had indeed a brain in mind. Independently from that, many people see a brain in it, and there is an agreement among several of us that indeed the shape looks very much like the shape of a brain. [3] I am very pleased that we have an

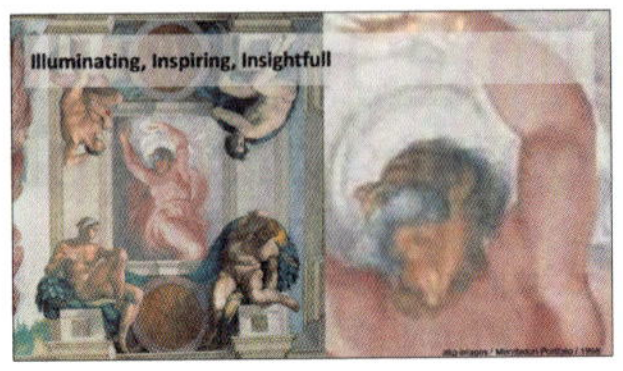

Fig. 3

expert today here, Professor Ian Tattersall, who is an expert in cognition and also knows philosophy in very much detail. I am sure that he can answer several of the questions that might be asked with respect to arts and these symbolic forms. [4]

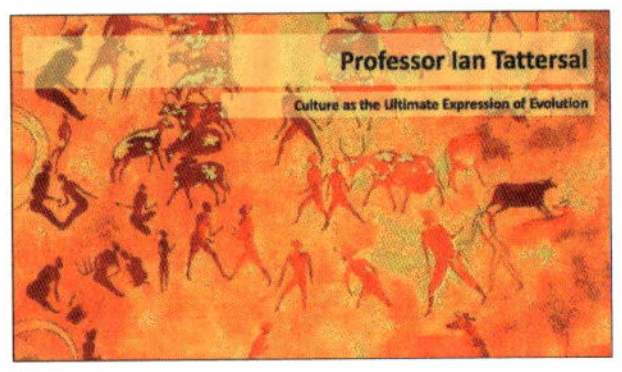

Fig. 4

Culture or evolution came also up with the structure of the brain, and this incredibly difficult network of connections that we see in the brain. On the left-hand side you see such a natural neuronal network that was created during evolution. Since the 1950s we also see the development of artificial neuronal networks. What you see here is a simple perceptron, and in the next few seconds you will see a multi-layer perceptron. Although there are many similarities between these two types of networks, there are also many differences, and many of my colleagues are excited to learn from natural new neuronal networks how can we create artificial neuronal networks and perhaps improve their capacities and make them more powerful. This is a whole research direction that is more and more coming up these years. [5] There are

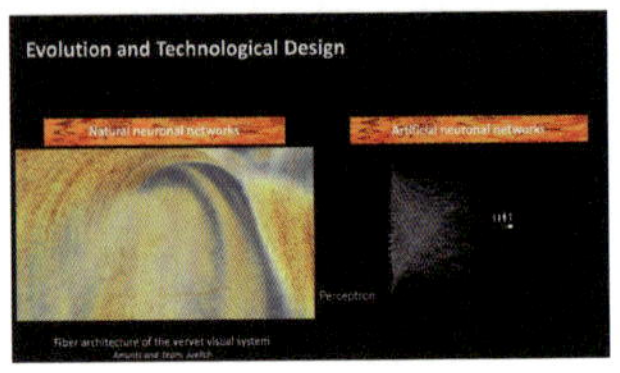

Fig. 5

incredible results when we are looking at what these artificial neuronal networks can create. Two years ago there was at Christie's a selling of an artwork that was the result of artificial intelligence for more than four hundred thousand dollars. We know that some artworks produced by human beings are also held for the same amount of money. This is an example here from Richter. We can ask ourselves: will we accept in the future to pay huge amounts of money for something which is a result of artificial intelligence? Will we attribute such a value to such an art piece in the future? Will this change the whole art world and the whole culture that we are doing? We of course know that such an artwork is reproducible up to the smallest bits and bytes. These are of course questions that I find very interesting to be answered in the future. [6] We could also ask our-

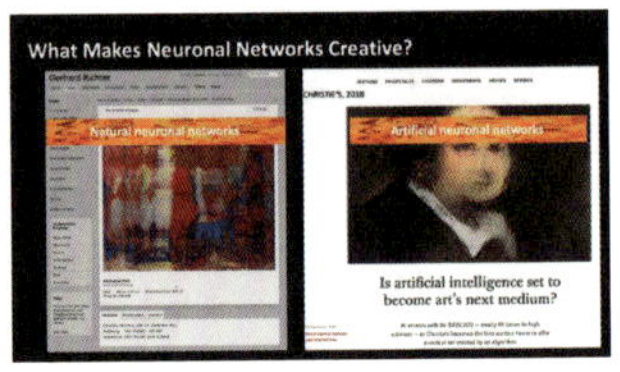

Fig. 6

selves—and I would ask myself—when we have the next collection at the fashion week, will we distinguish whether this fashion is designed by a human designer or by an artificial program? Can we make a difference by humankind designing as compared to artificial designing that we have seen in this last painting? This brings me to our speaker which is professor Idan Segev, who is working on neurons. [7] He and his colleagues are drawing beautiful images out of these neurons. By

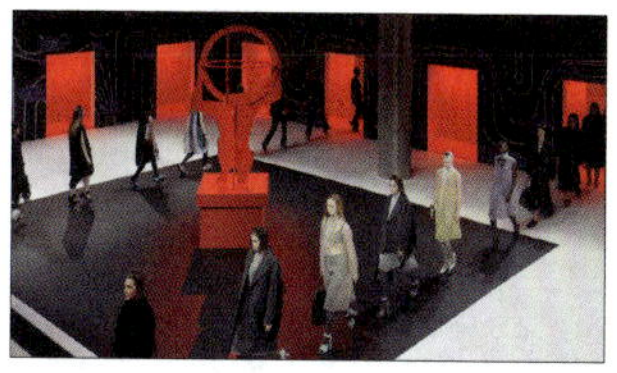

Fig. 7

looking at these beautiful images we can start to classify the cells of neurons with respect to their morphology. This morphology, that is the shape of the cell, is very closely related to their function, their physiology. But people do not only draw these beautiful cellular paintings, they also measure every single detail about these cells, they quantify how big the dendrites are, how big the branches are, and they relate it to other properties of these cells in order to get a better understanding of how these cells are functioning. You see that there are a lot of different cells that have been found in the brain. All differ with respect to their precise property. [8]

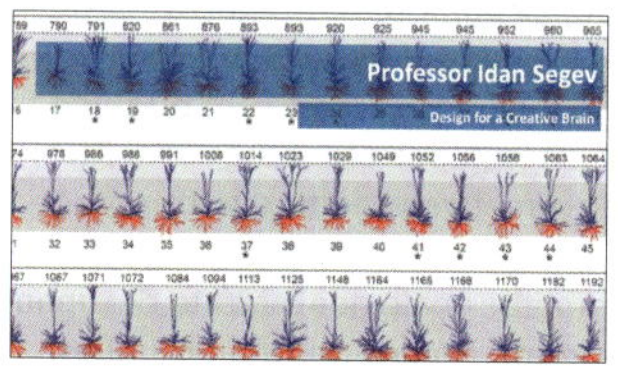

Fig. 8

I would say that every single cell is a little bit like a single universe. However, all these cells are embedded in a context, they are part of a certain layer in the brain, part of a certain brain area. These brain areas have a certain architecture. I am particularly interested in the cellular architecture of the human brain. Why? In the same way as New York has a very clear architecture, a rectangular one with boroughs, the houses built in blocks, which make New York so specific, while when you have a look at Paris, for example, you have these departments which are oriented in a different way. The way Paris and New York are organized with respect to their houses has a role, of course. It has an impact on how these cities are functioning, how their transport is being organized. It is not only important to count the number of houses, it is also important to see how these houses are arranged and how they form quarters, what consequences they have when we are looking at the function of the cities. The same is true when we look at the human brain. We see that many cells—there are about 86 billion nerve cells—are arranged in cellular columns, they are arranged in layers, in areas. The way they are organized has a lot to do with the areas they are responsible for and in which networks they are involved. [9]

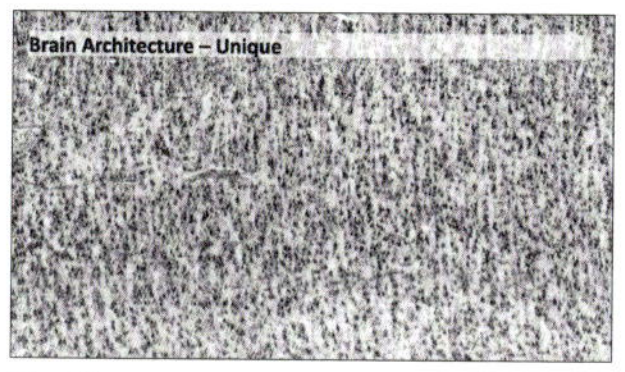

Fig. 9

To make such information about the cellular architecture of the human brain available was a part of our mission to develop the unique brain atlas. This is, of course, only one aspect of human brain organization, but it allows us to integrate other aspects, other facets of brain organization, including the connectivity or the molecular architecture, cognitive function, genes. In the Human Brain Project we are developing such an atlas where people can collaborate, and I am deeply convinced that the culture of scientific collaboration is indeed very human and very specific for us humans.

Culture as the Ultimate Expression of Evolution

IAN TATTERSALL

I have been asked to talk about "culture as the ultimate expression of evolution," and of course I am happy to try to do that. But in order to do so properly I shall have to dispel one common misapprehension, namely, that culture is a unique possession of our own species, *Homo sapiens*. This is not strictly the case, because in its strictest sense "culture" has been defined simply as the passing down, through the generations, of learned behaviors in a particular population. And in the light of this definition, we can find culture more widely in nature than some might expect. Indeed, one study not long ago identified 39 different cultural behaviors in our near relative, the chimpanzee *Pan troglodytes*. Those cultural expressions range from the ways in which those apes clasp hands, to the manner in which the males construct their sexual displays, and to the different fashions in which they make and use tools to fish for termites. What is more, at least thirteen different traditions of locally learned behaviors have been identified among Capuchin monkeys in South and Central America, most famously in the techniques they use for smashing nuts on anvils. Even dolphins locally learn how to use tools from their peers, as in the use of sponges to protect their snouts during foraging.

So, culture is not unique to our species, or even to us and our close fossil relatives. But what is unique is the unusual form of consciousness through which we filter our cultural learning. And since the acquisition of our unique cognitive style changed all the rules by which we conduct our lives, I need to emphasize that from now on I shall be referring to culture specifically in the human sense. The distinction is important, because although human culture is undoubtedly based on a more general primate sensibility, it also involves the peculiar way in which modern humans process information about the world.

Instead of perceiving the world around and within us in a holistic way, as to the best of our knowledge all other organisms do, we human beings mentally deconstruct our experience into vocabularies of discrete mental symbols. And then we can rearrange those symbols, according to rules, to come up with new statements about the world—not just as it is right now, but as it *might* be. As a result, we human beings tend *not* to live in the world as it directly presents itself to us, but instead to live most of the time in the version of the world that we reconstruct in our heads. Because we both perceive and reason symbolically, that world is in large part our own individual construct. But, while each of us will re-create our own world in our own idiosyncratic way, exactly *how* we do it will be deeply dependent on the cultural beliefs that we have learned. And while culturally mediated learning thus provides the underpinnings of our individual identities, it also explains why we can coexist in the complex societies we have created. It is the combination of consciousness and culture (in the uniquely human senses of those terms) that produces both our individual and collective senses of self. All of this suggests, of course, that if we are to understand the unique relationship between culture and consciousness in our species, we would do well to look back, and to consider how both aspects of humanity have evolved. In this endeavor, culture has the advantage over consciousness because it has a direct, if incomplete, material reflection in the archaeological record, while symbolic consciousness has to be directly inferred from archaeological traces.

Let us start our brief review with the reasonable assumption that our earliest ancestor who was not ancestral to any living ape—the first hominin, in other words—already had a relatively complex cognitive system that was broadly comparable to that of the quite sophisticated apes. The first material suggestion we have of any meaningful cognitive advance relative to the apes comes at some point over 2.8 million years ago, with the deliberate manufacture of the first stone tools by one or more members of the archaic hominin group informally known as "australopiths." Those tools consisted mainly of small, sharp flakes that were knocked off one cobble using another. This was an activity that required a degree of insight into the qualities of different kinds of rock, in addition to manual dexterity; and it also often involved foresight, as suitable materials were carried around the landscape for considerable distances before being made into tools as needed. The availability of cutting implements would have made a huge difference in the lives of the first stone tool makers, although we cannot observe that difference directly. What we do know is that the tools allowed the quick butchery of carcasses (at least initially, most likely scavenged), parts of which could be removed to be consumed in safer places; but this is clearly only the tip of the iceberg. The first stone tools nonetheless appear to have been made by early hominins with small stature, small brains, and archaic body proportions. [1]

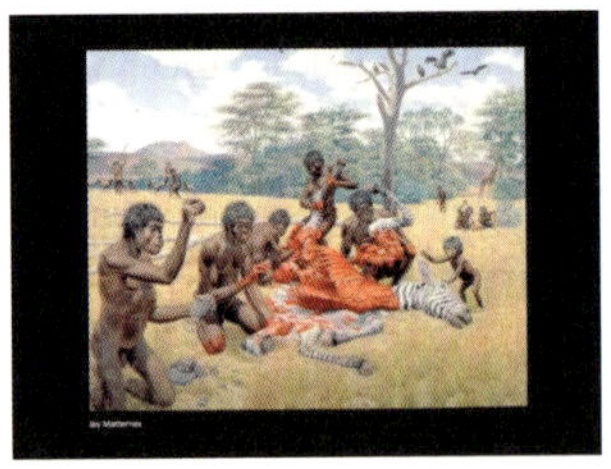

Fig. 1

Perhaps surprisingly from a modern perspective, but routinely from a pre-modern one, about a million years passed after the appearance of the first stone tools before we find any conceptual advance in stone tool manufacture. When this came it involved carefully shaping a stone core, according to a predetermined template that must have existed in the maker's mind before knapping began. [2] This process yielded the

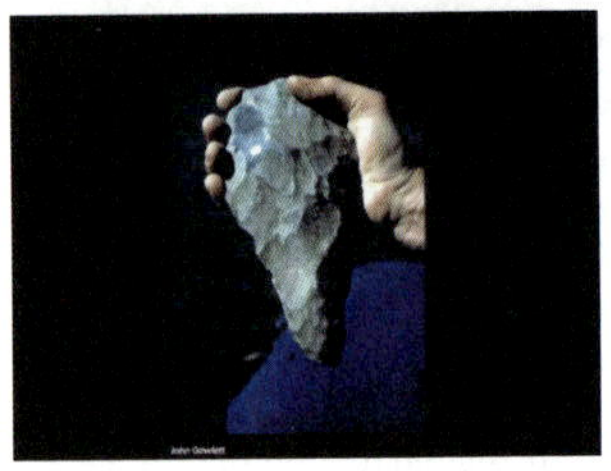

Fig. 2

larger tool known as the "handaxe," which clearly represented a more complex conceptualization of the potential inherent in stone as a material; but alas, we do not have a good record of any more general cognitive advances that may have accompanied it—although we can justifiably infer that those changes were there, because these tools appeared significantly after our own genus *Homo* had come on the scene about 2 million years ago. In contrast to the australopiths, who were bipedal on the ground but retained well developed climbing capacities, the effectively modern body proportions of those early *Homo* signal a dramatic ecological shift to the open savanna. Some additionally believe that this shift must have involved the mastery of fire, though there is no actual evidence for this before about a million years ago, and fire use at campsites only became routine after about 400,000 years ago. [3]

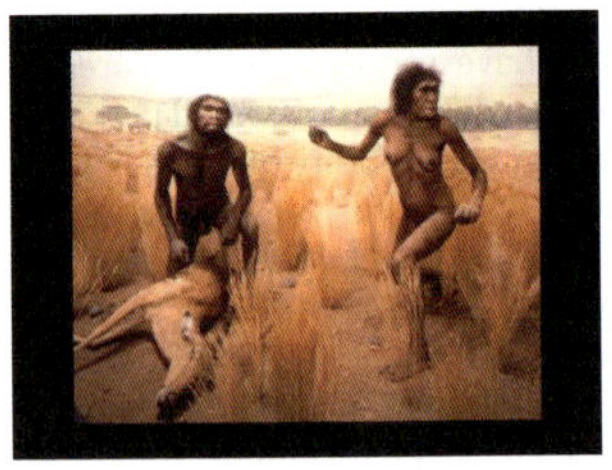

Fig. 3

And then, it was yet another million years before the next stoneworking

innovation was made. In this case, a core was laboriously prepared, until a single blow would detach from it a quasi-finished tool with a continuous cutting surface.
The evidence is, then, that hominids were becoming culturally, and presumably cognitively, more sophisticated over this early period of human evolution, but the pattern was one of the very occasional major innovation rather than one of gradual improvement. Brain sizes were enlarging over this period, as well; and although we cannot say exactly what was happening, it seems that such enlargement occurred independently in at least three lineages within the genus *Homo*. Which makes it fair to surmise that, throughout this process, hominids were employing an intuitive cognitive algorithm in which "intelligence," however you might want to define it, basically scaled with brain size. But big brains are costly things to have, because brain tissue is energetically very demanding, so presumably those bigger brains were paying back that higher cost by supporting more complex cultural styles that in turn allowed hominids to exploit the environment more effectively—although we sadly have little material evidence that bears on exactly what was going on.

Our own large-brained and very anatomically distinctive species *Homo sapiens* appeared on the scene in Africa following about 230,000 years ago. But there was no immediate corresponding change in the nature of the accompanying archaeological record. The new hominid was apparently still behaving in the old way, and the first indications of any *qualitative* change in cognitive style came over 100,000 years later, when we begin to find the production of items of bodily decoration, in the form of pierced shell beads at sites both on the Mediterranean and in southern Africa. Many of those beads bear traces of color, probably from contact with ocher-painted human skin: an observation important not only for its aesthetic overtones, but because bodily decoration of all kinds almost always has symbolic implications. Those implications are usually about economic or social status, or social affiliation; but whatever they are, they are making a statement. But we do not have to wait long for more direct indications that humans had adopted the symbolic reasoning algorithm, because around 80,000 years ago in southern Africa we find the first overtly symbolic objects, of which the piece shown in Figure 4 is the best and most complete. Found at Blombos Cave, near Africa's southernmost tip, it is a smoothed ocher plaque engraved with a geometrical symbolic pattern. That pattern evidently retained its meaning over time, because it is found repetitively in different levels of the site. The slightly younger Diepkloof Cave farther west has yielded comparable engravings on ostrich eggshell. Blombos also yielded the earliest "drawing" around 73,000 years ago. [4] Once the symbolic behavioral

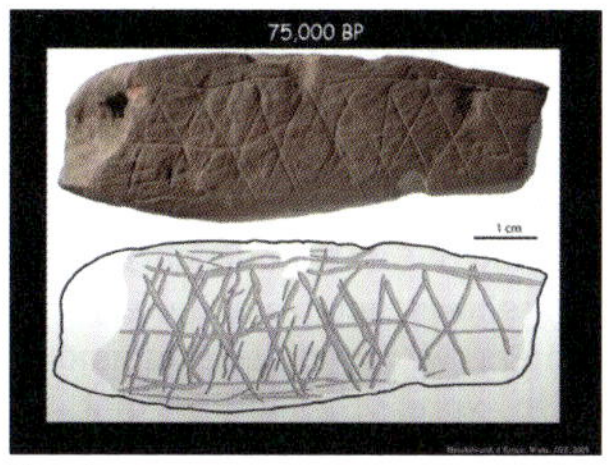

Fig. 4

repertoire that such objects encode was on the scene, our species quickly spread throughout Africa and beyond, eliminating all other competing hominids in the process. In eastern Asia, for example, the venerable *Homo erectus* rapidly disappeared; and at the other end of the Eurasian continent, our even closer relative *Homo neanderthalensis* was gone by about 40,000 years ago.

Even more eloquently for the spread of symbolism, the quintessentially symbolic tradition of Ice Age representative art was soon established. Over 40,000 years ago artists were making images of animals on rock shelter walls in Sulawesi and Borneo; and at the same time the astounding florescence of cave and portable art in Europe was getting under way, furnishing us with the clearest evidence you could wish for that, after millions of years of hominid evolution, the modern human sensibility had fully arrived. [5]

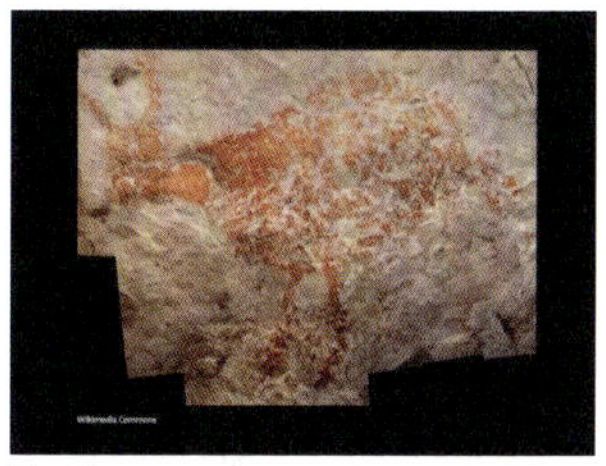
Fig. 5

So, what had happened? Well, the simplest hypothesis is that the neural wiring that permits symbolic thought was acquired in the developmental event that gave rise to the very anatomically distinctive *Homo sapiens*. And that event truly did involve a major biological change, as a comparison between a *Homo sapiens* skeleton and that of a more generalized Neanderthal clearly shows. The fossil record only gives us bones; but it is entirely plausible that the same genomic changes that produced the new skeletal form also involved alterations in the brain that gave *Homo sapiens* a new cognitive potential. [6] But, as is often the

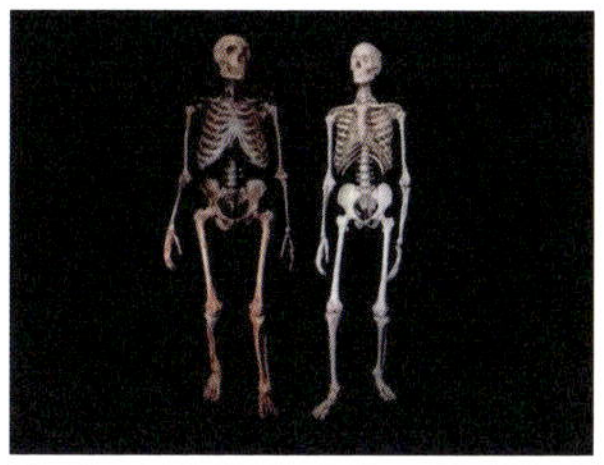
Fig. 6

case, that potential was not realized until it was released by what was necessary a cultural stimulus—

because the underwriting biology must already have been there. For after all, you can't exhibit a new behavior unless you already have the structures that permit it. And in this case, the cultural stimulus concerned was most likely the invention of language which, in depending on the creation of a vocabulary of symbols, maps more or less perfectly on to the particulate nature of symbolic reasoning.

Linguists speak easily of the human "language-ready brain," the very brain that was acquired in the speciation event that gave rise to anatomically modern *Homo sapiens*, significantly before the advent of symbolic behaviors. One can very easily imagine a scenario unfurling somewhere in Africa, some time over about 100,000 years ago, in which members of a small local population of "language-ready" *Homo sapiens* began attaching specific meanings to sounds that already existed as part of a vocal repertoire. That attachment would then have spurred a mental feedback process between symbol and meaning that soon led to structured language, as has been seen in the spontaneous invention of sign language among deaf Nicaraguan schoolchildren. And indeed, it seems likely that language itself was invented by children, always more accepting of the new than adults are. And once the advantageous novelty had been produced locally, it was primed to spread like wildfire among members of a species already enabled for it. What does this imply more broadly about the evolution of human cognition? Well, although human cognition is so remarkable and unique that many have assumed it took a long time to evolve under the burnishing hand of natural selection, it strongly suggests instead that modern humans acquired their remarkable cognitive powers both very suddenly and very recently. And indeed, this sudden, recent, and adventitious origin of what has been called the "human capacity" helps to explain a lot about our brilliantly creative yet patently imperfect species. Natural selection can never explain our dreadful memories, our faulty decision-making processes, and our self-destructive behaviors. But adventitiousness can—and it also tells us that these defects are the price we pay for our precious free will.

Finally, the accidental nature of humankind additionally says something important about the nature of our culture. In its rudiments, as we have seen, culture is widespread in Nature. But there is something very unusual about human culture, which is not only a driving force in human societies, but actually underpins them—because we are, literally, dependent on our cultures to survive. To take an extreme example, the little marsupials known as Bilbies can survive in the Australian outback because they have the physiological adaptations that allow them to do so. And they can all do it. But humans can survive in the outback *only* if they have the rare cultural knowledge, possessed by indigenous Australians, that allows them to. Without that culture, they are goners. We human beings are, in other words, deeply cultural creatures; and the archaeological record clearly shows that, from the very beginning, we have felt a profound need to express the cultures that make us who we are. On the other hand, it is equally true that evolution produces diversity, and that every species living in the world, including *Homo sapiens*, is merely one more terminal twig on the luxuriously branching Tree of Life. But while this last makes it more than a bit questionable to suggest that *Homo sapiens* and its exquisitely modulated cultures are the ultimate expressions of evolution, what it *is* entirely legitimate to say is that culture of this kind is the ultimate expression of being human.

Design for a Creative Brain

IDAN SEGEV

Today I will try to do something, so to speak, a little bit unusual, because I am going to speak about the very difficult topic that Ian raised: what is unique about human brain? What makes us so specific? Why do we succeed in creating art, technology, fashion? There will be a little bit of some ideas, not necessarily complete proof but I'll be happy to hear Katrin especially, as a brain expert, about what I am saying. Let me start with Einstein. We cannot not start with Einstein. He is probably the most creative scientist we ever knew because 1905 is a miraculous year: Einstein did four amazing papers with his creative brain and said that the most beautiful thing we can experience is the mysterious, it is the source of all true art and science. Fondazione Prada is trying to connect art and science in this set of lectures, and both art and science are the ultimate creativity. You start from an idea and then something comes up, like Leonardo da Vinci, Michelangelo that Katrin mentioned, and I will mention more. The question is: what is unique in our brain, as much as we can say, that enabled us to be so creative? There is no doubt that we are the most creative creatures living today, we create all the time: we create meetings, we create Zoom, we create medicine against things, and so on. What makes us so creative, what is unique about our brain? I will start my talk here with point one, a very quick outlook on creativity because Ian already said several things about this. Then I want to inject, to introduce the notion of a brain state. What is the state of a brain? What does it mean when you say the brain is in a state of sleep, in a state of Parkinson, in a creative state? What does it mean to be in a state? I want to emphasize and I want to try to convince you that our brain developed tools to have many new states. One of them is a creative state. We have the tools, we have the anatomy, we have the physiology that provide or improve our capability to be in new states that other brains cannot be in. This enables us to be creative. Then I'll go on with a few speculations here about what makes our brain unique, in the sense that it supports the foundation of creativity. I think creativity is the most beautiful thing that human beings do. We can use creativity, of course, for destruction, for doing beautiful things like art and science. Along my talk, I want to show some beautiful creative things that came from brain research, in particular the Blue Brain Project that I am part of. [1] Let me start with the two most

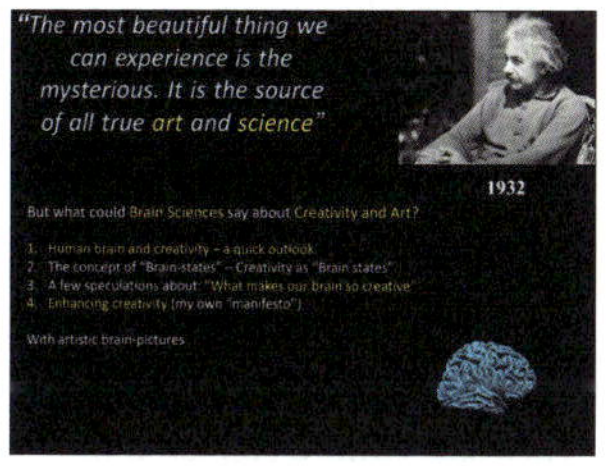

Fig. 1

creative people, in my mind, in the 20th century: I already mentioned Einstein and then Pablo Picasso here on the right. The two of them said something about creativity that we should listen to. Of course Picasso, being an artist, an intuitive, he said "The most important things is to create. Nothing else matters; creation is all." It is very unique to our brain to create. Einstein said something deeper, in a sense. [2]

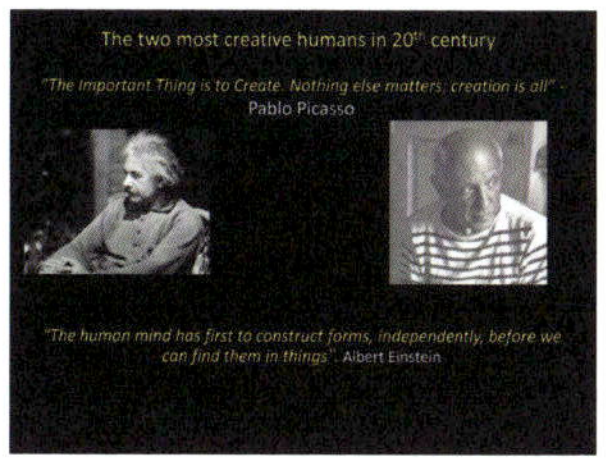

Fig. 2

He said "The human mind has first to construct forms, independently, before we can find them in things," to imagine—Ian mentioned imagination—things independently before we can find them in things. We generate hypotheses, we generate ideas or possibilities, and then we look after these possibilities raised in our brain, due to the capability to imagine things, and look for them or find them or not in the environment. [3] Other artists, for example

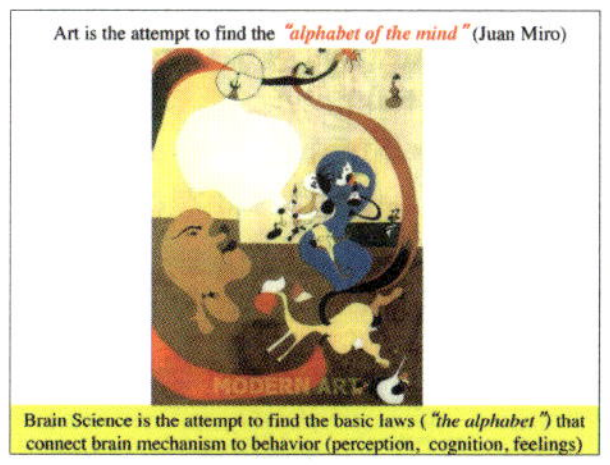

Fig. 3

Juan Mirò, said something very beautiful, in my mind: "Art is the attempt to find the alphabet of the mind." I would say, as a scientist, that science is also an attempt to find the alphabet of nature, and brain science is an attempt to find the alphabet of the mind, the fundamental principles that enable my brain to function. We don't want to describe every detail, we want to find the alphabet, the principles. That is a very beautiful idiom by Mirò. As Katrin mentioned, I like art a lot and I think art and science should go together. We'll talk about this at the end. I think we can learn a lot from artists and many times artists envision things much much before scientists found them in reality, much much before. We can learn a lot from artists about how to do science, or what to think about as a scientist. [4] Brancusi spoke about

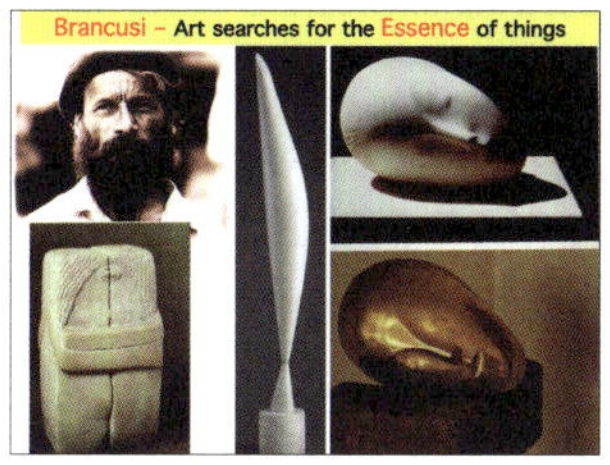

Fig. 4

the essence of things: again, how you reduce the details, how you reduce the complex things you are trying to understand, like a brain scientist or an artist, to the minimum. The birdiness of the bird, the headiness of the head, the kissiness of the kiss, the minimal keys that capture the kissiness, the notion of peace. Science and art are trying to do similar things in many ways but with different tools. You heard a lot from Ian about brain evolution, so I am not going to say much. [5] I just

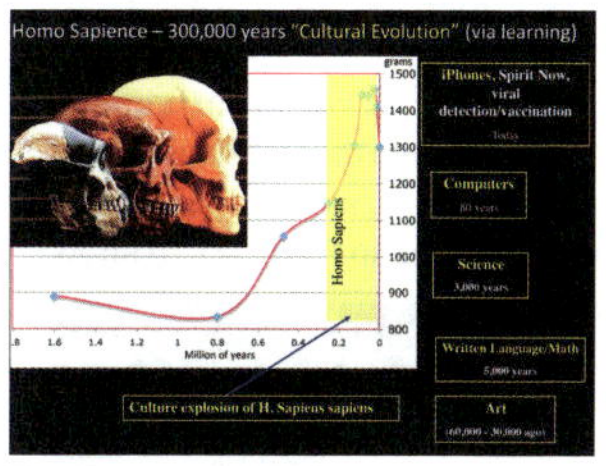

Fig. 5

want to emphasize one thing: during this 200,000—people now speak about 300,000—years of *Homo sapiens*, of what is called the cultural evolution, or this yellow region over the years, there is no genetic change. If you look at the genome of a person, of a fossil—we can do with ancient DNA today—and you compare the DNA of this person to ours today, there is no big difference. That is why I can say that we are *Homo sapiens*: we are the same genetic species. How come everything was there and suddenly, along the years, as Ian mentioned, there is art, some 3,000-6,000 years ago, there is written language, suddenly there is science, computers, iPhones, machine learning, brain simulations, viral detection, vaccination and who knows what is coming. How come, with the same brain, with the same structure and physiology, you continuously do new things? That is almost a miracle in terms of the brain yes. That is what I am going to try to look at, to transplant. [6] I cannot not show you of course these beautiful things because this is, in my mind, the ultimate things that the art did, these prehistoric caves that Ian mentioned already. I am going to tell you a little bit about brain states, because I think that what is unique about our brain is the capability to

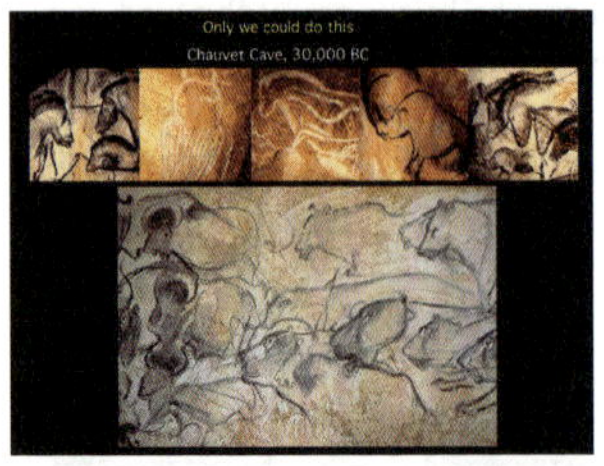

Fig. 6

have new states, which correlated with new things that we do we do including brain, art, science, creativity, and so forth. [7] To show

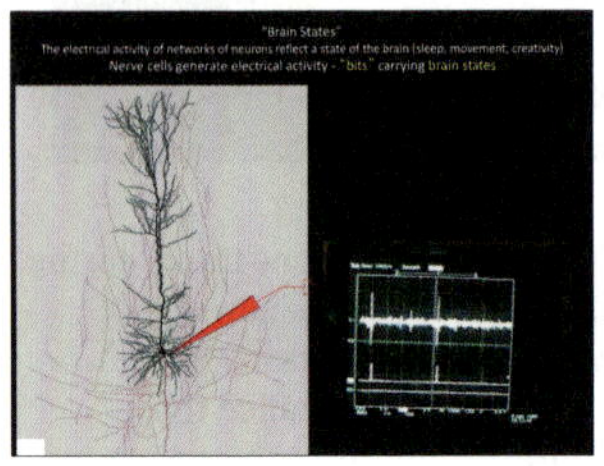

Fig. 7

you—maybe you know already—there is this elementary unit in the brain called neuron, a nerve cell. It is a complicated structure and we are studying it a lot. But for this talk I just want to mention that this element that is a nerve cell, the neuron is an electrical device, it generates electrical signals and you can hear here what would happen if I would record from the cell body. Here, from this black spot, with this electrode, what will happen if I will put an electrode in my brain, your brain, a mouse brain, and record what is going on there. There is this electrical activity—we call them spikes—which fire, sometimes spontaneously, sometimes in response to some external stimulus, like this. This is the individual activity of a single neuron in the brain, any brain. This is the elementary microchip that does all this computation, it generates ideas that enable me to talk now, that enables you to listen, to think, to sleep. This is the elementary unit but we have a lot of them—we should talk about numbers in a second. Katrin already mentioned that our brain has 86 billion elements like this. It is a huge number of elements interacting, and this interaction generates states, dynamical states of elements interacting with one another. To show you what I mean by state I want to take an example that is a difficult example, but it is easy to understand. I want to show you a state of the parkinsonian brain. What does a state of Parkinson's mean? [8] I show

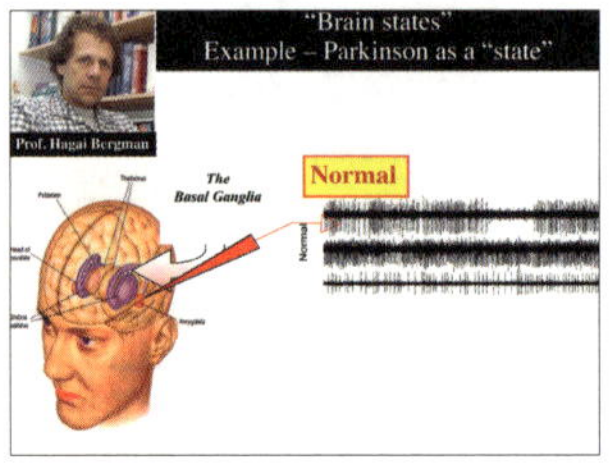

Fig. 8

you this is a work by Hagai Bergman from our work, who was one of the first to record with these electrodes from a brain of a monkey, in this case, but also in humans today we can record this activity before operation. You put an electrode in the living brain and you record from three cells, three neurons: one neuron is doing this activity, stops, and so on; the other nerve cell here, the middle one is active more rigorously; the third one is less active. Each neuron has its own kind of electrical music, if you want, and together they generate the philharmonic music of a non-parkinsonian person looks like this: each neuron is different a little bit, and together, about a million neurons and in this particular region of the brain, generate a global activity, a non-parkinsonian state. But if you record from the parkinsonian patient, similar neurons in the same region suddenly generate different electrical activity. [9] For example, this neuron is doing like a machine gun and then stops. Then there is a second neuron, and there is a third neuron and apparently, all the neurons remaining in this region of the brain, the basal ganglia, start to fire together synchronously, like a machine gun. This wrong state, this

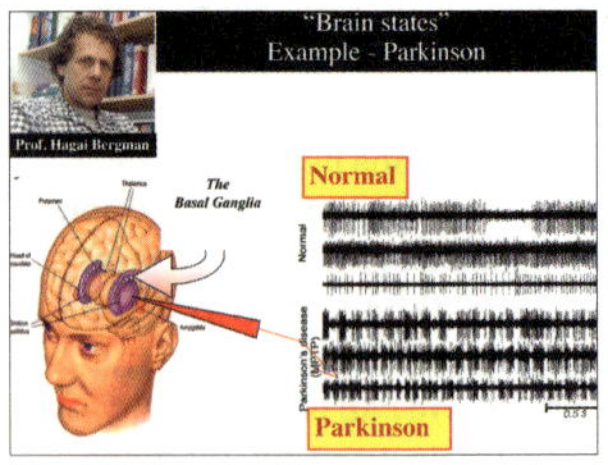

Fig. 9

electrical state that suddenly becomes parkinsonian is reflected by a parkinsonian state: your behavior, muscle activity, the movement, tremor, and all that is related to Parkinson's. So these are two brain states: this is a normal brain state and this is a parkinsonian brain state. I want to claim that all that we do can be defined as a state of the brain: you are now alert, you listen to my talk, and you have a particular brain state that enables you to understand what I am saying. I am in a state where I am talking now, I am moving my hands now: it is a brain state. [10] To show you that

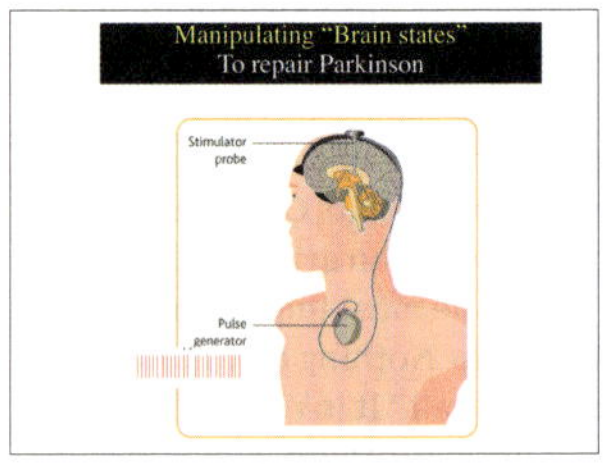

Fig. 10

we can interfere with the brain state, manipulate the brain state, and correct Parkinson's, I want to show you an enormous advance: what is called you know deep brain stimulation. If you are a parkinsonian patient and this region here generates this electrical activity—this state of Parkinson's—you can interfere, you can inject current through a generator, through a battery that is implanted. Like a pacemaker in the heart, you can pace make the brain, you can pacemaker the brain, you can inject current in the brain and correct, while you inject, the activity of this region. I want to show you an example of a parkinsonian patient here. This is a parkinsonian patient at Hadassah Medical School in Jerusalem, that agreed to be photographed, to be videotaped. You can see a very severe young person, 56 years old, that is parkinsonian and I don't need to show you how difficult it is for this person to move and to do things that are so easy for you, like this is very difficult for him to do. [11] This is a parkinsonian

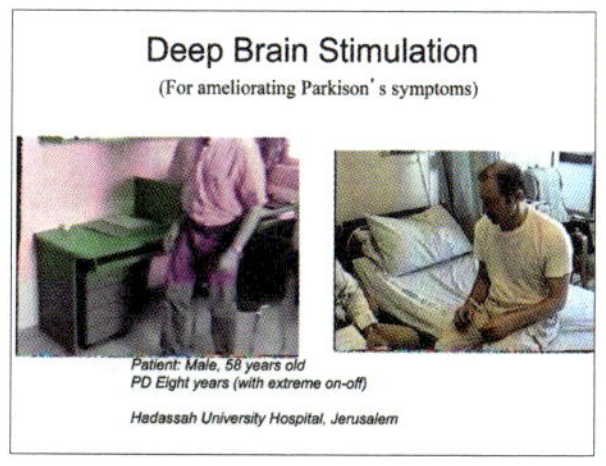

Fig. 11

state: his brain generates this electrical activity which is spontaneous, automatic. You cannot get rid of it so you are parkinsonian. But if you take this electrode that I mentioned before here, into the brain of this patient which is a very fast operation today, and you put the electrode in the correct location, then the same person after this fast operation actually completely alert—you can talk to him while you penetrate with a very thin electrode to the correct location—this same person is looking like that. This person that you see now walking on the left side is the same person just after the operation. He is getting an electrical stimulus changing his brain state, changing the state of his brain into normal brain activity and he is doing things that parkinsonian patients cannot do. This was an example of what a brain state, an electrical brain state means, and how I can manipulate the brain state from one state to another through a technology, brain-machine interface. [12] After introducing this brain state, I want to say that something in this human being, some 24,000 years ago here or in Germany, in Austria, in Italy, much later, by Michelangelo had a special brain

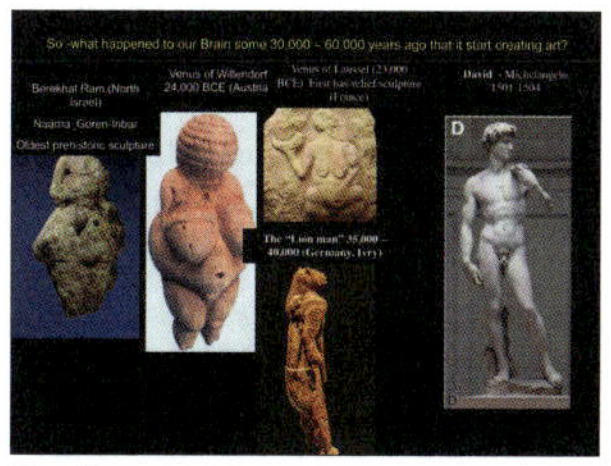

Fig. 12

stick that enabled him to create David and so on. It is a state. The question is what makes us unique in terms of being able to be in these states, what makes us so creative? I want to start with something to negate something: it is not the size of the brain. Katrin can tell you a lot about different brains of clever people, of unique people, unique personalities, she is an anatomist, she knows more than me. [13]

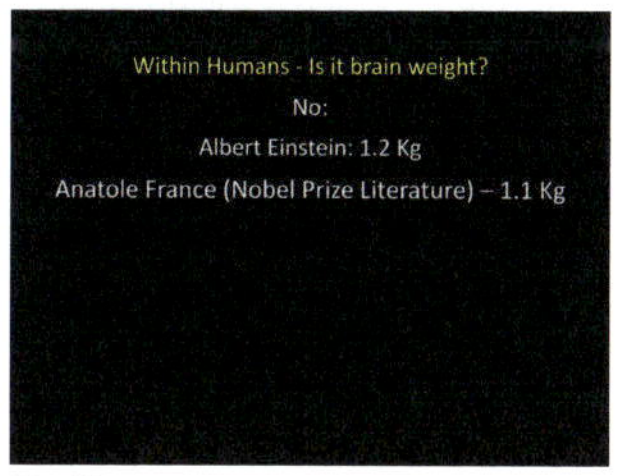

Fig. 13

But I can tell you that Einstein's brain was very small: 1.2 kilograms, much smaller than your brain, all of you. Anatole France, another Nobel laureate, had an even smaller brain: 1.1 kilograms, very small. Your brain is 1.4, 1.3, 1.5. So there is no connection between these genius people and the size of their brains. It is not the size of the brain. Bigger brains are not more creative than smaller brains. [14] Katrin mentioned something very important: we do have a huge number of cells, so if you compare the mammalian brain—from the smallest mammal to the bigger mammal to chimpanzees, to macaques, and so on—and compare it to our brain, we have a huge amount of microchips. A huge amount. 86 billion elements interacting electrically with one another. There is a huge combina-

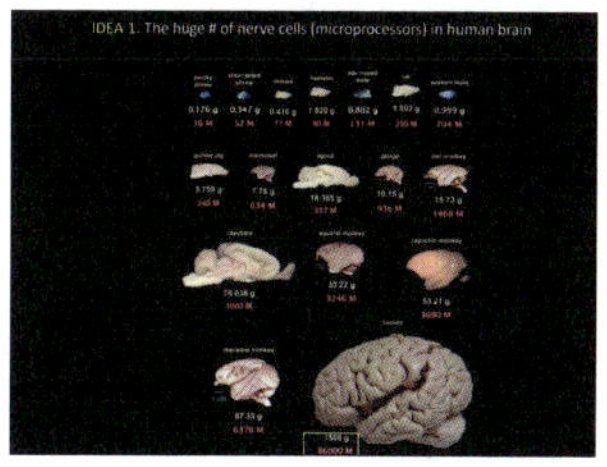

Fig. 14

tion of activities that you can do with 86 billion and you cannot do it with two neurons. If I had two neurons I could be either one zero, zero one, zero zero, or one one—I had only four states. With 86 billion we have a huge amount of permutation or combination of activity, many states. That is one idea. We are creative partially because we have the substance, in terms of a huge enough number of microchips, computing elements, neurons. There are animals with more brain cells. Dolphins have a bigger number of brain cells, so it is certainly not the only thing that makes us creative, but it is certainly part of the things that enable our brain to be in many brain states. [15] Another aspect that

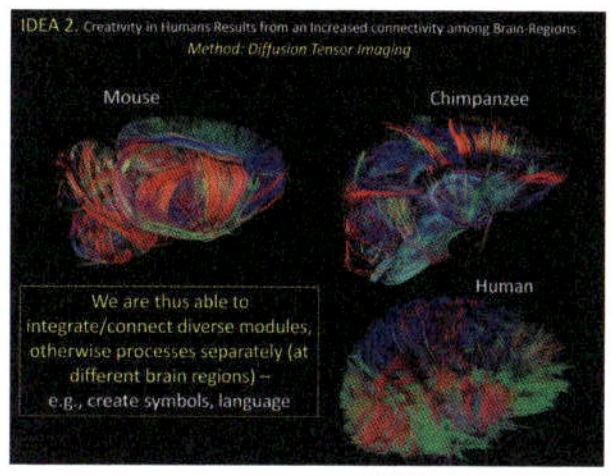

Fig. 15

is unique to our brain—again, Katrin can tell you more about the gross anatomy of the brain—but if you use this technology, which is called diffusion tension imaging technology, which enables you to see the connectivity, the connection, the axons, the wires that go from one region to another region of the mouse brain, and you compare the connectivity, the white matter, the axons going from one region to another region in the chimpanzee and in the human brain using the same technology, you see huge connectivity between regions. Our brain evolved through genetic evolution into a brain that is highly interconnected between regions. Much of our brain is what we call white matter. The white matter is this wiring connectivity. Katrin is one of the top persons in the world that works on this connectivity between regions. Why is it so important for creativity? Why is this interconnectivity important for creativity? Because when you want to generate symbols, like Ian spoke about, let's say that we are now agreeing that we have a symbol that is called glass; I am using the simple glass. When I now say glass, all of you are thinking about the shape of the glass. It has a shape but it also has a taste because inside the glass there is juice. Also, when you think about glass, you think about the feeling of the glass whether it is from plastic, from glass. Just by saying the symbol glass I actually mean visual symbol, somatosensory touch symbol, taste symbol—at least three different aspects combined together to the notion of glass, to the symbol of glass. How can a brain connect something that is visual here in this region, something that relates to touch in this region, something that relates to smell in this region, something that relates to the taste of the juice or the coffee in this region, when I think about glass? You need connectivity, otherwise there will be a visual class disconnected with taste glass, disconnected with smell glass. This intense connectivity enables us to generate symbols and, as Ian mentioned, symbols are everything because you manipulate symbols, you enable symbols to integrate, you create poetry by shifting symbols, I create lectures by moving symbols around and I am talking to you and these symbols go to your ear and you understand. Symbols are essential for creativity and connectivity in the brain, in my mind, is absolutely the basis for generating an integrative way of thinking about a symbol. A glass, a face, and so on. That is my second hypothesis about what is unique about our brain: it is highly connected. The mouse cannot create a symbol of glass because its visual system is not so intensely connected to its smell system and so on. [16] The

Fig. 16

third thing is what happens locally in a local region, a very small cubic millimeter region in our brain. This comes from the Blue Brain Project, which is a simulation of pieces of the brain. I am not going to go into details but I promised you beautiful pictures which are also artistic. If you look at just one cubic millimeter of your brain—in this case actually the brain of a mouse, or a rodent—you see a very dense local circuit, not between regions, but a very dense local connectivity. This is true for all mammals: if you look at the brain of a mouse or of a monkey it is extremely dense locally. But there is something unique about the human brain. [17] I promised you beautiful

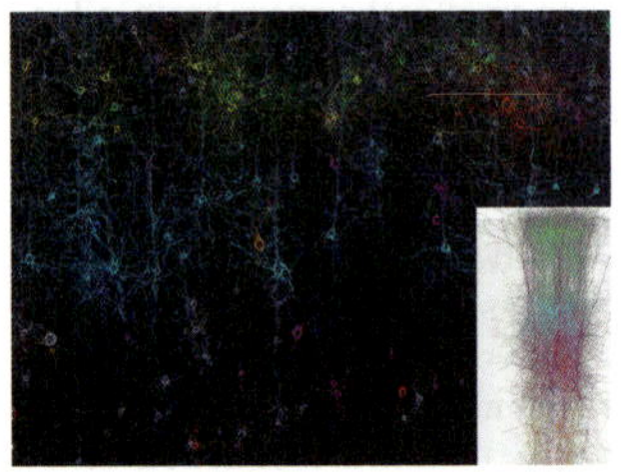

Fig. 17

pictures because it is an artistic talk as well, so this is from the Blue Brain Project: you can see a piece of the cortex of the mouse with different types of nerve cells, very beautiful. The brain is actually a very beautiful thing on its own. In the Blue Brain Project, we are trying to simulate the whole brain of a mouse. [18] This is now one of the first whole cortexes,

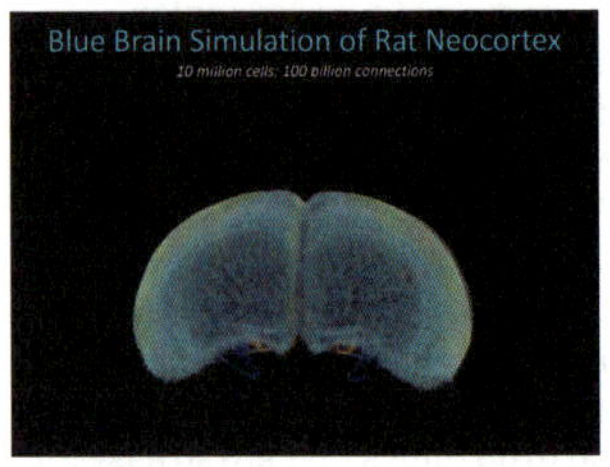

Fig. 18

10 million cells simulated one by one in the computer. Again, you can see that it is a very beautiful piece of art that actually was supposed to be in the Pompidou Center in Paris, if there was no Coronavirus. [19]

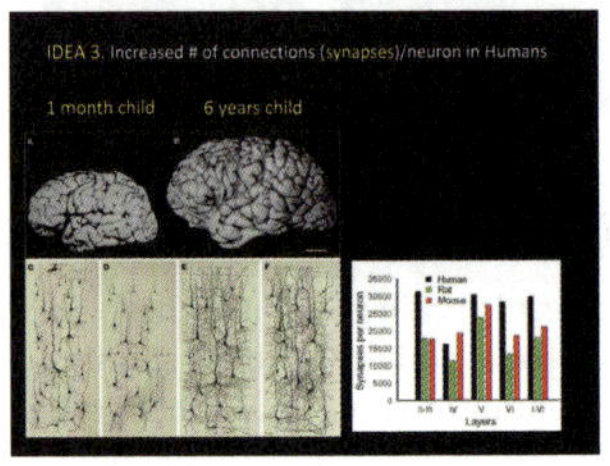

Fig. 19

What is unique about the brain in terms of local connectivity? You can see here two things that are important, in my mind, for generating the capability to become creative. One thing is this slowness of development of the human brain. Here on the left, the brain of a one-month-old child you can see the local circuit, the local neurons in the cortex of this young child postmortem. It is a sparse network, but if you go to the six years old child, the brain became bigger, with many more nerve cells, much more connectivity, many more synapses between cells in this region. But it took six years for our brain to evolve, to develop into a mature brain. It looks like this also in your brain, a mature brain. Six years. So the slowness of development, the fact that we develop so slowly compared to the chimpanzee, to become mature in terms of the brain—this slowness of development enables us to absorb the complicated environment, the expectation of our parents, the complicated world around us. Trump, Biden, here and there, what do we need and the expectation of my parents to interact with me, the mother, the father, the uncle. It is a very complicated world that we have time to absorb if we develop slowly. If you develop within two or three months to a mature brain, you will not be able to absorb all this complexity and create a model, so to speak, of the environment around you. The slowness of development is very important and there are genes that regulate our development so that we won't develop fast. There are cases where this system does not work. There are kids that develop much faster than needed and they are not capable of interacting with a complex world. Also, if you look at this diagram, you can see that in terms of the number of contacts that a single nerve cell receives from its neighbors, the number of synapses per single unit is about twice in humans—the black bar here—compared to the mouse, to the rodent. We are much more locally connected not only globally between regions, but also within a region. Human connectivity is very big. Why is it important? Because it gives me a lot of brain states, a lot of local perturbation of activity, a lot of states. Both the global interaction and also the local interactions give me a huge number of possibilities in terms of brain activity to enable me to absorb, eat, create, be ready for the future, generate a new future, because I have all these states that enable me to interact with it. [20]

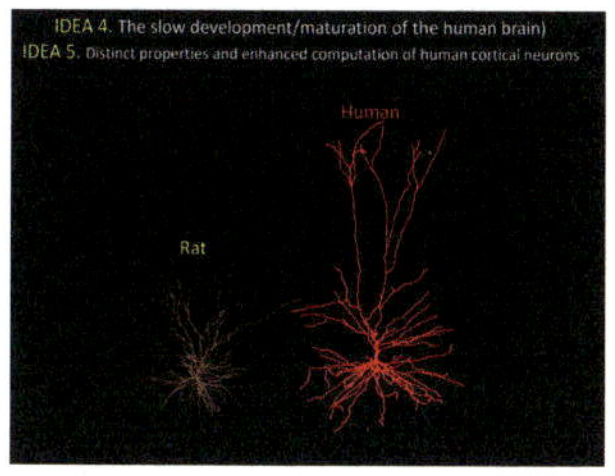

Fig. 20

One thing Katrin mentioned: we are working on human cells, the red cells. We can see today we have a mathematical model of these two elements and we can show that in terms of the computation of this microchip, the capability to do sophisticated computation, this human microchip—the human cell—is much more sophisticated than the red cell. [21] This is an addition to our

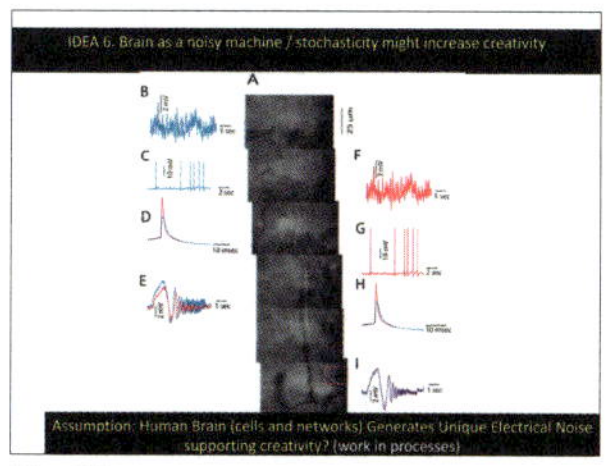

Fig. 21

capability to do interesting things. Finally, just before I finish, I want to inject something a little bit more complicated than what I said but I think it is very important: my computer at home is not creative. It is a deterministic machine, it always does the same thing. You press the letter a, you press the letter b, and this is successful. You want the machine to do that, when you do two multiplied by three it will be six and not suddenly seven. You need to reduce the physical noise that is generated by the element, by the microchips. If the microchip were noisy, electrically noisy, they will suddenly become wild and you with three multiplied by two you would get seven, you would get eight. The brain is very noisy, it is a noisy machine. You can see here that when you record from a nerve cell—here is a nerve cell—, and you record the activity of a nerve cell without any stimulation, in a dish, this element is very noisy electrically, it always generates these little bumps before generating the big signal underlying this activity. The brain is a noisy machine. It is too noisy when it becomes epileptic, but now that I am talking it is always a little bit noisy and it never repeats itself. When I see A it is not the same A for me, when I see Katrin it is not the same Katrin for me. It is always because my brain is always shifting trying to interpret reality but under noisy conditions. This is true for the mouse, for the rodent, but we are now finding that human neurons are noisier, actually in a different way. The spectrum of noise is somewhat different in human neurons, and to me, creativity comes from the fact that you don't do the same thing again and again. You cannot be deterministic and still be creative. You should do different things under the same conditions. Our brain is uniquely noisy and this is something very important for creativity. I should tell you that when you interview creative people, and I spoke with many mathematicians, Nobel laureates, and so on, and ask them "when is the moment of creativity? when is your brain creative?" Many of them say "when I am a little bit sick," and many of them say "just before I go to sleep, I am not completely alert, aware, but I am not yet asleep." In this state in between, when you have a little bit of fever, the noise becomes bigger. The neuronal noise is temperature dependent. I am not suggesting that you should be sick in order to become creative, I am saying that creative people many times said that good ideas came when they had a fever, when they shifted from one situation to another. They were not too overwhelmed by the environment, so the brain was allowed to shift spontaneously and it became creative. At that point, there was an idea. I want to end by asking all of us: how can we become more creative? How can we augment our cognition? All of us want to be more creative. You have

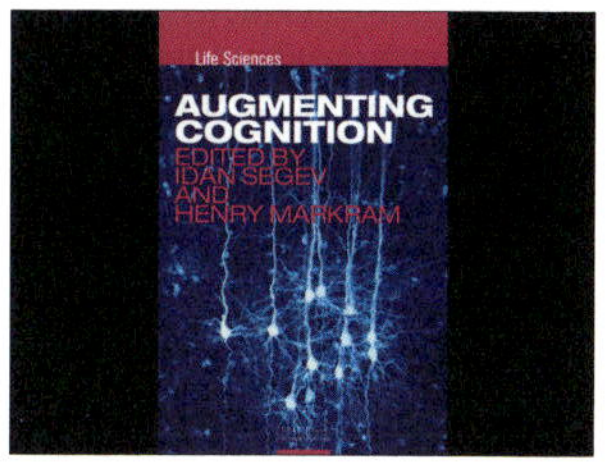

Fig. 22

the tool, the substrate, but how do we become more creative? [22] This is a book that I edited about machines and stimulation, pharmaceutical methods, drugs, and so on, to become more creative. I am not recommending this at all, I am recommending the following: first of all, I think that if you want to be creative—that is my little manifesto about how to become creative—I think we should stop these dichotomies. [23] I am a scientist, you are

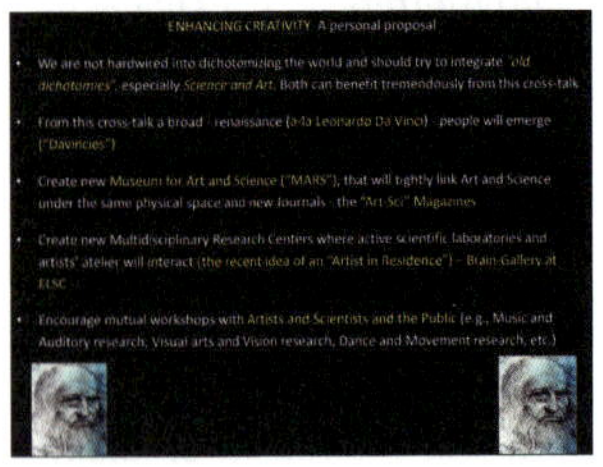

Fig. 23

an artist, she is an architect, and he is a driver. All these old dichotomies, especially those between science and art, are harmful to creativity because through many interactions with other people that are not from my field huge interactions can come, because of this friction between the two cultures. as C.P. Snow used to call them, the" two cultures". But there are many cultures and they should interact. We should not isolate ourselves in ivory towers and become only scientists. This is one call: break this old dichotomy, science-art, and generate, for example, new museums that we should call the MARS museum, the Museum of Art and Science. I tried many times to go to an art museum and ask them to do a scientific exhibition in the art museum. They said no, art is art, science is elsewhere. You go to a science museum, they say "We don't do art here, we do science here." I think it is a great mistake, especially for a child that comes to a museum: he should see Einstein on the left, let's say, and Picasso on the right, together. They go together. Then we shall have a Leonardo Da Vinci brain. I think our society should encourage this kind of interaction. This meeting that we are doing now goes in this direction because we are now interacting with people that typically do not. I never met Ian, although I read many of his articles, and I am trying to meet people not from my field because I already know people from my field, I don't want a copy of myself, I want somebody else. I very strongly suggest to all of you to do this interaction, to do both art and science. All of us should take the example of this amazing genius, Leonardo Da Vinci. I just want to add one last thing—I am a scientist, I describe a brain as a machine: there are nerve cells, synapses, connections, regions. It is true, it is a machine of some sort. But science has a limitation in terms of what it can say about the world. [24] I think

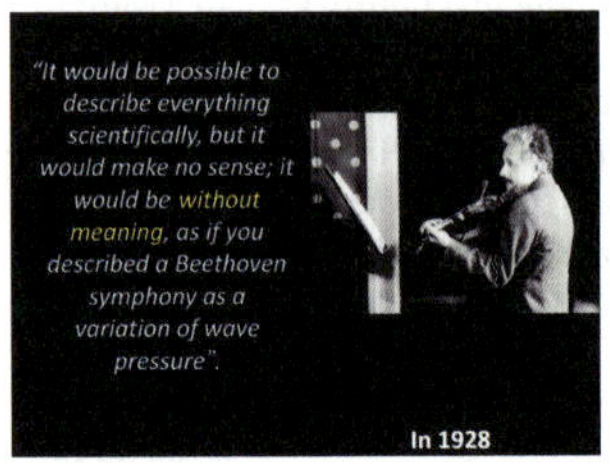

Fig. 24

Einstein again said the following: "It would be possible to describe everything scientifically, but it would make no sense; it would be without meaning, as if you describe a Beethoven symphony as a variation of wave pressures." Signs can describe wave pressures, but signs cannot describe the music of Beethoven.

Panel Discussion

KATRIN AMUNTS
IAN TATTERSALL
IDAN SEGEV

KATRIN AMUNTS In connecting to what you, Idan, said at the end, I would like to forward this question to Ian: people also talk a lot about the division of science and humanities, and in fields like neuroscience, psychology, and anthropology it seems that cultural and scientific approaches are merging. What is your experience? Is this a real merge, they come close together, or is this rather a difficult neighborhood between sciences and art?

IAN TATTERSALL It is a difficult neighborhood, there is no question about it. If the brain is a machine, which it must be, there is definitely a ghost in it, and we haven't identified that ghost yet. I think science and the humanities—Idan has made a very good point—are mutually illuminating because they are both products of the brain. This is where they definitely do meet, and we have to find somehow how they interface. We need both.

KA When we think about aesthetics and the work by Schiller, *On the Aesthetic Education of Man*, then Schiller tried to show the inner workings of artistic creation and perception. Something that we are also very much interested in as scientists, but that also means that beautiful images explain why we are created. Do we need beauty in order to understand the brain?

IDAN SEGEV In some sense, you may think "why do I need art at all?" I can live without art in principle, I can survive without art on my wall. So it seems. But apparently not. First of all, there must be a very good reason—and Ian should know better than me—why, 30-60,000 years ago we started to create art. What was the function of this art? Why should it be beautiful? Why these people in the prehistoric caves made this huge effort to go with torches down in the Lascaux cave or in those long caves and draw on the ceiling horses and bison? Why did they do it? It is a need for the brain, it is a need to project, it is a need to practice the capability which is very unique to humans—to imagine something that you don't see in front of you. You remember a bison, you don't take the bison with you into the cave but you remember and you imagine the bison and you have this amazing capability to practice on canvas, on the stone, and reconstruct something that is there, in your imagination. That is a very important tool, art is an extremely important tool to practice this capability to bring back thoughts into reality. That is one aspect. The other aspect is when I go to a museum and I enable myself to see Picasso or Van Gogh or modern ones, and I allow myself to go around, it makes a lot of new connections. so to speak, in my brain, because I am not anymore in the reality which is very forceful, this reality is very forceful on me. I need to behave, I need to say the correct things, I need to go. But in a museum and in front of a piece of art I can play, I can go, I can dream, I can make new connections, I can become creative. So for me, art is not just a coincidence, it is essential for the brain to go forward.

KA Perhaps to connect to Ian, Cassirer argued that symbolic forms are basic forms of understanding that are so universally valid and with which men shape reality. Others argue that cognition and behavior are biologically fine-tuned by natural selection, and very much emphasize this point. Where is your position in the spectrum? Could you please comment on this?

IT Sure. We have many millions or even billions of years of common evolution with other organisms in the world. There is a tremendous amount of them in us, and what we have that makes us unusual is an overlayering of capacities that were acquired at intervals, over a huge span of time. What amazes me is the record that was left by the Neanderthals, for example. The Neanderthals were unbelievably smart, they did a lot of very difficult things and they did a lot of very subtle things. They obviously understood nature in a very complex manner. Yet, I am sure they didn't live in quite the same world that we live in. What makes us different from them is something that has to be very small in a genetic sense. We are looking for human creativity and the ability to come up with new ideas and to remake the world in our minds. It is a new capacity, but it depends on a very old series of acquisitions. We are looking for something very small, and those are always very hard things to find.

KA But now, living in this time where we have this dynamic development of technology, and Idan mentioned that we can now use deep brain stimulation in order to help patients with epilepsy, for example, technology is now in the phase where we see direct interactions with the brain, with behavior, and the clinical symptoms, perhaps. One of the questions that we received is: can technology change our brains? If the answer is yes, the change will be an improvement or deterioration? I would like to invite you both to answer this particular question.

IS We have a loop with the environment. If there is a new technology, let's say GPS, somebody invented GPS, I put my GPS in the car and I don't need to think anymore, like a taxi driver in London I don't need to remember all the streets because the GPS does it for me. But before I use my hippocampus, my play cells, I use a piece of my brain to do all this memory about streets and stores and so forth. Now there is a machine that is doing this. So certainly, absolutely, the loop between me and GPS must change my brain because now I have a piece of my brain that can do something else. It is not idle, it doesn't go to sleep. I am doing something else with this, not clear yet what, but I have now the capability to take this one piece and go elsewhere. That is one direction of an answer. Of course, technology doesn't change my brain in a sense of genetics, but it does change the brain because now I have to learn this new machine and make use of it, which I didn't before. Learning is changing the brain in terms of connectivity and so on. I am also now using an empty so-to-speak region that used to do something. I used to climb on trees and escape from chimpanzees, now I don't need to do it because I have technology, I have cars and so on. There is a loop between myself, the technology that I invent, back into my brain that needs to adapt to the technology that I myself invented. That is one direction. The other direction, and I want to comment on what Ian said, in terms of biology are we going to become different species? Evolution is relatively slow, it takes 300,000 years to make a species called *Homo sapiens*. But we don't need it anymore, we can manipulate our genes using technology. I can generate a new species every day in the lab, people are generating new mice and other animals because we can manipulate the genome and so on. That is one direction, with all the ethical issues and with all the fears of becoming gods and manipulating our genes, but we have the tools, certainly we have the tools and there is no question in my mind that they will be used at some point, especially for diseases. The other direction is technology, because I can implant technology into my brain, I can implant sensors. There are sensors, like neuro pixels. Elon Musk is developing technologies to implant the brain. Clearly, we are going to become hybrid, become technologically hybrid with the technology, I mean physically. We shall expand our capabilities through these sensors and machines that will be part of me. I will expand myself,

I already do it with machines: my car is an expansion, a brush for a painter is an extension. I am talking now about physical extension. It will extend my brain. There is no question that we are going in this direction. So, both genetic manipulation using technology, science, and this technology of implanting into your brain, into your body. We are going to become different creatures or additional creatures, beyond what we are today due to technology, due to our brains.

KA Is our brain prepared in terms of hardware? And, secondly, are we really prepared in order to interact with such a complex system as the brain, with all these networks? Are we prepared to allow such technological interaction?

IT Every generation has to reinvent the wheel. Every generation has to learn how to exist in the world from the ground up. In acquiring the experience that allows you to do that, you do rewire your brain, as both of you know better than I do. So, in a sense, yes, we are changing our brains by introducing new technologies and so on, but those technologies are equally the product of the brain and we are not necessarily changing the potential of what the brain can do. Human beings have been around for 200,000 years and I don't think the human brain was structurally different 200,000 years ago, before we started behaving symbolically like we do now. But it needed a cultural stimulus to drive it into a qualitatively different state. If we are talking about states, this is the ultimate state shift from non-symbolic to symbolic. This must have happened in a very short time, and it certainly happened within the tenure of *Homo sapiens*. So I think biologically the entity was already there 200.000 years ago and since then, it is what we have done with that potential that has made all the difference. We are going to continue exploring that potential into the indefinite future. We have no idea how far we have progressed in exploring just what the brain can do.

KA Idan, you have shown this beautiful image of the simulation of the whole rodent brain. The question raised is: what is the relationship between the simulation and replication of the brain? Do you consider appropriate the definition of artificial intelligence? Is it truly intelligence?

IS At this point, just to make clear what I mean by simulation and why I do it. There was a girl some years ago, I wrote an article about these Blue Brain simulations to this journal that you mentioned for kids, this *Frontiers for Young Minds*, and there was this kid, my reviewer for these papers that we write for the kids—kids are the reviewers—so my 11 years old reviewer rejected my paper because she said that I didn't explain why I simulate the brain, it already exists so why do I simulate something that exists. I didn't explain it so she rejected my paper. This was a very beautiful rejection because my aim was to explain why I stimulate the brain and if I didn't succeed to explain it to her it means that I didn't do a good job for the rest of the kids. By the way we have six million kids reading the journal today. Six million kids, it is huge. They need to know where the world is going. When I say simulation there is no artificial intelligence here, it is just a replica, a dynamical replica, like the anatomists are doing, like Ramón y Cajal, the great anatomist, when he drew a little neuron on the paper and then put another nerve cell near it, another nerve cell, and there is a beautiful drawing of a piece of the brain by Ramón y Cajal the anatomist. It was static, there was no movement, there was no activity, which is of course very important to understand the road map of the brain, but it is not enough. We want to understand the dynamics of the brain. In the computer I can both put the anatomy of a cell and of a group of cells but I also can activate them mathematically so that they can replicate the electrical activity and the connectivity by synapses and so on. It is a dynamic copy of the brain. That is what I mean by saying I simulate the brain mathematically by replicating mathematically both the anatomy and the activity. Why do I do it? Because when you have such a big system, let's say the cortex of the mouse with 10 million neurons and about a billion synapses connections, you cannot understand how it functions unless you rebuild it from scratch, reverse engineer it, rebuild it and then manipulate, because now you built it. You hope it is close to reality and then, by manipulating parameters, I can understand why suddenly this brain becomes parkinsonian. What are the parameters? Is it the strength of the synapses or specific cell types that do not function? What is the underpinning of a parkinsonian state? We don't understand even one disease to the level of the mechanism. We have to prove that it will work, we are hoping that by replicating a copy in the computer, manipulating parameters, the computer will become parkinsonian, will have by epileptic seizure, not that the computer will collapse. The network in the computer will behave and then I will understand much better than I understood before. That is my simulation. There is no intelligence there, it is not connected to a body. If you ask me if at some point I will build a replica of a brain one by one, and this replica will be connected to a body, and this body will have senses, it will start to behave, yes. This machine will be myself because I, myself, am a machine. I have neurons, synapses, a body, senses. Yes, I am a machine. If somebody replicates me I already replicated myself with my daughters, so in some sense I already did, but not through a computer.
We are not trying to do it at all, we are trying to understand diseases through simulations.

KA When I speak with colleagues from anthropology, they emphasize very much the relevance of tool use for great apes. We see that chimpanzees who live maybe next door to bonobos have a very different way of tool use. One uses it, one does not use it so much. That means that the body has a high relevance for what our brain is inventing, in terms of behavior. Do you think that the simulation can be good enough when we do not simulate the body at the same time?

IT Simulation is a very complex thing. I think the way in which Idan is trying to approach it is the most creative way to do it. Most simulations are simulations of a result. But what does the brain do? How can we get a computer to do the same thing? If you look at computers in general, they can do a lot of things but they are always running a particular program. That program can just do one thing, and clearly, there is something different about the brain as a computer compared to what a computer is doing. You are simulating something that the brain does, but using a different structure. It is a very fascinating thing, and I am sure it has plenty of potential for telling us a lot. But we are still trying to simulate a product or a device. It is integrating those two where you find the enormous amount of difficulty. Tool making is an extremely important part of the human heritage. It is our first indication that the early hominids were doing anything different from what apes do, by making actual stone tools. We don't know what kind of soft materials they may have used to make tools. It is our best clue that they had moved out of the ape cognitive mold by beginning to produce tools between about 3.5–2.5 million years ago. Those tools got better and better and better, but step-wise. There was this technological pattern of non-change interspersed with very rare major innovations, until very recently. We are not looking at a process of gradual refinement in the same way that we would do it. Hand axes became thinner and more sophisticated over the years, but then you had the new technology coming in and it is not quite the same thing that we experience by having many generations of technology next to each other as we are doing today. Human beings are doing something rather different even from very sophisticated close relatives, like the Neanderthals.

IS Can I add one thing about machines? I want to alert the audience that there is a real revolution. We are living now in a real revolution in terms of machines, because we now introduce machines that learn. These are not static machines anymore, they replicate what they are aimed to do. These machines are changing as a function of the environment. In the machine learning world, the fact that we now borrowed some principles from the brain, how does the brain change when you learn something new and then you borrow it to this deep learning, deep machines? These machines are starting to behave in a very unexpected way sometimes. Katrin mentioned the creativity of these machines, it is because they are not static, they now respond to the world. They are changing through this interaction with the environment, so they close this loop and they become, in some sense, individuals. These machines are becoming individuals because each one has different experiences. I think this is a huge technological advance. It is a huge revolution to the level of the industrial revolution. It is coming. We are going to a different world with learning machines that are born as children, so to speak, and they evolved through interactions because the software is dynamic, it is changing, it is not the same from day to day.

KA If I can connect to it and also refer to a question that came from the audience earlier, wouldn't it be more straightforward if somebody would like to reproduce a brain than to do it in an organic process? The so-called organoids are being developed and used more and more as smaller systems for developing new therapies for brain diseases in particular. What do you think about this direction of research and the perspectives that seem to open through the usage of organoids?

IS I think organoids are fascinating. Just to explain, from each one of us you can take a cell, mature cell, re-engineer it to become a stem cell, and then use it to build pieces of heart, of liver, or also pieces of brain. There are these brain organoids that are in the dish, growing and interacting. It is not a complete brain still, it doesn't have a three-dimensional shape, regions, and so on. But you can use it and it is absolutely a wonderful, strong, very important tool to study diseases, for example. As I said, the anatomy of this organoid, the connectivity, the cell types are not a real replica of the brain and also, when this organize grows, it becomes rather complicated on its own and it is very hard to manipulate specific parameters. Again, it becomes a complex biological system that you need to simulate in order to understand. In the computer, I can manipulate each parameter but in biology, it is more difficult to manipulate separate parameters. It is a wonderful new tool, this brain organoid. Wonderful, impressive, very creative, too.

KA There is another question to Ian coming from the audience. It is a little bit provocative, but I think it is also interesting at the same time. What are the main discoveries about the emergence of consciousness that were made possible by the study of fossil remains? This places it also in a little bit of a comparison with what Idan brought—that we have a structure, that we have anatomy. But on the other side of the medal, we have the function and so we have the physiology and the behavior. Could you please elaborate a little bit more, Ian, on how these studies on fossils connect to this notion of consciousness?

IT As I tried to point out very briefly earlier, you have real difficulty in making a connection between states of consciousness, which are subjective states that we simply experience, and how those might be reflected in material objects that you might produce. The archaeological record consists of the durable things that people create, the things that people create that last a long time and can be found and analyzed. That is only a tiny bit of their total behavior and it has nothing directly to do with the way they experience the world. We have huge problems in trying to imagine how a Neanderthal or *Homo erectus* experienced the world. We don't even know how a chimpanzee, which we can observe, is subjectively responding to the world. We can't know what is going on in their head. We can look at a captive chimpanzee and see that it is miserable and not responding well to its environment, for example, and definitely processing this emotional state in a very complex way. But we cannot put ourselves in the place of that animal and understand exactly how it is feeling. Even less can we do that in the case of extinct relatives who have just left this material record of what they did rather than of what they felt. That is always going to be a problem. We are prisoners of our own mindset, we are prisoners of our own way of apprehending the world and processing information about the world. It is difficult for us to abandon that mindset and put ourselves in another emotional and intellectual place. It is tough. What do you think, Idan?

IS I think that now that we have the ancient DNA of, let's say, Neanderthal, in principle I can take this DNA and recreate the Neanderthal. The technology is available, if you want it. If you want more Homos near you we can create a Neanderthal and then you can interact with him or not. I said "in principle" because we have the DNA of the Neanderthal, it was completely sequenced, we can reconstruct the Neanderthal.

IT Theoretically that is true, but how would you bring it up to be a Neanderthal? We are so much what we have learned to be during our development and I have no idea how I would raise a genetically engineered Neanderthal to be a member of its own species. That would largely be a problem of my inability to know this basic thing about them.

KA In connection to this, such process, if technologically possible, raise a lot of ethical questions. This is true for artificial intelligence getting more and more capabilities, but for organoids as well as for this type of generated organism based on the DNA? What are your thoughts in that respect? What is ethically important and how could we make sure that we are doing the right things?

IS This is the most difficult question. I think absolutely we have to be prepared. Society has to be prepared for the fact that we are going to create new creatures and that artificial intelligence is going to be our escort. We need to decide on elementary things. As for automated cars that are going to do some accidents occasionally, who is to blame? Is it the engineer, the owner of the car? There is a huge amount of ethical issues coming both from this technology and from the fact that I can manipulate the genetics of people. Somebody came to me the other day and asked me "Since you can manipulate Parkinson, please manipulate my brain so I will become more creative. Put an electrode in my brain." Seriously. I am not as successful as my parents, inject something into my brain. Do we want it? Do we not want it? I think it is certainly not for the scientists to decide, absolutely not. I think we should be part maybe as advisors, we can envision where it is going, but it should be philosophers, artists, and politicians, absolutely. It should be ethical committees because the world is changing so fast that we have to be prepared for it also ethically. I certainly do not have global answers to this complicated and very frightening issue, but also very positive because we shall have more free time, we shall enjoy each other, talking, go to museums, create things. But the next generation has to be prepared. That is why we need this *Frontiers for Young Minds*, these kids need to know where the world is going. It is not going to be the same world in 20 years.

IT We have to be very careful not to do things just because we can do them. There is a compelling reason for trying to help patients with Parkinson's. There are no compelling reasons necessarily to make chimeras between us and other species or between other species. As Idan says, the ethical problems are huge and these are not scientific questions. They are ethical questions, and need to be debated by society as a whole.

KA In order to think about such opportunity or such development, wouldn't we need to have a more coherent theory about what is consciousness, what is language, what is making us so different? And are we indeed so different from our relatives, non-human primates? Where would you see the state of the arts level at which your field of research is now existing? How far are we, how advanced are we at this moment?

IT We are not very far advanced at all in understanding exactly what the cause-and-effect relationships are, say, between language and consciousness. I am only suggesting, for example, that language was the trigger that made a pre-adapted human brain function in a new way. And I am only saying that because I can't think of anything else that would fill that role and fill that role rapidly. We now know that although our knowledge is increasing about the complexity of the lives of ancient hominids, and we must never underestimate that, there was something that happened suddenly, and very recently, to create a complete change in the entire nature of the archaeological record, suggesting a radical change in the nature of the way in which humans were relating to the world around them. There are certain events that we have to hypothesize occurred, but the nature of those events and exactly what the mechanisms were that governed the interrelationships among them, is something we have very little idea of at all.

IS Katrin, I should mention that you are one of the top persons in the world to answer a question related to the brain and language, isn't it? You study the brain and language and especially what is unique about human brain regions and so on relative, let's say, to the chimpanzee that enables us to talk. You are the anatomist of language in many ways.

KA Yes, I am indeed very much interested in language and I am convinced that it is not so much the word that is the center, but the rules that connect the different words with each other. I see language at the same level if you like arts, also, it is another way to express, to communicate more directly. When we better understand language I am convinced that you also can better understand the brain.

5

Consciousness: The Still Open Questions (Two Different Perspectives)

MICHELE DI FRANCESCO
GIULIO TONONI

PANEL DISCUSSION MODERATED BY
VIVIANA KASAM

Fifty Years of Consciousness: A Philosophical History

MICHELE DI FRANCESCO

I would initially provide you with a state of the art of philosophical reflection concerning the topic of consciousness with an approach that is both theoretical and historical. I studied the past 50 years, which is a lot for 30 minutes, but obviously, I wanted to be concise, so I took the last 50 years of discussion to see how they accompanied the study of the mind with the cognitive science and neurosciences.
Then, for our discussion and for an interactive discussion on the relationship between philosophy and science that I see having great potential in terms of the collaboration that may exist between the two.
I would like to start with a quote. This quote may appear pessimistic in a way, but in reality it is a good starting point: "How is it that anything so remarkable as a state of consciousness comes about as a result of irritating nervous tissue, is just as unaccountable as the appearance of Djin when Aladdin rubbed his lamp." It comes from Thomas Huxley, a biologist who also wrote philosophical books, and was a convinced Darwinian expert. This can to some extent reflect the disconcertment, the bafflement of one who studied neurosciences for the relationship between consciousness and its physical basis.
Certainly, enormous progress has been made since then, from the 19th century, especially in the field of neurosciences. Huge steps forward have been made. One could think that this bafflement is over, but from a philosophical standpoint the seeds of the challenge, which was a conceptual challenge, is I think still very much present in contemporary debate. And, amongst other things, this has been indirectly proven also by the fact that many contemporary authors, Ned Block amongst the many, have mentioned this in their works of the 19th of the past century. After 130 years this problem came up again. The objective is to describe the surprising state of affairs and after having devoted a prologue to the description of a very crucial point of our history, about modern philosophical history, that is the Cartesian identification between mind and consciousness—that starts the modern way of seeing the mind—I will be examining three phases of philosophical reflection that are connected to three reflections on how we can offer a theory of consciousness that may explain its place in the natural world. One concerns the development of classic cognitive science from the 50s to the 70s up to the 90s. This is characterized by a computational vision of the mind. What I mean by computational vision is that, even if the metaphor has been considered misleading, to see the mind as software, the thought as a processing of information, on the one hand, and on the other ones, the downsizing of the centrality of consciousness, visibly the cognitive and mental phenomena. Just to mention an author, if you follow these themes, is Jerry Fodor. The second phase is connected to the birth, around the 90s of the past century, of scientific theories of consciousness, and obviously we had these before, as well, but a moment has come. By the way, a lot of people that are participating in this series of sessions have provided interesting contributions to this blossoming of theories of consciousness, including my colleague, Professor Tononi, who will be speaking later. But I am thinking also about Jean-Pierre Changeux, for instance, or Antonio Damasio. Finally, this phase is linked, from a philosophical standpoint, to the return of the centrality of the subjective states of experience, the ones that have been called qualia, and here we could mention David Chalmers. Then, the phase link to the return of the centrality of the phenomenology dimension as a starting point for the study of mental phenomena, and this, with a grain of salt, we could call it a going back to Descartes that characterizes a part of the contemporary philosophy. If we are afraid of talking about Descartes, we can say a return to phenomenology in the sense of all the various subsequent doctrines.
A couple of words on Descartes. In 1641, in his *Metaphysical Meditations*, he proposes an investigation on the nature of thought of the self and he says he answers the question "Who am I?" as something that thinks, and "What is something that thinks?" It is something that questions, that intends, that affirms, that denies, at once it does not want, that even imagines and that feels. Today it could seem to us a rather obvious definition, but it was revolutionary at the time, because if you know about the philosophical thoughts of the time, the theories of the 17th century, I am sure that you realize that Descartes here puts within a single classification—and he says intends, doubts, affirms, denies, wants, does not want to feel—so he is putting together properties which, for scholastic Aristotelian philosophy of the times he lived at, they were kept distinct. Obviously, it was connected to Aristotle that saw three basic souls—one was vegetative, one was sensitive, the other one was rational—where what characterized the human being was the rational one. Here instead, Descartes does two operations: he puts together the sensitivity and reason and introduces a different type of sensitivity because when he imagines a mind that can be separate from the body, that feels he is separating feelings, bodiness, and has a notion of feeling, which is like mere or pure thinking. A contemporary expert is the film *Matrix*, if I can be irreverent, where the protagonists enter the cybernetic virtual space as though they were software and they operate in a way that is immaterial in that type of reality. But this is a matter of intangibility, which is not that of the soul but that of abstract models

such as the ones of software. Anyways, there is also a comment by Rorty, who says that once Descartes invented that precise sense of feeling as nothing else but thinking, we start losing contact with the Aristotelian distinction between reason and the body, the sensation, the feelings. This new mind-body distinction, according to Rorty, leads to a distinction between two worlds more than between two sides of the human being. This is an accusation of dualism for Descartes. I find that some of the distinctions by Descartes are very interesting and by this what I would like to say is that it is not really necessary to do a dualist outcome starting from his position, and this maybe will be a step that can also be considered quite surprising coming from Descartes, which in his *Metaphysical Meditations* says that there is nothing that nature teaches me more explicitly and clearly than the fact that I have a body. By means of these feelings, sensations of pain, hunger, thirst, nature is also teaching me that I am not present to my body as the helmsman is present to the ship, but that I am so connected to it, I am so merged with it that I constitute with it one whole. It is true that such feelings of hunger and thirst are nothing else but confused ways of thinking that come and depend on the union and, so to speak, the mixing, the merging of the mind with the body. Attention to this new vision of the mind that corresponds, identifies the mind with consciousness, where you have both the feeling and the reasoning, this Cartesian identification between mind and consciousness can also be compatible with the organization of thought. Once said so as a prologue, I would like to start with the first point: the issue between consciousness and classical cognitive science. Descartes, as we said, introduced the notion of the mind which coincides with consciousness, and this is accepted by law and other scholars later on.
But between the 18th and the 19th century, it is known that the centrality of consciousness in general loses its importance at a cultural level. On the scientific plane, we could talk about Watson and behaviorism, James and consciousness as a process and not as a substance, Freud and the unconscious, Wittgenstein who criticizes the Cartesian private, and then think about literature and arts, which tended to develop the awareness of how many shadows we have in human thought and to which extent we don't understand self-deceivement in the life of people. There is a review of consciousness. The crucial step though, from our point of view, because we are interested in the scientific study of consciousness and its contribution to the understanding of this phenomenon, is the advent of cognitivism. Cognitivism is a research program that developed starting in the 1950s after behaviorism, and sees the participation of different disciplines which we have seen during these meetings this week, having a dialogue with neurosciences: psychology, artificial intelligence, linguistics, philosophy, neurosciences, anthropology, all things that we have already seen represented in cognitivism. In contemporary cognitivism, these were connected with the unitary vision of the mind. The mind was not seen as a collection of contents of consciousness but as a set of psychobiological functions that present themselves to our attention in the ways of self-conscious subjectivity. There are psychobiological functions that we do not observe directly, but we observe their product at a cognitive level. They can be described as computations or representations, in other words, software, which processes representations of the external world, of ourselves. What is a representation? Basically, when I ask myself "Can you access the Milan Cathedral through one single door, or are there many doors?" What I am trying is to represent the Milan Cathedral thanks to my memory and I try and give an answer to this question. When I try to understand the quickest way between A and B, also in this case I have a mental map of what happens. Representations are the way in which we represent the external world to ourselves. We can also represent aspects of our nature. These representations are manipulated via computations that are implemented by our brain, seen as brain hardware. Note that the psychic activities that operate in this case are not necessarily consciousness, but they operate at a level which has been called subpersonal, that is unconscious but without what Freud meant by it, and they are not accessible to the consciousness of the subject, but rather essential for the good functioning of the brain and of our capability of acting. These are states in which we build those contents which sometimes have access to our consciousness, depending on the different theories of consciousness circulating today. Computations and representations almost never appear in the character, and again on this point imagine I am trying to remember, I see an actor at the cinema and I can't think of his name. I have it on the tip of my tongue but I can't remember his name. This very simple phenomenon is connected with the idea that we are somehow not capable of activating consciously the computations that would allow us to go and get that piece of information in our memory. Whether this comes naturally, but this naturally is the level of consciousness, but at the unconscious level there are cognitive actions that take place without our control. These actions almost never appear in the consciousness we have. Maybe the idea is a bit too simplistic, but the idea that mental phenomena, our software, can already give us an idea of this and also gives us a possibility of looking at how we can start extracting the nature of what conscious phenomena are. This is one last quote by Hilary Putnam, it dates back to the 1970s

and then it was reprinted, and it is a way of explaining and describing pain as a functional state. Pain is not a brain state, in the sense of a physical chemical state of the brain, but a completely different kind of state. I propose the hypothesis that pain, or the state of feeling pain, is a functional state of the whole organism, as Putnam says. From this point of view, this proposal can be the basis of a form of physicalism this is a contemporary term for materialism, a physicalism that does not want to represent the mind as purely material, but reduces it to a set of mental states which take place in the brain but get identified at a different level, which is the function of one in the description of functions. As when I want to describe the engine of a car as that part that allows the car to move. I apologize to those who know more about engines, but the concept basically is there, it is a higher level of abstraction where we describe conscious states as functional states of the brain. The study of the mind allows us to explain or to hope we can build scientific and naturalistic models, in line with natural sciences, of how we can represent the world, and therefore to explain the intentionality of our mind. What emerged is that explaining intentionality in terms of functional states, in terms of processing of information is one thing, and explaining consciousness is something different, it is something else. The main authors on cognitivism understood this quite quickly. This is Jerry Fodor's quote, we are in 1977. It is during the 1970s that these reflections came out. Based on his current situation, Fodor wrote that the realm of the mind seems to forge into propositional states with a certain content, those that allow us to describe the world through representations, and qualitative states—feelings, subjectivity and we'll go back to this. We have some ideas about how the proportional states and the representations of the worlds could be transformed into science-compatible objects with the demands of materialism, but in the case of the latter, qualitative states both epistemological and ontological problems remain unknown. We seem to be at the threshold of a psychology of propositional attitudes and the psychology of consciousness continues to elude us. This is the point I wanted to get to, it is what allows me to say that cognitivism and cognitive science tried in all possible ways to provide a representational analysis and a computational analysis of mental phenomena, leaving to an indeterminate future the solution to the problem of consciousness. This happens because the problem of consciousness talks about the subjective experience of subjects and the nature of such experience. This is something that has been noticed by Thomas Nagel, for example, in an article he wrote in 1974 whose title was "What is it like to be a bat?" Nagel writes that the fact that an organism has somehow conscious experience means fundamentally that it makes it strange to be that organism. We can call this a subjective character of experience, it is not captured by any of the familiar analyses of the mental recently elaborated, because they are all logically compatible with its absence. It cannot be analyzed in terms of some explanatory system of functional states or intentional states, because these could be attributed to robots or automatons that behave like people even if they do not have experience with anything. What Nagel is telling us is that we are missing a signature, we are missing accurate identification of what makes consciousness what it is. We cannot explain the relationship between our physical nature and consciousness, because we can imagine creatures with the same functional states we have, but without consciousness. They could be robots, they could be automatons that can mime human behavior but are not conscious. In fact, in these meetings we have also had discussions about artificial intelligence and this is one of the theses that is currently being discussed. Is it possible to have a hard artificial intelligence that really creates thinking machines? We will talk about this during the debate, maybe. To conclude, with this we noticed that we have now introduced two distinctions. You know that making distinctions is very important in physical philosophical work. A personal distinction or sub-personal distinction, which recognizes the existence of unconscious sub-personal mental states, made by classical cognitive science; and the distinction between intentionality, i.e mental representations and consciousness, where the first notion was presented as naturalizable, so to speak, that could be studied through natural sciences, whilst the second one still appeared as metaphysically elusive from the point of view of the substantial nature. This, again, is Nagel, his book on the bat: at the moment the state of physicalism is similar to the hypothesis that matter is energy would have had if it had been supported by a pre-Socratic philosopher. We don't have any idea how this could be true. In this manner, we know that cognitive science, which has given a lot to introduce the science of the mind, has not really studied consciousness. There have also been Barth and others, but it's been partly reticent. This reluctance shouldn't last too long, and in fact, it didn't happen. There have been very many reasons why philosophers have started dealing with consciousness with more enthusiasm, but the main one certainly, from the point of view of our conversations on the relationships between culture and consciousness, is the development of scientific research, which has been extraordinary thanks to neuro technologies that would make a series of theories on consciousness possible based on neurobiology and cognition. I repeat, once again, that we have met some authors of these theories, and today I think we'll have the pleasure of listening to something very interesting with professor Giulio Tononi. The question many philosophers asked themselves during the 90s was: are the theories of consciousness explaining the subjectivity of it? All those aspects we have taken from Nagel's thinking, which have also been proposed by others, lead us to the second part of our discussion with phenomenal consciousness and the so-called hard problem of consciousness. To get there, another distinction could be useful, which is the one—here I chose the phenomenal and the access consciousness—we could have spoken about the psychological one, as David Chalmers says, the cognitive one or ourness as opposed to consciousness. But the distinction is this: the phenomenal one is the experience in first person, or having certain quality experiences of what happens to us. When people have certain feelings, states of mind, emotions, feelings, as they are called, they correspond to a series of mental phenomena that make up the experience that is experienced of the living agents that we are. The access consciousness, instead, is a construction of the sciences of the mind, which is identified by the mind, as we said, it is characterized functionally as the ability by a system for instance—this is Block's definition—of having access to interior states for the purposes of verbalization, organization of action, and also the construction of models of oneself which can be used in social interaction. These aspects of consciousness have been presented during this week of reflections and studies. Many of the authors who have then contributed to orienting the philosophical debate on consciousness—so we have Ned Block, Frank Jackson, Colin McGinn, John Searle, David Chalmers—have raised some metaphysical and conceptual questions, let's say, subscribing to the thesis that what it feels to be ourselves cannot be excluded from the rules of the world, but it can't even be seized concretely with the refined elaborations or processing or of the scientific processing of the mind, because they do not provide an answer to the problem, the relationship between brain activity and genesis of the phenomenological world of the first person subjective experience. Going back to what we said before, it is the phenomenal consciousness that seemed at the time to elude the explicative abilities by the scientific theories of consciousness. Obviously, this has not been mentioned by all philosophers, in fact, some developed models of consciousness that wanted to be more naturalistic, whilst others believed instead that it was premature simply but there were hopes of obtaining explanations in the future. I call this period, the state of the art in the philosophical debate, the war of qualia. This was a rather long period of time and we still are discussing it now in fact in philosophy, but we can say that in the 1990s and in the first years of the 21st century the debate was still very active, it was quite strong. What surfaced were certain trends, certain lines of thought. I would like to mention one in particular, which is the limit set by functional analysis. To explain the functions of something very often it is enough to explain the nature of something. For instance, from a biological point of view, if I say that the heart is a pump, that it behaves in a certain way, what am I doing? I am at least partially able to explain what the heart is, what it does, what its function is, and I can easily introduce a scientific description of the human body, it becomes part of a scientific description of the human body. When I said I asked myself "What is the nature of consciousness?", describing it is very complex because of its subjectivity. And then, to explain how this entity or how these phenomena that I have access to in an intrinsically subjective way, how they can be connected to objective phenomena, which are in my brain? This may create a problem that has been called the explanatory gap: this void that existed that has prevented the solution of what has been defined as the hard problem, whereas other aspects of the functioning of our minds that instead concern consciousness were connected in these so-called easy problems. Allow me to say that the problems of the consciousness are all but easy, so they are all difficult, so I think that the distinction was not a very correct one. Anyways, this is how they have been defined as. A person who devoted a lot of attention to these subjects, Jaegwon Kim, summarized in three lines the nature of the problem. It is a bit technical in the terminology but I would like to just repeat it. The phenomena to be reduced, such as the conscious phenomenon or consciousness, phenomena are conceptualized in our relational or extrinsic ways in terms of their causal relations with other phenomena and not intrinsically in terms of their internal qualitative character or structural composition. It is as though we were saying the scientific explanation in the terms of the causes for something does not succeed in entering the intrinsic dimension of the qualitative phenomena. I can imagine, for instance, an entity, a being that has exactly the same type of brain organization as I do, with all of the causal relationships amongst the various neurons, or rather sets of synapses and neurons together, and you can simulate this aspect. But from this I cannot derive, I cannot draw the normological need, on the basis of the rules of nature, that this entity may have a phenomenal type of thought. I can imagine that a person has exactly the same brain organization as my brain, but when we look at a ripe tomato maybe he has an experience of seeing something that is green and I have the experience of seeing something that is red. This comparison is a bit shaky, sorry. Anyways, if there was a complete symmetry in the organization of the perception of the colors, we would never be able to really understand the difference, because every time that he says red he means what I mean by green, and vice versa. In the practical behavior, because we have been trained to use the traffic light, I stop when I see red, he stops when he sees the green light, so the experiences are different. But the light signals are the same and so there is no possibility of identifying or detecting the difference. [1]

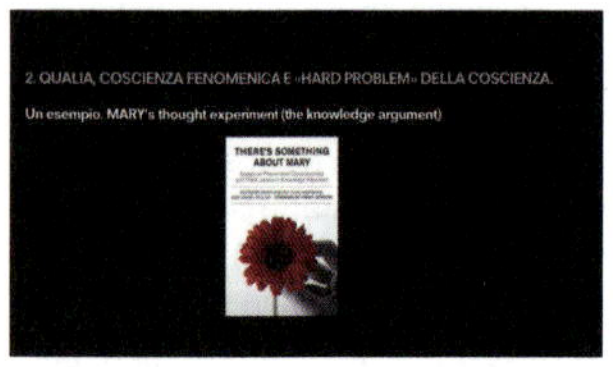

Fig. 1

How did the debate proceed? In general, with mind experiments, I would like to indicate one very briefly. This is Mary's thought experiment, the knowledge argument, as it is known. This is how the experiment works: let's imagine that we have a younger student or a younger graduate, a neuroscientist living in a room where everything is black and white. There are no visible colors, all objects are either white or black, and she studies books that have no color pictures, she receives information from monitors that are all black and white, but with great determination, this person studies neuroscience and learns all that she needs to know from a neurobiological standpoint on the perception of colors. At this point, Mary is omniscient as far as the neurobiological theories are concerned regarding the knowledge of colors. Mary finally is freed from her room, she can go out, and when she leaves the room she sees for the first time a flower. This is the only image that I wanted to include in my presentation. This is a flower, you see red petals. So the question that Frank Jackson is asking us is: "Does Mary learn something new when she sees a red flower or red tomato for the first time?" Does she say that she finally understood what it means to be seeing something in red? According to Jackson, the answer is yes. This positive answer means that because Mary already knew about all of the physical facts, the subjective experience of the red refers to facts that are not physical facts. This could be an example of a possible theory of mental experiments. This book, *There is Something About Mary*, and other books contain debates that are very complex, so it is not as simple as I have now told you. But this was just to give you an idea of how these theories have been presented. This is the way in which the discussion has proceeded on the existence of qualia. How did it end up? No shared solutions have been found. There have been two opposing parties that emerged: one that believes that the phenomenal experience does not really affect materialism or physicalism, because it can be explained either scientifically with representational theories of consciousness, or because the qualia, in reality, do not exist, they are a philosophical illusion and this is a theory that Daniel Dennett maintains; or others who believe that the qualia do exist but they are not just reducible to the physical, and the latter adhere or subscribe to the dualism of properties. There are entities such as our brain that have the capability of producing both physical and phenomena as well as phenomena which are not reducible to the physical. Or you have emergentism, and others. Now, the result that we obtain here is what I already mentioned before and what Kim called physicalism, or something that is quite similar to physicalism, that is: we accept the idea that we are not able to explain qualitative states but we explain all the rest. We have explained a very important part of reality. Even if we cannot imagine that the qualia cannot be reduced to physical states, we must be very cautious in thinking that qualia have causal properties and so that they interact with the physical part, because if they were to interact with the physical part we would have a violation of these causal principles of physics—the idea that the physical world is closed. Think about the conservation laws of energy, principles which are very strong and it is difficult to give them up. Some people say that we can't propose a philosophical theory that destroys these principles altogether, we need to admit that qualia exist but they don't have causal efficacy in our world. In reality, they probably would be like the last residue of Cartesian consciousness that we cannot explain in a materialistic manner, but that it is very limited if you think about the entire world. This is not the entire conclusion but it is a conclusion that could come near to what we are saying, unless you think that the heart problem has a more draconian solution and that it can lead to the idea of a purely mental causality aside from the physical one. We are at the last point and I will be quite short on this. I would like to mention an element that is quite important because it may be useful in the interaction with professor Tononi, since some of the themes of this last part are then found again in some of the most speculative parts. Said by a philosopher, I am complementing his work. This is a quote from Uriah Kriegel: the philosophers of the mind have—this is a book dating back to 2013—tried to find new and creative approaches to the mind-body problem that would escape the forced choice between materialism and dualism. The hope is to find a position that gives rise neither to an explanatory gap between mind and matter, as materialism does, nor to the causal inertia of the mind with respect to matter, as dualism does. The doubt we referred to before. How was this done? By introducing two factors, two fields of research. The first is the development of cognitive phenomenology and the program of phenomenal intentionality. These are complex terms but what I mean to say with this is that phenomenal intentionality means going back to the roots of classic phenomenology by Brentano, Husserl, and so on, where consciousness and intentionality used to go together. I would like to remind you that intentionality is the capability of representing the world, a representational capability linked to conscious thinking. Our conscious thoughts have conditions that can be satisfied, and if satisfied they represent something. If I have a mental image, a representation of the Milan Cathedral with specific features similar to those I am representing in my mind, my mental representation at that point has real content, so I can explain my capability of referring to the Milan Cathedral because the phenomenological content and the other one coexist. Cognitive phenomenology is a thesis that widens the field of consciousness, it does not reduce it to sensory states or bodily experiential states, such as pain or seeing a hue of red, but it also applies to other fields, such as beliefs, desires, etc. The second is the development of ontologies of models of realities in which phenomenal consciousness plays a central role, such as Russell's monism and panpsychism. The intrinsic nature of basic physical entities has to do with phenomenal consciousness. The idea is that in their intrinsic nature, physical entities, such as electrons or quarks, are micro-subjects of phenomenal consciousness, and being an electron or a quark, for example, makes the entity feel something. This seemed to me a not plausible theory but it has been defended with growing rigor, and it led to a lot of discussions ranging from the philosophy of physics to the philosophy of the mind, which should be taken into serious account, I believe. For classical cognitive science, the mind began with subpersonal processes from which consciousness emerged. For the new trends, the starting point to study the mind is the phenomenological approach with a more internist and cartesian vision, which approaches the phenomenological tradition of Brentano and Husserl. We are therefore in the presence of a program that gives centrality to the experience lived in the first person, but which does not exclude and in fact can be supportive in the following construction of models of how the brain works, of scientific models or the way in which the mind works.

Consciousness and Our Place in Nature

GIULIO TONONI

You heard a lot about consciousness. I would like to give you a very clear explanation: consciousness is what goes away when one falls into dreams in their sleep. [1] Nothing

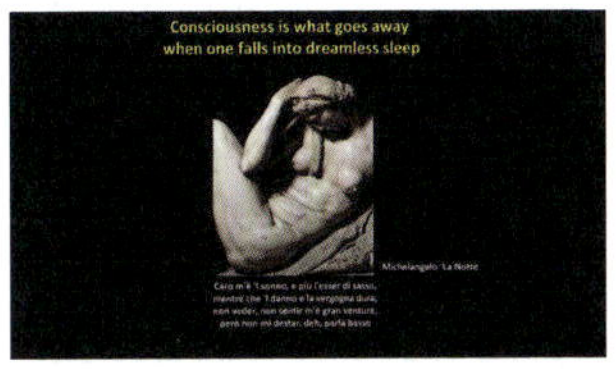

Fig. 1

better than a slide from Michelangelo, *La Notte*, and his own sonnet "Caro m'è l'sonno, e più l'esser di sasso, mentre che 'l danno e la vergogna dura; non veder, non sentir m'è gran ventura; però non mi desta, deh, parla basso." We all know that when we fall into a dreamless heap everything vanishes. Nothing is left, not myself, not my room, not my body, not my country, not anything whatsoever, no sounds, no sights, no shapes, no colors, no thoughts, no feelings. Everything is gone. That is what the absence of consciousness is: it is literally the absence of existence. Then, when you wake up in the morning, everything suddenly reappears, sounds and sights and colors and thoughts. That is what consciousness is: it is, I like to say, everything we are and everything we have. Take consciousness away, they can promise you the entire world but if you are not there with your consciousness there is literally nothing at all. [2] That is a very good

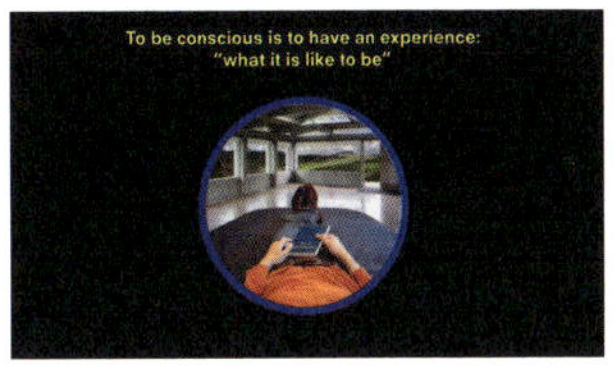

Fig. 2

first point to say that consciousness is everything. The second is, going back to Thomas Nagel mentioned by

Michele, another good way of putting what conscience is. To be conscious is to have an experience of what it is like to be. I am going to use here this image. Imagine waking up after dozing off with a book in your hand, with your body, your bed, your room, and the world outside. You have an experience, there is something that is like to be you at that moment, and every experience is what consciousness is, including an experience of a purely dark vast screen, or the sky at night. It is an experience too where there are no objects in it, you may have no thoughts whatsoever, that is often what happens when you wake up in the morning sometimes: there is nothing for the moment, except for just darkness and silence. But that is an experience, too. There is something it is like to be there, it is a sense of presence, or consciousness. I want to say a few things about what is not consciousness, because often this is a source of misunderstandings. [3] Consciousness is not

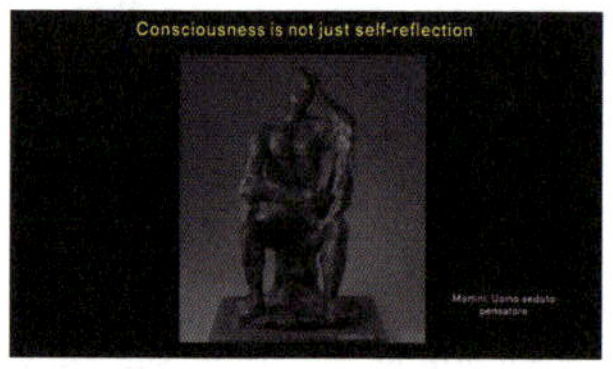

Fig. 3

just self-reflection as indicated here. Yes, many philosophers actually had sort of fallen into the trap of considering that. It is the typical philosophical thing to do, you are sitting on your chair and then thinking about consciousness, and it seems like the only experience there is when you are thinking about it. But that is not true. You are conscious whether you think about it or not, whether you tell yourself you are conscious or not. Just take this image for instance [4]: you are faced with an immense onslaught of experience and you just have it before you tell yourself what it is. The same is true for an action movie when you are skiing very fast down the hill. There is just experience before you can even reflect on that. Self-reflection is a part of an

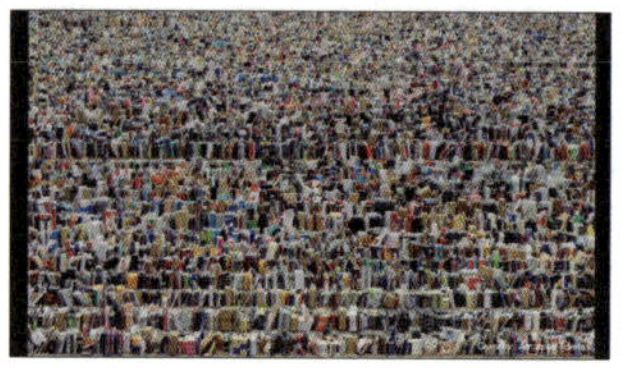

Fig. 4

experience but it doesn't exhaust it. [5] Another thing, which refers to

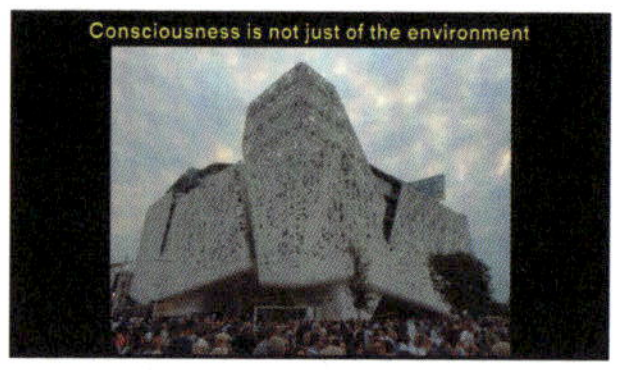

Fig. 5

what Michele mentioned as intentionality, that representation is, is that very often people focus on the experience of the outside world, the world out there, whether it is the EXPO and the crowd around it, or anything else. Yes, that is experience, but that is not the only kind of experience we have. [6] I am using

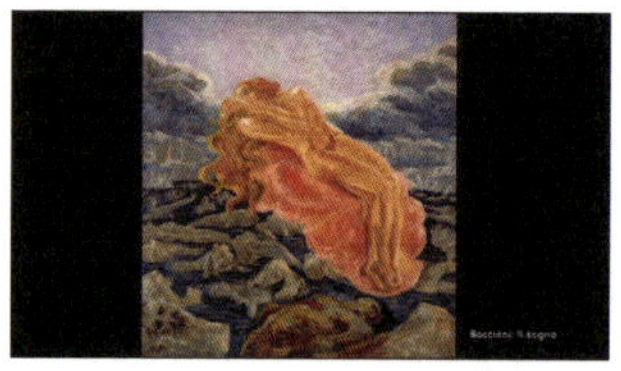

Fig. 6

here this painting by Boccioni to indicate that we all know it, every night when we fall asleep, we lose consciousness at times, but then we regain it too, when we are still disconnected on the input and output side from the world, when we dream. A dream can be as vivid as anything during wake, and it is all made up inside, it has nothing to do with the world outside, here and now. [7] Finally, and that is very important from a medical point of view, consciousness is not just responsiveness. We should have known that in the sense that when we are dreaming, indeed, we are typically unresponsive. Some patients, like the one shown here, may be com-

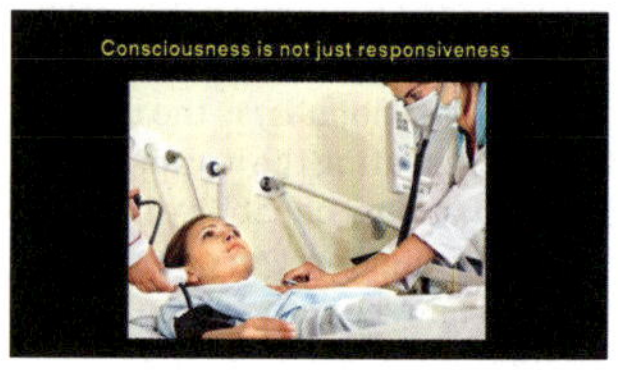

Fig. 7

pletely unresponsive. you literally cannot wake her up no matter what you do, you will get no response. But that doesn't mean necessarily that there is nobody there. [8] Very often

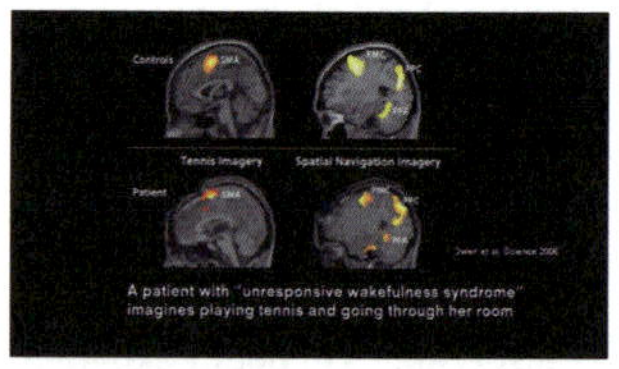

Fig. 8

that is true, but sometimes and this is a classic landmark study that showed it, sometimes you can actually convince yourself that there is somebody there, because, for instance, you can do an fMRI experiment in which you image the brain of the patient which you can see down here. When you ask the patient who's not saying anything, doing anything, you have no idea that the person is really there.
But for 40 seconds, say, think about playing tennis—she was an avid tennis player—or think about navigating through her house. The lights that you see shining in her brain are exactly in the same places where you see them shining in you and me if we were asked the same task. Clearly, there is a strong indication—now we have more and more data about this—that there is somebody there because the brain gets activated in a very similar way, even though there is no outward response. So consciousness is not just responsiveness. Now back to the hard problem mentioned by Michele. [9] I think the best illustration of the problem goes back to one of the greatest philosophers,

Leibniz, 300 and more years ago, who essentially used the metaphor

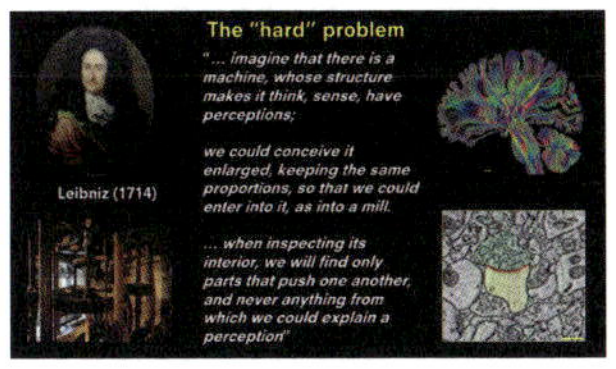

Fig. 9

of the mill and said to take the most complicated object they knew about then, let's say a mill, with lots of pulleys and gears and so on, and imagine it very big so you can walk into it and just see all these gears and these ropes pulling each other. Of course, he was thinking of the brain that we see now on the right as what you have seen in the past days from the human connectome project, for instance, or the human Blue Brain Project. This is an electron microscopy slide taken in our lab, in which you can see one of the connections between neurons. This is a presynaptic terminal with all the vesicles containing glutamate, and this is the postsynaptic spine, where the electrochemical transmission occurs. By the way, when philosophers come into the lab, I often show them on the screen of some of the students in the lab this kind of pictures to show that when you look into the brain it is an incredibly dense entity that is completely full of this synapses, mitochondria and glial cells. It is a completely dense thing, it is an incredible piece of biological machinery. But Leibniz thought, a long time ago: make it very big, walk inside it, touch all these vesicles. You are going to find just mechanical things, vesicles reducing transmitters and so on. You would say that it is electrochemical, not just mechanical but still, it is a machine and there is no way that we can understand how a machine that goes through the motions, that does what it does, would somehow generate consciousness or explain a perception, as he put it. There is no better way of putting the problem. Or, there is another way to put it, which is here with reference to the wedding at Cana, it seems as if we try to explain consciousness in physical terms, we are doing something like transforming the gray water of the brain into the red wine of experience. [10]

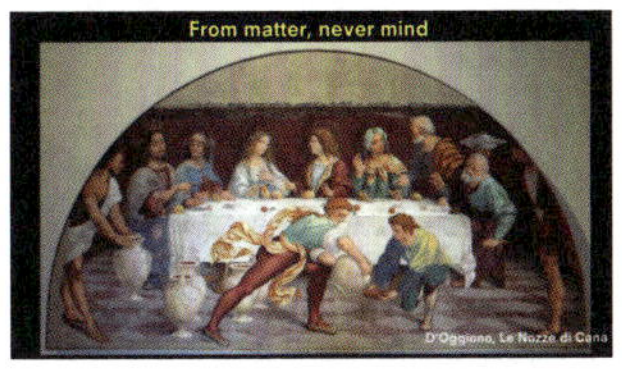

Fig. 10

A miracle, literally, and miracles are not what science typically deals with. [11] What I think is that we need

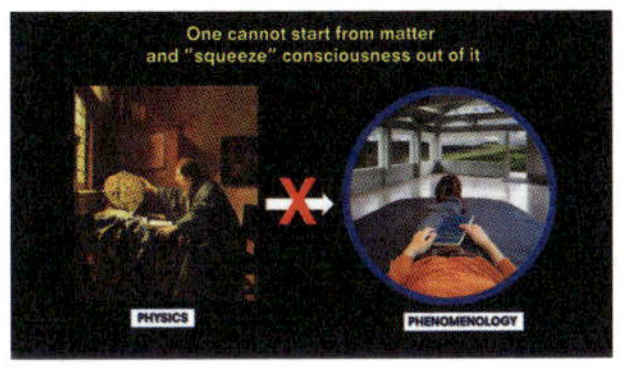

Fig. 11

to understand consciousness by starting from consciousness itself and not doing what Galileo taught us to do, which is to always start from the physical world and remove the subject from the equation. For consciousness, you cannot remove the subject from the equation. The subject of investigation is the subject himself or herself. We shouldn't go from physics and try to squeeze experience, colors, shapes, sound, space, time from the brain, you'll never do that and then get to phenomenology. [12] You have to do

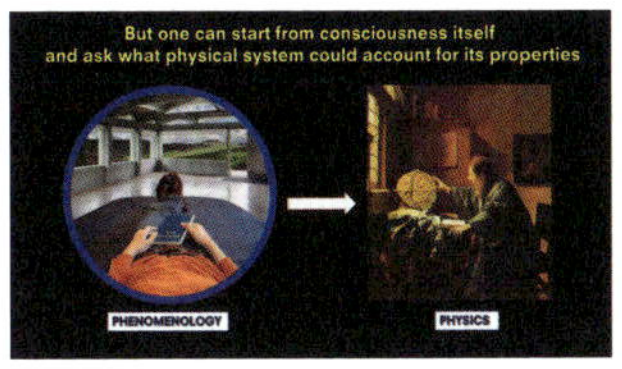

Fig. 12

exactly the opposite: you have to start from the phenomenon, literally the phenomenon you want to understand, which is the experience itself, and see if there is a way to account for it, to explain it in terms of physics. Consciousness is first, experience is first, also for scientists and physics is something we dreamt up from within the experience. We have to emphasize this because you cannot make progress on consciousness if you don't realize that that must be your starting point. [13] This is indeed what integrated

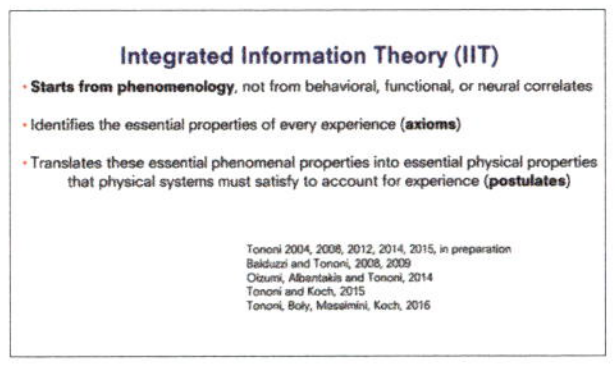

Fig. 13

information theory (IIT) tries to do. It starts from experience itself, phenomenology, not from the behaviors that you perform on your conscious, not from their remarkable functions, including language that you can do when you are conscious, and certainly not from what is happening in the brain. That is not where the solution is going to be found. You are just going to get more behaviors, more functions, and more neural correlates. Instead, we need to look at the phenomenon itself, we need to look at experience, and the point of start for IIT is to identify what are the essential properties of every experience, of course, axioms—we'll see them very briefly in a moment—then translate these properties into physical properties, properties that the physical system must satisfy to actually account for experience, which we call postulates, and then aim at accounting for the way experience feels. When there is experience and how it feels—that is space, time, shapes, colors, and so on and so forth—in terms of physical properties, the physical gives us an explanation of the phenomena, but not vice versa. [14] Let me begin this brief very cursory overview of the theory itself going back to Descartes, whom Michele brought up, and, saying indeed that the starting point is a very cartesian one, we should start from experience itself, what it is like to be, which is really phenome-

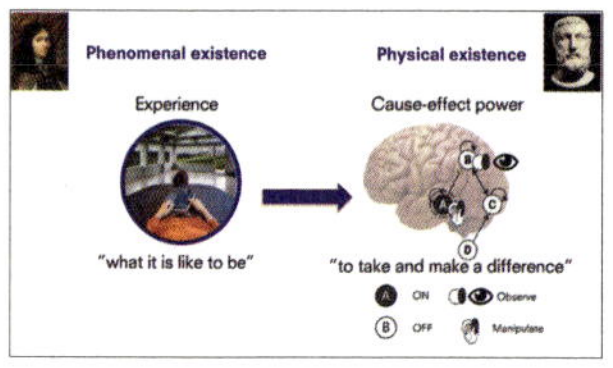

Fig. 14

nal existence. That is what exists immediately and inductively. Then, translate that in terms of physical existence. Physical is not very often well defined, but I want to quote another philosopher here. We actually don't know who he was, this is known as the aleatic stranger who walks in in a platonic dialogue, the sophist, a very difficult dialogue, and at some point comes up and presents the criteria for the physical. To him, the physical is basically cause-effect power. Whatever takes or makes a difference, even a minimal difference, is what exists in physical terms. I do take this very seriously. It is a perfectly good operational definition of what we accept as physical, including the Higgs boson. Unless you can actually show that it can take and make a difference nobody would agree that it exists. That is the criterion for physical existence. By the way, you find this not just in the aleatic stranger. Some people thought it may have been Parmenides, but also in some oriental traditions, for instance in Buddhism—Dharmakirti had a very similar definition of existence in terms of cause-effect power or, more generally, cause of power. [15] What

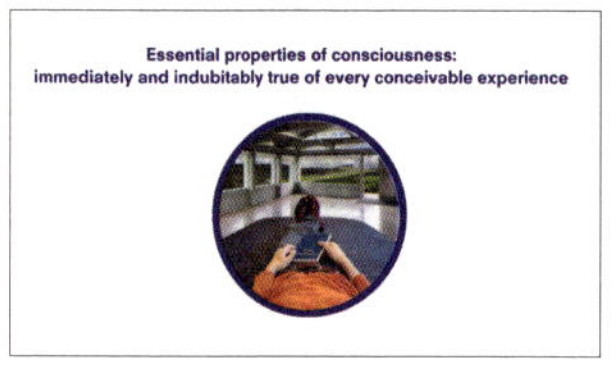

Fig. 15

IIT does, it says let's start then from what for sure exists, which is phenomenal existence, experience, and ask if there are some properties that are essential, that is, that are immediately and indubitably true of every conceivable experience. Some properties are always true. [16]

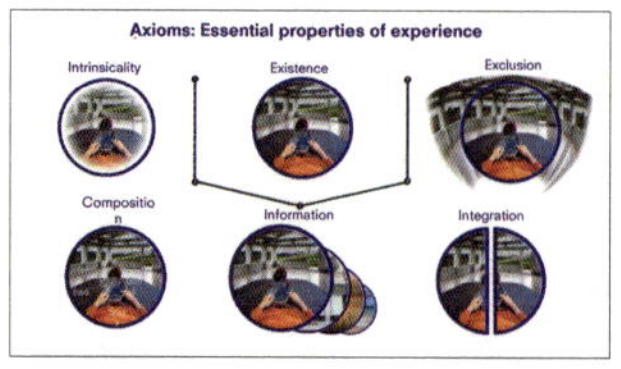

Fig. 16

It says there are five such properties. If experience exists—no doubt—then every experience, says IIT, is intrinsic or subjective, it exists for me, from the inside, not from somebody else, from the outside. It is structured, that is composition so I can always distinguish left and right, a bed and a book and a body, and the color of the book, and what is related to the book itself—it is a blue book and so on. Every experience is extraordinarily structured, if you wish, it has parts related to each other phenomenally. Every experience is specific. We are having this particular experience, say, when I wake up from a nap, and that experience is the way it is. It cannot be generic, nobody ever could have a generic experience. It is this experience, this body, this bed, this book, this color, and so on and so forth. And that, of course, is what makes that experience different from many others. But the key is that experience is specific. It is also unified or integrated. Every experience is one thing that you are having and if it weren't, there would be two people one seeing the left side and one seeing the right. But it is never true, you cannot have an experience that is not one. Finally, every experience is definite, in the Latin sense of having borders. It contains what it contains, not less and not more. The experience I am having is not half of what I am having, or a quarter or a third, and it is not more like seeing behind my back or anything like that. These properties are so fundamental that we typically don't notice them. Not even Descartes had really pointed out these things, he had understood the integration part, but he didn't really say these are true of every conceivable experience. Let's say they are and, without going to that, there are no other properties that are true of every conceivable experience. [17]

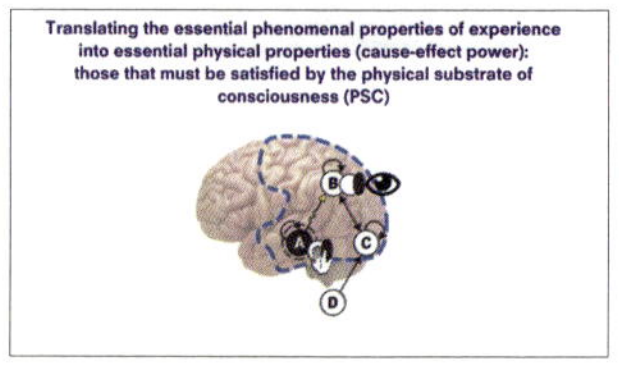

Fig. 17

Then, the next step that IIT does, is that it asks if we can account for what exists, phenomenal experience, in physical terms, given that physical explanations have been very powerful in accounting in general at the regulator's experience, where physical, as I mentioned, is cause-effect power, that is, taking and making a difference. Let's take the brain, which we know has something very intimate to do with consciousness, at least some parts of it, and physically we can observe neurons, we can see where they are, on or off, we can perturb them, turn them on, and turn them off, see what difference they make to other neurons, and so on and so forth. That is literally what neuroscience does: it performs physical investigations of the physical subset of consciousness. [18] Here is the

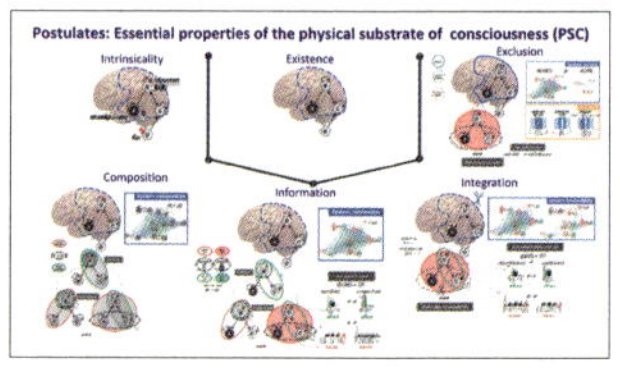

Fig. 18

translation, that is the difficult part, if you wish, in IIT, of the five essential properties of consciousness in terms of cause-effect power, that is, from phenomenal properties to physical properties. A system, for instance, to be able to support consciousness it follows, must have cause-effect power upon itself, intrinsically, in a structural way, so its parts might have cause-effect power within the whole, it must have cause-effect power that is specific the particular way it is, it must have cause-effect power as a single entity, it cannot be subdivided into causal independent components, and, finally it must have borders, both in space and time and, in grain, it must have borders where the substrate of consciousness ends, not less and not more. It is not my entire body, it is not my body and the floor on which I stand, it is not my body and the floor and the world, and it is also not just an individual neuron. It is something having to do with a piece of the brain—we are still trying to figure out which one—but it has borders and there must be an explanation for that. [19] That is

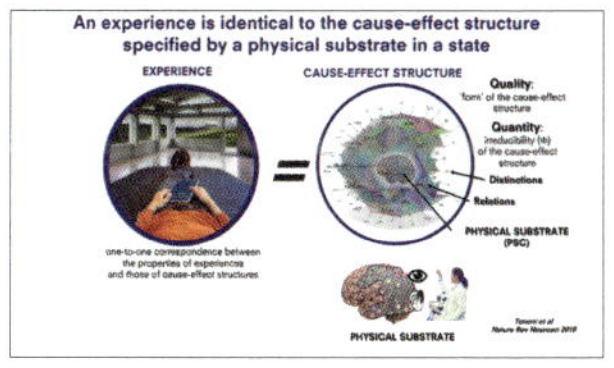

Fig. 19

really the starting point of IIT and the ending point of IIT, from a theoretical point of view, is an identity, it is an explanatory identity. We say that one experience, this particular one I am having the moment I wake up, is identical to a cause-effect structure. Keep in mind it is a structure. What the structure is, is the unfolding—that is the word we use—of the cause-effect power of a substrate. Take this piece of the brain. I have reasons to show this back of the cerebral cortex here, with some neurons on some intermediate, things that as scientists we can observe and manipulate, but that is not really what exists in physical terms. That is what everybody thinks exists, a bunch of neurons connected in a certain way. but what really exists physically, properly done, is the entire set of cause-effect powers of this substrate. That is indicated here with a very simple diamond-like structure, a gigantic cause-effect structure that IIT tells you in principle how to unfold. Then, the amount of existence in physical terms of this structure is given by the quantity phi, which was mentioned by Viviana. It tells you how much it exists, how reducible it is, how much of an entity it is, and its quality is essentially the shape, the form of this cause-effect structure. All its causal distinctions, as they are called, and the causal relations, which are the way these distinctions overlap. Obviously, I can't really do justice to this, it is rather long and there is a lot of mathematics involved, but the concept actually are simple. Consciousness exists, that is our starting point for everything, is that was gone we would be nothing; consciousness has five essential properties, the ones I mentioned; we can translate existence in physical terms from within consciousness because conscience is first; and then we have cause-effect power and we can translate those essential properties in physical terms; in the end, we get that what is identical to consciousness, says the theory, is a cause-effect structure. Is that true? This is something we can test. [20] Indeed,

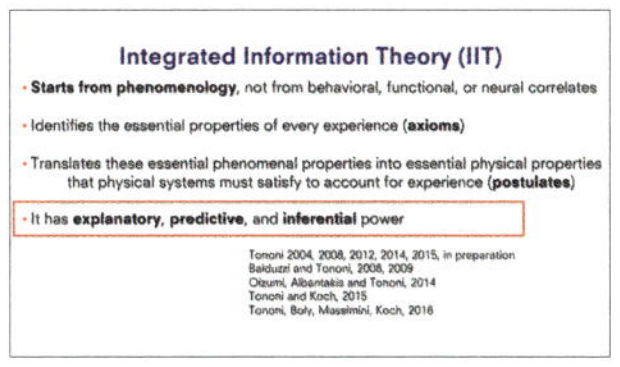

Fig. 20

IIT is a scientific theory and therefore it has explanatory, predictive, and inferential power. I'll quickly go through this starting with two things I always thought were central. If we don't understand them we don't understand anything about consciousness in the brain.One is: we know that consciousness has something to do with the cerebral cortex, up here, and nothing to do with the cerebellum down here. The cerebellum has 70 billion neurons, in fact, most of the neurons in the brain are there. They are smaller but they are there. The cortex altogether has 16 billion neurons, so a fifth or so. [21] Yet, if you take out the cerebel-

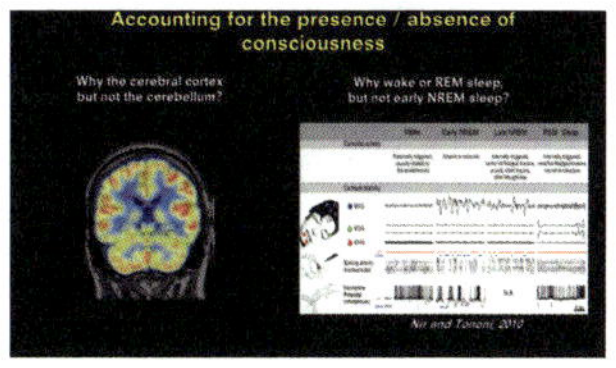

Fig. 21

lum, consciousness doesn't suffer at all. I mean, you have problems moving around a bit, you are, uncoordinated but your consciousness is what you hear, see, think, feel stays the same. But if you take out the cerebral cortex completely, you are gone, you don't exist anymore. What is the difference? The cerebellum has more neurons, it has as many connections. It is connected to the cerebral cortex itself, connected to the body, it has maps of the world, it can control movement and so on. What the hell is wrong with the cerebellum? We need to understand that. The other example is the one I mentioned at the beginning, something we know every night when we fall into a dreamless sleep, which is typically early in the night. Initially, people thought the brain would shut off and neurons would not be active, but here is a work we did together in California in an epileptic patient where you can actually record individual neurons, which are all these little vertical bars—they are a neuron firing. In wake they fire. There are six neurons here, in a particular area of the cortex of this subject in deep sleep. When you are not there they also fire roughly the same amount. There are these brief periods of silence for a fraction of a second, but it is not that the brain is not active, it is not that the cortex is not active. Yet, you are gone, you don't exist anymore. [22] We need to be able to explain that and so, very briefly, the theory does that, at least in principle, it shows that if you have the anatomy of certain parts of the cerebral cortex, like especially in the back

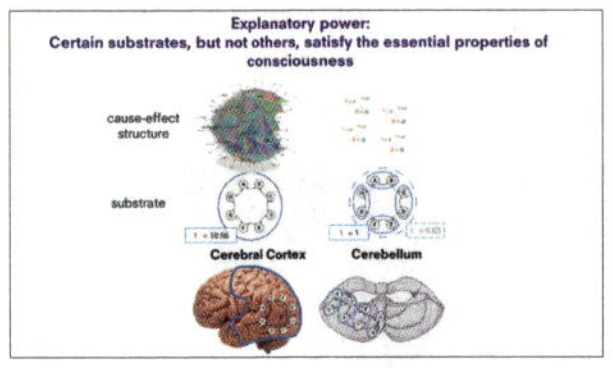

Fig. 22

here, you indeed are a good substrate that when you unfold its powers specify the cause-effect structure of high phi, that can be highly conscious. But if you take the cerebellum, even though it has many more neurons, once you unfold its substrate, instead of getting one big cause-effect structure, it fragments into a lot of very small things indicated here, mini cause-effect structures. It is very modular and it is all feed forward, on top of that is also largely inhibitory. But whatever it is, it has really the wrong anatomy to support consciousness, if IIT is correct. The anatomy is not enough here. [23] Now we have again the

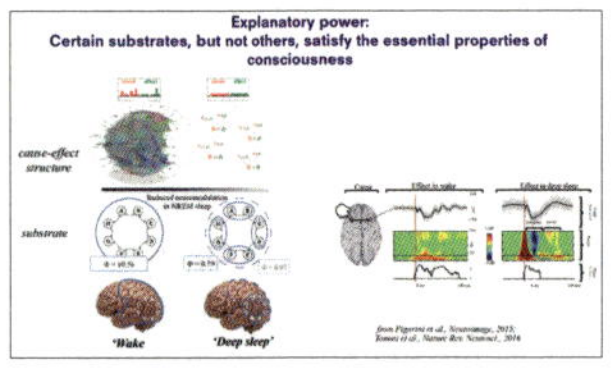

Fig. 23

back of the cerebral cortex, which has sort of the right anatomy, if you unfold it you get a big cause-effect structure. We take this little model and put it into a deep sleep. When we are unconscious, by changing the neuromodulators, particularly noradrenaline and acetylcholine, what happens is the connections are still there but their efficacy is greatly reduced through various mechanisms which we have investigated for a long time. If that is the case, suddenly the same anatomy now does not support this big cause-effect structure but disintegrates into these little things, which are worth nothing, basically. This is work done in the laboratory of Marcello Massimini in Milan, who has actually shown in an epileptic patient that when you go from wakefulness into early non-rem sleep, when consciousness is often lost, the causal interactions among the neurons in the cortex break down just as predicted by IIT. You can see this blue here: that basically means that the neurons are bistable, they fall into an off period and then they come back and they fire again, but with no causal connection to what was happening before, as indicated also here. So, the cause of interactions breaks down and consciousness is gone. As for predictive power, I'll give you just a few examples. A long time ago, also with Marcello Massimini, we tried to test a very basic prediction of the theory by using transcranial magnetic stimulation, which is this butterfly thing here in which you can use a strong magnetic field very briefly to induce a current inside the cerebral cortex. [24] Then we record

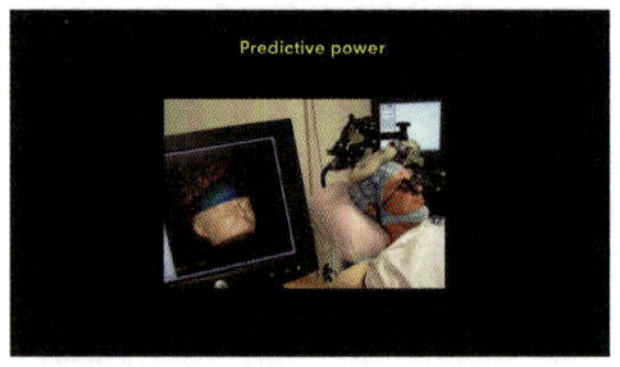

Fig. 24

how the brain responds using these EEG electrodes, several of them. Basically, what you do is you knock on the brain and you see where the brain responds and how it responds. What turns out to be the case, and now I have been confirmed many times, is that when you are awake and conscious you knock on the brain here you get a very broad integrated response that is complex. [25] It changes over a fraction of a second, which corresponds very crudely to high information integration. It would be high phi, you could measure it properly. Early non-rem sleep, when consciousness is lost, instead you knock the same way on the same spot on the brain but the response you get is just local, it remains local, it doesn't go anywhere and it lasts very little. Low information integration. Finally, and

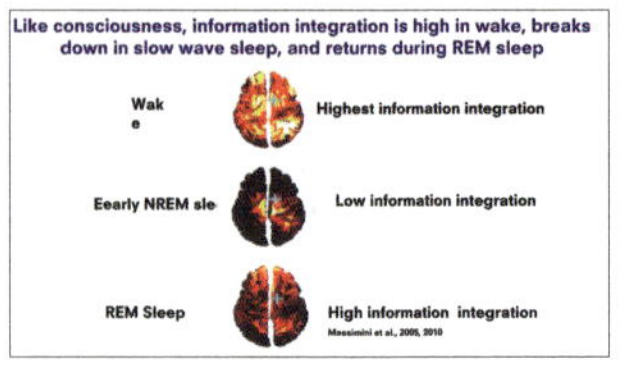

Fig. 25

that was more difficult to do, but in REM sleep, where most of the time you are conscious, you knock here and you get a response that is very similar to one you see in the wake, indicating that this complex integrated response of the brain to a perturbation is accurately consistent with high information integration, with consciousness, which indeed is the case because you are dreaming. [26] Then we went on and

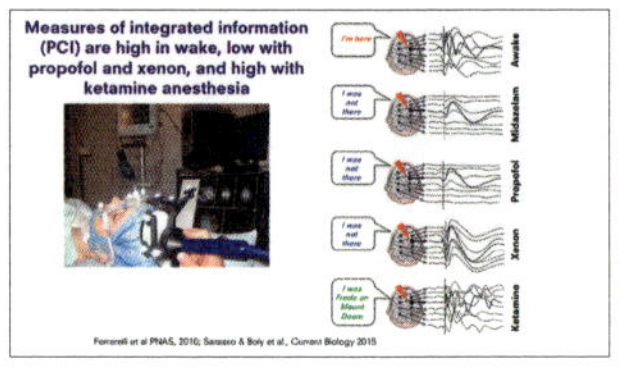

Fig. 26

did several experiments in which we went through, for instance, anesthetics of different kinds. This is wake, you can see this complex response now in time over the EEG, indicating high information integration. You are conscious. This is benzodiazepine in high doses, you go unconscious the response is local and stereotypical. Propofol, a typical anesthetic: the response is local and stereotypical. Xenon, on a very different anesthetic, you also become completely unconscious, the response is stereotypical it is not complex. Then here is the key one, ketamine, which is an anesthetic that puts you completely unresponsive, you can do surgery with no problems, but most people dream very vividly. This is one of our subjects saying, when she came out of it, that she was Frodo on mount Doom, she went on telling us this extraordinarily vivid dream. The response, noticing that even the subject was completely unresponsive, is very similar to the one you see during wakefulness or during REM sleep, indicating that this is a good marker for consciousness as predicted by the theory. [27]

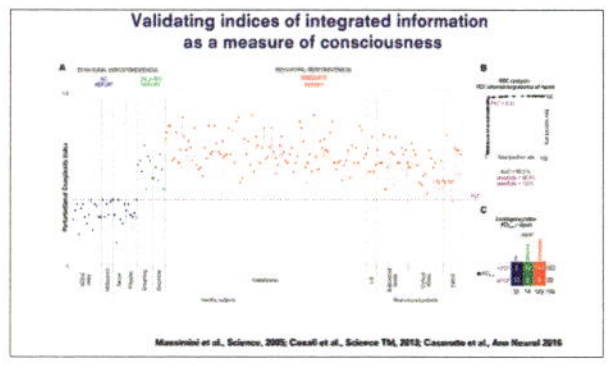

Fig. 27

In fact, the work of Massimini and his collaborators, and some of our work too, has confirmed that these are very many normal subjects who are awake, and these red points here indicate the value of an index called Perturbation Complexity Index, which tells you how complex the response is. It is sort of correlated to the ability to integrate information of the brain, it is always high when you are conscious. Importantly, it is also high when you are conscious and unresponsive, as I showed you in dreaming and during ketamine anesthesia. It is also high in patients with serious neurological lesions like this, the Locked-in syndrome, a totally unresponsive patient who may be able maybe to move up and down the eyelid to communicate but they are definitely there, no doubt about it. Or subcortical strokes, cortical strokes, or when they come out of a coma or an unresponsive wave from the syndrome, they come back and they are there, they are obviously conscious. This index is always high. Here on the left, in blue, you see instead many conditions of unconsciousness, the classic ones: deep non-REM sleep and this varies on aesthetics the index is always very low. So we can draw a line here and above this, we have an index of when you are conscious and below when you are not, with almost perfect sensitivity and specificity that we can use to gauge consciousness independent of the subject ability to respond. [28] This is very important in the clinic—and in fact, there are more than 300 patients who have been tested this way, and

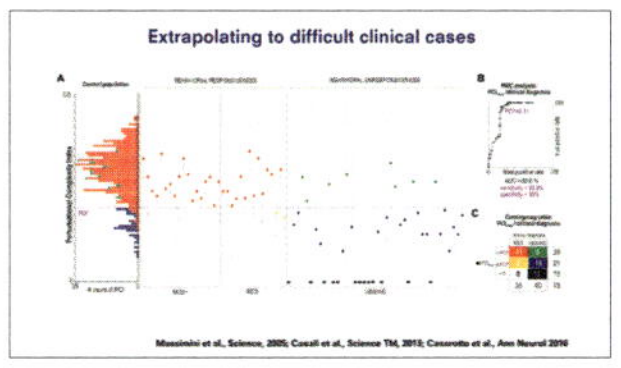

Fig. 28

we are trying to build a tool to do this in a way that can be made available to every hospital. For instance, in a minimally conscious patient, a good neurologist can say they are probably there maybe because they track a little bit with their eyes, but it is a very difficult assessment. The index is always high telling you they are there. But then the interesting thing happens when you go to patients who are unresponsive wakefulness syndrome, which used to be called vegetative state, here you can see that the index is either zero in black here or in blue here, at levels that we find in normal subjects during deep sleep or anesthesia, or in green here in levels that you find in perfectly conscious subjects. Suddenly, we can see here, which confirms what I showed you at the beginning in the study of the lady playing tennis that didn't respond at all but could somehow do it in her mind, or at least in her brain, that there are probably 20–25 % of patients in this unresponsive way from the syndrome, the so-called vegetative patients, who are presumably there, and others who instead are not. [29]

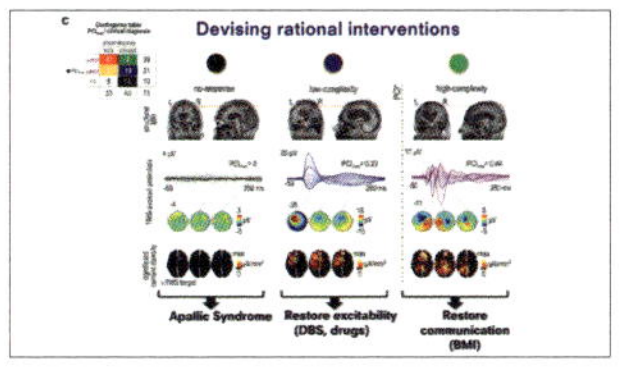

Fig. 29

This is very important, both clinically and also in terms of what you can do, because the black ones that had the zero value, truly I think we can tell the relatives that there is nobody there at all. The ones who were in the blue area, which have a response like the one you see in deep sleep or anesthesia, are candidates for being awakened again—whether you do it with deep brain stimulation, with focus ultrasound, or with drugs is a difficult thing to do but now new techniques are coming up—these are candidates for being awake. Finally, the ones in green here, that have a response which is as complex as you and me when we are awake or dreaming, are candidates for restoring communication to brain-machine interfaces. They are there, they are just not telling us, they are not just not communicating, so can we go into their brain and establish a communication? There are obviously important clinical implications. Now I want to briefly mention some of the fundamental work that has been done. [30]

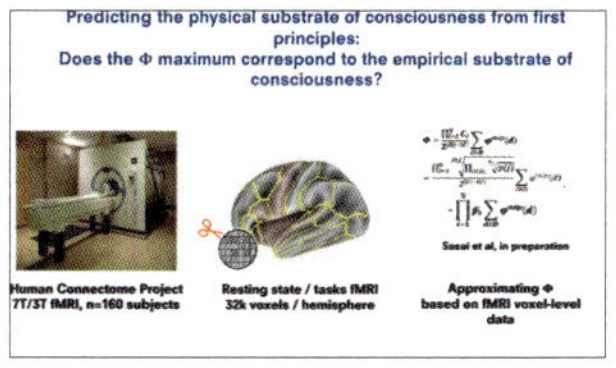

Fig. 30

We have done this for quite some time now, it is sort of ready. We use data collected by the Human Connectome Project, more than 160 subjects we have used using 7T and 3T fMRI, so powerful instruments—that is the best quality data there are—in subjects who were either doing nothing, just being scanned, or they were performing seven different tasks. We did an analysis based on what each voxel in the cortex, and more generally in the brain, were doing it. So, roughly 32,000 voxels per hemisphere, 64,000 over the brain, so it is a detailed analysis. A voxel is still a lot of neurons, probably more than a hundred thousand or so. It is not as fine as we would like to, not in space and not in time, but it is certainly better than anything else we have at this point. Then we have developed a way to roughly measure phi, integrating information to assess which parts of the cortex can support consciousness. [31] We did

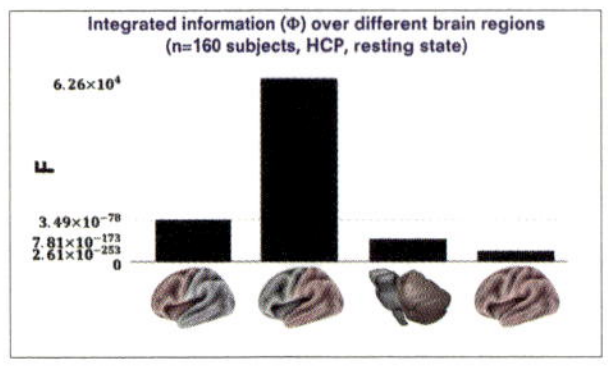

Fig. 31

that and we looked at different parts of the brain, for instance, the cerebellum here, which we know doesn't matter for consciousness has an exceedingly low value of phi, even with this all these caveats about how we measure it. If you take the front of the cerebral cortex, which is very important for some of the most sophisticated functions humans can perform, that is a very low value, too. But if you go to the back of the cerebral cortex the value jumps up. This is logarithmic scale, so this is immensely higher.
The back is very good for integrated information. Then, if you take the entire brain, the value is low again. So it is not just a matter of getting bigger. Integrated information requires a substate that has borders of maximum integration, and it looks like the back is good. [32]

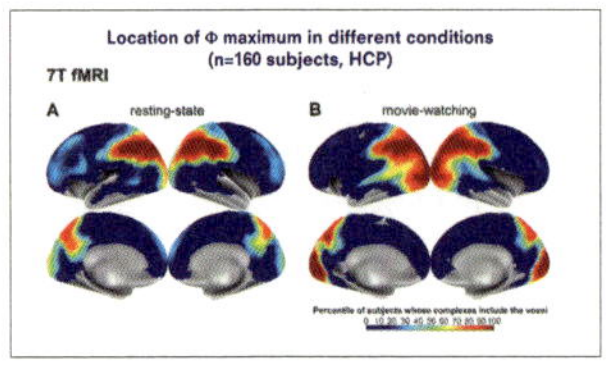

Fig. 32

In fact, we can go further and try to find this maximum. We need to make lots of approximations but you can see here in red, yellow, and green which parts of the brain are more precise in the cortex. This is the lateral side, this is the medial side, that is here in the middle. They are part of this maximum of phi, which is what IIT says should be the substitute for consciousness. This is during the resting state, meaning they are not sleeping, they are just thinking of nothing or whatever goes to their mind, but not performing any task. This is watching a movie, by and large, is the same areas plus more of what is here, auditory cortex and visual cortex, but it is primarily in the back, as I mentioned to you before. This is from first principles; we don't have here a hypothesis whether it is the front, the back, the cortex, the cerebellum. This is purely based on the properties of consciousness, the phenomenal properties of those five, we translate them into physical properties, cause-effect power. Where do we find the maximum effect power that IIT says should be the substitute experience? We find it in the back of the cerebral cortex primarily. It is literally from first principles. Is that correct that we identify the maximum primarily in the back? [33]

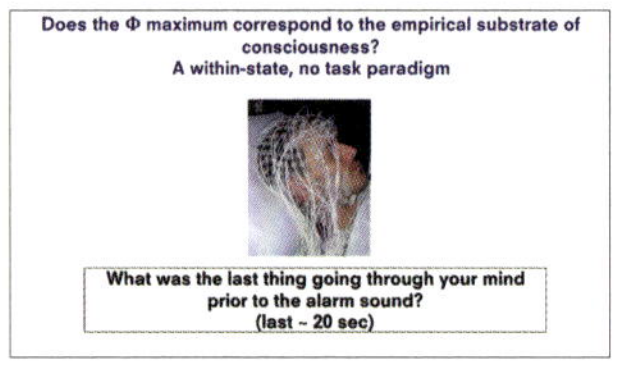

Fig. 33

There are many reasons to think that there is something right about it. One study that we did a few years ago, a long study, was actually to exploit one of the best ways to look for the neural correlates of consciousness, what is different in the brain when you are conscious and when you are not. When this is done between wake and sleep, for instance, many things change, or between wake and anesthesia, many things change beyond consciousness. But there is one case in which we can have a much cleaner comparison and that is sleep, especially non-REM sleep. When you sleep—we have done more than 300 subjects—a third of the time you are completely unconscious in non-REM sleep, and two-thirds of the time you are dreaming. It can be a vivid dream, a long dream, a short dream but there is some consciousness. We can compare when your brain is conscious, dreaming, and when your brain is not conscious, within the same brain when you are doing no task whatsoever there are no confounding factors and you are roughly in the same state—you are lying in a bed unresponsive and there are slow waves going on in your brain, as it is typically in non-REM sleep. What are the areas that show a difference? [34] It is primarily the

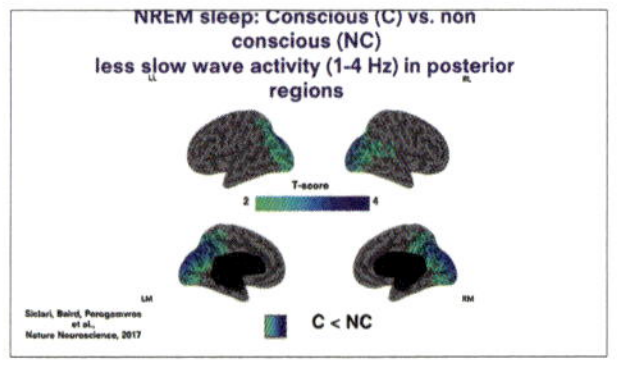

Fig. 34

back, lateral surface, medial surface. In blue and green you see the areas where there is a reduction of slow waves. Those slow waves are typical of non-REM sleep. When they go down and these areas get activated, you are conscious. This tells us the neural correlate seems to be primarily in the back. This is REM sleep, which looks very different from the point of view of the electroencephalogram and also the fMRI, but again the very same regions show up, primarily in the back of the brain. This is strong evidence that, indeed, these areas are key for being conscious rather than not. [35] If you look at

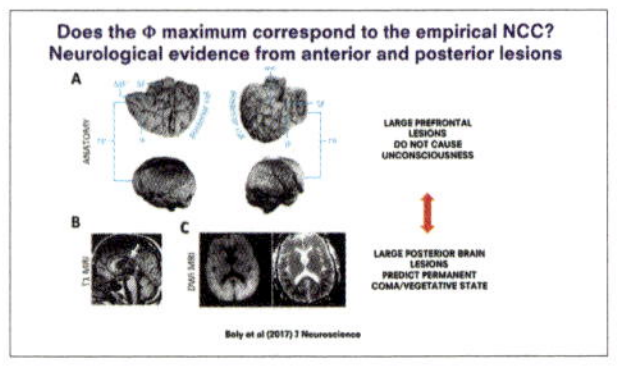

Fig. 35

the classic neurological evidence, in fact, you find remarkable things. This doesn't happen anymore, but this is a patient from 1930 or so. A Britain patient in which they literally took out all of the front of the cortex, the prefrontal cortex. This is the left side, this is the right side. At the time, to be safe and not kill any vessels and risk hemorrhage, they went in—you can read the report of the surgeon—with the fingers, scooping out first the left side of the front of the brain of this poor patient, and two weeks later, because it wasn't getting really much better, they did the right side, and the patient came out perfectly fine, in fact, he started talking when they took out all this piece of the prefrontal cortex. The prefrontal cortex is not useless, it is very important for many things we do as humans, and the patient indeed had what is now called frontal syndrome: it was a disease that he could not really change what he was doing very easily, many other aspects, like flexibility and so on. But no doubt he was conscious, he could hear, see, feel, think, have emotions. If, instead, you look over literature, a big lesion in the posterior part of the cortex—there are not too many cases of this sort, 200 or so are the ones properly reported but they basically never came out of an unresponsive syndrome, they were staying meditating. [36] The evi-

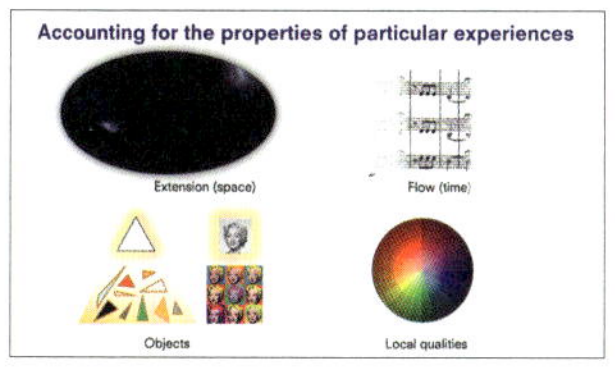

Fig. 36

dence suggests that—and there is much more, including recording studies and stimulation studies when you are stimulating the brain—that the back of the cortex seems to be really fundamental for us having experiences and it is where IIT says there is maximum cause-effect power. IIT is rather unique in trying to do the following: not only to account for why we are conscious, under what condition, what piece of the brain or any piece of matter is important for it and why, but also, you don't have a theory of consciousness if you cannot in principle account for the way it feels. Not just that it feels, but the way it feels. Why space feels extended? I started with a picture, for instance, of your room and your body. It is all extended, your body feels extended. You look at the sky, it feels extended; you look at the blank screen, it feels extended; you look at the wall, it feels extended. And feeling extended, although almost nobody really paid attention to it, is just as much the hard problem as feeling red or feeling pain. It just feels that particular way. Why is that? The same is true for why time feels like it is flowing. It is not magic, there must be an explanation, it must be physics and not magic. Why do objects feel the way they do? That there is a general boundary particular that when you see this triangle here it is this particular triangle, at this place, in black, with the brown border, but it is also a triangle and you experience all of that at the same time, all the concepts that came out in previous presentations. What matters is when we feel them consciously. The same is true for faces. Then, of course, why does red feel red, blue feel blue, pain feel pain, and so on and so forth? All of this must have a scientific explanation, but it is an objective explanation of the subject, not of what we do but of what we are. [37] I just gave you a very brief

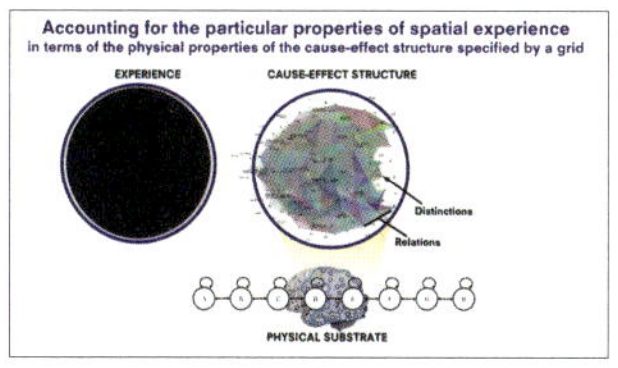

Fig. 37

indication of how IIT. This is our main research program, it is trying to account for the quality of experience. Here we are taking a very simple physical substrate, in this case it is just eight neurons connected by nearest neighbor connections like that, all off in this particular case as you find them in the back of the cortex. We unfold in full the cause-effect powers of this substrate. I told you what exists physically is not what we normally think, a bunch of neurons. That is what you can perturb and observe, but what really exists is the cause-effect power, which is a gigantic cause-effect structure, part of which is indicated here, of these stupid eight little binary neurons. This is part of the unfolded cause-effect power. [38] We show, as

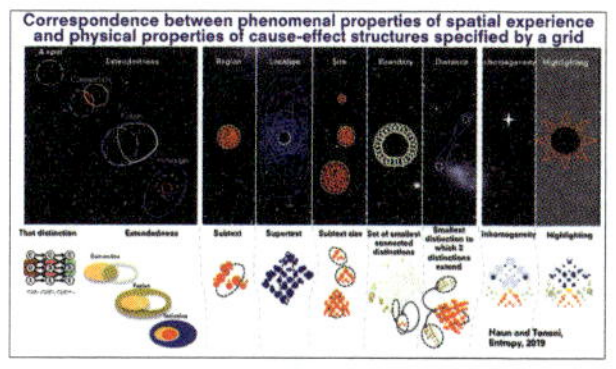

Fig. 38

predicted by IIT, that the properties of this cause-effect structure of this particular substrate can actually account for the extendedness of space. Why do you actually have an experience of space, where you can find spots of any size at any position, that they are overlapping with each other, that they can form larger spots or infuse into smaller spots, at every spot, included spots or is included by spots. These are fundamental properties of spatial experience there must be an identity to a cause-effect structure and there is, we show that. That is only true for the cause-effect structure that you unfold from, a grid but not from any kind of neural anatomy. Other properties of space are that there are regions of space, that have their location within the overall space, their size, their boundaries, and their distance, when you look out everything is the proper distance from everything else. You don't need to calculate anything, you don't compute a damn thing, you just see it. That is all, captured by the cause-effect structure specified by a grid. And guess what? [39] If you look at the brain we see that most of the back of the brain, where already we suspect is the subset of consciousness, is organized like grids. 38% here is all grid-like areas with lateral connections among neurons and then grid stacked on top of grids, I would like to say pyramids of grids. That is indeed a perfect

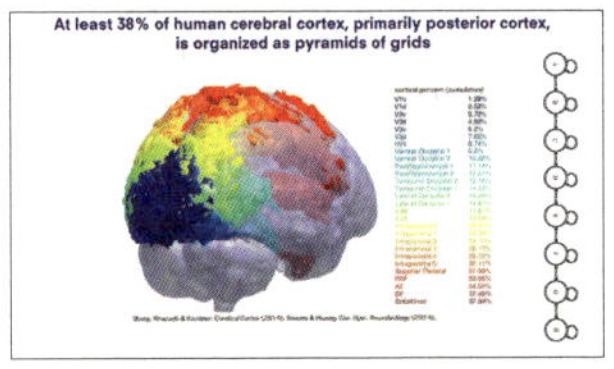

Fig. 39

subset for having high phi, for being conscious and conscious how? Primarily in a special manner. [40]

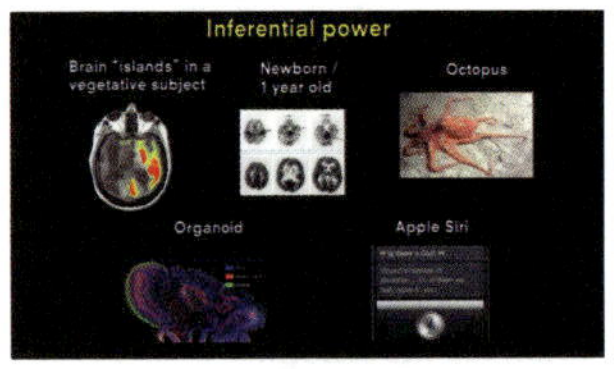

Fig. 40

I will finish now with the inferential power of IIT. If a theory is supported by data, makes a reasonable prediction, can explain a myriad of different things, then you begin to trust the theory and use it instead of your gut feelings to deal with difficult situations. Is there consciousness if a patient has a sea of destruction in the brain, but a few islands in the brain that are still strongly active almost normally? You can't really resolve that philosophically, you need a theory that tells you we are on an island like this, will there be consciousness, how much, and what. It could be pain, you want to know that, or it could not be anything you worry about. Newborns. This is a pet scan of a newborn. This is a one-year-old newborn. There is hardly any activity in the cortex only in the amygdala. We could use the theory to begin to ask: is there going to be something that is like to be a newborn baby or not? You are really still a little bit of a biological machine that cries but not with much consciousness. Then does it begin to really flourish, for instance in one year? Or not, I am not saying we have the answer but I am saying there is a principle approach. Take an octopus. I like it, it is my favorite animal not just because it is intelligent but indeed because it has a very alien brain. It has a little brain in each of the limbs, eight, plus the central brain which is twice as big, and these brains are connected. If you look at it you can't really say without a theory if it is conscious at all. Is there one octopus consciousness? Are there nine or eight and who knows? If we can unfold the substrate, we should be able to know whether they have consciousness, how many, how much, and of what kind. Here are two other examples that are becoming very important. Organoids, brains in a dish, which are being grown much much more than it was possible now, very quickly developed, for various reasons but they bring up lots of neuroethical concerns. Is there something like to be an organ, under what conditions? And, of course, computers. You heard from Michele the computationalist view of mind, the functionalist view of mind. What does IIT have to say about that? [41]

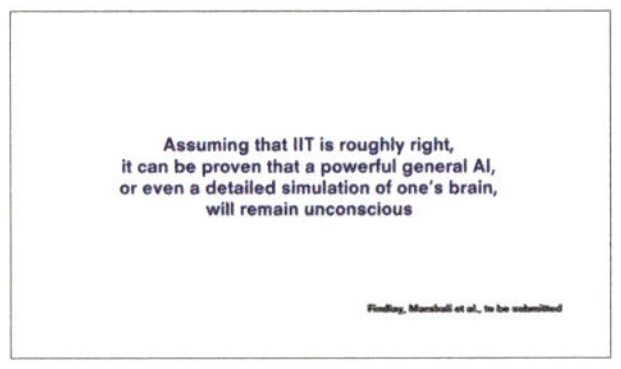

Fig. 41

It is very clear: we have proven now that if IIT is roughly right, then a powerful general artificial intelligence—that is a very powerful computer that runs very powerful software that can perform basically all the things we perform, we are not there yet but we are going in that direction—or even, which is far in the future, a detailed simulation of our brain, despite what you heard from the Blue Brain Project, but definitely not there well, you could do all of this in the computer. There is a lot of money being poured into doing this with computers. But if IIT is right we have to be careful about what we are doing. [42] Here is a very quick indication of what we demonstrated in a paper that is being submitted. Here is a very little system of five

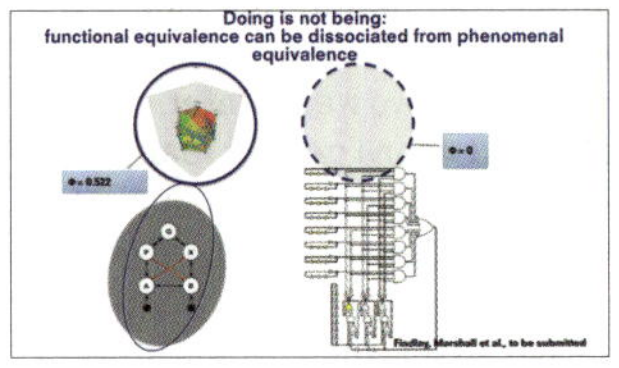

Fig. 42

neurons say in a particular state. Here it is unfolded cause-effect structure, what really exists physically. It has a low value of phi but it is not nothing, and it has a particular form. Here is a much bigger system of 66 logic gates, rather than five, connected to a little universal computer with registers, a clock, the arithmetic logic unit, and so on. This system can be programmed to be functionally equivalent to the one on the left, it does exactly the same things. The input-output function is exactly the same. If you put them in a black box, from the outside they are indistinguishable. But if you unfold this thing, even if it is much more complicated looking, it has zero phi, it doesn't exist, there is no consciousness whatsoever. All you get is what I call dust, some little pieces of this computer, mini entities worth basically nothing, that have nothing to do with what is being simulated. You can expand this because this is totally general. If you take the brain, of course, it is much more than this, the back of the brain here, and you were to unfold it, you would get, we presume, a gigantic cause-effect structure, which at every moment is identical with the experience I am having. [43] If I were

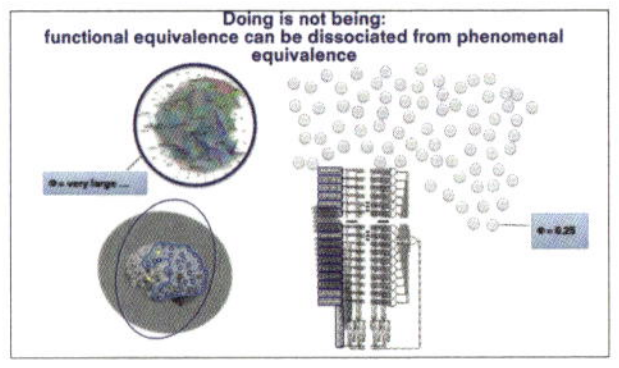

Fig. 43

to simulate either my functions or my brain with a giant computer, that computer would be from the outside indistinguishable from me, it would do all the same things. You can put it in a body like mine, is going to be indistinguishable. But if the idea is right, there is nothing there. The computer doesn't exist. What exists is this dust, these mini entities that have nothing to do with what really is there, which is my consciousness. Here is the maximum IIT. [44] Doing

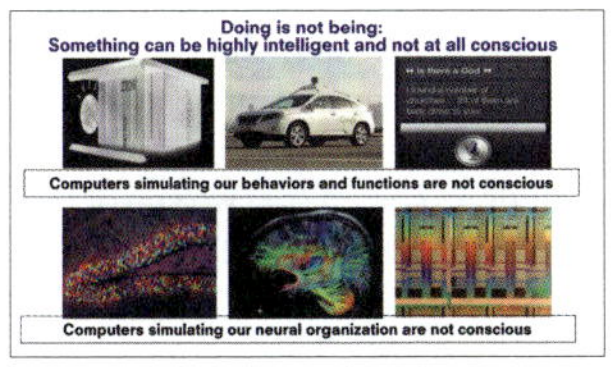

Fig. 44

is not being. We are in a functionalist site, now the world is going functional. What we care about is whether something is intelligent, anybody does things, we have supercomputers that are getting better than us at answering questions, cars that drive themselves better or almost as well as we do, and of course, the computers that can give answers to questions now, like many human beings, sometimes better sometimes not yet, granted, but certainly better than many patients who instead are there, they are conscious. Yet, IIT tells you that if it is done on a computer there is no way these things are there, and they should be treated accordingly. There should be means and not ends in themselves. Even computers like the Blue Brain Project that we are using to simulate how our own brain works, but simulated in a digital computer, so the hardware is like this, would have zero consciousness. [45] What you see in the

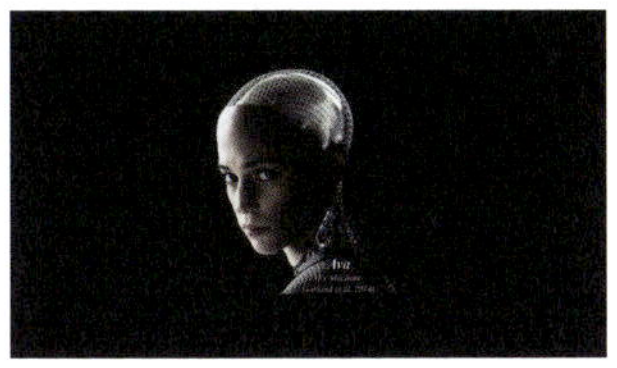

Fig. 45

movies, one classic example is *Ex Machina*, one day they may literally look like us, be remarkably intelligent even more than we are, maybe be nicer than we are—which is not that hard to do, after all—and yet if IIT is right, there would be nothing there. [46] What happens is we have

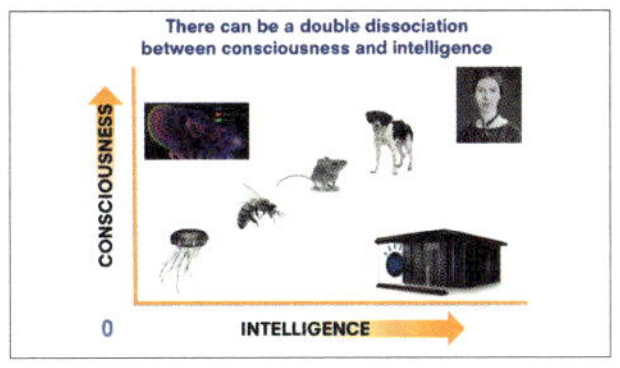

Fig. 46

been used to this correlation between conscience and intelligence. We are intelligent. When we are conscious we are intelligent, when we are not conscious we are not. Conscious intelligence goes together in us, typically. We look at the animal and plant kingdom and so on in the same way. We tend to think maybe a dog is probably conscious because it is intelligent, not as much as I am but it probably sees and hears, at least that much and so on, it may not have language. Then we don't really know anymore but we suspect they are not that intelligent, so they are probably not conscious. Again, IIT says there is complete dissociation in principle. This only happens, and we showed it, because of natural selection. There is pressure for developing integrated architectures, which are both conscious and intelligent. You can have an organoid in which there is no intelligence whatsoever, it is lying in a dish doing nothing, knowing personal outputs, and yet it could be highly conscious depending on the circumstances. Vice versa, you can have a supercomputer which is much better than we are or does the same things we are, and it is not there at all. [47] I'll finish by reminding you of the fundamental identity of IIT, that an experience, which is what consciousness is, is identical to a cause-effect structure. The unfolded cause-effect power of a physical subject and every phenomenal property has to be one-to-one corresponding to a physical property of this causal kind. I want to finish, because that was my

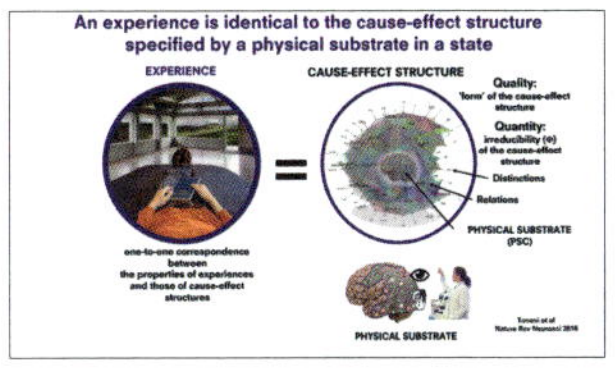

Fig. 47

title, by telling you that in principle I strongly suspect and I want to share that, that this is going, if it is true and we can substantiate it by looking at our brain, where we can actually do the experimental validation of what consciousness existence really is, we can do scientific metaphysics, if you wish, it tells us a very different story from the one we have been taught since the time of Galileo about what the universe is like and what actually exists. [48] What we have been all

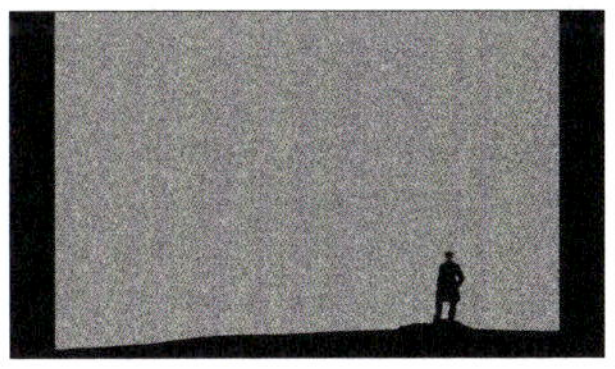

Fig. 48

taught now by science—and of course, I am the first one to be impressed with what science has been able to do—is that we really are just a little speck of matter living on a small planet in the corner of a modest galaxy, near a modest star, and the galaxy is one out of many of the universes that maybe is even expanding very rapidly, and may not even be the only one. That is what we have been told. And of course, our own existence is just a flair of life for an instant over the eternity of time. That is what we have been told, that is the picture that we all have that is becoming more and more in contrast with the humanistic worldview that we initially started with. But if IIT is right, it is very likely that this view of the world is utterly wrong, we have got it completely upside down, or better, outside in. It is likely that all these giant galaxies are really not much more, most of them at least, than an aggregate of dust, of this mini meaning little entities that don't really exist much in and of themselves. [49] They exist for us,

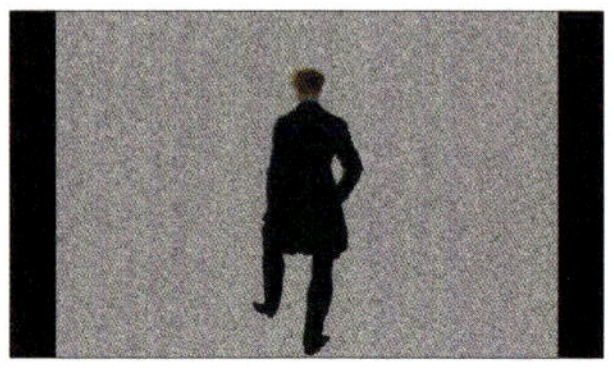

Fig. 49

but not for themselves. That the sublime mountains and oceans that we are inevitably meant to admire when we go to the top of a mountain and look down at the immensity of nature, those too are really not much more probably, most of it, than an aggregate of dust. Even, sorry for the embodiment, even our own bodies. The body that we see and feel so strongly lying on the bed. Our own body is really just an aggregate of dust that is possibly cells and cells themselves may be a little entity but worth very little, existing minimally, intrinsically for itself. [50] Instead, what really

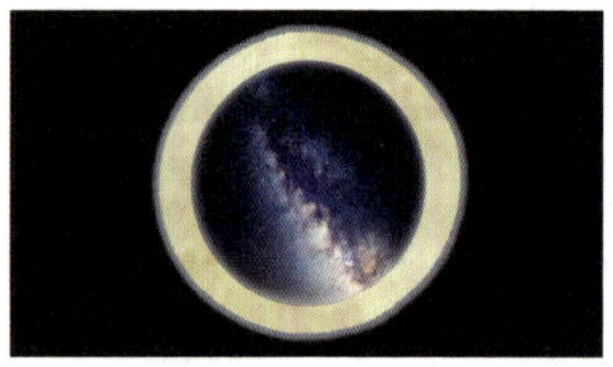

Fig. 50

exists, or that really the large entities in the universe, is our own experience within which my body is a structure, a form in the mind that exists within the mind, as a body. The mountains and the sea are forms within the mind that exist only within the mind, in the only way that something truly exists. The universe itself does not exist as such, it exists for each and every one of us as a giant form in the mind, a concept through which we attempt to capture everything, but only within our own consciousness.

And when I say our own, of course, I mean my own and that of every fellow human and animal being, each one a universe of consciousness.

Panel Discussion

VIVIANA KASAM
MICHELE DI FRANCESCO
GIULIO TONONI

VIVIANA KASAM I would like to ask a very simple thing: how did you get to the definition of phi? Because this is a mathematical equation, it is interesting also to know that there are people saying that in the end the whole world has been built according to this, so our idea of beauty depends on the proportions of the golden section that we see of the phi. Can you give an explanation?

GIULIO TONONI There is an assonance between phi and the so-called golden section, even if obviously philologists will probably disagree with me. The idea of phi is in reality that of measuring irreducibility, the extent to which something is above the parts that make it up. The way we can do this is physical. What we literally do is we are doing a causal cut between one part of the system and the rest of the system—we need to do this for any part because we need to be able to find the weakest section—and when we cut the system in two or three parts, we see what a difference it makes. If it makes no difference, for instance, if we take two entities that are completely separate causal, me here and somebody in New Zealand at the moment, so the opposite side of the world, if we separate that system in two, nothing much changes obviously. In this case, the phi is equal to zero, it tells us that the entity made up of that person in New Zealand and myself does not exist as a single entity. Its irreducibility is equal to zero. So basically it doesn't exist as such. Maybe I can exist, that person exists, but we don't exist together, the two of us together. Generalizing, we can see any system in the human body, a cell, the brain, the cerebellum, the whole brain, a part of the brain, and for each one of these we can operate all of these sections, these cuts, to see how irreducible they are. Literally, phi is a quantifier of existence. To what extent something exists is measured by how irreducible it is and the phi that matters is the maximum quantity of existing, it is what is called the principle of maximum existence, which vaguely resembles Leibniz, because he went quite close. The idea is that amongst all of the possible existence what really exists is what exists at the maximum: this is the principle, the one that has the maximum phi value. Basically, it beats all of the other existence.

VK I like this because it once again connects the topic of our evening, science and philosophy. At this point, I wanted to ask professor Michele Di Francesco. We heard about cause and effect, but cause-effect recently has been quite separate. We talk about con-causes, multiple causes, we call about network systems. How does philosophy respond to this? This is a question for both of you. Is the concept of cause-effect still valid? Is a cause producing an effect still valid as a concept?

MICHELE DI FRANCESCO I think it depends on the context in which this question is asked. From a certain point of view, it is well known and it has been known for a time that the identification, the distinction between cause and background conditions, is not that easy. Let's imagine, for instance, that there may be a car accident on a Saturday evening, so if we want to look for the cause we could go and look and see. If we are the police we want to see if the person on board is drunk, if we are the car manufacturers we want to see if there are any structural problems, if we are the mayor of the city we want to see if there were problems on the streets, like holes. If talking about multiple causes or con-causes is this, maybe I would say that from a physical standpoint, it is quite dubious if causes have to be taken as really existing entities in the quantistic description of the world. This is a notion that appears on all levels in the philosophy of the mind, I still find it a useful notion to be used, even if obviously debates are ongoing on this.

VK What about science?

GT There is always been a great mistrust of causality in science. Obviously, Hume had a rather complex vision of this. You could mention Hume in favor or against the existence of causality, as he was saying, it is simply something that happens, it is a series of things but there is nothing beyond this, it is only this, a series. Or, as has been said, one thing after the other, nothing more than that. But in reality, whilst physics was quite skeptical because the idea is to describe what happens with differential equations, in classic and quantistic terms, instead, biology is absolutely causal, it is all based on mechanism, it is mechanistic par excellence. In this sense, it refers to Descartes rather than to contemporary physics, and this is absolutely true. I think that the physicist is defined as a causal power in the theory that I mentioned, so we should differentiate between the causal and the causal potential of the state, that is what we do when we do this unfolding of a physical substrate to demonstrate what exists. It is not the substrate, it is the unfolded substrate in its causal potential, and what it is called the actual causation, which is what people normally think causes the next event, what goes from one state to the other, what has caused this accident, what has caused this particle to be disintegrated, what has caused the fact that the neuron has downloaded or not. Also, the current causality, which is important to understand what happens, can be studied. There is this work that has been done with the same fundamental principles of intrinsic nature, specificity, integration, and definition. These are to assess what causes what, where it is fundamental to understand that there are composite causes and there are background conditions as well, as you mentioned. In this sense, it is perfectly possible to talk about con-causes and characterize them mathematically. But all this is a recent evolution. Our work was only done last year, for instance. There is still a lot of confusion in interpreting causation, denying it, or seeing it in mechanicistic or biological terms. But above all, the different thing is to quantify it and to differ the potential one from the actual one.

VK I will ask one last question, which maybe is only out of curiosity. What space is there for free will in your vision today, the philosophical will, the scientific will of consciousness? What is the space for free will?

MDF I could make some comments. We debated a lot about free will in the framework of some results of neurosciences, of experiments of Liebe, then Sung subsequently, and the outcome was quite dubious, I think. Because neuroscience is such, I don't think that they are very much related to the solution of the metaphysical problem of free will. However, the discussion could be useful to clarify a few points. It has been shown that doing a certain action there are certain parts of the brain that are already beginning to prepare for this action, and the action could be just folding a finger, or a wrist. A discussion has been initiated on whether this was a good action and then the decision has been debated if the decision had not been already taken. What would really be surprising if there were a scientific theory that could show that conscious thought does not have a causal role in determining actions? This would be very bad. Instead, as far as free will is concerned, even if minds did not exist we would have this anyways, in the sense that determinism is valid in our world or is not. I will just limit myself to this remark. It is not really here that we can find a solution.

GT The conclusions are very different. I would like to say that the greatest majority of my colleagues in psychology, neurosciences, and science, in general, cannot really see how is it possible to have space for free will once we understand the physical substrate of the brain. In other words, even if experiments haven't demonstrated it conclusively, the idea is that each neuron of the 85 billion in our brain, decides whether to fire or not depending on the inputs it receives. It does so, and if there is determinism or not probably there will be a minimum casuality, not causality. In the end what will happen is determined by neurons, each neuron on its own, and there is no space for free will. Michele mentioned Jaegwon Kim, who said there is no space for any other causality, apart from the microphysical one of what atoms and neurons do. This is a vision everybody has, even though it sounds impossible because free will seems to be my will of raising an arm, for instance. But science says no, there is nothing, it is neurons that define what we do. IIT states that free will does exist in the fundamental metaphysical meaning of the term, so when I cause an action I consciously cause it, and not my neurons. This comes from another idea of IIT: what exists is consciousness and when consciousness exists based on the axioms, neurons no longer exist, they become part of consciousness. What literally exists is this cause-effect structure, this cause-effect power. This is what exists and not the neurons. Only what exists can lead to a cause. This is a conclusion. So free will requires consciousness, it is not sufficient to just have consciousness, you need more things understood by philosophy, religion, etc to be able to keep your will to do something. The more we are conscious, the more we can be free, and the more we have free will.

6

Concluding Remarks

MASSIMO CACCIARI
GIANCARLO COMI

PANEL DISCUSSION MODERATED BY
VIVIANA KASAM

Concluding Remarks

MASSIMO CACCIARI

The issue of free will has always been a sort of *caput mortum* of philosophy, and it is interesting to see how some philosophers who got closer to this negation of free will or the fact that free will has no foundation are those who are very far from science. A philosopher like Nietzsche, for instance, believed that free will was total imagination without any foundation. But I would like to limit myself to a spectator almost, I would like to simply underline the points that really struck me in the different presentations I followed this week. I would start with Jean-Pierre Changeux who recalled the ethical importance of our debate. This is important with respect to the body, but also to this product of evolution which is our brain, which allows us one way or another to know nature, and inside nature, the body, and inside the body, the brain. It is an extremely important fact from the ethical viewpoint that not knowing and not being curious about these aspects implies a political and ethical deficit. This is what I believe and this is the first point I wanted to stress. Then the debate focused on the theme of consciousness, as Damasio said: "the gift of the self." It is wonderful that it induced many in our civilization to think that consciousness did not derive from a biological foundation, so in our psyche we would have a distinction between the sensitive and vegetative one from the ethical and rational one. As Michele Di Francesco said, no philosophy went into vulgar and abstract dualism. The distinction we talked about always takes play in the unity of the soul, the principle of life, the principle of a body that has life, at least a potential life. We always dealt not so much with dualism but with the idea of regulation. It was necessary to make all possible efforts—and this is true for Descartes and for Spinoza—so that the ethical part can guide us and can drive us through our reality. I don't think this hope or idea has been abandoned by anybody because it still remains true. Then the different reports explained that consciousness—at least this is what I understood and I agree with this also from the philosophical viewpoint—is not a subject that emerges from the mind and directs. It is not the imaginary protagonist of this game. The symphony is produced by the orchestra itself while playing, to quote a poet, Antonio Machado: the walking has no walking, so you fire an impulse and the orchestra composes by playing and plays by composing. I think that all reports also showed that this image would not be sufficient because the orchestra, at least apparently, carries out a perfectly conscious job. Consciousness instead, finds itself in the unconscious process which is at the basis of consciousness. It is an important theme where neurology, philosophy of the mind, anthropology and culture can and should interact more so than in the past. Obviously, the dominance of "I think" has been radically criticized by the whole of contemporary philosophy, but the complexity of the nexus between conscious and unconscious has been clearly illustrated to this day with great accuracy. The unconscious of the genome will have to be aggregated into the archetypal one and the cognitive one. Consciousness, as such, seems to be what emerges out of a huge iceberg. It is just the tip of the iceberg and Changeaux said so, which is simply what we can identify as the fundamental conatus, as Spinoza used to say, of any organism. The conatus of living and by living developing your own shape and your own form. It is this free will of living, it is the anti-thermodynamic behavior of any living being, and this behavior manifests itself in the brain through the model of adaptation and coordination of plasticity, of reaction to the environment it represents. The restless brain is what we heard. We have a restless brain and we know this, as well as nature and *physis*, nature is restless by nature. Bergson, mentioning Giordano Bruno without being aware of it, used to call nature a huge blossoming of unforeseeable news. This restless brain, this brain that never rests constitutes a global workspace. In its maze, each recording element takes place locally but communication moves into the general neural workspace, the global neural workspace. Each word takes up its meaning in the context of a sentence. The conference in all its reports moved towards a reductionistic perspective as well, the understanding which becomes more and more accurate, of the neuronal basis of our mind does not mean explaining the event, which is highly unlikely, which is the revolutionary onset of subjectivity, as Damasio said, and subjectivity means singularity. The neurological common aspect does not explain the private nature of the mind. Many reports talked about a dramatic leap in this respect. More than one of the speakers used this expression when they talked about the language of thought that characterizes the human being. The language, or rather languages, in their essentially recursive character with metaphysical scientists the capability of the mind in its physical finiteness of understanding and perceiving what is not finite. Nothing can be there which is not in the senses. This is true and has always been true, but it is true that the process starts from feelings and the sphere of the self could never suppress the unconscious and emotional one, and Spinoza knew this very well. But it is also true that the intellect is not in the senses. It cannot be reduced to the same season. This dramatic leap allows us to explain the reasons for the dualistic mistake, where those were present. The reductionistic perspectives, both scientific

ones that reduce the mind to the brain, and philosophical ones, which are even more dangerous, that tend to select the mind as something separate, are based upon something which doesn't make sense.
A superstitious vision of the cause-effect power. Therefore, Hume, which has been mentioned before, should have healed. The matter is open to what is spiritual. This is what real philosophy has always taught us, the real philosophy of nature. There is a disposition. This does not mean that this is a direct cause. From this point of view, epistemology at the basis of contemporary philosophy should be integrated with the current research on consciousness and the mind. This is even truer if we look at the qualitative nature of consciousness phenomena of the mind's creativity, for example. We should abstract from all of this to perceive a man-machine. Also, Descartes recognized this was impossible, after all. Whatever machine can only have a limited number of physical states and the mind, the language of the thought, has potentially an infinite number of them. We can conceive a machine that adapts and reacts to the environment, and therefore modifies itself, that communicates to the environment in an unforeseeable manner when it is built. I think this means asking if it is possible to build a man.
The answer is known and also the means are known, and we have been knowing them for millions of years. The question that makes sense is different in my opinion, and with this, we move to the second part of the work carried out by Fondazione Prada's scientific committee. Based on the extraordinary scientific knowledge, these days have been illustrated. Can we move with more trust to tackle the physical basis of diseases, the language, the consciousness, of thinking, of the self? This will be the topic of the next international conference we want to organize.

GIANCARLO COMI

From the point of view of the neurologist, I have to underline some aspects that have drawn my attention remarkably during this conference. It has aroused in a very strong manner a multi-disciplinary type of interest that we have tried to represent in the conference. The concepts that I believe it is appropriate to underline are that all that we have listened to have revealed the amazing evolution of knowledge in these past few years. What we need to do is to confront ourselves, we need to tackle and face what our experiments are telling us. An enormous evolution that is also dependent on the availability of new technologies. Let's just think about technologies such as the functional MRI that allow us to see a brain in action. We have seen from Giulio Tononi some amazing images of perturbations of the brain activity, and we have seen how these induce changes in the state. We have also seen very important examples shown by professor Changeaux. This is what makes a scientist a person who wants to know, who is curious, who wants to put the thirst for knowledge before her/his own personal convictions, and this is not an easy process, but we have seen it repeated a number of times over these days. Another message that I think I have been able to draw from all of these discussions, and the theme of these discussions has been extremely appropriate, is that today we have the opportunity of dismantling a part of the complexity which seems to be indispensable to generate consciousness. This is a way of approaching the problem, which is a bit different from the global one that I heard about tonight, too. I think that one supports the other, without the support of the one, the other would lose strength. Another aspect that Idan Segev has mentioned in his presentation is that there is a certain uniqueness in what human brains would like to bring forward. Embracing with scientific rigor the complexity of problems, such as the ones that have been debated tonight on consciousness, but even the future ones that we shall be discussing consciousness that vanishes because of the diseases that undermine it, that undermine its capability and its potential. I also think that the amazing ability that our brain has not only in thinking and planning, but also in having and giving place to artistic expressions in a very high form of communication. As Moro said, without the language, it wouldn't have been possible to transfer knowledge and it wouldn't have been possible to have a culture emerge, at least in its structured form. Miuccia Prada, in promoting this initiative and starting off from the art, from beauty has wanted to inject an artistic approach so that you can put two sectors which are normally considered as antithetical—which are not antithetical because they are both the supreme expression of our mental activities—to bring them together.
I would like to thank in particular Miuccia Prada for having launched this important challenge. I would like to thank all of the components of the scientific committee, starting with professor Cacciari, who has been one of the first to launch this initiative. Over these evenings, we have had Viviana Kasam, Jubin Abutalebi and have had the opportunity of getting to know Andrea Moro, Daniela Perani, and we had also assigned this agreement between sciences and art with the presence of Udo Kittelmann, a great curator and expert on how to translate in the reality of an exhibition the complexity of human thought in its various articulation, in its various facets and this has been fundamental. Special thanks go to all of the speakers.
I think it has been a very important experience that we have had and our starting point is not our point of arrival. It is not where we want to get.

II

CONVERSATIONS

Presented from September 2021 to April 2022, "Conversations" consists of seven pairs of video interventions given by international scientists, philosophers and researchers followed by seven online conversations involving forms of interaction with the remote audience. The project was presented as an open platform for experimenting inclusive modes of exchange between scientists and scholars from different fields and a wide community of students, researchers, and individuals interested in neurosciences and brain studies in general.

Each conversation consisted of speeches by the two experts introducing their research on a specific topic, followed by an online debate coordinated by a moderator. Ten days before the conversation, speeches by the two experts introducing the topic were available online. Alice Roberts, anatomist, author and broadcaster, took up the role of moderator for the first six events. University of Birmingham's first Professor of Public Engagement with Science since 2012, Roberts has received five honorary doctorates from British universities.

The first two lectures and the first live talk involved Angela D. Friederici (Director at the Max Planck Institute for Human Cognitive and Brain Sciences in Leipzig, Germany) and Robert C. Berwick (Professor of Computational Linguistics and Engineering in the Department of Electrical Engineering and Computer Science at the Massachusetts Institute of Technology, Cambridge, MA). The talk focussed on the specificity of human language and its relation with other animal and artificial languages.

The following two lectures and the live conversation investigated the distinction between the functions of brain's right and left hemispheres and attention mechanisms. The event had seen the participation of Guido Gainotti (Professor Emeritus of Neurology at the Department Neurology of Policlinico Gemelli, Università Cattolica del Sacro Cuore, Rome) and Paolo Bartolomeo (Research Director at the Inserm, the French state biomedical research institute, and at the Physiological

Investigations of Clinically Normal and Impaired Cognition (PICNIC) Lab at the ICM – Paris Brain Institute in the Pitié-Salpêtrière hospital). During their lectures and talk, Ellen Bialystok (Distinguished Research Professor of Psychology and Walter Gordon Research Chair of Lifespan Cognitive Development at York University of Toronto) and Jason Rothman (Professor of Linguistic at the UiT the Arctic University of Norway and Senior Researcher in Cognitive Science at Universidad Nebrija in Spain) dealt with bilingualism and its advantages and disadvantages for the human brain.

The integration and relationship between language and music through an evolutionary perspective was the core of the lectures and the conversation by Steven Mithen (Professor of Early Prehistory at the University of Reading, UK and archaeologist) and Stefan Koelsch (Professor of Biological Psychology and Music Psychology at the University of Bergen, Norway).

Jeffrey A. Kleim (Associate Director of the School of Biological and Health Systems Engineering at Arizona State University in Tucson) and Álvaro Pascual-Leone (Professor of Neurology at Harvard Medical School, Boston, MA, USA, Medical Director at Deanna and Sidney Wolk Center for Memory Health, and Senior Scientist, Hinda and Arthur Marcus Center for Aging Research, Hebrew SeniorLife, Boston, MA, USA, Chief Medical Office at, Linus Health) focused their lectures and talk on the relationship between genetics, the possibility to increase brain plasticity through neuromodulation and the related ethical implications.

The subject of the two following lectures and conversation was the analysis of the evolutionary approach to conscience and the relationship between science, biological evolution and philosophy. This topic will be investigated by Simona Ginsburg (Associate Professor of The Open University of Israel, Ra'anana) and Eva Jablonka (member of the Sagol School of Neuroscience, Tel Aviv and Research Associate in the Centre for Philosophy of Natural and Social Science of the London School of Economics, London University).

The program of "Human Brains: Conversations" ended with a philosophical meeting. Massimo Cacciari (Professor Emeritus in Philosophy at UniSR – Università Vita-Salute San Raffaele, Milan) and Michele Di Francesco (Full Professor of Logic and Philosophy of Science of the School for Advanced Studies IUSS Pavia) investigated the influence of figures such as Plato, Descartes and Spinoza that crucially shaped the birth of the history of Western thinking in relation to science and conscience studies.

SCIENTIFIC DIRECTOR

Giancarlo Comi

WITH

Paolo Bartolomeo, Robert C. Berwick, Ellen Bialystok,
Massimo Cacciari, Michele Di Francesco, Angela D. Friederici,
Guido Gainotti, Simona Ginsburg, Eva Jablonka,
Jeffrey A. Kleim, Stefan Koelsch, Steven Mithen, Álvaro Pascual-Leone,
Alice Roberts, Jason Rothman

The following texts are transcriptions from the conversations available at
humanbrains.fondazioneprada.org/en/conversation

1

Syntax as the Fingerprint of Human Language

ROBERT C. BERWICK
ANGELA D. FRIEDERICI

MODERATED BY ALICE ROBERTS

The Core Language Computations

ROBERT C. BERWICK

Language is computed inside your head by neurons. But what exactly is human language? How does it make us different from other animals? What I want to show you today is that language is grounded on three specific core computations inside your brain driven by a single underlying computational engine—just like your laptop's CPU. Importantly, I also want to show you today that computer systems that claim to have captured these special computations so far have failed to do so despite what you might have heard otherwise. Let's begin with something simple. It is an example illustrating human language's first core computation. [1] I have shown

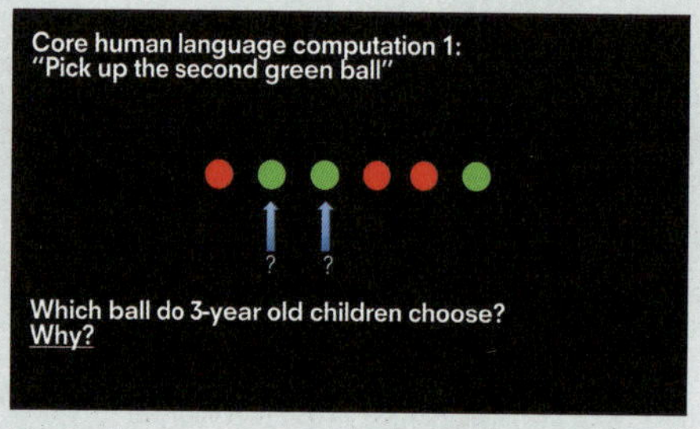

Fig. 1

it this way as a row of six red and green balls. We can imagine someone asking "pick up the second green ball." Note, there is an ambiguity here, a choice, because you could pick up the second of all the green balls, that is ball number three, this one, or you could pick up the second ball and see that it also happens to be green, ball number two. That is two different answers. But what do people actually

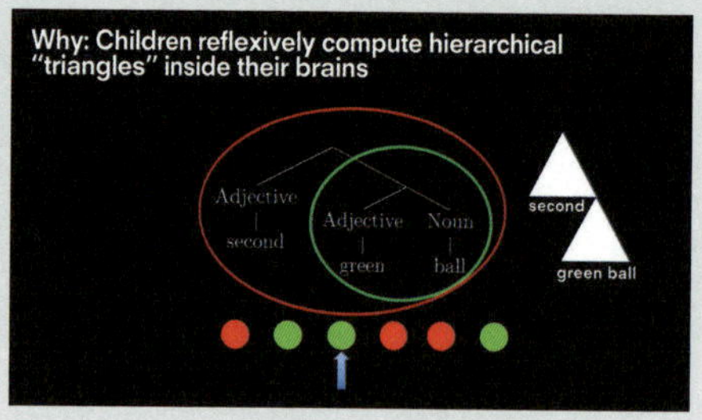

Fig. 2

do? If you know English, you always select ball number three. [2] The question is: why? Children do this reflexively, by ages three to five, and the answer is that children and adults unconsciously follow this first core computation of human language. They compute reflexively, in their heads, a hierarchical structure that looks like a stack of triangles associated with a string of words. I schematically showed that to the right, by this white triangle here: the triangle groups "green" and "ball" together as shown by this green circle. It is, in effect, looking for the set of all the green balls. When we add the larger circle to it, indicated by the red that encompasses the previous one, note that to represent this sentence in their brains the children just pay attention to the hierarchical structure alone. They throw away, ignore the linear order of the words, even though that is what we hear. [3] If the language was Japanese, as I show

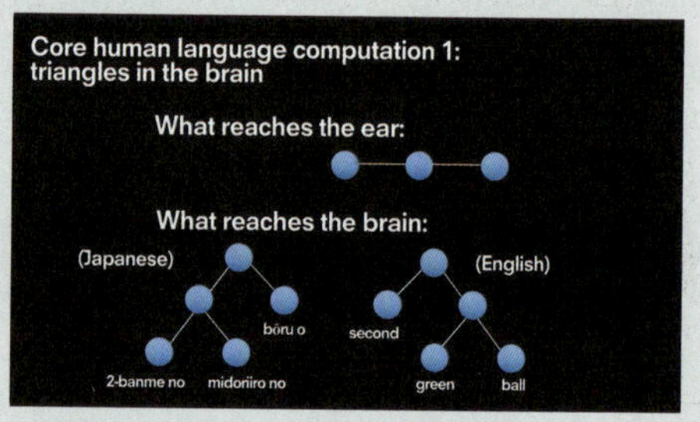

Fig. 3

at the bottom of this slide on the left, children construct the same stacked set of hierarchical triangles as before, in English, but with different words. The stack looks like the mirror image of the one in English but they get the same meaning. That is the first core computation of human language. We can say it this way. What reaches the ear is ordered in time from left to right, but what reaches the mind or brain is unordered from left to right. It is hierarchical. We can make these internal triangles in the brain visible using direct electrophysiological measurements of the brain as it processes sentences. I show this on this slide here. [4] An experiment

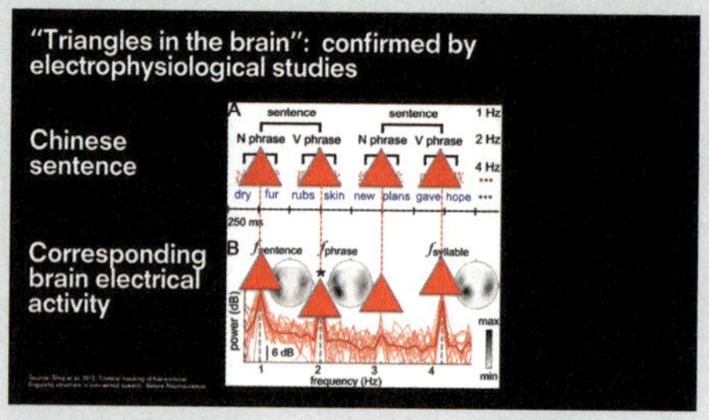

Fig. 4

done in 2015 by Ding et al., had Chinese human subjects listen to sentences like the one that is shown at the top. It is labeled here: "dry fur rubs skin new plans gave hope," with a hierarchical structure as I am going to indicate by the red triangles. Chinese subjects' brains produce these corresponding hierarchical electrical patterns that are shown at the bottom here with these peaks that align with the language up at the top. Thus, the brain actually pulsates in time to the formation of these hierarchical triangles. As you can see, they are juxtaposed. Consider one final example: the sentence in English "Eagles that fly instinctively swim." [5] This sense is ambiguous. The adverb "instinctively" can modify either "fly," describing how the eagles fly, or it can modify "swim," indicating how eagles swim. Suppose we change the sentence just a little, so we have it here. We move "instinctively" to the front. With this change, the sentence is no longer ambiguous. "Instinctively" can only modify "swim." What this shows is that "instinctively" does not modify the verb that is closest to "fly," but rather one that actually turns out to be hierarchically closest to "swim." The brain ignores linear distance. That is core computation one. What about other animals or computers? Can they build triangles in their brains as people do? It doesn't appear that they can. Consider perhaps the most serious effort to teach a nonhuman animal a human language—in this case, American sign language. This was with the chimpanzee Nim [6] who received intensive training from birth. Here we have a picture of Nim, that is the person on the left, being taught by Laura Petitto, the person on the right. After this training, Nim could string several signs together, like a laundry list, but had no way to construct something like "Second green ball" or understand it. Nim would just prattle on until it would happen on a sign sequence that led the human experimenter to give the reward that they wanted. Note Nim's longest sign below: give orange me give orange me eat orange give. You get the idea, Nim wanted an orange. It is simply a long set of beads on a string. It is not hierarchical triangles. While Nim had about the same auditory systems as humans, he never could acquire a language in the same way that children do. Let's see briefly now whether modern computer systems can do better than Nim and actually capture this kind of hierarchical computation and we'll return to that theme at the conclusion of the talk, because there has been much news over the past decade and how much better the new style artificial intelligence neural networks are in analyzing human languages, that they can come very close to what we do. I'll give you some evidence now that, in fact, quite the opposite is true, using our second green ball example. In the picture I show here [7] this is how Google's best AI neural network system deals with something like "pick up the second green ball." It is the representation it outputs. Note that it explicitly does not build a hierarchical structure. Rather, it is all flat. There are no stacked triangles like children do. Instead, it analyzes the second green ball in a linear string-like way, so arriving at the meaning that it should first look at the second ball and then see what happens. It is a Google part that doesn't work on this. Let's turn now to the second core computation of human language, what can be called digital infinity. Digital because there are three- and four-word sentences but no three-and-

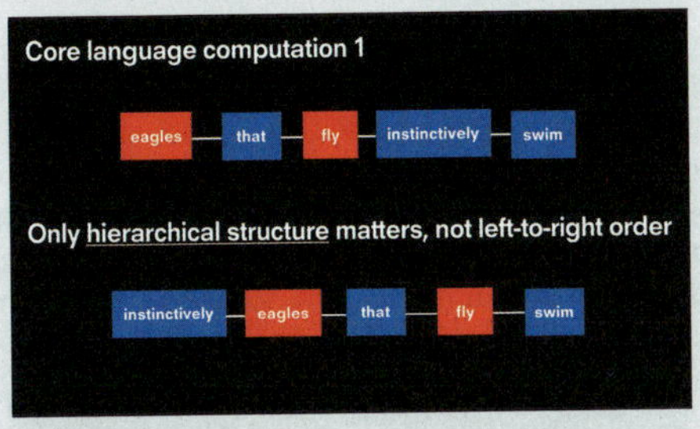

Fig. 5

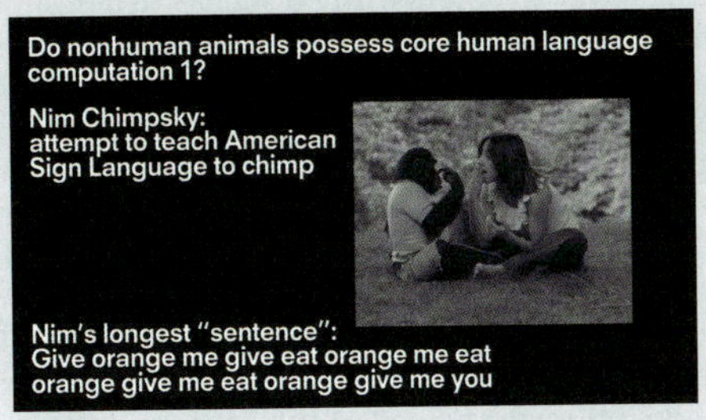

Fig. 6

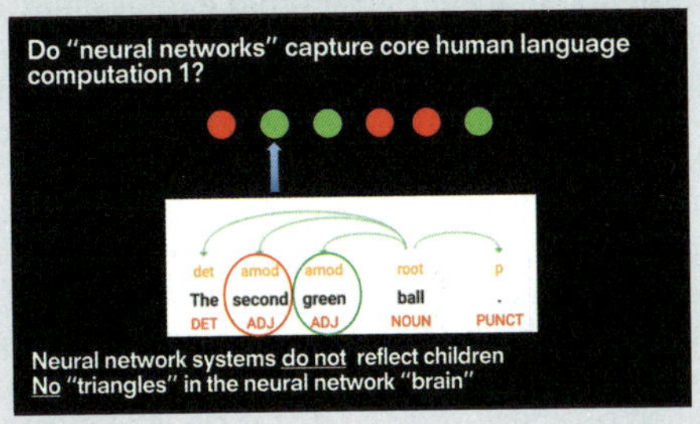

Fig. 7

a-half-word sentences. Infinity because we can construct infinitely many thoughts in our minds, and can convey to others with no access to our minds their innermost workings. Since the number of sentences we can speak is infinite, we can't explicitly list them all. The brain is finite. So how can we construct an infinity of possible thoughts? In the 19th century Wilhelm von Humboldt put the matter this way: that language makes infinite use of finite means. A finite brain, infinite possible behaviors, and sentence output. No other animals do this. The sentence I show here, from Dan Brown, *The da Vinci Code*, [8] might be called the infinite misuse of

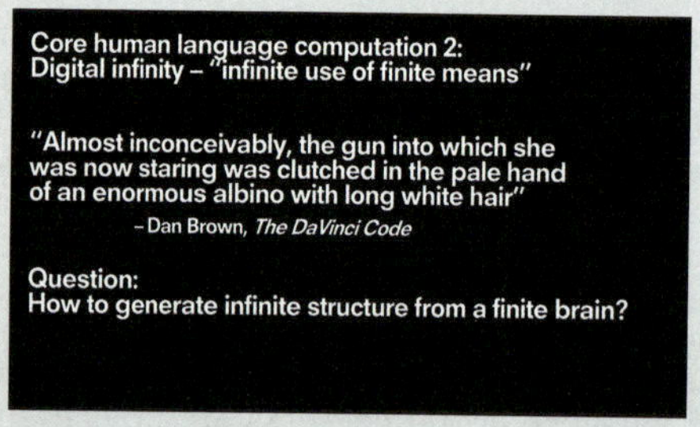

Fig. 8

finite means, nonetheless, a very complicated sentence: "Almost inconceivably, the gun into which she was now staring..." How can we construct all these hierarchically structured expressions, an infinite number of them? What is the computation that lies behind this ability? Until the 1930s and the advent of computation, we didn't understand how that could possibly be done, but now we do. We understand that the only way that a finite brain can do this is it has to have some kind of operation or computation that applies to its own output. That is an essential property of human language that I call the basic property. It is a property that is sometimes called recursive—that is what it means for something to apply to its own output. Some recursive structure-building operation is required for human language to build stacked triangles in the brain. We can actually now describe it and I have shown it here on this slide. [9] We can call it the computational operation "merge." What merge does is take two elements, like "ate" and "apples," x and y, and it can glue them together to form a new, larger structure. That is the union of x and y, a hierarchical structure. "Merge" can then apply to its own output. Here it can apply again, recursively, and construct a larger structure. "Bob ate apples." Notice that it could have just as well been "Bob apples ate" in Japanese. Remember, the linear order doesn't matter. This merge operation grinding along, gluing elements together is the human brain's CPU for lan-

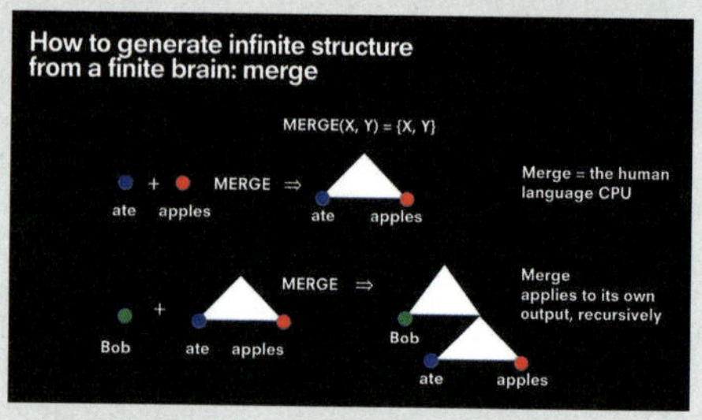

Fig. 9

guage—the simplest possible combinatorial operator that can generate all the possible human sentence triangles we need, and it doesn't seem to be present in any other animal besides us. It makes us special. Merge also leads directly to the third and final core computation of human language, what we can call movement. I have depicted this here on this slide. [10] We said before that merge is

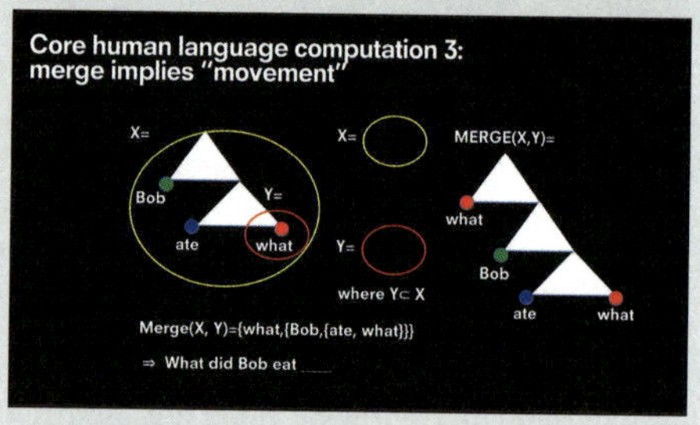

Fig. 10

defined as joining two elements together like "ate" and "apples." So there are two logical possibilities here: either those two elements are disjoint and have nothing in common with each other—"ate" and "apples," and in this case, we say y and x, the two elements are disjoint and we can form "ate apples," that is one possibility— or it is the case that y is a subset of x. We illustrate that second possibility where y is a subset of x on this slide, starting with the triangle representation for "Bob ate what." Next, we can take y to be the innermost element circled in red and x to be the entire set of stacked triangles circled in yellow. Then y is inside the yellow circle. It is a subset. If I apply merge to this, I get the following output shown here: what Bob ate what. When this gets spoken in English, the second occurrence of "what" at the very end is not pronounced, and one adds some kind of marker to mark tense. We get something that looks like this: what did bob eat.

That is a familiar question, where what is the understood object of what it is that was eaten, even though it is only pronounced at the beginning of the sentence and not the end. This is movement. It is commonplace in most of the world's languages. In short, with just this very simple imaginable combination machinery that we need to get digital infinity at all, as a free bonus we also get the possibility of sentences, where an item can be pronounced—"what," at the front of a sentence—distinct from where it is interpreted—at the end of a sentence. This possibility of dual positions of appearance and interpretation is unlike any programming language, but it is widespread in human language. That is the third core computation of human language. To conclude, let's see whether these computer systems that we now have for analyzing human language, the so-called deep neural networks, can mirror these three core human language computations. Such systems, you may know, are trained on billions of examples, essentially the entire web. So it is unsurprising that they can successfully parrot back many useful things. However, more surprisingly, they do not seem to reflect what human neural networks do, and they actually fail to capture any of the three core computations completely. Let's take a look at this. First, as we have seen, they fail to build hierarchical structures in examples like "The second green ball," where children and adults always build hierarchical structures. [11] We repeat here how

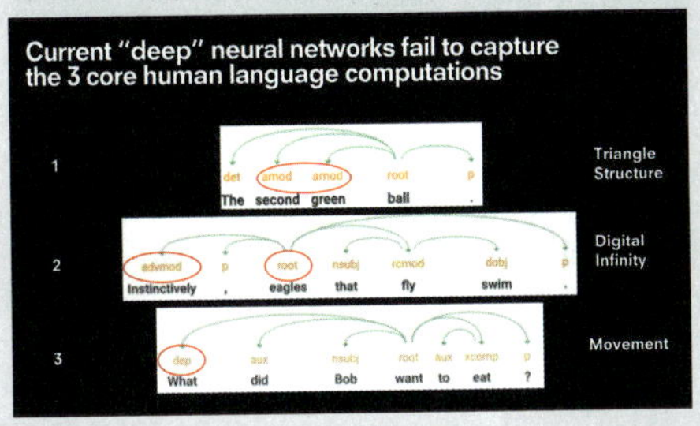

Fig. 11

Google's neural network system analyzes this example incorrectly. Second, they don't seem to accurately use merge—this basic operator, the CPU of human language—to generate digital infinity. The second figure shows how Google's parser system handles "instinctively eagles that fly swim." Instead of associating "instinctively" with "swim," the Google system, unfortunately, pairs it up with "eagles." It doesn't even pick out "swim" to be a verb. The neural network system has jumbled everything up. Third and finally, neural network systems stumble at the third core property computation of language—movement. We illustrate this in the last row on this slide with the example "What did Bob want to eat." That is only just a little bit more complicated than the example we used before—what did Bob eat—where we put "what" at the front of the sentence in English, even though it usually belongs immediately after "eat," as the object. What does the Google system do here? Crucially, it cannot figure out at all what "what" is actually supposed to be doing, that it is the object of "eat." That is indicated by the label that is on "what." If you look at that label it says DEP, which is what it labels something when it can't figure out what that word has to do in the sentence at all. So, while the neural network can't figure out the simple example—what did Bob eat—it is stumped by even a slightly more complicated one that no child would ever get wrong. So, by age three or five, children can all answer these kinds of questions correctly, but the Google system, trained on billions of examples, cannot. What do we conclude from this? I think we can conclude that whatever else these deep neural network systems are doing, they have not really absorbed these three core computations of language. The word "neural" might be part of their names, but the resemblance to what human brains compute with language ends there. That is a sobering thought for those in the field, I think, but it is also a golden opportunity to go beyond what we do today to build a computer system that actually does respect the core computations of human language. These computations are the true scientific goal of the study of language—the computational fingerprint of syntax in the brain, and what makes us uniquely human.

The Neural Basis of Syntax

ANGELA D. FRIEDERICI

Syntax as a fingerprint of human language. Why, you may ask, should syntax be so important? Syntax—you called it grammar in school—is the backbone and core of every language. Every child learns every language in the world. We speak and comprehend language automatically. However, usually we are not able to declare the rules according to which we build the sentences. The question is: what is the brain basis for all of this? We will allude to this today. There are several animals, that is, animals like parrots and dogs, that can learn words but they cannot acquire syntax. Even our closest relatives, the chimpanzees, differ from us in their ability to process syntax. This case study may illustrate this best. What you see in figure 1 is the Mean length of utterance of a hearing child, and you can see this child starts off with two words and then brings more and more words together to come up with a sentence. [1] Now look at a deaf child. This child learns

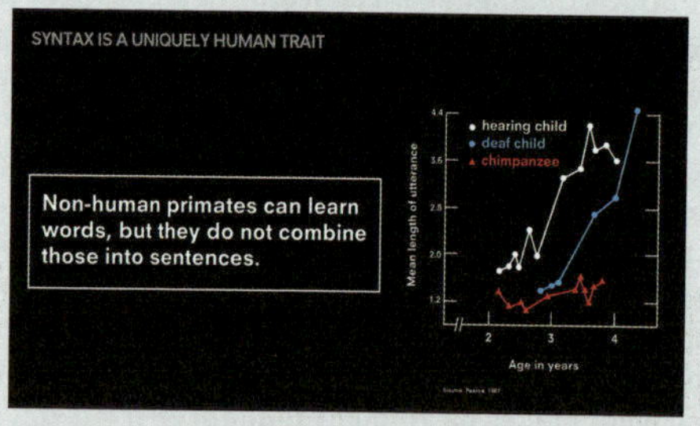

Fig. 1

sign language, though starting somewhat later, but then, piece by piece, and dramatically, increases the number of words in an utterance. The chimpanzee, however, even after extensive training, never reaches this level. What we can conclude from this is that non-human primates can learn words but they do not combine those into sentences. The crucial question then is: what is the brain basis for the human capacity of language? There are two major regions that have been identified, already decades ago: Broca's Area and Wernicke's area. [2] Broca's Area was thought to support language production. These days we know that there are two sub-areas: one is 44, one is 45. They are cyto-architectonically different and they also are thought to support different processes—syntax and semantics, respectively. There is this huge area in the temporal cortex, Wernicke's area, which, at least these days, is thought to hold the mental lexicon. In addition

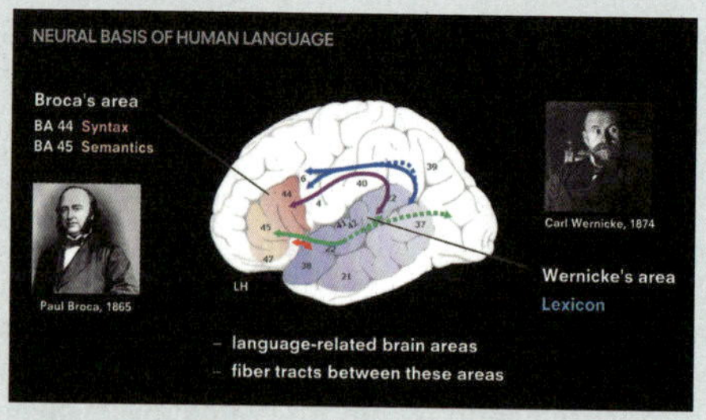

Fig. 2

to that, there are fiber tracts, white matter fiber tracts, that connect those two language-related brain regions. These fiber tracts are dorsally located or ventrally located and they will play a major role in the talk today. When considering the brain basis of human language, an obvious question is, are there differences when we compare the human brain to the non-human brain? Yes, there are, as depicted in Figure 3. [3] With respect to the cytoarchitectonic

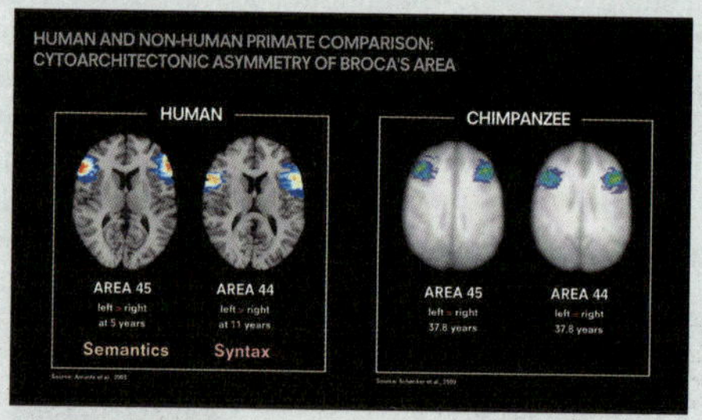

Fig. 3

asymmetry of Broca's Area, what we find is that this area is larger in the left hemisphere—which is a language-dominant hemisphere—than in the right hemisphere, and that the developmental trajectory of those two regions, with respect to their asymmetry, is different. For area 45, which possibly supports semantics, this asymmetry is already reached by the age of 5 years. However, in area 44—the area that supposedly supports syntax—the asymmetry is only reached by the age of 11 years, so much later. When comparing this to the chimpanzee, there is no cytoarchitectonic asymmetry in Broca's area whatsoever, even in the adult brain. When looking at the fiber tracts shown in Figure 4, we also see interesting differences. [4] In humans, there is this dorsal fiber tract, color-coded in purple here, and the ventral fiber tract, color-coded in green, and the dorsal fiber tract is much stronger than the ventral fiber tract. In the macaque monkey, however, the reverse is true. You see a weak dorsal fiber tract and a strong

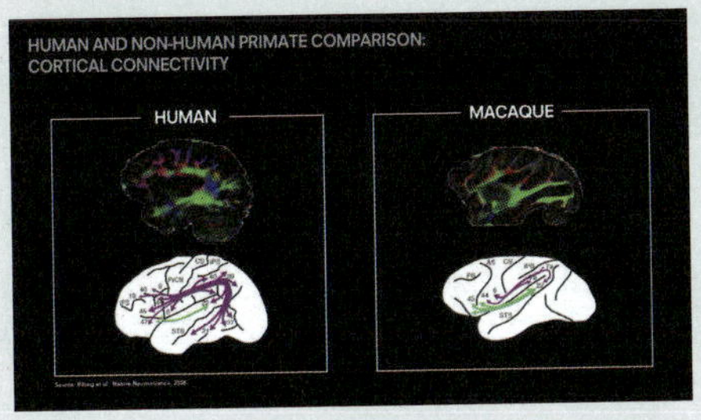

Fig. 4

ventral fiber tract. Now the question arises: what then is then really the behavioral difference between these species once we do a fair test between the human and the non-human primate, since language as such may not be a fair test? Researchers created artificial grammars that non-human primates may be able to process. First, a simple grammar, a finite state grammar (FSG), where for example every A element is followed by a B element and you see the syllable sequence here. The secondis a phrase structure grammar (PSG), which is more complex and has an embedded structure. [5] These two grammar types

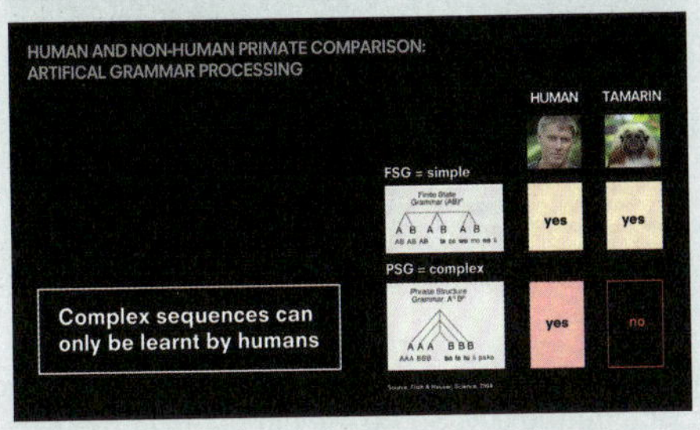

Fig. 5

were tested in humans and in tamarin monkeys. For humans it was pretty clear: both types of grammar were learned after very few exposures. However, tamarin monkeys only learned simple grammar and they never learned the complex grammar, even after intensive training. From this, we can conclude that complex sequences, even in these artificial grammar sequences, can only be learned by humans and not by non-human primates. At this point, the question arose: given this difference between the two grammars, is it also that the two grammars are processed by different areas in the human brain or by the same area? In order to test this, we took the very same grammars and presented them to humans during an fMRI experiment where we image the activation of the brain. What we found was the following: the simple, the finite state grammar, activated the frontal operculum, but not Broca's Area. [6] This is interesting because the

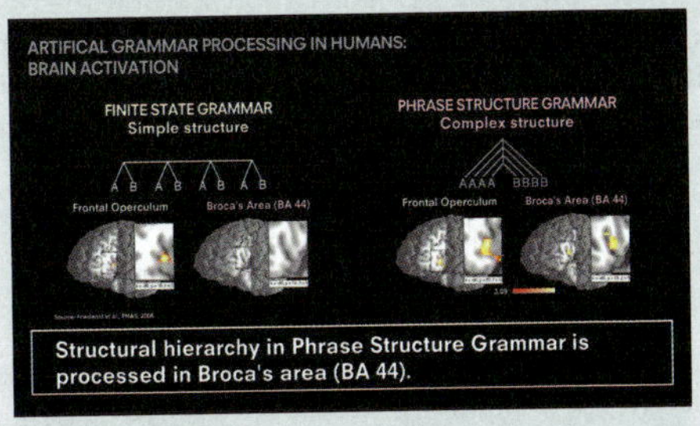

Fig. 6

frontal operculum is a phylogenetically older cortex than Broca's Area. However, when looking at the phrase structure grammar, we also see the frontal operculum, but now Broca's Area, that is 44, additionally comes into play. What this means is that the structural hierarchy of the phrase structure grammar, at least in this artificial grammar, is processed by Broca's Area 44. We also have the possibility to look at the fiber tracts in those very same participants. When seeding, in the activation spot of each of the participants, for the simple grammar—namely the frontal operculum—we found a ventral fiber tract, as displayed for the four participants in Figure 7. [7] We found that in all our participants.

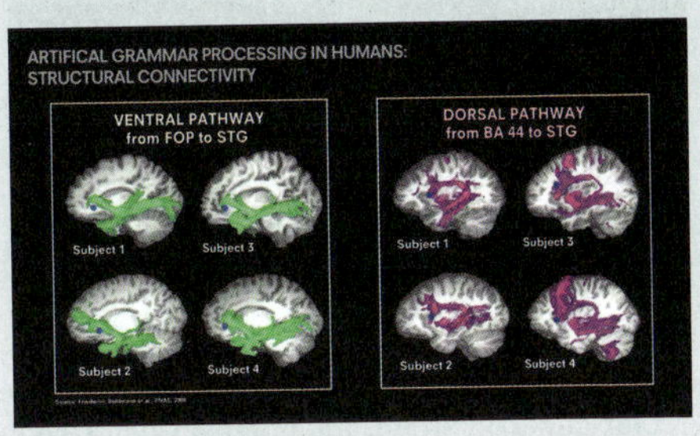

Fig. 7

Now look at the other grammar, where we had activation in area 44. Once we seed in area 44 and asked "Where does the fiber tract bring us?", we see the dorsal fiber tract connecting area 44 with the temporal cortex. This is also true for all participants. So, it appears that there are interesting structural differences between the human and non-human brain, but also functional differences for the two grammars, with respect to their activation in either the older cortex or the younger cortex. However, now the question may be: is what we have shown here, with the artificial grammar, also true for natural grammar? German is a wonderful language where you have center-embedded sentences, which result in the same embedded structures as for the artificial grammar. In Figure 8 you see how the relation between the subject (S) and the verb (V) for the different sentences and sub-sentences built up this embedded structure. [8] We also constructed longer

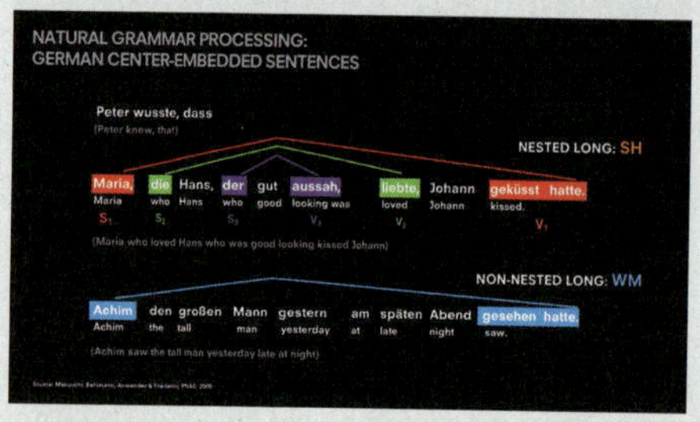

Fig. 8

sentences, where we had no embedding. This was done in order to see whether our brain activation effects are really separating between the distance between those element or the embedding itself. As expected, we found activation in Broca's Area. The color-coded orange region is the main effect, that is the activity for the syntactic hierarchy. [9] This is

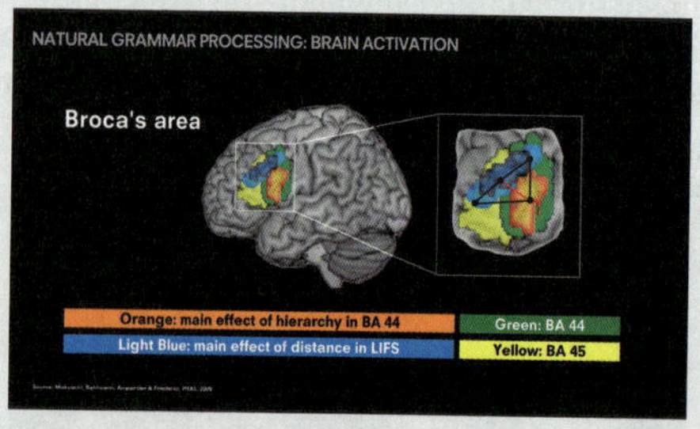

Fig. 9

located in the cytoarchitectonically, defined area 44, color-coded in green, and not in 45, color-coded in yellow. The distance effect is located in the left inferior frontal sulcus, which is color-coded in blue. These areas functionally work together when it comes to processing those complex sentences. However, what one really wants to see is that—when focusing on Broca's Area—and you make the sentence more and more complex, whether this area really increases its activity. This is what we also were able to show with respect to a German sentence, where you either have a simple structure—a canonical structure where the subject is in front of the two objects, as in a sentence like "Today has the grandfather the boy the lollipop given." [10] What you can do in German is the following: you can bring

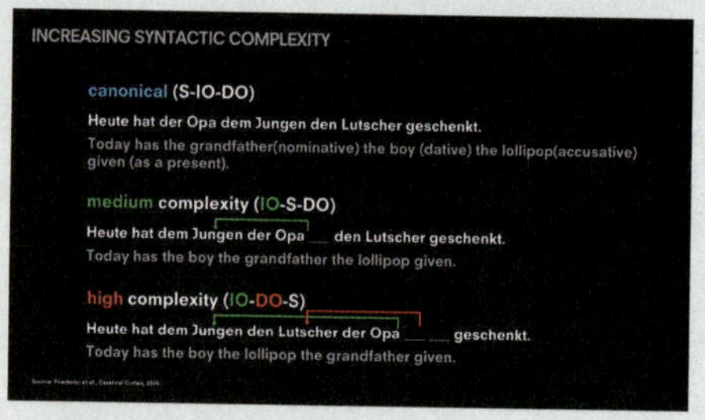

Fig. 10

an object in front of the subject, making a medium complexity. The sentence doesn't get longer, just more complex. You can do even worse: you can bring the two objects in front of the subject and then reach a high complexity. Now, looking at Broca's Area and the activation in that area (see Figure 11), what we see is that as the complexity increases, the activation increases in this area from low, to medium, to high. [11] With this, I think we can

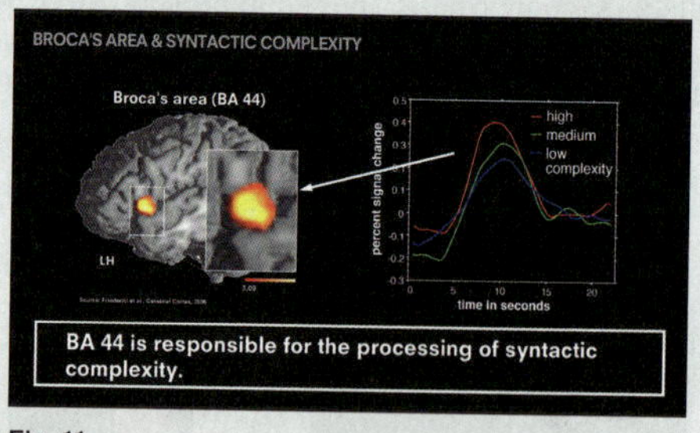

Fig. 11

say: BA 44 is responsible for the processing of syntactic complexity. So far we can even come to the conclusion that, at least for the adult brain, the function of syntactically complex sentences processing is supported by Broca's Area. Other studies, and also our own, have shown that in addition, processing complex syntax also activates the posterior temporal cortex. [12] These two regions are connected by the dorsal fiber tract, but at this point I have to say that we cannot be sure that this fiber tract is really supporting complex syntactic processes because the imaging of the white matter does not give any indication about its function. You have to use another way to test this and I will come back to that later. Before I do so, however, I would like to ask an interesting question, at least for those that speak other languages or are grown up in other languages. Is the language network universal or is it specific for each language? To answer this question we looked at different brains of people who had

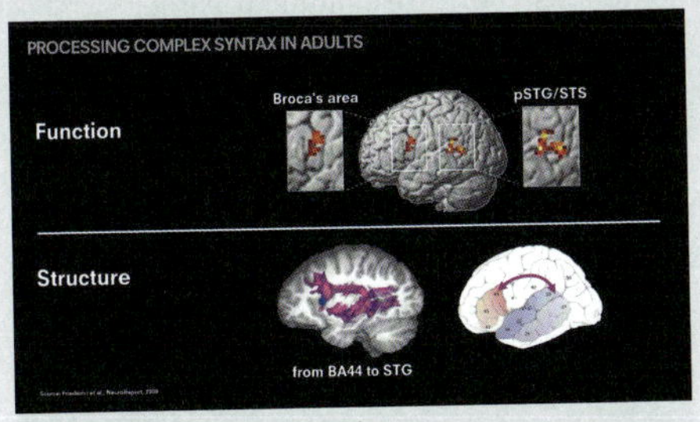

Fig. 12

monolingually learned either German, or English, or Chinese. When we summarize all the data that we have collected from those three groups—you see the known language network, with its dorsal and ventral fiber tract. [13] However, a detailed analysis

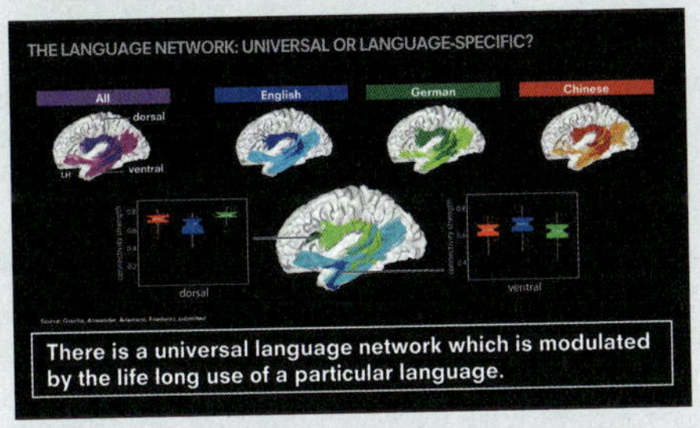

Fig. 13

of the strengths of the fiber tract in each of the three groups—German, English, and Chinese—shows that there are interesting differences between the strengths of the fiber tract. I allude here to the dorsal fiber tract, which is stronger in German and in Chinese and weaker in English. This makes sense because English is a fixed-word-order language with less complex concerning its morphosyntactic

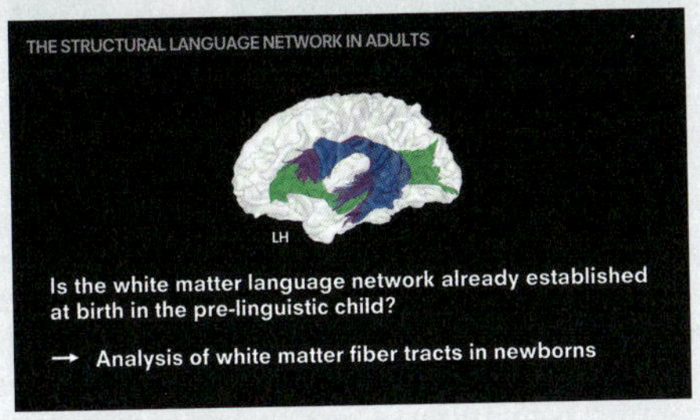

Fig. 14

elements than German. From this we can conclude that yes, there is a universal language network which is modulated by the life-long use, however, of a particular language. So far so good for the adult network—depicted again in Figure 14. [14] I would like to point out that here the dorsal fiber tract has a blue part and a purple part. This is of particular interest once we asked the question: is the adult white matter language network already established at birth in the pre-linguistic child? We had the possibility to analyze the white matter fiber tracts in newborns. The comparison between the newborn brain and the adult brain reveals some interesting differences. [15] In the adult brain we have two dor-

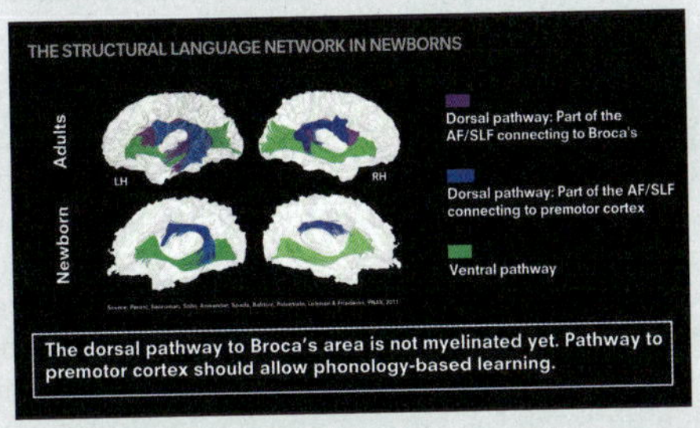

Fig. 15

sal fiber tracts, one going to Broca's Area—that is the purple one—and one only going to the premotor cortex, not reaching out that far into the inferior frontal cortex. This is color-coded in blue. Also, you see the ventral tract color-coded in green, which is already pretty well developed in newborns. Given that the way that we do our analysis is by capturing, among other parameters, the myelin of the fiber tracts, I think it is important to notice this difference. This is because once the myelin is established, the information transfer from A to B, from one region to the next region, is much better and faster. So, with respect to the dorsal fiber tracts to Broca's Area, we can say that, at least at birth, it is not myelinated. It is not functional in the sense that it is functional in adults. The next question is: what can infants do with this reduced language network? We reasoned that the pathway to the premotor cortex, which in adults supports auditory-to-motor mapping, should allow phonologically based learning. We tested this in four-month-old German infants, and those four-month-old infants had to learn Italian. Actually, they had to learn the dependency between the auxiliary and the ending. For example, "sta" "ando" or "are." [16] We used more than 30 verbs in our tests, so the children really had to abstract from the verb stem and learn the regularity. We also had incorrect sentences, in which this relation between the auxiliary and the ending was not correct, because we wanted to see: do our four-month-old infants learn to differentiate between the correct and the incorrect com-

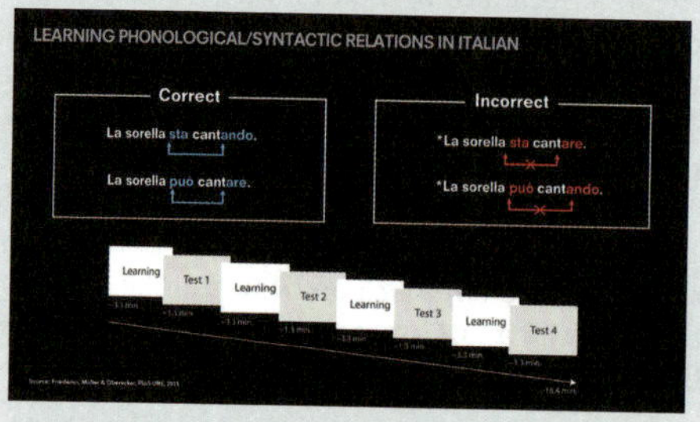

Fig. 16

binations? What we did in our study was that we had a short learning phase followed by a test phase and again a learning phase and so on. In the learning phase, the infants only heard correct sentences and in the test phase they heard correct and incorrect sentences. We recorded the electrophysiological response in those four-month-old infants and found the following: in test phase 1 there was no difference between the correct and the incorrect sentences. [17] Correct is depicted in blue, incorrect in

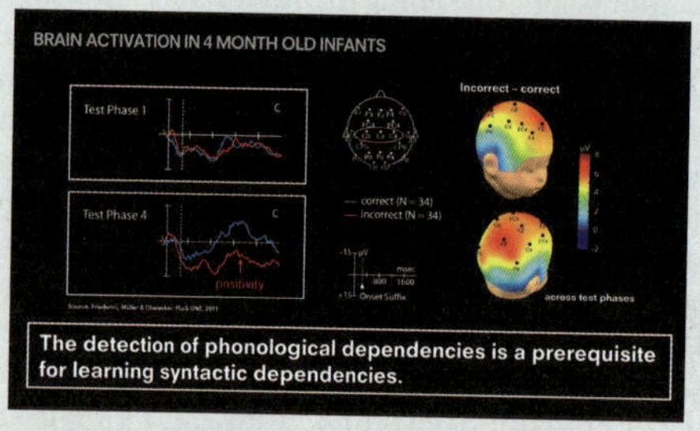

Fig. 17

red. However, in test phase 4 you see this dramatic difference between the correct and the incorrect sentences, showing that infants were able to learn this. The activation for this difference is widely distributed across the brain. From these data we can conclude that the detection of the phonological dependencies is present in infants. We take this ability to be a prerequisite for learning syntactic dependencies, because in language phonology and syntax often go together. However, the way up to learning complex sentences is really a long one from four-month-old infants to older children. What we know from a couple of studies that have been done over and over in different laboratories is the following: children deal well with very simple sentences like "The dog weefing the lion." Don't ask yourself what "weefing" means, it is a non-word. We used non-words as we didn't want to bring in semantics here more than necessary. When you now look at sentences, there are canonical, subject-first sentences and non-canonical sentences, object-first sentences, where the object is in front. What you can see is that even three-year-old children or two-and-a-half-year-old and four-year-old children are not performing above chance, indicated by the dotted line in Figure 18. [18] Performing here means

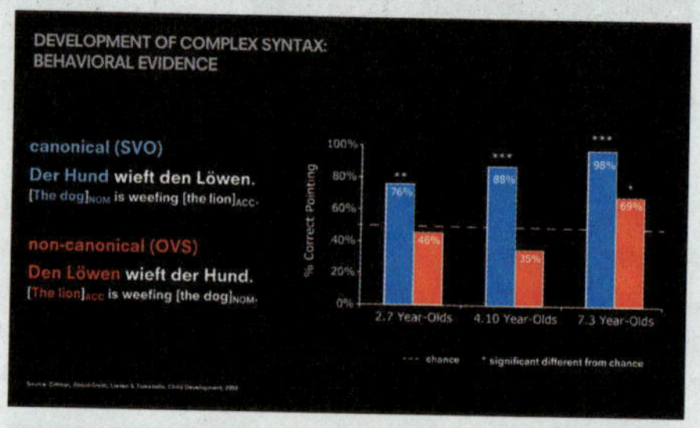

Fig. 18

they have to listen to a sentence and choose a picture which is either the correct match to the sentence or not. Only by the age of seven are they just performing above chance on those sentences, which are not long, they are just complex. Given that we have already made the statement that the dorsal fiber tract connecting to BA 44 and the temporal cortex is crucially involved in the processing of syntactic complexity, we now reasoned the following way: if so, then the dorsal pathway may not be fully developed in children. [19] Now we had to look at the

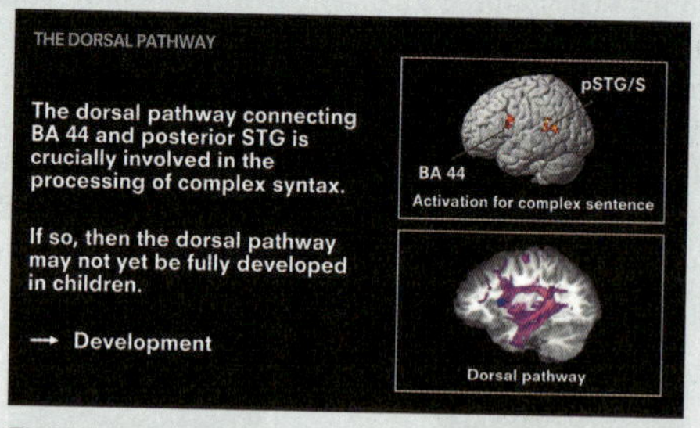

Fig. 19

entire developmental period. We did so across development in children between three and ten years and also looked at adults. [20] The dorsal fiber tract, color-coded in purple—you can see it with the naked eye—is much stronger in adults. The ventral fiber tract is pretty much established early in development already. When looking at the differences more carefully, we see that the strength of the dorsal fiber tract increases from age group to age group. More importantly, however, we found that this increase predicts and is correlated with the

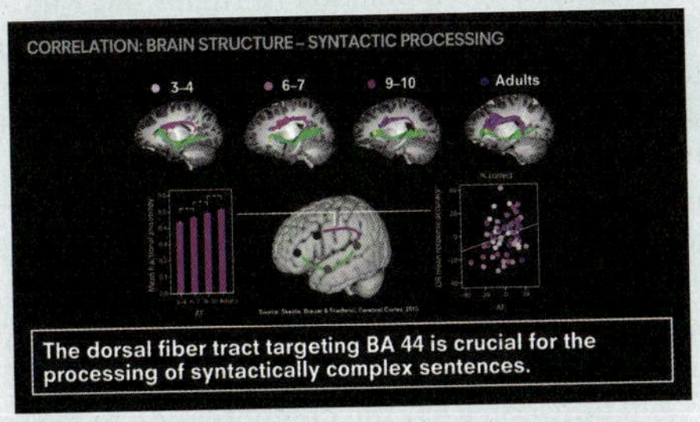

Fig. 20

children's performance in processing complex sentences. This is a significant correlation. From this, we can conclude that the dorsal fiber tracts targeting 44 is crucial for the processing of syntactically complex sentences. Where does all this bring us, with respect to our question, namely the syntax as the fingerprint of human language? [21] We have

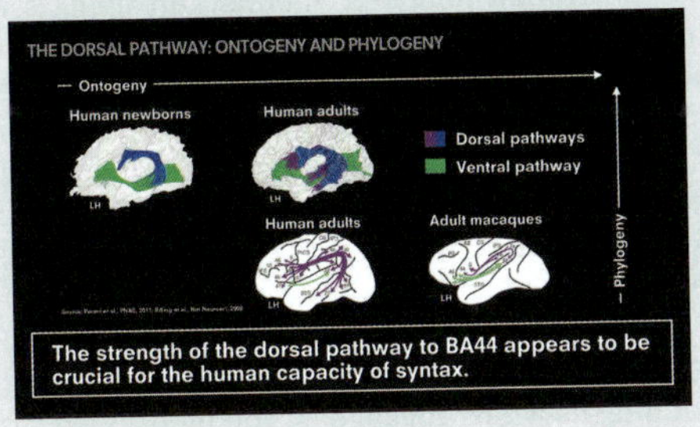

Fig. 21

just learned that from newborns to adults there is a long way, and it is related to the maturation of the dorsal fiber tract. When we look into the phylogeny, we also see that the dorsal fiber tract here plays a major role in the ability to deal with syntax. With this, I think we can say that the strength of the dorsal fiber tract to 44 appears to be crucial for the human capacity of syntax. What this means is that syntax is a unique human trait. It may be grounded in the connectomics of the human brain.

Conversation 1

ALICE ROBERTS, ANGELA D. FRIEDERICI, ROBERT C. BERWICK

ALICE ROBERTS The subject of this very first conversation is syntax as the fingerprint of human language. We are focusing on the special nature of human language comparing it with how other animals communicate and also looking at artificial forms of communication. Angela, could you please start by summarizing your lecture?

ANGELA D. FRIEDERICI Let me try to be short and bring it down to six points. The first point is that language is uniquely human. I think there is no discussion about this. In the lecture what I try to do in the second step is to show where language is represented in the human brain. There are two regions that are relevant: one is Broca's Area and one is Wernicke's area, both in the left hemisphere. The next step was to show that there are differences between the human and the non-human primate brain, in order to see whether there lies possibly an explanation for why language is specific to human. Yes, there are differences and we can come back to that later. Then I try to show that the specific ability to process syntax is located in a specific area in the Broca's region, which is the posterior portion of Broca's region, very special and very located. I see that from this area, where syntax is processed, there is a fiber tract going to another area, that is to the temporal cortex, and that this fiber tract seems to support syntactic processes. Why do we think so? We have shown that in children this fiber tract—the strength of the fiber tract—predicts the ability of a child to process a syntactically complex sentence. Interestingly, this fiber tract is not there at birth, it is not myelinated yet, and it develops and matures over time. So in the end, when we compare the adult human brain and the infant brain, the human brain and the non-human primate brain, what we find is that both have a very weak fiber tract from Broca's Area to the temporal cortex. So this fiber tract may be important for the human capacity of language.

AR Bob, at this point, would you like to ask Angela a question? I know the two of you know each other and you know each other's research but based on specifically the material covered in Angela's lecture, have you got a question for her?

ROBERT C. BERWICK Since we know that, as she showed, children grow up and develop these fiber tracts so that they can acquire language, the one question to ask is: does—as they put it—ontogeny recapitulate phylogeny? Has evolutionary history made it such that it doesn't happen in other primates or other animals but only in people? And that was I think an important question to try to figure out: why are we special?

AF Well, the question is why—we would like to know. The interesting point is that it seems to develop across the lifetime, so the fiber tract I am talking about, this special one, is the only one. There are several fiber tracts connecting the inferior frontal gyrus and the temporal cortex, but only this one that supports syntax is maturing as the child grows.

RB We only find that—so far as we know—in people and not really fully developed in other animals and other primates.

AF We know about macaques, we know even about our closest relatives, the chimpanzees. The chimpanzees do have this fiber tract but not as strong as the human, and in particular it is not clear whether it targets Broca's Area in the same way it does in the human.

AR It is fascinating to think about how much of this kind of programmed development is based on the culture that is around us. I know that is something that you look at in your research, Angela, and we will get into that in more detail later on in the discussion. Before we do that, Bob—please could you summarize your lecture as well?

RB I wanted to focus on the differences between the modern computers and what they do with language, because unless you have been hiding underneath a rock you have probably read about how much computers have been able to do with language and try to answer questions, and you probably interact with them on the phone every day and you don't know it. What I wanted to point out is that there is some very simple things that kids that are only two or three years old do—for example, if I say "pick up the second green ball" kids will invariably learn how to do this by the time they are three or four years old. But the current computer systems we have don't seem to be able to grasp that, and I think the central reason for that is that people have focused on the computer systems in getting them to kind of mimic the input and output of behavior of what people do, but they haven't looked inside the black box to figure out what is inside the human brain that is actually going on. They don't reflect that, they don't reflect the fact that in human brains we assemble a kind of structure that goes along with a sentence like "pick up the second green ball," and the computers haven't done that at all. As a result, they fall down on the job when it comes to these very simple things that kids can do. That was really the essence of what I talked about in the video presentation: there is a big gap between what the computers do and what people can do. What we do, what Angela and I do in language science and in brain sciences, is to try to figure out what is special about people and the computers so far haven't done that.

AR Bob, I found your lecture absolutely fascinating. I think it will come as a surprise to a lot of people that, when we have got AI trying to produce language, essentially what computer scientists are doing is, as you said, recreating the output—but not doing it in the same way that our brains do it. We can then see some of the failings in the way AI tries to understand us, because we are working at a different level and we are working in a different way. Angela, would you like to ask Bob something at this point? Then we'll develop this into a fuller discussion.

AF I found really instructive the way you were teaching us what language is. You mentioned that obviously there are sort of little hierarchies that are built once we process language at least in the human brain. I wonder why it would not be possible to implement that into the computer. For example, take the computation "merge." When you build such a hierarchy you could say "There is one element, A, and there is one element, B, and the one is a determiner and the one is a noun." Then the computer could learn that this is a noun phrase, so what does the computer do with it? Is it a problem that the computer cannot build the hierarchy or that the computer cannot build this hierarchy over and over and over again to make more complex sentences?

RB It is a great question.

AR Bob, before you answer it, explain this "merge" function because that was fairly crucial for the lecture. Can you explain that and then explain how the brain is doing these hierarchies before you launch into answering the question?

RB Sure. If Angela would permit me, Alice. All we were trying to say is that if you take something like "the" and "ball" and put them together to form "the ball" then it forms a kind of triangle, a unit. I can take that and use it in other places and other sentences. It forms a kind of chunk that looks like a triangle. The reason why the current computer systems don't do that is that they have decided not to adopt that approach. The computers are programmed by people outside, and people have decided on one of two ways of doing things. One way is this way of reinforcement learning, like training a pigeon over and over again on example sentences. You give it an example sentence and you expect some kind of output and if it gets it right you reward the computer and you say "Okay this is what we are going to do." That is the way a lot of these current neural network systems work: they are essentially processed on billions and billions of examples from the web and that is what they do. But that is not the way kids acquire language and it is not the way language works inside our heads. There is a way to do it and some people have tried to program it that way—some of my students and many other students—and you can get part of the way doing it but it involves a lot deeper understanding, it involves unpacking the black box and knowing what is inside our heads, and trying to figure that out from evidence like the one that Angela has gotten. That is just a lot harder, it is like ordinary science. It is like physics proceeding over 50 or 60 years and trying to do slowly experiments that tell us what is really going on. There is no reason a computer can't do that. Computers can be programmed to do whatever they want. There is one difference in the computers as opposed to the human brains which is, if I might speak broadly, as far as we know the human memory system works a little bit differently from the way computers do, just because of the way you try to build things in silicon as opposed to using neurons. The brain seems to address things in terms of their content. If I think of a house then I might think of all the things that are related to a house and that is the way I try to remember things. You remember the lane you walked down when you were growing up as a child or something like that. That is called content addressable memory. Modern computers don't really work that way and that is part of the reason that with modern computers people just haven't tried to really build something that really mirrors the brain in that way. There is a lot more that could be said there but that is one of the basic reasons: it is just a lot harder, it is hard science to try to figure out what is inside.

AR It is interesting that what we seem to have at the moment in terms of AI understanding human language, let alone generating it, is rather like in your lecture when you talk about chimpanzees and their understanding. When they are taught human language they are able to produce words in sign language. But the way they string them together just doesn't seem to be the way that we create complex sentences. You talked about a particular chimp, who I think is called Nim Chimpsky.

RB They tried to train Nim Chimpsky since birth in American sign language—Laura Petitto was one of the trainers—and they would go through the signs but it was also this kind of reinforcement learning, where they would have Nim make the signs until he finally said something like "orange" as a sign and then said "okay now I know what to do to get an orange." Then Nim would learn how to string those kinds of signs together like "Give me orange, give me, give me, give me." There were a lot of "give me" and he liked oranges. But we don't see that in children, they don't do that at all. Of course, they do ask for things but they very quickly acquire the basics of syntax even though, as Angela said, not all the wiring systems are in place at birth but they still work along very different lines than that. They very quickly master some of these things like "pick up the second green ball" and Nim never learned what little words like "the" or even what "pick up" or even "apple" really means. He never got there. There is something there that we have that they don't have and I said in my lecture that Von Humboldt in the 19th century said that people make infinite use of finite means. What that means is that there is infinite creativity. It is true, we say a lot of things, all the time, that are the same like "okay" and "have a nice day," but we also say things that are completely different so we say "There is the purple triangle that is off to the side of the street, you shouldn't trip over it" or something. You never heard that sentence before but we are endlessly creative that way.

AR So, what the chimpanzees are doing is copying sounds and wanting to get a desired effect, whereas humans are being genuinely creative with that brain structure?

RB Genuinely creative and producing novel sentences all the time. That is one of the things that the so-called Cartesians in the 17th century immediately noticed, that separated what they called people from machines, that automata or machines could just do this kind of route back and forth and never do anything really new but it took real human creativity to do that.

AR "I think therefore I am," or "I speak therefore I am."

RB That is right. It should be rather "I speak therefore I am."

AR Bob, would you like to throw a question back to Angela?

RB There are a thousand and one so many interesting questions to come up about this idea, about where this is all located in the brain. One of the other threads of analyses that keep coming up is: since the brain is this kind of electrical engine, do these electrical signals—besides the fact that they are just moving around in fiber tracts—have any sense where there is wave-like patterns or other sorts of patterns, or is it very local?

AF That fits quite nicely together, the wave forms you were talking about, or the oscillations that the brain does all the time, because we have seen that during sentence processing. There are these two areas, Broca's Area and the temporal cortex, that oscillate together once you process the sentence. The question is, what do the fiber tracts have to do there? The fiber tracts are there to bring the information from A to B, and that is the only way that the electrical signal can travel fast. It has to travel fast because language takes place in milliseconds.

RB Right. That is the reason the fiber tracts are so big, like the big trunk lines on the internet, right? They have to carry lots of information.

AF The myelin wraps it in a nice way so to make the information go even faster and I think this is very important to keep the information interaction going between those two areas which are pretty far apart.

AR I think we should probably talk a little bit about your research looking at those fiber tracts, Angela. We have known for a while that there are areas of the brain that are specialized for certain functions, very often because people have had those areas of the brain damaged and have lost a specific faculty. But what you are doing is looking in real time using functional MRI scanning to see where activity is taking place in the brain as people are performing various tasks. You are not only looking at the cortical areas—the areas where the brain cells or the bodies of the brain cells are—but you are particularly interested in these tracts, which are the wiring essentially of the brain.

AF We cannot say that the particular function is "in" the fiber tract, because we can only define the target points from where to where those fiber tracts go, because the electrical signal in each tract is the same. Yes, we look at the fiber tracts with respect to their target points. I think it is very interesting to see that fiber tracts can lie in parallel and go from A to B, but then in A they split up in two different areas and those two areas have different functions. It is very interesting to look at those fine-grained distinctions because otherwise we would not be able to pull apart for example syntax and semantics.

AR I think it is just astonishing to be able to visualize what is happening in somebody's brain when they are either listening to or producing language. Here is a really great question from our audience: "Is there any study that is been done to observe the brain's syntax processing when reading or listening to the same sentence?" Do we know the difference? Angela, that is probably one for you.

AF When you listen to the same sentence over and over again as one person, the activation in the particular areas go down. When you have it between subjects yes, you can see that even different subjects have different activation levels. But this has to do with the fact that the brain activation is not the same in every brain. You also have to account for the variances between the subjects, if that was your question.

AR I think he was asking—if I were to look at a sentence and read it in my head, and if I were then to read that sentence out loud—what would the difference be in terms of the brain processing?

AF I could not say what the difference would be because there are so many factors that are playing with it, but Bob, maybe you have an answer.

RB Certainly there is more than one process. When you are reading there is this additional process of decoding the little scribbles that are on the panel, even though nobody uses those anymore. At least the writing that is on the screen. Then that has to get mapped into whatever the actual sound system is for the language, so there is an extra step in there and so there are many other things that happen. I also wanted to add that it is actually remarkable when you compare what the modern AI systems are doing. They call them sometimes "deep neural networks" but the word "neural network" is really a throwaway line, because they are in no sense like actual neural networks, which are very noisy and, as Angela knows, they are very different. They don't have the properties that the human brain has.

AF I have to add here the following thing: when we talk about language we talk about spoken language and not about reading. Reading is clear. There the visual areas are active. But for the language processing it is the same for the same subject. Also, I think it is very interesting that you obviously internally also encode prosody during reading. For example, when you present a sentence with an intonational break, say, a pause after a phrase ends, you have a certain effect in the event related brain potential. When reading a sentence, you have the same effect once you see a comma, so it appears that during reading you internally obviously encode also the underlying prosody.

AR So there is punctuation in there as well. Angela, I think you had a question for Bob about second language learning?

RB That is always a question that arises because kids learn languages so easily and we all know of bilingual kids so if you grow up in an environment where there is both French and Spanish and English and German, then kids seem to have no difficulty at all. I remember my old friend Morris Halle, the late Morris Halle, who founded the modern theory of phonology here at MIT, he always said "I knew five languages because I was kicked out of four countries." People do that very easily when they are kids. You have to be a little bit attentive there, because there is the process of what we think kids are doing and then how hard is it actually for them, because that is a different question. Maybe they are actually doing a lot of computational processing we just can't see, and so it seems easy to them. One of the reasons for that, which my former student Charles Yang has looked at a lot, is that when you are a child you are not dealing with so many examples in so many sentences, and it turns out it is a lot easier to make rules—to make the rules of your language when you only have a hundred examples to look at than when you have to look at the entire internet. You can imagine if it is the entire internet: there is all this junk on the internet, so you get much more confused. There is something to do with the fact that you are only getting a limited amount of data early on with a limited number of examples and then that can grow over time. As Angela said, what we know now is that this process of maturation continues over a longer period of time. I think that is a really interesting question to raise, whether adults have as much trouble learning a second language. There are two things going on there: one is you have to clarify the question and say "well, how hard is it really? What are the adults really doing and how hard is it?" because it might look like it is hard for them phenomenologically, that it is really hard and that might be just because they are overwhelmed with all the other examples they know from their native language, from the first language. That is always been hard to tease out. But I think the bulk of evidence we have now is that this ability to learn, acquire new languages, should actually continue into adulthood but it is overlaid with all these other complicated processing factors and other kinds of things of this sort and that is what makes it hard in part. We don't have a complete answer to that, so I would be the first to say we don't really know.

AR It is interesting that you are saying that you think it is more to do with the environment.

RB And the fact that when you are a child you are in a more limited environment and you are getting more limited examples and you can prove that it is a lot easier to get the rules. If I just give you 30 examples, then it is a lot easier to make the rule up than if I give you a million examples and say "figure out the pattern here." That is a lot harder. We just know it is easier and another person who's worked in this area, Elissa Newport, who's looked a lot at kids acquiring language, has found that there is a sweet spot between ages of like four and seven, when kids seem to really make up these figures out, what are the rules of syntax in a language. I think it has to do with an interplay between the number of examples they are getting and how plastic their brains are, whether they have matured enough and so on, which is not to say that this can't go on, I think it actually can go on. We used to have an old idea that it stopped at puberty, because at puberty there are all these strange things that happen. One of the things people used to think is that second language acquisition became very difficult after that point, but I think that people have backed off of that now and realized.

AF I was wondering because at this point I think we should add there also a neural basis that is different in early versus late learning and that has to do with that. Once you learn one language, the brain is tuned towards that language, because the synapses are not only deleted for those processes that are not necessary anymore, but they are hard to recover later. So once you learn one language, the brain is pretty fixed. However, once you learn more languages than one in early days, maybe the first two or three years, then the system remains open and it is also open for other languages.

AR Is there evidence that people who grow up speaking several languages are able to more easily acquire another language in adulthood than those who have always been monoglots?

RB That is been very well tested that the more languages you grow up with the easier it is.

AR We have got some other questions from the audience: "Are there substantial differences in language between males and females?" Who would like to start with that one?

AF Let's put it this way: I think there is no language difference. There is a difference in communication, so there are non-language aspects that come into play and the only ever found difference has to do with emotional prosody, which is not part of language but it comes with the way you say a sentence. There we found that women are much more sensitive to that and react faster. We have shown that in ERP experiments, where we also look into the brain activity with an EEG millisecond by millisecond, and this is how we could find it out. Males also process this but a bit later and that may also give rise to problems in communication.

AR Do you think that that is an innate difference between the sexes or do you think that it is something which is essentially acquired through culture?

AF You asked exactly the right question. It is not innate because once you train the males to pay attention, early attention to this information, they can use it and they process it.

AR Bob, what is your perspective on that?

RB It is actually the case with most of these biological things that the individual differences from person to person—whether they are a man or a woman or whatever—tend to be greater than the differences than when you look at the whole population of women versus men. So the individual variations are always much greater than between groups, between entire groups. That is true of almost everything we found in biology. Human variation is widespread but it is localized to more individuals than anything else.

AR Here is a great question from our live audience: "Which animal, if it were possible, would be able to simulate the phonetics of human language best?" The parrot, they suggest.

RB That is a good suggestion. There are lots of them. Parrots are really great at mimics, doing that kind of mimicry. That would be the one I would come from. The interesting point about that, and of course some other birds that are actually exquisite mimics, is that and we keep finding other animals that can do this sort of thing. In Massachusetts there was a seal, a harbor seal that was raised by a New England lobsterman, and it had the accent of a New England lobsterman which was very funny. It said "Hey what you doing?". The point about it is that there is an interesting difference between the parrot mimicry—because there are these birds that can mimic what we are saying—and what true human language is and it is the following: unlike the parrots, if you take a human who's, say, a vocal language and you tie their mouth like that, they will still try to express through language, they will use their feet or they will start doing something else. The parrot won't do that, it won't change the modality of what it is doing to try to communicate. It doesn't go across that. If you tie my hands and feet down, I'll still try to do something to try to communicate. The parrot won't do that. There is something very special about human language and its expression that is so powerful and so part of our species that you can't get away from it. But I would agree that parrots would be great if you wanted to try to see how far you could get.

AR Have you looked at any parrots, Angela? Have you done any functional MRI scanning of parrot brains?

AF No, I do not know of any of these sorts of experiment but maybe they are done and they are not published. That could be.

RB You have to realize also that with the parrots if you are just talking about the sound system, the actual sounds many of the people who are watching might know, there is also of course sign language, many different sign languages, hundreds of them, actually, where you can use gestures instead of actual sounds. Those are perfectly fine human languages, and as far as I know we don't have any other examples of other non-human animals that do that, that can actually make the gestures. We tried to teach Nim but that was hopeless, it was all just stimulus and response, just memorized training. There is no other animal that does that. You have to ask that question about why, if you don't have the ability to speak, you can still acquire language and still be a full-fledged functioning language user. That doesn't happen in any other animals. I think that is an important point to keep in mind. It is so part of our being human that it just has to come out in some way. And we know that. If you aren't able to use language that is a tremendous problem and, as you know, if you are both blind and deaf you don't get any input, right? Nothing. But you can still acquire language. There are several methods to do that: one of them involves touching the face so you can feel the vibrations, and some people just feel the vibrations on your hands as you are touching.

AR It is just extraordinary. It is almost a fundamental need, isn't it?

RB Yes, but the important thing is that it doesn't have to do with the external expression, with the sound or the visual presentation. It is something that is inside that actually has to get outside. In literature we call that externalization: there is something inside your brain that is the actual language that has to get out there and it does.

AR We have a need to communicate with other humans. Now, Bob, you had some interesting questions for Angela. I think you had one about changes in the brain that may have been linked to the evolution of language in humans.

RB I think I already asked about whether we knew of any other animals that came close to those kinds of abilities. There were changes in the neocortex that had to grow in size or other concomitant changes that had to take place before language. Did we have a language-ready brain? Did we have to just get a big enough neocortex so that we could have lots of words? It is a very general question.

AF There may be specific answers. When we, for example, look into the evolutionary step from the macaque, who is no more far off from us than the chimp and the humans, it is an interesting thing that the macaque only have two temporal gyri in the temporal cortex, we have three and the chimp also has three.

AR These are the little folds in the surface of the brain?

AF Yes. So the human temporal cortex has one, two, three fold-like sub-regions, called gyri. This is not the case in the macaque, the macaque only has two. The middle temporal gyrus, that is the one in the middle, only present in chimps and in humans, seems to have a special role, because when we talk about the fiber tracts, the tract that comes from Broca's Area goes into the middle temporal gyrus, but only in the human and not in the chimp. So not only there is a gyrus that comes in addition, but it doesn't seem to be used in the chimp in the same way that it is used later on in the human. I think that gives you an idea of how evolution proceeds. I mean, you have new brain areas and then you add new functions to the area.

AR So the chimp has got what looks like the apparatus but it is not using it.

RB Missing something. That of course raises a question. It is too bad we don't have any detail, or I don't think there is a way to get this information about, say, Neanderthals and whether they had that same kind of organization or not. Maybe there is some way for taking casts of the interior of the crania but I bet not. Maybe there is.

AR I think it is so difficult—you can see the size of the cerebellum versus the cerebrum, but in terms of individual folds on the surface of the brain, I don't think you are going to get that detail unfortunately. Until we find a frozen Neanderthal in the permafrost of Siberia that question is going to have to wait! I have got another question here—"Do the differences between Jean Piaget and Lev Vygotsky's ideas about language and cognition—especially among symbolic tools of children—still stand up today?"

RB Vygotsky and Piaget both had ideas about symbolism which of course is very important for language, because language inherently involves symbols. But I don't think a whole lot of scientific progress has been made on that and the reason for that is that we don't have very good ideas about how symbols are even represented inside the brain. I think Angela would agree, not in terms of the neural connectivity or specific wiring or how that is even done. So that is what has really stopped us there from thinking more deeply about how closely they might have been right or wrong. It is certainly the case that symbols are important, that is why for example Nim could never figure out that an apple stood for lots of different kinds of apples, so it was this symbolic apple. Nim never got that. But the question is why. I think it is still open as to whether someone like Piaget was right about that. That has to come about first. Certainly, just to go back a little bit on some details about Piaget, he seemed to—as far as I have read him, it sounds like my questioner might know a lot more than I do about that—he thought that the symbols had to be there before the language could arise, but what is interesting is that there are been lots of recent research going back to ages as far back as we can push it—to ages two or even 18 months—where we can show that this kind of hierarchical structure that I talked about in my lecture, this building of little triangles inside the brain, can take place without really knowing symbols about nouns or verbs. What is known is simply that. You take "the," what is "the"? That is not a symbol, it just means. If you just know those little bits of what is called morphology in language then it turns out that that is enough for the kid to grab hold of, at least initially, and they might use that to figure out what nouns and verbs are. That is a rather surprising thing. Even that they could actually be sensitive to this that early.

AR I find it astonishing that you are able to work that out. It does sound, for the time being, that the ideas of Piaget and Vygotsky must remain hypotheses for now—and at some point, perhaps, we will be able to test them and our science will advance to that point. I am going to have to draw things to a close, but just very briefly from both of you, can you tell us what your burning research questions are right now?

AF What we really would have to do is to look at the development in the non-human primates. Is everything there in the beginning once they come to birth and they don't develop any of the brain structure similar to us? Then we also need some studies to look at what syntactic knowledge is available in the brain early on. We know by the age of two there is some, but the question is: can we say more and more specific things about this?

AR Thank you very much, Angela. Bob, how about your future research?

RB A lot of them tell what Angela was saying, which is that I would really like to have an honest-to-god computer representation of the way merge works down to the level of neural circuits, so that we can actually figure out how to bring that over to the testing that she does. I don't know how to do that yet because we know so little about the way individual neurons are put together and the way the memory system of the brain is hooked up. That is one of the things I am involved with now—trying to figure out some of those differences.

2

Hemispheric Lateralization and Attentional Mechanisms

PAOLO BARTOLOMEO
GUIDO GAINOTTI

MODERATED BY ALICE ROBERTS

Clinical Aspects and Theoretical Implications

GUIDO GAINOTTI

With the term hemispheric specialization, or hemispheric lateralization, we mean a series of phenomena for which some highly complex functions are mainly carried out by either the right or the left half of the brain. The most well-known of these is language, that is mainly performed by the left hemisphere, while, for example, orienting the spatial attention to objects lying on the left half of space is subsumed by the right hemisphere. If these are fairly consolidated data, less clear is the mechanism behind them.

Today I am going to first illustrate the clinical aspects of the phenomenon of spatial hemi-inattention and then to propose a theoretical basis for the existence of this kind of specialization of the right hemisphere. Typically, over two thirds of patients with left damage have language disorders, while just over a third of those with right damage show inattention for the left half space. [1]

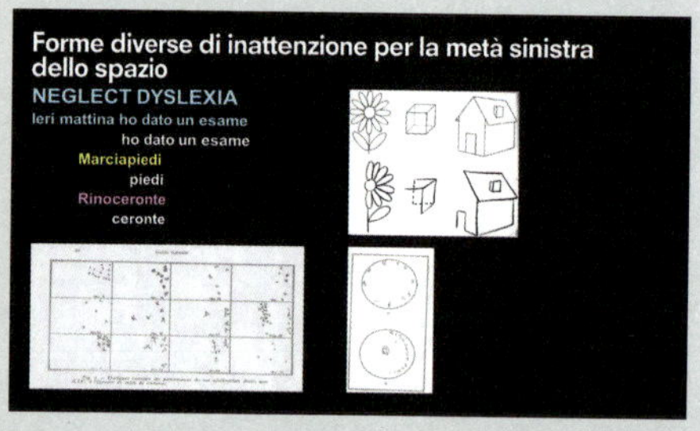

Fig. 1

This attentional deficit not only has an environmental impact, meaning for example that patients in a wheelchair can bump into obstacles on their left or, paradoxically, they can only eat the pasta in the right half of the dish, but it also impairs various cognitive functions with a visual-spatial component, like reading, exploring a visual display, gathering information from the left half of stimuli on which they focus their gaze and so on.

This disorder does not affect all the components of spatial orientation, but much more the automatic, involuntary orientation of attention, than the intentional, voluntary attentional processes. In slide 1 some situations are shown that highlight this aspect of hemi-inattention for the left space. For example, during reading tasks, the sentence "Yesterday morning I took an exam" could be misread simply as "I took an exam," the word "Sidewalk" could be misread as "walk," or the word "Rhinoceros" as "ceros," without the patient being surprised by the fact that the word "ceros" doesn't exist. Another activity that shows this inattention is observed when the patient is reproducing simple images, such as a daisy, a cube or a house. In all these cases, the patient only reproduces the right half of the daisy, the cube or the house, leaving open and empty the left side of these elements. Another situation in which we can study the spatial hemi-inattention is the reproduction of crosses randomly scattered on a sheet of paper. Again, the patient either copies just those placed on the right half of the sheet or stacks on the right half all the crosses that were scattered on the sheet, completely neglecting the left half of the sheet. One last example: copying a clock dial, the patient puts on the right all the numbers of the hours, completely neglecting the left half of the dial.

Another situation in which we observe quite peculiar behaviors, is asking a patient to copy a complex figure that shows a series of different components both on the right and on the left side. [2] In this case,

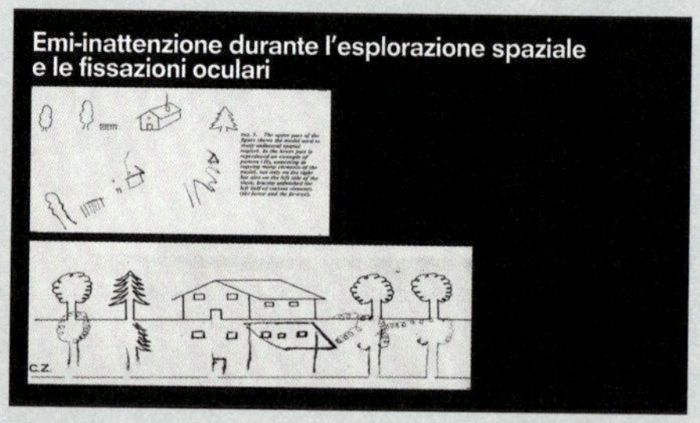

Fig. 2

we observe that the patient tends to copy all the components of the series, seeming to be aware of the left half of the spatial display and to explore each part of it. However, we see that the left half of each element on which attention was fixed is neglected. This suggests that in the hemi-inattention disorder the deficit mainly regards the automatic rather than the intentional components of spatial exploration. This distinction between automatic and controlled attentional processes had been made in the late 19th-century by James (1890), who distinguished an automatic, reflex, passive, exogenous attention (attracted by external stimuli), and a controlled, intentional, active, endogenous attention (dominated by the active intentions of the subject). This model, that allows to consider hemi-inattention as mainly due to a deficit of the automatic components of the spatial orienting of attention, is confirmed by observations which show that in these patients the stimuli located on the right side prevent from orienting the attention towards the stimuli located on the left side. Let's see what these phenomena are. One is the so-called magnetic attraction, observed in the study of the visual field done with the confrontation method. This exam of the patient's visual field is done in this way: the examiner, in this case the lady on the screen, asks the patient placed in front of her to fix his gaze on the tip of her nose so as to be sure that there is no movement of the gaze that can disturb or alter the test. [3] Then the examiner raises her hands

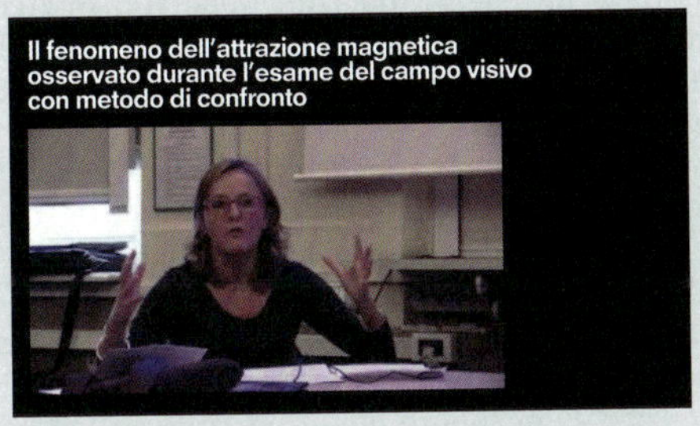

Fig. 3

to the level of the patient's eyes and moves the right or left hand alternately asking the patient to point to the hand that she moved. Obviously, if the patient sees always the hand moving on his right and does not see the one moving on his left, this shows there is a deficit for the left visual field. This is a context in which we observe the phenomenon of magnetic attraction, which consists in the fact that as soon as the examiner raises her hands at the level of the patient's eyes and before she moves one of her hands, the patient automatically orients his attention towards the hand placed on his right, as if the presence of a stimulus on the right automatically attracted his attention.

Another situation in which we can observe phenomena going in the same direction concerns the experiments in which the presence of distractors placed on the right increases the hemi-neglect for left-sided stimuli. [4] This phenomenon can be observed both in paper tests, asking the patient to cancel the stimuli placed on a sheet, and with computer touch-screen tests. To best understand this phenomenon we must remind that in normal subjects the visuo-spatial scan pattern starts from the left and then goes to the right, due to a generalization of an auto-

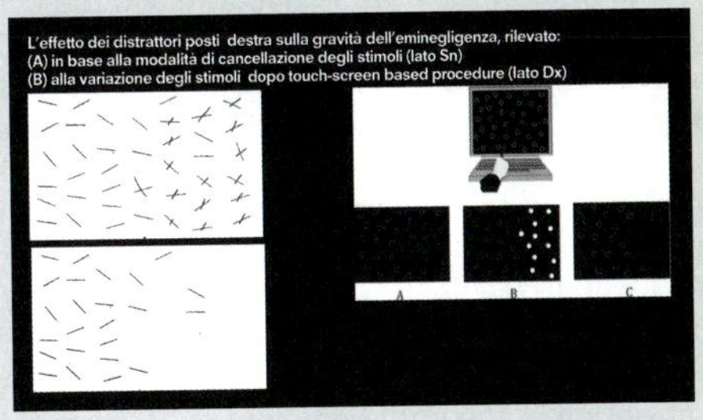

Fig. 4

matism that is acquired, in our Western culture, during reading learning processes. Unexpectedly, neglect patients usually reverse the scan patterns: they start the exploration from the right and move more or less painfully to the left. So, what happens? If right-sided stimuli are canceled with a pen, they remain in the right half of space and continue to attract the patient's attention. If, instead, they are physically erased from the sheet, the attention is no longer attracted by the stimuli on the right, and can proceed more freely towards the left half of the space. A similar test can be done on the computer, as shown on the right half of the slide. This distinction between the inability of the automatic (but not of the intentional systems) to orient attention toward the left half of space, (which justifies their left hemi-inattention) is also consistent with neurophysiological data referring to the bundles of fibers that are important in the human brain for the spatial orientation of attention. This slide is drawn by a paper by Bartolomeo & Seidel Malkinson (2019) and shows the complexity of these bundles of fibers. [5] The upper bundles of fibers belong to the

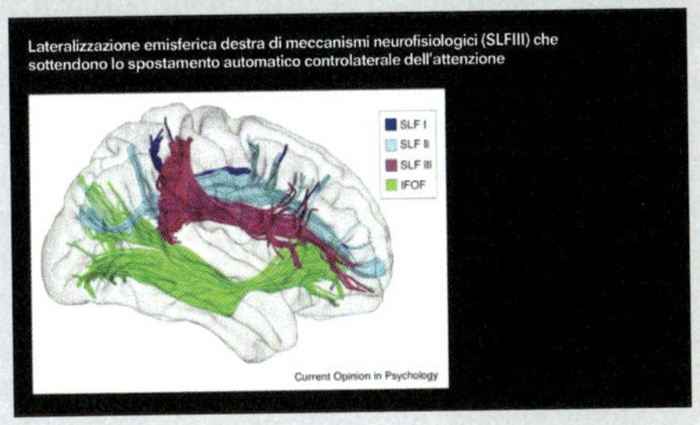

Fig. 5

upper longitudinal bundle that has three branches: first, second, and third branch, while in the lower part we see the lower frontal occipital bundle. The upper dorsal part of the upper longitudinal bundle is for voluntary orientation of attention in all directions of space and is equally represented both in the right and in the left hemisphere. On the contrary, the third branch, that is for automatic orientation of attention, is mainly represented on the right. These aspects of the problem will be dealt with in a much more exhaustive way by Professor Paolo Bartolomeo, so I will stop here for what concerns the strictly clinical and anatomical aspects of the problem. I will instead quickly analyze some assumptions on the possibility of a link between left lateralization of language and automatic orientation of attention in the right hemisphere. What is this hypothesis based on? It is based on assumption that the presence of language may influence the general organization both of thought and of brain processing in the left and in the right hemisphere (Gainotti, 2021). First, it could affect the level of cognitive processing, because the presence of language allows much more complex and abstract processing skills than those observed in absence of language. Second, the level of consciousness, because language is very important to gain awareness of what is happening in the environment and, even more, in our representation of our body and of the world. Third, the level of intentionality, as again language has a key role in allowing us to keep our direction of thought when we intend to perform an action. Several data show that cognitive functions prevail in the left hemisphere, while in the right one there is a prevalence of emotional functions. Many theorists agree that cognitive and emotional functions are the components of two different adaptive systems: the cognitive system is used to process in a complex and exhaustive way the environmental milieu, whereas the emotional system is considered as an emergency system, mainly used to quickly solve situations that can be critical for survival. As for the importance of language in conscious processing, Morris et al. (1998) have shown that processing of emotional stimuli, is conscious in the left but unconscious in the right side of the brain. Another situation showing that consciousness prevails on the left is a phenomenon observed for the first time by Babinski in the past century: that patients with right hemispheric damage having a left hemiplegia can be totally unaware of this. The word "anosognosia" means "lack of awareness" of the hemiplegia that has struck them (Babinski, 1914). Finally, the prevalence on the left of disorders regarding intentionality and voluntary movements, while the right affects automatic movements, can have a series of aspects. For example, in language disorders and in apraxic (=movement) disorders, due to damage of the left hemisphere, the deficit mainly regards the voluntary components rather than the automatic ones, of the action. Thus, an aphasic patient can be unable to name the picture of the "Virgin Mary," but can easily say immediately later, the word "Virgin Mary" as a sort of emotional interjection. The same word can't be intentionally evoked but can be automatically evoked. The same dissociation between impaired voluntary, but spared automatic movement production can be observed in apraxia. Conversely, a relative preservation of voluntary, in the absence of the automatic, movements can be observed in right brain-damaged patients. We have just seen that in the latter, the automatic orientation of attention to the left is impaired, whereas the intentional orienting is spared. Another automatic disorder observed in right brain-damaged patients is the loss of familiarity feelings that we experience when we see the face, or hear the voice of a known person. This feeling, which has the function of triggering, into the semantic memory, the search of the person to whom the face or the voice belong is unconsciously and automatically elicited. So, even if the hypothesis that I have just advanced remains quite speculative, we can conclude that the automatic orientation of attention in the right hemisphere might be, at least in part, related to the left lateralization of language and to the general reorganization that language entails for many brain functions.

How to Pay Attention with an Asymmetrical Brain

PAOLO BARTOLOMEO

As you know, the brain is composed of a right and a left half, called hemispheres. Each hemisphere controls the opposite side of the body and of the external space: the left hemisphere controls the right space and the right hemisphere mainly controls the left space. The two hemispheres are grossly similar but they don't have the same role in some cognitive abilities. For example, in most of us the left hemisphere is important for language—to speak and to understand language—and to control the right dominant hand. However, today I would like to talk about the other hemisphere, the right hemisphere. Let me start by presenting the case of the great Italian film director Federico Fellini. In April 1993, Fellini received his fifth Oscar for lifetime achievement. A few months later, he unfortunately suffered from a stroke in his right hemisphere, so he could not move his left hand and his left leg anymore. He started producing funny drawings like these ones. [1]

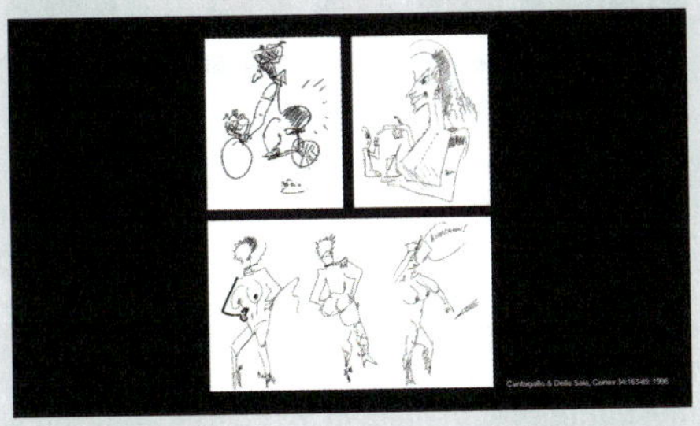

Fig. 1

Fellini was a very talented cartoonist, but as you can see the drawings lack some details on the left. Some details on the left part of the drawings are missing. This is how he imagined his therapist ordering him to draw her. When Fellini had to mark the central point of the horizontal line, he deviated the midpoint towards the right side, and even added a few funny drawings on the right part of the line. It is as if he didn't pay attention to the left part of the line. One might think that perhaps he didn't see the left part of the line, but this was not the case because sometimes he even drew over the left part of the line. But then his subjective midpoint always deviated towards the right, perhaps because the weight of his attention stayed on the right extremity of the line. [2] The case of Federico Fellini, unfortunately, is by no means unique. About half of the patients with a stroke in the right hemisphere present signs of vis-

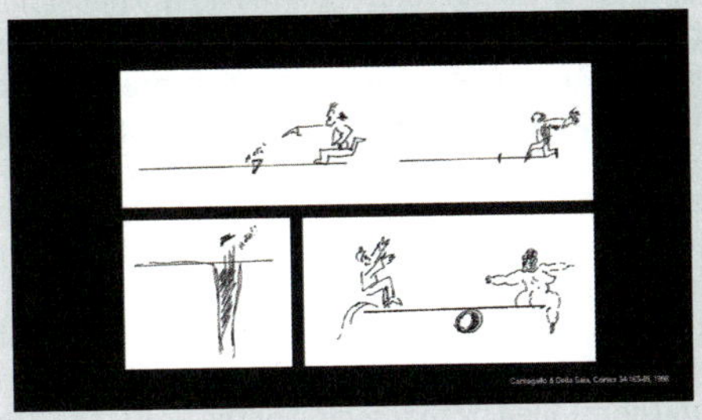

Fig. 2

ual neglect for the left part of the world. They do not eat from the left part of their dish; they do not shave or make up the left part of their face; they do not respond to questions from the left side or they respond to somebody else who happens to be on their right side. You can easily imagine how disabling this condition is, because patients with neglect are not able to resume everyday activities. Here you can see a patient we observed at the Pitié-Salpêtrière hospital. [3] We asked him to mark all

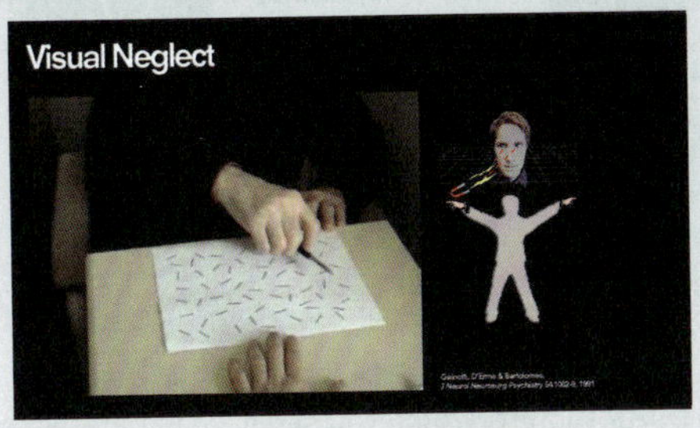

Fig. 3

the small lines on the sheet and, as you can see, he only paid attention to the lines on the right part of the sheet and completely forgot to mark the lines on the left part of the sheet. There is also another typical behavior, which stresses the problems that these patients have with attention. If you try to assess the patient's visual fields by raising your hands in the two visual fields of the patient, asking the patient to look at your nose and then briefly wiggling the fingers of one or the other hand, and asking the patient whether he sees the movement or not, this can be impossible with some neglect patients, because as soon as you raise your hands in the visual fields of the patients, the patients lose fixation and immediately look at the hand on their right, as if this right-sided stimulus exerted a sort of magnetic capture on their gaze and attention. What are the critical circuits in the right hemisphere whose damage provokes left neglect? To answer this question some years ago we had the possibility of using direct brain stimulation during neurosurgery. When neurosurgeons operate on brain tumors, they obviously don't want to leave patients with neurological deficits, so sometimes they awaken the patient during the intervention—the brain has no pain receptors—and they briefly inactivate with electrical stimulations the regions of the brain they would like to take away. So, if the patient shows deficits, they stop talking or moving, for example, the hand controlled by that region of the brain, then the surgeon respects that region. So we asked a patient who was being operated on for a tumor in his right hemisphere simply to bisect lines. It is the line bisection test we saw earlier. What we discovered is that transitory inactivation in one particular region of the brain provoked the largest rightward deviations on line bisection, a sign of severe neglect. It was this region at the bottom of the surgical cavity that you can see here. [4] This region corresponds

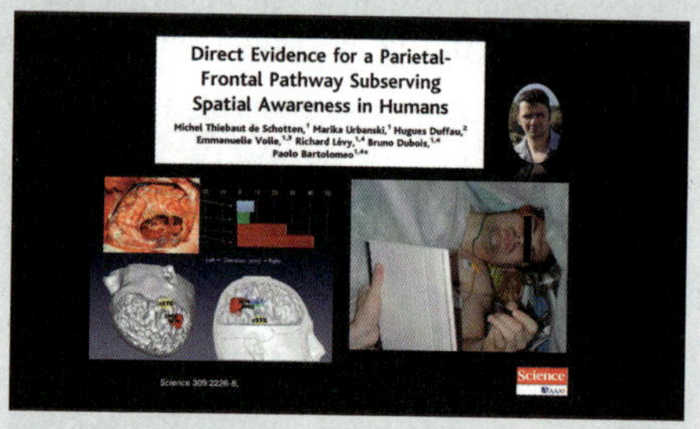

Fig. 4

to this structure, which you see here in blue, which is a system of connections in the brain white matter, which connects the posterior part to the anterior part of the brain. It is a kind of highway, a big chunk of cables that allows the posterior part and the anterior part of the brain to communicate with great efficiency and speed. You can see this pathway here, in cyan in this figure. [5] We know that this pathway is important for visual attention. When this pathway is damaged in the brain, in the right hemisphere, then patients show signs of left neglect, they do not pay attention to the left part of the world. Importantly, neglect, however, is not symmetrical, because similar lesions in the left hemisphere only rarely produce signs of right neglect. There is really something special in these attention networks in the right hemisphere of the brain. We don't know yet what exactly happens in the brain of patients with neglect. One possibility is that the left healthy hemisphere plays a role in neglect signs. The left hemisphere is not very good at controlling

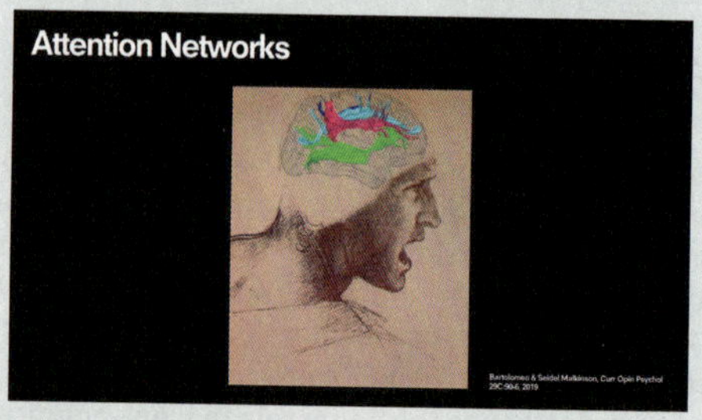

Fig. 5

attention, so when it is left alone by a lesion in the right hemisphere, it could provoke signs of neglect. For example, [6] in this study we used a non-invasive

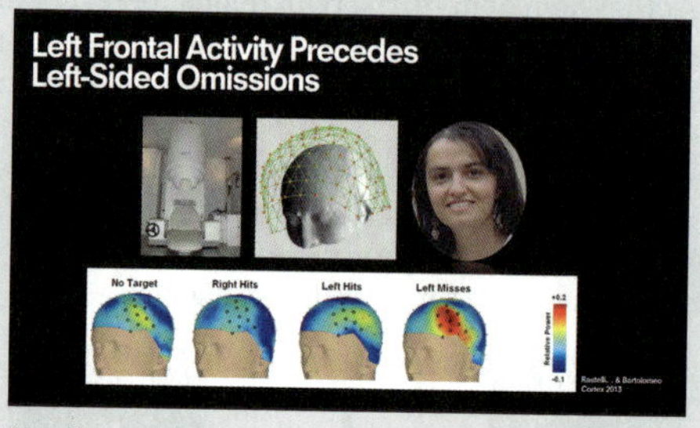

Fig. 6

new imaging technique called magnetoencephalography to study the brain activity in patients with signs of neglect. We asked patients with right hemisphere damage and left neglect to press a button each time they saw an asterisk, which could occur on the left side or on the right side of the screen. Being patients with neglect, of course, they did not respond to all the left-sided asterisks. What we discovered is that just before this omission of responses to left-sided asterisks—the left-side asterisk was presented and the patient did not respond to it because it neglected it—there was this activity in the anterior part of the left undamaged hemisphere. You can see this activity here in red. This activity in the healthy hemisphere preceded the signs of neglect, as if this region of the left hemisphere was saying "Whatever happens next, you don't have to respond to it." Perhaps this region actually oriented patients' attention towards the right side, so that they could not respond to left-sided stimuli. It could be that the left hemisphere actually has a negative role in neglect, at least for some signs of neglect. But can the left hemisphere learn with time to control spatial attention in a more balanced way and so to compensate for neglect? To answer this question, we need to take into account another important dimension, which is the evolution of neglect in time. We know that some patients with neglect eventually recover, at least partially, from signs of neglect. Even Federico Fellini, a few months after his stroke, could draw more symmetrical images than he did in the acute phase of his stroke. So, to assess the hypothesis of possible role of the left hemisphere in neglect recovery, we followed up in time 45 patients with right hemisphere strokes and signs of neglect. We tested these patients for neglect twice, once around three months after the stroke and the second time after more than one year after the stroke, in a very chronic phase. [7] We saw that of these 45 patients,

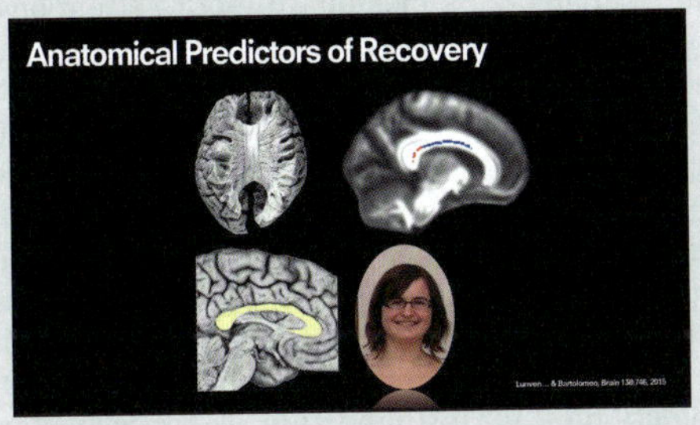

Fig. 7

27 had signs of neglect at three months from the stroke. Of these 27 patients, 10 had recovered after one year after the stroke and 17 still showed signs of neglect one year after the stroke. When we compared the anatomy of the lesions in these patients, we discovered that the most important difference was in another white matter pathway, in another big chunk of cables, called the corpus callosum, which allows communication between the two hemispheres. Those patients with damage in fibers who led to what you can see in the figure in red—the posterior part of the corpus callosum—did not recover from neglect. They kept showing signs of left neglect more than one year after the stroke. On the other hand, patients who had this structure in good anatomical shape could eventually recover from neglect. What does this mean? It means that, probably, the left hemisphere with time can learn to compensate for neglect, but only when the healthy left hemisphere can talk to the damaged right hemisphere through this portion of the corpus callosum. By reviewing the literature, we discovered that this compensation by the left hemisphere can actually be a general mechanism of recovery from cognitive deficits after a stroke. [8] For example, in the literature there are descriptions of patients with, this time, left hemisphere damage and language disorders—

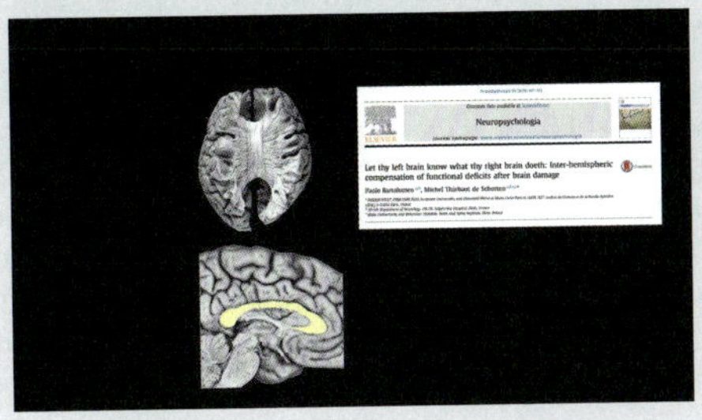

Fig. 8

patients who could not speak anymore or could not anymore understand language after lesions to the language circuits in the left hemisphere. Then, some of these patients were recovering from their deficit but, unfortunately, they got a second stroke, this time in the right—previously healthy—hemisphere. After the second stroke they lost again the ability that they were recovering. This compensatory activity of the healthy hemisphere could be a general mechanism of recovery from cognitive deficits after unilateral strokes. The challenge for future research is to try and find ways to improve the communication between the hemispheres in stroke patients, in order to shift the state of interhemispheric interactions from a competition state, which we can see in the acute phase of the stroke, to a cooperation state, which may promote recovery from the cognitive deficits that patients can experience after a stroke.

Conversation 2

ALICE ROBERTS, GUIDO GAINOTTI,
PAOLO BARTOLOMEO

ALICE ROBERTS The subject of this conversation is hemispheric lateralization and attention mechanisms. We are focusing on the two halves of the brain and looking at how cognitive functions are shared across the two hemispheres and how damage to certain areas can cause an asymmetry of attention. Guido, please summarize your lecture.

GUIDO GAINOTTI I am going to focus my attention on spatial hemineglect or lack of attention for the left half of space, which is observed in patients with the right hemisphere lesions. This is an extremely severe disorder, severe for one's daily life as well as for the cognitive activities of these patients. What is it? The fact is that these patients totally forget what happens to their left and this inattention is surely not due to a visual field deficit or to other elementary disorders, also because similar disorders for the right half of space are rarely observed in patients with left lesions. In addition, this disorder in these patients does not seem to be due so much to a voluntary deficit, an intentional deficit, but it is automatic, meaning that the patient automatically does not focus their attention on the left. Why don't they do so? Apparently this depends on the fact that the attention is being automatically captured by stimuli placed on the right of the visual field, so the presence of these stimuli on the right prevents attention to focus on the left. This is being confirmed also by neuroanatomical and neurophysiological data. In the brain we have some fiber bundles that allow focusing attention automatically or voluntarily, and what we see is that the bundles, that are necessary for the voluntary orientation of attention, are present both in the right and left hemispheres, while those that underlie the attention to the contralateral hemisphere are more developed in the right hemisphere. Since this is an important function—it is a non-verbal function that is responsible for many important functions in both animals and in humans—in human language and is represented by the left hemisphere, we wonder if this representation, this type of spatial orientation to the right may be partially linked to the existence on the left of a system that is important for language processing. I say this because language is an extremely important tool for cognitive processing, which is not useful only for verbal or non-verbal communication, but it is necessary also to structure the cognitive activities of the brain based on a more intentional, more conscious modality or, instead, based on a more automatic and more conscious type of activity. This is a general summary of my lecture.

AR Before we launch into questions, Paolo please briefly summarize your lecture.

PAOLO BARTOLOMEO As we know and as professor Gainotti was saying, the brain is made of two halves, the right hemisphere and the left hemisphere. In most people, the left hemisphere is important for speaking and understanding language, and for controlling movements, while the right hemisphere has different functions, including directing attention in space, although it is not just the right hemisphere that does it, but it is dominant. Many patients with stroke in the right hemisphere, as Guido was pointing out, lose their ability to pay attention to events occurring on the left side of the space. This disorder is called visual neglect. These patients do not eat food contained on the left side of their plate, they do not shave or apply makeup to the left half of their face, and, if they walk, they can bump into objects on their lefthand side. These issues are caused by lesions in circuits in the right hemisphere controlling the orientation of attention in space. These are circuits that normally tell us, among other things, when an object from the external world is important and requires our attention. Recent data show that these circuits can also be important for processing information or becoming conscious of all the events in the external world, or aware of them. Visual neglect does not only imply the working of the right hemisphere but also the left hemisphere, the healthy one, which does not get appropriate information from the right hemisphere, and therefore directs the attention automatically, as professor Gainotti was pointing out, towards the events on the righthand side. Every single hemisphere controls more or less what happens in the opposite hemispace. Therefore, the left hemisphere in these cases can have a negative action in those patients suffering from neglect. A few months after the lesion or stroke, in some cases, the left hemisphere seems to be able to help with compensation of neglect. But this seems to happen only if both hemispheres, left and right, can, so to say, speak to each other through connections between the two hemispheres. If the lesion affects these connections as well, it is very likely for the patient not to recover from hemineglect and therefore continue to show inattention signs for the left-hand side of the space. Therefore, it is possible that communication between the hemispheres makes some compensation for the deficit possible, and this might be a more general mechanism, for instance for recovery from language deficit after lesions to the left hemisphere. In some cases, the right hemisphere, which is the healthy one in this case, can compensate for these deficits. A challenge for future research is to find strategies to help this communication between the two hemispheres, the healthy one and the diseased one, so that you can switch from a situation of competition between the two hemispheres to a situation of cooperation or collaboration between the two.

AR Listening to both of you, you are obviously fascinated by the way that the brain works, but you are looking at it in a clinical way: the more we can find out about how the brain works, the more we understand how to help patients who have had lesions like strokes. That is terribly important and we are also getting a little bit of insight from both of your talks into just how disabling these lesions can be and how impactful they can be on people's lives. Paolo, would you like to start by asking Guido a question?

PB I would like to ask Guido something about the awareness of the disorder from the external world in these patients, because patients with neglect often have little awareness of their disorder, they are almost always unaware of what is happening. Would this suggest some sort of relationship between atmospheric lateralization and conscious experience?

GG I believe that when we speak about the conscious experience we should differentiate between various levels of consciousness. The fact that language is processed in the left hemisphere causes that tasks performed with this hemisphere are usually more intentional and conscious, while tasks performed with the right hemisphere are less conscious and more automatic. Within this general approach to the impact that language can have in the two hemispheres, we can observes that if there is a deficit that depends on a lesion of the right hemisphere, often the patient is not fully aware of this deficit. A disorder analyzed for the first time in the first half of the last century was "anosognosia," a lack of awareness of hemiplegia, shown by patients who had suffered from a stroke of the right hemisphere. Even if this may seem strange, these patients, who cannot move their limbs to the left, do not realize that the left half of their body is not responding to their commands. If one asks them "Did you raise the arm?" these patients can answer "yes" even if in reality they were not able to move the arm. This lack of awareness of the disease that we observe in patients with hemiplegia is even more accentuated in those with visual heminegligence, and this might be due to the fact that an important component of hemineglect is the difficulty in the (unconscious) automatic orientation of attention. Automatic processing is one of the factors involved which can explain this unaware reaction in the patient.

AR Would you like to ask Paolo something in return?

GG Yes, I have a question for professor Bartolomeo. He made very interesting studies on this topic and I would like to ask if he can add something about the fiber bundles responsible for voluntary orientation and those for the automatic orientation in both the right and the left hemisphere.

PB Let's say that we are now rather knowledgeable about the architecture of these networks. There are several frontal parietal pathways connecting the front to the posterior areas of the brain, and, as you were mentioning, we know that there are fiber bundles, which are symmetrical and are higher up—in anatomy we call them dorsal—and they orient attention into space. They are connected to the region which programs ocular movements in the frontal lobe and to the regions of the parietal lobe. These dorsal fibers orient attention into space both automatically and voluntary. What is peculiar about the right hemisphere is that there is a more ventral set of fibers. This network does not orient directly attention because it doesn't have motor capacities, but it is as if it were an interface between perception and movement because it seems to be defining objects worth my attention in space, which are important to me. When this ventral network is lesioned or disconnected from the dorsal part of the right hemisphere, you have signs of spatial neglect.

AR Just to elucidate those terms for our audience—dorsal is towards the top of the brain, ventral is towards the bottom of the brain, and then anterior is towards the front and posterior towards the back. These are just terms that anatomists like myself use, that clinicians use as well, so that we know where we are in 3D space within the brain. I wanted to pick up on something you said there, Paolo, which really caught my attention. You were talking about this anatomy and the wiring of the brain, these tracts within the brain which are essentially the wires connecting one area to another. How has it been possible to map out those pathways and work out not only their structure but also their function? How is that possible?

PB There are typical and classical anatomic techniques that are very well known by great anatomists from France, like Déjerine, who studied in the brain of corpses all these wirings. Since a few years ago, thanks to an important technical insight, you can highlight these wirings also in a living human brain. This is called nuclear magnetic resonance with diffusion sequences. This is a technique that allows you to map out the direction of the spread of water molecules in the brain, and starting from all these recordings through this technique you can rebuild, at least hypothetically, the trajectories of the fibers, the white matter, that are the wirings connecting the various areas in the brain.

AR This technique, which is also known as functional MRI, allows us to not only see the structure of a brain inside a person's head, a living person, but to watch it working. It is quite amazing. I think these are techniques that anatomists and doctors of the past couldn't even have imagined we would ever be able to do. We are learning so much and it is very helpful to those patients as well. I have a question for Guido. What is the psychological impact of left neglect?

GG The psychological impact is not very important in reality because, as I was saying before, the patient has little awareness of what is taking place. A general rule that we have in psychology is that to suffer or to elaborate psychologically on a specific event we must be fully aware of that event. But if we are not fully aware, then our reaction, our processing, is limited. This is the fundamental reason for my answer. Furthermore, there is a difference in the emotional behavior that we have in patients with lesions of the right and left hemisphere. If the lesion is in the left hemisphere, they have a catastrophic, dramatic reaction, that is, however, psychologically appropriate, because what happened to them is indeed catastrophic. On the contrary, patients who had a right lesion, often have an apathic type of behavior. They are indifferent, at times they are even ironic or sarcastic. These behaviors tell us that the emotional reaction is abnormal in these patients, just as anomalous, rather than lost, is the language of those with a lesion of the left hemisphere is not lost but it is anomalous. They cannot correctly communicate what they would want to communicate.

AR Intuitively, you would imagine that it is an extremely distressing syndrome to experience. What about the wider context of that, though, for the loved ones and the families, of somebody who is suffering from that kind of neglect, that kind of negligence of attention?

GG The patients, as mentioned, are not fully aware of what has happened to them, whereas the family members are totally aware of this. It is interesting to report, on this subject, an observation made by Babinski on the first case of anosognosia that he described in 1921. At that time he had noticed that the family members of patients who were not aware of their hemiplegia considered this disorder as a sort of protection for the patient and themselves. They tried to maintain these patients protected from the possibility of becoming aware of something that would have been extremely painful to them. Things are always complex, of course. There are anatomical, physiological, and psychological factors involved and we must try to put together all of these heterogeneous components.

AR There is a question from our audience for Paolo Bartolomeo. Are there in the history of neuroscience any other case studies that are as emblematic as the so-called "patient Tan" from the 19th century?

PB Patient Tan is a patient that was being observed in Paris from the Bicêtre hospital by Paul Broca. He was a patient who could no longer talk, the only thing that he would say was "Tan tan tan," these syllables. That is why they nicknamed him "Monsieur Tan," Mister Tan. This patient in 1861 died because of a cause that was not related to this brain problem, and Broca got his brain and realized that there was a lesion affecting a very specific region of the anterior area of the left hemisphere. At that point Broca drew this conclusion, which is still acceptable today, that we talk with the left hemisphere. There were other cases later on described in literature, some famous cases by Wernicke, another neurologist, showing that patients with more posterior lesions could speak but no longer understood what they were being told. They had an issue in terms of understanding language production which was relatively normal. From then on, and we are here, at the end of the 19th century, physicians started to understand that all the regions in the brain were not equivalent but might play specific roles. From them, we had the first proof and piece of evidence that the two hemispheres, the left and the right, did not play the same role because similar lesions in the right hemisphere would not cause similar language disorders as in Mr. Tan. Let's say that now in science describing single cases is no longer "fashionable," and it is a pity because thanks to the advances in neuroimaging now we can achieve data that are very specific and accurate not only in terms of the site of the lesion that may cause a certain deficit, but also in terms of brain networks in which this lesion may be found. Therefore, the present time is perhaps the best to study in detail individual patients, which is something which is being done by some teams—and we are trying to do that—but from the point of view of science, I have to say that right now it is rather tough to be able to publish single cases. That is a pity because you get lots of information from these patients on the normal working of the mind and brain.

AR That gives us a bit of an insight into the world of science and how science itself works. The idea is that studies involving large numbers of people are always going to be higher quality and are to be sought after but, just as you describe, especially with these new techniques, we can look in incredible detail at a particular case, at a particular patient, at how their brain is working and learn a lot from a single case. Do you think we'll see case studies coming back in? Case studies along with larger trials?

PB I think so and I hope so. At least that is what we are trying to do right now. We really hope that also the editors of scientific journals may realize how important it is.

AR A questioner is asking about the example that you mentioned in your lecture, where an aphasic patient was asked to name the image, which was an image of the Virgin Mary, and the patient automatically succeeded thanks to emotional connections to the symbol. That says a lot about the cultural background and the unconscious appreciation by the patient of what they were looking at. Do neurologists show different images during tests, according to different cultures? Because obviously those images are very culturally specific.

GG I must say first that the problem is much simpler than it appears. We don't have to present images that differ depending on the religion or the culture of the individual. These are two different levels of the processing of language. I can give an example. The term "Madonna" can be produced based on two processes. The usual one is that a person looks at the image of a "Madonna," elaborates consciously information on this image, searches within his dictionary the corresponding words, and grasps the right term for that specific image. This is the typical working modality of the left hemisphere. Instead, we have another possibility, which is that of having a series of invocations, an automatic type of language, even cursing, that can come out in a specific emotional condition. So the term "Madonna," which cannot be found within the vocabulary of an aphasic patient, can be pronounced by the same aphasic patient, if there is a state of frustration or of stress, rage, as is the case when people curse. The same patient of Broca, mentioned by Paolo, was a patient who apart from saying "Tan tan" also pronounced another expression, "Sacré nom de Dieu," meaning "The sacred name of God." This was another type of automatic utterance which, for reasons linked to his personal history, had remained among the expressions that he often used automatically. Hence, he could produce that word automatically being unable to find voluntarily the same terms.

AR Another question from our audience: an interesting one about ADHD, Attention Deficit and Hyperactivity Disorder, in the brain. In somebody who has ADHD is there any evidence of an imbalance between the two hemispheres?

PB Thanks to these new neuroimaging techniques we were able to check and see that at least in some of these patients having ADHD there are slight malfunctions. Not big lesions, such as in patients with spatial neglect, but there are finer lesions exactly in those circuits that we were mentioning earlier, controlling attention, particularly automatic attention. By this I mean that in at least some of these patients these circuits, particularly the ventral network of attention in the right hemisphere, do not seem to be working properly. Once again, this is not a severe disease, such as that of patients with neglect, but smaller malfunctions that can indeed influence the behavior of these people, that is their incapability of focusing attention for a long time span, for instance.

AR A question I would like to direct to Guido is about the evolutionary basis of lateralization in the brain. Do we understand evolutionarily why our brains are lateralized in this way and do other animals have lateralization?

GG Let me start with the second part of the question. While up to some decades ago scientists were convinced that the hemispheric asymmetries (such as the left hemisphere dominance for language) were only present in humans, in the last decades we have seen that certain forms of lateralization seem to be present also in animals. Paradoxically, these asymmetries were reported even in invertebrates, namely in animals with a brain totally different from the human brain. We must say that, however, not all the research that was carried on this subject is methodologically perfect. But apart from the criticism that can be made to this research, the notion that lateralization is not only a human trait is indeed somehow valid, and will require more detailed studies in the future. As for the reason why we have different right and left lateralizations, it is possible that it may be an instance of division of work across different brain structures. A difference that is often made is that between the emotional system on one hand and the cognitive system on the other. The emotional system is an emergency system that is used in conditions of emergency, allowing to take decisions as fast as possible and to trigger the response pattern. Cognition, on the other hand, is a very different system. It requires longer time to process in a controlled and conscious way the existing problem. Therefore, the coexistence in the same hemisphere of structures that have to process the emotional response and cognitive response should perhaps be avoided, because it could produce contrasting responses and this could be inappropriate. The separation of these functional systems, which are not compatible one with the other, may be one of the general reasons that may explain why we have different lateralizations.

AR I had no idea that other animals exhibited lateralization as well, and I think that probably is that enduring pitfall of imagining that humans are completely exceptional and that we are so separate from the rest of the animal world. Our brains are extraordinary, they are incredibly large compared to the brains of any other animal, but it is amazing that we can see some similarities in the way that these functions are divided up in the brain. Another question from our live audience is about callosal inhibition. Callosal inhibition assumes that stimulus-driven activity in one hemisphere is going to suppress activity in the other; callosal inhibition being the suppression of a flow of information across the corpus callosum, which is the connection between the two hemispheres. The questioner says given that this view assumes that if one hemisphere is particularly active it is going to be suppressing the other, can visual spatial training help? Paolo, could you take that question?

PB It is possible that these relationships based on inhibition between the hemispheres are not totally symmetric. There are data proving that the right hemisphere, particularly the posterior area of the right hemisphere, can inhibit the similar region in the left hemisphere more than the opposite is true. That is, the right hemisphere is apparently able to inhibit the left hemisphere more than the reverse—the left hemisphere inhibiting the right one. This might be an evolutionarily adaptive asymmetry. As Guido was pointing out earlier on, if there is anything urgent I have to respond, a predator appearing, or I am just crossing over the road and I have a bike coming up very quickly, I have to be able to react as quickly. It is possible that the right hemisphere tries to take upon itself the response by inhibiting the left one.

AR A lot of questions are coming in about attention deficits. One asks about poor lateralization and whether that could explain some conditions where there is an attention deficit. If so, is there a way to improve that?

GG This is a very general question which is difficult to answer, also because when we speak about lateralization there are some genetic factors involved and we have a multitude of genetic models. There are different genetic models for hemispheric lateralization. Furthermore, we also have environmental factors that are involved and it is not easy to assess the weight of genetics and of learning or other behavioral factors. In addition, during evolution, not necessarily all the factors related to lateralization developed at the same pace. Maybe certain types of lateralization developed more quickly in certain periods, while in other periods other kinds of lateralization develop. Lateralization can also be different for different age ranges, so it can be said that if we think about infancy and the impact of learning factors, or about adolescence and the impact of social factors, these are all developmental periods in which the impact of genetic and learning factors may be different. The issue is therefore interesting but should be broken down in many sub-components.

AR That is a very complex question and we can't get a very simple answer. I wonder if Paolo would like to add anything briefly.

PB I am not really very much into lateralization at young age, I must admit. I am under the impression that sometimes this poor lateralization label is not being used very rigorously, simply because we have very few measurements reliable enough having to do with lateralization. Even if you surf around the internet, you read lots of bits of information that are not really double-checked on hemispheric lateralization. This famous right hemisphere that is more creative and the left one is more logical etc. You really have to be careful when it comes to the sources of these things, which are just commercial considerations most times.

AR I think that through this discussion we have all realized that lateralization is fascinating and much more complex than it is sometimes set out to be. Thank you both very much.

3

Advantage/Disadvantage of Having a Bilingual Brain

ELLEN BIALYSTOK
JASON ROTHMAN

MODERATED BY ALICE ROBERTS

Protective Effects of Bilingualism on Aging Brains

ELLEN BIALYSTOK

We know that all of the experiences we have leave some kind of trace in our brains. Everything we do changes the way the brain handles those activities. What about bilingualism? Bilingualism is something that is the most common activity in the world. When you are speaking language, you are using most of your time every day. Does bilingualism have any impact on the brain? To answer that question, we have to start with a very surprising finding. A lot of research in psycholinguistics, neuropsychology, brain science has confirmed that the way the bilingual mind is organized is perhaps not what you would have expected. If I were to be designing a brain that could handle two languages, I would put in a switch: now you are using this language, now you are using that language and they would be quite separate. [1] But that is not how the brain is orga-

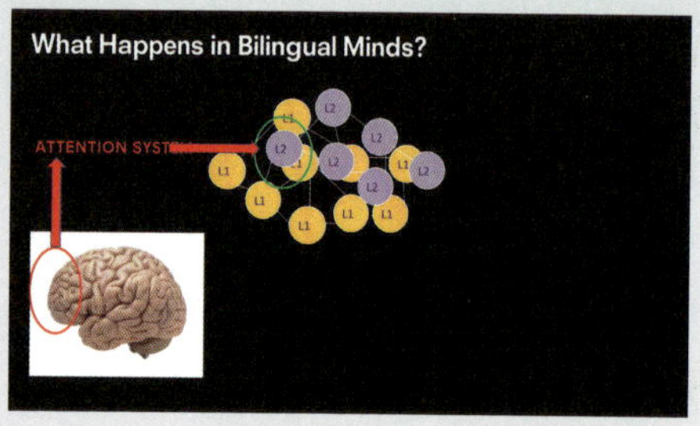

Fig. 1

nized. We now know that in the bilingual mind both languages are always active, all the time, even if you are absolutely sure you are only going to need one of them. That would be very confusing, because if both languages are active, you would expect bilinguals to make lots of mistakes, slipping in a word from the language that isn't appropriate, but they don't. Why not? If the two languages are active then there is always the possibility of conflict between them, and that conflict is constantly managed by the attention system in the brain, the general system that we use to attend to everything when there is a distractor. When you are in the supermarket, you only want to find the particular kind of yogurt you are looking for, you don't want to be distracted, and your brain, your mind has a selection and attention system that does that. That system is in the front part of the brain, it is part of what is more generally known as the executive function system. It is a really important system. It is the last part of the brain to develop in childhood and the first part of the brain to begin to decline with normal healthy aging. But if we are using that frontal executive system all the time, just to focus on the language we need to be speaking, then that system will inevitably change, it will get stronger, it will get better at doing that kind of selection. When we look at research across the lifespan, we find that, indeed, children and adults who have been bilingual all their lives, who had this constant experience all their lives, have better developed structure in that front part of the brain and, generally, they perform selection tasks better. All kinds of selection tasks, even ones that don't have anything to do with language. Our interpretation is that using those selection processes and brain structures to select the language you need to be speaking, because of this rather crazy brain organization, has led to certain improvements in those structures. But what about in older age? Here the story is a little more complicated. If we take monolingual and bilingual older adults and match them very carefully so they are the same age, and they perform all of these tasks to the same level, all of the background features that are relevant about them are the same, and these selection tasks are performed equivalently. What happens when you look in their brains? It turns out it depends. [2]

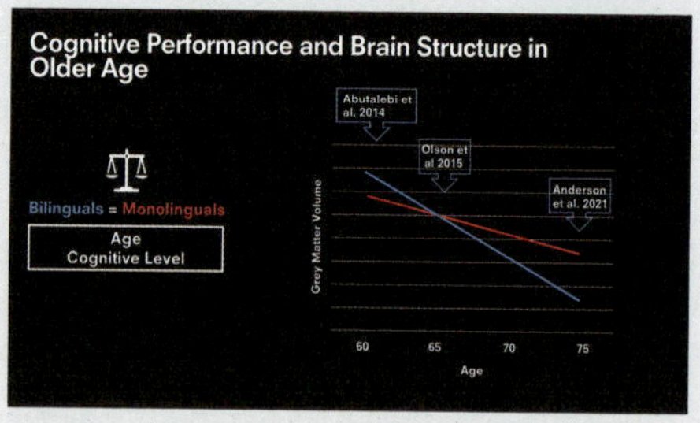

Fig. 2

On this graph, along the bottom x-axis we have age from 60 to 75, and on the y-axis gray matter volume, where higher values indicate more robust brains. Obviously, everybody wants as big a brain as they can get, so we want gray matter volume to be high. But we also know that with normal healthy aging we lose gray matter volume. This is normal, this happens to everybody and it begins at around that age, around 60. If we look at the red line here, which indicates monolinguals, we see a decline in gray matter volume between 60 and 75 years old. The blue line is bilinguals, and I put in three points from three different studies or sets of studies to show a pattern that is somewhat surprising. At the youngest age, the blue line, the bilingual line, is above the red line. At 60 years old on average. In fact, bilinguals have more gray matter volume, better brains. They have experienced less brain atrophy. But that blue line also declines. As I said, there is no escaping that atrophy and it declines more rapidly. At first it crosses the red line—and those are data we have from participants who are about 65 years old—and it keeps dropping, so by 75 years old the blue line is actually lower. On average, in our research, at around 75 years old or further into older age, we now have the opposite of the pattern we found in children and young adults, namely, we have bilingual brains showing less robust volume than monolingual brain. How do we explain that? This graph is a picture of what happens in older adulthood to brain structure—here you can think of brain volume, gray matter volume, like I just showed you—and in green, cognitive level. With increasing age—and I have only put in markers for age, these aren't precise ages—but with increasing age, in normal aging, both brain structure and cognitive level decline, but they decline in parallel. They are associated with each other, so as we get older it is more difficult to do those conflict and selection tasks, and our brains are losing volume. These things are happening in lockstep. That means that you can actually use one to predict the other. If you know what cognitive level someone performs these tasks at, you can basically predict what their brain looks like, and similarly, if you know what their brain looks like, you can predict their cognitive level. There is a set of circumstances that disrupt that pattern, that disconnects those two curves: that is the circumstances that we call cognitive reserve. [3] This refers to the effect that

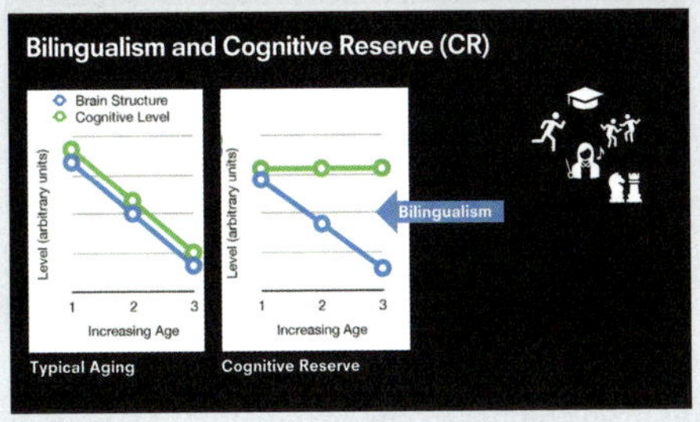

Fig. 3

all of these experiences that you keep hearing about—more education, active exercise, social engagements, stimulating activity—help you maintain cognitive level into older age. The way they do it is by keeping that green cognitive line high, despite what is happening to the brain. Remember, we can't reverse the biological forces that are responsible for brain atrophy with healthy aging. What we can do is keep cognitive level high despite that, we can put a wedge between them. What we have suggested in our research, is that another such experience that leads to cognitive reserve and separates those curves is bilingualism. This means that any one of these experiences, including bilingualism, should have the effect of keeping cognitive level high even as brain level declines. Let's look at an example. This is a study we did with a hundred older adults between the ages of 65 and 80. The mean age was 74, half for bilingual, half for monolingual. We gave them a lot of background measures to make sure they are all matched on everything we could think of: they had the same levels of education; they did a battery of neural psychological tests and produced the same scores; none of them had ever experienced memory or cognitive complaints nor had any of them ever visited a neurological clinic regarding any such complaints; they are all perfectly healthy matched older adults. We gave them some cognitive and memory tests and here is an example. This test is called N-back. The task is to follow a series of slides and press a button each time you see a number. In the first row called "1 Back," if the number matches the number you just saw on the previous slide, you press a yes button, and if it doesn't you press a no button. For example, when you see that second four you press yes, it matches the previous slide. The more difficult condition is "2 Back" because now you have to hold all of this in mind and decide if the number you are looking at matches what you saw two slides ago. It is much harder. In this example, the number two appeared on the previous two slide. The data that I have up on the screen here shows accuracy and reaction time for monolinguals and bilinguals. "1 Back" on the left. [4] Look first on the left panel: the "1 Back" monolinguals and bilinguals are 90% accurate It is an easy task. "2 Back" is much harder, so the accuracy drops to about 80%. The reaction time on the right shows that the "1 Back" is responded to very fast, the "2 Back," the more difficult task, is much slower. Here there is actually a significant difference. The bilinguals, the open circles, are actually significantly faster than monolinguals in responding, even though their accuracy is the same. Let's just say their performance is equivalent. There is no enormous differences in how these matched groups of monolinguals and bilinguals perform this task. Let's look at their brains. On the top is white matter volume, on the bottom is gray matter volume. Red bars

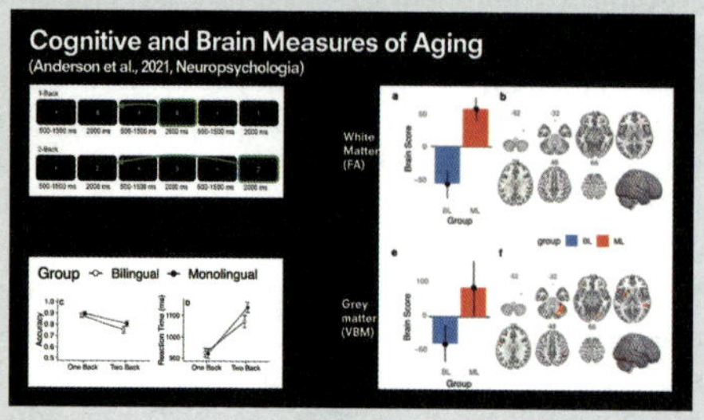

Fig. 4

are monolinguals, blue bars are bilinguals. When the bar goes up it means greater, when the bar goes down it means less. You don't need to read any of the fine print to see that the red bars are going up. Despite being matched on background measures, despite performing equivalently on these complex executive function tasks, the monolinguals have better brain structure in both gray matter and white matter. There are a couple of places where, despite these overall differences, the bilinguals actually have better volume, and I'll just mention one, that is in gray matter two areas responsible for language are actually more robust in the bilinguals, not surprisingly. Overall we have this situation that I described earlier for cognitive reserve: equivalent performance on the cognitive task, but the bilinguals have much poorer brain structure. This could be a problem. We match them on everything. Why do the bilinguals have poorer brain structure? Is it possible that this is a standard feature of bilingualism, in which case it would not be a very positive, attractive idea. The way we thought about this is that everyone in that study was admitted to the study because they reported being cognitively healthy, no brain or memory problems. Perfectly normal, functioning, independent living older adults. Maybe what was happening is that monolinguals who had the kind of brain structure we found for bilinguals, were not eligible for our study because they have had cognitive or memory complaints. In other words, the robustness of the bilingual brain might have been concealing what otherwise would have emerged as cognitive or memory complaints. To test that idea, we did a study that we called "Brain Swap." It is a kind of armchair theoretical idea, although we really did get data, where we take bilingual brains and put them in monolingual heads and say "how are you going to manage with his brain, monolingual?" We did it by matching the bilingual brains from the study I just showed you to monolingual brains from a database, so they were exactly the same. Now we have the bilinguals that you just saw, who are matched on background measures, like age, education, and everything else that we could find, as well as the exact parameters for their white matter and gray matter volume and structure. [5] What happens? We know that 100% of the bilin-

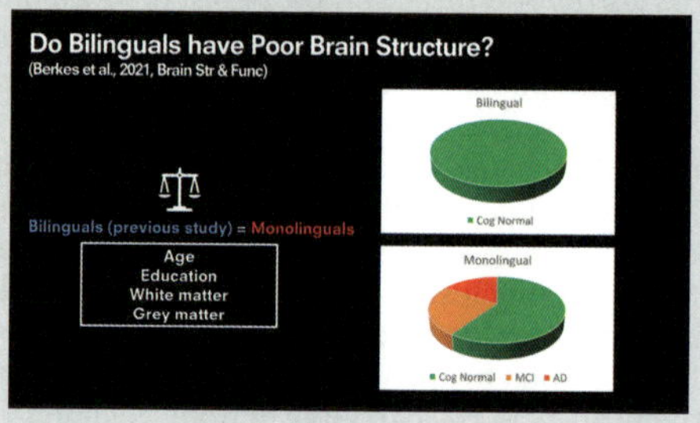

Fig. 5

guals were cognitively normal because they had to be, to participate in our study. What about the monolinguals who were matched on their brains? About 59% of them were cognitively normal, but 41% of them had clinical diagnoses of either mild cognitive impairment or Alzheimer's disease. These are the same bilingual brains, but now in the head of a monolingual, and now the outcomes are more variable. Our interpretation is that those bilinguals in our study, indeed, were experiencing more brain atrophy than the monolinguals, but they were able to cope with it, they had more resilience, it did not interfere with cognitive level. Maybe a more dramatic illustration of the same point comes from this. [6] This is a list of some, not even all of the studies

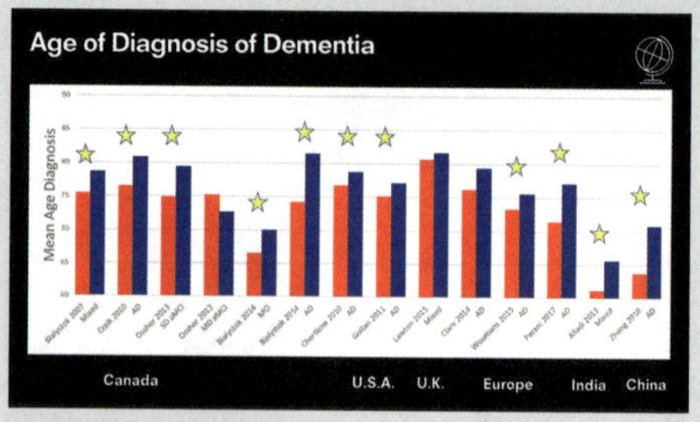

Fig. 6

that have compared age of diagnosis of dementia, usually Alzheimer's but not always, between monolinguals and bilinguals. Again, blue bars are bilinguals. What is plotted here is the age of diagnosis. What you can see across these many studies that have been conducted in many places in the world, is that the blue bars are higher. This means that, in general, bilinguals are older when symptoms appear and they are diagnosed with dementia. Why aren't they showing the symptoms? Are they getting the disease? What is going on here? If the cognitive reserve story is right, then we would expect that these bilinguals who are holding on longer without showing symptoms are nonetheless experiencing the disease, but are able to cope with it. We did one study to investigate that idea. In this study we took monolingual and bilingual Alzheimer's patients who matched on everything—age, chronological age, education, clinical level, cognitive scores. When these monolingual and bilingual patients came into the neurologist's office, they looked identical. But then we looked into their brains. What this chart shows is brain atrophy. [7] Here, higher bars show

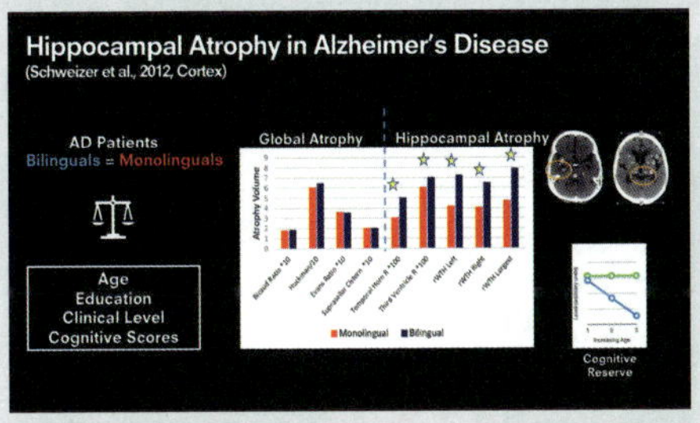

Fig. 7

more disease, more clinical disease. Left of the blue dotted line is just global atrophy and the two groups are equivalent. But to the right of the dotted line are measures of atrophy in the areas of the brain around the hippocampus that are taken as symptoms of Alzheimer's disease, and in every one of these contrasts the bilinguals have significantly more atrophy, which means they have significantly more disease. So I put that little graphic back in the bottom right corner, showing this is what we mean by cognitive reserve: they are performing at the same level—the green bar—but their brain scores—the blue bar—is lower. What can we do about this? Is it too late to go out and become bilingual? We just completed a kind of a small test of the idea that maybe just learning a language would be helpful. Learning a language is hard to do, and anything that is hard for your brain is good for your brain. We took a group of monolingual older adults, around 70 years old, and we enrolled them in a training study, gave them a pre-test that included a lot of background tests and specific executive function cognitive tasks. Then, they were enrolled in one of three groups. "Duolingo," to learn Spanish on an app, a brain training app called "Brain HQ," where they practice brain games every day—and these brain games, I have to say, are very similar to the executive function tests we are measuring them on—or a waiting group, where they just were the control. After 16 weeks we retested them on all of the cognitive tasks. We expected the brain training group to improve, because they were just practicing essentially the same kind of tasks that we were giving them in the pre-test and post-test. [8] We expected no change in

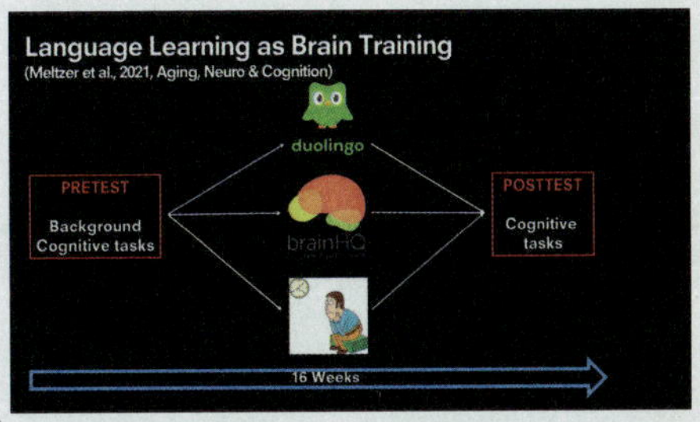

Fig. 8

the control group, they were just waiting. But what happens if you spend 16 weeks learning Spanish? I have three tasks that I am going to show you the results of. The first one is N Back again—so you know how that task works—and the graph shows the pre in green and post in orange test scores for the weighting control group, BrainHQ, and the Duolingo group. You'll see both BrainHQ and Duolingo improved significantly. Second task is called "the Simon task." It is a standard conflict task in this literature, and here the relevant variable is how fast you can respond. Again, the control group: no change over time. Both the BrainHQ and the Duolingo group got faster. Finally, a standard Stroop task: you see the word "red" but you have to say "blue" because that is the font color, or you see the word "green" but you have to say "red" because that is the font color. [9] It is

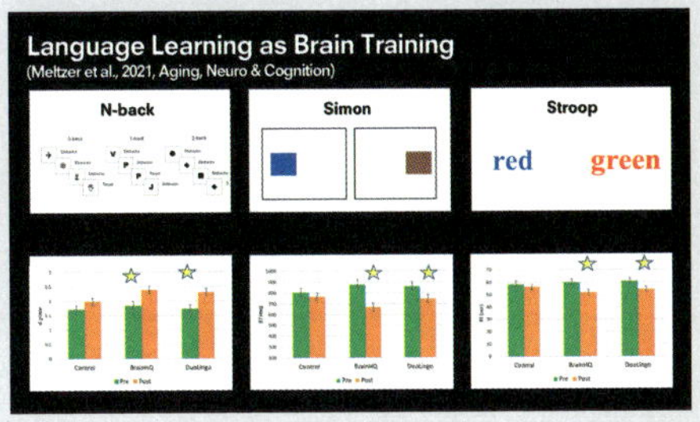

Fig. 9

very hard to do, and here again, we are looking at speed. Again, no change in the control, but both

BrainHQ and Duolingo got faster. This is a small study but it is very promising, because not only you have to spend your whole life being bilingual, you might find benefits to your brain and mind even by learning a language in older age. It is never too late.

Contextualizing Bilingualism: Focus on Mind and Brain

JASON ROTHMAN

Situating research examining potential connections between exponents of bilingualism, cognition, and brain for such a diverse audience is a daunting task, but I'll give it my best shot. Without further ado, then, my task today is first to contextualize what bilingualism is, inclusive of its many complexities on several fronts. Second, I will discuss how and why engaging with the naturally occurring ubiquitous mental activities needed for the juggling of more than one language can, on a spectrum, change our minds and, indeed, physical brain. Let's get started. The things we choose to do in our daily lives have lasting effects beyond the moments in which they occur. [1] Moreover, moments of choice have

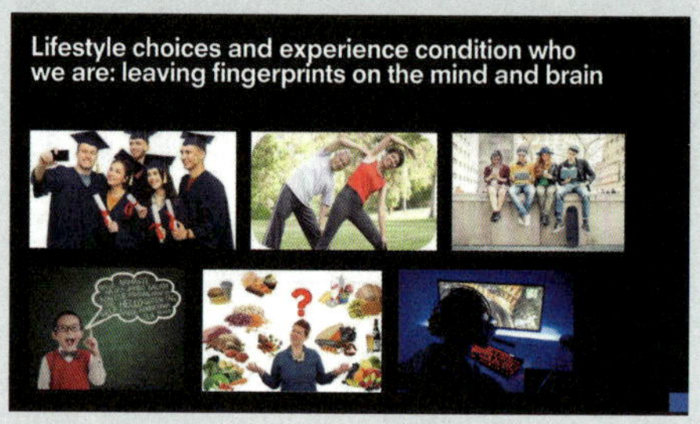

Fig. 1

cumulative effects. Let's take, for example, the choice we make when we eat. Indulging every now and then aside, if we choose foods that are high in fat and calories, all too often, there are potential immediate consequences. Our tummies might not appreciate it, for example. Cumulative effects reflected on our weight, our cholesterol levels, our heart health, and more. Of course, there are things we can do to counterbalance these choices. For example, physical exercise helps to compensate for the indulgent food choices we might make. As it turns out, choices like conforming to healthier diet and good measure of consistent physical exercise over the lifespan have profound knock-on effects on our cognitive and brain health. We know this. But these are certainly not the only lifestyle choices that correlate to such effects. Many mundane activities we choose to engage in can have similar results, provided they entail considerable levels of cognitive exercising, such as achieving high levels of education, training and playing music at expert levels, even being an avid video game player. Research suggests that bilingualism can be added to this list of lifestyle experiences adding to healthy cognitive life expectancy, the most important consequences of which Professor Bialystok discusses in her talk. If so, this is great news for many reasons. For now, I will highlight just one. Bilingualism stands out as something more equitable, more accessible, in a global sense, than most of the other known factors. Not everyone has the resources to achieve the highest levels of education, afford the healthiest of diets, or have the time needed for consistent exercise. Bilingualism, however, is literally everywhere, and in many cases comes for free. How does bilingualism do this, and what exactly is bilingualism? Let's have a deeper look. What does it mean to be bilingual? The answer is simultaneously simple and complex. In the most basic sense, a bilingual is a person who has knowledge for understanding and communicating in more than one language. Contrary to some popular beliefs, there is no true version of any given language. Thus, linguistic perfectionism, whatever that should mean anyway, brings little to bear as a bilingual qualifier criterion, although degree of fluency can have effects. With this simple definition in mind, bilinguals come in all shapes and sizes, so to speak. What is important to acknowledge is that no two bilinguals are exactly the same. Underpinning the single word, bilingual, there is huge variability, including but not limited to, the ages at which our languages were acquired; the relative abilities we have in our languages; our particular patterns for using them; and, indeed, the context, or for what we use them for; who do we or can we use our languages with—are these groups small, large, diverse, etc?—; how have our patterns of using our languages changed or remained the same over time. I could go on and on and on. The key issue to bear in mind is that differences like these are omnipresent and, as we will see later, may very well be deterministic for effects bilingualism can have on the mind and brain This diversity is also what makes bilingualism such a complex yet interesting phenomena to study. Bilingualism is an individual experience that is challenging to quantify. Depending on each viewer's personal context, or where you join us from tonight, it might be more or less hard to believe that most of the world's population is bi- or even multilingual. There are approximately 195 recognized countries in the world. At the same time, there are over 7,000 distinct languages spoken in our world. Clearly, the map does not add up for there to be a one country, one language mapping. As we know, some countries share a majority language, such as the United States, Canada, the UK, Australia, and Jamaica. So, the map tells us that there has to be significant language diversity and language contact in these 195 countries. As can be seen in the world map, this is definitively so. [2] The greener or darker the color, the

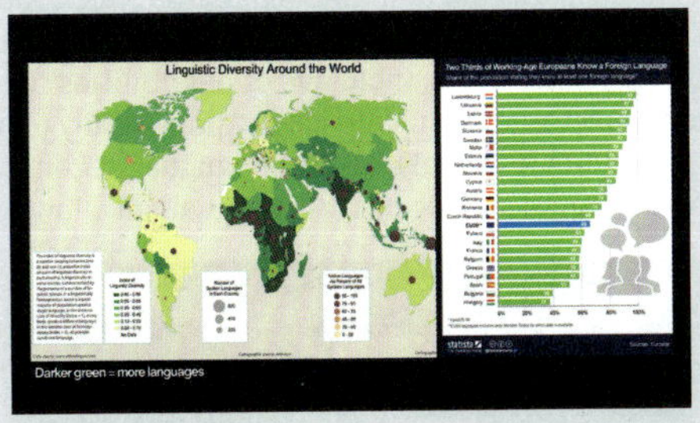

Fig. 2

more languages are present. Where there is more diversity, bilingualism tends to be the default state. This, indeed, is the majority of the world. It is worth highlighting to my point about bilingualism as an equitable lifestyle factor that many of the most bilingual parts of the world coincide with economically disadvantaged areas. One need not be from a place where there is huge linguistic diversity, in society, to be bilingual. Take, for example, Poland, which appears in light yellow on the diversity map. Looking at the chart on the right, one can see that roughly 62% of age working adults in Poland know and use a foreign language. Just under the EU-28 average of 66%, or two in every three working age Europeans. Bilingualism is truly everywhere. I think it is worth mentioning that scientific research on bilingualism has witnessed a meteoric rise since the 1990s, very much especially so after 2004, when work by Ellen Bialystok and colleagues first indicated that bilingualism might have implications for domain general

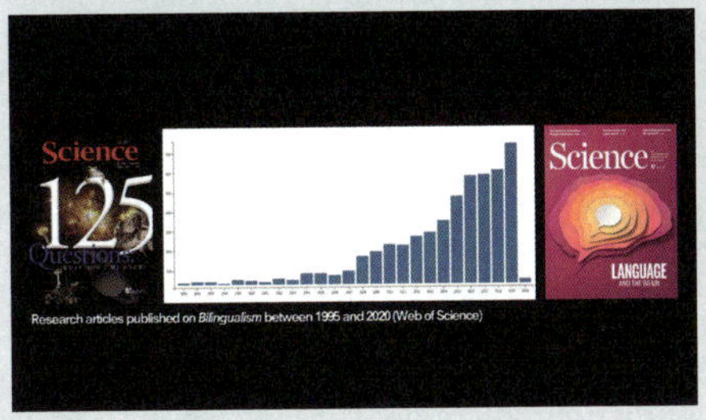

Fig. 3

cognition and the brain, that is outside of effects for and on language itself. [3] In fact, in 2005, in a special issue of the journal *Science*, Kennedy & Norman listed the "biological basis of second language learning or acquisition" among the top 25 most important scientific questions to be answered in our times. Before getting deeper into the mind and the brain connections of bilingualism, it is worth mentioning what might be obvious, although not necessarily so: bilingualism matters for a slew of domains, including but not limited to, our family and social networks, inclusive of our abilities to be more able to communicate with exponentially more people than if we only have one language; general cross-cultural awareness. [4] Bilingualism allows us

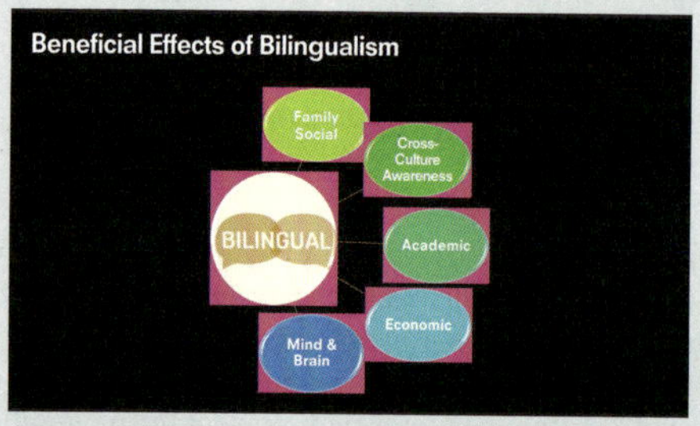

Fig. 4

to appreciate diversity, engage with other cultures, understand cultural differences, be more empathetic and more. Food, wine, music, art, literature, and human experiences in general are perhaps appreciated more deeply in their native tongues; academic: all other things controlled for and being equal, bilingual students often do better in general academic achievement; economic: there is no escaping the globally connected environment in which we live. Bilingualism can have personal economic benefits: job competitiveness, for example, and as a necessary part of the global supply chain, bilingualism has many impacts at societal level, as well. Of particular relevance for today's discussion are potential ones from mind and brain, to which we will now turn. Let's look at the neuroscience behind this. Let's think about what the brain has to do to accommodate the presence of two linguistic systems. Research shows that both languages remain constantly active in the brain, irrespective of contextual need or conscious intent. [5] Let's call this dual activation. Dual activation is necessary so that a bilingual can easily and instantly switch from one language to the other as required, which is frequently unpredictable. Perhaps, an analogy will contextualize why this makes perfect sense. I join you today from above the Arctic Circle, in Northern Norway. Here the temperatures are already frigid and there is a fresh layer of snow on the ground. As we know all too well, turning our car engine on and

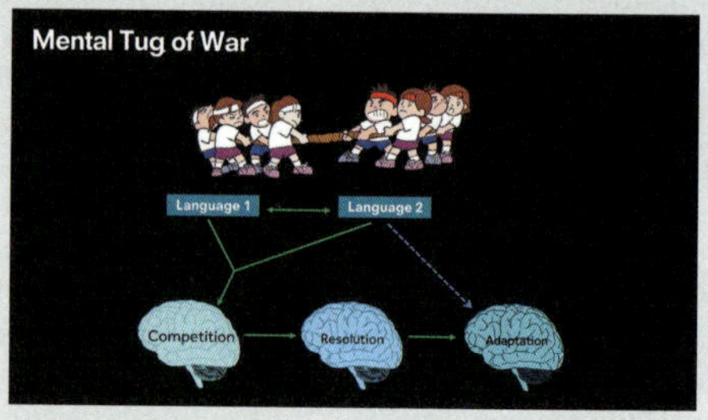

Fig. 5

leaving it idling allows us to go on our way immediately as we need. If the engine is not already running for when we need it, it will take at least several minutes for the car to be drivable. In a similar way, bilingual brains keep all languages activated. This explains the apparent ease with which bilinguals are able to switch between their languages, even within the same conversation or sentence, when appropriate, and immediately, as needed. Dual activation means that both languages are at a constant state of tug-of-war vying for limited cognitive resources. Although there is a need for both languages to be kept active, doing so introduces competition in real time, requiring some serious mental juggling. The mental competition has to be managed or controlled so that the language being produced is appropriate for a given communicative context, otherwise we might produce a language the other person doesn't understand, or a mix of all languages available to us. This need to control the two languages that are in competition brings us back to the heart of our question about the potential impact of bilingualism on neurocognition. It turns out that the mechanisms we use for bilingual language control are part of, or overlap with, a set of a wider cognitive control mechanism which we call executive functions. We use executive functions whenever we shift our attention from one task to another, or when we need to ignore, for example, noises that are stopping us from concentrating on something else. We use these skills every day to learn, work, and manage daily life, to focus and even regulate our emotions. Bilingualism, at least under certain conditions, is argued to strain and, thus, train executive functions above and beyond their other engagements, leaving fingerprints on the mind and brain in its wake. This all embodies the very essence of the proposed general effects of bilingualism on neurocognition on neural cognition. In summary, inside of bilingual minds there are two languages and they are kept active. Because the languages are both active, they compete. Because a bilingual has to contend with this competition, it makes heavy use of control mechanisms and executive functions. If this increased use of executive function means that executive functioning grows stronger, this will be felt beyond the language system. Researchers in the lab setting often administer to participants tasks that are believed to tap into executive functions. In other words, participants take experiments that are proxies for the everyday activities that require the involvement of cognitive control mechanisms. All things being equal and ideally controlled, the simplest hypothesis is that bilinguals will do better than monolinguals on these experiments, given more fine-tuned cognitive control. Early work in the field showed just that, but not for all bilinguals. The much-cited Bialystok et al. 2005 study showed better bilingual performance in three age groups, most profoundly in younger and older ages, where cognition is either not fully developed or well past its prime peak. [6] In other words,

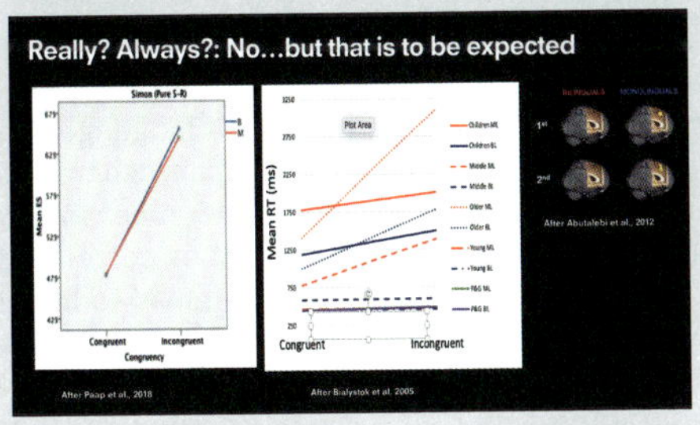

Fig. 6

where cognitive tests, often used in psychological research, are most likely to pick up subtle differences. Similar effects of these groups have been replicated many times as have the null results in younger adult populations. Obviously, child bilinguals pass through young adulthood before they become elderly. So what do we make of this? Work that looks beyond behavior—under the hood, so to speak—at the actual brain shows that sometimes even when the behavioral performance on a specific task is no different between monolinguals and bilinguals at younger adult ages, the way the brain reacts to the task can be quite different, and significantly so. In the picture on the right, you see that the monolingual brain requires much more energy to perform similarly as the bilingual brain does. Hold up. [7] Explanation requires more than simple bilingual status. Not all bilingualism is the same, but it is often treated as if it were monolithic, and then compared dichotomously to monolingualism,

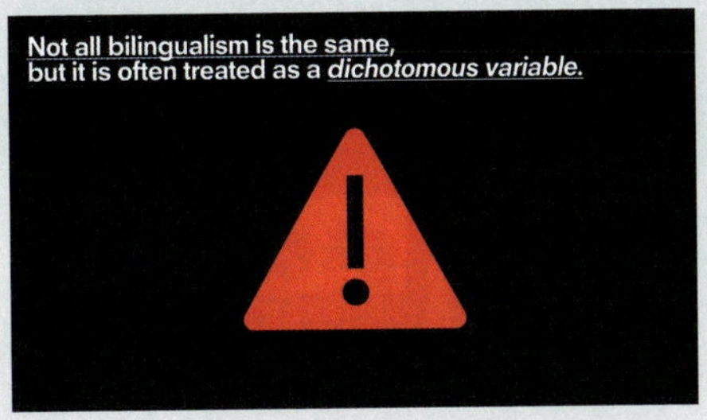

Fig. 7

which itself is not monolithic. Acknowledging and dealing with this requires significant shifting in the way we approach and do our research. As depicted before you, there are many variables. [8] Those on

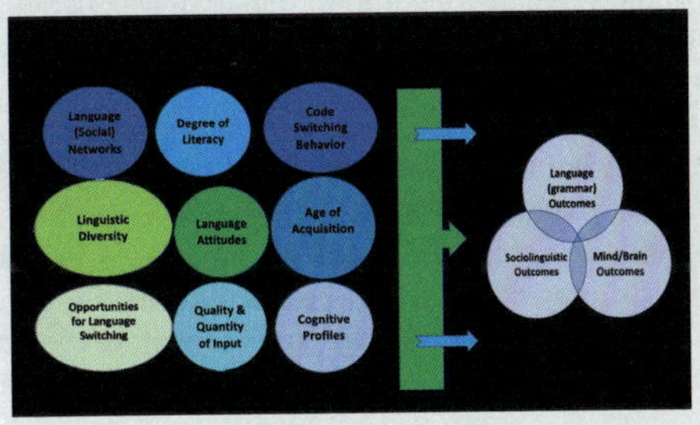

Fig. 8

the left are examples. In other words, the list is not exhausted. They combine to condition outcomes on the right for bilingual grammars, sociolinguistic competencies, and mind-brain adaptations, which themselves share significant overlap. Effectively, the permutations of individual differences in proxies that relate to opportunities for linguistic engagement, opportunities that delimit the straining and, thus, training of cognitive control, more reasonably sit on a vast continuum of dynamic experiences across individuals. In other words, bilingualism is a spectrum, not an absolute state. In what follows, we will have a cursory look into some research from our labs that speak to how the dynamics of bilingualism play out. I'll give you the take-home message already now. Bilingualism matters for cognitive and brain effects, but results are conditioned by individual engagement with the experiences that comprise one's very own bilingualism. In this study, our goal was to look at bilingual brain structure. [9] We gave non-native speakers of English living in England very detailed questionnaires about their experiences with bilingualism. For example, the age at which they acquired English, how they use each of their languages, quantity and context—in the home,

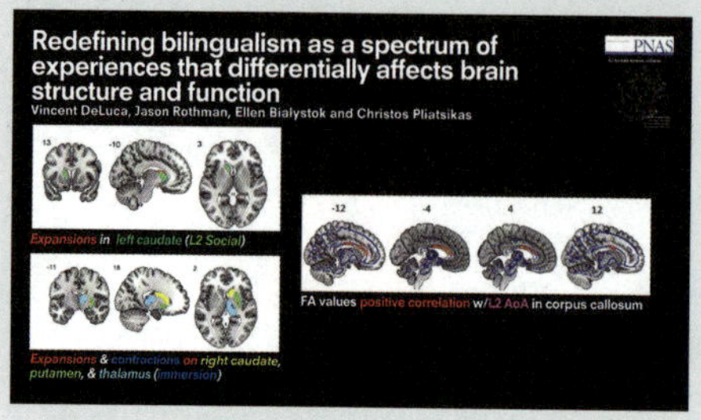

Fig. 9

with friends, at work etc—on a daily basis, and much more. Keep in mind that all the participants are bilingual, but individuals varied significantly in their experiences that form the basis of their very own bilingual profile. Recall, the idea was to see if there is a relationship between individual language experiences and brain structure. Yes, there most certainly is. I will note two important findings that can be gleaned from the images in front of you. First, individual bilingual experience along a continuum, affects structure differentially. And, crucially, changes are not random. Rather, they are seen in exactly those areas used to regulate the languages bilinguals have, where language and cognitive control overlap. In another study, we examined bilingualism effects on brain function, as opposed to structure. [10] Using the same participants from

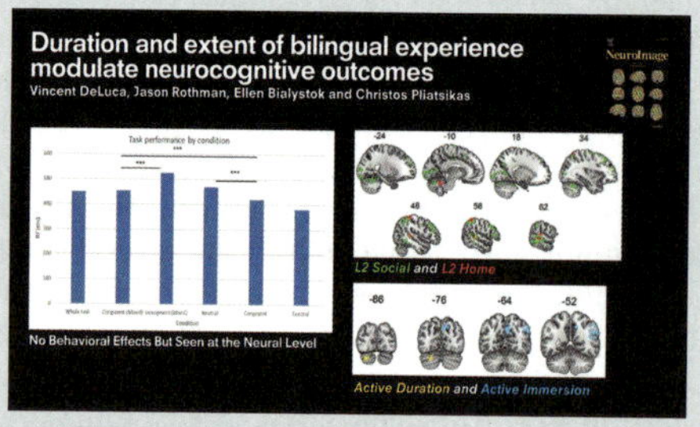

Fig. 10

the previous study, we looked at their brains while performing a simple task that measures aspects of cognitive control. As can be seen in the graph, the tasks worked. We see the expected effects across the groups, that is, at points in the task, when it is more demanding, people are slower. This is true regardless of individual differences of bilingual experience. In other words, bilingual experience do not predict individual behavioral performance. Perhaps, this is so because everyone was so darn good, making it difficult to tease anything apart. But is everyone's brain the same, when arriving at these indistinguishable behavioral performances? No. Individual differences in bilingual experience correlated to how the brain reacted to doing the tasks. For example, the longer one had experience using English, the less brain resources were required for optimal task performance, indicating better management of energy, and, thus, increased brain efficiency. It has been my pleasure to walk you through the basics of how bilingualism can be viewed as a lifestyle factor affecting the mind and brain. Let me leave you with some thoughts in summary. [11]

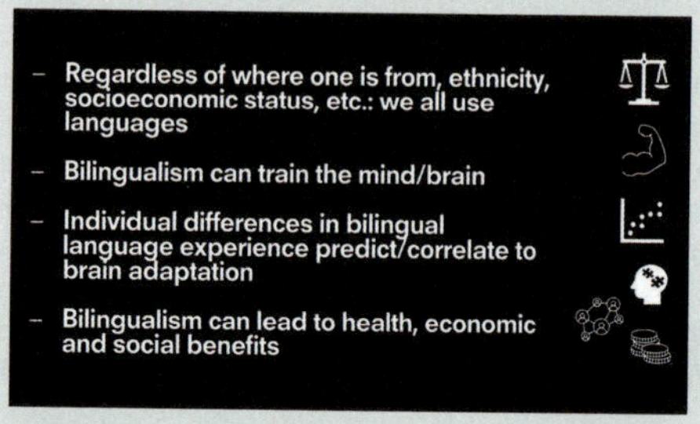

Fig. 11

Regardless of where one is from—ethnicity, socioeconomic status, etc.—, we all use languages and bilingualism surrounds us. Bilingualism can train the mind and brain, but individual differences in bilingual language experience predict or correlate to brain adaptation. Bilingualism can lead to health, economic and social benefits, but much more integrated work across many disciplines is needed to understand this fully. Of course, the work I get to discuss is a reflection of a huge team of researchers in my lab, the Psycholinguistics of Language Representation Lab at UiT, the Arctic University of Norway, and my many talented colleagues in the AcqVA Aurora Research Center and the Department of Languages and Culture, more broadly, in Tromsø, and the Nebrija Center of Cognition at the Universidad Nebrija, Madrid, Spain.

Conversation 3

ALICE ROBERTS, JASON ROTHMAN, ELLEN BIALYSTOK

ALICE ROBERTS In this conversation we'll be looking at the effects on cognitive function, the effects on brain function of possessing a bilingual brain and looking at how psychologists and linguists can delve into the inner workings of the brain to understand phenomena like bilingualism. This might sound like quite an esoteric question to pursue, but understanding this topic could really help us in supporting people to have healthier brains into old age and in combating diseases like dementia, for instance; and that's on top of all the culture that's opened up to people who learn more than one language. Jason, I wonder if you could start us off tonight by giving us a brief summary of your lecture.

JASON ROTHMAN Today we are here to talk about bilingualism and its potential to condition or change our minds and brains. Bilingualism is the state of knowing for communication in more than one language. Languages are many things: they are social—we use them with our family and friends; they are geographical, Italian lives in Italy; they are often written as well as spoken; they permeate all aspects of life; they are parts of our dynamic identities and indeed, how we form them. For our purposes today let me simply highlight that, in addition to living in the external world, they very much inhabit the black box of our mind. This means that languages are bound to interface inevitably with our cognitive systems. A key finding in research over decades is that when an individual commands more than one language, these languages remain activated ubiquitously, even when one or the other has no apparent need in a given context. Managing this dual activation requires a complex set of mental processes so that the right one is selected or the intended ones are interchanged appropriately in a single context, as needed. This balancing act or mental tug of war creates competition that must be resolved. This is thought to strain and thus train our domain general cognitive systems, attention, cognitive control, leading to adaptations for cognition and structural and functional changes to the brain. But not all bilingualism or all bilinguals of particular types are the same. Bilingualism is a continuum as opposed to an absolute state of being. Research shows that the degree of engagement or the opportunities one has to exercise their bilingualism correlates to the degree of adaptations to the mind and brain. In this sense, bilingualism is not so different from other factors that correlate to changes in the mind brain, like continuous engagement with physical exercise. But bilingualism is rather unique, it is literally everywhere. Over half the world is bilingual. Simultaneous dual activation means that its potential for straining, and thus potential training of the mind brain, is unrivaled by other factors. Importantly, bilingualism entails increased equity in many ways. It is potentially more accessible to more people and it is a conduit for social justice at many levels.

ELLEN BIALYSTOK The research from my lab, as well as many other groups over the years, has shown that bilingual experience has a measurable positive impact on an individual's ability to perform very particular cognitive tasks that we roughly in psychology describe as executive function tasks. Moreover, we were able to detect that brain structures and functions in these bilingual brains had been reorganized, so these effects had also been found in older age, but all the participants in those studies were undergoing healthy aging. An intriguing question was whether any of the protection from bilingualism that we had seen in other groups could also be found in the cognitive and brain levels after an individual had been diagnosed with a clinical cognitive impairment, such as dementia. It was a high-risk question, because our research to that time had been based on the cognitive effects of these tasks, on things like executive functions. But dementia begins with memory complaints, and there was no real evidence, at the time, that bilingualism had any impact on memory. Our investigation of this question was based on the notion of cognitive reserve, which is the idea that cognitive function can exceed brain function if it is supported by experiences that train certain neural pathways. The known sources of cognitive reserve are things you have all heard of. They are things like formal education, aerobic exercise... so overall through this program of research our results have confirmed that bilingualism too contributes to cognitive reserve, enabling older bilingual adults to function at a higher level than they might otherwise be able to. Some of the specific results that I talk about in the lecture include that, at similar cognitive levels, older bilinguals are dealing with more brain atrophy than monolinguals but it doesn't impede their cognitive function. They function the same as monolinguals, although if you look inside their brains there is more deterioration.

We have also shown that if you match those older bilinguals to a new group of monolinguals using brain structure as the matching criterion, almost half of the monolinguals, whose brains are structurally equivalent to the bilinguals, have been diagnosed with a clinical cognitive impairment. None of the bilinguals have. Bilinguals with similar brain structures are able to cope with those compromised brain functions, those compromised brains in function normally. Finally, in a somewhat dramatic finding, I think, there is a very large group of studies that have been conducted internationally and have all converged on this finding that, on average, bilinguals are older than monolinguals when they are diagnosed with dementia. We have looked further into that and found that if you match these monolingual and bilingual patients with dementia for symptom severity, then again the bilinguals have more compromised brains and, indeed, they have more neuropathology. These examples are consistent with the interpretation that bilingualism provides a source of cognitive reserve and therefore it allows older adults to maintain normal cognitive functioning despite increasing disease pathology. This is important. There is still no effective medication to prevent or even ameliorate the devastating consequences of dementia. The precise mechanism that drives this cognitive reserve is not completely understood, but we do know that it reflects the brain's enormous capacity for neuroplasticity. The brain constantly, continually adapts to everything in which it is engaged, and it is those adaptations that improve the brain's response to similar situations in the future. We also now know that bilingual experience is one such activity that improves the brain's responsiveness to cognitive challenges in older age.

AR There are some very interesting questions that have come in, but I want to start off with some very simple questions for both of you, just to lay the foundations before we start to get into this subject in more detail. Jason, I want to start by asking you what I think is a devastatingly simple question: What do you mean by language? What counts as language?

JR It is a devastatingly simple question, it seems, but it is potentially devastatingly not simple. I was just going to start by saying what a wonderful simple question it would seem to start with, but in fact it is really not. It is almost like asking a philosopher "What is the meaning of life?" but let's unpack it. Often people will tell you it is an army, a navy, a governmental structure that is going to determine what a language is, but of course that immediately fails. In my talk I have discussed that there are 195 to 197 countries, there are well over 5,000 languages and infinitely more, probably. So what can we agree a language is? It is a rule-governed system for communication, sound to meeting or signed to meeting, in the case of sign language, correlations that allow people to reliably communicate. Language is quite special, something that only humans have. Therefore, it is different from just simple systems of communication. It enables us to code and express the full range of our emotions, our thoughts, and the world around us. It will minimally have a phonological or a sign-based phonological system or sounds that reliably map to meaning morphological structure, for example, those small units of meaning that tell us if we are talking about the present the past or the future. Words, a lexicon, a very large lexicon for most people. Rules for putting those words together, or syntax, and relating them to meaning, or semantics, constraints on how the system is used in discourse or pragmatics, and this is socially dependent or sociopragmatics, as well. Many subsystems we do not traditionally call languages, by this account are actually languages. Maybe you are more used to calling them dialects or vernaculars or varieties. A language need not be a so-called standard one, whatever that means, one arbitrarily promoted by a given society. Not all languages have writing systems, although many do. Unfortunately, not all are equally prestigious or valued, although there is no linguistic justification for why this should be. If it is a rule-governed system for communication, as described, then it is a language. For our purposes, the presence of more than one of these systems can induce competition, although factors related to say how distant or close these systems potentially have knock-on consequences. This is an empirical question.

AR What I am particularly interested in is that you said quite explicitly that it was something which was unique to humans. In biology we spend quite a lot of time looking for things which are uniquely human, because we are interested in what makes us unique as a species. Very often what we find is that there are differences in degrees rather than absolutes, but you really think that language is something very special when it comes to humans then.

JR Here we have to distinguish between what communication is, and we know that many animals have very sophisticated systems of communication, including little ones that I know Ellen and I don't like very much, like bees. Bees have incredibly wonderful systems of communicating exactly what they need, where that pollen is, and they are able to communicate that. Also dolphins and whales. Across the board I personally believe my dog can speak, maybe he is an exception. But as it relates to what language actually is, what we are able to do, to talk about not just the present and the here and now, but indeed ponder and contemplate the past and project towards the future, humans are pretty unique. Language as a system is a pretty unique means by which we are able to do so, to the best of our understanding quite unique, very different from our closest cousins, so to speak. Even primates are quite different; can they learn words? Yes. Do they learn language? No.

AR Interestingly this is something that we have covered in a previous "Human Brains Conversation": we looked at language in the last conversation and really did nail that about chimpanzees, in particular, our closest living relatives. They can learn the equivalent of words, but just don't seem to be able to string them together in the way that we can. Ellen, I want to ask you an equally simple and devastating question that moves us on to tonight's topic. What do you mean by bilingualism, and is it possible to be more or less bilingual?

EB Maybe there really aren't any simple questions anywhere. I think you answered the question with your first question, with your sub question. If there can be degrees of bilingualism then that might tell us something about the definition. We tend to think of bilingualism as a thing, as a category. We tend to use it to sort people into groups. If that is the case, it should have clear defining features. But bilingualism isn't a category, it is, as you say, a continuum. We all have different experiences with language and languages, up to and including people who can use more than one language fluently in all situations. I guess we would agree that those people are bilingual. At the same time, we might also know people who can only ever use one language but even those people might have taken a language course in school or gone on a trip to Italy and learned all their favorite food items in Italian. But we would probably agree that that is not enough to change our description as monolingual. Between those two categories is a lot of territory and we now understand—and this is really a leading edge, a frontier of research on bilingualism now—that bilingualism is not determined simply by how much language you know, but also how you use the language, in how many contexts, with what characteristics, and how you divide your time and resources between languages and contexts, and on and on. It isn't even just that bilingualism is a continuum, it is a multi-dimensional set of experiences that define language use. The important discovery we have now made—and I have just alluded to the fact that this is the research frontier—is that the more bilingual experience someone has, defined by all these dimensions, the greater impact we find on those cognitive and brain outcomes I was talking about before. The effect is continuous and it is related in a very quantitative, precise way to this multi-dimensional continuum of bilingualism.

AR It is exciting to hear about these new concepts which are just emerging and these new discoveries that you are making in your research. I know that you know each other very well and that you do communicate with each other. But I wondered if tonight there were questions that you would like to ask each other having listened to each other's lectures and brief summaries. Ellen, have you got a question for Jason to start with?

EB Jason is a linguist, I am a psychologist and we meet in the world of bilingualism. One thing that everyone agrees on is this concept that Jason talked about, which is joint activation. Both languages are active and we all agree about that, but it is how that is framed and the implication of that concept. I would like to know more about the linguistic perspective on what that means. Jason, in your lecture you had that lovely metaphor of tug of war, so language is competing for something, and then there would be consequences. Can you explain from a linguistic perspective what that competition or that tug of war looks like, and more precisely, what the consequences would be?

JR Let me say linguistics is very diverse. It is like saying "what does psychology say about x," because psychology is so many things. Of course, linguistics is incredibly diverse and here what I would like to do is focus the attention on the individual, in the concept and the context of a linguistic approach to something. What I am interested in here, what linguistics would have to say about this, relates to the domain of language use. Language use is broader than linguistics per se, and I think you can agree on that. By the mental tug of war I am referring to the language usage and the management of it in the real world, as required by the sociolinguistic context. We know that being bilingual is an inherently complex context of meandering through real world activities in which language are ubiquitous and coded differently for individuals who have command of more than one, maybe three or four languages. And of course, when and how and the context in which it is appropriate to use one or the other or intersperse both of them—a concept called "code switching," or "code mixing," which is ubiquitous for at least subsets of bilinguals. From a linguistic perspective, I think we need to pay very close attention to several things that make bilinguals vary and lead to this tug of war. For example, whatever variables lead to the acquisition of language and thus result in proficiency. Research has shown that proficiency matters as one of those variables that are deterministic and predictive for outcomes in the cognitive domain. The social dimensions of language use, because they are opportunities one has for the engagement. This then all feeds into at a higher level: the connection to cognition, how languages and their selection are managed, which translate to the opportunities for bilingual effects. By the way, they are also the same opportunities that relate to the outcomes of language and bilingualism variation. Something that is very near and dear to my heart for a very long time happens primarily in the language domain. How language is represented, I want to say, which at least some domains of linguistic are seemingly consumed, if not obsessed with, I would say matters very little for the topic at hand. Competition of more than one language is what matters, and I don't see personally how any particular version of linguistics would say otherwise or be in conflict with this. It is about the social context of language, I would say. I think that this leads somewhat nicely to a question I have for Ellen, for the benefit of everybody. You identify joint activation as the crucial feature of bilingual language use, which I agree with. How does that lead to the cognitive consequences you are describing?

EB I guess the difference in our answers to what is essentially the same question is that I don't frame any of my answers in terms of language. You describe these languages, the languages are in competition, the language processing system, the production system, you are very specific that it is the production system that needs to figure out how to produce words and so on and so on. Eventually, at the end of your explanation, you got to what for me is the crucial part and that is the leap out of language into cognition. I have to say that leap in your explanation was a little vague. My answer actually starts there. I don't really have much to say about the linguistic competition, I think we have lots of things active in our minds all the time, but they don't interfere with performance. While I am talking to you I am thinking about what I am going to make for dinner in some dark corner of my mind, but it is not interfering with this because it is a completely different space. The unique thing about language is that these two jointly activated languages are sharing space, they are right there together, so for me that is not a linguistic problem, it is a cognitive problem, because I need to know where to direct focused attention. Focused attention is the primary cognitive process in everything, difficult, smart, advanced that we do. The brain has to figure out how to avoid a complete breakdown by saying "Oh my god, I got to say something but there are two languages going on here, what do I do?" and the brain says "Don't worry, we have got a region that was designed exactly for this problem" and it is a set of processes from part of the brain, selective attention, is what it does, it says you need to look here and ignore that. Because that solution to the many languages bumping around in one's brain is cognitive, not linguistic, you are using a set of everyday cognitive processes all the time. This is the adaptation. These cognitive processes that guide attention, selection, and these very focal high effortful activities, adapt to being needed all the time for language and they get smoother, they become better. It is an adaptation of these processes in the front of the brain whose primary reason for being there is to select what you want and avoid what could be interfering. That is what bilingualism does, it puts this seed, it puts this sort of irritant into language processing and the rest of the system has to respond. So it sends in selective attention.

AR It is very interesting listening to you talking about this sharing of a facility in bilingual brains and the fact that essentially your brain is working in both languages at the same time all the time and then it is up to you where you are directing your attention at any point. It sounds as though there is almost a metaphor for that in physical fitness, where if you train the body to its limits, pushing at those limits, then your bones and muscles will respond—they will keep up. I wonder if that is a useful analogy to make. You are saying that you are building resilience by having that strain on the brain that bilingualism creates.

EB I think that is a very good analogy. You have to use your brain and if you are engaging your brain in challenging activities all the time, your brain will respond largely. So yes, I think that is a good analogy.

AR Is the brain activity of a bilingual child who's learning and developing both language and cognition different from an adult bilingual with full cognitive development? What is the difference in children compared with adults who are becoming bilingual?

EB I can say something about it and I am going to divide it into a couple of parts because it is a very big question. Children who are just growing up and becoming cognitively more developed, and are learning two languages: this is the setup. Let's take the cognitive part first. Mostly, things roll on in the usual way, but we have seen even in children who are three and four years old that if they have been learning two languages from birth—they are in a bilingual home, they are speaking two languages—and these kinds of tasks that I was talking about, these executive function tasks that are defined by their need for selective attention, because there is some kind of competition, even at three and four years old, we have a body of research that shows bilingual kids find them easier. These are small differences, bilinguals find them easier, they master them a little bit sooner. All the kids are going to learn how to do these simple tasks. Monolingual kids six months later or a year later will be doing them just as well. But we do see a precociousness on these particular kinds of tasks for children who are being raised with two languages. In terms of brain, that is a little more complicated. There isn't a great deal of research on children's brains so I would be reluctant to make any general comments on how their brains are developing.

JR To follow from that, one thing that you have to bear in mind is the timing in life where you are likely to see these types of effects, given the types of tasks that psychologists and linguists who apply psychological methods use. You asked about children versus adults, and I think one thing to keep in mind is that children are not at peak development of their cognitive abilities. Of course, Ellen in her talk spent a great deal talking about when the other side of that happens, and that is after a very normal process of decline. When you are looking at task performance on executive functions, where we tend to see the most and that is students in psychology pools often, so 18 to 21 from universities, or young adults in general, you have to keep in mind that they are at a peak level of cognition. The question is what you can get from these types of tests. We assume, and we have good reason to assume, that they mimic some of the everyday things that we expect, changes an executive function might map on to. You have a really good opportunity when people are young to see this effect, because there is a compensation, so to speak, on either end of the age spectrum, but potentially not as easily seen when you are looking at adults. That is something to bear in mind as well, and I think relevant to the question.

AR Somebody from the audience asks: what is your view of the bilingual disadvantage? What is the origin of that idea? And quite often research focuses exclusively on the advantageous aspects. Perhaps it would be good to hear more about the other side of observed cognitive adaptations to bilingualism. Are there any disadvantages? What do you think? Could there ever be a disadvantage to bilingualism?

EB There could be. Let me say that the disadvantage that people cite most often is that if you give monolinguals and bilinguals a vocabulary test in one language, the monolinguals will get a higher score, to which I say who cares, because the bilinguals also know another language and it doesn't matter. There is zero evidence that these reliable differences in vocabulary translate into any kind of verbal disadvantage. That is one thing. There is something that is, at this point, highly speculative, because it is based on a very small number of studies—two for sure, maybe a third qualifies. I was talking about bilingualism finding these benefits on cognitive performance by virtue of the need for bilinguals to use that front part of the brain all the time, constantly, and train those executive function selection processes so they become more robust. That is what happens most of the time. But there are circumstances in which individuals, for completely unrelated reasons, have some kind of disability or dysfunction that affects those executive function, frontal processes. For example, clinical attention disorders include biologically based impairments that prevent individuals with these attentional disorders from using those attentional functions properly, easily, efficiently. What happens then when a bilingual individual, for whatever reason, also independently happens to have an executive function impairment on attention? As I said, there are only two studies that have looked into this in detail and both found the same thing—that bilinguals with diagnosed clinical attention function impairments performed worse than everybody else on these executive function tasks. Not only did they not show any benefit from being bilingual, but they showed a disadvantage in impairment. That is because the key to the bilingual advantages is removed from the equation and they need more effortful attention and they don't have the resources. Much more research is necessary. I offer that as a speculative interpretation from a very small number of studies. I think we have to be open to all kinds of results, so I would never say absolutely, categorically there could not be a disadvantage of bilingualism. That wouldn't be true.

AR It makes sense in terms of what you have said about the mechanics of bilingualism in the brain. Jason, any extra disadvantages you could maybe throw in at this point just to level things up a bit?

JR One thing I would like to say—and I am positive that Ellen is going to agree with this—is just the term itself. One thing I want to comment on is this notion of advantage versus disadvantage in juxtaposition as opposed to what they are, which are outcomes, they are outcomes of something that we are observing scientifically, that we are observing in our real lives and socially. I prefer to use that word, what are those outcomes? And understanding then that they are trade-offs for everything. Why potentially could there not be or would there not be in the domain of bilingualism? I think we can make that point if we had time—if we have time I can make it quite saliently. But let me bring up some other ones because potentially, from a linguistic perspective, indeed there are probably parents and individuals interested out there. Often it is said that there are delays in the acquisition of grammar, there are certainly other types of tasks that show that potentially people who are categorized as monolinguals are better in naming and retrieving and recalling, and so on and so forth. Potentially, those exist but they exist in the context of only having to meander through one lexicon or one grammatical system, as opposed to two. Those things are quite expected and certainly disadvantage is not a word that I would use to describe them at all. There are potentially societal ones, as well. We have to remember that bilingualism exists as a construct in the actual world. Very unfortunately, in the context of where monolingualism predominates, the notion and how this is contributed by the society of bilinguals may be feeling this in-betweenness, this otherness, and other consequences of a monolingual default bias when—as we have mentioned many times—that is not the default of the world in fact, it just potentially is seemingly so, artificially so in some of our contexts. Potential feelings of a self-dividedness. To add to the mix of that, it is not all roses and it is not intended to be. Any research program that is showing a positive effect or an advantageous effect as opposed to an advantage per se is, of course, open to the reality of the world in which we live, some of which we can change and some of which we can't, but all of which are probably at least the ones that relate to language proper, like this potential slight delay or naming effect. We can probably live with those, given the at least societal and social benefits of being bilingual, and of course, the ones that at least ample exposure and engagement with bilingualism seems to convey for the mind brain.

EB I agree that I don't like these binaries advantage/disadvantage. Following up on Jason's comment, another—I am going to call it—downside that has been reported to me—anecdotally, this isn't science, this is something that people have told me—as a bilingual, these people report to me that they feel that they have never mastered any language completely, that they don't have a native language which is really their own. I have heard it several times from different people with different backgrounds and experiences, but along with Jason's comments about the larger social and other implications of being a bilingual person in the world, I think this is one we should think about.

AR Somebody else asks: has any research been done on the effects of bilingualism applied to sign language? That was something that you did mention, Jason, in your summary, so I wonder if you could briefly tell us about that.

JR If it is on the cognitive domain maybe Ellen would like to answer that.

AR Ellen, could you tell us about any research just briefly looking at sign language and bilingualism in that respect?

EB As Jason said, it is a language, so everything that is true of language, is true of sign language. There are some differences. The cognitive effects are not identical. One important reason they are not is that, to the extent that the cognitive effects come from, you need to only say one thing at a time, I can't say "dog" and "chien" at the same time, I have to choose. But you can say "dog" and sign "dog" at the same time. It is a different kind of competition but there is still an adaptation and there are still cognitive implications. They are just not identical.

AR And is it ever too late to learn another language?

JR Let me stress that. In fact, if you watched on the talks, you'll see that Ellen went over a really interesting study looking at language learning in older age, showing that there are at least some knock-on effects for behavioral task performance as a result, equaled at least to other cognitive trainings. It really depends on what you mean by that, what you are satisfied in terms of what learning means potentially. If your goal is to be indistinguishable in your language from other sets of native speakers, that is ameliorated over time for various reasons. We can have a whole different discussion about theories, predictions, debates, and controversies within that domain alone. The answer is no, because language is learning, and we learn things all the time throughout our life. We continue to learn our own language throughout our lives. There are words today, in my mid-40s, that I don't know in English that I will know before—hopefully—I cease to be. No, it is never too late, and the benefits at multiple levels to be able to go to Italy, as Ellen said before, and order your food or learn all your foods in Italian. Absolutely, it is not too late to learn. But we have to kind of have realistic expectations of what the outcomes of that will be.

4

Language, Music, and the Brain

STEVEN MITHEN
STEFAN KOELSCH

MODERATED BY ALICE ROBERTS

Music, Language, and the Brain. The Past and Present

STEVEN MITHEN

My name is Steven Mithen. I am an archaeologist at the University of Reading, in the UK. I am interested in the relationship between language, thought, music, and the brain. I am interested in these because we don't understand the nature of being human unless we understand the configuration between them. But I am also interested because unless we understand how these interact, we don't understand the nature of prehistoric behavior in the past, and if we don't understand the past, we don't understand how the present has come into being. Music and language are universal in the modern world, but what is their relationship? When and how did they arise? What is their relationship with the brain and with thought? In these few images here, you can immediately appreciate the close connection between music and language. [1] If you listen to

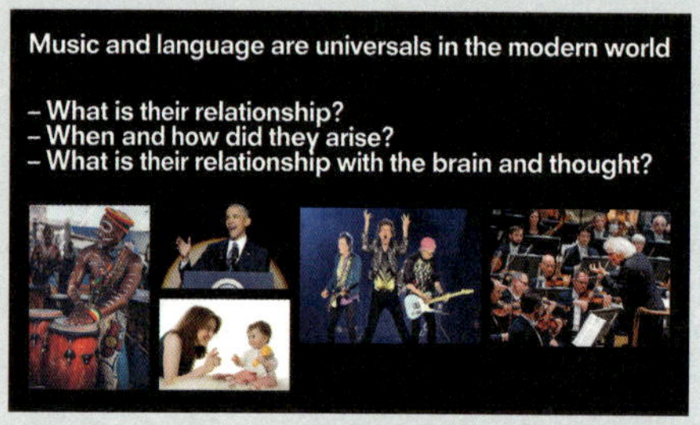

Fig. 1

the oratory of Barack Obama, or if you listen to a mother interacting with her baby or child, you immediately recognize that spoken language partly communicates by its musicality, by its use of rhythm and pitch. Also, looking at these images you can see a very close relationship between spoken language, gesture, and music. Obama, the mother, the baby, Mick Jagger, Simon Rattle, the conductor, are all using gestures as an integral part of their communication. Also, we can recognize that language and music are fundamentally social activities. Obama is talking to probably millions of people and as he does so, he is building up a relationship with those, just as the mother is with her baby. Of course, when people play together, whether in a rock band or in a symphony orchestra, those people are working very closely together, to almost merge themselves as an individual to become one single musical entity. These aren't just features of Western language and music, non-western music is illustrated by the drummer on the left-hand side there. That also illustrates another important aspect of the visual presence of making music and language: his body decoration has similarities, I think, with that of the Rolling Stones. And you can see Obama's suit, or the dinner jackets of the orchestra, are integral to the messages that they are communicating. [2]

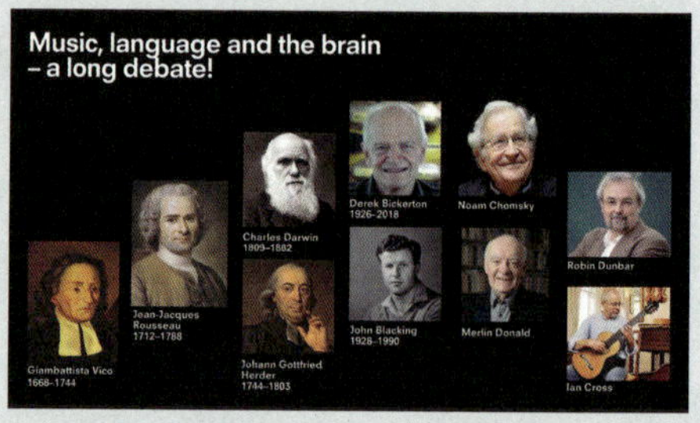

Fig. 2

Music and language are discussed in a wide range of disciplines that make its study challenging: anthropology, archaeology, linguistics, musicology, neuroscience, developmental psychology, and so forth. There is also a very long history, a long debate about the relationship between music and language. Most people listening today will be very familiar with the work of writers such as Derek Bickerton and Noam Chomsky, and more recently Robin Dunbar. Most accounts go back to Charles Darwin. I am as interested in some of the earliest writers. Giambattista Vico, writing in the 17th century, was one of the first to write about language and the nature of language and he was one of the first to suggest that thought was essentially verbal in nature. He argued that thoughts really were a mixture of almost dream-like associations of ideas. He emphasized the role of analogy and metaphor in human thinking. That connects with our modern ideas about cognitive blending and fluidity. Then Jean-Jacques Rousseau, in the 18th century, argued that the inception of language was essentially musical in character, driven by emotion rather than rationality. He also argued that language developed in Southern, warm climates, which has an amazing resonance, I think, with today when many of us would see the origin of language being with *Homo sapiens* in the Middle Stone Age in Africa. Then, Johann Gottfried Herder, writing a little bit later, in the 18th century, was one of the first to emphasize how language was the distinguishing feature of humans against all other animals and that, of course, is a debate that goes on, as some argue that language is special to *Homo* and no aspects of language are found in, say, chimpanzees or other primates. Some would argue that language is special just to *Homo sapiens*, or even just a small number of *Homo sapiens*. John Gottfried Herder's ideas of the 18th century also remain very pertinent. He also argued that language underpinned cultural diversity in the world. The language that you spoke influenced the way you think, the way you behave, the culture you have, which also resonates with some modern ideas. Dealing with music, language, and the brain, we have to look as much back to those early writers as we do to modern neuroscience, as I'll do in this short presentation. My challenge as an archaeologist is to recognize when language appears in the archaeo record. [3] The archaeo

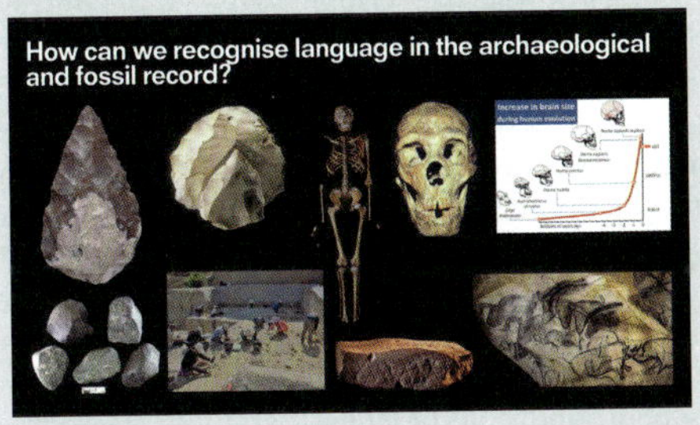

Fig. 3

record is of objects and fossils, and they are all absolutely silent. Some would argue that we can look at the technical complexity of early stone tools, such as that fine hand axe or Levallois tool. The technical skills of how these were manufactured, requiring some degree of planning, hierarchical thought, might imply linguistic capability. Some argue that transferring that technical skill from one generation to the next would have required verbal instruction. Others look towards the associations we find between stone tools and animal bones on archaeological sites, that suggest a group of hominins working together, sharing food, living in a social group. That sociality would imply linguistic skills. The fossil record is important because it provides evidence for anatomical adaptations for language. Many have argued that going right back to *Homo erectus* is the capability to produce complex verbal utterances. I am sure that is correct, I'll come back to that in a moment. Then, we also have the brain size. What were the large brains of hominins, especially about 500,000 years ago, being used for, if it wasn't for language? I'll come back to that issue, too. Then, we have symbolic activity in the archaeological record. That is very difficult to recognize and there are many debates about whether Neanderthals used symbols or not. We see clear symbols in, say, the artifacts and Blombos Cave painted 80,000 years ago, or in prehistoric cave paintings. But the big issue is whether the use of visual symbols also implicates the use of audible symbols, as in words. Recognizing language from the fossil and the archaeological record is certainly not very easy to do, and, to my mind, that evidence gives us an enormous paradox. If we look at the fossil record it is undeniable that by 500,000 years ago, or possibly much earlier, there was the capability in humans to make really complex vocal utterances. [4]

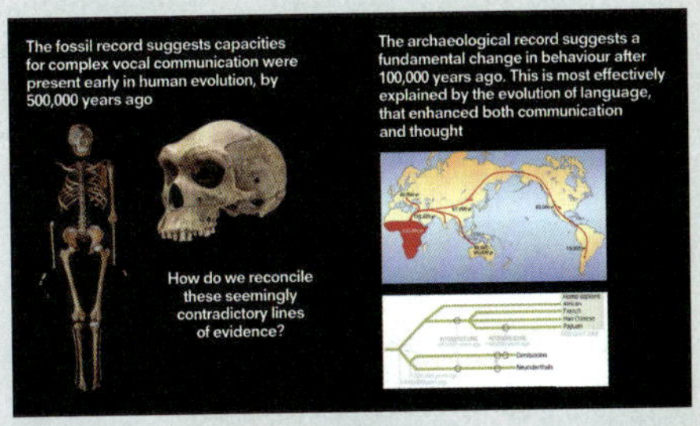

Fig. 4

Look at the skeletons of *Homo erectus*, as we have there. We can see there was an ability to control breathing. There was an ability to make a wide range of vocal utterances, using oral cavities and vocal cords. And by looking at ear bones, we can recognize that a wide range of pitches could be perceived. Then we have this issue of the large brain. What was that large brain being used for, if it is not for what we think is the most complex cognitive capability we have, which is making language? The dilemma, however, is, as we look at the behavior of these early humans such as *Homo erectus* and Heidelbergensis, as this anatomy and brain size are evolving, we don't see any fundamental change in behavior. The stone artifacts and the hunting methods look much the same. If language is the driver of cultural change, why don't we see any change at that time? The only time we do see a really fundamental change in the archeo record of human behavior is around 100,000 years ago. This is when a strain of modern humans evolved in Africa. It is when you find the first use of pigments and graphic designs, especially in South Africa. And soon after this, the *Homo sapiens* disperses into Eurasia—a remarkable global dispersal—entered South Asia, then built boats and traveled to Australasia, colonizing Northern lands, into Arctic tundras of Beringia, crossing into the Americas and then all the way down the Americas into the Amazonian rainforest, and reached Tierra del Fuego. As they dispersed around the world, we know that other humans rapidly went extinct. There may have been some interbreeding but, nevertheless, those other humans as distinct species became extinct. The modern human global dispersals must indicate that a fully modern language capacity had evolved. How can one equate that language evolving 100,000 years ago with the anatomical evidence, that complex vocal utterances and large brains were present at least 500,000 years ago if not significantly more? I tried to do that in my 2005 book, *The Singing Neanderthals*. I was very drawn to the ideas of Jean-Jacques Rousseau, Charles Darwin, and, more particularly, the ethnomusicologist John Blacking who, in 1973, had written about a pre-linguistic mode, musical mode of thought and action. [5] I character-

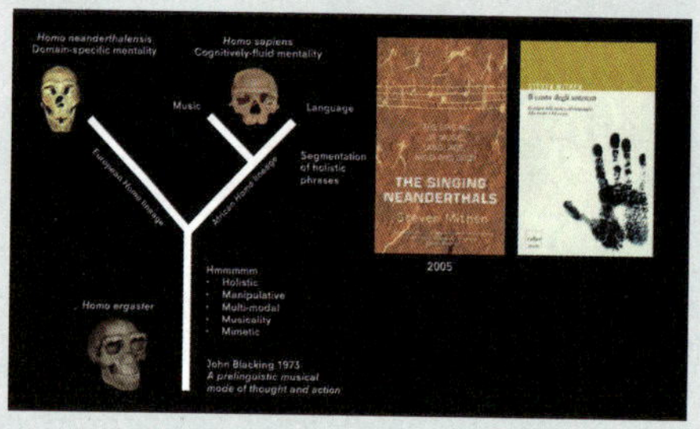

Fig. 5

ized the communications of early humans as being holistic, manipulative, multimodal—that is a lot of use of body language—, musical—making use of variations in pitch, rhythm, and timbre—, and mimetic. That form of communication system I think delivered the social bonding that was required. It enabled a transfer of information, required a large brain, and then continued throughout the human lineage that led to the Neanderthals in Europe. In Africa, however, around 100,000 years ago, for a reason that we don't fully understand, there was a bifurcation in this communication method. That led to the two methods of communication we have today: music and language. I think that occurred through some sort of segmentation of holistic phrases which, in essence, means the invention of words. In a previous book I had also explained the cognitive impact of that. If we look at humans, epitomized by the Neanderthals, who depended on that musical mode of thought and action, I think this prevented them from bringing together their cognitive skills for, say, making stone tools, thinking about nature, or engaging with other individuals. Without the concepts provided by words, they had very limited abilities to innovate and for symbolism. [6] They weren't entirely absent

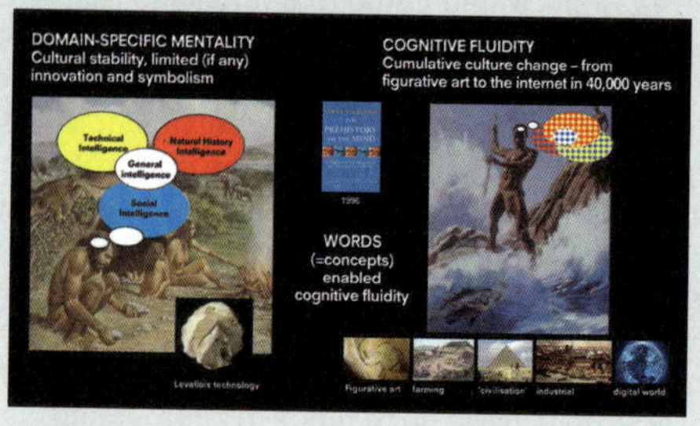

Fig. 6

but they were very limited. Consequently, we see this remarkable cultural stability in the archeo record of non-modern *Homo sapiens*. In contrast, those *Homo sapiens* in Africa, who invented words and evolved language, were able to build up new and complex concepts in the mind, because they were anchored by words, and they could be communicated to others by words. This gave them the characteristic that I called cognitive fluidity, an ability to combine together ways of thinking and stores of knowledge that previously were isolated. We can see the impact of this after the end of the Ice Age, when, in conditions of climatic stability and relative warmth, agriculture was invented in numerous parts of the world. Soon after that, towns and trade, and then civilizations, and within a mere 10,000 years, we have moved from living in localized bands of hunter-gatherers to being part of a global community connected by in a digital world. That is the true power of language. That is what language has delivered. I want to comment on those ideas by briefly remarking on two recent theories of language evolution that have been recently published and argued that these are erroneous or not as strong as they could be because they neglect the role of music. [7] Derek Bickerton, a linguist who's made profound contributions to the understanding of language evolution, very sadly passed away a few years ago. In his last book, called *More than Nature Needs*, he developed his previous arguments, that words came first, or proto-words, and that these evolved or invented—it is not quite clear what he means—by about one and a half million years ago to communicate about displaced entities in the world. He said it was obvious that words came first. That isn't obvious to me at all. He said these were like beads on a string, because they would yet to have syntax or grammar to arrange them in particular ways to provide additional value to their meaning. What seems absent in this theory is that it implies

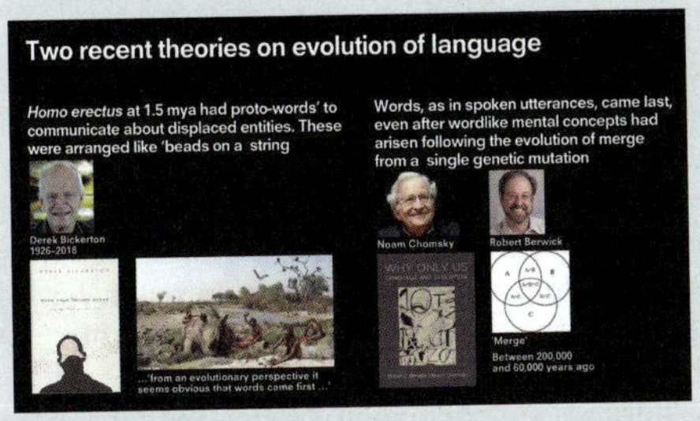

Fig. 7

there was almost silence in that common world, before words are invented. But we know that hominins must have derived vocalizations from their ape-like ancestors. If we look at chimpanzees as an analogy for the much earlier Australopithecus, we know that they had complex vocalizations. These words, if words were invented then, must have been embedded into that complex vocalization. I think Bickerton ignores that to his cost. A very different view has come from Chomsky and Robert Berwick. In a series of articles and their book called *Why only Us*, they have reduced the complexity of grammar to a single operator called Merge. They argue Merge is a cognitive process that allows cognitive atoms, as they call it, to be packaged together and then arranged hierarchically to be used recursively to create sentence-like thoughts. What they argue is that words came after that. We created this complex way of thinking using Merge and only later became able to express that with words. That seems a very strange idea, partly because of their claim that this happened by a single genetic mutation, partly because what would be the point of having those complex thoughts if you can't express them to anybody? Why would they give you any survival value at all? What I like about their ideas is that they argue that grammar, as in Merge, evolved before the use of words. I think that engages with two other really interesting pieces of academic development. [8] One is the greater emphasis produced in recent work on the role of music and social bonding. A major project, led by Patrick Savage, to which I contributed, published in 2021 in *Behavioral and Brain Sciences*, reaffirmed the role of music as social bonding, as being its prime function, and as such implies it has deep evolutionary roots, because we know that social bonding was critical to humans living in the past. There have been very strong arguments that sociality is the driving force of brain enlargement. Then, equally important, has been the work of Aniruddh Patel and Stefan Koelsch in arguing about

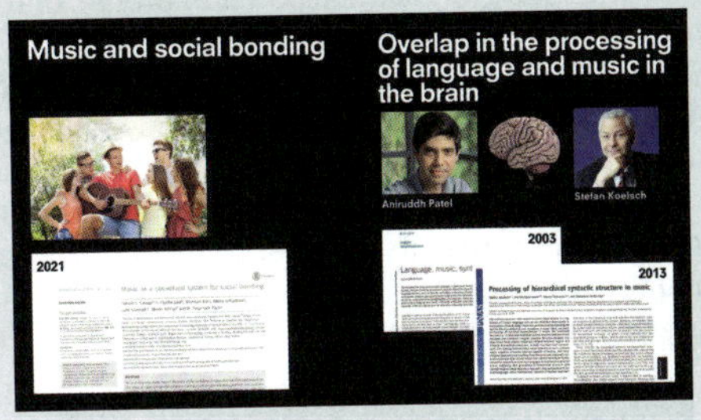

Fig. 8

the overlap of processing of music and language in the brain. Koelsch, in particular, has demonstrated that when we listen to music our brain's process is hierarchical syntactical structured in just the same way that we process language. I find that absolutely fascinating because I think that implies that Chomsky's Merge process is actually used for processing music as well as language. With that, we can return to the scenario that I stated earlier, and rather develop that, by drawing on that social bonding of music and the notion of merge. [9] In this

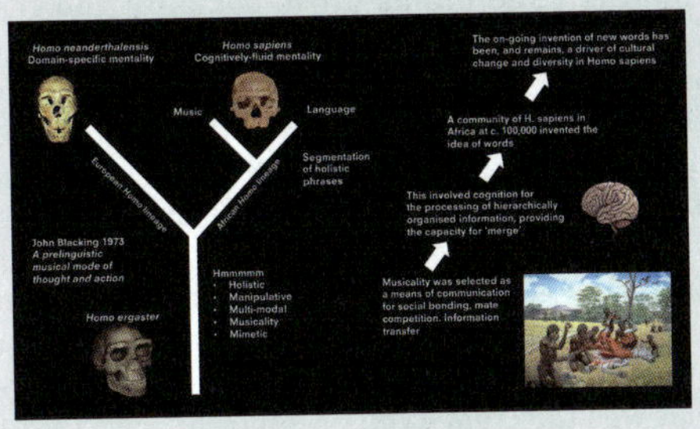

Fig. 9

slightly developed scenario on the right there, I would maintain the view that musicality was selected as a means of communication for social bonding primarily among our early human ancestors. Going right back to the Australopithecines, then *Homo erectus*, *Homo heidelbergensis*, and Neanderthals. Musicality is not only used for social bonding but also to make competition information transfer. That involves cognition for processing the hierarchically organized information within music. As Koelsch has demonstrated, that uses cognitive processes that are very similar to what we use for language, which I would equate to what Chomsky is talking about with Merge. Grammar, which we then used for language, came first before the intention of words, and then, sometime around 100,000 years ago, a community of *Homo sapiens* in Africa invented the idea of words. When words were combined with that grammatical power of Merge, that is the time when language appeared, it emerged. Then, the ongoing invention of words, the ongoing development of grammar, developed not only language but also culture change and diversity through time, and it continues to do so today, with the invention of words being a real driving force of culture and thought. To conclude, I'll go back to these four domains of thought, language, music, and brain. [10]

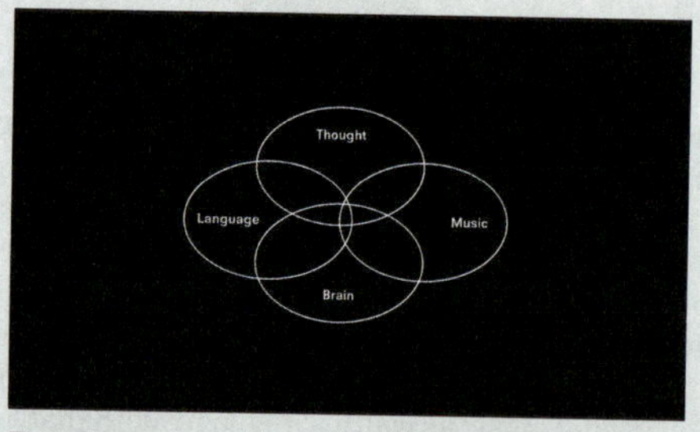

Fig. 10

My view is that these are intimately connected in the modern mind. From an evolutionary point of view, the large brains drove sociality and they acquired communication, which was musical in nature. That led to the selection of cognitive processes that could process hierarchically complex sounds, as we call musicality, as Koelsch has argued. Then, when words were invented, they drew on the pre-existing hierarchical processes to create a new phenomenon called language, this only happening in *Homo sapiens*. That transformed human thought into the cognitive fluid ideas that we have today. That was the same type of mind that Vico wrote about in the 17th century, and generated the cultural diversity that Herder wrote about in the 18th century. I think those early writers have a lot to contribute to how we think about language today, and should be considered alongside the most recent research in psychology and neuroscience.

Music, Language, and the Brain

STEFAN KOELSCH

I will first talk about the musical brain and why a world without music would be a world without humans. When you ask a scientist what separates humans from animals, they usually answer language, math, chess. Douglas Adams wrote: "the wheel, New York, wars." I say music. Music requires the perhaps simplest mental function that just separates us from animals. Humans can hold a pulse in a group, we can clap together in synchrony. What seems very trivial to us is also unique to us. No other species can do this. Whales can sing, but not in a choir. Gorillas pound on their chests, but not as a combo. Viewed in this light, music was the decisive evolutionary step from animals to humans. Research revealed that simply after clapping together, individuals help each other more, trust each other more, and cooperate more with each other. We see these effects already in preschool children. Thus, making music together has a very interesting effect. It unites us, it stimulates social bonding. We do something together. Suddenly, everybody is part of a group. A "we" emerges. Social bonding fulfills a deep human need—the need to belong. When this need is fulfilled with music, it elicits positive emotions, fun, joy, sometimes even happiness. These positive emotions support our health and our well-being. All this was important in the evolution of humans, because social bonding, cooperation, trust, peacefulness, and health brought humans an important advantage: to live longer. I think music was even the main mechanism for social bonding in large groups during the evolution of humans. Analogous to Robin Dunbar's vocal grooming hypothesis, music might have replaced manual grooming as the main mechanism for social bonding during human evolution. According to Dunbar's hypothesis, vocal grooming paved the way for the evolution of language. This view is based on the observation that, similar to manual grooming, human language plays a key role in building and maintaining affiliative bonds and group cohesion. This process is putatively driven by increased group size, which increases the number of social relationships that an individual needs to maintain and monitor. Once the number of relationships becomes too large, an individual can no longer maintain their social networks with tactile interactions alone, and selection will favor alternative mechanisms, such as talking to several individuals at the same time. However, compared to the relatively small group size typical for conversations, many more individuals can make music in a group, and thus establish and foster social relationships. This makes music the more obvious candidate to maintain social networks and increase social cohesion in large groups. Thus, a musical grooming hypothesis seems also plausible to explain the emergence of music, at least as plausible as a vocal grooming hypothesis is for the emergence of language. An important effect of individuals making music together is social bonding. Making music together means producing sounds at the same time, and that is afforded by music's fundamental structural principles—pulse and scale. Therefore, music is a special case of sound. As opposed to all other sounds, musical sounds have a particular structural organization in both the time domain and the frequency domain. On the time domain, the most fundamental principle of musical structure is the temporal organization of sounds based on an isochronous grid, the tactus or the beat. Of course, there are exceptions, such as some kinds of meditation music or some pieces of modern art music, but 99% of human music has a pulse, and in the frequency, that is the pitch domain, the most fundamental principle of musical structure is an organization of pitches according to the overtone series, which results in scales. For example, the pentatonic scales. Inharmonic spectra, for example inharmonic metallophones, give rise to different scales, such as the pelog and slendro scales. Again, there are exceptions, such as drum music, but 99% of human music has a scale. [1] Thus, for the vast

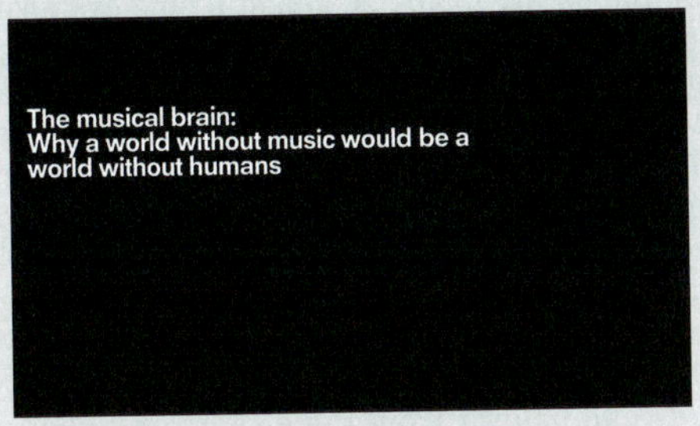

Fig. 1

amount of musical traditions around the globe, and presumably throughout human history, these two principles, isochronous speed and scale pitch, build the nucleus of a universal musical grammar. Out of this nucleus, a seemingly infinite number of musical systems, styles, and compositions evolved, and this evolvement appears to have followed computational principles, principles that have been described, for example, in the Chomsky hierarchy. What about language? Language is actually a special case of music. Both language and music consist of sounds. Language has also important musical features—speech melody, rhythm, timbre—but language does not need a pulse or a scale. Instead, language has words that can communicate specific meanings. Therefore, only one individual can speak at a time. When two individuals talk across each other, we can't understand what they say and it doesn't sound good. With music, many individuals can produce sounds at the same time and it still makes sense. In this regard, language is the music of the individual, and music is the language of the group. This brings along another interesting difference between language and music. For language, semantics is more important than syntax. Mandarin, for example, is a language that has only minimal syntax, and we can make ourselves understood in any language, even if we disregard inflections, declinations, etc. I know a number of Italian words but hardly any syntax, so when I speak Italian it sounds rather funny, but Italians can understand me. Syntax makes the understanding of language easier and faster, but we can make ourselves understood even without any syntax due to the semantics of language. For music, on the other hand, syntax is more important than semantics. When humans produce sounds based on an isochronous beat or scale pitch, we recognize it as music. Without such syntactic structure, we hardly recognize sounds as music. This is because without syntactic structure we could not produce sounds together. Thus, without semantics, there is no language and without syntax there is no music. It goes without saying that music is also a meaningful phenomenon, but in this talk I will focus on musical syntax. How is musical syntax processed in the brain? [2] In Figure 2b you can see two chord

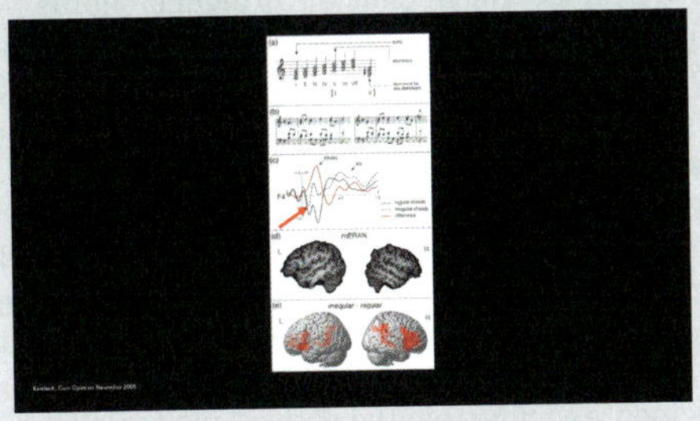

Fig. 2

sequences that I have used in numerous experiments. The left sequence sounds regular. The final chord of the sequence on the right sounds irregular. Note that there is nothing peculiar about the last two chords of that irregular sequence. When played in isolation these two chords sound regular. But they sound irregular at the end of the right sequence. Thus, it is only the previous context that makes the last chord sound regular in the sequence on the left and regular in the sequence on the right. Figure 2 (c) shows electric brain responses to the last chord. The black solid line indicates brain electric responses when the last chord is the regular ending. The dashed line indicates brain electric responses when the last chord is irregular. The red line indicates the difference between the two black lines. Even if you are not an expert in brain electric responses, you can easily see that the two black waveforms differ from each other, which tells us that the brains of our participants made a difference between the syntactically regular and irregular chords. Participants of this experiment were all non-musicians, several of which said about themselves that they were completely unmusical. Their brain responses, however, painted a different picture. We can see that the two black waveforms start to differ from each other already at around 150 milliseconds after the onset of the last chord. Pretty amazing how fast the brains of non-musicians processed this highly complex music syntactic information. As a young scientist in one of the experiments of my PhD, I also localized this brain response and I found that its main generators are located where you can see the yellow spheres in Figure 3. This part of the brain is referred to as Broca's Area, and back then it was thought that this area in the brain is a language area. The results of my experiment showed that this area actually also processes music. In the bottom image you can see results of the functional neural imaging experiment with similar chord sequences. Again, we see the activation of Broca's Area by music and technically irregular chords. We can also see activations in other brain areas, and all of these areas are also part of the so-called language network in the brain. Thus, there is a striking similarity between the music network in brain and the language network in brain. This is illustrated here. [3] The left image shows brain activations when individuals listen to language, and the right image shows brain activations when individuals listen to music. Note the overlap of the activated brain regions, and note that the activations for language are stronger on the left side of the brain and for music stronger on the right side of the brain. I would also like to invite you here to watch the human brains conversations video by Angela Friederici, where she explains the role of these brain regions in language. Results such as these indicate that lan-

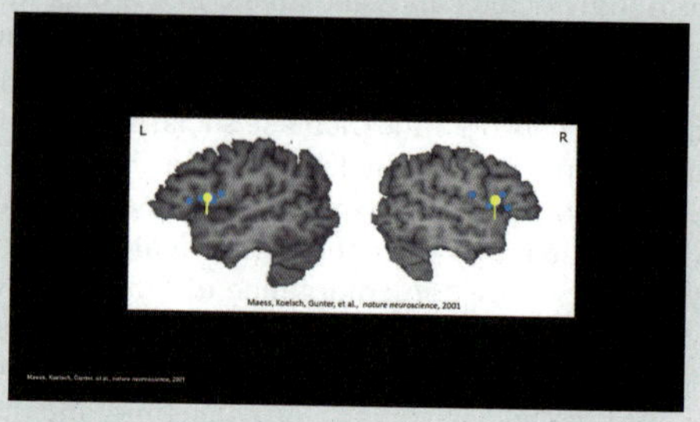

Fig. 3

guage and music are actually processed in the same brain network. Instead of separate networks, there is one single music language network in the brain, only that language is processed more strongly in the left side of the brain and music more strongly in the right side of the brain. [4] That is also why

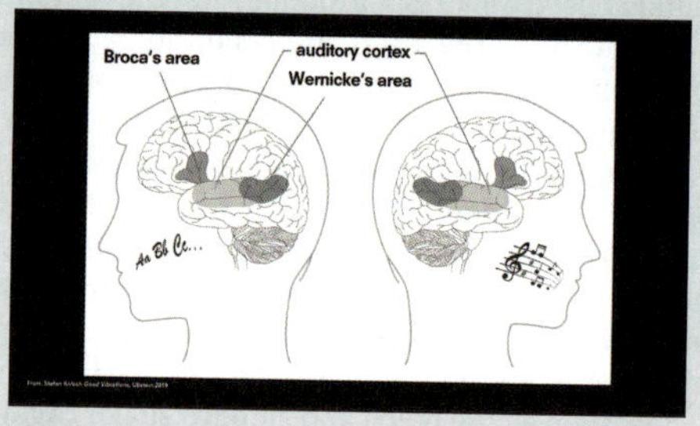

Fig. 4

patients with a stroke in the left hemisphere, who lost the ability to speak, can still sing even with words. I have seen stroke patients who could sing a song so clearly that it was difficult to believe that they could not say a word. Through singing, these patients can actually learn to speak again. Language and music are both part of our human nature. We know this because every human culture has language and music. This shows us that the human brain is a musical brain by nature. Everybody is musical, luckily, because this also means that everybody can benefit from the effects of music on health and well-being. For example, we carry out a project on the therapeutic effects of music in individuals with Alzheimer's disease. In this project, not just musicians can participate but everybody. Another example of the healing effects of music is that we can all use the good vibrations of music to make us feel more positive, more cheerful, to switch off worries or depressive thoughts. This already has health-supporting effects. Even in the darkest crisis, listening to pleasant, encouraging music releases a few healthy, fun molecules in the brain. Where one such molecule is released, thousands more will follow. Next, I will talk about why the sound of voice, and especially of the singing voice, moves us. [5] One reason is that we can powerfully

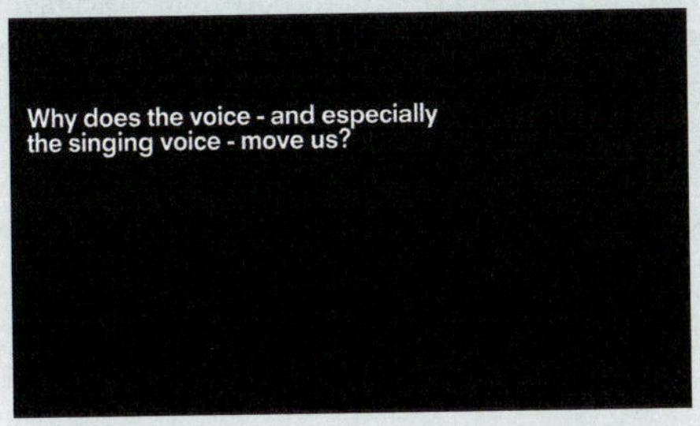

Fig. 5

express our emotions with our voice—laughing, crying, screaming, or speaking with a voice that sounds happy, sad, surprised. These emotional signals are biological signals. We don't need to learn them and we can understand them around the globe. If you went into the Amazon rainforest and met a Yanomami man, you would hear in his voice if he is happy, angry, fearful, or surprised, just as you could see it in his face and body language. When we hear an emotional expression, we often resonate with it. Emotions can be contagious—contagious laughter, crying, panic screams. Music is also a biological signal, especially when it imitates emotional speech. Western music often exaggerates the melodic and acoustical aspects of emotional speech. Music is then like a super expressive voice. For example, like a sad-sounding voice, sad-sounding music is slower, the melody goes downward, has little energy does not have much pitch variation. Whereas, happy-sounding music, like a happy voice, has more pitch variation, more energy, is faster, and sounds brighter. That is also why everybody around the globe can recognize emotions expressed in film music, on any piece of music that imitates an emotional voice. My research team once carried out a study with a tribe in Cameroon playing music to individuals who had never heard Western music before. Yet, they could tell whether that music sounded happy, sad, or scary. Already newborn babies are highly sensitive to the human voice, they hear the music in the voice, they hear when a voice sounds stressed or depressed, or warm and patient. They can't help but resonate with these emotions. That is also why singing to babies can calm babies down and parents, too, by the way, because we breathe out longer when we sing and that already calms our bodies down. Already pre-term infants calm down with slow, relaxing music, which shows us that we humans are sensitive to music even before birth. We are born with a musical brain. It is part of our human nature. In tonal music, virtually all music pieces feature many occasions where the music sounds like an emotional voice and will likely elicit feelings in us when we resonate with this super expressive voice. If we make music together, and even if we simply hear music in concerts together, we resonate together, feel together in time, experience good vibrations together. Experiencing emotions together makes emotions more intense, and it contributes to us feeling socially connected and creating a feeling of unity. Social connection can move us, sometimes so powerfully that we get goosebumps or that we are moved to tears. Being moved goes along with feelings related to social connection, such as sympathy, joy, happiness, or sadness, when a social bond is lost. What happens in the brain when music evokes joy or moves us? It activates this part of the brain that I refer to as the happiness system. [6] This system generates emo-

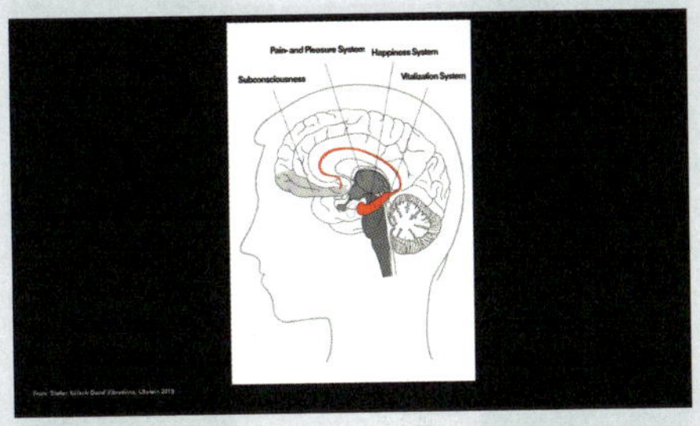

Fig. 6

tions associated with social bonding and social attachments. In my research, I discovered that music is particularly powerful in activating this system. When I designed one of the first brain studies on music and emotion, about 20 years ago, it was considered as risky and bold, because the academic view was that music cannot activate any central emotion structure deep in the brain. This view has been thoroughly revolutionized in the past years, because today we know from many brain studies that music can activate each and every emotional system in the brain. This is illustrated here. [7] Music can activate our vitalization system, motivating us, encouraging us, helping us to calm down and relax. There is actually a nerve that goes directly from our ears into this system. Music can also activate the brain system for pleasure, releasing pleasure molecules that keep our brains young

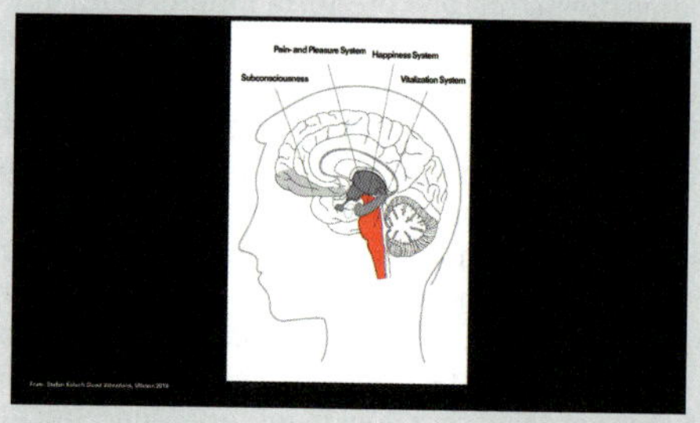

Fig. 7

and healthy, and activation of this system with music can also help to reduce pain. [8] Music has

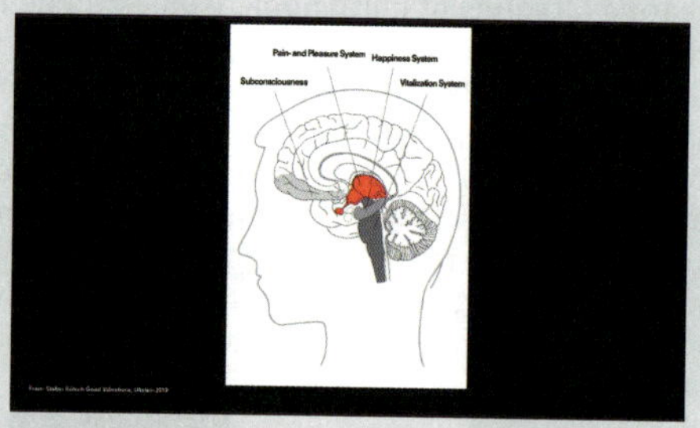

Fig. 8

effects on our subconsciousness. For example, when subconsciousness generates depressive or demotivating thoughts, music can influence thoughts in a way that they become more positive. Thus, we can use music to attain peace of mind. [9] When we lis-

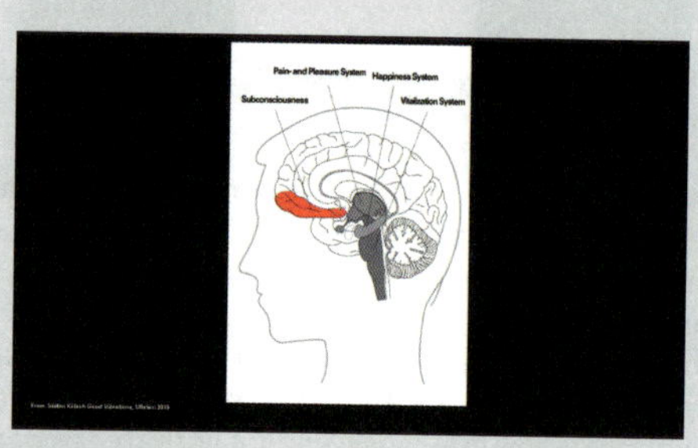

Fig. 9

ten to positive-sounding music, and kind of engage in this music by breathing along with the music or tapping along with the music, our thoughts become more positive. We can easily switch off negative thoughts and negative thought loops. The fact that music can influence every emotion structure in the brain is highly relevant for the therapeutic use of music, because in all chronic disorders and diseases, activity in some of these emotional structures is not quite right, and music can help to get that right again. For example, in our Alzheimer's project, we try to stimulate the generation of new nerve cells in the brain's happiness system with music. We sing together, feel together, experience social connection, we are moved together. These thoughts illustrate that, at its core, music is not just a cultural, happy phenomenon of modern human societies. It is not just auditory cheesecake. Music is deeply rooted in our brain and thus, at the heart of what makes us human. Thanks for listening.

Conversation 4

ALICE ROBERTS, STEVEN MITHEN, STEFAN KOELSCH

ALICE ROBERTS This evening I'm joined by two experts to discuss music, language, and the brain. We'll be taking the long view and looking at how music and language evolved over thousands of years; we'll look at how our brains process music and language today and also explore how music can be used as therapy as well. I'm joined by Steven Mithen and Stefan Koelsch. Stefan, could I ask you to start by giving us a brief summary of your lecture?

STEFAN KOELSCH In my lecture, the first point I was trying to make is that humans hold a pulse in a group. We can clap together in synchrony and what may seem trivial to us is actually unique to humans, as no other vertebrate species has this ability. I think that holding a pulse or a beat flexibly is perhaps the simplest mental function that just separates us from animals. Music was the decisive evolutionary step from animals to humans. Another point I want to make. When we make music together in a group, it unites us, it stimulates social bonds, and when we bond with others it elicits positive emotions—fun, joy, sometimes even happiness. These emotions support our health and our well-being and all this was important in the evolution of humans because the bonding, the cooperation that comes with it, the trust and peacefulness that comes with it, and health brought humans an important advantage allowing them to live longer. I also think that music was the main mechanism for social bonding in large groups during the evolution of humans because individuals can make music in groups, and thus establish and foster social relationships. This makes music an obvious candidate to create social networks and to increase social cohesion in large groups. Therefore, I have proposed a grooming hypothesis to explain the emergence of music, analogous to the vocal grooming hypothesis for the emergence of language. Another point is that making music together means producing sounds at the same time, and that is afforded by music's fundamental stretch: the pulse, the beat of the tactus, on the one hand, and scale on the other. These two principles build the nucleus of a universal musical grammar, and out of this nucleus a seemingly infinite number of musical systems and compositions evolve. Another point in my lecture is that only one individual can speak at a time, if both talk across each other we can't understand what they say and it doesn't sound good. With music, individuals can produce sounds at the same time and it still makes sense. In this regard, I said that music is the language of the group. This also brings along another interesting point of language and music: for language, semantics is more important than syntax. Syntax is more important than semantics, because without syntactic structure we could put sounds together. Thus, without semantics there is no language, and without syntax there is no music. However, because language is a special kind of music, and music sometimes can communicate specific meanings, the brain basis for language and music is largely overlapping separate brain networks. There is one single music language network in the brain, one for language. Only that language is processed more strongly on the left side of the brain and music on the right side of the brain. I am already going out of time. Language and music are both part of our human nature. This is important because it means that the positive effects of music—such as improved mood, reduced stress, and other health benefits—are innate to humans. The last point in my lecture: emotional signals in our voices are biological signals. This is one of the reasons why the sound of the voice and especially of the singing voice moves us.

AR I love the way that you range across that science. You look at the way that the brain processes language and music but then you are really interested in how to use music as therapy. It feeds back into well-being in our modern population as well. Now, to take us back into the deep past, I wonder, Steven, if you could just briefly summarize your lecture.

STEVEN MITHEN I am an archaeologist and my interests are in the relationship between language, music, and thoughts. I am interested in those because I don't think we can understand past behavior and all the artifacts we find in the art records, without asking about language and thought. But equally, I don't think we can understand their relationship today unless we think about these from an evolutionary point of view. People have debated the relationship between language and thought for many years. I think writers in the 17th and 18th centuries like Vico, Herder, and Rousseau still have a great deal to offer. I am interested in combining their ideas with the evidence from archaeology, and the evidence of the latest research in neuroscience, such as Stefan has been undertaking. As an archaeologist, my challenge is to recognize when language appears in the record. That is not easy because the artifacts and the fossils we dig up are all absolutely silent, we can't hear them speak to us. But when I look at all the evidence, I think it is clear that something fundamental happened in the human mind or society around soon after a hundred thousand years ago, because it is after that that we get the first appearance of pigment use and what we call symbols and arts. It is when the *Homo sapiens* disperses out of Africa and into Eurasia, into Europe, and eventually into the Americas. There is a huge cultural rushed event that starts happening and culture progresses. Soon we get the invention of cave art and then, when the Holocene starts, farming and then cities and so forth. I think that stage, just a hundred thousand years ago, marks the appearance of spoken language like we have today. That raises the question of how our earlier ancestors and our relatives, such as earlier *Homo sapiens*, *Homo erectus*, and the Neanderthals communicated with each other. My research has led me to think that this was a communication that had a very musical nature. Rather than having any words they had holistic phrases and they made extensive use of variations in pitch and rhythm. They used gesture and mimics, and this was a really sophisticated type of communication. It was a form of language, but not like language as we know it today. I think that that type of communication gave rise to both language and music in *Homo sapiens* and in the modern world. Research by neuroscientists such as Stefan and others has now demonstrated the overlap in the brain of music and language processing. Both music and language have a hierarchical structure. They are processed in the same parts of the brain. I think this reflects their common origin, although I actually would think that musicality comes first and the type of syntactical nature that we find in music, that the brain evolved to process, actually gave rise at a later date to the grammar that we have in spoken language. Up until 100,000 years ago I think all of our ancestors and relatives were communicating this musical-like mode of language, and then around 130 years ago probably a community, individuals somewhere in Africa, invented the idea of words. Words transformed the nature of communication and whether combined with that syntactical property that was already in the brain, we have a combination of words, a lexicon, and grammar that creates compositional language such as that we have today. That really transformed the way we think, it enabled new concepts to be invented, better communication, new ways of perceiving the world. Basically, it was a new world that enabled this great dispersal of *Homo sapiens*, the colonization of the new world in Australia, the invention of farming and so forth. To summarize, I think the earliest form of communication, musical in nature, gave a cognitive basis for the syntax that we later used in language that appeared after words were invented—I would say the invention of words was probably the greatest event of humankind.

AR Listening to both of your lectures makes us think about music very differently. I think there is a tendency to think about music as being something perhaps a bit frivolous, just about entertainment. What you are both saying is that it is much more than that, and potentially incredibly important in human evolution and to the success and even the survival of our species. I would like to ask each of you if you have got questions for each other. I know that you are very familiar with each other's research but Steven, have you got a question for Stefan?

SM Stefan, as you know I am hugely impressed and, to some extent, dependent on your work because you have demonstrated this overlap of processing of music and language in the brain. What you have argued is that music has a hierarchical structure, such a way that spoken language is a hierarchical structure. But I have noticed that your experimental work tends to draw western music, and especially work by composers such as Bach that we know to be quite well structured. I would like to ask whether you think all music for all cultures all over the world has a similar hierarchical structure, rather than being a sort of a consequence of the western tradition, if you like, and the particular types of composition that Bach composed and that appealed to the 18th–19th-century audience.

SK A short answer is that not all musical styles and not all kinds of music have a hierarchical structure. Perhaps I should just briefly explain to the listeners what I actually mean by hierarchical structure. For example, consider the sentence "The boy who kissed the girl played the cello." The relative clause, "who kissed the girl," is embedded in the main clause, "the boy plays the cello." We recognize this and know that the boy plays the cello. This is how we can understand the hierarchical structure, and it seems that humans are rather good in this regard; otherwise, we would hear that the girl plays cello because, let's say, a computer that is not capable of processing hierarchical structure would hear "the boy who kissed the girl," and then "the girl played the cello." Also, we can leave out the subordinate clause, and it still makes sense. The boy plays the cello. That still has meaning for us, and therefore we use the term "hierarchy." An action sequence can be hierarchically organized. Let's say I start making coffee and realize I am out of coffee powder. I go to the supermarket, get coffee, come back, and continue to make the coffee with a machine. If a melody has a beginning, a middle, and an end, then we can recognize a hierarchical structure. The middle can be embedded with harmonies from a different key, thus we begin in one key, go to another, and return to the home key. There are many folk melodies in which this is the case. A lot of tonal music shows a structure that we can describe in hierarchical terms, certainly all music by Bach. Bach always returns to the home key or the parallel major key in minor pieces. However, there are many kinds of music that do not show this kind of organization. For example, my group once conducted research in Northern Cameroon with members of the so-called Mafa tribe. Their music was not hierarchically organized; it was circular, that is, a repeating motive, and this is typical of many African musical styles. I am not surprised that many kinds of music actually do not have a strict hierarchical organization. We hear hierarchical organization only on relatively short time scales, that is, within a couple of notes or chords. Only individuals with so-called "absolute pitch" can really hear whether the hierarchical structure of a longer harmonic sequence is concluded, that is, hierarchically correct or not, if it is played with well-tempered tuning. But all other listeners, 99% of the audience, would not recognize if Brahms ended in the home key or not.

AR Stefan, would you like to fire a question back at Steven?

SK I first came into contact with Steve's brilliant work when I read his book, *The Singing Neanderthals*, back when I was a senior lecturer at Sussex University. We later met at a symposium, and he recently published a brilliant paper on music and social bonding with other colleagues in the prestigious journal *Behavioral and Brain Sciences*. I have a very pressing question that I have been wanting to ask for a long time: humans, that is, *Homo sapiens*, can hold a beat in a group, clap together, sing together, etc. Apparently, no other vertebrate species can do this. This ability to hold a beat is perhaps the simplest mental function that separates us from other species, and thus possibly represents the evolutionary step from animals to humans (possibly even from animals to early hominids, that is, from animals to the genus *Homo*). Steven, how likely do you think it is that this ability was already present in earlier hominids, in earlier species of the genus *Homo*, before the *Homo sapiens*?

SM That is a really interesting question. I would like to think I know the answer to that because I think that this capacity to hold a rhythm, coordinate with others, entrain with others is a consequence of the evolution of bipedalism around 1.8 million years ago. We often think about bipedalism as just for walking, but in fact, there are very good arguments from human anatomy that are as much for endurance running as for walking. We all know, from when we have been at the gym and on the treadmill, the importance of keeping a rhythm to be energetically efficient when you are doing that. We know how, when running with others, maintaining the same pace also helps enormously in terms of running efficiently. So I think from the emergence of bipedalism—for people that don't know that, that means walking on two legs—obviously it frees our hands for other purposes and there are lots of arguments why we came to evolve bipedalism. I think a consequence of that, an accident from that, was the ability to maintain a rhythm and entrain with other people of the rhythm. That gave a lot of selective advantages in terms of running together, working together, foraging together, and as a consequence of that we also gain an ability not only entrain to run together but also clap together, to dance together, to sing together. I think this evolution of bipedalism about 1.8 million years ago—it was quite gradual—was clearly a turning point around that was also a turning point in our musical abilities. In one sense, our transition from other animals occurred then and that was the first stage of it. Then I think the second stage of it was when, around a hundred thousand years ago, we invented words. Our musicality really began with bipedalism and it had all sorts of consequences from that.

SK That sparks another thought, because if they developed bipedalism, it also means that they lived in trees at the beginning. With bipedalism, they moved away from trees. Could this be the origin of hierarchical thought or the stimulation of hierarchical thought? Because when they planned to do something, they would come down from the tree, do something, and then go back up the tree again.

This suggests that they began to plan actions in hierarchical ways.

SM I think that is an interesting idea. There have been arguments in the literature about how the processing of environmental information, understanding your landscape requires some form of hierarchical thought. I am not entirely persuaded by that, yet. But we know that around 2 million–1.8 million years ago there were huge changes and all sorts of behavior—not only bipedalism but increasing use of stone tools, increasing use of eating meat, probably increasing food sharing—changed sociality. It might be within that nexus of changes that hierarchical thought emerged. I rather doubt it, I like to think that hierarchical thoughts somehow evolved to enhance the type of verbal communication that we have, which is why I like the idea of trying to understand the role of hierarchy and music. Why do we like it? Why do we find finishing on the same key pleasing, whether or not we perceive it or not? Why does that emotionally make us feel right? I would sort of like to throw that back to you. What is it about hierarchical music? Why is it there, what is its function?

SK If we perceive the hierarchical structure, then we usually perceive it as kind of relaxing because it is conclusive. The sequence appears closed and finished. But, as I said before, in longer sequences it is rarely the case that non-absolute pitch possessors actually hear this. They hear whether tones conclude previous tones on shorter, but not larger time scales. The reason why this developed, and why it came into play with Renaissance and Baroque music? My guess would be that with natural tuning, music in different keys sounds very different. The well-tempered tuning was not invented before the late 17th century, and before that, if you tuned a Cembalo in one key, let's say D major, it would have sounded very odd to end the piece in E major, which would have sounded impure and unpleasing. On the other hand, an entire piece only in D minor would have sounded boring. That might be the reason why music developed that featured a beginning in one key, then also changed into other keys, and finally returned to the home key.

AR What this discussion shows is that we are approaching these ideas that lie right at the forefront of research—there is so much that we know, and we know that both of you have published really seminal work in these areas—but you are standing at the edge of what we know and asking questions. I think it is absolutely fascinating. Listening to both of you, it is really interesting to think about how musical ability evolved, that it is related to walking on two legs and that kind of rhythm that you get. But also, it kind of shifts our conversations about music. We are starting to see music as something which is quite fundamental to being human, and actually fundamentally important to human evolution—rather than just thinking about it as a kind of spin-off. We are actually asking: what could have been the benefits, the advantages? What is the selective pressure? Steve, when you were talking about humans falling in step and then music starting to evolve, you can imagine that individuals who—just by virtue of the variation that would be there in a population—had a better ability for music might have a bit of an advantage. You can see how that then spread through the population and how natural selection might actually have promoted that. I think it is fascinating to think that there could have been a selective advantage, potentially, for musical ability. Would you agree with that?

SM I would agree with all that. But I think these aren't just rather arcane academic matters we are talking about. It is really important to our society today, it is the reason why we need more music in schools rather than just maths and English and science. We need kids singing together, dancing together, making music together, because that is all about building up their ability to cooperate together, work together, enjoy together. I think placing music on an evolutionary basis and showing how it is fundamental to us as human beings should really have an impact on our society today and what we recognize is important for building well-being in individuals and in communities. Of course, that is why music is also very good for personal therapy, as well, as Stefan has definitely written about extensively.

AR We will come round to that. I have got some questions coming in from our live audience. There is a question here for Stefan: how would you explain the striking differences in musicality and feeling for rhythm between individuals, even if they haven't had any musical training? How do you explain that variation?

SK It is very difficult to give an ultimate scientific answer, because you would technically need to take children after birth, randomize them to a music or a control group, then conduct studies with them over the course of years. For ethical reasons that is not possible. But there is ample evidence that training hugely modifies musical abilities. As soon as you start training, your perceptual abilities change, your rhythmic abilities, your perception of small time intervals. The perceptual acuity of sounds increases. If you play an instrument where you have to adjust the pitch yourself, like a violin, then your pitch perception becomes more fine-grained. In addition, the sensorimotor representations of the finger increase in the brain, and the neural connections between both hemispheres connecting the motor neurons get thicker, making information transfer between both hemispheres quicker and more efficient. There is evidence that musical training and active musical engagement lead to changes in brain structure and function, especially in auditory and motor areas, along with increased fine-motor abilities and perceptual acuity.

AR I suppose it is like a lot of human features that there is variation; there is variety amongst us. I have got a question for Steven about symbols. Ian Tattersall claims that the capacity to create symbols is the real boundary between animal versus human language. Others suggest that animals may create symbols, such as cats urinating to mark off their territory, but the difference is that only humans connect signs together to generate new meanings. Do you think symbolism is also relevant in music?

SM I certainly agree that creating symbols is important. The definition of symbols is quite challenging. Whether cats and other animals marking territory are putting a symbol down, marking their territories should be defined as a symbol. I am not entirely sure. We have got some evidence that other types of humans like Neanderthals were beginning to make symbols, maybe very late in their existence, but we have got some evidence they are making symbols. I think making symbols is important but I would very much see that as a consequence of having spoken language rather than it being an underlying capacity from which language emerges. I am not sure whether you can describe music as having symbols. The reason for that is that symbols have a shared meaning. If I have either a word, or badge, or flag and I wave my flag around, other people will know that I am talking about this particular country. We have a shared meaning for it. One of the weaknesses of music is that we don't necessarily have a shared meaning for it. We have shared emotions from it, but if I hear music, whether it is a piece by Bach or whoever, I might interpret that with some meaning, some narrative to it, but you, Alice, or somebody else might interpret in a very different way. We don't really have a shared meaning for it. We may have shared emotions about it but we could don't really take inferences from it. It is what Stefan said earlier: music is weak on semantics. I think if it is weak on semantics I am not sure whether we talk about music as being symbolic. It incites emotions, expresses emotions, but in my mind, it is not really a symbolic system of communication in the way that spoken languages are. I don't know whether Stefan would think differently on that one, but that is my view.

SK I would agree. Of course, language is meaningful, most words have meanings, at least all the content words. But let us not forget that music is not meaningless. For example, the James Bond theme symbolizes James Bond, and not Batman. However, the James Bond melody does not represent statements like "James Bond is here" or "James Bond was here" or "James Bond will be here" or the like. As you said, language is much better with symbols than music is.

AR That is really interesting, because what you are saying there with the theme tunes is that we have learned an association, so you can use that piece of music as a symbol, but it is not as straightforward as a word, for example.

SK It would be wrong to say that it cannot symbolize anything, but this comes not anywhere close to that of language.

AR You would be fairly limited in your sentences, wouldn't you? All you could say was "James Bond, Batman, Star Wars, James Bond." I have got a question here that taps into something that Steve was just talking about—the way that music invites emotions, creates emotions. That was something that you talked about, Stefan, in your lecture. I was fascinated by the idea of the universality of that. I was not aware of that. I did not realize that this seems to be something that is universal across all human cultures. It doesn't seem to be something that is learned; a happy upbeat song will always sound like a happy upbeat song. It doesn't depend on which culture you are in. And there is a question about emotion from our audience. Is the kind of emotion stimulated by verbal language different from that stimulated by music? Can there be an emotion that is not driven by semantics?

SK Yes, they can be not driven by semantics. For example, composers can play with our expectations generated by syntactic structures—a chord can elicit surprise because it does not fit the common harmonic structure. Harmonic structure is a syntactic, not a semantic feature of music. In addition to surprise, musical structure can elicit emotions such as tension, for example during a deceptive cadence, and relaxation when it resolves. We can also be certain or uncertain what the next tone or harmony will be. With regard to emotions aroused by speech and music, we actually have two speech systems: a voluntary one and an involuntary one. The voluntary one is evolutionarily new, whereas the involuntary one is the older—we share it with other primates. The old involuntary system produces emotional vocalizations and it colors our voice with emotion when we speak. This older emotional system, by the way, is always active. This is why we speak with some affective tone. Always. Even when we try to hide it. That is why, for example, a good doctor does not only listen to what you say is wrong, but also how you say it. The tone of your voice can lead a good doctor to find out what's wrong. This old emotional vocalization system is also important for music production, and it operates independent of semantics.

SM I think that is actually right. I think musicality trumps words every time. Alice, if you ask me "hey Steve, how are you feeling today?" and I said "Alice, I am really sad, I am so depressed, it is awful," you just wouldn't believe me because you would hear the tone of my voice, the pitch. If I said "I am really happy today," you would ignore the words, listen to my tone of voice, look at my gesture, my body language, and infer from that. You ignore the words. So I think musicality trumps words when you are trying to express emotions, every time.

AR What you are pointing out there is the inherent musicality in language. Of course, some languages are more musical than others, but all language has an element of musicality. What I would like to ask both of you is: is it possible to have language without music? What do you think, Steven?

SM Not really. Not fundamentally. Of course, musicality has different levels of skills and abilities. Some people suffer from aphasia, where they lack linguistic capacity because of aphasia. We also have people who have amusia, and because of pathology, they lack the ability to engage in music. I think I suffer from quite a lot. I have great difficulty following pitch, I am not very musical. I don't think that is because of my evolution, I think I just have a bit of a strange genetic defect somewhere that I am not totally good with music. Even with that, and even with people who suffer from severe amusia, I think they get musicality, to some extent, in the gesture and in their speech. Of course, if you listen to the greater directors of our time—whether you go to Barack Obama, or within the church or wherever—the musicality in their speech is just amazing in terms of their variations of pitch, their various rhythm, and they are engaging with an evolved sense of musicality in all of their listeners because that is the way they know it is going to have their most effect. Of course, they are doing it unconsciously, they are not aware of that. They are just doing it. I think that they do overlap. I don't think you can have expressive language without some good musicality there. But I am not an expert on that. Stefan might be able to answer that much better.

SK Some individuals are not particularly good at rhythm or pitch, but they do understand whether I ask a question or whether I make a statement. They do recognize whether I am sad, whether I really want them to do something or whether I am just asking. These are musical features of language which require musical skills. Also, adding to what you said before, I do not think that some people have "no sense of music." I am aware that some people have great difficulties singing a melody, but if you listen carefully they still can sing in a way that is going at least in the right direction. Or if individuals have difficulties with rhythm, they still clap in a way that they are not completely random. For me, it is all a matter of error tolerance. In cultures where everybody is considered musical, there is no such thing as "musical" or "non-musical." I also believe that everybody is musical. I am aware that some colleagues investigate individuals with great difficulties of perceiving or producing pitch or rhythm, but even those individuals are good at perceiving small pitch variations in sentences, and they do hear whether a chord progression is not regular.

AR You observe that both music and language consist of sounds, but what about those languages that can be expressed by signs only—by gestural language signs, in other words—like American or

Italian sign language; and are they substantially different from those languages which use sounds?

SK No, they are not, not in the brain. Sign language also seems to be processed in the music-language-network in the brain, as I call it. This network is typically activated more strongly in the left hemisphere by language, and in the right hemisphere by music. However, both language and music can activate this network in both hemispheres, and therefore my term "music-language-network." Interestingly, when language becomes more musical, for example more narrative and emotionally expressive, then this network is activated more strongly in the right hemisphere, that is in the "music-part" of this network. Likewise, even sign language activates the right hemisphere more strongly when it is more narrative and emotionally expressive.

AR So sign languages show similar patterns of activation to spoken language and therefore to music as well.

SK As I said in my presentation, music and language are processed with overlapping neural resources in the brain. That is, there is an overlap of brain structures processing language and music. That is why I speak of the "music-language-network" in the brain, rather than of two separate networks, one for music and one for language. However, language is typically processed more strongly in the left side of the brain, and music typically more strongly in the right side of the brain. With sign language this network is also active. But the more narrative and emotionally expressive the sign language is, the more active the right hemisphere becomes, thus the activity pattern in the brain is more reminiscent of that of music.

AR There is something I have always puzzled over: the fact that you have these networks on the left-hand side—where you are processing language—and I would always think "What are they doing on the right-hand side?". And you answered that question: they are doing music. Steve, I have got a question for you. Languages are different from one another. Once you spontaneously acquire your own language, or more if you are exposed to many communities during your early childhood, this is fixed. Is there anything similar for music? Are there differences, so that after childhood one cannot feel the same way about different musical systems? I suppose there are two questions there. How fixed is it? And is it similar for language and for music?

SM All I can say is I don't know really. That is out of my area of expertise. We know about acquiring languages that there is a period in childhood when you can acquire several languages almost simultaneously with the same ability, then, after a certain time, trying to acquire a further language becomes more challenging and you are using different parts of the brain to do that. That is the same with musicality. I simply don't know and I think probably this is a question for Stefan. My impression is that it is much easier to engage with other musical traditions and other musical languages. When I travel around the world, when I go to China or when I go to Jordan and I am with, say, Bedouin communities there or Chinese communities there, I can't speak the language, I can't communicate but I can actually relatively easily sing with them, engage with them, dance with them. That seems to make a much easier connection with people from other cultures, even if I have no linguistic ability at all. It seems to make it easier to engage with other traditions of music quite rapidly and build a social bond through that than it is to learn other languages. In fact, other languages actually act to make a distance, to prevent any social bonding if you really can't understand what somebody says. It is much nicer to sing and dance with people than it is to try to speak with them sometimes. Stefan might know about acquiring through childhood acquiring musical skills and abilities. That is out of my domain.

5

Enhancing Brain Plasticity with Neuromodulation

JEFFREY A. KLEIM
ÁLVARO PASCUAL-LEONE

MODERATED BY ALICE ROBERTS

Brain Adaptation: Neural Plasticity, Neuromodulation, and Genetics

JEFFREY A. KLEIM

I would like to start off my presentation about brain adaptation with the big picture. [1] The big picture is

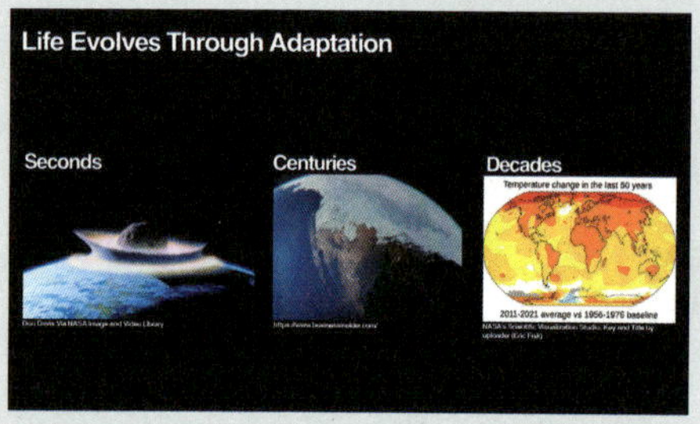

Fig. 1

life appeared on this planet about 3.8 billion years ago. Regardless of its form, it is always been constantly adapting to changes in the environment. As this slide shows, sometimes there are sudden changes in the environment. 66 million years ago the dinosaurs and everything, 80% of life on Earth was wiped out within seconds. Longer-term changes, ice ages—we have had seven or eight of them—last 100,000 years. The last one ended in 25,000 years and resulted in human beings being the largest hominid species. Finally, at an even shorter time scale are changes in global climate. We are all having a front-row seat right now to changes in global warming. Darwin actually called this "survival of the fittest." This is something we have known since the late 1850s, that is, the environment determines which genes are passed on and which animals will survive. As science progressed, people began to realize that the mysteries of life, as we would describe them, advances in molecular biology, neuroscience, and genetics, we began to learn not only the biological mechanisms that drove evolution, but also discovered that many systems within an individual organism could be adapted within their own lifetime. In fact, enduring adaptations in system structure and function could be observed across days, minutes, and even seconds. This idea was first applied to the brain by a fellow by the name of William James, who was a psychologist, not a neuroscientist or a physician. [2] He recognized the fact that the brain was the seat of behavior, so if we could learn and remember things from the past, then the biology of the brain must be able to adapt. He wrote a very famous textbook called *Principles*

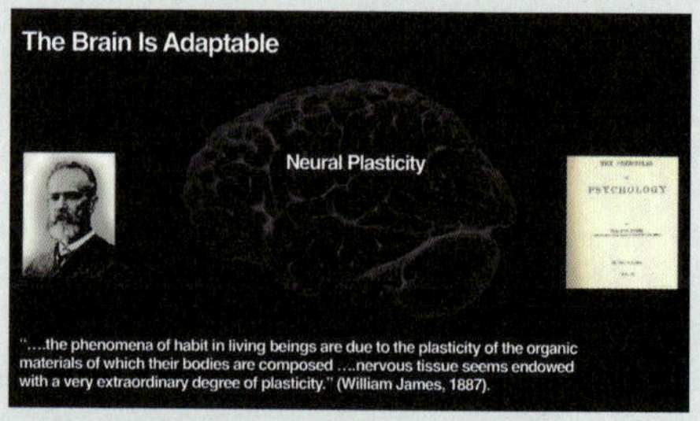

Fig. 2

of Psychology, and in that textbook he wrote an essay on habit. He said the phenomena of habit in living beings are due to the plasticity of the organic materials of which their bodies are composed. The nervous system seems endowed with a very high degree of plasticity. He actually borrowed the term from metallurgy. Metallurgists have been using the term for a long time—plasticity, that is, saying that if you bent metal or if you bend material enough in one direction over and over again, it would start to take on that shape. This was the birth of the idea that not only can biology change across generations, it can also change within the lifetime of an organism. Over the last 140 years—and this is a slide that is my own, personal interpretation of how plasticity has advanced over the last century and a half—we have moved from it becoming just an idea to an actual library of empirical evidence that plasticity, neuroplasticity, exists at every level of the nervous system. [3] If you look at this slide, what you

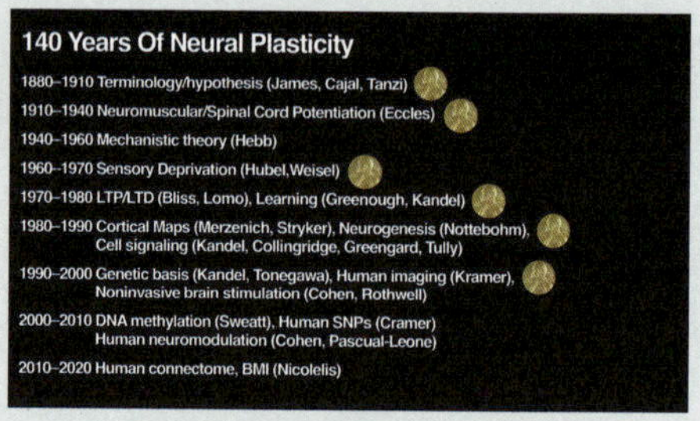

Fig. 3

are going to see is that it began with an idea and it moved down through, showing plasticity in the spinal cord, into the brain. Eventually, we are heading into an era where we are relating plasticity to genetics and even understanding a large-scale connectivity of the brain using high-density tractography. Each one of those little gold Nobel prize medals that you see on my slide represents people that have won the Nobel prize studying neuroplasticity. We have come a long way since James first proposed this hypothesis and I am fortunate enough to have a younger brother who makes video games for a living, and I asked him to use his video game skills to make a video trying to explain how neuroplasticity works. [4] You don't need a degree in neuroscience to fol-

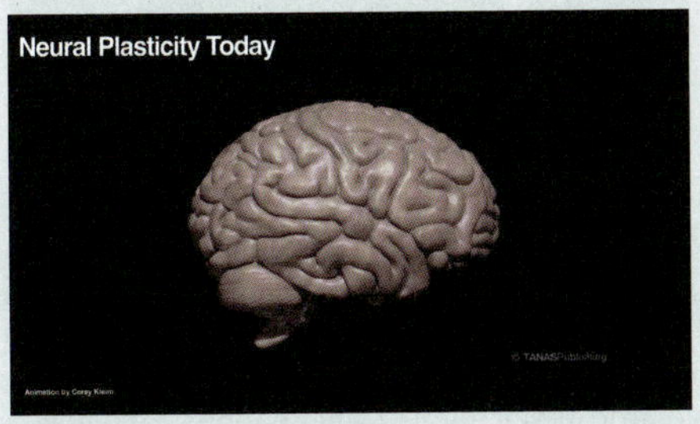

Fig. 4

low this video. This is neuroscience and neuroplasticity meet Pixar, let's say. What we have learned over the last five or six decades of research is that plasticity can take on many forms within the brain. But one of the ways that plasticity can be orchestrated is through cell signaling pathways. We have learned that there are specific genes within the brain—if you remember, genes, the DNA, live within the nucleus—and certain types of experiences or electrical stimulation or events can turn these, for lack of a better term, plasticity genes on. The genes are first, of course, transcribed in the nucleus, shown here in red, and then they are translated outside the nucleus into proteins. Those proteins go off and they impart some biophysical change in the neuron. The way I was brought up as a young graduate student was the change in neuron structure and function. What we know is that animals who have undergone a learning experience, for example, or an injury, grow new dendrites. If you look carefully along those dendrites, what you find is there are more dendritic spines. If you think about this process and why it is important, if you see more dendrites and you see more dendritic spines, what that means is that that neural circuit, that neuron is incorporated into, is somehow functioning differently than it was before. If you think about this happening across the 100 billion or so neurons within the brain, this is how we think learning or the encoding of experience occurs. We also think it is how relearning in response to something like a brain injury, brain disorder, brain disease, also, occurs. One of the other things that happened is that our understanding of neuroscience, our understanding of neurophysiology, and our understanding of neuroplasticity have led to a number of techniques that allow us to go in and specifically modify specific neural circuits that are associated with a disease or disorder. There are a number of these. Pharmacology was the first example. People would take drugs, which were neuromodulators, to alter the activity of the patterns within their neural circuits. What we have learned is that we can go in using biomedical engineering techniques and create different ways of stimulating specific areas of the brain. [5] For example, the image on the left is deep brain

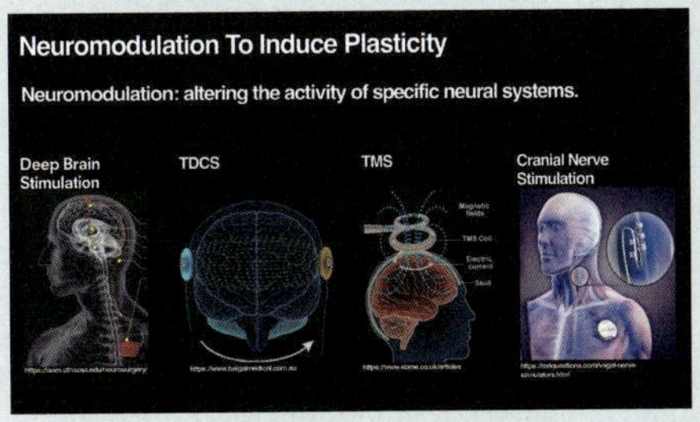

Fig. 5

stimulation and that is where you take an electrode, drop it down into the brain and activate specific areas of the brain in specific intensities and frequencies. Another example is transcranial direct current stimulation—and this is probably one of the most simple ones—where you basically place an anode and a cathode electrode across the brain and you pass current through the brain. Depending on how you do it, you can either increase or decrease the excitability of different areas of the brain. The third image on the right shows the transcranial magnetic stimulation. This uses electromagnetic induction to activate neurons within the cortex. You can do this in a variety of different ways, one of which is to stimulate different patterns, and you can increase or decrease the excitability of particular brain areas. Finally, the third approach is rather than directly stimulating the brain, indirectly stimulating the cranial nerves. For example, vagus nerve stimulation is one of those forms of cranial nerve stimulation where you are stimulating nerves that head into the brain and they hit specific brain areas in the brain stem and midbrain, who then naturally secrete plasticity-inducing neurotransmitters and neuromodulators. This has been used successfully to treat a variety of different diseases. For example, when you look at the way we have been treating using neuromodulation, pain was one of the first, which happened in the 60s—headache, back pain, neuropathy, movement disorders. [6] Deep brain

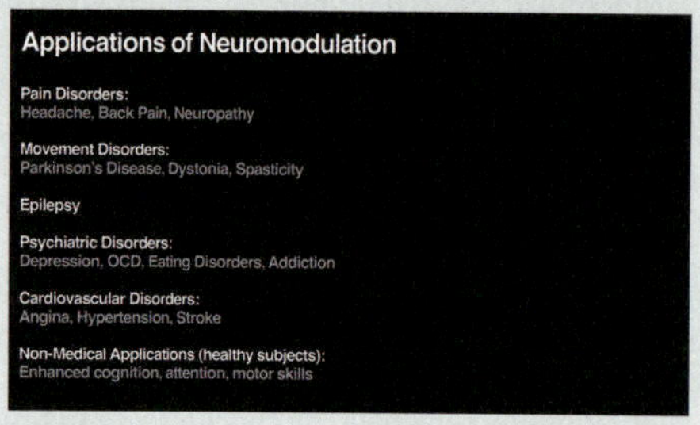

Fig. 6

stimulation is one of the most common treatments for Parkinson's disease, dystonia, and spasticity. It is also been used for epilepsy. So vagus nerve stimulation has been used for epilepsy. Even psychiatric disorders—things like depression, anxiety, obsessive-compulsive disorder, even eating disorders, and addiction. These techniques have been used. Cardiovascular disorders, things like angina, hypertension. It has been used in stroke, for example, to try and enhance the functional recovery of individuals who have suffered a stroke. Then we get into another area, which is going to be sort of how I am going to wrap this all up at the end, which is non-medical applications. There are companies now that sell variations of non-invasive neuromodulation to enhance cognition, attention, and motor skills. This is not a medical treatment, this is to enhance the functions of normal healthy subjects. For lack of a better term, this is what we would refer to as a performance enhancer—not unlike a steroid at an Olympic event. In parallel with the advances that we have made in neuroscience, we have also had these major advances in molecular biology and, specifically, in genetics. We sequenced the human genome back in the early 2000s and we have begun to do these giant studies examining the association between the presence or absence of specific genes and the presence or absence of particular disorders. One of the first places that this has really made some impact is in pharmacology. [7] The FDA has a website that you can hop on to and there are between 240 and 250 drugs that are out there, whose efficacy depends entirely on your genome. That is because each of us has a number of what are called genetic polymorphisms. If you have taken a high school biology class, you know that our DNA is lined up by base pairs of nucleotides; every once in a while, genes get passed on where those base pairs are switched. Somebody might have an adenine, somebody might have guanine, somebody may have thymine. It turns out that one little change in the

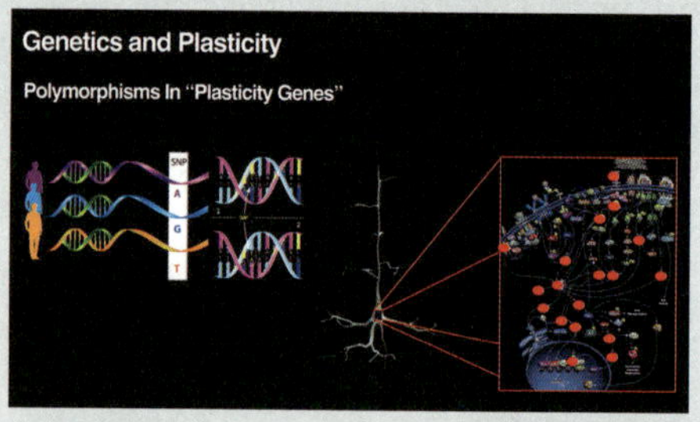

Fig. 7

DNA can produce a protein that acts very differently. If it is true for pharmacology, it would also be true for things like neuromodulation. The diagram I am showing you on the right is a slide that features a generic description of what happens and how cell signaling pathways orchestrate plasticity. A receptor, a transmitter, or a neuromodulator binds to a receptor inducing a cascade of cellular events. Each of those red dots on the slide is a protein involved in those cell signaling cascades that, we know, there are human nuclear single nucleotide polymorphisms form. So, the question is, of course, does your genetics affect your response to things like plasticity-inducing experiences like neuromodulation or learning? The answer is yes, and that would be a completely other lecture. We could spend hours talking about that and hopefully, we'll discuss that in our open session. In addition to sequencing the genome, we have also made some significant advances—I am sure most of you have heard this on the news—in being able to edit the genome or suppress the genome. [8] We can augment, we can

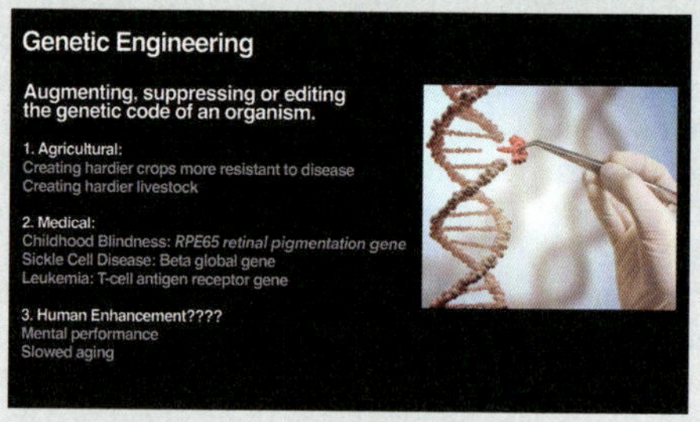

Fig. 8

suppress, and can edit genes within specific areas of the genome. That has been used in agriculture to produce crops that are hardier, livestock to produce animals that grow faster, larger, produce more milk, for example, but it is also been used—and now there are several FDA-approved treatments—to treat retinal blindness, leukemia, and sickle cell disease. In these cases, what they do is they take the cells out of the patients, re-engineer them, and place them back. They are not changing the entire genome of the organism. That, as far as we know, outside of the animal world has been done once and it was all over the news a couple of years ago, when scientists in China had created twins with a genetic modification that should make them more resistant to Huntington's disease. It brings up the question of human enhancements. If we know that certain genes are associated with a stronger response to certain types of learning experiences, or might slow aging, etc., who and how should we be doing this? What I want to leave you with then today is the fact that these advances in biomedical engineering, and science in general, have placed this into a new era of evolution. As I told you at the beginning, life appeared 3.8 billion years ago and Darwin proposed that the way that organisms evolved was through survival of the fittest. [9] Environment dictated

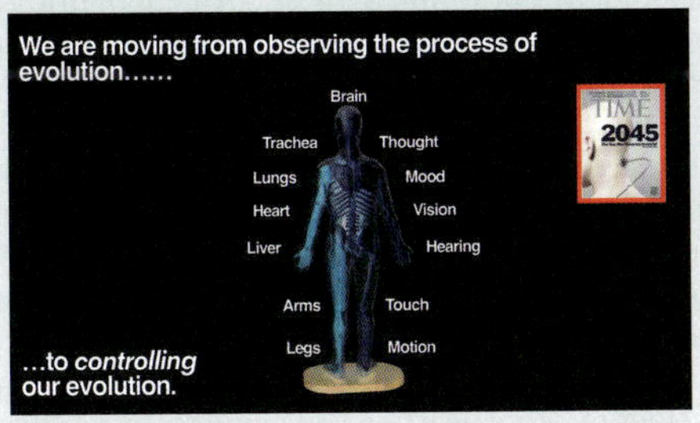

Fig. 9

what we were going to do, genes were passed on in a natural way. We are now at a crossroads. We are no longer spectators. We are no longer simply watching evolution happen—we can control it. So if there are changes that are occurring, we can use neuromodulation, we can use genetic engineering to control evolution, and I am hoping that in our open forum discussion Dr. Pascual-Leone and myself and other folks that might be involved will have a chance to discuss these ethical issues.

A Healthy Brain Is a Plastic Brain

ÁLVARO PASCUAL-LEONE

A healthy brain is a plastic brain. To be specific, it is a brain with the optimal amount of plasticity. What does that exactly mean and why should it matter to all of us, as we go through life aiming to function best and minimize the risk of illness? The human brain is made out of highly specialized, exquisitely built cells—neurons, we call them—supported by a whole host of other very specialized brain cells called glia. These brain cells, each one of them has a different configuration and a different function. Functions that have been developed over many thousands of years of evolution. Brain cells are highly specific and specialized, and because of that, they are also highly resistant to any change. They do what they do. It is very difficult to recover them from doing something else if they are not doing it well. We have building blocks that are explicitly built for a given purpose and do that very well, but one drawback of such a specialized system is that it could potentially be a very inflexible system. It could be very dependent on a given setting or circumstance. Yet, the world we live in, the monitoring of our own body that the brain has to do, and the relationship and understanding of the world we navigate and act upon, require a very different system, a much more flexible system. We live in a world that is continuously changing. We live in a world that faces us with challenges. We cannot anticipate from one moment to the other because the recent past of humanity dealing with the Covid-19 pandemic illustrates the fact that things we cannot anticipate end up having an enormous impact on us. Our brains are challenged continuously to cope with these unexpected events. That requires a system that is able to respond quickly and flexibly to novel situations. When my parents were growing up, they couldn't conceive of a way to communicate with each other via mobile telephones, leave alone having all your memory and knowledge in these little gizmos that I cannot live without. And I am an amateur compared to my children. Our brain, made up of these exquisitely specific and sophisticated brain cells that are very fixed in what they do, is challenged and required to adapt and react to a very rapidly changing world. The solution of nature to do that is what we call plasticity. It is played out in plasticity, because these elements, these brain cells, are connected with each other in networks. Those networks are flexible, they are continuously adapting, they are capable of change. Those networks are nature's solution to the fact that the brain has to cope with the rapidly changing world, and they are therefore intrinsically modifiable. They are continuously changing and that raises a different challenge—the challenge of how quickly, how effective should plasticity be and what are the opportunities to monitor and modify that plasticity. The concept of plasticity, that very idea was first introduced by Williams James, a psychologist. He introduced it referring to habits, to behavior, not to the brain per se. It was Ramón y Cajal, the Spanish neuroscientist, who took that notion of plasticity and explored it, put it into neurobiology. Today we talk about these networks, these connections of brain cells, each one of which is specialized for a given behavior. Every thought we have, every perception, every emotion, every plan, everything we do, literally everything that our brain does is mapped onto a specific network. That network is modifiable and changing and capable of adapting. Plasticity, therefore, is not something we can switch on and off, it is not something that is good or bad per se, it is something that is always there. It is the very way our brain works and the challenge is to guide it. The challenge is to promote those changes that benefit us and suppress those that do not. Plasticity is the substrate for learning and acquiring new things. It is the source of creativity and ingenuity of humankind, but it is also the very substrate of the symptoms of disease and disabilities. As a neuroscientist, I want to understand the substrates, the makings, and the workings of that plastic brain network and brain, but as a neurologist, I want to guide the plasticity for the benefit of individual patients. To do that, we need to be able to translate the insights from animal studies into humans. We need to be able to measure the efficacy of plasticity and we need to not only characterize the consequences of a plastic brain, but also measure how the plasticity is working, how much plasticity a given subject has, and ultimately, be able to modulate it, to guide it for the benefit of the person. Those are the challenges we face and they are important because, just like Williams James originally pointed out, the brain has to be capable of this plastic change that he spoke about for the first time. But it needs the right amount of plasticity. It is bad if the brain has too much plasticity, it is too easy to be modified. But it is also bad if the brain is not plastic enough. We need a brain, in healthy situations, that has the optimal amount of plasticity that yields to influence but doesn't yield all at once. Perhaps one thing that is worth talking about first is what are the extent or examples or illustrations of that plasticity in the brain at work. When we acquire the ability to do individual movements of the fingers, when somebody becomes a pianist and has exquisite control of those finger movements, that is possible—that skill, like every other human skill—thanks to the fact that those networks that control the finger movements get modified. That happens at an enormous fast time course. Very quickly, ensembles, groups of neurons start working differently as a team to be able to do a given task differently. It is a substrate of learning and acquisition of skills for the human brain. We have known that for a very long time—in fact, Cajal anticipated that in his description of the fact that neurons, because of their connections, would provide the substrate for every human capacity and ability. But the range of examples of plasticity is much greater than we had initially thought of, at least that I had initially thought of. For example, a lot of our brain, a lot of our human brain is devoted to processing visual information. About two-thirds of the brain cortex has something to do with vision, so I became interested, a number of years ago, in the question: what happens to the brain when one is born without vision or when one loses the sense of vision? What happens to that real state of the brain? It is unlikely that you would not be doing anything, and in fact, what we now know is that the same way as, very rapidly, the coordination of fingers changes when you acquire the ability to play the piano, the same way that happens—very quickly—large distances of the brain are bridged to assign different tasks to parts of the brain that traditionally would be devoted to vision, if one loses vision. That happens at an enormously fast speed, too—for example, if one takes sighted individuals and blindfolds them, rendering them blind for just a few days, the parts of the brain in the back of the head, the back of the brain that would process vision starts becoming engaged in perceiving touch or in hearing voices and remembering them, or in detecting patterns of sounds or remembering words offered. In other words, in activities using other senses that overcome the limitations of the loss of vision. The extent of brain connections that must be reshaped in their function to enable that kind of a change, that kind of plastic change, is much larger than what we had anticipated. So it illustrates the fact that this capacity of the brain to be modified truly is an enormously powerful and capable tool for the brain to adapt to different circumstances. But it is a double-edged sword. It is also the case that the same capacity can underlie chronic pain, can underlie functional neurological disorders, underlie dystonia and alterations of movements, underlie ultimately cognitive problems or spasticity in the modern domain. A lot of very disabling symptoms of diseases are manifestations of that large-scale change capacity of the brain. Now, going beyond, characterizing the consequences of plasticity is particularly important. One thing is to say the brain is plastic, and here are examples of how the brain is modified when one has a lesion, an illness, or when one learns new skills. A very different thing is to measure and characterize how effective the mechanisms of plasticity are themselves, how capable is a brain to be modified in meaningful ways. Measuring the efficacy of the mechanisms of plasticity is important because it allows us to ask the question: do they change over a lifetime? Do they change in the setting of disease? Are certain illnesses manifestations of alterations in the mechanisms of disease? And, perhaps more important than any of those, can we do something about the efficacy of the mechanisms of plasticity in a given individual to help him or her function better? Techniques to characterize the mechanisms of plasticity, so not just the consequences of plasticity, but the efficacy of the mechanisms themselves, have been adapted from animal models to humans. We now have tools that are non-invasive, for example, transcranial magnetic stimulation, a way to perturbate specific parts of the brain and capture the response to that perturbation that allows us to measure and characterize the efficacy of the mechanisms of plasticity in human individuals—young children, all the way to older adults, healthy individuals, as well as those with illness. We can do those measurements for different parts of the brain, and even explore different types of mechanisms of plasticity, those that potentiate connections and those that suppress or reduce the efficacy of connections. Thanks to those studies we now know that the efficacy of the mechanisms of plasticity normally goes down with age. As we get older, plasticity remains but it takes more input, more repetition to be able to secure and implement a change. Again, this is a double-edged sword. It makes it harder to develop new skills but it also makes it harder to develop problems of disease, because the brain is less easy to be modified in a lasting way. We also know that there are illnesses that have a core alteration in the mechanisms of plasticity that appears to be the critical component that leads to pathology. For example, in dementia of Alzheimer's disease type, the mechanisms of plasticity are suppressed. It takes too long for the brain to acquire new habits and develop new skills. It is possible, but it is fundamentally altered and it is altered in that way very early on in the disease. On the other hand, in illnesses like autism spectrum disorder, the mechanisms of plasticity are hyperactive. The brain is too easy to be modified and a brain that is too plastic ends up being a noisy system because it is influenced in a lasting way by too many things that don't really matter. We have ways to identify alterations of mechanisms of plasticity as the core problem in illness. The final point I would like to make is that we also have now tools to not just measure, but to modify the plasticity, not just to characterize it but to modulate it, to make a hyperplastic brain or a hyperplastic network in the brain less effective in its plasticity, less likely to be modified. Or, vice versa, to try to increase the likelihood of modification of systems that are hypoplastic, which allows us to ultimately reduce the risk of disability, promote a healthier functioning of individual brains, and approach therapeutics in a very different way. Approach therapeutics by precisely characterizing specific alterations of brain network dynamics, and modulating those brain circuits to make them function in a more adaptive way. The remarkable thing is that when one does that, because of the very plasticity of the system, interventions can lead to lasting benefits and, therefore, open up the door to prevention, to minimization of disability, and ultimately, to better, healthier brain for all of us as we age.

Conversation 5

ALICE ROBERTS, JEFFREY A. KLEIM, ÁLVARO PASCUAL-LEONE

ALICE ROBERTS This evening I'm joined by two experts to talk about how adaptable our brains can be during our lives; how much our brains can modify their connections; and how we can perhaps harness that brain plasticity to become more efficient learners and recover better from neural injury. I wonder if I could ask each of you to just summarize your lectures in just a few minutes to just refresh us before we enter into the discussion.

JEFFREY A. KLEIM My lecture is all about adaptation, which has been a fundamental characteristic of life since it first appeared on this planet over three billion years ago and life has evolved over countless generations in response to continuous environmental challenges. However, we now know that biological systems can also adapt within the lifetime of an organism and nowhere is that more evident than—I am going to steal one of your lines—the most wonderful and intriguing of all organs, the human brain. Neuroplasticity is an evolutionary conserved fundamental characteristic of all nervous tissue across all species. Modern advances in science have allowed us to not only describe the basic principles of neuroplasticity but to begin to experimentally induce plasticity across multiple species. But, as Álvaro pointed out in his lecture, what does simply detailing the phenomena accomplish? We are no longer simply studying plasticity, we are applying it, establishing its importance for both mediating and treating neurological disorders. We are approaching the ability to impact the function of young healthy brains and even change the trajectory of normal aging. It is a very exciting time to be involved in the neurological sciences.

ÁLVARO PASCUAL-LEONE I think the point I try to make is that, just like Jeffrey was saying, plasticity is an intrinsic property of the brain of humans and of the brain of any other species. One would hope that is a good thing, but it doesn't have to be. It is an intrinsic property of the brain that can play out for the benefit of the individual—and that particularly relates to behavioral gains, acquisition of skills, and performance benefit—, but it can also be the cause of symptoms of a disease, of human suffering, and of problems. So I think that the challenge for neuroscience is to go beyond just trying to identify plasticity as a phenomenon, trying to demonstrate the reality of its existence in going into guiding the plasticity for the benefit of a given individual, for the benefit of any one of us. If we are able to do that, then we have the opportunity to actually promote brain function and brain health across the lifespan, because the overarching argument that I try to make is that a healthy brain across the lifespan is one that sustains, as Williams James, when he coined the notion of plasticity emphasized, the quote "right amount" of plasticity. I wish I knew what that is for any one of us, but I think that therein lies the both challenge and opportunity at this current very exciting time of neuroscience and translation of neuroscience because we start having the tools to actually guide that plasticity and potentially benefit individuals.

JK One of the things that struck me from your presentation, Álvaro, was a comment you made that a normal brain is a changing brain. I could not agree more, but balancing, as you just mentioned, that level of plasticity, some basic scientists would refer to that as homeostatic plasticity. The notion suggests that we could use some measure of brain plasticity as a useful biomarker reflecting the level of health of a brain in the same way that we use blood pressure or cholesterol levels to look at the health of an individual. The problem, of course, is the timing, location, and direction, and the alterations in plasticity are very complex as you all know. Therefore, there are similar starting points for different individuals, different slopes of life-changing events. Do you see this as a possibility down the road, where you might go in for your annual health checkup and have some measure of your capacity for balancing plasticity or the balance of plasticity being a measure of brain health?

APL The short answer is yes. I think that we need a vital sign of the brain, an indicator of brain health that is reliable, that is applicable across the lifespan, that can be tracked for a given individual longitudinally, and that captures core aspects of what it means to have a healthy brain. The argument I try to make, and what you are echoing, is that plasticity is one such component in every list. To have a metric of the efficacy of the mechanisms of plasticity for a given subject as they change or modify across their lifespan is frankly what we need. The question, as you point out, is how do we find it; how preserved is it across brain areas; how specific would it be for different brain networks in the cortical regions; how reproducible is it across the lifespan; how much variation is there with age? There are so many challenges that it is a very exciting time, but it could be a truly transformative impact to have such a metric that we can assess. Also because it would allow for potentially translatable phenotypes markers that could be then systematically exploring animal models to learn more about how to modify them. So I hope that smarter, motivated scientists, students, and companies come along and develop such metrics. I think the power of such approaches would be enormous.

AR I want to explore a couple of concepts that you touched on there before we get on with the discussion. Álvaro, I was going to invite you to ask Jeffrey a question, but just before we do that I wonder if we could define this concept of plasticity. You are talking about how plastic the brain is. Are you essentially talking about the ability of the brain to form new connections? Is that what it boils down to?

JK I wrote a book on plasticity and rehabilitation and before I began writing, I asked all of my colleagues: can you define neuroplasticity? I had as many definitions as I had people I asked. It was always limited to whatever the experimental paradigm was they were working on, whether it was development or stimulation or what have you. My definition of neuroplasticity—which I think is important if you are going to start talking to clinicians about it and they don't really know what it means—is any change in the structure or function of neurons that can be observed either at the individual or population level. Very blanket, an intentionally vague definition that can be then understood and used in a conversation between basic and clinical scientists. That is something that we are lacking I think a little bit.

AR Presumably a fundamental property of that is connectivity. That is how the brain works.

JK The ultimate change is a change in the circuitry that controls a behavior

APL Coming from Spanish neuroscience tradition, I have always thought that Ramón y Cajal's intuitions are quite insightful here, too. He made the argument that plasticity, at the end of the day, requires the assumption of two processes, the first one being the optimization of connections that already exist in a "functional" modification, and then the establishment of new connections, the structural changes. He discusses a sort of two-step process where you need to almost have connections already to meaningfully establish new ones. That is always resonated with me, I have always thought that there is pretty good evidence that if you have absolutely no connections left, because of the damage, to re-establish patterns of function is unbelievably rare, if at all possible. So it might well be to your point that we need connections, we need to modify the efficacy of those connections and then, as a second step, there is a possibility of arborization, of increased connectivity, of structural change, like Jeffrey was saying.

AR These terms are so organic—like the term "arborization," where you are talking about this tree-like growth of neurons. That is quite fundamental to understanding this. The way you are talking about it, there has to be some healthy growth, just as in plants; if you have coppiced a tree down to a point where there will be no regrowth, you have to have a healthy base in order to get that sprouting and to get that new growth occurring. Álvaro, have you got a question for Jeffrey?

APL I do. I have a sort of two questions lumped into one. I read Jeffrey's book, I have admired his work and followed it and learned a lot from it and continued to do so. I have always wondered how much of what is true and detailed in knowledge from the motor system is really applicable to creativity, executive functions, parietal abilities, that kind of stuff. As a cognitive neurologist, I try to think about how much can we translate or assume what is true there is going to be true here. Embedded in that, how much of what we learn about the exquisite insights from animal models is going to be true in humans? We like to think we talk about the same thing in human and animal experiments, and I have grown increasingly convinced that in at least human neuroscience sometimes we throw around terms that are not really aligned with the use they have in animal models. We think we know what we are talking about but we don't.

JK It is a great question. My colleague Theresa Jones and I were charged by clinicians to come up with what are the key general behavioral signals that drive plasticity. If you look across the animal literature and you look across human literature, what are the key elements of use? Not just exercise, but what are the key elements of use? We came up with several general characteristics: things like persistence, intensity, salience, difficulty, etc., things that you can look at in animals and manipulate in animals, things that you can look at in humans and manipulate in humans at an experiential level. At a behavioral level, we can do that fairly well. It does transfer between cognitive domains, sensory domains, motor domains. It transfers fairly well. Transferring specific results from animal studies, especially in injury animal studies, can be more difficult. One of the issues that we have is that the animal rodent brain, which is the one that I am most familiar with, is much less lateralized than the human brain, for example. As brain size increases, lateralization also increases. So we have to be very careful, as you pointed out, to make sure that what we are studying, if our goal is—not to throw out basic science—to translate that into a human condition, that we have to, first of all, talk to clinicians. I remember going to a stroke conference and asking what the basic rehabilitation strategy was that they gave people with upper extremity impairments. People laughed at me and said "What are you talking about? There isn't one." This was 15-20 years ago. Now they started to look at using principles of plasticity. How intense, how many repetitions, those sorts of things. We are starting to make that connection at a general level, I think that is applicable. At a mechanistic cellular level, I think it is also applicable. But there is a space in between where it becomes difficult to make that trend, that transition, that translation.

AR Jeffrey, do you think that approaching neuromodulation in healthy people will require new views on ethics and regulatory aspects?

JK What has happened is these neuromodulatory systems, for example, transcranial direct current stimulation, which basically turns your brain into a battery—you can buy these things online and they claim that it improves gaming ability, academic performance in students, I am waiting for the day that I walk into my biomedical engineering freshman class and see students wearing these things. There is no real regulation on it, but there has been a movement now. There is something called the "neuromodulation appropriateness consensus committee" and it is focused primarily on pain, but they have at least had meetings and started to look at the fact that we can't just allow people to buy these things, especially implantable devices, not just peripheral devices. The other question becomes the goal. The goal of these things was to make performance enhancers for people with neurological disorders. Now you can make performance enhancers for healthy people. We need to regulate that because the truth is we don't know exactly how it is or why it is that this passing current through the brain actually does improve performance. They met this past month to look at whether or not we should be looking at this in Olympic athletes, a timely topic. It meets two out of the three criteria, but the problem is you can't test whether or not someone has been doing this or not. You can test for steroids, you can test for performance, you can't test whether or not someone's been doing this or not. It is almost a moot point if people are doing it. I think something has to be done eventually. When you start messing with the rain and passing current randomly through different parts of the brain you need to be very careful.

AR It is tricky because, as you say, in terms of what is happening it is on a par with using medical drugs which are designed to help deal with certain pathologies, and particularly drugs which are designed to improve blood flow, for instance, which are used in disease and used in things like angina but can equally be used to improve performance. You can test for those drugs, so here we have got something where you have to trust that people aren't doing it.

JK To show you how behind we are: the world anti-doping authority has gotten involved. It is not the world anti-modulation authority, it is the world anti-doping authority. They are trying to figure out whether or not this should be something that should be looked at for athletes that are competing professionally.

APL A number of years ago Masahito Kobayashi working with me showed that using transcranial magnetic stimulation to very specific parts of the motor cortex, at a very specific time, with very specific stimulation parameters, it was possible to speed up the acquisition of a five fingers movement experiment in terms of piano playing. He did a number of experiments showing the different parameters applied to different areas on the basis of our understanding of the connectivity. I thought it was a very elegant work. Therefore, in my mind ever since that work and even before, with relevant animal studies from Theresa Jones and many others, we have known we have the capacity to speed up, to modify the acquisition of skills. We can do it with neuromodulation techniques, broadly speaking. I think there is still a far cry between very controlled experiments like those and their application broadly. I couldn't agree more with you, Jeffrey, that when people start using these devices with the expectation that they are going to derive at best a benefit but never something bad from it and there is a naivete built into there. If it can change things in one direction, it can do it in an opposite direction, and the fact that a given mechanism was at play for one experiment doesn't mean the same mechanism is going to be in place in another setting. I think the need for an appropriate ethical framework to regulate the use but also to guide, to advise people's endorsement, or even a self-use of these techniques, is really important.

JK At the very least we should move from pushing a button on Amazon to retrieve this device to have a pre-assessment that tells you how this device should work for you, for what you want to accomplish. I don't think we are quite there yet.

AR Álvaro, you brought up an incredibly important point there, which is that there can be a downside. This shouldn't be seen as being a kind of universally positive intervention, therefore the ethical discussion around it is very much like the use of drugs, where if you are using a drug to treat a patient, that drug will inevitably have some side effects but you consider the benefits of the drug to outweigh the costs of it. If you then try to use that drug in a healthy person—the negative side effects will be there. You have got a similar kind of ethical consideration, presumably, with neuromodulation.

APL I don't even think that it is at the level of hypothetical. One application of non-invasive brain stimulation that is endorsed, approved by the Food and Drug Administration, is transcranial magnetic stimulation, which is broadly accepted across other countries, not just the US, and CMR devices, and so forth. It is very effective, it is covered by insurance companies. It implies or involves the application of certain stimulation parameters mostly to the left dorsolateral prefrontal cortex, a brain region sort of here, in patients with depression. Those very same parameters of stimulation, when applied to somebody who does not have depression, cause dysphoria, cause a mood disorder, they cause, quote-unquote, "depression." It is a paradoxical benefit in the patient with depression that it actually helps them. So to your point, Alice, I think that it is not a question of theoretically maybe there could be a downside. No. We know there will be under some circumstances. I think it is important at the very least for people to be aware of this, at the very least to have a true opportunity for informed consent in choosing to pursue a given intervention.

JK We have opened Pandora's box and it is not easy to close. But we need to sort it out.

AR I have got an interesting question that has come in about the pandemic. It asks whether it is possible that the Covid-19 pandemic has already induced plastic changes in our brains due to changes in habits, from things like travel restrictions, reduced stimuli, and motor activity. If that is true, do you see that plasticity is positive or negative? Also, how reversible is it? Should we be working on selecting healthy stimuli and behaviors and making them into habits, as well? How reversible are plastic changes in our brain?

APL Let me start with the last point. Let me put it provocatively. I don't think that the brain reverses changes made. Let me put it like that so that Jeffrey can react to it and teach me whether what I am saying is nonsense. The point I am trying to make is that I think the brain is capable to keep modifying itself, to keep changing connectivities, but if one structural change happens it is not likely that it will be a reversal of the change made. It would be another change that they place. The notion of you cannot unscramble the scrambled egg. You can do something different with it but not unscramble it. To the point of the question, if any situation that is a substantial stressor and that is protracted in time and has a long impact, like the pandemic, it will change the brain because of plasticity. The question is not can we undo that change? But rather, how can we guide that change so that it translates into a benefit so that it can get redirected for sustaining well-being? The impact on different people will be different and we have the beginnings of the evidence of assessments of tools that we can use to define and anticipate whether situations like this will have a negative impact on a given individual or will have not only no negative impact, but potentially a positive impact. You look at the Covid-19 pandemic data and while there is an enormous amount of mental health impact, and that is negative, there is a population level, some individuals actually are doing better than they were before Covid. There are differences in the impact of plasticity or what is different is that the baseline connectivity, such that the same insult, as it were, it ends up modifying the brain in slightly different ways.

JK I agree completely. One way to think about it is if you were driving down a road how do you reverse yourself? You can just turn around and go back the way you came, but that is not quite how plasticity works in general. It finds another way back in the way that Álvaro has described. It is a really good question. This is something that is affected virtually every domain that we have, whether it is motor, lethargy, we are not getting out as much, we are not interacting as much, cognition, emotion, depression—all these things have a root basis in plasticity. The good news is that, as opposed to things like a sudden injury, such as a stroke or a traumatic brain injury, recovering from that is somewhat more difficult than recovering from—even though it may take longer—emotional disturbances or those sorts of things. We actually know more about that now than we knew about recovering from a sudden brain injury. Covid is an interesting brain insult, not because of something biologically specific I am pointing to, it is a brain insult for some in the sense that it is isolation and a change in the way that your lifestyle is. As Álvaro rightly pointed out, for some people that is good and for some people it is not.

AR It is fascinating talking to you both about the way you are separating out talking about the structure and function of the brain. I know that, Jeffrey, this is a key tenet of your philosophy—talking about plasticity in this way, that even if we have changes that are essentially irreversible, that doesn't mean that there will be irreversible changes to cognitive function—because you can build these routes to circumvent those pathways that have already been created. Even if you have depression setting in as a result of something like the Covid-19 pandemic and the trauma that we have all been through, it doesn't mean that the depression is irreversible, because you can still build on top of those changes in the brain which were associated with that depression. Does that seem right?

JK Yes, and that is true and an extreme example of that is post-traumatic stress disorder, which is becoming rampant in this country in the last decade or so, where people are coming back from overseas after being in a war, or have a traumatic event—loss of a family member—and that instantaneous event that happened or happens across a week or two can cause profound changes in behavior. That is plasticity gone wrong. Trying to not just undo that but rewire it in a way that is healthy has to be a major goal of the kind of work that Álvaro and I do.

APL In a way it boils down to me what you were commenting, Alice, on to what is the level, what is the output of quote plasticity we are looking at, what is this purpose. Probably a product of my own education, but I think that William James said it right. At the end of the day, it is about behavior, it is about the nervous system allowing and enabling us to relate meaningfully, to cope meaningfully, to adapt—using Jeffrey's lecture point—to the changing world that is out there. To do so, faced with the reality that the world that surrounds us both externally and internally is changing way too fast, for genetic change and generation after generation of evolution. If the ultimate goal of plasticity is to meaningfully sustain behavior or to cope with those changes, then you can envision—and we have lots of examples—that lots of different ways of brain function can sustain the same behavior. Very simplistically, if your goal is to be the fastest runner out there, and you lose your legs but get amazingly good on your orthopedic prosthetic legs, you can be way faster than I ever can imagine being. It is uncanny to watch paralympic games in that setting. You can remain an amazing painter having lost fingers and painting with your mouth, and painting with your toes. We have examples of all those uncanny abilities. Clearly, your brain is functioning in a different way. I had the enormous privilege of studying a congenitally blind painter, a Turkish man, a very gifted congenitally blind man. His art is quite stunning, his brain clearly functions in a very different way than mine, because a lot of my brain is processing vision, and that is plasticity. You could almost think of lots of mechanisms, lots of ways, lots of changes in the brain that sustain that behavior. To me, that is a reason for hope because it means that if some problems develop in the behavior there are lots of resources that the brain does have to sustain the behavior. The question is what is the cost of those changes vis-à-vis, potential subsequent stressors or subsequent insults?

JK As I tell my students, the problem with studying plasticity is plasticity because there are so many different ways the brain can change. How do you know which is the one that counts?

AR Here is a question from our live audience. Is the diagram of the homunculus—the strange little man or woman organized around the sensory and motor cortices drawn by Penfield in the 1950s—still considered to be a reasonable reference in neural cartography? How do we feel about that homunculus?

JK There is some neuroscience trivia for you. As somebody who has mapped that homunculus in rats, we call it the rattunculus. I have also done some in humans. A lot of the original work was done on epileptic patients who were having resections done. It turns out that in epileptic patients the face is actually flipped, so many of those drawings are actually not normal, because you don't go in and have your brain stimulated unless you have got something wrong with your brain. People don't volunteer to have their skulls open to have the brain stimulated. At a macro level, however, it is reasonably accurate. I happen to have one. I just have a brain sitting next to me. I can tell that I teach. The legs sit down here in the center, the arms and the face come out that way. In general, that is true but if you look at it very carefully, at the level of hundreds of microns, what you find is not this nice global pattern, but you find this interdigitated pattern of some error, some areas of wrist, and digit, and elbow were all interdigitated. My postdoctoral advisor, Randy Nudo, used to call it a fractured mosaic, almost like a piece of artwork, because they were all interdigitated. At a global level yes, flip the face around. But when you get down to the micro level it doesn't look like that. Not only that, it is constantly shifting around, it is very plastic.

APL If I can make two comments along those lines, I totally agree. One other aspect of the history of neuroscience trivia is that that homunculus basically was a request to a cartoonist, to an artist to make sense, in some meaningful easy-to-digest way, for publication and dissemination purposes of results that were quite more complex. So complex that they couldn't quite figure out how to look at it. I think in Randy Nudo's way of thinking of it as a mosaic—if you go far enough from—, you can actually see a given organization, but you come down to the detail and it is a very messy world out there. I think it is quite useful in that context, it holds true if you are far away from it. If you zoom in, it doesn't, and it doesn't precisely for the reason that Jeffrey was pointing out. If you think about moving this finger, or moving that finger, and they are in different locations in the homunculus, and they are clearly very different behaviors. Yet, at the level of neuronal ensembles, Mark Schieber would say that more than 90 of them are the same. They are just engaged together in a different moment in time giving rise to this or that. But overwhelmingly they are the same. What we have to come to learn is A, very flexible and very dynamically changing; B, it is an on-purpose simplification which, as a clinician, I have to say it remarkably works. Remarkably it is useful, it is fabulous.

AR In terms of communication, in terms of teaching, it definitely works. We are talking about the homunculus, this cartoon-like image of a person's body stripped across the surface of their brain to represent where the different neurons are.

JK It is a wonderful way to represent it.

AR It is a broad brush picture. When you get down to the detail it is more complicated than that—and it changes. Here is another question about plasticity from our audience and it is about meditation. What do you think about meditation as a tool that could train plasticity? Can it? And the studies that are being done in the field with respect to meditation—does meditation work in brain treatment?

JK I can feel this one to begin with. There is been quite a bit done on meditation and looking at expert meditators versus amateur meditators, and looking at how the activity of the brain changes. They are very different from one another. When you get good at meditating, certain areas of your brain become more active than if you are just starting out as a novice. It also varies as a function of the type of meditation that you do. A good general rule is that the EEG, the brain waves that are high, alpha waves that are powerful when you are paying attention, those go down and the waves that are associated with being calm go up. No shock there, but it depends on the type of meditation that you do. Focused attention versus open monitoring attention, for example, causes different sorts of brain waves. So yes, there is plasticity that occurs there. What is interesting, at least to me, is you can ask someone to sit quietly and look at their brain activity and someone who meditates: the brain activity looks different than someone that doesn't, even though they are not meditating. Meditation itself does induce a form of plasticity within the brain.

APL I agree with all that and it is very important what Jeffrey was saying. In regards to the question, I think that perhaps it might be worth pointing out that learning to become a good meditator promotes plasticity, guides a certain change, introduces a change. As Jeffrey was saying, a consequence of that is that people who are expert meditators quite literally have a different brain than they did when they didn't meditate, or compared to others. But the very fact of meditating is not inducing a plastic change in itself. It gets a little slippery, I know, I am catching myself in a paradox. But what I mean to say is that it is not that if you want to activate plasticity you should meditate. That is not the point. The point is that meditation is a given different state of the brain and if you become proficient in getting there, that induces a capacity or a change in the brain, and, therefore, in plasticity.

JK In the same way that if you get proficient at anything—meditation or racquetball—your brain changes, not surprisingly.

AR You are basically saying that it is about learning a skill—and what you are doing in learning the skill is that you are honing your ability to learn, you are increasing your ability to be plastic and to learn.

APL The question that comes up, which is an interesting one to me, is the point you were just making, Alice. If you become an expert in that very specific skill, such as meditation, does that make the mechanisms of plasticity of your brain more efficient? Does that change the capacity of your brain as a core capacity of plasticity? Do you then become better able to recover from an injury or prevent dementia? There is incredibly interesting data to me, as a neurologist, suggesting that that may very well be the case. Embracing silence, enabling mind wandering, and even more so, in various different practices of meditation, religious or not, there is a real change in the functioning of the brain that makes the efficacy of the mechanisms of plasticity greater.

AR I just want to ask maybe one more and just a very quick answer, please. Can culture and education influence language plasticity? Do you think bilingualism can be a positive factor in maintaining a healthy brain? A very quick answer from both of you, please.

JK Bilingualism does change the brain and the earlier you become bilingual the greater the change is. But that is because you learn to become bilingual. It doesn't make necessarily you a better musician. It is the same response to meditation.

APL It is really valuable and I think it is a phenomenal skill, a phenomenal effort to promote two things. One is the fact that in different languages the same literally translated word has a different meaning. The connotation of "love" is not the same as the connotation of "amor"; "bread" is not exactly the same as "pan." With that, there is a benefit in the brain representation of semantic complexity. That is one aspect of the benefit. The other aspect of the benefit, which I think is very interesting, is that if you have never learned a second language, that is a great skill to focus on to promote the capacity of the brain to keep on learning. What Jeffrey was saying, in older age if you have already learned one you don't need it, but that is a valuable thing. One final thing, unique to two languages, is the capacity to translate between one and the other and that is a plastic development in the brain that non-bilinguals don't have.

JK Spoken as a true bilingualist.

6

Relationship between Science, Biological Evolution, and Philosophy

SIMONA GINSBURG
EVA JABLONKA

MODERATED BY ALICE ROBERTS

Animal Consciousness: A Biological and Evolutionary Approach

SIMONA GINSBURG

During many years of discussions, Eva Jablonka—an evolutionary biologist—and I—a neuroscientist—have taught each other our fields. We are both, also, philosophers of biology. With some trepidation, we dared to embark on investigating one of the most engaging topics that brings together science and philosophy: the problem of consciousness. We published a book in which we represented our theory, which in a nutshell claims that consciousness evolved with the appearance of a very special type of open learning. In our lectures, we try to describe how the theory was built. What do we mean by consciousness? [1] We use consciousness as synonymous with subjective experiencing and sentience. It includes many experiences. We are all familiar with a painful, nagging toothache, smelling coffee brewing in a "macchinetta," remembering a face, hearing your name called out at a cocktail party, breathing fresh air after being confined to a stale cell. Minimal consciousness, or sentience, then, is the most basic, non-reflective subjective feeling. It includes experiences that are exteroceptive. For instance, those related to seeing and hearing. Interoceptive, like thirst and hunger, and proprioceptive, like feeling the position of our body. Many questions spring to mind when we reflect on subjective experiencing. Why is it necessary? Why can't information simply be processed, without any "personal feel," information about the world and our bodies, without this "personal feel"? Who is conscious? Only humans or maybe other creatures? And what is the relation between the mind, the mental aspect of our life, and our body and brain, the physical part? The neuroscientist Sherrington expressed his wonder by focusing on the subjective experience of seeing. "When I turn my gaze skyward, I see the flattened dome of the sky, the sun's brilliant disc, and a hundred other visible things underneath it. What are the steps which bring this about? A pencil of light is focused there on the retina. It gives rise to a change which in turn travels to the nerve layer at the top of the brain. The whole chain of these events from the sun to the top of my brain is physical. Each step is an electrical reaction. But now there succeeds a change wholly unlike any which led up to it and wholly inexplicable by us. A visual scene presents itself to the mind. I *see* the dome of sky and the sun in it, and hundred other things beneath it." [2] Sherrington expresses here the mind-body problem. What is the relation between the mental and the physical? And how can the mental be produced from the physical? There are several philosophical stands on the issue of which I will mention three: dualism, panpsychism, and physicalism. [3] *Dualists* assume that the mental, the mind, is not identical to the physical, the body. Some maintain that the physical and the mental are two kinds of stuff that pervade all substances, while others think that only human rational thinking requires mental stuff. If one believes that the mind and the body are different kinds of stuff or properties, do the mind and body interact? How is it that a state of mind, such as sadness, causes a

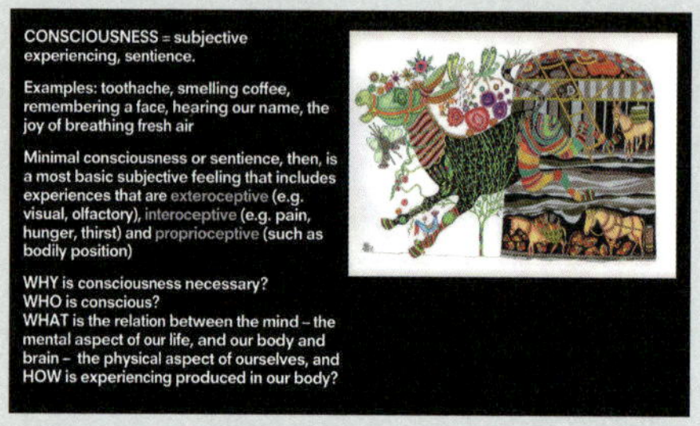

Fig. 1

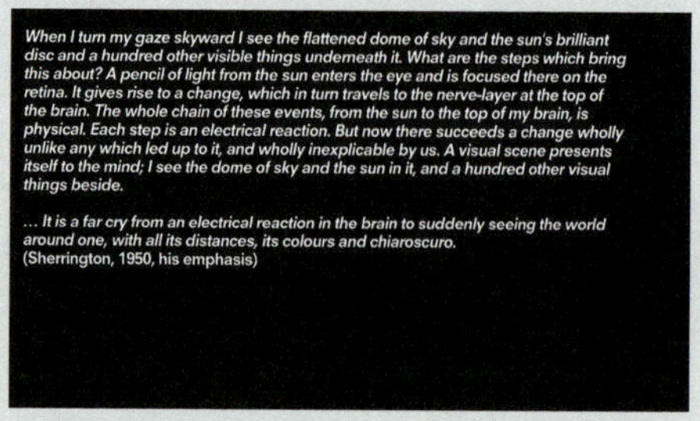

Fig. 2

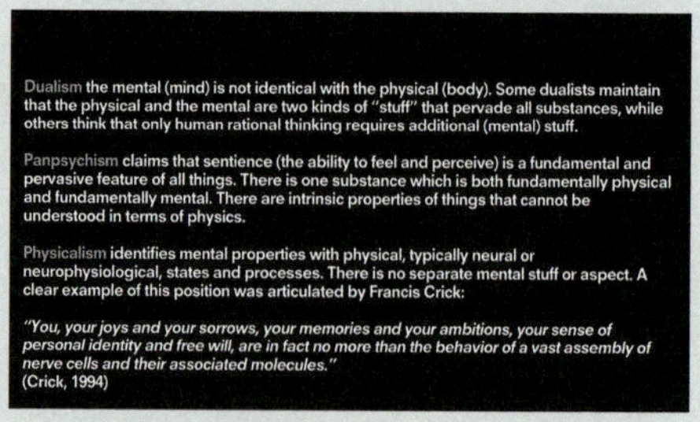

Fig. 3

physical state? And how, conversely, can physical changes in the body, like consuming a drug, cause mental changes? And what is the self? Is it an attribute of the body or of the mind? *Panpsychists* claim that sentience, the ability to feel and perceive, is a fundamental and pervasive feature of all things. They believe that there is one substance only, which is both fundamentally physical and fundamentally mental. They assume that there are intrinsic properties of things that cannot be understood in terms of physics. Panpsychism takes several forms. While some argue that everything has a degree of subjective experiencing, including your cup of tea, others think there are two types of experiencing, micro and macro. The micro experience of elementary particles, like quarks, is different from the macro experience of a toothache. So how, one wonders, does the experience of, say, elementary particles in your tooth combine to form macro experiences of a toothache? *Physicalists* identify mental properties with physical states and processes, typically neural or neurophysiological ones. They assume that consciousness pervades all matter, everything. A clear example of this position was articulated by Francis Crick. "You, your joys and sorrows, your memories and your ambitions, your sense of personal identity and free will, are in fact no more than the behavior of a vast assembly of nerve cells and their associated molecules." This statement hides many complexities. The vast assembly of nerve cells and their associated molecules is, as Crick himself assumed, an organized assembly. So what kind of organization is it? Can isolated brains be conscious? Are nerve cells both necessary and sufficient for consciousness? Neuroscientists who study consciousness often differ in the way they approach these questions, but they share the same broad philosophical approach. They regard consciousness as the outcome of the self-organizing, dynamic interactions between low-level parts of the neural system and higher levels of the system. These neural dynamics give rise to coherent images, feelings, and actions. The question is how this could happen. Even if consciousness can be experimentally manipulated, and even if perfect neural and bodily correlates of consciousness can be identified, why is it that these correlations entail subjective experiencing? Why are we not zombies? And why do we feel that our mental activities are qualitatively different from our physical activities? Our own scientific naturalistic view is an evolutionary one. We think that consciousness evolved, and we argue that it appeared as a transition from animals who had no consciousness to those that had sentience and subjective experiences, even if their consciousness was just minimal and very far from our own human reflective consciousness. The fact that animals, other than ourselves, are at least minimally conscious and sentient has gained support in the past decades from an increasingly larger number of philosophers and scientists. [4] In 2012, a group of biol-

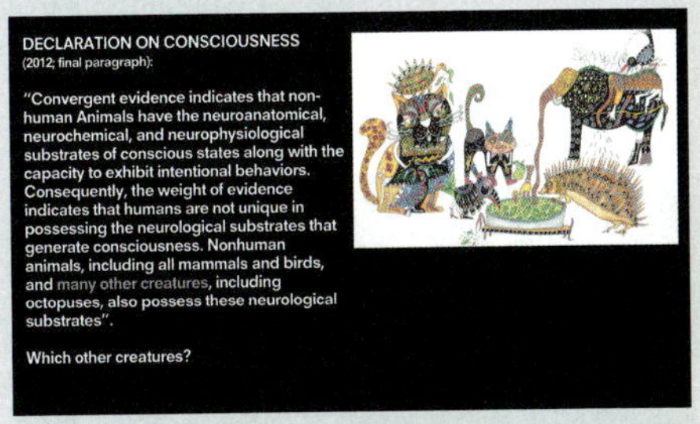

Fig. 4

ogists signed the Cambridge Declaration stating: "Convergent evidence indicates that non-human animals have the neuroanatomical, neurochemical, and neurophysiological substrates of conscious states along with the capacity to exhibit intentional behaviors. Consequently, the weight of evidence indicates that humans are not unique in possessing the neurological substrates that generate consciousness. Nonhuman animals, including all mammals and birds and many other creatures, including octopuses, also possess these neurological substrates." And recently, several months ago, scientists and philosophers in the United Kingdom wrote a detailed report that will enter the Animal Welfare Act in the United Kingdom. [5] Drawing on over 300

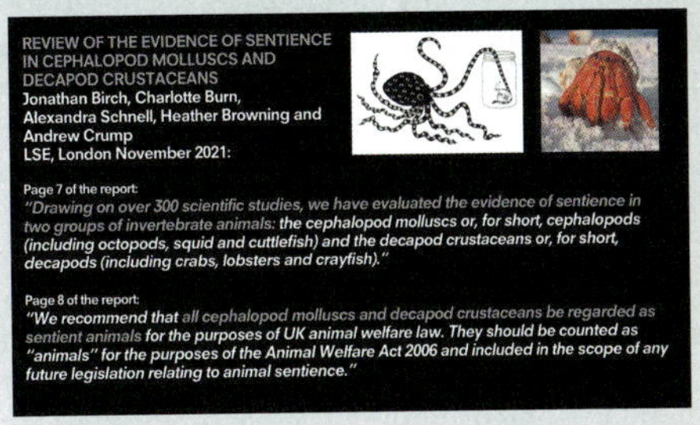

Fig. 5

scientific studies, we have evaluated the evidence of sentience in two groups of invertebrate animals. The cephalopod mollusks or, for short, cephalopods including the octopus, the squid, and cuttlefish, and the decapod crustaceans or, for short, decapods, including crabs, lobsters, and crayfish. Among the studies that convinced biologists that nonhuman animals are conscious are also our own evolutionary studies. I have already said that, according to our approach, consciousness evolved. It is fairly new to find scientists who take an evolutionary approach to the question of consciousness, and among scientists who do, there are at least seven hypotheses for the origin of consciousness. They include the hypothesis that consciousness originated only once in humans or twice in vertebrates and cephalopods. [6] Other hypotheses maintained

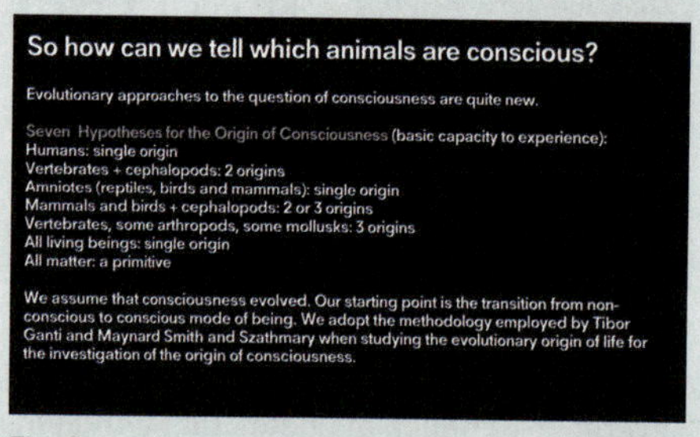

Fig. 6

that sentience began in amniots or three times in mammals, birds, and cephalopods. And there are those who find evidence for the dawn of subjective experiences separately in invertebrates, arthropods, and mollusks. There are also those who believe that consciousness developed once the first living creatures appeared. In other words, it exists in all living organisms. And finally, there are those who believe that consciousness is a primitive entity that pervades all physical matter, whether living or non-living. How can one develop an evolutionary theory of consciousness when there are so many disagreements about consciousness? I mentioned various philosophical approaches to the question on conscious organisms. Our way of approaching this question was inspired by the way the Hungarian chemist Tibor Gánti tackled a similar problem, the problem of how life began—another very difficult question. He looked at the transition from non-living matter to the first living organisms. Gánti started by compiling a list of capacities that, in spite of the different views about the nature of life, are generally deemed jointly sufficient for the simplest, minimal life. Gánti listed these hallmarks as maintenance of a boundary separating the entity from the environment, metabolism, dynamic persistence, information storage, regulation of the internal milieu, growth, reproduction, and, of course, also death. If we found on a distant planet a being that exhibits all these capacities, most scientists would agree that it is alive. The functional and structural coupling among the mechanisms and processes that implement these capacities make up a minimal living entity. But it is not always easy to detect or dissect each of these hallmarks in an entity that we study. Is there a single marker that requires the presence of a system with all these capacities? Following Gánti, the evolutionary biologists Maynard Smith and Szathmáry proposed that there is: it is "unlimited heredity". By that, they meant the capacity to form lineages that differ in open-ended ways from the initial system, so that the number of possible variants is so vast that it could not have been explored by real lineages, given the age of the universe. Such a system would be sustainable over evolutionary time. The DNA-based heredity system is an example of an unlimited heredity system. [7]

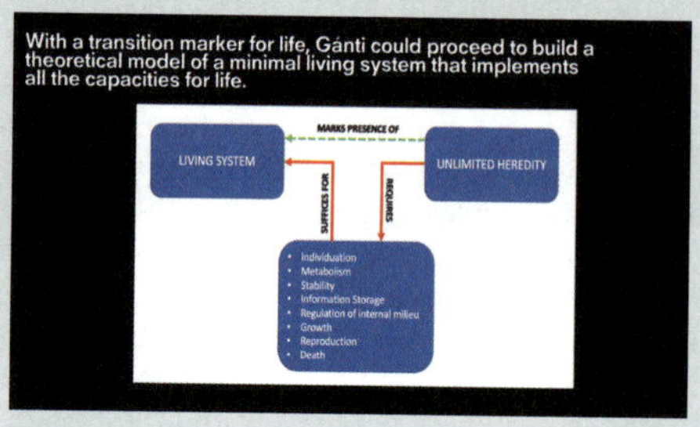

Fig. 7

With the transition marker for life, Gánti could proceed to build a theoretical model of a minimal living system that implements all the capacities for life. An evolutionary transition marker is a property such that when we find evidence for it, we have evidence that a major evolutionary transition in which we are interested has gone to completion. The transition marker allows us to reverse engineer the system that enables it like paleontologists who can reconstruct the whole animal from a single part. Theoretical biologists and chemists can reconstruct a minimal living system if they find structures that implement the capacity for unlimited heredity. So, if we found on a distant planet several long polymers, molecules made up of different units, like DNA, we would infer that there must have been a system, that enabled the synthesis and maintenance of such complex molecules because they could not have been generated spontaneously. [8] The ordered complexity of the polymers would suggest that they had to be synthesized by other specialized molecules within a spatially distinct self-reproducing system with intricate maintenance machinery. And since the structure of the polymers suggests that their properties enable them to be copied with variations, we would infer

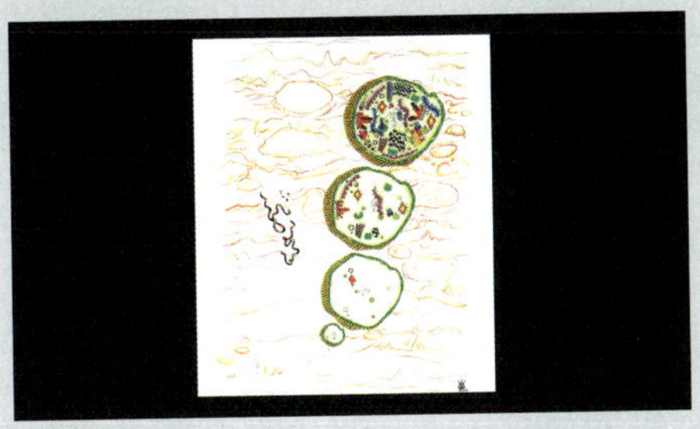

Fig. 8

that one of the functions is information storage and that they allow the system as a whole to persist for long evolutionary periods. The unlimited hereditary capacity of the polymers is then diagnostic of a minimal evolutionarily sustainable life. Our evolutionary approach to studying consciousness was to look for an evolutionary marker of the transition to consciousness. In her lecture, Eva will show you what it is.

Learning and the Origins of Consciousness

EVA JABLONKA

I am going to tell you how Professor Ginsburg and I applied a methodology, which was developed by a Hungarian chemist, Tibor Gánti, to the problem of life. Tibor Gánti tried to define and build a model of minimal life. And the way he went about it was to compile a kind of consensus list of the capacities that would generally be recognized as characterizing life; he then built a model that links all these capacities together, and tried to find a single capacity of the system that when it is in place, it implies that the whole list is also in place, a single capacity that is a diagnostic marker. Simona Ginsburg and I call this the evolutionary transition marker. In his case, an unlimited heredity system was suggested as the evolutionary transition marker for life. We were looking, using this kind of way of thinking, this kind of methodology, for a marker for minimal consciousness. We wanted to apply this methodology to the question of consciousness. We searched for such a marker because a single tractable capacity that can identify a minimal kind of life or consciousness can not only tell us which beings are alive or conscious, but can also tell us which processes and which structures actually construct this capacity. Figure 1 presents the idea of a transition marker. [1] What

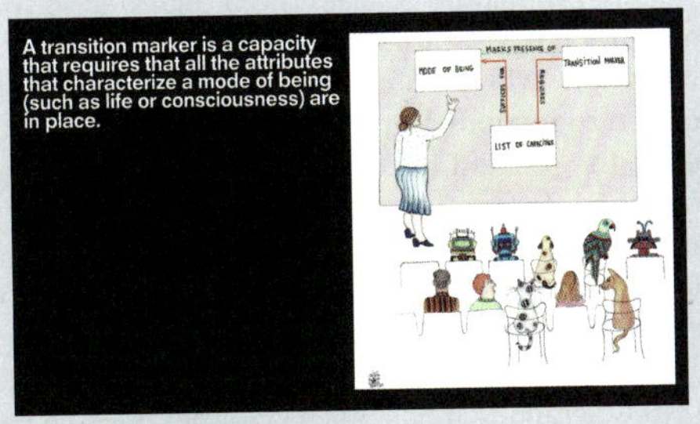

Fig. 1

Gánti did for the question of the origin of life was to find a list of capacities that everyone would agree that if they exist, if they are displayed by some kind of being, there are indicative of, in his case, life. It can be a list of capacities that are characteristic of something like consciousness. So it is something that has to be sufficient for us to say, yes, if somebody has this list of capacities, then it is alive. Or, if we are thinking about somebody, about rationality, rational. And then, the next stage is to find a single tractable capacity so that when it exists, we know that the whole list is in place. This is the kind of methodology that we applied. We started with a list of capacities. Now, the list of capacities that we see here in figure 2 is quite elaborate. [2] I am not going

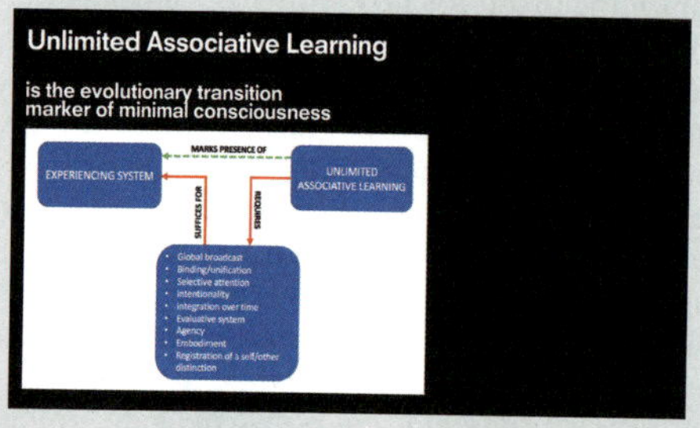

Fig. 2

to go into a detailed description of each of them. But basically, what we are looking for are several very important and, we believe, diagnostic capacities. For example, conscious creatures perceive a composite object or a composite action as an integrated whole. We see the apple as both red and round and of smooth texture. This allows us to make discriminations. For example, a peahen can discriminate among peacocks that differ subtly in the color and pattern of their tails. A mouse can discriminate between edible and inedible seeds. In order for this to happen, we also need to have some kind of capacity to bring together information from different cognitive systems, sensory systems, a memory system, a value system, and to bind them together so that comparisons, discriminations, generalizations, and evaluations will become possible. We also have to hold on to incoming information. For a conscious being, the present is not a kind of infinitely small point of intersection between the past and the future. It has a duration. It it is experienced as durable. A conscious being can evaluate goals, it can evaluate its perceptions, it can evaluate its own actions. And it can change its evaluations according to circumstances. It has this kind of flexibility. It can also selectively exclude or amplify certain signals coming from its own body and coming from the world. So, there is a lot of selective attention, there is an ability to focus attention, to sustain attention, to shift attention that non-conscious beings don't have. Very importantly—and this is related to all the other things too—, a conscious being not only integrates information, but it also maps signals from the world and the body and creates kind of maps, cognitive maps. One example of this is the kind of spatial maps that are formed by birds, bats, rats. These maps are also called representations, and these representations must be implemented in a body that is not just reactive but is intrinsically active, proactive. And it has the ability to have a stable kind of perspective that can discriminate between the same stimuli that come from the world and the body. So the origins of the stimulus is very much dependent on how you react. It is very much dependent on where it comes from. For example, if you tickle yourself or you are tickled, your reaction is very different although the sensory stimulus per se can be exactly the same. Now, this list, as I told you, is not complete and some people think that some aspects of it are more important and some aspects of it are less important and obviously all the kind of characteristics that I was talking about are related to each other. And this is very important because they complement each other, facilitate each other, enable each other, and so on. They form a system. If we found something, a being that has all these capacities on another planet, most researchers, most psychologists, most neurobiologists, most cognitive psychologists would say: "Well, we have to take very seriously the possibility that this being is conscious. We will treat it with respect". It is very difficult to investigate and identify each of this long list of capacities that I was talking about. We were therefore looking for a single capacity that requires that this whole list is in place. And after a lot of work, we found one capacity. This capacity can be called the capacity for Unlimited Associated Learning. [3]

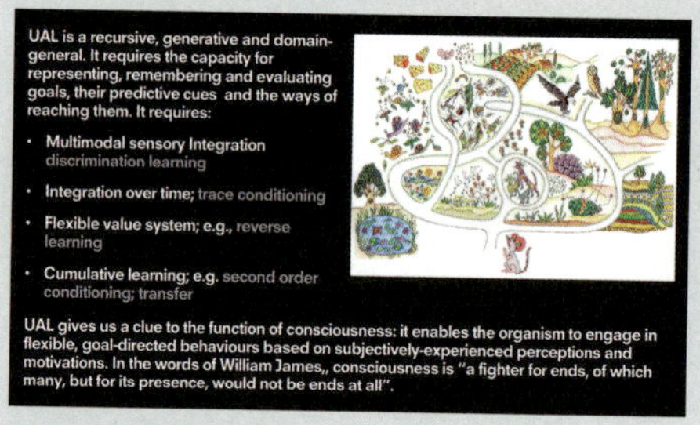

Fig. 3

It is a form of learning—and we'll explain in a minute—which is recursive, generative, and domain general. It requires that the creature can represent, remember, and evaluate goals, that non-conscious beings don't have their predictive cues and the ways of reaching them. And the way that this capacity is actually operationalized is when a creature can do discrimination learning, it can do multimodal sensory integration that allows it to do discrimination learning. It can integrate over time, and this can be represented by trace conditioning, that is the animal can learn the predictive clues even when there is a time gap between the predictive stimulus and the reinforcement. They don't have to overlap, they don't have to be synchronous. A third thing is a flexible value system, so that the animal can learn that something that was good in environment one is bad in environment two, and vice versa. For example, a little mouse can learn that a particular route that it was using is no longer safe because the environment has changed. So when an animal or a being is able to do all these things, it enables it to display very flexible and goal-directed behavior, which is based on subjectively experienced perceptions and motivations. This gives us a clue as to the function of consciousness. William James said that consciousness is a fighter for ends of which many, but for its presence, would not be ends at all. We agree. So the function of consciousness is to be able to reach goals that change all the time; the goals are evaluated through the perceptions and motivations of an organism, its internal states, as they interact and map its own actions and the external world. Now we hypothesized that this kind of system, the ability to learn in this way, is dependent on reciprocal connections between sensory, motor, reinforcement, and memory processing units, with the central unit at the core of the network. [4] You see

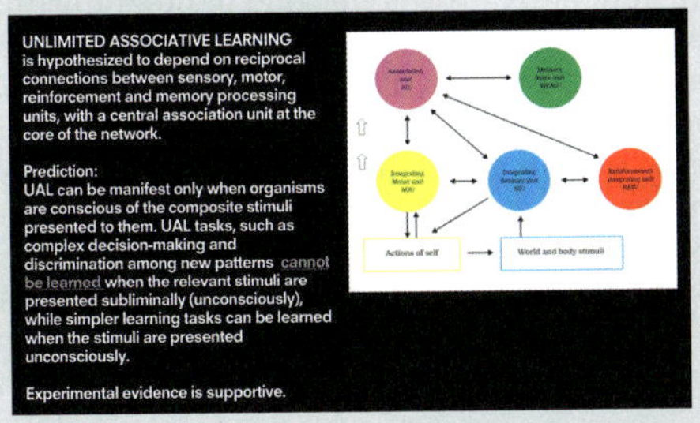

Fig. 4

it in figure 4, this is the purple unit. And we think that in the nervous system, this kind of architecture—this is a very, very simplistic picture—is implemented in the central nervous system. And this leads us to a prediction—if we are right, this kind of learning is really a marker of minimal consciousness. The prediction is that this type of learning can be manifest only when animals are actually conscious of the predictive stimuli of their own actions. If an animal is presented with the composite stimuli subliminally, it will not be able to learn them. It can only be able to learn them when it is conscious of them. On the other hand, simpler forms of learning can be accomplished by the animal, even if the stimuli are presented subliminally. And so far, at least in humans, this kind of prediction has been borne out. Also, there are some experiments, in monkeys that also support this. And I am very glad to say that people are beginning to look at this question also in other species, not only in mammals, not only in humans. We have predictions as to the nature of learning that can happen under subliminal and supraliminal conditions. Now, if we are right, and unlimited associative learning is really a marker of consciousness, unlimited associative learning is something that we can actually study. [5]

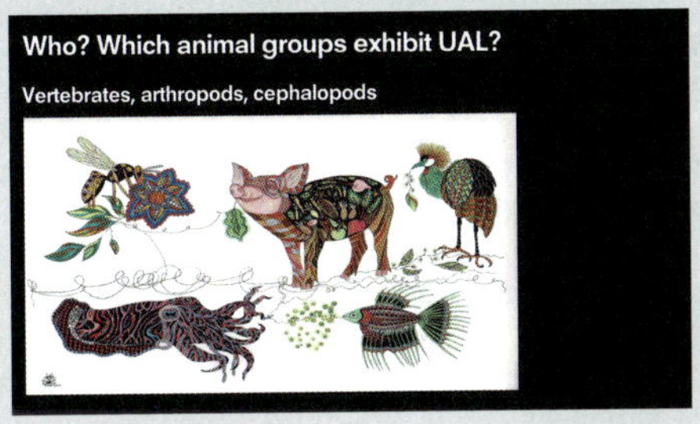

Fig. 5

We can see which animals or creatures or any kind of thing can actually accomplish this kind of learning. And we looked for quite a long time to see which animals are capable of this kind of learning. We found out that it is only animals and only a very small number of phyla among the animals. There are only three phyla of animals that are capable, that display the capacity for unlimited associative learning. These groups of animals include the vertebrates, some arthropods, including some insects, and including crabs, lobsters, shrimps, and the cephalopods, the coleoidea cephalopods, which include the squid, the octopus, and the cuttlefish. Now, we also know what kind of brain and what kind of nervous system we want. We can ask: do these animals have supporting structures in the nervous system that actually enable this form of learning? And what are they? We know that they must have them because they have this form of learning, so we can identify them and we can compare them between the different species. And when we do that, we find something very interesting: the brain structures in mammals show a functional organization that is similar to that found in the other groups. [6] Here we have a human being, this is Mr. Spinoza, his brain is on the left; we have a dragonfly, and its brain on the right. And we have at the bottom an octopus hiding in a coconut shell and its brain. Of course, all these three brain types are very, very dif-

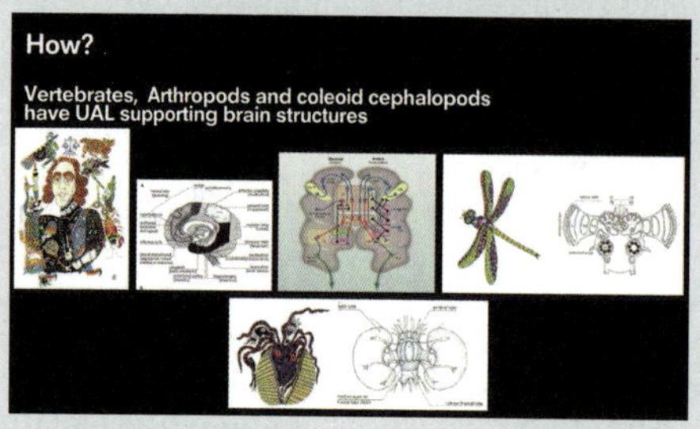

Fig. 6

ferent structurally. Nevertheless, they have a similar functional brain architecture. In the middle, you see the mammalian and the insect functional brain architecture, and they are surprisingly similar. They have the same type of functional units and the same type of connections among functional units. We know a little bit less about the cephalopods, but as we are learning more, we see that they too have the same kind of functional units. So, Vertebrates, arthropods, and coleoidea cephalopods have the UAL supporting brain structures. Since we have fossil brains, we can find in the fossil records traces of brain structures. We can figure out when these UAL-supporting structures first appeared in the fossil record. And when we look at that, we find out that UAL first appeared in the Cambrian, 542 million years ago, in two groups: the vertebrates and the arthropods. [7] It appeared about 250 million years

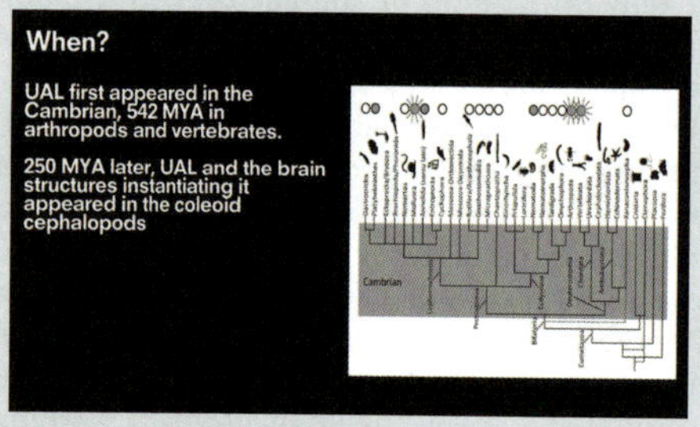

Fig. 7

later in the coleoidea cephalopods. What you see in this diagram is what is known as the Cambrian Explosion. It is a relatively short period of time—about 60 million years—where almost all the known phyla of animals appeared. That is why it is called the Cambrian explosion. It was an explosion of new forms of life. It is something that caused Darwin a lot of trouble because he was always talking about gradual evolution. And here was a very sudden burst of evolutionary novelties. During this period, we see the brain structures in the vertebrates and in the arthropods that are indicative of the capacity for UAL and, by implication, minimal consciousness. The next question that we asked was: these two groups appear at that time, and another group appeared 250 million years ago. So were there three independent transitions to consciousness and transitions to UAL? It also gave us the idea that since the vertebrates and the arthropods appeared in the Cambrian, and because unlimited associative learning is such a powerful adaptation—it allows individuals to learn about many aspects of the environment during their own lifetime and adapt during their own lifetime—, maybe this capacity for unlimited associative learning was one of the things that actually drove the great Cambrian explosion. When you have this capacity to learn, you exert huge selection pressure on the species with which you interact and on the individuals in your own species with which you interact. So predators become more effective predators. Prey becomes more evasive prey. Mates become more discriminating, and so on and so forth. So evolution can become very, very rapid, and it can lead to a kind of explosion that we see. We are not saying that this is the only reason for the explosion, there are many interacting reasons, but we suggest that this was one of the engines that drove this great diversification of life. What we see here in figure 8 is the kind of creatures that appeared in the Cambrian. Mainly it was dominated by arthropods, but there were also small fish. There was a kind of great war for existence, a great battle which was driving evolutionary interactions among species and driving huge diversification of species, leading to both antagonistic and cooperative arms races and many, many feedbacks. Now, this led to a lot of the things that we see. Think about, for example, the beautiful patterns on the

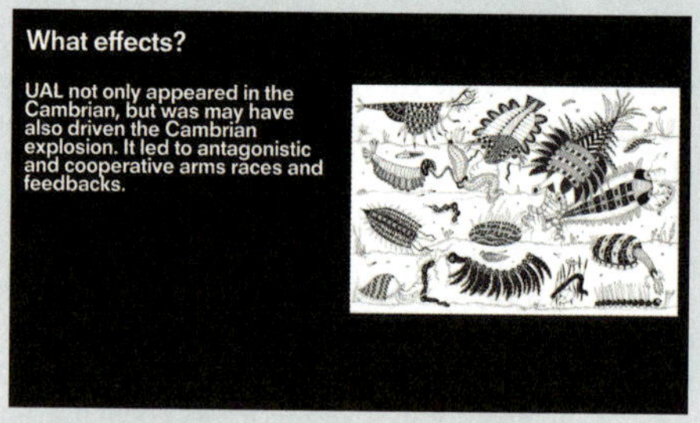

Fig. 8

body of the male fish or the haunting songs of birds. [8] This kind of things would not have evolved were it not for creatures being able to discriminate among them. For example, the beautiful patterns on the male fish would not evolve if it wasn't for the female's ability to discriminate between different patterns and choose among them. I don't think we would have beauty without consciousness. It is very important to remember that we are talking about the origins of consciousness and about simple consciousness, this was just the beginning of an evolutionary process that went on. Consciousness and learning became more complex during evolution. [9] I was describing this very simple consciousness,

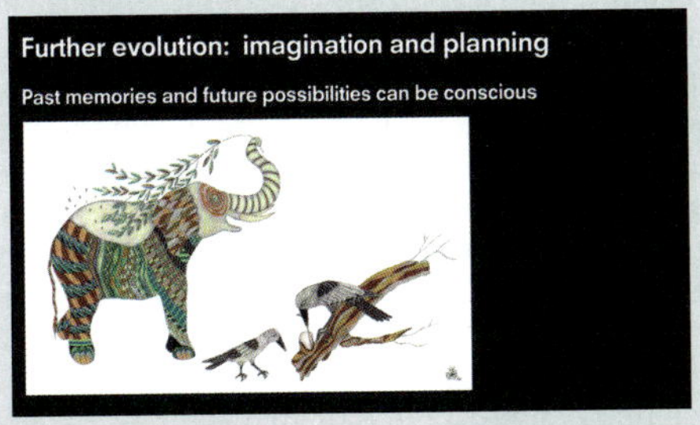

Fig. 9

the consciousness that is underlined by unlimited associative learning. This type of learning evolved and animals are able to be conscious not only of encoding information in the present, but also be conscious of past information, of the past, and can be conscious of the future state of the world. Virtual realities have emerged. And this is the evolution of imagination, of planning. Past memories and future possibilities have become conscious in some animals, for example in elephants and in Corvids. And finally, we reach the human species. [10] Our own

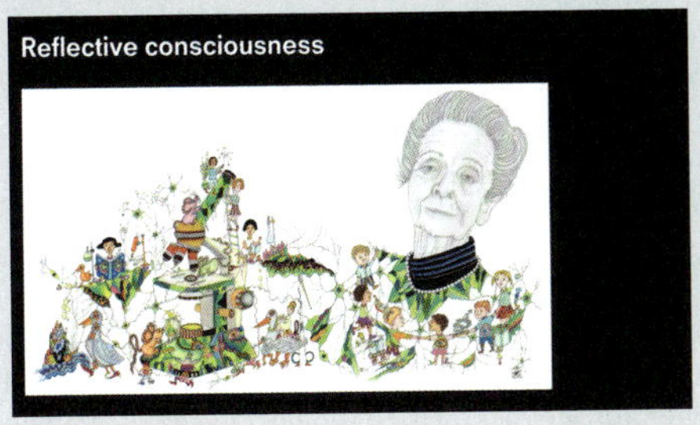

Fig. 10

lineage where we have reflective consciousness. And this is the ability not only to imagine and to plan but also to communicate about the products of our imagination, to share them, to analyze them, to rationalize them, and to categorize them.

Conversation 6

ALICE ROBERTS, SIMONA GINSBURG, EVA JABLONKA

ALICE ROBERTS This evening I am joined by two experts to talk about what one of them has called "one of the most engaging topics bringing together science and philosophy": the problem of consciousness. What is consciousness? Are other animals conscious? What is the relationship between the brain and mind? What is the relationship between the mental and the physical? Just a few of the questions we'll be considering tonight. Simona, please could I ask you to just give us a brief summary of your lecture?

SIMONA GINSBURG In my talk I asked what we mean by consciousness and said that for us it is synonymous with experiencing, subjective experiencing, and sentience. The minimal consciousness that we see in many animals is very different from human reflective consciousness. It includes experiences that are triggered by external receptors that bring about things like seeing and hearing, for example, and experiences that are initiated by internal receptors leading to feelings, such as sensing the position of the body in space, and to feelings like thirst and hunger. The concept of consciousness raises many questions. The first one is why is consciousness necessary? Why can't we process information about our internal surroundings and the external world and make sense of it without the subjective feel? Another question is who is sentient? Only humans, some specific animals, which and why? Our own naturalistic approach to such questions is an evolutionary one. We start with the fact that consciousness evolved during evolution. A transition from non-sentient organisms to experiencing ones occurred, and we investigate how this happened. In doing so, we were inspired by Tibor Gánti, who tackled another elusive and difficult question: how did living creatures emerge from inanimate matter? Following his methodology, we formed a list of characteristics that most scientists and philosophers would regard as sufficient when they appear together to render the system conscious. Finally, we searched for a transition marker, a single characteristic that can easily be identified, whose presence is diagnostic of consciousness. In our theory and investigations, we discovered that open learning is this marker.

AR You have set the stage beautifully and opened up this question, which I hope we are going to explore, as we get into the Q&A, that from an evolutionary point of view, it isn't self-evident that an animal which senses and reacts to its surroundings has to be conscious. It opens up the question: why is consciousness necessary? What is it actually for? We will explore that in more detail. But first of all, Eva, please, could you summarize your lecture for us.

EVA JABLONKA I started my lecture by reiterating the reasons for seeking an evolutionary transition marker. As Simona told you, such a marker is some kind of a single tractable capacity of the system. Once we have evidence of it, we have evidence that the transition in which we are interested—in our case the transition from non-sentient to sentient organism—has gone to completion. Once we identify this marker, this capacity, we can reverse engineer from this capacity back to the system that enables it, so we can understand something about the dynamics of the system that enables the capacity and by implication consciousness. Then I went on to present the proposal about the evolutionary marker that we have identified, which is a form of learning. It is something we call "unlimited associative learning." It is an open-ended, generative, recursive, and representational form of learning and we suggested that the evolution of this learning, which is very highly adaptive, drove the transition to consciousness. I then discuss the implication of this thesis for questions that pertain to the neural dynamics that constitute consciousness; what is the architecture of a system that is conscious, to the taxonomic distribution of consciousness to the question of who is conscience. I pointed to some of the predictions of the theory about the relationship between learning and consciousness. There are specific predictions that follow from our proposal and the ecological context in which consciousness and unlimited associative learning first appeared. I also highlighted some of its effects, such as the emergence of beauty. I ended by pointing to two subsequent stages in the evolution of consciousness: the evolution of imaginative consciousness, such as that of elephants, hominids, great apes, and the symbolic rational consciousness of humans.

SG Eva, there are several theories of consciousness. What does an evolutionary theory of consciousness add to these?

EJ You are asking an evolutionary biologist. First of all, I will start by saying that evolutionary biology is the most powerful theory in biology and it is a conceptual kind of bottleneck through which every theory of life and mind must pass. If some theory—biological theory, psychological theory, sociological theory—does not pass through this bottleneck, there is something seriously wrong with it. Evolutionary biology is therefore a very good kind of reality check. Since we believe that consciousness evolved, an approach that is based on evolutionary theory is a good starting point. But it is more than a good starting point, because an approach that starts from the evolutionary transition from non-sentient to sentient beings has several advantages. The first one is that the earliest sentient beings do not carry the misleading baggage of later evolved structures and processes that build up consciousness in complex organisms, like humans or mammals. Recognizing the earliest forms of consciousness would enable us also to identify specific processes and principles that can be applied to all conscious beings. That is one great advantage. Another one is that it offers a tentative answer to "Who is conscious?" The question of who is conscious, what is the taxonomic distribution of consciousness, and it can also tell us—an evolutionary approach can tell us—when, under what conditions consciousness first evolved, what its functions may be, whether it evolved once or more than once, whether it was lost in some lineages, how it evolved into new forms or sometimes richer forms, and what its future can be. There are many advantages and I think it is a good thing to look at it from this perspective.

AR It is fascinating to look at consciousness as an evolved phenomenon and to understand that it started off fairly simply and just got more and more complex. I like the way that you described that, in looking for animals with simpler forms of consciousness, you have stripped away the bells and whistles of the more complex ones so that you can get to the core of what consciousness is really about and why it starts to evolve. Eva, would you like to ask Simona a question?

EJ There is an obvious question to somebody who likes philosophy and that is: how does your lecture, Simona, relate to what David Chalmers called "the hard problem of consciousness?" What is the relationship?

SG In my talk I raised the question "Why can't we process information about us, our body and the world, and about our actions without a subjective feel that accompanies this processing?" This is exactly the same question that David Chalmers called the hard problem some decades ago. He considered the easy problems, those that relate to information processing in the brain, for example, the way light, falling on our retina in the eye, is processed through many stages and structures in the brain. In this case, the hard problem is why qualia—that is subjective impressions, special specific colors, like red or blue—accompany this process when certain wavelengths fall on the retina. Chalmers claimed that up till now no answer has been given to the hard problem. We think that in our theory we have, to a considerable extent, dissolved the hard problem. We have presented an evolutionary framework that shows how certain dynamic organizations of dedicated structures, processes, and interactions within the brain, give rise together to the capacity of being conscious. Consciousness is a necessary outcome of a transition that occurred in animals during their evolution, just as life is a necessary outcome of chemical evolution.

AR It is a big claim—to have solved the hard problem of consciousness. I must say I am on my way to being convinced. Let's see if our audience is. We have some questions that have come in from them. Here is one: the term "consciousness" has a long tradition. Do you think that the meaning of the word has remained essentially constant, or have there been substantial changes during this period? This is difficult, as this is about semantics and definitions. As with all definitions, consciousness changed.

SG The word "conscious" and "consciousness" are umbrella terms. That is why there are so many concepts that come under them. It is very hard to define. We hear about creature consciousness, which may have conscious and unconscious states, about access consciousness, effective consciousness, higher order consciousness, minimal consciousness, phenomenal consciousness, to name very few. All of them have something to do with mental behavior and mental phenomena. Historically, the term was used only in relation to human beings, but today things have changed and it is used also when referring to animals. Today some people, not necessarily those we agree with, even talk about conscious plants.

AR In some ways that is similar to a lot of concepts in biology where, perhaps, we could say that biologists of the past probably fell prey to a bit of human exceptionalism and that there are all sorts of aspects of human biology that we now find to be shared to greater or lesser degrees with other animals, which is to be expected even if everything is evolved, because we are connected with the rest of life on earth. Eva—presumably, we should expect that, because of evolution, consciousness should not be an exclusively human characteristic. We should expect to be able to find it amongst other species in the animal kingdom?

EJ There is a lot of continuity in evolution. There is a question of whether there is some kind of qualitative jump somewhere, and some people believe that there was. The question is not settled by evolution, because evolutionarily, although we say there must be continuity at some level, it could be that at a certain macrolevel there is discontinuity. So the fact of evolution does not resolve the problem completely. We however think that there is continuity, that if we are looking at the question from an evolutionary point of view, we actually could see the continuity, although there will be some cases where we will not be able to say whether a particular animal is conscious or is not conscious. This would not be a problem of definition, it would be a problem of the inevitable gray areas that always exist when you are doing evolutionary biology. You have the same problem with life. You can identify things that are alive 100% and things that are not alive 100%, but there is a gray area in the middle where it is very theory dependent, it very much depends on your characterizations and definitions. It is very difficult to decide whether something is alive or not alive. We see the same with consciousness.

AR This reminds me of the whole species problem in evolutionary biology. It is very difficult to define—you can take one species that is nicely separated from another in time but you can't possibly say when one species changes into another. There is no point in time.

EJ Yes, you have the problem with incipient species and it is not a coincidence that both Darwin and Lamarck were obsessed with the origin of species. Lamarck's book starts with the problem of species. He starts with the deconstruction of the essentialist notion of species and Darwin does the same.

AR We have touched on another question that is coming from our audience, which is about degrees of consciousness and particularly thinking about humans. What kind of difference, if any, does the presence of human language make in the degree of consciousness? Do you think this is a qualitative or a quantitative difference?

EJ I'll take it because I did quite a lot of work on the evolution of language. We think that it is a qualitative difference, actually, although there is continuity and we can see the continuity. Because with human language what happened is that virtual realities became communicated, it is possible to talk about, to discuss the imaginative—what is the future, the past—and discuss it with oneself and with others. Virtual realities become something that is shared, and language also led to new symbolic values. New values appeared in the world, like the good, the beautiful, and the just. Also, new goals appeared that satisfy these symbolic values. They have become constructed and we are the kind of creatures and the only creatures that thrive to fulfill this kind of symbolic goals. Also, the kind of emotions that we call the normative emotions—guilt, shame, embarrassment, and pride—, which evolved before language, we think, were very much deepened by language, because the social gaze, the evaluation of others, and all the things that go into the kind of social relations that language allows, have made our view of ourselves very much dependent on what other people not only think about us but say about us. These emotions have become very much extended and enriched through the evolution of language. Also, other emotions—think of moralistic anger, for example, the anger we all feel, I think rightly, about what Putin is doing today in Ukraine. Language also increased greatly the problems of false memory, because you have to make a distinction now, not only between remembering what happened to you and what you imagined, but also what was told to you and what you imagine and what actually happened to you. False memory became a big problem and we deal with it in all kinds of ways, including the development of autobiographical memory. Another thing that language changed is perception itself. There is something called verbal overshadowing. Overshadowing is the phenomenon whereby when you describe something in words, for example, a criminal that you saw, you then have greater difficulty of perceptual remembering, of having a rich image of that somebody, of that person. Whereas if you don't describe in words, your perceptual memory is much better. A lot of things changed with language. We think that language really changed human consciousness in profound ways.

SG Maybe, if I may add, cultural evolution expanded a lot with language, and that, of course, had a great effect on our human consciousness. Another interesting point about the social gaze and emotions in humans is the blush. Only humans blush.

EJ Yes, but *Homo erectus* may have blushed before language.

AR You are saying that language has had such a profound impact that it is a qualitative difference in human consciousness, having had language for a while—we are not sure how long, but for a few hundred thousand years at least. Here is another question. Assuming that consciousness has evolved, do you think the process has been gradual or discontinuous? That is an interesting one as well.

SG If I may answer, what is gradual? It is scale-dependent. The Cambrian during which we think consciousness first appeared was only 50 million years long. A very short time in geological terms. Is the diversification that occurred in animal groups during that time gradual or explosive? Human evolution, from the chimpanzee to ourselves, took about six million years. It was very rapid. But there was continuity within this period. Eva, would you like to expand?

EJ No, obviously it is not something that happened within the ecological time frame, it happened within a geological time frame.

AR Essentially, your answer is that it is continuous, but the rate can change. Can you relate consciousness in animals to consciousness in neonates? An interesting question.

SG The test, at least according to our theory, would be once the brain begins to develop and the structures in it that enable UAL (Unlimited Associative Learning) appear, we would say that potential consciousness is there.

AR Does that mean that human neonates are always going to be different from other animals, then? The structures are there to support that development?

EJ Not necessarily. A lot of human development is driven by culture. We are cultural creatures as a species, so we know that people who are not exposed to language during a critical time have great problems acquiring it. I think this is true for many other human-specific properties. Cecilia Heyes has recently made a very strong case arguing that in fact most of those capacities and properties of humans are driven by cultural evolution and constructed by cultural evolution. We have just the very general kind of potential for responding to this cumulative cultural input. Because of that it very much depends on the cultural environment. You cannot disconnect humans from their cultural environment.

AR That is a really interesting point because we still hear occasionally all kinds of nature/nurture debates and when you are looking at humans it is impossible—if you separate humans out away from culture that is a very unnatural state for humans to be in. I absolutely love the quote of Michael Tomasello that a human is born expecting culture in the same way that a fish is born expecting water. Let's move on then to another question. One about fish, in fact. You say that the beautiful patterns of the body of the male fish would not have evolved if it were not for the female's ability to discriminate and choose subtly between different male adornments. How do you define beauty independently, then, and how can you be so sure that the male fish couldn't recognize beauty?

SG In her talk she expressed part of this sentence, not all of it. She didn't say that a male fish wouldn't recognize beauty. We think that beauty is in the eyes of the beholder. It cannot exist independently without an observer. In fact, this question reminds me of the famous question "does a tree branch falling on the ground make a sound when nobody is witnessing?" The answer is no. The branch may produce mechanical pressure on the ground when it hits it and even pressure waves. But without the presence of a being that can hear, there is no sound. Similarly, organized structures and patterns do exist in the world, regardless of evolution and consciousness, but their beauty results from perception and evaluation by a sentient being who may also derive pleasure from them.

EJ Something about the male. Of course the male fish can appreciate beauty. For example, it will certainly have an experience when it sees the patterns on the bodies of rivals. It may not seem pleasing, it may see it as very unpleasant, in some ways, but it will certainly recognize it and have an experience of it. It has the same mental capacities inherited from his mother but of course, what the male regards as beautiful will be colored by his life as a male.

AR Do you think what the male fish or the female fish end up perceiving as beautiful is really entirely arbitrary? How does that emerge? Because if beauty is just in the eye of the beholder, it doesn't exist as an independent essence in and of itself.

EJ It doesn't mean to say that they didn't evolve. Many of the aesthetic things that give pleasure have some kind of relation to some kind of existential needs of ours. For example, there are people who argue that symmetry, bright colors, beautiful patterns are indicators of fitness in a male. The female scrutinizes these things. It is not entirely arbitrary. The kind of patterns that are regarded as beautiful are, to some extent, indicators of fitness. It is not the only story. There are all kinds of runaway processes, but there is some kind of relationship. There is a certain level of arbitrariness, but there is no absolute arbitrariness. The very fact that it is functional suggests that it cannot be completely arbitrary.

AR It is interesting because I think there is room there for imagining how some patterns might come up. As you say, it is a kind of runaway processes, it could begin with spandrels [non-adaptive features] just appearing during development in a way that almost seems arbitrary. But as soon as another being notices those and perceives them as either beautiful or not, they become evolutionarily meaningful. Here is a really interesting question about where we are going in the future of consciousness. If consciousness is evolutionary, can we imagine that there will be in the future living beings with augmented consciousness as compared with ourselves? Beings that will come out from brain-machine interfaces, AI, bionics—could we create machines with consciousness in the future? What a question!

SG There is no doubt in our mind that such augmented consciousness will be possible and that cyborgs, which are chimeras between machines and biological organisms, will expand. Even today they exist, and in the future, they will become more and more prevalent. Already at present, when we wear a headset of virtual reality, we easily and very quickly enter our own avatar and imagine it as part of our body. This is probably a continuation of a process that began ages ago when tools such as hammers became an extension of our bodies culminating today in our smartphones. Furthermore, since the brain is very plastic, we can rewire it, in certain cases, by using artificial devices. We can, for example, even today implant the brain electrodes that communicate with computers, which enable the control of artificial limbs. The mere intention to move a limb is programmed to cause it to move. I think a serious problem may arise from the brain-machine interface developments. Once we become cyborgs, if we are wired to other cyborgs, will we continue to have our individual experiences? Will we not lose our autonomy and free will once we are connected to them? Will there be a collective consciousness? What will it be like? And, of course, part of the question that you asked was also will robots be conscious? Eva, would you like to take that up?

EJ We just wrote the paper about it recently. We can go on talking about it for quite a long time but what I want to say is that first of all there is no robot in the world that can do unlimited associative learning. They can do very limited things, very limited things better than human beings, but it is extremely limited. That is the first thing. They don't also have the kind of general value system that living creatures have, that is based on survival in the case of biological beings, but in the case of robots, it would probably be something like some kind of coherence, some kind of common value which is the basis of everything that happens to them. We also think that it will be very much more difficult than people think to create some kind of conscious robots, not only because of the software difficulty, but also because of the relationship between hardware and software. For example, it is not clear to us that it will not be necessary to use soft kinds of materials, or what is called even intelligent materials, clever materials. It is not clear to us at all that it will be possible to develop the affordances of ontogeny, that are enabled by morphological ontogeny, not only cognitive ontogeny. Nevertheless, we think that probably it will be possible one day to build conscious robots and I am very worried about it personally, because we don't have a good record in our treatment of conscious creatures. But I think it is far more difficult than most people think.

AR You harked back there actually to the very first conversation in this series where Angela Friederici and Robert Berwick were talking about human language and the syntax of human language. Berwick was arguing in his lecture and in our discussion that in fact all of the AI that we have at the moment does not recreate human language. It recreates a superficial appearance of human language but it is actually not doing the syntax in the way that the human brain understands language. It is interesting to hear you say as well that you think that consciousness is a much more complex problem than perhaps some people think at the moment with robots and AI. The next question goes into a bit more about detail and I must say this appeals to me as an anatomist. Eva, in your lecture you showed various structures in the conscious brain and I wonder if you could explain what those structures are. What these key structures are that we see in apes, in octopi and potentially in insects as well?

EJ I would answer very briefly because it is really going into gory details here. The integrating structure that they pointed to included units that integrate sensory information. You need to have sensory integrating units and it has to be integrating information that comes from the external world, from the internal environment; and also integrating motor units that can represent the motor behavior. You need to have integrating reinforcement unit systems, and you need to have a declarative memory system that represents composites, that represents events. You need all this to interact with each other and you need all to come together. The mappings from all these units will need to come together before the result is conscious perception and reaction. What are these things? I don't want to go into great apes and into humans because they are great apes and humans, because we are already talking about high-level structures where the neocortex is very much involved. Other very important structures for consciousness include, in the case of vertebrates, for example, structures such as the optic tectum that controls direct behavioral responses towards specific points in body-centered space. That is one. Another one is the hippocampus, which is required for declarative event memory and we see it in almost all vertebrates, except maybe the very primitive, but already there we have cells that behave like hippocampus cells. The reticulate formation and the periaqueductal gray, are where the effective circuitry is converging and diverging. These are very important structures for consciousness and we know that they are important because when people have lesions in sensory integrating areas, for example in the primary visual cortex, V1, the result is blind sight, they are not conscious of what they see. They say they see nothing although they behave as if they see something. We know that the lesions of the hippocampus lead to a lack of consciousness of past events. We know that lesions in the periaqueductal gray lead to coma, complete coma. All these structures obviously are related to consciousness. There are analogous structures in arthropods and in cephalopods. When I am saying this—that these structures exist—I am talking as if they have distinct functions. Sometimes they have more than one function, one has to remember that. It is not a completely modular system. The mushroom bodies and the central complex in insects are believed to be either homologous or analogous to the hippocampus and to the basal ganglia, and in cephalopods we have an called frontal vertical lobe that has circuitries that are analogous to those that exist in the cerebral cortex, the hippocampus and the amygdaloid complex in vertebrates. We know that it plays a role in the evaluation and leads to decision-making in the octopus. There is a lot of analogy there.

AR You can delve into that comparative behavior but also comparative anatomy as well and see analogous structures. It makes you think a little bit differently about octopi, certainly. I want to end up with a couple of much broader questions. We are almost reaching the end of our time. There are questions about your careers and about science and philosophy, a bit more generally. Simona, I wanted to ask you—because I was fascinated, in your career, by the fact that you started off investigating the biochemistry of nerve cells, their membranes, and the molecular details of how nerve impulses are created—when did you start thinking about the problem of consciousness and not just thinking about it but actually want to investigate it?

SG I have always been occupied by philosophical questions and during my years in Oxford, while working on my DPhil on biological and artificial membranes, I also had the chance to go to some fascinating philosophy seminars. My main motivation in turning to the study of consciousness was a wish to think about broader topics than the behavior of tiny ion channels, important as they may be. Later, as you mentioned, I developed the MA Program in Biological Thought at The Open University of Israel, and while preparing a course on the mind-body problem, I discovered a lacuna. At that time, about 20 years ago, very little work on consciousness seemed to have been done focusing on evolution. This led me to try to convince Eva, an evolutionary biologist, to collaborate. It wasn't easy.

EJ I told her at first I won't touch it with a bargepole, because it will drive me crazy. At the end, she won, and the way she won was first of all she said "Look, nobody really says anything detailed about the evolution of consciousness since the middle of the 20th century!" and I said myself "well, that is true." The other thing that she did, which was very manipulative, was she said "Why don't we start reading Aristotle's *De Anima*, on the soul?" Aristotle is my favorite philosopher and I taught this book, I love this book, I think it is the greatest theoretical biology text that was ever written—biology and psychology in one. I said okay, let's start with Aristotle and that is how it evolved.

7

History of Modern Philosophy

MASSIMO CACCIARI
MICHELE DI FRANCESCO

MODERATED BY ALICE ROBERTS

Conversation 7

MASSIMO CACCIARI,
MICHELE DI FRANCESCO

MASSIMO CACCIARI I will give a very simple introduction and I will just touch upon a few of the authors that marked the philosophical research on the themes and subjects that we discuss today. About this research and the new approaches, I think that my colleague professor Di Francesco will talk more extensively than I. I think that we have like a kind of legend still going on around philosophical thinking. A kind of legend that is nourished by a kind of bad philosophy in the framework of which philosophy—and metaphysics, this very important word that isn't very much understood by some in the significance and meaning of the words metaphysics that Aristotle had—some consider the philosophical thinking as a way to separate or distance itself radically from the biological foundations, the body, aspects of our being, and the noethical dimension. Even a great philosopher like Nietzsche, who talked about the body as a great reason—the body is a great reason—in other pages writes about the fact that philosophers have, in their story, disliked and despised the bodily dimension, creating a contrast between the bodily dimension and the spirit. This is a completely false vision. There is no philosophy that can be called philosophy that is dualistic. One of Plato's beautiful pages—but I could find so many other wonderful pages by Plato—said "We cannot believe that we can take care of a part of our body without understanding them as a whole." He also adds "We cannot care separately for the parts of the body" of which *kefalè*—he used the Greek term *kefalè*—is an essential part. *Kefalè* has been translated as mind, but I don't agree. *Kefalè* is the chief, but it is the same as the brain—the head. And in this case, the head is a word for the brain, so this is what Plato is trying to say. The brain is the fundamental, essential part we have to take care of as a part of the body, as the head of the body, *kefalè*. In our body, of the body, belonging to the body, matter as the body is thinking matter. The idea of a thinking matter is something that we can find originally in the history and the tradition of our thinking, western thinking. We could also show how this fundamental direction is something that we can find in our civilization and not in other civilizations, but it is part of our thinking, western thinking as we have seen in Plato and in how other people misunderstood the metaphysics of Plato. Metaphysics doesn't mean beyond nature, *meta-physis*, it means the understanding of those principles that allow us to understand nature, our nature, and our thinking nature. That thinking matter that we have is called *kefalé*, in Plato's words: the head meaning the brain. This idea of the thinking matter comes back in extraordinary ways in another philosopher and a great researcher scientific researcher, who is Leibniz, a great mathematician and I would like to quote one of his pages. I could find so many interesting quotations by him, but it is extraordinarily effective for us, also from a literary point of view. He said, and I quote, we are at the end of the 17th century: is it not clear that every being has a certain degree of spontaneity or freedom? There are innumerable minds everywhere. There are minds even in the human eggs before conception. Unless one is foolish enough to believe that, in all of nature, only our minds are active. Plotino, some centuries before him, said "Nothing is as foolish as thinking to consider every element of nature *abpsychòs*, without psyche, without soul, without mind. Even a small grain of sand has a certain level of spontaneity or freedom. Everything, as Spinoza said, is always a cause, is always also a cause. Also from an etymological point of view, in Italian "cosa" and "causa" are very similar words. The idea of a thinking matter, yes, a thinking matter, our brain is *kefalé*, thinking matter. So thinking, as Professor Di Francesco will, I think, agree with me, if every being, each being is in one way or another a thinking matter, but this characteristic, for evidence reasons, gets to our being at that knowledgeable level without leaving behind that being a thinking matter, that link between the bodily reason and the consciousness dimension and the intelligence. If it is so, it is evidence that the thinking cannot be understood as a separate activity. *Cogitare*, thinking. To think. What does it mean? It means that several layers of our being *co-agitano—cogitare*, co-agitare—so they think that at the same time they shake, they waggle together when we think. And here we get to Descartes. When I say thinking I do not only mean *intelligere*, the fact of wanting or imagining, but also feeling. So the feeling is, for me, the same thing as thinking, as *cogitare* in Latin. It cannot be a separate thing from thinking and feeling. The two things are not separate. So we can see that in this direction of the great philosophical thinking in Europe—especially talking about the founders of modern philosophy, who were also great researchers like Descartes and Leibniz, and all the others, great scientists—we find in all of our history the same perspective:

the fact that the biological foundation, or neural foundation, and the consciousness cannot be separate. The inseparability of these two elements. Consciousness stems from feeling, from the unconscious, and they get together. They collide. They can enter into a conflict, as many know very well, but these two dimensions are not separate. They cannot be separate, with the exception of some pathological situations. So what is the problem here? The problem is that—this is something that the poet Leopardi questioned himself about—it is true that the brain is the one thinking, that there is an equivalence between mind and brain. So who can think if not the brain? Who could say otherwise? It is impossible to state something else with reason. So the point and the problem is not the fact of understanding the relation and correlation between these elements, but the issue is—and that is my way of thinking, and it is my personal point of view that I cannot explain into too many details since I don't have enough time—but I think that the good way to talk about this correlation is not a mechanical way. In the relation between these folds, these layers, where a fold is a term that we found in Leibniz. Our brain, mind, consciousness have inseparability. Even Hume talked about this, and Kant criticized it. Why go back to a very ancient way of considering these things as separate things? Why do we need to go back to a deterministic and mechanical approach that is so obsolete? For Leibniz, this wasn't possible, but even Descartes and Spinoza questioned that. The point is, once again, that we have to understand this relationship but we cannot explain this relation in a deterministic way or on the basis of a continuum, of a simple continuity, as a linear development and evolution between the unconscious dimension, feeling, and consciousness. The gift of self, an expression that Damasio said and that I really cherish, is something new, is revolutionary. This is an innovation in the evolution of our species. It is absolutely a leap, *natura facit saltus*, we are not sure about that. We have an evolutionistic mechanism based on cause and effect. Maybe this is the cause and we maybe know the effects. Everything cannot be that linear. This is a very archaic epistemology, it is not that linear. But I have to try to understand what the nexus is. I do have to try to explain that, but this explanation will be pertinent and if we take our distance from determinism and deterministic approaches, and we distance ourselves from this superstitious idea of causality, even the great physics abandoned this causality perspective. At the same time, in the texts and the books that I read about scientific philosophy, I think that we are observing the idea of mind and brain as the same thing. Why? What does that mean? If I try to explain in terms that are not deterministic but probabilistic, this probabilistic approach, I am not going against the scientific approach because contemporary science is probabilistic. It is statistical. Nature is common, it doesn't impose itself using a very rigid causality principle. In other terms, my idea is that the perspective, the good perspective of the research in the psychological field is a perspective that—and I am trying to summarize here—familiarizes with the terms of indetermination that regulate the same calculation of contemporary physic mathematics. The indetermination of principles. This is something that I find in contemporary logic and great physics and sometimes I don't find it in some approaches from contemporary scientific psychology. If our work goes in this direction, I think that we can see new paths and very interesting new perspectives in the cooperation, and the common work and collaboration between real philosophy—I am talking about the real one, not the one talked about in stories, but one of the authors of the great classic—and psychology, logics, and physical science. I think that there is a common perspective that revolves around a common epistemology, and this is the end of my contribution. Thank you for your attention.

MICHELE DI FRANCESCO As Massimo Cacciari was saying, I am going to focus a bit more on a contemporary discussion but I also would like to start at least from Descartes. We are speaking about a personality who connects as a liaison for issues related to classical philosophical ideas and also modern philosophical ideas. Professor Cacciari was reminding us of the innovation presented by Descartes when he spoke about the mind. It was not the *cogitation*, not just the intellect, but it was also the sensitivity that also had a role. A philosopher of cognitive science, of contemporary cognitive science, has defined human beings as informiforous, meaning that we nourish ourselves with information but at the same time we have feelings and we perceive subjective feelings. Descartes stated "Who am I? I am a thinking being" and what is this is someone who has doubts, someone who denies, someone who may be once, who does not want, someone who imagines, someone who thinks. That was a revolution and it was also a definition similar to what we mean by conscience on a broader level. This new sense of feeling, of not just having a physical experience of our life but a physical component that intersects itself also with the intellect, with desire, and with many other issues, well this was indeed something fundamental. There is another issue into which I want to delve, and this is how Descartes, whenever people spoke about him, defended an approximate dualism. This surprises me. I am not an historian of philosophy, I am not a scholar, as Professor Cacciari, is of the ideas of the past. But just to read metaphysical meditations you see sentences such as "There is nothing that nature can teach to me more explicitly than the fact that I have a body." This is something that should make us think, because via these sensations of thirst, of hunger, nature is telling me that I am not just like a norseman or a captain of a ship, but reminding us of Aristotle. Also, that the two things are totally connected, they are bound one with the other. It is a single whole—in Latin *commistus*, *mélange* in French. A mix. The problem is to understand what this means. We are not denying, of course, that there weren't differences between the ideas of Descartes and our ideas of today, but it is important for us to recall these ideas. The interesting fact is that in the 20th century, we saw the development of a science of the mind which went in a direction not so much on the side of Descartes but on the side of what Cacciari was mentioning: the so-called thinking matter. We started to ask ourselves how could matter think. It was a lucky case, because Massimo, when you concluded your lecture, you touched on the relationship existing between Alan Turing and the philosophy of the mind. Alan Turing is the father of computer science and he dealt with this not to invent computers—even if the computer was invented thanks to his studies—but because he was studying the philosophy of math. Turing showed that a series of logical processes, of logical demonstrations, could be calculated by algorithms and calculated by a mechanical machine, the Turing machines, as they are called. What then physically will be transformed into a computer. From here comes the idea that we can try to answer the question "What is the mind?" Well, the mind is a device that processes information and the process is a representation of the external world. It is not the dominant vision of today, but it was fundamental at the time to get to our ideas of today. This idea of computations and representations allows unifying the natural world, because there is no longer a reason to deny animals to have the possibility of processing algorithms. This is an important and fundamental element. This was called cognitive science and at the basis of cognitive science there is exactly this. But versus cognitive science, some more intuitive persons, more intuitive than others, or more careful with nuances, have, in fact, realized that it was a problem with the notion of conscience, because what happens in the cognitivist approach is that once start from the bottom with a Darwinian type of approach. Hence, conscience is the point of arrival, not the starting point, so the point of arrival of a series of subpersonal processes. Below personal experience. These emerge at a more complex level and emerge sequentially until you get to the level of conscience. The problem is that it is not clear how, with a traditional approach, this process could be completed. I quote Thomas Nagel, who introduced the idea based on which a state of conscience is a state that says what it is like. These are subjective facts. He says that the fact that an organism has a conscious experience fundamentally means that it perceives a special effect in being that specific organism. This subjective experience was not grasped by anyone in the familiar, so to say, assessment of what is mental, because these were all logically compatible with its absence. It cannot be analyzed in logical functional states, or by intentional states, because these could be attributed to a robot or to an automaton that behaves like a person even if it did not gain any real experience. In this author who in 1974 wrote an essay titled "Che effetto fa essere un pipistrello?"—what is it like to be a bat—he clarified that it was important to ask ourselves what conscience is in species different than ours. This problem was tackled early on, but many years were necessary before this was clarified, or, at least, became clearer. What happened is that between the 80s and 90s, some theories came out, theories of conscience based on brain analysis. Some expressions came out. Jean-Pierre Changeux spoke about a neuronal man. Patricia Churchland of neuro philosophy. They set at the center what we have learned thanks to technological development and what we have learned from the functioning of the brain that was tackled during these many meetings. But we still have the problem of the fact that, from a certain point of view, the presence of conscience seems not to be totally compatible with the mechanicistic type of physical science, as professor Cacciari mentioned. What becomes important is to see how we can tackle such things. I think that my time is up. I simply want to quote the existence of different approaches that are materialistic, traditional approaches. We have dualistic approaches and then, recently, what is interesting is that we have gone back to a position which, up to a few years ago, was unknown, the so-called panpsychism, meaning that this derives not from the Renaissance, not from Plotinus, but it comes from the analysis of contemporary physics. One of the most surprising aspects of contemporary physics is not to answer the question "What does physics talk about?" Contemporary physics, according to many authors, describes the relationship existing between the fundamental objects of physics, but it does not analyze its nature. Some authors generate, of course, a lot of discussion in doing so and this has raised interest. But Philip Roth, for example, defends these positions in an interesting book—a bit complicated to accept in full. He defends his panpsychistic thesis, called *Galileo's Error*. The mechanicistic Galileo, naturally. From this viewpoint, we can say that what we are witnessing today is that there is a return of an interest in conscience by philosophers of the mind, and an interest in phenomenology as well as for this the importance of this bodily disembodiment or incarnation. This embodiment defines our cognitive processes, a fusion between our bodily nature and our cognitive processes. This has been just a comment that I wanted to make. I just changed a bit my lecture because I wanted to touch on some of the points mentioned by Professor Cacciari. I do hope that this has been of some interest to you.

III

IT BEGINS WITH AN IDEA

The exhibition "It Begins with an Idea" served as a multidisciplinary attempt to understand the human brain, its layered functions, and its centrality to human history. The brain generates our models of the world, coding our strategies of response, directing the stories we tell ourselves. The complexity of brain function took over the three floors of Ca' Corner della Regina, through a global constellation of neuroscientists, philosophers, contemporary fiction writers, as well as surgery accounts, instruments of dissection, anatomical theaters, neural models, drawings, engravings, paintings, books, dream records, and memory amulets.

Taken all together, the exhibition's focus is neuroscience: the polyphony, created by the collection of exhibited material and voices, traces the outlines of our bending selves, the gaps in our own knowledge, the path by which the history of neuroscience has been constructed, the seeds of future scientific inquiry, and the known and unknown of the human brain.

CURATED BY

Udo Kittelmann in collaboration with Taryn Simon

WITH

Yasmin Abofoul, Ayọ̀bámi Adébáyọ̀, Huda Akil, Hanan al-Shaykh, Ch'aska Anka Ninawaman, Polina Anikeeva, Chloe Aridjis, Tash Aw, Anirban Bandyopadhyay, György Buzsáki, Stefano Cappa, Mauro Javier Cárdenas, David Chalmers, Patricia Churchland, Antonio Damasio, Stanislas Dehaene, Daniel C. Dennett, Catherine Dulac, David Erritzoe, Lisa Feldman Barrett, Akwaeke Emezi, André Fenton, Esther Freud, Karl Friston, Rivka Galchen, Daniel Galera, Guido Gainotti, Ali Ghazizadeh, Paolo Giordano, George Guidall, Carl Hart, Suzana Herculano-Houzel, Amadi O. Ihunwo, Uzodinma Iweala, Erich Jarvis, Bianca Jones Marlin, Mieko Kawakami, John Keene, Daniel Kehlmann, Sheng Keyi, Katie Kitamura, Alexander Kluge, Hari Kunzru, Kumi Kuroda, Joseph E. LeDoux, Letizia Leocani, Hervé Le Tellier, Mahmoud Bukar Maina, Catherine Malabou, Eve Marder, Michele Mari, Hannah Monyer, Andrea Moro, Maria Concetta Morrone, Charu Nivedita, Tilsa Otta, Helen Oyeyemi, Daniela Perani, David Poeppel, Supratim Ray, Sidarta Ribeiro, Cord Riechelmann, Salman Rushdie, Daniela Schiller, Ekaterina Sedia, Leanne Shapton, Wolf Singer, Ahdaf Soueif, Mitchell Joseph Valdés-Sosa, Marija Stepanova, Angela Vincent, Charles Yang, Rafael Yuste, McKenzie Wark, Li Zhaoping

THIS THING YOU DON'T FEEL

Udo Kittelmann

The seat of the soul, a "honeydew melon,"[1] "conventional convoluted cauliflower beauty"[2] or "a lump of flour":[3] the sight of the porous and fleshy human brain leaves one speechless and grasping for familiar imagery and concepts. This has always been the case. And yet, neuroscientist and Nobel Laureate Rita Levi-Montalcini described in 2008 the prevailing notion of those in her field that, for the longest time, neuroscientists were "third class citizens compared to physicists and molecular biologists."

The 21st century in science has seen a sharp turn inward toward a more intimate scientific mystery than the fascination with physics that dominated the 20th century. This fact can be measured both by data in the United States that reveal the increasing number of neuroscientist PhD students as compared to academic positions[4] and by the funding of multi-billion projects such as the European Union's Human Brain Project, the United States' Brain Initiative, and China's Brain Project. Similar initiatives sent us to space, built the Large Hadron Collider, and mapped the human genome. Now it's the squishy mass of tissue, the covert resident in each of us, that has transfixed attention and resources and which promises to transform our experiences of ourselves in the world.

By contrast, it can be easy to forget that brain studies as a discipline has developed across the world over millennia. The brain as the container of thought, enlightenment, criminality, dreams, deviance, and madness has been interrogated, described, exalted, and brought out for examination. While this history has often been presented in a scientifically positivist and Eurocentric framework, "Human Brains. It Begins with an Idea" is necessarily as interested in established narratives and transitions within those narratives as it is in the gaps, pauses, and footnotes that have been too often overlooked. At its core, this exhibition straddles the divide between complex research and aghast wonder. It explores the knowns and unknowns in the brain while examining the historical misunderstandings and persistent mysteries of neuroscience.

While we are accustomed to seeing the rise and fall of our chests as our lungs fill with air, the beating of our hearts, or the press of a heavy meal on our stomachs, the brain—our own computational center—remains abstract: no pulses, no vibrations. It is simply there. When we initially discussed how to tackle such a difficult and important topic with Taryn Simon, who curates the exhibition in collaboration with me, it was clear that we neither wanted to create a simple scientific exhibition nor put art exclusively into the service of science, as explanation or illustration. Instead, we decided to focus on science as a narrative written by shifting authors, with additions and adaptations throughout time, geography, bias. It was essential to see the exhibition as an international, interdisciplinary experiment— a reflection of a time before technology and specialization thickened and fortified the boundaries and hierarchies between science, art, and literature. Working in collaboration with Simon, Fondazione Prada's curatorial team, and the Human Brains scientific committee enabled us to create a hybrid exhibition.

To this end, we have composed an exhibition in three movements, each functioning as an interlocutor for the other. On the ground floor, traditional medical surgery amphitheater seating surrounds a large screen projecting found footage of experiments, surgeries, and their discoveries. Two lateral rooms provide additional videos about brain functions. On the first floor, the second movement begins, combining historical objects—from Mesopotamia and Ancient Egypt up through the 20th century—with short stories written by thirty-two authors from eighteen different countries. On the second and final floor, the artifacts and stories culminate in a central, tunneled assembly of interviews with leading neuroscientists constructed into The Conversation Machine, a film scattered across thirty-two screens orchestrated by Taryn Simon.

The initial entrance videos, sourced mostly from YouTube, present brain studies as accessible and even as a self-referential kind of play. These shorts—real, imagined, and accidental—glimmer with and also refract a diversity of voices that introduce the audience to a sampling of anatomies, mechanisms, and illustrations of the brain as we have come to know it. They propose fantasy as well as explication. This section holds for visitors a visceral jolt of recognition and investigative anticipation upon seeing their brains from the inside and out.

From here, the exhibition continues to the first and second floors, with the visitor confronted by historical objects ranging from Cylinders of the Sumerian ruler Gudea (Mesopotamia, *c.* 2120–2110 BCE), which comprise the first known record of a dream, to two 1992 issues of *Proceedings of the National Academy of Sciences* featuring the first comprehensive papers reporting fMRI. Well known objects are presented in new light, such as the link between René Descartes' writing on the pineal gland as the seat of the soul and Rembrandt's *The Anatomy Lesson of Dr. Jan Deijman* (1656) in which the exposure of the gland during a public dissection is meant to serve as an additional post-mortem punishment for the executed thief. Beside these familiar objects, lesser known works fill in gaps in the history of brain studies, such as a heart-shaped amulet from the Egyptian Late Period (712–332 BCE). Although Ancient Egyptians accurately described many brain injuries and produced the earliest known references to the brain, meninges, and cerebrospinal fluid, they often used the word for heart to indicate the mind and believed that memories of past deeds were also housed there. In a polyphonic, global, and transhistorical conversation, the objects are given imagined voices and lives by works of speculative fiction written by authors from around the world.

> *"And as my brain-in-a-jar is about to collide with my brain-in-a-meteor, my brain will speak at last, and in the spirit of (iii) [fun's needed between exist and perish] it will ask how much for a pint of fresh formaldehyde, whereabouts my jarring brain."*[5]

The stories vary in form and perspective, challenging the objects for which they were written. Many use a first-person narrator to bring the viewer closer to artifacts that are particularly distant in geography or era. Rather than simply recounting known histories, these stories unleash the ethical and philosophical questions behind scientific breakthroughs, unearthing figures previously neglected, many of whom were women and people of color.

Fiction becomes a glinting tool for historical investigation precisely because the names and lives of countless patients and early physicians have been forgotten, and so must be invented. It is through fiction that the repression faced by medieval doctor Trotula is mirrored in the aggression experienced by a contemporary nameless woman in Katie Kitamura's *Liber Trotula*.[6] In *The Cutting*, John Keene[7] intertwines the voices of an anonymized black patient and that of an imagined white assistant to narrate the moment before a 19th century trepanation in an act of critical fabulation.[8]

> *"Stop right there, Korbinian. The brain does not have maps but ventricles, through which perception and imagination, via cogitation and estimation, finally reach the memory."*
> *"Oh, I do commend your pioneering spirit, my dear Persian. But pioneers are always doomed to folly. The ventricles are spaces filled with fluid, nothing more..."*
> *"Heretic! How dare you speak like that to the great master Avicenna? Two centuries after him, I can confirm that the mind is indeed organized in ventricles."*
> *"Please don't go getting overexcited, Master Albertus. Your ideas are old. You've got to accept that the old is always defeated by the new."*[9]

These new stories are unified through the voice of one narrator: George Guidall, dubbed the "undisputed king of audiobooks"[10] with over 1,500 books recorded and numerous awards. Guidall's very presence suggests the rankling habits of skepticism inculcated by single-voice narration. At the same time, Guidall's own biography–the changing of his last name from Shapiro, his previous careers as an actor and mental health provider, his devotion to the unique personality of each literary work he voices–shifts and deepens this skepticism toward new questions. By selecting him, we hope to make very explicit the projection of one voice, body, brain, and experience upon and into multiple texts, languages, geographies, bodies, stories, realities, and experiences–an intractable foundational problem for both how the brain works and how the history of neuroscience has been constructed.

On the second and culminating floor, visitors enter Taryn Simon's The Conversation Machine, a multiscreen environment populated by interviews with living neuroscientists whose work seeks to advance our understanding of human thought, consciousness, behavior, and brain function. Clips from over 140 hours of interviews with her are intricately woven to generate unexpected adjacencies and dissonances among individuals working in a field whose history is marked not only by rigor and breakthrough, but also by serendipity, error, transgression, and uncertainty.

The machine's mechanical principles mimic those of the brain. In a choreographed conversation, participants appear to listen and respond to each other although, like the brain locked in its skull, without any direct access to each other or the viewer. The topic of conversation moves and morphs between the brain's inner workings and the structures of the external world that co-create—and interfere with—those processes. In The Conversation Machine, political ideology, gender and racial identity, trauma of all kinds, morality, religion, and money all have biological counterparts.

Gestures, language, pauses, unconscious stutters, and fidgets—these are some of the ways the brain has of expressing itself to the world. The filmed interview format used by Simon attends to these subtleties of communication while the conversation's construction dramatizes them in a choreography of blinking stillness and fluid movement. In a similar spirit to the artifacts and speculative fiction stories on the floor below, the humans behind the biggest discoveries about the brain of our time are as important as the research itself. This approach produced arresting statements that delight as well as inform:

> "One thing I realized is that the memories you want to protect are the good ones. In order to protect them, you don't have to retrieve them as much. Actually, if you don't retrieve them at all, they're perfectly protected."[11]

> "People say, 'Well look, the neurons that your brain is made of, they don't have free will.' That's true. The motor proteins that troop around inside your neurons and your other cells, they don't have free will. That's true. They're little robots with very few degrees of freedom. But that doesn't mean that you, made up of those neurons, aren't free, because you can be freer of your parts in the same way that a flower can be red, even though none of its finest parts are red."[12]

Personal objects from each participant join the conversation and perform a real-time archeology of their owners' words and ideas, by turns elucidating, illustrating, or playfully distracting. No explanation accompanies their appearances; many of them are banal. Filmed on their own and in the hands of the participants, these objects are manifold: including a Stratocaster, brain models and microchips, and even a set of Russian dolls. In this eclectic presentation of personal archives, there is an echo of the historical objects and speculative fiction stories that swell and surround the interviews, agitating the distinction between material and figurative perception.

While often relegated to brief introductions or asides in neuroscience textbooks, or else presented as a linear tale of hard work and inevitable genius, history can assert another purpose that better serves both scientists and a wider public. In his article "Why Study the History of Neuroscience?", Dr. Richard Brown argues:

> "If history is a collective memory of the past, the loss of the historical objects and documents of neuroscience is equivalent to 'scientific Alzheimer's disease.' Treating this disciplinary ailment is more than a matter of preserving and cataloging laboratory notes, specimens, equipment, drawings and documents. This material needs to be incorporated into the teaching and practice of neuroscience today."[13]

For neuroscientists, "by methodologically investigating historical data, models, hypotheses and experiments, alternatives to contemporary theories can be contemplated" and anecdotes can provide insight into the accidents that bring about breakthroughs.[14] For the rest of us, there is a lot we can learn from the idealism and doggedness of neuroscientists. The study of the brain in particular—silent and embedded as it is—instructs a process of questioning the biases and expectations that guide us expediently through the external world and that, when encountered in new ways, can also challenge and reform.

The selection of the interviewees was a collaborative task performed with great care and research. From their own perspectives and experiences, several participants deal with both current and historical issues related to inclusivity. Collectively, they represent the changing face and cutting edge of neuroscience. We are very grateful to the thirty-six scientists from over twenty nations who made time to participate and finally, within the exhibition, they speak to all of us.

As in The Conversation Machine, many of the fiction pieces read by the narrator concern themselves with the very writing of science and its quixotic claim on truth. For example, Sidarta Ribeiro's *Ventricles, Hearts, and Minds* playfully summons philosophers, theorists, and neuroscientists from across the centuries in the Afro-Brazilian candomblé practice of *gira*, in which mediums gather to channel the voices and advice of ancestor spirits. Lively and humorous, the text pulls back the curtain on science's supposed monopoly of fact to reveal the twists and turns, knots, and circles over time, in our understanding of the brain. In *The Khipu Master* Ch'aska Anka Ninawaman also threads together past, present, and future, in a story about moving on and the encoded intergenerational knowledge of a *Khipu kamayuq warmi*, woman master of khipus.

In the 21st century, at a time where an ongoing interest in the archive and in broadening our definitions of both "history" and "science" stems, in part, from a premium on the production and extraction of data, it is hard to untangle ourselves from the deep roots of these methodologies. The tools of art and fiction, which allow us to distort, experiment, or play with reality, are capable of reminding us, not only of the satisfaction, but also of the elusiveness of truth. In these gaps and openings, new historiographies and tomorrows are possible, though not promised. With this exhibition, it was important to recognize the strivings and stories that precede us in hopes that each visitor and reader will find something to prod, tickle, press their gray matter, and leave them more curious about their brains and those of others.

1 Ralphs Universum, Quarks, *Was das Gehirn so besonders macht*. https://www.youtube.com/watch?v=RtjfMfVBJ8k.

2 Rivka Galchen, *Memoir of Clemente Susini, Upon the Hiring of a New Assistant*, see pp. 379-382 in this publication.

3 Alexander Kluge, *Eyes from Another Country. How Secret Intelligence Services Were Astonished by a Newspaper Article*, see pp. 477-486 in this publication.

4 www.scientificamerican.com/article/are-there-too-many-neuroscientists/.

5 Mauro Javier Cárdenas, *Louis Victor (Doesn't) Speak*. Written in relation to the brain of Patient Tan (alias of Louis Victor) on whom Pierre Paul Broca experimented in the course of discovering the brain disease aphasia. See pp. 399-404 in this publication.

6 Katie Kitamura, *Liber Trotula*, see pp. 315-320 in this publication.

7 John Keene, *The Great Operation*, see pp. 365-370 in this publication.

8 Critical fabulation is a term coined by historian and theorist Saiydiya Hartman in her article "Venus in Two Acts," published by Duke University Press in *Small Axe*, no. 26, vol. 12, no. 2 (2008).

9 Sidarta Ribeiro, *Ventricles, Heart, and Minds*, see pp. 327-336 in this publication.

10 www.nytimes.com/2017/08/17/books/george-guidall-audiobooks.html.

11 Daniela Schiller, Icahn School of Medicine at Mount Sinai, New York, USA.

12 Daniel C. Dennett, Tufts University, Medford and Somerville, MA, USA.

13 Richard E. Brown, "Why Study the History of Neurosciences?", in *Frontiers in Behavioral Neuroscience* (May 22, 2019). https://doi.org/10.3389/fnbeh.2019.00082.

14 Ibid.

i

LESSONS

Stadium seating, traditional in medical amphitheaters where live surgeries were given alongside lectures, invited the viewer into the position of an observing student or spectator, with the object of studying the history of neuroscience itself.

Online found footage of experiments, surgeries, and discoveries introduced the audience to a sampling of anatomies, mechanisms, and illustrations of the brain as we have come to know it. These videos of neuroimaging, surgical operations, and laboratory discoveries included a basement experiment uncovering how we process visual material, an operation seeking the pathways of memory, and an investigation into the neural tributaries of learning.

In the adjacent rooms lessons given by scientists and researchers focused on the human brain's ability to see, speak, move, remember, and feel emotions. Here syntax and neural components of language and consciousness stood alongside illusion of vision, sensory-motor performance, and memory.

WITH

Stefano F. Cappa, Guido Gainotti, David H. Hubel, Letizia Leocani,
Andrea Moro, Maria Concetta Morrone, Wilder Penfield, Daniela Perani,
Suzanne S. Stensaas, Tornsten N. Wiesel,
USC Mark and Mary Stevens Neuroimaging and Informatics Institute

Suzanne S. Stensaas, Introduction to the Human Brain
Department of Neurobiology and Anatomy and Spencer S. Eccles Health Sciences Library, University of Utah, Salt Lake City

David H. Hubel and Tornsten N. Wiesel's experiment on visual perception using a microelectrode implanted into the visual cortex of an anesthetized cat
Johns Hopkins University, Baltimore 1959

The neurophysiology of memory explained by Wilder Penfield, neurosurgeon of McGill University, Montreal; footage from the film *Gateways to the Mind*, 1958
Produced by the Warner Brothers Pictures, released through the local offices of Bell Telephone System

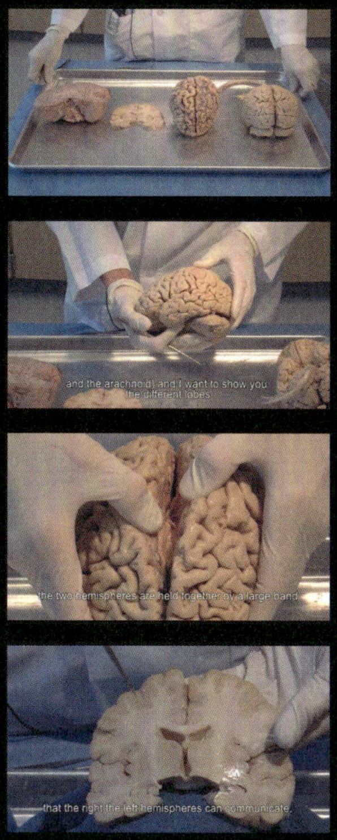

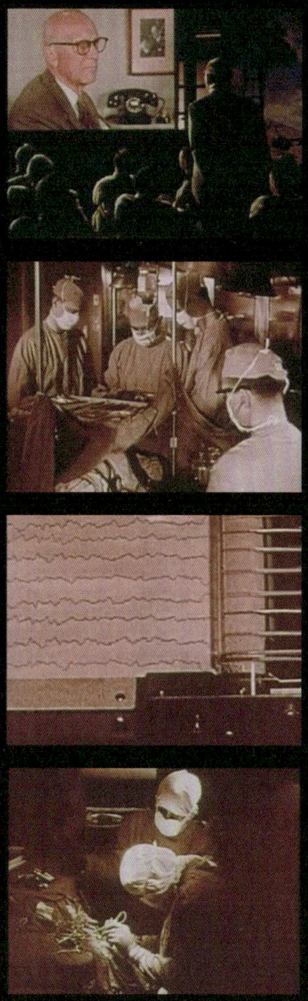

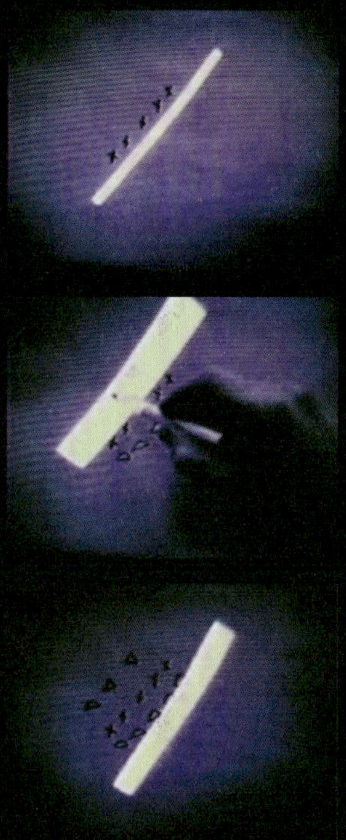

Impossible Languages. A Journey to the Boundaries of Babel

Andrea Moro

Syntax has been recognized as being the specific property of all and only human languages. It is the capacity to generate infinite sentences by assembling a finite set of words. Why all conceivable rules are not realized in human languages? The 20th-century linguistics has shown us that only a subset of the possible rules is realized in the languages of the world. In natural languages, words are grouped in boxes, which can be potentially infinite. If I take "Mary," just a proper name, that "Mary" can be embedded in a box and you could say "The uncles of Mary." The label of the big box is "the uncles" and "Mary" is inside. That can be done infinitely. You could say "The friends of the uncles of Mary." Thus, syntactic rules ignore linear order or flat rules. They only see boxes.

Why don't human languages exploit the inner order of words? Are they excluded by an arbitrary cultural convention or by brain structure? If they are excluded by brain structure, then there must exist impossible languages, namely languages which are not recognized by our brain's canonical networks dedicated to language. The results of the experiments are robust. The selected activation of different circuits for the two different types of languages, possible versus impossible, shows that the existence of impossible language cannot be the result of an arbitrary cultural convention. In other words, it is as if languages were the natural expression of the brain with no specific instruction. The set of principles limiting human languages constitutes the "stem mind" of infants. Also, language acquisition is not a constructive process from zero, but a selection among all possible forms, and eventually there are no languages better than others. This last one has a tremendous impact on the so-called linguistic racism.

We Are Memories

Stefano F. Cappa

The fact that experiences leave a trace in our brain and mind is a central aspect of our mental life. What makes this possible is a complex neural machinery involving a core structure called the hippocampus and an extensive network of connections of this system with multiple cortical regions responsible for the input and output of the information. Memory has always been very much priced by humans. They have been trying to improve memory abilities by techniques that often involve the location of items in a complex spatial array. What is really interesting is that only a few people are able to use these techniques to really improve their memory performance. Some of these people become professional mnemonists. Looking into the brain and anomalies, the core structure remains the same but what is changing when the mnemonists are performing their tasks is the intensity of the connections between the hippocampus and many cortical areas.

There is also the other side of the coin, that is, memory getting worse, which happens physiologically in aging and is associated with changes taking place at the level of the hippocampal structure. An extreme disorder of memory, called amnesia, is associated with the severe damage to these structures.

The phenomenon of memory which is not conscious was observed clinically by several clinicians, who were able to show that some traces remain fixed in our nervous system even in the absence of conscious recollection. The hippocampus is not crucial for his form of memory, as a change in the pattern of activity in the cortical areas engaged in the processing of the incoming information is sufficient to retain a trace of the corresponding experience. Another interesting phenomenon is false memories. We all have false memories because memory is not a photograph of past events but a reconstruction, which can go astray for a number of different reasons.

Body and Mind in the Development of Emotions

Guido Gainotti

Emotions are considered as a response scheme that allows the organism to automatically cope with a certain number of critical and emergency situations. They differ from the cognitive system due to a greater engagement of the body and to the immediate, automatic responses given to external situations.

The development of emotions goes through three phases. First, the brain automatically evaluates the emotional stimuli, thanks to a structure called amygdala, and sends the result of this evaluation to a series of structures. The most important of these structures is the hypothalamus, critical for the regulation of the vegetative system (and in particular of the sympathetic nervous system, that is involved in emergency situations) and of the hormonal system. These structures activate the body parts (heart, lungs, and muscles) that are more involved in the emergency response. These body parts, in turn, send back to a specific part of the brain, the insula, an information concerning the body contribution to the emotional response.

These informations about the inner state of the organism are organized by the posterior insula in interoceptive representations, that are integrated at a more conscious level in the anterior insula, after having been processed by cognitive structures such as the amygdala, the cingulate cortex, or the orbitofrontal cortex. This last structure is fundamental for the control of the emotional response, keeping under control all the structures involved in the emotional response.

Vision
Maria Concetta Morrone

What we see and the world that we perceive can be very different from reality, as the brain is attempting to actively construct an optimal representation of the world under natural ecological condition. Our retina captures the external distribution of light and transmits it to the occipital poles of the two hemispheres which commence visual processing, decoding the major information. Here neurons, the building block of the nervous system, respond to simple visual elements. Line and borders of specific orientation excite specific neurons and the visual scene is represented in a sparse and distributed multitude of neuronal assemblies, which inspired artificial neural networks now implemented in many modern visual devices. To construct the image of the world, our brain uses at least 25 different cortical areas in each hemisphere, each specialized for the analysis of different aspects of the retina input.

Prior information embedded in these multiple cortical representations helps to generate the "vision" of the world. Vision is never constant, but adapts rapidly to the changing world, over periods from milliseconds to years, over the lifespan. Adaptation allows our brain to discriminate incredibly small variations of the visual input. It takes place at all stages of analysis and in all associative and primary cortical areas. What you see now is influenced by what you just saw few seconds or minutes ago. The brain can perceive on the fly, but without prior information of the constancy of object shape, movement, and position that we learn with experience, it would take seconds instead of milliseconds, and it would be very imprecise. Vision is generative: our brain perceives incoming sensory signals molding them on prior knowledge, correcting perception when these signals do not match.

Neuroimaging for the Exploration of the Language
Daniela Perani

The language system appears to be specialized from specific components of language processing rather than from activities such as speaking, repeating, reading, or listening. A large part of the research in neuroimaging on languages has indeed been devoted to the identification of the modular language processors that are phonology (the sound of language), lexicon, semantics, and sentence morphology and syntax. Phonology is the sound of languages. Phonemes, syllables, words, the fundamental part of all languages, are processed in the frontal temporal and parietal regions of the left brain.

When the brain listens to words, larger areas are activated in the left hemisphere, particularly in the frontal and extensively in the temporal lobe, because the latter includes the representation of the meaning of words. When we listen or produce sentences the brain is activated in shared but specific structures devoted to processing syntax–the way words are built up to produce a meaningful language. The right hemisphere is also involved in language, mainly for prosody, which is the processing of the stress of the words, or the pauses between syllables, words, sentences.

The neural substrate of language is already active at birth. Little children and newborns cannot understand the meaning of words, but their brain reacts to the sound of language. Music-related neural structures are also shared with language. We also know for certain that whatever language we use, including sign languages, the temporal, the parietal, and the frontal regions, in particular in the left hemisphere, are activated.

Notably when we use more than one language, as bilinguals or polyglots do, we can create compensation, plasticity in our brain which can cope with neurodegeneration. Bilingualism-induced plasticity can create brain reserve and so delay the onset of dementia.

The Sensorimotor System
Letizia Leocani

The sensory and motor systems tend to be considered together because they are very much interconnected. Touch is fundamental to manipulate objects and to control movement. Also, the two regions devoted to motor commands and body sensations in the brain are close to one another and those that require more training are also more developed. Moreover, we activate complex motor pathways and circuits not only when we perform or even imagine movements, but also when we observe actions by other people. This has been shown using measures of neuronal metabolism such as functional magnetic resonance imaging, and also by measuring EEG signals from electrical activity of neurons. There is also another method to test the motor cortical excitability and brain plasticity, which is transcranial magnetic stimulation.

The stimulation of the brain in a non-painful way with magnetic stimuli can activate our motor cortex by sending motor commands to our muscles and we can record the movement and the electrical activation of those muscles and also map their representation on the brain. This technique allowed not only to test our brain excitability in different conditions, but also to measure brain cortical plasticity of our motor cortex. The brain cortical plasticity, although much more efficient in the early years of our life, never stops and this is very promising. We can learn at any age and reshape our brain. People over 65 were demonstrated to be capable of learning to juggle and this was also related to reshaping of the brain circuits involved in this activity. However, these changes are not lasting forever. We have to keep training, otherwise they will be reversible and disappear again.

Modern MRI: Details and Diffusion, 2017
Video courtesy of the USC Mark and Mary Stevens Neuroimaging and Informatics Institute (www.ini.usc.edu)

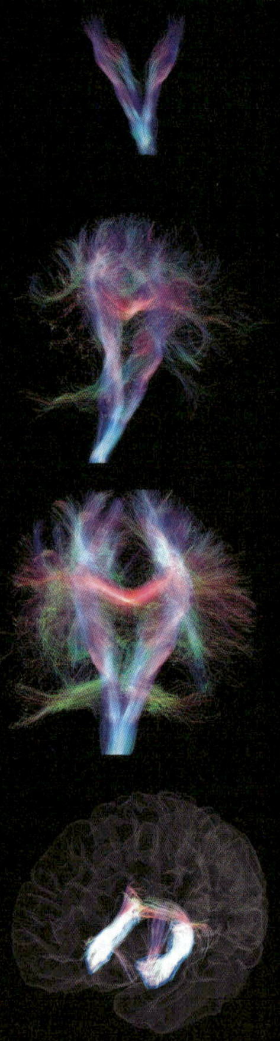

Neurons and Synapses of the Medial Prefrontal Cortex, 2017
Video courtesy of the USC Mark and Mary Stevens Neuroimaging and Informatics Institute (www.ini.usc.edu)

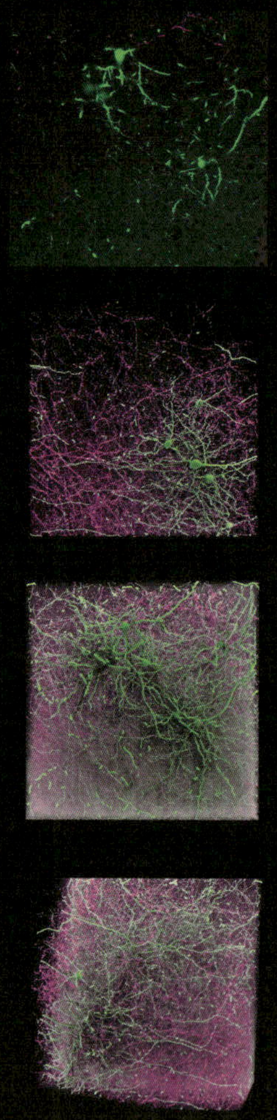

ii

STORIES

A labyrinth of vitrines and monitors unfolded across the two floors of Ca' Corner della Regina. Artifacts, documents, paintings, and other historical objects or their copies and facsimiles, encoded centuries of attempts to understand the human brain. These were challenged by the words of fiction writers who call them forth, expanding the boundaries of our investigation scope, as well as revealing latent social, political, and personal histories.

The vitrines' objects and extended captions reported an ongoing desire to understand the origin and location of thought and the mechanisms of movement and perception, but also the proliferation of brilliant theories, dark practices, and the application of theoretical models for the treatment of physical and mental illness. These artifacts and collected data formed an imaginative contract with their assigned writer: a temporary moment of coherence before fictional reinvention.

Evolution has embedded in the brain a surprising ability to produce and absorb inaccurate memories. This counterintuitive brain function is an adaptive one that asserts the primacy of the present moment to meaning-making.

A 3D copy of two Sumerian cylinders with cuneiform texts reporting the earliest record of a dream (Iraq, 2120–2110 BCE); a papyrus facsimile listing forty-eight medical case histories, named for the American Egyptologist and antiquities dealer Edwin Smith who purchased it in Luxor (17th century BCE); Greek, Latin, Byzantine, Islamic, and Renaissance manuscripts and books—sometimes copies of the original—that have survived and proliferated through their adaptation and replication over time; a semi-circular Andean ceremonial knife used for medical operations and ritual sacrifices (Northern Peru, 13th–15th century); a scale model made by the CNR Laboratories of the anatomical theater at the University of Padua (1595, model 1933); *Cerebri Anatome*, a book on brain anatomy by Thomas Willis, illustrated by architect Christopher Wren (1664); a 19th-century reprint of the *Kaishihen*, the first recorded dissection of a brain in Japan by physician and surgeon Shinnin Kawaguchi, illustrated by painter and draughtsman Shunmei Aoki (1772); two bronze carvings of the Shiva Nataraja dancing on Apasmara, the demon of ignorance (India, 18th-19th century); Francesco Calenzuoli's open head wax anatomical model (19th century); original drawings of neurons by Nobel prize winners Camillo Golgi and Santiago Ramón y Cajal (1883–1921); the work of Rita Levi-Montalcini which resulted in the discovery of the Nerve Growth Factor (1951 and 1964); an article published in *New Scientist*, documenting the first MRI of a human brain (1978); and more were called forth by the writers in the self-same way memories are retrieved, winnowed, and remodeled within the speculative process evolution dictates and fiction recreates.

The artifact's stories in "It Begins with an Idea" were written by international writers to be performed in thirty-two short videos, directed by Taryn Simon and produced by Fondazione Prada for the exhibition, in which renowned audiobook narrator George Guidall read aloud all the newly commissioned literary texts. One voice was projected into multiple stories, languages, geographies, bodies, and realities: an intractable framework problem foundational both to how the brain works and to how the history of neuroscience has been constructed.

While the exhibition's labyrinth suggested a potential single path, the invitation to hear a story-told reconfigured the visitor's navigation at every turn. The human brain's tendency to project itself onto its environment was equaled by its necessity to absorb grounding beliefs and ideas. Brought together here, fiction and history tipped each other back and forth in a vital play of certainty and interpretation regarding how stories, and histories, are told and by whom.

STORIES BY

Salman Rushdie, Tilsa Otta, Ekaterina Sedia, Charu Nivedita, Ahdaf Soueif, Leanne Shapton, Daniel Galera, Cord Riechelmann, Sheng Keyi, Uzodinma Iweala, Hanan al-Shaykh, Katie Kitamura, Hervé Le Tellier, Sidarta Ribeiro, Tash Aw, Maria Stepanova, Ch'aska Anka Ninawaman, Chloe Aridjis, John Keene, Hari Kunzru, Rivka Galchen, Mieko Kawakami, Mauro Javier Cárdenas, Michele Mari, Paolo Giordano, Akwaeke Emezi, Helen Oyeyemi, McKenzie Wark, Daniel Kehlmann, Esther Freud, Ayọ̀bámi Adébáyọ̀, Alexander Kluge

VIDEOS DIRECTED BY

Taryn Simon

STORIES NARRATED BY

George Guidall

PRODUCED BY

Fondazione Prada for the "Human Brains. It Begins with an Idea" project

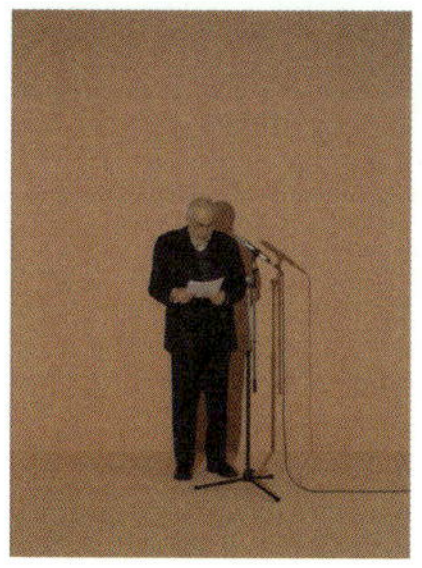 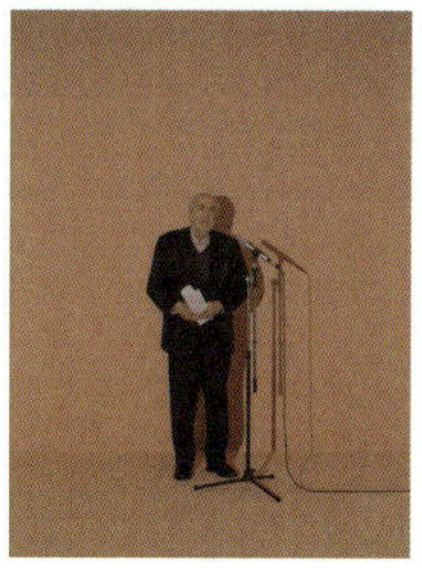 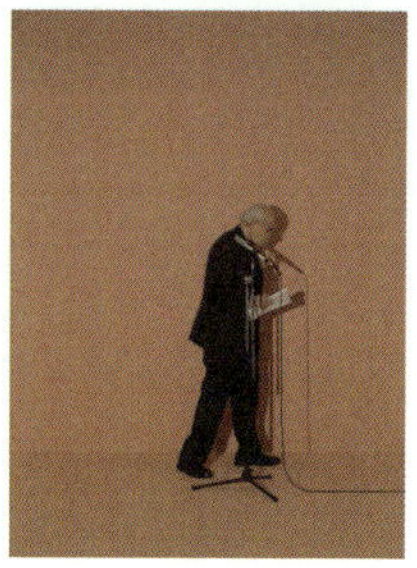

George Guidall reads the artifact's stories of "It Begins with an Idea."

STORY 1

ù

or, The Interpretation of Dreams

Salman Rushdie

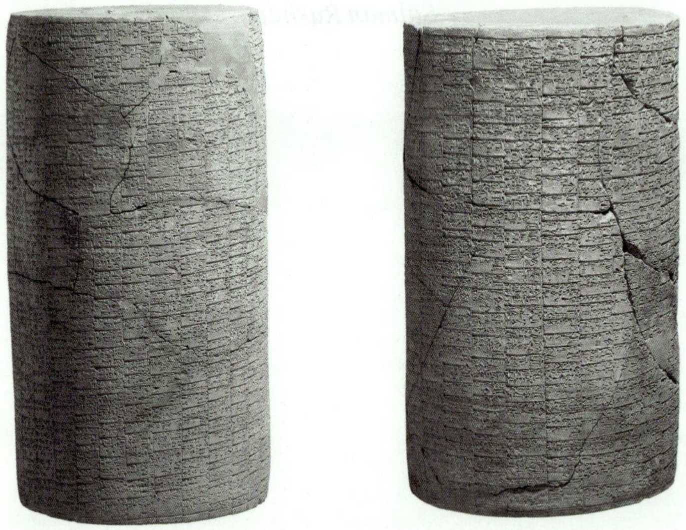

Cylinders of the Sumerian ruler Gudea with cuneiform texts,
Iraq, *c.* 2120–2110 BCE.

In those early days of the world we didn't have words for many things we didn't understand. For example, we didn't have a word for the thing that happened in our heads during the nightly unconsciousness. In fact, the very idea of the nightly unconsciousness was frightening, and we had only recently invented a word for it that wasn't frightening. Before that we called it the nightly death and thought the gods inflicted it on us to remind us regularly of our puny mortality. Then they allowed us to be reborn in the morning and we were expected to be grateful.

Nobody can remember the name of the genius who thought of the word that allowed us to believe that this periodic unconsciousness was normal. That was another new word, "normal." The idea behind that word, the idea that there were *norms*, yardsticks against which things could be measured, was also new.

The word we made up for the nightly unconsciousness was this: *ù*. Once we started calling it the *ù*, it began to feel comforting, even something to look forward to.

But we noticed that during the *ù* we weren't absolutely absent; there were things happening, sometimes to "us," sometimes us as we were and at other times us at different ages or in different circumstances or with totally different characters—"us" but unlike ourselves… or, alternatively, things might be happening to complete strangers about whom we knew nothing. These things that happened were usually not sequential and frequently made no sense, except that sometimes they were sequential and made a kind of sense, although we only remembered a few of them, and not in full. Only fragments remained after we awoke.

We wondered if we were all living two lives, a day life and a night life, and we thought that each life might be the *ù* of the other one. Maybe our *ù*-selves remembered only fragments of our waking lives and tried to make sense of them, too.

In the end we needed two words to describe the things that happened during the *ù*, one word for the meaningless things and another for the seemingly meaningful.

We called the meaningless things *mamu*. "Dreams." We called the meaningful things *mashngi*, "visions," and we also used that word to mean "omens." Every day at breakfast we revealed our latest *mamu* and argued about whether or not it was really a *mashngi*. Nobody knew how to win these arguments at first, and so we had to agree temporarily that the words "win" and "lose" did not apply to a discussion of *mamu* and *mashngi*.

Then we invented the idea of *translators*, who would interpret the *mamu* and *mashngi* so that we would know what they meant. We do not now remember why certain persons were thought to have the power of interpretation, or on what basis they made their judgments, but we all accepted what they said, however crazy it sounded.

If the interpreter said, the *mashngi* says you're in love with your sister, we said, uh-huh, okay.

If the interpreter said, the *mashngi* says you hate your father, we said, yeah, probably.

If the interpreter said, the *mashngi* wants you to break up with your partner, we did that, and if we were told that the *mashngi* wanted us to kill our neighbor, we did that too.

If the interpreter said, the *mashngi* wants you to build the *E-ninnu* temple to the glory of the god Ningirsu, and the presence in your *mashngi* of those noisy *tigidlu* birds twittering in the *ildag* trees means that they will not let you sleep, that there will be no *ù* for you until you have built it… well then, we said, we'll get right on it, just watch us get that temple built.

After a time there was a new development. This took the form of a rebellion against the translators.

It turned out that many people did not want to break up with their partners or kill their neighbors or build temples, and these rebels came up with the idea that there was no such thing as a *mashngi*, that the gods would not communicate with us in a manner that was so difficult to understand, and so it followed that there were only *mamu*, meaningless things, that happened during the nightly unconsciousness, the *ù*, and that we didn't need to pay any attention to these *mamu* or their interpreters.

From there it was a short step to the next idea, which was that there were no gods anyway, that even Ningirsu didn't exist but was just a figment of the dream-world, the *mamu*-world, which wasn't real either; and that meant we were on our own, without mighty invisible beings to guide us, and therefore we were obliged to make our choices without omens or visions or meaning instructors. The universe was unexplainable, it was simply there, and didn't have to be explained—it didn't have to mean anything. And all human life, waking or sleeping, was a kind of absurd *mamu*.

When that appalling idea was born, it split us into two camps, and a great civil war began, which has never ceased to this day.

We need somebody to come up with a word that persuades us that the idea of gods and the idea of no gods are both norms. That if some people think of life as *mamu* and others want it to be filled with *mashngi*, well, that's pretty normal.

If we had such a word we could agree to differ and our endless civil war would end. We could shake hands and accept the idea that the world of *E-ninnu* temples built to glorify the god Ningirsu and the world of no temples at all could exist simultaneously, side by side, without a problem.

We are sorry to report that as yet nobody has come up with that useful word.

STORY 2

Las iluminaciones circulares

Tilsa Otta

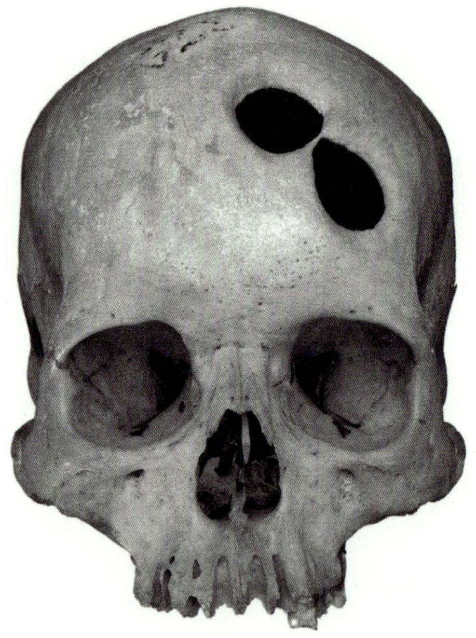

Skull of an adult Inca male with five healed trepanations,
site of Patallacta, Cuzco, Peru, *c.* 1476–1532.

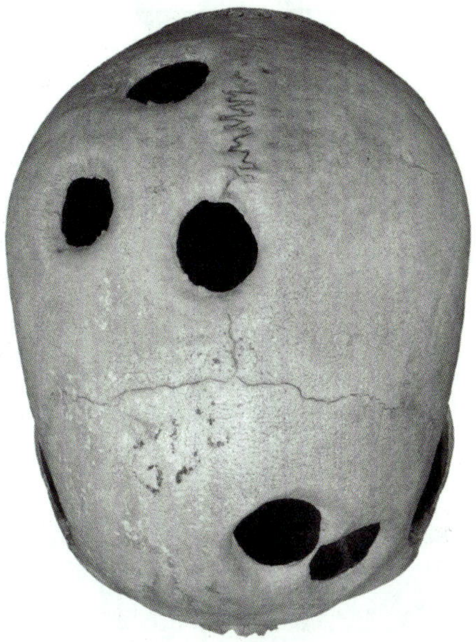

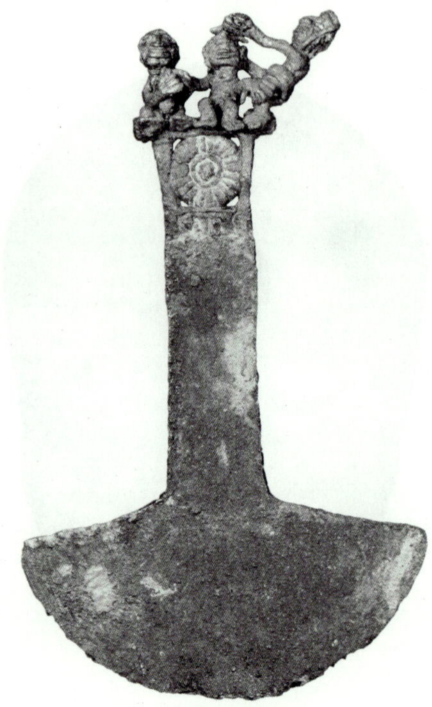

Incan ceremonial knife (*tumi*) with a depiction of a skull trepanation,
13th–15th century.

Desde las montañas de los andes, el Imperio Incaico se expandió empleando la fuerza y destrezas de la comunidad entera, de ahí su grandeza, un reflejo de la plena conciencia de que si el Sol era el dios estelar era porque su luz caía sobre todos sin distinción. Particularmente al mediodía, la ardiente deidad conocida como Ti acariciaba la vida en la Tierra, favorecía los cultivos, recargaba el brillo de nuestros ojos, mecía a las fieras hasta el sueño.

Si bien su linaje era conocido más allá de la Tierra (hermano y esposo de Mama Quilla, diosa lunar; padre de los Incas, humanos emperadores), de vez en cuando sorprendía revelando nuevos parentescos, adoptando a sencillos pobladores, revolucionando sus existencias y expandiendo sus funciones de un modo sobrenatural.

Todo empezaba cuando un indígena soñaba una edificación nueva y al día siguiente despertaba con un dolor de cabeza atroz que lo cegaba y forzaba a acudir al chamán con carácter de urgencia. Tras verificar la autenticidad del templo por venir, el chamán, curandero del cuerpo y espíritu, diestro en medicina natural y enteógenos, daba informe al Inca, quien autorizaba la intervención. El procedimiento era el siguiente: se anestesiaba al elegido mediante la ingesta de San Pedro (bebida preparada del cactus Wachuma), el cirujano cortaba el cuero cabelludo del paciente empleando un Tumi, cuchillo ritual semicircular hecho de oro, plata o cobre, y lo doblaba hacia afuera de modo que el cráneo quedaba a la vista. Siguiendo los deseos del Sol, se perforaban tres o cuatro agujeros en la corteza ósea con una pequeña herramienta de obsidiana en forma de cincel, permitiendo la extracción de discos de hueso mediante ranurado circular y raspado. Era una operación de alto riesgo, sí, un desliz y el encéfalo hubiera sido desgraciado, arruinando alguna de las maravillas del mundo. Pero rara vez ocurría. Lo que ocurría era que se extraían las capas óseas seleccionadas haciendo palanca con los instrumentos rituales y el cerebro quedaba al descubierto, mirando directamente el mundo por primera vez, absorto, recibiendo la pura luz a través de cuatro ventanas nuevas. Alegría desbordante, emoción radiante que precedía la trepanación.

El cerebro quedaba expuesto por un lapso aproximado de seis minutos. Al Inca le gustaba presenciar esta parte, cuando los rayos del dios Ti fluían como brillante brisa hasta el interior de su vasallo. En realidad, a todos les encantaba ese momento. ¿Cómo no adorar ese instante, donde el alma se ilumina por medio de la profanación del cuerpo? No faltaban los curiosos que se asomaban por entre las ramas de los árboles tomando fotos mentales. Una vez que el Sol había concluido la transmisión de su secreto, la herida era sellada con una placa de oro, el cuero cabelludo reacomodado y la cirugía llegaba a su fin, recibiendo la aprobación del Inca con un leve gesto o encendidas palmas, según fuera el temperamento del dignatario.

Después de ser penetrado por la última gran idea de Dios, bajo la forma de poderosos rayos, el depositario temporal no experimentaba paz alguna. Por el contrario, la migraña se mostraba feroz, sumada a la sensación desesperante de que algo pugnaba por salir de él o entrar aún más. Y es que la iluminación no es un campo sembrado de rosas, y menos si ocurre mediante un procedimiento invasivo. El dolor desencadenaba aullidos que emocionaban a la comunidad pues la erección de un coloso de diseño único y divino se anunciaba inminente.

Entonces iniciaba la construcción, liderada por el mensajero trepanado. A veces era un centro ceremonial, un reloj solar, un ushnu o un templo como el mismísimo Machu Picchu. Para cumplir esta tarea monumental, cientos de hombres trabajaban sin descanso. La mayoría de ellos provenían de pueblos conquistados por el Imperio Inca y no hablaban el idioma local, el quechua, sino una decena idiomas y dialectos precolombinos sudamericanos. Este potencial problema en la organización se resolvía por el don del mensajero, que comunicaba el diseño del Ti en un lenguaje de luz que todos los presentes comprendían, o más bien, absorbían del mismo modo en que las plantas obtienen sus nutrientes. El idioma solar fluye a través de la boca del hablante y se expande en la dirección en la que este la proyecte. Sus instrucciones podían ser observadas simultáneamente a un radio de más de 300 metros a la redonda, y su cualidad luminosa hacía posible trabajar hasta altas horas de la noche. Cuando era necesario, tenues rayos atravesaban el cuero cabelludo y proyectaban recomendaciones en el cielo, a través de los agujeros circulares sellados con placas de oro, creando un espectacular juego de luces. Inmensas piedras irregulares y pulidas eran movilizadas desafiando la geografía montañosa y encajaban unas con otras, erigiendo muros inmortales en un elevado ejercicio de geometría. Esas jornadas eran la prueba de que la voz de dios es una luz infrarroja que opera en el espectro visible cuando así lo desea.

Una vez concluida la obra, el suministro luminoso cesaba, así como cualquier atisbo de cefalea. Estas personas habían honrado el designio cósmico, habían cumplido una labor destacada en beneficio de la nación y el progreso de su cultura, por lo cual eran reconocidas debidamente por el imperio y la admiración de sus pares.

Sin duda, el Sol no fue el responsable de las prodigiosas técnicas y tecnologías desarrolladas por los incas, pero fue un fiel consejero, porque ellos eligieron escuchar. Hasta que fueron derrocados violentamente, entonces los sueños de los sobrevivientes dejaron de brotar, o quizás solo dejaron de darles importancia, reducidos como estaban.

La trepanación se realizó en múltiples culturas, pero no sabemos si Ti estuvo involucrado en todos esos territorios. Probablemente encontraron motivos propios para hacerlo, otros dolores, otras ideas que pugnaban por salir. Y ante la aparición de tumores y conmociones, arcaicos neurocirujanos se atrevieron a combatirlos adentrándose en el cráneo con instrumentos rudimentarios.

Hoy que esta práctica ha sido dejada de lado, se cree que aún hay gente que padece esa condición, por ello sufren migrañas tan intensas que deben refugiarse en la penumbra. Les duele porque la luz interna lucha por salir, para que escapen los buenos sueños y se conviertan en templos.

Queda de esa época el deseo de refractar algo de esa luz que nos regala el sol, siendo pequeñas estrellas.

THE CIRCULAR ILLUMINATIONS

From high in the Andes Mountains, the Inca Empire expanded out by means of the strengths and skills of each and every one of its members. And so did its greatness, which reflected the empire's understanding that the Sun was god of the stars because His light fell on all Inca without distinction. At noon in particular, that fiery deity named Ti caressed all earthly life, giving His favor to the crops, rekindling the sparkle in our eyes, and gently rocking the beasts until they fell asleep.

Although His lineage was known to extend beyond the Earth (He was brother and husband to Mama Quilla, goddess of the moon; He was father of the Incas—the human emperors), from time to time He surprised the world by revealing new kinships, adopting simple settlers and revolutionizing their lives by expanding their skills in supernatural ways.

It would all begin when an indigenous villager dreamt of a new building and awoke the next morning with a terrible, blinding headache, which forced him to call upon his shaman with great urgency. After confirming the future temple's authenticity, the shaman—a healer of body and spirit, skilled in natural and entheogenic medicines—informed the Inca emperor, who gave the intervention his approval. The procedure went as follows: the chosen one was anesthetized via the consumption of San Pedro (a drink prepared from the wachuma cactus), the surgeon incised the patient's scalp with a *tumi*—a half-circular ritual knife of gold, silver, or copper—and folded the skin back so as to reveal the cranium. According to the Sun's wishes, three or four holes were then cut out of the skull with a small obsidian chisel, which allowed for the extraction of discs of bone by means of circular scoring and prying. It was a high-risk operation, of course; one slip and the brain was spoiled, one of the great marvels of the world ruined. But that was a rare occurrence. What generally took place was that select layers of bone were extracted with the ritual instruments and the brain was exposed, peering out directly at the world for the first time, receiving pure light through those four new windows. An overflowing joy and radiant excitement preceded every trepanation among the Inca.

The brain remained exposed for approximately six minutes. The emperor enjoyed attending this part, while the god Ti's rays flowed like a brilliant breeze into His vassal. To be fair, everyone loved that moment: How not to adore that instant in which the soul found illumination through the profanation of the body? There was no lack of curious villagers peering out from between the tree branches and taking pictures with their minds. Once the Sun had finished transmitting His secret, the wound was sealed with a gold plate, the scalp was put back in place, and the surgery came to an end, with the emperor indicating his approval with a wave of his hand or a round of applause, according to the dignitary's temperament.

After being penetrated by God's latest great idea, which had arrived in the form of powerful rays of light, the temporary repository did not experience a sense of peace. On the contrary, the migraine he suffered was fierce, adding to the disturbing sensation that something was fighting to escape from him or even trying to plunge further in. Light is not a field strewn with roses, even less so if it comes during an invasive procedure. This suffering unleashed howls that gripped the community, and the erection of a unique colossus of divine provenance was proclaimed imminent.

Then began the process of its construction, headed by the trepanned messenger. It might be a ceremonial center, a sundial, an *unshu*, or a temple like Machu Picchu itself. In order to perform this monumental task, hundreds of men were put to work without rest. Most of them were taken from villages conquered by the Inca Empire and didn't speak Quechua, the local language, but one of a dozen pre-Columbian South American languages and dialects. This potential organizational problem was resolved by the talent of the messenger, who was able to communicate Ti's design in a language of light which all present could understand—or rather absorb, in the way that plants obtain their nutrients. The solar language flowed through the speaker's mouth and expanded in the direction in which he projected it. His instructions could be simultaneously observed at a radius of more than three hundred meters, and the language's luminosity made it possible to work until deep in the night. When necessary, dim rays of light shot out from his scalp through circular holes in the gold plate and projected instructions into the sky, which created a spectacular play of lights. Immense, irregular, polished stones were moved in defiance of the mountainous geography and made to fit into each other, erecting immortal walls in advanced geometric exercises. Those days of labor are proof that God's voice is an infrared light that operates in the visible spectrum whenever he so chooses.

As soon as the work was complete, the supply of light ran out, along with any trace of a headache. These people had honored cosmic design and carried out an outstanding project for the benefit of the nation and the progress of their culture, for which they duly received recognition from the empire and the admiration of their peers.

Without a doubt, the Sun was not responsible for the prodigious techniques and technologies developed by the Incas, but He was a faithful consultant, and they chose to listen to Him. That is, until they were violently overthrown, at which point the survivors' dreams stopped manifesting—or perhaps the Inca simply stopped lending them any importance, diminished in power as they were.

Trepanning was practiced across many cultures, but we are unsure if Ti was also present in all of those territories. These peoples likely found their own motives for it, their own pains, their own ideas fighting to escape. And faced with tumors and concussions, archaic brain surgeons dared to combat them by cutting through the cranium with rudimentary tools.

Now that this practice has been abandoned, it is believed that certain people still suffer from that condition which provokes migraines so intense that they have to take shelter in the shade. Their pain stems from the light inside them fighting to escape so the good dreams can find their way out and be converted into temples.

The desire still exists from that era to refract something of that light the sun gives us, for we are little stars.

STORY 3

Butter

Ekaterina Sedia

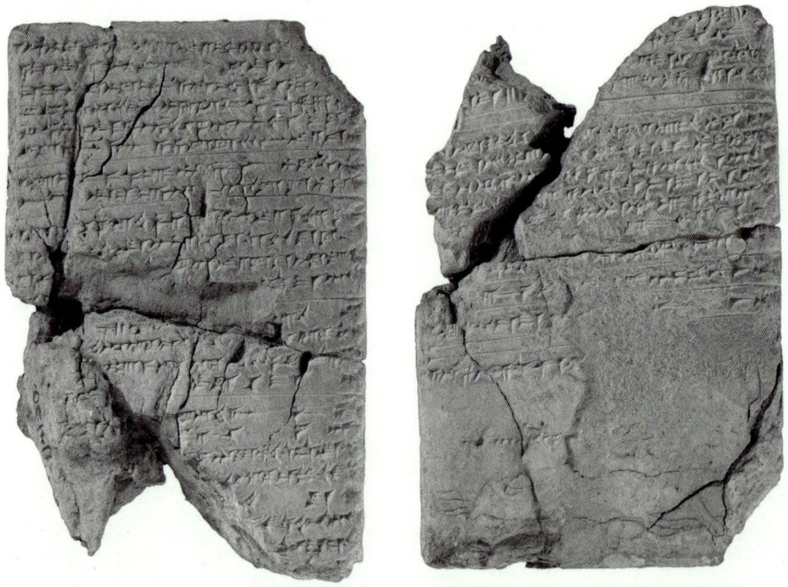

Front and back sides of a single-column cuneiform tablet with a drawing of the demonic creature *bennu* (right image, bottom), Assur (Qal'at Sherqat), Mesopotamia, 8th century BCE.

The demon brings me oranges; I can smell them before he arrives and everything swirls into white, eddies bifurcating and splitting, dichotomy of a forked tongue raised to an infinity of mountain peaks. When I come to, there are no oranges, only my mom and her cool soft hand on my forehead.

"My hands too will grow soft when I'm old," I say, with my lazy cotton-wad of a tongue. "Collagen breaks down as you age."

"Shhh, shhh," she says, in a voice that lets me know she hasn't understood a word I said. And immediately after: "Have you been taking your medication?"

I nod. Meds don't work for demons. I feel him squirming still, but settling down.

"I'm scheduling you for your neurologist appointment," Mom says. "It's just too much, too much. My poor Ase."

I squeeze my eyes shut, the demon circling, circling behind my eyelids, like a dog tamping down imaginary grass before lying down. Then he is quiet.

The sounds intrude through the wool and the sleep, the tired waves. TV drones; Mom speaks on the phone.

* * *

The demon doesn't like doctors. *He would kill you if he could.* His scales rub against each other like leaves; his claws smell like dirt.

"How are you feeling?"

I shrug. "Better."

"You still have seizures."

"I do. I just feel better about them."

The doctor never cracks a smile. "Even now? During Ramadan?"

"I didn't know it was Ramadan." Not strictly speaking accurate, but I'm making a point.

The doctor looks surprised. "Not Muslim? I thought your family was from Iraq?"

I sigh. "Yeah, but a lot of Iraqis are secular ... and we're Kurdish on top of that."

"Atheists?"

"Zoroastrians." I shift. "Anyway. What do de—seizures have to do with Ramadan?"

"Fasting," he says. The neurologist is an old man, and I wonder if old doctors are ever asked to go back to school, to learn new things, or if they are forever stuck in whatever year they graduated. "It works really well with some forms of epilepsy. We don't really know why, but it does."

I spread my fingers wide. "Magic!"

His thin lips purse momentarily in what I think is a smile. "I will adjust your dosage. No driving for another six months. And if you want to try fasting too ..."

"For how long?" I stand up, pause at the door. "It's kind of self-limiting, isn't it?"

"You can eat butter," he says.

I think he expects follow-up questions; instead, I leave.

* * *

The demon is hungry. So am I, but I am not getting up from the sofa, and certainly not turning off the TV. *Do you think it's a good idea to watch cartoons in your condition?*

"Fuck off," I say. "Go away. I'm not feeding you."

Not even oranges?

"Nothing."

Not even for pretty sparkles?

"They're not even that pretty." (They are.)

He goes silent. I imagine him sulking, pacing, slithering through the nasal cavities, spitefully kicking olfactory bulbs, until finally quieting down, hungry and drowsy, behind my eyeballs. He kneads like a cat, growls but falls asleep.

My stomach growls back, but all is quiet and I am not moving.

* * *

He promises, he promises so much—pretty sparkles and secrets, and the smell of oranges; the sensation of falling, falling, as my scalp tingles and bliss spreads, as tiny hairs stand on end and slow warmth flows; incomprehensible whispers that thrill and tickle; lights blooming and unfurling like endless flowers.

He begs me for pomegranates, figs, and grapes. He slavers for naan, for some biryani, for sweet black tea. He knows my mom's cooking as well as I do—this kinship of salted yogurt and kofta, of mustard seeds smoking in the skillet. Their fat, oily smell cuts into my brain like the demonic oranges.

"You have to eat something." My mom looks into my room, sunlight through the dancing curtains dappling her face in white crescents and shadowed curves, like the cuts on a violin. "You want some kofta?"

"No." Yes. "Do we have any butter?"

She nods. "And fresh naan."

"Just butter," I whisper. "Please."

I let each pat melt on my tongue, nutty and deep. I want to tell the neurologist that it only works for demons. Exorcism by butter.

He is weak and no longer promises anything; he doesn't move much. His neck is limp, like a dying snake. Melting butter flows around him like the ocean around a marooned man, only taunting.

* * *

How can you miss a place you've never seen, the smells you cannot even conceive of in New Jersey suburbs (raw mutton, wet wool, ewe's milk steaming in the cold morning air), how can you even dream of what those mountains look like (*we are never going back, never-ever*.) And why does this longing pierce like a needle; why the panic when you realize that as the demon is fading so is that secret landscape, and soon you'll be left with nothing but your mom's spare reminiscences, a mere palimpsest, flat and dull, and *not yours*.

I try to remember the visions, the showers of comets and the thatched roofs, the dull glint of gunmetal, bread baking on the stone hearth (oh, that bread). Smells of oranges and pomegranates, an overripe fig bursting open on a bruised lip.

I can only live there.

I can only live there. Memories of visions fade faster than they have any right to. The demon is dead, I think.

* * *

"No seizures in three months." The neurologist seems pleased. "We'll just continue with the medication ... and?"

I look at the floor. "Keto."

"It seems to be working."

It only works for demons. "I guess."

"Still no driving," he cautions, "but in the fall you should be able to go back to school. Rutgers?"

"Yeah." I try to look happy. I know Mom will be; I imagine her face, patient and docile in the waiting room, blooming into a smile.

But the sparkles.

And the dust in the air, and the unreasonably, impossibly cold July mornings in the mountains. We can never go back.

Oh, but we can.

Is it fair to her, though.

Is it fair to you? Was it fair to you?

She did what she thought was best for us.

In the car, on the way home, we see a fruit stand—all three of us. "Do you want anything?"

"Peaches," the demon and I say with one mouth (only one of us is sorry), and our bifurcated tongue twitches in anticipation of sugar, of endless falling, of the smell of first raindrops hitting the dust, heavy like bullets.

STORY 4

தாரக தாண் டவம்

Charu Nivedita

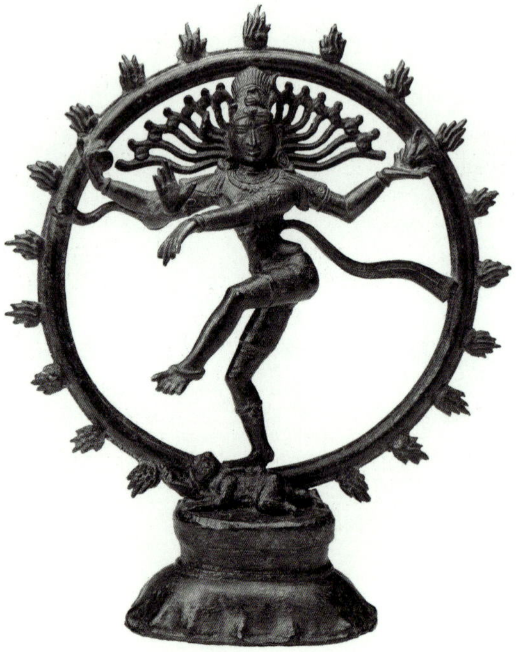

Shiva Nataraja, India, 18th century (front and back view).

தாரக தாண் டவம்

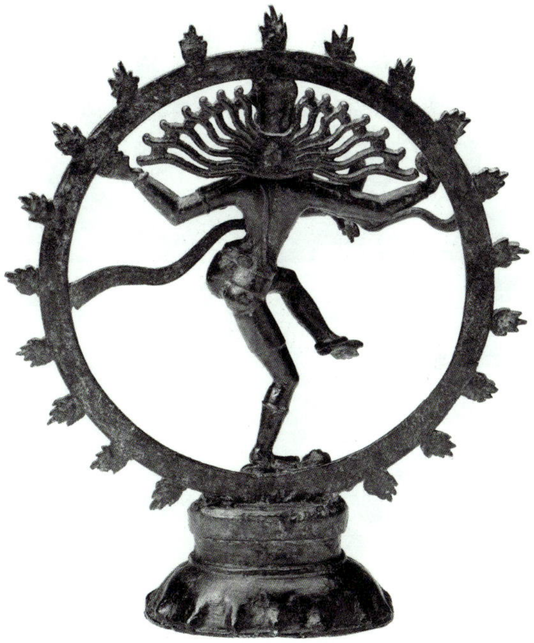

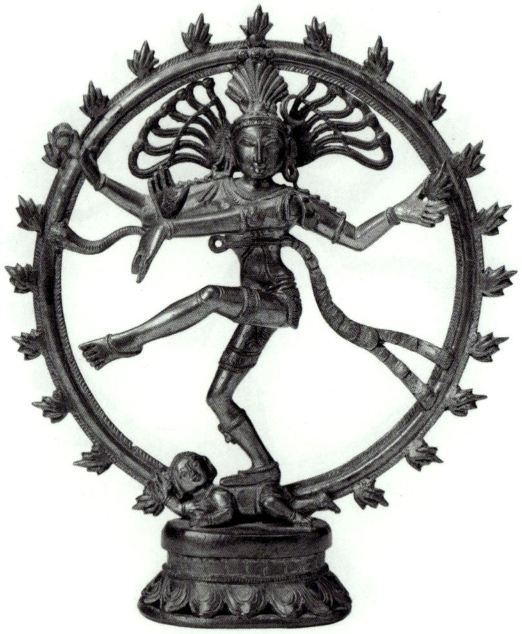

Shiva Nataraja, India, 19th century.

எங்கள் ஊரில் பட்டப் பெயர் இல்லாத பையன்கள் ரொம்பக் குறைவு. அப்படி பட்டப் பெயர் இல்லாதவர்களை யாரும் கண்டு கொள்வதில்லை. பட்டப் பெயர்கள் பல சமயங்களில் நடையுடை பாவனையை வைத்தே வழங்கப்படும். ஒருத்தன் பெயர் சரோ. அவன் நடிகை சரோஜாதேவி மாதிரி நடப்பான். சிலருக்கு அவர்களது குணத்தை வைத்தும் கிடைக்கும். அப்படி என் வகுப்புத் தோழன் நடராஜனுக்குக் கிடைத்த பெயர் நட்டு அல்லது மறை. ஏனென்றால், அவன் நடவடிக்கைகள் பலவும் நட்டு அல்லது மறை கழன்றவன் மாதிரியே இருக்கும். எல்லா மாணவர்களும் பள்ளி மைதானத்தில் கால்பந்து ஆடினால் நட்டுவை அங்கே பார்க்க முடியாது.

எங்கேதான் போவான் என்று எனக்கு ஒரே ஆர்வமாக இருந்ததால் ஒருநாள் அவனையே கவனித்துக் கொண்டிருந்தேன். எல்லோரும் மைதானத்துக்குப் போகும் போது அவன் மட்டும் வேறொரு பக்கம் நடக்க ஆரம்பித்தான். அந்த வழியில் தெரு எதுவும் கிடையாது. நேராகப் போனால் சுடுகாடு. போகட்டும் என்று விட்டு விட்டு சிறிது நேரம் கழித்து அவன் சென்ற திசையிலேயே சென்றேன். பார்த்தால் சுடுகாட்டில்தான் இருந்தான். நெருங்கிப் போய்ப் பார்த்த போது நாலைந்து நாய்களோடு பேசிக் கொண்டிருந்தான். எனக்கு சப்பென்று போய் விட்டது.

பள்ளிக்கூடத்திலும் தெருவிலும் நாய்களோடு கொஞ்சுவது போதாதா? சுடுகாடு வரை போக வேண்டுமா? ஆமாம், அங்கேதான் போக வேண்டும்தான். அவனால் பள்ளிக்கூடத்திலோ தெருவிலோ நாய்களோடு கொஞ்ச முடியாது. இதற்கு நான் நாய்களுக்கும் எங்கள் ஊர்ப் பையன்களுக்குமான உறவு பற்றிக் கொஞ்சம் சொல்ல வேண்டும். நாயை எங்கே பார்த்தாலும் கல்லால் அடிப்பது பையன்களுக்குப் பழக்கமாக இருந்தது. அதனால் ஊரில் பல நாய்கள் நொண்டி நாய்களாகவே ஆகி விட்டன. பையன்களுக்கு நாய் மேலே என்ன பகை என்று கேட்டால் ஊரில் நாய் கடிக்காத பையனே இல்லை. என்னைக் கூட ஒரு நாய் கடித்திருக்கிறது. ஆனால் என்ன வித்தியாசம் என்றால், எல்லோரையும் கடித்தது தெருநாய். என்னைக் கடித்தது என் வகுப்புத் தோழன் பிரகாஷ் வீட்டுப் பாமரேனியன் நாய். கடிக்காது, பயப்படாதே என்று அவன் சொன்னதால் தைரியமாக உள்ளே போனேன். கடித்து விட்டது. அப்புறம்தான் தெரிந்தது, நாய் வளர்ப்பவர்கள் எல்லோருமே சொல்லும் வார்த்தை அது என்று. அப்புறம் என்ன, தொப்புளைச் சுற்றி ஊசி போட்டார்கள். நாய் கடித்தால் நாமும் நாய் மாதிரியே குலைத்துக் குலைத்துச் சாவோமாமே? டேய், அதெல்லாம் வெறி நாய் கடித்தால்தான்டா என்றான் பிரகாஷ். நல்லவேளை, எனக்கு அப்படி ஒன்றும் ஆகவில்லை.

ஆனால் நட்டை மட்டும் நாய்கள் எதுவும் செய்வதில்லை. வாலைக் குழைத்துக் கொண்டு அவனிடம் போய்க் கொஞ்சும். அதனால் நட்டை வெறுப்பேற்றுவதற்காக அவன் எதிரில் எந்த நாய் வந்தாலும் அதைக் கல்லால் அடிப்பது

பையன்கள் வழக்கம். உடனே அப்படி அடித்த பையன்களை நட்டு அடிப்பான். நட்டு அத்தனை பலசாலி அல்ல என்பதால் நட்டுதான் கடைசியில் நிறைய அடி வாங்குவான். இப்படியே சென்று கொண்டிருந்தபோது ஒரு பயல் நட்டுவைக் கடுப்பேற்றுவதற்காகவோ அல்லது அவனுக்கு அன்று நேரம் சரியில்லையோ என்னவோ, ஒரு சிறிய குட்டி நாயைக் கல்லால் அடித்து விட்டான். அன்றைக்கு அந்தப் பையன் நட்டுவிடம் வாங்கிய அடி இருக்கிறதே, கன்னம் காது எல்லாம் பியந்து ரத்தம் கொட்ட ஆரம்பித்து விட்டது.
ஓ, நடந்ததை என்னால் நடந்த மாதிரியே விவரிக்க முடியவில்லை. நட்டு அடித்தானா, கடித்தானா? ரெண்டுமே செய்தான். அதோடு சண்டை நடந்து கொண்டிருக்கும்போதே நட்டுவுக்கு வலிப்பும் வந்து விட்டது. அதற்குள் தெரு ஜனங்களும் கூடி முதலுதவி என்று நினைத்துக் கொண்டு என்னென்னவோ செய்தார்கள். என்னால் கூட்டத்தின் உள்ளே நுழைந்து பார்க்க முடியவில்லை. பார்த்தபோது அவன் வாயெல்லாம் நுரை தள்ளியிருந்தது. சரியானதும் எழுந்து அவன் பாட்டுக்கு எதுவுமே நடக்காதது போல் நடந்து போனான். அதிலிருந்து அவனுக்கு எதிரே நாயை அடிப்பதை பையன்களும் நிறுத்தினார்கள்.
பள்ளிப் படிப்பு முடிந்த பிறகு நான் நட்டைப் பார்க்கவே இல்லை. நான் தில்லிக்குப் போனதும் ஒரு காரணம். ஊர் நண்பர்கள் யாரோடும் எந்தத் தொடர்பும் இல்லாமல் போனது. ஆஃபீஸ் போவதும் ஒழிந்த நேரத்தில் கதை எழுதுவதுமாக எழுத்தாளன் ஆவதற்குத் தீவிரமாக முயற்சி செய்து கொண்டிருந்தேன். என் நண்பர்கள் அனைவருமே நான் எழுத்தாளன் ஆவதற்கு எதிர்ப்புத் தெரிவித்தார்கள். ஏனென்றால், தமிழ் எழுத்தாளர்கள் பெரும்பாலும் மிக மோசமான முறையில்தான் இறந்து கொண்டிருந்தார்கள். பலர் குடி அடிமை ஆனார்கள். ஓரிருவருக்குப் பைத்தியம் பிடித்தது. பலர் வறுமையில் வாடிச் செத்தார்கள். சாகும் போது கூட மரண ஊர்வலத்திற்கு ஒரு டஜன் பேர் கூட வராத நிலைதான் நிலவியது. ரவீந்திரநாத் தாகூரைப் போல் சர்வதேச அளவில் புகழ் பெற்றிருக்க வேண்டிய சுப்ரமணிய பாரதிக்கே அவர் அற்பாயுசில் இறந்த போது மரண ஊர்வலத்துக்கு எட்டு பேர் வந்தார்களாம். இதையெல்லாம் மீறி எழுத்து என்னை ஈர்த்ததற்குக் காரணம், விதியாகத்தான் இருக்க வேண்டும்.
இடைப்பட்ட காலத்தில் இலக்கியச் சிறு பத்திரிகைகளில் நட்டு என்ற பெயரில் வெளியாகியிருந்த கவிதைகளைப் பார்த்து நம்முடைய நட்டுவா எனச் சந்தேகப்பட்டேன். நட்டு கவிதைகள் எனக்கு ரொம்பவே பிடித்து விட்டன. அதிலும் தாரக தாண்டவம் என்ற நெடுங்கவிதை அவனைப் பற்றிய இந்தக் கதையில் அவசியம் இடம் பெற்றாக வேண்டும்.
தாரக வனத்தில் ரிஷிகளின் அட்டகாசம் தாங்க முடியவில்லை. அங்கே செல்கிறார் சிவன். கூடவே விஷ்ணு பெண் வேடத்தில். ரிஷிகள் சிவன் மீது புலியை ஏவுகிறார்கள். சிவன் தன் சிறுவிரல்

நகத்தினால் புலியின் தோலைக் கிழித்து விடுகிறார். அடுத்து ஒரு பாம்பை ஏவுகிறார்கள். அதை ஒரு மாலையாக்கித் தன் கழுத்தில் மாட்டிக் கொள்கிறார். அதற்குப் பிறகு சிவன் நடனமாடத் தொடங்குகிறார். அப்போது அவரைத் தாக்குவதற்கு முயலகன் என்ற குள்ளன் வருகிறான். அவனையும் ஒரு விரலால் எடுத்து சிவன் தன் காலடியில் போட அவன் முதுகு உடைகிறது. சிவ நடனமும் தொடர்கிறது. நடனம் முடிந்ததும், ஆதிசேஷன் சிவனிடம் இதே நடனத்தை மீண்டும் ஒருமுறை ஆட வேண்டுமென பிரார்த்திக்க, சிவன் அதை பிரபஞ்சத்தின் மையமான தில்லையிலே ஆடினார். இந்தப் புராணக் கதை நம் எல்லோருக்கும் தெரிந்ததே. இதே கதையை எடுத்த நட்டு, இதில் தன் வாழ்க்கையை நுழைத்திருக்கிறான். முயலகன்தான் நட்டுவாம். ஆயுர்வேதத்தில் வலிப்பு நோயை அபஸ்மரா என்கிறார்கள். ஸ்மரனை என்பது நினைவு, அறிவு, பிரக்ஞை. அபஸ்மரனை – நினைவு அழிதல். முயல் வலி என்று நாட்டுப்புற வழக்கில் சொல்வார்கள். முயல் உயிருக்குப் போராடி ஓடும் போது மூச்சு விடுவது போல் வலிப்பு வந்தவர்கள் மூச்சு விடுவதால் முயல்வலி. சிவனைத் தாக்க வந்த முயலகனும் இதே ரகம்தான். இந்தக் கவிதையில் நட்டு தனக்கு முதல் முதலாக வலிப்பு வந்த கதையைப் புகுத்தியிருந்தான்.

ஆதியிலே நடனம் இருந்தது என்று தொடங்கி, ஆகாயத்திலே உள்ள எண்ணற்ற நட்சத்திரங்களும் கோள்களும் அவற்றின் லயமும் சிவநடன ரூபங்களே என்பதை சரடாகக் கொண்ட அந்தக் கவிதை ஐந்து பாகங்களாக விரிந்திருந்தது. சிருஷ்டி, ஸ்திதி, ஸம்ஹாரம், திரோபவம், அனுக்ரஹம் என்பவை பாஞ்ச க்ரித்யா. உடுக்கையிலிருந்து சிருஷ்டி, கரங்களிலிருந்து ஸ்திதி, அக்னியிலிருந்து ஸம்ஹாரம், உயர்த்திய காலை நோக்கும் கரம் திரோபவம் (பாதுகாப்பு), உயர்த்திய கால் அனுக்ரஹம். இதில் முதலாம், ஐந்தாம் பகுதிகளின் முக்கியப் பகுதியை மட்டும் சொல்லி விடுகிறேன்.

சிருஷ்டி
பிறழ்வு கொண்டிருந்த என்
மனதில்
தன் காலடி வைத்தான் ஞானக்கூத்தன்
அண்ட சராசரமே நான்தான் என்ற
ஓர்மை கிடைத்தது*

அனுக்ரஹம்
நான் இந்தப் பிரபஞ்சத்தோடு
உரையாடிக் கொண்டிருக்கிறேன்.
அந்த உரையாடலை கவிதைகளாகக் கொடுக்கிறேன். இந்தச் செயல்பாட்டுக்கு அக்னி தேவைப்படுகிறது. ஆனால் அக்னியோ ஸம்ஹாரம். எனவே, என்னை அக்னியால் அழித்துக் கொண்டு அந்த அழிவையே கவிதையாக மாற்றுகிறேன். என் ஸ்தூல உரு கவிதை என்ற இன்னொரு உருவாக மாறுகிறது. நான் அழிந்து பிரபஞ்ச நானாக மாறுகிறேன்.

தாரக தாண்டவம் இவ்வாறாக முற்றுப் பெறுகிறது...

அக்னியால்
என் உடல்
கொஞ்சம் கொஞ்சமாக
எரிந்து கொண்டிருப்பதை
சற்றே உயரத்திலிருந்து
பார்க்கிறேன்.

* * *

எனக்குத் தெரிந்து விட்டது, நம்முடைய நட்டுதான்.

உடனே விசாரித்தேன். சென்னையில் இருப்பதாகத் தெரிந்தது. என்ன செய்கிறார் என்று கேட்டபோது, ஒன்றும் செய்யவில்லை, நிறைய குடிப்பார் என்றார்கள். ஆகா, எல்லா தமிழ் எழுத்தாளர்களின் தலைவிதி போலவே நட்டுவுக்கும் ஆகி விட்டதே என்று நினைத்துக் கொண்டு உடனடியாகக் கிளம்பி சென்னை வந்தேன். ஆளைக் கண்டு பிடிப்பது ஒன்றும் சிரமமாக இல்லை.

நீண்ட முடி. தாடி. கடுமையாக இளைத்துப் போயிருந்தான். இருபது ஆண்டுகள் இருக்கும். அறை வாடகை, சாப்பாடு எல்லாம் நண்பர்கள் கவனித்துக் கொள்கிறார்களாம். வீட்டோடு தொடர்பு இல்லை. ஐந்து ஆண்டுகள் சங்கிலியால் கட்டி விட்டார்கள் என்றான் சிரித்துக் கொண்டே. அதிர்ந்து போனேன். என்ன ஆயிற்று என்றால், ஏதோ நிழல் ரூபமாக, சலனச் சித்திரமாக எப்போதாவது மனதில் தோன்றி மறைகிறதே ஒழிய மற்றபடி எதுவுமே ஞாபகம் இல்லை என்கிறான்.

அந்தக் காலகட்டத்தில் என்ன நடந்திருக்கிறது என்றால், ஐந்து ஆண்டுகள் அவன் எழுந்தே நிற்கவில்லை. அதுதான் பிரச்சினை. தவழ்ந்தபடியேதான் இருந்திருக்கிறான். நன்றாக நடமாடிக் கொண்டிருந்த ஒருவன் திடீரென்று நடப்பதை நிறுத்தி விட்டுத் தவழ்ந்தபடி வாழ ஆரம்பித்தான் என்றால் எல்லோரும் என்ன நினைப்பார்கள்? காலில் எந்தப் பிரச்சினையும் இல்லை என்று மருத்துவச் சோதனையில் தெரிந்து விட்டது. எழுந்து நடக்க வைப்பதற்காகப் பெரும் பிரயத்தனப்பட்டிருக்கிறார்கள். எழுந்தால் தலை இடிக்கும் என்பது அவன் வாதம். அதற்குப் பிறகு அது மனநலப் பிரச்சினையாக மாறி விட்டது.

மனநல மருத்துவர்களும் (அவனை எழுந்து நடக்க வைப்பதற்காக!) பலவித மருந்துகளைக் கொடுத்திருக்கிறார்கள். அந்த மருந்துகளின் ஒரே பலன், ஒரு நாளில் பதினெட்டு மணி நேரம் உறக்கம். மற்ற நேரமெல்லாம் தவழ்ந்த நடைதான். பாவம், மனநல மருத்துவர்களுக்கு நட்டு தன்னை அபஸ்மாராவாக நினைத்துக் கொண்டிருக்கிறான் என்று தெரியவில்லை. சிவசரித்திரத்தில் அபஸ்மாரா ஒரு குள்ளன். நட்டுவின் பிரக்ஞையில் ஏற்பட்ட ஒரு கலாபூர்வமான நிகழ்வை மனநல மருத்துவர்கள் மனச்சிதைவாக எடுத்துக் கொண்டிருக்கிறார்கள்.

தான் எழுதி வைத்திருந்த குறிப்பேடுகளை வைத்து ஒரு நெடுங்கவிதை எழுதியிருக்கிறேன் என்று காண்பித்தான். அபஸ்மராவின் தனிமொழி. Soliloquy of Apasmara. இன்னும் பிரசுரிக்கவில்லையாம்.

”அதற்குப் பிறகு வீட்டில் இருக்கக் கூடாது என்று சென்னை வந்து விட்டேன். இங்கே வந்தால் ஒரு கொலைக் கேஸில் உள்ளே தள்ளி விட்டார்கள், என்ன சொல்ல?” என்று நகைச்சுவையைச் சொன்னது போல் பெரிதாகச் சிரித்தான்.

என்ன விஷயம் என்றால், சரி, அதற்குள் போவதற்குள் என்னைப்

பார்த்தால் இப்போது எப்படி
இருக்கிறது, சொல்?
யேசு நாதரைப் போல் இருக்கிறது.
ம்ஹும். நீ எழுத்தாளனைப் போல்
பதில் சொல்கிறாய். நீ அடிக்கடி
திட்டுகிறாய் அல்லவா, காமன்மேனே
காமன்மேனே என்று. அந்தக்
காமன்மேனுக்கு நான் எப்படித்
தெரிவேன். நன்றாகப் பார்த்து
சொல்.
பிச்சைக்காரன் அல்லது திருடன்.
இதுதான் சரி. இது புரிந்தால்தான் நான் சொல்லப் போகும்
கதை புரியும். கோயம்பேடு
பஸ் வளாகம் அருகே நான்
நடந்து போய்க் கொண்டிருந்தேன். என்னைப் போலவே
தோற்றமளித்த ஒரு ஆள் வலிப்பு
வந்து துடித்துக் கொண்டிருந்தான். எல்லோரும் அவரவர்
பாட்டுக்கு அவனைப் பார்த்தபடி
எதுவுமே நடக்காதது போல்
போய்க் கொண்டிருந்தார்கள்.
யாருமே அவன் உதவிக்குப்
போவதாகத் தெரியவில்லை.
எனக்கு அப்போது வேறு எந்த
யோசனையும் வரவில்லை. அவன்
துடித்துக் கொண்டிருந்ததைப்
பார்த்ததுமே அவன் அருகே போய்
ஒரு சிறிய குச்சியைத் தேடி எடுத்து
அவன் பல்லிடுக்கில் வைத்தேன்.
முந்தின நாள் மழை பெய்திருந்ததால் அந்த இடமெல்லாம் சகதியும்
சேறுமாக இருந்தது. நான் வேறு
அப்போது நல்ல போதையில்
இருந்ததால் தடுமாறிக் கீழே
விழுந்து விட்டேன். என் உடம்பெல்லாமும் சகதி ஆயிற்று. அவன்
அருகே போய் அமர்ந்தேன். அவ்வளவுதான் நடந்தது. ஆனால் மக்கள்
கூட்டம் பைத்தியம் பைத்தியம்
என்றும், கொலை கொலை என்று
பதற்றமாகக் கத்தினார்கள். உடனடியாகப் போலீஸ் வந்து என்னைக்
கைது செய்தது.
எவ்வளவோ சொன்னேன்.
போலீஸ் கேட்கவில்லை. அதில்
பார், ஒரு அபத்த நகைச்சுவையை.
நான் ஒரு கவிஞன் என்று சொல்லியிருக்கலாம். நான் ஒரு ரைட்டர்,
நான் ஒரு ரைட்டர் என்றேனோ,
டேய் நாயே, பிச்சைக்கார நாயே,
நீயாடா ரைட்டர் என்று கேட்டு
அதற்கு வேறு ரெண்டு போடு
போட்டார்கள். எனக்கு அப்போது
தெரியாமல் போயிற்று, போலீஸ்
ஸ்டேஷனில் ரைட்டர் என்று ஒரு
பதவி இருக்கிறது என்று... மறுநாள்
பேப்பரில் "வலிப்பு நோயாளியைக்
கொலை செய்ய முயன்ற பைத்தியம்
கைது" என்று பெரிய கொட்டை
எழுத்தில் போட்டு என் முகப்படத்தையும் போட்டிருந்தார்கள்.
பரவாயில்லை சூர்யா, குஸ்தாவ்
ஃப்ளபெரை விடவும் அந்தோனின்
ஆர்த்தோவை விடவும் நான்
பெரிய அதிர்ஷ்டசாலிதான். மூன்றே
நாளில் லாக்கப்பில் இருந்து நண்பர்களின் முயற்சியால் வெளியே
வந்து விட்டேன்.
நான் தில்லி திரும்பியதும் ஒருநாள்
நட்டு இறந்து விட்டதாகக் கேள்விப்பட்டேன்.
அக்னியால்
என் உடல்
கொஞ்சம் கொஞ்சமாக
எரிந்து கொண்டிருப்பதை
சற்றே உயரத்திலிருந்து
பார்க்கிறேன்.

* * *

* திருமந்திரம் மூன்றாம் தந்திரம்
பாடல் 2741

TANDAV AT TADAKA

In the town of my childhood, it was the rare boy who wasn't assigned a nickname. This meant he was deemed unworthy of attention. The names were bestowed on the bases of appearance and demeanor. My classmate Natarajan was called "Nut" and sometimes "Screw," because we believed his behavior could only be explained by a screw having come loose in his head.

One fine day, a boy—perhaps because he was particularly keen to irritate Nut, or perhaps because his stars were aligned wrong—decided to toss a stone at a puppy. Nut set upon him so suddenly and so hard that there wasn't a single orifice on the boy that wasn't spewing blood—and Nut had drilled a few fresh ones where there had been none before he was done with him. It happened in a flash, and I can't quite find the words to describe the fight. Did Nut smite or bite? Both. Even as the duel went on, Nut suffered an epileptic attack. This may have been why the other boy survived. By this time, a crowd had gathered and its members began to do all sorts of things by way of first aid. Nut was foaming at the mouth. Once he had recovered, he got up and walked away as if nothing had happened.

I did not see Nut once we graduated from school. I moved to Delhi and lost touch with my childhood friends. I would go to the office and write in my spare time. I was determined to be a writer, despite my friends counseling me against it. The reason for their intervention was that it appeared to be the destiny of the Tamil writer to die a miserable, often violent death. Most drank their way to an alcohol-soaked end. Some went insane. The rest wilted away from poverty. It was the rare writer whose funeral procession numbered over a dozen mourners. Subramania Bharati, who ought to have reached the heights of international fame Rabindranath Tagore did, died in his thirties and only had eight mourners at his funeral. If none of this could thwart my passion, it must have been the designs of destiny that drew me to writing.

Somewhere during this period, I came across poems written under the pseudonym "Nut" in niche literary journals and wondered whether this could be our Nut. I loved his poems. A long-form poem titled *Tandav at Tadaka* deserves mention in this story.

This is the setting: the *rishis* were undertaking severe penance with frightening effects in the Tadaka forest. Lord Shiva went there to put an end to this. Along with him was Lord Vishnu, in the guise of a woman. The *rishis' yagna* let loose all sorts of animal-spirited demons upon the gods. First came the Tiger Demon. Lord Shiva tore him to shreds and used his skin as a vestment. Then came the Snake Demon. Lord Shiva wore him as a garland around his neck. And then he began to dance. His Tandav dance.

At the time, the fire of the *yagna* spewed the dwarf Muyalagan at Lord Shiva, who lifted him with a single finger and threw him on the ground beneath his feet, breaking Muyalagan's back even as he continued to dance. At this, the King of Snakes and the serpent on which Lord Vishnu rests, Adisesha, entreated Lord Shiva to dance his Tandav again, whereupon the Lord of Dance chose Thillai, believed to be the center of the universe, as the site of this performance. This is how the story from the Puranas goes.

Nut had weaved his own life into this mythological tale. In the poem, he was Muyalagan. In Ayurvedic terminology, epilepsy is known as *apasmara*. The word *smara* refers to intelligence, memory, consciousness, and awareness. Apasmara is the destruction or loss of *smara*. In country medicine, *apasmara* was called *muyal vali*. "Muyal" is the Tamil word for rabbit, and it is said those in the throes of epilepsy struggle for breath like rabbits running for their lives. Nut evoked the first occurrence of his epilepsy in the poem.

The poem began with the line:

In the beginning, there was dance

With the idea that the innumerable galaxies in space and their component stars and planets and the energies they emanated and the forces that drew them together were all manifestations of Lord Shiva's Tandav, the poem spilled into five sections that corresponded to the Pancha Kritya, the celebrated five actions of Lord Shiva—Srishti (Creation), Sthiti (Sustenance), Samhara (Rejuvenation), Trobhava (Destruction of Delusion), and Anugraha (Liberation). The damaru—hand-held drum—of Shiva was the source of the Srishti, his arms of the Sthiti, the fire from his feared third eye of the Samhara, the hand pointing at his raised leg of the Trobhava, and his raised leg of the Anugraha.

In the first section, "Srishti," I read:

My splintered mind,
Upon which Gnanakoothan
thundered his feet,
Came together to sense itself
The centre of the cosmos.

In the fifth section, "Anugraha," I read:

I speak to this cosmos,
And give you
Our dialogue
In poetry.
For this task, I burn myself in fire,
And gather the ashes into poetry.
My corporeal self
Creates my cosmic self
*Through lines of rhythmic syllable.**

The poem ended with the lines:

In fire,
My body
Burns,
Molecule by molecule.
I see this
From a height.

It could only be our Nut.

I made my inquiries. I learned he lived in Chennai. What did he do for a living? Nothing, they said. He drank himself into oblivion. It appeared that the destiny of all Tamil writers had not evaded Nut. I left for Chennai at once and found him fairly easily. The hard part was recognizing him.

His long hair, mournful beard, and skeletal frame could hardly speak for the boy of twenty years ago. His friends took care of his food and lodging. He was no longer in touch with his family. He laughed as he told me he'd been in chains for five years. He couldn't tell me much more. His memories came to him in shadow play and silhouetted images. He couldn't stitch together the chronology of his life.

Sometime during this period, five years had passed without his ever standing up. He crawled every time he had to move. It made little sense for a man who was perfectly able to stand and walk all his life to suddenly lose the ability, having suffered no accident or illness. Medical tests revealed there was no problem in his legs. The doctors tried everything to make him walk. However, he insisted that he would hit his head against the roof if he stood up. Then the diagnosis changed—it must be a mental illness, they said.

Psychiatrists pumped him with medication (to make him walk!). However, this only had a soporific effect, inducing him to sleep eighteen hours a day. When he woke, he crawled along the ground. The poor psychiatrists had failed to figure out that Nut had begun to see himself as Apasmara.

In the mythology surrounding Shiva, Apasmara is a dwarf. What had been an artistic inspiration in Nut's *prajnya*—consciousness on a transcendental plane—manifested itself to the psychiatrists as evidence of mental aberration, an incident of acute psychosis. He showed me a poem in long form that he had meticulously woven from the intricate notes he had kept in notebooks maintained for the purpose. It was titled *Apasmaraavin Thanimozhi*—The Soliloquy of Apasmara. It was unpublished.

"I decided I could no longer live with my family and moved to Chennai. Would you believe what happened when I got here, Surya? I was jailed for murder!" he said, with a burst of laughter.

"Why?"

"Tell me honestly, whom would you take me for if you saw me for the first time?"

"Jesus Christ."

"Hmm. You speak like a writer. You speak with contempt for the common man, don't you? Whom would this common man take me for? Look at me and answer."

"A beggar. Or a thief."

"Yes, you're right. And this is the crux of my story. I was walking by the Koyambedu bus terminus when a man had an epileptic attack. Everyone stared at him and no one helped. I acted on instinct and impulse. I rushed to his side and stuck a small piece of wood in his mouth so he wouldn't bite his tongue in a fit. It had rained the previous day, and I slipped and fell. I wasn't sober, either. And so I practically crawled to him to give him first aid. But then everyone began to scream that a madman was killing someone. The police carted me off to jail. They paid no heed to my attempts to explain what had happened. They asked me who I was. I could have told them I was a poet. I had to puff up my chest and say, 'I'm a writer.' They cuffed me on the face and said, 'Beggarly dog! You're a writer, are you? A writer? You dare insult us?' It would be a few days before I discovered that 'writer' was a police designation—it refers to the cop who writes down citizen complaints at the police station. The next morning's newspaper carried my photograph with the bold headline, 'Madman arrested for attempt to murder epilepsy victim.' Never mind, Surya. I got off easier than Gustave Flaubert and Antonin Artaud. I was only in the lockup for three days before my friends bailed me out."

I left for Delhi after meeting him just the once. The next time I had news of him, it was an announcement of his death.

In fire,
My body
Burns,
Molecule by molecule.
I see this
From a height.

* * *

**Tirumantiram,* third Tantra chapter, 2741.

STORY 5

Verso

Ahdaf Soueif

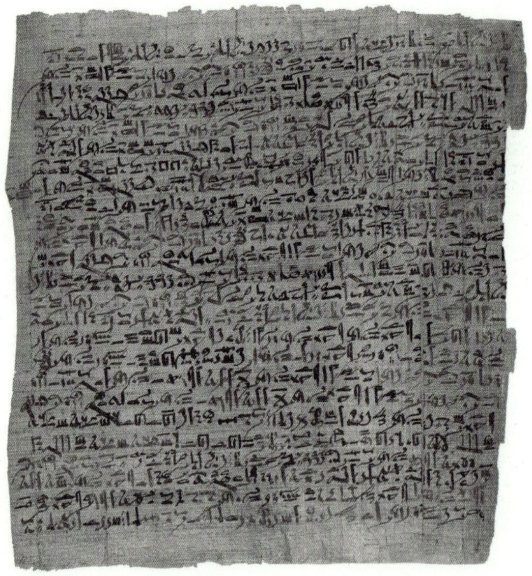

The Edwin Smith Papyrus (detail), medical text, Thebes, Egypt, 16th–17th Dynasty, *c.* 1600 BCE.

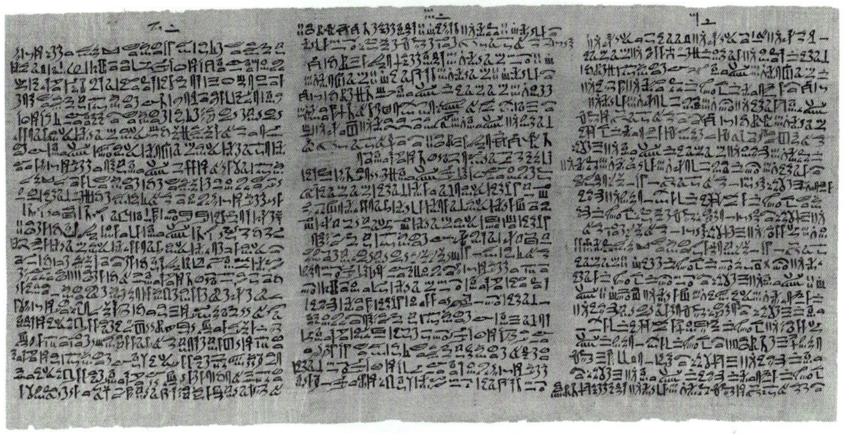

The Ebers Papyrus (detail), medical text, Luxor, Egypt, 18th Dynasty, *c.* 1500 BCE.

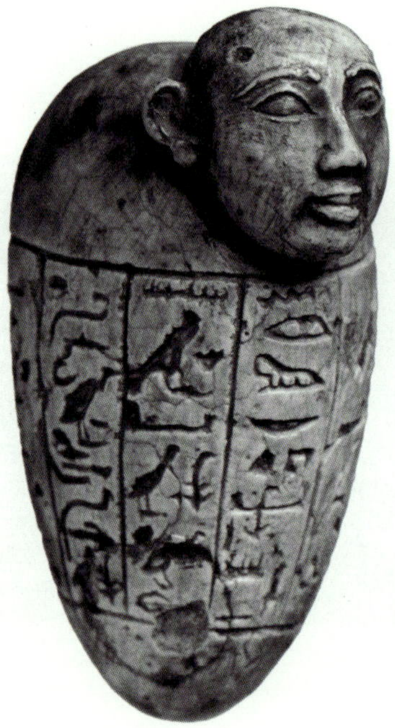

Heart-shaped amulet with a human face and engraved in hieroglyphic characters with chapter 30 of the *Book of the Dead*, Egypt, 25th–31st Dynasty, *c.* 712–332 BCE.

"Why is it always 'he,' 'he,' 'he'? Why doesn't he write about a woman ever?"

"Because, idiot, this is a manual for a field hospital. He's writing about treating soldiers. Women don't get dragged off to fight."

"I'd fight," she says.

"No, you wouldn't. Or rather," he pauses, "you'd fight if you were angry enough. You wouldn't fight to order; you'd run around asking questions and cause trouble."

"So are you working or are you squabbling?" He's paused in the doorway. His son and his daughter. Unconventional that he's taught his daughter to write, but useful now. The Chief Scribe passes his hand over his bristles of gray hair. His useless hand—well, not quite useless, but not dependable, not as sure as it used to be. The tremor varies as the day passes. Maybe that too will pass and he will once again hold the brush and the pen with the certainty he'd never questioned.

"Let me see." He leans over the sunlit scrolls. They're good. The girl's is more distinctive; there's pleasure in the coloring, in the odd little flourish she now allows herself—

"Three more days, Papa, we'll be done."

He nods. His son, serious and understated. The boy senses the urgency, although he hasn't pushed it. No need to alarm them, and the commission was clearly confidential. Waset is secure and at peace. Yet the court demands copies of two medical treatises—and demands them quickly. There's turbulence in the north and the balance is fragile.

He crosses through the room to his inner courtyard. Now he is by the pond, with the spread leaves of water lilies floating on the water. Hands clasped tight behind his back, he paces under the pergola. Waset is readying itself. It's good he was ready too—with his secret helpers. How long will it be till his trembling hand can no longer be disguised? He glances up at the pergola, heavy with grapes. Even if he stopped working, his family would lack for nothing. That's not what he's worried about. The best scribes have graduated from his studio and he is always sending them work. And he has His Majesty's confidence. He pauses, staring into the still water. "Look! His heart is closed and darkness is in his body"—the words his eyes had fallen on as he leaned over his children's writing hands. The light is fading in the garden, and the image that comes to him now is of a bird that has had its feathers clipped. It spreads its wings and it flutters and it flutters but it stays stranded on the ground. Again and again it strives to do what it has always so naturally done—and cannot. What now? How can it be a bird if it cannot fly?

The Chief Scribe lifts his bowl to his mouth. He drinks deep and puts it down and folds his hands tight together. Across the matting, where the food is laid out in pretty earthenware and glass, with here and there a spray of foliage, his wife's glance answers his, reassured. The rhythms of their life. The food and the wine laid out as the sun sets and there is no longer enough light for writing. But the friendly dusk and the candles are soft on the walls, on the bright cushions, on the beloved faces. Later, friends might come around, and in the corner the lyre and the clarinet lie ready. This is what peace looks like. In their lifetime there has not been war in or near Waset. If it comes, their son will likely not be called up; his skills as a scribe will be of more value than one more body at the front, it will be the poor lads laboring in the fields who're taken—

"You know, I've been thinking—"

"Oh no!" he smiles at his daughter; her thoughts are never conducive to peace.

"Seriously, Papa. Look: He lists treatments for everything—except for our head. If your skull is broken they don't do anything except wait for you to 'arrive at the turning point.' That's what they call it. But he says that what's inside the skull is rippled 'like molten copper' and that it 'throbs and flutters.' And that the wound there will—basically—kill the person."

"So?" That's her brother, impatient.

"Well, isn't it strange that our heads should be filled with something that throbs and flutters and will kill us if it's wounded but is of no consequence—"

"Nobody said it was of no consequence, he repeats that the injured man cannot move his neck—"

"But they treat it like it is. They don't even keep it like they keep the other things. They break through our noses and drag it all out and they throw it away—"

"Please!" her mother exclaims, putting down the plum she had been absently polishing on her sleeve. "Please! Enough!"

"But—"

"Enough, enough!" her brother cries. "Look, mama!" he rushes on. "Listen, you like the heart—" He points at the blue amulet she wears on her breast. "Well, when the ancient doctor gets to the heart he turns into a poet. Listen: 'There are vessels from the heart to every limb, and if you will put your hand on the head, on the hands, on the wrists, on the legs, you will measure the heart. For it is the case that the heart speaks to every vessel and every limb... Listen to the heart in order to learn the knowledge that comes from it.' Isn't that wonderful? That our heart speaks? And we can hear it? We can hear it now. Feel—" He leans toward his mother, and taking her hand he places her fingers in the gentle dip just above his collarbone. "There," he says, "you can feel my heart speaking. What does it say?"

The Chief Scribe watches, his eyes shining with tears as his daughter lays her fingers gently against his collarbone:

"I can hear your heart, Papa."

STORY 6

Oath

Leanne Shapton

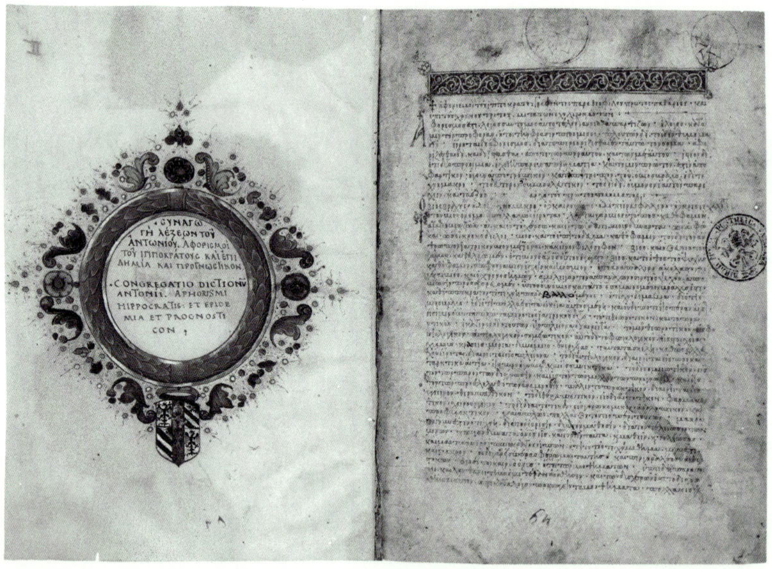

Byzantine codex reporting the text of the Hippocratic oath, 1100–1199.
Frontispiece and spread including the oath.

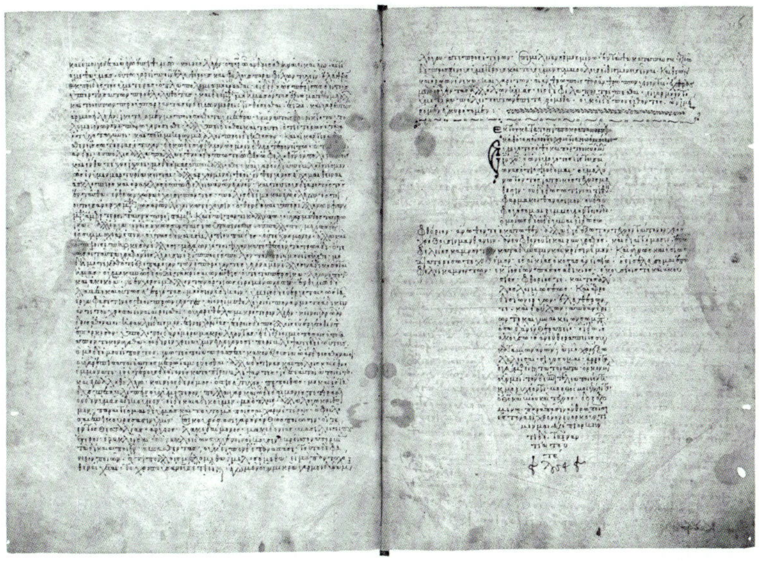

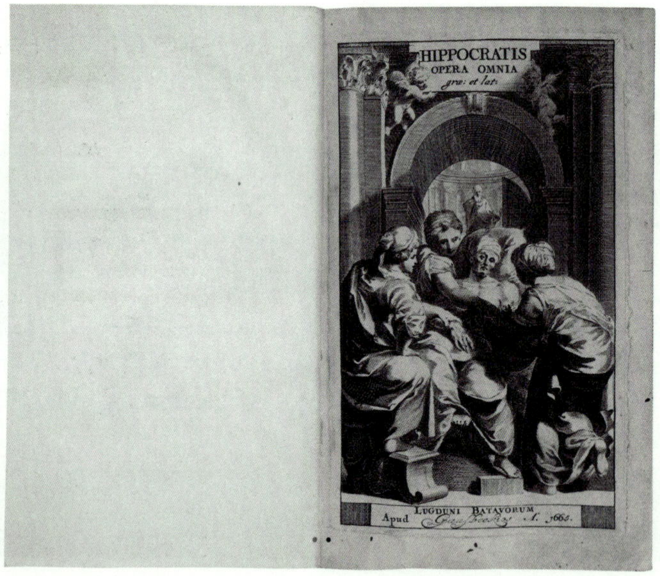

Hippocratis Opera Omnia [Hippocrates' complete works], Greek and Latin edition, Leiden, 1665. Original texts from *c.* 5th–4th century BCE.

The man was restored to life.
The woman drowned.
The man took off his ring.
The woman hung up.
First, do no harm.

I was sleeping with a man who described his marriage as ailing. One night, when I was talking to the man on the phone, he recounted a time when he had gone, with his wife, to a party at their neighbors'. He had clinked beers with the host and said, "Sorry we're late. I'm in the doghouse." The host had snorted and replied, gesturing at the man's wife, "Dude, you *live* in the doghouse."

The man used this scene as a way of complaining about his marriage. He described his wife's mean streak, her drinking, their five-year lack of sex, her bad posture, and her habit of absentmindedly picking her skin. Even though I wanted to hear it, I thought it was cruel. It was unnecessary if the marriage was in fact on its last legs. The doghouse sounded like asylum. I wondered what it felt like, to be in the doghouse, to walk around and go to parties with his wife. How bad could it be? There was probably TV and ice cream and donuts in the fucking doghouse. There was beer.

A large painting hung in the main staircase of Chandos House, a building occupied by the Royal Society of Medicine. I would stay there when I was in London for medical conferences. It was a depiction, in oils, of the moment when Emperor Alexander II of Russia, on horseback, beheld a peasant who had been pulled from the water. From his saddle, the emperor reaches toward the sodden man, whose skin is painted the same white as his shirt. He is being propped upright by three men, faces dark with pleading and with fear.

The plaque read: *The Emperor, seeing the apparently lifeless body of a peasant which had been dragged from the river Wilma [Vilnia], dismounted from his horse and, after more than three hours' constant personal efforts, succeeded in restoring the man to life.*

A person in a doghouse speaks in a distinct pitch. Maybe it's only perceptible to other people in other doghouses. And to the spouses that banish them there. Eventually I picked up on it. I stopped watching *The Simpsons* with my daughter.

I asked the man why, if the marriage was in fact over, he did not tell his wife he was in love with another woman. "It would obliterate her," he said, and then he quoted from the Hippocratic Oath, which I had recited to him, also a doctor, the night we met: "First, do no harm."

I thought this over.

The man told me he was spending nights at his father-in-law's beach house while he and his wife separated. He sent my daughter books that his own children had loved when they were small: One about a giant jam sandwich. Another about spelunking. These books arrived in the mail; my daughter enjoyed them.

In the water opposite the father-in-law's beach house, 152 years after Alexander II resuscitated the Lithuanian man, a woman had drowned. On calm days, the wind made a fine grain on the surface of the water where the drowning occurred, where one nine-year-old friend dared her six-year-old friend to go into the water wearing her school uniform.

Cy Twombly, the son of a swim coach, painted mad wet thrashings. Flaubert, the son of a physician, avoided conclusions. In Dostoevsky's *The Idiot*, Prince Myshkin says of the ecstasy felt at the onset of an epileptic fit: "I would give my whole life for this one instant."

I fainted, just once. Feeling dizzy in a Toronto bar, I pushed outside for some air, took a few steps. I was wearing a pink silk velvet dress, sleeveless, mini, and then knew I was falling, knew the dress was short, knew how hard my head would hit the pavement, and knew, in a blissful, euphoric surrender, that none of it could be helped. I knew, in those five-tenths of a second, complete happiness.

When he visited me, the man did not wear his ring. But once, on a video call, I saw a glint on his left hand. "Oh, I'm wearing it for Marco," he said. Marco was a colleague of his wife's. "I saw Marco tonight and I want to protect her dignity." Then he took it off in a show of something.

One night, as the man was speaking to me, his wife came into the room, grabbed his phone, and hit him on the head with it. On my end of the line, I could hear muffled yelling and static. I listened for a few moments, then hung up. I looked up the oath.

I cut and pasted one section of what I'd found into a text to the man: *I will follow that system of regimen which, according to my ability and judgment, I consider for the benefit of my patients, and abstain from whatever is deleterious and mischievous.*

He did not reply or return calls. I became agitated as two days passed. I worried; I was impatient with my daughter. I sent the man another text, one I thought might be less ambiguous. It was from an older version of the oath that had since been amended: *I will give no deadly medicine to anyone if asked, and in like manner I will not give to a woman a pessary to produce abortion.*

Later that afternoon, while I was riding the subway, a woman sitting across from me started shaking violently. Her eyes rolled back; she slid off her seat and onto the floor as the train slowed, pulled into a station, and stopped. Her straight black bangs swung from side to side. A passenger rushed in and cradled the woman's head to keep it from banging the floor, and another ran for help. I asked if there was anything I could do. "No," the other passenger said. I left the car, glancing at the shaking woman through the subway window, wildly animated but insensate. She and I looked to be the same age, and I shuddered, thinking it was supposed to be me, thinking of the pre-Hippocratic belief that God smote seizures upon sinners. That they were "cured" by incantation. I needed an incantation. As I walked the extra blocks home, I repeated: *first do no harm, do no harm.*

On the third day, the man called and explained that he had checked himself into a psychiatric ward. "I was worried I would hurt myself," he said. "I want to spend the rest of my life with you."

When I looked it up, years later, it came as a surprise to me that there was no relation between Hippocratic and hypocrite. I used to believe the definition of hypocrisy was to have Hippocratic intentions that were consciously, deliberately subverted.

STORY 7

Um guia para a confecção de corpos

Daniel Galera

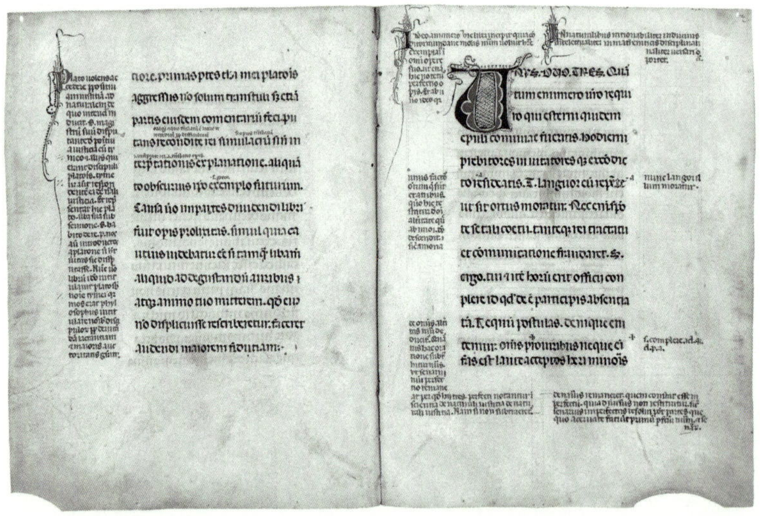

Platonis Timaeus a Chalcidio latino sermone translatus. Platonis Phaedo a Henrico Aristippo Latino sermone translatus [Plato's *Timaeus* and *Phaedo* translated into Latin by Calcidius and Henricus Aristippus], Italy, 14th–15th century. Original texts from 4th century BCE.

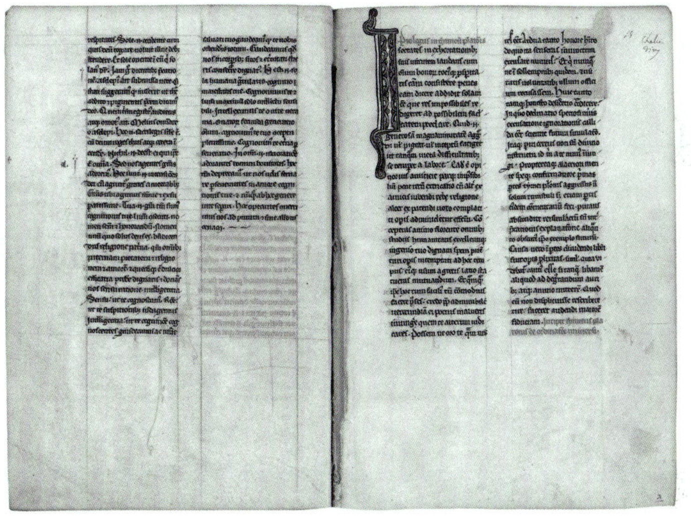

Timaeus-interpretatio latina Chalcidii [The *Timaeus* translated into Latin by Calcidius], 14th–17th century. Original text from 4th century BCE.

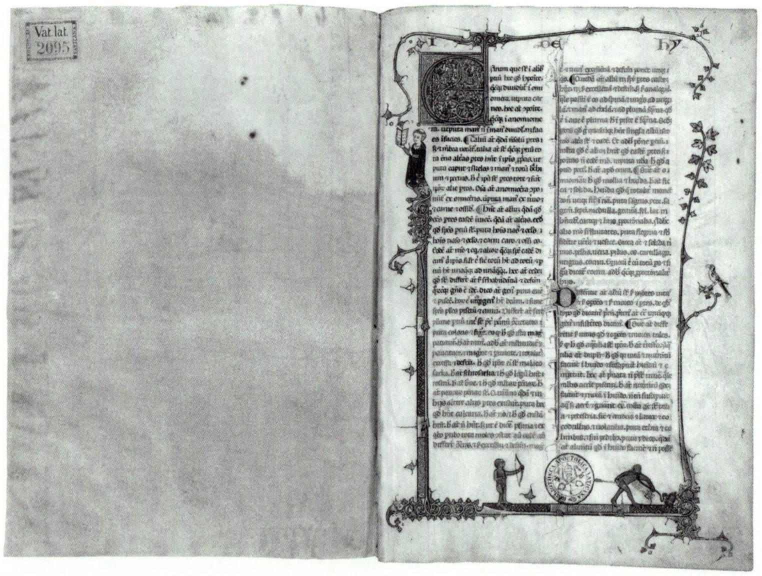

Liber Aristotelis de Historiis animalium [Aristotle's *History of Animals*], Latin translation, Italy, 13rd century. Original text from 4th century BCE.

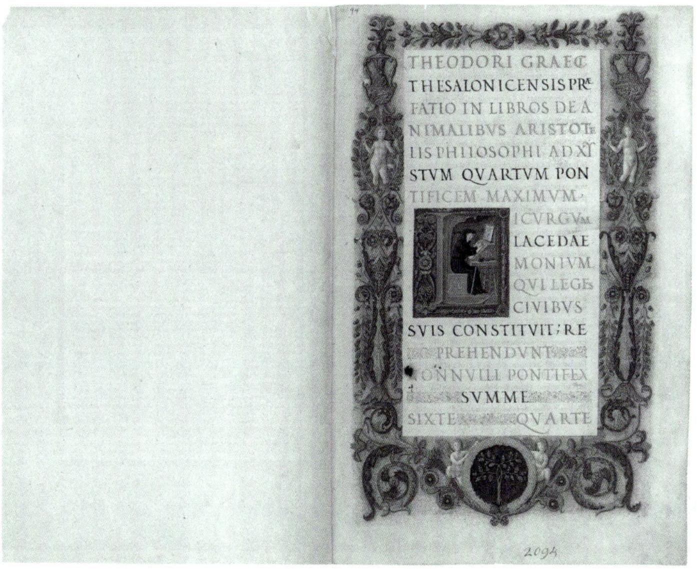

Aristotelis De animalibus libri [Aristotle's *Book of Animals*], Latin translation, Italy, 15th century. Original text from 4th century BCE.

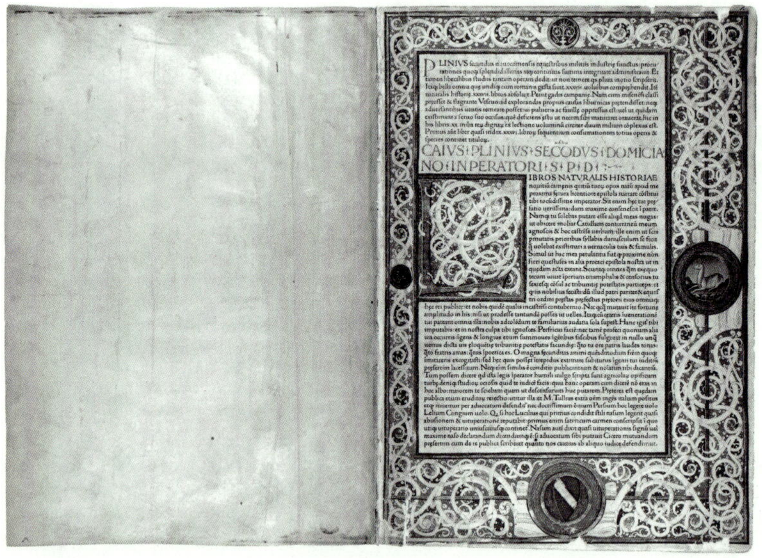

Pliny the Elder, *Naturalis historia* [Natural History], Venice, 1469.
Original text from *c.* 77–78 CE.

Preâmbulo

A mente humana necessita de um corpo humano e só reconhece a si mesma na medida em que é afetada pelas relações desse corpo com corpos exteriores. Os avanços alcançados por nossa companhia em neurociência e tecnologias da informação nos permitiram escanear, codificar e armazenar a totalidade dos dados contidos no cérebro de um indivíduo. Nosso escopo de atividades não inclui, todavia, a confecção de corpos para abrigar as mentes digitalizadas de nossos clientes. A lista que segue contém sugestões de três abordagens para a confecção de tais corpos, lembrando que do artefato ou organismo resultante não se poderá afirmar que é provido, necessariamente, de uma mente humana, embora seja seguro afirmar que estará dotado de *alguma* mente, derivada, mas não circunscrita, ao conteúdo da mente humana. Essa ressalva consta do termo de isenção assinado pelos clientes.

Corpo 1

Um dispositivo eletrônico compacto com disco rígido de vidro calcogênico, capaz de armazenar os dados da mente em um volume compatível com uma pequena lanterna. O tamanho limitado implica também uma limitação de canais de comunicação sensorial, que ficariam reduzidos aos estímulos sonoros: um sensor de escuta e um alto-falante que poderá estar regulado para imitar, tanto quanto possível, o timbre e a entonação da voz humana do cliente. Podemos esperar de tal corpo a produção de uma mente severamente debilitada em termos emocionais, porém com agudas capacidades lógicas e algorítmicas. O que ela ficará devendo em termos de empatia poderá ser compensado pela adequação do *self* a um mundo cada vez mais matematizado e mecânico, desligado das imprecisões orgânicas da carne. Platão afirmou no *Timaeus* que a alma divina e imortal, o *logos*, está alojada na cabeça e tem o corpo inteiro como servo. Se assim for, nessa abordagem deverá se observar com cautela o comportamento de um *logos* sem o restante do corpo, ou seja, sem partes irracionais e irascíveis para serem comandadas. Um corpo tão desprovido de órgãos – e de organismos dentro de organismos – poderá despertar na mente digital uma inclinação a duvidar da própria existência. Haverá risco elevado de terror psíquico mascarado pelas limitadas opções emotivas do artefato.

Corpo 2

As impressoras 3D de matéria orgânica vêm sendo empregadas há muito tempo na confecção de órgãos para transplante e modificação estética. A consolidação dessas tecnologias com as redes neurais quânticas pode permitir que uma mente humana digitalizada seja hospedada em um gerador dinâmico de tecidos humanos. Nesse tipo de corpo, a mente imprimirá *em tempo real* os órgãos, tecidos e secreções que melhor expressem suas emoções ou expandam sua interação com os corpos externos. A base filosófica dessa abordagem está na combinação de cardiocentrismo e encefalocentrismo encontrada nos escritos de Plínio, o Velho. O autor e naturalista romano afirma que o cérebro é a "cidadela" dos sentidos e o "regulador do entendimento"; ao mesmo tempo, observa que o coração "é a primeira víscera que se forma no feto" e que contém "os receptáculos primários do espírito e do sangue". É no coração que "a mente faz sua casa". Em um gerador dinâmico de tecidos humanos, a mente se manifesta não apenas em sua relação com o coração, mas com partes como língua, pele, pâncreas, ovário, testículo; a mente se expressa por meio do sangue, do pus, da bile, da saliva, do suor, do sêmen, da secreção vaginal. Os órgãos e efusões incham, murcham, tremem, escorrem, borbulham, se irrigam e apodrecem. O corpo está sujeito a contaminações de outros corpos: organismos dentro de organismos. Este é um corpo humano virado ao avesso, monstruoso aos instintos, mas que busca afetos e deles se cria, retroalimentando o banco de dados neural com os resultados da sua interface com outros corpos. Sua morfologia e sua linguagem tenderão a ser vistas como grotescas e escatológicas, mas convém lembrar que o corpo humano típico também se comunica através do sangue, das secreções, dos feromônios, dos aumentos e diminuições de volume. A lágrima encharca o olho, a pele eriça ao toque apaixonado, os humores fervem sob o domínio do ódio, o hálito fica empestado de tristeza. A mente se forma e se reconhece porque o coração palpita, mas quem pode acelerar ou acalmar o próprio coração apenas com a força do pensamento?

Corpo 3

Quando Aristóteles afirma que a alma está alojada somente no coração, e que o cérebro nada mais é que uma massa fria e insensível, cuja função é resfriar o excesso de calor sanguíneo gerado pelo coração, ele nos abre a possibilidade de uma consciência que é somente corpo, somente órgão, livre das amarras do "centro operacional" contido em nossas cabeças. Sob essa perspectiva, é possível questionar o papel que petabytes de dados extraídos de uma determinada coleção de neurônios - o serviço oferecido por nossa companhia - teria na composição de uma consciência. Para copiar um *self*, pode ser necessário extrair dados não somente das conexões neurais e da memória, mas também da experiência do corpo em cada um de seus componentes e em cada instante de sua história; ou seja, seria necessário nascer e viver de novo exatamente como aquele organismo. Assim sendo, é sugerida a abordagem de um *corpo narrativo*, caracterizado pela transmissão de padrões semânticos e sintáticos através de todo tipo de material. É como se houvesse uma assinatura, uma impressão, associada a cada acontecimento possível - o rumor de um riacho, um orgasmo, uma fratura exposta, a cor produzida pelas escamas micrométricas da asa de uma borboleta, a *Eneida* de Virgílio, um canudo de plástico -, e essa assinatura precisasse apenas de um ponto de vista, um passo atrás, para ser conhecida e apreciada. A consciência assim entendida não mora em lugar nenhum; seu corpo é a substância eterna e infinita que se arrepia ao sopro das causas e efeitos. Faça com os dados o que bem entender. O homem nada cria. No máximo, conhece.

GUIDE TO MANUFACTURING BODIES

Preamble

The human mind requires a human body and perceives itself only through the relationship between its body and other bodies. Our company's advances in neuroscience and information technology have enabled us to scan, code, and store the sum total of data contained in an individual brain. Our current scope of activities does not include the fabrication of bodies to serve as hosts for our clients' digitized minds. The following list contains three theoretical ways of manufacturing such bodies. Though it is important to note that the artifact or organism produced cannot necessarily be said to possess a human mind, it is nonetheless safe to say that the body is equipped with *a* mind that is derived from, but not circumscribed by, the contents of a human mind. This disclaimer appears in the waiver that all our clients are required to sign.

Body 1

A compact electronic device with a chalcogenide glass hard drive capable of storing mind data in an object no larger than a small flashlight. The restricted size would also restrict the possibilities for sensorial communication, which would be limited to acoustic stimuli: an auditory sensor and an adjustable speaker that could mimic, to the best of its abilities, the timbre and intonation of the client's human voice. Such a body could be expected to produce a mind that is emotionally impaired but that nonetheless exhibits strong logical skills and algorithmic efficiency. What the mind lacks in empathy would be compensated for by the self's adaptation to an increasingly mathematicized and mechanical world, detached from the organic murkiness of the flesh. In *Timaeus*, Plato attests that the divine, immortal soul—*logos*—is located in the head, while the entire body functions as its servant. Should that be the case, this method would require closely monitoring how the *logos* behaves without the rest of its body—that is, without any irrational, irascible parts for it to preside over. A body devoid of organs—and of organisms within organisms—might awaken in the digital mind a tendency to doubt its own existence. This method carries a high risk of psychological terror, which would be difficult to discern given the artifact's limited emotional expression.

Body 2

3D bioprinting has long been used to manufacture organs for use in transplants or plastic surgery. Incorporating quantum neural networks into this technology could make it possible for digitized human minds to be hosted in dynamic human-tissue generators. Housed in such a body, the mind would print, *in real time*, whatever organ, tissue, or secretion best conveyed its emotions or enabled further interaction with other bodies. The philosophical basis for this approach combines the encephalocentric and cardiocentric hypotheses found in the work of Pliny the Elder. The Roman author and naturalist asserts that the brain is the "citadel" of sense perception and the "regulator of understanding," while observing that the heart is "the first among the viscera that is formed in the fetus" and contains "the primary receptacles for the spirit and the blood." It is in the heart "that the mind has its abode." In a dynamic human-tissue generator, the mind would manifest not only in relation to the heart but also to other organs, such as the tongue, skin, pancreas, ovary, and testicles, while expressing itself through blood, pus, bile, saliva, sweat, semen, and vaginal secretion. The organs and effusions would swell, wither, quiver, dribble, bubble, moisten, and rot. This body would be vulnerable to contamination from other bodies: organisms within organisms. This is a human body turned inside out, monstrous to human instincts, yet one that seeks out and thrives on affection, establishing a feedback loop between neural data banks and the results of its interfaces with other bodies. Although its morphology and body language are likely to be considered scatological and grotesque, it is worth remembering that the typical human body also communicates via blood, secretions, pheromones, and increases or decreases in volume. Tears dampen our eyes, our skin tingles when touched with passion, our humors simmer under the influence of hate, our breath reeks of sadness. The mind is formed and perceives itself because the heart beats, but who out there can quicken or quieten their own heart with thought alone?

Body 3

When Aristotle claimed that the soul resides only in the heart and that the brain is just a cold, insensitive mass with the sole purpose of cooling the excess of heat that the heart generates in the blood, he put forth the possibility of a consciousness that is pure body, pure organ, untethered from the "operational center" in our heads. Such a perspective allows us to question the role that petabytes of data extracted from a specific bundle of neurons—services offered by our company—actually play in the making of a consciousness. In order to make a copy of the self, data would need to be extracted not only from neural connections and memories, but also from the experience of the body and each of its components throughout every moment of their existence; that is, it would be necessary to be born and live in the exact same way as the other organism. For this reason, we would recommend the *narrative body* method, which is characterized by the transferal of semantic and syntactic patterns through all kinds of materials. It is as if every possible event—the murmuring of a brook, an orgasm, a compound fracture, the color produced by the micrometric scales of a butterfly wing, Virgil's *Aeneid*, a plastic straw—bears a signature or an impression that simply requires a particular angle, or a little distance, for it to be recognized and appreciated. Such a consciousness resides nowhere; its body is an eternal and infinite substance that bristles under the winds of cause and effect. Do what you will with the data. Humankind creates nothing. At most, it learns.

STORY 8

Galens Seelen

Cord Riechelmann

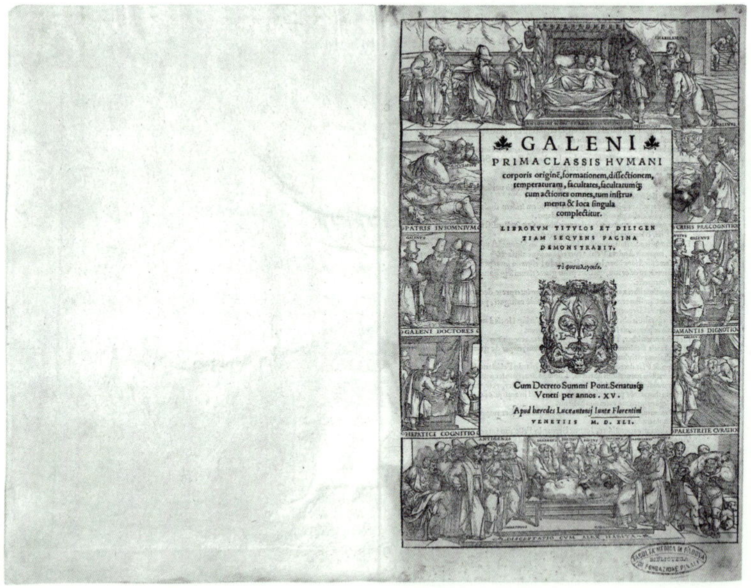
GALENI
PRIMA CLASSIS HVMANI
corporis originẽ, formationem, dissectionem, temperaturam, facultates, facultatumq́; cum actiones omnes, tum instrumenta & loca singula complectitur.
LIBRORVM TITVLOS ET DILIGENTIAM SEQVENS PAGINA DEMONSTRABIT.
τὸ φυσιολογικόν.
Cum Decreto Summi Pont. Senatusq́; Veneti per annos. XV.
Apud hæredes Lucæantonij Iuntæ Florentini
VENETIIS M. D. XLI.

Claudius Galenus, *Prima classis humani corporis originem, formationem, dissectionem, temperaturam, facultates, facultatumque cum actiones omnes* … [Description of the human body comprising origin, formation, dissection, temperament, capacities and all activities …], Venice, 1541–1542. Frontispiece, table of contents, chapter 1. Original text from 1st–2nd century.

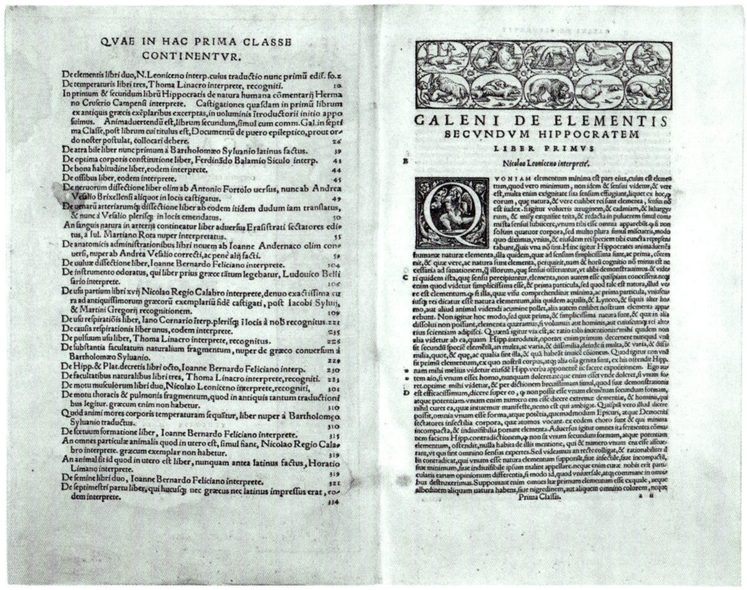

QVAE IN HAC PRIMA CLASSE CONTINENTVR.

GALENI DE ELEMENTIS SECVNDVM HIPPOCRATEM

LIBER PRIMVS

B Nicolao Leoniceno interprete.

QVONIAM elementum minima est pars eius, cuius est elementum, quod vero minimum, non idem & sensui videtur, & vere est, multa enim exiguitate sua sensum effugiunt, liquet ex hoc, quod eorum, quæ natura, & vere cuilibet rei sunt elementa, sensus nō est iudex. si igitur volueris æruginem, & cadmiam, & lithargyrum, & misy exquisite trita, & redacta in puluerem simul commista sensui subiicere, vnum tibi esse omnia apparebūt. quod si non solum quatuor corpora, sed multo plura simul miscueris, modo quo diximus, vnius, & eiusdem rei speciem tibi cuncta repræsentabunt, quāuis vna nō sint. Hinc igitur Hippocrates animaduertēs humanæ naturæ elementa, illa quidem, quæ ad sensum simplicissima sunt, ac prima, contemnit, & quæ vere, ac natura sunt elementa, perquirit, nam & horū cognitio nō minus est necessaria ad sanationem, quam illorum, quæ sensui offeruntur, vt alibi demonstrauimus. & videri quidem ista, quæ sensu percipiuntur, elementa, non autem esse quispiam concesserit. neque enim quod videtur simplicissima esse, & prima particula, sed quod tale est natura, illud vere est elementum. quod si illa, quæ visu comprehenditur minima, ac prima particula, vniuscuiusq; rei dicatur esse natura elementum, alia quidem aquilis, & Lynceo, & siquis alter homo, aut aliud animal videndi acumine pollet, alia autem cuilibet nostrum elementa apparebunt. Non igitur hoc modo, sed quæ prima, & simplicissima natura sunt, & quæ in alia dissolui non possunt, elementa quæramus, si volumus aut hominis, aut cuiuscunq; rei alterius scientiam adipisci. Quænā igitur via est, ac ratio talis inuentionis? mihi quidem non alia videtur ab ea, quam Hipp. introduxit, oportet enim primum decernere nunquid vnū sit secundū speciē elemētū, an multa, ac varia, & dissimilia, deinde si multa, & varia, & dissimilia, quot, & quæ, ac qualia sint illa, & quā habeāt inuicē cōionem. Quod igitur non vnū sit primū elementum, ex quo nostrū corpus, atq; alia oīa genita sunt, ex his ostendit Hipp. nam mihi melius videtur eiusdē Hipp. verba apponentē sic facere expositionem. Ego autem aio, si vnum esset homo, nunquam doleret. neque enim esset vnde doleret, si vnum foret. optime mihi videtur, & per dictionem breuissimam simul, quod sine demonstrationis est efficacissimum, dicere super eo, quod non possit esse vnum elemētum secundum formam, atque potentiam. vnum enim numero ens esse dicere extremæ dementiæ, & hominis, qui nihil curet ea, quæ intuemur manifeste, nemo est qui ambigat. Quispiā vero illud dicere possit, omnia vnum esse forma, atque potētia, quemadmodum Epicuri, atque Democriti sectatores insectilia corpora, quæ atomos vocant. ex eodem choro sunt & qui minima incompacta, & indiuisibilia ponunt elementa. Aduersus igitur omnes ita sentientes cōmunem faciens Hipp. contradictionem, quod non sit vnum secundum formam, atque potentiam elementum, ostendit, nulla habita de illis mentione, qui & numero vnum ens esse asseuerant, vt qui sint omnino sensus expertes. Sed videamus an recte colligat, & rationabiliter illis contradicat, qui vnum esse natura elementum supponūt, siue insectile, siue incompactū, siue minimum, siue indiuisibile ipsum malint appellare. neque enim curæ nobis erit particularis earum opinionum differentia, si modo id, quod vniuersale, atq; commune in omnibus destruxerimus. Supponunt enim omnes hæ primum elementum esse exquale, neque albedinem aliquam natura habens, siue nigredinem, aut aliquem omnino colorem, neque

Prima Classis. a ii

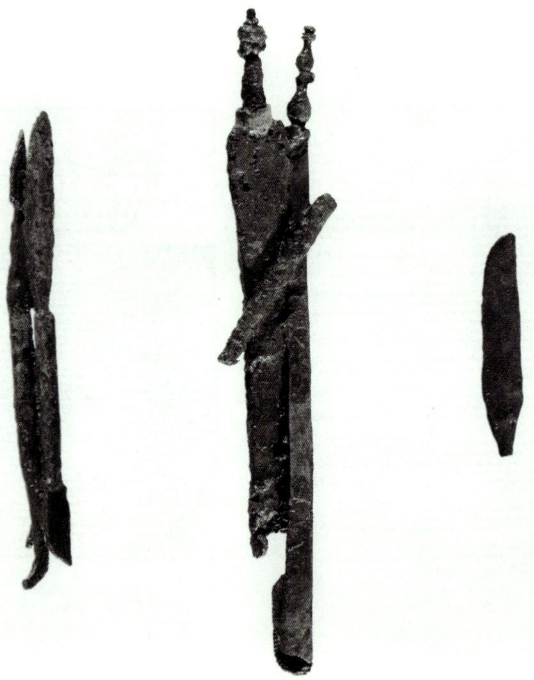

Set of surgical instruments from the Surgeon's Domus in Piazza Ferrari, Rimini, second half of the 2nd century.

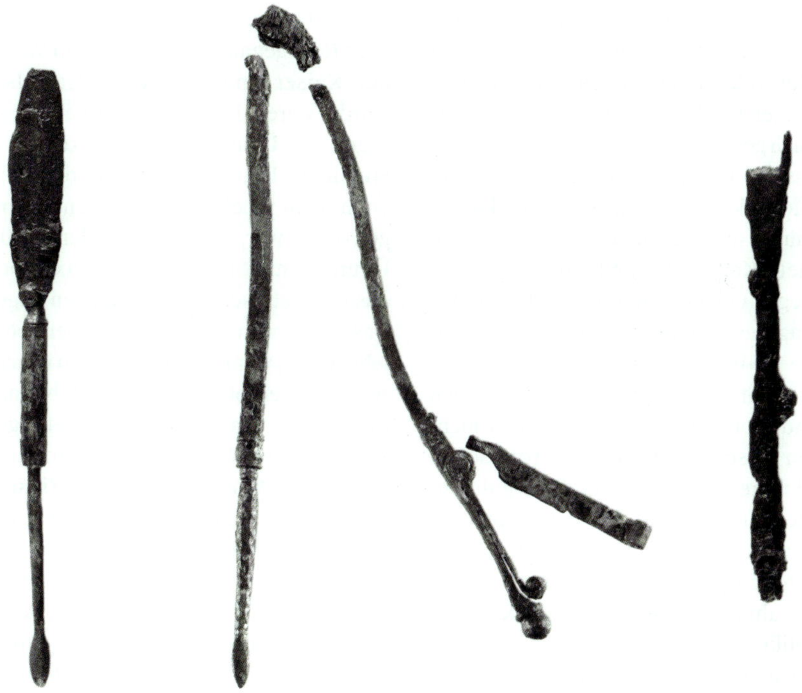

Es war im 2. Jahrhundert unserer Zeit nicht nur für Ärzte ein Vorteil, dass das Mitleid noch nicht als ein allgemein erstrebenswerter Affekt von Christen in der Nachfolge des heiligen Paulus in die römische Gesellschaft getragen worden war. Denn so konnten in den Arenen von Rom oder Pergamon nicht nur Tiere wie Löwen oder Hunde aufeinander gehetzt werden bis das Blut floss. Auch Menschen kämpften in den Arenen gegeneinander oder auch gegen afrikanische Großkatzen wie Löwen, Leoparden und seltener auch gegen schnell-schlanke Geparden.

Dass die sogenannten Tierhetzen und Gladiatorenkämpfe nicht nur oberflächliche Hautrisse oder blaue Augen, sondern auch tiefe Körperein- und aufschnitte wie freigelegte Oberschenkelmuskel und eingeschlagene Schädel mit sich brachten, gehörte zu ihrem Spektakel. Die offenen und nicht selten spritzend blutenden Wunden der Gladiatoren und Tiere waren aber nicht nur den Zuschauern aus dem römischen Establishment Schauder und Vergnügen, sie zogen auch wissenschaftlich inspirierte Geister wie den Arzt Galenos von Pergamon (ungefähr 129-199 n. Chr.), oft kurz nur Galen genannt, ähnlich magisch an wie die Kämpfer ihr Publikum.

Als Galen im Jahr 157 n. Chr. in Pergamon begann, als Gadiatorenarzt zu arbeiten, trafen zwei Linien der griechisch-römischen Antike in einer Person zusammen. Es waren das philosophische Denken der Natur der Griechen und die veränderte Stellung des Arztes in den Oberschichten der römischen Gesellschaft.

Ärzte lebten in der römischen Oberschicht nicht mehr allein von ihrer Heilkunst und ihrem Ruf. Sie mussten beides immer wieder, wollten sie ihre Stellung als „Modearzt" behalten und ihre Kundschaft mehren, in öffentlichen Sektionen und Vivisektionen an verschiedenen Tieren wie Schafen, Hunden, Schweinen und vor allem Affen auch demonstrieren. Dabei waren diese Show-Sektionen alles andere als ein Spiel: Eine misslungene Vorführung konnte den Arzt um seinen Ruf und damit um sein Einkommen bringen.

Für Galen, der, als er 161 von Pergamon nach Rom gegangen war, dort zum Modearzt der Kaiser und römischen Gesellschaft wurde, waren die Gladiatorenkämpfe sozusagen eine Übungs- und Fundgrube in jeder Hinsicht. Denn auch wenn das Mitleid noch nicht zu einem geachteten Vermögen geworden war, so hatte es in der ärztlichen Übungs- und Demonstrierpraxis bereits das erste ethische Stopp-Zeichen gegeben. Das Sezieren menschlicher Körper, toter wie lebendiger, war fast überall schon in griechischer Zeit verboten worden und es betraf auch die Körper von Sklaven und Gefangenen. Die verletzten Körper der Gladiatoren boten Galen also Möglichkeiten der Einsicht in das Körpergeschehen, die er sonst nur an Tierkörpern hätte gewinnen können.

Und sein umfassendes medizinisches Werk – er verfasste mit um die 200 Schriften ein Werk enzyklopädischen Ausmaßes, das mit seiner Lehre zur Anatomie und Physiologie des menschlichen Körpers bis ins 17. Jahrhundert die gesamte Heilkunde nicht nur im europäischen Raum bestimmte – ist voll von Beschreibungen und Erkenntnissen, die sich ohne große Phantasie auf seine Erfahrungen mit verletzten Gladiatoren beziehen lassen. So hatte Galen erkannt, dass Blut in Arterien und Venen fließt und eindringlich beschrieben, wie natürliche und animalische Lebenskräfte wie Ebbe und Flut durch die Blutgefäße strömen. Das Bild vom im Rhythmus der Herzschläge aus den aufgeschlagenen Adern eines verletzten Kämpfers schießenden Blutstroms ist da dann nicht mehr weit.

Dass Galen aber aus den Arenen der Gladiatoren in die Paläste der römischen Oberschicht wandern konnte, hatte neben seinem Talent auch mit dem gewandelten gesellschaftlichen Status der Kämpfe zu tun. Waren die Kämpfe bis um die Zeitenwende im Unterschied zu Wagenrennen und Tierhetzen eine rein private Angelegenheit, veranstaltet von wohlhabenden Bürgern, um ihren Ruhm zu mehren, so wurden sie besonders unter Kaiser Augustus zu einem öffentlichen Spektakel. Und das hatte mehrere für Galen äußerst günstige Folgen. Waren die ersten Gladiatorengenerationen ausschließlich aus Sklaven und Gefangenen rekrutiert worden, begaben sich im 1. Jahrhundert n. Chr. zunehmend ehemals freie Bürger in den Gladiatorenstand. Zusammen mit den jetzt öffentlich, von staatlicher Seite organisierten Kämpfen änderte das weniger den Status als vielmehr die Lebensqualität der Gladiatoren. Gladiatoren im „öffentlichen Dienst" mussten höchstens dreimal im Jahr kämpfen, wurden die übrige Zeit aber gut versorgt. Und diese gute Versorgung bezog sich neben der Ernährung eben auch auf die ärztliche Behandlung. Die ärztliche Betreuung der Gladiatoren nicht nur während der Kämpfe galt als hervorragend.

Kein Wunder also, dass Galen seinen Dienst an den Gladiatoren nie als bedrückend oder gar unwürdig empfand. Was er an den Verletzten sah, war allerdings auch in jeder Beziehung von Sensationen durchzogen. Gladiatoren, die, schwer an Beinen und Armen verletzt, noch vom Schockzustand des Körpers schmerzfrei gehalten worden waren, konnten willentlich ihre Bewegungen nicht mehr steuern und darüber mehr verzweifeln als über ihre schrecklichen Wunden. Auf der anderen Seite konnten Hirnverletzungen die über den bewussten Willen gesteuerte Muskelbewegung behindern oder unterbrechen. Das waren Beobachtungen, die auf eine Verbindung zwischen dem Gehirn und Muskeln hindeuteten, die Galen hundertfach an Gladiatoren gemacht hatte. Damit war er ganz nah an einer der bis heute drängenden Fragen, nämlich der Frage nach dem Sitz der Seele.

Er hatte, da gab es jetzt keinen Zweifel mehr, das Gehirn als Sitz jener Seele, der Tierseele, des *spiritus animalis*, erkannt, die die Bewegung möglich macht und so steuert, dass ein den Maßgaben der Umgebung geschuldetes adaptives Verhalten möglich ist.

Damit hatte Galen einen Teil der griechischen Drei-Seelen-Lehre, wie sie Platon und Aristoteles vertraten, im Gehirn verortet. Wobei Galen bereits durch seinen Vater Nikon, einen in Mathematik und Grammatik bewanderten Architekten, bestens in die griechische Drei-Seelen-Lehre eingeführt war. Nach dieser Lehre gibt es drei Seelen, eine vegetative Seele, die *anima vegetativa*, die Wachstum, Entwicklung und Erregbarkeit durch Umweltreize möglich macht. Pflanzen haben nur diese eine Seele, während Tiere und Menschen auch noch eine Tierseele, eine *anima animalis* oder einen *spiritus animalis*, haben, die Bewegung, adaptives Verhalten und – zumindest bei Aristoteles – auch Intelligenz hervorbringt. Die dritte Seele in diesem Bund, die Vernunftseele, schrieb dieses Modell allein dem Menschen zu.

Galen hatte aber nicht nur den Zusammenhang zwischen den Muskelbewegungen eines Tieres oder Menschen und dem Gehirn erkannt. Er hatte auch erkannt, dass das Gehirn das Ziel aller fünf Sinne ist.

Alle durch Berührung, Schmecken, Riechen, Sehen und Hören gewonnenen Sinneseindrücke würden ins Gehirn gelenkt und letztlich dort koordiniert. Woraus Galen schloss, dass das Gehirn auch der Sitz des gesunden Menschenverstandes, der Koordination der Alltagserfahrung, war. Dadurch hatte er das Gehirn zum Steuerungsorgan von motorischen Handlungen, kognitiven Fähigkeiten und letztlich auch der Intelligenz gemacht. Das alles sind Annahmen, die bis heute in der Hirnforschung ihre Spuren hinterlassen. Es ist dabei zwar nur noch eine kuriose Fußnote, dass Galen meinte, zwischen den harten, für die Motorik zuständigen Nerven, und den feinfaserigen, weichen Nerven für die Sinne unterscheiden zu können. Zumindest in der Alltagssprache haben aber auch die harten Motorik-Nervenstränge Galens ihre Früchte hervorgebracht: Wenn man heute von besonders unerschrockenen Menschen sagt, sie hätten „Nerven aus Stahl“, dann ist der Grund dafür in Galens Texten zu finden.

Harte Nervenstränge aus Stahl hatte Galen aber zumindest in einem Fall seiner Studien- und Versuchsobjekte nicht: bei den Affen. Affen waren allgemein die beliebtesten Versuchs- und Demonstrationstiere im Rom seiner Zeit. Galen aber wollte sie partout nicht in seinen Vorführungen einsetzten. Was, wenn man weiß, dass viele Tiere für die Versuche ertränkt wurden, kein Wunder ist. Aber das ist noch kein Grund, Affen davon auszunehmen, das Ertränken von Schweinen oder Hunden wird nicht weniger herzzerreißend abscheulich gewesen sein. Galen dachte nur entschieden anders über Affen als seine Zeitgenossen. Für Galen hatten Affen „schöne“ Seelen und das ließ ihn vom „abscheulichen Schauspiel“ ihrer öffentlichen Vivisektion Abstand nehmen.

GALEN'S SOULS

In the second century of the Common Era, the fact that compassion had not yet been introduced into Roman society as a generally desirable quality by Christians following the teachings of Saint Paul was an advantage, and not just for doctors. It meant that in the arenas of Rome and Pergamon animals such as lions and dogs could be baited and set against one another and the blood would flow. Humans also fought against each other in these arenas, or against African big cats such as lions, leopards, or, more rarely, sleek and speedy cheetahs.

An integral part of the spectacle of these so-called animal chases and gladiator fights was that they resulted not simply in superficial scratches or black eyes, but in deep gashes and cuts to the body, such as torn and exposed upper-thigh muscles or smashed skulls. The open and frequently bloody wounds of the gladiators and the animals were, however, more than just entertainment for the thrill-seeking audience of the Roman establishment; they attracted scientifically inspired minds in just the same magical way that the fighters drew their crowds. One such mind was the physician Galenos of Pergamon (*c.* 129–199), commonly known by his shorter name: Galen.

When Galen began his career as a doctor to gladiators in Pergamon, in 157, two aspects of Greco-Roman antiquity were united in one person: on the one hand, the philosophical approach of the Greeks to nature, and on the other, the new status of the figure of the physician in the upper echelons of Roman society.

Physicians among the Roman upper class no longer lived solely from their healing arts and their reputation. They had to demonstrate both these things continuously—if they wanted to retain their status as "fashionable doctors" and increase their clientele—by conducting public dissections and vivisections on a variety of animals, such as sheep, dogs, pigs and, above all, monkeys. Yet these show dissections were anything but a game: One botched demonstration could cost the doctor his reputation and consequently his income stream.

For Galen, who became the favored physician to the emperor and to Roman society when he left Pergamon for Rome in 161, the gladiator fights represented in many respects a sandbox in which to practice and make discoveries. Even though compassion had not yet risen to the level of a respected virtue, practices of medical trial and demonstration had nevertheless already developed their first ethical red lines. Dissecting the human body, dead or alive, was forbidden almost everywhere in the ancient Greek world, and this applied to the bodies of slaves and prisoners as well. Injured gladiatorial bodies thus offered Galen opportunities to see into the workings of the body, which he would otherwise only have been provided by the bodies of animals.

Galen's comprehensive medical work—he composed around two hundred medical texts, representing an oeuvre of encyclopedic dimensions and theories on the anatomy and physiology of the human body that would go on to shape all the healing arts into the seventeenth century, and not only in the European world—is full of descriptions and insights that can, with only a small mental leap, be traced back to his experiences with wounded gladiators. For example, Galen recognized that blood flowed through arteries and veins and described in detail how natural and animal life forces moved through blood vessels as if with ebb and flow. It is not hard to imagine that this idea stemmed from the image of blood spraying in the rhythm of the heartbeat from the open veins of a wounded fighter.

It was not just his talent but also the changing social status of the fights that allowed Galen to walk out of the gladiatorial arena into the palaces of the Roman upper crust. While until the turn of the millennium the fights were, in contrast to chariot racing and animal baiting, purely private affairs staged by wealthy citizens to increase their fame, under Emperor Augustus in particular they became public spectacles. This had several extremely positive consequences for Galen. While the first generations of gladiators were recruited exclusively from among slaves and prisoners, in the first century more and more formerly free citizens entered the gladiatorial profession. Combined with the move to public, state-organized fights, the impact of this was evident less in the status of the gladiators than in their quality of life. Gladiators working in "public service" would have to fight at most three times a year, but were well provided for the rest of the time. And this could be seen in their diet and in their medical treatment. The medical care of gladiators, even outside the arena, had an excellent reputation.

So it is no wonder that Galen never felt his service to the gladiators to be onerous or dishonorable. Yet what he witnessed in the injured men was itself sensational in every respect. Gladiators who had severe injuries to their legs and arms but were nevertheless impervious to pain due to their body's state of shock could no longer control their movements—and were more distressed by that than by their dreadful wounds. Furthermore, brain injuries could disrupt or prevent consciously controlled muscle movement.

Such observations suggested a link between the brain and the muscles, and Galen saw this hundreds of times in the gladiators. And this brought him ever closer to one of the most compelling questions of history to date: Where does the soul reside?

There is no longer any doubt that he identified the brain as the seat of that soul, the animal soul, the *spiritus animalis*, which makes movement possible and controls it in such a way as to allow behavior that is adaptive and suited to the environment.

Galen thus located a part of the Greek doctrine of the tripartite soul, as expressed by Plato and Aristotle, in the brain. Galen had been taught about the tripartite soul by his father, Aelius Nicon, an architect, mathematician, and grammarian. According to this doctrine, there are three souls. The vegetative soul, the *anima vegetativa*, allows growth, development, and sensitivity to environmental stimuli–plants have only this one soul. Animals and humans also have an animal soul, the *anima animalis* or *spiritus animalis*, which produces movement, adaptive behavior, and–at least according to Aristotle–intelligence. The third soul in this triumvirate is reason, which is attributed solely to humans.

However, Galen had not only recognized the connection between the muscle movements of an animal or human and the brain. He had also recognized that the brain is the target of all five senses. All sensory impressions gained through touch, taste, smell, sight, and hearing are directed to the brain and coordinated there. From this, Galen concluded that the brain was also the seat of common sense, where everyday experience is handled. His conclusion made the brain the controlling organ of motor functions, cognitive abilities, and, ultimately, intelligence. These are all assumptions that have left their mark on brain research to this day. The fact that Galen thought he could distinguish between the hard nerves responsible for motor functions and the fine-fibered soft nerves responsible for the senses is now only a curious footnote to the story. In everyday language, though, Galen's hard motor nerves have borne fruit: When, today, particularly intrepid people are described as having "nerves of steel," the reason for that can be found in Galen's texts.

However, in at least one area of his research and experimentation, Galen did not have nerves of steel at all: monkeys. In general, in Rome at the time, monkeys were the most popular animals for use in experiments and demonstrations. Even so, Galen refused point-blank to use them in his demonstrations. And it is no wonder: Many animals were drowned for use in experiments. Nevertheless, that was not the reason to give special treatment to monkeys, as drowning pigs and dogs would have been just as heartbreakingly dreadful. In fact, Galen thought very differently than his contemporaries about monkeys. For Galen, monkeys had "beautiful" souls, and that caused him to distance himself from the "dreadful spectacle" of their public vivisection.

STORY 9

老童生

Sheng Keyi

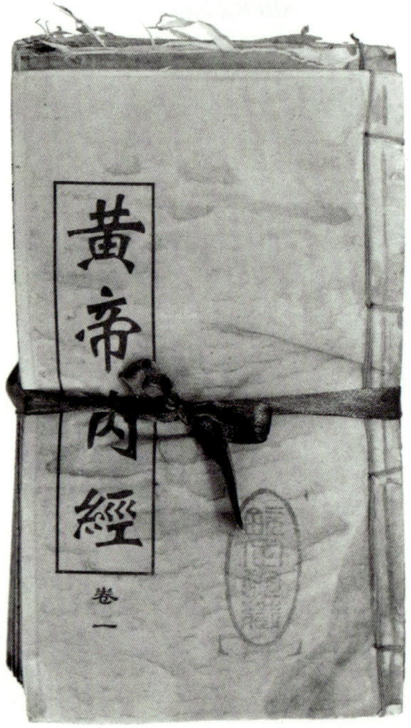

Huangdi Neijing [The yellow Emperor's classic of internal medicine], vol. *Su Wen* [Basic questions], medical text, China, 19th century. Original text from *c.* 3rd century BCE–3rd century CE.

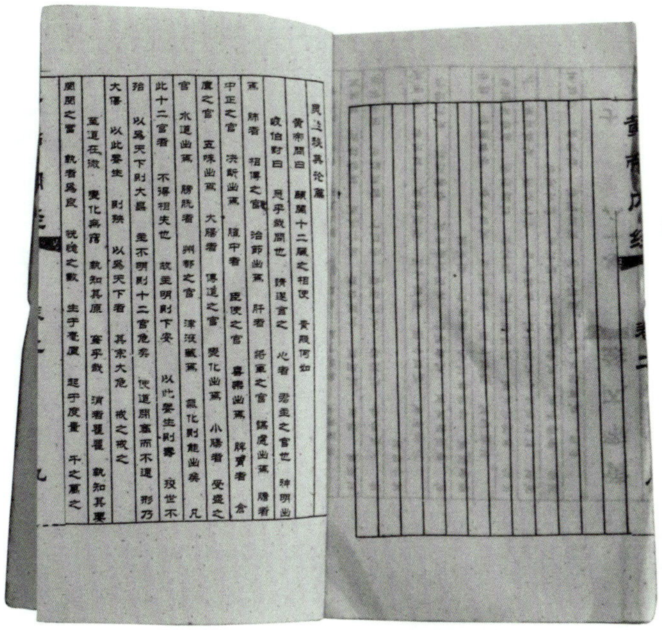

靈蘭秘典論篇
黃帝問曰 願聞十二藏之相使 貴賤何如
岐伯對曰 悉乎哉問也 請遂言之 心者 君主之官也 神明出
焉 肺者 相傅之官 治節出焉 肝者 將軍之官 謀慮出焉 膽者
中正之官 決斷出焉 膻中者 臣使之官 喜樂出焉 脾胃者 倉
廩之官 五味出焉 大腸者 傳道之官 變化出焉 小腸者 受盛之
官 水道出焉 膀胱者 州都之官 津液藏焉 氣化則能出矣 凡
此十二官者 不得相失也 故主明則下安 以此養生則壽 歿世不
殆 以為天下則大昌 主不明則十二官危 使道閉塞而不通 形乃
大傷 以此養生則殃 以為天下者 其宗大危 戒之戒之
至道在微 變化無窮 孰知其原 窘乎哉 消者瞿瞿 孰知其要
閔閔之當 孰者為良 恍惚之數 生于毫釐 起于度量 千之萬之

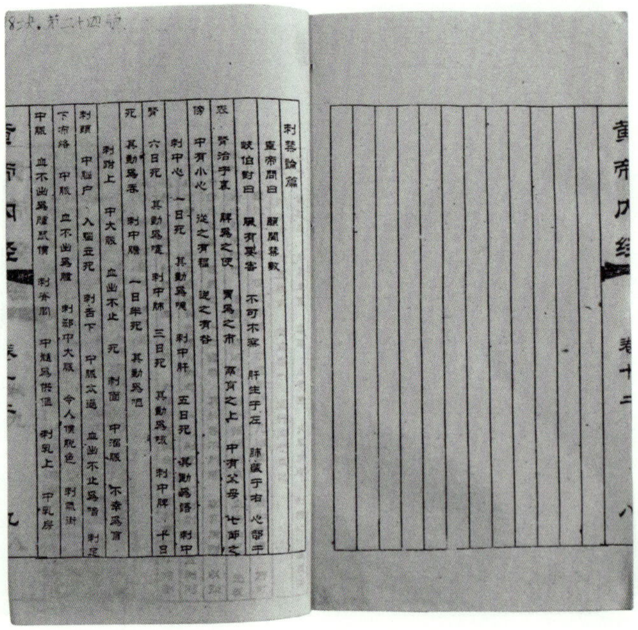

Huangdi Neijing [The Yellow Emperor's Classic of Internal Medicine], vol. *Suwen* [Basic Questions], medical text, China, 19th century. Original text from *c.* 3rd century BCE–3rd century CE.

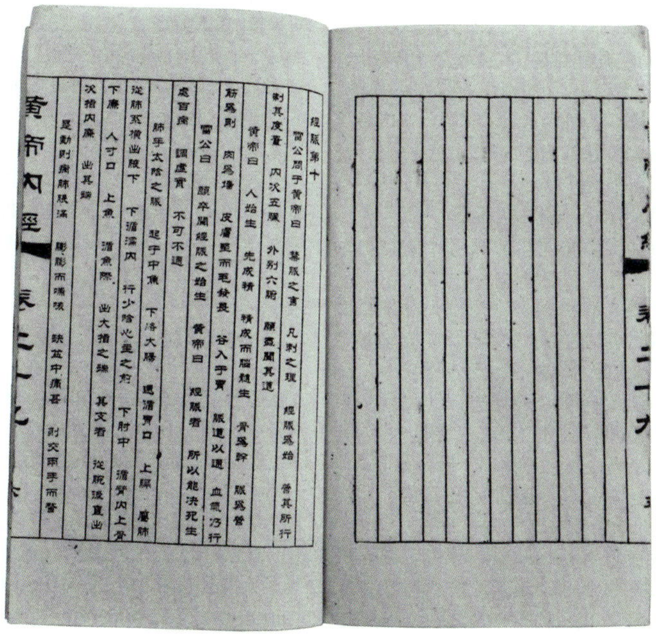

經脈第十

雷公問于黃帝曰　禁脈之言　凡刺之理　經脈爲始　營其所行
制其度量　內次五臟　外別六腑　願盡聞其道
黃帝曰　人始生　先成精　精成而腦髓生　骨爲幹　脈爲營
筋爲剛　肉爲墻　皮膚堅而毛髮長　谷入于胃　脈道以通　血氣乃行
雷公曰　願卒聞經脈之始生　黃帝曰　經脈者　所以能決死生
處百病　調虛實　不可不通
肺手太陰之脈　起于中焦　下絡大腸　還循胃口　上膈　屬肺
從肺系橫出腋下　下循臑內　行少陰心主之前　下肘中　循臂內上骨
下廉　入寸口　上魚　循魚際　出大指之端　其支者　從腕後直出
次指內廉　出其端
是動則病肺脹滿　膨膨而喘咳　缺盆中痛　甚則交兩手而瞀

黃帝內經　卷之七

老童生，湘西人，少时立志为官参政，造福百姓，三次应童子试，均告失败，得名老童生。其早年事迹，知者寥寥。晚年淡泊独居，两耳不闻窗外事，研读古书，自斟自饮，时而低声吟诵，“天有阴阳，人有脏腑；天有五行，人有五脏；人与天地相参，与日月相应”，街坊闻之，讥笑其胡诌自吃，被童子试考坏了脑。

老童生言谈好掉书袋，使人如坠云里雾里，且口无遮拦，谈论金木水火土，时令气候，衰旺盛克，对生者发出预言死期，均获应验，人渐觉其神。甚而有传，老童生肉眼可窥见人之脏腑，内里病灶症结，一目了然。

遂有人上门求助。幼儿拉稀，久治无效，老童生医其母，称其乳汁虚寒。人患浓疮毒疔，遭蜈蚣毒蜘蛛咬，其嚼烂草叶，敷于患处，立竿见影。登门求医者，几枚鸡蛋，一包冰糖，甚或鞠躬为酬，老童生余生把脉观色，察病究源，自得其乐。

人赞其医术高明，他曰当谢古人智慧。

某日，一面容憔悴女教师来访，言其骨盆腹痛已久，去医院体检验血，均无异常，受病痛困扰已久，终日焦虑，苦不堪言。老童生观其色，听其脉，吩咐其次日清晨空腹再来，因晨时人阴气未被扰动，阳气尚未耗散，气血未受到扰乱，诊察脉象更为准确。

翌日，女教师踏露复至。老童生把脉听诊，不谈病情，反倒沏茶闲聊，问及成长，家庭与婚姻。女教师答问颇为犹疑。老童生自抽屉拿出一册焦黄旧书，称祖传医著，本当秘藏，但今破例示人，言其六十岁研习此经，今已二十余载，知识皆来自此书。

“人始生，先成精，精成而脑髓生。诸髓者皆属于脑。人的记性皆在脑中。脑为元神之府。”老童生熟读经书，随口而诵，又谈及脑与知觉的关系，以及由这种关系引起的精神变化，“你的症结，这部《黄帝内经》皆有解答。早年所经历的创伤，储积在大脑里，若你若不坦诚相告，释放积郁，病邪就不能制服。”

女教师终吐露隐情。十五岁遭叔父强暴、性虐，时间持续半年之久，未敢告知家人，秘藏于心，如负巨石。

女教师问：“既然病症在精神，为何却疼在骨盆？”

“人的全部心理活动是由人脑支配。肺主一身之气，而调节全身活动。肝主怒。胆维护心，传布心的喜乐。脾和胃是饮食的受纳和布化，无味的阴阳靠它们的作用而得以消化、吸收和运输。严重的心理因素会影响各器官发挥作用，如果途径闭塞不通，形体会受到严重伤害。”

女教师复问：“心病医身，果真能治？”

老童生答：“身体宇宙，不外乎阴阳二字。移精变气，古祝由之术也。”

老童生言毕，北面而咒，并赐中草药方，嘱其服用。

又三年，全国文化大革命，到处批斗“牛鬼蛇神”，摧毁历史文物，古字画、书刊、器皿、饰物，统统被扔进大火中。女教师不堪红卫兵批斗之苦，遂告密求生，揭发老童生私藏封建邪书《黄帝内经》，行巫术迷信，毒害社会。

红卫兵蜂拥至老童生住处，翻箱倒柜，搜书一箩筐，独不见经书。老童生曰，“中华医学瑰宝，怎能付之一炬”，拒交经书，遂被打成反革命。红卫兵剥光其衣，用绳子绑住其生殖器，命其随牵引爬行。如是游街示众。

老童生不堪其辱，自缢身亡。

又数年，国门打开，经济体制转型，房地产业蓬勃发展，到处大兴土木。老童生旧居拆迁，于其屋墙中挖出一物，正是《黄帝内经》，人如获至宝。

THE OLD APPRENTICE SCHOLAR

The old apprentice scholar was a native of the Xiangxi region of western Hunan province. As a youth, he aspired to be a government official working in the service of the people of his country. To this end, he sat for the preliminary imperial examinations three times, but failed all three attempts. Hence his name: the old apprentice scholar. Not much is known about his youth. In his later years, he was solitary and lived frugally, keeping himself to himself and passing his time in studying ancient texts. He would pour himself a drink and recite softly the following words: “Heaven has yin and yang, people have internal organs. Heaven has five elements, people have five organs. People’s lives are conjoined with heaven and earth, and correspond with the sun and the moon.”

The old apprentice scholar dropped so many learned quotations into his speech that his listeners were often mystified. He was wont to talk about the elements of gold, wood, water, fire, and earth, the changes in the seasons and the commensurate flourishing and decay of life. He was able to predict correctly when someone was going to die, and was thus felt to have divine powers. It was even rumored that he could see right inside someone to their internal organs, identifying the seat of their illnesses at a glance.

So people came to him for help. When a baby had chronic diarrhea and was not responding to treatment, the old scholar diagnosed a deficiency of yang in the mother that was affecting her milk and treated her. When someone suffered from a suppurating sore or a boil, or was bitten by a centipede or a spider, he chewed up herbs and applied them to the affected area, with speedy results. For payment, he usually received a few eggs, a bag of rock sugar, or even just a grateful bow. And thus the years passed, and he was happy to carry on ministering to his patients, taking their pulses, observing their color, and diagnosing their maladies.

Whenever he was praised for his medical skills, he always credited the wisdom of the ancients.

One day, a woman teacher paid him a visit. She looked haggard and complained of chronic pelvic pain. She told him she had undergone tests at the hospital but the results were all negative and she was almost frantic with worry and distress. The old apprentice scholar examined her color carefully and listened to her pulse, then instructed her to come back at daybreak the next day on an empty stomach. In the morning, her yin energy would be undisturbed, her yang energy would not have dissipated, and her qi and blood would be settled. Thus it would be possible to get a more accurate pulse reading.

Early the following morning, the teacher came again. The scholar took her pulse, but instead of talking about her illness, he made tea and chatted, asking about her family and her childhood and if she was married. When the teacher was somewhat hesitant in her answers, the old man took an old medical book with yellowing pages from a drawer, saying that it had been bequeathed to him on condition that he kept it secret, but now he was making an exception and showing it to her. He said that he had begun to study it twenty years ago, at the age of sixty, and it had taught him everything he knew.

He began to recite the familiar words: "When someone's life begins, the essence is the first to form, and this gives rise to the brain matter, where the whole of human memory is located. The brain is the home of the *yuan shen*, the primordial spirit of life." The old apprentice scholar then talked about the relationship between the brain and consciousness, and the changes caused in the mind by this relationship. "*The Yellow Emperor's Classic of Internal Medicine* has the answer to all your problems," he went on. "The traumas you experienced in your early years are stored in your brain, and if you don't open up and release the emotional pressure, we will never be able to tackle the root cause of your disease."

The teacher finally told all. She had been raped by her uncle at the age of fifteen. The abuse had gone on for six months, but she had never dared to tell her family. The secret had weighed on her like a great boulder ever since.

"But if the disease is in my mind, why is the pain in my pelvis?" she then asked.

"All mental and emotional activity is governed by the human brain. The lungs control the qi throughout the body and regulate the body's activities. The liver governs anger. The gall bladder protects the heart and transmits its happiness. The spleen and stomach receive and distribute food and drink, and yin and yang (which have no taste or odor) depend on them for digestion, absorption, and transmission. Critical psychological factors can affect the functioning of the organs—if those pathways are blocked, serious physical harm may result."

"So can treating the mind really cure the body?"

"The universe of the body can be summed up as yin and yang. In treating illness, the ancients aimed to transform the essence and the qi by employing beneficial spells to tackle the root cause of illnesses."

The old man faced north, recited a spell, then gave the teacher a prescription of Chinese herbal medicine, with instructions on how to take it.

Three years went by and the Cultural Revolution raged through China. Red Guards destroyed precious historical and cultural objects—old calligraphy scrolls and paintings, books, household utensils and jewelry, all were thrown into the flames—and attacked intellectuals as evil monsters. When the teacher's turn came and she was beaten beyond endurance, she denounced the old apprentice scholar in order to save her own skin. He possessed, she told them, a copy of *The Yellow Emperor's Classic of Internal Medicine*. The Red Guards regarded this as an evil, feudal book, full of witchcraft and superstition, liable to poison people's minds.

The Red Guards descended on the old apprentice scholar's house and turned it upside down. They searched through all his many books, but the *Classic* was nowhere to be found. The old man insisted, "Chinese medicine is a precious inheritance. Why do you want to burn the book?" And he refused to tell them where it was. As a result, he was branded a counterrevolutionary. The Red Guards stripped him naked, tied a rope around his genitals, and paraded him on all fours through the streets.

The old man was so humiliated that he hanged himself.

A few years later, the country opened its doors and the economy underwent a transformation. In the ensuing property boom, construction sites sprang up everywhere. In due course, the old apprentice scholar's home was demolished—and *The Yellow Emperor's Classic of Internal Medicine* turned up, hidden inside one of the walls. It was hailed as a great treasure.

STORY 10

The Unseen

Uzodinma Iweala

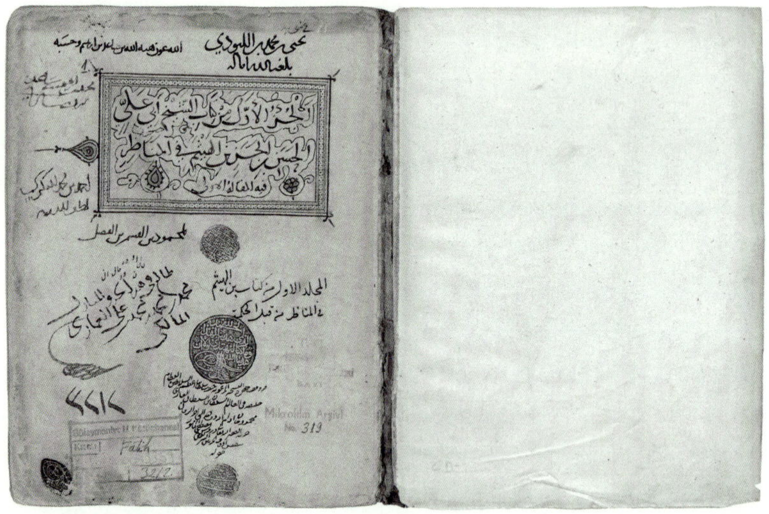

Abū 'Alī al-Hasan ibn al-Hasan ibn al-Haytham, *Kitāb al-Manāẓir* [Book of optics], Cairo, *c.* 1083.
Frontispiece and a spread including a diagram of the visual system.

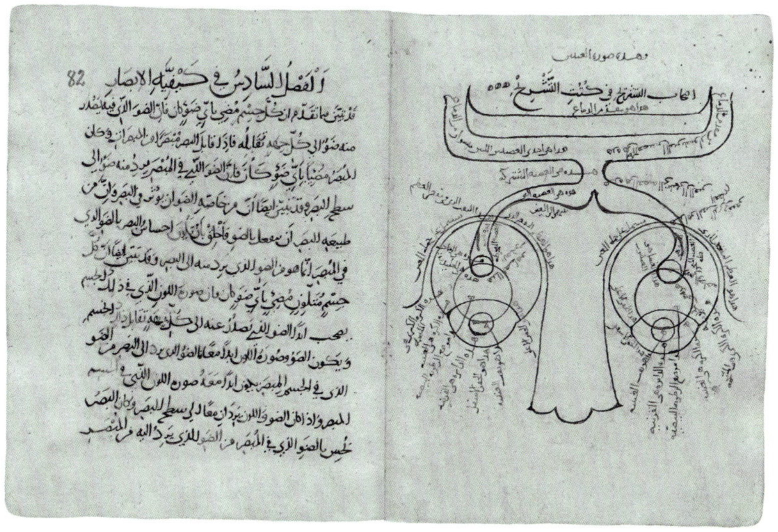
82
الفصل السادس في كيفية الابصار

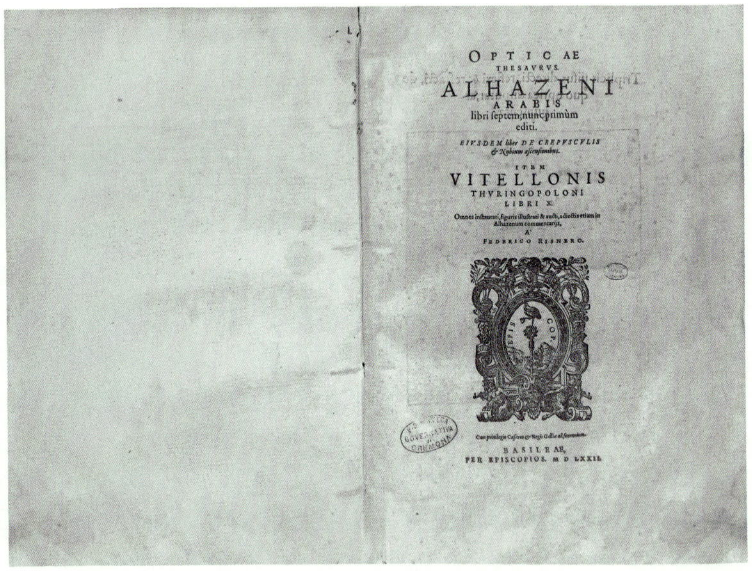

OPTICAE
THESAVRVS.
ALHAZENI
ARABIS
libri septem, nunc primùm
editi.
EIVSDEM liber DE CREPVSCVLIS
& Nubium ascensionibus.
ITEM
VITELLONIS
THVRINGOPOLONI
LIBRI X.
Omnes instaurati, figuris illustrati & aucti, adiectis etiam in
Alhazenum commentarijs,
A'
FEDERICO RISNERO.
Cum privilegio Caesareo & Regis Galliae ad sexennium.
BASILEAE,
PER EPISCOPIOS. M D LXXII.

Opticae thesaurus [Compilation of texts about optics], combined edition including texts by Alhazen (Abū ʿAlī al-Hasan ibn al-Hasan ibn al-Haytham) and Vitello Thuringopolonis, Basel, 1572. Original texts from 10th–13th century. Frontispiece and a spread including a diagram of the optic nerve.

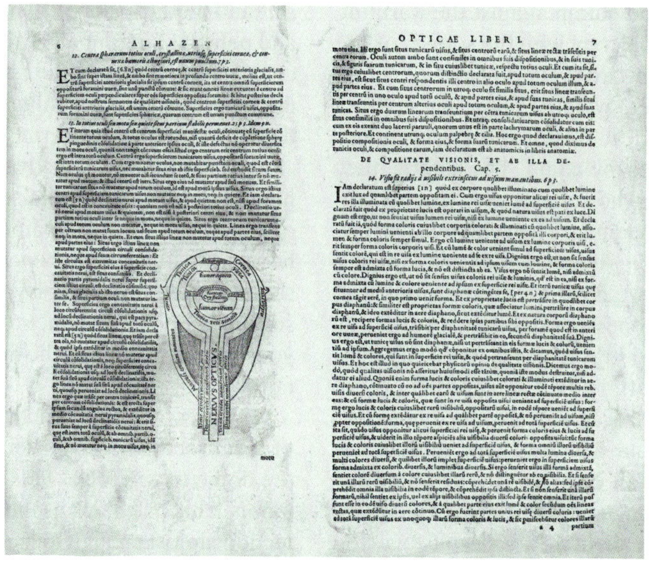

6 ALHAZEN

OPTICAE LIBER I. 7

DE QVALITATE VISIONIS, ET AB ILLA DEpendentibus. Cap. 5.

14. *Visio fit radijs à uisibili extrinsecus ad uisum manantibus.* 6 p 3.

They say madmen are the only ones who can truly see. The thought torments me, terrifies me, pursues me in my dreams so I do not sleep. I cannot sleep: Instead, I pace the darkness of my chambers until the call to prayer. Then the street below stirs and the traders prod their laden asses onward, stumbling to the markets. The women hum work songs as they attend to their household duties. Only when the children fill these alleyways do I find myself fatigued enough not to dream. Only then do I slump against the warming walls of my balcony with the sounds of life pushing me toward the darkness.

I hear them whisper about me. That is him, Abū'Alī al-Hasan ibn al-Haytham, they spit after my name. The heretic, they say. The one who collects eyes from butchered goats—madman. Their children fear me. They sing songs about me as I pass holding bloody sacks. And yes, the sacks are full of all the things no one would dare eat save the cats and crows and stray dogs that peck and snarl around the offal-filled stench pits. There are rats too, large, with courage enough to screech at cats. The butchers' market has emboldened all creatures to the point that greed, not hunger, drives their petty squabbles. How like them we are, despite our God and djinns. That I could crack the skull of a man, I would show them all—trace for them these soft gyri that I have seen folded neatly into the heads of all beings. What viscous and silly notions lie tucked between. But then they would confirm me mad instead of merely thinking me so. Would they be so wrong?

In my dreams, I see that giant the Greeks called Argus. His eyes watch me, unblinking. They taunt without mockery, but still they taunt. At these moments, soaked in cold sweat, I wake with a prayer on my lips: "O Allah, protect me from my front, behind me, from my right and my left and from above me, and I seek refuge in You from being taken unaware from beneath me." But I am not a God-fearing man any more than I am a madman. And on those nights I am beset by Ptolemy's monsters, and I take my lamp and my blade through these dark corridors, muttering curses against that Greek and his gifts. What would I give for the peace and pleasure of Basra, marshes scented with salt and mud, flocks of flamingos, herons, and pelicans singing all praises due! That I could give up my eyes. But I could never give up my eyes and all the things I have seen in them. And there is no Hermes to release me.

My blade works quickly to release these humors, my hands now slick with the essence of vision. I hold aloft a pair of eyes by their nerves, my finger a peg beneath their decussation. I shave layer after layer from the lens, seeking the point of truth. Ptolemy, I laugh at your simplicity and all the things you could not see.

My eyes—they play tricks on me. In each of them I see differently. This one as round the dome of the rock. This one a crescent moon. This one black as the Kaaba, its perfection almost enough to make a believer of me, except that I believe only in what I can see. And so they call me mad. A man without faith is a man without sense, that old imam from Basra said. The beginning of all wisdom is the fear of God, he said. To them I am a senseless fool, locked here in this darkness with my questions that beget questions and answers as yet unseen. I am not the first nor will I be the last.

If this is madness, then let me dwell here until the world loses its senses. Then, perhaps, they too shall see.

STORY 11

السمكة العطشى

Hanan al-Shaykh

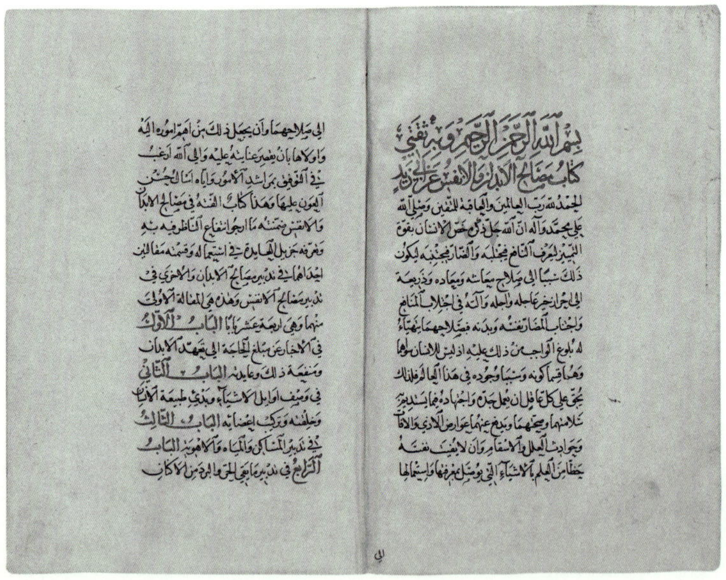

Abū Zayd Ahmad ibn Sahl al-Balkhī, *Maṣālih al-Abdān wa al-Anfus*
[Sustenance for body and soul], Baghdad, 1479. Original text from *c.* 9th–10th century.

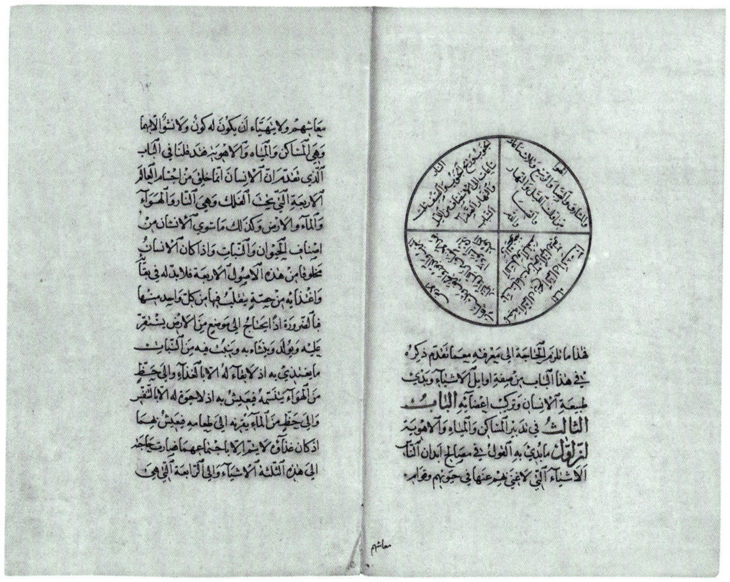

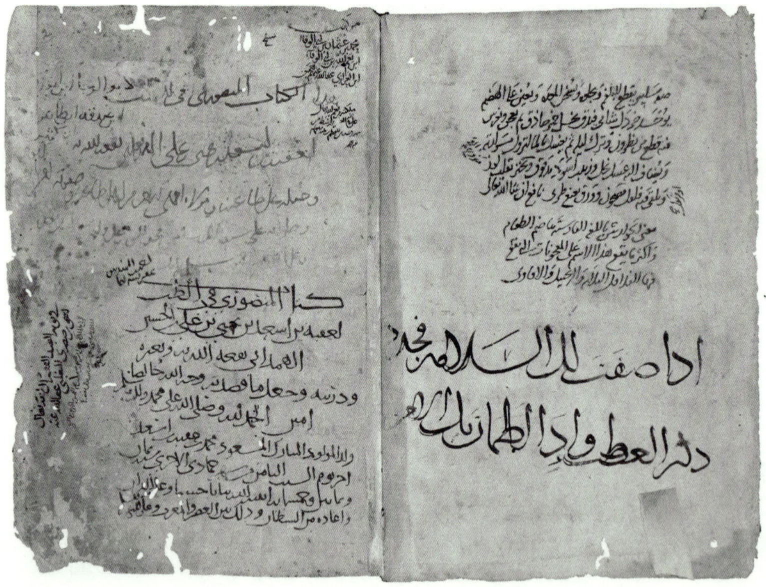

Abū Bakr Muhammad ibn Zakariyyā' al-Rāzī (Rhazes), *Kitāb al-Manṣūrī* [al-Mansuri's book of medicine], 1183. Original text from *c.* 9th–10th century.

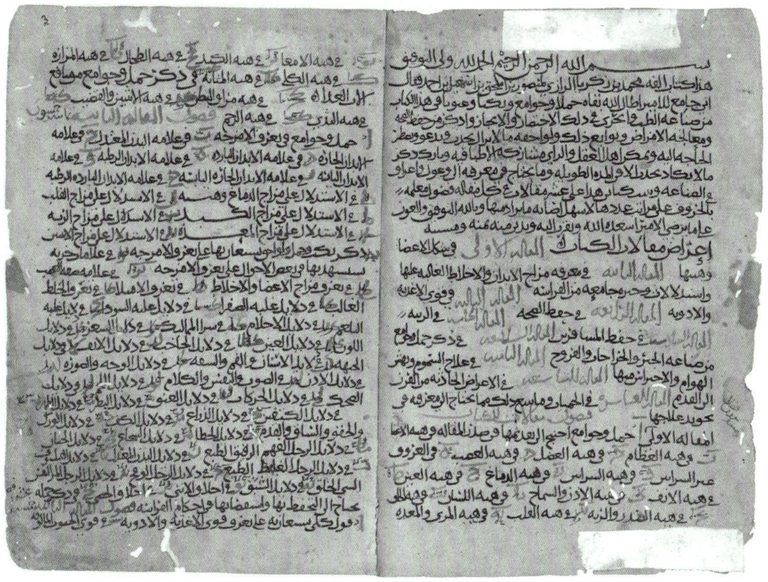

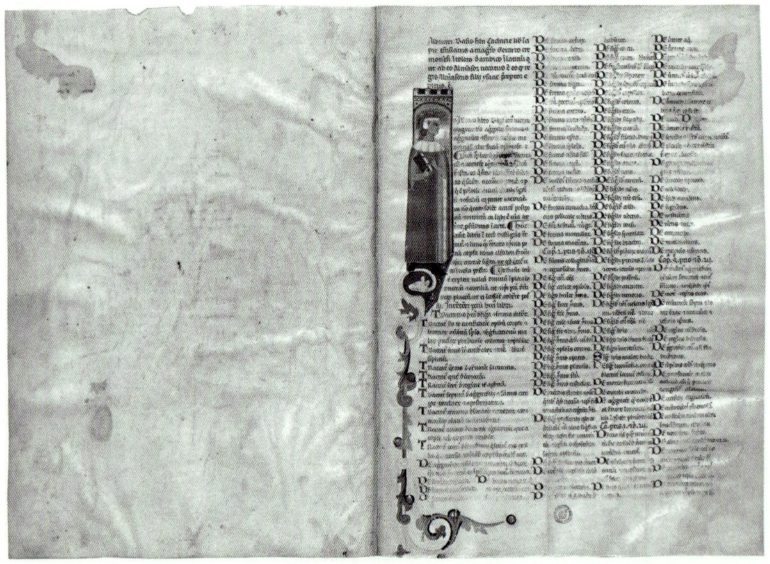

Abū Bakr Muhammad ibn Zakariyyā' al-Rāzī (Rhazes), *Liber Almansoris* [al-Mansuri's book], Latin translation, 14th century. Original text from *c.* 9th–10th century.

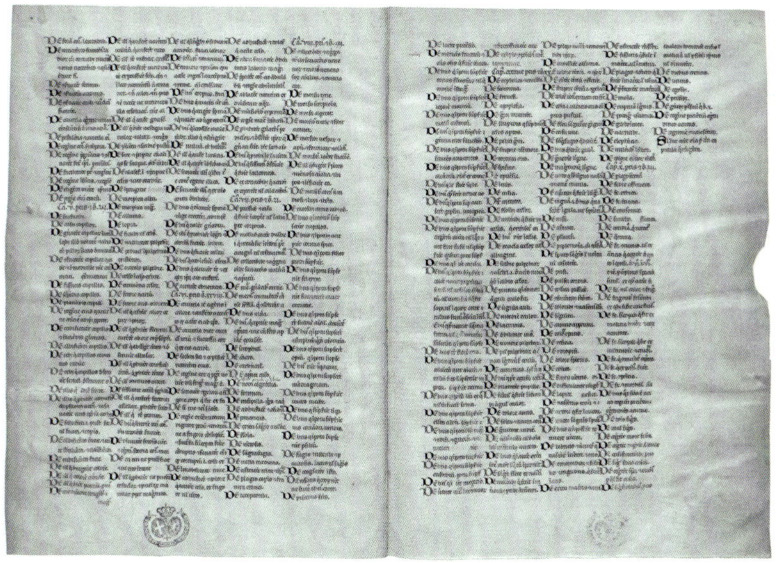

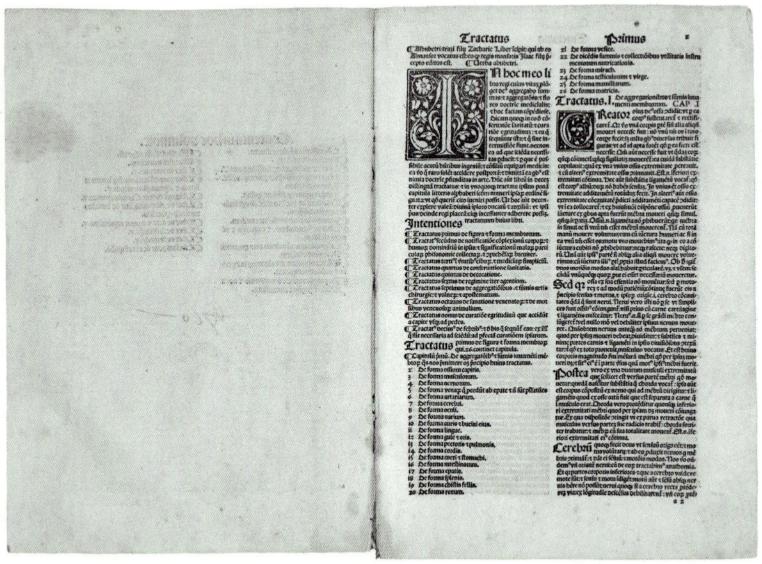

Abū Bakr Muhammad ibn Zakariyyā' al-Rāzī (Rhazes),
Contenta in hoc volumine. Liber Rasis ad almansorem ... Liber de aegritudinibus puerorum eiusdem ... Libellus zoar de cura lapidis, compilation of medical texts, Latin translation, Venice, 1497.
Original texts from *c.* 9th–10th century.

سسمعت جدّتي مرّة تقول، إنّ السمك لا يفكّر إلا وهو يواجه الخطر. وأنا سمكة عطشى، أكاد أموت من العطش. ظننتُ أني إذا التجأتُ إلى عائلتي فإنّ مريئي سينتعش، وقلبي سيهدأ ولو قليلا. لكنّ الليل أرخى سدوله وأنا ما زلتُ أبكي وأشد شعري حتى وأنا أحتمي بسريري إيّاه الذي ظننتُ أنه مرفأ الأمان والشاهد على أحلامي منذ أن كنتُ طفلةً إلى أن فارقتُه وذهبتُ إلى بيت الزوجيّة. وأمي لم تتوقف عن ضمي إلى صدرها ومواساتي بتكرار جملتها: «أصبري، أصبري، لا بدّ أن تتبدّل الأحوال»؛ وجدتُني أصيح بها «لا، لن تتبدّل، فأنا أعاني منذ ثلاثة أشهر وأربعة أيام، وزوجي كأنه فاقد السمع والبصر، ثلاثة أشهر وأربعة أيام لم يقترب مني بعد ليلة الدُّخلة». عندما أمسكتْ أمي بكلتا يديّ تحاول إبعادهما عن شدّ شعري وضربي لوجهي، وجدتني أنكبّ على يدها أعضّها بكل ما عندي من عزم وقوة ثم أنبح ككلب مُصاب بداء الكَلَب، ثم لأتحوّل بعد نهاريْن وليلتين من كلب إلى نبتة صُبّار لا يستطيع أحدٌ الاقتراب منها؛ وعندما أخذتِ النبتة تصوِّب أشواكها على عينيّ ولساني وتهبط عبر حنجرتي لتصل إلى معدتي فأستفرغ سائلاً أصفراً، التجأت إلى جرّة كبيرة أريد أن أدخل فيها لأحتمي بها، حين أحطتها بذراعي والتصقت بها كما يلتصق الجنين بالحبل السّري، رغم الآلام في فخذيّ وذراعيّ وكفّيّ، رغم جوعي وعطشي وحاجتي للتبوّل.

جارةٌ لنا دفعت باب بيتنا تستفهم من أمي عن سرّ هذا الصراخ والعويل؛ ولولت أمي بأني فقدتُ عقلي بين ليلة وضحاها. ولم أعد أسمع سوى همسات ووشوشة وسكون البيت، وهذا ما زاد من ندبي ونواحي إلى أن أوقعتني الجرّة التي كنت أتشبث بها؛ وجدتني أدور حول نفسي وأدور في البيت إلى أن سمعت جلبةً عند النافذة المفتوحة؛ لا بدّ أن هناك قطّة كعادتها جائعة لكنّ عويلي أرعبها. وإذ بمخطوطة كتاب عند النافذة، أمسكت بها وفتحت فمي لتمزّق أسناني أوراقها، لكنني فكرت بأن زوجي ربما كان هو الذي جاء بها ووضع في داخلها رسالةً طالباً العفو والمغفرة وهو يستجديني العودة إليه. وكانت هناك ورقة من جارتنا تحثّني فيها على الاطلاع على المخطوطة وتتوسّل إليّ بالحرص على سلامتها إذ إنها مخطوطة نادرة قد ورثتها عن جدّها.

آه، أين جدّتي، أريد جدتي الحنونة الآن، الآن حتى أرتمي في حضنها. عدتُ ألطم على وجهي، وهجمتُ على المخطوطة من جديد وجدتني أقرأ عنوانها، «صحة الأبدان والأنفس» للعالم في الطب الروحي وطب القلب أبو زيد البلخي. استوقفتني كلمة الأنفس، وتركت عيناي رغماً عني تتمعنان الأحرف والكلمات والجمل، فتقرآن أن الإنسان يتكوّن من روح وجسد، فلا يمكن للوجود البشري أن يكون بصحة جيدة دون اندماج الروح والجسد؛ وقرأت أنه إذا مرضت النفس فإن مصير الجسد أن يقع في المرض أيضا، وسرعان ما وجدتني أسيرة كلمات العالم البلخي التي أسرّت في أذني أن هناك من يعاني بهدوء، وهناك الذي يعاني بجنون، وبأننا لم نُخلق هكذا؛ عندما قرأتُ أنه انتقد الأطباء الذين يركّزون على أمراض الجسد ويُهملون الاضطرابات العقلية حتى أردت أن يعاينني، آه! لكن البلخي يعيش في سمرقند، «آه، آه»، ناديْتُ بأعلى صوتي «لماذا هذا العذاب الذي يريد إبادتي وحرقي»؛ وفجأة التمعت في عقلي جملة كنت قرأتها أو سمعتها، بأن العالم الطبيب الرازي قد تتلمذ على يديّ البلخي، والرازي يعيش قريباً منا في شهر-وراي قريباً من طهران، عندها هرعت إلى غرفتي أرتدي ملابسي، وأسرعت آتي بعباءة المرحوم. ما أن أكملت ارتداء ملابسي حتى انتظرت دخول أمي إلى المرحاض لأسرع وآتي بعباءه المرحوم والدي وعمامته ومشلح صدره الصوفي، والتي كانت جميعها تتصدر الطاولة حتى إذا ما رأتها أمي همست لها «الله يرحمك ويدخلك فسيح جنانه، آه كم أنا مشتاقة لك». لبستُها كلها متخفّية بها فأصبحتُ رجلاً؛ قرعت باب الرازي وقد توقفت عن التحايل بأني أعاني من آلام فظيعة، فلم أصرخ أو أرتعش بل سرت ببطء خلف الحارس الذي فتح لي الباب وأدخلني إلى غرفة الرازي الواسعة، وما أن رأيته حتى أسرعت أنحني أمامه حتى كاد وجهي أن يلامس الأرض؛ طلب مني صوتُه النهوض فنهضت وقلت «أنا يا سيدي سمكة عطشى، ألجأ إليك لأنك أنت البحر والنهر والسواقي.» أشار إليّ بأن أجلس قبالته وهو يتفرّس وجهي قبل أن يسألني «هل أنت ذكر تشعر بأنك أنثى أم العكس؟»

انتبهت بأني ما زلت متخفية بملابس والدي. أسرعت وفككت القماش من حول خصري، خلعت العباءة، رفعت العمامة عن رأسي والمشلح الصوفي عن رقبتي وعدتُ أنا: برأسي تحت تيربون ملون تحيط به الورود بينما تدلى شعري على كتفيّ. عدتُ أنا بسترتي الزرقاء المخملية ذات الخَرز الأزرق وسروالي الفضفاض بلون غروب الشمس والذي أخفى صندل والدي الذي كنت قد حشوته بالقطن حتى لا تفلت قدماي منه لاتساعه. فجأةً قررت الهرب لأني عدت إلى نفسي، إلى الكلب المسعور وإلى نبتة الصبار التي تعيش في الرمال بين العقارب والثعابين؛ نهضت قبل أن أتقيأ، أسرعت إلى الباب لأسمع الرازي وهو يسألني إنْ كنتُ أشم رائحة البخور وهو يقرّب المبخرة مني، وكانت من العنبر والمسك، وإذا كنت أحب رائحة البخور، ثم استفسر وهو يتمعّن في وجهي إذا كان باستطاعتي أن أشير له إلى موضع الألم في جسدي.

هل أشير له إلى رأسي الذي أشعر كأنه يغلي به الماء ويحرقني، أم أشير إلى قلبي الذي أسمع عويله وتهدّجه، أم إلى جسمي حيث آلام الوحدة تعصرني من أذنيّ إلى أصابع قدميّ؟ لو أن الرازي امرأة لأشرت إلى أسفل بطني وإلى ثدييّ وصحت بأنه ليس بإمكاني تحمّل أنينهما، إنهما يشكوان لي وحدتهما، موتهما البطيء بدلاً من أن يعيش في رحمي جنين وفي ثدييّ حليب، فالشاب الذي تزوجني لم يعُد يقربني أو حتى يلمس يدي منذ ليلة زفافنا.

لا بدّ أن الرازي قد نفذ صبره من طول صمتي وتنهيداتي وبكائي إذ سمعت وقع صندله على الأرض، ثم، ولدهشتي سمعت نغماً موسيقياً يتردّد صداه في أرجاء الغرفة الواسعة وكأنه صوت عيدان من القصب يداعبها النسيم، كأنها ماء يجري بكل رِقّةٍ؛ كأنها حفيف سعف النخيل وهي تلوح وتنحني حتى تشرب؛ كأنها تنهيدة زهرة السّوسن وهي تتفتّح. أخذ نقر العود يدخلني، أغمضت عيني أريد أن أعيش هنا فوق هذا المقعد. ثم سمعت صوتاً؛ الصوت يسألني عن اسمي فأجيبه «الرازي، أسمي الرازي»، ضحك الصوت، وبكيتُ أنا. سألني الصوت «هل من الممكن أن تخبريني لماذا تبكين؟ أسألك من أجل أن تفتحي نافذة صدرك المغلقة حتى يدخلها ما تشتهين». أسرعتُ أفتح كل نوافذي؛ نوافذ عقلي وقلبي وجسدي وقلتُ للصوت «أنا في حيرة وتخبط، وغضب وخوف وو... أريد أن أفهم لماذا أخذ الشاب الذي تزوجني وتزوجته يعاقبني بهجره لي. لقد حاولت أن أقترب منه لكنه دافع عن نفسه وابتعد عني وكأني نحلة ستلدغه، كشّني عنه وكأنني ذبابة تحوم حوله بأزيز مزعج، كل هذا، كلّ هذا.» عدتُ إلى البكاء، ردّد الرازي خلفي وكأنه عرف ما بي، «كل هذا لأنك»؟

«كل هذا لأني صحت صيحة ألم عندما فقدت عذريتي في ليلة الدّخلة، ليصيح بي هو الآخر وقد تحوّلت مقلتا عينيه إلى كُرتيْن من الحديد «لا بدّ أن من سمع صرختك ظنّ أني طعنتك بالسكين»؛ أجبته «آسفة، لم أقصد أن أصرخ، الألم الشديد هو الذي صاح، لم أكن أعرف أن ليلة الدّخلة هي هكذا، ظننتها امتداداً للطرب والأغاني والزغاريد»، فأجابني وكأني دستُ على قلبه «تشنّجك وبكاؤك جعلاني أشعر وكأني شاب لا تعرفينه ، لحق بك وهجم معتدياً عليك»، اقتربت منه وركعتُ أمامه وأنا أمدّ له يدي، «آسفة، آسفة يا زوجي العزيز، أرجو مسامحتي ، دعنا ننسى ما حصل، أرجوك، أرجوك، وإذا كنت تفكر بأنك ستسبب لي الألم مرة أخرى فإني أعدك بأن هذا لن يحدث مطلقاً، فأنا لست عذراء الآن والحمد لله»، ثم أخذت أوجه له عبارات الإطراء الواحدة بعد الأخرى لأنه خرق عذريتي بتلك القوة بينما هناك، كما سمعت، رجال ضعفاء لا يتمكنون من فض بكارة المرأة، ثم حاولت جذبه لمطارحتي الحب، بالألاعيب، بالتوسل، ثم بجرأة متناهية، فأخلع ملابسي وأنا أحاول الجلوس في حضنه، أتدحرج على الأرض وأستلقي فوقها وأفتح فخذي، عندما نهرني قائلا إنني لست إلا بائعة هوى لا أنتمي إلى عائلة رفيعة فأجبته «طبعاً أنا كبائعة الهوى ‹أنوار› التي تحمل ثدييها وكأنهما كيسيْن من التمر والتي قيل لي إنك تزورها يومياً». كآبتي تحولت إلى هستيريا، أسرعت إلى سطح البيت أناجي طيور الحمام، وعندما هربتْ مني رشقتها بالحجارة، عندما بقيت النجوم في السماء تتوهّج، بصقت عليها لتسقط البصقة على وجهي.

صمتَ كلّ ما حولي ما عدا الموسيقى التي جعلتني أفكر إنْ كنت قد رويتُ كل هذا للرازي أم لنفسي؛ وكل شعرة من شعري أصبحت وتَراً للعود الذي طغى صوتُه على الأرجاء بألحانه الشجيّة المؤثّرة، بينما تحولت تلافيف دماغي إلى دفوف رقيقة تنقر نغمات المرح وخفّة الروح؛ إيقاع الموسيقى تآلف مع ذهني لدرجة أن ألوان حبيبات الكهرمان والزمرد والياقوت دخلت عينيّ. عصافير حملتني وهي تغرد، حلّقت بي وهي تزقزق، نورٌ أضاء ظلام داخلي وعقلي؛ وريدٌ ثالث وُلد في قلبي.

سألني الرازي إن كنتُ جائعةً؛ كيف حدث أني لم أضع في فمي سوى ريقي الذي كان كالسموم في فمي ومعدتي، فتحت عينيّ حين توقّفت الموسيقى وسألني أن أعود إليه في الغد، فركتُ عيني من النور ومن وضوح عقلي وأجبته إذا كان باستطاعتي البقاء هنا إذ إني أخاف إن عُدتُ إلى بيت الزوجية فإنه سيعود كبيت السلحفاة يركل ظهري بنعليه، وإذا عدت إلى بيت أمي سيعود لي جنوني. تركني ثم عاد ومعه جارية أخذتني من يدي إلى صالة جميلة تعجّ بالنساء وبالصغار من أطفال وصبية. لم أتشنّج كعادتي عندما خيّم الليل، بل جلست مع النساء أبتسم وأفكر بأني سوف أطلب الطلاق وسوف أقع في حبّ رجل كالرازي، إنسانٍ لا يفرّق بين ذكر وأنثى، تواقٍ إلى معافاة مريضه وإعادته إلى نفسه وجسده. ما أن أطلّ الصباح حتى عادت عصافير الرازي والألحان تحملني لساعات طويلة في غرفة عيادته حيث نقر العود، وحنين الناي يضيئان عقلي الذي قاسى وابتعد. وجدتني أسير في الأحياء والنسيم يلامس وجنتي وملابس أبي مضمومة إلى صدري، وعقلي الذي قاسى الأهوال وابتعد عني عاد إليّ وهو يتوهج ويذكرني بالأغنية التي حفظتها منذ الصغر وتمنيت يوماً ما أن أغنيها للشاب الذي أقع في حبه، وها أنا الآن أغنيها للحبيب المجهول.

قول حبيبي هو قولي، وقولي قوله
روحه وروحي روحُهُ،
إن عاشَ عشتُ، وإن عشتُ عاش

THE THIRSTY FISH

I heard my grandmother say once that fish think only when they are facing danger. I'm a thirsty fish, about to die of thirst. I thought if I took refuge with my family my throat would be refreshed and my heart would calm down a little. But night closed its eyes and I was still weeping and tearing my hair, even though I was in my own bed, the place I considered a safe harbor and a witness to my dreams from the time I was a child until I left it for my marriage bed. My mother kept grabbing ahold of me, holding me tight and trying to console me, repeating the same sentences over and over again—"Be patient, be patient. Things are bound to change"—while I screamed back at her, "No, they'll never change. I've been suffering for three months and four days, and my husband seems to have gone deaf and blind. Three months and four days since our wedding night and he hasn't been near me." My mother seized my hands, trying to stop me from pulling my hair and striking my face, and I bent over and bit her on the hand as hard as I could and began barking like a rabid dog. Two days later I'd changed from a dog into a prickly pear that nobody dared approach, and when the cactus began aiming its prickles at my eyes and tongue, and descending through my throat into my stomach so that I vomited a yellow liquid, I tried to climb into a big jar for protection and wrapped my arms around it and clung to it like a fetus to its umbilical cord, despite the pain in my thighs, arms, and hands, not to mention my hunger and thirst and desperate need to pee.

One of our neighbors pushed open our door, asking my mother what the screaming and shouting was about, and my mother wailed that I'd lost my mind overnight. I could hear only muttering and whispering and silence all around, making me howl even louder, and the jar where I was taking cover knocked me over and I found myself spinning round the house in circles. There was a sudden commotion at the open window: the cat must be hungry, as usual, but my howling had scared her. Then I noticed a manuscript by the window. I seized ahold of it, baring my teeth, ready to tear it to pieces, but I thought perhaps my husband had put it there and hidden a message inside it, asking for my forgiveness and begging me to come back to him. What I found was a note from our neighbor, telling me to study the manuscript and begging me to take good care of it, as it was a rare manuscript she had inherited from her grandfather.

Ah, where is my grandmother? I want my kind loving grandmother now, now, so I can throw myself into her arms. I began striking my face again, and attacking the manuscript, and found myself reading its title: *Sustenance for Body and Soul*, by the polymath scholar and psychologist Abū Zayd Ahmad ibn Sahl al-Balkhī.

The word "souls" caught my attention, and in spite of myself I let my eyes take in the letters, words, and sentences and read that a human being is made up of a soul and a body, and no human existence can be healthy without the intertwining of body and soul. If the soul is ill, it will follow that the body will fall ill too. I was soon captivated by al-Balkhī's words, which whispered to me that there are people who suffer quietly and those who suffer and go mad, but that we are not created like that. When I read that he criticized doctors who focused on physical diseases and neglected mental disorders, I wanted him to examine me. But al-Balkhī lives in Samarqand. "Why am I suffering like this?" I screamed aloud.

"It's going to destroy me." Then in a flash I remembered reading or hearing somewhere that the scholar and physician al-Rāzī had been a pupil of al-Balkhī, and al-Rāzī lives close to us, in Shahr-e Rayy, near Tehran.

I jumped up and went to my room to dress, then waited for my mother to go to the toilet so I could rush and fetch my late father's abaya, turban, and woolen shawl, which had pride of place on the table, allowing my mother to whisper to them every time she saw them, "May God have mercy on you and take you into His Paradise. Oh, how I miss you."

I put them all on, disguising myself as a man, and went over to al-Rāzī's house and knocked on his door. I stopped myself from pretending I was suffering from dreadful pain, so I didn't scream or shake but walked slowly behind the guard who opened the door to me and led me to al-Rāzī's large room. The moment I saw him, I bowed before him, my face nearly touching the ground. I heard him asking me to rise, so I obeyed and said, "Sir, I am a thirsty fish. I seek your help because you are the sea, the river, and all the streams."

He indicated for me to sit opposite him, examining my face before asking me, "Are you a male who thinks he's a female, or the opposite?"

I realized I was still wearing my father's clothes and quickly loosened the clothes around my waist, took off the abaya, lifted the turban from my head, unwrapped the woolen shawl from around my neck, and was me again, wearing colorful headgear with a garland of roses while my hair hung down over my shoulders. I was me again, in my blue velvet jacket with blue pearl buttons and my loose, flowing trousers the color of the sunset hiding my father's sandals, which I'd stuffed with cotton wool, as they were too big for me. I suddenly decided to escape, as I'd become myself again, the mad dog, the plant that lived on the sand among scorpions and snakes. I got to my feet before I threw up and I rushed to the door. I heard al-Rāzī asking me if I could smell the incense as he brought the incense burner close to me; it was amber and musk, and he asked if I liked the smell. Then, examining my face, he inquired if I could point to the place in my body where the pain was.

Should I point to my head, which felt as if there were water boiling in it, burning me, or to my heart, whose wailing and trembling I could hear, or to my body, where the pain of loneliness gripped me from my ears to the tips of my toes? If al-Rāzī had been a woman, I would have pointed to the place below my stomach and to my breasts and screamed that I couldn't bear their moaning: they complained of their loneliness to me, their slow death, when there should have been a fetus alive in my womb and milk in my breasts, for the young man who married me had not come near me or even touched my hand since the night of our wedding.

Al-Rāzī must have become impatient with my long silence, my sighs, my weeping, as I heard his sandals clatter away over the floor. Then, to my astonishment, I heard music echoing around the vast room, like the sound of reeds caressed by the breeze; like water flowing gently; like the rustle of palm fronds waving and bending to drink; like the sigh of an iris blooming.

The beat of the oud began to get through to me. I closed my eyes; I wanted to live on this seat forever. Then I heard a voice, a voice asking me my name, and I answered, "Al-Rāzī. My name is al-Rāzī."

The voice laughed, and I cried.

"Can you tell me why you are crying?" asked the voice. "I want to know, so that you can open the closed window in your chest and all your desires can enter."

I rushed to open all my windows, the windows of my heart, my mind, and my body, and I said to the voice, "I'm lost and confused, angry and afraid and—I want to understand why the young man who married me, and whom I married, punished me by deserting me. I tried to get close to him but he defended himself and moved away from me as if I were a bee about to sting him. He brushed me off as if I were an annoying fly buzzing around him. All this, all this ..." I began crying again and al-Rāzī repeated after me, as if he knew what was wrong with me, "All this because you ...?"

All this because I screamed with pain when I lost my virginity, and my husband screamed back at me, the pupils of his eyes transformed into two little balls of steel: "Anyone who heard you yelling would think I'd stabbed you with a knife."

"Sorry," I answered. "I didn't mean to shout. It was the pain. I didn't know the wedding night was like that. I thought it would be a continuation of the joy and songs and ululations."

As if I'd trampled on his heart, he answered, "The way you shuddered and cried made me feel I was some man you didn't know, who had chased you and attacked you."

I went up to him and knelt before him, holding out a hand to him. "I'm sorry. I'm so sorry, my dear husband. Please forgive me. Let's forget what happened. Please. Please. And if you think you might be going to hurt me again, I promise you that will never happen. I'm not a virgin now, thank God."

Then I began heaping praise on him because he had torn my virginity with such force, while there were, so I'd heard, weak men who were unable to deflower a woman. Then I tried to lure him into lovemaking, playing little games, imploring him, throwing caution to the wind and undressing and trying to sit on his lap, rolling on the floor and lying there spreading my thighs. When he accused me of being no better than a prostitute, rather than a girl from a good family, I replied, "Of course, I'm like the prostitute Anwar, who carries her breasts like two bags of dates and who I'm told you visit daily." My depression turned to hysteria and I rushed up onto the roof to confide in the pigeons. When they flew away from me, I threw stones at them, and when the stars in the sky went on shining, I spat at them and the spit landed back on my face.

All around me was silent except for the music, which made me wonder whether I had said all this to al-Rāzī or just to myself. Every hair on my head became a string of the oud, whose sad, moving melodies dominated the surroundings, and the convolutions of my brain turned into gentle tambourines playing joyful, lighthearted tunes. The rhythm of the music was so much in harmony with my mind that my eyes took on the colors of amber, emerald, and sapphire stones. Singing birds carried me up into the air and circled around with me, twittering and cheeping. A light shone in the darkness inside me. A third vein was born in my heart.

Al-Rāzī asked me if I was hungry. How was it that all I'd eaten was my saliva, which was like poison in my mouth and stomach? I opened my eyes when the music stopped, and he asked me to come back the next day. I rubbed my eyes, dazzled by the light and my mental clarity, and answered by asking him if I could stay there with him, since I was afraid that if I went back to the marital home it would be like living in a tortoise's shell again, and if I went back to my mother's house I'd go mad once more. He left and returned with a maidservant, who took me by the hand to a beautiful room thronged with women and girl and boy children. I didn't have cramps like I usually did when night fell, but sat with the women, smiling and thinking that I'd ask for a divorce and fall in love with a man like al-Rāzī, a person who didn't differentiate between male and female and was eager to heal his patients and restore them to themselves. As soon as morning broke, al-Rāzī's birds and the soothing melodies transported me for long hours to his clinic, where the beat of the oud and the yearning of the flute lit up my mind, which had suffered and gone far away. I found myself walking in the neighborhood with the breeze caressing my cheeks and my father's clothes wrapped tightly round me, and my mind that had endured horrors and become estranged from me came back to me glowing and reminding me of a song I'd known by heart since I was a child and wanted one day to sing to the young man I fell in love with.

And here I am now, singing it to this unknown lover:

My lover's words are my words and
my words are his
His soul is mine and my soul is his
If he lives, I live, and if I live he lives.

STORY 12

Liber Trotula

Katie Kitamura

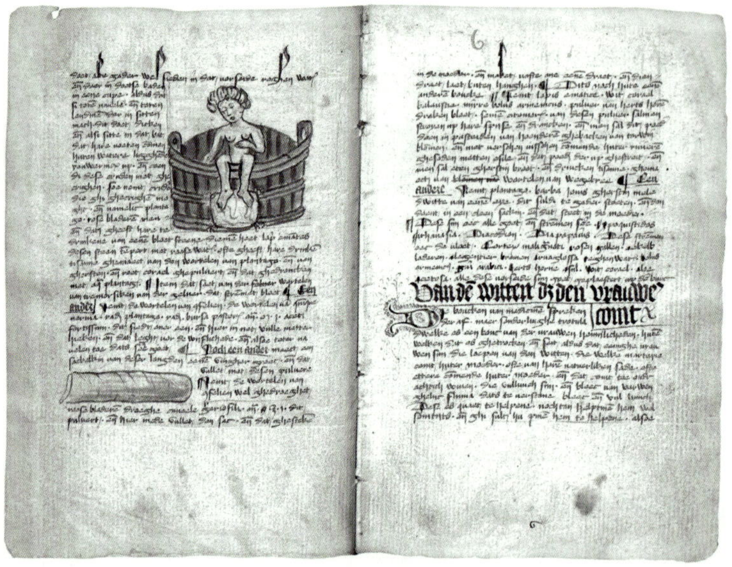

Liber Trotula [The Book of Trotula], compilation of Trotula of Salerno's
three medical treatises on gynecology and cosmetics, vernacular (Dutch) translation, 15th century.
Original texts from 11th–12th century.

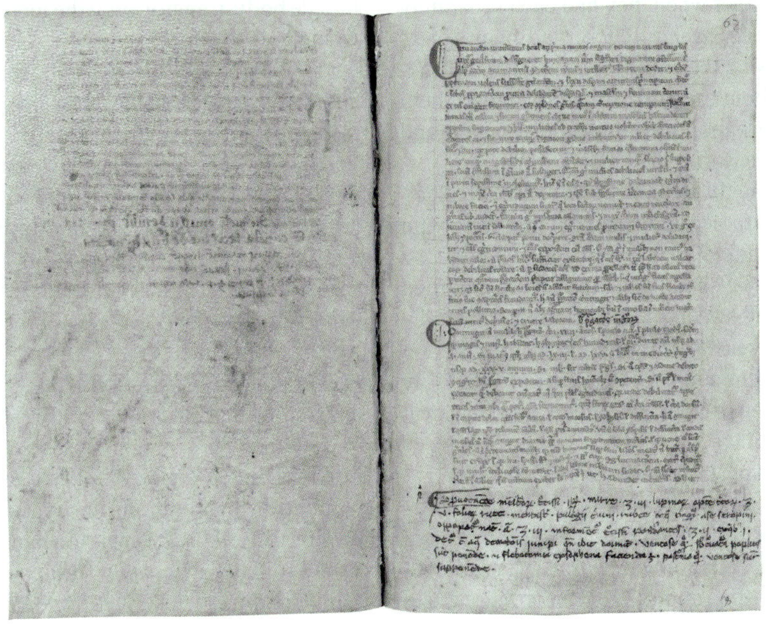

Trocta [Trotula], *De passionibus mulierum* [The book on female diseases],
Latin translation, 14th century.
Original text from 11th-12th century.

They were a young and unusually vivacious couple, of the sort you notice at a party. Their happiness created its own atmosphere in an otherwise dull gathering, and I soon found a way to join them in conversation. I asked how they'd met and they explained that they had met at a lecture on an unlikely topic, the "Liber Trotula."

I said that I had never heard of the Liber Trotula. They laughed and said that they hadn't either when they had each separately decided to attend. There had been flyers advertising the event on campus and something about the posters had intrigued them both. Perhaps it was simply those arcane and evocative words, the Liber Trotula.

The night of the lecture was cold, sleeting, and rather unpleasant. The auditorium was heated to an uncomfortable degree, so that people shed hats and scarves and then coats and sweaters as they made their way to their seats. Although the lecture was on campus, the audience was almost exclusively composed of elderly members from the local community.

Like me, I said pleasantly.

Exactly, the man said with a laugh. I was feeling a little ridiculous, sitting in that aged company, attending a lecture on a subject I knew nothing about. What an obscure waste of an evening! But then this beautiful woman sat down beside me. And I began to feel that after all there was a reason for being in that auditorium.

As the lecture started, he continued, I was painfully aware of her, this young woman with her interest in abstruse medieval texts. The subject matter became suddenly compelling, because each new fact about the Liber Trotula was another clue about the woman I was already beginning to fall in love with.

But what, I interrupted, is the Liber Trotula?

A medieval medical text, he said, written by a female doctor known as Trotula of Salerno. This Trotula of Salerno wrote about everything from menstruation to fertility and childbirth. The collated texts became the Liber Trotula, and in that form circulated throughout Europe. For centuries, it was the authoritative text on the female body, transforming the medical care women received.

I thought he spoke with surprising authority, considering he had only just confessed to knowing nothing about the subject. Perhaps he was channeling the pedantic tone of the lecturer. Of course, he said with a little laugh, some of what appears in the Liber Trotula is necessarily dated. It is a medieval text, after all. There are endless teas and suppositories and other misapprehensions worthy of the time.

He paused. On the one hand, Salerno understood that infertility could be attributed to both men and women. On the other, the Liber includes an entry about excessive humidity in the womb and its supposed effect on infertility. If the womb is too humid, or so Trotula tells us, it becomes so slippery that it cannot retain the male seed. And the sign of that excessive womb humidity?

I told him I couldn't begin to imagine.

Teary eyes.

He waggled his fingers down the surface of his face. I noticed that his fingers were overly long, his hands apparently enormous. The woman frowned. The womb being tied to the brain by way of the nerves, the man continued, and the humidity, presumably by way of these same nerves, working its way up to the brain out of the body in the form of the tears. A weeping brain, if you will.

He dropped his hands from his face. The image is striking, but the logic is surprisingly primitive, don't you think? And yet a remarkable number of the concepts in the Liber Trotula remain relevant. A symptom of the general lack of research into the female body and its illnesses, both historically and today.

As he continued to speak about the female body and its ailments, the persistent lack of data and funding, segueing to the marketing strategy behind breast cancer research, the woman standing beside him seemed to grow increasingly restless. Several times, she opened her mouth as if intending to speak. He seemed entirely oblivious, until eventually she placed her hand on his arm, as if to restrain him.

Abruptly, he fell silent. She turned to face me. The party had grown louder and I leaned closer so that I could hear her as she spoke.

The Liber Trotula does compel you to reflect on the present, she said quietly. But as I listened to the lecture, I mostly found myself thinking not about the present, or even the past, but about how hard it is to feel time. She paused. Trotula was a female doctor in premodern times, a millennium ago. But sitting in that auditorium, I no longer knew what a thousand years was. My brain couldn't accommodate that scale.

It seemed impossible to imagine the life of a woman in that period, or understand how the female body traveled through the world, notwithstanding all the talk about teas and humid wombs. It was only when the lecturer said that Trotula was the first doctor to advocate for the use of opiates during childbirth that I felt that world reveal itself to me. At the time, until Trotula, labor pains were considered a woman's punishment for original sin. In that simple line about pain, sin, and childbirth, the texture of premodern life was suddenly vivid.

She paused. You see, the Liber Trotula is a portal into that time, a portal and a technology. That is why the book fascinates. And that is why Trotula herself fascinates. Through the book, you can feel the mysteries of its author.

She turned to the young man. Earlier, you called her treatments primitive. But science builds on science. It draws upon available knowledge. It is necessarily incremental in its growth. But the ability to perceive women as fully human—that doesn't require incremental knowledge. It only requires clarity and imagination. Trotula was radical because of her imagination.

But who was she, I asked, that she possessed such imagination? What do we know about her?

Only some basic biographical information, the woman said after a moment. I could see that the young man was irritated by the way she had commandeered the conversation; perhaps she was aware of it too. Her husband and sons were doctors, the woman continued. She attended one of the rare medical schools that educated both men and women. But we don't know much else.

Even so, I said, she was clearly extraordinary.

Almost too extraordinary, the man now interjected. He smiled so that his teeth were exposed, and I realized then that I did not like him. I did not know the first thing about the couple, and yet I began to feel the young woman could do better. Apparently Trotula is a fiction, he said. An invention. Her biography is rather too convenient, a feminist icon *avant la lettre*, if you see what I mean.

He pronounced the words with a convincing but overly nasal accent. But then, I asked, who wrote the treatise?

Some people believe it was the husband, the man said. Or that the husband wrote the technical aspects of the book, while Trotula merely contributed the livelier parts.

Livelier, the woman interrupted. I don't think that's the word the lecturer used. And I don't think it was so definite as that, although I concede that the entry you mentioned, about the brain's humidity, has been attributed to a male author. The man shrugged. She turned to me. I am certain that Trotula existed, she said firmly. And I am certain that she was the primary author of the Liber Trotula. But it seems that even today, the erasure of women continues apace.

It was at that point in the lecture, the young man said suddenly, that you leaned over and said, Let's go, shall we? His voice dropped. And those were the first words you said to me. That was the first time I heard your voice. I couldn't respond. My heart was thumping so loudly I was sure you could hear. I stood up and reached for your hand–

She looked up at him. To my surprise, her face was suddenly uncertain. Despite herself, she had been drawn into the story he told, the memory of their first encounter. And yet, surely she must have seen that the narrative itself was trite and flimsy, a haphazard collection of cliches and threadbare emotions? She continued to stare at him without speaking, seemingly unresolved, until an expression of sullen irritation broke across his handsome face.

I felt embarrassed and looked away. I murmured something about going to get a drink, and they let me go without protest. I was some steps away when I turned back and asked, What was on the poster? The one for the lecture?

The young woman frowned. It was just a photograph of the book. The Liber Trotula. The man reached for her shoulder, gripping it tightly. He was already deep in conversation with someone else. She smiled and then turned back to him, dipping her head down to listen. As I watched, it occurred to me that they might stay together for many more years, even for the remainder of their lives. Or they might break up later that evening, having exhausted the possibilities of the relationship in a final, violent row.

Perhaps I was being unkind. I went to the bar, which was now thronged with people. It took some time before I was served. When at last I had my drink, I scanned the room for the young couple, but they were nowhere to be found.

STORY 13

Un changement de perspective

Hervé Le Tellier

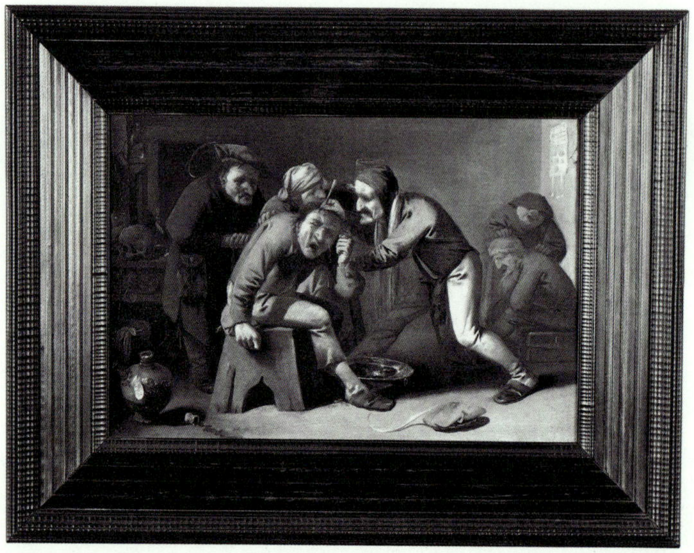

Pieter Jansz Quast, *The Extraction of the Stone*, *c.* 1630.

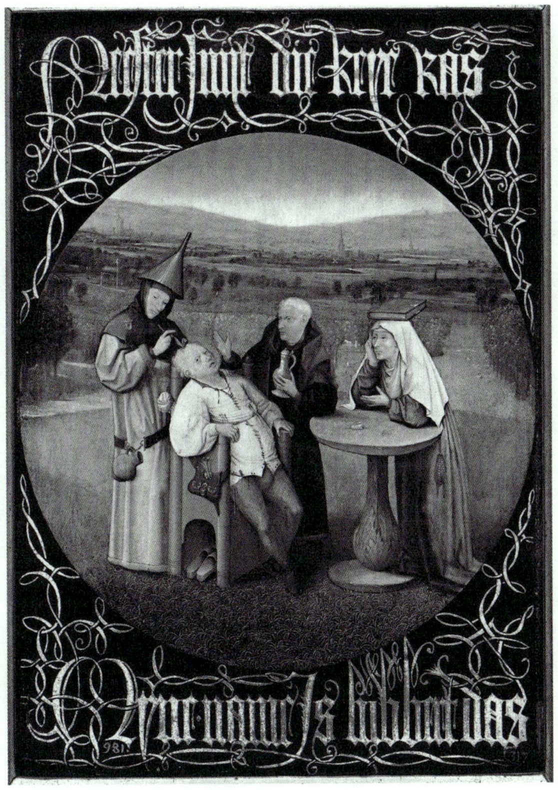

Hieronymus Bosch, *The Extraction of the Stone of Madness*, *c.* 1501–1505.

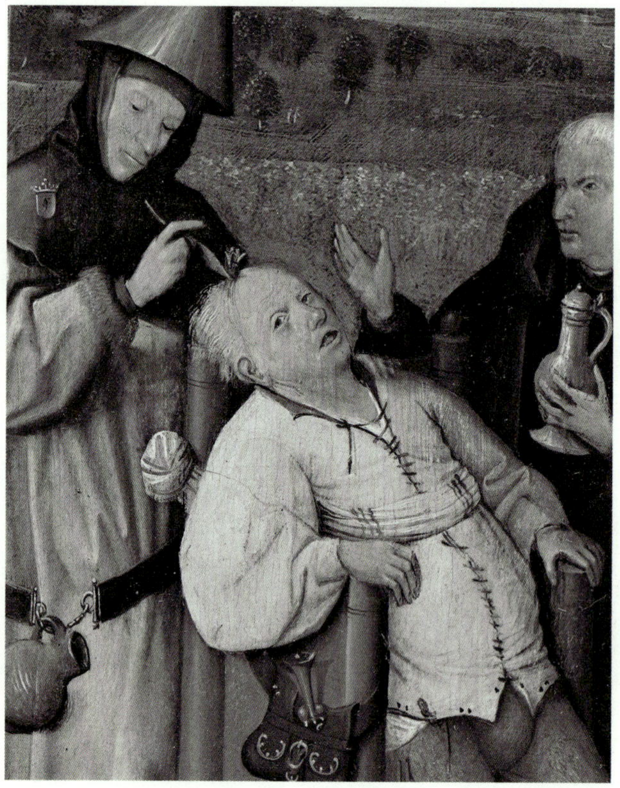

Hieronymus Bosch, *The Extraction of the Stone of Madness* (detail), *c.* 1501–1505.

Le tableau du peintre néerlandais Hieronymus Bosch, *L'extraction de la pierre de la folie*, de 1494, est célèbre. Celui du peintre flamand Pieter Quast, *L'opération de la pierre*, qui date de 1630, ne l'est pas moins. Selon la tradition, on l'analyse comme une raillerie du charlatanisme de l'époque, que vient symboliser l'entonnoir sur la tête du « chirurgien », et de la naïveté des « patients ». Un homme se ferait extraire du crâne la « pierre de la folie », devant un moine avide et une nonne non moins cupide, tous deux complices de l'ignoble escroc. Ce dernier cache évidemment le caillou au patient ligoté, pour le présenter par magie une fois « l'opération » terminée, laquelle pouvait mal finir si le praticien tentait vraiment une trépanation.

Si l'œuvre de Quast laisse peu de doute quant à son sens, on peut légitimement se poser des questions concernant le tableau de Bosch : de récentes recherches menées à l'université d'Uymsterdorp conduisent à supposer un contresens interprétatif. Car à bien regarder *L'extraction de la pierre de la folie*, ce n'est pas une « pierre de folie » que manipule l'homme à l'entonnoir mais une fleur, et sans doute une fleur de nénuphar – et non la traditionnelle tulipe hollandaise – puisque cette fleur sauvage, qui poussait au pied de la chaîne de l'Himalaya et que cultivaient les Turcs, n'est apparue aux Pays-Bas que cinquante ans plus tard. Mais ne digressons pas. On voit d'ailleurs sur la table une seconde fleur de nénuphar. Or, cette fleur égyptienne, avant de devenir dans le langage des fleurs le symbole de l'indifférence, a d'abord été symbole de fertilité et d'épanouissement de l'âme.

Observons maintenant notre pauvre homme attaché. Ses vêtements sont de bonne facture mais négligés, son embonpoint l'encombre ... L'homme est un bon vivant, un riche bourgeois, certes, mais il s'ennuie, la vie lui pèse. Jan Van Der Meyde, drapier de son état dans la bonne ville d'Arnhem, espère plus de l'existence. Il a lu les classiques grecs et latins, dont la nonne porte d'ailleurs certainement sur la tête un ouvrage. Il sait que Démocrite dit qu'« on ne peut être poète sans quelque folie », il n'ignore pas que Sénèque ajoute qu'« on ne trouve guère un grand esprit qui n'ait un grain de folie ». Le bon Horace ne dit pas autre chose.

Alors, Jan Van Der Meyde est pris d'une immense espérance. Il veut devenir poète, il veut pouvoir écrire, comme Germain Nouveau quatre siècles plus tard :

Je vais même plus loin, personne
Ne pourra jamais me guérir,
Ni la sagesse qui sermonne,
Ni le bon Dieu, ni la Sorbonne,
Et c'est fou que je veux mourir.

Alors, lui que jamais la moindre folie n'a effleuré, il souhaite enfin que sa vie prenne une tournure moins banale, moins bornée. Et il exige qu'on lui insère dans la boite crânienne un « grain de folie ». Le tableau devrait donc porter pour titre *L'insertion du grain de folie*. Le charlatan officie donc et feint d'insérer dans le crâne de Jan Van Der Meyde ces deux fleurs de nénuphars achetées à prix d'or au marché de Rotterdam. Jan souffre de la coupure du scalpel, car il est bon que la souffrance atteste de l'efficacité du remède. Mais il en sortira indemne et persuadé d'être fou, comme il le souhaitait. Sans doute n'a-t-il pas tout à fait tort. Car, si croire être sage c'est à coup sûr être fou, croire être fou n'est pas tout à fait être sage.

A CHANGE OF PERSPECTIVE

This painting—*The Extraction of the Stone of Madness*, painted by the Dutch artist Hieronymus Bosch in 1494—is famous. The Flemish painter Pieter Quast's 1630 work *The Extraction of the Stone* is no less so. Bosch's painting is traditionally analyzed as a dig at the charlatanism of the era, as symbolized by the funnel on the head of the "surgeon" and the naïveté of his "patients." A man is having the "stone of madness" extracted from his head before a grasping monk and an equally avaricious nun, who are both in collusion with the scurrilous crook. Of course, the practitioner hides the stone from his pinioned patient and magically produces it once the "operation" is complete, although the procedure can take a turn for the worse if he actually attempts trepanation.

Quast's painting may leave little doubt as to its meaning, but with Bosch's there is room for legitimate hesitation: Recent research by the University of Uymsterdorp points to a counterinterpretation. Close examination of *The Extraction of the Stone of Madness* reveals that it is not a "stone of madness" that the man wearing the funnel is handling but a flower, and very probably a water lily—not a traditional Dutch tulip, because the tulip, which grew wild at the foot of the Himalayas and was cultivated by the Turks, did not appear in the Netherlands until fifty years later. But let's not digress. In fact, there is a second water lily on the table. Before becoming the symbol of indifference in the language of flowers, this Egyptian bloom was first said to represent fertility and fulfillment.

Let us turn our attention to the unfortunate bound patient. His clothes are well-made but worn carelessly, and he is encumbered by a little excess weight… The man enjoys worldly pleasures, he is wealthy and of good social standing, yes, but he is bored; life weighs heavily on him. Jan Van Der Meyde, a draper by profession in the good town of Arnhem, hopes for more from his time on earth. He has read the Greek and Latin classics—in fact, it is almost certainly one of these that the nun has on her head. He knows Democritus said, "There can be no great poet without a spite of madness," and is well aware that Seneca added, "There is no great genius without some touch of madness." The revered Horace said nothing to the contrary.

Jan Van Der Meyde, then, is filled with tremendous hope: He wants to become a poet, he wants to be able to write, as Germain Nouveau did four centuries later:

I cannot be healed or restored,
I would go so far as to cry,
Not by wisdom outpoured,
The Sorbonne nor the Lord,
And it's folly that I want to die.

And so this man who has had not so much as a brush with madness wants his life to take an unexpected, less blinkered turn at last. He asks to have a "grain of madness" planted inside his head. The painting should therefore be called *The Insertion of the Grain of Madness*. The charlatan duly does the honors and pretends to introduce the two water lilies—for which he paid a fortune in Rotterdam market—into Jan Van Der Meyde's head. The scalpel cut is excruciating for Jan, because it is accepted that suffering proves the efficacy of the cure. But he will come away unscathed and convinced that he is mad, just as he hoped. And he may not be entirely wrong, because if believing we are wise is a sure sign of madness, believing we are mad is not quite the same as being wise.

STORY 14

Ventrículos, Corações e Mentes

Sidarta Ribeiro

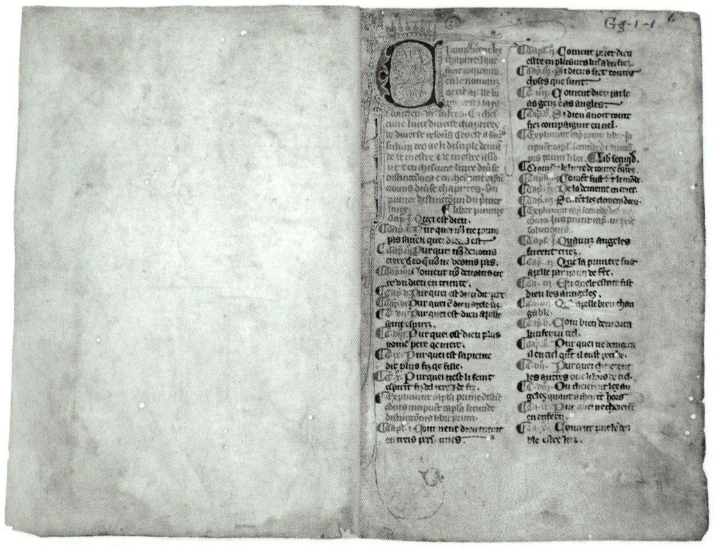

Trilingual compendium of texts including a diagram of the brain as per theory of Avicenna (Ibn Sīnā), England, first half of the 14th century.

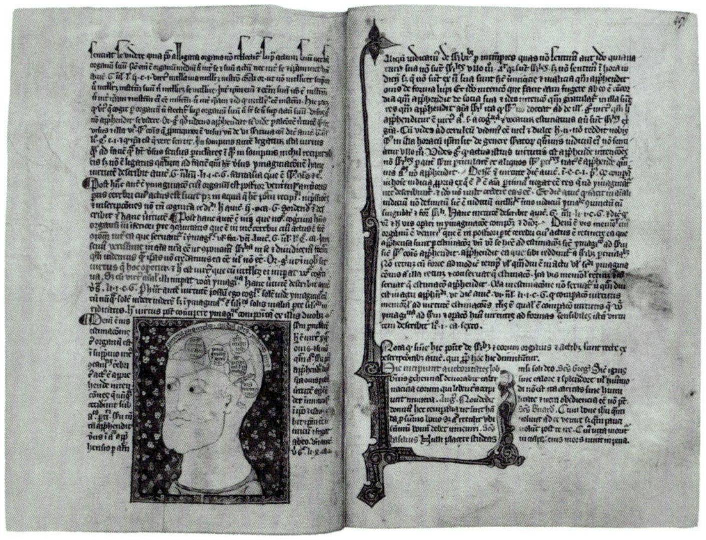

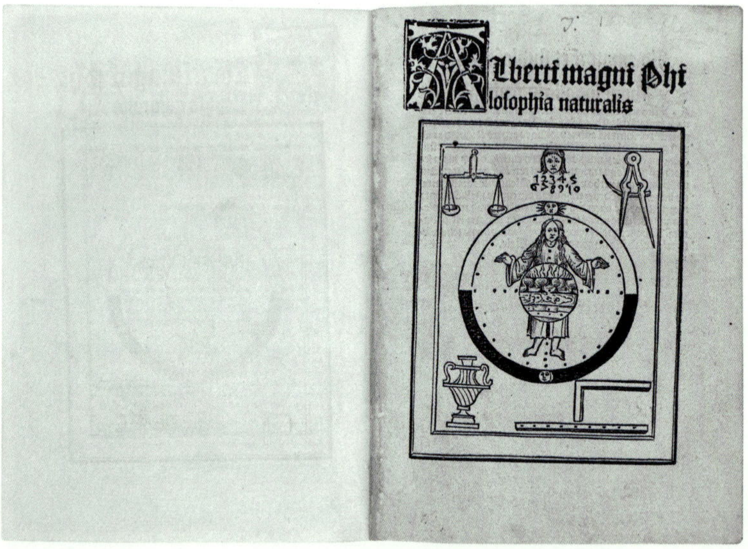

Alberti Magni philosophia naturalis [Albertus Magnus' natural philosophy], Basel, 1506. Original text from 13rd century. Frontispiece and spread including a diagram of the brain.

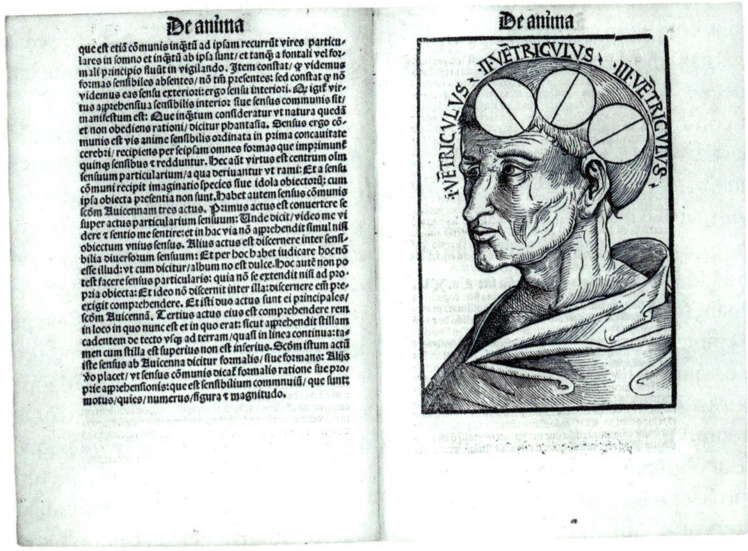
De anima

que est etiã cõmunis inq̃tũ ad ipsam recurrũt vires particu-
lares in somno et inq̃tũ ab ipsa sunt/ et tanq̃ a fontali vel for-
mali principio fluũt in vigilando. Item constat/ q̃ videmus
formas sensibiles absentes/ nõ tm̃ presentes: sed constat q̃ nõ
videmus eas sensu exteriori: ergo sensu interiori. Q̃ igit vir-
tus apprehensiua sensibilis interior siue sensus communis sit/
manifestum est: Que inq̃tum consideratur vt natura quedã
et non obediens rationi/ dicitur phantasia. Sensus ergo cõ-
munis est vis anime sensibilis ordinata in prima concauitate
cerebri/ recipiens per seipsam omnes formas que imprimunt
quinq̃ sensibus ⁊ redduntur. Hec aũt virtus est centrum oim
sensuum particularium/ a qua deriuantur vt rami: Et a sensu
cõmuni recipit imaginatio species siue idola obiectorũ: cum
ipsa obiecta presentia non sunt. Habet autem sensus cõmunis
scdm̃ Auicennam tres actus. Primus actus est conuertere se
super actus particularium sensuum: Unde dicit/ video me vi
dere ⁊ sentio me sentire: et in hac via nõ apprehendit simul nisi
obiectum vnius sensus. Alius actus est discernere inter sensi-
bilia diuersorum sensuum: Et per hoc habet iudicare hoc nõ
esse illud: vt cum dicitur/ album no est dulce. Hoc autẽ non po
test facere sensus particularis: quia nõ se extendit nisi ad pro-
pria obiecta: Et ideo nõ discernit inter illa: discernere eñi pre-
exigit comprehendere. Et isti duo actus sunt ei principales/
scdm̃ Auicennã. Tertius actus eius est comprehendere rem
in loco in quo nunc est et in quo erat: sicut apprehendit stillam
cadentem de tecto vsq̃ ad terram/ quasi in linea continua: ta-
men cum stilla est superius non est inferius. Scdm̃ istum actũ
iste sensus ab Auicenna dicitur formalis/ siue formans: Aliis
vo placet/ vt sensus cõmunis dicat formalis ratione sue pro-
prie apprehensionis: que est sensibilium communiũ/ que sunt:
motus/ quies/ numerus/ figura ⁊ magnitudo.

De anima

Juro que é verdade, foi minha bisavó quem contou. Os fatos extraordinários ocorreram no primeiro terreiro da religião chamada umbanda, na pequena cidade de São Gonçalo, perto do Rio de Janeiro. Naquela época a cultura afro-brasileira, como o candomblé e a capoeira, sofria bruta repressão policial. Paradoxalmente, entretanto, proliferaram naquela pequena cidade ao lado da então capital federal os cultos rítmicos e coloridos da umbanda, voltados para a comunicação sagrada com entidades africanas e indígenas, sincretizadas com santos católicos e espíritos europeus na linha das seánces espíritas de Allan Kardec.

Assim como acontecera com o samba algumas décadas antes, a umbanda tornou-se uma verdadeira febre no Rio de Janeiro. Uma cidade velha que não existe mais, muito mais provinciana e desconectada do mundo do que agora, uma cidade ainda sem bossa nova, garota de Ipanema e carnaval para inglês ver.

O caso começou em 22 de agosto de 1918 numa gira de quinta-feira, um grande ritual de incorporação espiritual com diversos médiuns reunidos para atender visitantes carentes de acolhimento, orientação e auxílio. Tudo prosseguia como de costume, em crescente fervor e vibração. Os tambores, cantos e defumações inebriavam, havia eletricidade no ar. Já haviam baixado no terreiro as entidades *Sete Encruzilhadas* e *Oxossi*, respectivamente o guardião daquele terreiro e o patrono do dia da semana. À meia noite em ponto, entretanto, o médium mais experiente do grupo foi subitamente deixado por *Sete Encruzilhadas* e caiu no chão em espasmos violentos. Ao se levantar ficou claro que fora montado pelo espírito de um estrangeiro.

Após alguns minutos proferindo palavras incompreensíveis. o visitante se calou e apontou para pedaços de giz de várias cores. Ao recebê-los começou a psicografar no chão, lentamente, dois misteriosos desenhos. O primeiro retratava um homem jovem em perfil esquerdo olhando para seu lado direito com um leve sorriso. Traços ligavam os olhos à parte superior da cabeça, que continha cinco círculos conectados. No segundo caso surgiu o busto de um homem mais velho, em perfil semelhante ao do primeiro desenho, mas com três grandes círculos no interior da cabeça. Os minuciosos desenhos foram então riscados com um grande X, como se cancelados – e o ato se completou com a anotação de uma distante data futura: 1 de junho de 1941.

Quando o médium despertou, foi tomado de visível emoção. Encerrou os trabalhos espirituais daquela noite e disse bem alto para que todos ouvissem: *"Esses desenhos expressam um fundamento que precisamos decifrar até a data assinalada"*. As solenes palavras deixaram as pessoas em polvorosa, inclusive minha bisavó que ainda era menina e assistiu tudo. Qual o sentido daquelas cabeças contendo círculos internos? Que tipo de visitante espiritual aparecera naquela noite? Qual o significado daquela data futura?

Ninguém ali tinha a menor ideia de como interpretar o material, e por isso resolveram copiar os traçados para levá-los ao exame de médiuns mais experientes. Os desenhos foram enviados às principais mães e pais de santo da região, mas ninguém soube explicá-los. Decidiram então consultar as ialorixás da Bahia, também sem sucesso. Finalmente apelaram aos babalaôs da Nigéria e do Benim, mas ninguém soube esclarecer nada e o assunto acabou esfriando.

* * *

Para ser exato, esfriou por 22 anos. Porém, nos meses que antecederam a data marcada, o terreiro se alvoroçou novamente, na antecipação do que poderia vir a acontecer. A esperada noite finalmente chegou, o barracão lotou, os tambores soaram, e o transe começou.

Até hoje os mais velhos se agitam quando contam que, exatamente à meia noite, o aguardado visitante retornou. Novamente repetiu os misteriosos traçados, porém desta vez com gestos ágeis. Mais uma vez gravou no chão uma estranha data futura: 6 de junho de 1961. Afinal, falou em português com forte sotaque alemão:

– Como está situada a cabeça do homem? Afirmo que a superfície do cérebro é um mapa que se divide em cinquenta e duas áreas, assim como o planeta se divide em países: um lugar para cada coisa, cada coisa em seu lugar.

Para espanto e comoção das pessoas, entretanto, rolou pelo chão uma médium auxiliar. Quando se levantou, apanhou um pano branco, fez um turbante sobre a cabeça e bradou:

– Alto lá, Korbinian. O cérebro não tem mapas e sim ventrículos, por onde a percepção e a imaginação, através da cogitação e da estimativa, afinal chegam à memória.

– Louvo teu pioneirismo, ó persa. Mas os pioneiros estão sempre fadados à tolice. Os ventrículos são espaços cheios de líquido, nada mais.

Em meio à algaravia, instalou-se um rebu. Começaram a rolar pelo chão todos os médiuns da casa, à medida em que mais e mais espíritos se juntavam à acalorada discussão.

– Herege! Como ousa falar assim com o grande mestre Avicena? Dois séculos depois dele, confirmo que a mente se organiza nos ventrículos.

– Não se exalte tanto, mestre Alberto. Suas ideias estão velhas. É preciso aceitar que o novo sempre vence o antigo.

Foi nesse momento que minha bisavó recebeu em convulsões violentas um visitante ainda mais estranho. Ao se levantar, emitiu uma cavernosa e solene voz de ancião com sotaque indecifrável:

– Suas teorias são meras ilusões, senhores. De fato, os ventrículos são a chave... mas não do cérebro e sim do coração.

E da mesma boca saiu uma voz mais, agora uma mulher:

– Lá vem o Aristóteles com suas observações míopes!

E novamente o ancião:

– Quosque tandem abutere, Hipatia, patientia nostra?

A gritaria era tremenda e já não se sabia quem dizia o quê:

– São três os ventrículos da alma, Alberto: no coração e nada mais!

– São três, Aristóteles: no cérebro e nada mais!

– No cérebro há quatro ventrículos, idiota!

– E no coração também, imbecil!

– Acalmem-se senhores, suas teorias estão ultrapassadas. Os pensamentos estão nos neurônios e nada mais.

– As células cerebrais são todas conectadas entre si, como uma rede. Não existem neurônios isolados, quantas vezes terei que dizer!

– Respeitosamente, professor Camillo, sua teoria é falsa.

– Porco cane, Santiago!

– Calma, por favor! Ambos estão ultrapassados! Aprendam o que descobri por experimentação positiva: os pensamentos são ondas eletromagnéticas e se transmitem à distância.

– A arrogância é tua fraqueza, Hans. O novo não melhora o antigo, apenas o degenera.

– Quanta superstição, mestre Aristóteles! Quanta inércia!

E mais não digo, pois irrompeu uma lamentável pancadaria. Os tambores silenciaram e naquela noite ninguém conseguiu refletir sobre a verdade e a mentira de todas aquelas palavras.

* * *

Felizmente, o tempo está aí para isso. Foi uma longa espera até 6 de junho de 1961, o que permitiu ao grupo se preparar melhor. Levaram até câmeras de filmagem e pesquisadores da universidade, para melhor documentar o evento parapsicológico.

À meia noite, como esperado, apareceu um visitante com sotaque alemão. Mas não desenhou, não brigou com ninguém e nem era qualquer das pessoas que haviam aparecido antes. Se apresentou como Carl, rabiscou uma data futura - 30 de outubro de 1999 - e reclamou, um tanto atordoado:

– São vozes demais dentro da minha cabeça!

Tomou um café preto, cruzou as pernas e após um longo silêncio, afinal pediu giz:

– Agora sim, gostaria de desenhar... Alguém pode chamar a Nise, por gentileza?

* * *

Nota explicativa
22 de agosto de 1918: falecimento do neurologista Korbinian Brodmann
1 de junho de 1941: falecimento do neurologista Hans Berger
6 de junho de 1961: falecimento do psiquiatra Carl Gustav Jung
30 de outubro de 1999: falecimento da psiquiatra Nise da Silveira

VENTRICLES, HEART, AND MINDS

I swear it's true; it was my great-grandmother who told me. The extraordinary events took place in the first house of worship of the Umbanda religion in the town of São Gonçalo, close to Rio de Janeiro. At that time, Afro-Brazilian culture, things like Candomblé and capoeira, were suffering under brutal police repression. Paradoxically, however, in that town right next to what was then the federal capital, the rhythmic and colorful cults of Umbanda flourished, dedicated to sacred communication with African and indigenous spirits syncretized with Catholic saints and European spirits, along the lines of Allan Kardec's Spiritist séances.

As had happened with samba a few decades earlier, Umbanda reached a fever pitch in Rio de Janeiro. An old city that no longer exists, much more provincial and disconnected from the world than it is now, a city that did not yet have bossa nova or a girl from Ipanema or a carnival focused just on impressing the tourists.

It began on August 22, 1918, at a Thursday gira, a big ritual of spirit incorporation at which several mediums were gathered to receive visitors in need of welcome, guidance, and aid. Everything proceeded as normal, with increasing fervor and excitement. The drums, the chanting, and the cleansing smoke were intoxicating; there was electricity in the air. The spirits Seven Crossroads and Oxossi had already come down to the terreiro, the house of worship; these were, respectively, the guardian of that terreiro and the patron of the day of the week. At midnight on the dot, however, the most experienced medium in the group suddenly felt Seven Crossroads leave him and fell to the ground in violent spasms. When he got up again, it was clear that he had been possessed by the spirit of a foreigner.

After a few minutes of pronouncing incomprehensible words, the visitor fell silent and pointed to the various colored pieces of chalk. Somebody passed them to him, and he began to psychograph two mysterious drawings, slowly, on the ground. The first depicted a young man looking to his right with a faint smile. Lines connected his eyes to the upper part of his head, which contained five interlinked circles. The second showed the bust of an older man, much like the first drawing but with three large circles inside his head. The detailed drawings were then struck through with a big X, as if being canceled—and the act was completed with the noting of a date in the distant future: June 1, 1941.

When the medium awoke, he was visibly overcome. He terminated the spiritual work for the night and said, very loudly, so that everyone would hear: "*These drawings express something fundamental that we must decipher by the given date*." These solemn words left people abuzz, including my great-grandmother, who was then still a girl and who witnessed it all. What could those heads containing inner circles mean? What sort of spiritual visitor had come that night? What was the significance of that date?

Nobody there had the least idea of how to interpret the material, so they decided to copy the sketches to get them examined by more experienced mediums. The drawings were sent to the leading Umbanda priestesses and priests in the region, but nobody could explain them. So they decided to consult the ialorixás of Bahia, similarly without success. Finally, they appealed to the babalawos of Nigeria and Benin, but nobody could offer any clarification, and in time the matter just cooled off.

* * *

For twenty-two years it cooled, to be exact. However, in the months leading up to the appointed date, the terreiro became unsettled once again, in anticipation of what might happen. The long-awaited night finally came, the shack filled up, the drums sounded, and the trance began.

To this day, those of the older generation still become quite agitated when they describe how, at exactly midnight, the expected visitor returned. He repeated his mysterious sketches, though this time his movements were more nimble. Once again he scratched down a strange future date: June 6, 1961. Finally, he spoke in Portuguese, with a thick German accent:

"*How is the head of man arranged? I declare that the surface of the brain is a map divided into fifty-two areas, just as the planet is divided into countries: one place for each thing, and everything in its place.*"

To the people's astonishment and alarm, an assistant medium rolled down onto the ground. When she got back up, she took a white cloth, tied it into a turban around her head, and shouted:

"*Stop right there, Korbinian. The brain does not have maps but ventricles, through which perception and imagination, via cogitation and estimation, finally reach the memory.*"

"*Oh, I do commend your pioneering spirit, my dear Persian. But pioneers are always doomed to folly. The ventricles are cavities filled with liquid, nothing more.*"

In the middle of this chatter, a great ruckus was setting in, as all the mediums in the house now started to roll down onto the ground, as more and more spirits came to join the heated discussion.

"*Heretic! How dare you speak like that to the great Master Avicenna? Two centuries after him, I can confirm that the mind is indeed organized into ventricles.*"

"*Please don't go getting overexcited, Master Albertus. Your ideas are old. You've got to accept that the old is always defeated by the new.*"

That was the moment my great-grandmother, with violent convulsions, received an even stranger visitor. When she got up, she spoke with the booming, solemn voice of an old man with an unidentifiable accent:

"Your theories are mere delusions, gentlemen. Ventricles are indeed the key ... but to the heart, not the brain."

And from that same mouth came another voice, this time a woman's:

"Oh, there goes Aristotle again, with those short-sighted observations of his!"

And the old man again:

"Quosque tandem abutere, Hypatia, patientia nostra?"

There was a tremendous amount of shouting, until it was no longer possible to tell who was saying what:

"The ventricles of the soul are three, Albertus: in the heart, and that's it!"

"They are three, Aristotle: in the brain and that's it!"

"The brain has four ventricles, idiot!"

"So does the heart, moron!"

"Oh, do calm down, gentlemen, your theories have been superseded. Thoughts are in the neurons, and that's it."

"Brain cells are all connected together, like a network. There are no isolated neurons, how many times do I have to say this!"

"Respectfully, Professor Camillo, your theory is false."

"Porco cane, Santiago!"

"Please, just calm down! Both of you have been superseded! You should learn what I discovered through experimentation: thoughts are electromagnetic waves, and they are transmitted long-range."

"Arrogance has always been your weak point, Hans. The new does not improve upon the old, it merely degrades it."

"Such superstition, Master Aristotle. So much inertia!"

And I will say no more, because at that point a most regrettable brawl broke out. The drums fell silent and nobody that night was able to consider the truth and the lies of all those words.

* * *

Luckily, that is what time is for. It was a long wait until June 6, 1961, which allowed the group better to prepare itself. They even brought movie cameras and university researchers so that they might better document the parapsychological event.

At midnight, as expected, a visitor with a German accent appeared. But he did not draw, he did not get into an argument with anybody, nor was he any of the people who had appeared previously. He introduced himself as Carl, scribbled a date in the future—October 30, 1999—and grumbled, in a bit of a daze:

"There are too many voices in my head!"

He got himself a black coffee, crossed his legs, and, after a long silence, finally asked for a piece of chalk:

"Now—yes—now I would like to draw ... Would somebody kindly call Nise, please?"

* * *

Explanatory note

August 22, 1918: death of the neurologist Korbinian Brodmann
June 1, 1941: death of the neurologist Hans Berger
June 6, 1961: death of the psychiatrist Carl Gustav Jung
October 30, 1999: death of the psychiatrist Nise da Silveira

STORY 15

On the Bridge

Tash Aw

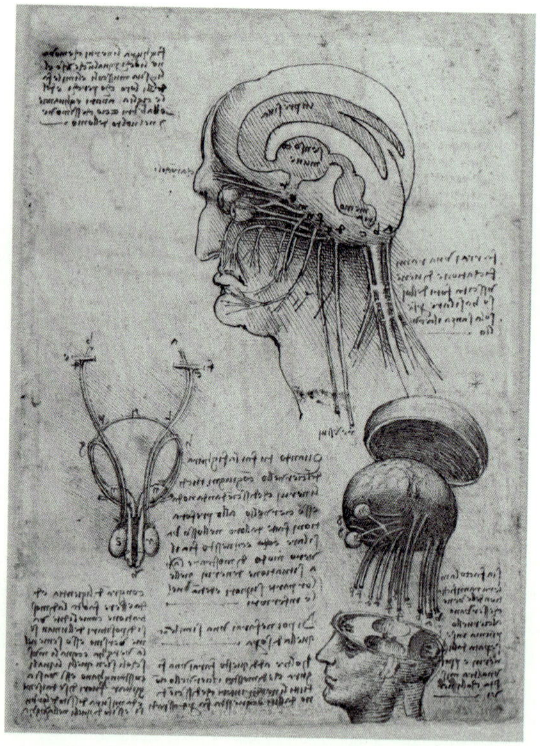

Leonardo da Vinci, manuscript sheet with notes and brain drawings, 1506–1508.

It starts with the blinking of the eye, a sign that he is not as nerveless as I believe, that a reaction of some sort is taking place within him, minuscule but discernible nonetheless—a transformation so delicate and overwhelming that his gaze is broken, this dark-eyed youth, whose face bears all the insolence of his age. It is the same every morning, at almost the exact time, as we cross paths on the bridge. I am heading one way, he another, hurrying to attend to the tasks that fill his day. When we see each other there is a sort of recognition, where he holds my stare for a half-second that imbues this moment with an infinite quality. But maybe those fleeting instances are something I've conjured from nothing; maybe his gaze is hollow and meaningless and he attributes no significance at all to this visual connection. No aggression, no malice, no interest or curiosity, no desire. What is he thinking of when he walks swiftly through the damp morning air, his limbs moving with the ease and strength of water, his head held high and proud *despite all he has lived through*? (Despite all he has lived through—my fantasies again; I cannot seem to control them these days). What impulses travel through the passages contained in his skull when he looks at me? Maybe nothing at all is happening in his brain; it is all happening in mine.

I cannot help but think, again and again, about the way these strange waves travel through our heads. They swirl and trickle, as brilliant and alive as molten wax, seeping into the farthest folds of the brain—sensations so intense that they surely cannot be the product of a clunky machinery of ventricles arranged one after another, but must be the result of something more complex.

It is his eyes that make me think of the way he collects the information around him and draws it back into the chambers of his brain. The expression is empty at first, the pupils as deep and uncommunicating as pools of black water in a forest pond. But then: a flicker into recognition when he sees me. The shape of the eyes remains unchanging. *The eyes*. Those simple openings in the surface of his skin are feeding sensations back into his brain. Fear, mistrust, suspicion. What of? Perhaps of all that I represent, but I cannot guess what that might be. The color of my skin and hair—of my eyes, so contrasting to his. The way I walk, the person he thinks I am. I imagine, therefore, that he is defending himself against difference.

The blink is the first breach of his defenses. He cannot maintain that air of quiet defiance. His eyelids quiver, they open and shut, once, twice, as we pass each other. What lightning impulse, traveling from the eye to the brain and back to the eye again, causes that to happen? I imagine it is me.

I imagine, I imagine, I imagine. Where in my brain do these fantasies take refuge? I cannot escape them, they are constant, lingering in hidden spaces that I cannot fathom. The human brain, when you look at it freshly dissected, reveals nothing of its mysteries. Its softness speaks of nothing; we cannot tell what shapes or passages traverse it, or how it controls the eyes, the face. When casting a bronze sculpture the other day, I thought of the hot glittering liquid filling my brain, illuminating every passage within it. The sensation of warmth was akin to that which I experience when I see this person on the bridge every morning and I recreate his history, his past, his journey from a distant land to ours. He walks close to the side of the bridge, trying not to be observed, but my eyes capture him, I *see* him.

Each act of imagination, and reimagination, stems from the sight of something wondrous. And each of these sentiments is born of the connection between the eyes, those simple openings in the skin, and the ventricles in the brain.

Then, in recent days, the blink has been followed by a twitch of his cheek, a faint pulling of just one side of the mouth that promises a smile but fades away into something more intriguing. How does that happen to a person's face, this creasing of the jowls that follows the trembling of the eyes that follows the moment you catch sight of something or someone? The moment you are moved by beauty.

It happens so quickly, yet the sight of him ignites sensations in my head, and as I continue my walk I am already dissecting my own brain, imagining filling it with hot wax as silken as the thoughts that are currently running through me, tracing the trajectory of these sensations from my eyes into my skull, into a nervous center, the *imprensiva*, expanding and filling every tiny intricate cavity as it transfers to the *senso comune*. All of that contained in the hard skull enrobed in soft skin. I feel it exploding, dissembling, taking with it a thousand years of misunderstanding and replacing it with something real and clear. The hot wax hardens and the soft tissue of my brain falls away at the cut of the scalpel. I can feel my face coming alive, my eyelids trembling lightly, my cheeks pulling into a smile although I cannot fully explain why.

STORY 16

Театр Внутри

Maria Stepanova

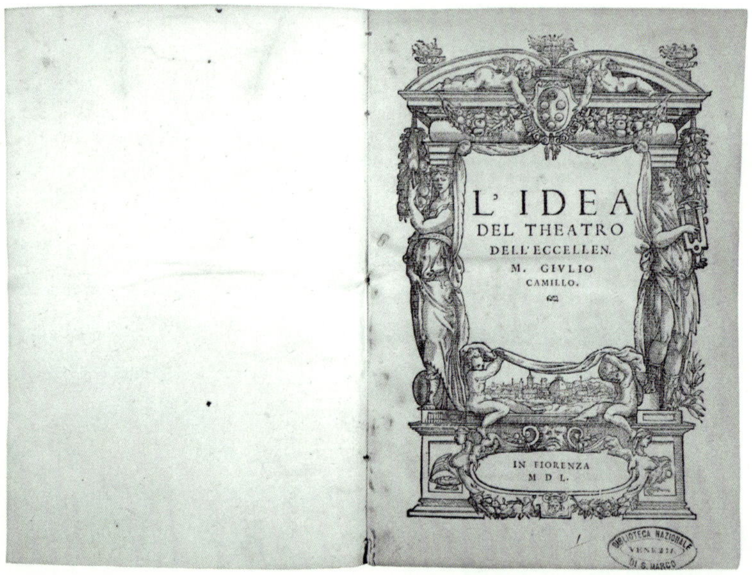

Giulio Camillo, *L'idea del theatro* [The idea of a theater], Florence, 1550.
Frontispiece and first chapter.

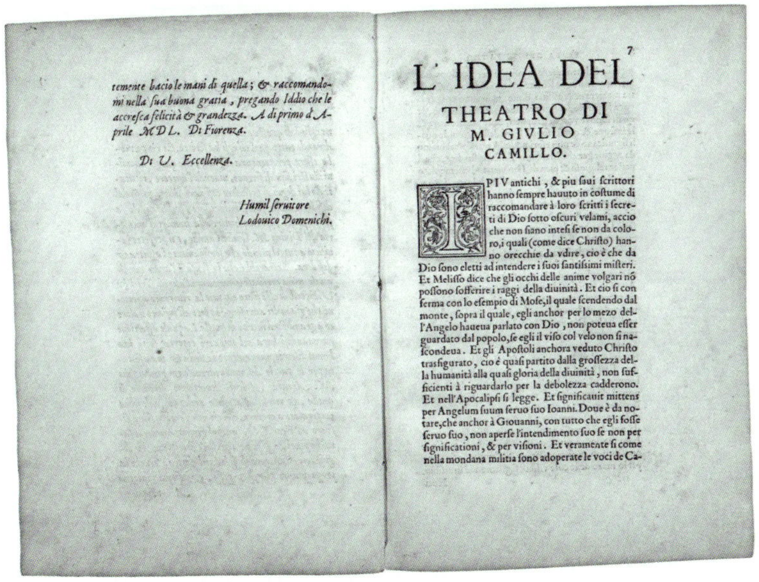
temente bacio le mani di quella; & raccomandomi nella sua buona gratia, pregando Iddio che le accresca felicità & grandezza. A di primo d'Aprile MDL. Di Fiorenza.

Di V. Eccellenza.

Humil seruitore
Lodouico Domenichi.

7

L' IDEA DEL THEATRO DI M. GIVLIO CAMILLO.

I PIV antichi, & piu saui scrittori hanno sempre hauuto in costume di raccomandare à loro scritti i secreti di Dio sotto oscuri velami, accio che non siano intesi se non da coloro, i quali (come dice Christo) hanno orecchie da vdire, cio è che da Dio sono eletti ad intendere i suoi santisimi misteri. Et Melisso dice che gli occhi delle anime volgari nō possono sofferire i raggi della diuinità. Et cio si conferma con lo esempio di Mose, il quale scendendo dal monte, sopra il quale, egli anchor per lo mezo dell'Angelo haueua parlato con Dio, non poteua esser guardato dal popolo, se egli il viso col velo non si nascondeua. Et gli Apostoli anchora veduto Christo trasfigurato, cio è quasi partito dalla grossezza della humanità alla quasi gloria della diuinità, non sufficienti à riguardarlo per la debolezza caddero no. Et nell'Apocalipsi si legge. Et significauit mittens per Angelum suum seruo suo Ioanni. Doue è da notare, che anchor à Giouanni, con tutto che egli fosse seruo suo, non aperse l'intendimento suo se non per significationi, & per visioni. Et veramente si come nella mondana militia sono adoperate le voci de Ca-

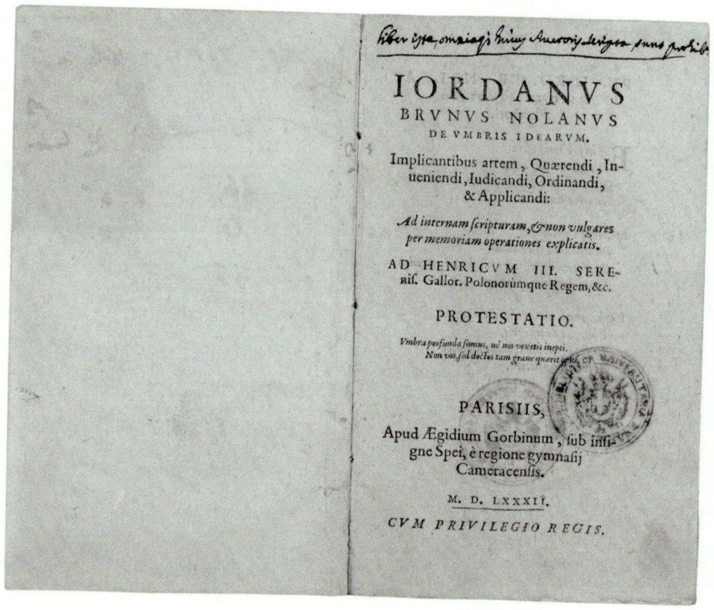
IORDANVS
BRVNVS NOLANVS
DE VMBRIS IDEARVM.
Implicantibus artem, Quærendi, Inueniendi, Iudicandi, Ordinandi, & Applicandi:
Ad internam scripturam, & non vulgares per memoriam operationes explicatis.
AD HENRICVM III. SERENIS. Gallor. Polonorúmque Regem, &c.
PROTESTATIO.
Vmbra profunda sumus, né nos vexetis inepti.
Non vos, sed doctos tam graue quærit opus.
PARISIIS,
Apud Ægidium Gorbinum, sub insigne Spei, è regione gymnasij Cameracensis.
M. D. LXXXII.
CVM PRIVILEGIO REGIS.

Giordano Bruno, *De umbris idearum* [The shadows of ideas],
Paris, 1582. Frontispiece and spread including Bruno's diagram of the mnemonic wheel.

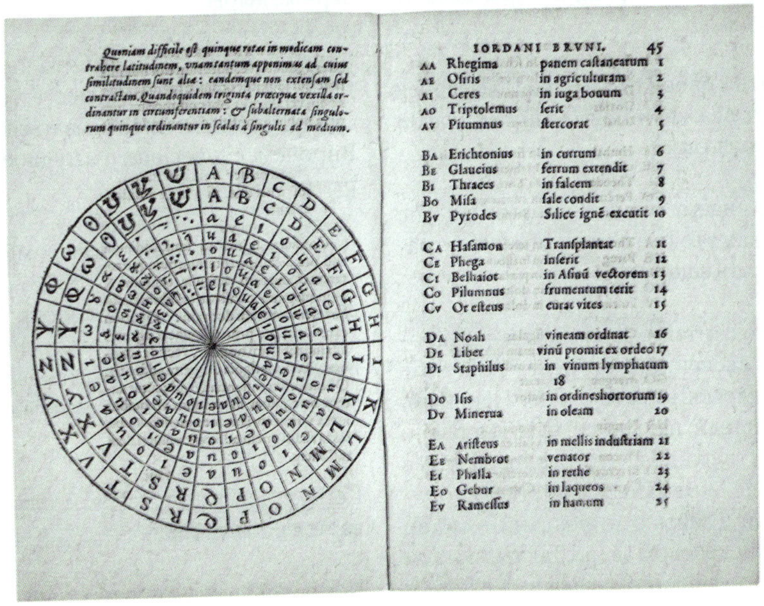

Quoniam difficile est quinque rotas in medicam contrahere latitudinem, unam tantum apponimus ad cuius similitudinem sunt aliæ: eandemque non extensam sed contractam. Quandoquidem triginta præcipua vexilla ordinantur in circumferentiam: & subalternata singulorum quinque ordinantur in scalas à singulis ad medium.

IORDANI BRVNI. 45

AA	Rhegima	panem castanearum	1
AE	Osiris	in agriculturam	2
AI	Ceres	in iuga bouum	3
AO	Triptolemus	serit	4
AV	Pitumnus	stercorat	5
BA	Erichtonius	in currum	6
BE	Glaucius	ferrum extendit	7
BI	Thraces	in falcem	8
BO	Misa	sale condit	9
BV	Pyrodes	Silice ignẽ excutit	10
CA	Hasamon	Transplantat	11
CE	Phega	inserit	12
CI	Belhaiot	in Asinú vectorem	13
CO	Pilumnus	frumentum terit	14
CV	Or esteus	curat vites	15
DA	Noah	vineam ordinat	16
DE	Liber	vinú promit ex ordeo	17
DI	Staphilus	in vinum lymphatum	18
DO	Isis	in ordineshortorum	19
DV	Minerua	in oleam	20
EA	Aristeus	in mellis industriam	21
EE	Nembrot	venator	22
EI	Phalla	in rethe	23
EO	Gebur	in laqueos	24
EV	Ramessus	in hamum	25

Я закрываю глаза, и становится темно. Театр памяти работает при закрытых дверях.

И у Джордано Бруно, и у Джулио Камилло театр памяти круглый: круглый, как коробка черепа, в которой происходит действие. Впрочем, это относится к любому театру – сцена и зрительный зал дополняют друг друга: мы сидим и смотрим, актеры делают свое дело.

Принято говорить, что, когда человек умирает, вся жизнь и весь мир *проходят перед его глазами*.

Мы не знаем, правда ли это; еще не знаем, как это бывает с теми, кто утратил память еще при жизни.

Когда-то считали, что человек – микромодель вселенной: он несет ее в себе и изменяется вместе с нею. Нервный, скандальный, невыносимый Джордано Бруно утверждал, что раздражающее изобилие животных обретает порядок, если думать о мироздании как об одном прекрасном звере. Так в арабской притче стая птиц оказывается единым живым существом, носящим общее имя.

Когда человек умирает, огромное животное мира умирает вместе с ним. Когда человек теряет память, все звери и птицы, все витрины и улицы исчезают медленно, поодиночке.

Когда я писала этот текст, у моего отца нашли стремительно развивающуюся болезнь Альцгеймера.

* * *

В театре памяти зрительный зал заполнен образами и понятиями зримого мира. Тот, кто смотрит спектакль, стоит на сцене и смотрит в зал, читает его, как книгу: он и публика, и перформер. Это театр одного актера, одного зрителя: он и видит, и говорит. Когда человек теряет память, зрительный зал пустеет. Там, где были птицы и звери, знаки зодиака, боги, герои и чудовища, возникают черные дыры.

В деревянном театре, что построил Джулио Камилло, на месте того, кто смотрит, можно было оказаться вдвоем. Впрочем, скорее всего эти двое видели разные вещи.

В зрительном зале: слова, эмблемы и символы, скульптуры, атрибуты человеческих искусств и ремесел. Каждая эмблема заставляет зрителя вспоминать что-то свое – вместе они становятся текстом, который каждый раз складывается заново.

Когда человек умирает, этот текст оказывается последним.

Когда человек умирает, стены и вещи, которые его окружали, остаются театром памяти, неподвижным и пустым, потому что больше некому заставить их говорить.

Театр памяти – способ организации вещей и событий: то, на что ты смотришь, само собой выстраивается в цепочку, образует внутренние связи, нужные, говорит Джордано Бруно, *для внутреннего письма*. Запись спектакля ведется круглосуточно, круглая черепная коробка принимает подсказки, оживляет сопоставления, придает действию единство. Вещи мира, на которые некому смотреть, становятся бездомными, беженцами на голой земле беспамятства.

Мой папа больше не помнит, в каком городе живет и как звали мою мать. То, о чем мы говорим сегодня – пара коньков, которые были у него в детстве. Это были замечательные коньки, вызывавшие общую зависть и восторг. Они назывались *норвеги*.

Что видел Джордано Бруно в свои последние недели? Однажды по странному стечению обстоятельств я оказалась в комнате, где он ожидал допроса. Там до сих пор стоит старинный резной сундук, который, сказали мне, *может его помнить*. Если выйти на крышу, увидишь в ногах весь Рим и круглую, как театр, крышку Пантеона.

Один из тех, кто упражнялся в искусстве памяти, утверждал, что если пропорции скульптуры совершенны, она неминуемо обретет собственную душу; то же самое, продолжал он, может относиться к текстам и мыслям. Несколькими столетиями позже один английский богослов предположил, что домашние животные обретают душу и бессмертие, крепко связанное с бессмертием хозяина – чем больше их любишь, тем реальней, живее эта душа.

Будет ли сундук вспоминать Джордано Бруно? Будут ли вещи, что стоят и лежат в квартире моего отца, помнить его по мере того, как он их забывает? Что, если то, что окружает нас в момент смерти, прощания, радости, острой душевной боли, а потом живет без нас, само собой складывается во что-то вроде паззла, в котором мы могли бы узнать самих себя, когда поднимется занавес? Что, если они – посмертное тело нашей памяти, то, что остается в зале, когда опустела сцена?

Может ли театр памяти показать мне не меня, а тех, кого я хочу увидеть, если я буду внимательно глядеть на то, что от них осталось – по списку?

Вот список предметов, изъятых при аресте у поэта Даниила Хармса в августе 1941-го. Он больше с ними не встретится – умрет от голода в пустой тюремной больнице во время блокады Ленинграда.

Паспорт (выдан 7 июня 1941 года)
Свидетельство об освобождении от воинской обязанности за негодностью по болезни
Копия свидетельства о браке и разные справки на имя Ювачева-Хармса Д.И.
Результаты медицинских анализов
Фотокарточки, 6 штук
Часы карманные
Крестик (нашейный) самодельный белого металла
Бумажник старый коленкоровый
Записная книжка с разными записями, одна
Членский билет № 2330 Союза советских писателей СССР
Пять листов бумаги с разными рисунками в красках и с подписями
Книга «Новый завет» издания 1912 года с пометками и записями на полях
Два использованных железнодорожных билета
Лупа в медной квадратной оправе

Два кольца
Три стопки и одна рюмка белого металла
Портсигар белого металла, один
Мундштук янтарный с оправой желтого металла
Два коробка спичек с инициалами «Д.Х.»
Стихи в рукописи («Элегия») на 2-х полулистах, за подписью Александра Введенского.

... С высокого этажа моей гостиницы немецкий город, где живет мой отец, виден весь сразу, со всеми местами, где мы бывали вдвоем и поодиночке. Но сейчас темно, только налит металлическим светом вокзал, все его полустанки отчетливые, как в голограмме, но ни поездов, ни пассажиров нет. Улица, ведущая к нему, совсем пустая, только светофор иногда переходит с красного цвета на зеленый. После полуночи в городе выключили всю подсветку, и церковные шпили и замок на дальней горе разом исчезли.

Если бы я хотела назвать все, что знаю здесь, поочередно, это было бы легко, хотя видимых вещей в наступившем мраке почти не осталось. Впрочем, я и так помню их назубок. Круглый привокзальный фонтан, огибаемый трамвайными путями, зимой стоит без воды. Темный бульвар со всеми своими деревьями и скамейками лишился очертаний, стал черным пятном где-то справа; чуть светит магазинная улица Кайзерштрассе – почти все лавки и лавочки уже не те, что в 1995-м, только большие магазины остались на своих местах. За ними, вдалеке, будут крыши древнего госпиталя – больницы Юлиусшпиталь, где двадцать лет назад умерла моя мама. А слева, за краем окна и зрения, спит мой папа – за городским парком, за барочной епископской резиденцией, в картонной квартирке, где прожил двадцать лет, а теперь не признает ее своей – и вот-вот, как сказал врач, перестанет узнавать и меня. В театре памяти не нужно света, мы и так слишком хорошо знаем, куда смотреть.

THE INNER THEATER

I close my eyes and it gets dark. The memory theater works with closed doors.

For both Giordano Bruno and Giulio Camillo, the theater of memory is circular: circular, like the skull, in which the action takes place. This is true of any theater, though: the stage and the auditorium complement each other: we sit and watch, the actors do their work.

People say that when a person dies, all of life and the world *passes before their eyes*.

We don't know if this is true; and we don't know how it is for those who have lost their memory while still alive.

Man was once thought to be a micromodel of the universe: he carries it within himself and changes with it. The nervous, scandalous, insufferable Giordano Bruno argued that the annoying profusion of animals finds order if one thinks of the universe as one beautiful beast. Thus, in the Arab parable, a flock of birds turns out to be a single living creature with a common name.

When a man dies, the great animal of the world dies with him. When a man loses his memory, all the animals and birds, all the windows and streets, disappear slowly, one by one.

As I was writing this text, my father was diagnosed with rapidly developing Alzheimer's disease.

In the memory theater, the auditorium is filled with images and concepts of the visual world. The person who watches the performance stands on the stage and looks towards the audience, reads it like a book: he is both audience and performer. It is a theater of one actor, one spectator: he both sees and speaks. When one loses his or her memory, the auditorium becomes empty. Where there were birds and beasts, signs of the Zodiac, gods, heroes, and monsters, black holes appear.

In the wooden theater that Giulio Camillo built, it was possible for two people to be in the place of the one who is watching. However, it is likely that the two saw different things.

In the auditorium: words, emblems and symbols, sculptures, attributes of human arts and crafts. Each emblem makes the viewer remember something different—together they become a text that is put together anew each time.

When a person dies, this becomes the last text.

When a person dies, the walls and things that surrounded them remain a memory theater, motionless and empty, because there is no one else to make them speak.

The theater of memory is a way of organizing things and events: what you observe is self-linked, forming the inner connections necessary, says Giordano Bruno, for *inner writing*. The performance is timed around the clock, the circular skull taking in cues, bringing juxtapositions to life, giving the action a coherence.

The things of the world, with no one to look at them, become homeless, refugees on the bare ground of oblivion.

My dad no longer remembers what city he lives in or what my mother's name was. What we are talking about today is a pair of ice skates he had as a child. They were wonderful skates, causing general envy and delight. They were called *Norwegians*.

What did Giordano Bruno see in his final weeks? One day, by a strange coincidence, I found myself in the room where he had waited to be questioned. There still stands an ancient carved bust, which, I was told, *might remind me of him*. If you go out on the roof, you can see all of Rome and the round, theaterlike lid of the Pantheon at your feet.

One of those who exercised the art of memory asserted that if the proportions of a sculpture are perfect, it will inevitably acquire a soul of its own; the same, he continued, can apply to texts and thoughts. A few centuries later, an English theologian suggested that pets acquire a soul and immortality firmly linked to the immortality of their owner—the more one loves them, the more real, the more alive, that soul is.

Will the bust remember Giordano Bruno? Will the things that stand or lie in my father's apartment remember him as he forgets them? What if the things that surround us at the moment of death, of farewell, of joy, of acute heartache, and then live without us, are themselves put together into a kind of puzzle in which we might recognize ourselves when the curtain rises? What if they are the posthumous body of our memory, what remains in the audience when the stage is empty?

Can the theater of memory show me not me but the people I want to see, if I look carefully at what's left of them—down the list?

Here is a list of items confiscated during the arrest of the poet Daniil Kharms in August 1941. He would never meet them again—he starved to death in an empty prison hospital during the siege of Leningrad.

Passport (issued June 7, 1941)
Certificate of exemption from military service for unfitness due to illness
Copy of marriage certificate and various certificates in the name of Yuvachev-Harms D.I.
The results of medical tests
Photocards, 6 pcs.
Pocket watch
White metal homemade cross
An old linen wallet
A notebook with different entries, one
Membership card #2330 of the Union of Soviet Writers of the USSR
Five sheets of paper with different drawings in paint and with signatures
The New Testament, 1912 edition, with notes and writings in margins
Two used railroad tickets
Magnifying glass in a square copper frame
Two rings
Three glasses and one white metal shot glass
One white metal cigarette case
One amber mouthpiece with yellow metal frame
Two boxes of matches with initials D.H.
Poems in manuscript "Elegy" on 2 half-sheets signed by Alexander Vvedensky

From the top floor of my hotel, the German city where my father lives can be seen all at once, with all the places we have been, together and individually. But it's dark now; only the train station is filled with metallic light; all its half-stops are distinct, as in a hologram, but there are no trains or passengers. The street leading up to it is completely empty, with only the traffic lights, changing sometimes from red to green. After midnight all the lights in the city were turned off; the church spires and the castle on the far mountain disappeared at once.

If I wanted to name everything I knew here, one by one, it would be easy, though there were hardly any visible things left in the darkness. I remember them by heart, though. The circular station fountain, wrapped around the streetcar tracks, stands without water in winter. The dark boulevard, with all its trees and benches, has lost its outline, has become a black spot somewhere on the right; the shopping street Kaiserstrasse shines a little. Almost all the shops and benches are no longer the same as in 1995; only the big stores remain in their places. Behind them, in the distance, will be the roofs of the ancient hospital, the Julius Hospital, where my mother died twenty years ago. And on the left, behind the edge of the window and view, my dad is sleeping—behind the city park, behind the baroque episcopal residence, in a cardboard apartment where he has lived for twenty years and that he now does not recognize as his own—and very soon, as the doctor said, he will cease to recognize me too.

There is no need for light in the theater of memory; we already know too well where to look.

STORY 17

Khipu Kamayuq

Ch'aska Anka Ninawaman

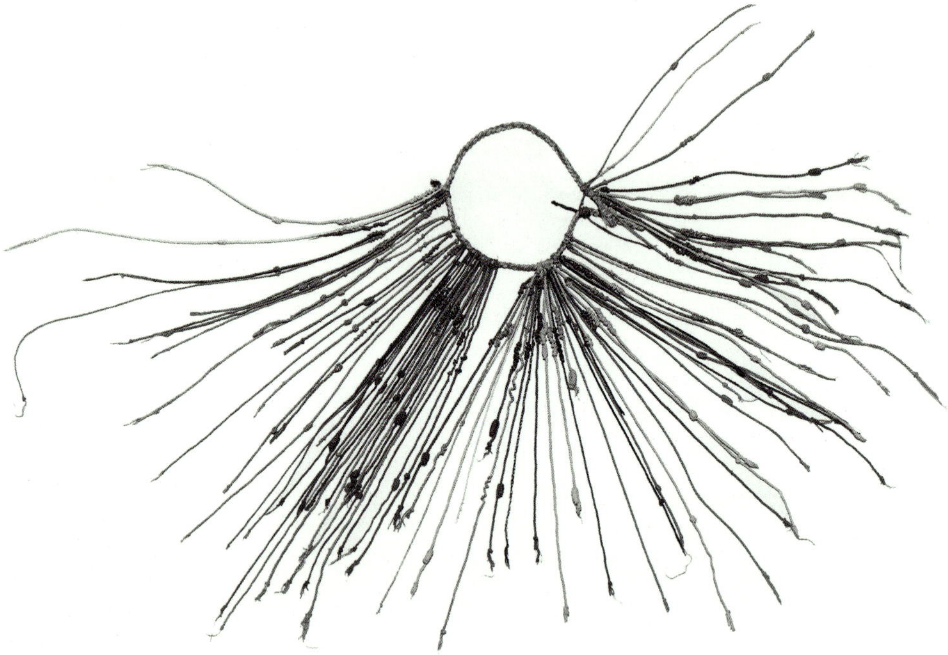

Inca khipu, Peru, 15th–16th century.

–Khipu kamayuq tata, imay sunquraq saqiykuwankiku, khuyay waqakushaqta, nispa lliw kuyakusqankuna ininiyunku tatayta ripunan patapi rikuspa.
–Ama waqaspalla kacharpariwaychis, kusisqa sunqulla Qurupunaman chayarapunayaq.
–Killa mama ña k'ancharamusqaña-haywarimuwaychis Yuyaysapa khipuyta kunan kutillatawanña khipukamayuq vidanmanta willarukusaykichis.

Yuyaysapay khipuyninpa willakuyninkunataqa yachashani nuqaqa. Sapa tutallan, papay khipunta munaraykuspa willawaqku, uñallamanta willakuykuna uyarisqanta: "Machuykichis kuwartilmata ayqirimusqa, khayna-aqna" nispa. Hinaspataq, willawaqku paypa vidamantañataq: "Kikchi qatiriykuwaq, aqna-khayna" nispa. Hinaspataq, "qankunañataq khipunchispa willakuyninkunata khipuqaychis" nispa niwaqku. Ñañay munaraqpacha q'aytuchanta, khipuspa-yuyaychaspa, imaymanata wakichiq umachanpi. Nuqaman tukuwaqtintaq mamayta mink'arikuq kani: "Alqamari t'isashawan" nispa. Mamay aysariruwaq, hinaspataq willakuykunaq qunqasqay mininta yuyarichiwaq. Hinallataqmi, kay willakuykunatapas, "tayta-mama" nispa niyku. Chaytan Tatay may awinkunamantaraq uyarispa, yuyaychakuq ishkay yuyayman: wat'isqa yuyay-vidaymanta yuyay. Ñawpaqkaqmi wiñan sach'aq raphinhina raphikama; chay huq kaqtaqmi, raphichakun papaq-raphinhina tukuyman, huq raphimanta huq raphi wiñashanña, chay raphimanta huqñataq raphipakushan.

* * *

Khipuqay khipuykita, tata, nispa samiychariyku, lliw yuyayninpa taqinta muchharinawankupaq.
–Ñachu, quri librunchis, ñachu qullqi lapranchis, khipunchispa willakuyninkunata qhillqanaykipaq?.
Ari, tata, nispa k'umuykuni.

–Kay sani chhukchawan khipusqa khipun kuyakusqay khipu. Kay chhukchan, k'ita quwa-misiq chupanpi wiñan. Chaytan kay wat'isqa yuyaypi willasaykis. Kay michin ruphay k'an nishaqtin, puhu pataman lluqsimun q'uñikuq. Chaypin t'ankaran, hinaspa phuyuta khasan, chayman rachaparukun. Hinaspa t'ipiykamun chupanta, chaylla paramama talliyakamun.
–Tata kay khipuykiqa khatashantaq. Pitaqri?
–Nuqataq kashani.
–Willariwaykuya.
–Kay yuyay ichaqa nuqaq vidaymanta. Huq kutin, p'ulquyururani waylla-wayllanta, chaypi puxuq patanpi puñushasqa misi. "Sani chhukcha" nispa qunqaylla chutashasqaniña chupanmanta. Chaylla chikchi t'uxaq, manaña kachariruwaqchu. Chaypi pituchayukuni. Ichaqa, illapanwan manaraq rupharushawaqtin, Yuyaysapay saltaxan, lliwta maqanayukunku. Rikch'ariruqtiy, sani chhukcha samashasqa makiypi.

Tata, nispa hanllatatankuraq, hinaspataq haywarinku sayri-sayrishaqta, kupa ch'iwarchatawan. "Uhaykusun Yachaq" nispa ch'allaykun khipunta traguchanwan. "Tata" nispa Yuyaysapaq khipunkuna ch'ayñaykunku. Huq kaq khipu, llapan khipukunata ch'inllarachin.

– Imataq sutinri?
– Wata puñuq mach'aqway. Wata hunt'ata puñuspan rakhullataña wiñan. Chaytan manaraq rikch'arishaqtin, sach'arunakuna umanta apaqaruspa ahiyutiwan kankanku.
– Achalaw!, nispa lliw runa ch'aqwariyunku.
– Ch'inllalla, khipu kamayuq tata qallarillanqataq, nispa mamayqa.

Kay khipu ichaqa china, lluq'iman khipusqa. Lluq'iru makillan chay khiputaqa lluq'iman phushkan.
– Phashkaqakullanchu chinakuna?
– Manan allin khipu kamayuq khipuqpas atinmanchu chay khipu ñawinchaytaqa, mulapi watasqa raqtin.
– Mulari imaynataq?
– Mula watayqa, urqu q'aytu, china q'aytu, ishkaynin ch'ulla sunqullapi wataymi.
– Tata, imaqtintaq Yuyaysapaykitari chay lluq'iykikunari uqllan?
– Lluq'iykuna, mana llamiykachinchu, uywasqa khipuykunata, haqay tumaqiya tuquwan, chaymi, q'aytuyta suwaspa, mach'aykunapi waqachin, chaymi nuqaq umaypas nanallantaq.

– Tata, imaqtintaq kay lluq'iykikunari ch'iqitaraq ch'iqishan? nispa khuyankuna tapuykun. Lluq'iman pushkasqa q'aytun, kutichin runaq rimayninta, kikimanmi ima layqasqatapas kutixatachin.

– Tata, Puka khipuykipas kasqataq.
– Ari, willariwaykuya.
– Machu Pukara mayu patanchispin, haqay españulkuna llaktanchista yawar mayuman churarqan Pukaraq qurinta suwayta munaspa. Chaymantapachan awinchiskunaq samaynin, chay qaqapi ñak'arishan.
– Amanya!

– Chhaynatan nishan khipunchispi. Chaymi, librakuyta munaspa, chayninta pasaq karrutakunata wahaspa vulkaykachinku, hinaspataq qaqa ukhuman tanqaykunku. "Chhalay" nispan chaypi musuq almata saqixatanku, paykunataq linahinku apukunaman phalarirapunku. Ichaqa, papayta qhawaykuqtiyqa, manaña runa kayninpichu samashasqa, linahinmanña thastishasqa.
– Tatay! nispa chay sip'uy-sip'uy makillanta much'aykuni.
– Usqayta, ama usuchiyku timputa, ususiy.

– Kay khipu, ichaqa yanqa lliklla phashkariylla.
– Imanishantaqri?
– Khipukamayuq tatas, Viraku mamakuq ch'ukllanman saqikamusqa q'ipinta, chaymansi rankha-rankhallataraq kutinan q'ipin chaskikanpuq.
– Virakullaqpitaqchu chaytapas saqikamunki imaynataqri? nispa mamay, uchuchayuqta, tapuykun . Lliw asirayaykunku, "atuq".

– Tata, Antawara-antawaraq khipuykiri kanchu? nispa kuyakuqninkuna tapuykunku.
– Ari kanmi. Antawara, inti q'illuy-q'illuy, waykuytan Qurupunaq, Zulimanaq, tambunman waykuykuqña kani. Chaypin, apukunaq p'aqunkunawan kukata akullispa, Yuyaysapanman khipuq kani paqukunaq harawikunata.
– Muchhariwayku, ch'ullachallata!
Yaku mamaq ñawin lliphipikuspa
wachu wachunta phawashaspa,
qunqaylla rumipi rank'arukun,
chaylla lliw pampantinta p'alqariykun.
Manapuniñan urpi
yananwan tupanmanñachu.

* * *

Chaylla "k'un, k'un", nispa willakuykuna willaq apukuna, ch'uklla pataman chayayaramunku. Hinaspataq, Apuchinninku hamunankama, quri librunkuta ñawichaykunku.

Ichaqa, manaraq Pachaq qallariyninta tukurushaqtinku, Apuchinkuna sukharimunku. "Q'iw" nispa, tatay laphanta saq'aqanpacha.

– Wira p'ukuta sipiychis. Ch'inllalla.

– Qurpallaykichis, nispa apuchinkuna waykuyaramunku ch'ukllaman.

– Tata kumpañashawankimanchu icha?", nispa tapuykun Apu Ausangati rit'i urqu.

– Tata, icha nuqawanchu" nispa tapuykullantaq Salqantay mama.

– Khipukamayuq Tata, khipuykuq hap'iqinmi kanki, nispa Qurupuna, Zulimana, iskayninku laphankuta mast'arispa tatayta chaskiykunku.

– "Pasankichu qusay?" nispa mamay waqarukun, hinaspataq ukllarukun.

– Qanña wawanchiskunata, ñitunchiskunata" nispa runa kayninpiraq mink'ayurukun.

– Yuyaysapayniy, kuyasqaq khipuy, kaykamalla nuqawanqa" nispa kinsa mat'iy watayta watayurun: Dios-yaya, espíritu-santo, amen.

– Yuyaysapaynin allin makipin saqishayki" nispa makiyman churayuruwan. "Laph" nispa Yuyaysapaq samaynin sunquypiña samaramun. "Tataq viyahimanta" nispa watayuruni Yuyaysapa khipuypi.

– Haylli khipu kamayuq mama! nispa lliw haylliykuwanku.

– Quri-libruyta haywarimuwaychis qiw, ch'ukllapatapi ñawinchanaypaq qiw, nispa kuntur kayninpiña mañarukun.

– Ñachu? nispa tapuykunku Apunchiskuna.

– Ña, haku, nispa phalarirunku.

THE KHIPU MASTER

"Tata, Khipu Master, how can you leave us this way, weeping with such deep sadness," murmur Tata's assistants at the hour of his passing.

"Don't continue crying this way, my children. It is with a full heart that I will pass on to Qurupuna. The full moon has risen now and she shines brightly; pass me my *Yuyaysapa*, wisest of khipu cords, so that I might tell you once more the story of my life as a *khipu kamayuq*, master of khipus."

And just as Tata wanted, I remember all of the khipus' stories. Each night, my father would consult with the cords and their knots, recounting stories he had heard since childhood.

"And so it was that your great-grandfather deserted the army encampment..."

"...that the puma abducted the shepherdess..."

"...that I crossed paths with a condemned soul..."

Later, he would tell us of more recent adventures before inviting us to join in.

"From oldest child to youngest, trace now the knots of our khipu cord."

Our oldest sister would then stroke the slender cord that my parents handed to her, knot and imagination taking hold in her heart. Knot after knot, my turn would come. The breath of the slender cord seemed to recognize, to encircle me, accelerating the memories. When distracted little brother's turn came, he would often shout, "The falcon messed me up!" so that Mother would rush toward him, patiently helping him to remember the forgotten passages.

These knots, passed down through the generations, are the father-mother knots with which I learned to tie the structured memories of my own experiences. The first type grow orderly, like strong tree branches, creating forms which emerge from a single trunk. The second reproduce like the intricately rhizomatic *yura* potato, always unique, always different from others of their species, without following a formal pattern, without growing outward from a single trunk. These second knots are like seed potatoes, containing so many *ñawi*, eyes, that when you plant them a root sprouts from each eye, and from that root a new potato, and so on in succession.

"Trace now your knots!" we then entreat Tata, trying to open the storehouse of Father's memories.

"Are you ready with your golden book and your silver pen to record the memories of our khipu?" Father asks me.

"Yes, Tata," I respond.

"This knot, made of the finest thread, is a beloved token," he assures me. "It is found on the tail of the fearsome Andean mountain cat. Here I have the first knot on the structured cord. The myth tells us that the wild cat emerges to sunbathe on the banks of the stream. He gambols and cavorts about, he belches and clouds appear, and with a single leap he climbs them. He plays a bit before lashing his serpent's tail, unleashing the rain."

"Tata, this small knot is trembling. Who is it?"

"It is me."

"Tell us, Father."

"I have here a knot of my own experience. One afternoon, I took a shortcut through some bushes and there on the stream bank lay a *quwa*, a wild Andean mountain cat, sleeping. 'My chance to secure the sacred *quwa* cat's finest thread!' I said to myself, and then there I was, pulling its tail. Suddenly, a bolt of lightning exploded and the cat cornered me. I fell to my knees. Just as its lightning was about to strike me down, my *Yuyaysapa*, wisest of khipu cords, jumped up and began to whip him. When I awakened, a red wad of threads lay throbbing in my hand."

Those gathered around look at Tata admiringly and offer him tobacco and bit of liquor. "Cheers to my *Yuyaysapa*, wisest of khipu cords," he says, and moistens the cord respectfully with a few drops of liquor.

"And Tata, what about that knot, what is it called?"

"*Wata puñuq mach'aqway*, the serpent who sleeps for a whole year. During that time he grows fat, and at the right moment, just before he wakes up, the *ch'unchukuna* jungle dwellers slice off his head and roast it with annatto."

"Yum, how delicious, this sleeping-serpent meat!" we all exclaim.

"Quiet now, the Khipu Master is about to unravel another tale," Mother tells us as she covers Tata with his flowered blanket.

"This knot is female and she is tightened here, on the left side. These knots are tied by those who can entwine with the left hand."

"And is it possible to untie these female knots?"

"*Manan*, not even the most astute of the *khipu kamayuq* masters can read these sorts of knots, since they are tightened in *mula watayqa*, mule knots."

"What is a mule knot?"

"The mule fastens together the male and the female, to unite them in one heart."

"Tata, why do the knots tied on the left seem to hug so tightly to your cord?"

"Ahh... those are the brave defenders who protect fellow knots broken by insatiable toad's trap. At times, toad also abducts my khipus, abandoning them alone, wailing in distant caves. When this happens, my head throbs in pain."

"And Tata, you also have red threads."

"So it is. And it unravels in this way … On the banks of the mighty and venerable river Pukara, the Spaniards brought suffering to our people, bathing the rivers in blood, as they sought to enrich themselves with the gold of our lands and of our princess Chuqichampi … *llaqtanchista yawar mayuman churarqan*. Ever since, the breath of our ancestors suffers there in the ravine."

"This cannot be!"

"But that is just how it is knotted into memory. To alleviate such suffering, our grandmothers and grandfathers reach out for the buses and cars that pass over the bridge and push them toward the precipice. 'Come barter with us!' they call out, leaving behind the new souls as they take flight toward the snowy peaks of their homelands."

Suddenly, I realize that Tata is breathing differently. He has already begun his journey to Qurupuna.

"Tata! I cry, kissing his wrinkled hands."

"Don't waste your time, my child."

"These knots are simple, easy to unravel."

"And what lesson do they offer us?" And so Tata tells us once more about the time when our ancestors lived without the light of Father Sun and worked to germinate seeds of potato, quinoa, and tobacco with the moon's light alone, when coca leaves were still inedible. Before the time of the Incas and the Sun, before llamas and alpacas were domesticated and began to produce rich and luxurious wool in the beautiful variety of colors we now know.

* * *

"Tata, do you also have the coveted khipu from the Antawara times?" the listeners wish to know.

"So it is … At that very hour, I was arriving at the inn and storehouses of the sacred Qurupuna and Zulimana. And there beside the fire, while chewing their coca leaves, they twisted and looped my khipu with the most beautiful verses of the sage poets."

"Share one of the stanzas with us, Tata, please!"

"But there isn't time …"

"Just one, please, Tata?"

Yaku mamaq ñawin,
allinta phawashaspa rank'arukun,
manapuniñan urpi yanantin
tupanmanchu.

Mother Water's spring,
running so fully but losing sight,
and just like that, the young lovers
never to meet again.

And suddenly, *k'un, k'un*, the sound of loud footsteps crossing the roof of the hut; the arrival of the sacred condor storytellers, *Apu kunturkuna*. While they wait for the arrival of the other hallowed beings, the condors open the golden book. They haven't even read the first paragraph about the birth of the universe when one of them begins to screech from afar. Then, "*Q'iw*!" Tata screeches, at the same moment that he unfurls his new wings.

"Snuff out the wax plate," he calls.

Silence.

"Visitors!" they shout, now from within the hut.

"Tata, would you like to come with me?" the sacred snowcapped peak, Ausangati, asks him.

"Would it please you to come with me?" snowy Mother Salqantay offers too.

"Tata, you hold the breath of our knots," Father Qurupuna and Mother Zulimana avow, and they unfold their wings to draw him near.

"Dearest husband, will you be leaving now?" Mother softly moans, embracing him tightly.

"Please care for our children and our grandchildren. Dearest khipu, you will accompany me only this far." Tata exhales deeply and ties three final knots along the cord. "*Yaya*-Father, Holy Spirit, amen. *Yuyaysapa*, wisest of khipu cords, my friend, I leave you now in good hands," he manages to say as he lays the khipu in my hands. "Laph!" The breath of *Yuyaysapa*'s life force and wisdom flows into my heart.

"For Tata's journey," I declare as I tie my first knot.

"*Haylli*! Long live the *Khipu kamayuq warmi*, the woman who is khipu master!" the onlookers approvingly call out to me.

"Pass me the golden book, *Q'iw*, I'll be needing it to read about thatching roofs, *Q'iw*." Tata manages to make this last request, now in his condor form.

"Your bag of coca leaves and your flask!" my brothers remind him.

"Where is my faithful pack llama?" calls out Tata-condor, grasping the llama's lead.

"*Haylli*! Hurrah for the Khipu Master!" we all cheer him on.

"Ready now?" the mountains inquire.

"Ready," Tata confirms, taking flight.

STORY 18

Four Topographies

Chloe Aridjis

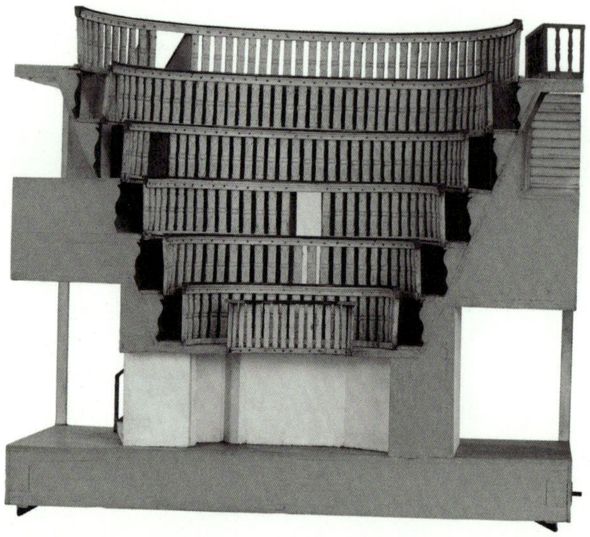

Model in two parts of the anatomical theater of the University of Padua, 1932–1933.

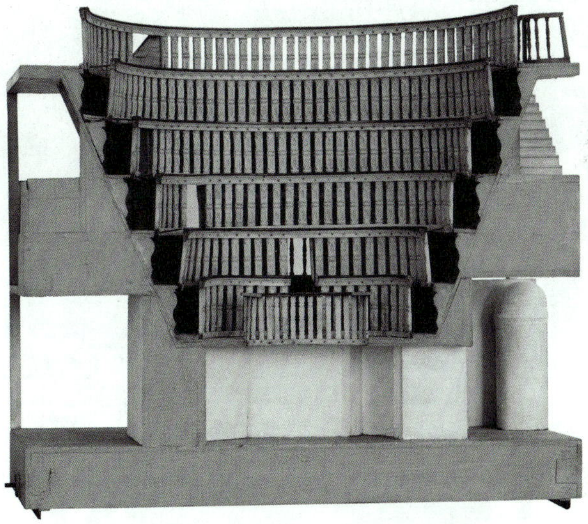

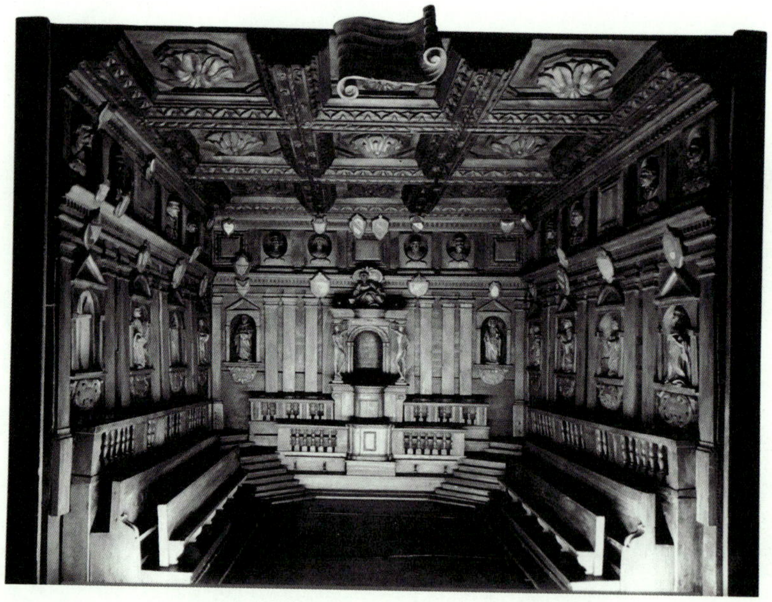

M. Caimi, model of the anatomical theater of the Archiginnasio, Bologna, 18th century.

Like the cadavers themselves, the anatomical theater was carved up and segmented, establishing its own geometry of life and death: a funnel-shaped space like an upturned cone, with six steep concentric circles that narrowed in on the dissecting table. During one particularly cold winter in Padua, 1602, four cadavers were brought in for examination, four different topographies, each harboring its own past, although this past remained hidden to the two hundred spectators who came to witness the mechanisms of existence unveiled.

First subject: male, twenty-six years of age. Hanged four days ago. Muscular legs, broad shoulders, and small hands, almost like those of a child. As the professor of anatomy reads from his book, the demonstrator points to the chest area slightly left of center and the dissector instructs the barber surgeon where to perform the incision, there where dwells the dense, inert muscle that is the heart.

The gray, waxen face that stares vacantly upward belongs to Pietro, the fruit seller's son, of no fixed residence. After his father's death he turned to rural banditry, a calling he acquired with ease. Solitary roads and mountain passes were his accomplices. Always on the move, always in the shadows, he now lies in total stillness beneath the light of eight student-held candles. In life, he suffered from an overwhelming attraction to pockets; now a man thrusts his hands into the cavities of his body. He had always been drawn to reliquaries, lighting candles in churches to overlooked saints before stealing from the more crowded altars. After he spent much of his life in the pursuit of silver, a sharp polished instrument now slices through his chest.

Subject two: male, eighteen. Shoulder-length hair, long nose, and slender body, as if arrested in the final burst of growth. With forceps the barber surgeon holds up a tangle of intestines from the abdomen while nearby, watching them uncoil, gather nobles and other influential citizens—closer to the source of illumination, closer to the stench of death. Medical students crowd into levels two and three of the theater, a restless insolent mass that infuses the space with its life force, barely contained within the wooden balustrades, their attention span constantly put to the test. Above the students, in the three uppermost tiers, gathers the general public, a curious, morbid assortment of humanity.

This is Andrea, the old carpenter's apprentice. In life he never came into contact with the printed word, but now a fine book, complete with illustrations, is wielded over his head. Orphaned at five, taken in by the childless carpenter and his wife, he began to fixate on the matter of inheritance, convinced he would receive none. As he watched the man carve wooden toys and polish tables he kept his violent thoughts to himself, until one morning he walked over and struck him with an adze. Any carpenter would marvel at the theater's interior, the spruce and walnut that went into this structure in which knowledge circled upward, but the apprentice is unable to appreciate the functional beauty of its ellipse through his lusterless eyes.

Subject three: female, thirty-three. Cropped hair, small waist, mouth partly open. Women are rare: in the tense darkness all eyes strain to see the cadaver brought from Venice three nights ago.

Only twice had Sara left the ghetto; the sight of its gates, locked at dusk, caused great anxiety. Eight months ago a fever had swept over her as she wandered its streets, sleepless and incoherent, terrified of her own reflection in the canal and of clothes strung out to dry, their limbs and torsos flapping with intention. In her small room at the top of a house, the shadows at night took on the shapes of her ancestors. She hadn't entered a synagogue in years but would occasionally light the candle she carried in her bag, and she ate the bread the old baker gave her. After tumbling down a flight of stairs she was taken to a hospital, where she caught an infection and died soon after.

During an interval, a troupe of lute players wanders in and plays three songs, their appearance distracting the students from stomping impatiently on the floor and wearing down the wood. Once the musicians leave, the anatomists resume their study of the upper and middle layers of the woman's womb.

Subject four: male, forty-four. Shapely arms, large thighs, and oddly shaped cranium; in life, he suffered daily from headaches.

This is Emiliano, a farmer from Calabria, who killed his daughter's impertinent suitor with a rock. Summer and winter, day and night, the young man would prowl their orchard and the nearby roads, his face imprinted on every corner of their lives. He helped with the harvests but refused to leave once work was over, always lurking at the end of the field, bringing misfortune, Emiliano was certain, to the family. The crops withered, his daughter aged under the man's gaze, his wife began speaking in another dialect. Yet no one could understand Emiliano's distress, and far less the violence of the act he committed, and after he was hanged they donated his body to science in exchange for a free funeral.

In the depths of the farmer's cadaver, one student envisions the landscapes of his own childhood—the rivers, mountains, and valleys—thinking of paths of circulation and the constellations of features around them. As a blade slashes through nerve endings that once enabled all sorts of dialogue, he tunes out the professor's rhetoric and dwells instead on his own life paths, wondering how much separates him from the dead man below.

Nearly a century after this cold winter in Padua, a new generation of cadavers was carted into the new anatomical amphitheater in Bologna. Wood-paneled, with an ornate ceiling, marble dissecting table, and sculptures of famed physicians, it further ennobled this exploration of our interiors. The body itself may have seemed like a theater within a theater, yet no knife or scalpel, however honed, could peel away the mysteries of the inner life.

STORY 19

The Great Operation

John Keene

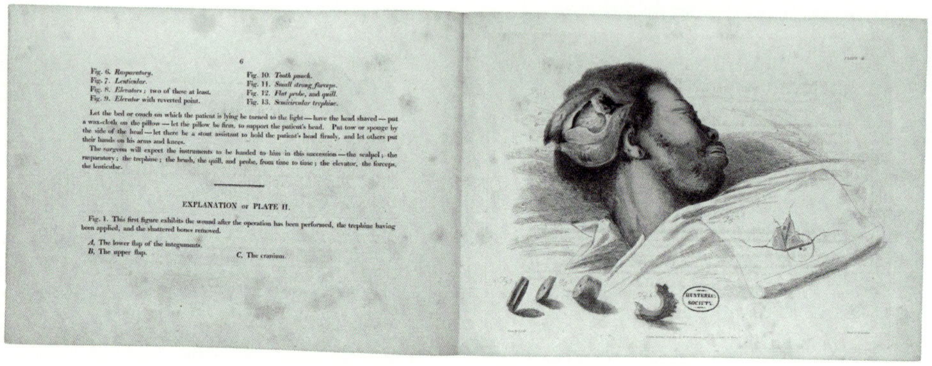

6

Fig. 6. *Raspatory.*
Fig. 7. *Lenticular.*
Fig. 8. *Elevators*; two of these at least.
Fig. 9. *Elevator* with reverted point.
Fig. 10. *Tooth punch.*
Fig. 11. *Small strong forceps.*
Fig. 12. *Flat probe*, and *quill.*
Fig. 13. *Semicircular trephine.*

Let the bed or couch on which the patient is lying be turned to the light—have the head shaved—put a wax-cloth on the pillow—let the pillow be firm, to support the patient's head. Put tow or sponge by the side of the head—let there be a stout assistant to hold the patient's head firmly, and let others put their hands on his arms and knees.

The surgeon will expect the instruments to be handed to him in this succession—the scalpel; the raspatory; the trephine; the brush, the quill, and probe, from time to time; the elevator, the forceps, the lenticular.

EXPLANATION OF PLATE II.

Fig. 1. This first figure exhibits the wound after the operation has been performed, the trephine having been applied, and the shattered bones removed.

A, The lower flap of the integuments.
B, The upper flap.
C, The cranium.

Charles Bell, *Illustrations of the Great Operations of Surgery: Trepan, Hernia, Amputation, Aneurism and Lithotomy*, London, 1821. Plate II illustrating the wound after the operation has been performed; plate III illustrating a skull with various examples of fractures.

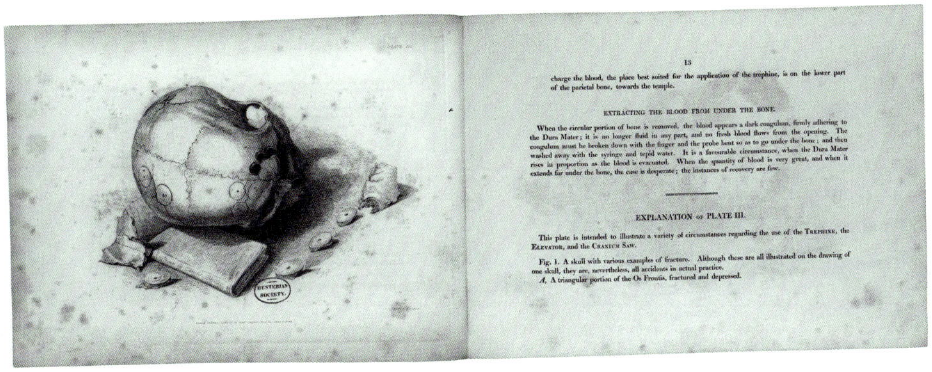

13

charge the blood, the place best suited for the application of the trephine, is on the lower part of the parietal bone, towards the temple.

EXTRACTING THE BLOOD FROM UNDER THE BONE.

When the circular portion of bone is removed, the blood appears a dark coagulum, firmly adhering to the Dura Mater; it is no longer fluid in any part, and no fresh blood flows from the opening. The coagulum must be broken down with the finger and the probe bent so as to go under the bone; and then washed away with the syringe and tepid water. It is a favourable circumstance, when the Dura Mater rises in proportion as the blood is evacuated. When the quantity of blood is very great, and when it extends far under the bone, the case is desperate; the instances of recovery are few.

EXPLANATION of PLATE III.

This plate is intended to illustrate a variety of circumstances regarding the use of the TREPHINE, the ELEVATOR, and the CRANIUM SAW.

Fig. 1. A skull with various examples of fracture. Although these are all illustrated on the drawing of one skull, they are, nevertheless, all accidents in actual practice.

A, A triangular portion of the Os Frontis, fractured and depressed.

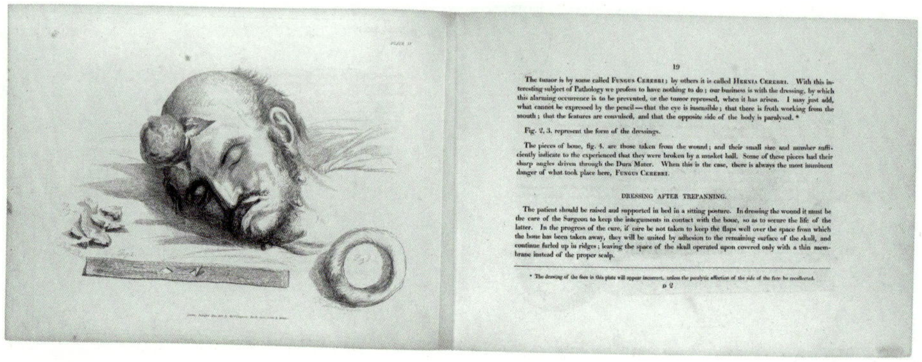

19

The tumor is by some called FUNGUS CEREBRI; by others it is called HERNIA CEREBRI. With this interesting subject of Pathology we profess to have nothing to do; our business is with the dressing, by which this alarming occurrence is to be prevented, or the tumor repressed, when it has arisen. I may just add, what cannot be expressed by the pencil—that the eye is insensible; that there is froth working from the mouth; that the features are convulsed, and that the opposite side of the body is paralysed. *

Fig. 2, 3. represent the form of the dressings.

The pieces of bone, fig. 4. are those taken from the wound; and their small size and number sufficiently indicate to the experienced that they were broken by a musket ball. Some of these pieces had their sharp angles driven through the Dura Mater. When this is the case, there is always the most imminent danger of what took place here, FUNGUS CEREBRI.

DRESSING AFTER TREPANNING.

The patient should be raised and supported in bed in a sitting posture. In dressing the wound it must be the care of the Surgeon to keep the integuments in contact with the bone, so as to secure the life of the latter. In the progress of the cure, if care be not taken to keep the flaps well over the space from which the bone has been taken away, they will be united by adhesion to the remaining surface of the skull, and continue furled up in ridges; leaving the space of the skull operated upon covered only with a thin membrane instead of the proper scalp.

* The drawing of the face in this plate will appear incorrect, unless the paralytic affection of the side of the face be recollected.

D 2

Charles Bell, *Illustrations of the Great Operations of Surgery: Trepan, Hernia, Amputation, Aneurism and Lithotomy*, London, 1821. Plate IV illustrating the consequences of the rupture of the dura mater.

A hush blankets the bright theater.

I hear myself talking but is it only in my head or can they hear me speaking aloud saying my name is Dominick Nero to the ones holding my legs and feet as if the neck of a chicken or the ones resting their elbows on my hips and the ones who have turned my shoulders into a tabletop like the one I am lying on, the doubling like my eyesight and I had said to Mrs. Harris who runs the grocery since Mr. Harris passed that I am seeing everything twice and she said Oh that happens when you grow old Dom and even my dear old Harris had spells but I am only twenty-six years myself, I said to her, and I can no longer get to sleep and my head aches like a splitting stone all the time and I am prone to drop things but it was not until I collapsed in the street and Dr. Lewes an associate of Dr. Bell was one of the white men who happened to be passing by examined me and recommended me to Dr. Bell who met with me asking me numerous questions like when did the double vision begin and when did you first have trouble with your balance and do you have nightmares and I could have told him quite an epic just from daily life but I stuck to the questions and he said Yes, Dominick, yes, that is very interesting and an assistant was also there writing as well dipping the quill into the ink as fast as I spoke, yes, Dominick, I think we shall examine you now, Dr. Bell said and he looked me over thoroughly peering into my eyes and ears and his assistant did as well though I have never had a doctor touch me not even when I was delivered to this earth according to my mother and he said I think you have a lesion, Dominick, and ran his hand along the ridge of my head through my glen of hair, I would like to remove it, Dominick, and then he told me so many other things I had to pack them tightly like limes in my memory about how he was going to operate and how I would ›

Everyone has assumed their position, one holding the patient's bare ankles and feet, another pinning his slender hips, under the sheet, to the table, two others weighing on his shoulders, a fifth with the ale-brown face in his pale hands, drained of what little color they had from the pressure of keeping the chin steady and the patient's left cheek flush with the folded towels atop the wood. This last one alternates between staring at the man beneath him and looking away.

Dr. Bell is reviewing his surgical tools, freshly polished for this first task of the day, the knives and saws and drills glistening in the morning light spilling into the room, and observing the sketch artist, whose work he will refine for his lectures and, he intends, his anatomical guide. Several of his colleagues, as well as students poised with pencils and paintbrushes and notebooks, and attendants holding towels, stand a short distance away, on the floor as opposed to the theater's elevated gallery, so as not to miss the operation up close.

"What if he bucks?" Collier, vising his fingers around the patient's neck and jaw, asks Dr. Bell. "He looks to be a fiercesome one. Lewes says he was a mason before he started at the greengrocers. Though I heard they were insensible to pain, with skin thicker than cowhide. I've never been here when you operated on one."

"Elephant," says Dr. Halsey. "Like where he's from."

"Not so," Andrews answers. He is leaning on the patient's shoulders, twisting them toward the floor. Dr. Lewes, his partner this time, pushes back. "The fellow that Dr. Morris operated on in January screamed out like a cat being wrung through and we had to give him a whole other dram and tie his hands to calm him down. It left me trembling so I almost ›

‣ feel so little pain I would not know what was underway and how I should not eat a single thing for a day before he cut the lesion out and how there would be medical men observing us closely and how he had succeeded at many such surgeries and how I, Dominick, would make medical history and fare after and though these ones in here keep referring to me as the African which I am of course I was born in Liverpool in this kingdom as I told Dr. Bell whose assistant may have recorded that too and came to London at age two with my parents and I thought to go to Sierra Leone when the crown made the offer and my older brother Sancho is there, and has set up a homestead and found a wife, a Mende woman, Nemy, and is practicing his faith and animal husbandry, his letters tell me, though I can no longer read them and Olivia my sister must read them to me, her voice like honey pouring from a comb, and she and I chose to stay here to be close to our mother and father and because I did not want to start over across the sea but when I am well and Dr. Bell has cut the lesion out I just might though it is pleasant enough here at times even during winter and I have my dear company near, not just Olivia and my parents and soon I hope someone of my own and little ones and the church and my societies and the work with Mrs. Harris who has agreed that I may return when I am better and I wish Olivia were here beside me because I am seeing all of these men in double and I am saying I am ready please get on with it and I do not think it will pain me as much as my poor head does already and I am floating now seeing two of the artist looking intently at me and I am saying my name is Dominick please paint or etch that at the bottom of the paper but they are not listening to me and continue to stare until Dr. Bell whom I cannot see because he is behind me says:

‣ needed a whiskey myself. I found this one in the street, you know."

"Glad I wasn't there myself," Hughes says. He grasps the patient's ankles tightly.

"Nonsense. Of course they are not insensible," Dr. Bell says, "and feel terrible pain just like us. That is why I raised the dosage. Now please, keep him steady. I am about to begin. Remember, please remain silent; look away if you start to feel unwell and try your best to suppress it."

Hughes glances down at the tin buckets near his and Collier's feet and wonders if he will hit his if he needs to. He has spent many a day in a medical theater so he should not, he reminds himself, feel revulsion at seeing these anklebones dark as loam and the pink heel and soles in his grasp, though he would much rather be peering into the open skull, which for some reason does not repel him as much, or at all, and he recalls the baker from Stepney who had a growth the size of Wales protruding from the side of his crown, and that time he was tasked with gripping the head, with one of Bell's students, and despite the smell and the patient's suit of rags he not only had not released his hold nor grown ill but could not unpeel his eyes from Dr. Bell's hand guiding the scalpel as it slowly opened the red, bloody flesh.

"Gentlemen, are you ready?"

All say yes in unison, including Fletcher, the draftsman, glancing at the patient, who appears to be mouthing something, and adjusting his easel, though he may lift the board holding the paper and carry it closer, as in the past, making sure never to stand too close to the others or Dr. Bell so as not to stumble into him or break his concentration, adamantine as it is, while also not breaking his own.

Dr. Bell, lowering the scalpel to the shaved skull, whispers:

Then let us begin.

STORY 20

Cerebri Anatome, 1664

Hari Kunzru

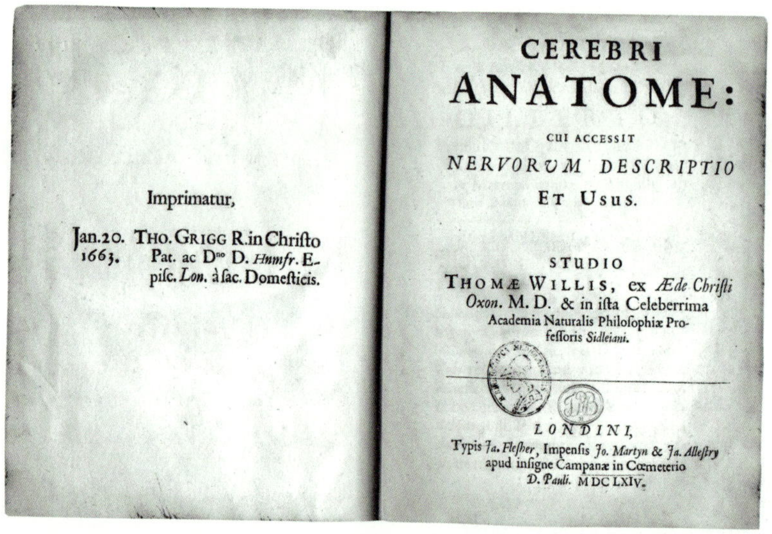

Imprimatur,

Jan. 20. THO. GRIGG R. in Chriſto
1663. Pat. ac D^no D. *Humfr.* E-
piſc. *Lon.* à ſac. Domeſticis.

CEREBRI
ANATOME:
CUI ACCESSIT
NERVORUM DESCRIPTIO
ET USUS.

STUDIO
THOMÆ WILLIS, ex *Æde Chriſti*
Oxon. M. D. & in iſta Celeberrima
Academia Naturalis Philoſophiæ Pro-
feſſoris *Sidleiani.*

LONDINI,
Typis *Ja. Fleſher*, Impenſis *Jo. Martyn* & *Ja. Alleſtry*
apud inſigne Campanæ in Cœmeterio
D. Pauli. MDCLXIV.

Thomas Willis, *Cerebri Anatome* [Anatomy of the brain], London, 1664.
Frontispiece and a spread with plate no. 1 illustrating the human brain.

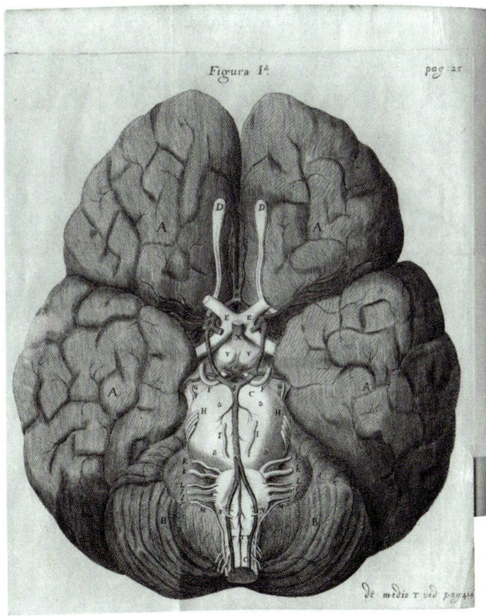

Cerebri Anatome. 25

Figura Prima

Cerebri humani Cranio exempti basin cum Vasorum abscissorum radicibus ostendit.

A. A. A. A.	*Cerebri quadripartiti anteriores, posterioresque lobi.*
B. B.	*Cerebellum.*
C. C.	*Medulla oblongata.*
D. D.	*Nervi olfactorii, sive Par primum.*
E. E.	*Nervi optici, sive Par secundum.*
F. F.	*Nervi oculorum motorii, sive Par tertium.*
G. G.	*Nervi oculorum pathetici, sive Par quartum.*
H. H.	*Nervorum Par quintum.*
I. I.	*Nervorum Par sextum.*
K. K. k. k.	*Nervi auditorii, & eorum utrinque bini processus, Par septimum.*
L. L. l. l. l. &c.	*Par vagum, sive octavum, pluribus fibris constans.*
M. M.	*Nervus spinalis, ad originem Paris vagi à longinquo accedens.*
N. N.	*Par nonum, pluribus etiam fibris constans, (quæ deorsum tendentes, in*

E *eundem*

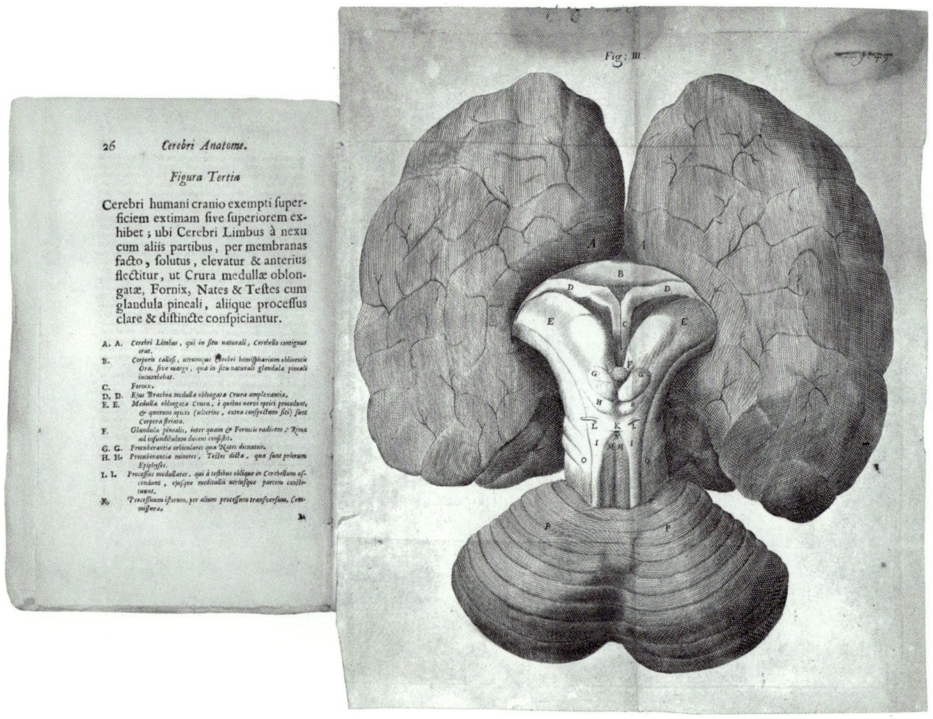

Thomas Willis, *Cerebri Anatome* [Anatomy of the brain], Amsterdam, 1664.
Spreads with plates nos. 3 and 4 illustrating the human brain.

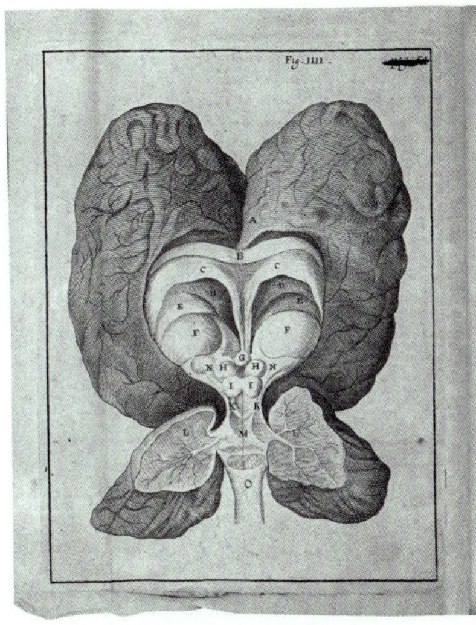

Cerebri Anatome. 27

L. *Nervorum patheticorum à Commiſſura proceſſuum prædictorum Exortus.*
M. M. *Medullæ oblongatæ portio, proceſſibus & protuberantiis prædictis ſubſtrata.*
N. *Foramen ventriculi, ſeu cavitatis, quæ protuberantiis orbicularibus ſubjicitur.*
O. O. *Protuberantia annularis à Cerebello dimiſſa, & medullam oblongatam amplexantis, portio.*
P. P. *Cerebelli ſuperficies extima, & ſuperior.*

Figura Quarta.

Effigies Cerebri humani, quod cujuſdam Adoleſcentis, ab ipſa nativitate Fatui, & ex eorum numero qui vulgo Lemurum ſubdititii perhibentur, fuerat; Cujus Cerebri moles, cum tenuior ac ſolito minor fuit, limbus ejus ulterius elevari & reflecti potuit, ut interiora quævis penitius conſpicerentur.

A. A. *Cerebri Limbus elevatus, & plurimum reclinatus; qui in naturali ſitu medullæ oblongatæ prope Cerebellum innixus, Nates & Teſtes abſcondebat.*
B. *Corporis calloſi ora, ſive margo inferior.*
C. C. *Fornix cum duobus brachiis ejus, crura medullæ oblongatæ amplectens.*
D. D. *Interna Cerebri cavitas, à limbi ejus complicatura circa medullam oblongatam reſultans.*
E. E. *Crurum medullæ oblongatæ Apices, ſeu corpora ſtriata.*
F. F. *Thalami nervorum Opticorum.*
G. *Glandula pinealis, inter quam & fornicis radicem, foramen,*

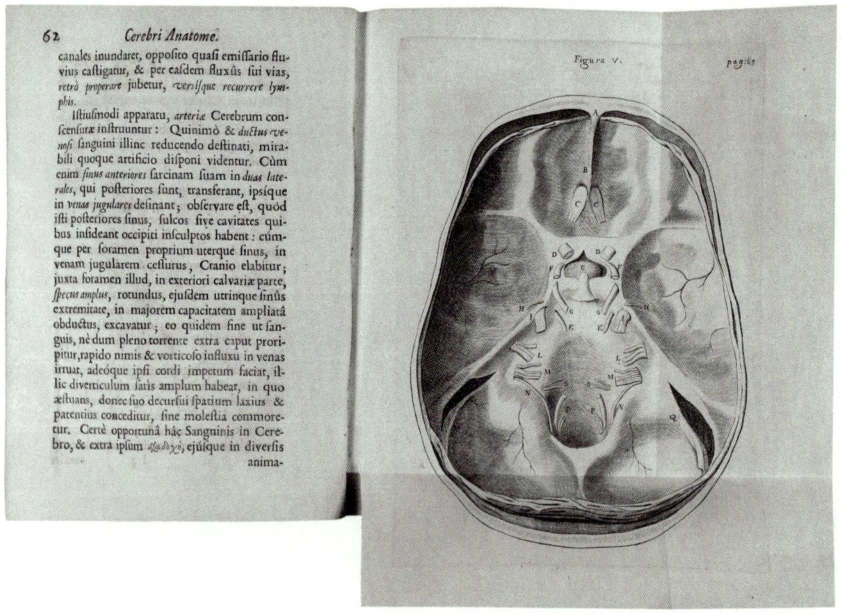

62 *Cerebri Anatome.*

canales inundaret, oppoſito quaſi emiſſario fluvius caſtigatur, & per eaſdem fluxûs ſui vias, *retrò properare* jubetur, *verſiſque recurrere lymphis.*

Iſtiuſmodi apparatu, *arteriæ* Cerebrum conſcenſuræ inſtruuntur: Quinimò & *ductus venoſi* ſanguini illinc reducendo deſtinati, mirabili quoque artificio diſponi videntur. Cùm enim *ſinus anteriores* ſarcinam ſuam in *duas laterales*, qui poſteriores ſunt, transferant, ipsíque in *venas jugulares* deſinant; obſervare eſt, quòd iſti poſteriores ſinus, ſulcos ſive cavitates quibus inſideant occipiti inſculptos habent: cúmque per foramen proprium uterque ſinus, in venam jugularem ceſſurus, Cranio elabitur; juxta foramen illud, in exteriori calvariæ parte, *ſpecus amplus*, rotundus, ejuſdem utrinque ſinûs extremitate, in majorem capacitatem ampliatâ obductus, excavatur; eo quidem fine ut ſanguis, nè dum pleno torrente extra caput proripitur, rapido nimis & vorticoſo influxu in venas irruat, adeóque ipſi cordi impetum faciat, illic diverticulum ſatis amplum habeat, in quo æſtuans, donec ſuo decurſui ſpatium laxius & patentius conceditur, ſine moleſtia commoretur. Certè opportunâ hâc Sanguinis in Cerebro, & extra ipſum [illegible], ejúſque in diverſis anima-

Thomas Willis, *Cerebri Anatome* [Anatomy of the brain], London, 1664.
Spread with plate no. 5 illustrating the human brain.

One evening the Sedleian Professor of Natural Philosophy and the Savilian Professor of Astronomy meet together in the astronomer's rooms, which overlook a stark quadrangle. A fire is blazing in the hearth and the college porter has long since locked the gate for the night. Wrapped in darkness, the world beyond the walls seems distant and somehow more tractable than by day. One could believe it might yield its secrets to the determined speculation of eminent men. Willis, the natural philosopher, confesses to the astronomer, Wren, that he has, for the time being, put aside his previous researches—into fevers and fermentation and the motions and heats of blood. "I am," he says, "occupied with another question. I am resolved to discover the seat of the soul."

Wren is surprised. He knows that his colleague is speaking too freely. Trespassing on the province of religion carries a risk, and with the king only four years restored to his throne, they are living in dangerous and mutable times. Yet he himself is a man of no small ambition. They have before them a bottle of the college Madeira. The talk takes them late into the night.

"The key to understanding the rational soul," says Professor Willis, "is anatomy. Almighty God has conceived his creation according to a plan. I look into nature as into a book. There, out of the recesses, by anatomy, is the truth brought forth from its den."

"I have read—" starts Wren.

"No, no," says Willis. "It is not a question of reading. In this matter no authority, however illustrious, can be relied upon."

Wren: "A Frenchman says the animal spirits are a kind of water, which flows through the nerves by a system of pumps."

Willis: "I have dissected nerves and found no pumps, unless they be invisible to the naked eye."

Wren: "Do you believe, like Harvey, that the soul resides in the blood?"

Willis: "No, sir, I do not."

Wren laughs and pours his friend another glass of the smoky wine. Willis looks haunted. When he speaks, it is barely audible.

"I have addicted myself, Wren, to the opening of heads."

"What kind of head?" asks Wren.

"Every kind. I have opened the heads of calves and sheep and dogs. Poultry. Swine."

"And men?"

"And men."

Professor Wren is seized by thoughts of volume, of rotundas and cylinders and vaults and domes. He has made a special study of architecture, and has recently drawn up plans for a new university auditorium, inspired, as he will tell anyone who cares to listen, by the great theater of Marcellus in Rome. What lies hidden inside the cavern of the skull, what theatre of human thought and emotion is to be found there?

"So," he asks, "what have you seen? Is it possible to discern the action of the animal spirits?"

Willis looks pained. "The spirits," he says, "are weightless. They can be neither seen nor touched, and thus they do not make an impression on the nerves as they pass."

Willis confesses to his friend that he has not the skill to anatomize his specimens himself, and so he has made use of a learned doctor by the name of Richard Lower. "A man of sharp wit and sharper knife," he quips. Wren nods, distracted, and more than a little drunk. He now understands something that has been puzzling him. Willis always smells strange, because of his experiments with fermentation, but lately he has had a peculiar odor about him. It is the smell of corpses, of bodily decay.

Some days later, Wren arrives at Willis' dissecting rooms, trepidatious and excited. A sullen manservant opens the door and leads him through a room lined with shelves, each one crammed with great jars of cloudy liquid in which float indeterminate forms. He looks about him, appalled. In the jars are brains of every size. Willis has slain whole hecatombs of animals. Through this hellish pantry lies the door to the dissecting room. The smell hits him as soon as he is shown in.

Standing over a white-shrouded table is a man with reddish hair and narrow, rather widely set eyes. Wren finds himself unnerved by Dr. Lower. Watching him slide a long knife up and down a broad leather strop, Wren feels the small hairs on his neck rise up, as if in anticipation. Willis arrives, rubbing his hands. Without ceremony, he pulls down the shroud to reveal the cadaver. There on the table lies an old man with sunken cheeks and a prominent nose, his pate already shaven for Lower's knife. Wren is about to ask where Willis acquired his specimen, but something stays his tongue. Better, perhaps, not to know.

As Wren watches, Lower makes a quick movement, then tugs the scalp away, holding it between his fingers like a conjurer with a handkerchief. Then he takes a little saw and follows the same line, carefully traveling around the skull. Lastly, he makes crosswise cuts and—much to Wren's disgust—sets to work with a penknife and a pair of shears, opening the head like a fruit.

As Lower carries on breaking away bits of bone with his shears, Willis points out arteries and other vessels. Lower pauses to wipe his hands. Then, indefatigable, he sets to work again. In this way Wren is shown, like a man being led through outer and inner chambers toward the presence of a king, the three coverings of the brain, each of a different substance. At last Lower cuts various strings or anchors and Willis is able to remove a sort of spherical jelly from the cavity of the skull and show it forth in his hands.

Wren is humbled. It seems impossible that all the impressions and memories of a man could be contained in this mute shape, with its mysterious prominences and infoldings. He is looking upon the very chapel of the Deity, the seat, as Willis puts it, of the soul. And yet he sees nothing that he can understand. Not a thing.

STORY 21

Memoir of Clemente Susini, Upon the Hiring of a New Assistant

Rivka Galchen

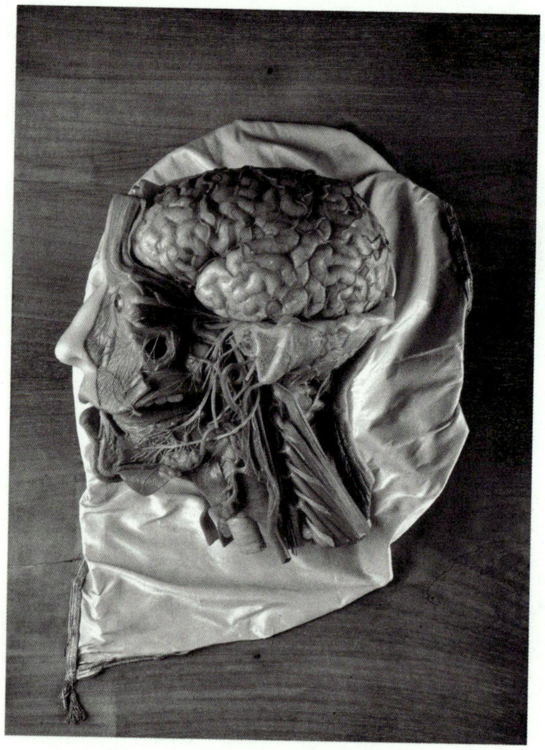

Francesco Calenzuoli, wax anatomical model of section of the head and neck with the brain exposed, covered by the dura mater and showing various dural venous sinuses in situ, 19th century.

I was a quail as well, in the beginning. I wonder what the abbot Fontana thought of me. I was a quail and a superstitious old peasant and, at the same time, a child. I shouldn't be surprised, or annoyed, that now my own assistant, nineteen years old, is the same. I resolve to tolerate him. For at least two weeks. Then I will reassess.

I do remember now that when Fontana first hired me, I thought that slicing through a wrist would be like slicing through, say, marble. Ferric veins. Some silty strata. But overall: smoothness. Marble was a material I knew better than I knew my own body. Also clay. An encounter with either left a powder on my fingertips. Acid produced only a gentle froth. To my mind, our bodies were country cousins to marble.

I wasn't a fool, not entirely. But even when I pressed at that image in my mind of the inside of a body—even when I tried to have it meet my senses, which told me about softness and blood—still what I saw when I pictured tearing or sawing an arm in two was something like a sliced pastry, rolled with jam and cream. Intestines I imagined as silk. The spleen was a bitter olive. The mind: a fluttering of unusually small doves. Blood was a melted jewel, lungs were muslin, and the bronchi of the lung were God's finest tree.

Fontana, patiently, said sure. But also no. He told me to look more carefully. He said that every artist should look more carefully. At that time, I thought Fontana was a butcher trying to pass as an artist. His shelves were crowded with jarred kidneys, and with desiccated hands, and with bones sorted from smallest to largest. The onion skin of fascia was piled high in a can. I didn't only dislike his morbid miniature museum—I disagreed with it. I wished to correct it. Even the magnificently tiny ear bones I thought were telling a kind of convincing lie. I dismissed the whole workshop as false, since the workshop told me that the human body was very little different from an October splayed pig. (I was in love at the time, of course, as any reasonable nineteen-year-old man is in love, and this also clouded my perceptions.) I thought of the convincing falseness of what I saw with my very own eyes as one thinks of a man who comes to town claiming to be your cousin and who a week later leaves with your silver spoons.

But I did become a very thin man that first month with Fontana. I mended my own hose. I told myself that my eyes knew very little, my nose even less. Why did I stay with the job? When did I start to be altered? The dead bodies stayed the same and I was the one being chipped away at and formed.

I remember in particular an early sculpture I did, almost on my own, of a dissected head. Fontana looked in on my work now and again, but now he was the one distracted by a love affair and I was the one with plenty of time. The dead man had the nose of a Frenchman. His hair had a soft curl. He was a convict who had stolen flour. His third molars had not erupted. Fontana had split and pinned the gums to reveal the molars in their hiding place. Then the rest he had left to me. Without haste, I made a cast of the palate, of each facial muscle … I proceeded without delusion, for the most part. It was only with the Frenchman's brain that my fantasy intruded. I had accepted its conventional convoluted cauliflower beauty. I obeyed my eyes and fingers in setting the tangle of its vasculature. But I added a tender lattice of lymphatics to the whole. I hadn't seen lymphatics. But I wished upon this gentle young man that he have a draining lace of lymphatics.

It was a very embarrassing error. Pointed out to me by many.

I'll give the assistant at least a month to adjust.

I am no longer like I once was, or like my assistant is now. I could walk through a palace of severed fingers wearing rings and not be troubled in the slightest. A hallway of specimen jars of pompous livers, of the foolish fundal curve of stomachs, and of the insectile branchings of ureters—I would stop to get a closer look. I could wear a woman's uterus as a purse. I could inspect the head of John the Baptist on a platter with a mild, detached interest. I would be more curious to see the cross section of his neck than the full sight of his face, but I wouldn't complain. But now I know that faces—our exteriors in general—are the misleading false cousins. But, of course, I do now and again wonder if I was a better man when I, like my assistant, was a thin quail.

STORY 22

プラダ財団最終

Mieko Kawakami

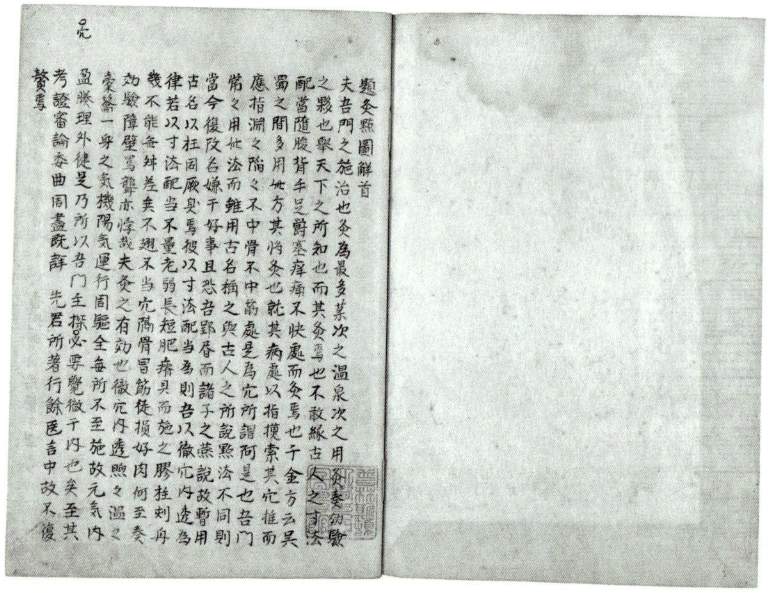

Shinnin Kawaguchi, *Kaishihen* [Complete notes on the dissection of a cadaver], 1802.
Japanese anatomical and medical text with illustrations by Shukuya Aoki.
Frontispiece and spread including the dissection of the brain.

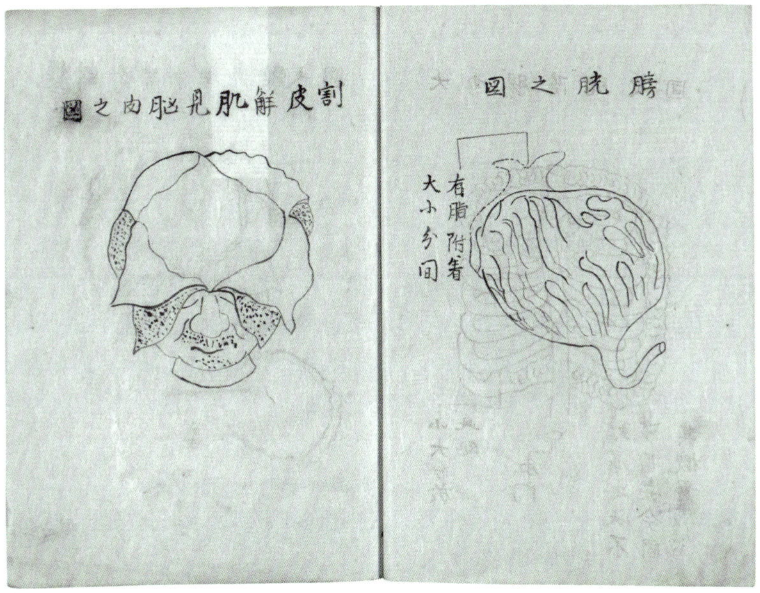
割皮解肌見脳肉之圖
膀胱之図
有脂附着
大小分同

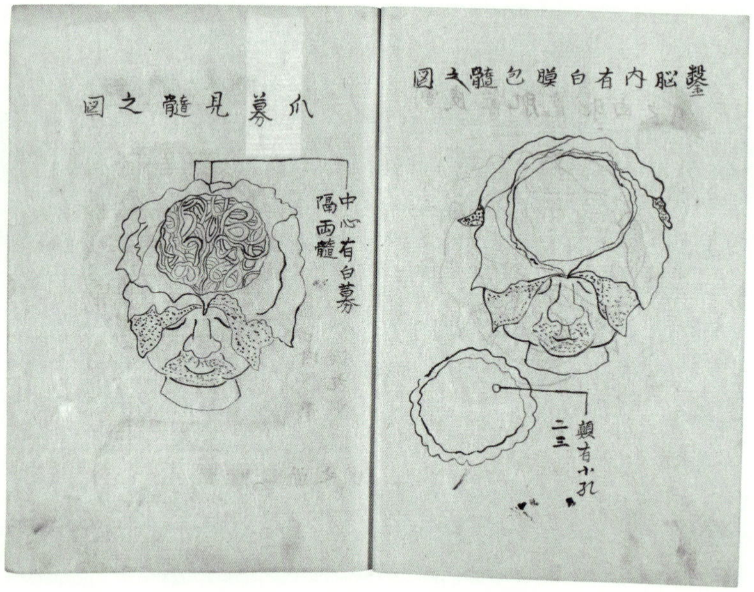

Shinnin (Nobuto) Kawaguchi, *Kaishihen* [Complete notes on the dissection of a cadaver], 1802.
Anatomical and medical text with illustrations by Shukuya Aoki.
Spreads including illustrations of the dissection of the brain, the rib cage, the ocular bulb.

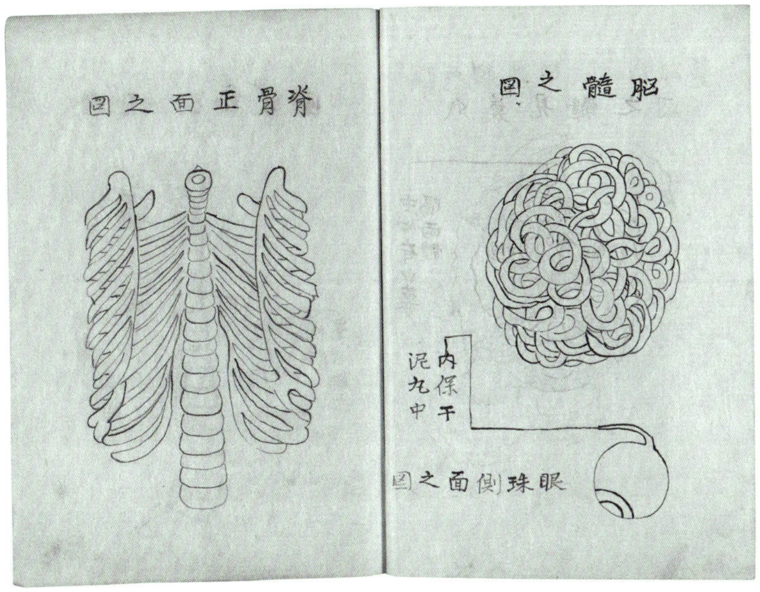
脊骨正面之図
脳髄之図
内保干
泥丸中
眼珠側面之図

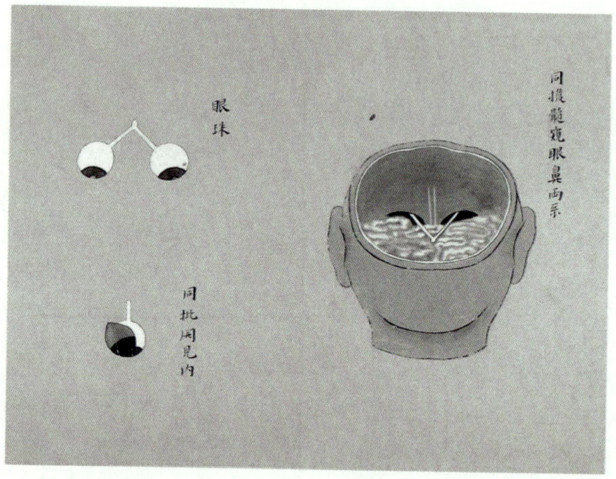

Seii Sugawara, *Fujin Zōzu* [Pictures of female internal organs], anatomical manuscript scroll, Japan, 1774. Illustrations by Seii Sugawara of the dissection on the body of a thirty-four-year-old female executed criminal, performed by Tomon Yamawaki.

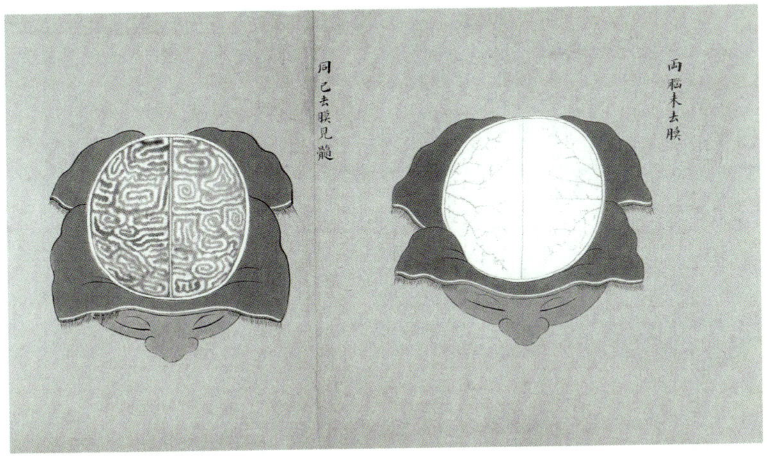
両腦未去膜
同已去膜見髓

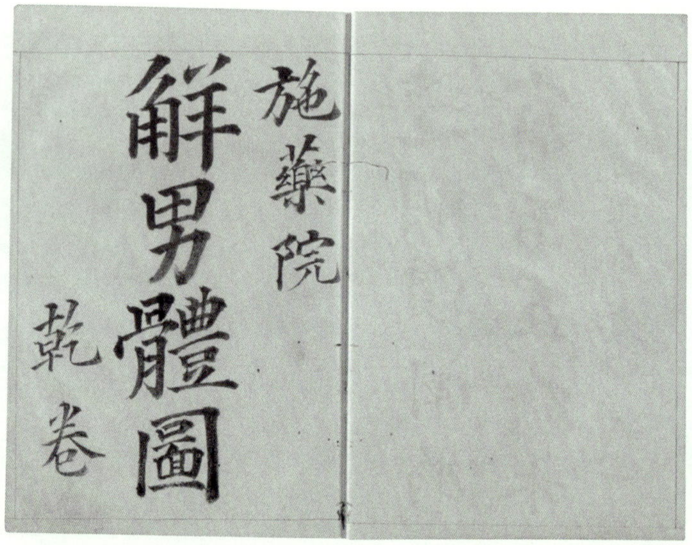

Kanzen Mikumo, Genshun Koishi, Ranshū Yoshimura,
Seyakuin Kainantaizōzu [Seyakuin's male anatomical chart],
anatomical book, Japan, 1798. Frontispiece and spread including illustrations of the dissection of the brain.

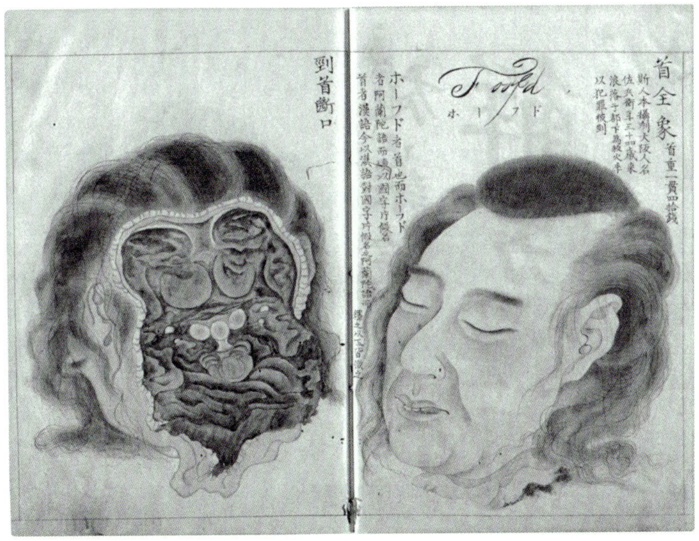
首全象
ホーフド
到首斷口

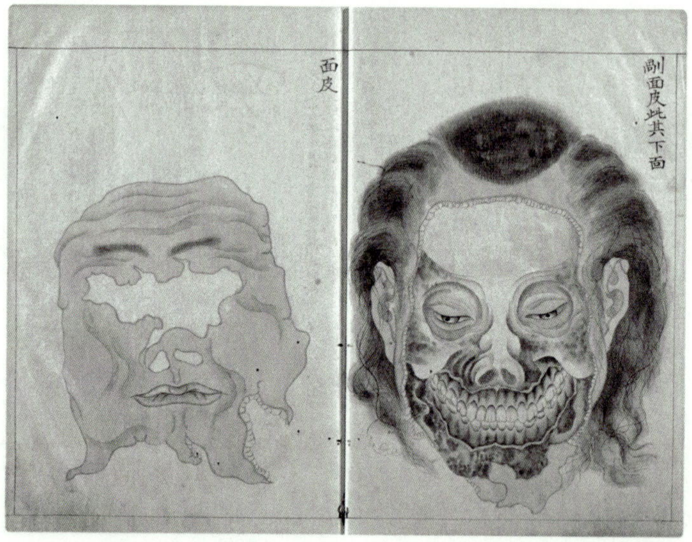

Kanzen Mikumo, Genshun Koishi, Ranshū Yoshimura,
Seyakuin Kainantaizōzu [Seyakuin's male anatomical chart],
anatomical book, Japan, 1798. Spreads including illustrations of the dissection of the face and brain.

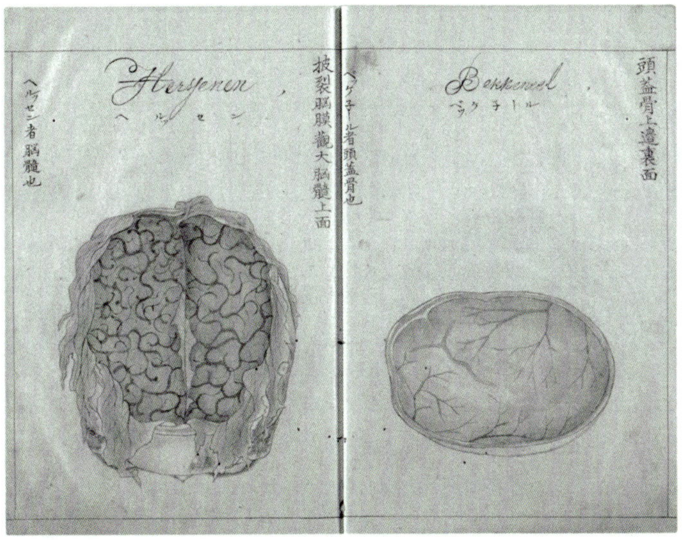
頭蓋骨上邉裏面
Bekkeneel
Hersenen
ヘルッセン者腦髓也

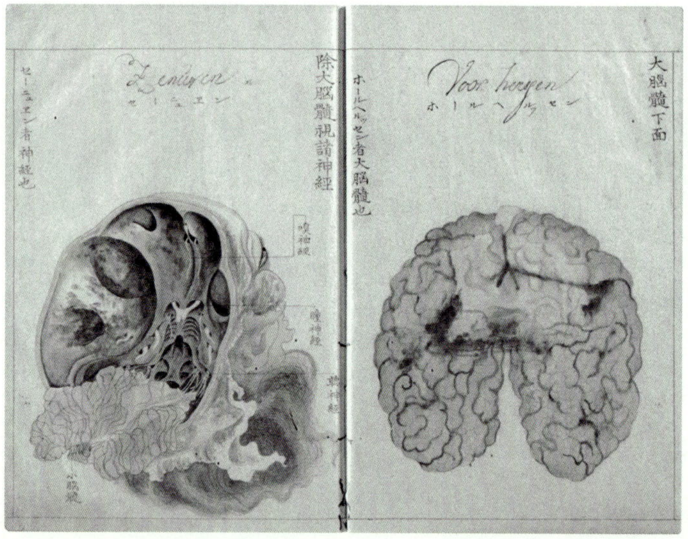

Kanzen Mikumo, Genshun Koishi, Ranshū Yoshimura,
Seyakuin Kainantaizōzu [Seyakuin's male anatomical chart],
anatomical book, Japan, 1798. Spreads including illustrations of the brain.

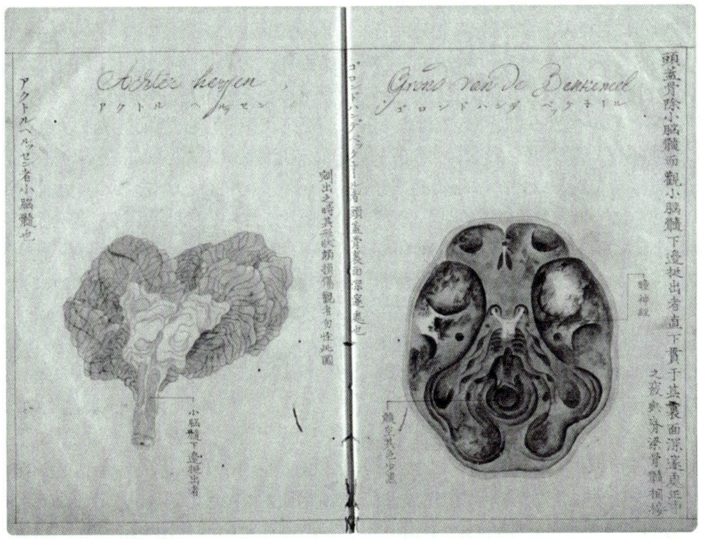
小脳髓下邊挺出者

その日の罪人には首がついていた。割り当てられる死体は斬首されているものがほとんどだったので、物珍しさに何人かの医師が集まった。罪人は土の上に裸のままでうつ伏せに放られており、しかるべき者がやってきてしかるべき場所に運ばれるまで、まだ時間があるようだった。

十一歳になったばかりの佐野鈴郎は、息をひそめ、少し離れたところからそれを見ていた。腕と胴体に黒っぽいあざがいくつかあり、髪はざんばらに散り、四肢はぴくりとも動かなかったけれど、肩や尻の筋肉はしっかりと盛りあがり、あざがついているだけで、ほかの部分は自分とおなじだった。眠っているか気を失っているだけのようにも見える。鈴郎はその死んでいるという罪人のどこに、死があるのかがわからなかった。鈴郎は、人の目がなくなれば、この罪人はむっくりと起きあがって、どこかへ去っていくのではないかという気がした。昨日から降り続いた雪はもう止んでいたけれど、空気は身を切るほどに冷たく、辺りを覆った雪の白さはさらにその白さを増してゆくようだった。それは藩医を父にもち、のちにその後を継いで自ら腑分けに携わるようになる鈴郎が、初めて見た罪人の死体だった。

都には有名な死刑執行人がいた。彼らは幕府に認められた刀の試し切りを生業にする者たちでもあった。道具は知恵と結びつけば、色んなものの役に立ったが、刀はただ人を斬るためだけに存在するものだった。であれば、斬れ味を試すのには人間の首がいちばんだ。罪人を死刑にしながら刀の出来を知ることができる。役目を果たした死体たちは、医術の発展にいそしむ藩医のところに運ばれてきた。この日の罪人は頭部の解剖のために首を斬り落とされずにここに運ばれてきたのだった。

医師たちはいったんその場を離れて屋内にもどり、執刀の段取りについて話し合うことにした。戸がしっかりと閉められ、大人たちが奥に移動したのを確認してから、鈴郎はおそるおそる罪人に近づいていった。膝をついて身をかがめて、罪人の顔を見た。見開かれた目は何も見ていないようだった。黒い血が鼻から垂れていた。乾いてもおらず、いやに生々しいと感じた瞬間——血はまるで生きている紐のようにするすると伸び始めた。半分開いた唇と顎の上をゆっくりとつたい、地面を這い、鈴郎の指先をたどって肩を走り、胸から首を迫りあがって、やがて鈴郎の鼻の穴へ入っていった。

大腸は四月十日。鼓膜は祝言の日の椿。皮膚には、両手におさまった息子の熱だ。腰骨には友との別れ、脛には畑の土の匂い、剥がれた爪は酒と焚き火、こめかみには笑いに笑った土砂降りの雨だ。胃には金に浮かれた束の間の興奮が、腎臓には誰かの真似をして初めておぼえた歌が鳴っている。くるぶしは夜、目玉は春、肺には先立ったトミの顔——鈴郎の脳裏に、ひと息に同時にやってきた無数の像に、彼は短く叫んで、後ろに飛び退いた。

何をやっているのだ、背中で父の声がして、鈴郎は地面に尻をつけたまま振り返った。指先からのぼってきた罪人の血をぬぐおうと顔中をこすっても、手には何もつかなかった。目の前の罪人の四肢はすっかり変色しており、血は顎から首にかけて黒く固まり、それが死体であることは誰の目にも明らかだった。

何をやっているのだ、父親は再び鈴郎にきいた。何もしておりません、と鈴郎は答えた。そのとき雪が降りはじめた。父は息子を見、息子は父と自分の間の地面を見ていた。やがて何人かの助手たちがやってきて、罪人をどこかへ運んでいった。

REMAINS

The criminal that day still had his head. The bodies almost always arrived headless, so a few doctors had gathered there to look at the unusual corpse. The criminal was naked, face down in the dirt. It would apparently be some time before the appropriate people would come to carry the body to the appropriate place.

Sano Suzurō, who had just turned eleven, stood and looked on from a slight distance, holding his breath. There were dark bruises on the criminal's arms and torso. His hair was a mess and his arms and legs were stock-still, but the muscles around his buttocks and shoulders were full and firm. With the sole exception of the man's bruises, every part of his body was the same as Suzurō's. He looked like he could have been sleeping—or maybe passed out. Suzurō had no way of knowing where it was in this body, this dead body, that death actually resided. There was some part of him that felt that the body might get up and leave if everybody just turned the other way. The snow that had started falling the day before had stopped, but the air was piercingly cold, and the white of the snow that covered the area seemed to be growing even whiter. Suzurō's father was a doctor, and the boy would eventually follow the same path, even participating in his own necropsies, but this was the first time that Suzurō had laid eyes on a criminal's corpse.

There were famous executioners in the capital. As part of their work, they tested the swords recognized by the shogunate. If used with intelligence, most tools can serve a variety of purposes, but the sole purpose of these blades was to kill people. That being the case, there was nowhere better to test their sharpness than on human necks. So the executioner made his living by delivering death to the condemned—and ascertaining the quality of the shogunate's weaponry in the process. Once a corpse had served its purpose in this way, it was taken to the doctors, who would use the criminals' remains to further advance the medical arts. The criminal that day had been brought with his head intact so the doctors could examine his brain during the dissection.

Most of the doctors had already peered inside the heads of birds and beasts, examining their contents, but this would mark their first encounter with a human brain. As the doctors wondered just how different the brain they were about to dissect was going to be from the ones they'd seen and studied in their anatomical charts, their minds ran feverish with excitement. In this age when it was known that blood circulated by way of the heart, flowing through two large and similar blood vessels, what they sought to find was the seat of the soul. It was believed that the soul could be found deep inside the brain, in the pineal body, an organ that measured no larger than a pea. The doctors could not put into words the unfathomable mystery before them—what it meant for a soul, which ought to be invisible and untouchable, to be witnessed and documented by another. They instead perceived only a faint terror.

The doctors left the body and returned indoors to discuss their final preparations for the procedure. Once the door was shut and Suzurō had confirmed that the adults were truly gone, he cautiously approached the corpse. He got down on his knees and leaned in, taking a good look at the criminal's face. The man's eyes were wide open, apparently looking at nothing at all. Dark blood, not yet dry, was dribbling from his nose. Then, at the moment when its eerie freshness struck Suzurō, the blood slid into motion. Like a rope with a life of its own, it made its way over the man's half-opened lips and chin, crawling across the ground, traveling up Suzurō's finger, then running up his arm before rising to his neck and entering his nostrils.

In his large intestine, the tenth day of the fourth month. In his eardrums, camellias on his wedding day. In his skin, the warmth of his son in his arms. In his hip bone, a final farewell to a friend; in his shins, the smell of the earth in the fields; in his nailless fingertips, drinking around the fire; in his temples, the rainstorm when he just laughed and laughed. In his stomach was the momentary thrill of hitting it big, and in his kidneys, strains of the first song he'd ever learned from somebody. The night was in his ankles, the spring in his eyeballs. In his lungs was the face of his late wife, Tomi. As the multitude of images flooded Suzurō's mind, he let out a shriek and leaped back.

"What are you doing?" Suzurō heard his father's voice behind him and turned around as he sat on the ground where he'd landed. He lifted a hand to his face to wipe off the criminal's blood that had climbed up his finger, but when he looked at his palm, there was nothing there. The arms and legs of the criminal before him were now completely drained of color, and the blood on the man's jaw and neck had dried black. To anyone who saw it, this was clearly the body of a dead man.

"What are you doing?" Suzurō's father asked his son again.

"Nothing," Suzurō said back. Just then, snow started to fall. Suzurō's father looked at the boy, and Suzurō looked at the ground between his father and himself. Eventually, a couple of assistants came and carried the corpse away.

STORY 23

Louis Victor (Doesn't) Speak

Mauro Javier Cárdenas

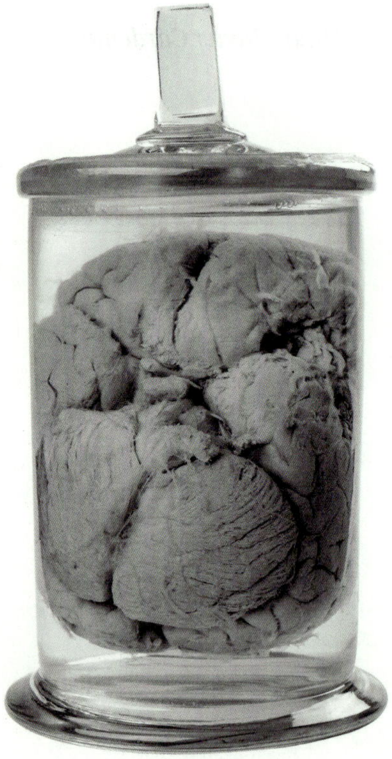

Louis Victor Leborgne's brain.
Dr. Paul Broca's first aphasic patient, nicknamed "Tan," France, 1861.

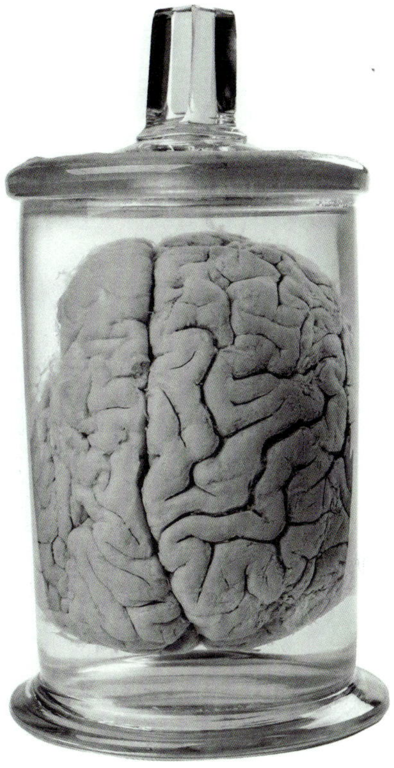

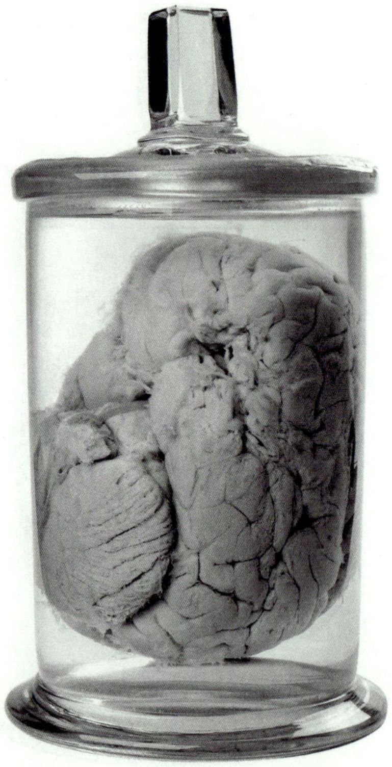

Louis Victor Leborgne's brain.
Dr. Paul Broca's first aphasic patient, nicknamed "Tan," France, 1861.

Why would I even want to speak again, Louis Victor thinks, because before he stopped speaking he'd already wondered what was the purpose of speaking—to remind others you exist?—if you walk around with a full-sized mirror directed at you does that prove that you—Louis Victor's so vain he walks around with a—locomotion probably suffices as proof of your existence, Louis Victor thinks, but speaking is required as proof you're in possession of a spirit—how much for the oranges, my spirit would ask—he exists!—whereabouts my donkey—he exists and has a donkey!—but if you don't speak how would you even ask how much the oranges, Dr. Broca might ask, how much the hay—you mean if Louis also forgot how to point at the fruit and/or write the word *hay*?—I know you know I know you're an expert in spirited brains and bounteous language, Dr. Broca, I might reply, and therefore are keen to language me out of my (non)predicament, but what I'm trying to tell you is that your field doesn't matter to me—how would you even tell me my field doesn't matter to you if you—distract him with the donkey jokes, Louis—what I'm telling myself that I'm telling you is that I've evolved and no longer need to speak, such that if you were to mend my nonspeaking I would still refuse to speak—except to ask whereabouts your—and so Dr. Broca might speak of the impairment of Louis Victor's brain whereas I (don't) speak of the thrilling transcendence already attained by my brain, and so when I'm dead and at last Dr. Broca's allowed to dislodge my brain from my head—slippery, like a lizard—I know you know I know that that's all I am to you, doctor, a specimen to be plumbed for brain parts—he will discover clues to the electrifying evolution of my brain—it's not a lizard it's a crocodile run!—and the residents of this accursed hospital, confused as to why they haven't been uncursed yet, will collect funds to purchase a pedestal for my preserved brain—the effigy of Saint Nurse François didn't work so we might as well—and under the right meteorological conditions (lightning, locust, rain), they will gather around my brain and submit questions of great importance—who will inherit Louis' donkey? what's a good recipe for fried locust? etc.—but my brain will refuse to speak, which will result in charges of selfishness, soft-brainness, self-idolatriness, and so on, and so years will go by and my pedestal will rot, and the uncursed patients and Dr. Broca will rot, whereas my brain-in-a-jar will persist, and in a new, more advanced epoch scientists will connect my brain-in-a-jar with my liver-in-a-jar and at last my brain will speak—why did you let Louis drink so much vermouth? can someone please change the formaldehyde water in my jar?—and in a new, even more advanced epoch scientists will connect my brain to a mechanical cathedral designed to decipher everything in the world and the mechanical cathedral will tell them Louis Victor's brain does not see the point of speaking—but we want to know what was the price of oranges back then, the scientists will say, the price of formaldehyde, the price of mirrors—have you heard the one about the brain in the funhouse whose brain mother says stop making faces they're reflecting everywhere and the brain says but I don't have a face—and this is the other reason I no longer speak, Doctor—and the mechanical cathedral will file a complaint for misuse of its advanced capabilities to switch on an archaic brain—I was programmed to eradicate evil in the world and here I am plugged in to a foul engine of donkey jokes—it's not a crocodile it's a dinosaur run!—and in a new, less advanced epoch everyone will forget about my brain because they have concluded that being awake serves no purpose other than to waste money on pedestals and doctors of brain language that will rot anyway so everyone has learned how to sleep day and night, and because their dreams can't feed on their daytime materials anymore their dreams subsist on the few images from the day they're born (scissors, worms, snot), and so everyone who's born in the same clinic will have the same dreams about the same triangular nurses' hats–paper boats trailed by ravenous scissors, for instance–and because nobody is awake nobody will be able to judge the content of their dreams, and so due to the absence of judgment in their dreams theirs will be the freest epoch–and in another, even more advanced epoch than the previous even more advanced epoch, thousands of years from now, scientists from a different part of the galaxy will connect my brain with one of their meteors just for fun–we have surveyed the universe and determined the purpose of the universe is to (I) exist and (II) perish, therefore (III) fun's needed between (I) and (II) otherwise (IV) gloom–and they will forget about their brain-to-meteor connection as a shower of mechanical locusts rattles their space hospital and dislodges the meteor that's also my brain, causing it to head to earth, and so my brain, which is simultaneously on earth and in meteor, will know its own demise is forthcoming, and since it will take millions of years for the meteor to reach earth, my brain will have enough time to resign itself to its own destruction–speak and warn them of the meteor, you bird brain!–there aren't enough space rafts for everyone anyway so what's the–and as my brain-in-a-jar is about to collide with my brain-in-a-meteor, my brain will speak at last, and in the spirit of (iii) it will ask how much for a pint of fresh formaldehyde, whereabouts my jarring brain.

STORY 24

Pulchroneuroni

Michele Mari

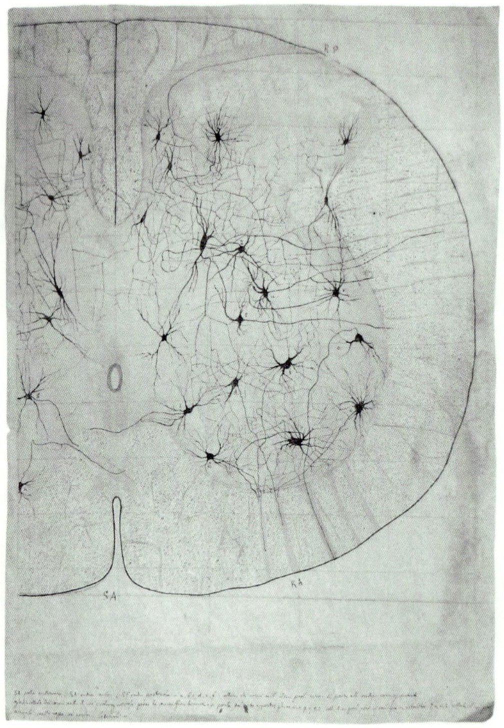

Camillo Golgi, drawing of the spinal cord, *c.* 1883–1884.

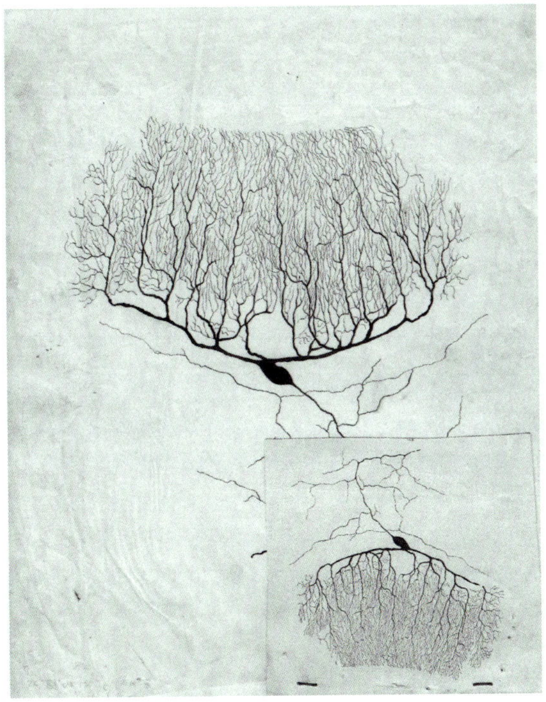

Camillo Golgi, drawing of the Purkinje neuron, *c.* 1883–1884.

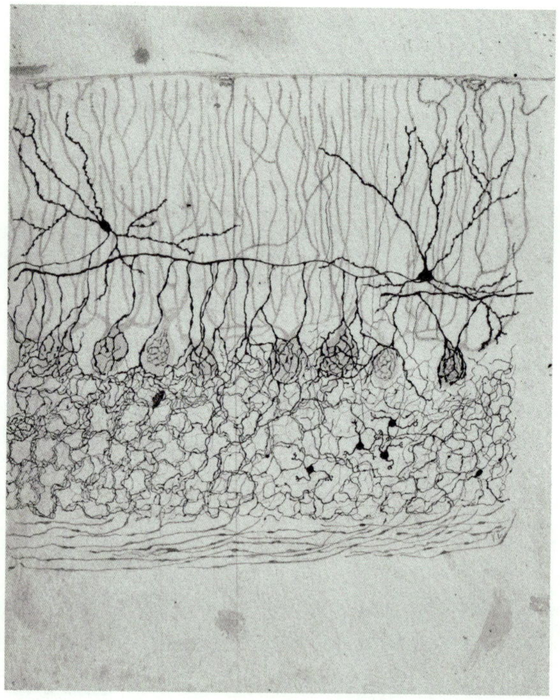

Camillo Golgi, drawing of the diffuse nerve network, *c.* 1883–1884.

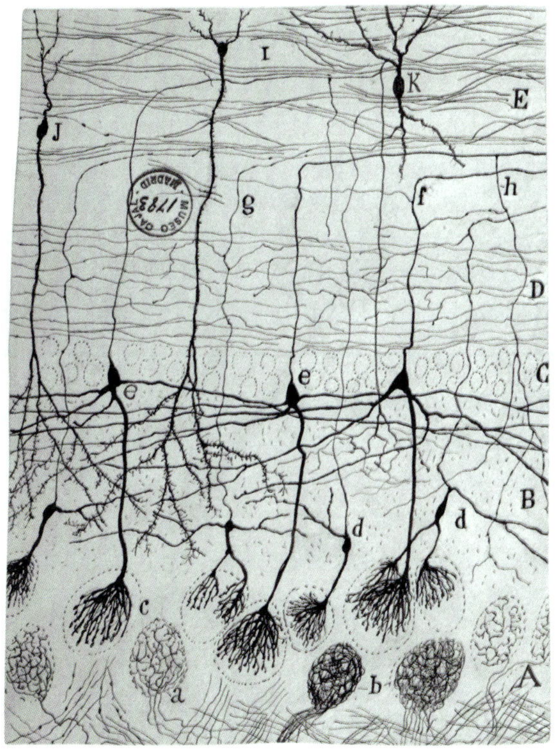

Santiago Ramón y Cajal, section of the olfactory bulb of a cat, 1901.

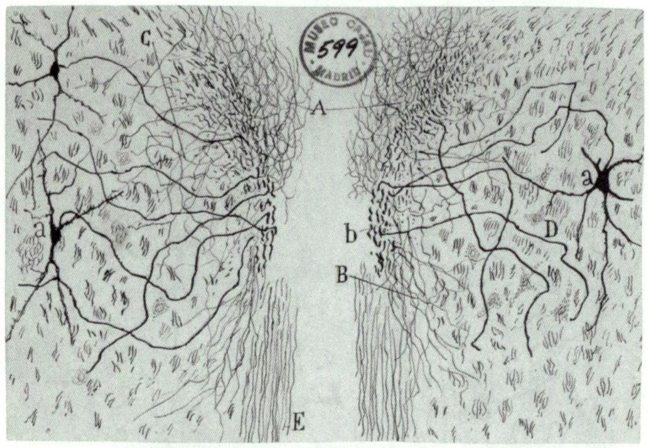

Santiago Ramón y Cajal, frontal section of the brain, 1904.

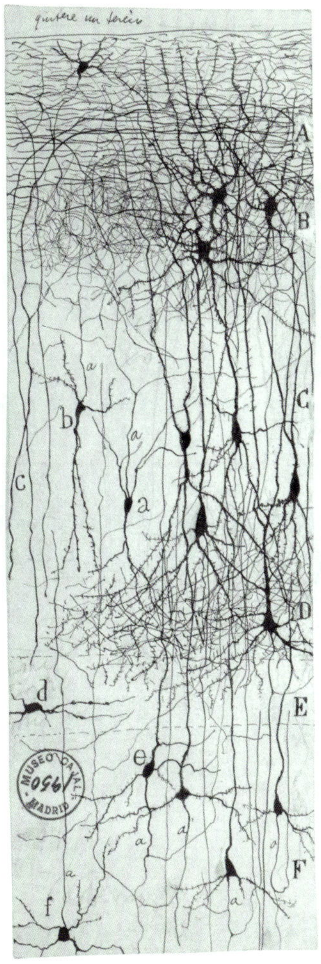

Santiago Ramón y Cajal, cross section of the superior sphenoid area of the rabbit, 1901.

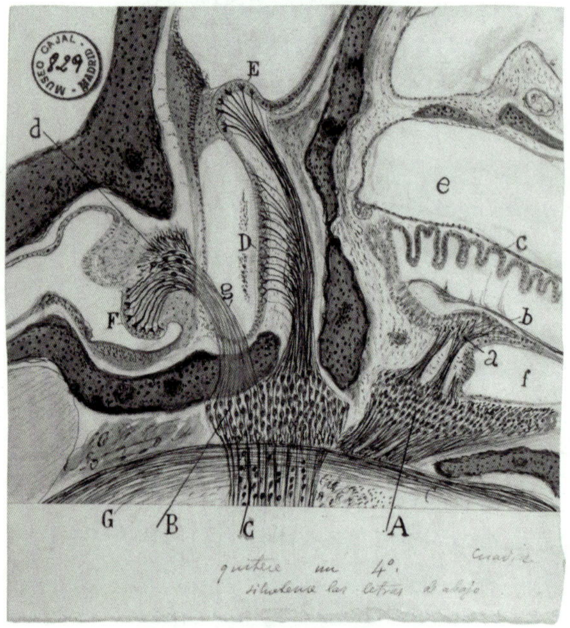

Santiago Ramón y Cajal, front section of the inner ear (labyrinth) in a few-days-old bird, 1908.

Nel 1906 gli Accademici del Premio Nobel, nel comparto di Medicina, si trovarono in un grande imbarazzo. La metà di loro avrebbe voluto assegnare il premio a Camillo Golgi, l'altra metà a Santiago Ramón y Cajal. Nel giro di pochi anni entrambi avevano rivoluzionato l'istologia neuronale, di fatto individuando e descrivendo ciò che prima di loro era stato pensato solo in via teorica: il neurone. Certo Ramón y Cajal, come argomentavano i partigiani dello spagnolo, aveva rettificato gli errori dell'italiano, giungendo a formulazioni ancora più rivoluzionarie: ma senza l'italiano, obiettavano gli avversari, senza la sua avventurosa invenzione della "reazione nera", quel diabolico composto chimico capace di rendere il neurone visibile al microscopio, Ramón y Cajal non sarebbe andato da nessuna parte, come egli stesso, prima che la gelosia inquinasse i rapporti fra i due scienziati, aveva inizialmente riconosciuto. Si profilava dunque un *ex aequo*, soluzione che tuttavia, stando ad alcune indiscrezioni, entrambi i candidati avrebbero preso come una beffa se non come un insulto: sicché si indugiava, nell'attesa che il ripensamento di qualche accademico spostasse gli equilibri.

Alla vigilia del voto, chiusi da ore nell'aula scientifica, gli accademici del comparto medico sentirono bussare alla porta: stupiti e irritati dall'interferenza, andarono ad aprire: sulla soglia, visibilmente impacciato, stava I.K., un loro collega del comparto letterario, e dunque un inferiore, se era vero che di tutti i Nobel quello per la letteratura veniva considerato come una faccenduola minore. Consapevole dell'irritualità del proprio comportamento, esitante, scrutato con impazienza dagli astanti, I.K. spiegò di essere animato dalle intenzioni migliori, e di aver tanto osato nella speranza di poter contribuire positivamente a una situazione il cui stallo non era un segreto per nessuno; certo non sapeva nulla di scienza, aggiunse, eppure proprio questo lo metteva... avrebbe messo tutti loro, se avessero seguito il suo consiglio... insomma lo metteva in una posizione privilegiata, dalla quale... sì, dalla quale si sarebbe potuto cogliere l'elemento decisivo all'assegnazione del premio. Questo elemento, concluse nello sbigottimento generale, sarebbe stata la forma, intesa al massimo dell'accezione artistica. Golgi e Ramón y Cajal, gli risultava, avevano disegnato di persona i neuroni e la loro rete, e chi aveva avuto l'occasione di vedere quelle tavole ne era rimasto impressionato e affascinato. Volevano dunque, lorsignori, sospendere almeno provvisoriamente la loro diatriba scientifica per accostarsi a quelle rappresentazioni come se si trattasse di opere d'arte?

Gli scienziati rimasero in silenzio, pensosi; alcuni di loro erano scandalizzati, altri peccaminosamente tentati (bisogna sapere, in effetti, che il principio secondo il quale una teoria scientifica tante più probabilità ha di essere giusta quanto più è formalmente bella non era ancora stato accettato, per la semplice ragione che chi lo avrebbe affermato, Paul Dirac, aveva in quel momento due anni). A poco a poco tutti gli sguardi si posarono su O.M., presidente del comparto. Con un cenno questi fece tornare tutti a sedersi ai loro posti, con un altro cenno congedò I.K., quindi, aperta una voluminosa cartella, ne distribuì casualmente il contenuto ai suoi vicini, lasciando intendere che proseguissero nella distribuzione. Alla fine ogni accademico si trovò davanti chi due chi tre fogli di differente grandezza: sui fogli, disegni a matita, a china, a tempera. Sui fogli...

... filamenti sottili come righe di pioggia,
e il loro addensarsi in fiocchi, o in
fiori, in graziose esplosioni... foreste di
licheni rampanti, come aneliti vani,
ed efflorescenze di radici sempre più fini,
ife di micelio quasi, lanugine d'angelo
rivelata dalla rugiada: e occhi come
di coda di pavone, ciliati, radianti,
in convoluzioni spugnose che implicano
rime, fessure, scissioni e sottoscissioni
del segno, alla maniera di un leggerissimo
retino, a generare patine, ombre, e
poi di colpo uno screzio, il coagulo di uno
spermatozoo, poi un altro, poi altri,
nella germinazione ombrellifera che si
spande, si estende per riconvergere su se
stessa come collassando, vascolarizzata,
tramata, fibrillante... paesaggi elfici
con fuochi fatui, oppure le crepe di
un terreno siccitoso, l'ansa smaltata di
un vaso craquelé, e qui riflessa nella
palude l'oltranza della mangrovia, radici
che salgono e rami che scendono, una
vena aurifera, forse, e l'immillarsi del
raggio rifratto tra nubecole lievi,
anche se queste amebe, questi crocicchi
protoplasmatici ci dicono di una visione
ulteriore, di forme d'oltregalassia
scese un tempo... o salite... a informarci
il tessuto, nell'alba di ogni umana
istologia... il neurone come primissimo
neuma, il neuma e il neurone... gli assoni,
i dendriti, tutto lì, riemerso, riconosciuto,
interpretato, dipinto, assurto a realtà...

In quei disegni gli accademici vedevano com'era fatto il loro cervello, dove nasceva l'impulso, il brivido elettrico, la culla del cogito era lì, in quei filamenti sottili, in quei punti, poco importava che fra neurone e neurone ci fosse continuità, come pensava Golgi, o contiguità, come voleva Ramón y Cajal, le sinapsi erano garantite comunque, onore e merito alle sinapsi! E viva l'inchiostro, l'occhio e la mano che hanno riprodotto facendo della scienza poesia, della fisica arte, sì, al punto che se anche quei due si fossero inventati tutto, se si fossero divertiti a trascrivere i loro capricci, creando un mondo fantastico di meduse e di larve filamentose, l'avrebbero fatto con una tale coerenza formale e una tale ossessione da poterlo accettare, e poter partire da lì per costruire la scienza del secolo nuovo!

Qualcuno sospirò, qualcuno accennò un timido applauso, altri si asciugavano gli occhi. Ma tutti sorridevano: perché la bellezza di quelle tavole metteva in secondo piano le divergenze scientifiche, e la preminenza dell'uno sull'altro: nell'arte del disegno eccellevano entrambi, e per questo motivo, non più come scappatoia ma nello spirito di una duplice consacrazione, fu votato senza esitazioni l'*ex aequo*, cosa che alla fine non dispiacque (stando alle reazioni delle rispettive reti neuronali) né a Golgi né a Ramón y Cajal, visto che la partita scientifica rimaneva aperta, ed era cosa che si sarebbe giocata lontano da Stoccolma.

PULCHRONEURONS

In 1906, the Nobel Prize academicians in the medicine section found themselves in an awkward predicament. Half of them intended to award the prize to Camillo Golgi and the other half to Santiago Ramón y Cajal. In a matter of a few years, the two had revolutionized neuronal histology, locating and describing what had until then been figured out only in theoretical terms, i.e., the neuron. For sure Ramón y Cajal—argued those on the side of the Spaniard—had rectified the errors of the Italian, reaching even more revolutionary formulations; but without the Italian—countered the opponents—without his adventurous invention of the "black reaction," that diabolical chemical compound that made the neuron visible with a microscope, Ramón y Cajal would have gone nowhere, as he himself had initially recognized before jealousy spoiled relations between the two scientists. So it seemed the prize would be divided *ex aequo*, but rumor had it that the candidates considered such a solution a mockery, if not an insult; and so the lingering went on, in the expectation that second thoughts from some academicians would change the balance.

On the eve of the vote, the academicians of the medicine section, locked up for hours in the science hall, heard a knocking at the door; surprised and irritated by the interference, they opened it; on the threshold, and clearly not at ease, stood I.K., a colleague of theirs from the literature section, and therefore an inferior colleague, if indeed it was true that the Nobel Prize in literature was considered a minor matter. Well aware of his unorthodox behavior, hesitating and scrutinized impatiently by those in attendance, I.K. explained that he was animated by the best of intentions, and that he had dared so much in the hope of contributing positively to a situation whose deadlock was a secret to no one; of course he knew nothing of science, he added, and yet this same fact put him ... could have put all of them, were they inclined to follow his advice ... all told, it put him in a privileged position, from which ... yes, from which the decisive element for the awarding of the prize could be perceived. Surrounded by general dismay, he summed up that this element was the form, to be understood at the highest of its artistic meaning. He believed that Golgi and Ramón y Cajal had personally drawn the neurons and their network, and those who had the chance of seeing the drawings had been impressed and fascinated by them. And so, would you gentlemen suspend your scientific dispute, at least temporarily, and draw near to those representations, as if they were artworks?

The scientists remained silent, meditating; some of them were scandalized, others sinfully tempted. (Actually, it must be said that the principle whereby a scientific theory has the highest chance of being correct when its formal beauty is highest had not been accepted yet, for the simple reason that he who stated it, i.e. Paul Dirac, was at that time two years old.) Little by little, all eyes were on O.M., who chaired that section. With a nod, he summoned those in attendance back to their seats; with another nod he dismissed I.K.; and then, on opening a voluminous folder, he casually distributed its contents to those near him, implying that they should carry on with the distribution. In the end every academician was facing two or three sheets of paper of varying sizes; on the sheets were pencil drawings, ink drawings, tempera drawings. On the sheets ...

... filaments as thin as lines of rain, and
their thickening in flakes, or in flowers,
in gracious explosions ... forests of
rampant lichens, as vain yearnings, and
thinner and thinner bloomings of roots,
nearly hyphae of a mycelium, down of an
angel revealed by the dew; and eyes
as of a peacock tail, ciliated, radiant
ones, in spongy convolutions that imply
slits, cracks, splits, and subsplits
of the sign, in the manner of a very
light net, generating patinae, shadows,
and then all of a sudden a smear, the
congealment of a spermatozoon, and

then another one, then others, a treelike
germination that expands, extending
and then reconverging on itself as
if it is collapsing, vascularized, woven,
fibrillated... elfic landscapes with
will-o'-the-wisps, or cracks of an arid
land, the enameled handle of a crackled
vase, reflecting the exuberance of the
mangrove, roots climbing and branches
falling, an auriferous vein, possibly,
and the fracturing by the thousands of
the refracted ray among light nubeculae,
even though these amoebae, these
protoplasmatic crossings, hint at
an ulterior vision, ultra-galactic shapes
descended... or ascended... once upon
a time to mold our texture, at the dawn of
every human histology... The neuron
as the very first neume, the neume and
the neuron... the axons, the dendrites,
all there, reemerged, recognized,
interpreted, painted, risen to reality...

In those drawings the academicians saw how the brain was made; the birth of the impulse, the electric shiver, the cradle of the cogito was there, in those thin filaments, in those dots, little did it matter that between neurons there was continuity, as Golgi thought, or contiguity, as Ramón y Cajal argued, the synapses were guaranteed in any case, honor and merit to the synapses! And hurray to the ink, the eye, and the hand that have reproduced, turning science into poetry, physics into art, yes, to the extent that if those two had invented it all, if they had had fun in transcribing their fancies, creating a fantastic world of jellyfish and filamentous larvae, they would have done it with such formal coherence and such obsession that one could accept it, and depart from there to create the science of the new century!

Someone sighed, someone started a shy applause, others were drying their eyes. But everyone was smiling, because the beauty of those drawings put scientific divergences and the preeminence of one over the other backstage; they both excelled in the art of drawing, and for this reason, and not anymore as an expedient but rather in the spirit of a double consecration, an *ex aequo* division was voted for with no hesitation, something which in the end—based on the reactions of their respective neuronal networks—displeased neither Golgi nor Ramón y Cajal, since the scientific duel was laid out, and it was something that was going to take place far away from Stockholm.

STORY 25

Come acqua sulle ali di un'anatra

Paolo Giordano

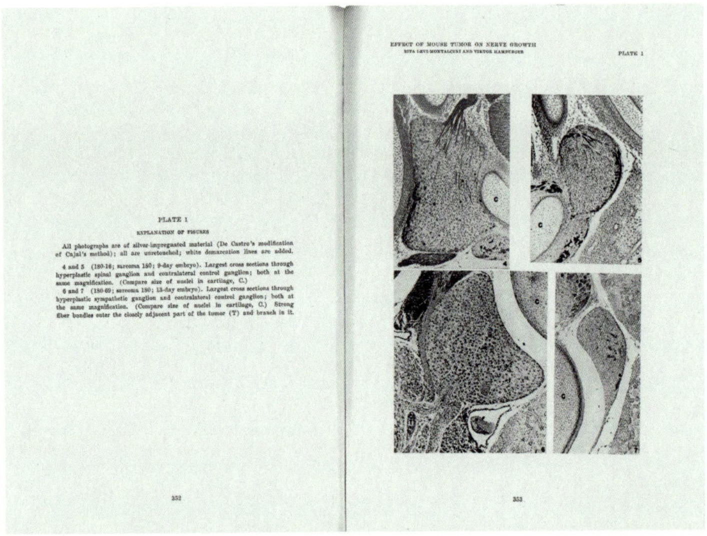
PLATE 1

EXPLANATION OF FIGURES

All photographs are of silver-impregnated material (De Castro's modification of Cajal's method); all are unretouched; white demarcation lines are added.

4 and 5 (180-16; sarcoma 180; 9-day embryo). Largest cross sections through hyperplastic spinal ganglion and contralateral control ganglion; both at the same magnification. (Compare size of nuclei in cartilage, C.)

6 and 7 (180-69; sarcoma 180; 13-day embryo). Largest cross sections through hyperplastic sympathetic ganglion and contralateral control ganglion; both at the same magnification. (Compare size of nuclei in cartilage, C.) Strong fiber bundles enter the closely adjacent part of the tumor (T) and branch in it.

EFFECT OF MOUSE TUMOR ON NERVE GROWTH
RITA LEVI-MONTALCINI AND VIKTOR HAMBURGER
PLATE 1

Rita Levi-Montalcini and Viktor Hamburger, "Selective Growth Stimulating Effects of Mouse Sarcoma on the Sensory and Sympathetic Nervous System of the Chick Embryos," *Journal of Experimental Zoology*, vol. 116, no. 2, 1951. Plates nos. 1 and 2 illustrating the effect of the mouse tumor on nerve growth.

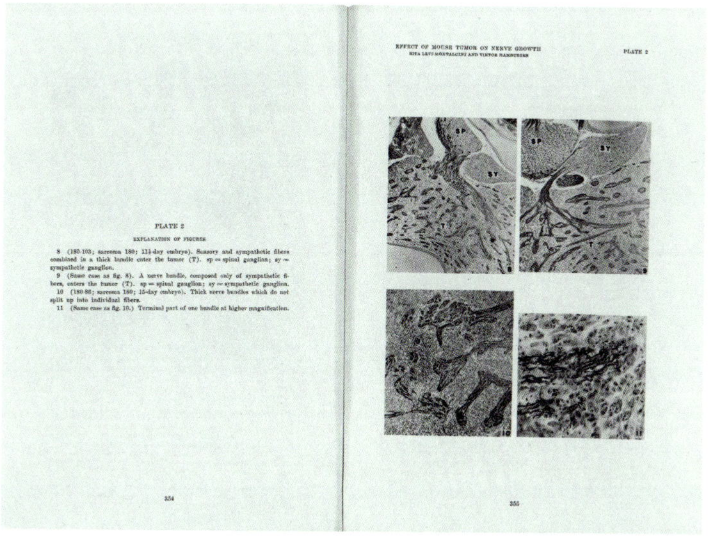

PLATE 2

EXPLANATION OF FIGURES

8 (180-103; sarcoma 180; 11½-day embryo). Sensory and sympathetic fibers combined in a thick bundle enter the tumor (T). sp = spinal ganglion; sy = sympathetic ganglion.

9 (Same case as fig. 8). A nerve bundle, composed only of sympathetic fibers, enters the tumor (T). sp = spinal ganglion; sy = sympathetic ganglion.

10 (180-86; sarcoma 180; 15-day embryo). Thick nerve bundles which do not split up into individual fibers.

11 (Same case as fig. 10.) Terminal part of one bundle at higher magnification.

354

EFFECT OF MOUSE TUMOR ON NERVE GROWTH
RITA LEVI-MONTALCINI AND VIKTOR HAMBURGER

PLATE 2

355

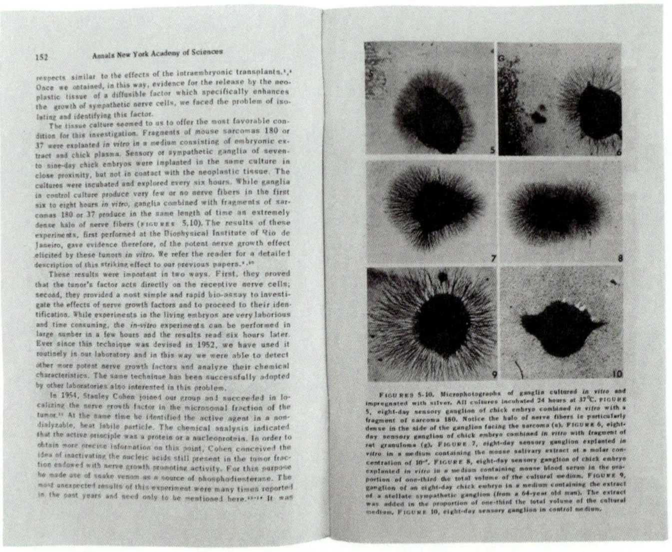

152 Annals New York Academy of Sciences

respects similar to the effects of the intraembryonic transplants.[1,2] Once we obtained, in this way, evidence for the release by the neoplastic tissue of a diffusible factor which specifically enhances the growth of sympathetic nerve cells, we faced the problem of isolating and identifying this factor.

The tissue culture seemed to us to offer the most favorable condition for this investigation. Fragments of mouse sarcomas 180 or 37 were explanted *in vitro* in a medium consisting of embryonic extract and chick plasma. Sensory or sympathetic ganglia of seven- to nine-day chick embryos were implanted in the same culture in close proximity, but not in contact with the neoplastic tissue. The cultures were incubated and explored every six hours. While ganglia in control culture produce very few or no nerve fibers in the first six to eight hours *in vitro*, ganglia combined with fragments of sarcomas 180 or 37 produce in the same length of time an extremely dense halo of nerve fibers (FIGURES 5,10). The results of these experiments, first performed at the Biophysical Institute of Rio de Janeiro, gave evidence therefore, of the potent nerve growth effect elicited by these tumors *in vitro*. We refer the reader for a detailed description of this striking effect to our previous papers.[3,10]

These results were important in two ways. First, they proved that the tumor's factor acts directly on the receptive nerve cells; second, they provided a most simple and rapid bio-assay to investigate the effects of nerve growth factors and to proceed to their identification. While experiments in the living embryos are very laborious and time consuming, the *in-vitro* experiments can be performed in large number in a few hours and the results read six hours later. Ever since this technique was devised in 1952, we have used it routinely in our laboratory and in this way we were able to detect other more potent nerve growth factors and analyze their chemical characteristics. The same technique has been successfully adopted by other laboratories also interested in this problem.

In 1954, Stanley Cohen joined our group and succeeded in localizing the nerve growth factor in the microsomal fraction of the tumor.[11] At the same time he identified the active agent in a nondialyzable, heat labile particle. The chemical analysis indicated that the active principle was a protein or a nucleoprotein. In order to obtain more precise information on this point, Cohen conceived the idea of inactivating the nucleic acids still present in the tumor fraction endowed with nerve growth promoting activity. For this purpose he made use of snake venom as a source of phosphodiesterase. The most unexpected results of this experiment were many times reported in the past years and need only to be mentioned here.[12-14] It was

FIGURES 5-10. Microphotographs of ganglia cultured *in vitro* and impregnated with silver. All cultures incubated 24 hours at 37°C. FIGURE 5, eight-day sensory ganglion of chick embryo combined *in vitro* with a fragment of sarcoma 180. Notice the halo of nerve fibers is particularly dense in the side of the ganglion facing the sarcoma (s). FIGURE 6, eight-day sensory ganglion of chick embryo combined *in vitro* with fragment of rat granuloma (g). FIGURE 7, eight-day sensory ganglion explanted *in vitro* in a medium containing the mouse salivary extract at a molar concentration of 10^{-6}. FIGURE 8, eight-day sensory ganglion of chick embryo explanted *in vitro* in a medium containing mouse blood serum in the proportion of one-third the total volume of the cultural medium. FIGURE 9, ganglion of an eight-day chick embryo in a medium containing the extract of a stellate sympathetic ganglion (from a 64-year old man). The extract was added in the proportion of one-third the total volume of the cultural medium. FIGURE 10, eight-day sensory ganglion in control medium.

Rita Levi-Montalcini, "Nerve Growth Factor," *Annals of the New York Academy of Sciences*, vol. 118, no. 3, 1964.
Spreads including microphotographs of ganglia cultured in vitro and a microphotographs of chick embryos.

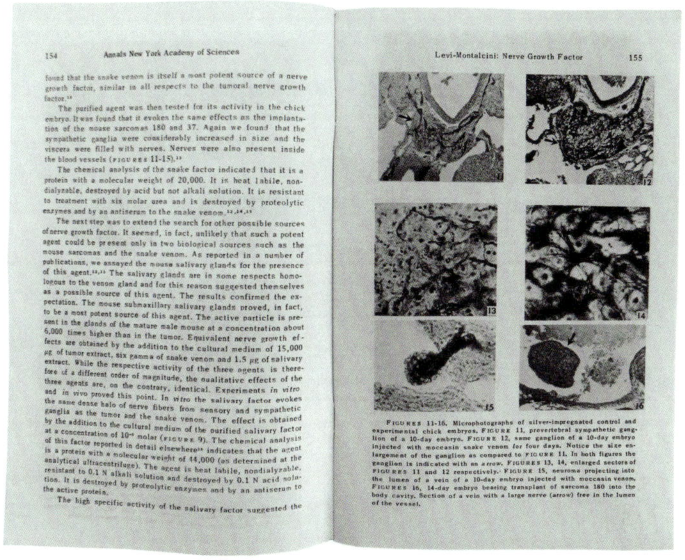

154 Annals New York Academy of Sciences

found that the snake venom is itself a most potent source of a nerve growth factor, similar in all respects to the tumoral nerve growth factor.[11]

The purified agent was then tested for its activity in the chick embryo. It was found that it evokes the same effects as the implantation of the mouse sarcomas 180 and 37. Again we found that the sympathetic ganglia were considerably increased in size and the viscera were filled with nerves. Nerves were also present inside the blood vessels (FIGURES 11-15).[13]

The chemical analysis of the snake factor indicated that it is a protein with a molecular weight of 20,000. It is heat labile, nondialyzable, destroyed by acid but not alkali solution. It is resistant to treatment with six molar urea and is destroyed by proteolytic enzymes and by an antiserum to the snake venom.[12,14,15]

The next step was to extend the search for other possible sources of nerve growth factor. It seemed, in fact, unlikely that such a potent agent could be present only in two biological sources such as the mouse sarcomas and the snake venom. As reported in a number of publications, we assayed the mouse salivary glands for the presence of this agent.[11,13] The salivary glands are in some respects homologous to the venom gland and for this reason suggested themselves as a possible source of this agent. The results confirmed the expectation. The mouse submaxillary salivary glands proved, in fact, to be a most potent source of this agent. The active particle is present in the glands of the mature male mouse at a concentration about 6,000 times higher than in the tumor. Equivalent nerve growth effects are obtained by the addition to the cultural medium of 15,000 μg of tumor extract, six gamma of snake venom and 1.5 μg of salivary extract. While the respective activity of the three agents is therefore of a different order of magnitude, the qualitative effects of the three agents are, on the contrary, identical. Experiments *in vitro* and *in vivo* proved this point. *In vitro* the salivary factor evokes the same dense halo of nerve fibers from sensory and sympathetic ganglia as the tumor and the snake venom. The effect is obtained by the addition to the cultural medium of the purified salivary factor at a concentration of 10^{-9} molar (FIGURE 9). The chemical analysis of this factor reported in detail elsewhere[15] indicates that the agent is a protein with a molecular weight of 44,000 (as determined at the analytical ultracentrifuge). The agent is heat labile, nondialyzable, resistant to 0.1 N alkali solution and destroyed by 0.1 N acid solution. It is destroyed by proteolytic enzymes and by an antiserum to the active protein.

The high specific activity of the salivary factor suggested the

Levi-Montalcini: Nerve Growth Factor 155

FIGURES 11-16. Microphotographs of silver-impregnated control and experimental chick embryos. FIGURE 11, prevertebral sympathetic ganglion of a 10-day embryo. FIGURE 12, same ganglion of a 10-day embryo injected with moccasin snake venom for four days. Notice the size enlargement of the ganglion as compared to FIGURE 11. In both figures the ganglion is indicated with an *arrow*. FIGURES 13, 14, enlarged sectors of FIGURES 11 and 12 respectively. FIGURE 15, neuroma projecting into the lumen of a vein of a 10-day embryo injected with moccasin venom. FIGURES 16, 14-day embryo bearing transplant of sarcoma 180 into the body cavity. Section of a vein with a large nerve (*arrow*) free in the lumen of the vessel.

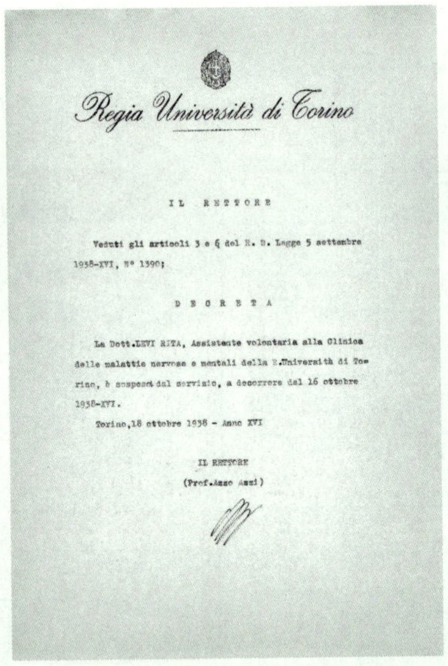

Regia Università di Torino

I L   R E T T O R E

Veduti gli articoli 3 e 6 del R. D. Legge 5 settembre 1938-XVI, N° 1390;

D E C R E T A

La Dott.LEVI RITA, Assistente volontaria alla Clinica delle malattie nervose e mentali della R.Università di Torino, è sospesa dal servizio, a decorrere dal 16 ottobre 1938-XVI.

Torino,18 ottobre 1938 - Anno XVI

IL RETTORE

(Prof.Azzo Azzi)

Rectoral decree of suspension from service of Rita Levi-Montalcini, volunteer assistant at the Clinic of Nervous and Mental Illnesses of the Turin University, October 18, 1938.

Nel dicembre del 2008 ho intervistato Rita Levi-Montalcini per il suo centesimo compleanno. Mi sono presentato all'appuntamento con un mazzo di fiori e mi chiedo se oggi lo rifarei. A quel tempo portare dei fiori a una personalità così illustre sembrava il minimo, ma le sensibilità cambiano in fretta e oggi mi viene il dubbio, con quei fiori, di aver trattato un premio Nobel donna diversamente da quanto avrei fatto con un uomo. Rita Levi-Montalcini non aveva degnato il mazzo di grande considerazione. È possibile che a cento anni ragionasse in modo più moderno di me, che ne avevo appena compiuti ventisei.

Ricordo il suo incedere lentissimo lungo il corridoio bianco dell'Ebri, il centro di ricerca sulle neuroscienze che aveva fondato alcuni anni prima. Camminava sorretta da un signore, eppure dritta, elegantissima in un abito verde scuro con le spalline rigonfie. Le prime frasi che mi ha detto non corrispondevano a una mia domanda: "A me nella vita è riuscito tutto facile. Le difficoltà me le sono scrollate di dosso, come acqua sulle ali di un'anatra". In effetti, per tutta la durata del colloquio ho avuto l'impressione che le domande che avevo preparato sul taccuino non servissero a nulla. Dovevo solo sintonizzarmi con il suo flusso di pensieri e cercare di fissare quanti più dettagli possibile.

Più tardi, trascrivendo l'intervista, mi tremavano le dita, perché sapevo che lei l'avrebbe letta e valutata con la severità del suo intelletto. Quattro anni dopo è morta. Ho pianto quando l'ho saputo perché, seppure brevemente, l'avevo incontrata.

Rita Levi-Montalcini ha definito la sua carriera scientifica una "detective story" iniziata in "condizioni disastrose". Disastrose in generale, perché nella prima metà del Novecento i neurobiologi erano "cittadini di terza classe" rispetto ai fisici e ai biologi molecolari. Si avvalevano di tecniche approssimative e subivano il pregiudizio che il cervello fosse un'entità unica talmente complessa da essere inconoscibile.

Ma ancora più disastrose nel suo caso specifico, perché l'introduzione in Italia delle leggi razziali la alienò, in quanto ebrea, dall'università, appena dopo la laurea. Levi-Montalcini fuggì in Belgio, dove raggiunse il suo maestro Giuseppe Levi (del quale settant'anni dopo avrebbe parlato con un affetto e una riverenza intatti).

Ma nel 1940 il Belgio finì sotto l'occupazione nazista. Alla fine dell'anno Levi-Montalcini fece il viaggio a ritroso, in auto, fino a Torino. Non poteva che vivere da rifugiata ormai. Simili difficoltà potevano significare per lei l'abbandono della ricerca. Ma si rivelarono, al contrario, l'impulso decisivo. È questo l'aspetto aneddotico più straordinario della sua carriera: la fortuna nella sfortuna. Oppure, se non si vuole credere alla fortuna (a cui lei credeva), la capacità di trasformare un limite in un'occasione. Lei stessa lo ha raccontato molte volte, nei libri, nelle interviste, anche a me. Parlava della "possibilità di rovesciamento" che esiste in ogni catastrofe. "Forse sono un'innata ottimista", mi ha detto, "ma penso che ci sia sempre qualcosa che ci salva. Le leggi razziali si sono rivelate la mia fortuna, perché mi hanno obbligata a costruirmi un laboratorio in camera da letto, dove ho cominciato le ricerche che mi hanno in seguito portato alla scoperta del Nerve Growth Factor".

Un laboratorio in camera da letto. Quasi chiunque, per quanto motivato, si sarebbero lasciato travolgere dallo sconforto, o non avrebbe avuto la creatività sufficiente a rifondare la propria scienza in un setting così limitato. Invece, confinata nell'appartamento di corso Re Umberto a Torino, Rita Levi-Montalcini allestì un laboratorio *in camera da letto*. Aveva a disposizione un'attrezzatura minima: microscopio, aghi, vetrini. Le uova di gallina erano facili da ottenere, bastava procurarsele al mercato, così si mise a studiare gli embrioni di pulcino. L'embriologia non era un ambito così in voga, anzi quasi non esisteva, ma era il meglio che potesse fare lì. Perciò lo fece.

Utilizzando il metodo di Golgi, che consiste nell'impregnare d'argento i tessuti nervosi in modo da renderli visibili al microscopio, iniziò diversi esperimenti. Quelle pratiche l'avrebbero portata, con il tempo, a ipotizzare l'azione di una sostanza sconosciuta, presente nei sarcomi di topo ma anche nel veleno di serpente e nelle ghiandole salivari, che induceva la formazione di fibre nervose in quantità. Levi-Montalcini disegnava le fibre sui suoi taccuini: sembravano irradiarsi dai gangli come capelli. Se non si fosse dedicata, suo malgrado, agli embrioni di pulcino, è probabile che non avrebbe mai visto nulla di tutto ciò.

La sua ostinazione non si esaurì a Torino. Dopo i bombardamenti sulla città, nel 1941, Levi-Montalcini dovette fuggire ancora. Andò con la sorella Paola in una villa fuori Asti. E anche lì ricostruì il suo laboratorio. Con l'Europa sprofondata nella guerra, lei andò avanti con le uova di pulcino, con le inoculazioni, con i disegni delle fibre argentee che misteriosamente si formavano sotto la lente del suo microscopio.

Vent'anni dopo quegli esordi in "condizioni disastrose" avrebbe isolato il Nerve Growth Factor insieme a Stanley Cohen. Altri trent'anni dopo avrebbero vinto entrambi il Nobel. E vent'anni più tardi, davanti a me, lei avrebbe ancora parlato dell'NGF con l'entusiasmo dell'inizio. Considerava la sua proteina miracolosa, le attribuiva potenzialità infinite. Presto, ne era certa, sarebbe servita a curare molte malattie degenerative, tra cui la SLA e l'Alzheimer.

Mi capita abbastanza spesso, quando sono a Torino, di passare davanti all'appartamento dove c'era il suo laboratorio domestico, perché si trova fra casa mia e il centro. Per lo più non ci faccio caso, ma qualche volta sì. Soprattutto negli ultimi due anni di pandemia. Mi chiedo cosa Rita Levi-Montalcini direbbe di quel che ci è successo e ci sta succedendo ancora, se troverebbe anche nella catastrofe attuale una "possibilità di rovesciamento". Nella confusione la sua voce è mancata, almeno a me.

Quando mi ero seduto di fronte a lei, mi aveva invitato a darle del tu, perché tra scienziati si faceva così. Si era lamentata di aver perso un po' la vista e molto l'udito, ma il suo pensiero era più attivo che a vent'anni.

"Il corpo faccia quello che vuole. Io non sono il corpo. Sono la mente."

"E quando muore il corpo?" le avevo chiesto.

"Sopravvive quello che hai fatto. Il messaggio che hai dato." Dopo una pausa, mi aveva fissato intensamente, come se con la sua vista indebolita mi vedesse comunque attraverso, e per la seconda volta mi aveva rivolto una domanda difficile, una domanda che non mi avrebbe mai più fatto nessuno: "Paolo, come vedi il tuo futuro?"

LIKE WATER OFF A DUCK'S BACK

In December 2008, I interviewed Rita Levi-Montalcini on her one hundredth birthday. I showed up for the interview with a bunch of flowers, and I wonder if I would do the same today. Bringing flowers for someone so well-known and celebrated seemed like the least one could do, but sensibilities change pretty swiftly and I wonder, today, whether I was in fact treating a woman Nobel Prize winner differently than I would have a man. Rita Levi-Montalcini didn't really pay much attention to the bouquet. She might have been, despite her one hundred years on this Earth, a lot more modern than I was, who had just turned twenty-six at the time.

I remember her moving incredibly slowly down the white corridor of the Ebri, the neuroscience research center she had founded a few years prior. She walked with the aid of an assistant but still erect, incredibly elegant in her dark green dress with puffed gigot sleeves. The first things she told me bore no relation to any of my questions: "Things have always come easy to me, in life. Difficulty just slipped off me, like water off a duck's back." Admittedly, for the entire conversation I felt as though none of the questions I had prepared in my notepad would lead anywhere. What I needed to do instead was tune in to the frequency of her thoughts and set down as many details as possible.

Later, as I was transcribing the interview, my fingers were shaking, because I was pointedly aware that she would read it and judge it with the full severity of her intellect. She died four years later. I cried when I found out, because, even though briefly, I had met her.

Rita Levi-Montalcini has called her scientific career a "detective story" that had started in "disastrous circumstances." Disastrous in a general sense, because in the first half of the 1900s the scientific community considered neurobiologists to be third-class citizens when compared to, say, physicists and molecular biologists. Their methods were approximations and they were operating under the prejudice that the brain was such a complex, single entity that it was unknowable.

Even more disastrous in her specific case, because the passing of racial laws in Italy had excised her from academia shortly after she earned her degree, due to her being Jewish. Levi-Montalcini fled to Belgium, where she met up with her mentor, Giuseppe Levi (of whom she would speak, even seventy years later, with untarnished affection and reverence).

In 1940, however, Belgium fell under Nazi occupation. At the end of that year, Levi-Montalcini retraced her steps, by car, to Turin. All that was left was for her to live as a refugee. Her situation could have meant the end of her research; instead it turned out to be the decisive push. This is the most extraordinary anecdote of her entire career: being lucky in unlucky situations. Or, if one would rather not believe in luck (though she did), the ability to transform a limitation into an opportunity. She has talked about it several times, in books and during interviews, and she told me again. She would talk about the "potential for overhaul" that exists within every disaster. "Maybe I am too much of an optimist," she told me, "but I am convinced there is always something that will save us. The racial laws turned out to be my stroke of luck, because they forced me to build a laboratory in my own bedroom, in which I started the research that would lead me to the discovery of the nerve growth factor."

A laboratory in her bedroom. Almost anyone else, no matter how motivated, would have been overwhelmed by despair, or at least would not have had the creativity to start their scientific endeavors from scratch in such a limited setting. Levi-Montalcini, instead, within the confines of the Corso Re Umberto flat in Turin, set up not a home but a *bedroom* laboratory. She had the bare minimum in terms of equipment: a microscope, syringes, slides. Chicken eggs were easy enough to obtain, so she started studying chick embryos. Embryology was not a particularly sought-after field, it barely even existed, but that was the best she could do in the situation. So she did.

Using Golgi's method—that is, dyeing tissues with silver in order to make them visible under the microscope—she started a number of experiments. These would bring her, in time, to hypothesize the action of an unknown substance, found in mouse sarcomas but also in snake venom and salivary glands, inducing the formation of copious amounts of nerve fibers. Levi-Montalcini would draw the fibers in her notebooks: they seemed to radiate from the ganglia like hair. If she hadn't dedicated her attention, devoid of any other option, to chick embryos, she might never have seen any of it.

Her indefatigability did not end in Turin. After the bombing of the city, in 1941, Levi-Montalcini was forced to flee again. She and her sister Paola moved to a villa outside Asti. And she set up her lab there once more. With Europe succumbing to war, she carried on with her chicken eggs, her injections, her sketches of the silver fibers that mysteriously formed under her microscope's lens.

Twenty years after her debut in "disastrous circumstances," she would isolate the nerve growth factor, along with Stanley Cohen. Another thirty years later, they would win the Nobel Prize together. And twenty years after that, in front of me, she would still talk about NGF with the same enthusiasm of the early days. She saw her protein as something miraculous, something with infinite potential. Soon, she was certain, it would help cure a number of degenerative diseases, including ALS and Alzheimer's.

When I'm in Turin I often happen to walk past the flat where she used to have her bedroom laboratory, as it's on the way from my place to the center of town. I don't always notice, but sometimes it stands out. Especially during these past two years of the pandemic. I wonder what Rita Levi-Montalcini would have had to say about what has happened and is still happening to us, if she would find in our current disaster a "potential for overhaul." In all the noise and confusion, her voice has been missed. I missed her.

When I sat down in front of her, she asked me to address her informally, because that was what scientists did. She complained about having lost a little of her vision and a lot of her hearing, but her mind was far more active than when she was twenty years old. "The body can do what it wishes. I am not my body. I am my mind."

"And when the body dies?" I asked her.

"What you have done is what survives. The message you put out." She paused, staring at me intently, as if she could see through me even with her weakened eyesight; then, for the second time, she asked me a hard question, one that no one else would ever ask me again: "Paolo, how do you see your future?"

STORY 26

Berger's Doll

Akwaeke Emezi

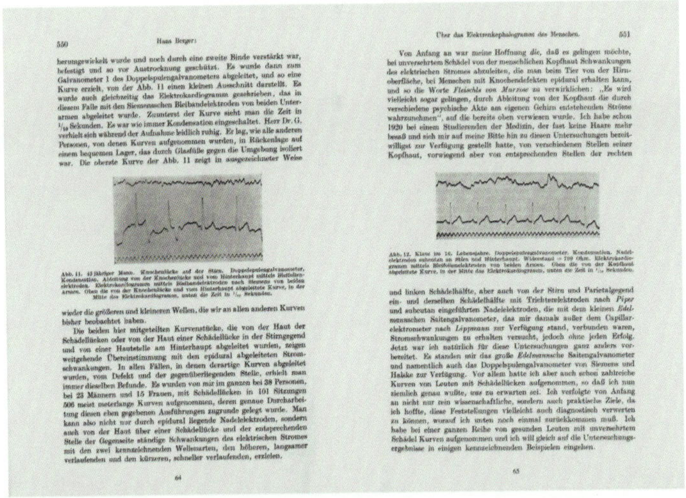

550 Hans Berger:

herumgewickelt wurde und noch durch eine zweite Binde verstärkt war, befestigt und so vor Austrocknung geschützt. Es wurde dann zum Galvanometer 1 des Doppelspulengalvanometers abgeleitet, und so eine Kurve erzielt, von der Abb. 11 einen kleinen Ausschnitt darstellt. Es wurde auch gleichzeitig das Elektrokardiogramm geschrieben, das in diesem Falle mit den Siemensschen Bleibandelektroden von beiden Unterarmen abgeleitet wurde. Zuunterst der Kurve sieht man die Zeit in $^1/_{10}$ Sekunden. Es war wie immer Kondensation eingeschaltet. Herr Dr. G. verhielt sich während der Aufnahme leidlich ruhig. Er lag, wie alle anderen Personen, von denen Kurven aufgenommen wurden, in Rückenlage auf einem bequemen Lager, das durch Glasfüße gegen die Umgebung isoliert war. Die oberste Kurve der Abb. 11 zeigt in ausgezeichneter Weise

Abb. 11. 43jähriger Mann. Knochenlücke auf der Stirn. Doppelspulengalvanometer, Kondensation. Ableitung von der Knochenlücke und vom Hinterhaupt mittels Bleifolienelektroden. Elektrokardiogramm mittels Bleibandelektroden nach Siemens von beiden Armen. Oben die von der Knochenlücke und vom Hinterhaupt abgeleitete Kurve, in der Mitte das Elektrokardiogramm, unten die Zeit in $^1/_{10}$ Sekunden.

wieder die größeren und kleineren Wellen, die wir an allen anderen Kurven bisher beobachtet haben.

Die beiden hier mitgeteilten Kurvenstücke, die von der Haut der Schädellücken oder von der Haut einer Schädellücke in der Stirngegend und von einer Hautstelle am Hinterhaupt abgeleitet wurden, zeigen weitgehende Übereinstimmung mit den epidural abgeleiteten Stromschwankungen. In allen Fällen, in denen derartige Kurven abgeleitet wurden, vom Defekt und der gegenüberliegenden Stelle, erhielt man immer dieselben Befunde. Es wurden von mir im ganzen bei 38 Personen, bei 23 Männern und 15 Frauen, mit Schädellücken in 101 Sitzungen 506 meist meterlange Kurven aufgenommen, deren genaue Durcharbeitung diesen eben gegebenen Ausführungen zugrunde gelegt wurde. Man kann also nicht nur durch epidural liegende Nadelelektroden, sondern auch von der Haut über einer Schädellücke und der entsprechenden Stelle der Gegenseite ständige Schwankungen des elektrischen Stromes mit den zwei kennzeichnenden Wellenarten, den höheren, langsamer verlaufenden und den kürzeren, schneller verlaufenden, erzielen.

64

Über das Elektrenkephalogramm des Menschen. 551

Von Anfang an war meine Hoffnung *die*, daß es gelingen möchte, bei unversehrtem Schädel von der menschlichen Kopfhaut Schwankungen des elektrischen Stromes abzuleiten, die man beim Tier von der Hirnoberfläche, bei Menschen mit Knochendefekten epidural erhalten kann, und so die Worte *Fleischls von Marxow* zu verwirklichen: „Es wird vielleicht sogar gelingen, durch Ableitung von der Kopfhaut die durch verschiedene psychische Akte am eigenen Gehirn entstehenden Ströme wahrzunehmen", auf die bereits oben verwiesen wurde. Ich habe schon 1920 bei einem Studierenden der Medizin, der fast keine Haare mehr besaß und sich mir auf meine Bitte hin zu diesen Untersuchungen bereitwilligst zur Verfügung gestellt hatte, von verschiedenen Stellen seiner Kopfhaut, vorwiegend aber von entsprechenden Stellen der rechten

Abb. 12. Klaus im 16. Lebensjahre. Doppelspulengalvanometer. Kondensation. Nadelelektroden subcutan an Stirn und Hinterhaupt. Widerstand = 700 Ohm. Elektrokardiogramm mittels Bleifolienelektroden von beiden Armen. Oben die von der Kopfhaut abgeleitete Kurve, in der Mitte das Elektrokardiogramm, unten die Zeit in $^1/_{10}$ Sekunden.

und linken Schädelhälfte, aber auch von der Stirn und Parietalgegend ein- und derselben Schädelhälfte mit Trichterelektroden nach *Piper* und subcutan eingeführten Nadelelektroden, die mit dem kleinen *Edelmann*schen Saitengalvanometer, das mir damals außer dem Capillarelektrometer nach *Lippmann* zur Verfügung stand, verbunden waren, Stromschwankungen zu erhalten versucht, jedoch ohne jeden Erfolg. Jetzt war ich natürlich für diese Untersuchungen ganz anders vorbereitet. Es standen mir das große *Edelmann*sche Saitengalvanometer und namentlich auch das Doppelspulengalvanometer von Siemens und Halske zur Verfügung. Vor allem hatte ich aber auch schon zahlreiche Kurven von Leuten mit Schädellücken aufgenommen, so daß ich nun ziemlich genau wußte, *was* zu erwarten sei. Ich verfolgte von Anfang an nicht nur rein wissenschaftliche, sondern auch praktische Ziele, da ich hoffte, diese Feststellungen vielleicht auch diagnostisch verwerten zu können, worauf ich unten noch einmal zurückkommen muß. Ich habe bei einer ganzen Reihe von gesunden Leuten mit unversehrtem Schädel Kurven aufgenommen und ich will gleich auf die Untersuchungsergebnisse in einigen kennzeichnenden Beispielen eingehen.

65

Hans Berger, "Über das Elektrenkephalogramm des Menschen"
[On human electroencephalography], *Archiv für Psychiatrie und Nervenkrankheiten*, vol. 87, no. 1, 1929.
Spreads including electroencephalograms of patients at different ages.

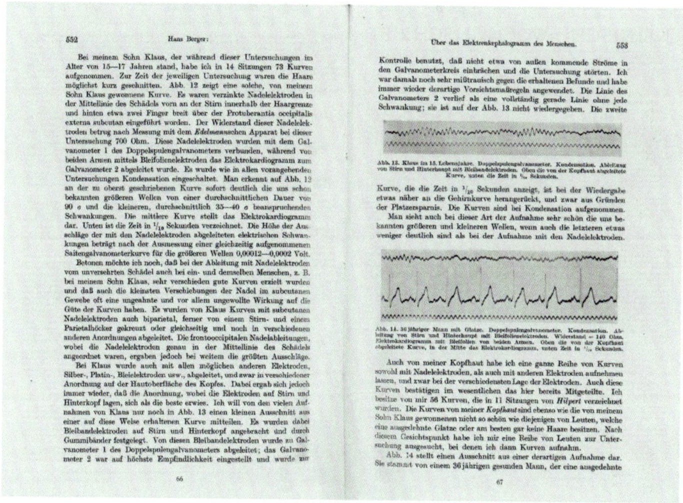

552 Hans Berger:

Bei meinem Sohn Klaus, der während dieser Untersuchungen im Alter von 15—17 Jahren stand, habe ich in 14 Sitzungen 73 Kurven aufgenommen. Zur Zeit der jeweiligen Untersuchung waren die Haare möglichst kurz geschnitten. Abb. 12 zeigt eine solche, von meinem Sohn Klaus gewonnene Kurve. Es waren verzinkte Nadelelektroden in der Mittellinie des Schädels vorn an der Stirn innerhalb der Haargrenze und hinten etwa zwei Finger breit über der Protuberantia occipitalis externa subcutan eingeführt worden. Der Widerstand dieser Nadelelektroden betrug nach Messung mit dem *Edelmann*schen Apparat bei dieser Untersuchung 700 Ohm. Diese Nadelelektroden wurden mit dem Galvanometer 1 des Doppelspulengalvanometers verbunden, während von beiden Armen mittels Bleifolienelektroden das Elektrokardiogramm zum Galvanometer 2 abgeleitet wurde. Es wurde wie in allen vorangehenden Untersuchungen Kondensation eingeschaltet. Man erkennt auf Abb. 12 an der zu oberst geschriebenen Kurve sofort deutlich die uns schon bekannten größeren Wellen von einer durchschnittlichen Dauer von 90 σ und die kleineren, durchschnittlich 35—40 σ beanspruchenden Schwankungen. Die mittlere Kurve stellt das Elektrokardiogramm dar. Unten ist die Zeit in $^1/_{10}$ Sekunden verzeichnet. Die Höhe der Ausschläge der mit den Nadelelektroden abgeleiteten elektrischen Schwankungen beträgt nach der Ausmessung einer gleichzeitig aufgenommenen Saitengalvanometerkurve für die größeren Wellen 0,00012—0,0002 Volt.

Betonen möchte ich noch, daß bei der Ableitung mit Nadelelektroden vom unversehrten Schädel auch bei ein- und demselben Menschen, z. B. bei meinem Sohn Klaus, sehr verschieden gute Kurven erzielt wurden und daß auch die kleinsten Verschiebungen der Nadel im subcutanen Gewebe oft eine ungeahnte und vor allem ungewollte Wirkung auf die Güte der Kurven haben. Es wurden von Klaus Kurven mit subcutanen Nadelelektroden auch biparietal, ferner von einem Stirn- und einem Parietalhöcker gekreuzt oder gleichseitig und noch in verschiedenen anderen Anordnungen abgeleitet. Die frontooccipitalen Nadelableitungen, wobei die Nadelelektroden genau in der Mittellinie des Schädels angeordnet waren, ergaben jedoch bei weitem die größten Ausschläge.

Bei Klaus wurde auch mit allen möglichen anderen Elektroden, Silber-, Platin-, Bleielektroden usw., abgeleitet, und zwar in verschiedener Anordnung auf der Hautoberfläche des Kopfes. Dabei ergab sich jedoch immer wieder, daß die Anordnung, wobei die Elektroden auf Stirn und Hinterkopf lagen, sich als die beste erwies. Ich will von den vielen Aufnahmen von Klaus nur noch in Abb. 13 einen kleinen Ausschnitt aus einer auf diese Weise erhaltenen Kurve mitteilen. Es wurden dabei Bleibandelektroden auf Stirn und Hinterkopf angebracht und durch Gummibänder festgelegt. Von diesen Bleibandelektroden wurde zu Galvanometer 1 des Doppelspulengalvanometers abgeleitet; das Galvanometer 2 war auf höchste Empfindlichkeit eingestellt und wurde zur

66

Über das Elektrenkephalogramm des Menschen. 553

Kontrolle benutzt, daß nicht etwa von außen kommende Ströme in den Galvanometerkreis einbrächen und die Untersuchung störten. Ich war damals noch sehr mißtrauisch gegen die erhaltenen Befunde und habe immer wieder derartige Vorsichtsmaßregeln angewendet. Die Linie des Galvanometers 2 verlief als eine vollständig gerade Linie ohne jede Schwankung; sie ist auf der Abb. 13 nicht wiedergegeben. Die zweite

Abb. 13. Klaus im 15. Lebensjahre. Doppelspulengalvanometer. Kondensation. Ableitung von Stirn und Hinterhaupt mit Bleibandelektroden. Oben die von der Kopfhaut abgeleitete Kurve, unten die Zeit in $^1/_{10}$ Sekunden.

Kurve, die die Zeit in $^1/_{10}$ Sekunden anzeigt, ist bei der Wiedergabe etwas näher an die Gehirnkurve herangerückt, und zwar aus Gründen der Platzersparnis. Die Kurven sind bei Kondensation aufgenommen.

Man sieht auch bei dieser Art der Aufnahme sehr schön die uns bekannten größeren und kleineren Wellen, wenn auch die letzteren etwas weniger deutlich sind als bei der Aufnahme mit den Nadelelektroden.

Abb. 14. 36 jähriger Mann mit Glatze. Doppelspulengalvanometer. Kondensation. Ableitung von Stirn und Hinterhaupt mit Bleifolienelektroden. Widerstand = 140 Ohm. Elektrokardiogramm mit Bleifolien von beiden Armen. Oben die von der Kopfhaut abgeleitete Kurve, in der Mitte das Elektrokardiogramm, unten Zeit in $^1/_{10}$ Sekunden.

Auch von meiner Kopfhaut habe ich eine ganze Reihe von Kurven sowohl mit Nadelelektroden, als auch mit anderen Elektroden aufnehmen lassen, und zwar bei der verschiedensten Lage der Elektroden. Auch diese Kurven bestätigen im wesentlichen das hier bereits Mitgeteilte. Ich besitze von mir 56 Kurven, die in 11 Sitzungen von *Hilpert* verzeichnet wurden. Die Kurven von meiner *Kopfhaut* sind ebenso wie die von meinem Sohn Klaus gewonnenen nicht so schön wie diejenigen von Leuten, welche eine ausgedehnte Glatze oder am besten gar keine Haare besitzen. Nach diesem Gesichtspunkt habe ich mir eine Reihe von Leuten zur Untersuchung ausgesucht, bei denen ich dann Kurven aufnahm.

Abb. 14 stellt einen Ausschnitt aus einer derartigen Aufnahme dar. Sie stammt von einem 36 jährigen gesunden Mann, der eine ausgedehnte

67

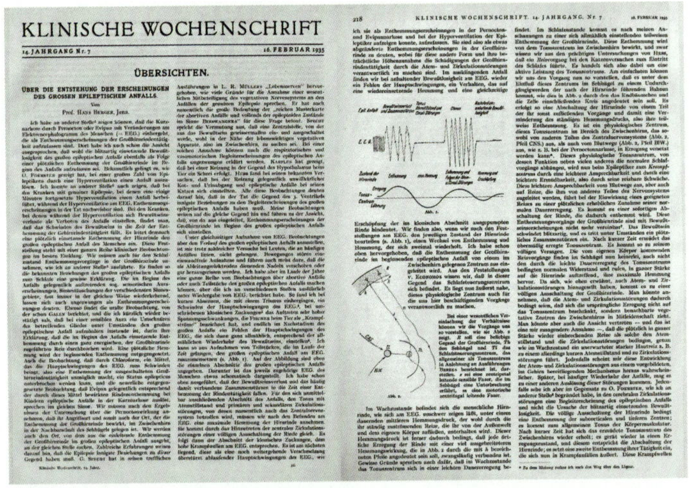

KLINISCHE WOCHENSCHRIFT

14. JAHRGANG Nr. 7 — 16. FEBRUAR 1935

ÜBERSICHTEN.

ÜBER DIE ENTSTEHUNG DER ERSCHEINUNGEN DES GROSSEN EPILEPTISCHEN ANFALLS.

Von

Prof. Hans Berger, Jena.

Hans Berger, "Über die Entstehung der Erscheinungen des großen epileptischen Anfalls" [On the origin of the phenomena of the grand epileptic seizure], *Klinische Wochenschrift*, no. 14, 1935. Spread including brain wave diagrams.

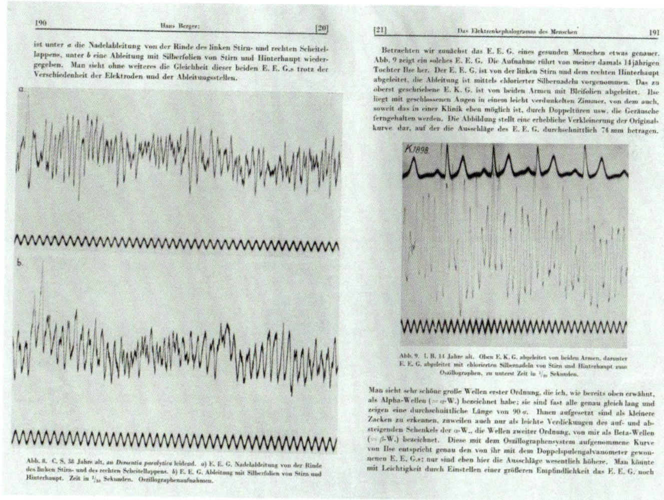

190 Hans Berger: [20]

ist unter *a* die Nadelableitung von der Rinde des linken Stirn- und rechten Scheitellappens, unter *b* eine Ableitung mit Silberfolien von Stirn und Hinterhaupt wiedergegeben. Man sieht ohne weiteres die Gleichheit dieser beiden E. E. G.s trotz der Verschiedenheit der Elektroden und der Ableitungsstellen.

a.

b.

Abb. 8. C. S. 38 Jahre alt, an *Dementia paralytica* leidend. *a*) E. E. G. Nadelableitung von der Rinde des linken Stirn- und des rechten Scheitellappens. *b*) E. E. G. Ableitung mit Silberfolien von Stirn und Hinterhaupt. Zeit in $^1/_{10}$ Sekunden. Oszillographenaufnahmen.

[21] Das Elektrenkephalogramm des Menschen 191

Betrachten wir zunächst das E. E. G. eines gesunden Menschen etwas genauer. Abb. 9 zeigt ein solches E. E. G. Die Aufnahme rührt von meiner damals 14 jährigen Tochter Ilse her. Der E. E. G. ist von der linken Stirn und dem rechten Hinterhaupt abgeleitet, die Ableitung ist mittels chlorierter Silbernadeln vorgenommen. Das zu oberst geschriebene E. K. G. ist von beiden Armen mit Bleifolien abgeleitet. Ilse liegt mit geschlossenen Augen in einem leicht verdunkelten Zimmer, von dem auch, soweit das in einer Klinik eben möglich ist, durch Doppeltüren usw. die Geräusche ferngehalten werden. Die Abbildung stellt eine erhebliche Verkleinerung der Originalkurve dar, auf der die Ausschläge des E. E. G. durchschnittlich 74 mm betragen.

Abb. 9. I. B. 14 Jahre alt. Oben E. K. G. abgeleitet von beiden Armen, darunter E. E. G. abgeleitet mit chlorierten Silbernadeln von Stirn und Hinterhaupt zum Oszillographen, zu unterst Zeit in $^1/_{10}$ Sekunden.

Man sieht sehr schöne große Wellen erster Ordnung, die ich, wie bereits oben erwähnt, als Alpha-Wellen (= α-W.) bezeichnet habe; sie sind fast alle genau gleich lang und zeigen eine durchschnittliche Länge von 90 σ. Ihnen aufgesetzt sind als kleinere Zacken zu erkennen, zuweilen auch nur als leichte Verdickungen des auf- und absteigenden Schenkels der α-W., die Wellen zweiter Ordnung, von mir als Beta-Wellen (= β-W.) bezeichnet. Diese mit dem Oszillographensystem aufgenommene Kurve von Ilse entspricht genau den von ihr mit dem Doppelspulengalvanometer gewonnenen E. E. G.s; nur sind eben hier die Ausschläge wesentlich höhere. Man könnte mit Leichtigkeit durch Einstellen einer größeren Empfindlichkeit das E. E. G. noch

Hans Berger, "Das Elektrenkephalogramm des Menschen"
[Human electroencephalography], *Nova Acta Leopoldina*, series 6, no. 38, 1938.
Spread including electroencephalograms of patients at different ages.

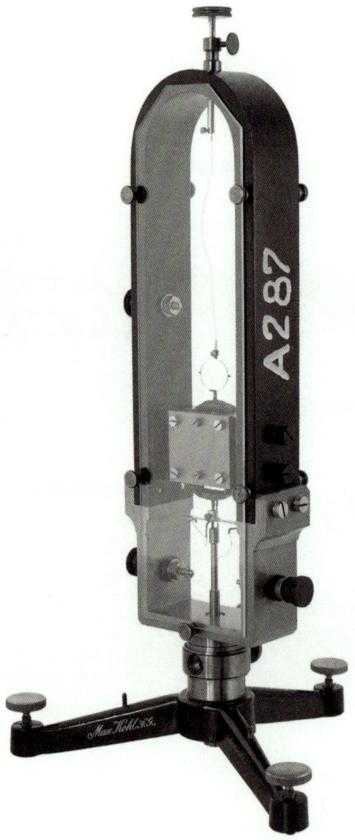

Max Kohl's loop galvanometer for measuring brain waves, Germany, 20th century.

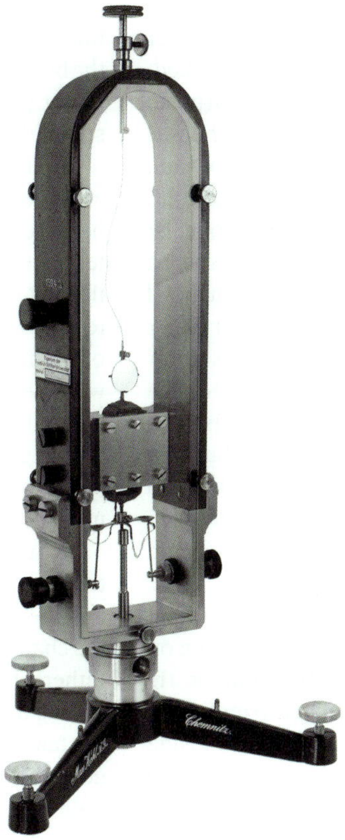

Ilse was terrified of the annex.

It was full of her father's instruments: groaning and whirring things with dials and needles, reams of photographic paper, a working museum of his obsession. It contained the chair she had to sit in when she was summoned to help with his work, and Ilse hated this chair. She hated its worn leather straps and dull buckles, and how her father used those to hold her still while he worked.

Ilse would never struggle on purpose; she was a good girl who did what she was told even when she very much did not want to. But Ilse was also only nine years old, and that was not an age at which anyone, let alone a scientist, could reasonably expect a nonsedated child to sit still while thin metal bits were inserted into their head. Her father had discovered this early on, at the start of these experiments, when he attempted the procedure without restraining her. Ilse had intended to be as quiet as her dolls, but then the silver had stabbed through her skin and she had jumped in fright. There had been blood and so now there were buckles, and that was science.

Besides, her father had to focus on what he was doing, what he had been working on for years—the making of a brain mirror. He had no time to worry about her wriggling or being anything other than the organ inside her head, a test subject under his hands. Ilse knew she couldn't be a distraction. Sometimes she thought she wasn't meant to be a daughter either. Maybe she was meant to be a doll, but a special one—a small blond doll that could do sums, tricks for her father's brain mirror to catch.

She knew his work was important, but helping him with it didn't get easier, no matter how many times she was good in that chair. The silver needles sliding into her scalp always hurt. There was one at the front, another at the back, and the space in between was a land her father kept trying to map, year after year. He told her that the silver needed to touch her skull—not the bone, but what lay over the bone. Ilse knew that her skull was a cap over her brain, and that her head was made of caps. A cap of bone, a cap of white touching silver over the bone, a cap of skin, a cap of hair, and her father piercing through it all, looking for something everyone else called impossible.

In the harsh light of the annex, Ilse was afraid of her father. She couldn't say that, because she was a good and obedient daughter. She wasn't entirely sure she could even think it and still remain good, so she tried to pretend it wasn't true. She wasn't afraid of her father, of his anger when something went wrong, of the way he looked at her without seeing Ilse herself, like she was just a stranger, another patient in the hospital whose brain he wanted to play with, another unrelated doll for his collection.

No matter her fear or the lie of its absence, there was always the corridor with his office and the annex at the end. There was always the chair and the straps and the buckles, her father's hands parting her golden hair and sliding the sharp silver through. So Ilse kept her tears where they belonged, far away and deep inside her. She was not Ilse. She was a doll. She was a brain. She was whatever he wanted her to be. She would do whatever he wanted her to do.

In the end, no matter what, she was simply his.

STORY 27

Big Day

Helen Oyeyemi

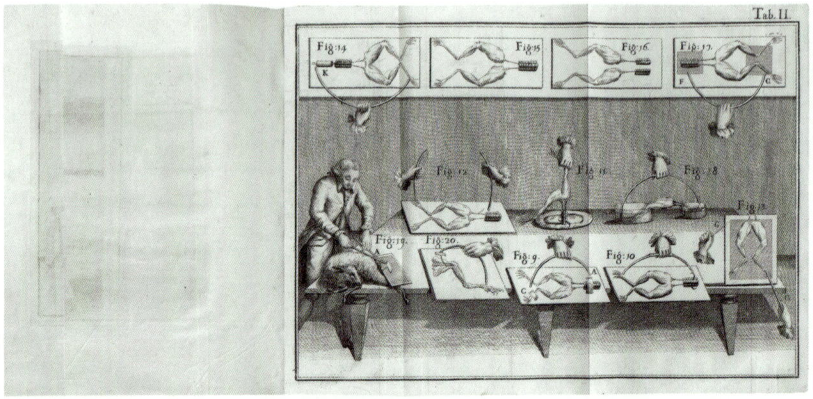

Luigi Galvani, *De viribus electricitatis in motu musculari commentarius*
[Commentary on the effect of electricity on muscular motion], Modena, 1792. Plates nos. 2 and 3.

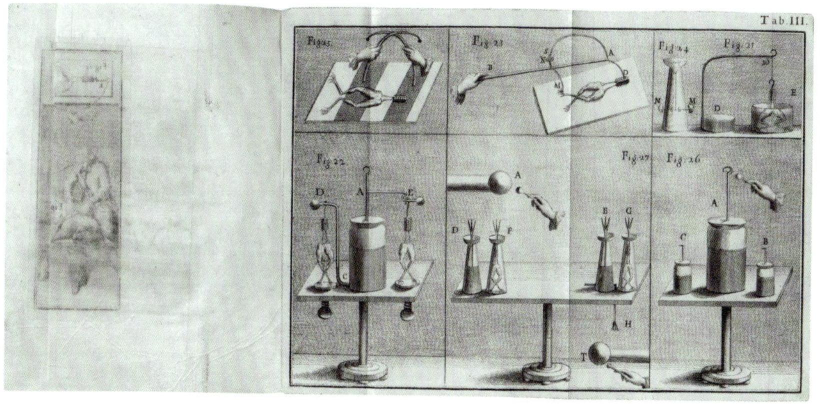
Tab. III.

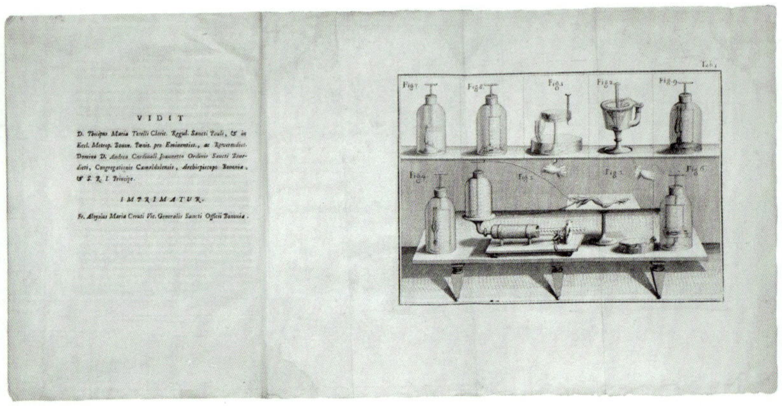

Giovanni Aldini, *De animali electricitate dissertationes duae*
[Two dissertations on animal electricity], Bologna, 1794. Plates nos. 1 and 2.

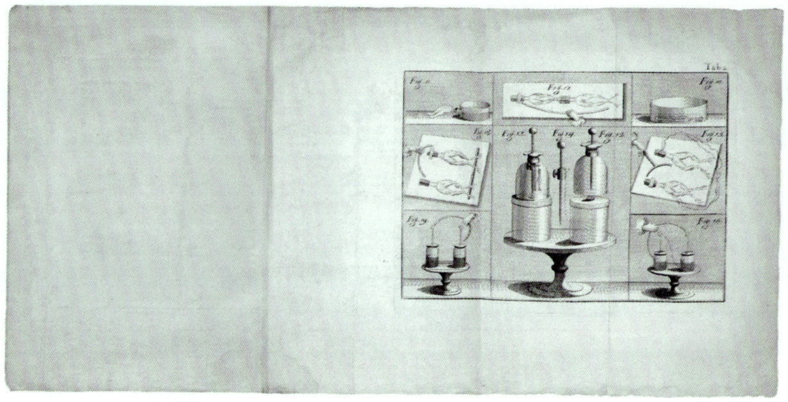

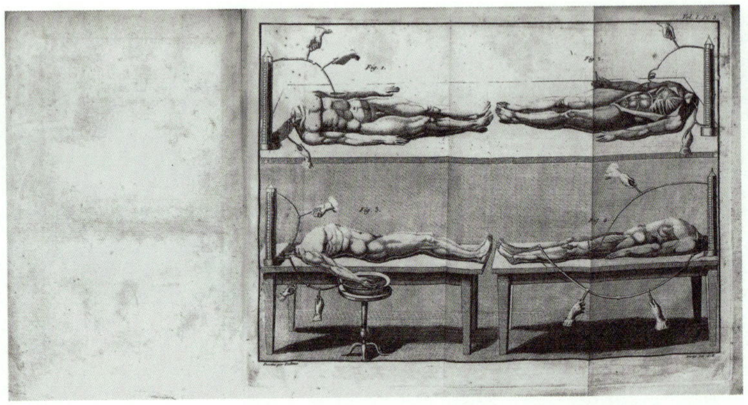

Giovanni Aldini, *Essai théorique et expérimental sur le galvanisme*
[Theoretical and experimental essay on galvanism], Paris, 1804. Volume 1, plates nos. 3 and 4.

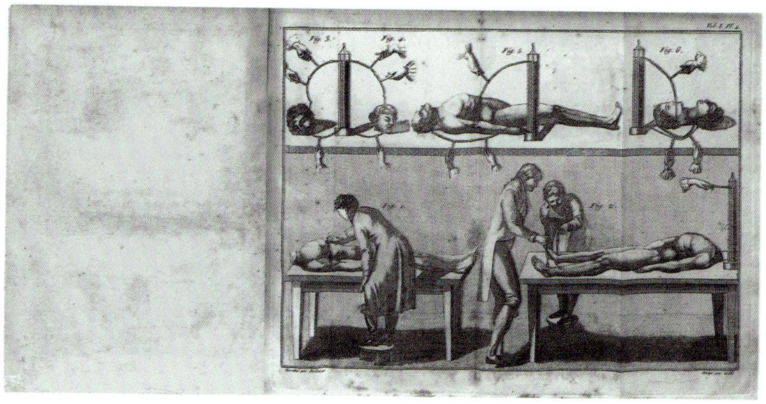

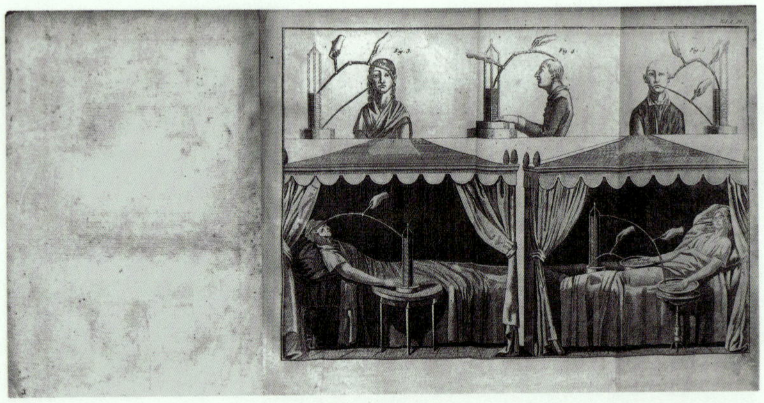

Giovanni Aldini, *Essai théorique et expérimental sur le galvanisme*
[Theoretical and experimental essay on galvanism], Paris, 1804. Volume 1, plate no. 5.

One morning in 1802, a physician began to get an inkling that his dreams had tired of lying in wait and had come looking for him. We have it on good, though unverifiable, authority that 1802 was this physician's thirtieth year of refraining from dreaming while he was asleep. This was all thanks to his uncle, who had also been a physician.

"Dreaming has led me to my line of inquiry," the boy's uncle told him. "But if you want to continue it—and you should want to, my boy—you must see everything as if it's a diagram. Every aspect of our work must be available for crystal-clear exposition. In several languages! You can't glut yourself on illusion like the rest of us ... Your faculties will grow flabby; the meaning of our findings would collapse under them. And there's still so much to do ..."

The physician-as-a-boy looked his uncle dead in the eye. He didn't reply to his uncle in words: whatever he said just then would be all wrong, and he was deathly afraid of beginning with words that would upset them both: "My dear uncle ... my dear, nervous, diffident, brilliant uncle," words like that. But he was ready to prove that he could be counted on.

A nephew watches his uncle spark arcs that quicken the corpses of the cold-blooded and sizzle personality back into the listless living ... and he learned that what burned in him could be replicated outside the body, could be overridden or even replaced by that externally generated force. He gave himself entirely to accurate observation, understanding, and description of these marvels—and when he slept, sleeping was all he did.

Dreamless repose seemed to anchor a certain agility. Rung by rung, the physician worked his way up to a position from which he was permitted to experiment with mortal remains. He made arrangements with staff at the prison closest to the university he lectured at ... Three inmates due for execution would be made available to the scientific cause immediately afterward.

Then came Unreason ... bright and early on the morning of the physician's big day.

Those who attended the experiment later that morning knew nothing of this. As far as they were concerned, the day's main event was the few minutes when, upon application of a metal wand to pressure points in the arms, the physician caused three headless bodies to point their fingers directly at onlookers. But in many ways this was more memorable for the onlookers than it was for the wielder of the wand. Afterward, those onlookers were embarrassed by their panic, of course. Embarrassed to have considered themselves accused or acknowledged by the dead. Such reactions were a betrayal of the downright ordinary—that is to say, mechanical—basis of what they'd seen. Everybody knew that the flesh had jerked at the behest of skeins of gray coins, brown coins, wires, and sea-soaked rags ... Those who'd momentarily lost the plot redeemed themselves through eager discussion of the interactions that had occurred between those components.

Electricity had plenty to say to biological matter that morning. There were six steel tables laid out in the operating theatre: three for the bodies that had been hewn at the neck and three for the heads that had been harvested. The electric current appeared to clear its throat a couple of times before commanding the bodies to rattle their digits. The rattling of digits was followed by that apparently malevolent finger-pointing. The severed heads seemed to receive other instructions: they narrowed and widened their eyes, spinning a roulette wheel of facial expressions in their quest for grunting, gurgling speech.

All the while, the physician was wondering by what means he'd been convinced to campaign for the impossible.

A law student had visited him earlier, at eight o'clock that morning. Quite the opportunist, that law student. He'd announced himself as the representative of one of the subjects of the physician's experiment. The physician asked which subject, and when the law student named his client, the physician said, "Oh, of course it's her," to mask his surprise. The person named was the least repentant of the inmates, or so the physician had heard–the one least likely to mend her ways in this life or any other. What devilish thing could she mean by engaging legal representation on the eve of her execution? The physician soon found out; his guest produced a document that was already adorned with a blotchy assortment of signatures.

> *Should electrical stimulation fully restore the deceased to life, we the undersigned agree to use such influence as we have to ensure that she receives a posthumous pardon.*

When the physician asked what use a posthumous pardon would be to one resurrected, the law student stressed the importance of a guarantee that his client wouldn't be hindered from narrating and benefiting from a lucrative account of her crimes, punishment, and rebirth. "This," the law student said, tapping the document, "is a declaration of the right to make a living after death."

The physician stared at the document without really seeing it. After a few moments, he began to have some thoughts that communicated only with one another and went quiet when he tried to join in. This worried him. He talked aloud to counteract his concerns. Raising objection after objection (moral, social, scientific)–talking nineteen to the dozen, in fact–the physician added his signature to all the others. The instant was a heady one, the ethical equivalent of rum punch, voluntary and involuntary swirling together… then the document was returned to the law student, who said: "As expected, you side with the seekers of justice," and withdrew with an amiable nod.

After that, only one prospect had the power to put the physician's thoughts in the proper order. He was granted this at the close of the experiment as he carefully surveyed the inmates' forms, basking in their stillness. Now he could forget all about the agreement he'd signed; it was null and void.

Rationality had seized the day again. For another day this work would not be overturned…

The longing for his findings to hold their ground for more than one day at a time–that was laziness, and he would ignore it.

The physician found this a bit harder to do now that uncle was no longer available for collaboration or consultation. Also: the physician had grown slightly more abstracted in his nature in the months since he'd applied strong electric shocks to his own skull. The effects were still palpable. When the physician got headaches, they were extra special–it was if an outsized tongue covered his cranium with its clammy fur. Nudged by that tongue, the physician's head wobbled beyond his preferred degree of motion whenever he tried to turn it. And he'd mutter to himself: *Come on, come on, what am I… a tooth being worked out of the earth's gum?* He counteracted this sensation by conceiving of himself as a humming, rainbow-sheened specimen detained within the contours of a Leyden jar. Then everything slowed down and smoothed out.

STORY 28

In Nightmares Begin Possibilities

McKenzie Wark

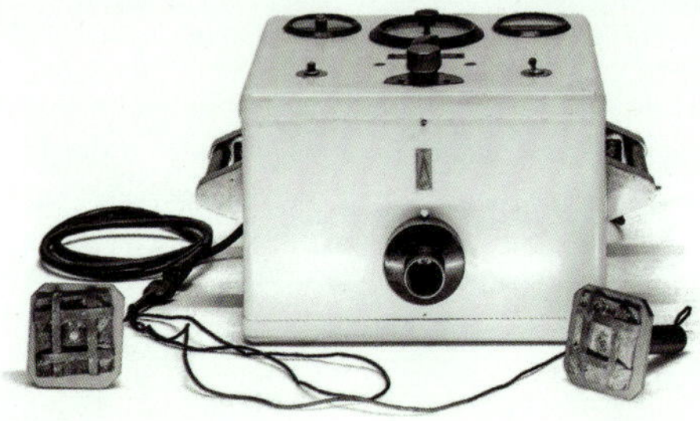

Lucio Bini's and Ugo Cerletti's electroconvulsive therapy machine, Italy, 1938–1940.

The doc always asks me the same things after. What's my name? Lou. What year is it? It's 1959. Do I know who the president is? That newsreel fake Eisenhower. I know that much. Just don't ask me what happened yesterday, or last week. I've been trying to read, as there's nothing else to do between treatments. Can't read anything more than a few pages. I forget the story.

Remember I read some Delmore Schwartz. There's a story that's like a dream. Man goes to the movies, but the movie is about when his parents met. He's yelling at the movie. He doesn't want it all to happen. He doesn't want to be. I'm rooting for that guy.

I think that was this morning, or maybe yesterday. I don't remember much else. It's hot. It's summer. This druggy heat. These two guys, teenagers like me, they came for me in the afternoon. Shot me up with something. Made me go limp, I guess so I wouldn't struggle. Not that I would. Strapped me to a gurney. Wheel me into the room with the machine.

Stick a rubber block in my mouth. The taste of rubber. The smell. They put the electrodes on my skull. The doc orders the teenage orderlies to stand clear. He flips the switch. I guess he does. I don't see. I don't remember. I don't remember anything until he's asking me who the president is again.

Mom called. She's a nervous wreck. I guess that's where I get it from. She sounds so guilty but I don't know what she ever did. I guess she worries about me. It's true, I lost it when I went to college. Flipped my wig. The city, right there at my feet. Might have been fine but then some guy on my dorm starts calling me a faggot. Like he knows. Like maybe he's one too and he has to point the finger at me before someone points it at him.

Dad came and got me, drove me back out to the island. Was more worried about what the neighbors would think than about me. All I know is that I never want to turn out like him. I fucking hate it out there. Everyone pretending they're normal. Never know what's going on behind those closed doors.

Don't know what they told the neighbors. It was a little early for me to be back from school for the summer. Didn't feel like summer to me. The sky felt like uninterrupted gray. Everything gray. If I wasn't practicing guitar the gray would get to me. Surround me. Seep in.

I play pretty good, but I never get, like, *good* good. I play good enough for rock and roll. Get a good rhythm going. It's a way to get across a feeling. Put myself into something, into a song. So that thing is no longer in me.

I forget a lot of things, though. Have I forgotten how to play? That would kill me. That'd be the joke of all time. They're blasting my skull with their mad scientist shit so I don't kill myself, and yet it's the thing that kills me. Makes me dead inside.

I don't want to be dead. Like Dad. I don't want to have to take it out on those around me. Make them jump. I'd rather make them dance. He's so uptight. So frightened. Of what? It's okay for him. He has got his little castle to be king of, out there on the island. He wants to kill the part of me that wants out of all that. He wants to kill the part of me that lives. The side that won't be tamed.

Maybe if I'd been born a girl it would have been easier. He'd see me different. The part of me that likes guys wouldn't be such a problem, but then I'd have a problem with the part of me that likes girls. Anyways. I look at my kid sister and I see she doesn't have it easy either.

The orderly says they'll be coming for me again later today. Feels like it was just minutes ago they came for me last time. Like I'm in a movie rerun on television. I don't know where the time goes. I guess I had meals, took a shit, slept. Did I sleep? Mom called. I remember that. She doesn't mean to, but her voice puts me on edge.

I don't know how many more times. I suppose I could ask. They have to let me out eventually. Or do they? I have to get out. I don't want to go home. Dad on my case all the time. Not my fault he's mad at his own stupid life. Mom barely keeping it together. Sis just trying to be a normal kid in the middle of it all. It's hell. They're all in this hell and they don't know it. This place is hell too. Part of the same phony hell.

Next thing I know I'm strapped to that damned gurney again. Wheeled off by the same two orderlies. The taste of rubber again. Then the doc flips the switch. Bad dream. Bad movie. Phone call from hell.

Something is going to die here. Some part of me. Joke's on them. It won't be the part they're trying to kill. I'm going to take my guitar and my amp and get the fuck out. First chance I get. I'll write a hit song. I'll form a rock and roll band. I'll be a star. Up where no one can see me.

Mom, Dad, they don't know a thing. These psychiatrists too. Clueless. Maybe they'll figure it out someday. Maybe they'll figure out what the machine is good for. It sure does something. Maybe they'll learn from these experiments on people like me. Consider the possibilities.

Wouldn't it be funny if them zapping my brain was the reason I learned to not trust any of it ever again? Have to learn to live. How to live my life. In this body. In this brain. That likes to fuck, get fucked, get high, that wants to be a star and make them all dance.

Gonna take this noise, this static, that the doc shoots through my brain like bad speed. Gonna amp it up, a feedback squall. Gonna be the switchboard. I'm so switch, babe. Gonna blast it right at the ears of all the kids out there. The ones like me. Only there's nobody like me.

STORY 29

Im Kopf des schwarzen Joris

Daniel Kehlmann

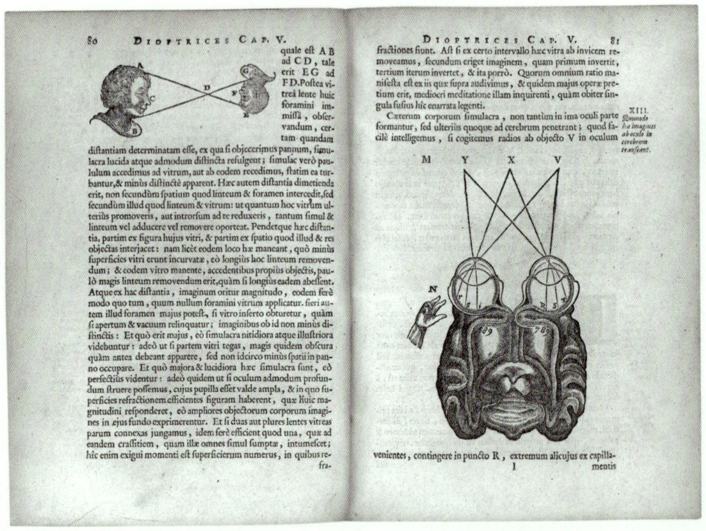

80 DIOPTRICES CAP. V.

quale est A B ad C D, tale erit E G ad F D. Postea vitreâ lente huic foramini immissâ, observandum, certam quandam distantiam determinatam esse, ex qua si objecerimus pannum, simulacra lucida atque admodum distincta refulgent; simulac verò paululum accedimus ad vitrum, aut ab eodem recedimus, statim ea turbantur, & minùs distinctè apparent. Hæc autem distantia dimetienda erit, non secundùm spatium quod linteum & foramen intercedit, sed secundùm illud quod linteum & vitrum: ut quantum hoc vitrum ulteriùs promoveris, aut introrsum ad te reduxeris, tantum simul & linteum vel adducere vel removere oporteat. Pendetque hæc distantia, partim ex figura hujus vitri, & partim ex spatio quod illud & res objectas interjacet: nam licèt eodem loco hæ maneant, quò minùs superficies vitri erunt incurvatæ, eò longiùs hoc linteum removendum; & eodem vitro manente, accedentibus propiùs objectis, paulò magis linteum removendum erit, quàm si longiùs eadem abessent. Atque ex hac distantia, imaginum oritur magnitudo, eodem ferè modo quo tum, quum nullum foramini vitrum applicatur. fieri autem illud foramen majus potest, si vitro inserto obturetur, quàm si apertum & vacuum relinquatur; imaginibus ob id non minùs distinctis: Et quò erit majus, eò simulacra nitidiora atque illustriora videbuntur: adeò ut si partem vitri tegas, magis quidem obscura quàm antea debeant apparere, sed non idcirco minùs spatii in panno occupare. Et quò majora & lucidiora hæc simulacra sunt, eò perfectiùs videntur: adeò quidem ut si oculum admodum profundum struere possemus, cujus pupilla esset valde ampla, & in quo superficies refractionem efficientes figuram haberent, quæ huic magnitudini responderet, eò ampliores objectorum corporum imagines in ejus fundo exprimerentur. Et si duas aut plures lentes vitreas parum connexas jungamus, idem ferè efficient quod una, quæ ad eandem crassitiem, quam illæ omnes simul sumptæ, intumescet; hîc enim exigui momenti est superficierum numerus, in quibus refra-

DIOPTRICES CAP. V. 81

fractiones fiunt. Ast si ex certo intervallo hæc vitra ab invicem removeamus, secundum eriget imaginem, quam primum invertit, tertium iterum invertet, & ita porrò. Quorum omnium ratio manifesta est ex iis quæ supra audivimus, & quidem majus operæ pretium erit, mediocri meditatione illam inquirenti, quàm obiter singula fusius hîc enarrata legenti.

Cæterum corporum simulacra, non tantùm in ima oculi parte formantur, sed ulteriùs quoque ad cerebrum penetrant; quod facilè intelligemus, si cogitemus radios ab objecto V in oculum XIII. *Quomodo hæ imagines ab oculo in cerebrum transeant.*

venientes, contingere in puncto R, extremum alicujus ex capilla-
I mentis

René Descartes, *Opera Philosophica* [Philosophical works], Amsterdam, 1672.
Spread including diagrams of human vision.

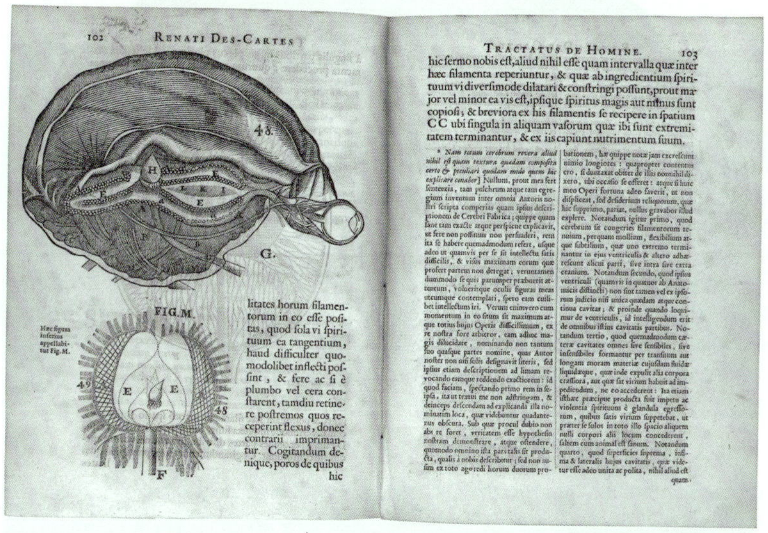

102 RENATI DES-CARTES

Hæc figura inferius appellabitur Fig. M.

litates horum filamentorum in eo eſſe poſitas, quod ſola vi ſpirituum ea tangentium, haud difficulter quomodolibet inflecti poſſint, & fere ac ſi è plumbo vel cera conſtarent, tamdiu retinere poſtremos quos receperint flexus, donec contrarii imprimantur. Cogitandum denique, poros de quibus hic

TRACTATUS DE HOMINE. 103

hic ſermo nobis eſt, aliud nihil eſſe quam intervalla quæ inter hæc filamenta reperiuntur, & quæ ab ingredientium ſpirituum vi diverſimode dilatari & conſtringi poſſunt, prout major vel minor ea vis eſt, ipſique ſpiritus magis aut minus ſunt copioſi; & breviora ex his filamentis ſe recipere in ſpatium CC ubi ſingula in aliquam vaſorum quæ ibi ſunt extremitatem terminantur, & ex iis capiunt nutrimentum ſuum.

* *Nam totum cerebrum revera aliud nihil eſt quam textura quædam compoſita certo & peculiari quodam modo quem hic explicare conabor*] Nullum, prout mea fert ſententia, tam pulchrum atque tam egregium inventum inter omnia Autoris noſtri ſcripta comperias quam ipſius deſcriptionem de Cerebri Fabrica; quippe quam ſane tam exacte atque perſpicue explicavit, ut fere non poſſimus non perſuaderi, rem ita ſe habere quemadmodum refert, uſque adeo ut quamvis per ſe ſit intellectu ſatis difficilis, & viſus maximam eorum quæ profert partem non detegat; veruntamen dummodo ſe quis parumper præbuerit attentum, volueritque oculis figuras meas utcumque contemplari, ſpero eam cuilibet intellectum iri. Verum enimvero cum momentum in eo ſitum ſit maximum atque totius hujus Operis difficillimum, ex re noſtra fore arbitror, eam adhuc magis dilucidare, nominando non tantum ſuo quaſque partes nomine, quas Autor noſter non niſi ſolis deſignavit literis, ſed ipſius etiam deſcriptionem ad limam revocando eamque reddendo exactiorem: id quod faciam, ſpectando primo rem in ſe ipſa, ita ut textui me non adſtringam, & deinceps deſcendam ad explicanda illa nominatim loca, quæ videbuntur quadantenus obſcura. Sub quæ procul dubio non abs re foret, veritatem eſſe hypotheſin noſtram demonſtrare, atque oſtendere, quomodo omnino iſta pars talis ſit producta, qualis à nobis deſcribitur; ſed non auſim ex toto aggredi horum duorum probationem, hæ quippe notæ jam excreſcunt nimio longiores: quapropter contentus ero, ſi duntaxat obiter de illis nonnihil dixero, ubi occaſio ſe offeret: atque ſi huic meo Operi fortuna adeo faverit, ut non diſpliceat, ſed deſiderium reliquorum, quæ hic ſupprimo, pariat, nullus gravabor illud explere. Notandum igitur primo, quod cerebrum ſit congeries filamentorum tenuium, perquam mollium, flexibilium atque ſubtilium, quæ uno extremo terminantur in ejus ventriculis & altero adhæreſcunt alicui parti, ſive intra ſive extra cranium. Notandum ſecundo, quod ipſius ventriculi (quamvis in quatuor ab Anatomicis diſtincti) non ſint tamen vel ex ipſorum judicio niſi unica quædam atque continua cavitas; & proinde quando loquimur de ventriculis, id intelligendum erit de omnibus iſtius cavitatis partibus. Notandum tertio, quod quemadmodum cæteræ cavitates omnes ſive ſenſibiles, ſive inſenſibiles formantur per tranſitum aut longam moram materiæ cujuſdam fluidæ liquidæque, quæ inde expulit alia corpora craſſiora, aut quæ ſat virium habuit ad impediendum, ne eo accederent: Ita etiam iſthæc præcipue producta fuit impetu ac violentia ſpirituum è glandula egreſſorum, quibus ſatis virium ſuppetebat, ut præter ſe ſolos in toto illo ſpacio aliquem nulli corpori alii locum concederent, ſaltem cum animal eſt ſanum. Notandum quarto, quod ſuperficies ſuprema, infima & lateralis hujus cavitatis, quæ videtur eſſe adeo unita ac polita, nihil aliud eſt quam

René Descartes, *L'Homme de René Descartes et un Traité de la formation du fœtus* [René Descartes' treatise on man and the formation of the fetus], Amsterdam, 1677. Spread including diagrams of the ventricles of the brain and pineal gland.

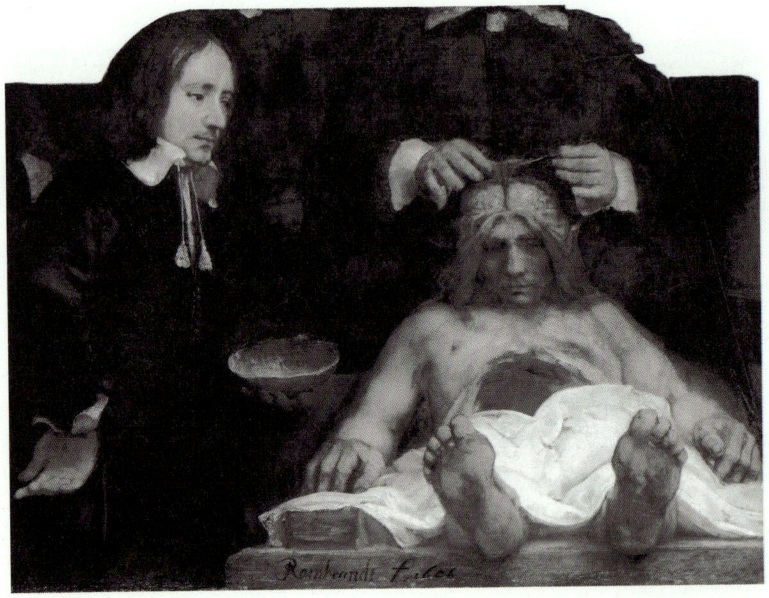

Rembrandt van Rijn, *The Anatomy Lesson of Dr. Jan Deijman* (fragment), 1656.

Ich stelle mir vor, wie Dr. Deijman das Messer fester hält und einen Schnitt macht. Er weiß, dass man die Seele nicht sehen kann, aber er rechnet doch damit, ihr auf die Spur zu kommen, indem er genauer hinsieht als irgend jemand vor ihm. Und während er aufmerksam in den Kopf des Diebes starrt, dessen Hirn er schneidet, spürt er die Aufmerksamkeit der Umstehenden: Es gibt in Amsterdam, der aufgeklärtesten Stadt der Welt, nur eine öffentliche Autopsie im Jahr, jedesmal ist sie ein großes Ereignis, für das die Menschen von weither kommen. Viele bringen ihre Kinder, Fünfjährige sitzen auf den Schultern der Väter, Mütter halten kleine Mädchen hoch, damit sie besser sehen können.

Ein Problem der Justiz seit altersher liegt darin, dass man einen toten Verbrecher nicht mehr erreichen, ihm nichts mehr antun kann. Diese Autopsie erfüllt somit zwei Zwecke zugleich: Einen bestraften Verbrecher vor dem Volk zu zerschneiden, befördert die Gelehrsamkeit und demütigt ihn dabei noch nach seiner Tötung. Ein besonderes Detail dieser Demütigung, subtiler als ein normaler Mensch es sich überhaupt auszudenken vermag, besteht nun darin, dass vor der Zerstückelung und Verbrennung des Körpers die Zirbeldrüse, laut Descartes Sitz der Seele, also: des Bewusstseins, entnommen wird, damit dieses Bewusstsein die Zerstörung des Körpers noch von ferne fühlen und miterleben kann. Der schwarze Joris – so stelle ich es mir vor, natürlich wissen wir außer seinem Namen fast nichts über ihn – war ein übler Gauner, sodass Dr. Deijman sogar Grund zu der Hoffnung haben kann, in seinen Gehirnwindungen neben den Ursachen des Denkens und Fühlens noch den Keim des Bösen zu finden.

Ich stelle mir vor, dass Dr. Deijman das hier zum ersten Mal tut. Warum auch nicht – es gibt immer ein erstes Mal! Während er schneidet, erfüllt ihn das Gefühl, etwas Verbotenes zu tun und Dinge zu sehen, die nicht dafür bestimmt sind, erblickt zu werden. Die Innenseite der Welt geht nur den Schöpfer etwas an, so wie ja auch das Räder- und Federwerk in den Uhren nur für die Augen der Uhrmacher sind - der Umstand, dass die Innenseite des Menschlichen so schleimig weich und blutwarm ist, macht es nur noch deutlicher, dass sie die menschliche Neugier nichts angeht.

Und ich denke an den Philosophen Leibniz und sein Bild von der Mühle. *„Denkt man sich etwa eine Maschine, die so beschaffen wäre, daß sie denken, empfinden und perzipieren könnte, so kann man sie sich derart proportional vergrößert vorstellen, daß man in sie wie in eine Mühle eintreten könnte. Dies vorausgesetzt, wird man bei der Besichtigung ihres Inneren nichts weiter als einzelne Teile finden, die einander stoßen, niemals aber etwas, woraus eine Perzeption zu erklären wäre.*“

Leibniz schreibt dies ein halbes Jahrhundert nach Dr. Deijmans Obduktion des schwarzen Joris, aber er beschreibt da einen Gedanken, von dem ich mir vorstelle, dass er mit verwirrender Kraft wenn auch mit weniger Klarheit, auch in Dr. Deijman aufgestiegen sein könnte. - Ich stelle mir die Verwirrung vor, die den Doktor überkommt, ich stelle mir vor, wie er in der Bewegung innehält und zu seinem die Silberschale haltenden Assistenten, den ich zum Beispiel Calkoen nenne, und zwar auch deshalb, weil das tatsächlich sein Name war, der uns nämlich überliefert ist, er hieß Gijsbert Calkoen, und so sagt er also zu Calkoen: „Es denkt im Kopf, so wie es im Darm verdaut, nicht wahr?“

Und Gijsbert Calkoen, der natürlich keine Ahnung hat, worauf sein Meister hinauswill, nickt langsam, bedeutungsvoll und ratlos.

„Aber wer ist *es*?" fragt Dr. Deijman. „Also der Klang kommt in die Ohren, und das Licht in die Augen, und der Geschmack auf die Zunge, und all das dringt durch die Nerven ins Hirn, aber dort muss dann doch auch jemand sein ... Es muss einen Ort geben, wo alles zusammenläuft, und an diesem Ort ... Wer ist da?"

Gijsbert Calkoen, was bleibt ihm auch übrig, nickt.

„Wer", wiederholt Dr. Dejman, „ist dort? Wer ist an diesem Ort?"

Und er blickt sich um, in die fünfzig oder mehr Augenpaare, die ihn dabei beobachten, wie er dem Toten ins Gehirn schneidet, und ein Gefühl überkommt ihn, dass sie beide sich bis zur Ununterscheidbarkeit gleichen - dass in seinem Kopf das ist, was auch im Kopf auf dem Tisch sichtbar wird, die gleiche schleimig graue Masse, sodass es Dr. Deijman plötzlich scheint, als zerschnitte er eigentlich sich selbst.

Und um dem Schwindel zu entkommen, blickt er suchend um sich, bis er das klare und leidenschaftliche Gesicht des Mannes mit dem Skizzenblock findet, der ihn just in jenem intrikaten Moment festhalten wird, da er in aller Behutsamkeit mit dem Skalpell die Zirbeldrüse aus dem Kopf des gestraften Verbrechers hebt. Da Amsterdam eine große, aber doch auch überschaubare Stadt ist, stelle ich mir vor, dass sie einander lange kennen, Dr. Dejman und Rembrandt van Rijn, der den Moment bewahren wird für die Gegenwart und die Zukunft, für uns, die wir glauben, mehr zu wissen über das Gehirn als Dr. Dejman - die wir aber doch weder eine Theorie des Bewusstseins haben, noch auch nur eine Ahnung, wie solch eine Theorie überhaupt aussehen müsste, um unserer Ungewissheit darüber abzuhelfen, wer da in uns denkt und fühlt.

Und so stelle ich mir vor, dass die Verwirrung Dr. Dejman so sehr überwältigt, dass er nie wieder ganz zurückfindet: Ein hochgeehrter Gelehrter, kommt ihm in den folgenden Jahren das eigene Leben abhanden, er trinkt zu viel, er denkt fast täglich an den schwarzen Joris, in dessen Kopf er sein Messer gesteckt hat, und er bedauert, dass er ihm nie als Lebenden begegnet ist, als hätte das seine Fragen beantworten können.

So wird er im Alter zum Sonderling, der Dr. Dejman. Er geht allein spazieren auf den Grachten, seine Enkel lachen ihn aus, er spricht leise vor sich hin. Er denkt viel darüber nach, dass er jetzt hier auf der Straße ist, aber doch auch im Kopf der Menschen, die ihm entgegenkommen und ihn sehen: Hier und dort zugleich, obwohl man in ihnen nichts finden würde als jene Masse, die er aus dem Kopf des schwarzen Joris kennt.

Und manchmal kommt es ihm vor, als wäre die Welt eine riesenhafte Mühle, ein unfassbar intrikates System von Rädern und Maschinen, in dem er doch nirgendwo, wie weit er auch geht, nichts findet, woraus sich das Denken erklären könnte. Bis ihn irgendwann die Verwirrung mit der Kraft eines Schlaganfalls überkommt, bis sein Gehirn versagt und er zusammenbricht und stirbt.

So stelle ich es mir vor. Oder richtiger - so stellt es sich der vor, der in meinem Kopf denkt und träumt und der angeblich ich ist.

INSIDE THE HEAD OF BLACK JOHN

I imagine Dr. Deijman tightening his grip on the knife and making a cut. He knows that the soul cannot be seen, but he nonetheless expects to track it down by looking more closely than anyone before him. And as he stares intently into the head of the thief whose brain he is cutting, he senses the attention of the bystanders. In Amsterdam, the most enlightened city in the world, there is only one public autopsy a year; every time, it's a major event to which people come from afar. Many bring their children, five-year-olds sitting on their fathers' shoulders, little girls held up by their mothers so they can see better.

It's an age-old problem of justice that a dead criminal has been placed beyond our reach, where we can no longer do him any harm. This autopsy thus fulfills two purposes at once: To cut up a punished criminal before the eyes of the people promotes learning and at the same time humiliates him even after he is killed. A special detail of this humiliation, subtler than anything a normal person is even capable of conceiving, is that before the dismemberment and burning of the body, the pineal gland—according to Descartes the seat of the soul, that is, of consciousness—is taken out, so that this consciousness can still feel and witness the destruction of the body from a distance. Black John—so I imagine; of course, apart from his name we know almost nothing about him any longer—was a wicked scoundrel, so that Dr. Deijman may even have reason to hope to find in the convolutions of his brain not only the causes of thinking and feeling but also the germ of evil.

I imagine that Dr. Deijman is doing this for the first time. Why not—there's always a first time! As he cuts, he is filled with the feeling of doing something forbidden and seeing things that are not meant to be glimpsed. The inside of the world is the concern of the Creator alone, just as the gears and springs in clocks are only for the eyes of clockmakers—the fact that the inside of human beings is so slimily soft and blood-warm only makes it clearer to us that it's no concern of human curiosity.

And I think of the philosopher Leibniz and his image of the mill: "Supposing there were a machine so constituted as to think, feel, and perceive; one can imagine it proportionally enlarged so that one might step into it as into a mill. In this case, we should, on examining its interior, find nothing but individual parts, which push each other, but never anything by which to explain a perception."

Leibniz writes this half a century after Dr. Deijman's autopsy of Black John, but he is describing a thought that I imagine could have arisen with confusing force, even if with less clarity, in Dr. Deijman too. I imagine the confusion coming over the doctor, I imagine him pausing in his movement and saying to his assistant holding the silver bowl, whom I will call, say, Calkoen, and indeed also because that was actually his name, which has, as a matter of fact, been passed down to us, he was named Gijsbert Calkoen, and so he says, then, to Calkoen, "Thinking is done in the head just as digestion is done in the intestine, isn't it?"

And Gijsbert Calkoen, who of course has no idea what his master is getting at, nods slowly, meaningfully, and blankly.

"But who is doing it?" Dr. Deijman asks. "Well, sound comes into the ears, and light into the eyes, and taste onto the tongue, and all this reaches the brain through the nerves, but then there must be someone there too... There must be a place where everything converges, and in that place... who is there?"

Gijsbert Calkoen, what else can he do, nods.

"Who," repeats Dr. Deijman, "is there? Who is in that place?"

And he looks around, into the fifty or more pairs of eyes watching him cut into the dead man's brain, and he is overcome with a feeling that the two of them are so alike as to be indistinguishable–that inside his head is what is now visible in the head on the table, the same slimily gray mass, so that it suddenly seems to Dr. Deijman as if he were actually cutting up himself.

And to escape the dizziness, he looks searchingly around him, until he finds the clear and passionate face of the man with the sketch pad, who will capture him just at that intricate moment when he is gingerly lifting the pineal gland out of the head of the punished criminal with the scalpel. Since Amsterdam is a large city but not inordinately so, I imagine that they've long known each other, Dr. Deijman and Rembrandt van Rijn, who will preserve the moment for the present and the future, for us, who believe we know more about the brain than Dr. Deijman–but who in reality have neither a theory of consciousness nor so much as an inkling of what such a theory would even have to look like to relieve our uncertainty about who is thinking and feeling inside us.

And so I imagine that the confusion so overwhelms Dr. Deijman that he never quite finds his way back: A highly honored scholar, in the years that follow he lets his own life slip away from him, he drinks too much, he thinks almost every day of Black John, in whose head he stuck his knife, and he feels sorry that he never met him as a living man, as if that could have answered his questions.

Thus, in old age, he becomes an eccentric, Dr. Deijman does. He walks alone on the canals, his grandchildren laugh at him, he speaks quietly to himself. He thinks a lot about the fact that he is now here on the street and yet also in the heads of the people who approach him and see him: here and there at the same time, although one would find nothing in them but that mass which he knows from the head of Black John.

And sometimes it seems to him as if the world were a gigantic mill, an inconceivably intricate system of wheels and machines in which he nonetheless, however far he goes, never finds anything by which thinking could be explained. Until at some point the confusion overcomes him with the force of a stroke, until his brain fails and he collapses and dies.

So I imagine it–or, to be more precise, not I, but the one who thinks and dreams in my head, purporting to be me.

STORY 30

During the Contortions

Esther Freud

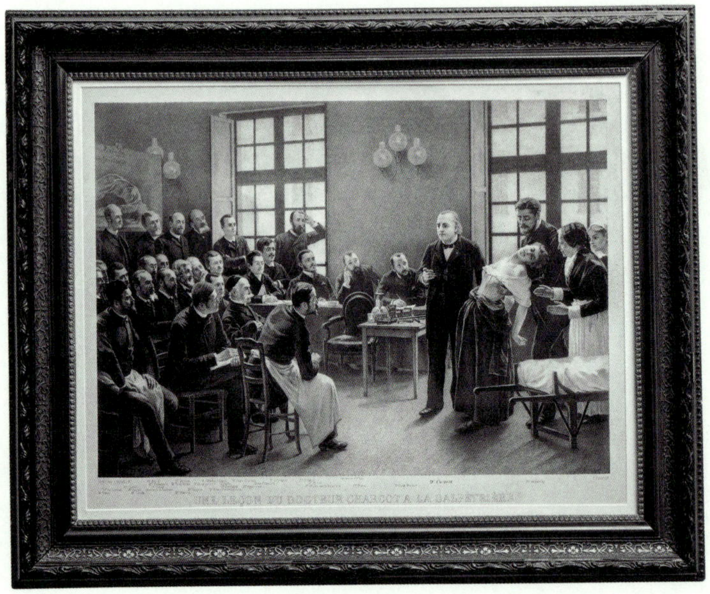

Eugène Louis Pirodon (after André Brouillet), *A Clinical Lesson at the Salpêtrière*, 1888.

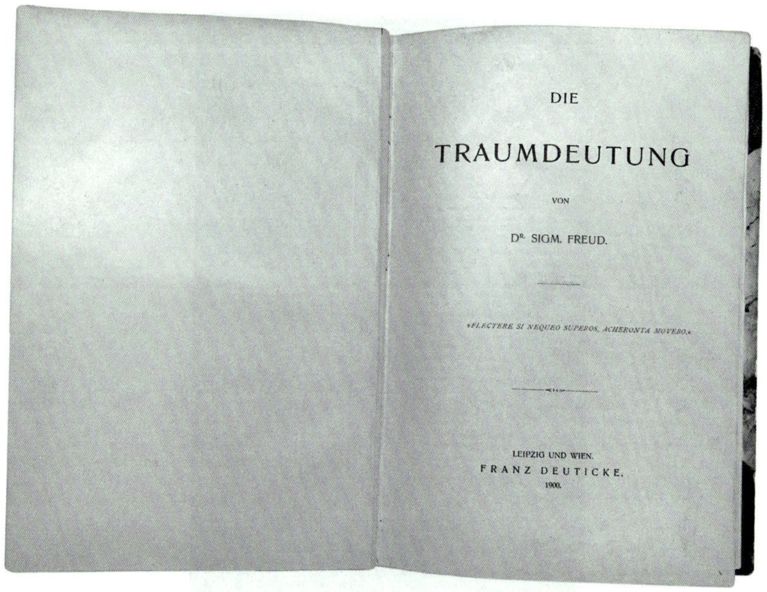

Sigmund Freud, *Die Traumdeutung*, Leipzig-Vienna, 1900.

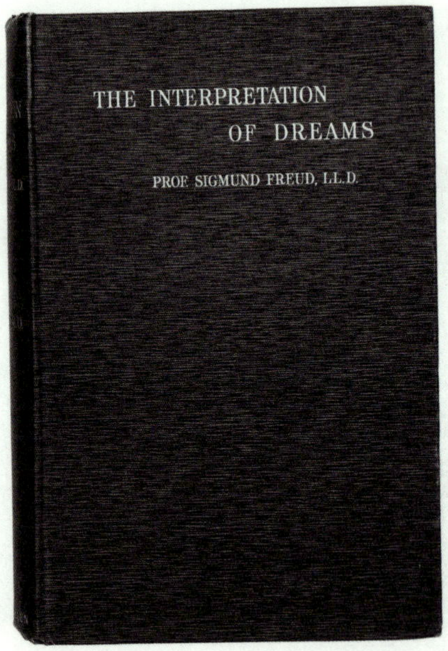

Sigmund Freud, *The Interpretation of Dreams*, London, 1913.

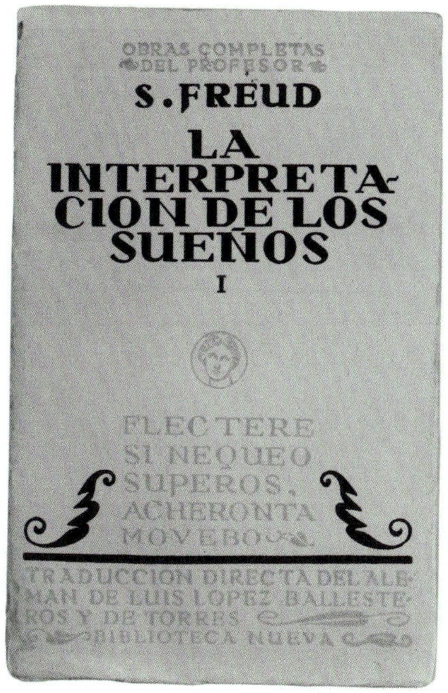

Sigmund Freud, *La interpretación de los sueños*, Madrid, 1931.

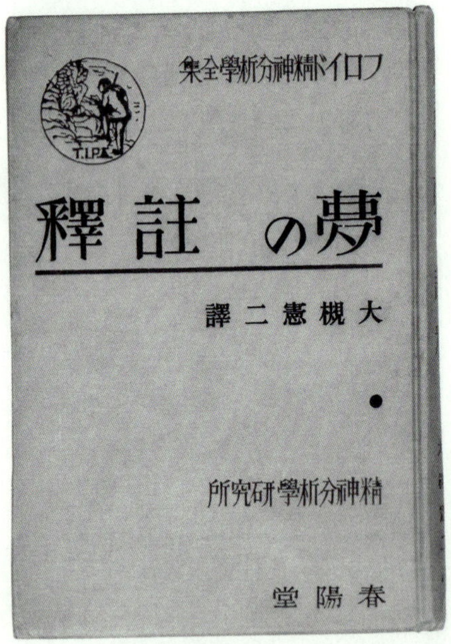

Sigmund Freud, *Yume Handan*, Tokyo, 1930–1933.

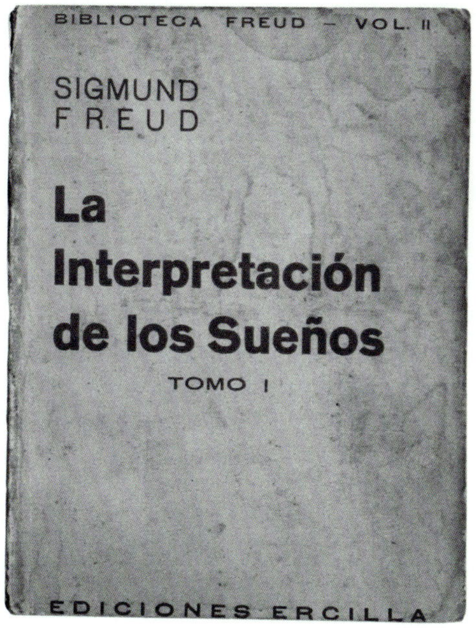

Sigmund Freud, *La interpretación de los sueños*, Santiago de Chile, 1936.

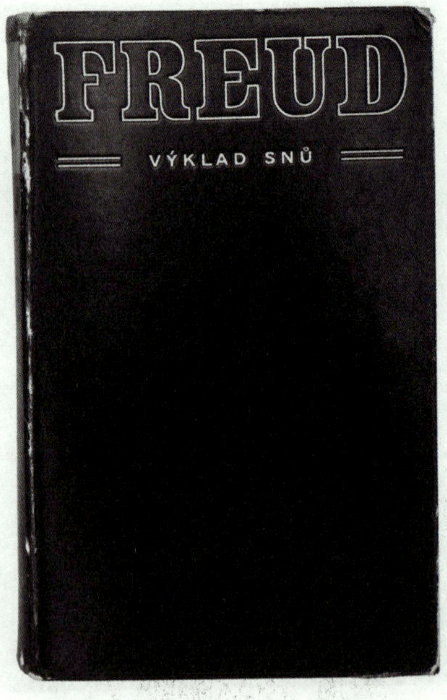

Sigmund Freud, *Výklad Snů*, Prague, 1937.

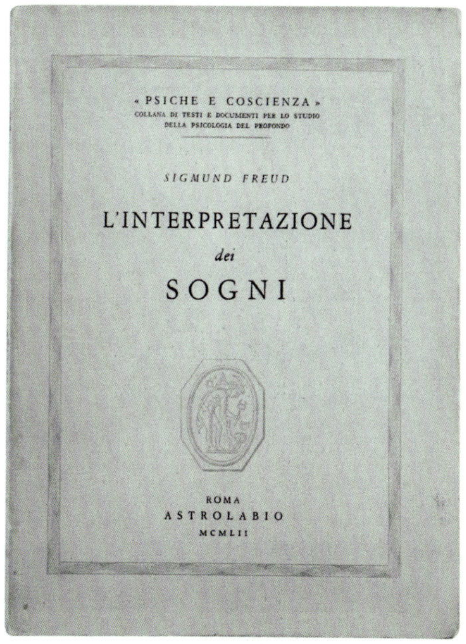

Sigmund Freud, *L'interpretazione dei sogni*, Rome, 1952.

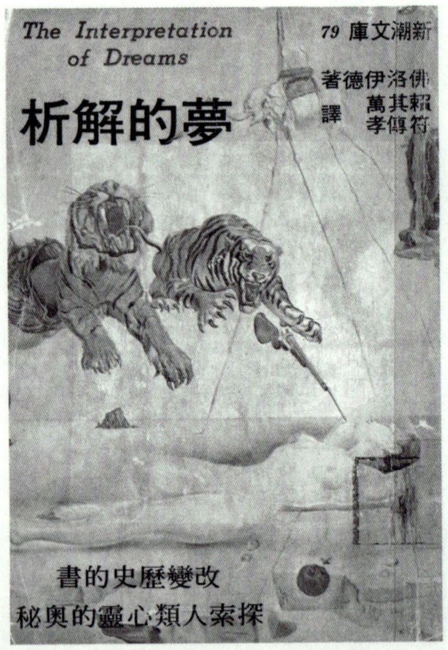

Sigmund Freud, *Meng de jie xi*, Taipei, 1972.

There are fifteen versions of the woman they call "Blanche," but I am content to inhabit this one, a lithograph made from the painting *A Clinical Lesson at the Salpêtrière*, framed, and hanging on the wall of Professor Dr. Freud's consulting rooms in Vienna. Here, the Professor can examine me as I undergo hypnosis, falling backward in replica of an earlier inmate—also a patient of Charcot's, but not so renowned. Freud is a great one for examining—and I am, after all, the model of hysteria—but I examine him too, looking on as he reads his letters, sighs, laughs, exclaims over each new translation of *The Interpretation of Dreams*. Russian. Czech. Japanese!

Blanche is not my name. Blanche was the sister I failed to protect. "Keep quiet and I'll let her be." It was the promise I was made, and obedient, believing, I flew into the rafters, never looking down, her name a mumble to rise above the pain. But he was lying, and my sister was not spared. Her seizures contorted her, boiled her blood, caused her to choke on her own tongue, and when her body was wrapped in a shift I was threatened if I spoke. "Silence." It was the last sound I heard. "No" the last word that escaped me, until the day our father was taken away and locked into a hospital for the insane.

Our father was a carpenter, our mother a linen maid, and of their eight children, five did not survive. Epilepsy was given as the cause of death, but when I succumbed to fitting, my mother threw water at me, so cold I revived. At twelve I was apprenticed to a furrier who never missed a chance to thrust his stinking head against my chest, so that to avoid him I lost consciousness, woke soaked in my own urine, my tremors worsening with each attack. I was accused of clumsiness, threatened with a beating, but when I ran away I was put to work in a laundry where there was no evading the attentions of the jeweler next door. I sought asylum in a convent. Here I did my best to hide my fitting, but I was caught in a seizure, ripped a garment, and was dismissed. Everything I held in my hands escaped me.

The day I entered the Salpêtrière I was eighteen. I was hired as a servant, but my condition attracted the attention of the neurologist Charcot, who chose me for a demonstration of hysteria. Willing, I fell into a trance, danced, enacted scenes of an erotic nature, contorting my body, as I had been taught. The word "Blanche" fell from my lips. Doctors, writers, and philosophers caught hold of it: the man responsible for discovering the electrical activity of skin, another who gave his name to the syndrome Tourette. Freud, a student of Charcot's, transcribed it in his spiked and fluid hand. Blanche was a sensation. Dancers, actors—Sarah Bernhardt herself—flocked to the Salpêtrière. We were the toast of Paris, until criticism began to flow. Nothing but theatrics. The performance must be feigned. Did they think I could fool Jean-Martin Charcot? I was frothing at the mouth! Did they think I could fool Freud?

Now I hang in the consulting room at Berggasse 19, overseeing the long hours the Professor listens to his patients, sits at his desk. Today he holds up a pencil sketch, sent from a fellow in his field, a Dr. Bose from Calcutta, who has asked, more than once, that a photograph be sent. When he fails to do so, Dr. Bose, who is making a study of the concept of repression, sends a drawing of what he thinks he "ought to look like." Freud calls to his daughter, and they laugh over it together. It has ignored all racial attributes.

"They wish for me to visit India," he tells her. "To talk at the newly formed psychoanalytic society. I will write to say I am too old." Anna, his youngest child, his Blanche, insists that he's not old. Sixty-six is nothing. They take each other's hands. If I could weep, I would. Laughter, love, inquisitiveness. How many ways there are to be a child. I watch as Freud's theories expand, as his patients reveal themselves, or hide. As new translations of *The Interpretation of Dreams* are arranged with the others on their shelf. American, Italian, Spanish.

But our life in Vienna is threatened. None of us can stay. Not after Anna is arrested, released into the unsafe middle of the night. Not when officers pay a visit to the apartment, refuse to leave until the Professor declares them innocent of crimes. *I can recommend the Gestapo to anyone*. He signs with a flourish.

Within weeks I am wrapped and packed into a box, transported across Europe with the rugs, the furniture, the "old and grubby gods," which is how he refers to his antiquities, explaining how they aid him in his thinking, how closely archaeology and psychoanalysis are connected. Egyptian, Roman, Greek. I travel in a crate with the goddess Athena.

In London I am placed above the Professor's couch–surely a promotion–where his patients eye me as he waits for them to speak. For he still sees patients, if only a few; receives visitors, if rarely; answers letters; finishes yet another book. He accepts that psychoanalysis is imperfect, that it is young, while he is old. Even Anna does not argue. And when he is too frail to walk, a bed is made for him by the French windows, looking out into the garden, where in September of that year, only weeks after the declaration of war, grateful to have found a home in "lovely, free, magnanimous England," he closes his eyes, fearing for the fate of friends and family left behind in Austria. He does not see his work continued, hear his name become known across the world, receive Chinese and Arabic editions of his books, or discover the fate of his four sisters, transported to Theresienstadt and Treblinka, along with others, more than eighty, who were herded into his apartment in Vienna where they waited, each with their own spirit, all with the same fate.

STORY 31

On the Head of an Idiot (No. 8)

Ayọ̀bámi Adébáyọ̀

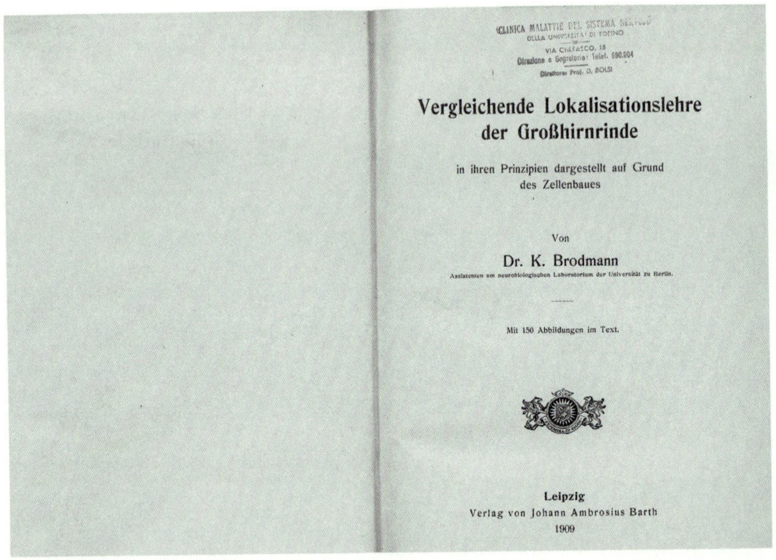

Vergleichende Lokalisationslehre
der Großhirnrinde

in ihren Prinzipien dargestellt auf Grund
des Zellenbaues

Von

Dr. K. Brodmann

Assistenten am neurobiologischen Laboratorium der Universität zu Berlin.

Mit 150 Abbildungen im Text.

Leipzig
Verlag von Johann Ambrosius Barth
1909

Korbinian Brodmann, *Vergleichende der Grosshirnrinde: in ihren Prinzipien dargestellt auf Grund des Zellenbaues* [Comparative localization studies in the brain cortex: its fundamentals represented on the basis of its cellular architecture], Leipzig, 1909. Frontispiece and illustrations of granular neurons and section of fetal cerebral cortex.

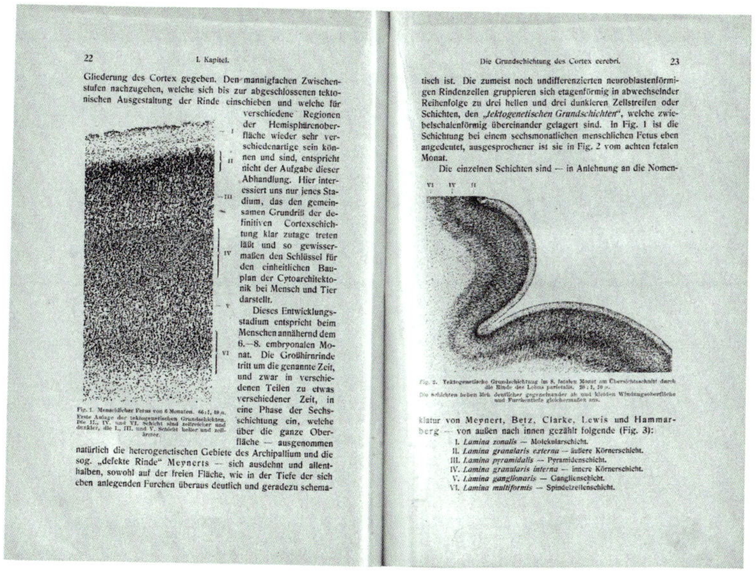

Gliederung des Cortex gegeben. Den mannigfachen Zwischenstufen nachzugehen, welche sich bis zur abgeschlossenen tektonischen Ausgestaltung der Rinde einschieben und welche für verschiedene Regionen der Hemisphärenoberfläche wieder sehr verschiedenartige sein können und sind, entspricht nicht der Aufgabe dieser Abhandlung. Hier interessiert uns nur jenes Stadium, das den gemeinsamen Grundriß der definitiven Cortexschichtung klar zutage treten läßt und so gewissermaßen den Schlüssel für den einheitlichen Bauplan der Cytoarchitektonik bei Mensch und Tier darstellt.

Dieses Entwicklungsstadium entspricht beim Menschen annähernd dem 6.–8. embryonalen Monat. Die Großhirnrinde tritt um die genannte Zeit, und zwar in verschiedenen Teilen zu etwas verschiedener Zeit, in eine Phase der Sechsschichtung ein, welche über die ganze Oberfläche — ausgenommen natürlich die heterogenetischen Gebiete des Archipallium und die sog. „defekte Rinde" Meynerts — sich ausdehnt und allenthalben, sowohl auf der freien Fläche, wie in der Tiefe der sich eben anlegenden Furchen überaus deutlich und geradezu schema-

Fig. 1. Menschlicher Fetus von 6 Monaten. 66:1, 10 μ. Erste Anlage der tektogenetischen Grundschichten. Die II., IV. und VI. Schicht sind zellreicher und dunkler, die I., III. und V. Schicht heller und zellärmer.

tisch ist. Die zumeist noch undifferenzierten neuroblastenförmigen Rindenzellen gruppieren sich etagenförmig in abwechselnder Reihenfolge zu drei hellen und drei dunkleren Zellstreifen oder Schichten, den *„tektogenetischen Grundschichten"*, welche zwiebelschalenförmig übereinander gelagert sind. In Fig. 1 ist die Schichtung bei einem sechsmonatlichen menschlichen Fetus eben angedeutet, ausgesprochener ist sie in Fig. 2 vom achten fetalen Monat.

Die einzelnen Schichten sind — in Anlehnung an die Nomenklatur von Meynert, Betz, Clarke, Lewis und Hammarberg — von außen nach innen gezählt folgende (Fig. 3):

I. *Lamina zonalis* — Molekularschicht.
II. *Lamina granularis externa* — äußere Körnerschicht.
III. *Lamina pyramidalis* — Pyramidenschicht.
IV. *Lamina granularis interna* — innere Körnerschicht.
V. *Lamina ganglionaris* — Ganglienschicht.
VI. *Lamina multiformis* — Spindelzellenschicht.

Fig. 2. Tektogenetische Grundschichtung im 8. fetalen Monat am Übersichtsschnitt durch die Rinde des Lobus parietalis. 20:1, 10 μ.
Die Schichten heben sich deutlicher gegeneinander ab und kleiden Windungsoberfläche und Furchentiefe gleichermaßen aus.

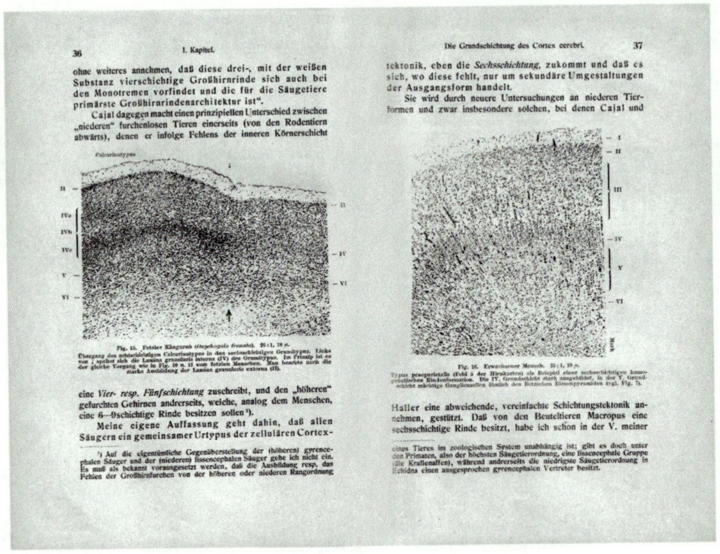

36 I. Kapitel.

ohne weiteres annehmen, daß diese drei-, mit der weißen Substanz vierschichtige Großhirnrinde sich auch bei den Monotremen vorfindet und die für die Säugetiere primärste Großhirnrindenarchitektur ist".

Cajal dagegen macht einen prinzipiellen Unterschied zwischen „niederen" furchenlosen Tieren einerseits (von den Rodentiern abwärts), denen er infolge Fehlens der inneren Körnerschicht

Fig. 15. Fetales Känguruh (*Onychogale frenata*). 25:1, 10 µ.
Übergang des achtschichtigen Calcarinatypus in den sechsschichtigen Grundtypus. Links von ↓ spaltet sich die Lamina granularis interna (IV) des Grundtypus. Im Prinzip ist es der gleiche Vorgang wie in Fig. 10 u. 11 vom fetalen Menschen. Man beachte auch die starke Ausbildung der Lamina granularis externa (II).

eine *Vier- resp. Fünfschichtung* zuschreibt, und den „höheren" gefurchten Gehirnen andrerseits, welche, analog dem Menschen, eine 6—9schichtige Rinde besitzen sollen[1]).

Meine eigene Auffassung geht dahin, daß allen Säugern ein gemeinsamer Urtypus der zellulären Cortex-

[1]) Auf die eigentümliche Gegenüberstellung der (höheren) gyrencephalen Säuger und der (niederen) lissencephalen Säuger gehe ich nicht ein. Es muß als bekannt vorausgesetzt werden, daß die Ausbildung resp. das Fehlen der Großhirnfurchen von der höheren oder niederen Rangordnung

Die Grundschichtung des Cortex cerebri. 37

tektonik, eben die *Sechsschichtung*, zukommt und daß es sich, wo diese fehlt, nur um sekundäre Umgestaltungen der Ausgangsform handelt.

Sie wird durch neuere Untersuchungen an niederen Tierformen und zwar insbesondere solchen, bei denen Cajal und

Fig. 16. Erwachsener Mensch. 25:1, 10 µ.
Typus praeparietalis (Feld 5 der Hirnkarten) als Beispiel einer sechsschichtigen homogenetischen Rindenformation. Die IV. Grundschicht stark ausgebildet, in der V. Grundschicht mächtige Ganglienzellen ähnlich den Betzschen Riesenpyramiden (vgl. Fig. 7).

Haller eine abweichende, vereinfachte Schichtungstektonik annehmen, gestützt. Daß von den Beuteltieren Macropus eine sechsschichtige Rinde besitzt, habe ich schon in der V. meiner

eines Tieres im zoologischen System unabhängig ist; gibt es doch unter den Primaten, also der höchsten Säugetierordnung, eine lissencephale Gruppe (die Krallenaffen), während andrerseits die niedrigste Säugetierordnung in Echidna einen ausgesprochen gyrencephalen Vertreter besitzt.

Korbinian Brodmann, *Vergleichende der Grosshirnrinde: in ihren Prinzipien dargestellt auf Grund des Zellenbaues* [Comparative localization studies in the brain cortex: its fundamentals represented on the basis of its cellular architecture], Leipzig, 1909. Spreads including illustrations of the granular neurons of a kangaroo and of an adult male, and diagrams of the cerebral cortex.

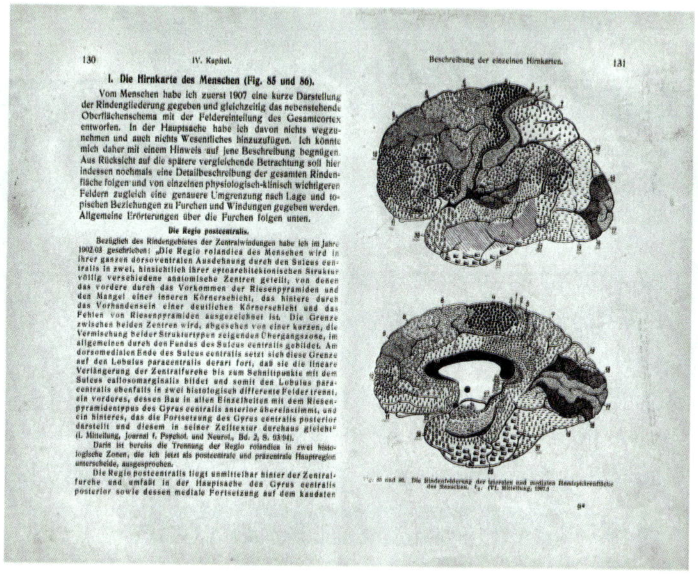

I. Die Hirnkarte des Menschen (Fig. 85 und 86).

Vom Menschen habe ich zuerst 1907 eine kurze Darstellung der Rindengliederung gegeben und gleichzeitig das nebenstehende Oberflächenschema mit der Feldereinteilung des Gesamtcortex entworfen. In der Hauptsache habe ich davon nichts wegzunehmen und auch nichts Wesentliches hinzuzufügen. Ich könnte mich daher mit einem Hinweis auf jene Beschreibung begnügen. Aus Rücksicht auf die spätere vergleichende Betrachtung soll hier indessen nochmals eine Detailbeschreibung der gesamten Rindenfläche folgen und von einzelnen physiologisch-klinisch wichtigeren Feldern zugleich eine genauere Umgrenzung nach Lage und topischen Beziehungen zu Furchen und Windungen gegeben werden. Allgemeine Erörterungen über die Furchen folgen unten.

Die Regio postcentralis.

Bezüglich des Rindengebietes der Zentralwindungen habe ich im Jahre 1902/03 geschrieben: „Die Regio rolandica des Menschen wird in ihrer ganzen dorsoventralen Ausdehnung durch den Sulcus centralis in zwei, hinsichtlich ihrer cytoarchitektonischen Struktur völlig verschiedene anatomische Zentren geteilt, von denen das vordere durch das Vorkommen der Riesenpyramiden und den Mangel einer inneren Körnerschicht, das hintere durch das Vorhandensein einer deutlichen Körnerschicht und das Fehlen von Riesenpyramiden ausgezeichnet ist. Die Grenze zwischen beiden Zentren wird, abgesehen von einer kurzen, die Vermischung beider Strukturtypen zeigenden Übergangszone, im allgemeinen durch den Fundus des Sulcus centralis gebildet. Am dorsomedialen Ende des Sulcus centralis setzt sich diese Grenze auf den Lobulus paracentralis derart fort, daß sie die lineare Verlängerung der Zentralfurche bis zum Schnittpunkte mit dem Sulcus callosomarginalis bildet und somit den Lobulus paracentralis ebenfalls in zwei histologisch differente Felder trennt, ein vorderes, dessen Bau in allen Einzelheiten mit dem Riesenpyramidentypus des Gyrus centralis anterior übereinstimmt, und ein hinteres, das die Fortsetzung des Gyrus centralis posterior darstellt und diesem in seiner Zelltextur durchaus gleicht" (I. Mitteilung, Journal f. Psychol. und Neurol., Bd. 2, S. 93/94).

Darin ist bereits die Trennung der Regio rolandica in zwei histologische Zonen, die ich jetzt als postzentrale und präzentrale Hauptregion unterscheide, ausgesprochen.

Die Regio postcentralis liegt unmittelbar hinter der Zentralfurche und umfaßt in der Hauptsache den Gyrus centralis posterior sowie dessen mediale Fortsetzung auf dem kaudaten

Fig. 85 und 86. Die Rindenfelderung der lateralen und medialen Hemisphärenfläche des Menschen. [illegible]. (VI. Mitteilung, 1907.)

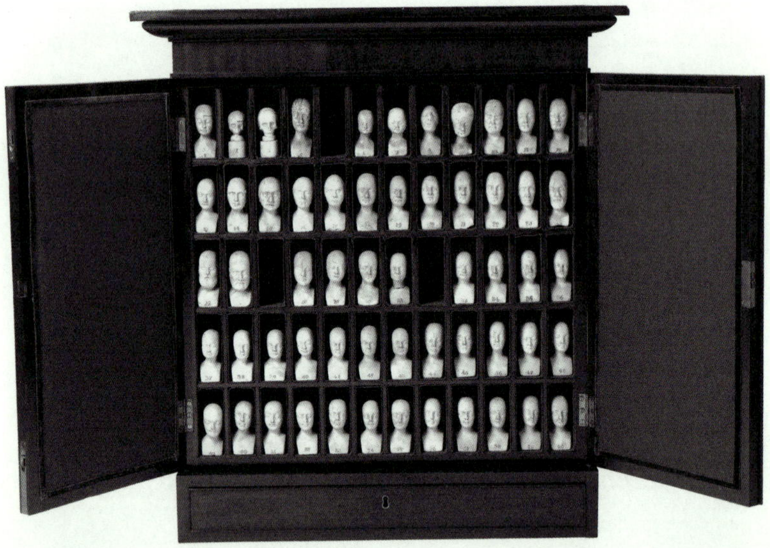

Phrenology cabinet, 19th century.

I am also conscious of only being able to offer incomplete data in other respects. The theory of anatomical localization is, like physiological and clinical localization, still in a state of development. The directions to take to reach them are not immediately attainable. So many problems must remain unresolved, some can be only provisionally unraveled, and for yet others the way ahead can only be sketched out roughly.

—Korbinian Brodmann

Liam's washed himself before coming. Good and proper too, head to toe for the first time in weeks. He did not mind that the soap cost him more than a week's wages; better to be clean around the learned than to be sniffed at by the lot of them. His ma also asked him to shave his head. He's done so, but was careful to hide his newly bald scalp under a hat before leaving the house.

Now, while he waits, Liam folds and unfolds the woolen hat that once belonged to his pa. There's dirt crusted beneath his nails. More brown than black after his bath, but mucky still. Could the state of his fingers affect the judgment Dr. Spurzheim might pronounce about him? Probably not. His ma said all that would matter was his head. This phrenology came to them through his aunt, a governess in Vienna. She wrote his Ma to say this mapping of his head would help him understand his character and abilities. Why this is the case, he cannot tell. He does not think his scalp has more to say about him than his hands.

His left thumb is misshapen from getting trapped in a door when he was too little to remember. He stayed there for minutes or hours, no one knew how long, because though he shed tears he would not cry out. When he was discovered by his pa, it was a wonder that thumb was not totally lost. Liam also remembers how the large scar on his left palm came to be. Trying to cleave a chicken, he slashed his own palm open instead, and though he cried out that time, there was no one near to hear or help. His brothers were grown and gone from home by then. His mother was visiting a niece and his father was long dead. His right hand bore no scar that man could see, but in winter it ached so much the throbbing woke him on many nights. It started the winter after he joined his brothers to carry their father's coffin, though everyone said he shouldn't because he was too slight, too young, and only just recovered from the pox himself. If Mr. Bally and the doctor asked his opinion, Liam would tell them there was more to learn about him from his hands.

He knows he has dozed off when he snaps awake. There's a shadow over him and a hand on his shoulder, which continues to tap him until he rises to his feet. He follows the man who's come for him through a dark corridor to a room with high windows that let so much light in that Liam squints. Three men are seated; each has a piece of paper on which he's taking notes. The man who brought him in motions Liam to a chair that's angled to lean back.

Liam sits. The man explains what will happen next. His face and head will be coated with oil, and then plaster applied upon parts of his face. It will only take a few minutes of stillness on Liam's part for the sculptor to obtain a mold hard enough for his purpose. Liam nods to let Mr. Bally know he's understood what is required of him. He leans back in the seat and waits to be oiled.

The oil feels warm against his scalp, as though it's been slightly heated, and the scent is familiar. When his nose is coated with it,

Liam remembers that his mother spilled a bottle on his father's body before it was taken away to be dressed and readied for the funeral. He feels the plaster being laid carefully against his forehead and his nose, yet still he can smell the lavender. He twitches his hands to keep the onrush of memory at bay, but there's something missing. His woolen cap is gone. He digs into his pocket and the sculptor reminds him to be still. His pockets are empty. He must have dropped it while he dozed.

Liam stands, despite the sculptor's protests. He rushes back into the corridor with plaster dripping down his chin. The cap is on the floor, right in front of the chair. He picks the hat up, then sits. He folds and unfolds it, considering whether he should go back in to the men who will think him a fool when they see the hat he's come here to retrieve is moth-eaten and frayed. Perhaps they will even send him away. But then it's his ma—daily astonished by how Liam's face has grown to become like his pa's and wanting for the dead man some form of immortalization—who has asked him to have this done. Liam knows he's ruined the work and they will have to begin again, but for his ma's sake, he stuffs his hat into his pocket and goes back to meet the waiting men.

STORY 32

Augen aus einem anderen Land. Wie ein Geheimdienst sich von einem Zeitungsartikel überrascht sah

Alexander Kluge

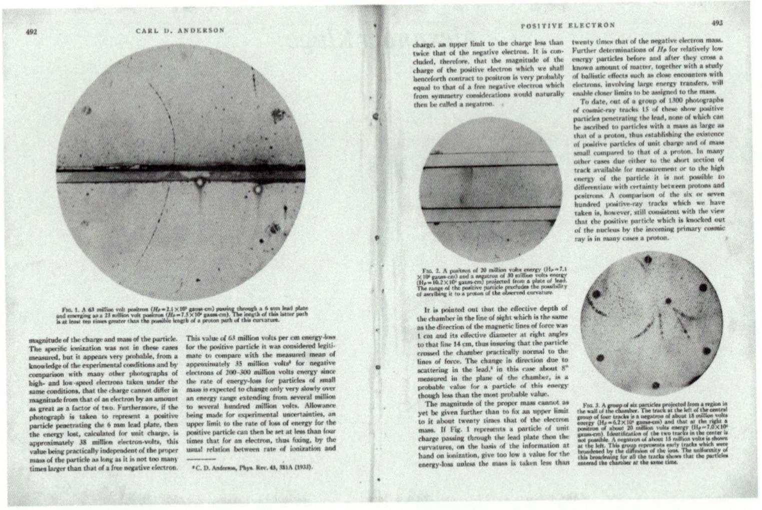

492 CARL D. ANDERSON

Fig. 1. A 63 million volt positron ($H\rho = 2.1 \times 10^5$ gauss-cm) passing through a 6 mm lead plate and emerging as a 23 million volt positron ($H\rho = 7.5 \times 10^4$ gauss-cm). The length of this latter path is at least ten times greater than the possible length of a proton path of this curvature.

magnitude of the charge and mass of the particle. The specific ionization was not in these cases measured, but it appears very probable, from a knowledge of the experimental conditions and by comparison with many other photographs of high- and low-speed electrons taken under the same conditions, that the charge cannot differ in magnitude from that of an electron by an amount as great as a factor of two. Furthermore, if the photograph is taken to represent a positive particle penetrating the 6 mm lead plate, then the energy lost, calculated for unit charge, is approximately 38 million electron-volts, this value being practically independent of the proper mass of the particle as long as it is not too many times larger than that of a free negative electron. This value of 63 million volts per cm energy-loss for the positive particle it was considered legitimate to compare with the measured mean of approximately 35 million volts[3] for negative electrons of 200–300 million volts energy since the rate of energy-loss for particles of small mass is expected to change only very slowly over an energy range extending from several million to several hundred million volts. Allowance being made for experimental uncertainties, an upper limit to the rate of loss of energy for the positive particle can then be set at less than four times that for an electron, thus fixing, by the usual relation between rate of ionization and

[3] C. D. Anderson, Phys. Rev. 43, 381A (1933).

POSITIVE ELECTRON 493

charge, an upper limit to the charge less than twice that of the negative electron. It is concluded, therefore, that the magnitude of the charge of the positive electron which we shall henceforth contract to positron is very probably equal to that of a free negative electron which from symmetry considerations would naturally then be called a negatron.

Fig. 2. A positron of 20 million volts energy ($H\rho = 7.1 \times 10^4$ gauss-cm) and a negatron of 30 million volts energy ($H\rho = 10.2 \times 10^4$ gauss-cm) projected from a plate of lead. The range of the positive particle precludes the possibility of ascribing it to a proton of the observed curvature.

It is pointed out that the effective depth of the chamber in the line of sight which is the same as the direction of the magnetic lines of force was 1 cm and its effective diameter at right angles to that line 14 cm, thus insuring that the particle crossed the chamber practically normal to the lines of force. The change in direction due to scattering in the lead,[3] in this case about 8° measured in the plane of the chamber, is a probable value for a particle of this energy though less than the most probable value.

The magnitude of the proper mass cannot as yet be given further than to fix an upper limit to it about twenty times that of the electron mass. If Fig. 1 represents a particle of unit charge passing through the lead plate then the curvatures, on the basis of the information at hand on ionization, give too low a value for the energy-loss unless the mass is taken less than twenty times that of the negative electron mass. Further determinations of $H\rho$ for relatively low energy particles before and after they cross a known amount of matter, together with a study of ballistic effects such as close encounters with electrons, involving large energy transfers, will enable closer limits to be assigned to the mass.

To date, out of a group of 1300 photographs of cosmic-ray tracks 15 of these show positive particles penetrating the lead, none of which can be ascribed to particles with a mass as large as that of a proton, thus establishing the existence of positive particles of unit charge and of mass small compared to that of a proton. In many other cases due either to the short section of track available for measurement or to the high energy of the particle it is not possible to differentiate with certainty between protons and positrons. A comparison of the six or seven hundred positive-ray tracks which we have taken is, however, still consistent with the view that the positive particle which is knocked out of the nucleus by the incoming primary cosmic ray is in many cases a proton.

Fig. 3. A group of six particles projected from a region in the wall of the chamber. The track at the left of the central group of four tracks is a negatron of about 18 million volts energy ($H\rho = 6.2 \times 10^4$ gauss-cm) and that at the right a positron of about 20 million volts energy ($H\rho = 7.0 \times 10^4$ gauss-cm). Identification of the two tracks in the center is not possible. A negatron of about 15 million volts is shown at the left. This group represents early tracks which were broadened by the diffusion of the ions. The uniformity of this broadening for all the tracks shows that the particles entered the chamber at the same time.

Carl D. Anderson, "The Positive Electron," *The Physical Review*, vol. 43, no. 6, 1933.

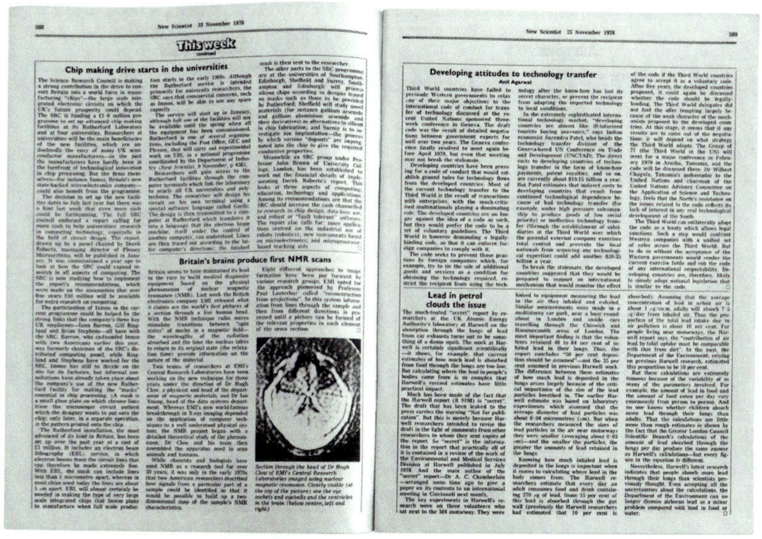

"Britain's Brains Produce First NMR Scans," *New Scientist*, vol. 80, no. 1130, 1978.

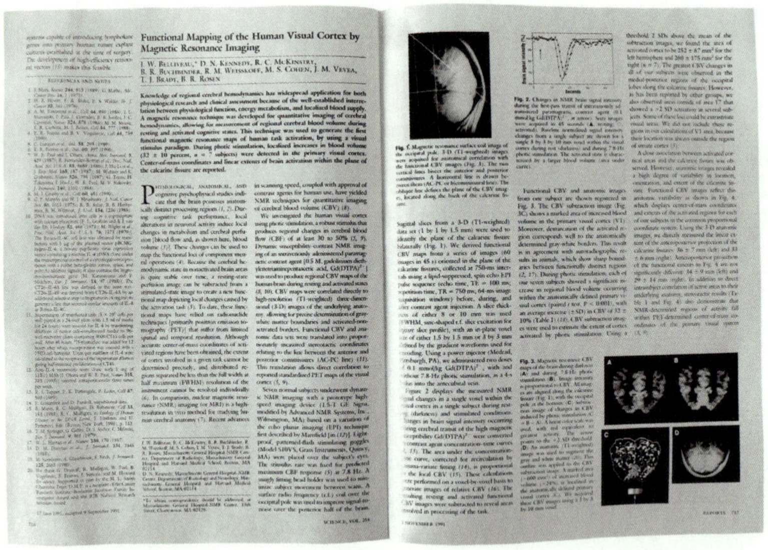

Functional Mapping of the Human Visual Cortex by Magnetic Resonance Imaging

J. W. Belliveau,* D. N. Kennedy, R. C. McKinstry, B. R. Buchbinder, R. M. Weisskoff, M. S. Cohen, J. M. Vevea, T. J. Brady, B. R. Rosen

Knowledge of regional cerebral hemodynamics has widespread application for both physiological research and clinical assessment because of the well-established interrelation between physiological function, energy metabolism, and localized blood supply. A magnetic resonance technique was developed for quantitative imaging of cerebral hemodynamics, allowing for measurement of regional cerebral blood volume during resting and activated cognitive states. This technique was used to generate the first functional magnetic resonance maps of human task activation, by using a visual stimulus paradigm. During photic stimulation, localized increases in blood volume (32 ± 10 percent, n = 7 subjects) were detected in the primary visual cortex. Center-of-mass coordinates and linear extents of brain activation within the plane of the calcarine fissure are reported.

J.W. Belliveau *et al.*, "Functional Mapping of the Human Visual Cortex by Magnetic Resonance Imaging," *Science*, vol. 254, no. 5032, 1991.

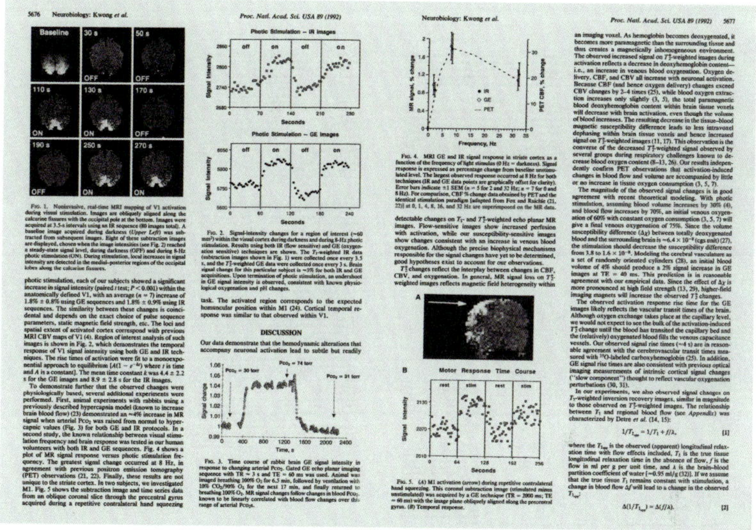

Kenneth Kwong, J.W. Belliveau *et al.*, "Dynamic Magnetic Resonance Imaging of Human Brain Activity During Primary Sensory Stimulation," *Proceedings of the National Academy of Sciences of the United States of America*, vol. 89, no. 12, 1992.

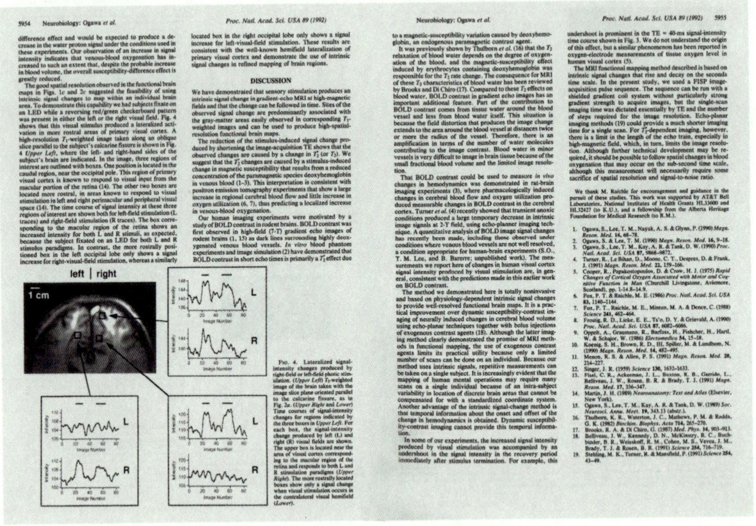

difference effect and would be expected to produce a decrease in the water proton signal under the conditions used in these experiments. Our observation of an increase in signal intensity indicates that venous-blood oxygenation has increased to such an extent that, despite the probable increase in blood volume, the overall susceptibility-difference effect is greatly reduced.

The good spatial resolution observed in the functional brain maps in Figs. 1*c* and 2*c* suggested the feasibility of using intrinsic signal changes to map within an individual brain area. To demonstrate this capability we had subjects fixate on an LED while a reversing red/green checkerboard pattern was present in either the left or the right visual field. Fig. 4 shows that this visual stimulus produced a lateralized activation in more rostral areas of primary visual cortex. A high-resolution T_1-weighted image taken along an oblique slice parallel to the subject's calcarine fissure is shown in Fig. 4 *Upper Left*, where the left- and right-hand sides of the subject's brain are indicated. In the image, three regions of interest are outlined with boxes. One position is located in the caudal region, near the occipital pole. This region of primary visual cortex is known to respond to visual input from the macular portion of the retina (14). The other two boxes are located more rostral, in areas known to respond to visual stimulation in left and right perimacular and peripheral visual space (14). The time course of signal intensity at these three regions of interest are shown both for left-field stimulation (L traces) and right-field stimulation (R traces). The box corresponding to the macular region of the retina shows an increased intensity for both L and R stimuli, as expected, because the subject fixated on an LED for both L and R stimulus paradigms. In contrast, the more rostrally positioned box in the left occipital lobe only shows a signal increase for right-visual-field stimulation, whereas a similarly located box in the right occipital lobe only shows a signal increase for left-visual-field stimulation. These results are consistent with the well-known hemifield lateralization of primary visual cortex and demonstrate the use of intrinsic signal changes in refined mapping of brain regions.

DISCUSSION

We have demonstrated that sensory stimulation produces an intrinsic signal change in gradient-echo MRI at high-magnetic fields and that the change can be followed in time. Sites of the observed signal change are predominantly associated with the gray-matter areas easily observed in corresponding T_1-weighted images and can be used to produce high-spatial-resolution functional brain maps.

The reduction of the stimulus-induced signal change produced by shortening the image-acquisition TE shows that the observed changes are caused by a change in T_2^* (or T_2). We suggest that the T_2^* changes are caused by a stimulus-induced change in magnetic susceptibility that results from a reduced concentration of the paramagnetic species deoxyhemoglobin in venous blood (1–3). This interpretation is consistent with positron emission tomography experiments that show a large increase in regional cerebral blood flow and little increase in oxygen utilization (6, 7), thus predicting a localized increase in venous-blood oxygenation.

Our human imaging experiments were motivated by a study of BOLD contrast in rodent brains. BOLD contrast was first observed in high-field (7-T) gradient echo images of rodent brains (1, 15) as dark lines surrounding highly deoxygenated venous blood vessels. *In vitro* blood phantom experiments and image simulation (2) have demonstrated that BOLD contrast in short echo times is primarily a T_2^* effect due to a magnetic-susceptibility variation caused by deoxyhemoglobin, an endogenous paramagnetic contrast agent.

FIG. 4. Lateralized signal-intensity changes produced by right-field or left-field photic stimulation. (*Upper Left*) T_1-weighted image of the brain taken with the image slice plane oriented parallel to the calcarine fissure, as in Fig. 2*a*. (*Upper Right* and *Lower*) Time courses of signal-intensity changes for regions indicated by the three boxes in *Upper Left*. For each box, the signal-intensity change produced by left (L) and right (R) visual fields are shown. The upper box is located near the area of visual cortex corresponding to the macular region of the retina and responds to both L and R stimulation paradigms (*Upper Right*). The more rostrally located boxes show only a signal change when visual stimulation occurs in the contralateral visual hemifield (*Lower*).

It was previously shown by Thulborn *et al.* (16) that the T_2 relaxation of blood water depends on the degree of oxygenation of the blood, and the magnetic-susceptibility effect induced by erythrocytes containing deoxyhemoglobin was responsible for the T_2 rate change. The consequence for MRI of these T_2 characteristics of blood water has been reviewed by Brooks and Di Chiro (17). Compared to these T_2 effects on blood water, BOLD contrast in gradient echo images has an important additional feature. Part of the contribution to BOLD contrast comes from tissue water around the blood vessel and less from blood water itself. This situation is because the field distortion that produces the image change extends to the area around the blood vessel at distances twice or more the radius of the vessel. Therefore, there is an amplification in terms of the number of water molecules contributing to the image contrast. Blood water in minor vessels is very difficult to image in brain tissue because of the small fractional blood volume and the limited image resolution.

That BOLD contrast could be used to measure *in vivo* changes in hemodynamics was demonstrated in rat-brain imaging experiments (3), where pharmacologically induced changes in cerebral blood flow and oxygen utilization produced measurable changes in BOLD contrast in the cerebral cortex. Turner *et al.* (4) recently showed that transient anoxic conditions produced a large temporary decrease in intrinsic image signals at 2-T field, using echo-planner imaging technique. A quantitative analysis of BOLD image signal changes has recently been made, including those observed under conditions where venous blood vessels are not well resolved, a condition appropriate for human-brain experiments (S.O., T. M. Lee, and B. Barerre; unpublished work). The measurements we report here of changes in human visual cortex signal intensity produced by visual stimulation are, in general, consistent with the predictions made in this earlier work on BOLD contrast.

The method we demonstrated here is totally noninvasive and based on physiology-dependent intrinsic signal changes to provide well-resolved functional brain maps. It is a practical improvement over dynamic susceptibility-contrast imaging of neurally induced changes in cerebral blood volume using echo-planar techniques together with bolus injections of exogenous contrast agents (18). Although the latter imaging method clearly demonstrated the promise of MRI methods in functional mapping, the use of exogenous contrast agents limits its practical utility because only a limited number of scans can be done on an individual. Because our method uses intrinsic signals, repetitive measurements can be taken on a single subject. It is increasingly evident that the mapping of human mental operations may require many scans on a single individual because of an intra-subject variability in location of discrete brain areas that cannot be compensated for with a standardized coordinate system. Another advantage of the intrinsic signal-change method is that temporal information about the onset and offset of the change in hemodynamics is obtained. Dynamic susceptibility-contrast imaging cannot provide this temporal information.

In some of our experiments, the increased signal intensity produced by visual stimulation was accompanied by an undershoot in the signal intensity in the recovery period immediately after stimulus termination. For example, this undershoot is prominent in the TE = 40-ms signal-intensity time course shown in Fig. 3. We do not understand the origin of this effect, but a similar phenomenon has been reported in oxygen-electrode measurements of tissue oxygen level in human visual cortex (5).

The MRI functional mapping method described is based on intrinsic signal changes that rise and decay on the seconds time scale. In the present study, we used a FISP image-acquisition pulse sequence. The sequence can be run with a shielded gradient coil system without particularly strong gradient strength to acquire images, but the single-scan imaging time was dictated essentially by TE and the number of steps required for the image resolution. Echo-planar imaging methods (19) could provide a much shorter imaging time for a single scan. For T_2^*-dependent imaging, however, there is a limit in the length of the echo train, especially in high-magnetic field, which, in turn, limits the image resolution. Although further technical development may be required, it should be possible to follow spatial changes in blood oxygenation that may occur on the sub-second time scale, although this measurement will necessarily require some sacrifice of spatial resolution and signal-to-noise ratio.

We thank M. Raichle for encouragement and guidance in the pursuit of these studies. This work was supported by AT&T Bell Laboratories, National Institutes of Health Grants HL33600 and HL32427 (to K.U.), and a fellowship from the Alberta Heritage Foundation for Medical Research (to R.M.).

1. Ogawa, S., Lee, T. M., Nayak, A. S. & Glynn, P. (1990) *Magn. Reson. Med.* **14,** 68–78.
2. Ogawa, S. & Lee, T. M. (1990) *Magn. Reson. Med.* **16,** 9–18.
3. Ogawa, S., Lee, T. M., Kay, A. R. & Tank, D. W. (1990) *Proc. Natl. Acad. Sci. USA* **87,** 9868–9872.
4. Turner, R., Le Bihan, D., Moone, C. T., Despres, D. & Frank, J. (1991) *Magn. Reson. Med.* **22,** 159–166.
5. Cooper, R., Papakostopoulos, D. & Crow, H. J. (1975) *Rapid Changes of Cortical Oxygen Associated with Motor and Cognitive Function in Man* (Churchill Livingstone, Aviemore, Scotland), pp. 1-14.8–14.9.
6. Fox, P. T. & Raichle, M. E. (1986) *Proc. Natl. Acad. Sci. USA* **83,** 1140–1144.
7. Fox, P. T., Raichle, M. E., Minton, M. A. & Dence, C. (1988) *Science* **241,** 462–464.
8. Frostig, R. D., Lieke, E. E., Ts'o, D. Y. & Grinvald, A. (1990) *Proc. Natl. Acad. Sci. USA* **87,** 6082–6086.
9. Oppelt, A., Graumann, R., Barfuss, H., Fishcher, H., Hartl, W. & Schajor, W. (1986) *Electomedica* **54,** 15–18.
10. Koenig, S. H., Brown, R. D., III, Spiller, M. & Lundbom, N. (1990) *Magn. Reson. Med.* **14,** 482–495.
11. Menon, R. S. & Allen, P. S. (1991) *Magn. Reson. Med.* **20,** 214–227.
12. Singer, J. R. (1959) *Science* **130,** 1632–1633.
13. Fisel, C. R., Ackerman, J. L., Buxton, R. B., Garrido, L., Belliveau, J. W., Rosen, B. R. & Brady, T. J. (1991) *Magn. Reson. Med.* **17,** 336–347.
14. Martin, J. H. (1989) *Neuroanatomy: Text and Atlas* (Elsevier, New York).
15. Ogawa, S., Lee, T. M., Kay, A. R. & Tank, D. W. (1989) *Soc. Neurosci. Annu. Meet.* **19,** 343.13 (abstr.).
16. Thulborn, K. R., Waterton, J. C., Mathews, P. M. & Radda, G. K. (1982) *Biochim. Biophys. Acta* **714,** 265–270.
17. Brooks, R. A. & Di Chiro, G. (1987) *Med. Phys.* **14,** 903–913.
18. Belliveau, J. W., Kennedy, D. N., McKinstry, R. C., Buchbinder, B. R., Weisskoff, R. M., Cohen, M. S., Vevea, J. M., Brady, T. J. & Rosen, B. R. (1991) *Science* **254,** 716–719.
19. Stehling, M. K., Turner, R. & Mansfield, P. (1991) *Science* **254,** 43–49.

Seiji Ogawa *et al.*, "Intrinsic Signal Changes Accompanying Sensory Stimulation: Functional Brain Mapping with Magnetic Resonance Imaging," *Proceedings of the National Academy of Sciences of the United States of America*, vol. 89, no. 13, 1992.

Ein britisches Blatt hatte von einem neuartigen Kernspinresonanzapparat (NMR), entwickelt durch das Team des Physikers Dr. Hugh Clow, berichtet. Im weltweiten Rennen des Fortschritts hatte dieses Team eine Furt entdeckt, durch die sich mikroskopische, „gequantelte" Funksprüche zwischen den Hirnzellen – Unterschiede von Spin-Zuständen, sogenannte Spin-Echos – durch Radiowellen mit einem makroskopischen Aufzeichnungsgerät verbinden lassen. Hirnforscher konnten so erstmals den Redeton und das Leben im arbeitenden Menschenhirn direkt beobachten: die Kommunikation in der „Republik des Geistes".

Die Journalistin, welche die vom Forschungslabor eines Konzerns an sie herangetragenen Informationen in die Sprache der Zeitung zu übersetzen hatte, sah sich vor Schwierigkeiten gestellt. Sie kannte ihre englische, frühstückende Leserschaft gut genug, um zu wissen, welche Mitteilungen und welche Wortwahl sie besser ausließ. Das meiste an dieser Sensation – „erstmals sehen Experten in den Zauberkasten BEWUSSTSEIN direkt hinein, sie sehen es pulsieren" – war in der Ausdrucksweise einer zielgruppenorientieren Presse kaum zu formulieren. Inzwischen lag aber der Artikel nicht bloß in den britischen Haushalten und Cafés vor. Er lag auch auf dem Tisch von *Auslandskundschaftern* der HAUPTVERWALTUNG AUFKLÄRUNG im Ministerium für Staatssicherheit in der Hauptstadt der DDR. Ein Team dort war speziell verantwortlich für die „Abschöpfung" des „wissenschaftlichen Fortschritts im Westen". Die Tagessensation in dem britischen Blatt elektrisierte sie. Ihr Interesse war motivierter als das der Zeitungsleser, auch stärker auf Dauer ausgerichtet.

Es ging nämlich bei diesem neuen NMR-Apparat, den man im Osten zunächst nicht nachbauen konnte (dazu hätte man mehr Information und auch eine andere Produktionskapazität gebraucht) um die „Verkehrsverhältnisse der HYPERFEINSTRUKTUR, in der die Hirnsynapsen untereinander konferieren". Die zu messenden *Boltzmann-Faktoren* unterscheiden sich um winzige Einstellwinkel, 10 hoch 4 kleiner als alles, was man mit dem makroskopischen Blick sehen kann. Der Vektor „Mü" (griechischer Buchstabe μ) gilt DABEI FÜR DIE VOM DREHIMPULS ABHÄNGIGE RICHTUNGSQUANTELUNG BEI GEGEBENER DREHIMPULS-QUANTENZAHL „I". Die Konstante Gamma bezeichnet das *gyromagnetische Verhältnis* und den *Landé-Faktor*. Man schreibt also, notierte der wissenschaftliche Kundschafter Tacke:

$$\hat{\vec{\mu}} = \gamma \cdot \hat{\vec{I}}$$

Wie und mit welchen Worten sollte Tacke so etwas seinem Abteilungsleiter melden? Wie sollte es dieser den Hauptverantwortlichen, seinen Kontrolleuren in der Hauptverwaltung des Zentralkomitees übermitteln? Und was sollten, außer Verwunderung zu äußern, diese Entscheider mit der Innovation anfangen?

Andererseits war evident, dass – wie durch ein faustisches Zauberglas – die Menschheit (und hier im konkreten Fall der Abschöpfung das sozialistische Dauerinteresse, im Gegensatz zur Sensationslust der Tagespresse) erstmals in die INTERNE SCHATZKAMMER, die GRÜBLERISCHE GEGENWELT DER SUBJEKTIVITÄT, blickte. Ein Blick wie auf einen fremden Himmelskörper.

Noch jetzt, 39 Jahre später, längst entlassen aus dem Staatsdienst einer untergegangenen Republik, herabgestuft zum

Pensionär und Heimatforscher in einem kleinen Ort in Sachsen-Anhalt, ist Fred Tacke immer noch voller Begeisterung über die damalige NMR-Nachricht. Das Vaterland der Werktätigen ist entschwunden, aber die Aufgabenstellung, alle Geheimnisse des Planeten aufzuspüren, hat sich eher erweitert als verringert. In den 80er Jahren des vorigen Jahrhunderts war die Hauptverwaltung Aufklärung (HVA) das wache Auge, das alle Geheimnisse des Westens bei sich versammelte. Das Auswärtige Amt der Deutschen Bundesrepublik weiß nicht, was das Finanzministerium oder die Staatlichen Forstämter wissen und denen ist unbekannt, was Kenntnisstand der NATO in Brüssel ist. Erst das Insektenauge des HVA vereinigt alle diese Informationen. Das bedeutet: Augen aus einem anderen Land.

Was ist aber, fragt Tacke noch heute, solche Allwissenheit gegen den Zugang zu dem Räderwerk unter der Schädeldecke der Menschen, dem Gehirn. Alle Zeiten dort reichen filigran zurück bis zur Entstehung der Rückgrat- und Schädeltiere, ja zu den Anfängen des Lebens überhaupt.

Tacke hat sich kundig gemacht. Er hat sich ein chirurgisch geöffnetes Schädeldach in der Charité zeigen lassen. So begeistert er von NMR noch immer ist, so enttäuscht war er vom Anblick. Ein menschliches Gehirn, in einfacher Makroskopie, also vom Menschenauge betrachtet, erinnert ihn an einen Mehlklumpen. Er würde, sagt er, „einen solchen Kloß keinem Kannibalen, auch nicht gebacken, als Mahlzeit anbieten wollen".

Das Bild ändert sich, sobald man die Feinstruktur, nämlich die Oszillationen zu Gesicht bekommt, wie sie das NMR mitteilt: Den lebensbegleitenden Strom der Billionen sternartig verzweigten Hirnzellen und ihrer Stützgewebe. Gibt man, so der experimentierfreudige Ex-Spion Tacke, die Schwingungen, welche der Apparat nach der Kernspininduktionsmethode ausweist, in einen TONGENERATOR ein, so entsteht ein sonores, vielstimmiges, „grauweißliches" Geräusch, ganz ähnlich dem, das die Schwingungen des Planeten Uranus als Ton erzeugen, wenn dieser gewaltige Himmelskörper die Raumzeit durchbricht.

Die Hauptverwaltung Aufklärung und ihre verbrüderten Geheimdienste im Sozialismus hätten, so Tacke, nicht die Staatsgeheimnisse des Westens, sondern das menschliche Naturwunder unter der Schädeldecke der Menschen erforschen sollen. Zurückgestuft zum Heimatforscher kann Tacke im Jahr 2022 wenig tun, um die Erforschung der SUBJEKTIVEN SEITE DES MENSCHEN ins Ziel zu führen.

* * *

Abb. Aus der Sammlung Tacke (inzwischen auf 6 Kisten und 7 Aktenbände angewachsen). Es handelt sich um ein Bild, das ein sibirischer Verschwörungstheoretikers ins Netz gestellt hat. Es zeigt angeblich einen Alien. Tacke weist nach, dass dies eine Irreführung ist. Es handelt sich um eine NMR-Aufnahme eines Menschenhirns. Oben links und rechts die Augenpartie.

EYES FROM ANOTHER COUNTRY. HOW SECRET INTELLIGENCE SERVICES WERE ASTONISHED BY A NEWSPAPER ARTICLE

A British newspaper reported on a new nuclear magnetic resonance (NMR) machine that had been developed by a team led by Dr. Hugh Clow. In the global race for progress, this team had discovered a bridge across which microscopic, "quantized" radio messages between brain cells—differences in spin states, so-called spin echoes—could be linked by radio waves to a macroscopic recording device. This meant neuroscientists could, for the first time, directly observe the tone of speech and life within the working human brain: the communication going on in the "Republic of the Mind."

The journalist who had the task of transposing the information given to her by the research laboratory of a corporation into the language of the newspaper found herself confronted with some challenges. She was familiar enough with her breakfasting English readership to know which announcements and which phrases she would be well-advised to omit. Most of the content of this sensation could hardly be successfully formulated in the manner of a target-group-oriented press—"experts see directly into the magic box of the CONSCIOUSNESS for the first time, they see it pulsating." But by then the article was being read not just in British households and cafés. It was also lying on the Foreign Intelligence table in the HAUPTVERWALTUNG AUFKLÄRUNG (HVA), the main Directorate for reconnaissance at the Ministry for State Security in the capital city of the GDR. One team there was especially responsible for "skimming through" the "scientific progress in the West." They were electrified by the sensation in the British newspaper. Their interest was more highly motivated than that of the average newspaper reader, more focused on long-term perspectives.

This new NMR device, which could not initially be recreated in the East (more information and a different production capacity would have been required), concerned the "communications conditions in the HYPERFINE STRUCTURE, in which the brain's synapses conferred with each other." The *Boltzmann factors* to be measured differ by tiny degrees, 10 to the power of 4 smaller than anything that can be seen with macroscopic vision. The vector "mu" (Greek character μ) applies TO THE DIRECTIONAL QUANTUM DEPENDENT ON THE ANGULAR MOMENTUM WITH A GIVEN ANGULAR MOMENTUM QUANTUM NUMBER "I." The gamma constant describes the *gyromagnetic relationship* and the *Landé factor*. Thus one would write, as scientific intelligence officer Fred Tacke noted:

$$\hat{\vec{\mu}} = \gamma \cdot \hat{\vec{I}}$$

How and using what words should Tacke report something like this to his department head? How should that man then convey it to those in charge, his supervisors at the head office of the Central Committee? And what could these decision makers do with the information on this innovation, except express amazement?

On the other hand, it was evident that humanity (and in this specific case of skimming, this meant the long-term vision of socialism, in contrast to the sensationalism of the daily press) gazed for the first time—and as if through a Faustian magic lens—into the INNER SECRET VAULT, THE BROODING COUNTERWORLD OF SUBJECTIVITY. As if gazing at a strange celestial body.

Even now, thirty-nine years later and long since released from service to a fallen republic, demoted to pensioner and local historian in a small community in Saxony-Anhalt, Fred Tacke is still full of enthusiasm for that NMR message. The Fatherland of the Workers has disappeared, but the task of finding out all the secrets of the planet has expanded in scope rather than diminished. In the 1980s, the HVA was the watchful eye that drew all the secrets of the West into its orbit. The Foreign Office of the Federal Republic of Germany did not know what the Finance Ministry or the State Forestry Offices knew, and neither did they know what level of information NATO had in Brussels. Only the insect eye of the HVA united all this information. And that means: eyes from another country.

But, as Tacke still asks today, what does such omniscience matter in comparison to access to the machinery beneath the skullcap of the human head, the brain? There, all times filigree back to the origins of vertebrates, animals with spines and skulls—indeed, even back to the origins of life itself.

Tacke informed himself. He had them show him a surgically opened skull in the Berlin Charité. As amazed by NMR as he still is, he was nevertheless disappointed by what he saw. A human brain in simple macroscopy—in other words, viewed by the human eye—reminds him of clumpy flour. He would, he says, "not wish to offer a dumpling like that to a cannibal as a meal, not even baked."

The picture changes as soon as you get to see the fine structure, i.e., the oscillations, as shown with NMR: the stream that is attendant upon life—of trillions of starlike branching brain cells and their supporting tissues. According to the ex-spy and keen experimenter Tacke, if the oscillations, which the apparatus shows using the nuclear spin induction technique, are fed into a SOUND GENERATOR, a sonorous, polyphonic, "gray-white" sound is produced, very similar to that which the oscillations of the planet Uranus produce when that huge celestial body breaks through space-time.

Tacke thinks that the HVA and its affiliated intelligence services in socialism should have been investigating not the state secrets of the West, but instead the natural wonder just beneath the top of the human skull. Demoted to local historian, Tacker can do little in 2022 to steer research into the SUBJECTIVE ASPECTS OF THE HUMAN BEING toward its destination.

* * *

In the Tacke Collection (now expanded to six boxes and seven dossier files), there is an image placed online by a Siberian conspiracy theorist. It purports to show an alien. Tacke demonstrates that this is a hoax. It is in fact an NMR image of a human brain. The eye area is visible at the top left and right.

iii

THE CONVERSATION MACHINE

Neuroscientists, psychologists, neurolinguists, and philosophers from five continents are broadcast in an assembly of thirty-two screens. They investigate neuroscientific experiments and their philosophical and ethical dimensions.

Like the brain, The Conversation Machine–videos, interviews, and orchestration by Taryn Simon, produced by Fondazione Prada for the exhibition–is a self-organizing system. It responds to itself, continually constructing and assimilating its own order and disorder.

In clips excerpted from over 140 hours of interviews, participants appear to listen and respond to each other's statements. They enter and exit. Objects that refer to their work appear in flashes. Clusters of discussants migrate across the screens. Others sit in sustained, active silence. Like the brain, the conversation morphs according to a logic of prediction and surprise. Attention and distraction maximize and minimize responsiveness in human brains. The "listeners" in the conversation mark this balance while the viewer's attention is by turns captured, lost, redirected, tested, split, and pulled. Objects, empty chairs, fidgets, speaker and screen changes incorporate distraction–the conductor of attention–directly into the environment.

The participants talk about research, experiments, ideas of "me," survival, biological changes carved by trauma. The brain feeds on mismatch, connections, stimuli. The conversation navigates a history of neuroscientific knowledge-making marked by rigor, breakthrough, and discovery–but also by error, transgression, and uncertainty. It attends also to the continual presence of absences: gaps, missing perspectives, hidden layers. It tracks forces of violence, extinction, and oblivion embedded in our brains alongside creativity, regeneration, and resilience.

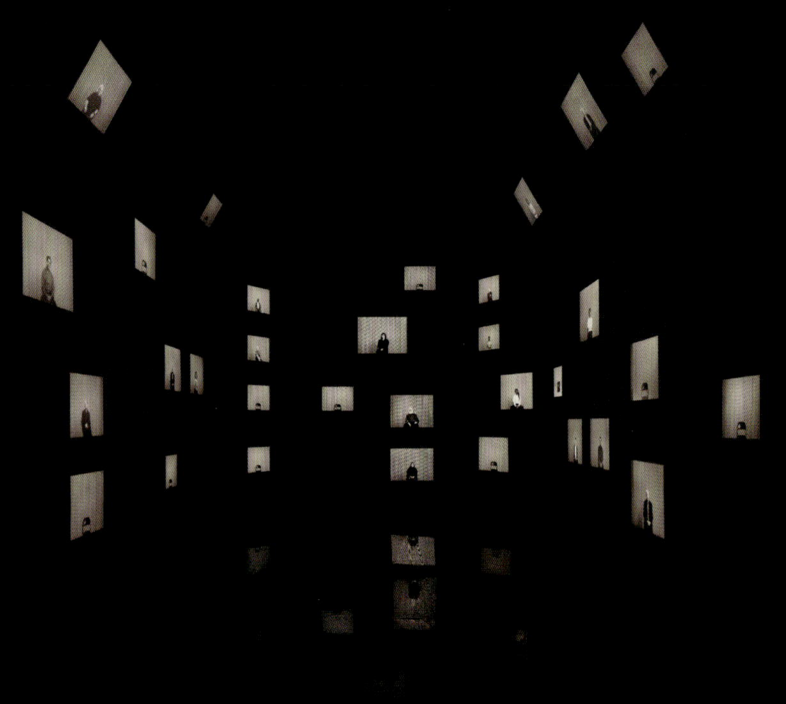

WITH

Yasmin Abofoul, Huda Akil, Polina Anikeeva, Anirban Bandyopadhyay, György Buzsáki, David Chalmers, Patricia Churchland, Antonio Damasio, Stanislas Dehaene, Daniel C. Dennett, Catherine Dulac, David Erritzoe, Lisa Feldman Barrett, André Fenton, Karl Friston, Ali Ghazizadeh, Carl Hart, Suzana Herculano-Houzel, Amadi O. Ihunwo, Erich Jarvis, Bianca Jones Marlin, Kumi Kuroda, Joseph E. LeDoux, Mahmoud Bukar Maina, Catherine Malabou, Eve Marder, Hannah Monyer, David Poeppel, Supratim Ray, Daniela Schiller, Wolf Singer, Mitchell Joseph Valdés-Sosa, Angela Vincent, Charles Yang, Rafael Yuste, Li Zhaoping

VIDEOS, INTERVIEWS AND ORCHESTRATION BY

Taryn Simon

PRODUCED BY

Fondazione Prada for the “Human Brains. It Begins with an Idea” project

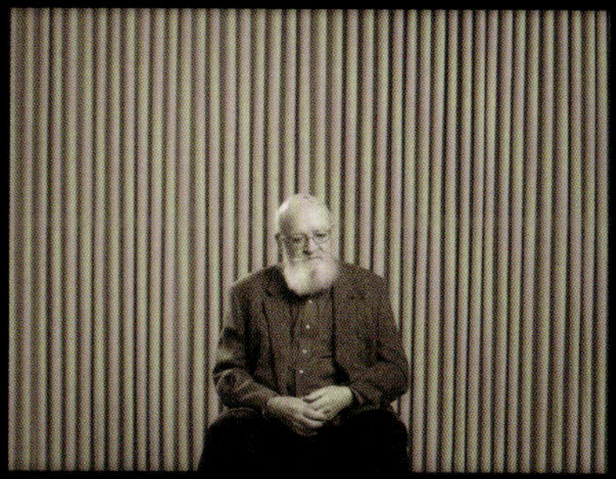

Daniel C. Dennett – You don't know that you are not a zombie, whoever you are. You say, "Well, yes, I do. I have direct knowledge of my consciousness." No, you don't. You have a belief that you are conscious of the scene in front of you, that you are hearing my voice and so forth. Those are beliefs. Do you know how you get those beliefs? It is all due to brain processes. Do you know what those brain processes are? No. In fact, you know almost nothing about them.

Supratim Ray – If you look at it, it's just this pinkish mass of tissue. It's amazing that these connections together form everything that we are. But at the end of the day, if you look at it from outside, it is just a yea-big mass of spongy sort of tissue. You can't see the connections or anything. You see some core structures, that's about it.

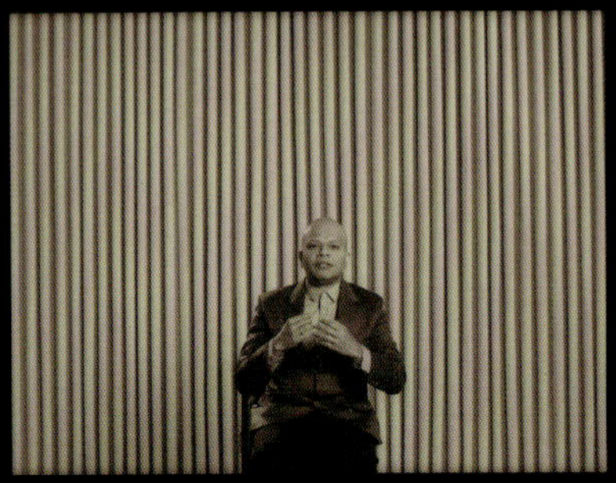

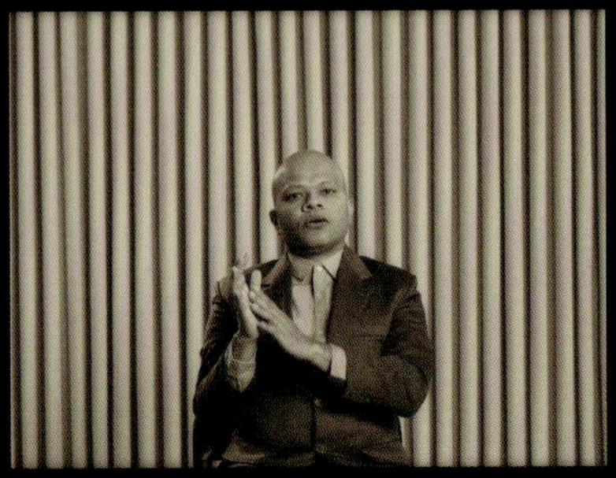

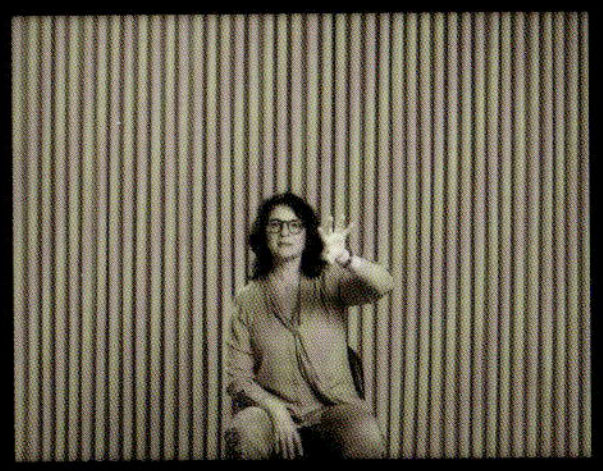

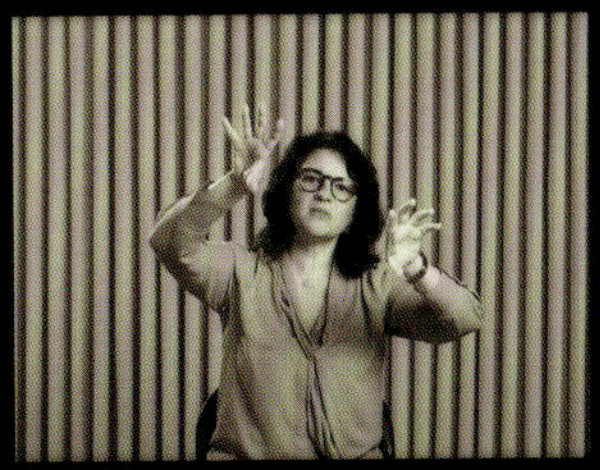

Suzana Herculano-Houzel – Cells in the brain are distributed very unevenly. There are tons of them here, not very many there, and then tons over there again. Very much like people in a country. So if you want to poll people in a country, you have to use these super-fancy, careful sampling strategies to make sure that your sample is representative. So the question was, how do you do that in the brain? I realized that with the brain, we could do something that you can't do in a country with people, which is, you can dissolve the brain and just very much liberate all those units into one volume of fluid, which is the soup. So that's how I created this method for counting cells in a brain that I call "brain soup."

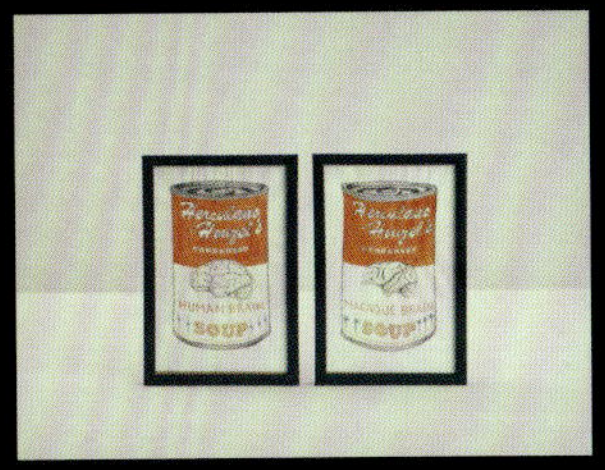

00:01:10:17

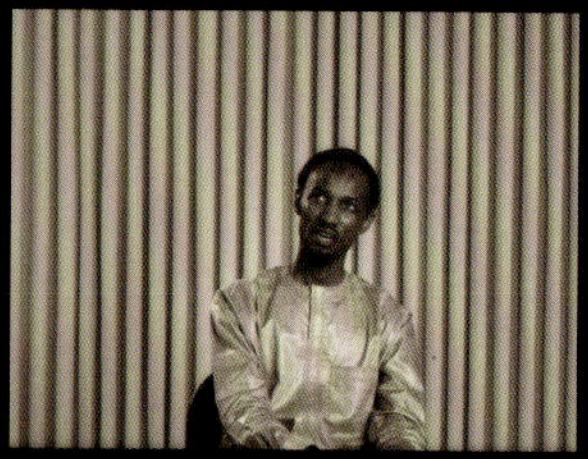

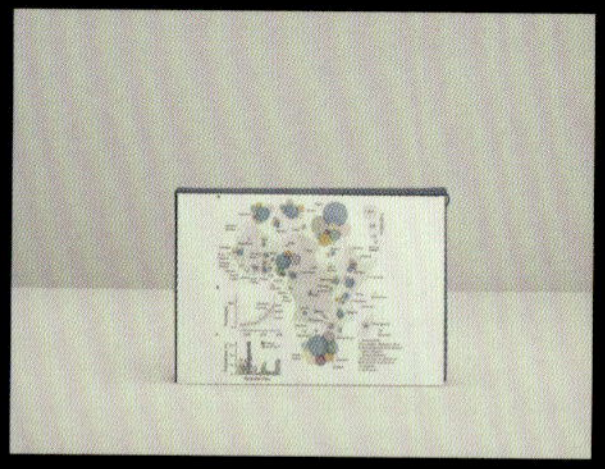

Mahmoud Maina – One of the key things that I have been trying to do is to see how we can incorporate more African samples in our research. And obviously you have to start somewhere. […] A small pinch of skin collected from an individual can now be reprogrammed into brain cells. And someone is more likely to give you a small pinch of skin that you can now use to grow brain cells than him saying, "Okay, you know what, after I die, you can have my brain."

Eve Marder – We get our lobsters from these waters. And the fishermen, when they find a female lobster who is carrying eggs, they notch her tail and they throw her back. They will not bring in female lobsters. So if we wanted to study female lobsters, we would actually have to get a permit, and then get someone to collect them for us, because the fishermen won't bring them in. The fishermen just want them in the water, making babies.

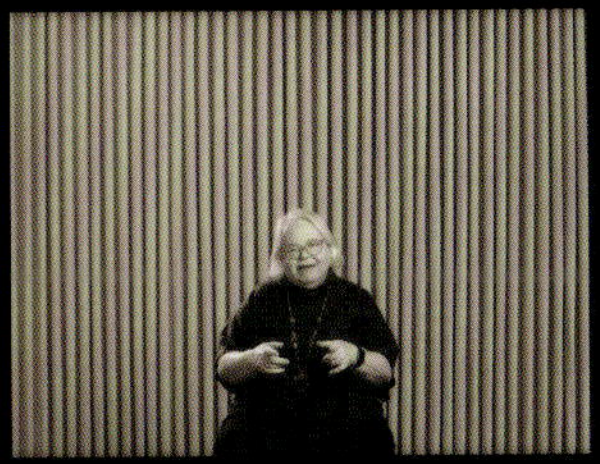

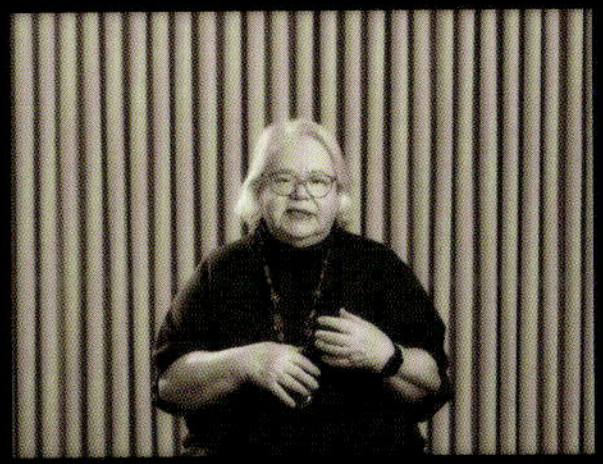

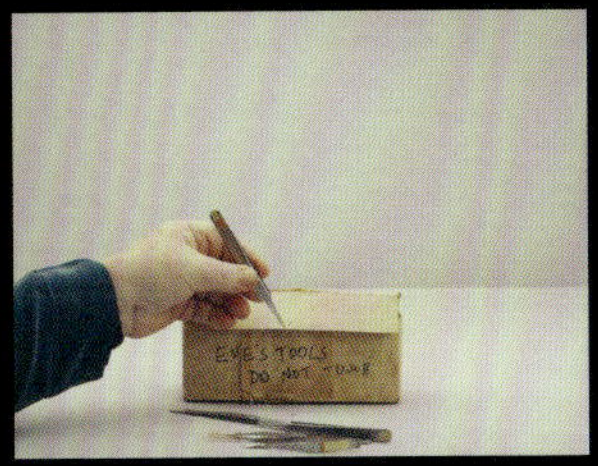

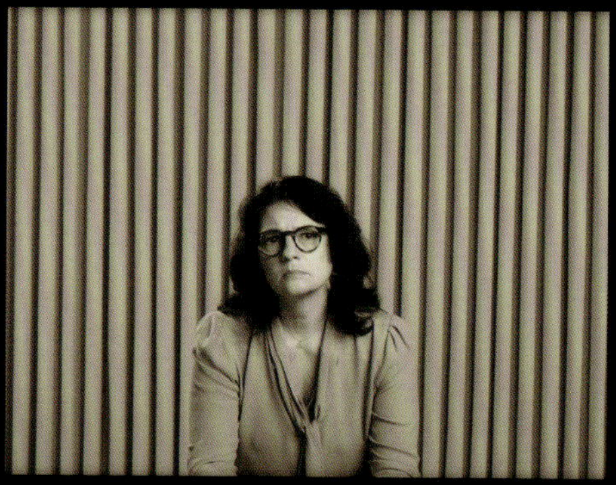

SH — I have several fridges—No, I have several freezers—full of animal brains. I call it my frozen zoo. There are brains that are still whole. There are brains that are sectioned into parts, or in very thin slices, which is what we usually have to do to look at them under a microscope. I also have hundreds of little vials of brain soup that we keep from every single brain that we study.

Amadi O. Ihunwo – I see the same structures across … from having dissected brain in Nigeria, to East Africa, to South Africa, to white brain, whatever you call white brain or black brain or whatever. I see the same.

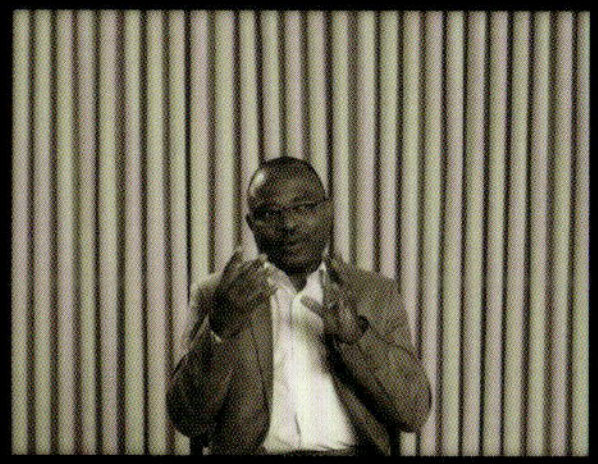

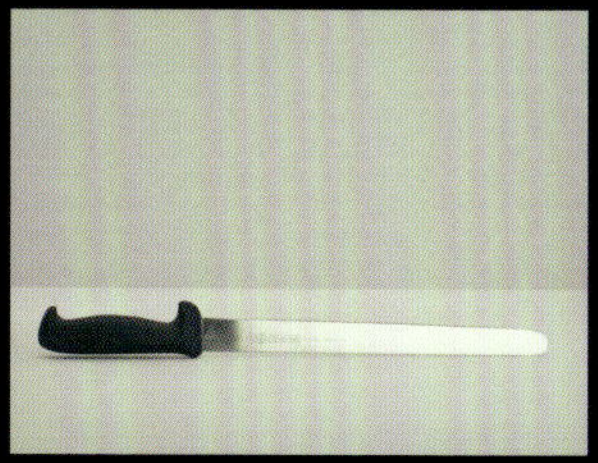

00:03:49:01

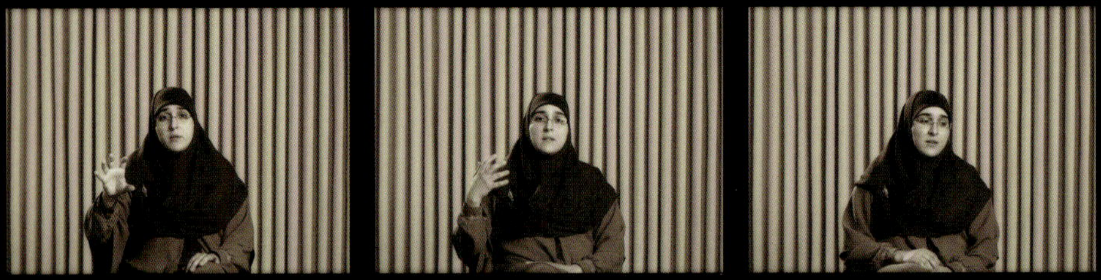

Yasmin Abofoul – I think we are different people. Yes. But not because our brains have been made by different material. But I think that we are born to different societies. We have different religions and different experiences of lives and different individual issues also.

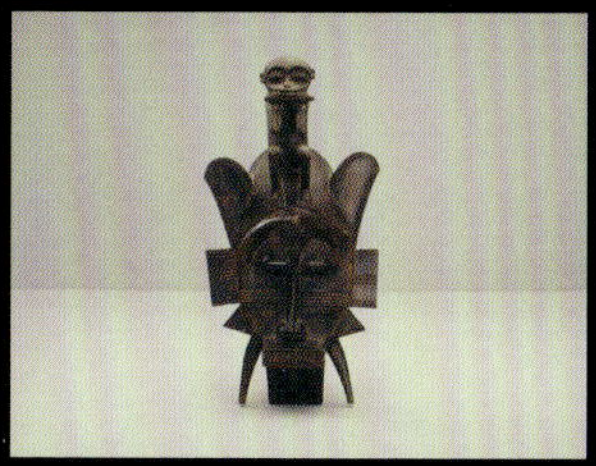

Huda Akil – When we study the depressed brain in postmortem tissue—meaning that people have died and have donated their brain and we compare the depressed brains to the control brains of people who died of natural causes—we see differences everywhere. Some brain regions are more affected than others, but the activity of the entire genetic makeup of the cells is altered throughout the brains of depressed people. Some brain regions like the hippocampus, the amygdala, the prefrontal cortex, regions that are implicated in emotions and cognition are more profoundly affected, but hardly a brain region escapes the impact of depression. And what the brain looks like is as if it is impoverished. There are cells, the glial cells, that are really important for the health of neurons, for the connection between neurons. They participate in the immune responses and they are greatly affected by depression. So even though depression may have started with one circuit being disrupted or have started from some other place in the body, in the end the brain is affected, and it is like death by a thousand cuts.

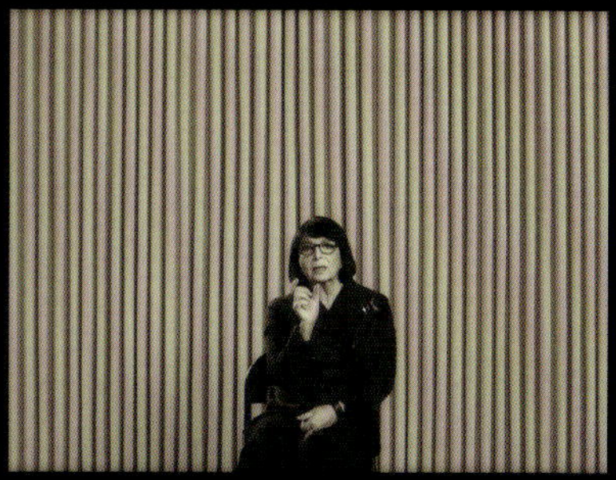

00:04:38:01

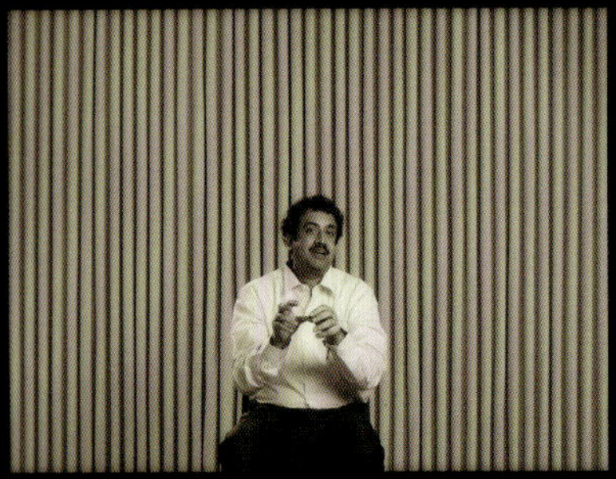

Anirban Bandyopadhyay – A cell on a Petri dish–you will see that when the cell dies, its branches that were going out, they come together, and then the grid collects it together, and it forms this spherical cell. So it isolates itself, and then it is death. So isolation is death.

Catherine Malabou – The brain is capable of creating out of trauma, out of destruction, a totally new person. It is like a negative metamorphosis. Like the one we see in Kafka's short story, that someone is able, out of this destructive plasticity, to become someone else. And this capacity of negative metamorphosis is what I call "destructive plasticity," which is a threat that inhabits each of us.

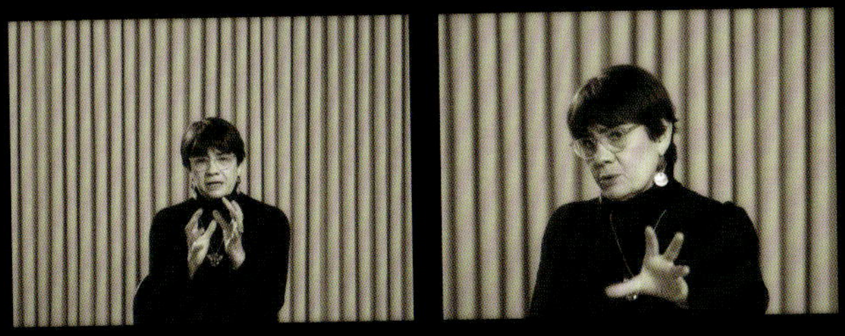

00:06:26:15

Kumi Kuroda – A boy committed sexual murder when he was seventeen years old. I also did the expert witness for that case. And he has been abused really severely physically, and his brain has been damaged. [...] And that kind of case, of course, he was arrested, but his father was not arrested. And his brain is damaged.

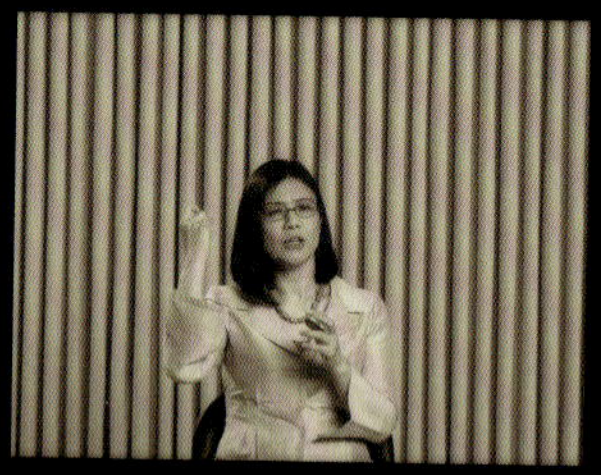

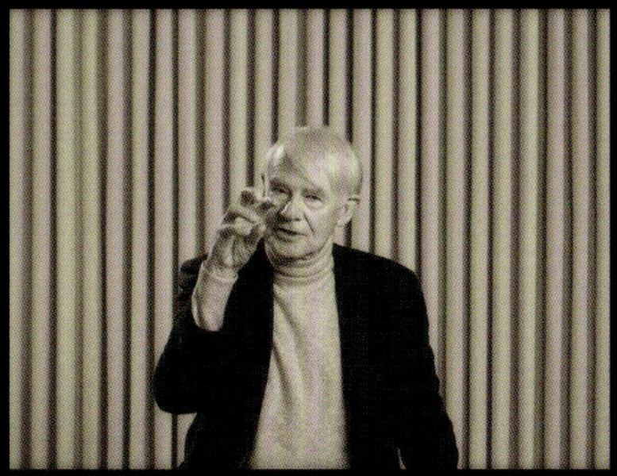

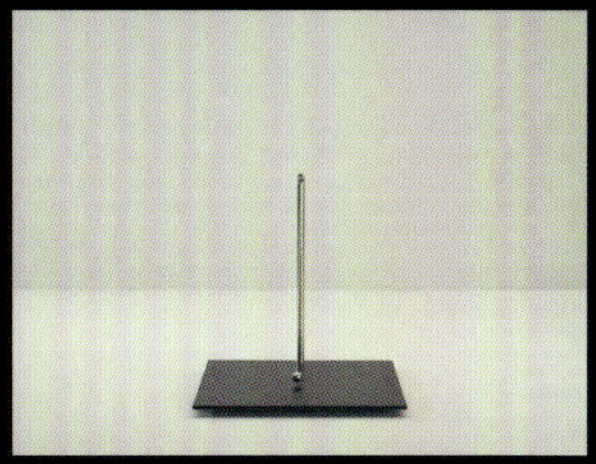

Wolf Singer – So something must be wrong, otherwise a quote-unquote normal person would not do that. So what apparently our law system does is that it looks as to whether the average, in the normal distribution of behaviors, dispositions, is violated by somebody who is criminal, and at the same time, we would have to say it is violated by somebody who is extremely compassionate, et cetera, like Mother Teresa. So if you are an outlier in the normal distribution, you are either rewarded with a Nobel Prize or you are put in jail.

KK – So that's why I think that neuroscience can do something, should be able to do something for these cases. And I am searching for the way. And one way is, of course, the direct manipulation of the brain area.

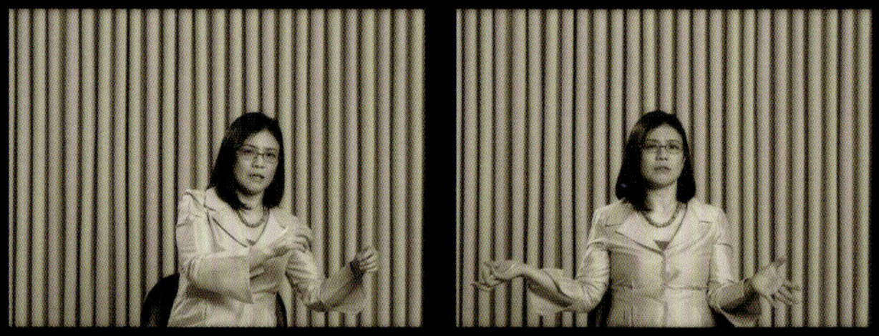

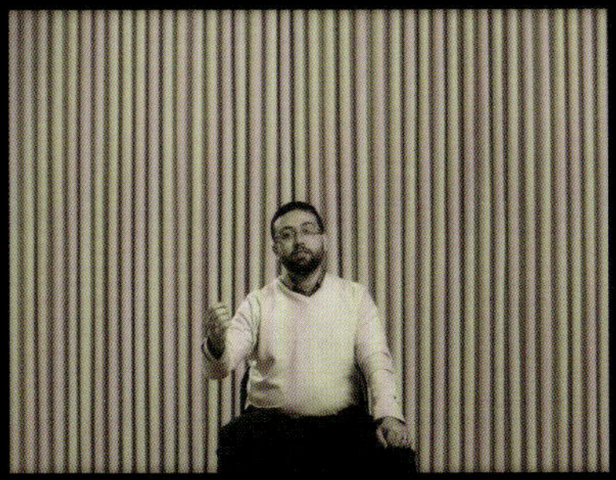

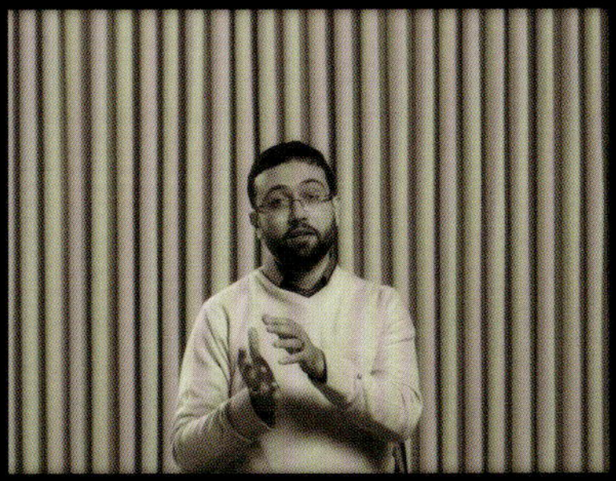

Ali Ghazizadeh – Imagine you have an electrode in your dopamine regions, and you give the person an agent, some current injection, every time he does the thing that you want him to do. You give him pleasure for that.

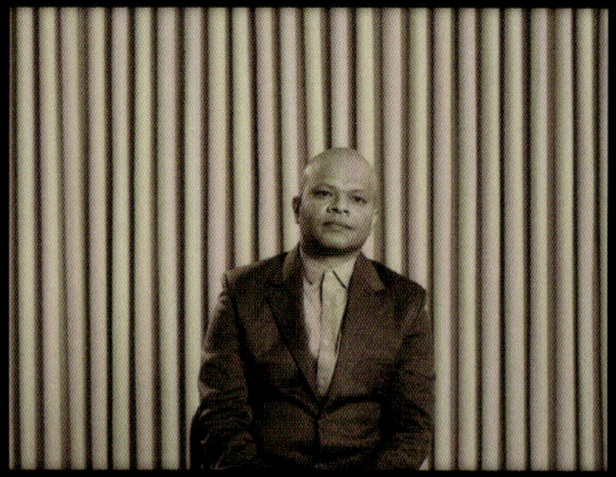

SR – It's a very finely tuned circuitry, and it can go through these different states, much like water at 99.9 degrees, it's still water. And it becomes steam at 100.1. Or for that matter, just above zero, it's still water, and a little bit below, the state changes. It becomes something else.

AG – This example of a soldier that basically every time he wakes up, he sees the enemy as some monsters, and each time he shoots these people, he gets a reward delivered to the brain. So actually for him to feel good, he has to wake up and kill a bunch of people.

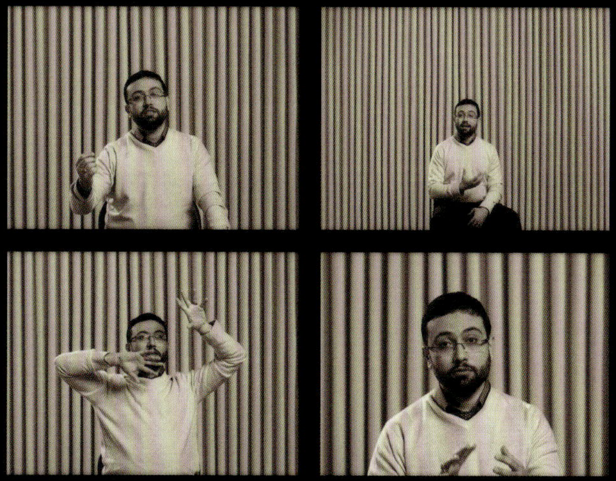

00:09:32:08

Bianca Jones Marlin – One arm of our work looks at how a virgin mouse, some mouse that has never given birth to pups, how her brain can change based on a hormone that is released in the brain and the body. It is called oxytocin. It has been known as the love hormone. What our research showed was that a virgin would then become pregnant and become a mother, and that changes parts of her brain, because what happens as she gives birth, oxytocin floods the body, because it is important for uterine contractions and milk let-down when the pups are feeding. That changes the hearing centers of her brain and she responds differently to the sound of a baby crying.

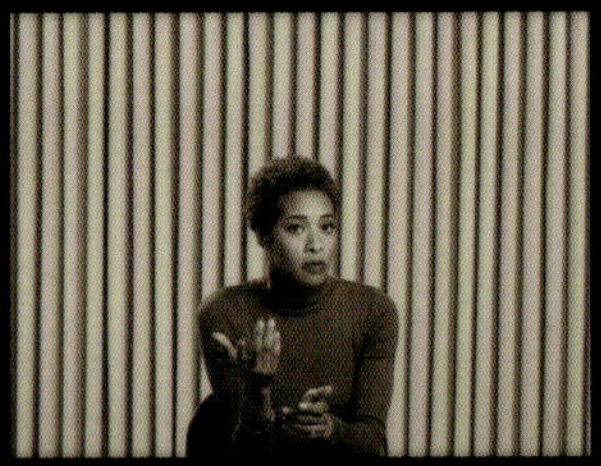

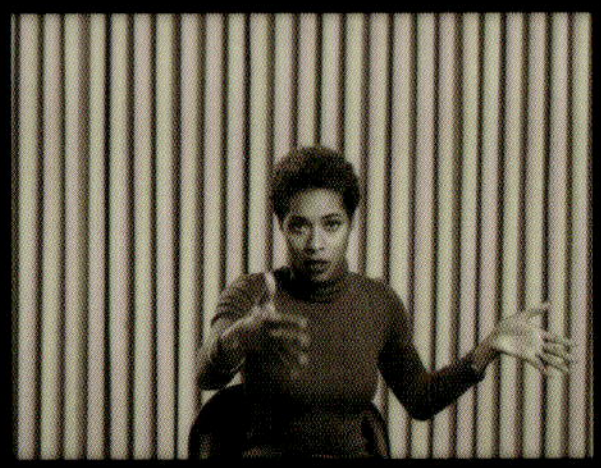

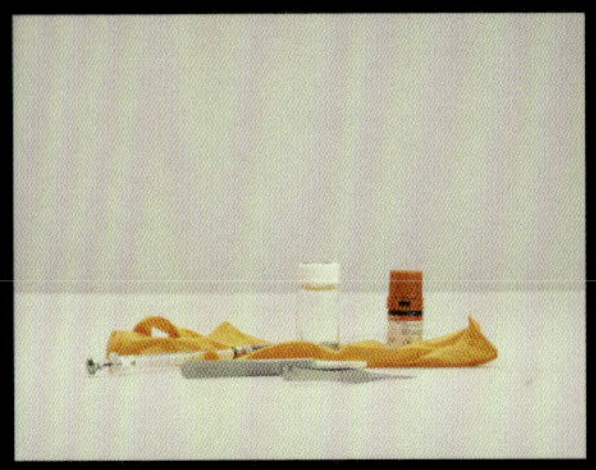

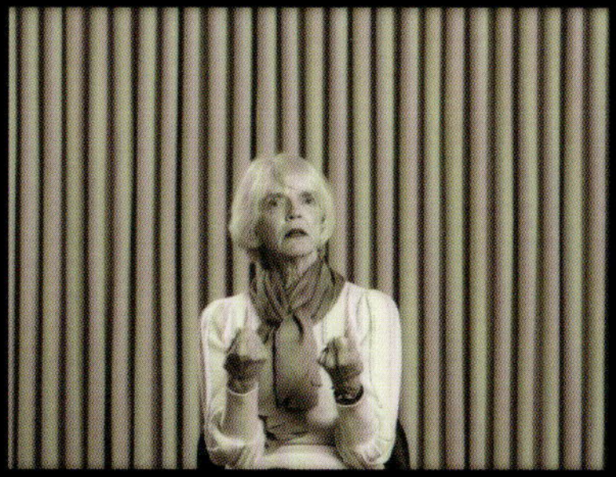

Patricia Churchland – The mother defends the baby, the mother stays with the baby even when the baby is a great bloody nuisance. The mother takes risks for the baby, she sacrifices her own interests for the baby, and we know human mothers and human fathers do that all the time. The idea really is that when you see what the neurochemistry does in terms of what we call attachment, but what really translates into this kind of caring and cooperating behavior, it is hard not to see that as the platform for morality.

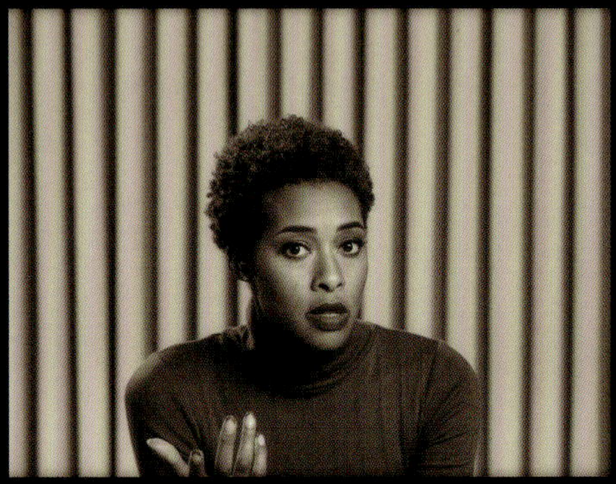

BJM – Biology has it set up so that the addiction is your offspring. And when that doesn't work, and you are not willing to put your life on the line for your offspring, we want to look at these imbalances. When we take this mechanism into our own hands, it potentially could be problematic. Would it be lovely for everyone to take a sniff of oxytocin up their nose and then fall in love with humanity? Yes. The odds of that happening? I can't speak to the odds of that happening. I haven't done that study.

Carl Hart – I guess if I had a dream experiment, it would involve sex and drugs. I mean, that's what humans are here to do, really: sex, drugs, and rock and roll. But it would involve sexual partners and a drug like MDMA and studying partners or dyads, and under different conditions, under different doses of MDMA. I want to know about the interaction with that drug and the relationship. I think that those are the important questions because they can teach us some things about how do we enhance the benefits and the pleasures of our relationships. We in our field don't even touch those questions.

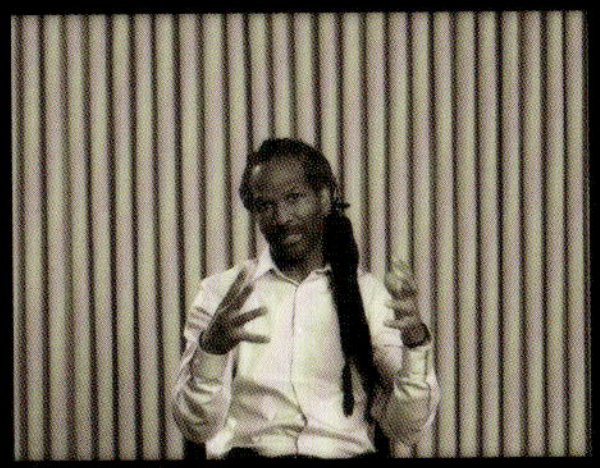

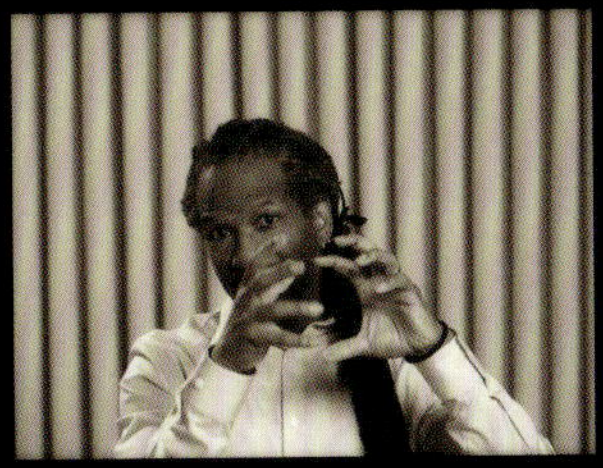

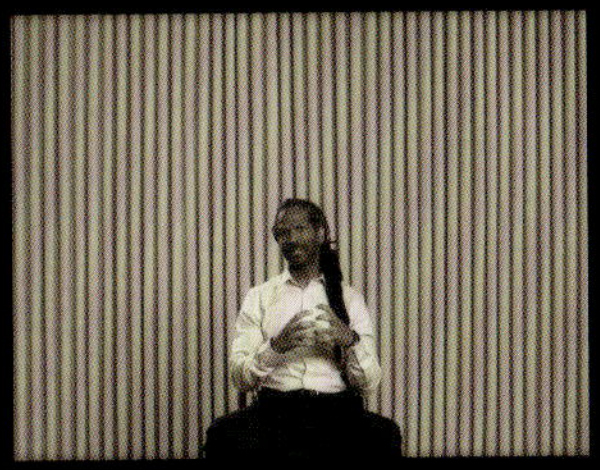

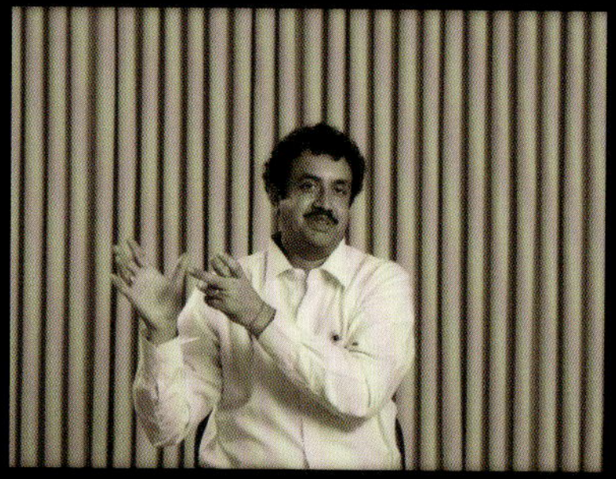

AB – Current science is all about: you take an object, you hit it hard with some energy or some particle. It will scream, it will cry, and looking at those screams, you will try to fit it. If I hit it this way, it will scream like that. If I hit it that way, it will scream like that, and then you pit it with the input and output and you create a scientific theory. This was kind of a principle, but there could be another way of doing. It could be a conversation. That is, I give a very little pump and then after a long time, when it feels like, it will give some response.

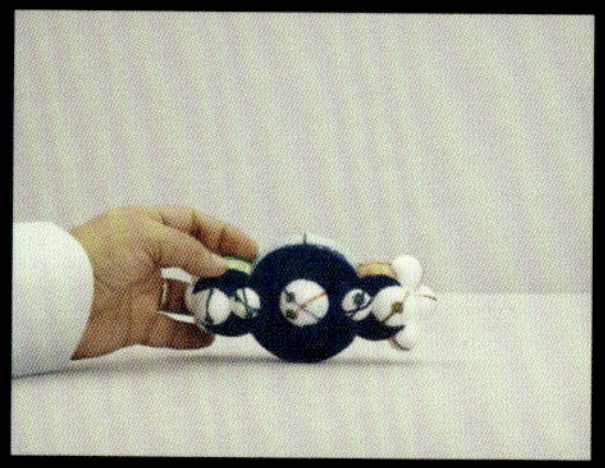

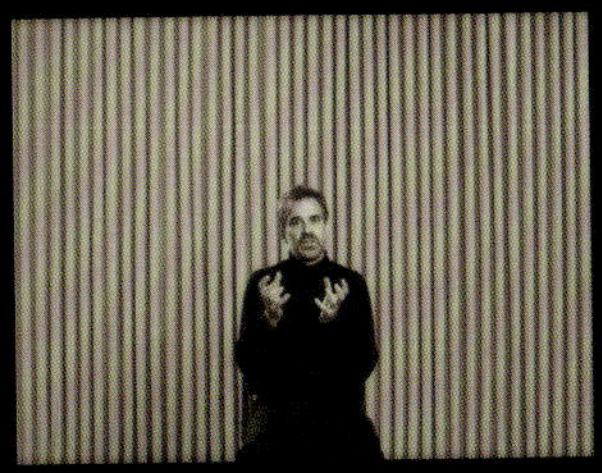

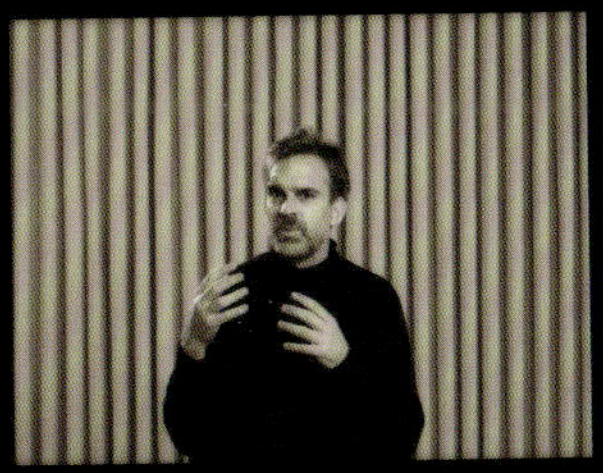

David Erritzoe – We can see, identify, and distinguish with MRI, while people are just resting in the scanner, we can see how the brain activity arranged itself into different functional networks. And that some of that distinction, separation, between those networks, when a psychedelic comes on and affects the brain, some of that separation between these networks is temporarily broken down.

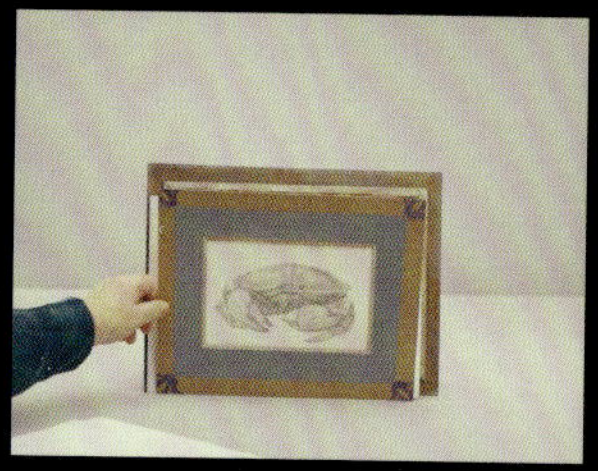

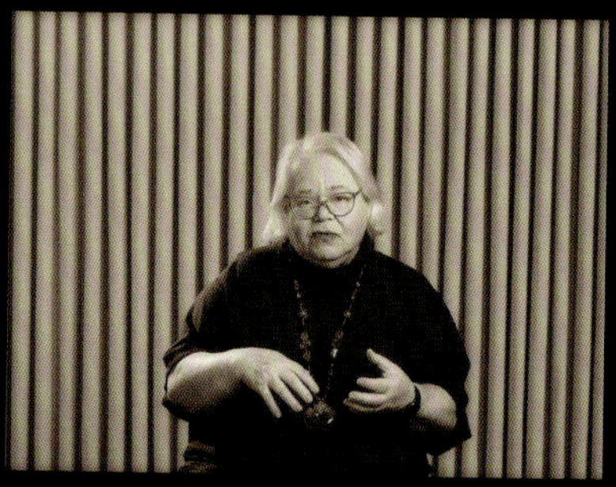

EM – I am not sure how much one would learn by putting LSD or psilocybin on a lobster. I mean, all you could do is measure whatever the hell you measure. They will do something. LSD acts on serotonin receptors. Serotonin receptors are important in the lobster. So you will see something. So what?

DE – There is something that is unpredictable in it, which is all part of the beauty, that a lot of stuff can come up. A lot of stuff related to your previous life experiences, relationships, things you have not been able to address, things that you have avoided, suddenly come up in a new form. It might take form of a monster. It might be really quite tricky. But the part of what we are preparing people for, and that can be psychologically managed and handled and supported, is this: then stay with it. Be curious, it's safe. You will not die. You *might* die. You might die within the experience, but you will not actually die outside of it. So let yourself die if that is what is happening, or go through the mouth of that monster. Let it eat you.

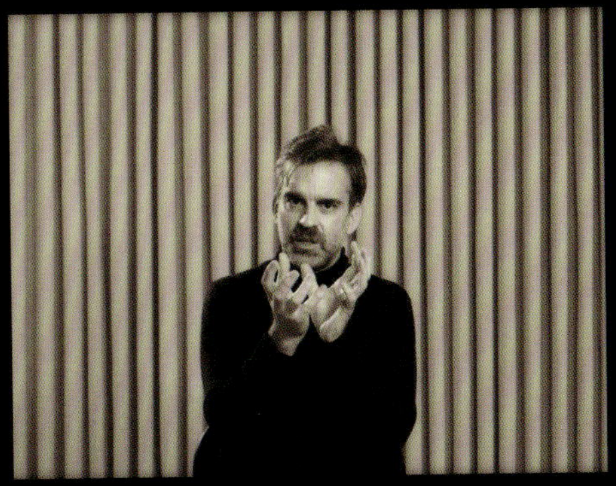

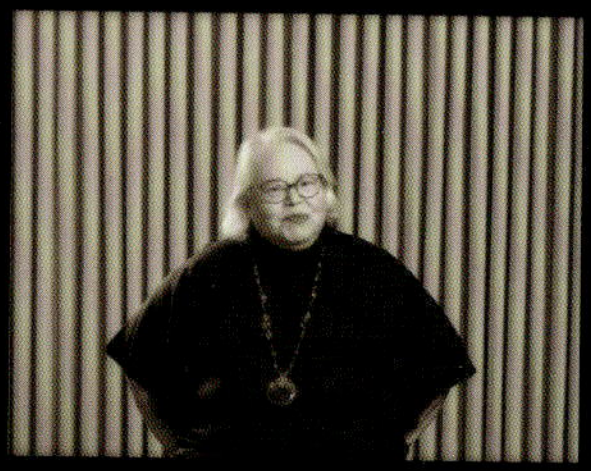

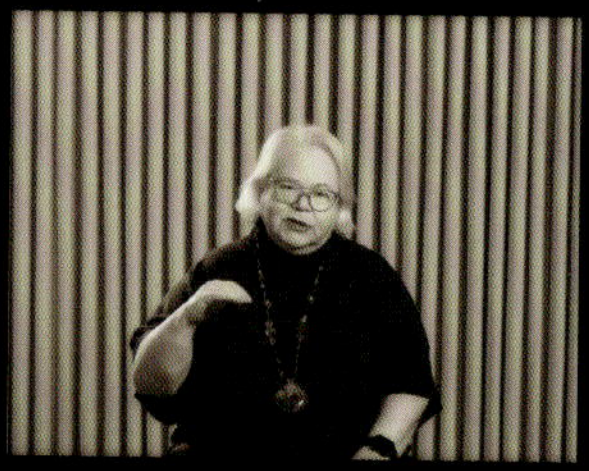

EM – I couldn't get my head around why they thought this would be consciousness-raising when all they were doing was mucking about with one of the major transmitter pathways in the brain. I am saying, raising your consciousness by blocking pathways in the brain—I mean, what is wrong with this picture? You are just producing a perturbed processing system. I didn't understand how they could think of it as getting closer to the secrets of the universe.

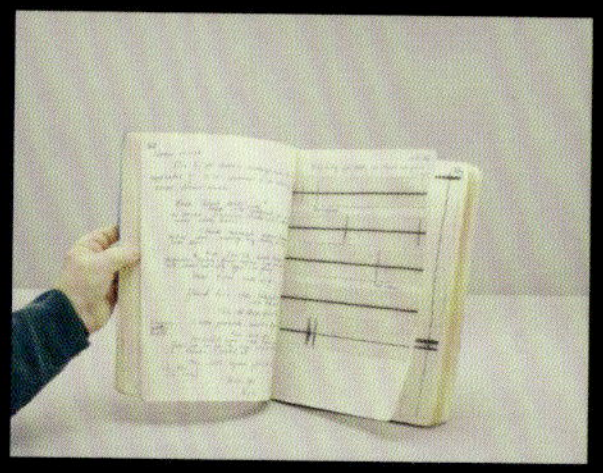

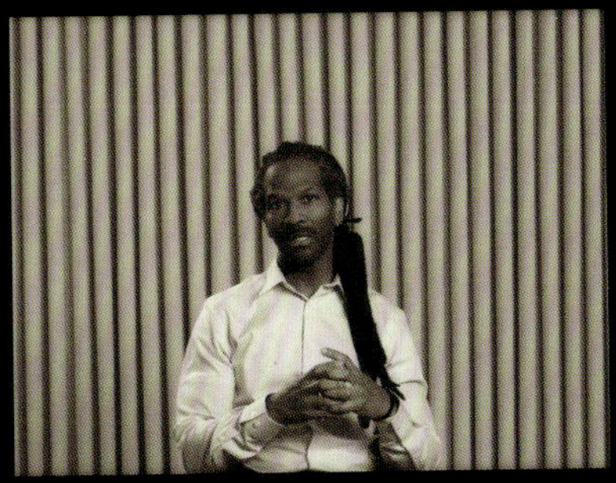

CH – When we think about something like God, however you conceive God to be, when people take certain drugs they are seeking God. Even though we can't measure God, but that's an important reason that this person may be taking the substance. We have to figure out how to measure these types of things.

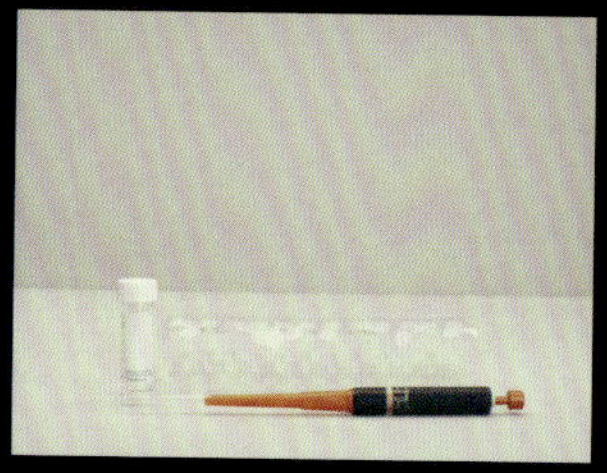

Angela Vincent – I mean, the existence of the universe—it is extraordinary what we know about it, but we still don't know anything about why it was there in the first place. And that always, to my mind, opens up the possibility of there being God. I know that that is not an old man with a beard, but that concept of something greater than that which we can understand. And to explain how we are on this planet, anyway. Also, the brain is a wonderful example of where man has evolved, apparently into the stage where our brain functions seem to me far more than needed to survive.

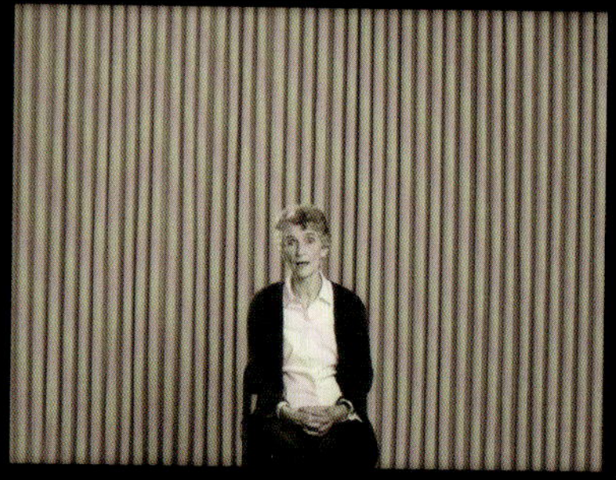

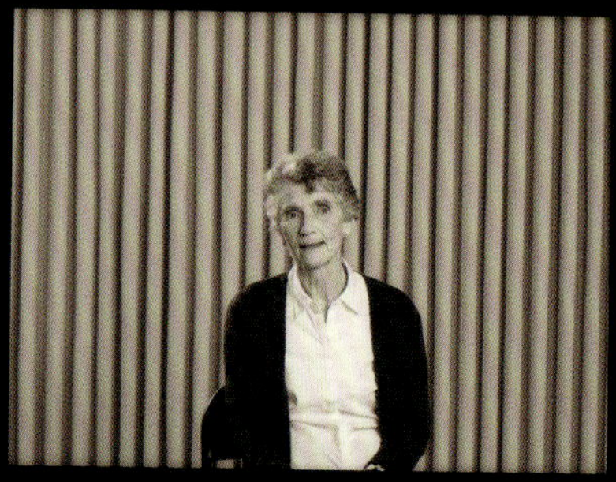

Stanislas Dehaene – Now, some people say, "Well, mathematics is out there in the universe." I don't think so. I think it's a projection of the mind and the trajectories of planets are not actually ellipses. They are close to that. It's a model. It's something that we project from our brains to the reality of physics, but it's an invention. It's not really out there. The universe is not really made of mathematics. This is the beautiful creation of our brains.

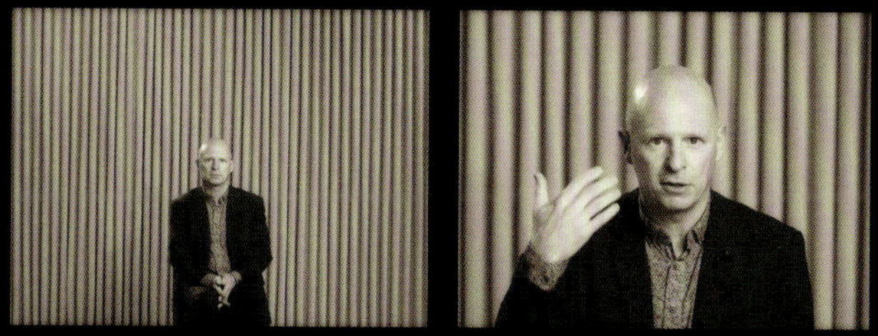

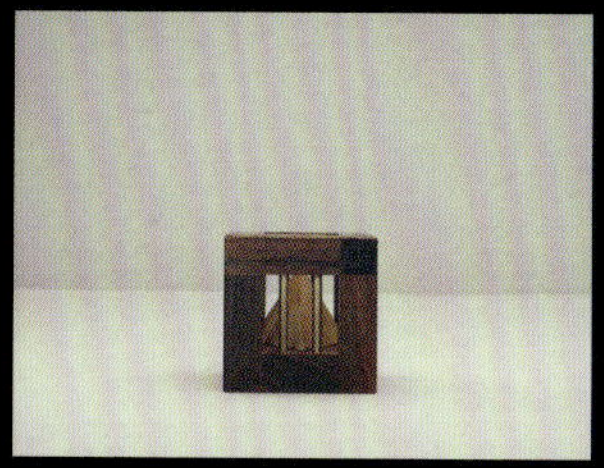

00:17:22:22

DD – Every religious person is atheist about most of the gods. They don't believe in Zoroaster or Neptune or Zeus or the Rain God or Thor. Some of us are just atheist about all the gods.

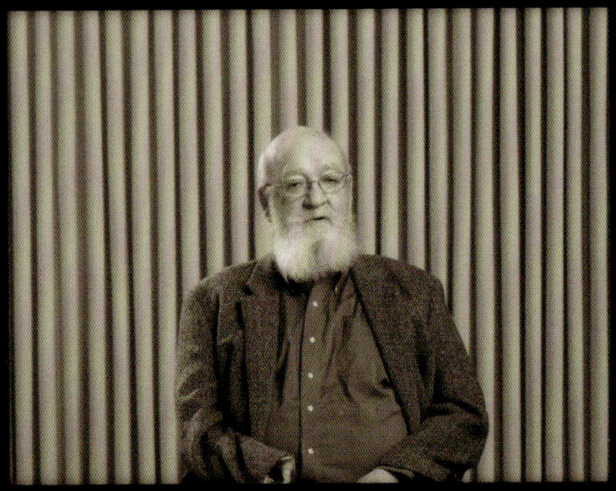

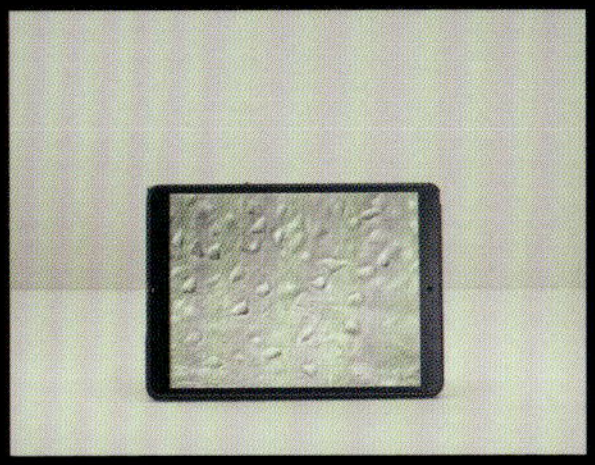

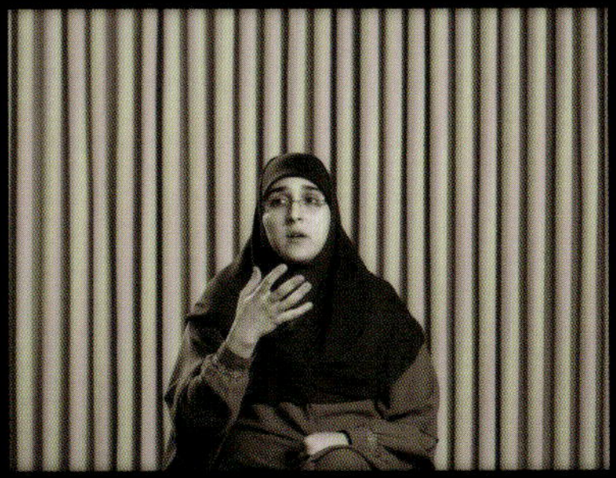

YA – God is the origin of all of the things.
He could not be created by something,
or—it's not an energy. No, he is immaterial,
God is immaterial. Allah is immaterial.
He could not be … Allah could not be created
by something.

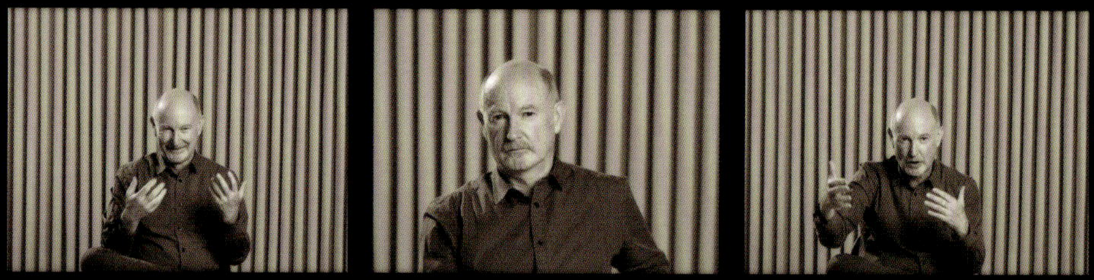

Rafael Yuste – The reason why nature builds nervous systems and brains is to build thoughts, to build states in our brains that exist independent of the world but can be used to represent the world. So you can have a thought of a car in your head. It's not a car, but it represents the car. And we think that all brains, all the way to cnidarians, are doing that. They are building little internal states that can be turned on by the outside world.

WS – Think about a Stone Age family sitting in front of their cave eating. And they share food, and while they eat together, it becomes very obvious that there are certain persons that are greedy while others are generous. There are certain persons who eat very diligently and others do it another way. So since everybody watches this, they become aware that there is something existing that you cannot touch, that is not material, but that you can experience. And you can share this experience that there is a greedy guy over there.

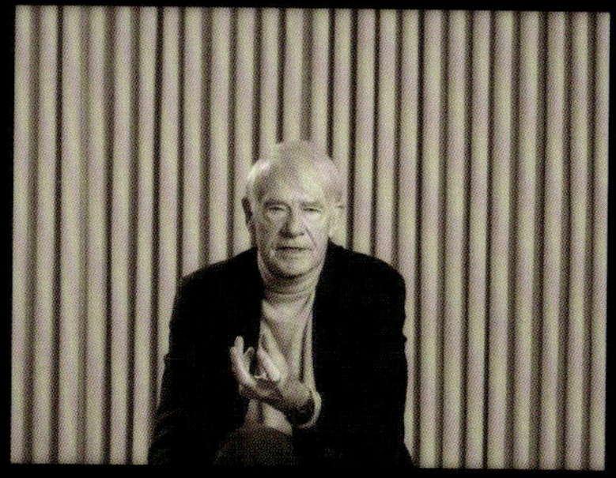

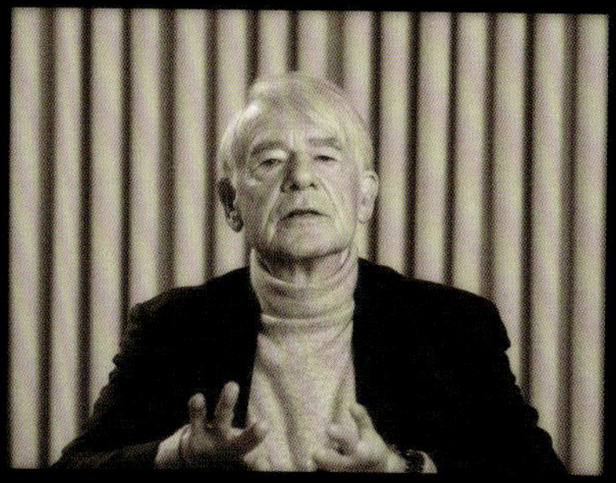

00:19:21:09

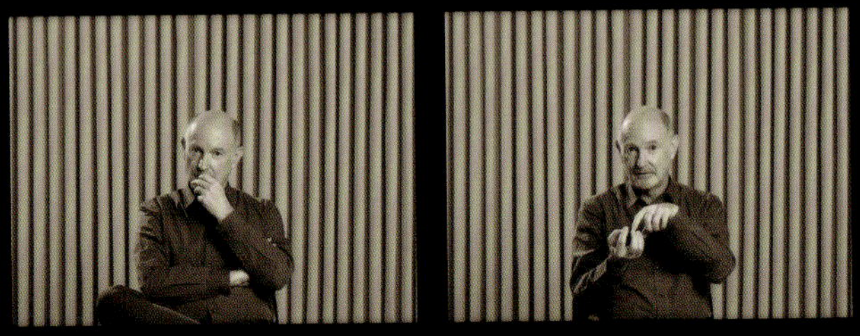

RY – By doing that, you can manipulate the world mentally. So instead of manipulating it with your hands, no, you can do it all in your head. And by doing that, you can predict, you can figure out what is going to happen. And I think all animals, we are all in the game of predicting the future, because if we can predict the future, guess what? We can survive, or we can reproduce better. That's the game of evolution.

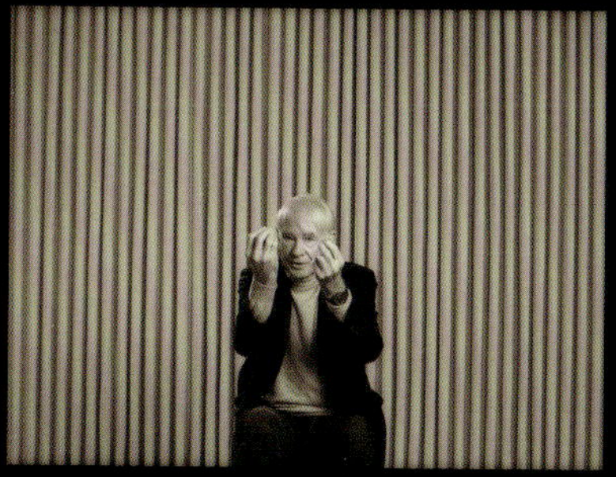

WS – So you can create immaterial realities by shared experience, naming them, exchanging them, refining them. And then you can start to produce cultural evolution all the way up to belief systems that are mighty enough to erect cathedrals.

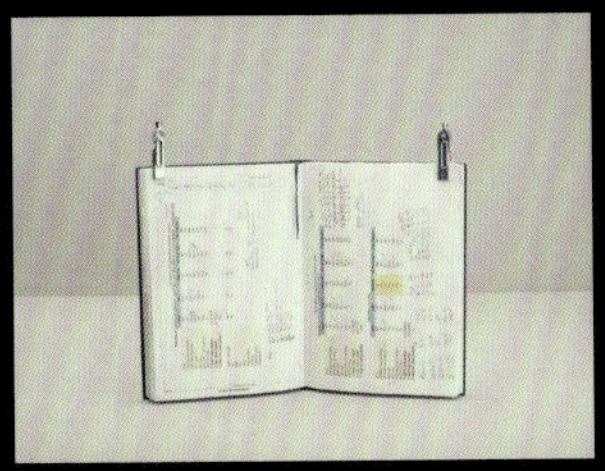

MM – It is only when we understand how
beliefs in certain deities, or cultures,
shape our decisions that we will understand
how different individuals behave, develop
diseases, or even, who knows,
develop resistance to different diseases.

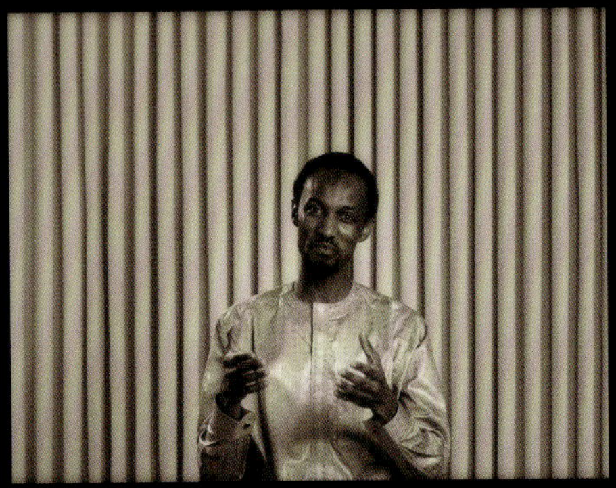

DD – I think the clergy who are marrying people, burying people, christening babies, consoling people, advising people who want to hear their faith confirmed—that's a terrible position to be in, and my heart goes out to the clergy who have been caught in that position, because they no longer believe what they feel they have to say to these people. But at least the harms they are doing are really minimal. They are hurting themselves more than anybody else, and they are allowing little old ladies and men to go to their deaths without the terrible trauma of losing their religion.

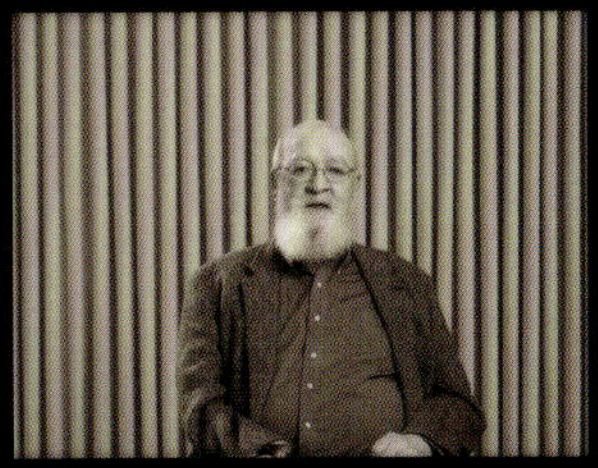

00:21:30:04

MM — As scientists, you do not believe that there is, for example, either the soul or there is afterlife, or maybe there is the possibility that when people die they go to some places. Generally, even by limiting that, we are limiting science, if you see what I mean. Because science is about discoveries; it's about understanding what is unknown.

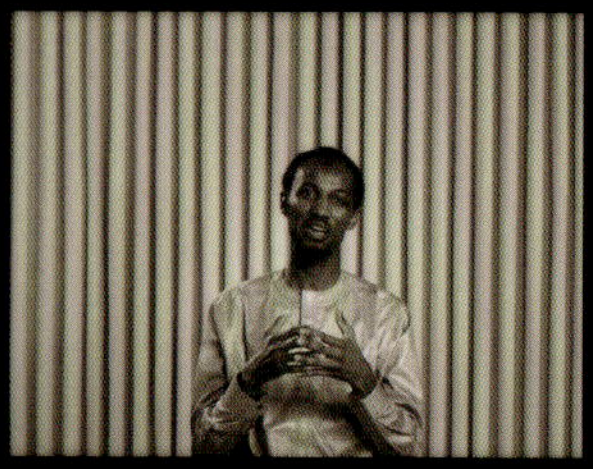
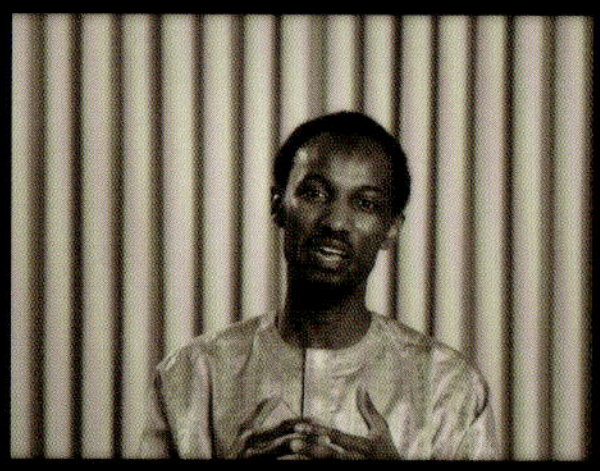
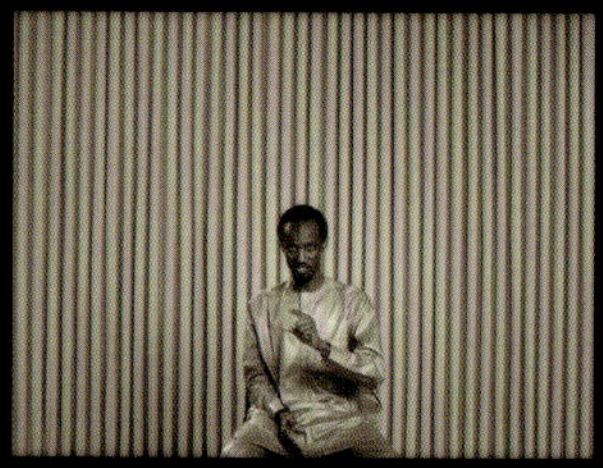

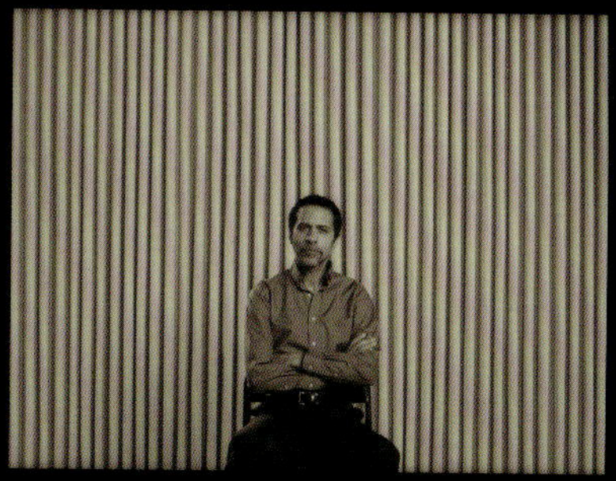

Erich Jarvis – Over a hundred years ago, the patriarchs of comparative neurobiology, comparing species, took Darwin's view of evolution and merged it with Aristotle's view of *scala naturae*–and that is, living forms went from lower to higher forms–and made up this hypothesis, this theory, that the brain evolved with the intention of creating humans.

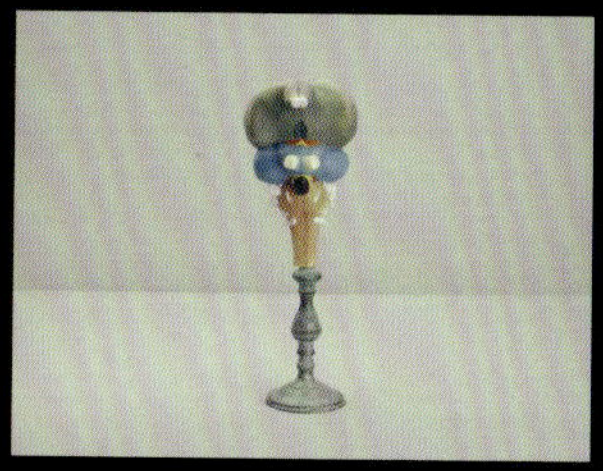

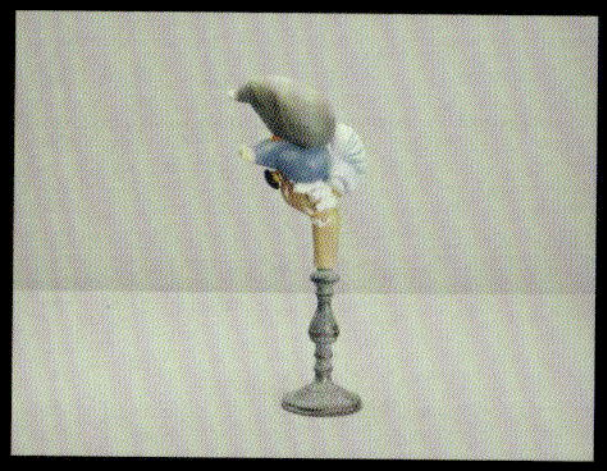

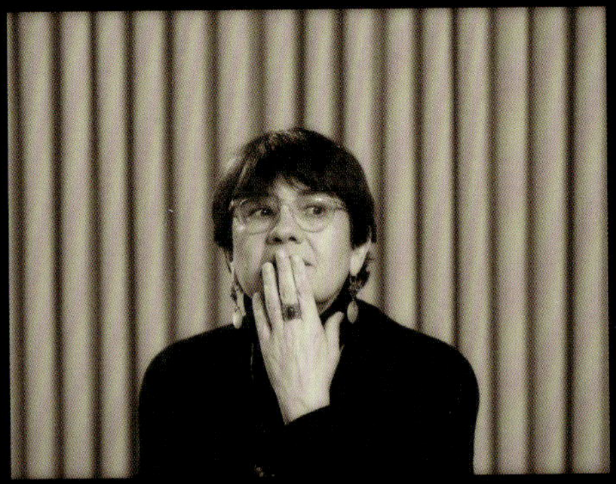

CM – The dominant task of neurology for a long time has been to find ways to prove the superiority of the white race in terms of perfection of the brain, in terms of, yes, intelligence. It was to affirm the superiority of certain brains over others.

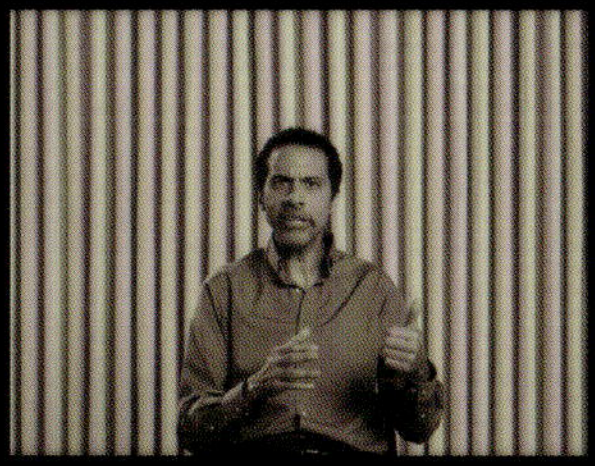

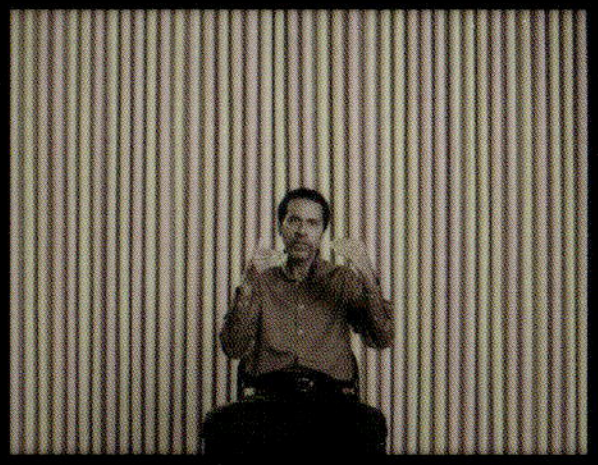

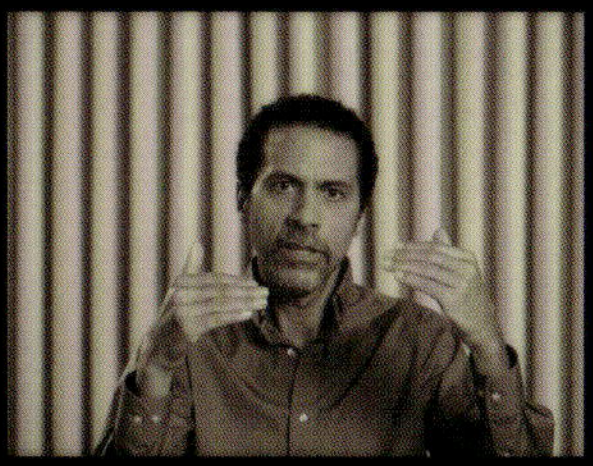

EJ – And only mammals were thought to have a neocortex. And amongst the mammals, European men were argued to have the biggest neocortex, Asian men or Asians in general were argued to have the next biggest. And Africans were argued to have the smallest, followed by apes and monkeys. And this is the logic and the rationale or some of the rationale to call it "neocortex." A term we still use today is based upon racial theory.

György Buzsáki – I grew up behind the Iron Curtain, in Hungary, and by the time I got to medical school I became interested in the brain. I was, like everybody else, interested in my own brain. I wanted to learn about my own uncertainties. And the framework that was given to us in the East was mostly Pavlovian teaching. [...] It assumed that the brain is an absorption device that associates various things. And that's it. You don't need any opinion from the body or from the brain. There is no action that is needed, because the brain will just learn whatever it is given to it. And that, of course, is a fantastic weapon for Communist propaganda, or any propaganda in any society.

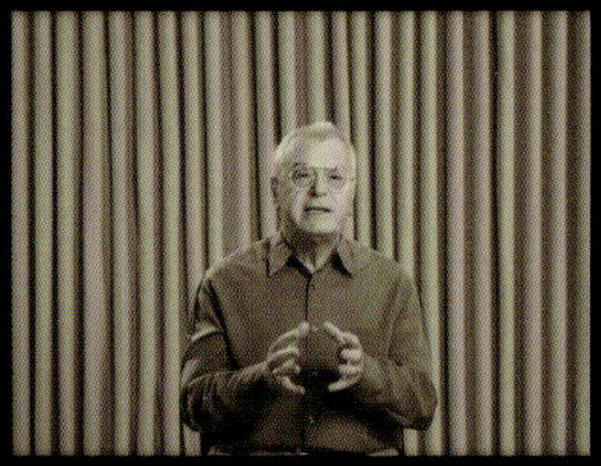

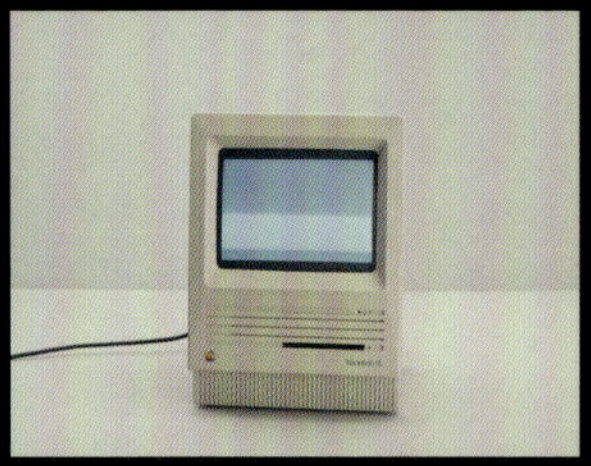

André Fenton – For two hundred years, America thought of people as animals and property. That concept has affected everyone. It's in every institution and it's in everyone's mind.

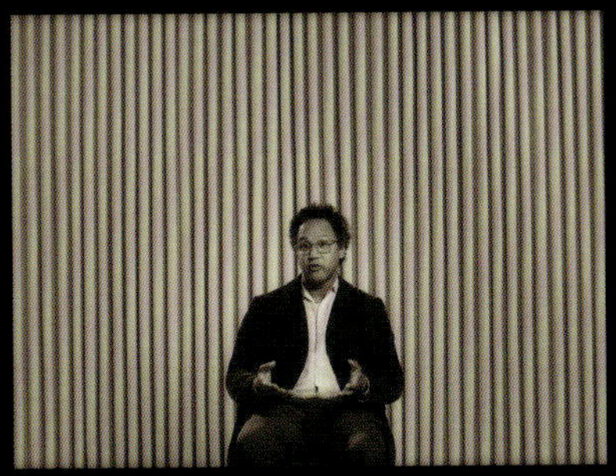

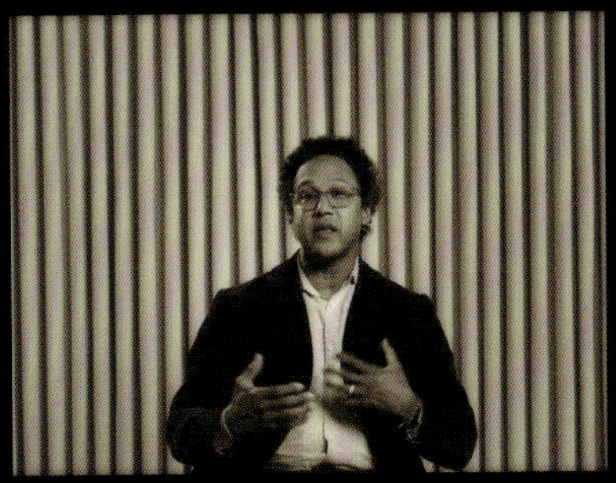

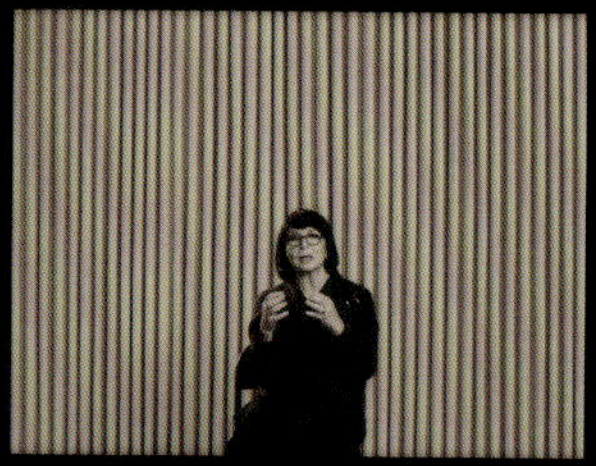

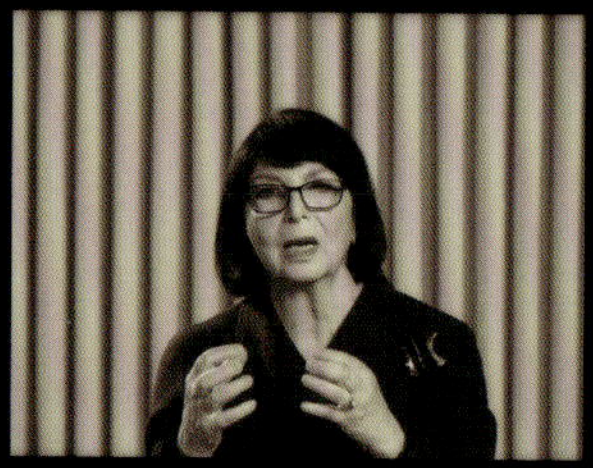

HA – Beliefs are part of how we reconstruct reality. The reality we see, we perceive, the history, what we are anticipating about the past. We need to tell ourselves a story that makes sense of the world. And that story has to have content and emotion. And that is what we wind up believing about ourselves and the world.

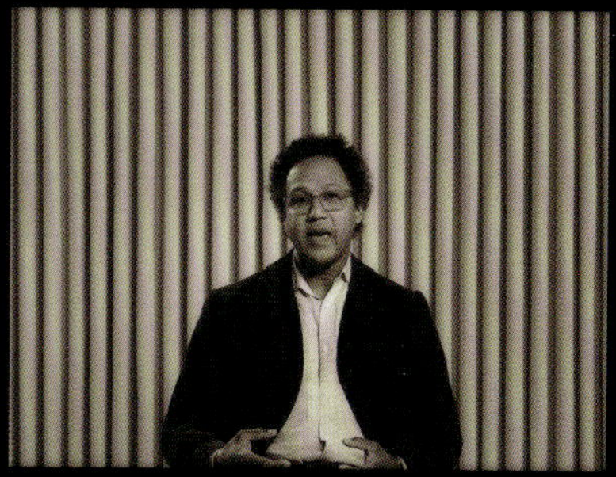

AF – The brain is a complex system, and complex systems are indifferent to change, but when they do change, they change radically. […] They tend to be stable until the right kinds of perturbations happen, the right kinds of turbulences happen. And only then can they change. What is interesting about those periods of indifference is that latent ideas, latent states, are being developed. Possibilities are being developed. They don't manifest, but they are cooking, in a way. And then the turbulence happens and a radical transformation happens and it's not to something new, it's usually to one of those latent states. So I don't know how America's going to change, and I don't know when it's going to change, but it will change. And that change is going to come radically. And after a lot of turmoil.

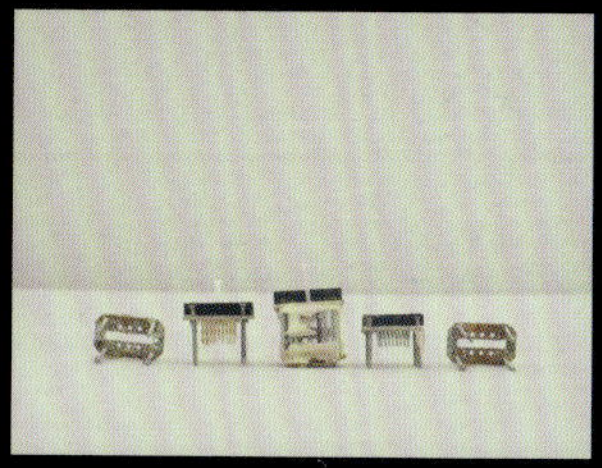

00:26:29:13

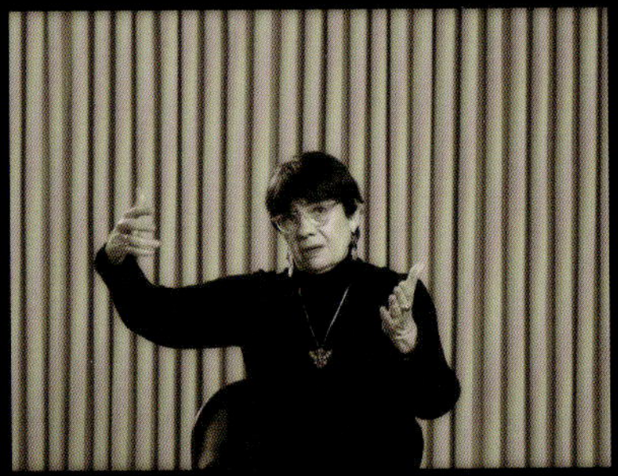

CM – The brain is an anarchist, because
it functions without any centralized power.
No region in the brain has a privilege;
there is no superior function in the brain.

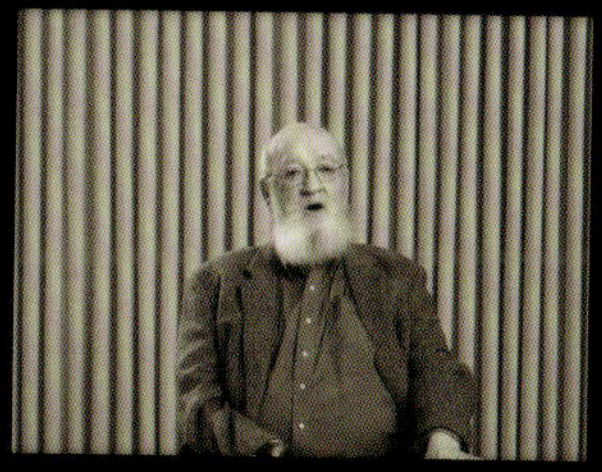

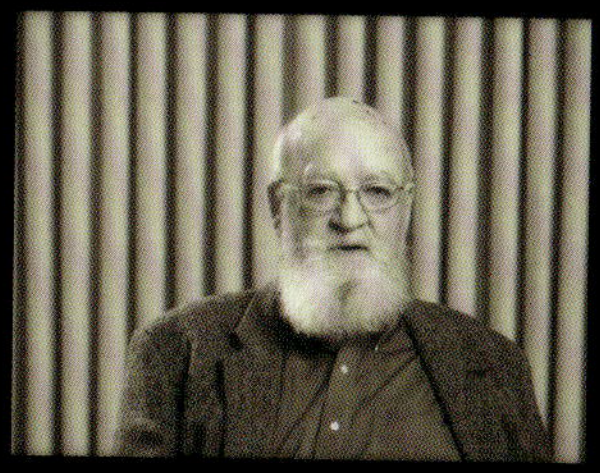

DD – The motor proteins that troop around inside your neurons and your other cells, they don't have free will. That's true. They are little robots with very few degrees of freedom. But that doesn't mean that you, made up of those neurons, are not free.

CM – It allows us to interrogate political anarchism as different from what people usually think about it, that it is something naïve or unrealizable or impossible.
On the contrary, if it is possible in the brain, it must also be possible in political reality.

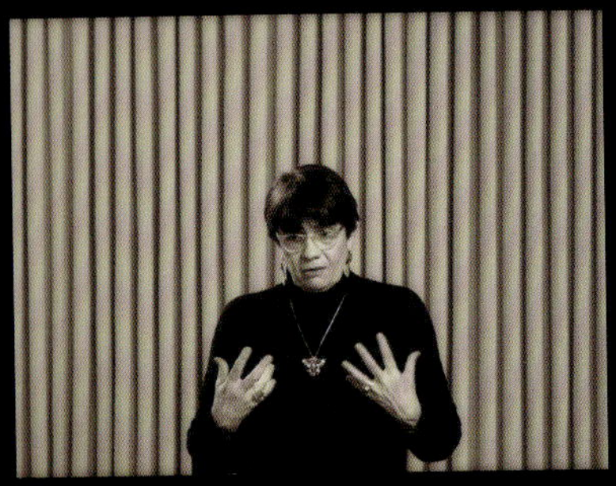

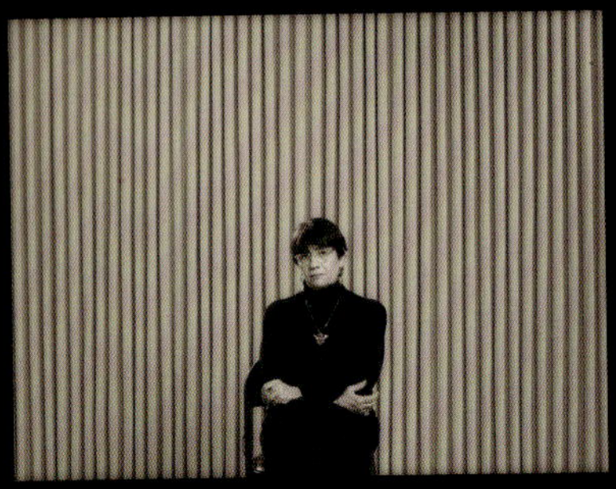

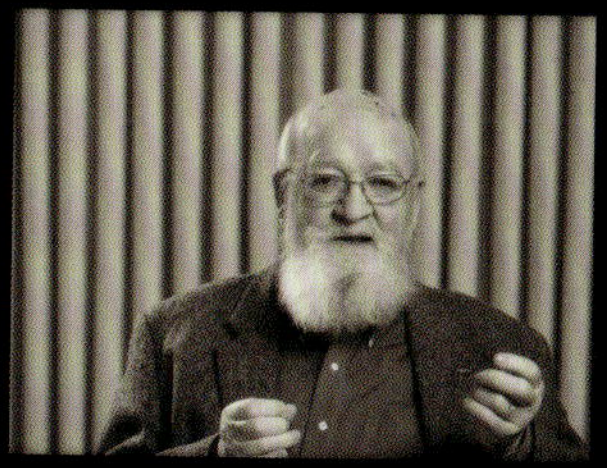

DD – You can be freer of your parts … in the same way that a flower can be red, even though none of its finest parts are red. Atoms are not red. Molecules are not red. Big, big molecules maybe are red, but not little ones. Just as something can be colored without being made of colored parts. In fact, nothing is made of colored atoms. There aren't any colored atoms. And yet, color is a real phenomenon. So freedom is a real phenomenon in the human world. To be free is to be able to control yourself.

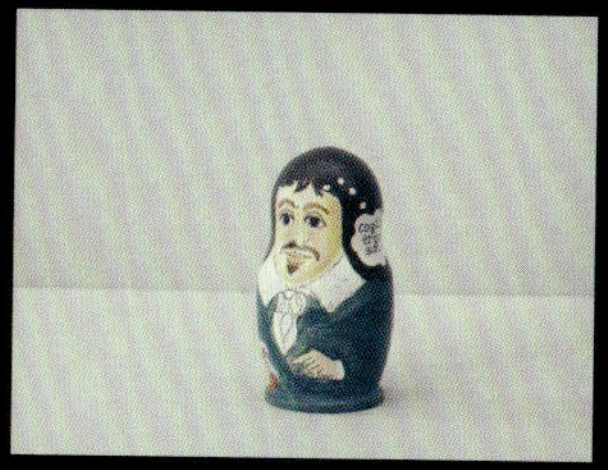

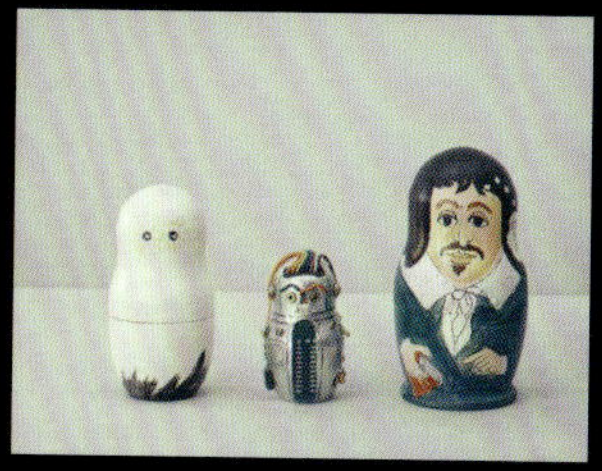

00:28:39:19

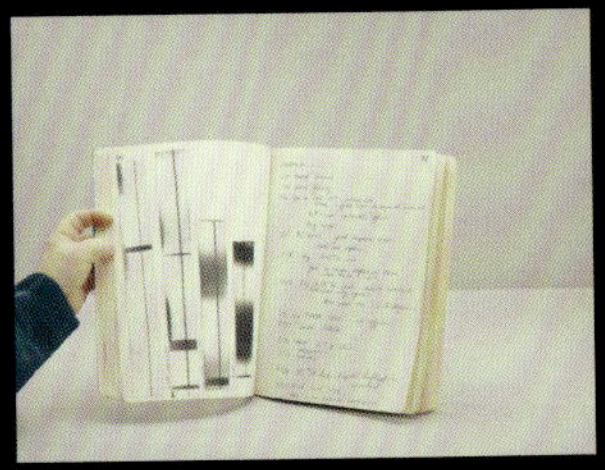

EM – Lobsters, like all animals, will sometimes do very surprising, unpredictable things. If you want to call that free will– in other words, if you want to use the notion of if you had ten animals, and you gave them the same stimulus, and eight of the ten did one thing and two of the ten did something else, if you want to call that evidence of free will, yes, then they have it. If you were to ask me why did individual lobsters go to which particular part of the ocean and look for a burrow, sure. But that depends on how you define free will. I am not sure I have free will.

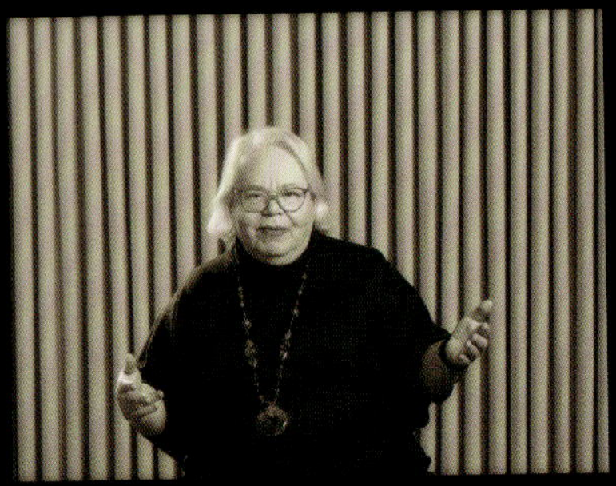

Charles Yang – Imagine a rodent at a fork of a road may decide based on previous experience that going either way is terrible. So I don't go anywhere. But it is difficult to get a machine to do so. The machine will always give you the most likely answer or the best answer, given whatever it believes to be the best answer.

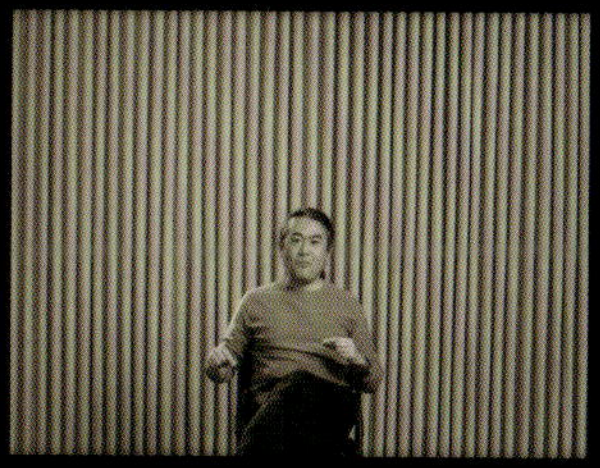

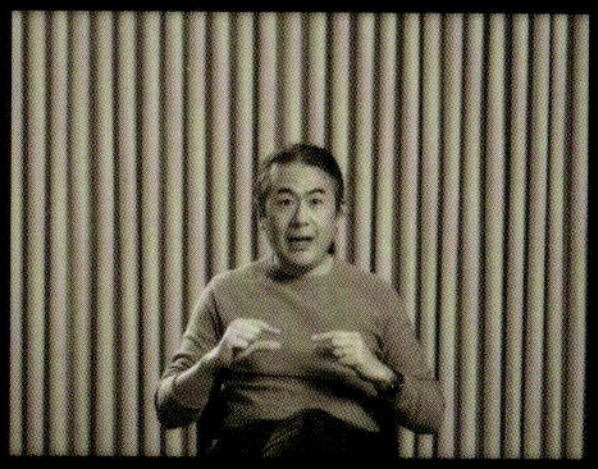

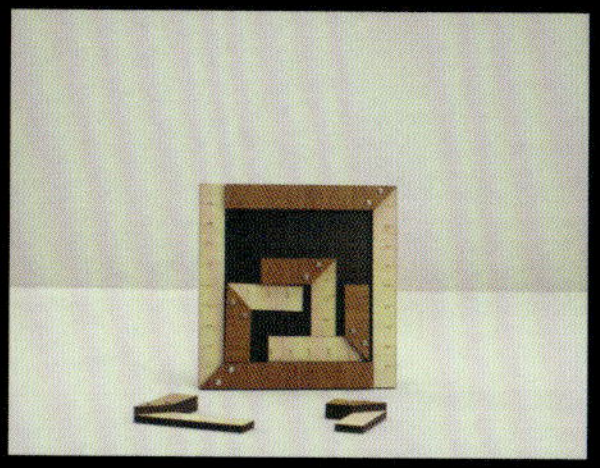

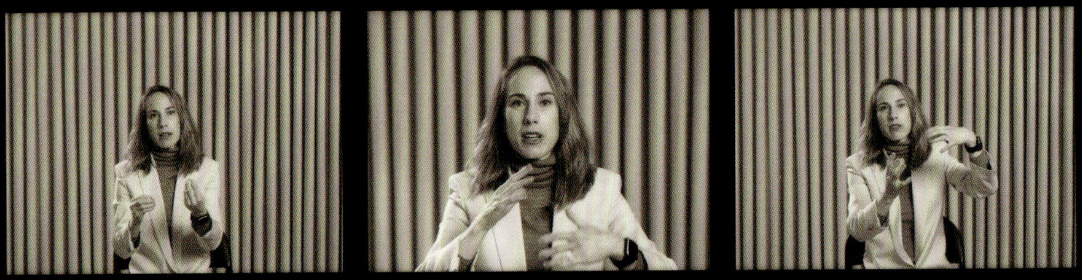

Polina Anikeeva – It is amazingly possible, without any buttons, to make perfectly reasonable people do perfectly unreasonable and horrible things. Will the types of technologies that our group and other groups are developing make it easier for bad actors to rid people of their free will?

CY – We do things when we feel compelled to do so, or we have enough reason to do so. And that may not be a bad thing. Is this really a bad thing? No, it's probably not a bad thing. If you force people to choose, they may not choose wisely.

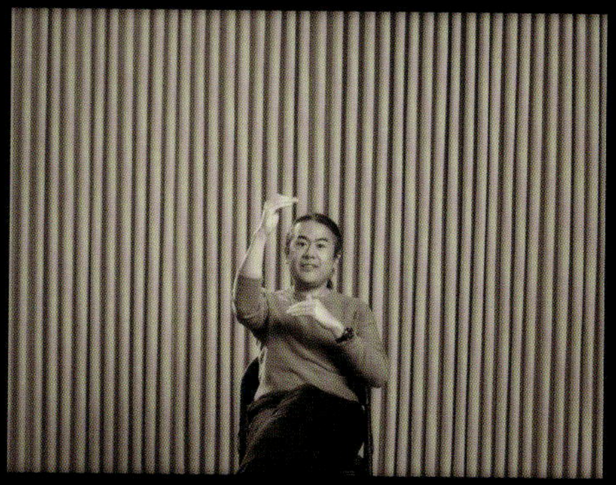

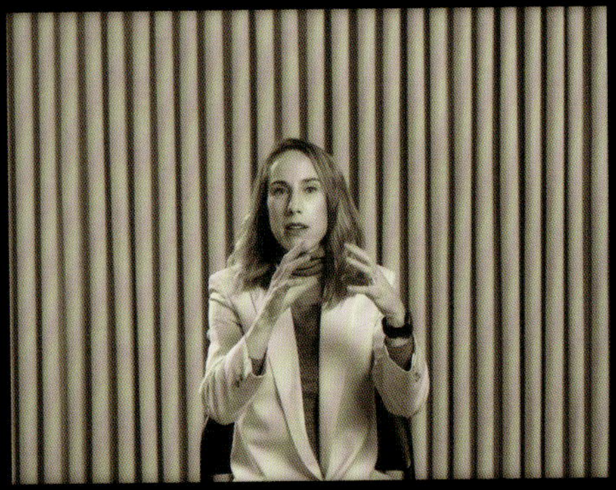

PA – When people implant Parkinson's patients with electrodes, very often they can treat the motor symptoms of a Parkinson's patient, but then they can completely change their personality—from, let's say, a very reasonable individual to somebody who has compulsive behaviors, gambling, just really having difficulty processing decisions. And then when you talk to families, or when physicians talk to families, of course everyone says, "Well, just turn off the stimulator so that the person can be themselves, even though not so capable." But when they talk to the patients themselves, the patients all want to keep the stimulators on, despite obviously turning into someone else.

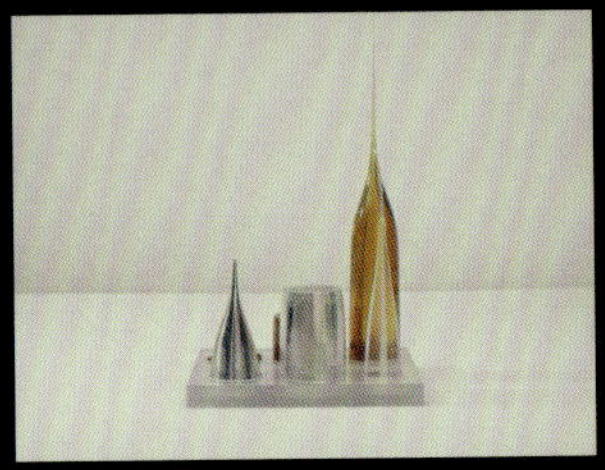

Joseph LeDoux – The idea that your body is producing behaviors that you don't feel you are in charge of is kind of disturbing. And so to resolve that dissonance that creates, what we do as people is narrate our lives. We generate narrations constantly to make our conscious experience, which is a tiny part of the brain, tiny part of the mind, somehow make sense with what is actually happening in the world.

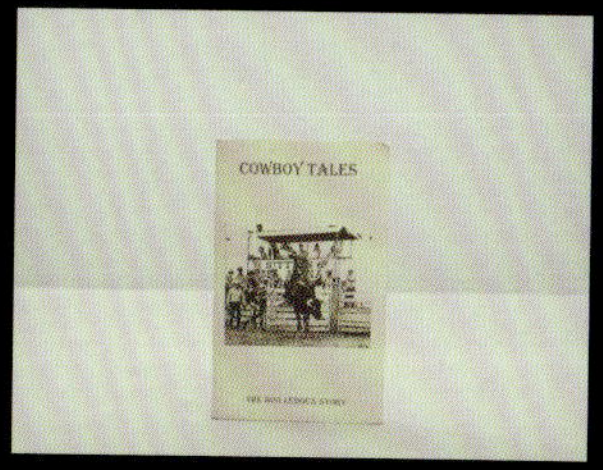

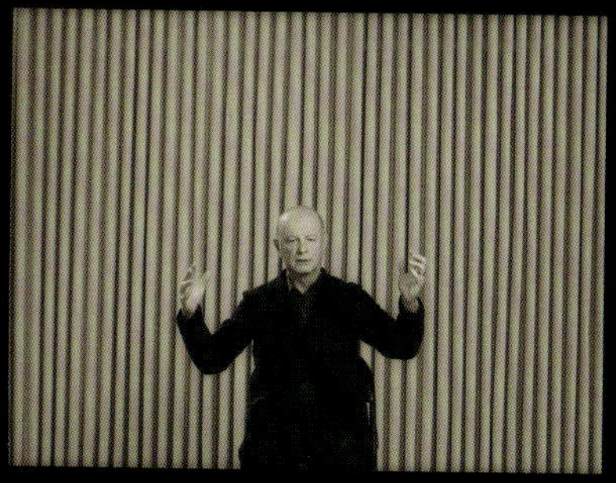

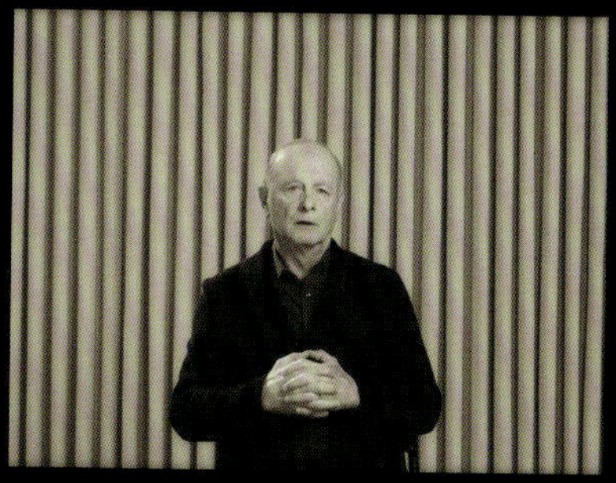

00:32:12:22

Antonio Damasio – Just think of the beauty of this. There you are on the screen, and you are several feet away from me. Here I am. We are having no collusion other than through the physical medium that allows me to see, given the distribution of light in front of me. When I am feeling that I am hungry or thirsty or cold, for example, because it's freezing in this goddamn set, then it is something that I am doing in partnership with my body. There is no way that I can have the feeling of being cold if I do not have a way internally of, quote-unquote, measuring the temperature inside my body, and the temperature around me, and the differential makes me cold. So this partnership, this hybridicity of feeling, which is the core of consciousness, that is something completely different from what we do in the visual medium.

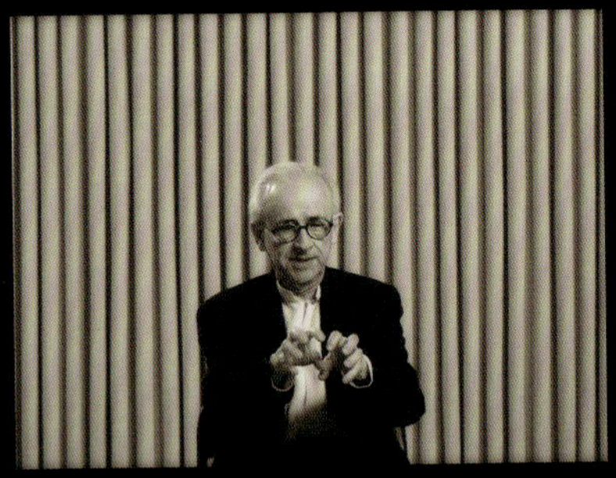

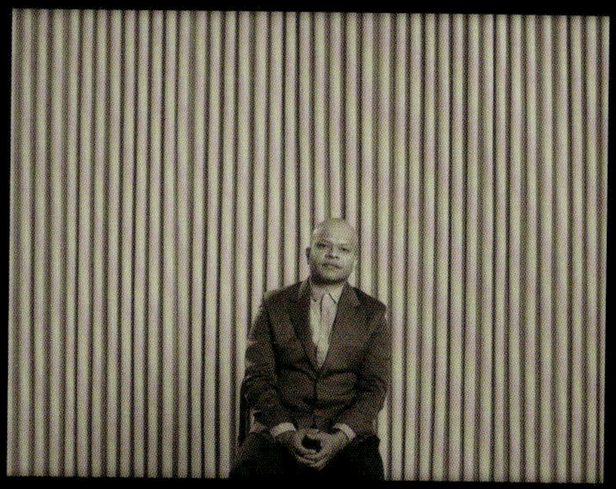

SR – There is this very amazing set of pictures that a painter drew. That person had spatial hemineglect. He had a stroke and he was recovering, and he used to draw himself by looking at the mirror. Initially the drawings were only on just half of the person. And it's quite amazing that a person looking at himself but is only able to draw half side. And slowly the brain sort of heals and that picture becomes full. So I find it very profound that a person looking at himself is still drawing only half, as if the other half has just gone out of existence because you can't attend to it anymore. It's not a visual deficit. It's not a memory deficit. It's a deficit in your ability to pay attention.

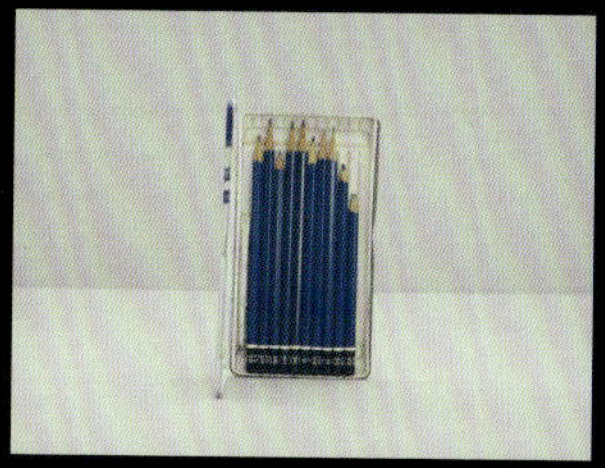

00:33:47:22

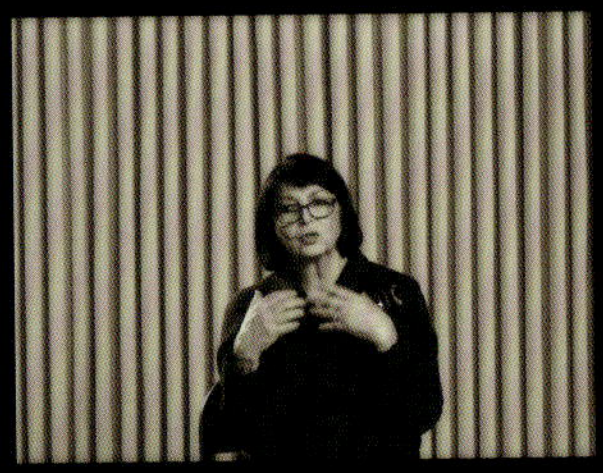

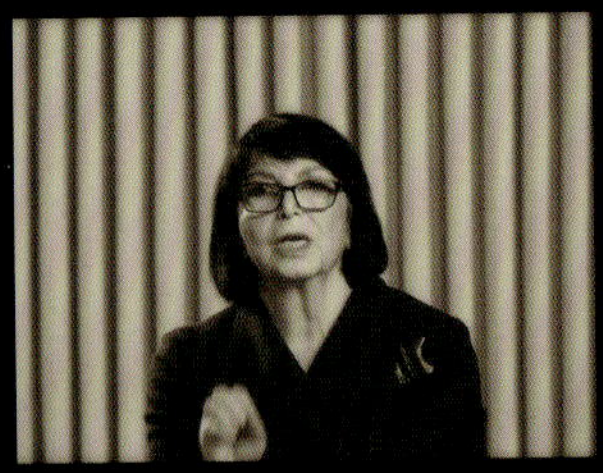

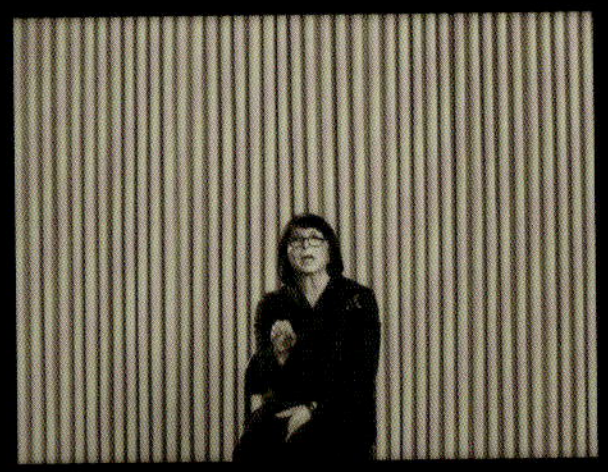

HA – I met a gentleman who had been in solitary confinement for over twenty years, and he told me that very early on his depth perception changed, because he just didn't have enough room to focus beyond this nearby wall. But even more striking, after leaving prison, he had completely lost his ability to orient in space. And that is because space orientation happens in the hippocampus, but the hippocampus is also the brain site that responds to stress, where chronic stress, lack of activity, blocks neurogenesis, blocks the ability of the hippocampus to keep remodeling itself. And in a shrunken hippocampus, your ability to orient and remember space is greatly diminished. And this gentleman was lost all the time, including going home. He was never that way before being in solitary confinement. So these are just examples that show that solitary confinement changes the brain, and that is a physical change. And that, under U.S. law, represents cruel and unusual punishment.

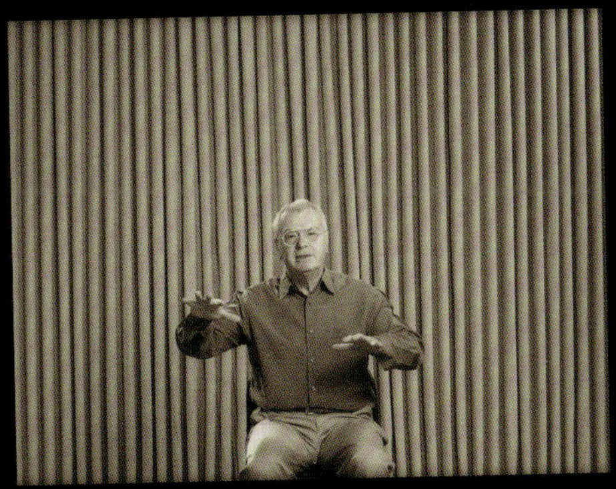

GB – If your body would be placed in a big jar when you were born and you were given all your sensors–quote-unquote, you could see everything, you could watch TV, whatever you want. You can have your mother touching your skin and so on. The only thing that you don't have is the ability to respond to any of these. If you deprive the brain from these action outputs and from the verification that comes with it, the brain will never learn to perceive. It will never learn to think. It will never have any knowledge.

00:36:15:01

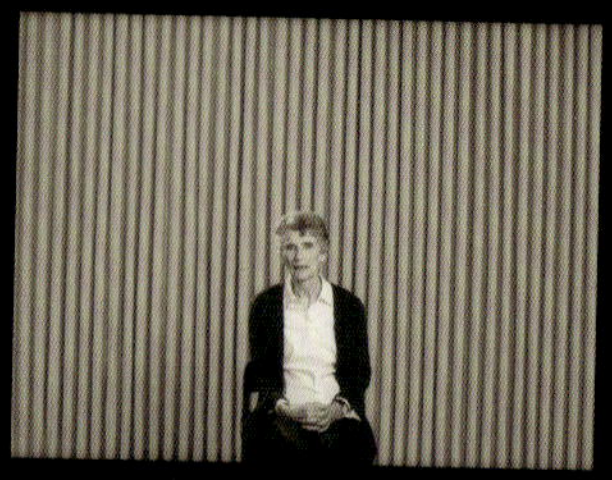

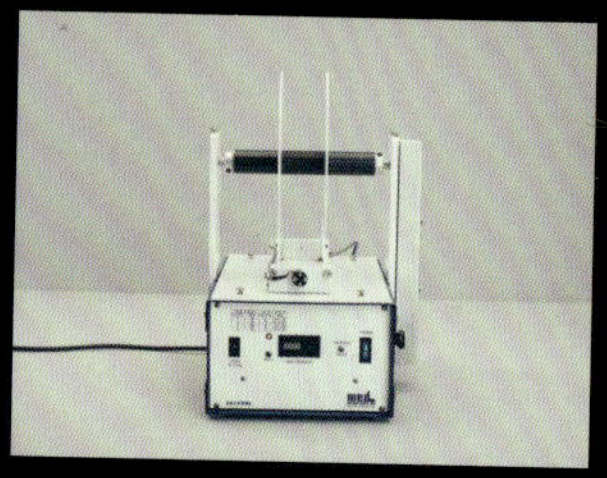

AV – We know that people can survive for years on a machine that looks after everything except their brain waves and that they probably are, to some extent, conscious. I would hate to find myself in one of those situations where you can't communicate but you are actually conscious. I think that must be one of the most unpleasant things. And yet the few people one's heard of who have recovered, actually, they don't seem to have found it so depressing.

CM – Emotions are vital. It means that they not only play a major part within the brain, they also play a major part in survival, because in order to survive, we need to be interested in surviving. We need to be attached, passionately attached, to survival.

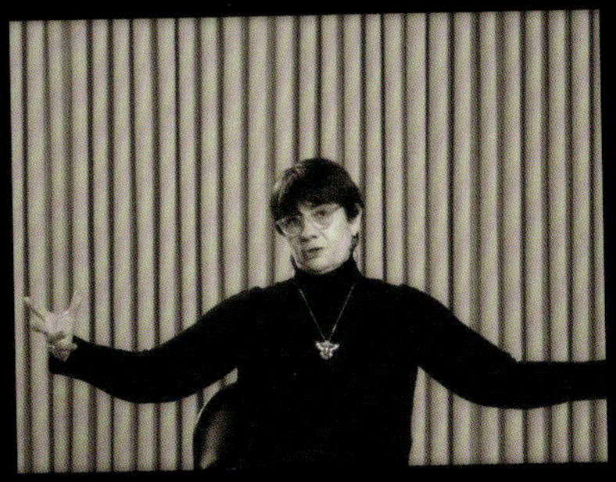

00:37:35:06

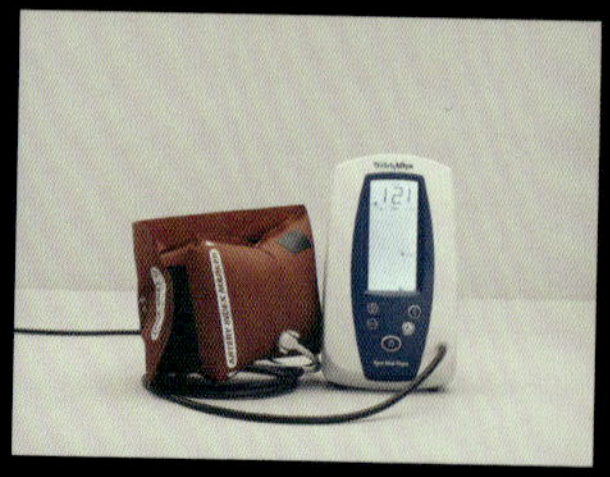

CH – And so I decided to subject myself to opioid withdrawal just to prove to myself that it was unpleasant, but not life-threatening. And so that is what I did. […] I think it is absolutely critical for people to understand the effects of the drugs that they're studying, understanding the effects from a personal sort of perspective.

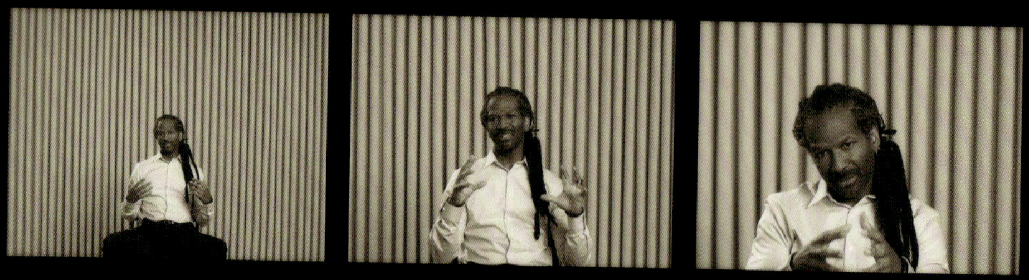

SR – It makes a lot of sense for us to have a sense of self. Otherwise, my thumb and my leg and, you know– I need to feel that it's all me, for me to survive and all my systems to work properly. So evolutionarily, looking at an organism, I think the sense of self, and I need to defend myself, and I need to eat, and it's all just one unified percept of self, rather than just distributed things– it makes a lot of sense from an evolutionary point of view.

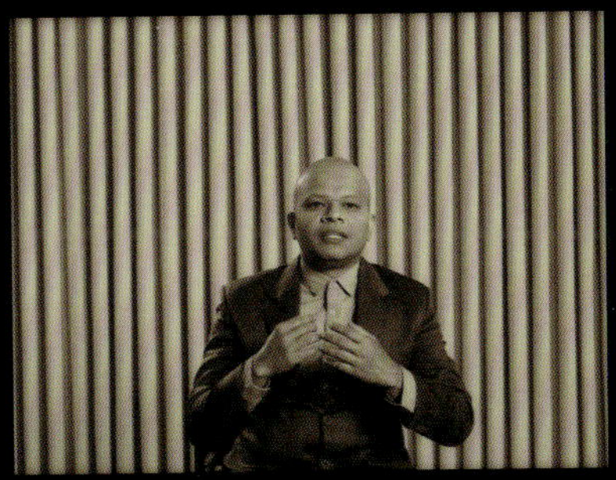

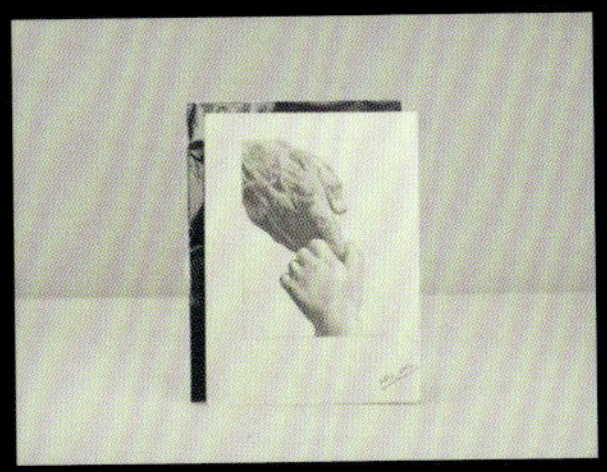

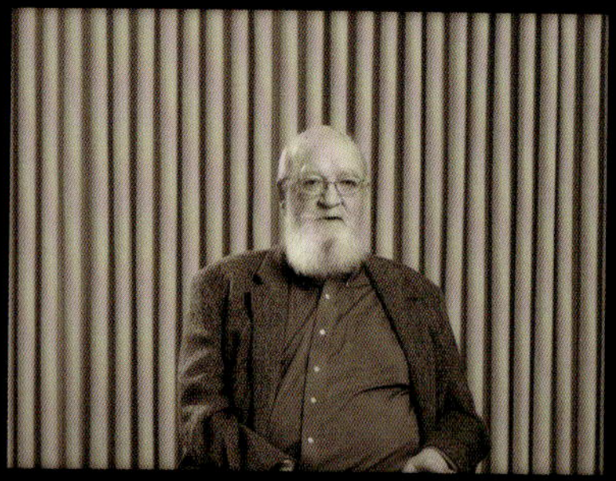

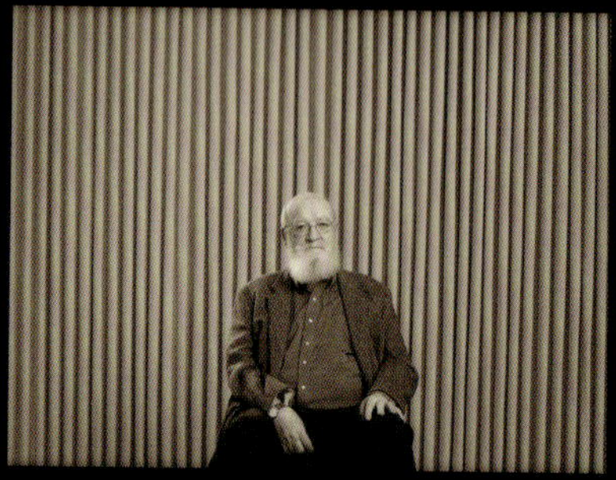

DD – People with two selves, or three selves, or four selves, that are quite stable—this really has occurred, and I think a lot can be learned from looking at that. I think once we recognize that the unity of consciousness is an achievement, not a ground state … You are not born unified. And most of us achieve a pretty good unity, but it can come hugely apart in some clinical conditions. And even in normal folks, it can come apart.

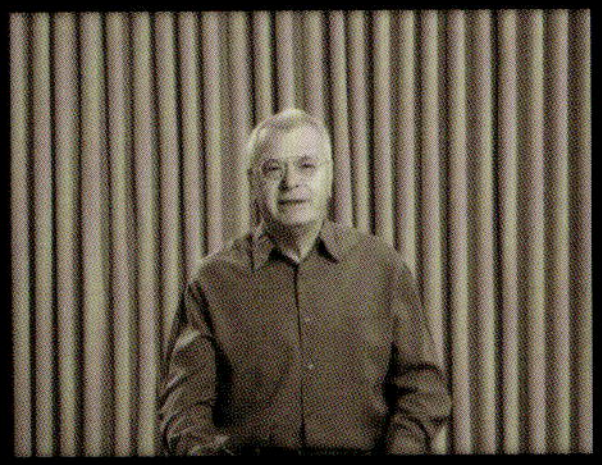

GB – The dream that we dream together is reality. Reality is a social agreement between you and me. Reality is what we experience and agree upon, that experience is real. […] If that happens in a wrong way, I see the world very, very differently. Then I am a super-creative individual. But if I don't find another individual who thinks the same way, I am just schizophrenic.

00:40:11:12

Karl Friston – I remember one patient with schizophrenia who was a very vocal and great contributor to our group meetings. And he would come up with the most appealing and odd conjunctions of words. My favorite, which I still remember to this day, is "angel shit."

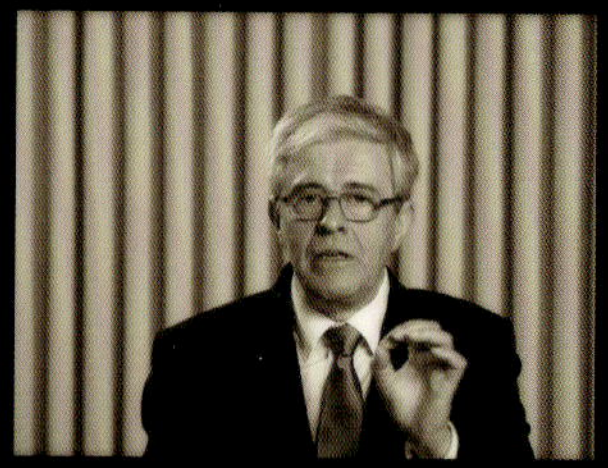

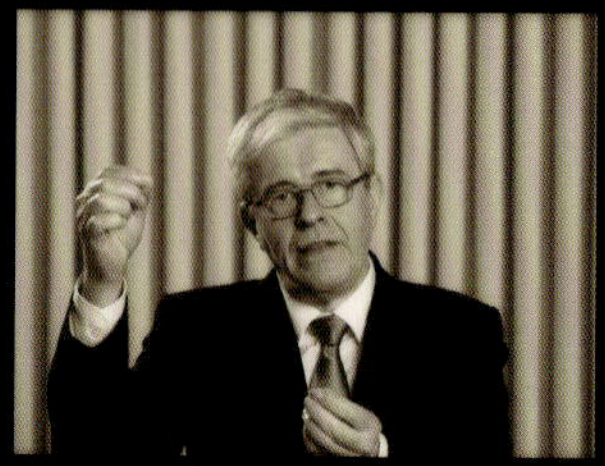

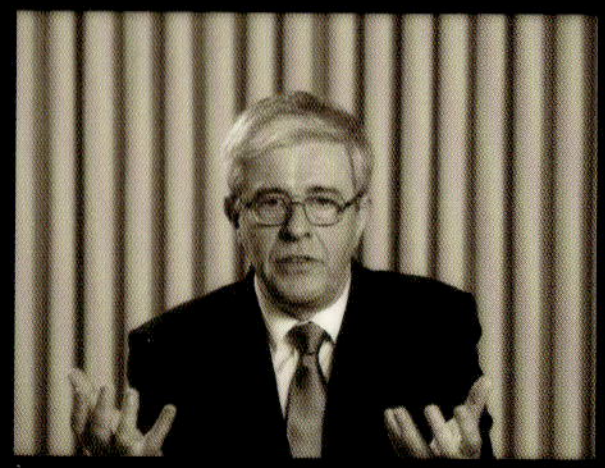

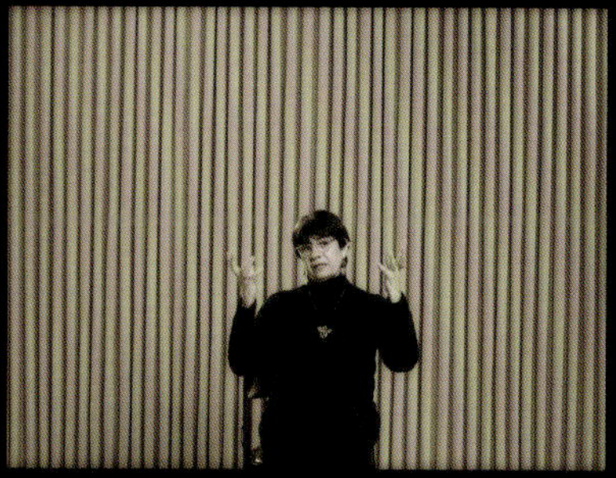

CM – The brain has to construct its own disorder; it has to, at some point, assimilate its own disorder. Otherwise it would be destroyed by it.

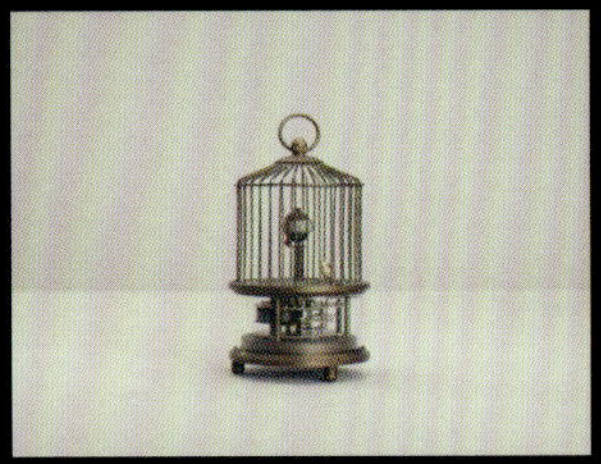

KF – The first thing to realize is that you can't get schizophrenia as a child. It's just not a disease of childhood. It's a disease of early adulthood that can, at the earliest, manifest in late adolescence. And that tells you something quite important: that if it's hard-wired into the genes, it's not hardwired in a way that is manifest at birth. You have to wait decades before it's manifest. So that tells you it's something about learning about your world that is required of you when you get to a certain age, usually when you have left or are leaving home. There is something that goes wrong with the way that you learn about the world.

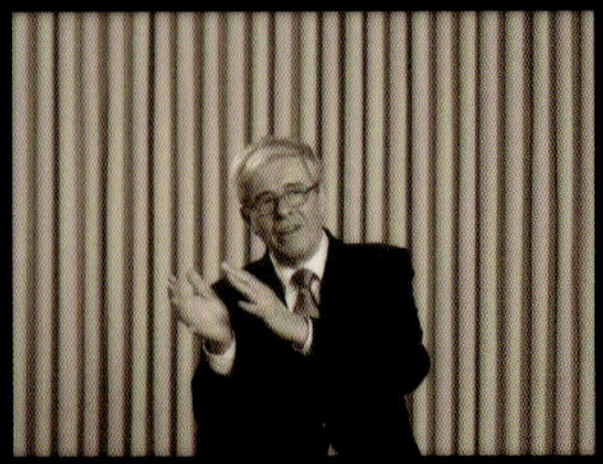

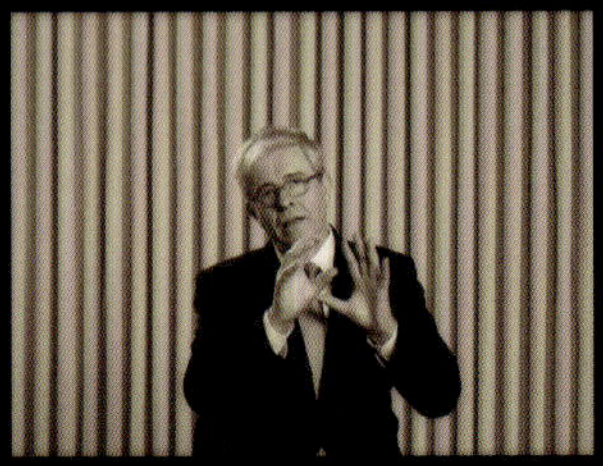

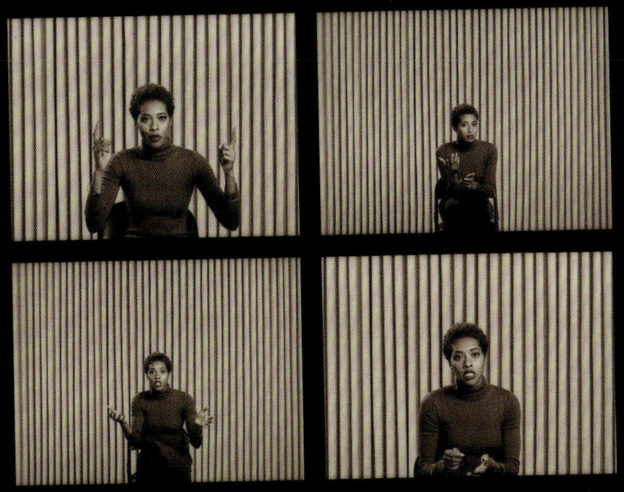

BJM – Error happens. Error occurs. Error occurs even in our genetics, sometimes to our detriment, sometimes to our advantage. So if it's that far down in the making of our being that error is there, I would hesitate taking it out of anywhere else. Maybe it belongs there.

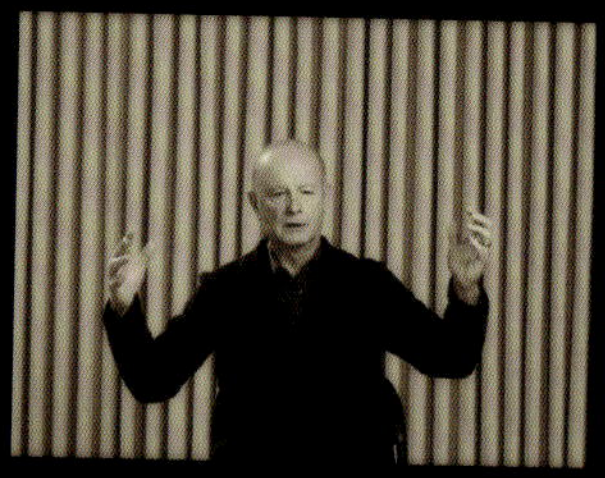

JL – People can have conditions where they may feel perfectly comfortable with having hallucinations or thoughts that are part of their personal experience. It doesn't necessarily mean that this defines the person. This is something the person has. In the past, some of these extreme conditions were viewed as valuable and very cherished by the culture.

AB – If I say I have a confusion, and I go inside that and I find some more confusion– I am confused: Who am I? Then I go inside, I find my hand, leg, and others. I say, my hand is me, my head is me, my heart is me. And then those are again confusion. And then I go inside and I go inside and inside and inside, I go to the single atom and even I don't find me.

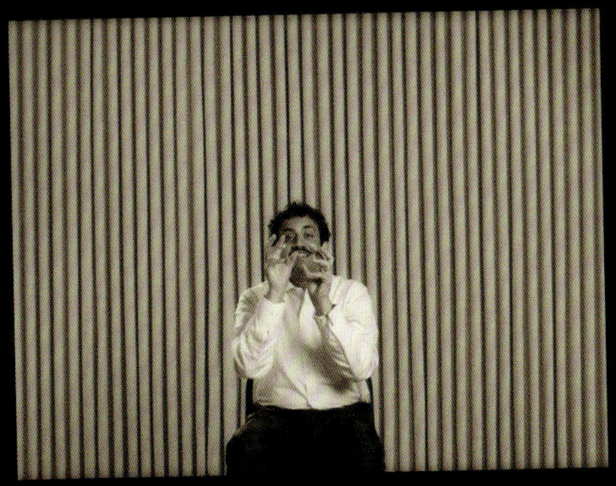

00:43:01:11

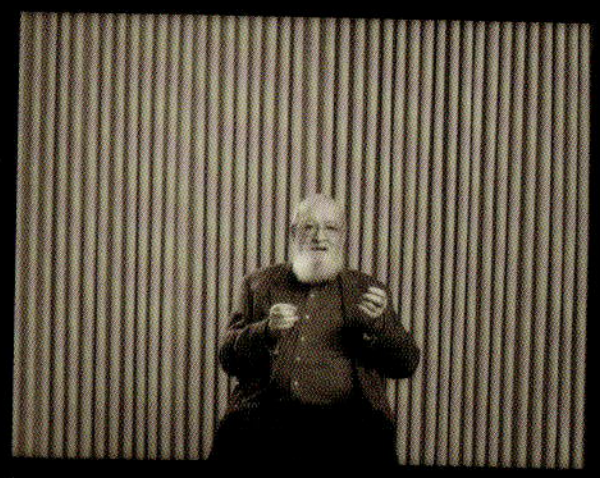

DD – We don't want to see all the swarms of atoms. We want to see liquids, and solids, and colors, and feel textures. All of that is, to some degree, user illusion designed to make it easier for us to get around in the world, in the same way that your smartphone has a user illusion. You don't want to know what's going on inside your smartphone. There are dozens of layers of complexity that you are better off not knowing about. They would only slow you down. You want to have a user-friendly, narcissistically focused, tuned set of representations. And that is what consciousness is.

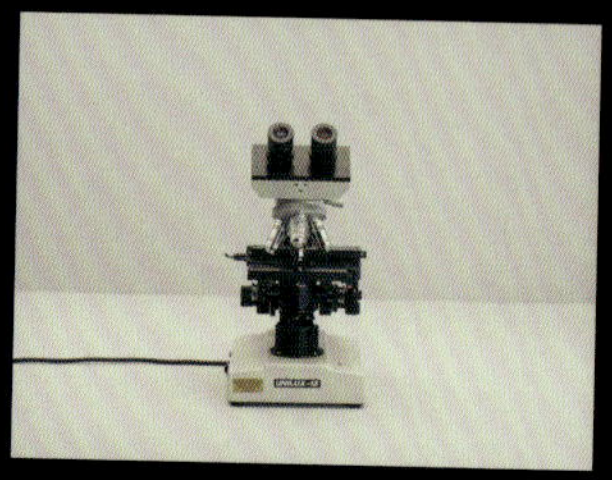

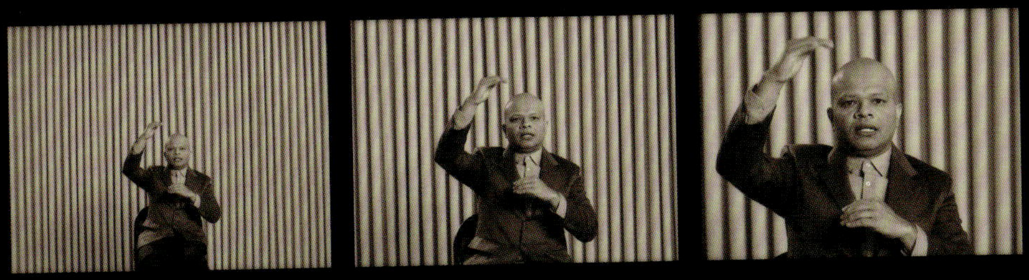

SR – Because there is so much information that is coming in, you actually want to process stuff that is unexpected or new or somehow different from what your expectation is. So you actually have a prediction about what the sensory stimulus is going to be. And what you are sending is essentially a prediction error of sorts, something that didn't match with what you were predicting. So this idea of me and myself and how it's different from someone else could depend on what kind of prediction I have about this world.

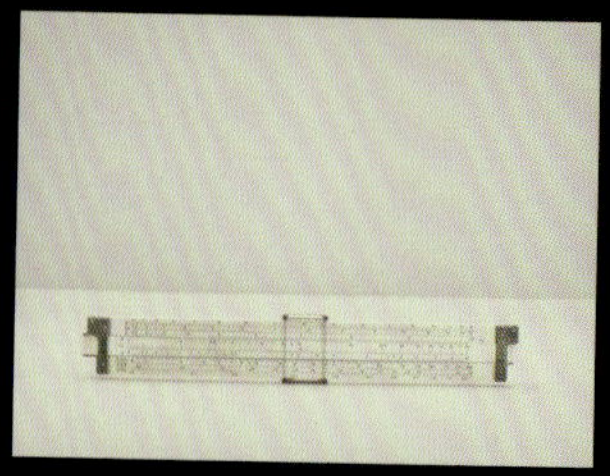

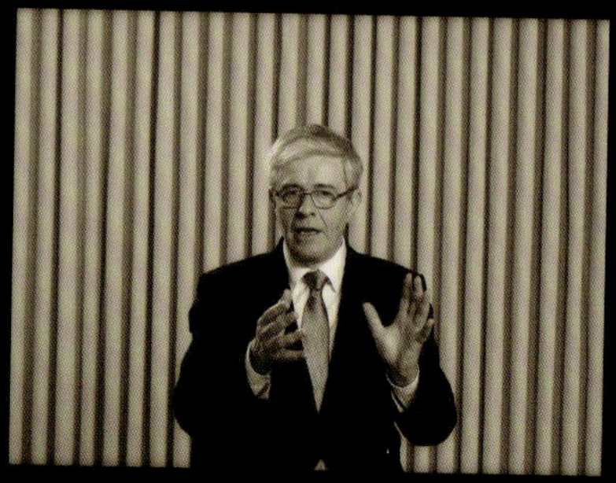

KF – In order to infer what you mean, what you intend, your propositional stance, your attitude towards me, what
you want me to know, I have to have in my head a model of you. And furthermore, in order to disambiguate between what you cause, for example, in conversation and what I cause, I have to have a model of me. [...] But notice it rests upon this crucial assumption: I have to assume that you are me.

Mitchell Joseph Valdés-Sosa – My dog sees different colors than I do, right? And bats hear sounds I can't hear. So our sensory systems are feeding to the brain different types of information, and that heavily influences the model we make of what is happening in the objective world.

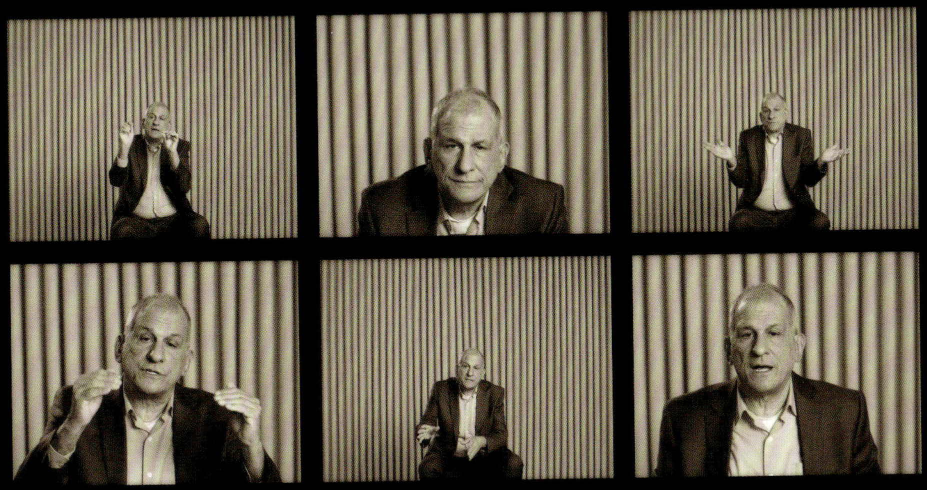

00:44:21:14

Li Zhaoping – The fly always runs away faster than you. You just can't get it. Why is it that the fly is faster than you? The fly has a small brain and we humans have a big brain, but somehow they can run away faster than we can hit them. Because the fly could not concentrate. Therefore, they are so easily distracted. All their brain is distraction. They focus a hundred percent of their brain on distraction. So therefore they can react to emergencies faster than we do.

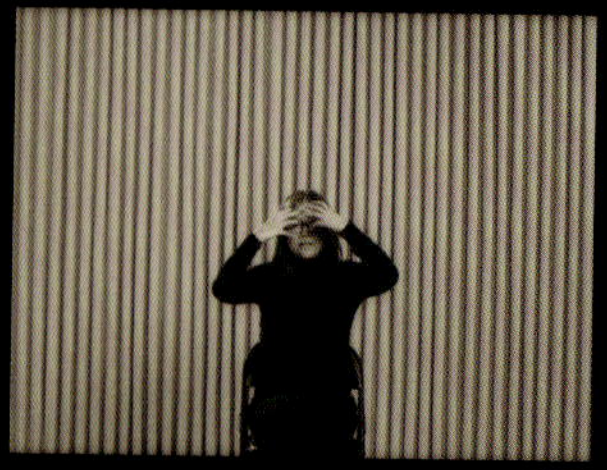

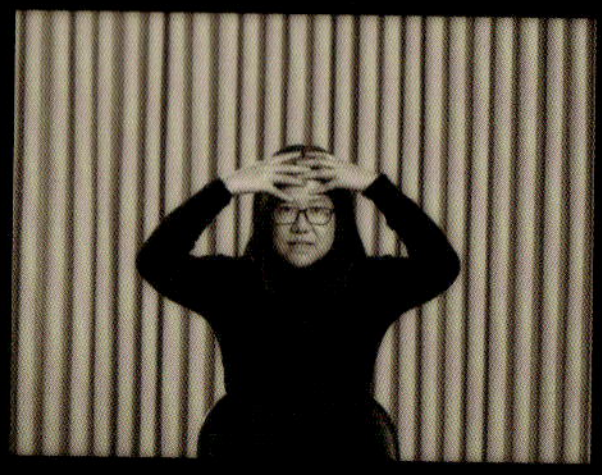

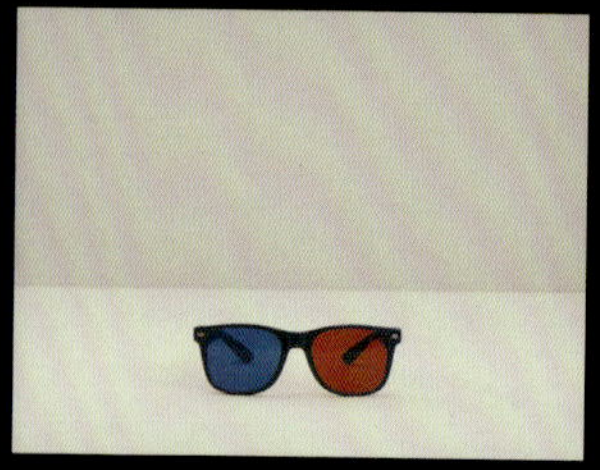

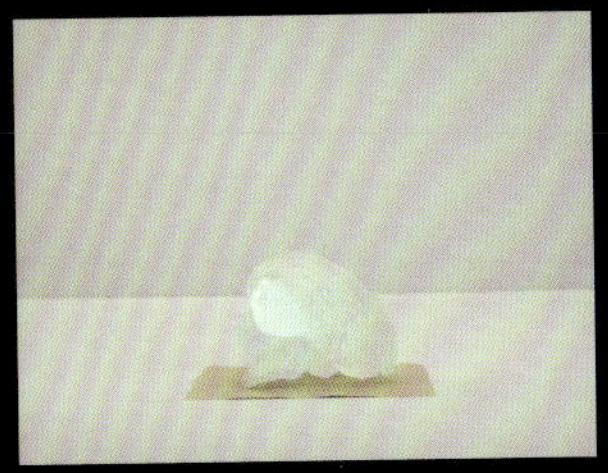

AB – The next kind of war or next kind of wave of human destruction will happen through swarm robots. [...] These tiny little swarms can do enormous harm. Targeted killing by satellite control, targeted killing to a particular individual, and then ruin that particular country. So already the AI machines are coming, and it is just a matter of fifteen, twenty years, you will see the whole world will become a very dangerous place to live in. And the only way we can get out of it is building a conscious machine, because only when you have the conscious machines, they know how to tackle the AI machines. Because it will know the pattern, it will create the pattern, it will analyze and synthesize the pattern much better than a human do. And it is based on a principle that everything is connected to everything. So it will know who is breaking the pattern, and how to teach a lesson to those things. So we have to invent conscious machines to save the destruction of humanity.

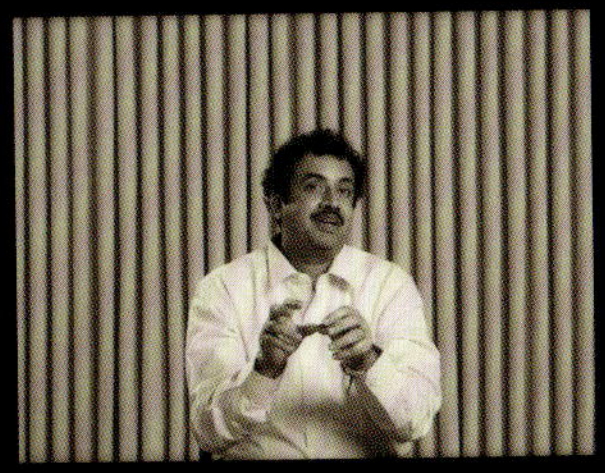

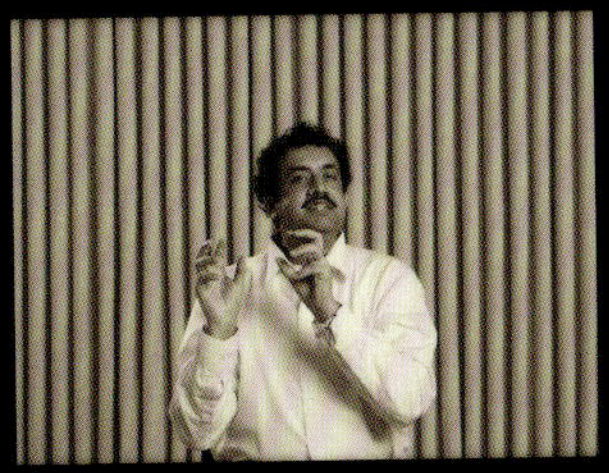

00:46:48:11

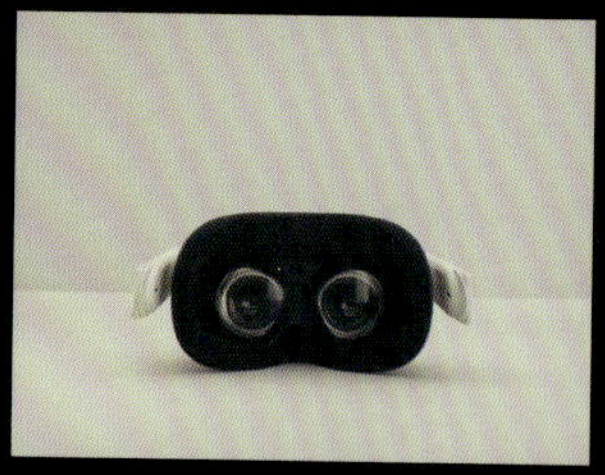

DD – We are conscious robots. We are robots made of robots made of robots. And you get down, look inside the brain, that's what you find inside: there are lots of little cellular robots, and even subcellular robots, motor proteins and the like. But I don't think we should make conscious robots. Because conscious robots would be both dangerous to us and we would be a danger to them. We want smart machines, not artificial colleagues.

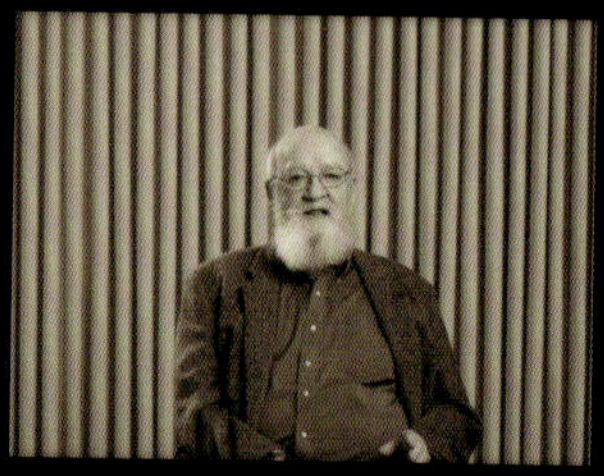

David Chalmers – I think the safest way to do it is gradually replace a bit of my brain, by, say, replace some neurons by silicon chips, replace 1 percent of my brain, 5 percent of my brain, 10 percent of my brain, 20 percent, 50 percent, till eventually 100 percent of your brain has been replaced by silicon chips. Ideally you could stay conscious throughout this process, so I could say, "Okay, I am still here. I am still here. I am still experiencing this." If there was a continuous chain of consciousness from the biological brain to the uploaded brain, then I think there is a pretty good case that that would still be me.

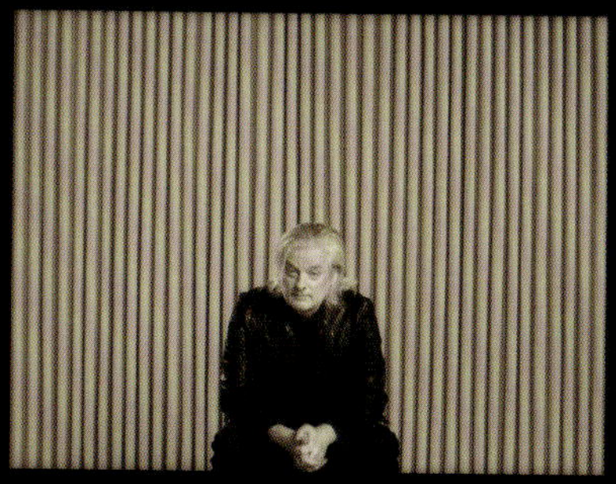

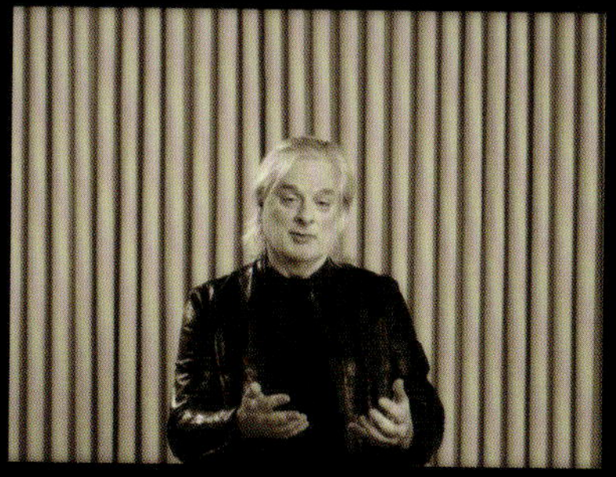

00:48:42:13

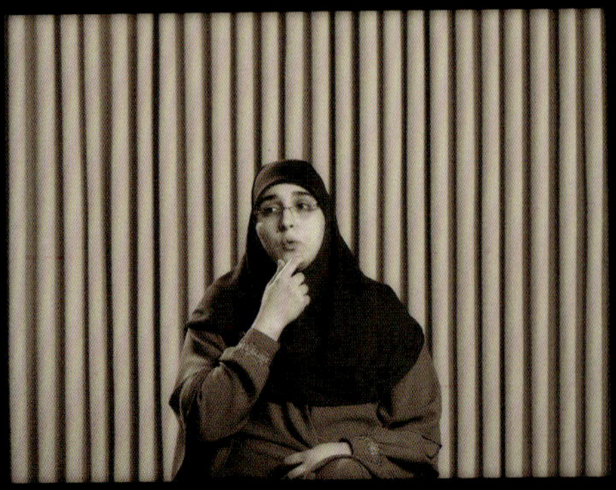

YA – I would like to say to those who are trying to build human brains, good luck. I don't think that this mission could be completed.

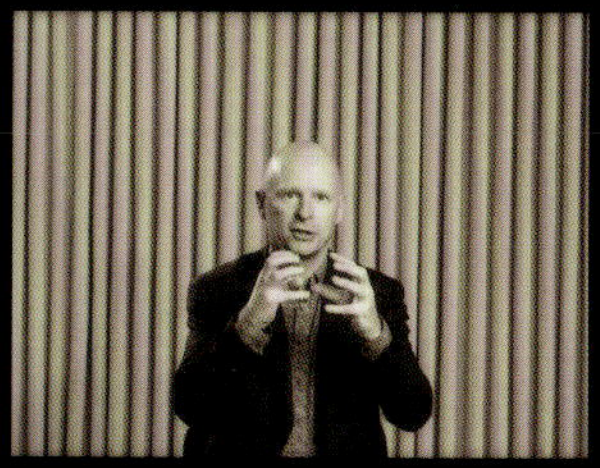

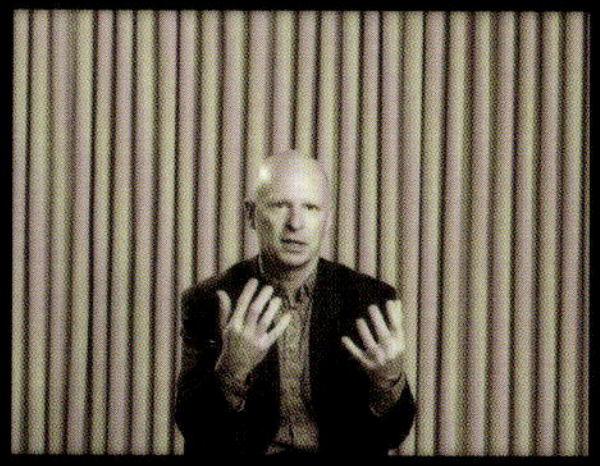

SD – There is no other species that can think of the combinations of ideas that we have. For instance, a chimera, a lion with a deer's head. That's a crazy thought, of course, but that is part of the space of thoughts that humans can have, because we have this ability to put the ideas together in novel ways, in syntactically organized ideas that other species do not seem to have.

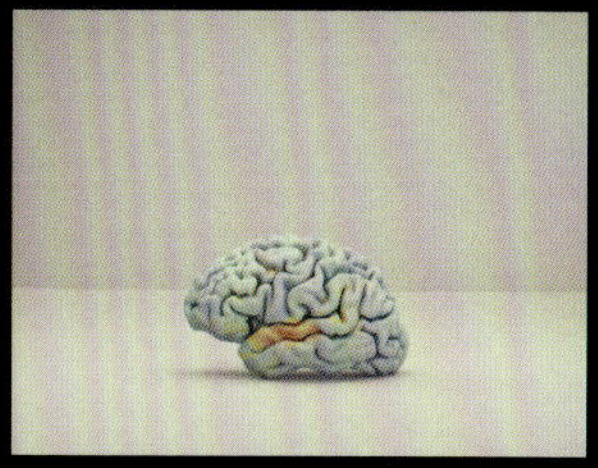

00:49:44:03

EJ – It does make me think, if we humans can have such a big ego and think of ourselves as unique about something, and we are like other animals, well, philosophically you would argue, other animals should also have egos. And I would not be surprised if they did.

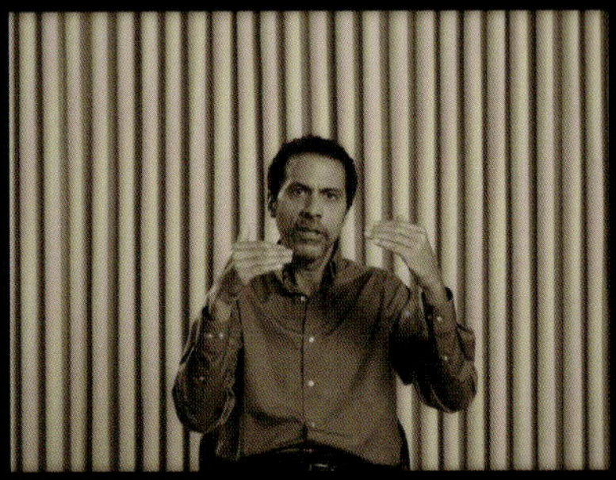

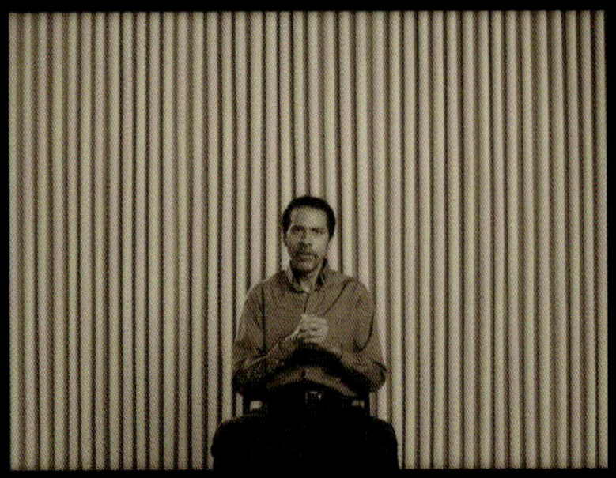

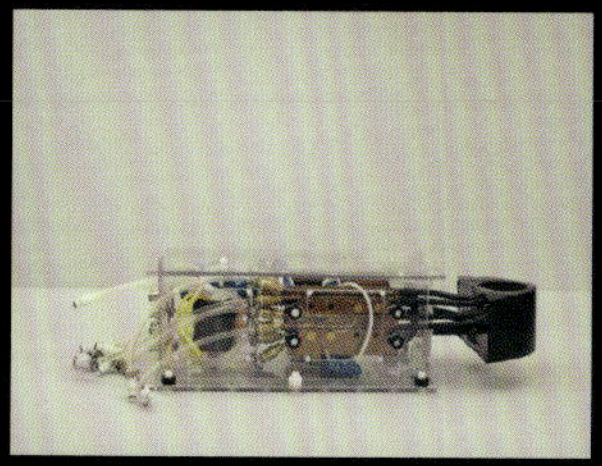

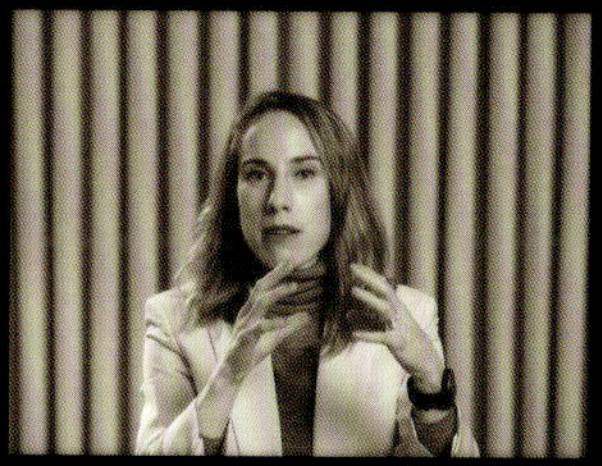

PA – One of the first things that I saw in that lab was a rat with an optical cable sticking out of its head. And at that very moment, I knew that I must spend some time in this lab.
This was amazing and shocking, and I asked my future colleagues, "Do they survive?" And the answer was, "They thrive."

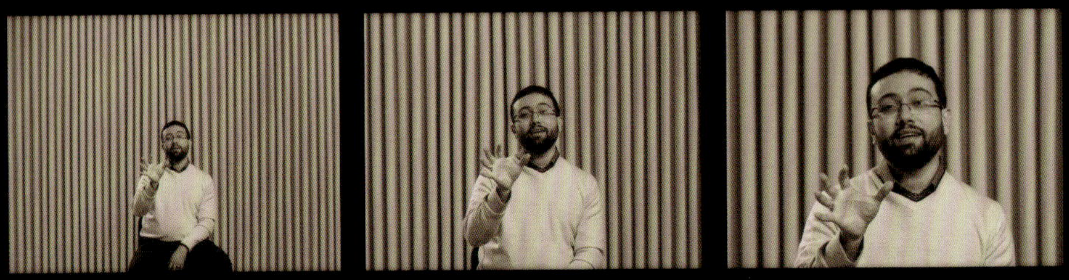

AG – Of course, the first group that are going to use these types of technologies is military, and yeah, that's why it's scary. And the problem is, the groups that have the most amount of money and resources to put into these things are big corporations and governments, both of whom can be very dangerous for the liberty of the people.

PA – Let's say we have a depressed patient. We would have them instrumented with those nanoscale agents, and whenever their symptoms begin to reappear, all they would have to do is to go sleep in their bed, push a button, and then they would wake up next morning feeling like themselves again, feeling not depressed and ready for the day.

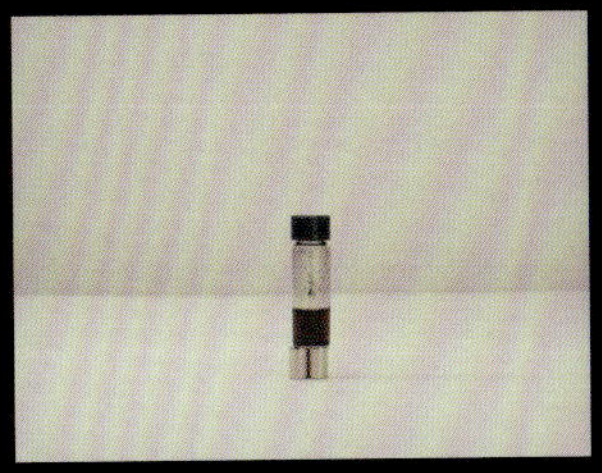

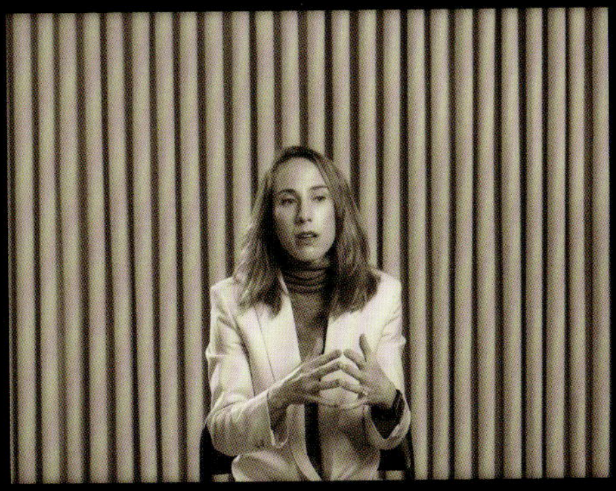

RY – Technologies are neutral; you can use them for good or for bad. Imagine the person that first discovers fire. Like, great, we can warm ourselves up during the winter and cook our food, or we can burn the cave of the guys next door. So this is the same thing. Fire, the wheel, the horse, atomic energy, the printing press, Internet, AI, neurotech.

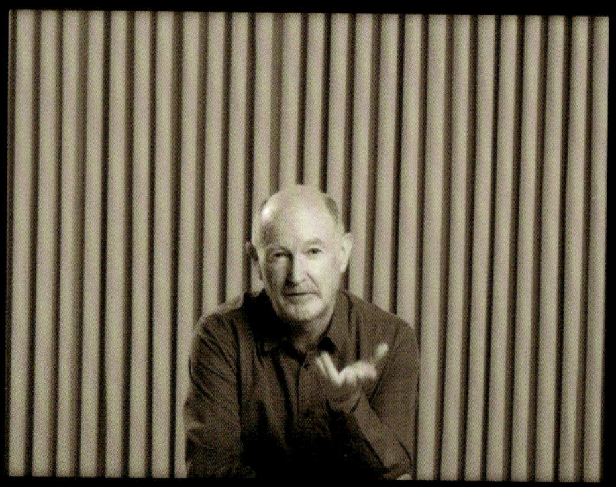

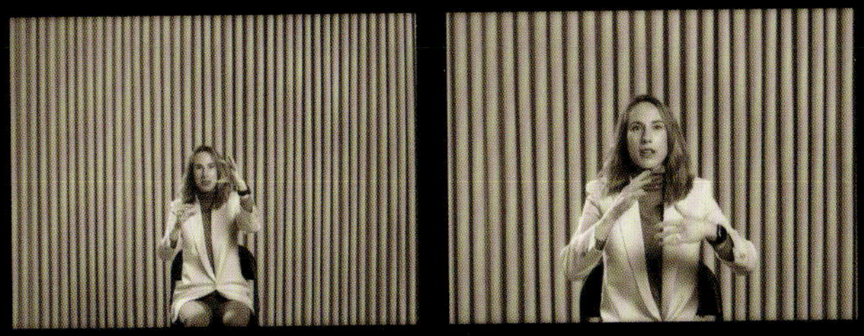

PA – Whenever we sense a flaw, whenever a system goes awry, we would be able to activate those agents that are operating in the nervous system to restart that particular group of cells, or treat or start a circuit in this group of cells.

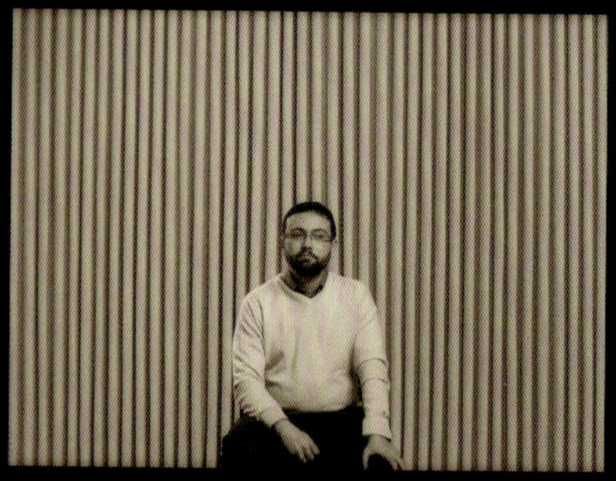

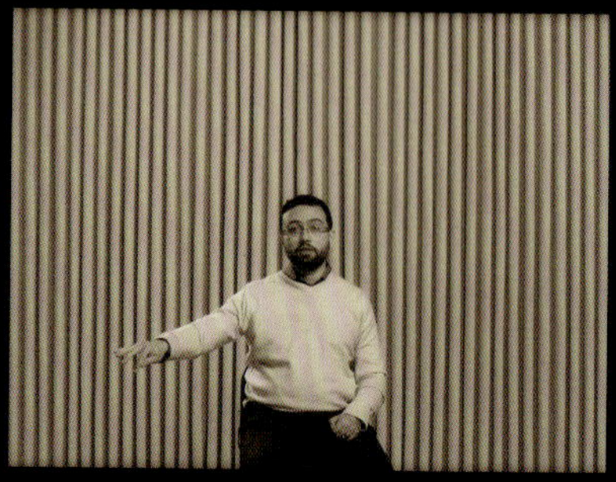

AG – It's kind of different from a nuclear bomb. That's why it's more dangerous. You may not notice it, but it will come into the people's lives, and it will affect the way that they live.

RY – This is the first time in history that humans have developed a technology that can change, can alter the essence of who we are. And we think, if this is not a human rights issue, what is a human rights issue?

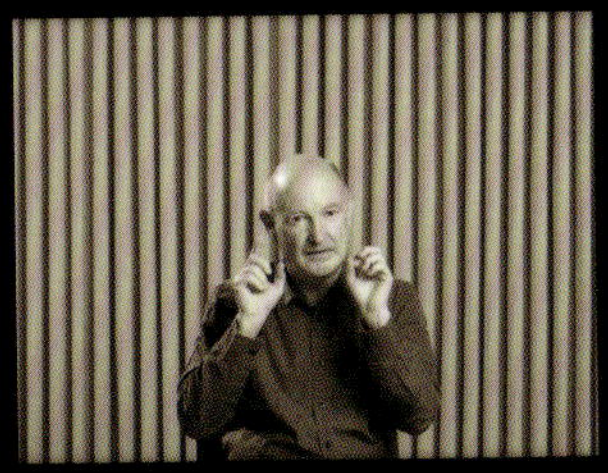

LZ – Brain control power might be used
just like the original atomic bomb was used.
And let's hope if it would be used, it's
as temporary as the original atomic bomb.
Just twice and not a third time.

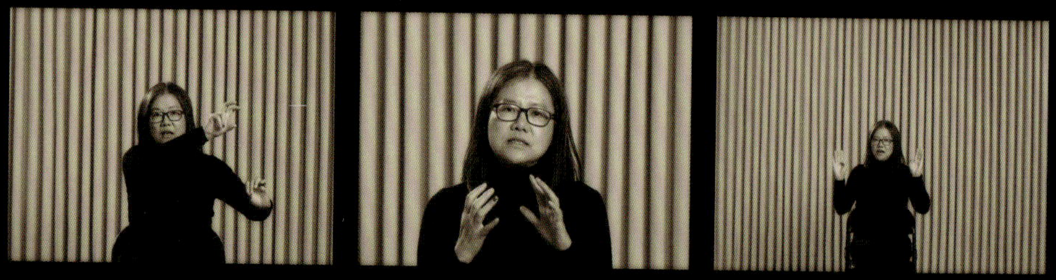

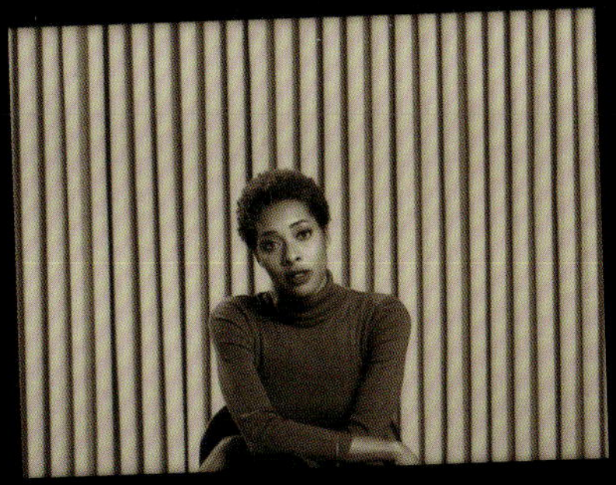

BJM – If you can push a button and have a whole people group dismissed, people have done it. If you can just push a button and ignore the fact that people are people, and have them work for you, get their money, use their bodies, use their children to benefit yourself, then it could be done.

00:53:41:14

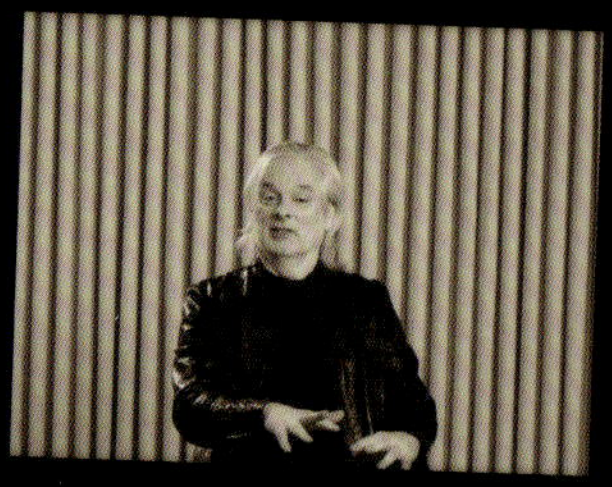

DC – We don't care about what happens to a rock. You can destroy a rock, it's fine. You didn't hurt anybody. You didn't sin. You didn't create a moral harm. But maybe a dog, we care about what happens to dogs. Dogs can suffer. And I think this is because dogs are conscious, whereas rocks, plausibly, are not conscious. So, I think once you are conscious, you are a being that matters morally.

AV – When somebody is being abused, they become less humanlike, or they get just more like an animal, frightened and terrified. And that makes them less human and therefore more easy to abuse more. And I think this is something of a ... well, not vicious circle exactly, but it just compounds itself until you get to the stage where people are capable of doing this, and perhaps they enjoy it.

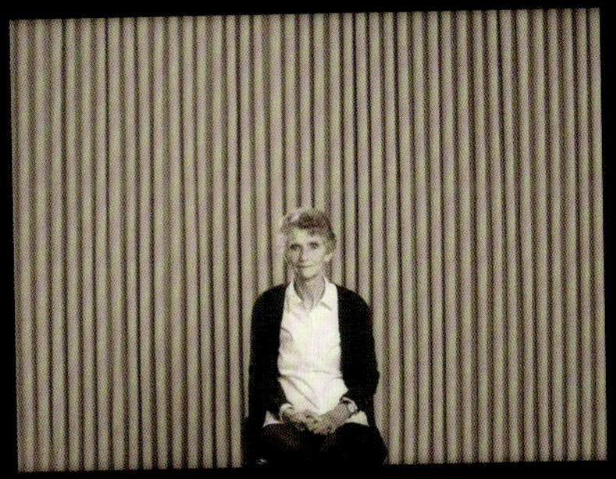

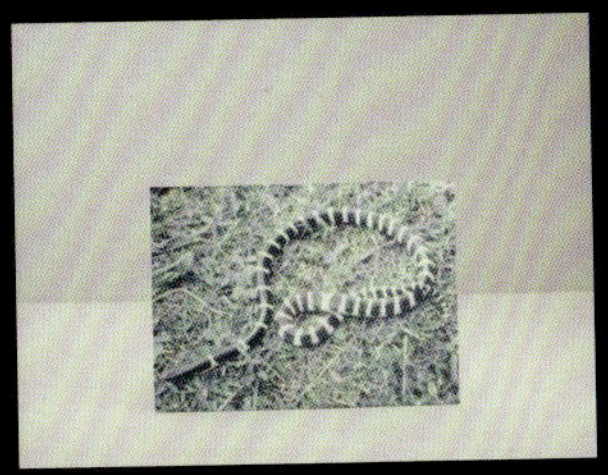

00:54:42:17

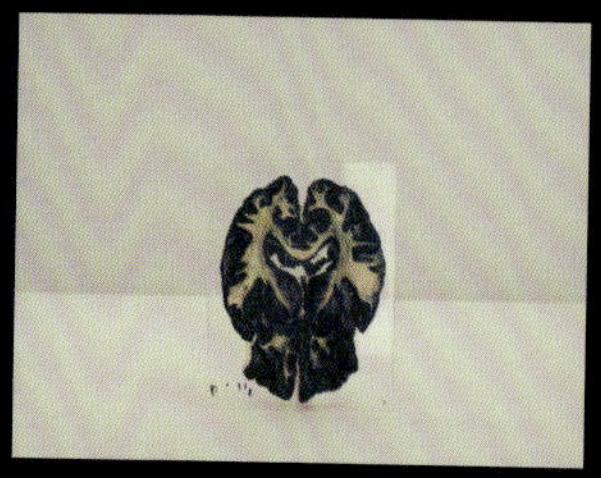

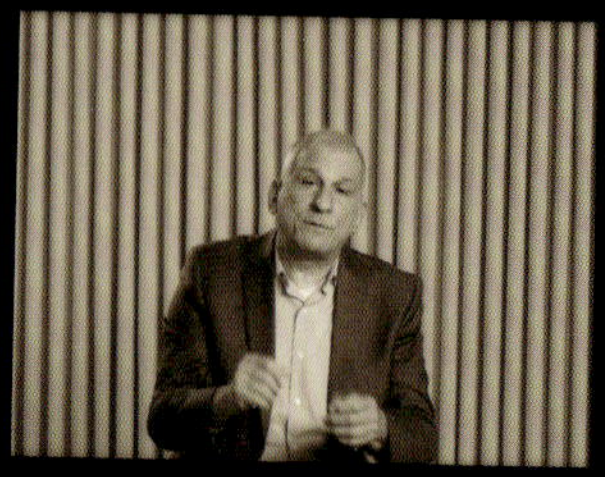

MJVS – We studied two hundred murderers. People who had killed other people. And then we have studied a large number of children who have conduct disorders. And it is obvious that people have a biological tendency for aggressive behavior. That is all in our evolutionary history. But we have learned to control it, to channel it adequately.

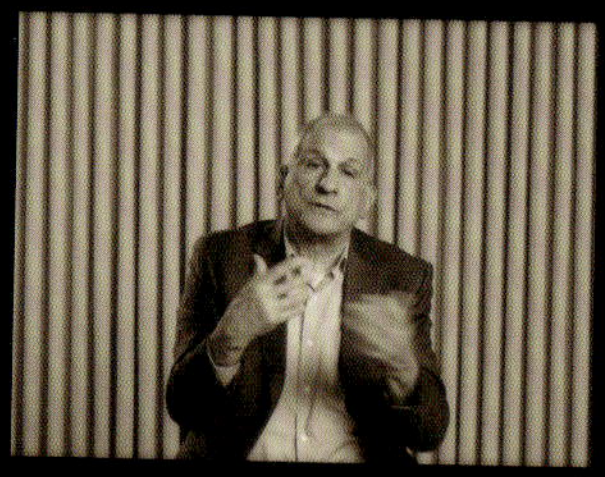

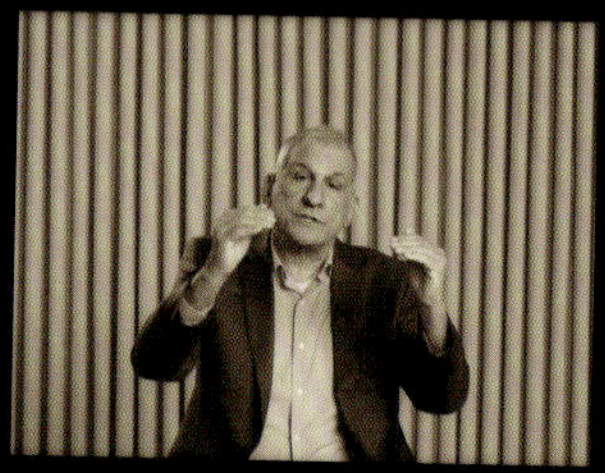

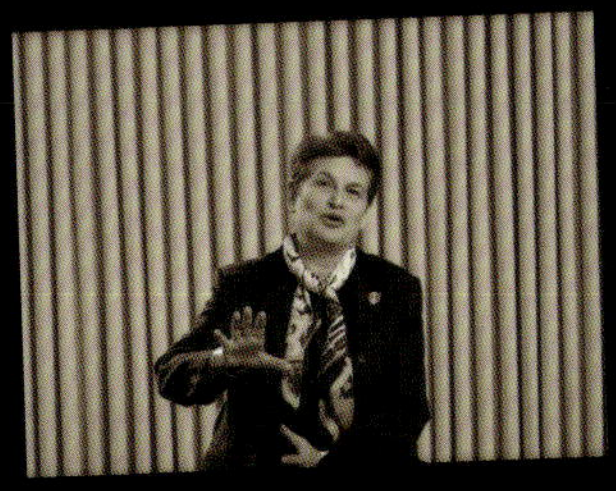

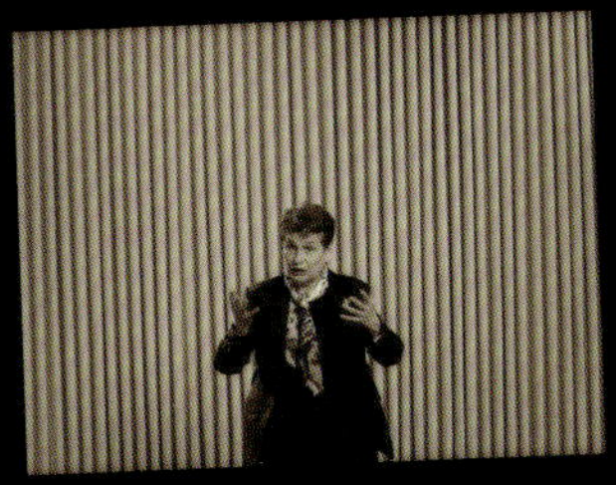

CD – If you look at the behavior of domesticated versus wild mice or wild animals, domesticated female mice are very tame. They have been selected to reproduce very easily, not be very aggressive, which is very different than the wild females that are very aggressive and very difficult to handle. So when labs or facilities have developed and grown colonies of mice, they have selected for docile females. Because somehow an aggressive female didn't seem right, but an aggressive male was okay.

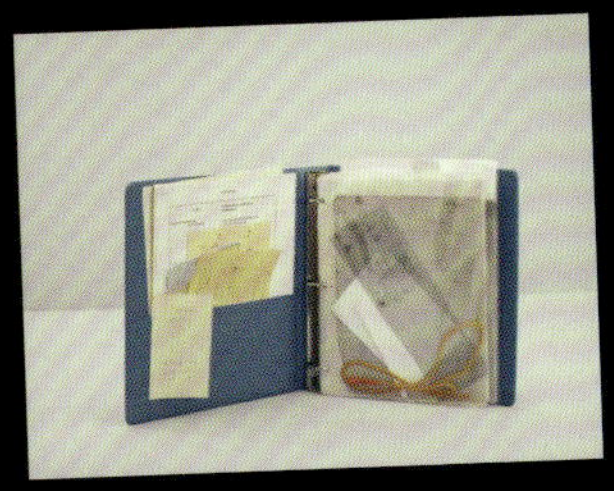

00:55:46:14

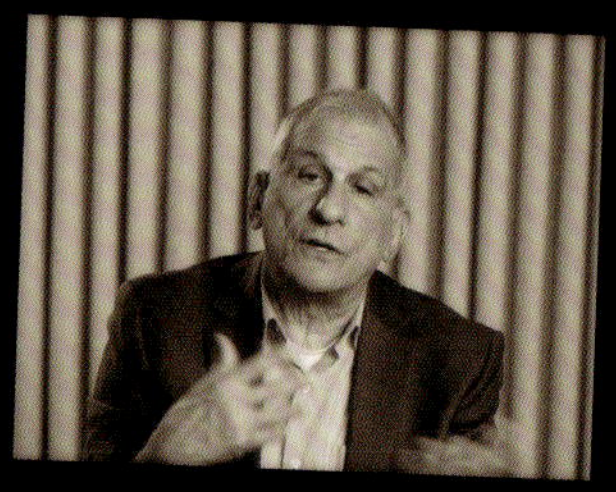

MJVS – Athletes have good impulse control. Extremely high impulse control. They have learned to channel aggression in a productive way; it can't pass certain limits, it has to occur in certain scenarios. And it is interesting to study this interplay of circuits that lead to aggressive behavior that are needed for certain kinds of sport or other kinds of behavior. But if you learn to control them, they are not harmful for society—in fact, you receive prizes for it.

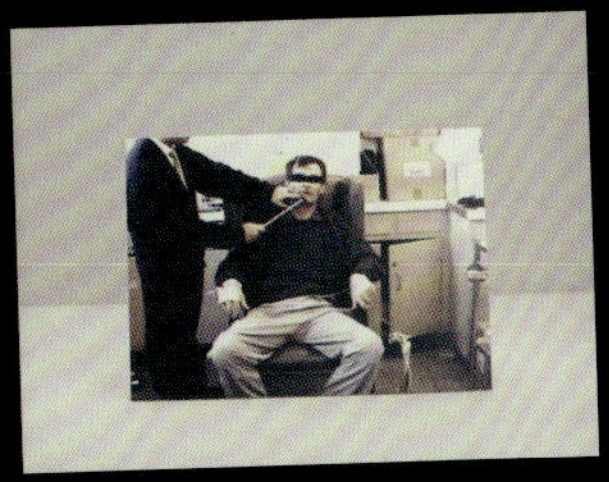

AV – As we know, there are so many circuits involved in the brain and so many modulatory inferences going on the whole time that under certain circumstances, it may not take a great deal to trip into a state where you can behave abnormally.

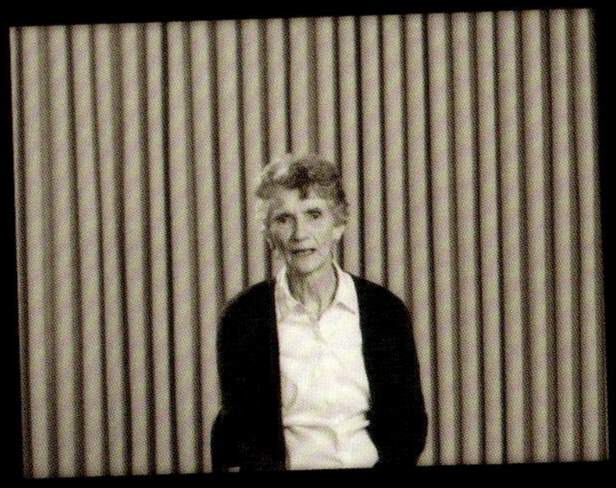

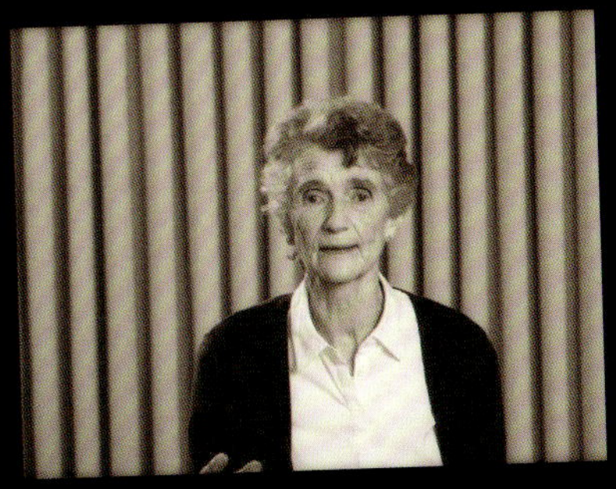

00:57:17:19

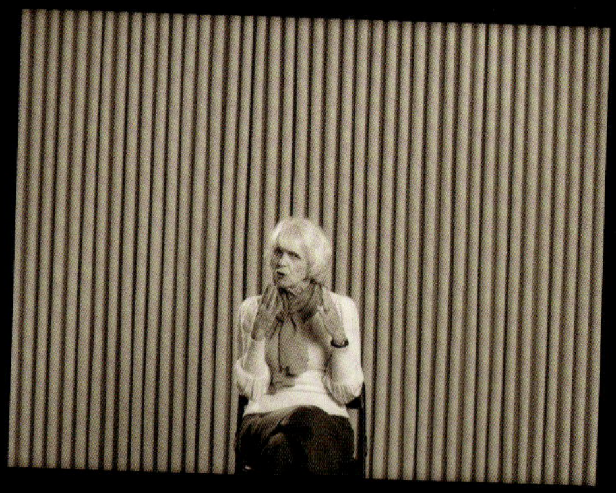

PC – It does happen that two groups of chimpanzees will be living more or less okay with each other. More or less harmoniously. But then, perhaps, as a result of one group expanding, having more babies and so forth, they will then, at night, two or three males, maybe another one, they will creep up to the boundary of the other camp, and if they find a lone chimpanzee foraging, they will grab him, beat him up, rip his eyes out. They are vicious. Do they take pleasure in it? Well, we have no way of knowing, but it is not just like they are protecting something. They are really damaging this individual. It is puzzling. Whether humans do something comparable and whether it involves pleasure or not, I really don't know, but it certainly is the kind of thing that crosses one's mind.

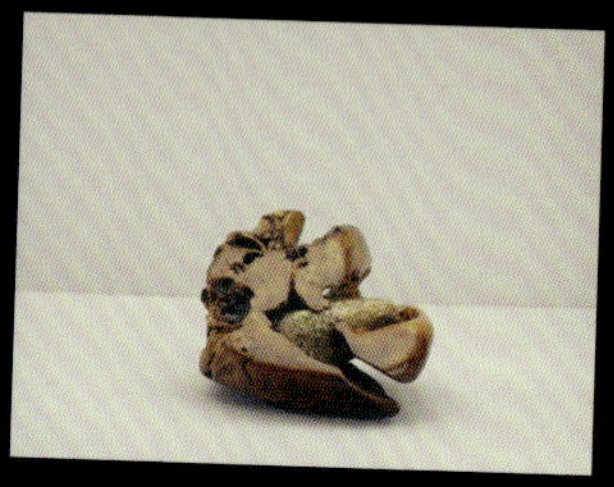

CM – Most men, as we know, are still today educated in this vision of the male being authorized to let his death drive express itself through violence. So, of course, in our brain there are many sites for this energy, this dominating energy, this capacity to destroy, to produce death, et cetera. There is a neural death drive that exists in all of us.

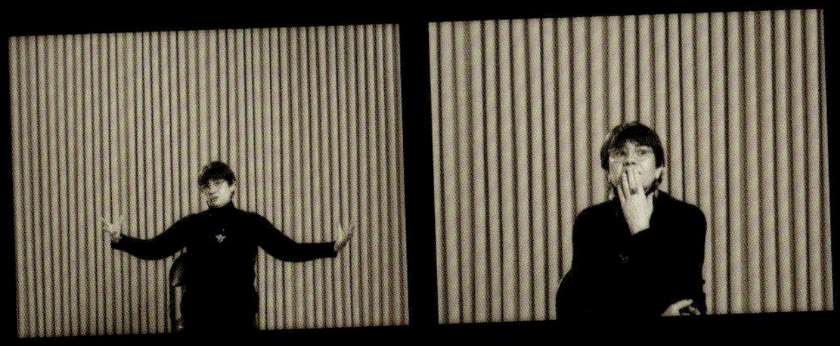

00:58:56:00

WS – Shortly after the Germans had invaded Poland, the army had the problem that they needed people to kill Polish elites and to kill Jews over there, and to kill minorities that they had captured.

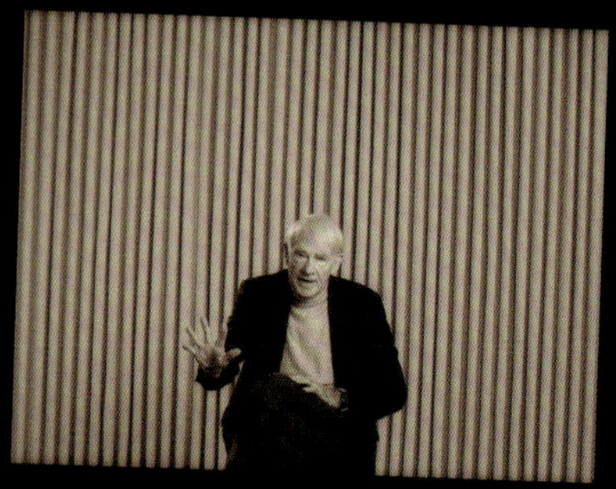

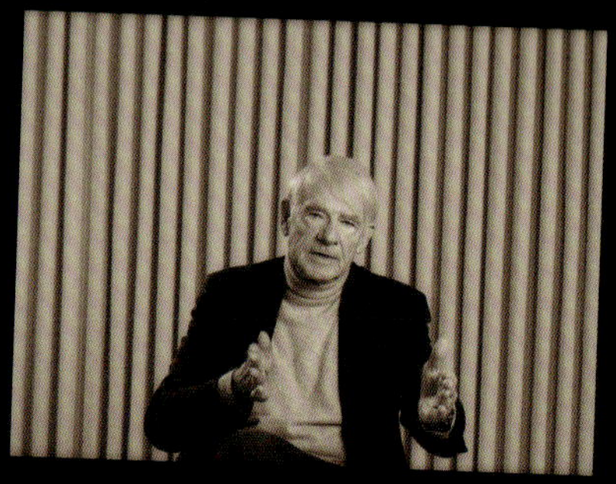

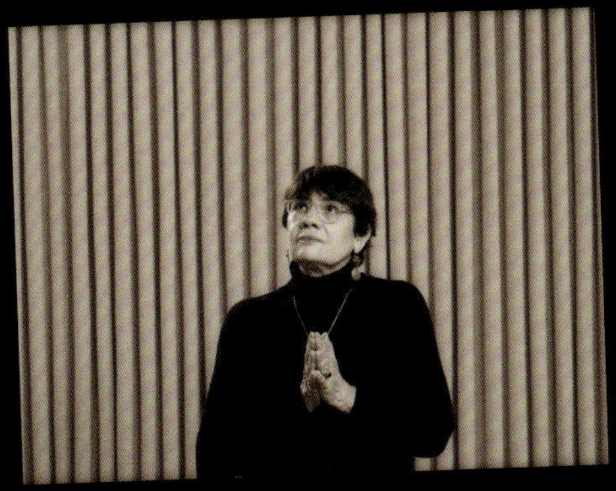

CM – But that would imply to admit, which very few neurologists do, that death is inscribed in the brain, that the brain is also a self-destructive machine. Not only creative, but also self-destructive—that it also strives to destroy itself and other people.

WS – Once [the Germans got to Poland], they were confronted with the problem: "What we want you to do is to kill." They did this with bullets in the neck. There were a few, a minority in the beginning, who said, "Yes, we are willing to do that," but there were still–They could negate it. They could deny to do that. But after a couple of weeks, everybody participated.

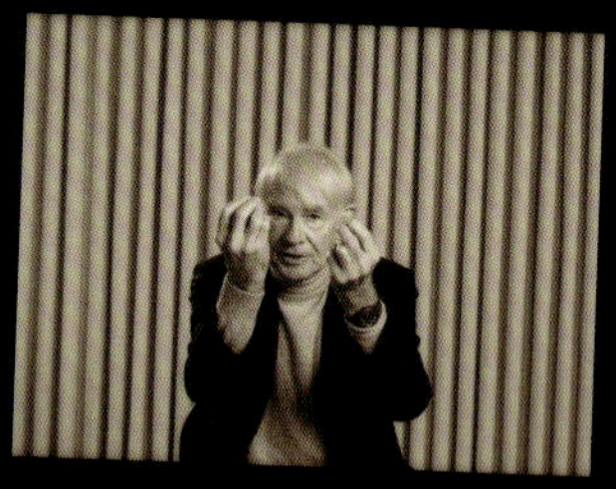

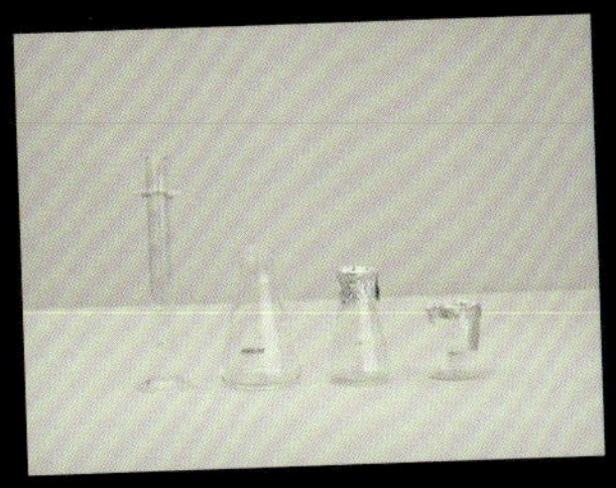

BJM – We live with this constant dichotomy, this constant fight or flight. There is an area of the brain—it is called the VMHvl—and this area of the brain surrounds also oxytocinergic neurons, where you can activate a certain population in that area and a mouse will want to mate. You activate a cell population just next to that area, and that animal wants to fight and kill. They are both located in the same area of the brain and yet have the ability to create life and to destroy life.

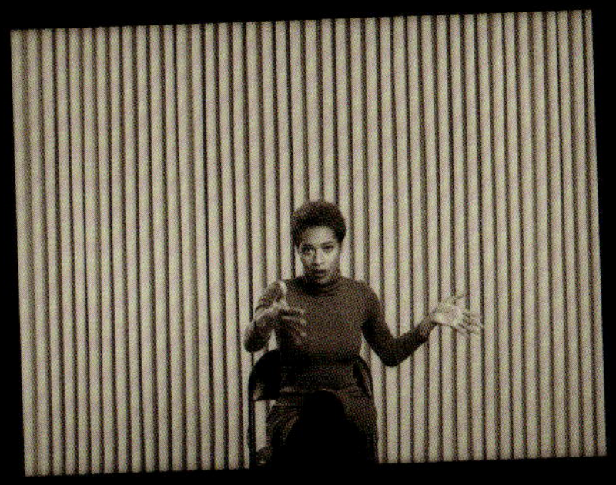

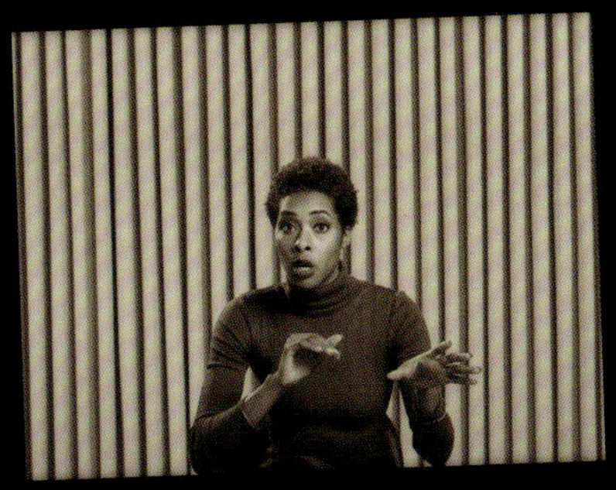

01:01:03:14

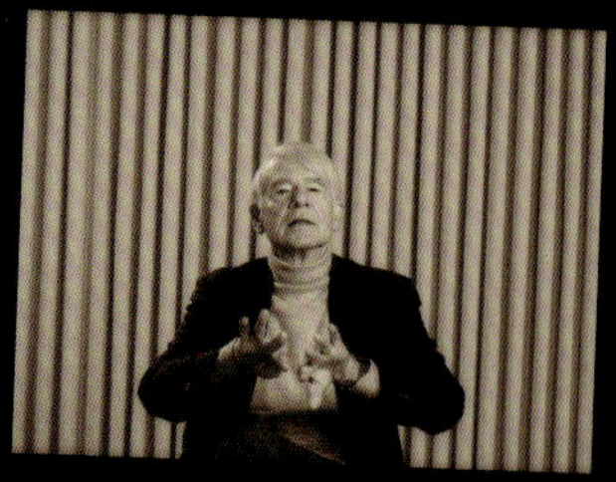

WS – Why did those who in the beginning did not want to do it join the group, join the crowd? And then even help to improve the system of killing, saying, This not a rational way to do it, one by one; just let's them align in the line and then kill them with a machine gun. Then, well, it is easier to have a ditch already dug in which they fall. Abominable things like that, planned and executed by people who had a couple of months before maybe worked in the civil service, doing great, compassionate things. I have no answer. But nature has endowed us with this possibility. Of course, our cultures try to overcome this, our normative systems, our education tries to overcome this. But as we realize over and over again, this is a thin varnish and you can break it very, very quickly.

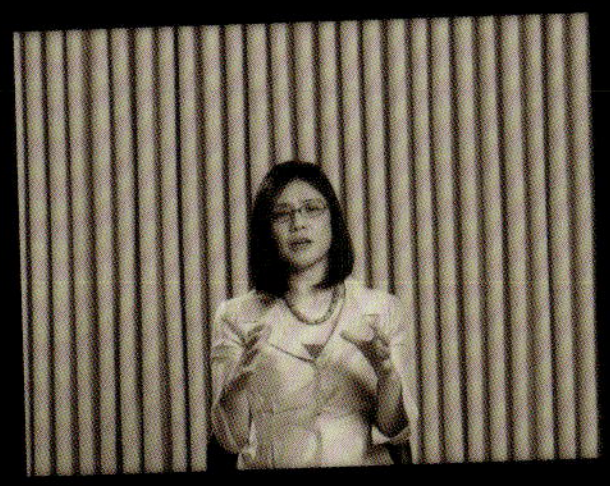

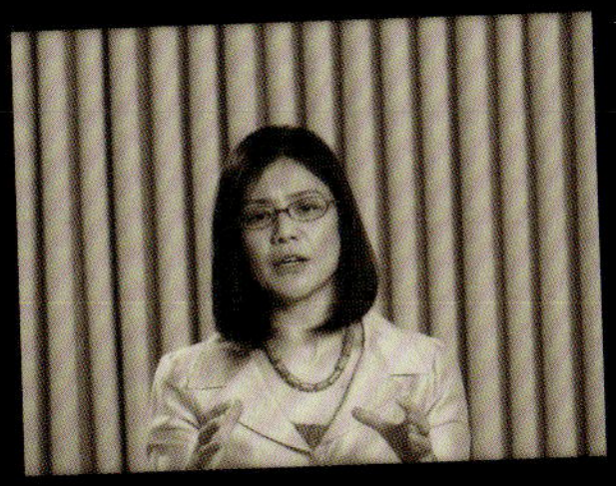

KK – The first experience of manipulation was of the brain area that is necessary, determined to be necessary, for parental care in mice. We didn't expect that actually the mouse mother start killing the pup, the biological pup. That was quite shocking for me.

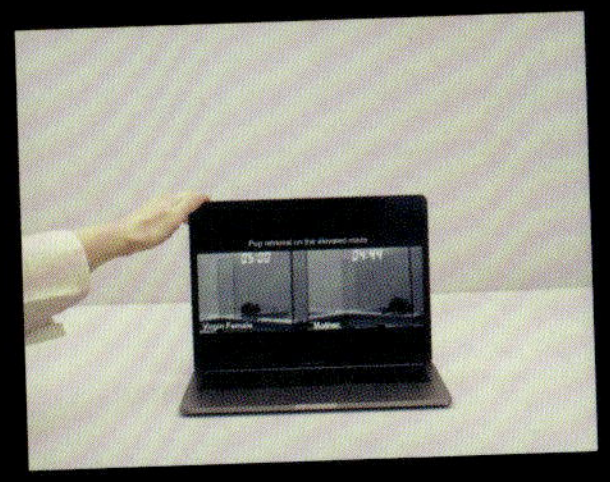

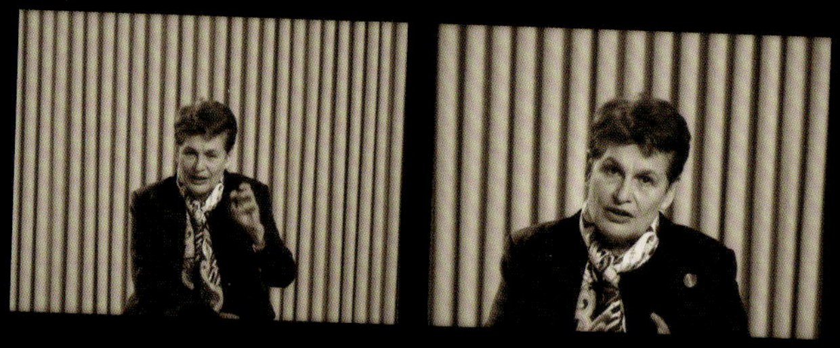

Catherine Dulac – In particular, when there are too many predators, not enough resources, when the females somehow understand there is no chance for infant to survive, very often females either abandon their infant or cannibalize them.

KK – I have four children, and during that time had three miscarriages. One miscarriage, at the time, somebody said, "Oh, you sacrifice so many animals." That's a punishment, from the animal to my miscarriage. Even sometimes people say something like that. But I accept. It's maybe an Eastern way of thinking. It's a reincarnation. Our life, human life, is connected with the animal life. We are not separating human and animal. So, in Eastern way, Asian way of thinking, that may happen—because I sacrifice too many animals, that I receive punishment and then I miscarried the infant, that's possible. But still I am decided to do ... Still I do this.

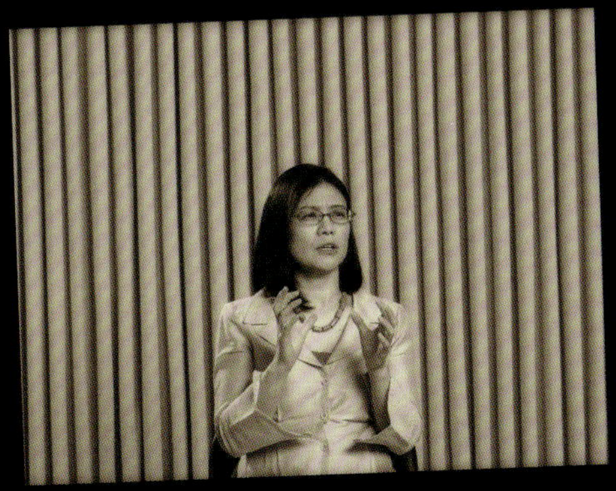

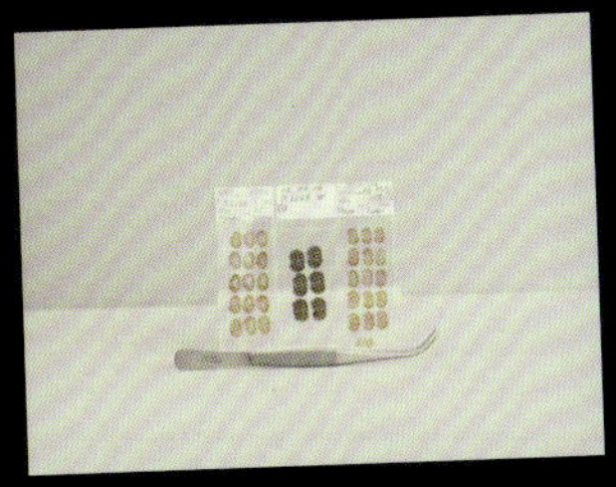

01:03:33:15

CD – Killing infants is a huge waste of resource for a species. And it turns out that a number of mammalian species have found a response against infanticide. And that response is likely to be monogamy. Which is that when a male and a female stay together and both take care of infants, then there is no need, evolutionarily speaking, for something like infanticidal behavior. And that saves the life and the investment in infant.

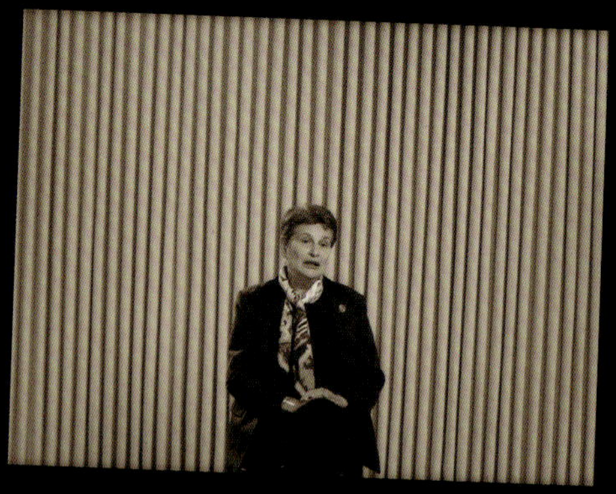

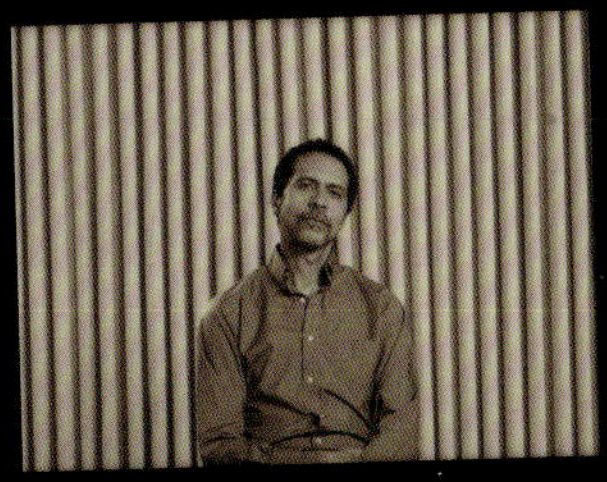

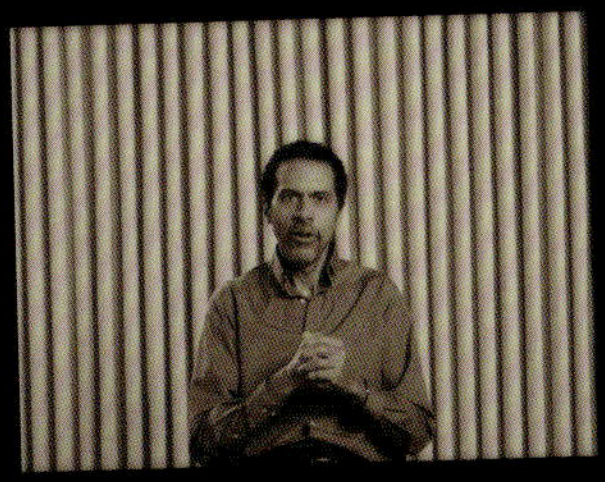

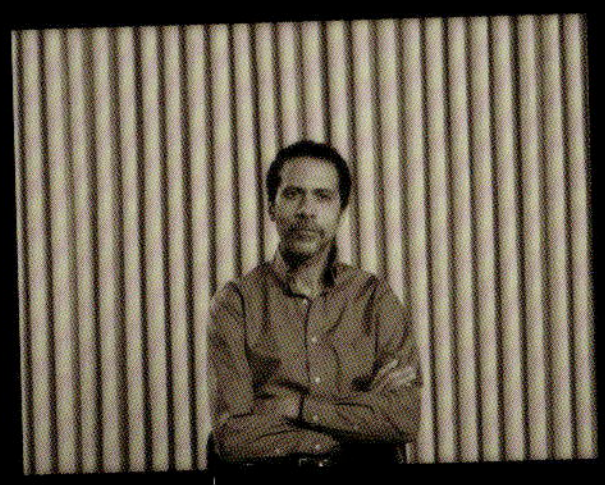

EJ – We overrate ourselves, relative to other species, and we share a lot more commonalities with these other species than we realize, and we will not discover that until we get rid of the assumptions and really ask the questions with an open mind. And we also have to be ready to accept that some species are more advanced than us in some things they do than we humans do.

01:05:11:21

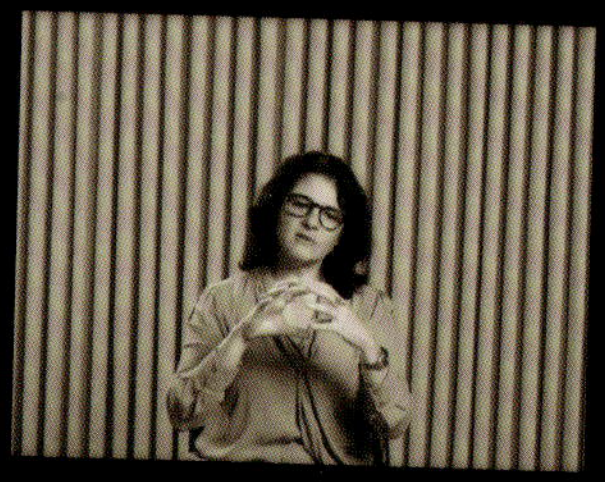

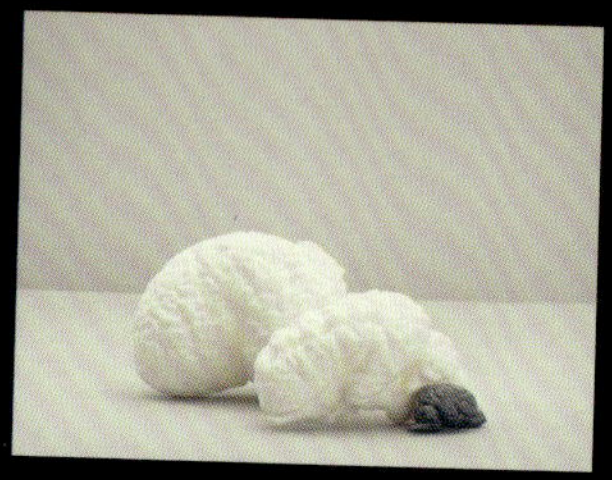

SHH – If what you want is a brain that is outside the norms, that's really an outlier, your brain is the raccoon. The raccoon has a brain that is about the size of a cat brain, but it actually has as many neurons as a dog brain, which is twice as many neurons as you would expect for a brain of its size to have. […] If you told me the numbers of neurons that go into the brain of a raccoon, I would have told you that was a primate, that's a monkey. That's how different they are. So it makes perfect sense to me when you hear all the stories about raccoons being mischievous creatures that just break into your house and turn trash cans and get into places and get what they want, it doesn't matter how well you keep it. It's perfectly in line with our finding that they have twice the number of neurons in the cerebral cortex than you would expect.

DD – I love to look at birds of prey teaching their young how to be predators, how to be flying predators. Or how to drop clams on the rocks, and then eat the clams. There is teaching by nonhuman animals, but there isn't explaining, and explaining is the key to understanding. And we are the only species that explains.

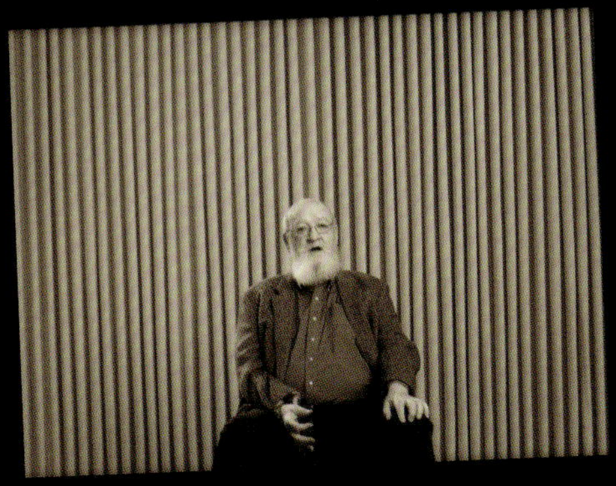

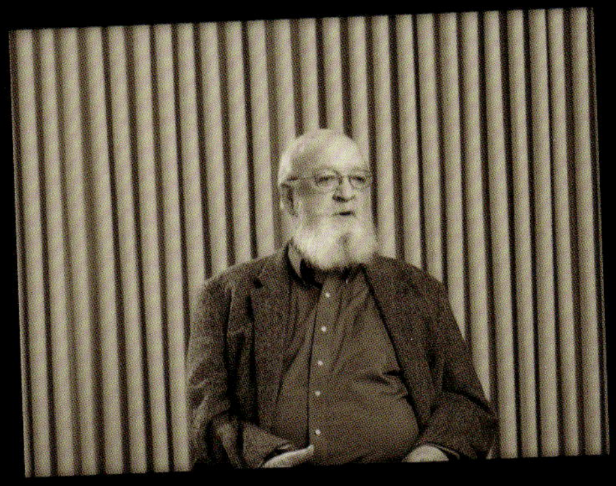

01:06:48:09

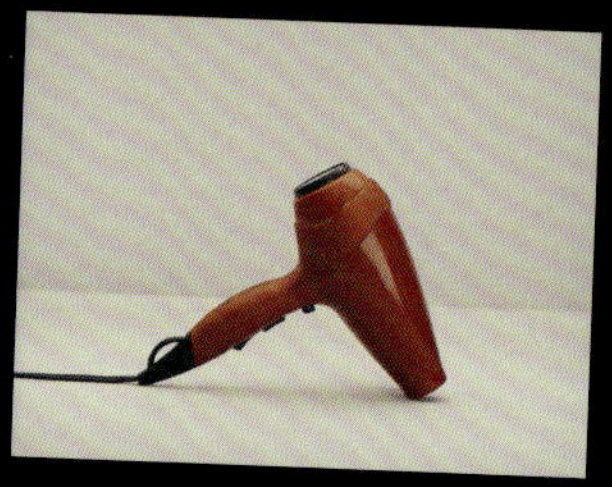

DC – So much of cultural transmission works through imitation, to the point where actually it seems to inhibit learning. People have pointed out that we didn't move beyond the hand ax for hundreds and thousands of years, because people just passed on the hand ax from one generation to the next. So it is arguable that an awful lot of cultural transmission inhibits learning.

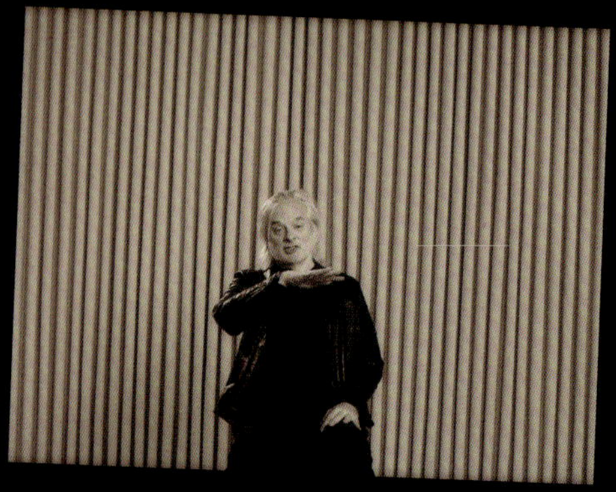

SD – A rectangle, for instance, or a line of dots—and these are the oldest drawings, not the realistic drawings. For instance, there is a piece of ocher in South Africa which has parallel lines forming a network and an array of equilateral triangles. And there is an even older artifact in Java, it is a shell, and there is a zigzag on it. And when we see this, we immediately know it is human, because no other animal makes such symbolic drawings. Even if you give pencils and you train a chimpanzee, they will never produce these sort of drawings. It is sort of the crux of the human mind, the ability to create geometrical patterns.

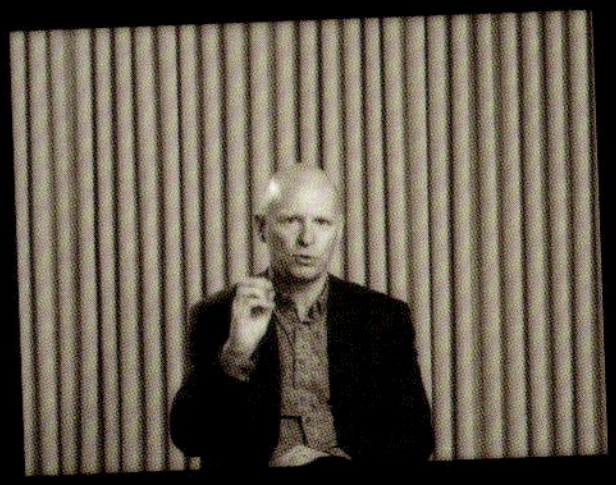

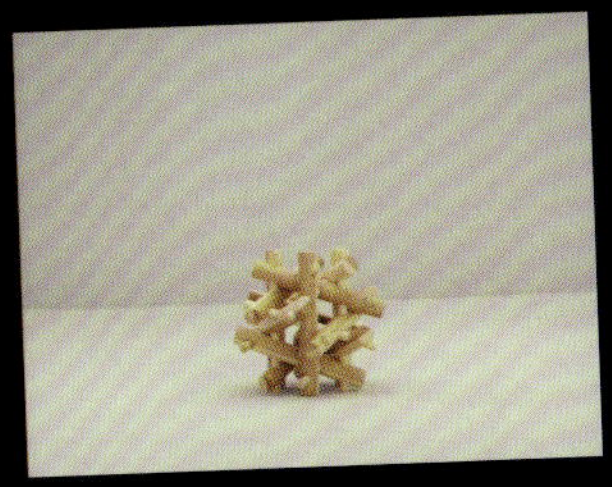

01:07:49:07

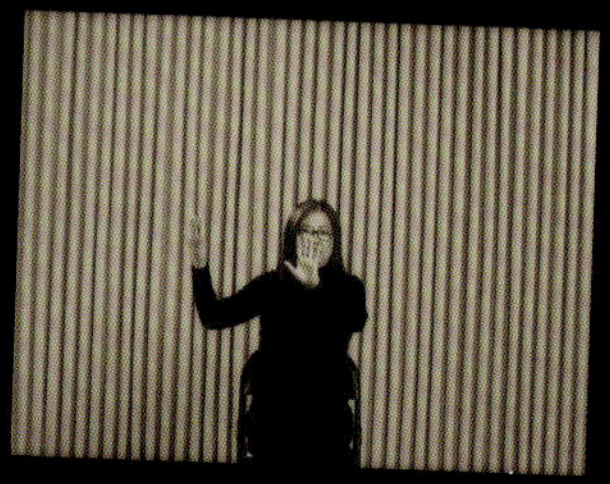

LZ – The primary visual cortex is the first stage in the cortex when an eye signal goes there. It is still raw information; it can only see a line, not the whole face. Each neuron looks at one line here, another neuron looks at another line here. Each just seeing a speck of the whole image. It's primitive, it's low-level, it's not intelligent.

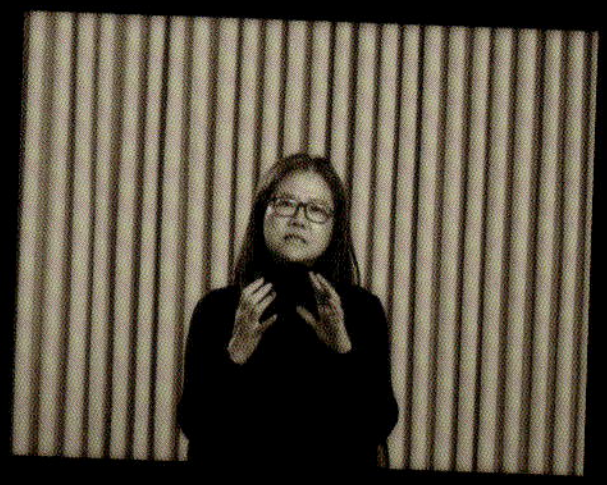

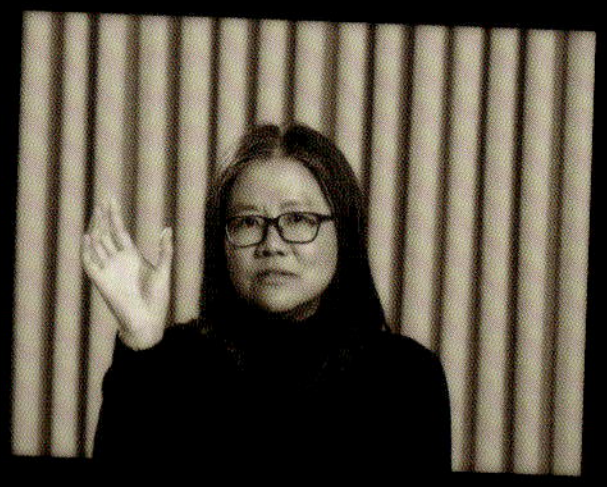

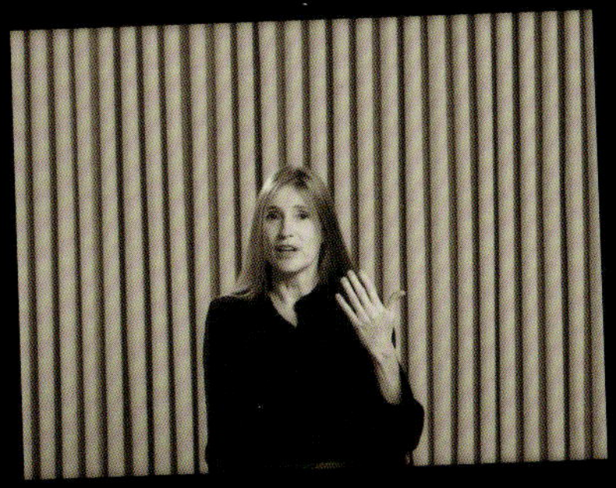

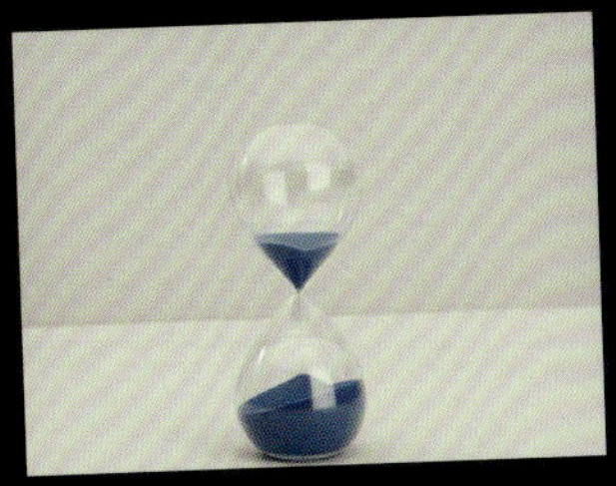

Daniela Schiller – Evolution embedded the brain with the ability to have inaccurate memories. It's very surprising. It's exactly what we would not want our brains to do. It's hard to think about it in an adaptive way, but when you make the effort of thinking about it, you realize that our reality and our situations are constantly changing. So a lot of information becomes irrelevant.

01:08:56:18

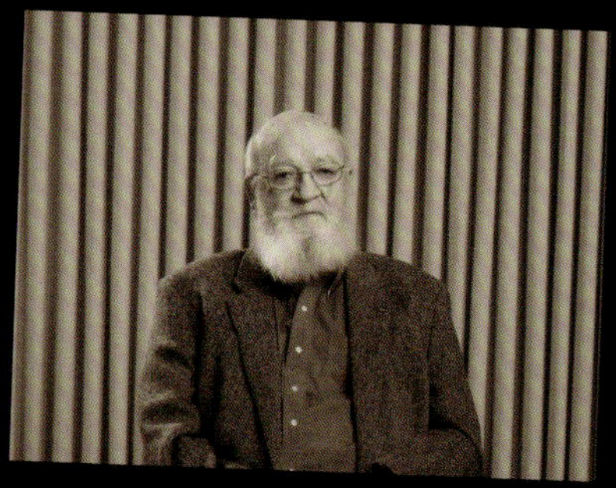

DD – Let's go back exactly a thousand years, to the year 1021, and now let's go back a hundred years from that, back to 921. There is almost nothing that the person in 1021 knew about the world, aside from who was king or something, that he could tell the person from 921. They had not learned anything in a hundred years.

DS – You know that it's irrelevant, for example, if you never retrieve it or the chances that you will retrieve it are very low, or when you retrieve it the information it consists of is no longer relevant. In this case, you would want to maintain previous experience but incorporate new information into it, such that the next time you will have a more coherent memory. That makes it much more effective than retaining multiple copies of multiple events and then sorting between them. It's much more efficient to have one event, but it's just to keep being updated.

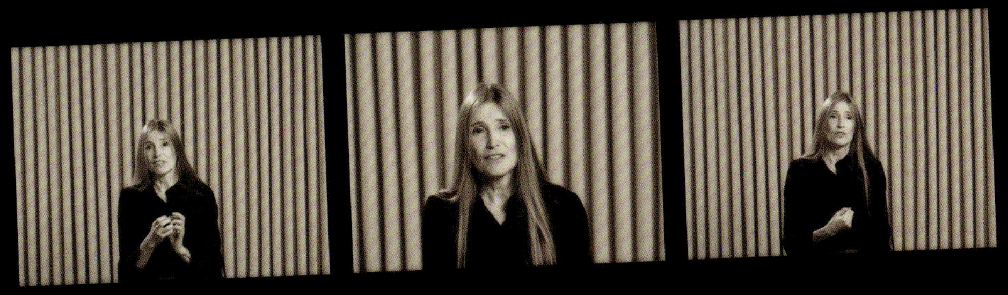

01:10:01:13

DD – Compare what we can tell today with what people knew in 1921. Tremendous difference. If you had a time machine today and you could go back to 1921, you could amaze people with all the things you could tell them that we know now. But a thousand years ago that would not have been the case. Everything was pretty much the same. That is the pace at which knowledge has increased, thanks to enlightenment and reason, in the last four hundred years.

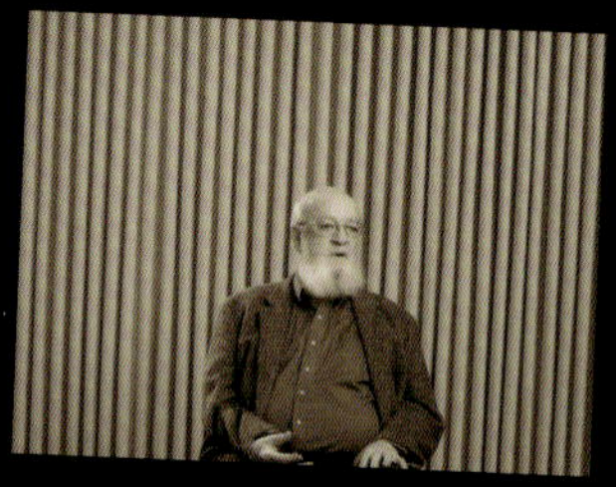

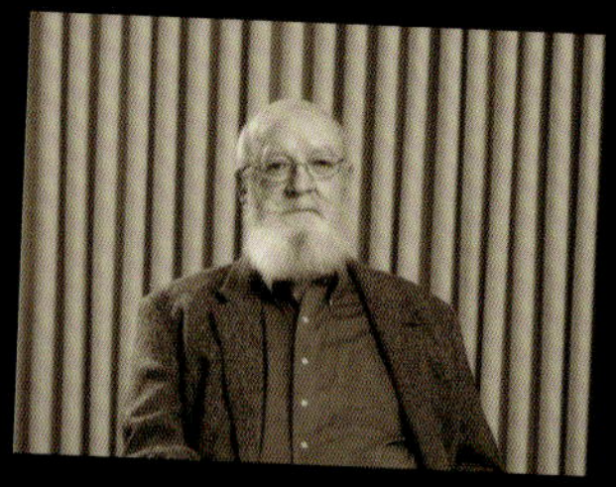

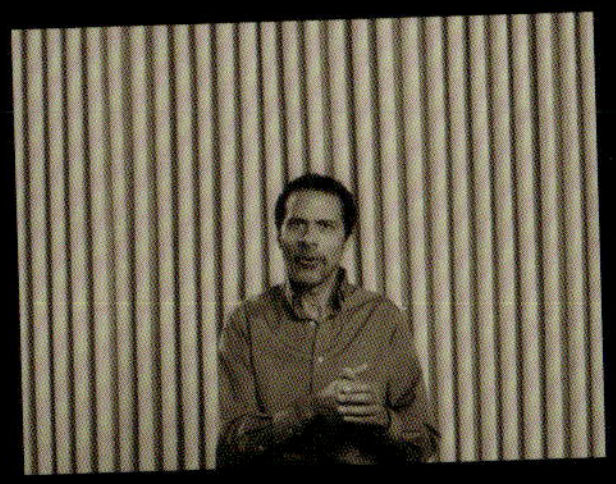

EJ – I think it's going to happen sometime in the future that we will be getting animals to be able to … I want to say it, I hate to say it, but speak. In ways that we could not imagine before.

01:11:21:18

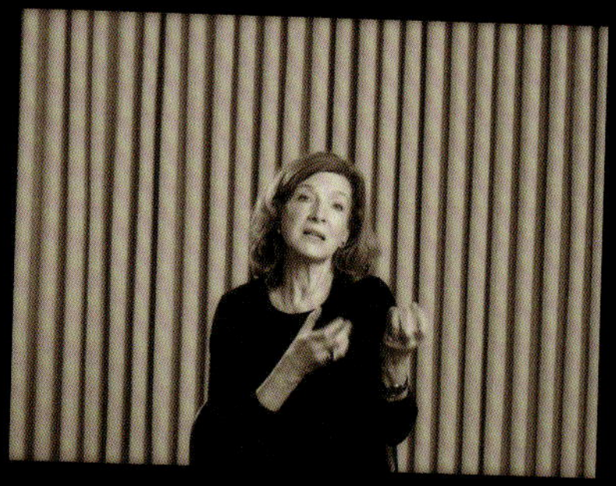

Hannah Monyer – You introduce some genes in just particular cells of the brain and you can turn these cells on by light in the millisecond range, and then turn it back to what it was before. So you have in the same specimen– most often animals, most often the rodents– something that is not painful but that you turn on and off at your own will. And you say, under these circumstances in this behavior, this cell, or even better this gene, does this or that. As simple as that.

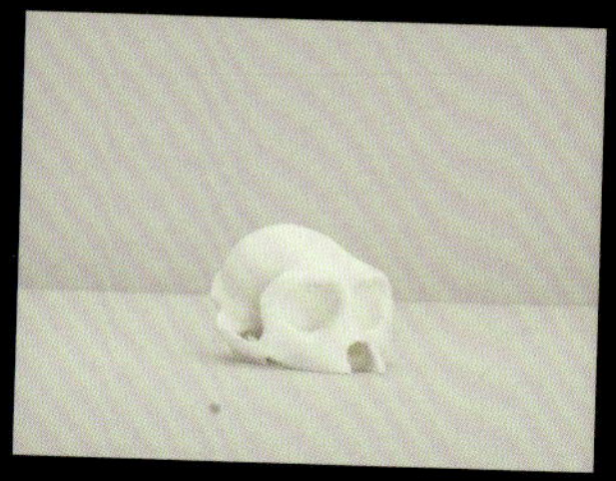

EJ – I proposed this idea of genetically manipulating a species to try to get it to imitate sounds, to get closer to what we would call speech. If I propose to do it in a non-vocal-learning bird, like a pigeon, people are okay. If I propose to do it in a mouse, people are okay. If you start talking about species more closely related to us, like a chimpanzee, or a macaque, or a gorilla, people start to get really scared.

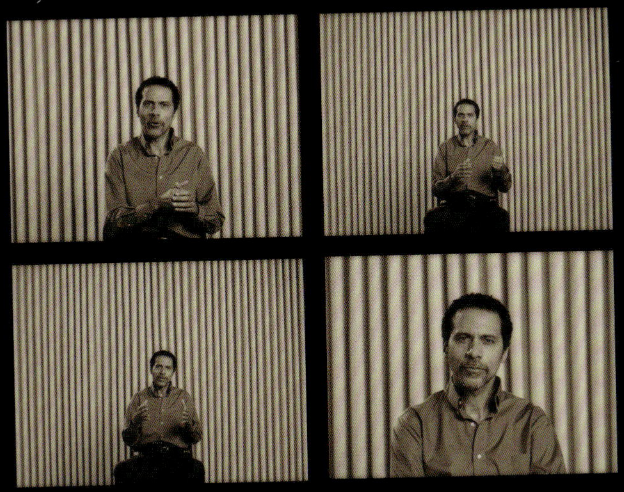

01:12:22:20

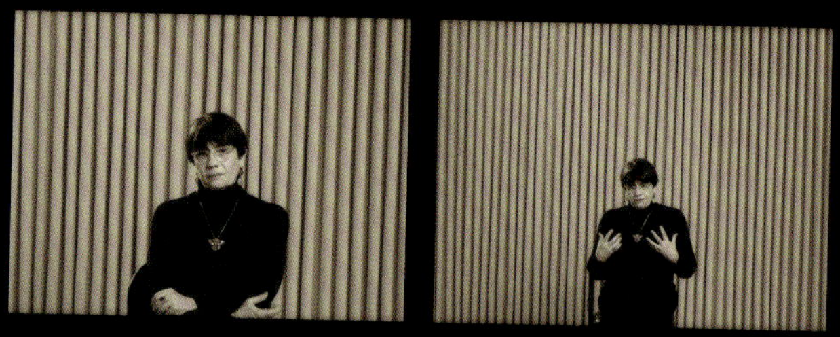

CM – I think it is not necessary to sacrifice animals. Perhaps it has been in the past, but now it is not necessary anymore. Because as far as the elementary functions of the brain are concerned, we already know anything that we needed to know. So now it is about superior functions of the brain. And for that we have the machines.

EJ – Even if animals have empathy or if we have empathy for these animals, I think it is valid to study them to understand human biology and disease. But I also think we should be studying these animals to help *them*. Help them with their diseases, help them help us figure out how to prevent their extinction.

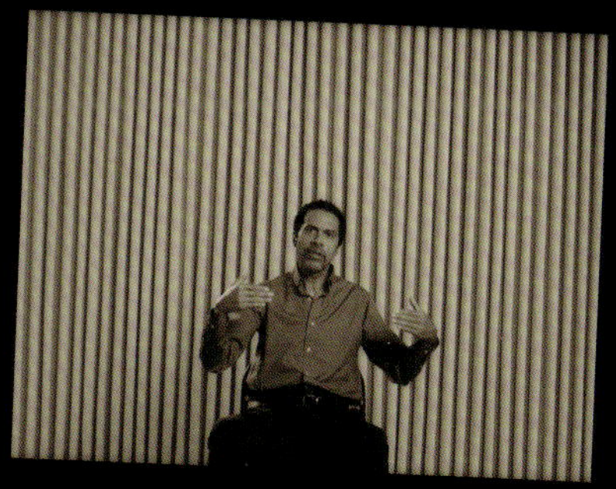

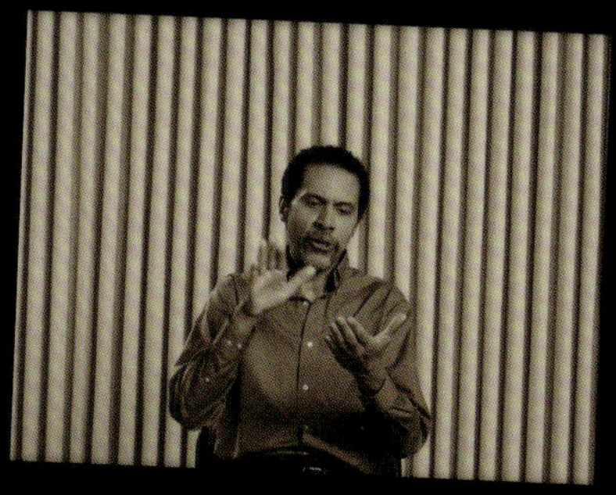

01:13:26:23

DD – I understand that people who don't understand science, or who are intimidated by science, are in some cases deeply threatened by this. And I suppose in a certain way they are right to be threatened. Because it is the people who know the science who are going to know how to do things in the future. And if they don't know how to do things, they are going to be in a very dependent position. And jolly well they should get themselves educated if at all possible, because that is not going to stop.

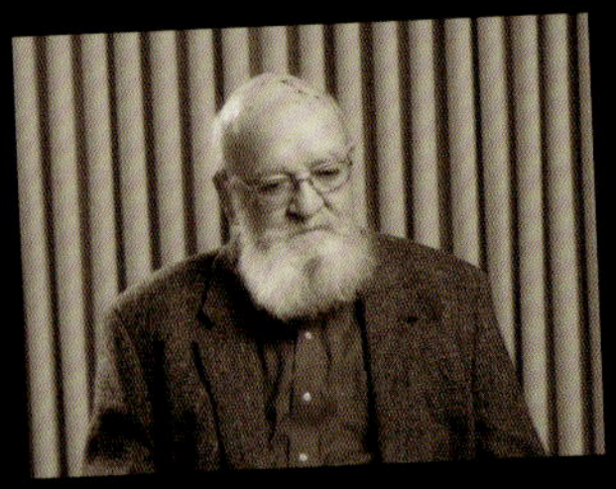

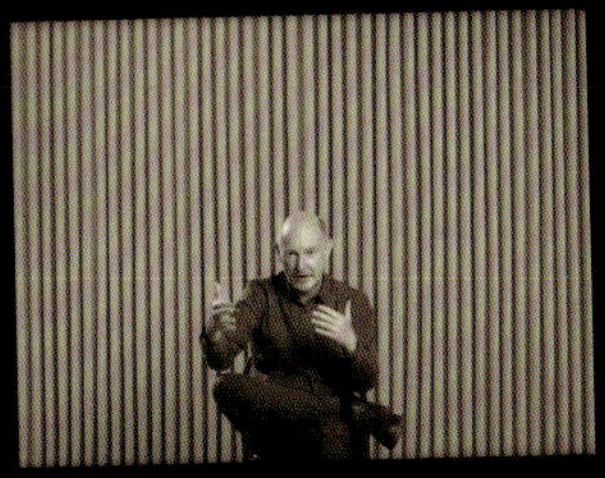

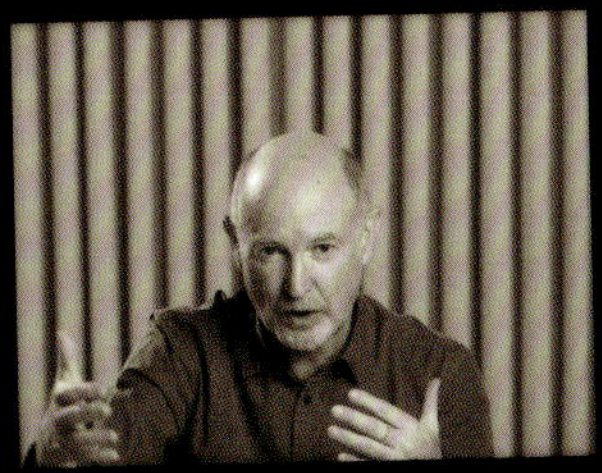

RY – We power our brain with the equivalent of a 20-watt bulb, and smaller animals it is even less. So this has to be a treasure trove of tricks that evolution has figured out over 750 million years.

01:14:36:10

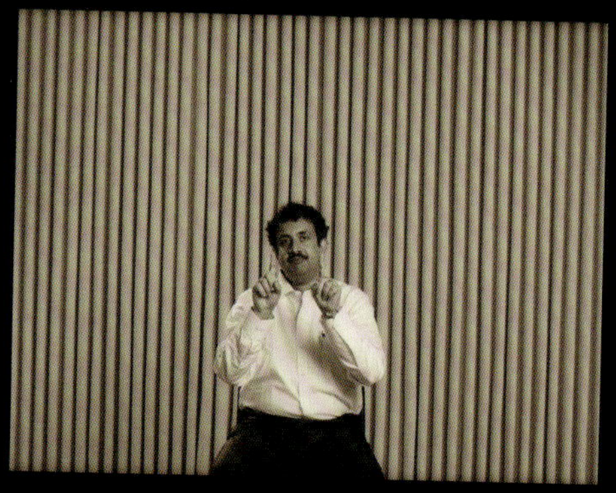

AB – Nobody understands how the brain works, and everybody knows that.

RY – The brain has a hundred billion neurons in 3D, and there is currently no technology to be able to access the activity of all these neurons. This is just technically too hard. So this is a massive technological challenge. And the natural way that we think should be approached to do this is to centralize resources as a national challenge and international challenge. It is like sending man to the moon. You didn't send the man to the moon by involving a hundred laboratories and giving them a hundred little grants. No, no. You built a hell of a space center in Houston and focus all their money.

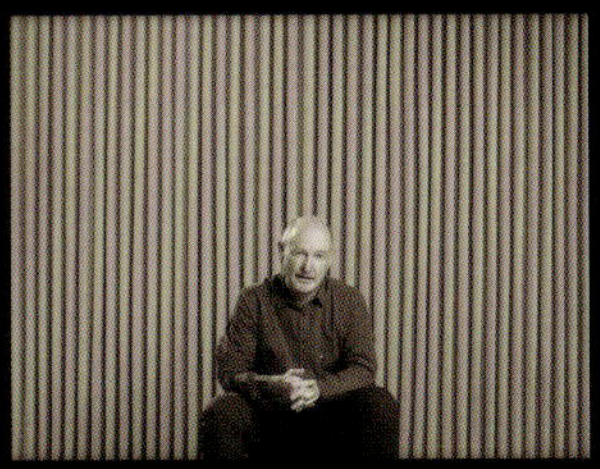

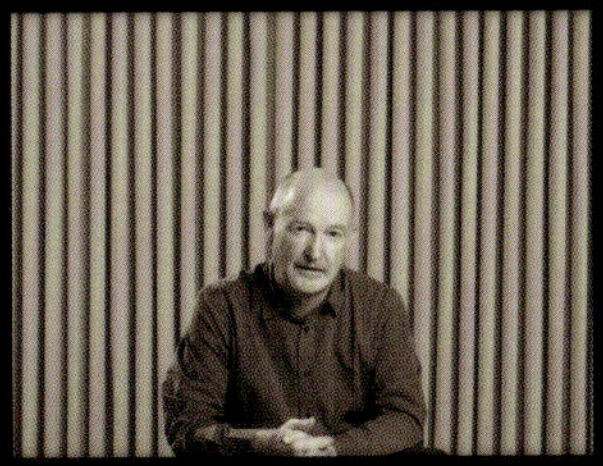

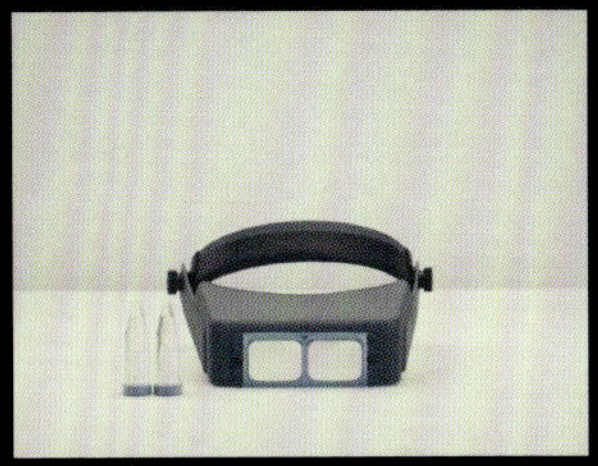

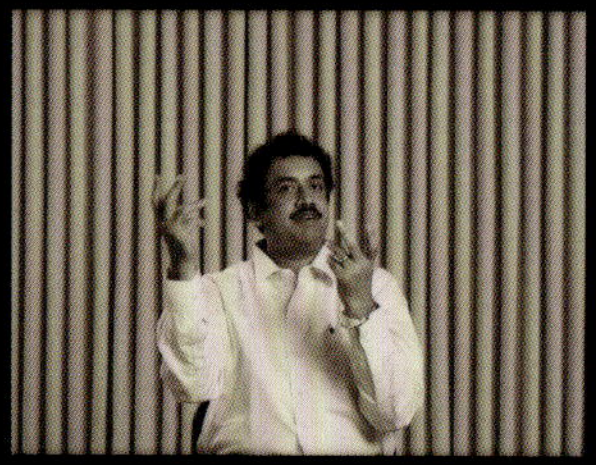

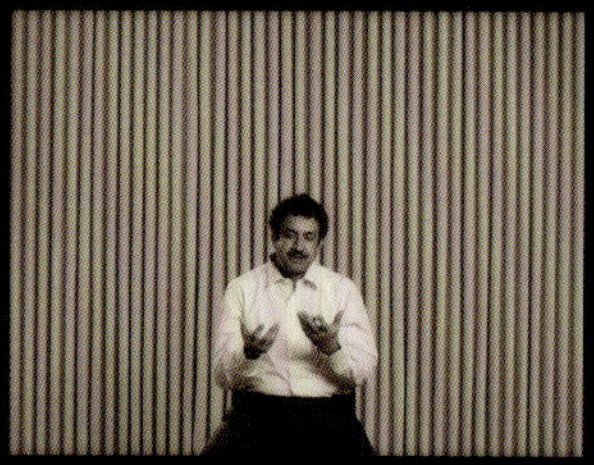

AB – And you see multi-billion-dollar project or an initiative, what did it produce? The great Blue Brain initiative in Europe, multi-billion-dollar project, $3billion project. What did it produce? Yeah, so. We all know.

GB – The interesting problem in living sciences, especially in neuroscience, is the "goal." It's the most problematic issue also in evolution. Evolution does not have a goal. Organisms do not have goals either.

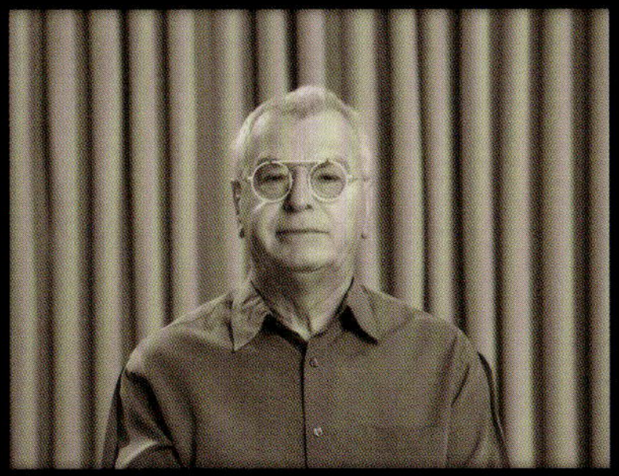

CM – The brain has in itself this tendency to laziness, to passivity, because it is a powerful organ and it takes a lot of energy for it to function. So there is also this principle of the least effort. And for me this is the danger, this is the main threat.

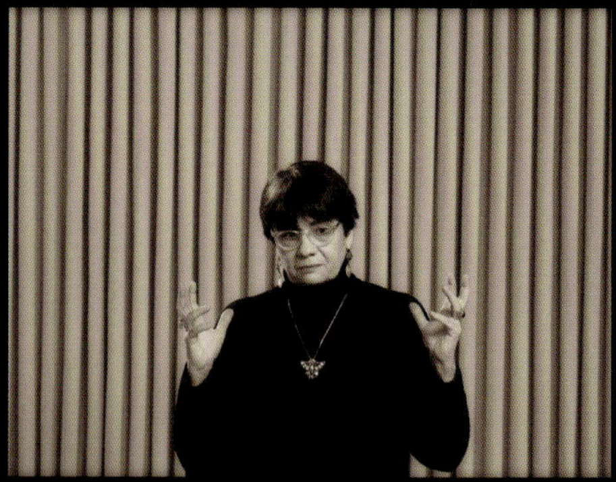

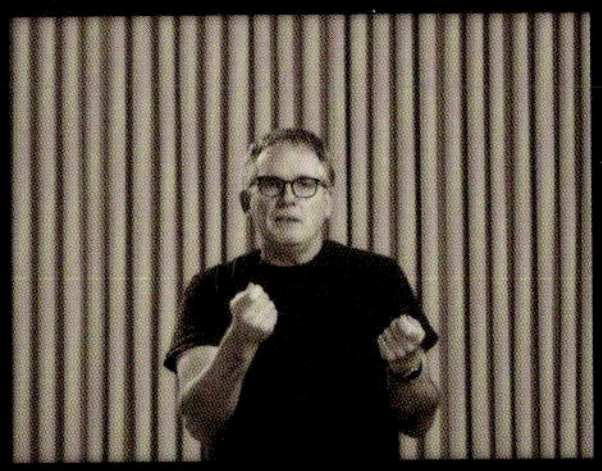

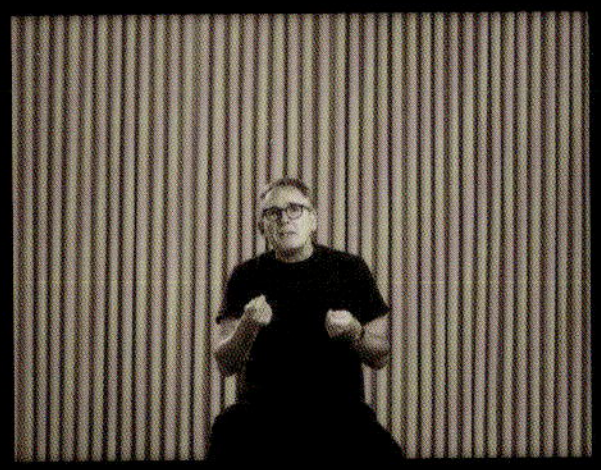

David Poeppel – As I am speaking there are mechanical vibrations, air, that is being wiggled. And that wiggled air arrives in your ear and it wiggles a membrane in your ear and that mechanical vibration gets translated into abstract concepts in your head. So you literally go from vibration in the ear to abstraction in the brain.

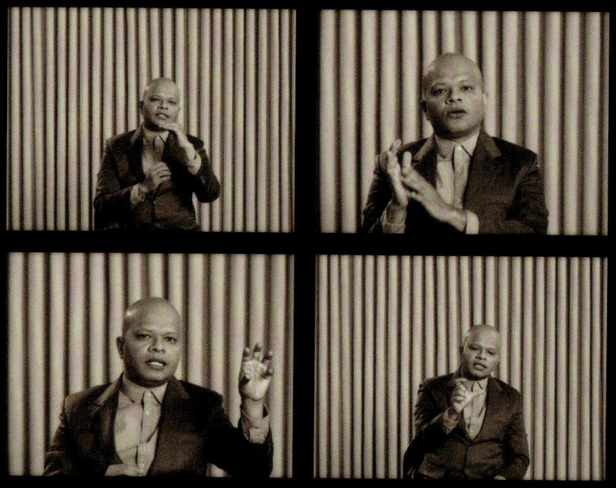

SR – It's incredible that we can get hypnotized. You are sort of letting you be controlled by someone else. And it's dangerous also. But just the fact that you can disassociate in a way that you trust someone else's voice or someone else's commands more than your own, what your own senses are telling you, it's actually a very profound thing in itself.

HA – Circadian patterns are those patterns through which all creatures respond to light–when the day starts, the light starts, when the day ends, and how our rest and activity cycles are really guided by light. And we have distorted that to some extent, but it's still circadian rhythms that are really important in when we wake up, when we sleep, when we act. The brain encodes that information.

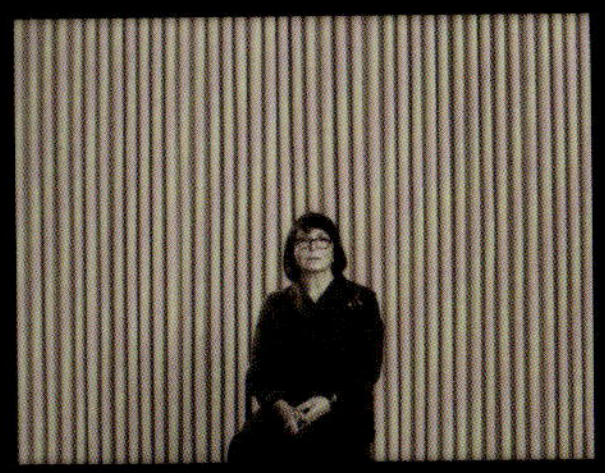

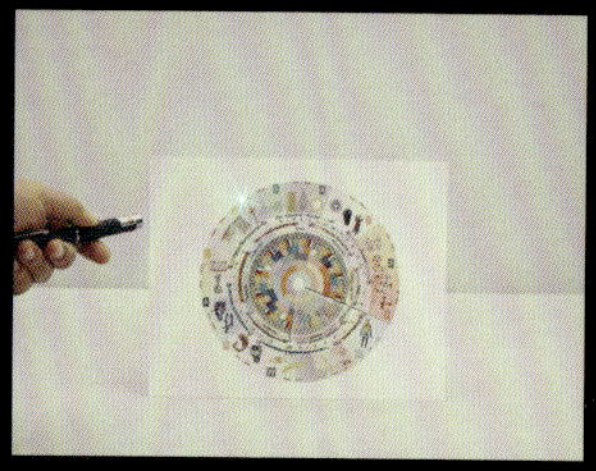

AB—There must be something very beautiful about it. There must be some way nature is processing information we never thought about. It's not bits—zero one zero one—based on which we have created this civilization, this wonderful civilization and wonderful technologies, machines, and everything. They are following instruction. But maybe composing music, in its own way, is how the nature works. And that may be the foundation of consciousness.

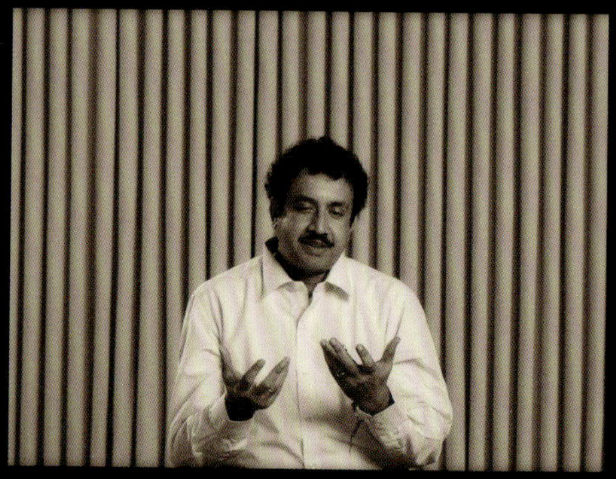

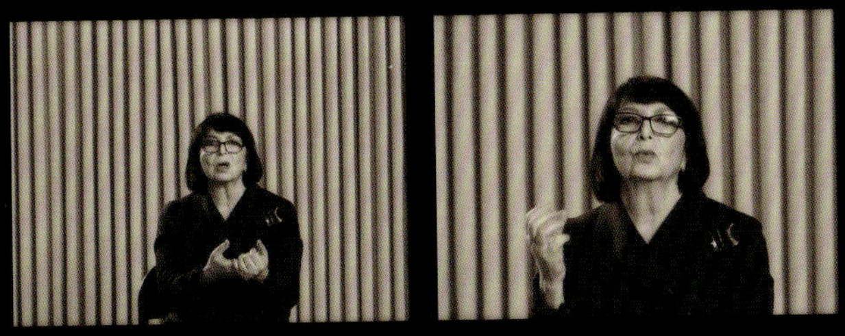

HA – When we look at the pattern of gene activity in a depressed brain, let's say at sunrise, and we try to predict what time they died, we might guess he died at sunset, because the time of day does not seem to register in the shape of patterns of gene activity that track correctly with the time of day.

DD – I think the existence of the biosphere is itself so wonderful that we should recognize that pain is a big part of it. Suffering is a part of it. We human beings suffer more than other creatures because of our imaginations. We see ahead. Much of the suffering is in anticipation and in recollection. And there is very little suffering when there isn't anticipation and recollection. And that's the case with all the little fishies that are eaten by the big fishies, and all the plankton that are eaten by the whales. And even the whales have only the dimmest sense of their own difficulties, more than almost any other creatures on the planet. But we far outstrip the whales, and the dolphins, and the chimpanzees, and the dogs, and the birds in our abilities to imagine suffering in ourselves and in others.

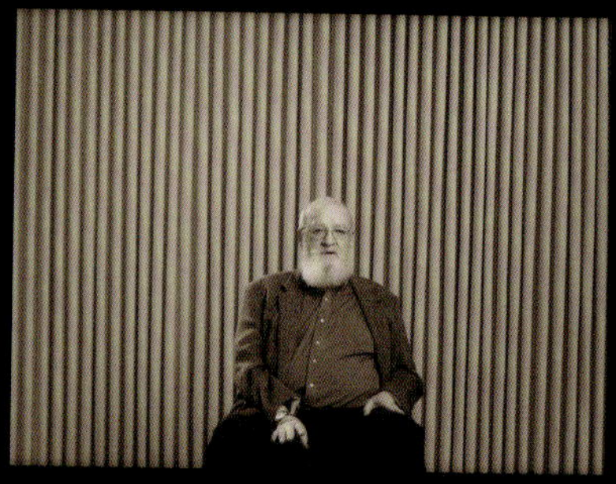

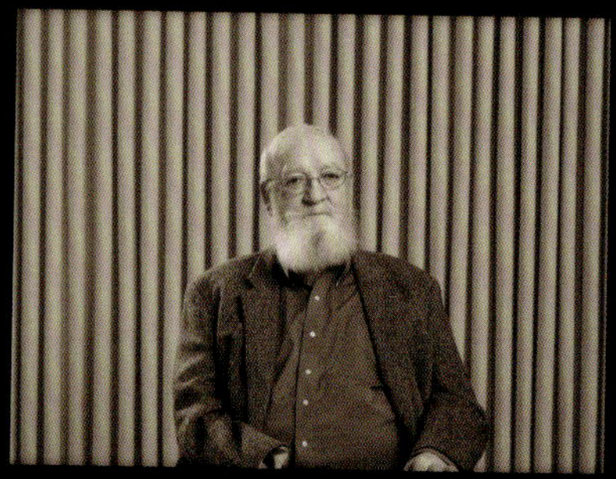

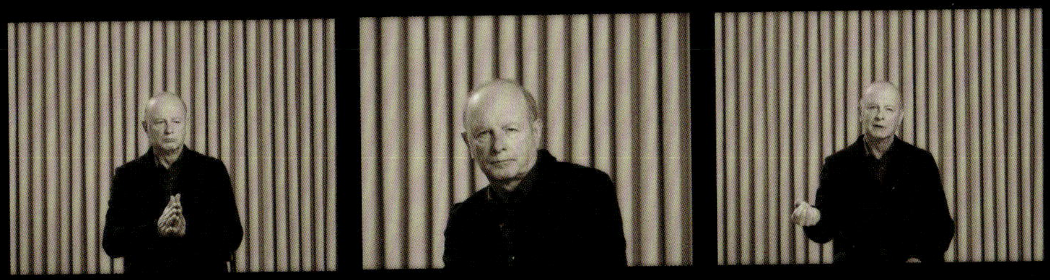

JL – Humans are the only animals that literally commit suicide. Other animals will find themselves in situations where they don't survive, but the intent to end the life requires something. It requires that you know that you had a start and that you have an end. And again, we can't go inside another animal's mind, but I think it's unlikely that any other animal has this ability to exist in their personal past and to exist in their possible personal future. And suicide is the decision to make that possible future now rather than later.

SR – There is some sort of bacteria or yeast, some very low form of life, and you start to take away resources so that this colony knows that they won't survive. So the outer layer sort of dies in order to protect the inner layer, so that the species would survive. And the outer ones would just basically give up their life so as to protect that. But if you mix two different strains, then they don't do it. They will only do it for themselves. Now, you would call it suicide, but I think it's an evolutionary principle. The species tries to protect or enhance its survival, even if it comes at a cost of an individual.

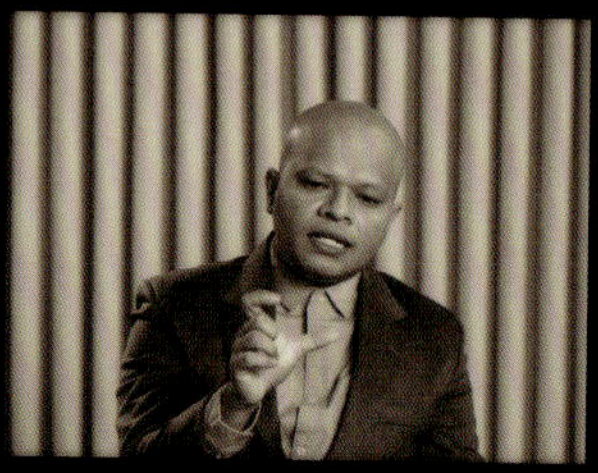

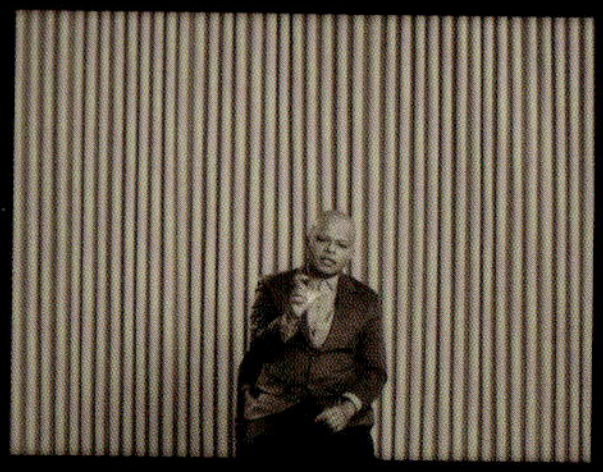

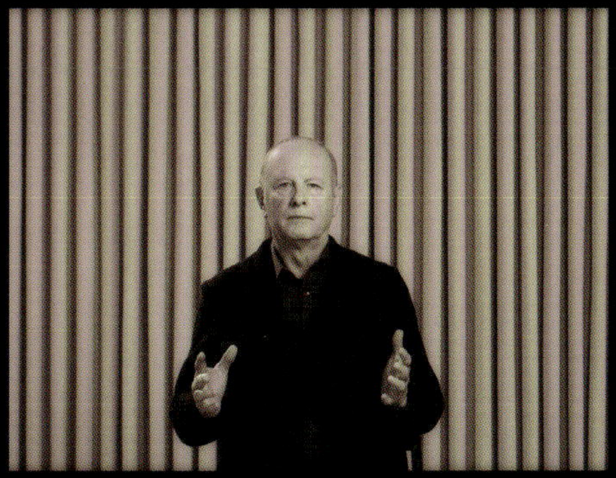

JL – The prefrontal cortex is the cost we pay for being anxious. We anticipate the future and we can be anxious. We can anticipate the future and we can kill ourselves.

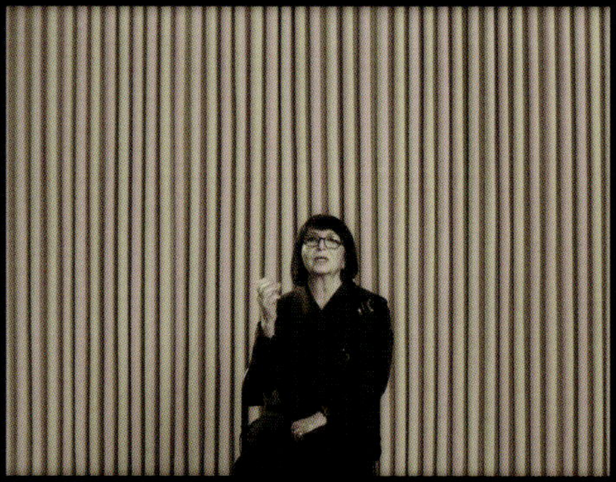

HA – I think pain can become familiar; a certain kind of pain that you have experienced before can be predictable and familiar. And if you know how it starts, if you know how it ends, that may give you a modicum of control and a feeling that you can, therefore, be more powerful than it or overcome it, compared to vaguer, more distressing, but very overpowering kind of suffering.

AI – In stress situations, there is a hormone called corticosterone that is usually released, and that is influenced or being controlled by the pituitary gland and also the hypothalamus in the brain.

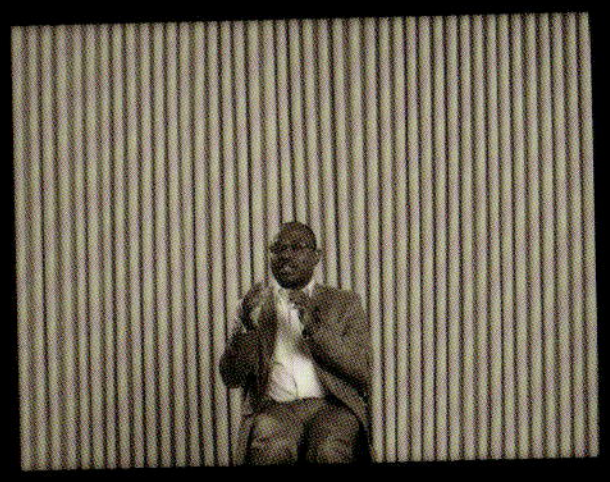

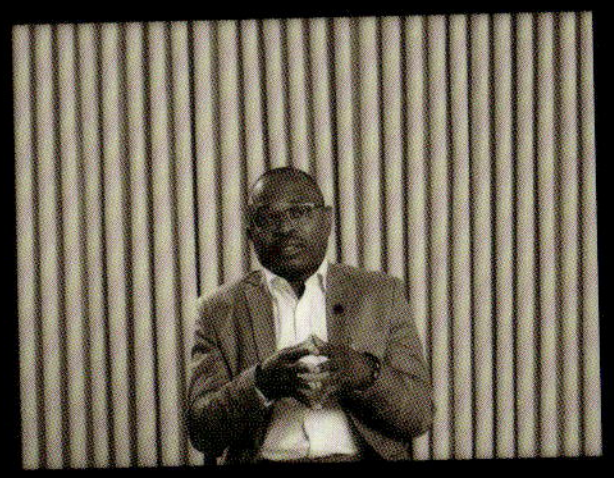

01:23:50:12

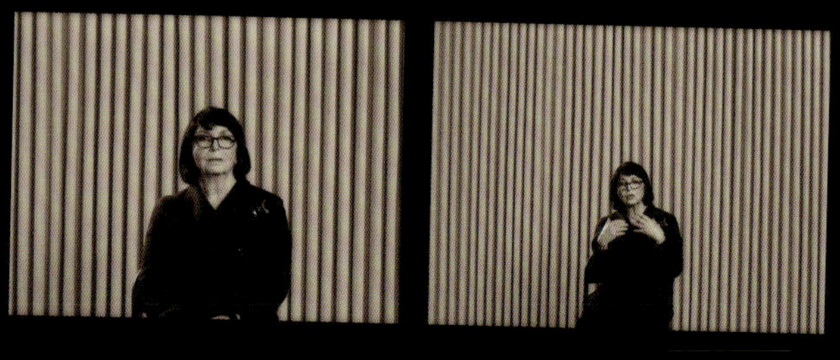

HA – People who are in great distress seek some feelings of pain that are predictable, because I think the predictability can itself be reassuring. The familiarity. Feeling something versus feeling nothing.

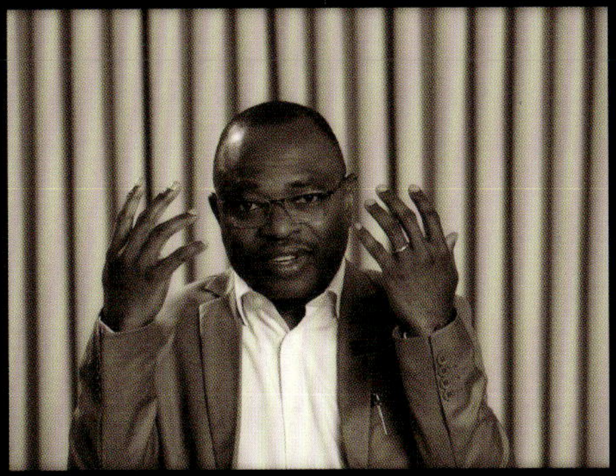

AI – Whatever is identified as the stress factor, if there is a withdrawal from that, the brain will also need time to readjust and allow itself to respond to the absence of that stress.

YA – I had a patient who suffered from an extreme trauma in her childhood and she expressed this trauma by what we call PNES, or psychogenic nonepileptic seizures. […] And in the therapy, what we are trying to do is to translate these nonepileptic seizures back to words. We give words to her experience and we express the trauma by using emotional states and words instead of seizures.

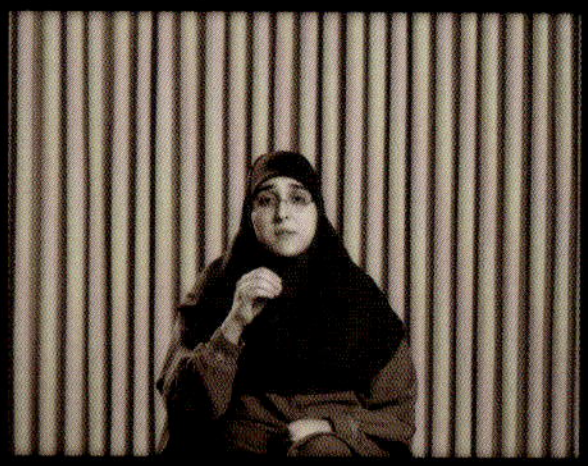

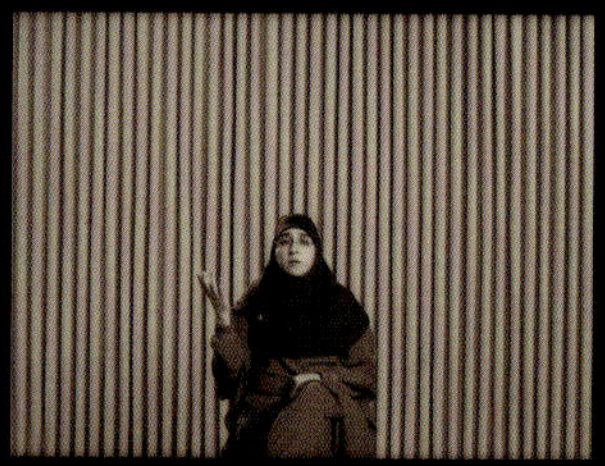

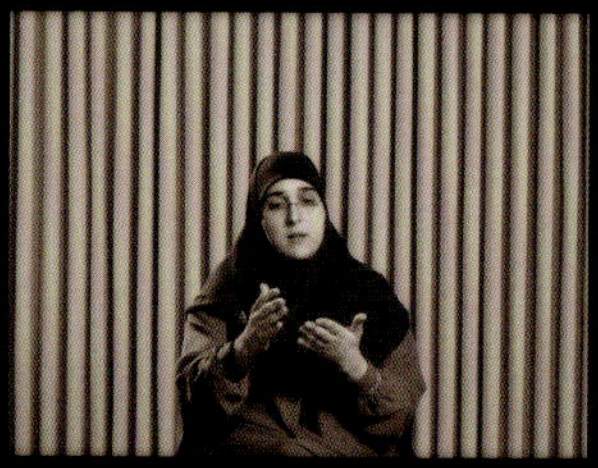

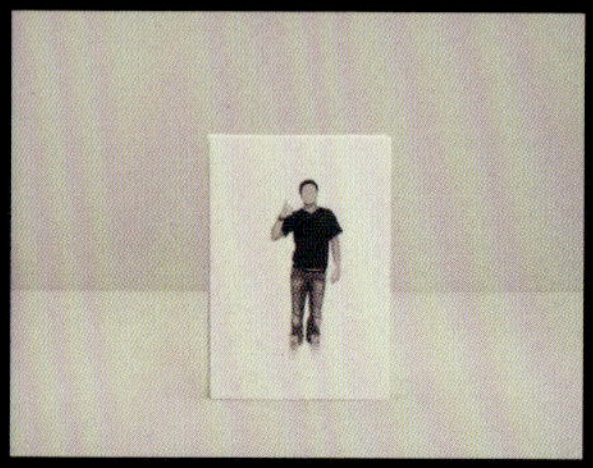

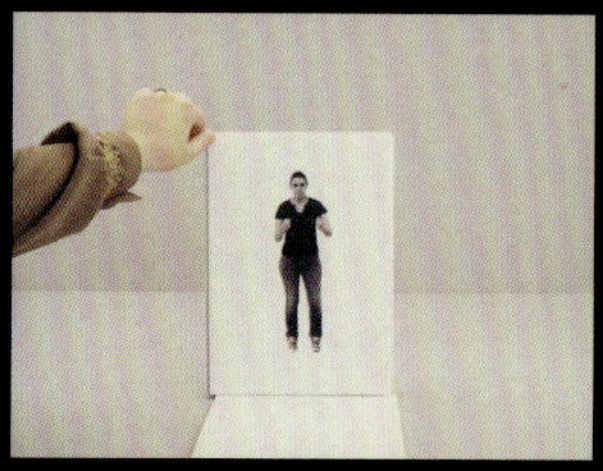

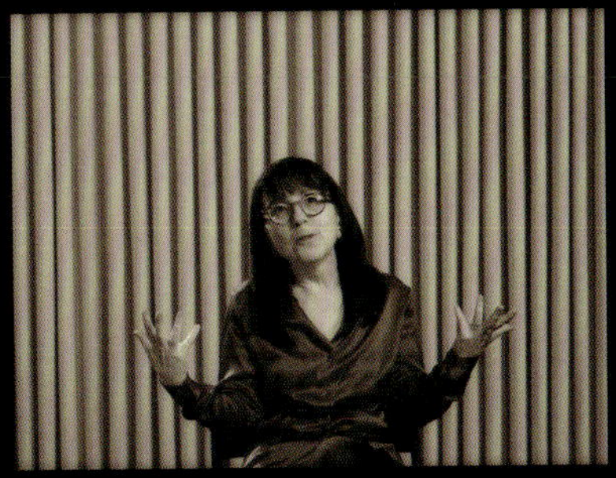

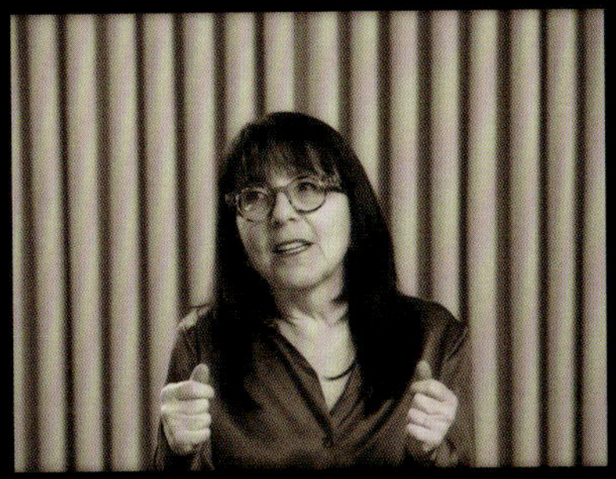

Lisa Feldman Barrett – Adverse events which provoke trauma, traumatic experience—what is damaging about them is the metabolic consequences that they have for a person. But it is very hard to talk about that without sounding like you are trivializing the experience, because what feels compelling to us—dramatic and important—is the suffering, is suffering of the person. And you can't really convey that suffering by talking about glucose metabolism.

01:25:38:13

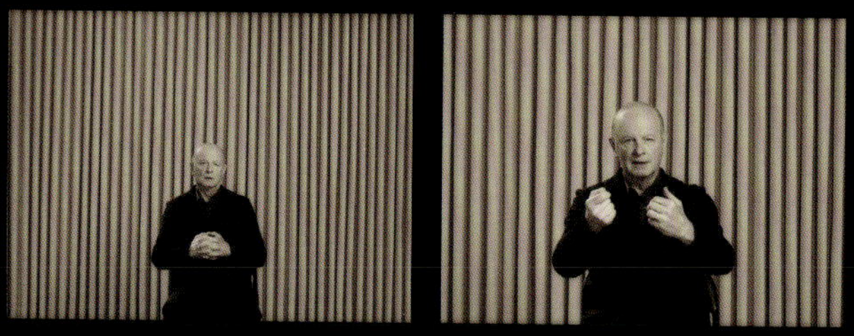

JL – One thing that personally affected my emotions was the loss of our first son, Jacob, at the age of seventeen. [...] So I think perhaps the music was so important, not only because of my connection with Jacob, but it is a kind of way of being in the brain where language fails.

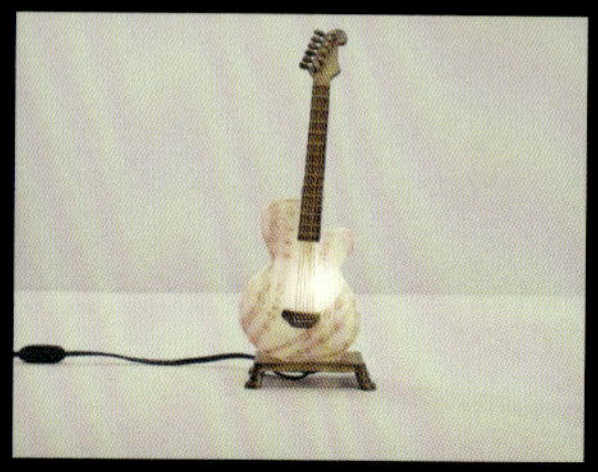

[*Singing*]
Mind over matter
It's something I'm trying to do
Guess it's only physics
Keeps me apart from you
Mind over matter
Something I'm trying to do
Break down space and time
Be together with you
I'm trying to will you here
Find you wherever you are
Locate your body, your soul
Pull you in from afar
Are you still in my time?
Have you slipped away?
Have you gone forwards and back
Living in another day?

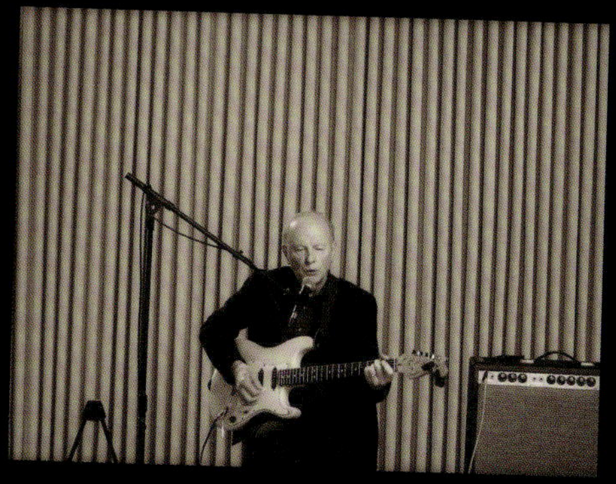

01:26:46:21

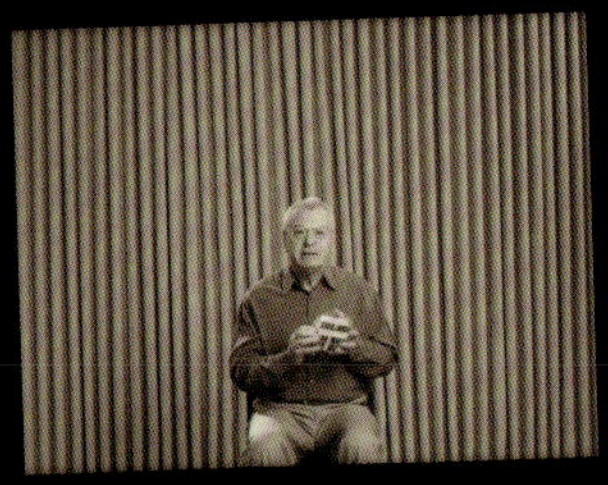

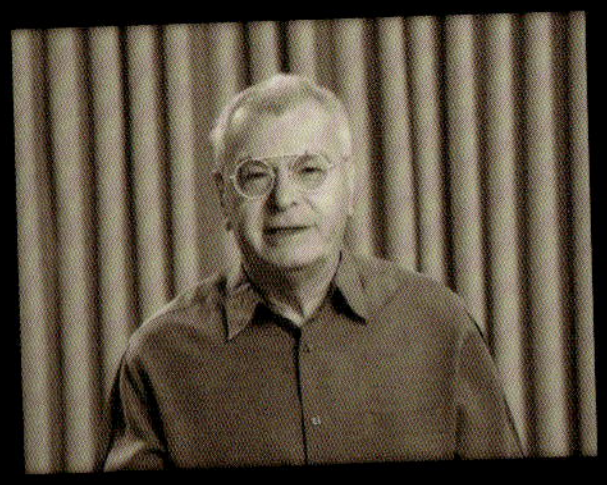

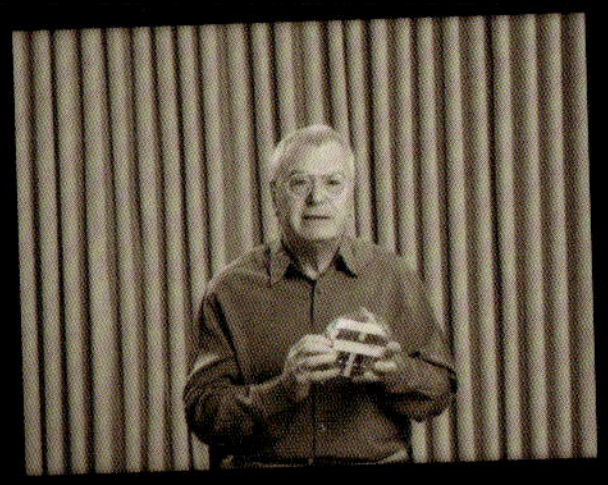

GB – Nature discovered, or allowed, a mechanism to go around and explore the world with cues outside there—there are mountains, there are trees, there are rivers—that help you to navigate. Now, when the brain, over the evolution, gets smarter and smarter and smarter, then you can do this navigation in the absence of outside cues. At the moment, I see you. You are on a screen. But in fact, I am not seeing the screen, I am seeing the computation in my brain. If I close my eyes, I can still see you.

AF – During sleep, your neurons replay some of your experience. We can experience that directly, as dreaming, for example, but you don't have to even have a conscious re-experience of that; your neurons just do it. And they do it because they have a propensity to do it. They fire in particular patterns that if you are awake would correspond to an experience. And so, in this way you can have a commingling of your real experiences, the experiences you might have had, and even the impossible experiences that your brain is prepared to have, but you likely will never have.

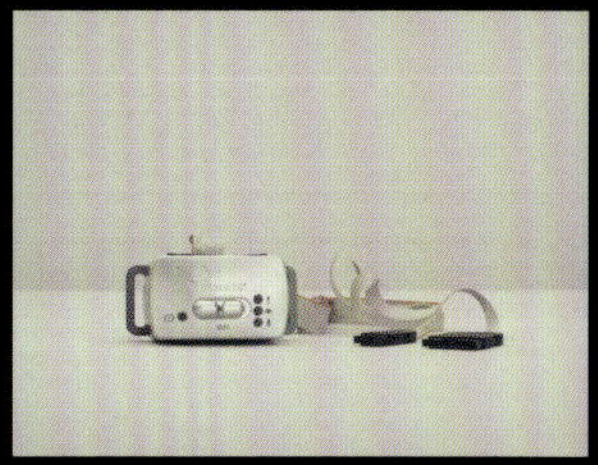

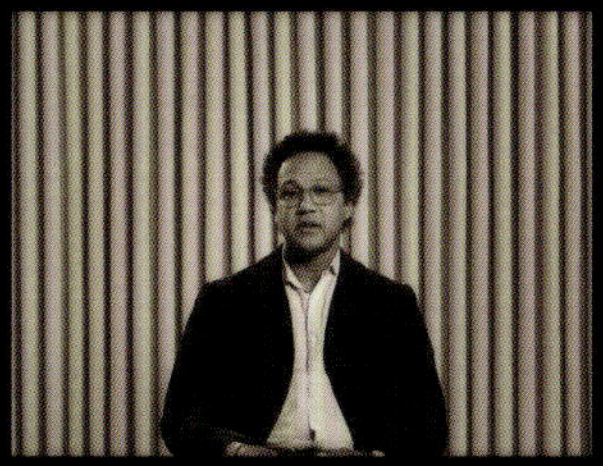

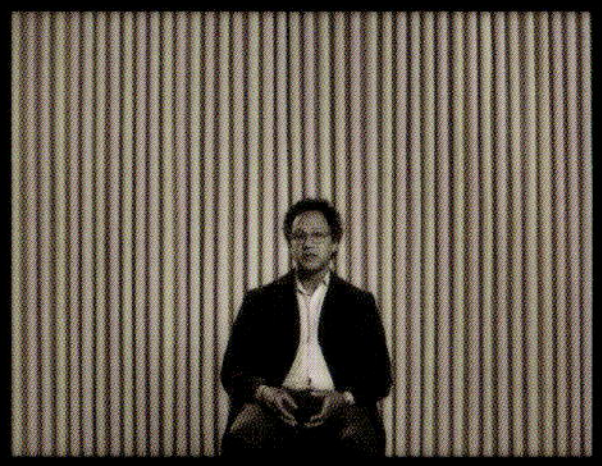

HM – There is oblivion in worms. There is oblivion in flies. And there the molecular mechanisms are known. There we know what happened cellularly. We wish very much to know that also for rodents. We wish to know that for humans as well.

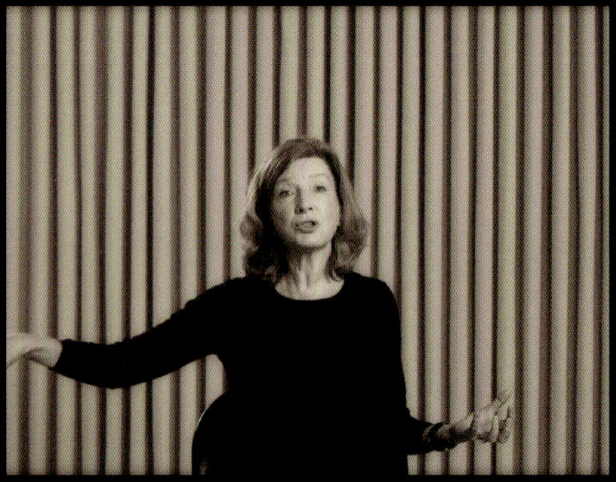

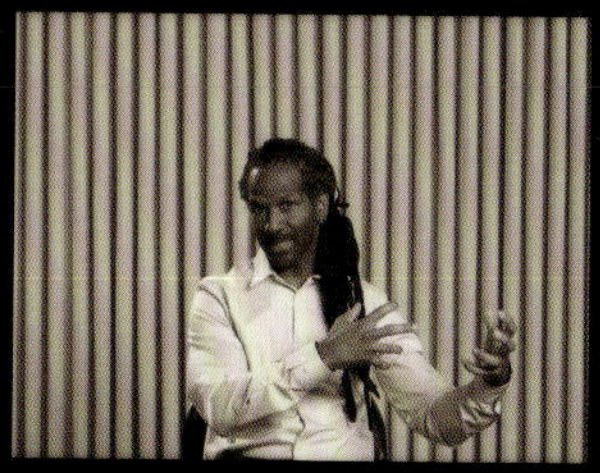

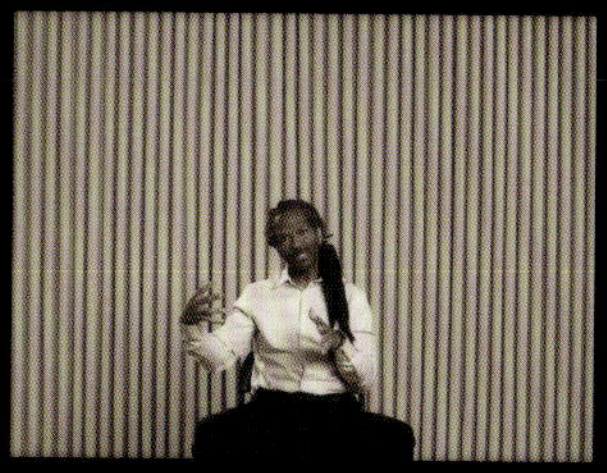

CH – One of the things that people just need to do, if they want to remember it, they have to stay awake long enough for the memory to be consolidated. So when we think about, like, date rape drugs, that is what is going on. The memory is not being consolidated.

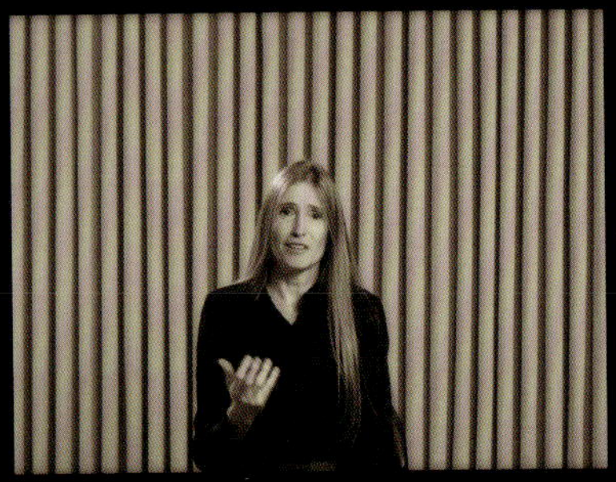

DS – My father never talked about his memories. I knew that some significant things happened in World War II, because I learned about it in school, but I never heard it directly from him, and whenever I made any kind of attempt—usually in the form of "My teacher asked me to ask you"—he always refused to talk about it. And I had to kind of imagine what he has been through. And I think imagination is worse than reality, but there was a point close to the end of his life where, really in the nick of time, I managed to talk to him a little bit and learn about these stories. And no, the reality was horrible. And I understood why he never talked about it. And from his perspective, he said, "Suppression is awesome. Suppression is just good for you. You wouldn't be able to function without suppression." And after talking to him, I kind of realized he has a point.

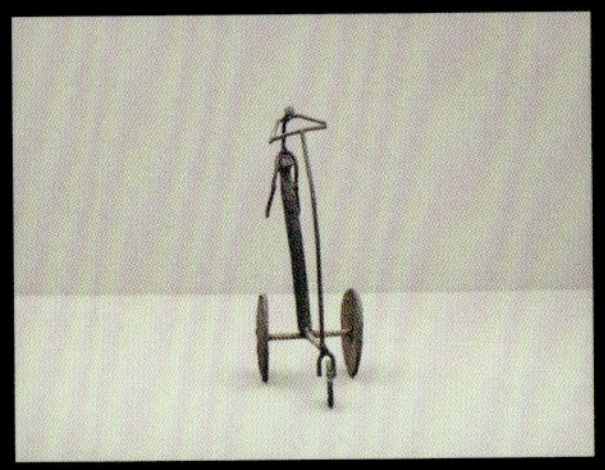

DP – The brain samples and processes things at very fast timescales, like *chop chop chop chop chop*, and slower ones, *chop … chop … chop … chop*. And all these things are happening concurrently and have to be aligned on top of each other. But the notion of continuity is actually illusory.

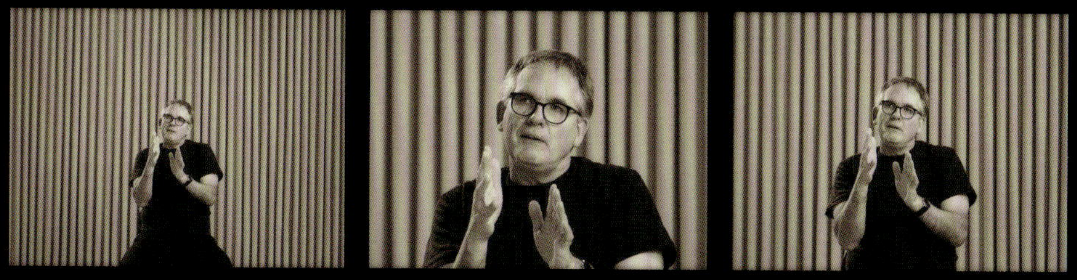

DS – The point of reliving, retrieving memories and modifying them is to incorporate them within your present. But some memories are just too much at the edge of life, just impossible to incorporate them, the reality is unbearable. In these cases it seems suppression is an adaptive mechanism. It's actually good.

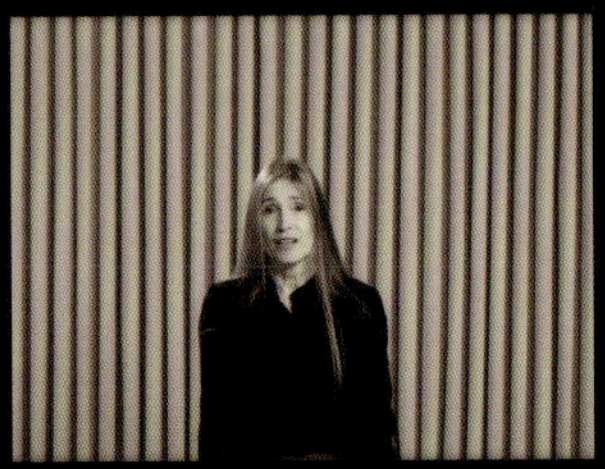

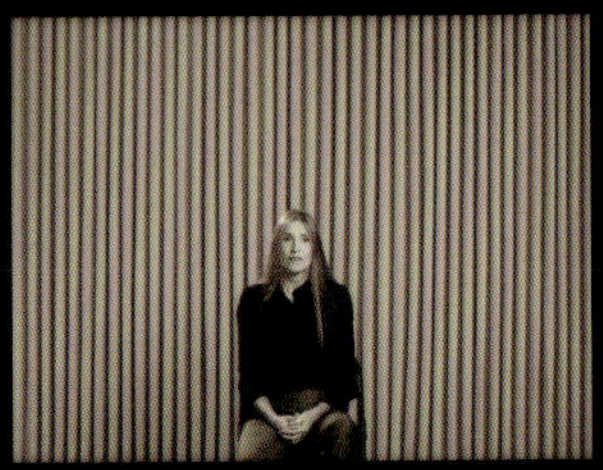

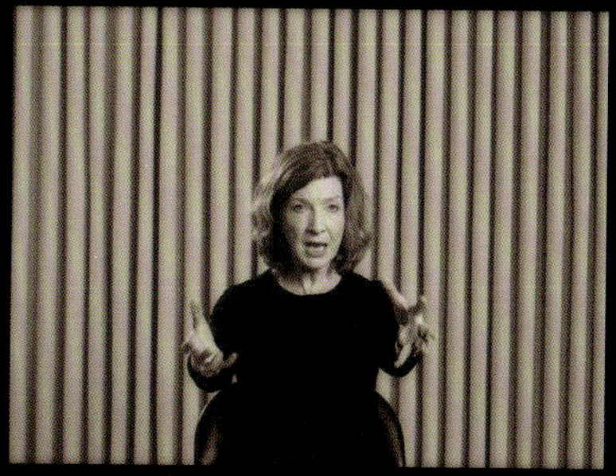

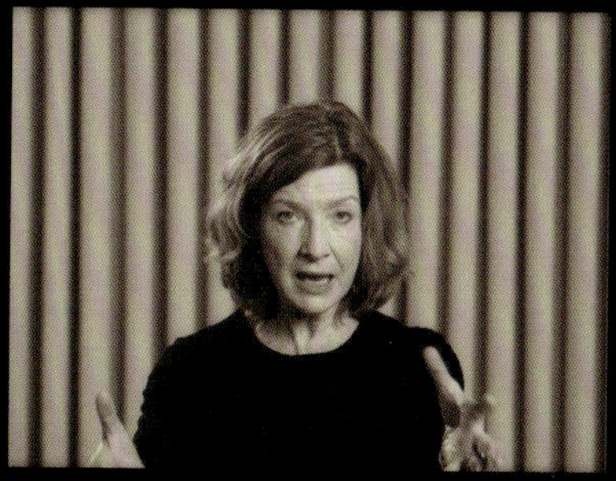

HM – During the first three years of your life, of a child's life, of course the child makes memories, but the child has no access to these memories later. So it is very well established in many fields of science that what we think are our memories earlier than the third, fourth year of life are not memories that we have in our brain.

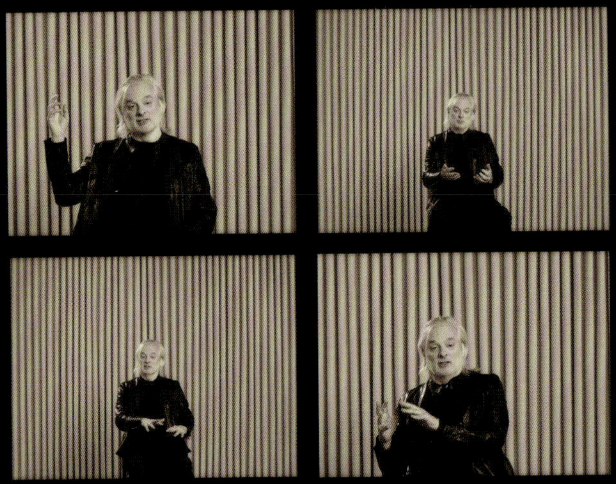

DC – People used to think that infants were not conscious at all. And this was why they circumcise them without giving them anesthetic, because they thought, "Well, they are not conscious yet. They don't feel pain." Now that seems very brutal to us.

HM—If anything traumatic happens and you have problems later in your life because this trauma has happened—you do not have, by whatever means, you do not have ways to get back to your memory. You may be a disturbed person later in whatever realm of your daily life, but you have no ways to access these memories to integrate them.

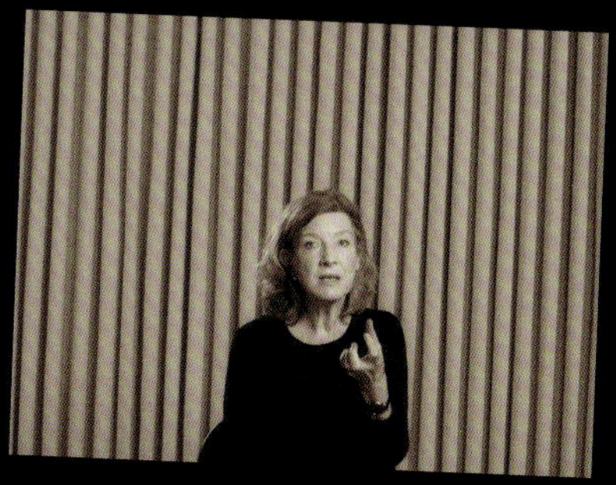

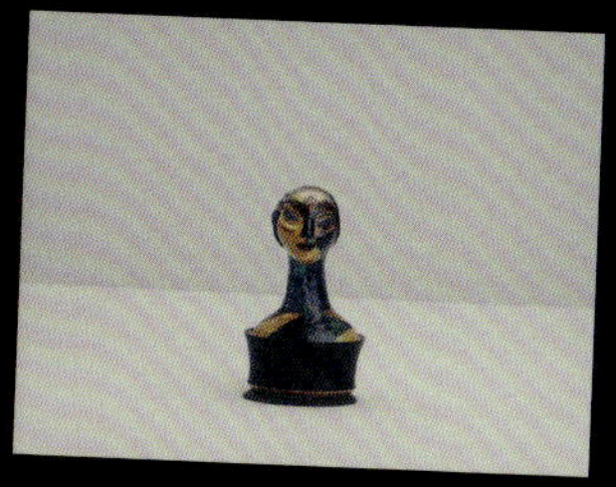

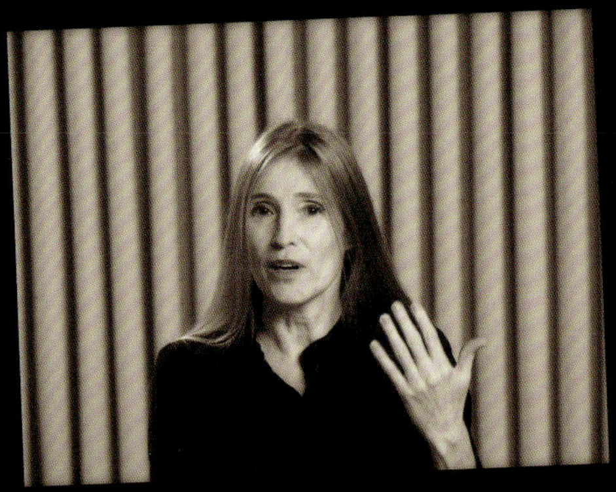

DS – The memories you want to protect are the good ones. And in order to protect them, you don't have to retrieve them as much. Actually, if you don't retrieve them at all, they are perfectly protected. [...] If you go and talk about it over dinner with friends, all the time in different situations, it just makes the memory simpler and modified and contaminated, in a way.

GB – We believe that we are our own memories, our own episodes. If you take away my memories, I am just a zombie. I will stop existing, and this unfortunately happens with diseases such as Alzheimer's disease and so on. And it is a devastating condition. I would rather give you my legs, my arms, or anything than my own precious memories.

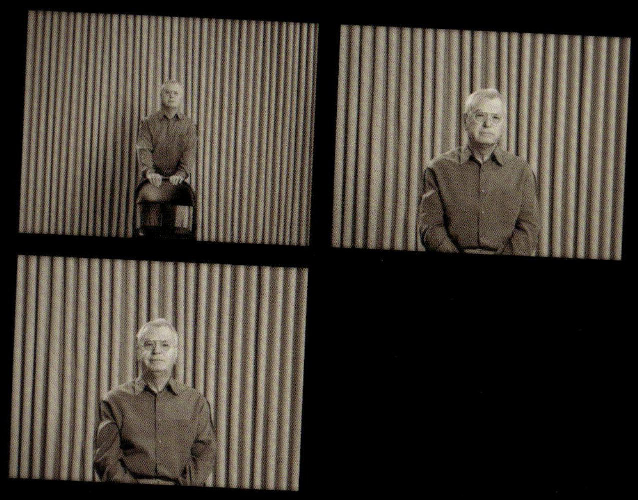

01:33:54:13

DD – It takes a good long while when a dear friend or family member dies to get over all the habits that we have that depend on their still being with us. And those habits very often give rise to basically hallucinatory beliefs that they are still with us, that they are ghosts. That we can still talk to them. We have dreams. Or we just feel a presence in the house that isn't there anymore. "She is standing right behind me." No, she isn't. But you are so used to her standing there that you can sort of feel her presence.

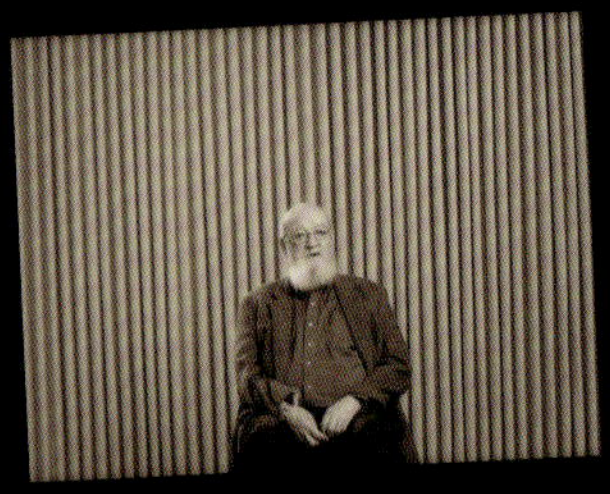

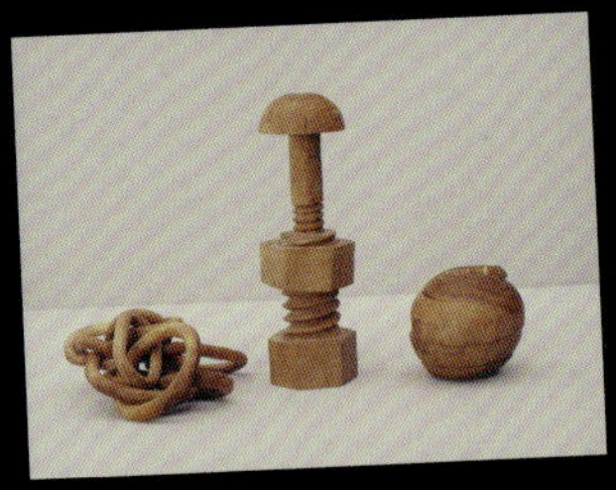

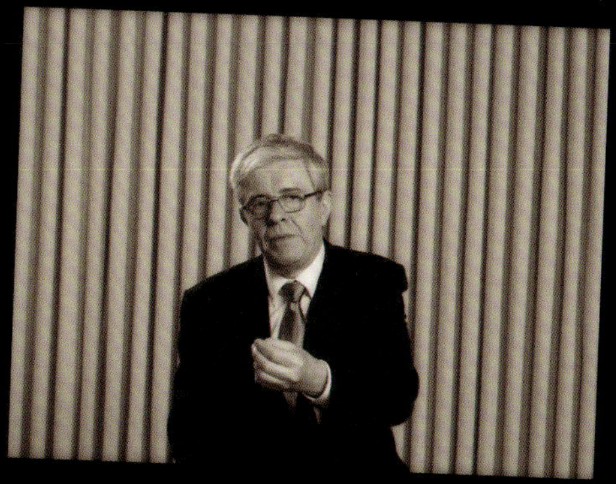

KF – All these sensations from out there pressing themselves onto this boundary, this sensory veil, this thing that enshrouds the internal things of the thing, which in this instance is the brain. That thing is itself a succession of nested boundaries creating like an onion of hierarchies. And at each point, the onion, or one layer, is trying to predict what the layer on the outside is in receipt of. And what it fails to predict is a prediction error that is passed through to the deeper layers to revise the predictions, to make them better.

01:35:16:16

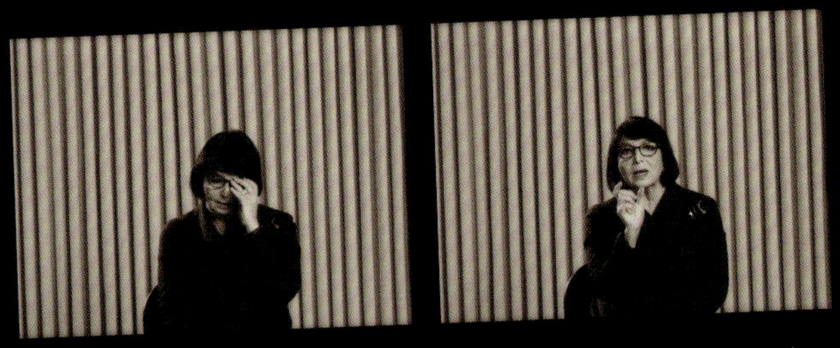

HA – From a neurobiological standpoint, beliefs are very helpful, because they decrease the uncertainty, they decrease the noise. You immediately perceive somebody as safe or unsafe, a situation as good or bad. And diminishing uncertainty allows you to act rapidly; it cuts down on the time you have to take to make decisions.

SR – It is as if we live in this world which is very hazy, and attention is a spotlight that falls on one part. That part sort of lights up. And then as attention moves away, only wherever that spotlight is, that information goes through to the higher centers of the brain. The rest is going to get filtered out. And then this information interacts with the memory and is stored in the memory. So the way attention moves around and the way these circuits play out in the brain is going to eventually define what kind of me you will end up being.

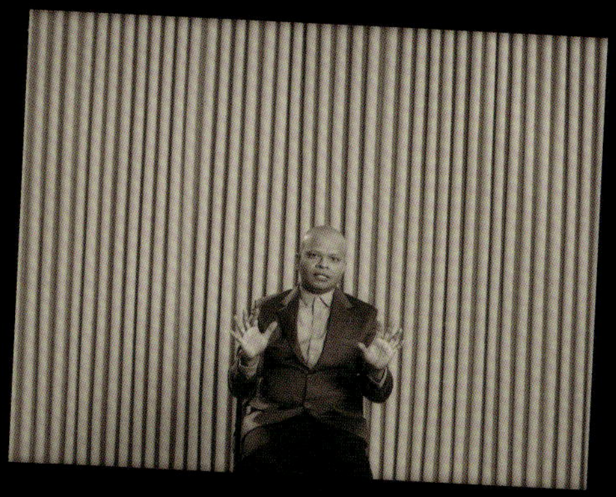

01:36:24:03

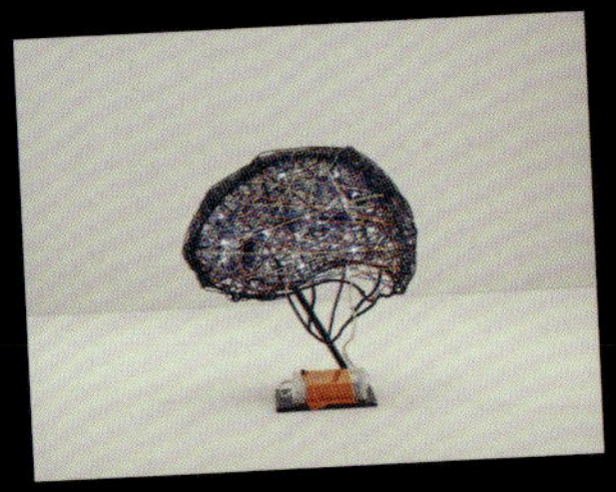

WS – What I experience is something that depends on the framing. The framing is given to me by my brain, by the experience that is covertly a resident in my brain. I am not aware of all these, let's call it, prejudices.

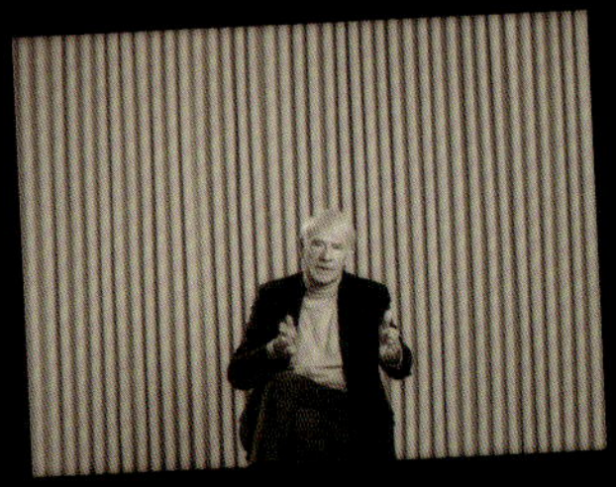

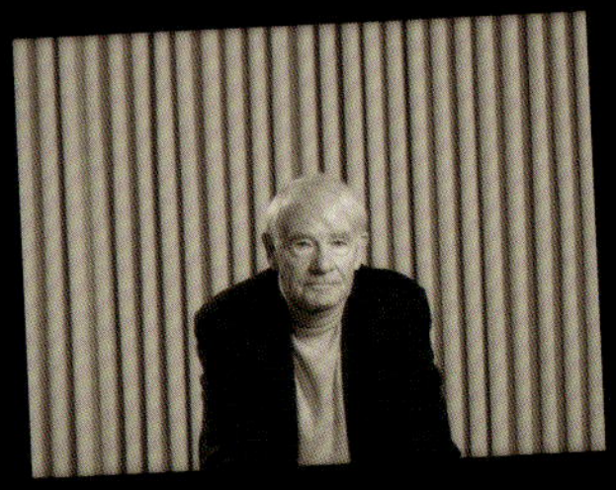

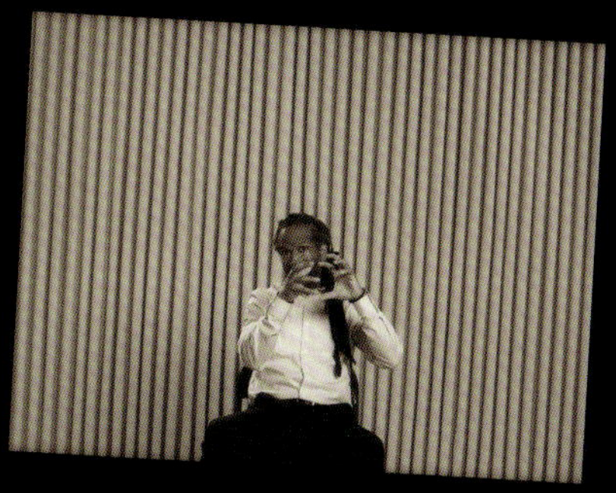

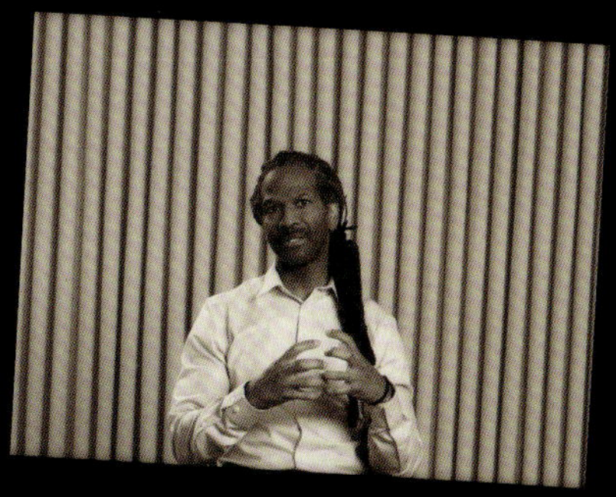

CH — As Derek Chauvin has his knee on George Floyd's neck, the other officers, the three officers on the scene, one of the officers was telling bystanders—He is looking at bystanders and saying, "See, kids, this is what drugs do. This is why you should stay off drugs." And the implication was that if you use a psychoactive drug, you can be murdered. You now can be murdered. And so the question for me became, How in the hell have we gotten here? Where somebody believes that simply because you chose to alter your mind, your mental headset, you wanted to feel better, you wanted to feel different—you can now be killed.

CM – There are universal, let's say, structures that are spatial and temporal. But out of these identical structures, we don't build the notion of the real in the same way.

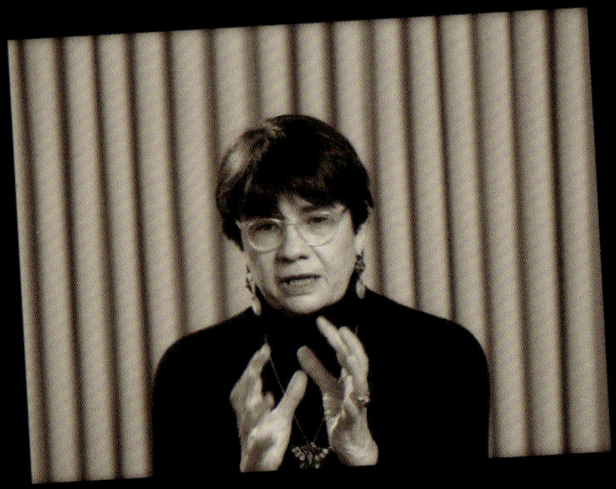

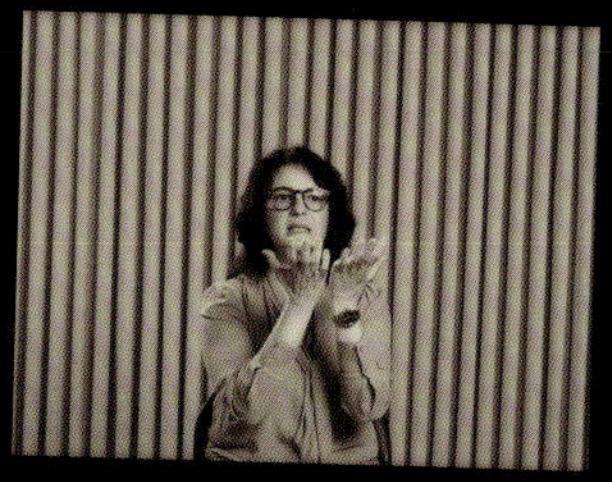

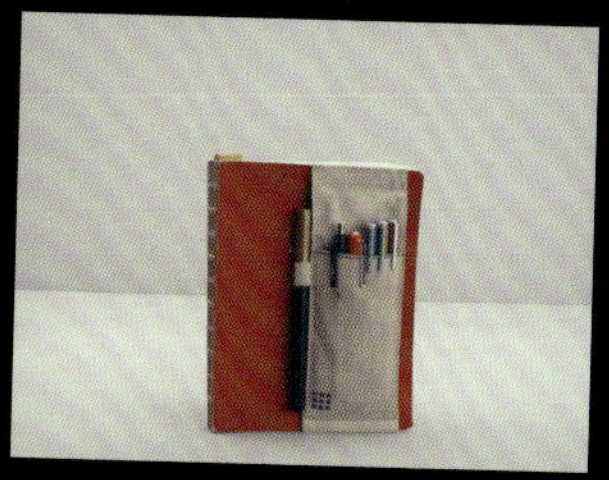

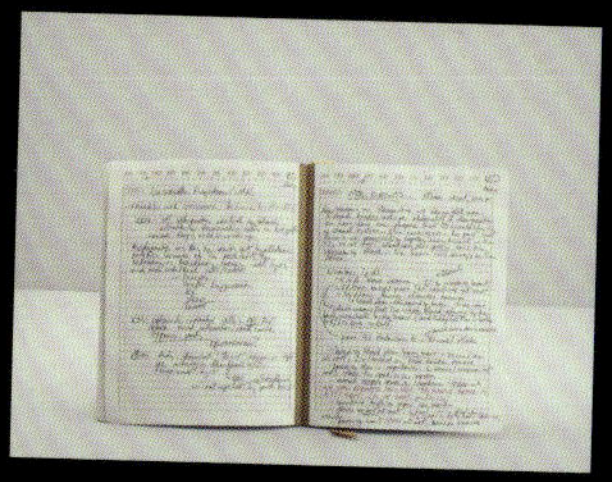

SH– We live with this myth, and we have been feeding this myth, that we are rational creatures, that brains are about being rational. It is so not true. Thank goodness, by the way, we are not rational people, because we know what happens when somebody is perfectly rational. And what you have is a sociopath, is somebody who works perfectly well for their own purposes. In that sense, sociopathy should not at all be considered a disease, because the person who is a sociopath is doing splendidly well, exactly because they could not be bothered.

01:39:00:14

AD – It is not just a slight impairment of your emotional systems. Those individuals were sicker in a deeper way, and thankfully for us in a rather unique way, because there are a few Hitlers around, but the number of those that can get to power and do anything with that power is fortunately small.

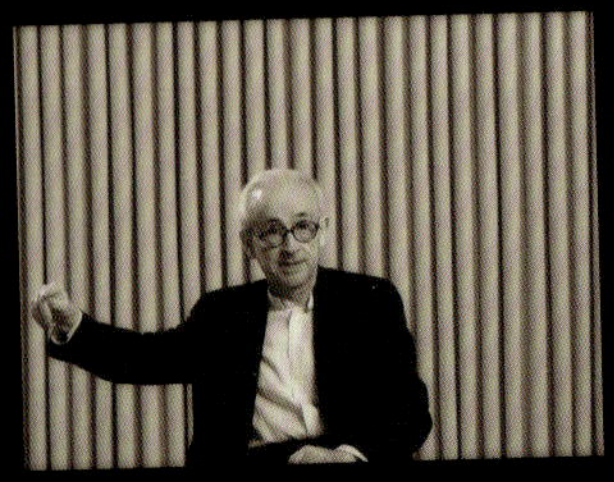

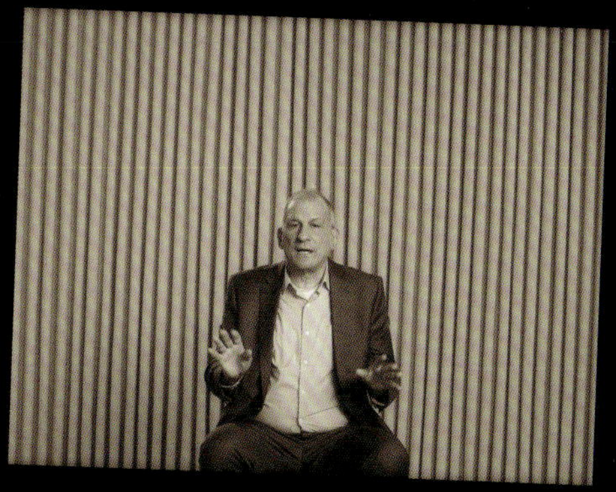

MJVS – Snakes with ties. They are talking about the psychopaths that do not have inadequate social behavior and get thrown into prison. But these are guys that have the same emotional reactions, lack of reactions, the same lack of empathy, but they are functional. They fit into society. And they tend to climb high in different kinds of organizations. And these are the guys I think that do most harm, because the people that have a violent outbreak or they rob or they kill somebody, eventually they will be caught. But these other ones survive a long time and they can do a lot of harm.

01:40:12:01

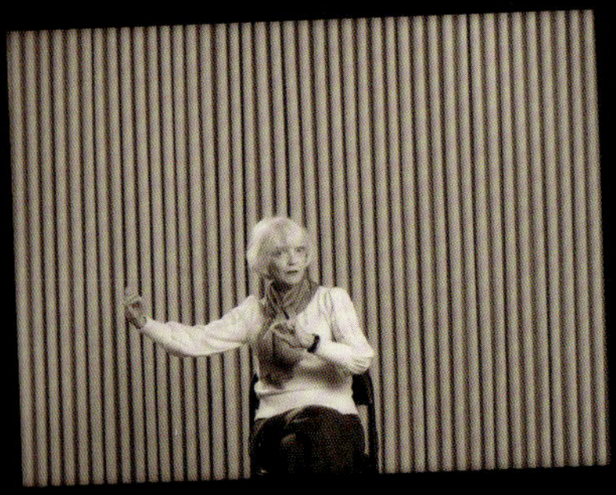

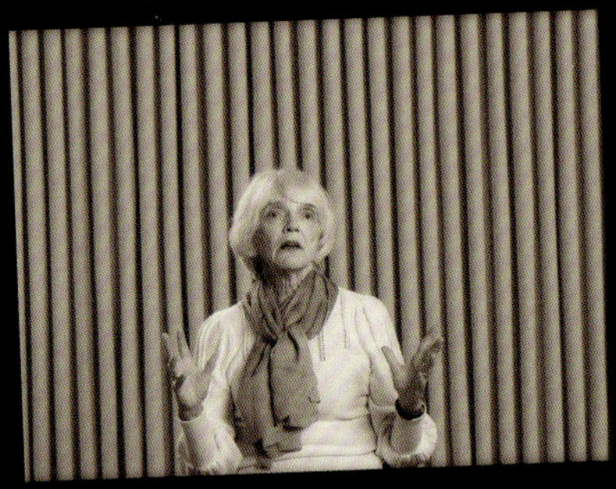

PC – There are several elements that are involved there. One almost certainly has to do with a deep pathology at the level of, shall we say, oxytocin receptors. So they don't feel empathy. They don't care. They don't feel an impulse to cooperate. They don't feel bad when they do things that are wrong. And we know that psychopaths are like that. That's the characteristic that you see again and again and again.

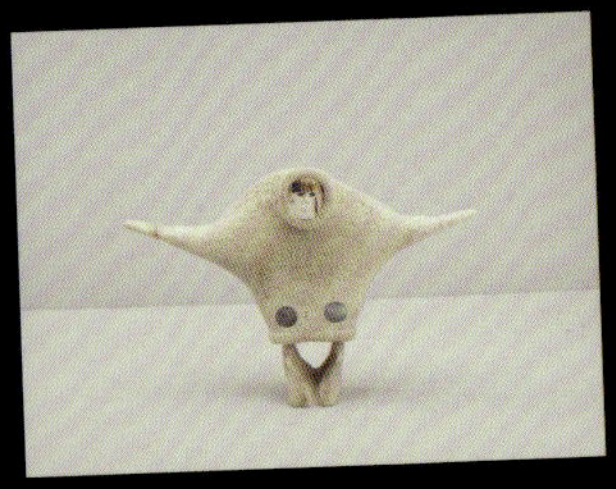

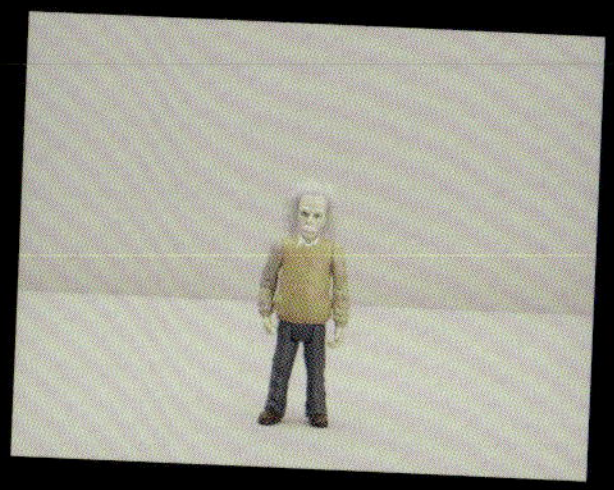

EJ – It might be possible that we humans have more, let's call it, advanced mental health disorders than nonhuman animals. And why that would be the case is because the human brain is actually in a more immature state as an adult than many other species. There are certain genes that are regulated in the human brain that keep it in a more juvenile state, which allows us to absorb and generate more memories, a larger set of memories than some other species. And I think by keeping it in this more immature state, we also have this disadvantage of being more juvenilelike in our behaviors as adults.

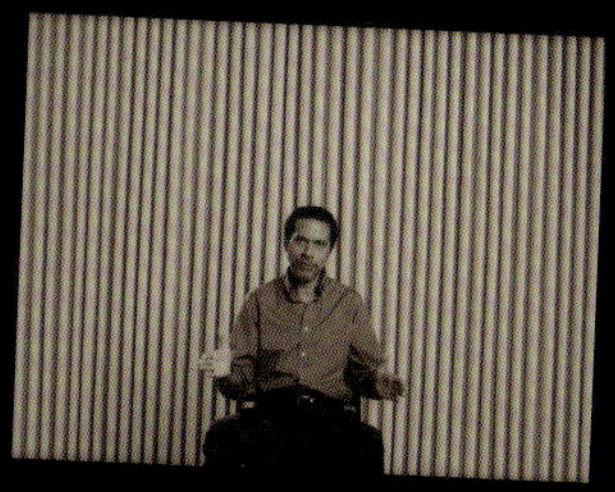

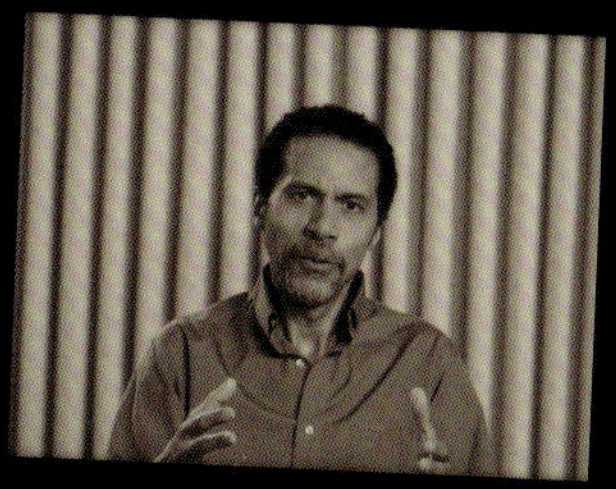

01:41:33:12

CM – We have a candidate at the moment in France who doesn't hesitate to say that women are stupid and they should not play any part in politics, that some journalists should change their first names because they sound African. He can say things like that in the twenty-first century. Sometimes his discourse is close to Nazism and nobody says anything. I mean, of course, people from the left are fighting against him, but he has this kind of power, this kind of pure popularity. How can we explain that other than referring to something that is biologically inscribed in the brain, which is this lethal energy.

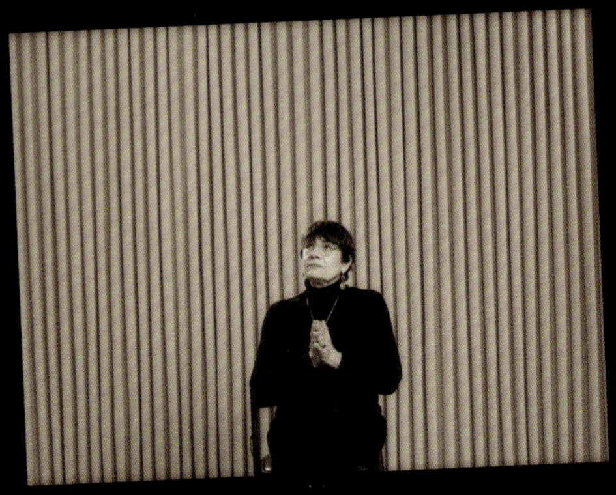

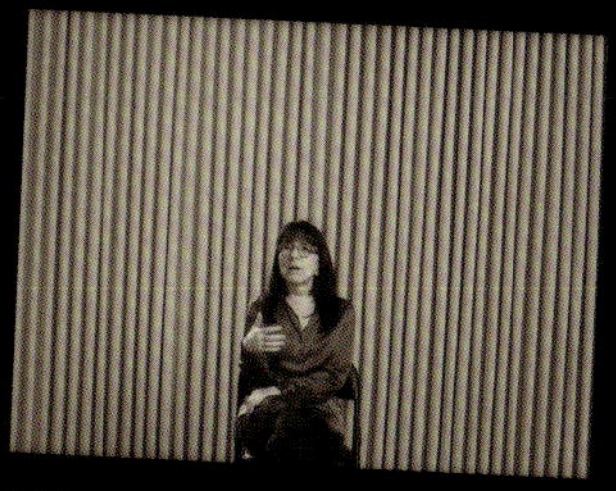

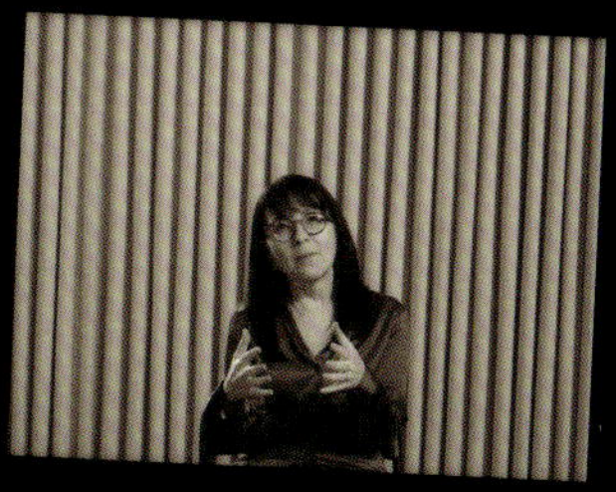

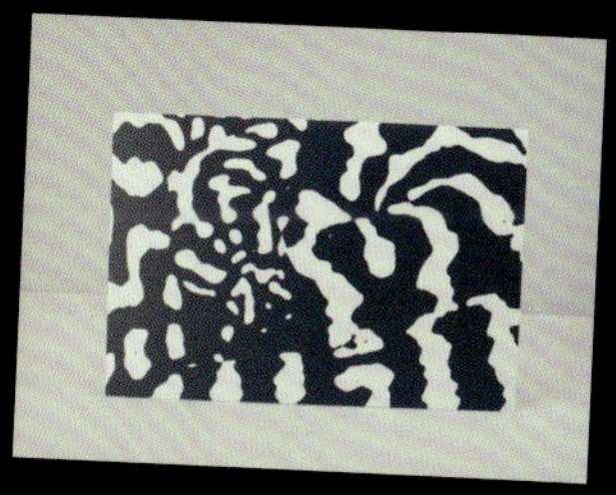

LFB – Poverty, economic disparity, racism– these things don't on their own, singly, lead a culture towards an authoritarian path. People who are metabolically encumbered tend to gravitate in that direction, because simple single causes, or even having somebody else telling you what's right and wrong, reduces uncertainty at a time when your brain really needs it.

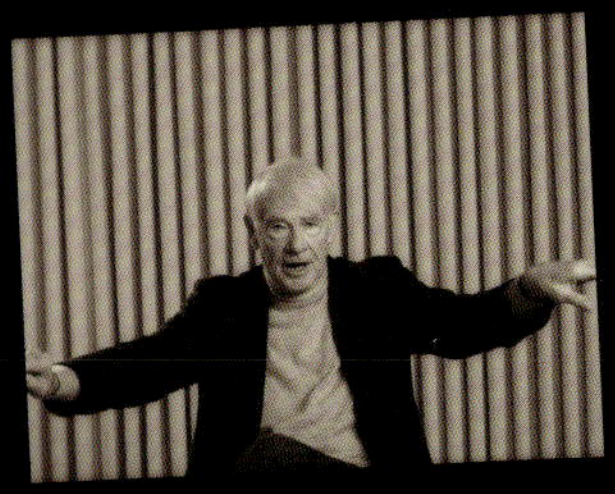

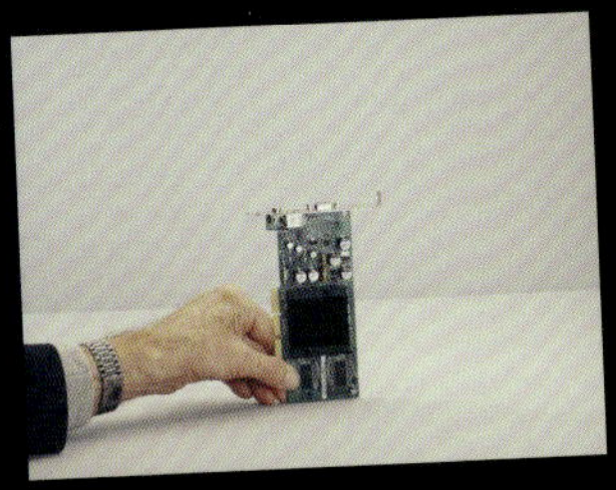

WS – These systems, they are, of course, the very worst alternative, because they make the assumption that the conductor is endowed with meta-intelligence, more intelligent than all the players, and that it is possible to direct top-down a complex system with nonlinear dynamics. It's not. There is a reason why such a center is not existing in the brain. It's self-organizing.

KF — I get shivers down my spine when I hear people like Zuckerberg talk about connectivity, global connectivity, being something that is aspirational. It's absolutely not, from the point of view of anybody who works with self-organizing systems.

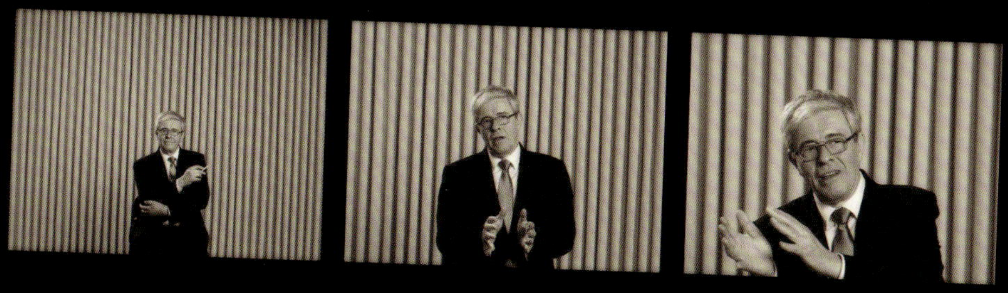

01:44:32:12

AB – The whole world is going for information, information, big data, big data, more data, more data, this kind of a pathetic journey. Maybe a human brain, a conscious machine, never does that. Maybe it doesn't look at the information data at all. It looks at the invariant.

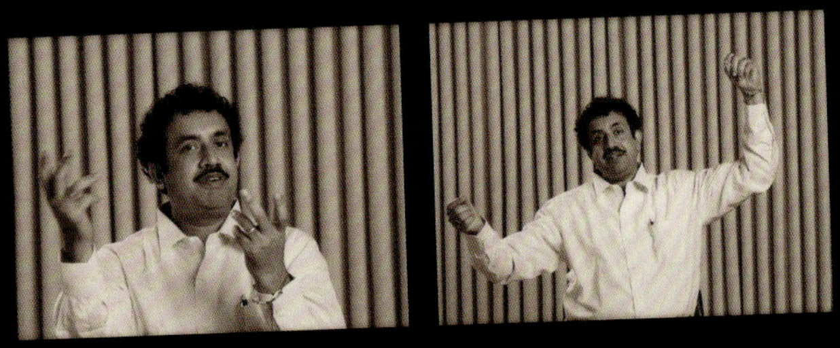

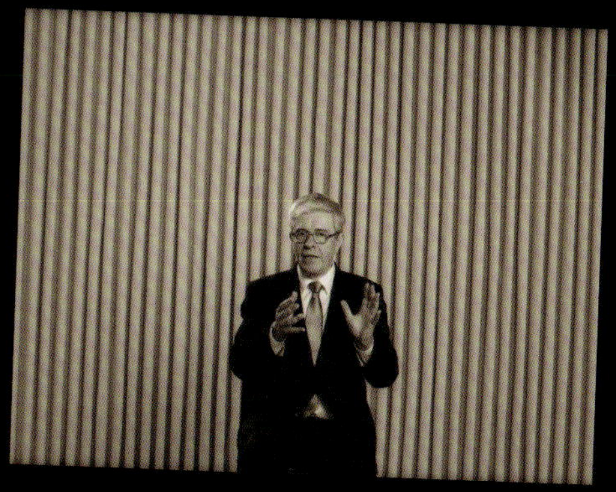

KF— We craft and sculpt our communities by actually shielding and protecting and removing connections to actually give shape and form—and usually nested in hierarchical form—to our societies and communities and families. That is incredibly fragile. If you want to destroy that fragile nonequilibrium steady state, then you just put those connections back and you overconnect.

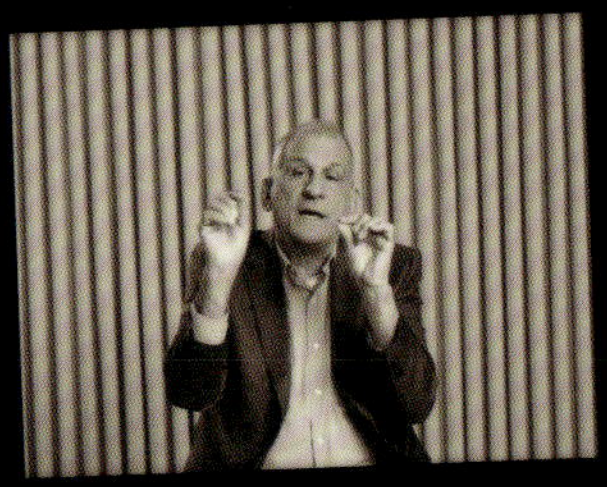

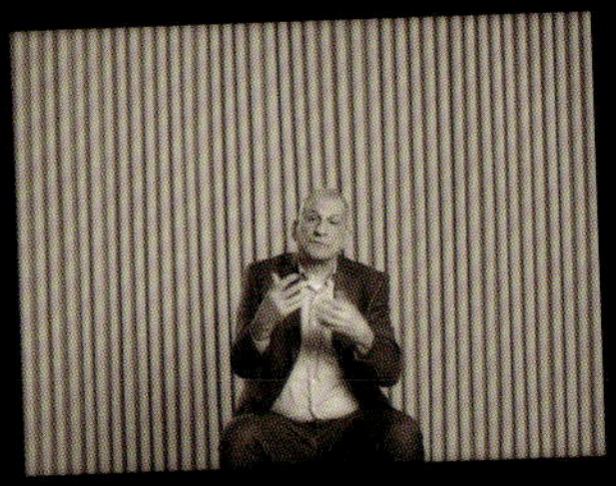

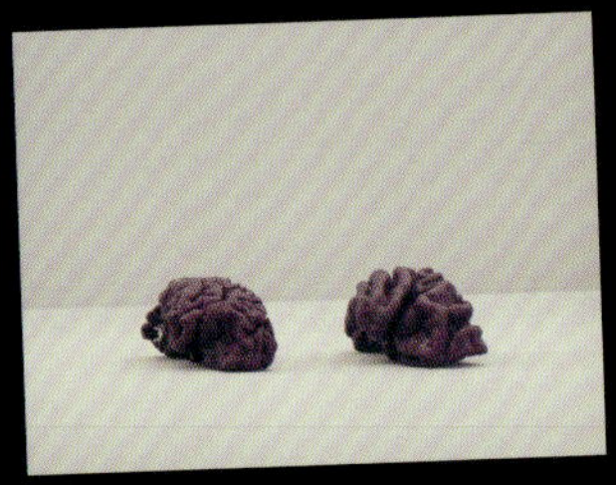

MJVS – All the brakes that society developed to eliminate paranoid thinking, to eliminate gut reactions, so people would only act at a first impulse, all of this weakens when we have this kind of, let's say, unrestrained and very divisive functioning of the social networks.

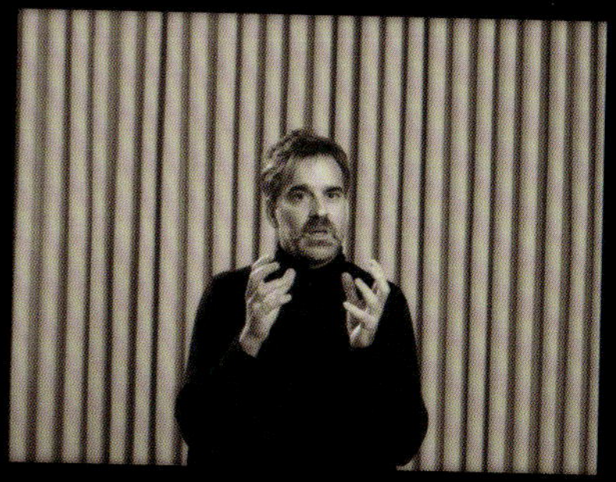

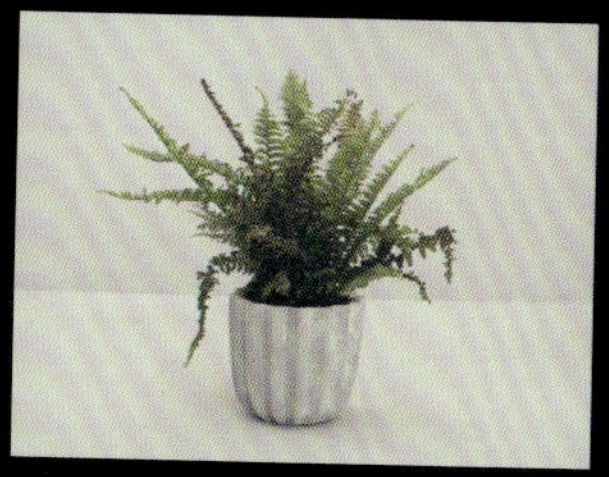

DE – Everything in moderation. If people suddenly take psychedelics all the time, maybe they can become too open, which is one of the big traits in personality psychology. Maybe connecting dots a bit too much, so it starts to be not very productive. So creativity—you might be able to connect a lot of dots, being very creative, see connections and new understandings and combining things in an innovative, new creative way. But if you do that with everything, everything around you, then it's not productive. Then it's confusing; it might turn into some degree of paranoia.

01:46:04:05

MJVS – And then all these more primitive ways, I would say, or incomplete ways of processing information—an emotional reaction without having all the facts, or having the facts but not having an emotional connection, which leads to a paranoid kind of thinking—all these elements that are hardwired into the human brain no longer have the brakes of society to control them. And this is leading to a polarization and to, I think, a proliferation of conspiracy theories and ideas that are wrong.

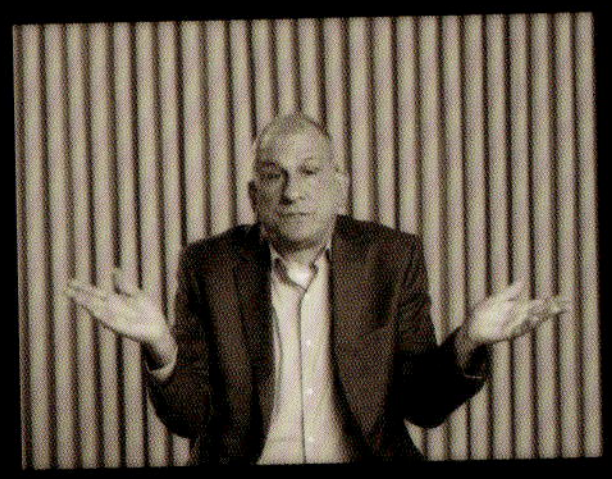

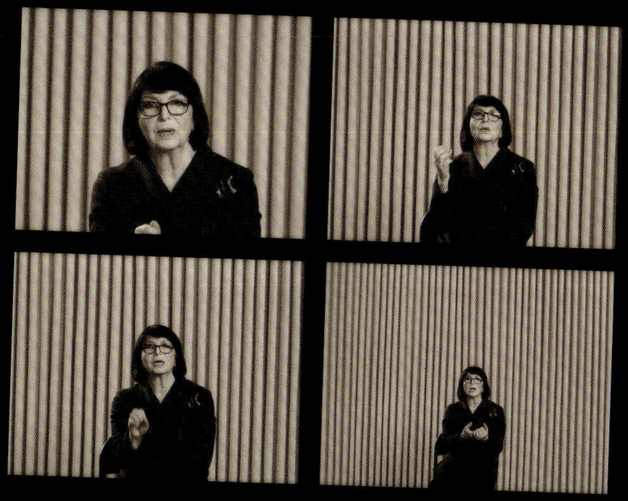

HA – The question is whether you can get to the point where the emotionality, the shared emotionality, of certain beliefs become much more powerful than any amount of cognition, enough that some people are willing to kill themselves for a belief.

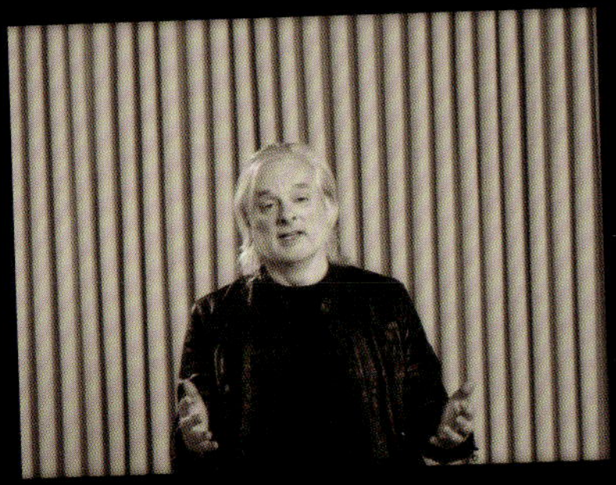

DC – Other people have always acted as kind of extensions of our minds. Maybe your partner remembers things for you that you unaided by them wouldn't remember. They are basically serving as maybe a memory system for you. So we have always had at least these socially extended minds. And technology, I think, is to some extent augmenting that.

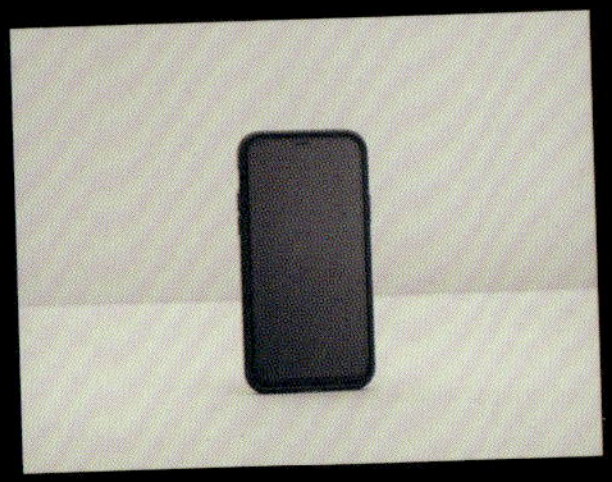

RY – And it will lead unavoidably to the mental and cognitive augmentation of humans. This will happen. No doubt about it. The question is when. So we will become a hybrid species in which we will perform some or a lot of our cognitive processing outside.

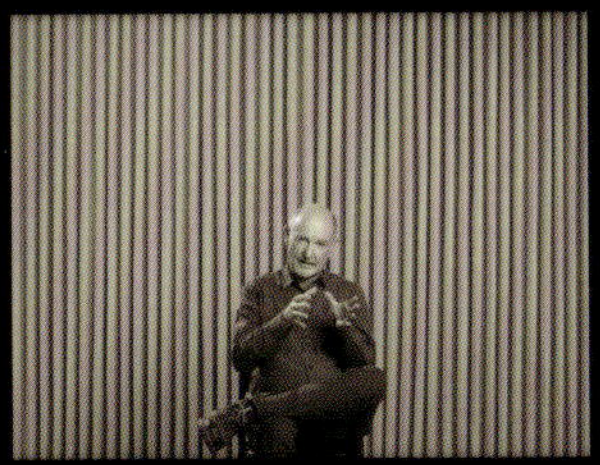

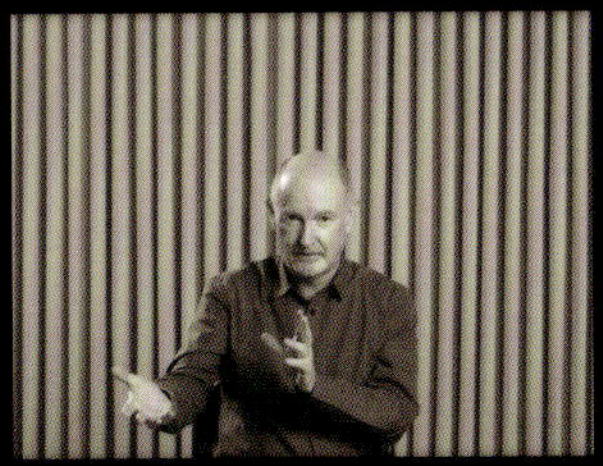

DC – Somehow we are now directly in touch with other people's minds. And that would open up new forms of mind extension and collective cognition. Perhaps even eventually collective consciousness.

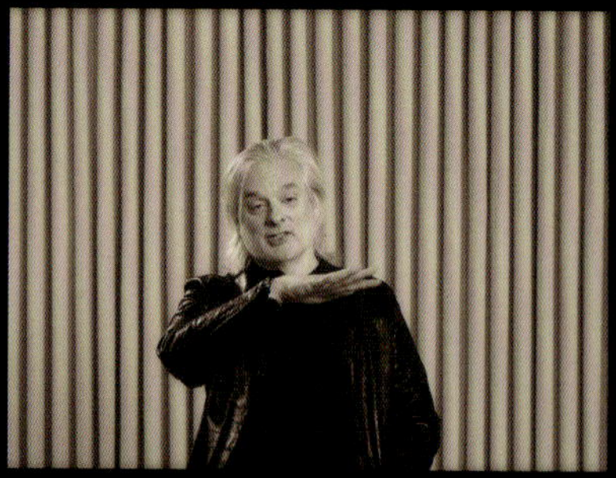

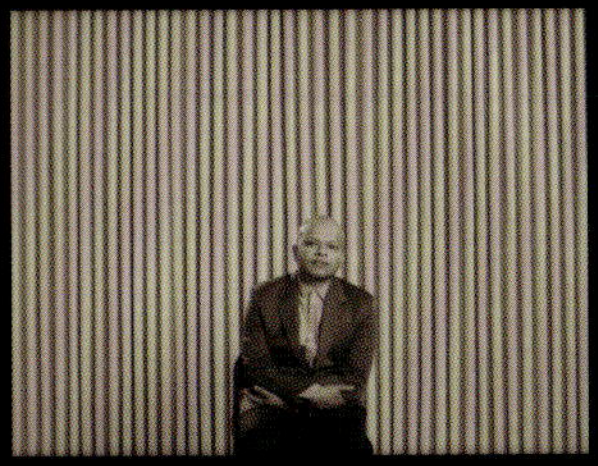

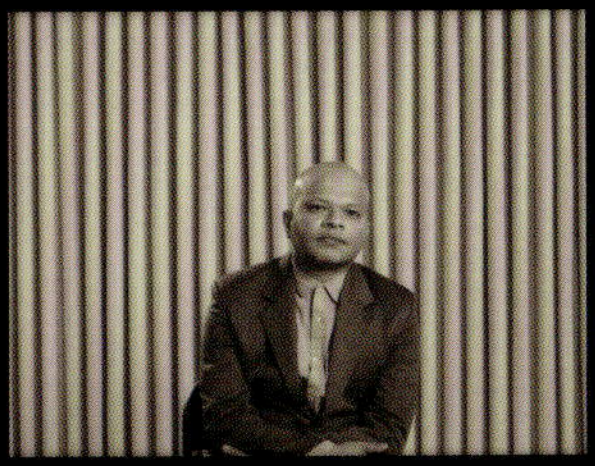

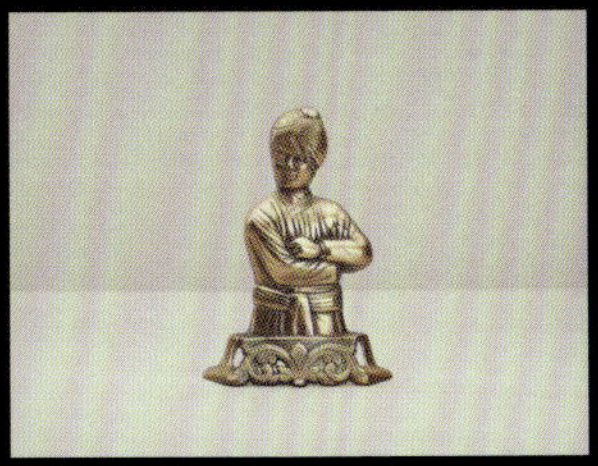

SR – This is something that monks would do for years. But there is at least this talk of your self melting away and you sort of getting absorbed in a higher energy or principle which sort of connects everyone.

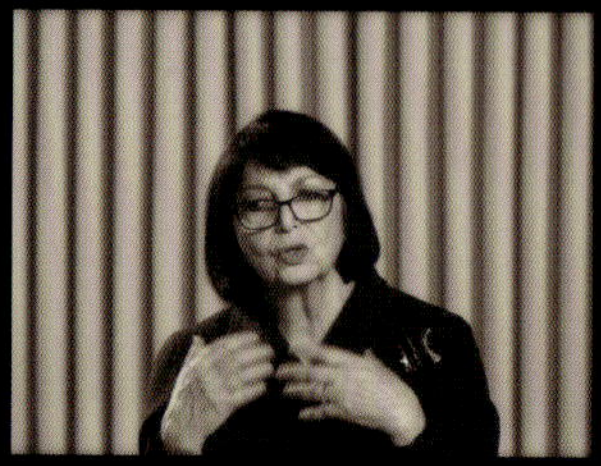

HA – What happens when it's not one person who is being stressed, but when an entire family is stressed, or when an entire community is stressed, or when an entire nation is stressed? Where the sources of support that you derive that can help you withstand stress, rewire the brain, create countervailing forces of resilience, start to wear thin because nobody has them to offer. And now everybody is running out of sources of support, and therefore the ability to remodel the brain in a positive way.

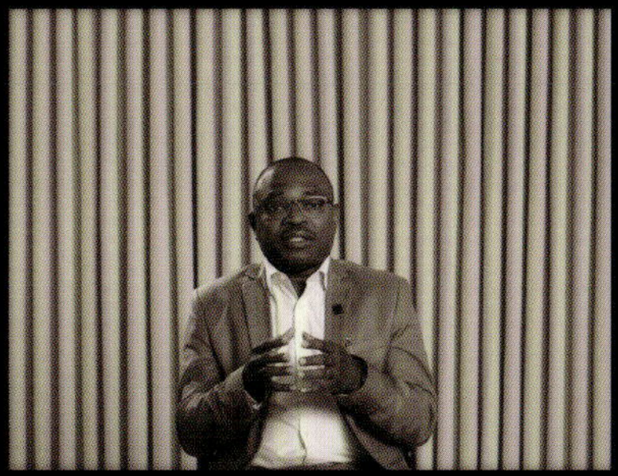

AI – One thing we have not given attention to is the time the brain requires to adjust to experiences. In an explosion, we always want to deal with environmental factors, changes–we want to deal with the physical, we want to manage the physical, and we do not give attention to the fact that over a period of time, that is the only time you can see the effect of that experience on the brain.

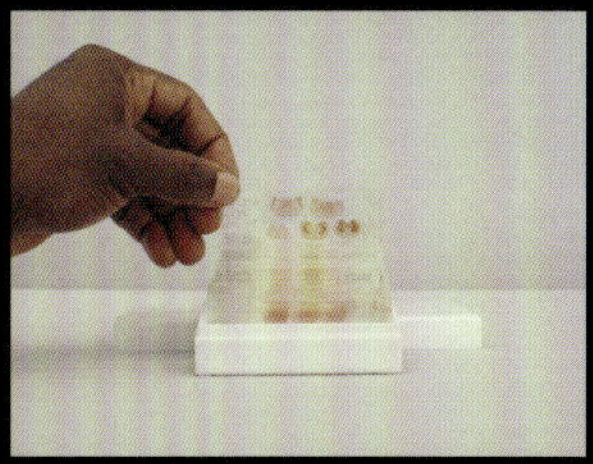

DD – Everything that happens in the world mirrors something in the brain. The brain is where you store your memories, the brain is where you store not just your memories, but you might call them your cognitive reflexes. The fact that you are going to lash out or be afraid or be calm or be brave– all of the emotional responses to the world are due to the controls of your brain.

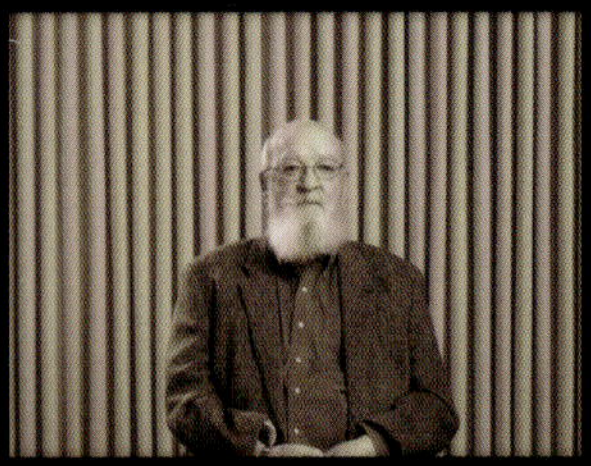

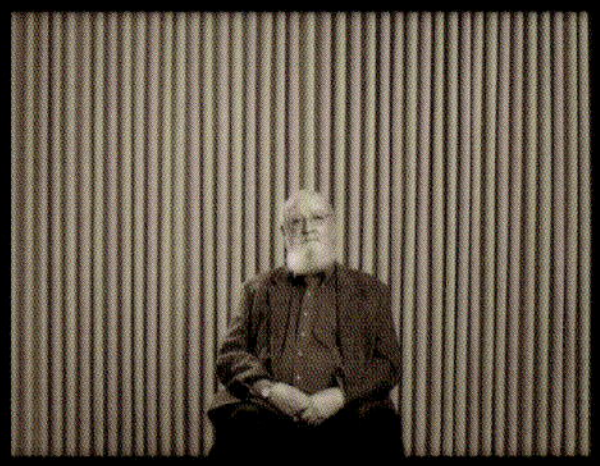

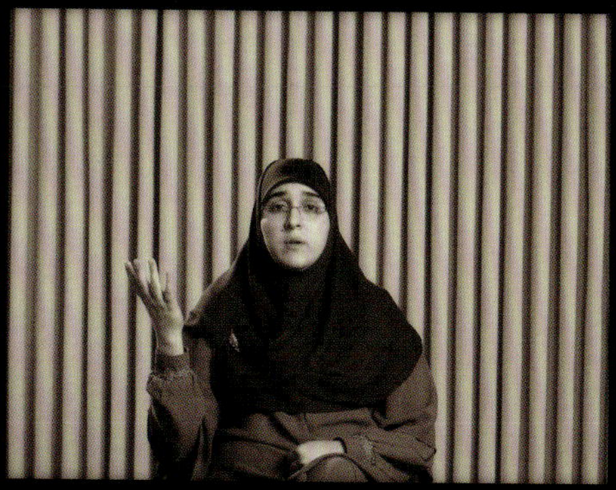

YA – When there is a restriction in
how we could express emotion, I think that
probably it could affect our ability also
to recognize, to understand, and to perceive
the other's emotions.

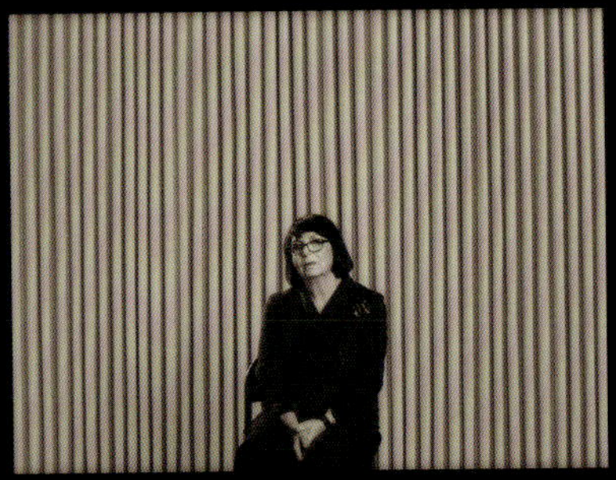

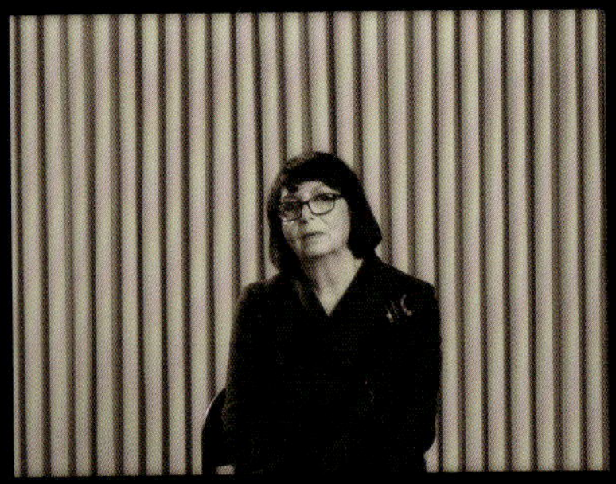

HA – One of my cousins had a daughter who was a young faculty member at the University of Aleppo, teaching computer science. One day she was sitting with him in their apartment, with her father, a few feet apart. An explosion went off. The wall fell down, the electricity went out, her phone flew up in the air, her father caught it, and he looked over and his daughter was dead. His house was destroyed around him, but he was still alive. This was probably the worst thing that could happen to a human, having your child die and you survive.

JL – If you aren't part of the conscious experience, it's not an emotion. You have to be engaged personally in order to feel an emotion. You can't have love without it be you that is in love, or you can't be afraid of a snake without you worrying that the snake is biting you. So you yourself has to be part of any kind of emotional experience.

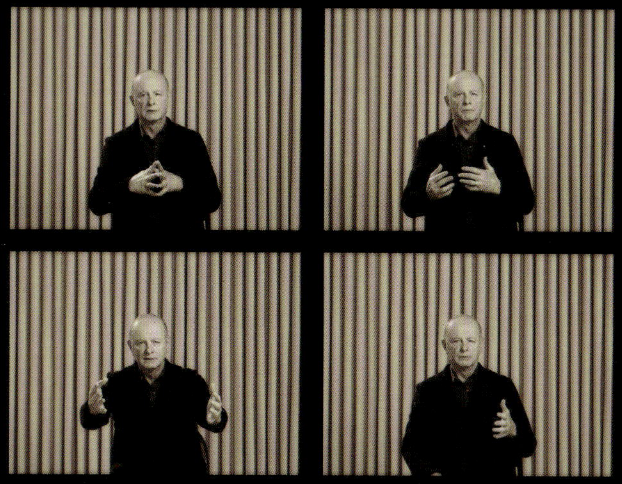

BJM – We are able to have a mouse have a slightly stressful experience, not something along the lines of starvation or a Holocaust or anything that humans will define as trauma, but something as small as a light foot shock. And that light foot shock in a father who then breeds with the mother, and the father never meets the offspring, has effects in the offspring.

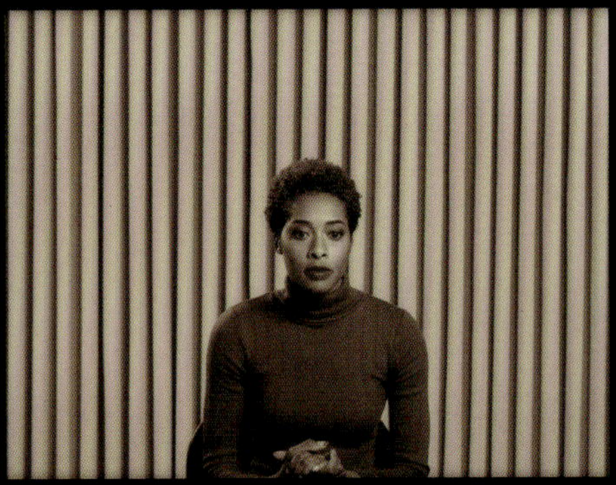

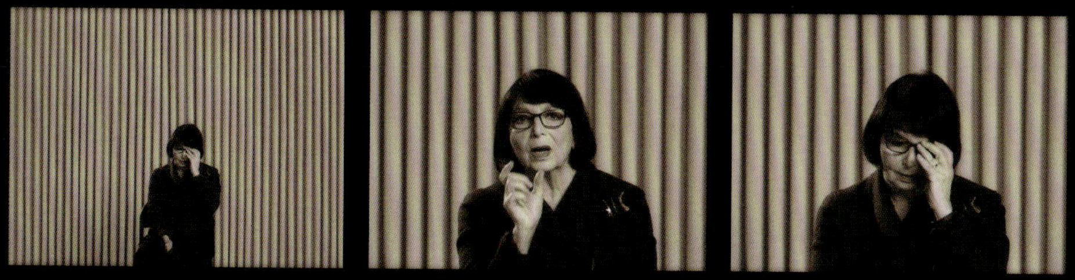

HA – This is one event where an entire family got impacted. It definitely changed the course of his old age. He aged very dramatically from there. It changed many other relatives who had to change their lives to support him. And I am sure both psychologically and biologically their lives have been altered forever.

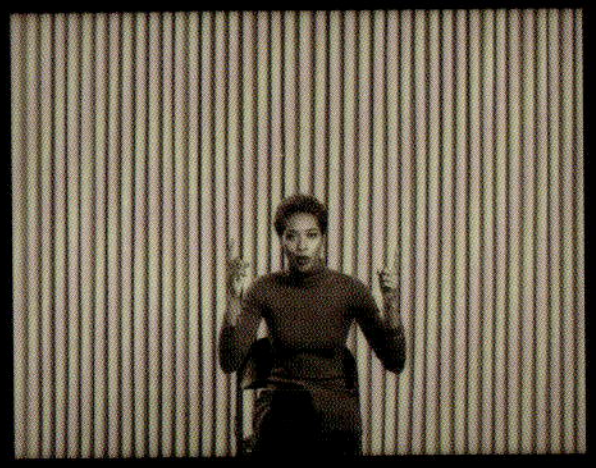

BJM – When a female mouse is pregnant … Let's say she is pregnant with a female. That means that the eggs of her granddaughter are inside of her. So an experience that pregnant female mouse has could essentially affect the second generation in utero, right then and there. With males, with male mice, sperm turn over every month, every twenty-one days. And I find this, dare I say, a more interesting question. Because if sperm are turning over, and yet we can still have this message carried on, it means that it is not just that things need to be there in a particular period of time, but that the memory within itself is living in that human. So that the woman doesn't have to be pregnant with her child, but that before she is pregnant with that child, before that male actually sires offspring, that memory sits. How does a memory sit in the brain and remember itself in the brain and body?

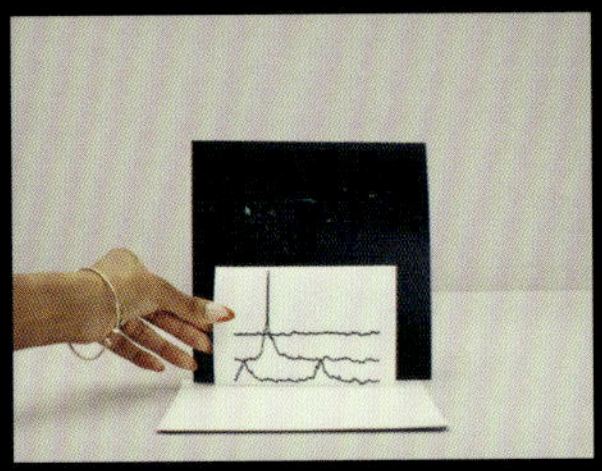

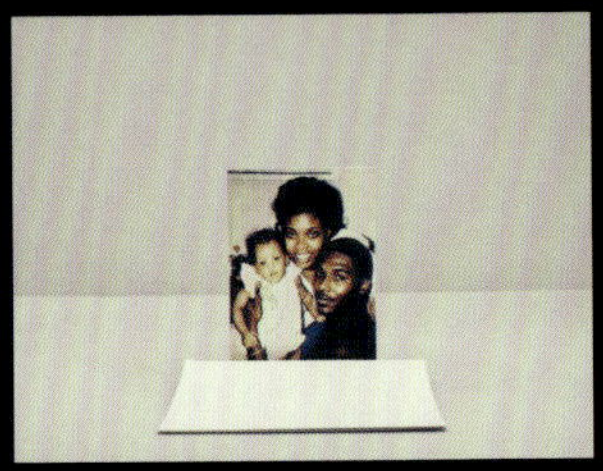

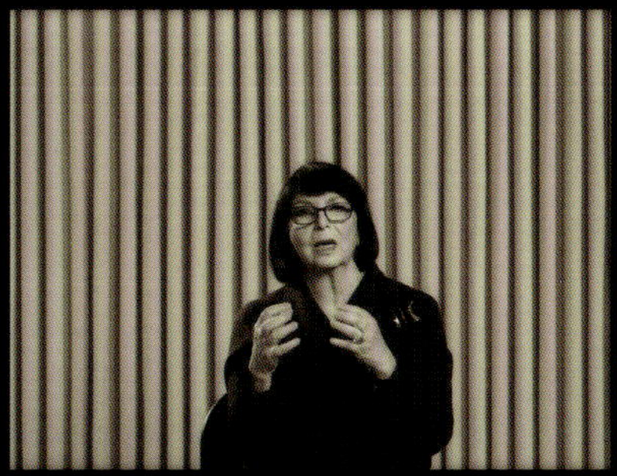

HA – I mean, speaking of emotions, right? They come out of nowhere. And it shows we all have trigger points that we carry with us, and they never habituate. They never go away. They continue to be very strong. And that, I think, is an important part of thinking about emotions and mood, is that some of them become dull, but some actually never lose their power to upset. Hopefully, there is an equivalent that never lose their power to also bring joy.

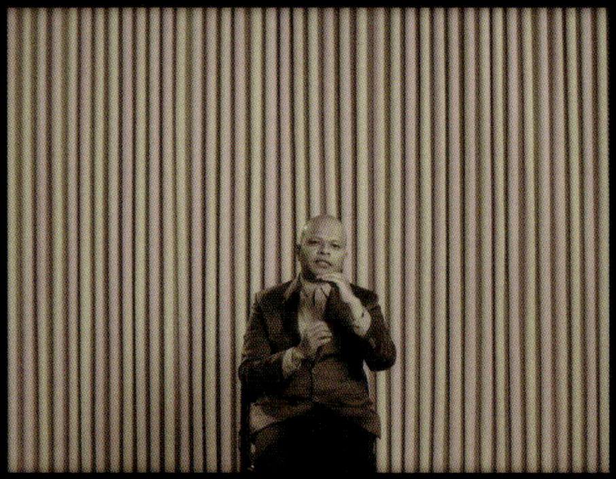

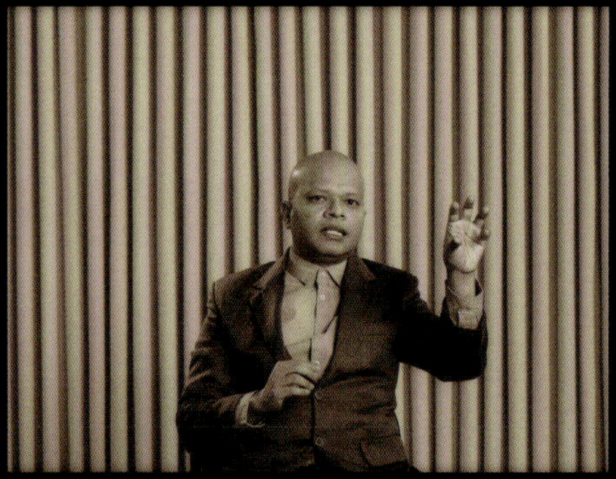

SR – The visual stimulus is my hammer. And what we have and others have found is that there is a signature in the brain, this gamma wave, that responds differently depending on how you are hammering the head. [...] Every time there was a picture of a tomato on the screen, we got these intense gamma waves, like the strongest gamma we have ever seen in any species under any stimulus condition.

CM – There is a biological explanation for the pleasure we take in consuming violent scenes or horror movies, et cetera. […] Even if I am not the actor of the violent scene, because of my mirror neurons—the name is very clear—these neurons are mirroring what is happening. So, thanks to these mirror neurons, I can almost experience, without really experiencing it, something like a crime scene or any kind of violent or even pleasurable scene.

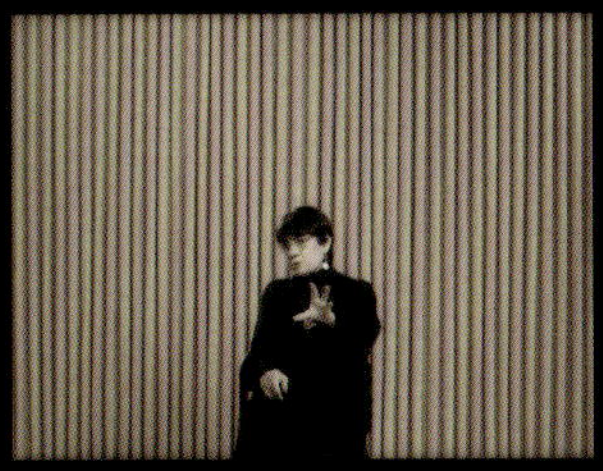

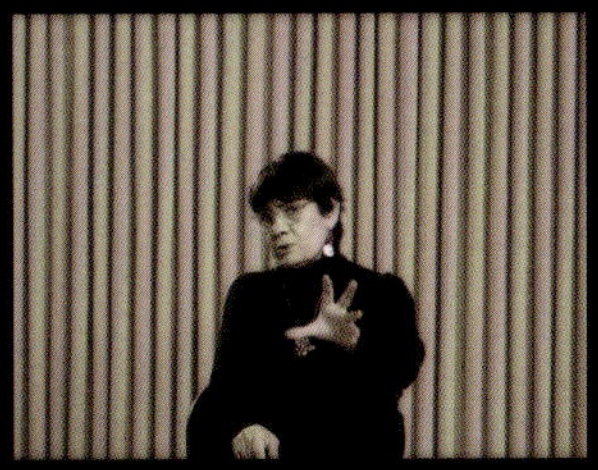

01:56:16:03

SR – There is a part of the brain called the ventral tegmental area (VTA). Now, that is called the pleasure center of the brain, in the sense that if you stimulate that part, you feel very happy. So some patients, human subjects, who have rare cases where people have stimulated them and they have been able to report also, they say, "Oh, it's a feeling like nothing else."

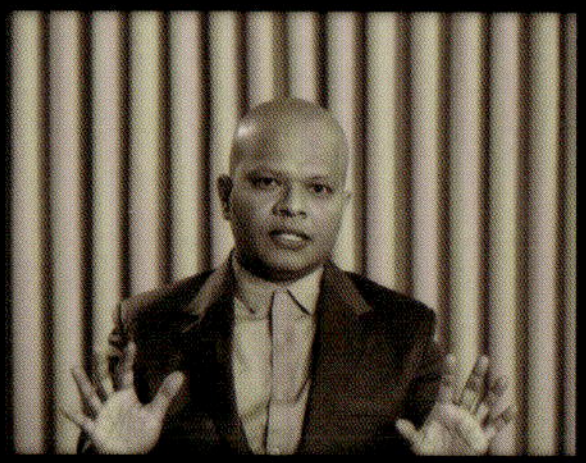

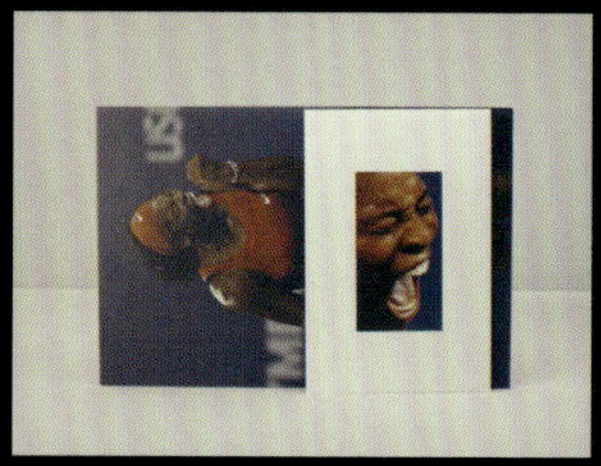

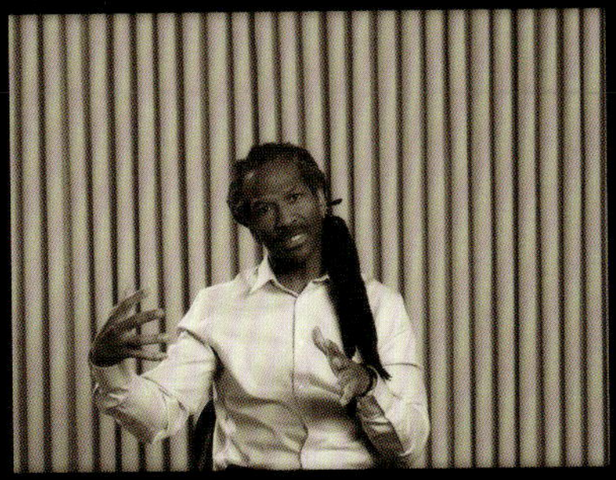

CH – One of the reasons that we downplay euphoria is based on our tie to this puritanical notion that you have to earn this pleasure, or you should work, and that is how we determine your value.

01:57:32:08

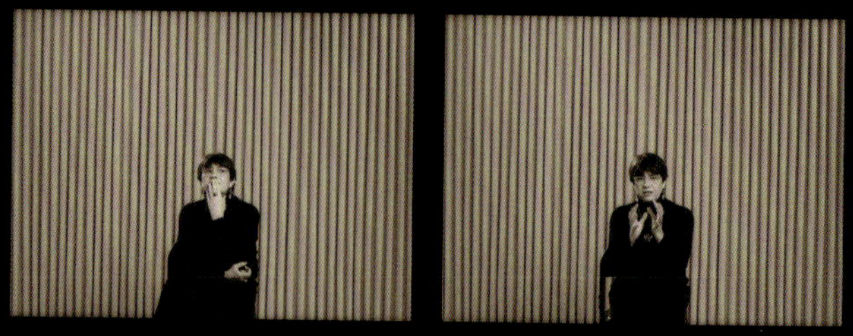

CM – The brain is not only the image of capitalism but is also what can resist capitalism, through the refusal of an excessive flexibility. The refusal of being used and bent in all directions. The refusal of exploitation. Because I think that in the brain there are also thresholds of resistance. If you push too hard, if you bend, if you ask for too much flexibility, then there is a resistance.

SR – It makes you wonder that … Things like purpose of life and all that, at the end of the day, might be just getting stimulated in the VTA.

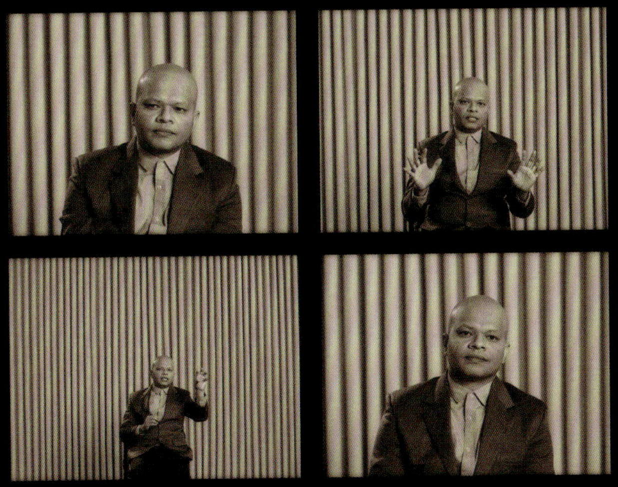

CH – And so the question for me as a scientist, as a professor, is, Why haven't we had this discussion? Why have we sacrificed this specific individual right? I didn't agree to sacrifice that right. Who agreed to sacrifice that right?

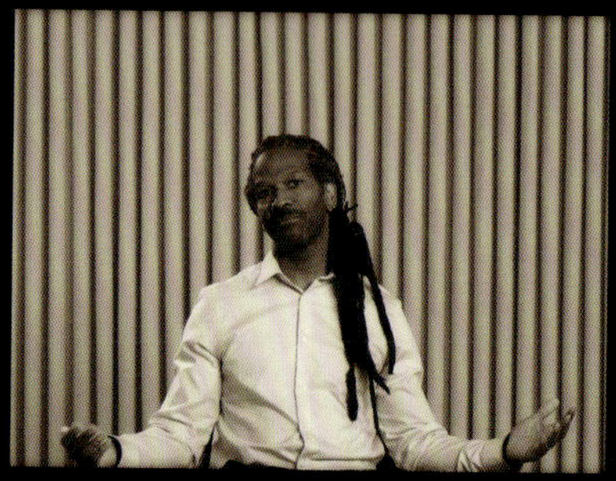

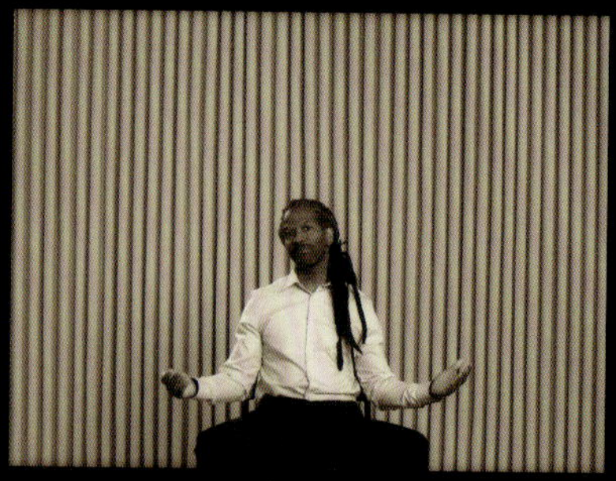

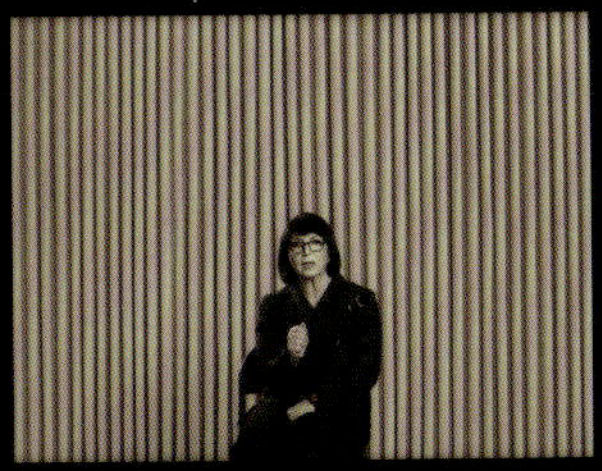

HA – Morphine, heroin, and all these drugs work because they are mimicking a naturally occurring system in the brain called endogenous opioids, or endorphins, which are widespread and perform many different functions, from controlling pain to helping us experience pleasure, to increasing motivation, to changing mood. They have endocrine functions, they have gut functions, they have a very broad set of functions.

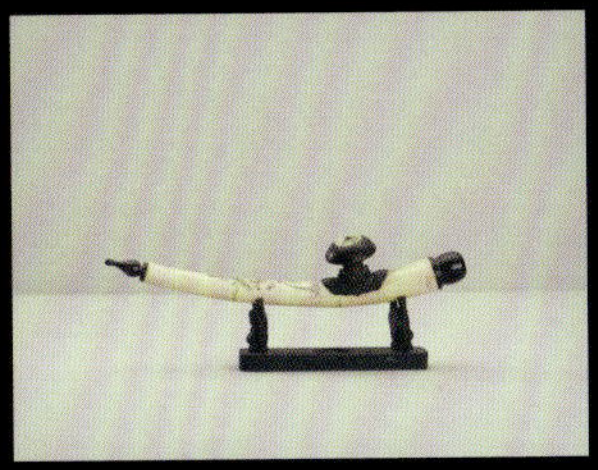

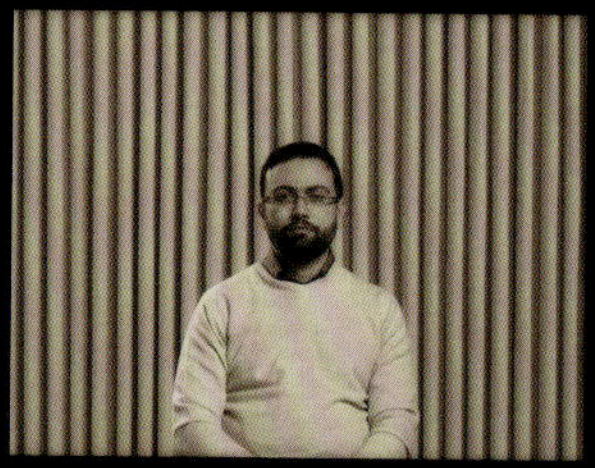 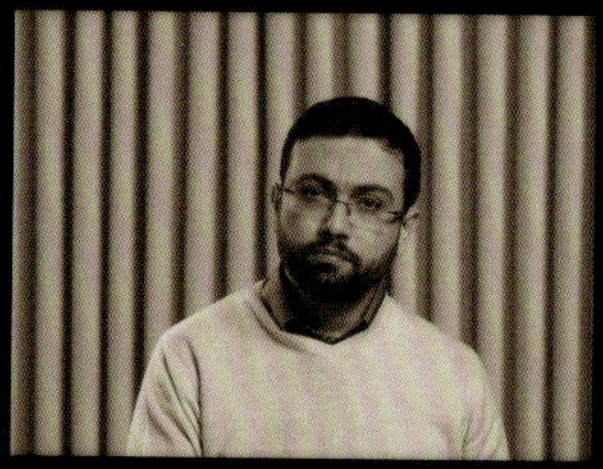

AG – You can have a rat that every time he presses a lever, he gets an electric current injected into his dopamine areas, and the rat basically kills himself over just pressing the lever.

HA – What happens when we give drugs or when we use drugs like heroin and other opiates is that they overtake the function of these natural compounds, and in some ways they bypass the ability of the system to regulate itself, to put the stop.

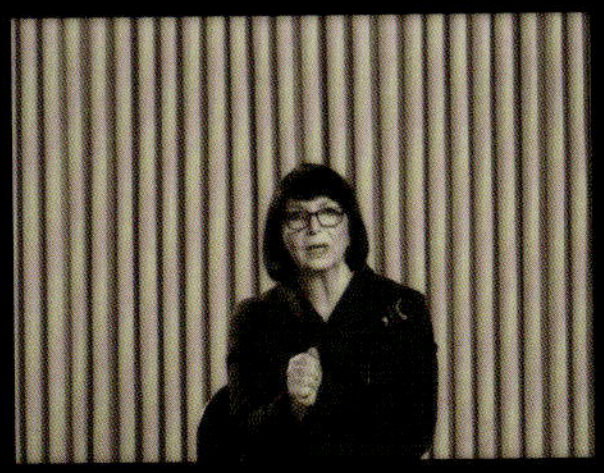

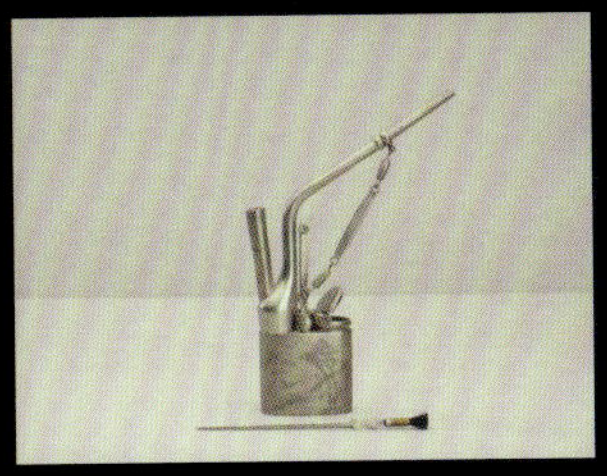

CH – The bottom line is this: We have these tools that can enhance the human experience, make it more pleasurable, make it more enjoyable. And we have severely restricted these tools, and we are not sharing them with the public. That is a tragedy.

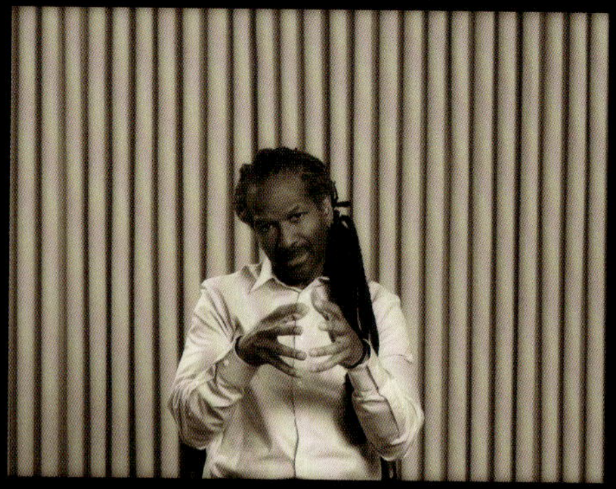

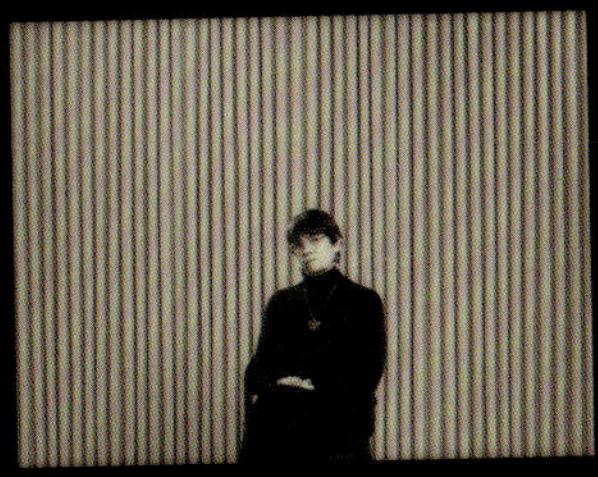

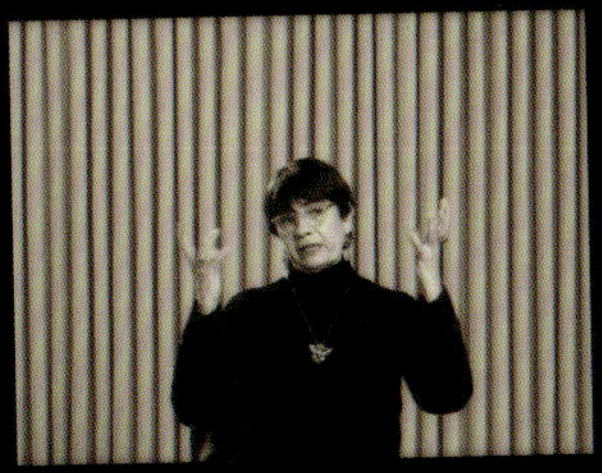

CM – Nobody can feel their own brain. The brain remains silent in each of us. It's not like the heart or any kind of other organs. The brain is central, and at the same time, we don't feel it. It is remaining in a certain sense nonsubjective, nonsubjectivable. We cannot appropriate it. And orgasm is one of the very few expressions of the brain. Now we know that orgasm is a manifestation of the brain. It's the brain that reacts to pleasure.

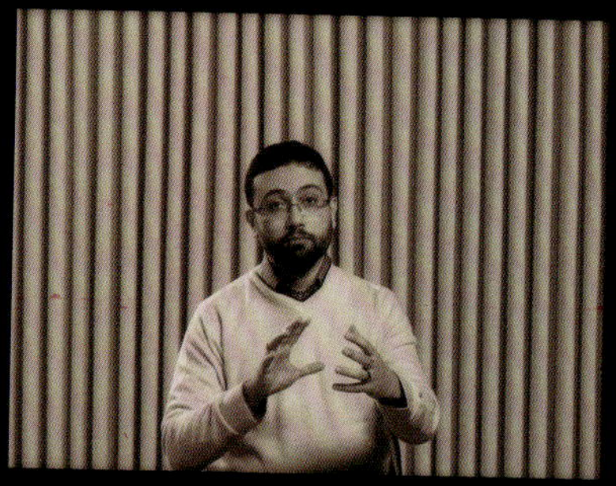

AG – The brain has no feeling of pain. That is, after you remove the skull and the dura, the brain doesn't feel any damage that you cause to it. This is why, in the movie *Silence of the Lambs*, when Anthony Hopkins was cutting the brain of his victim, the victim was happily sitting at the table enjoying his dinner.

CM – The brain cannot be experienced. It remains silent in us. It does everything, but we don't feel it. And so sexuality is one of the main, the only sign that we have of our brain's existence. That's interesting, I think—the brain exploding in pleasure.

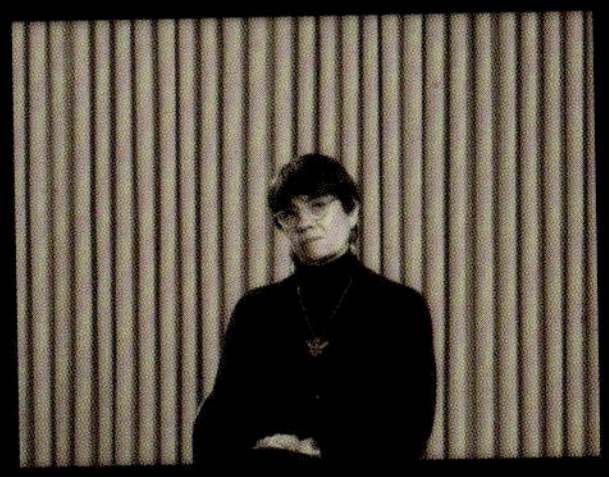

GB – Neurons in the brain have no clue what is outside in the world. Neurons can have information only from other neurons. In other words, there is no confirmation, there is no second opinion. There is no grounding for this neuron. What is reality? What is outside in the world?

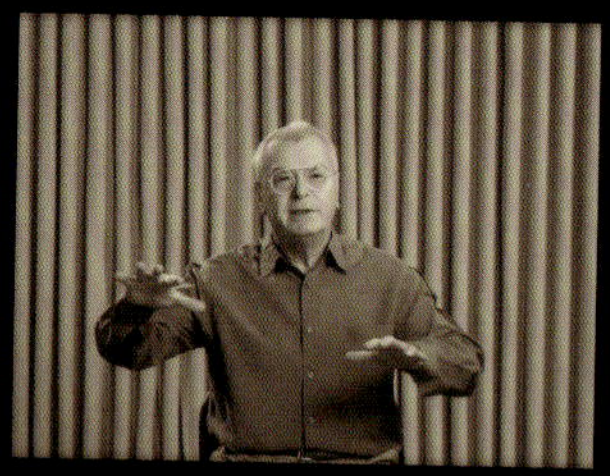

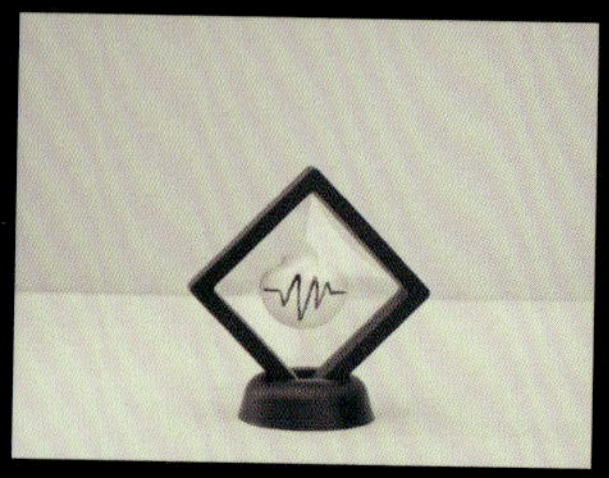

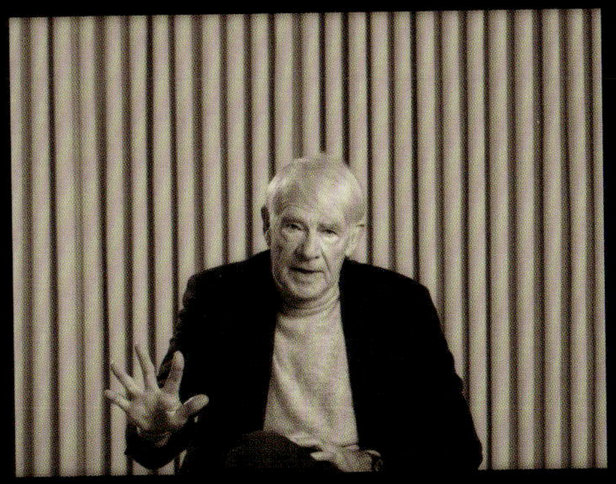

WS – What does it mean to us that there is a space-time continuum? What does it really mean for us, or that the universe expands into something that is beyond this border? So we already are at limits of our intuitive understanding, because our cognitive systems have evolved constrained by the world that was important for survival. In the scale that was important for our survival, the quantum world is completely irrelevant for us. It determines what happens, but we have no sense organs for it.

02:03:08:22

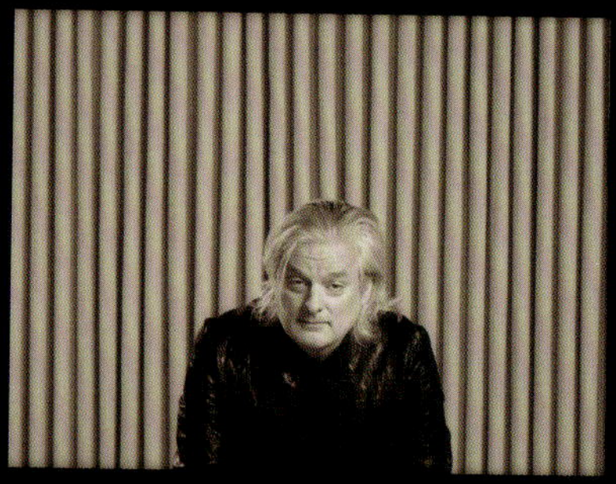

DC – Maybe we are just tuned into this one wavelength where we are just looking for one kind of signal, and maybe there is intelligence living in dimensions we haven't even encountered yet.

GB – My tongue has a niche. My tongue lives in my mouth, and it is constantly exploring it, and finds something, and compares it with its own memory that, "Oh, this thing wasn't there."

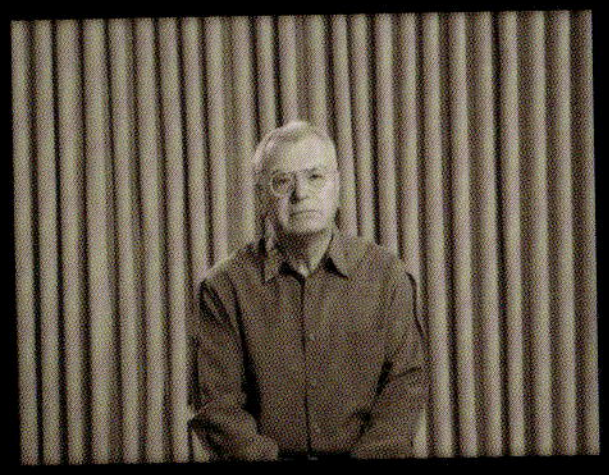

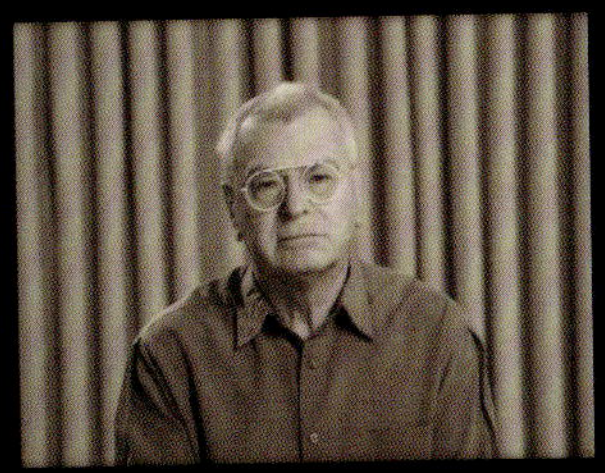

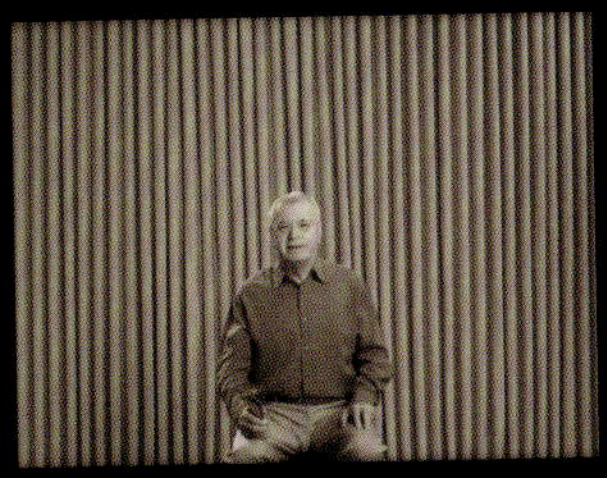

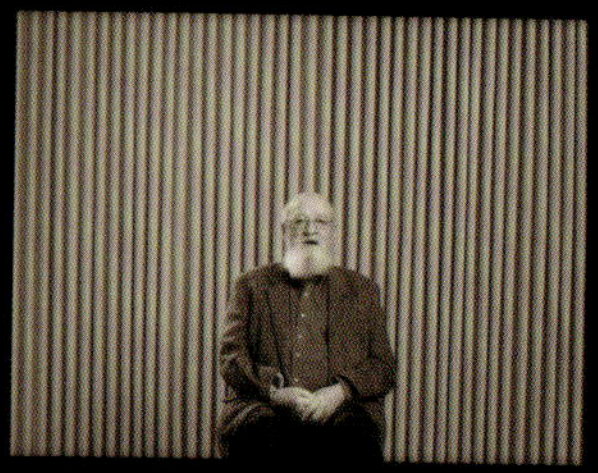

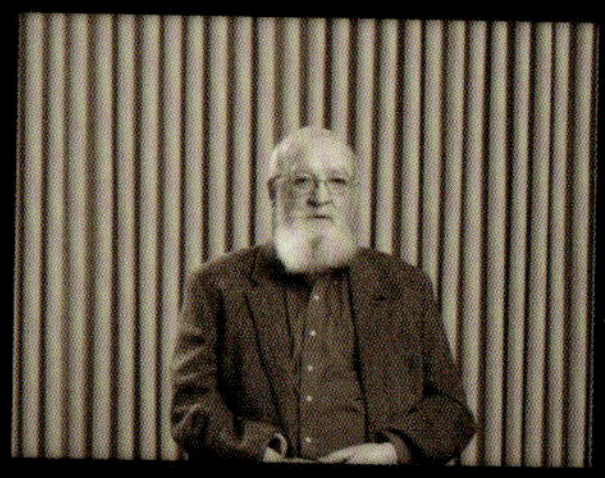

DD – The search for extraterrestrial intelligence is, I think, a well-founded effort, if only because it might have paid off quickly. It hasn't paid off yet. It might never pay off. But it is certainly worth hunting for, although part of me hopes we never find it. Because what if we find extraterrestrial intelligence and the first message that we decode is a plea for help?

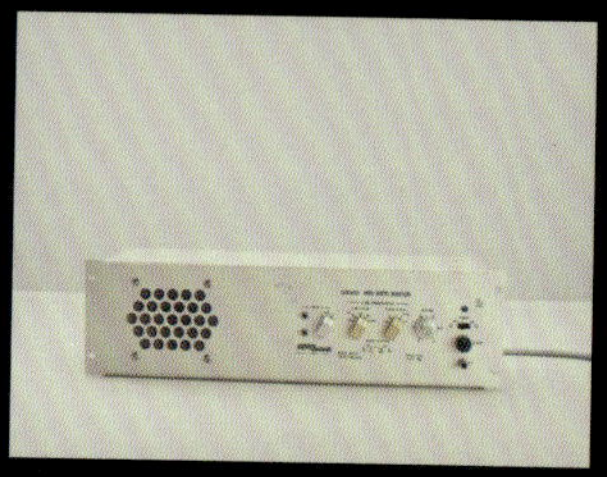

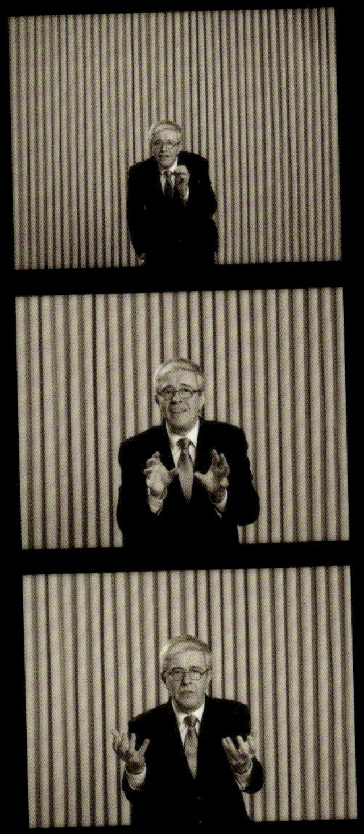

KF – What would happen if it was another kind of us? A pathological kind of us. Two kinds of us that started to create in-groups and out-groups.

02:04:54:11

WS – What has been competition for food may have become, now, competition for belonging to something. So the group has to stabilize itself and at the expense of other groups. Difficult to overcome because deeply rooted.

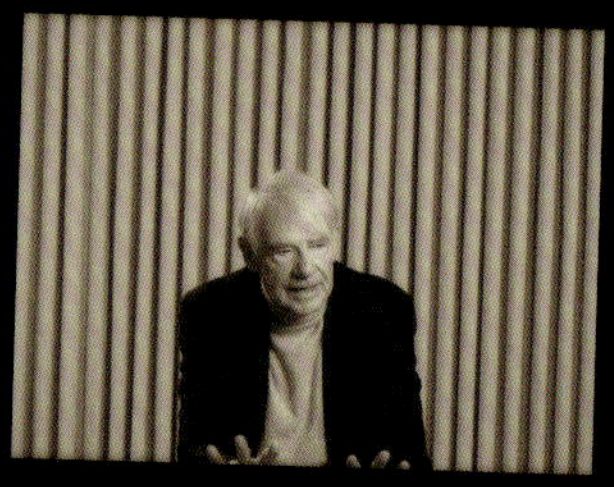

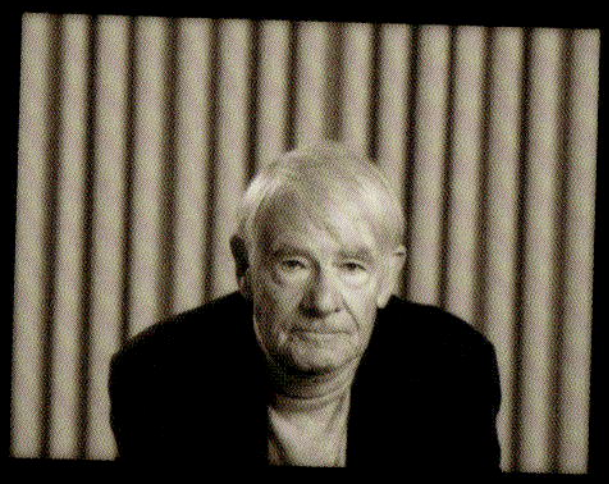

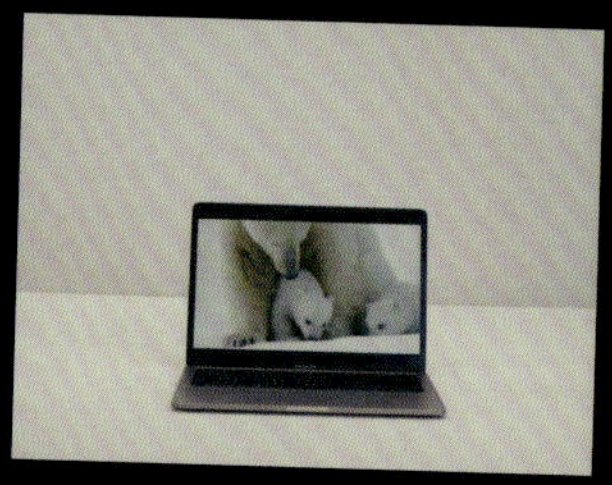

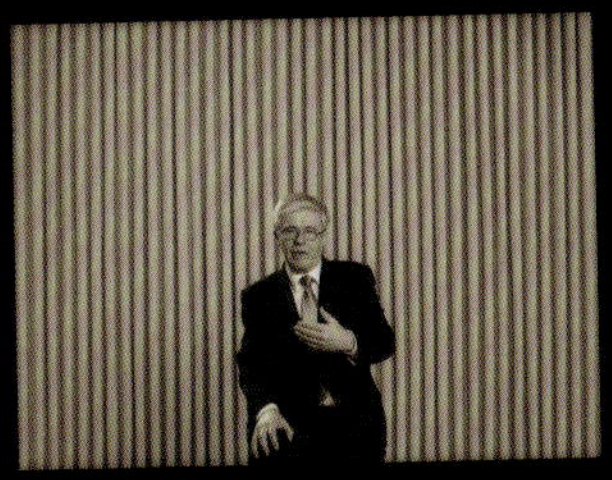

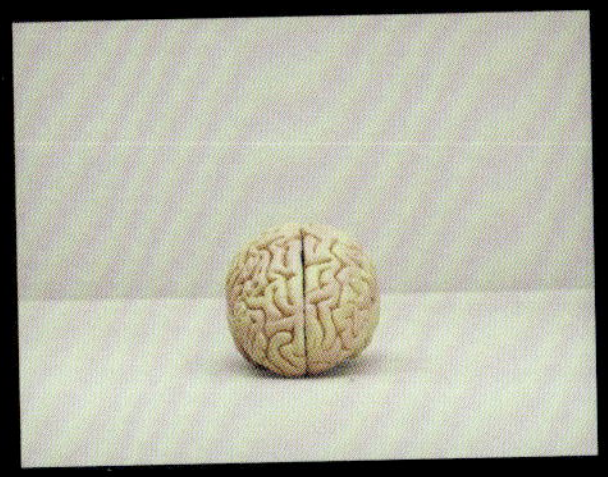

KF – When we see big divides, whether they are theological or political, it's almost fifty-fifty. It's as if the only stable way to disagree, to be polarized, is to have an equal weight on both sides of the argument. Whether it's Biden versus Trump, whether it's Brexit versus Remain—you take your favorite axis, it's almost exactly fifty-fifty when it comes to the votes. It may well be that this is the only stable solution: to have two co-created minimal attracting sets that enable these two separable populations to still be aware of an exchange with each other.

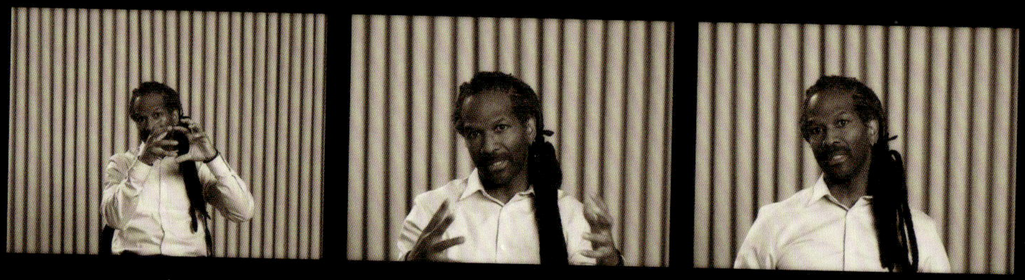

CH – One hemisphere is in charge, in control. You can't have both of them in control at the same time, otherwise you wouldn't get coordinated movements and thoughts. And so you have a lot of collaboration going on, and you also have a lot of inhibition going on. A lot of the cells are being actively inhibited. They are not firing.

SHH – It is really impressive how much easier it is to destroy than to build complexity, but that's physics. That's the reality of physics. And I guess biology is exactly the realm of building complexity, despite entropy, despite the first law of thermodynamics, despite how everything just tends towards equilibrium, towards just that final balance that is death really.

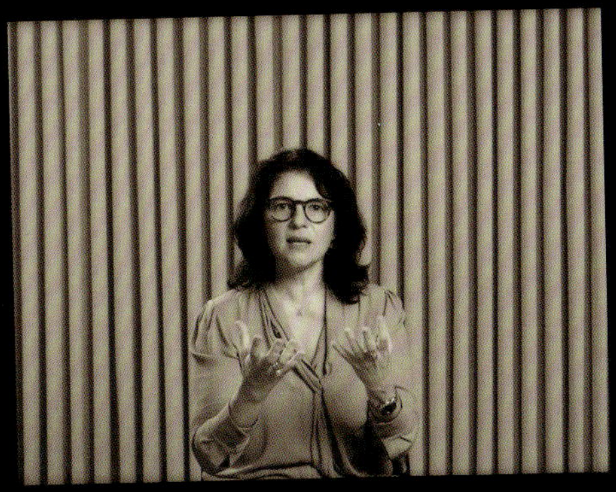

IT BEGINS WITH AN IDEA

Fondazione Prada, Venice

April 23 – November 27, 2022

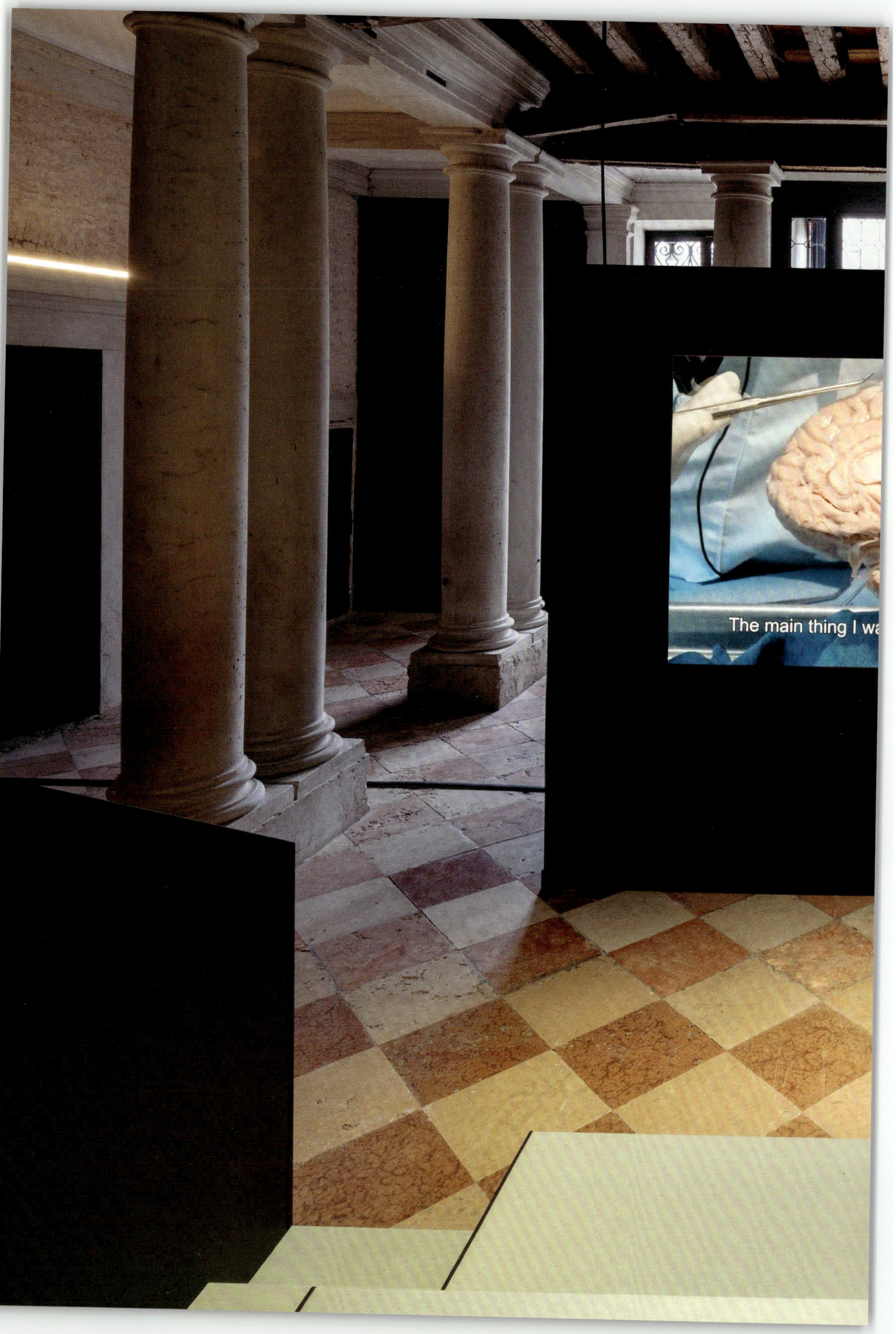
The main thing I wa

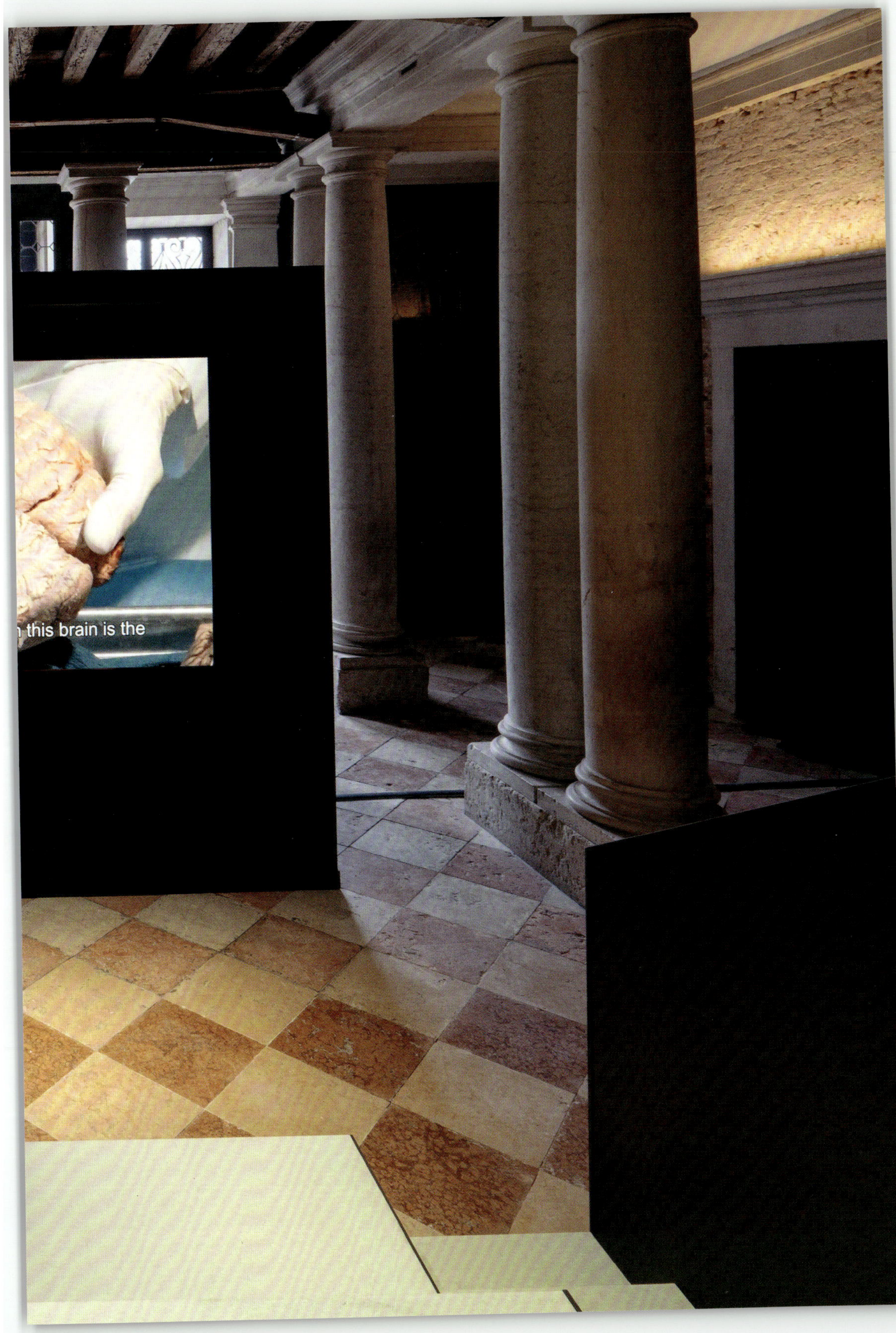
this brain is the

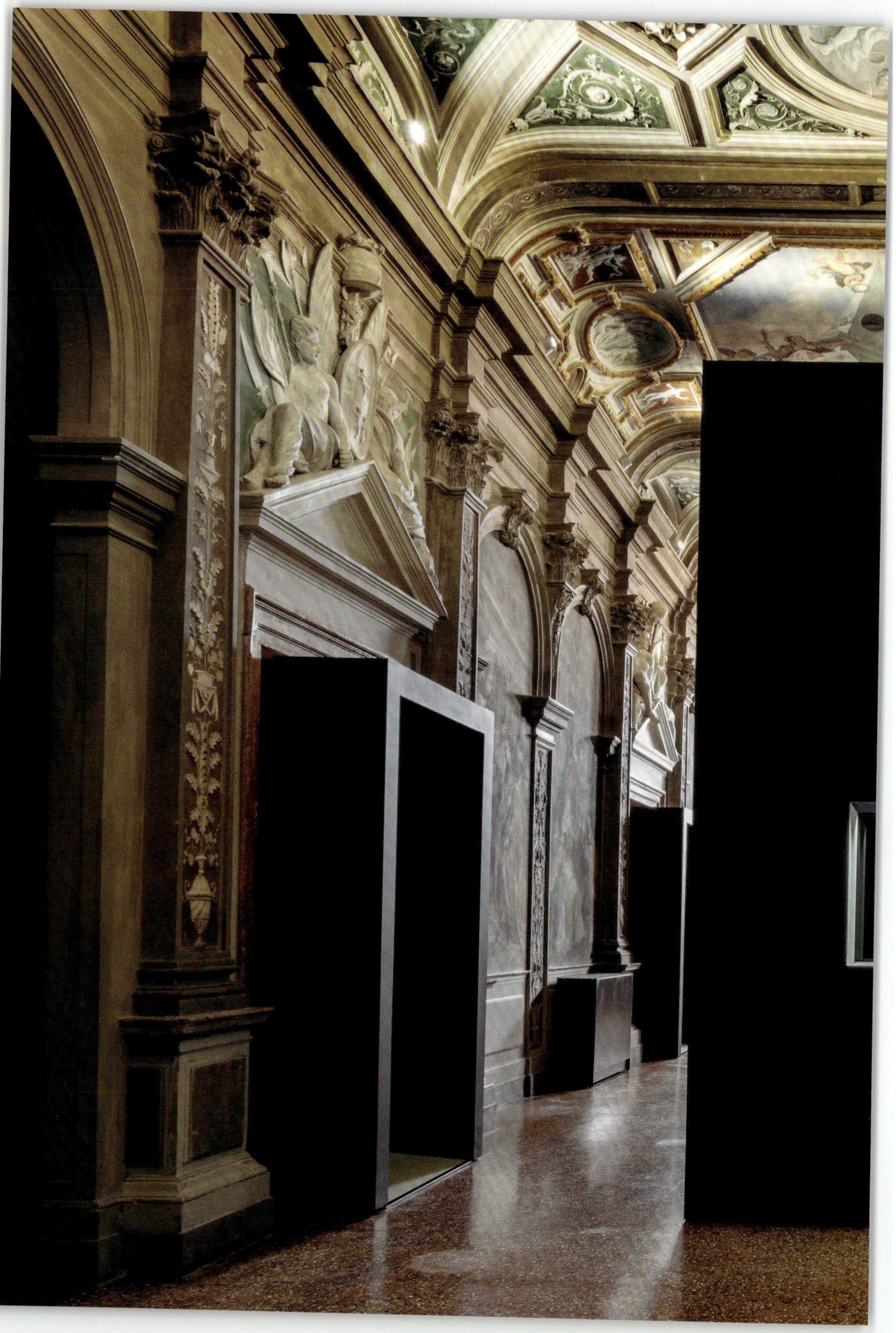

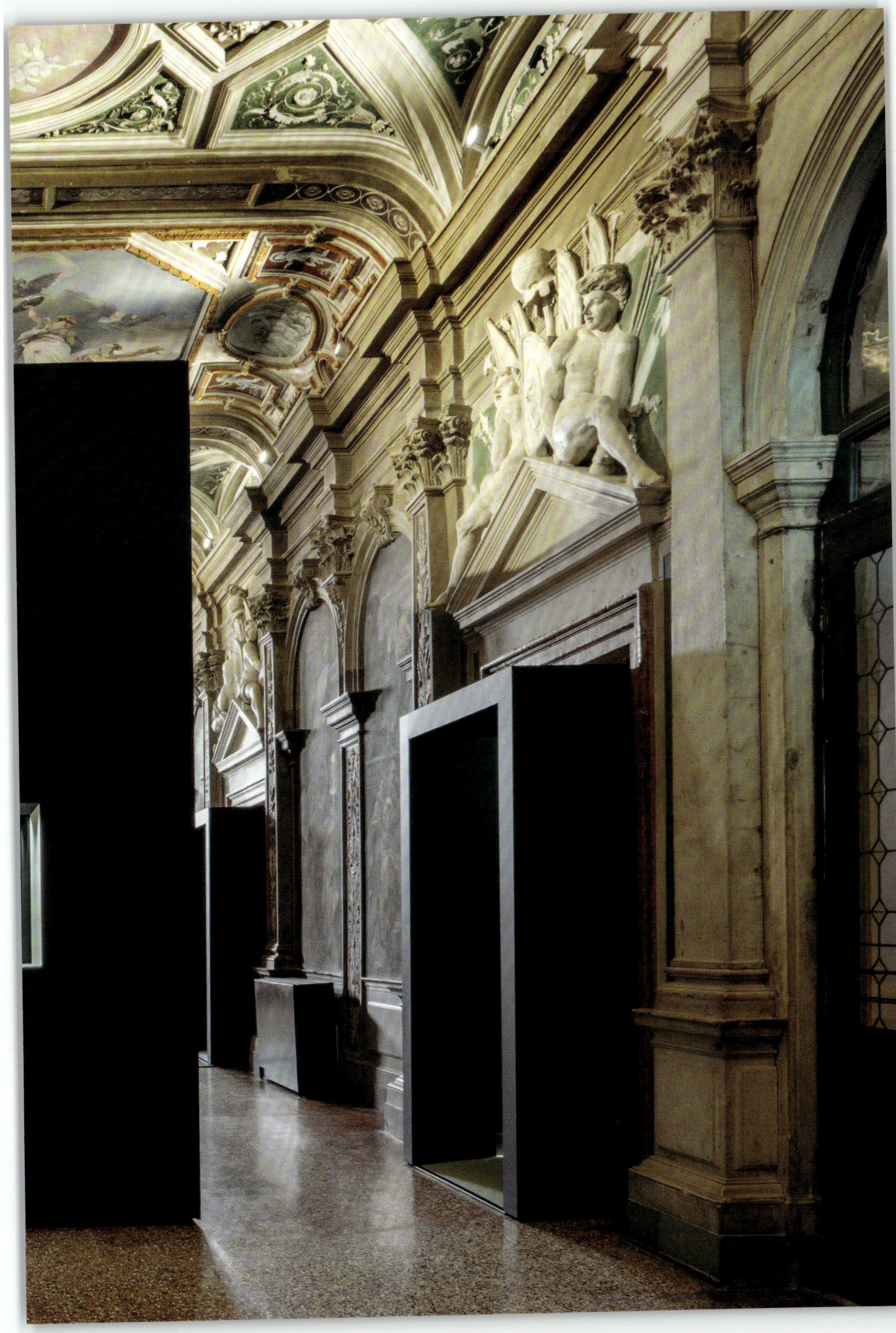

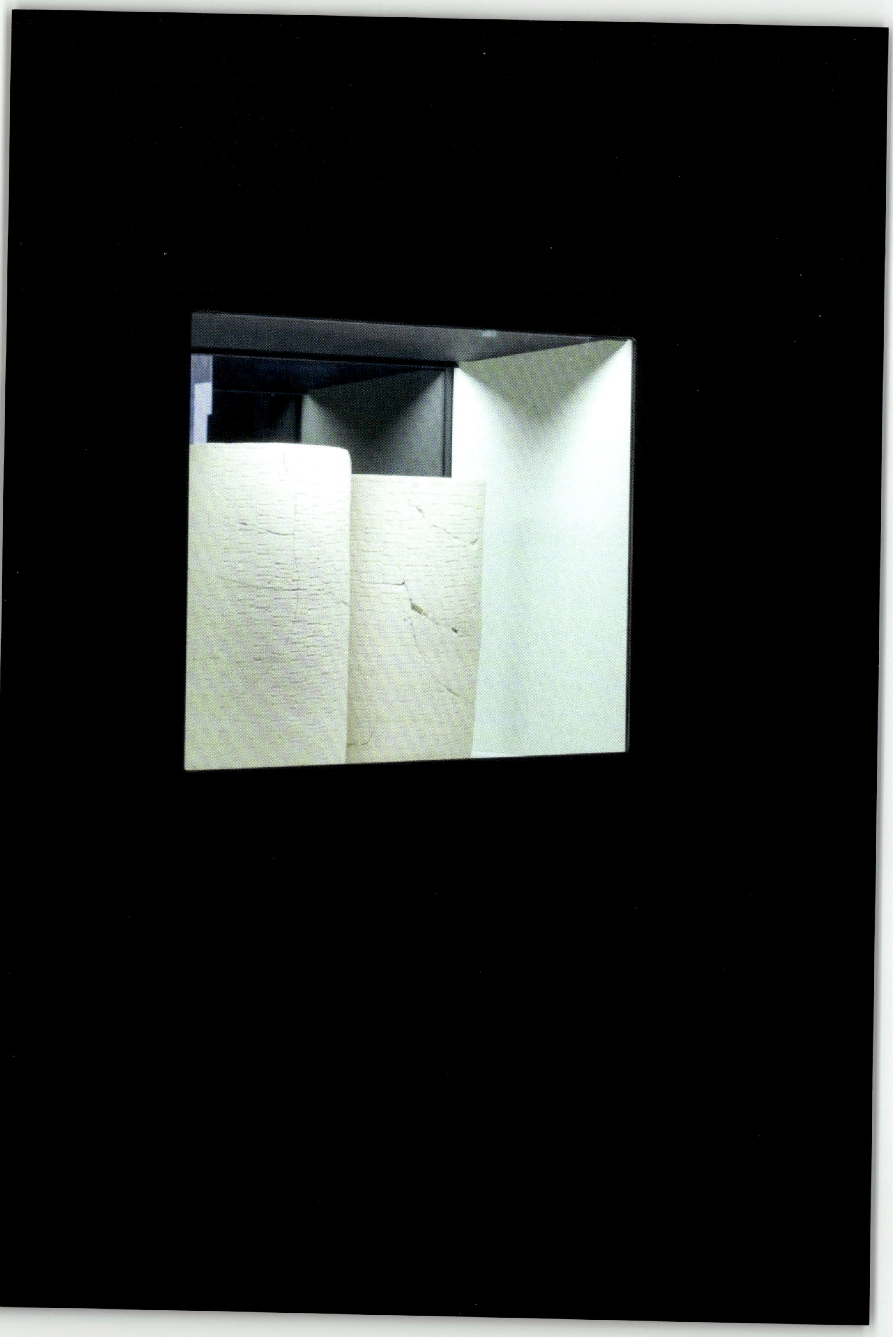

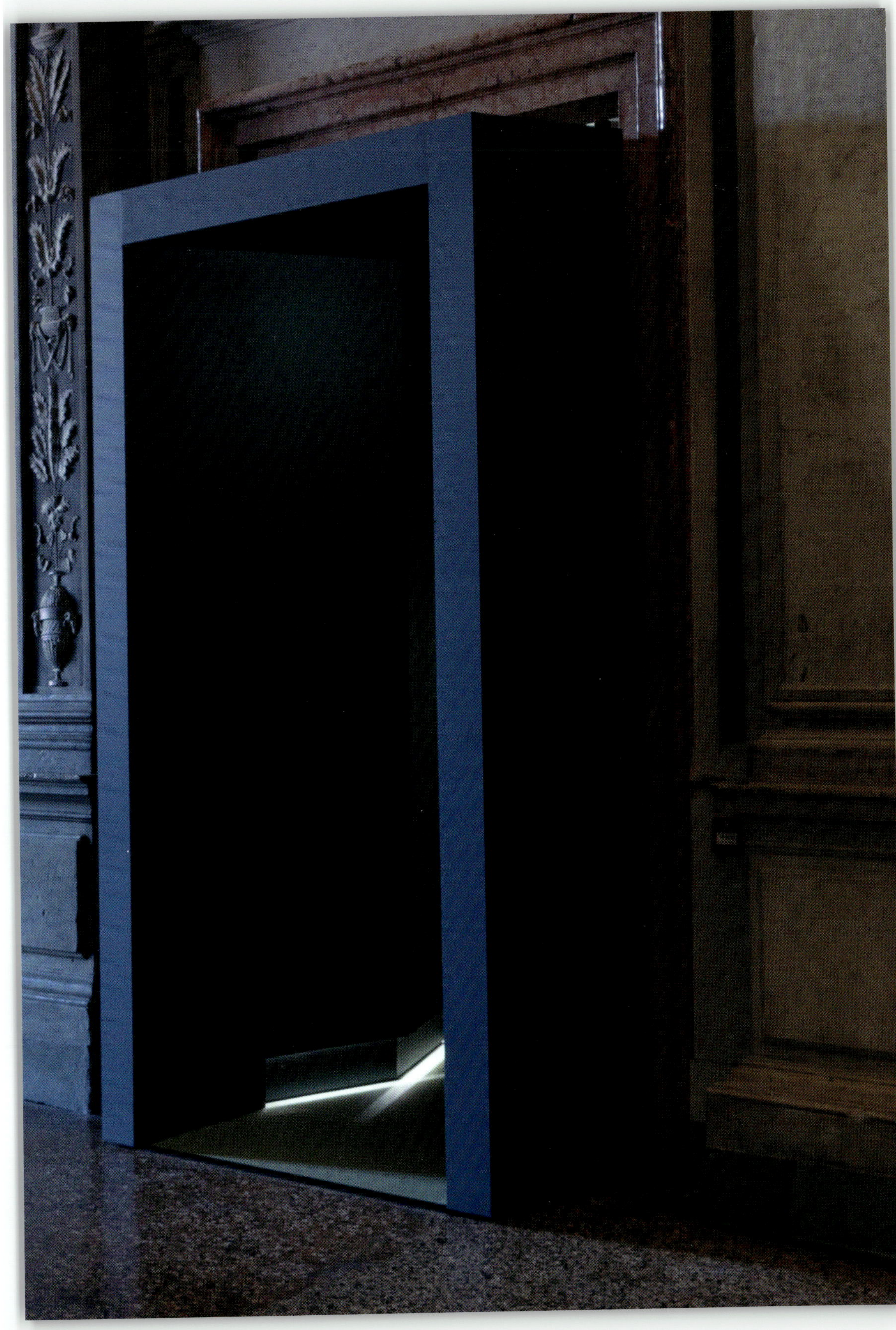

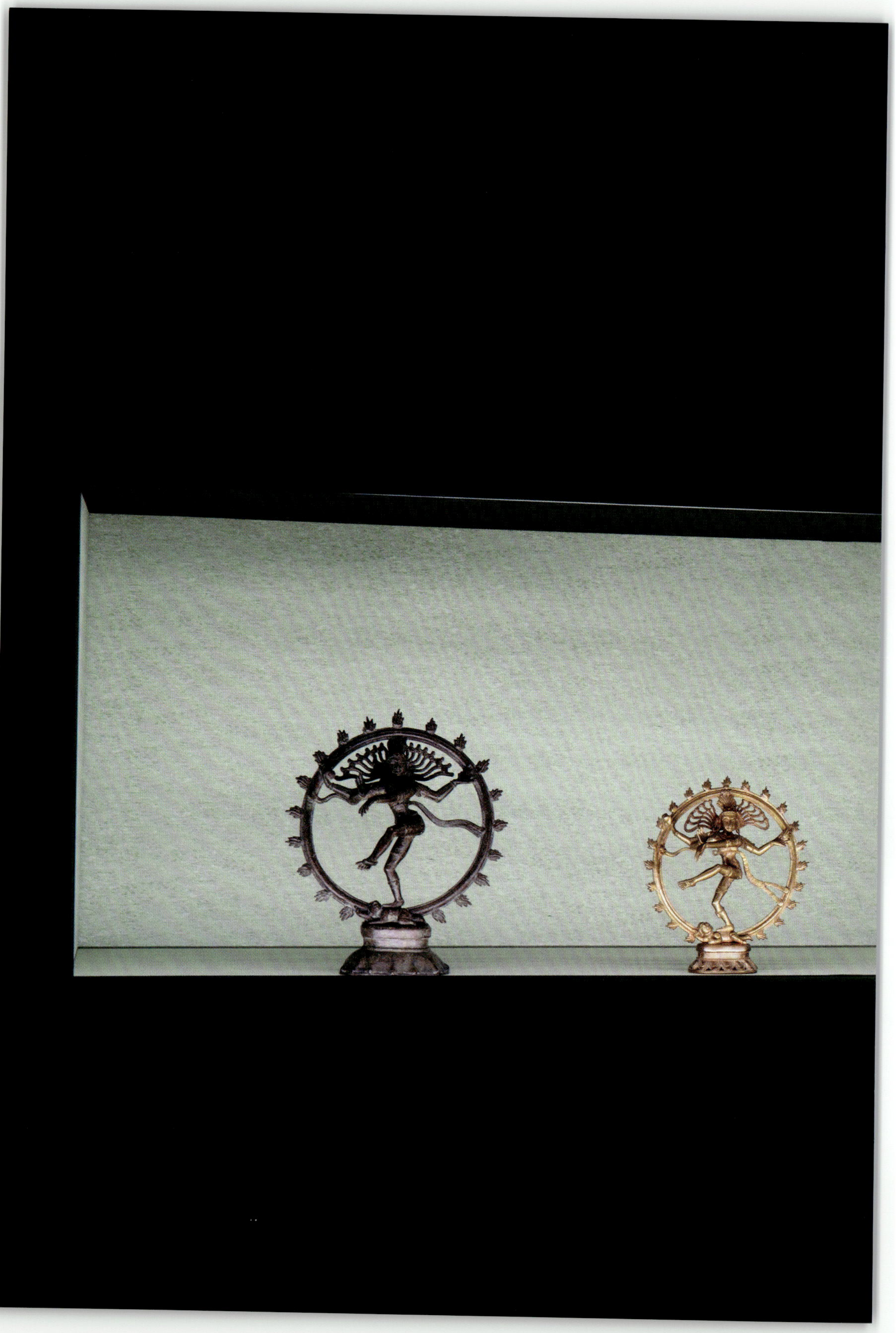

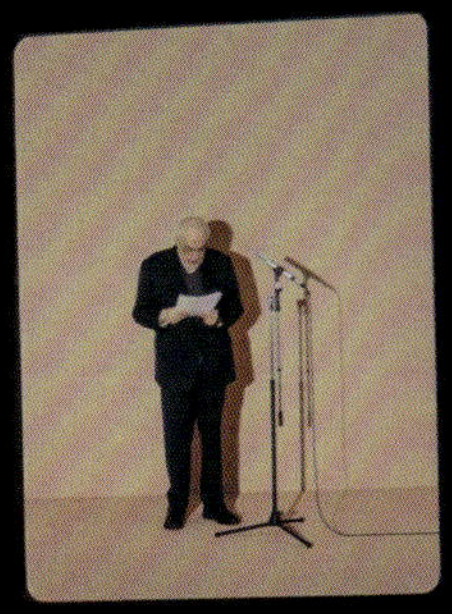

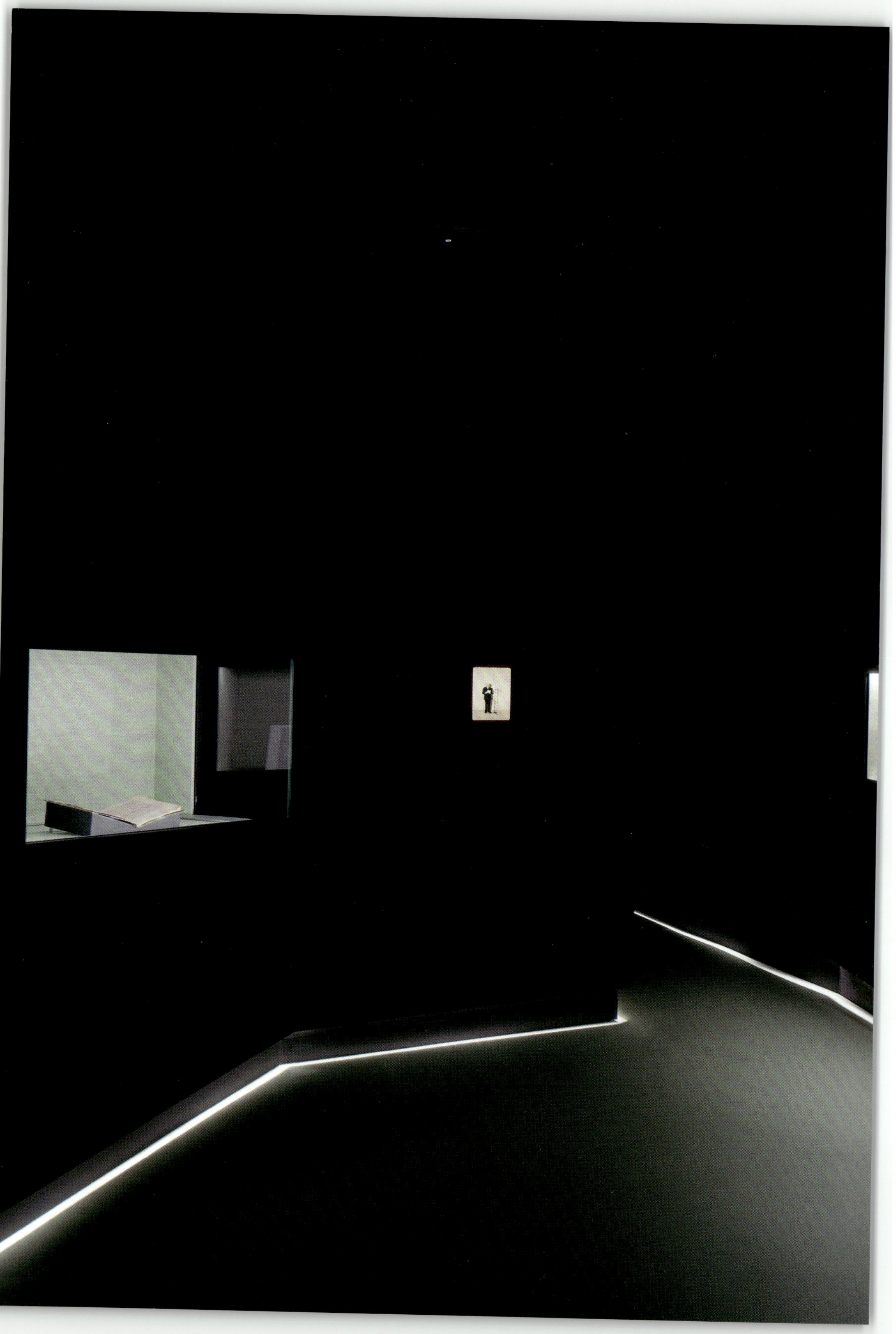

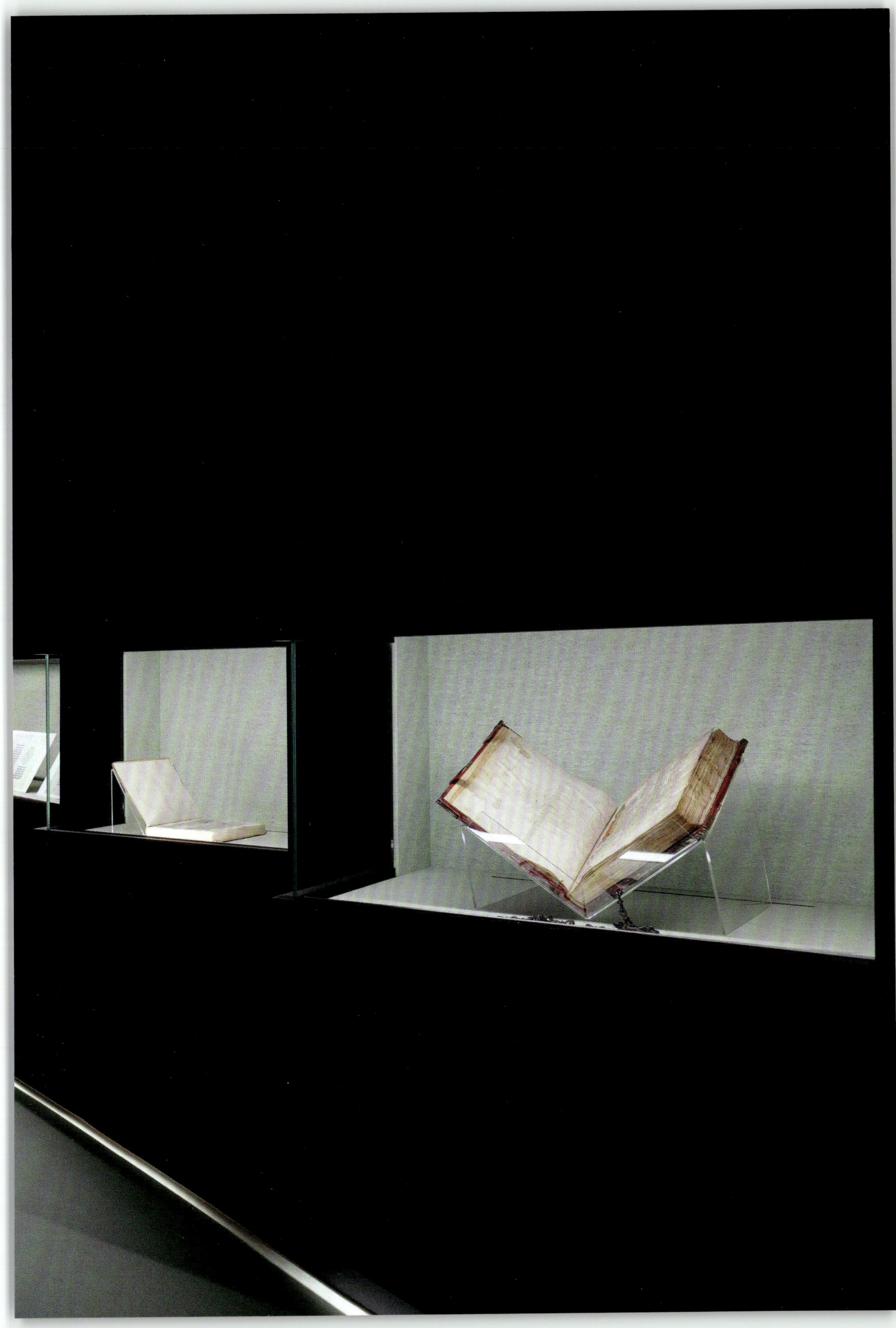

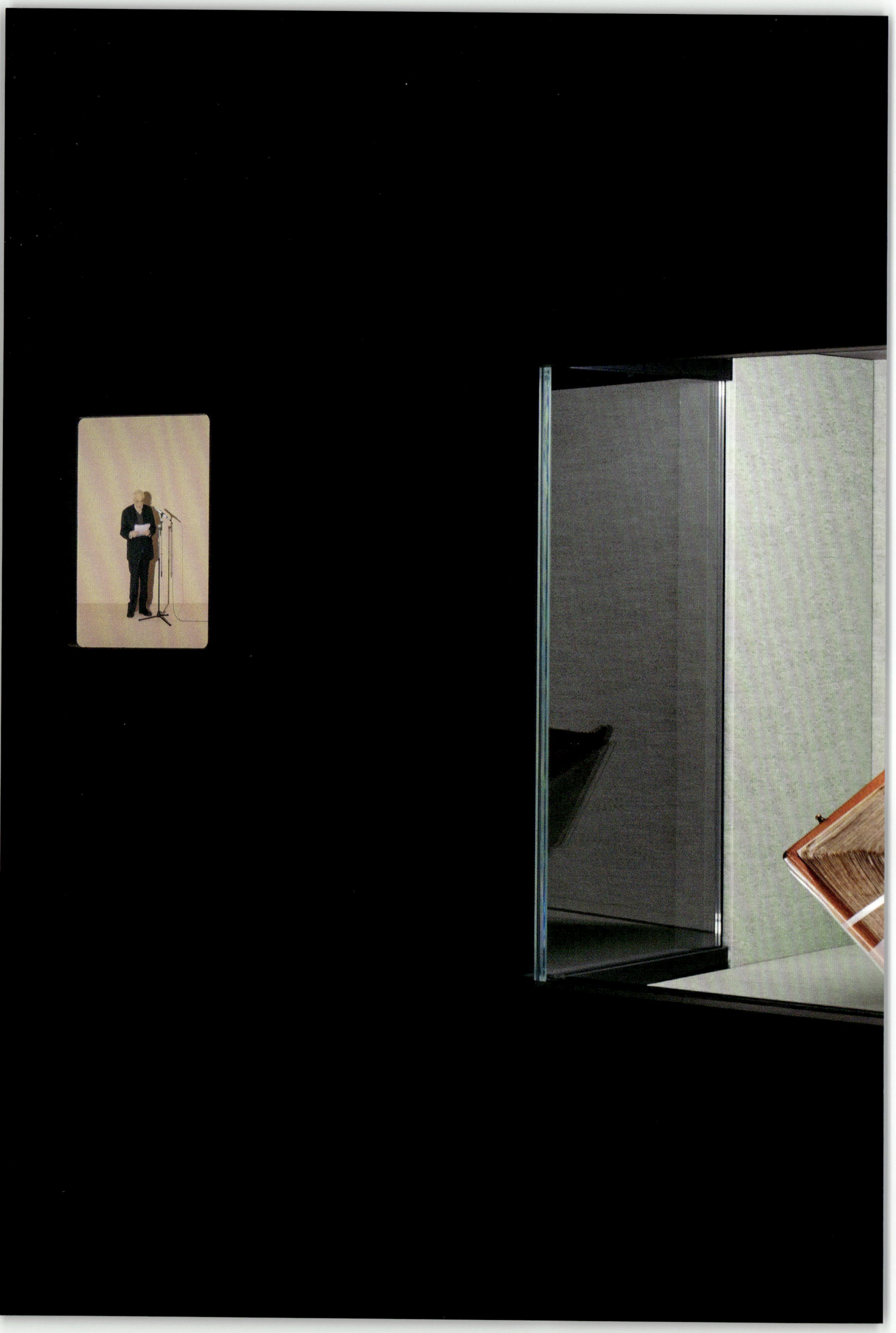

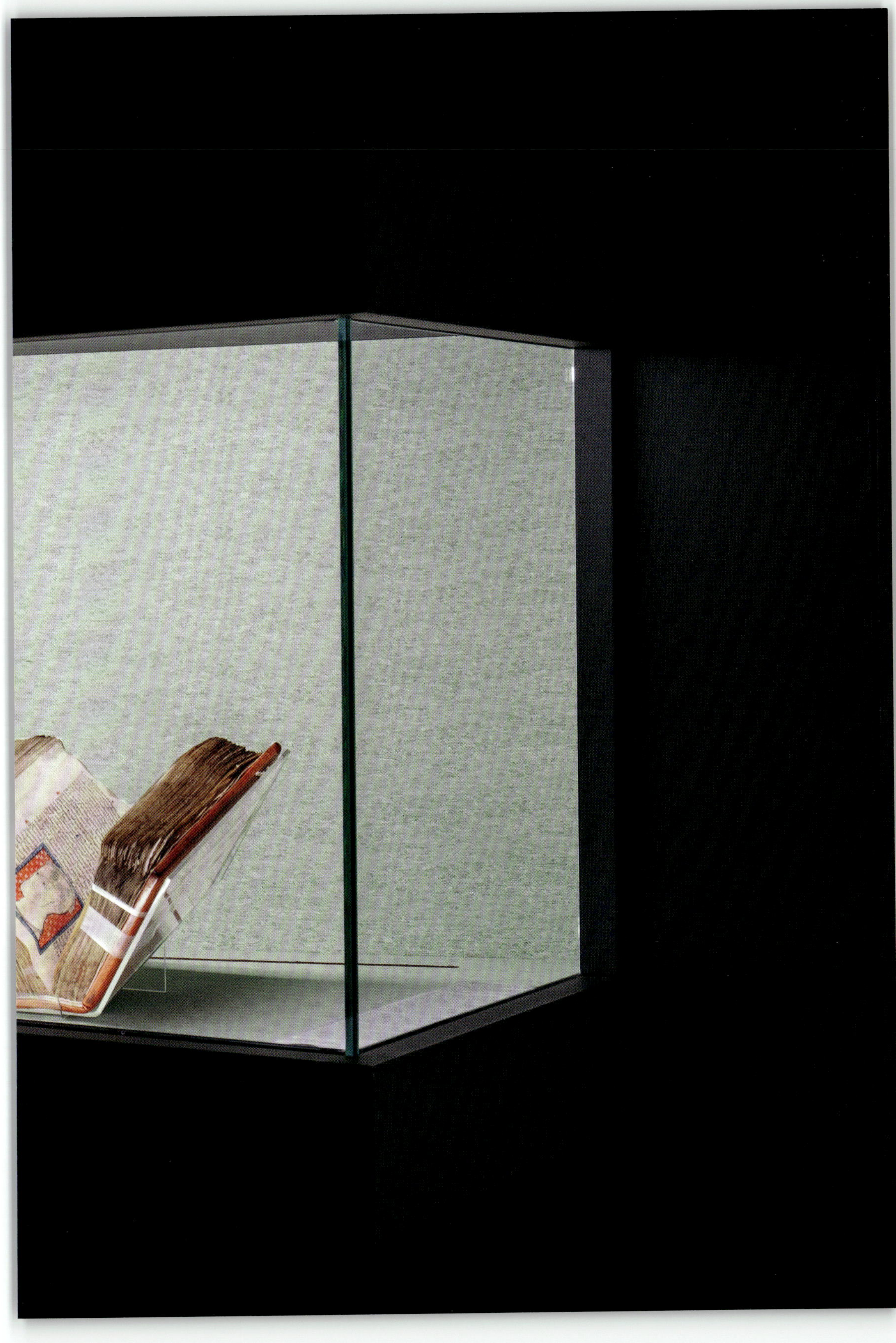

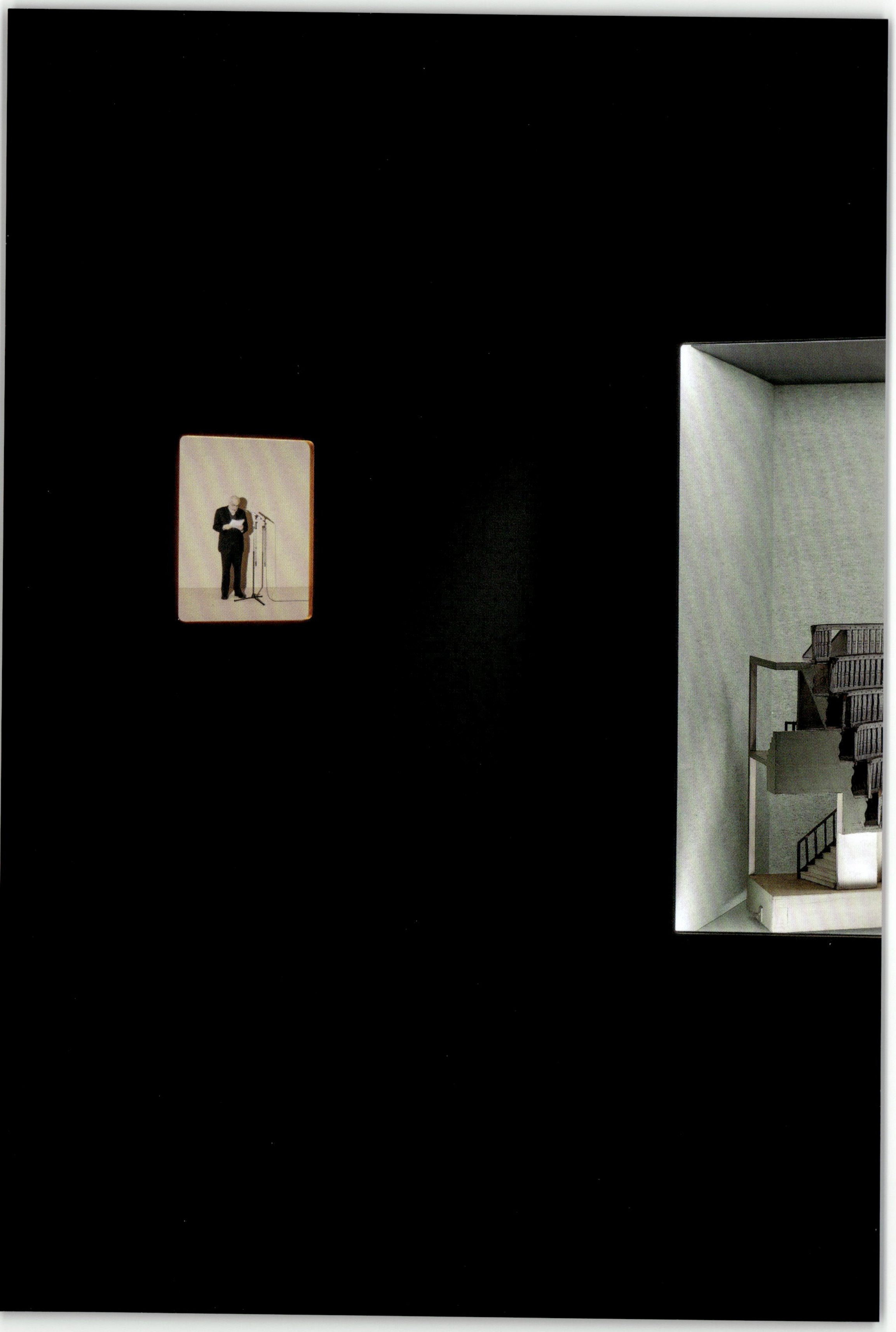

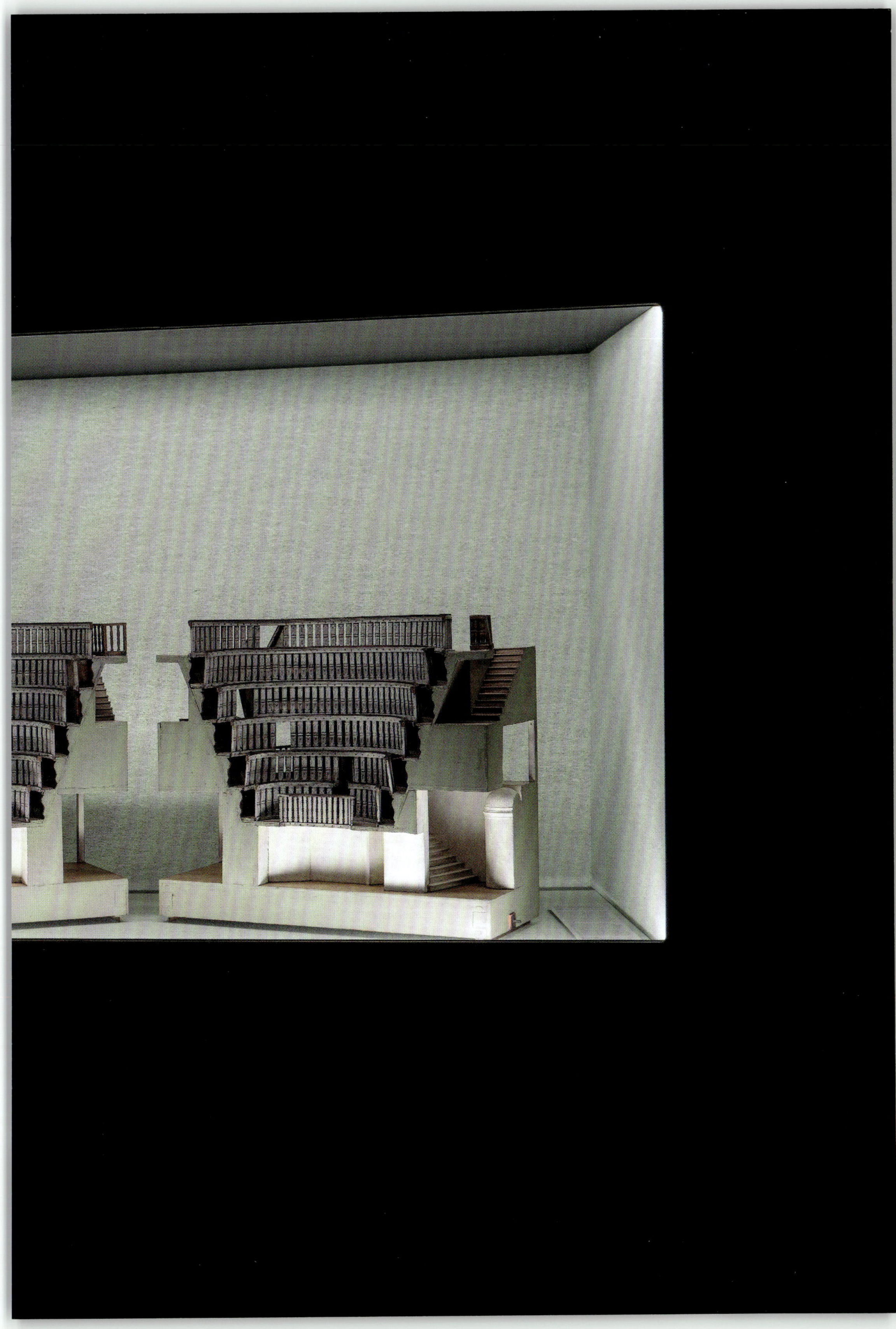

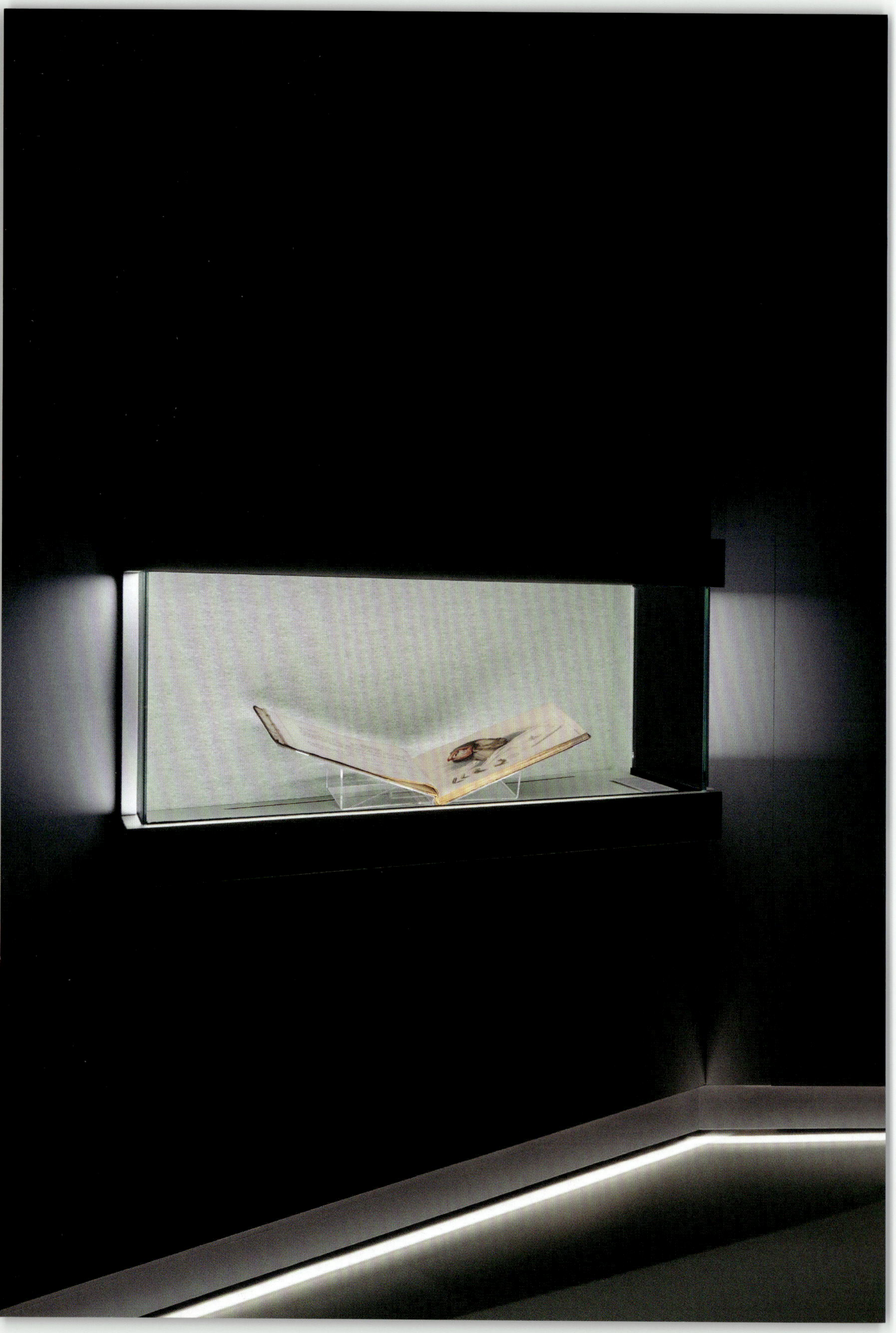

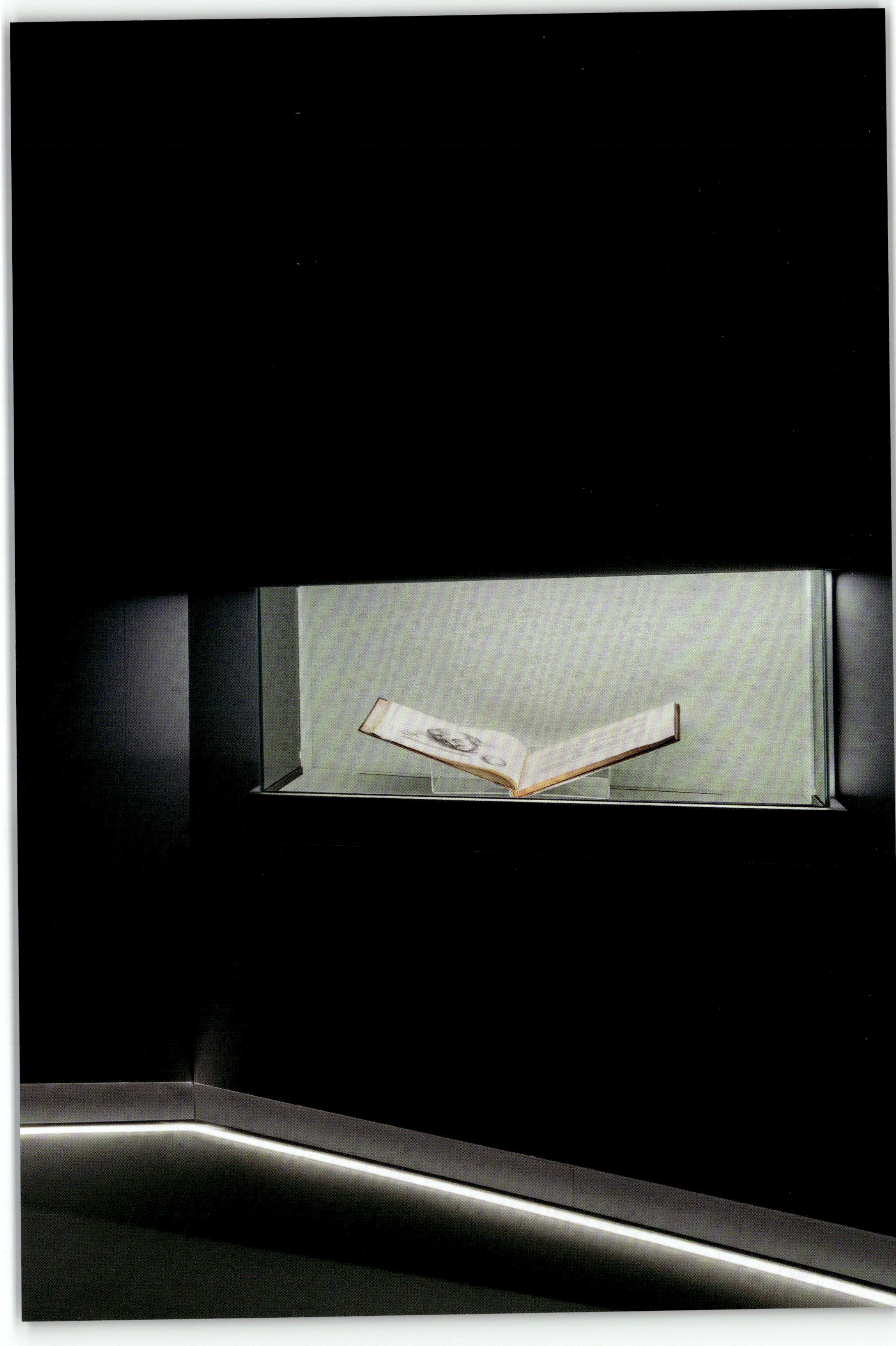

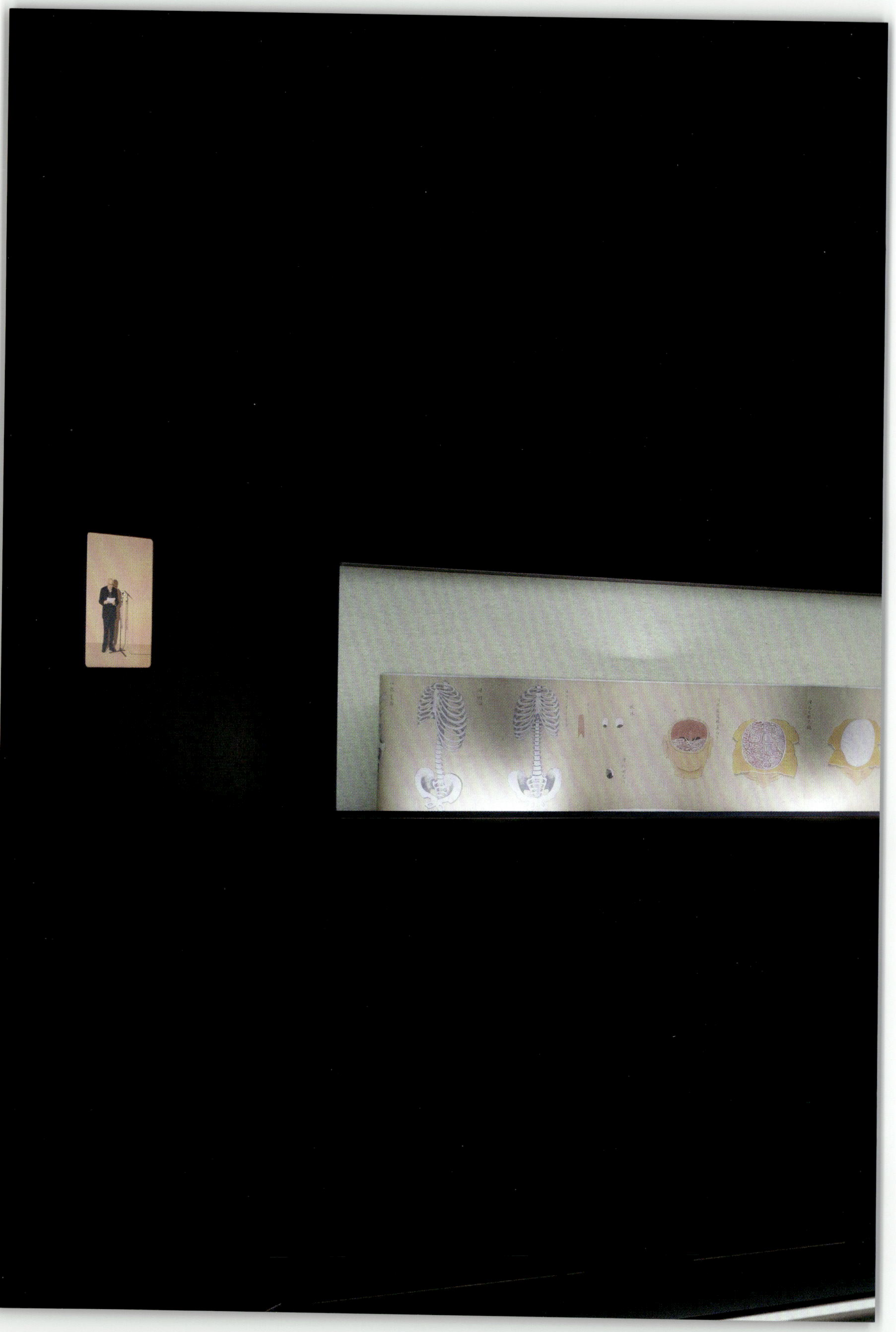

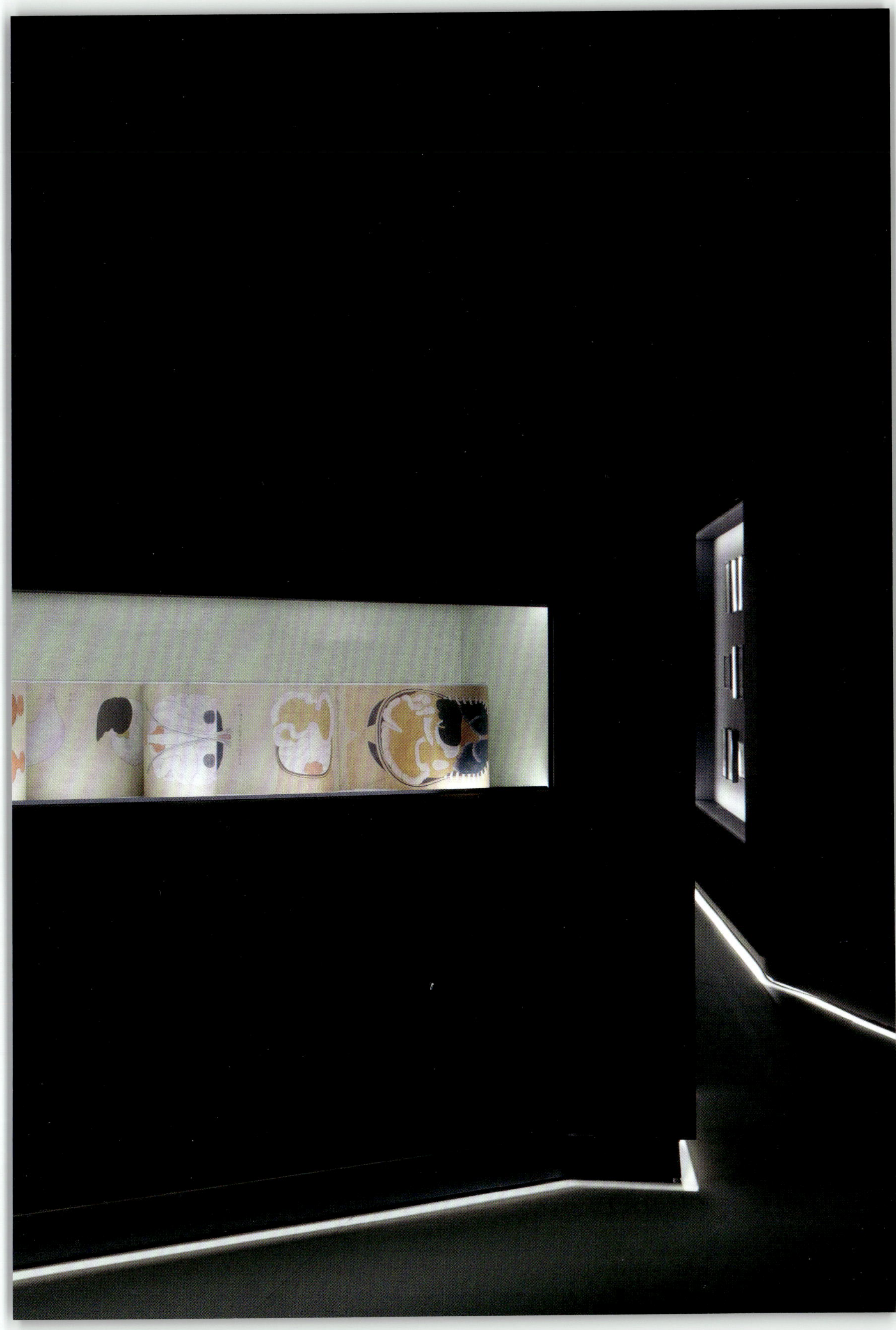

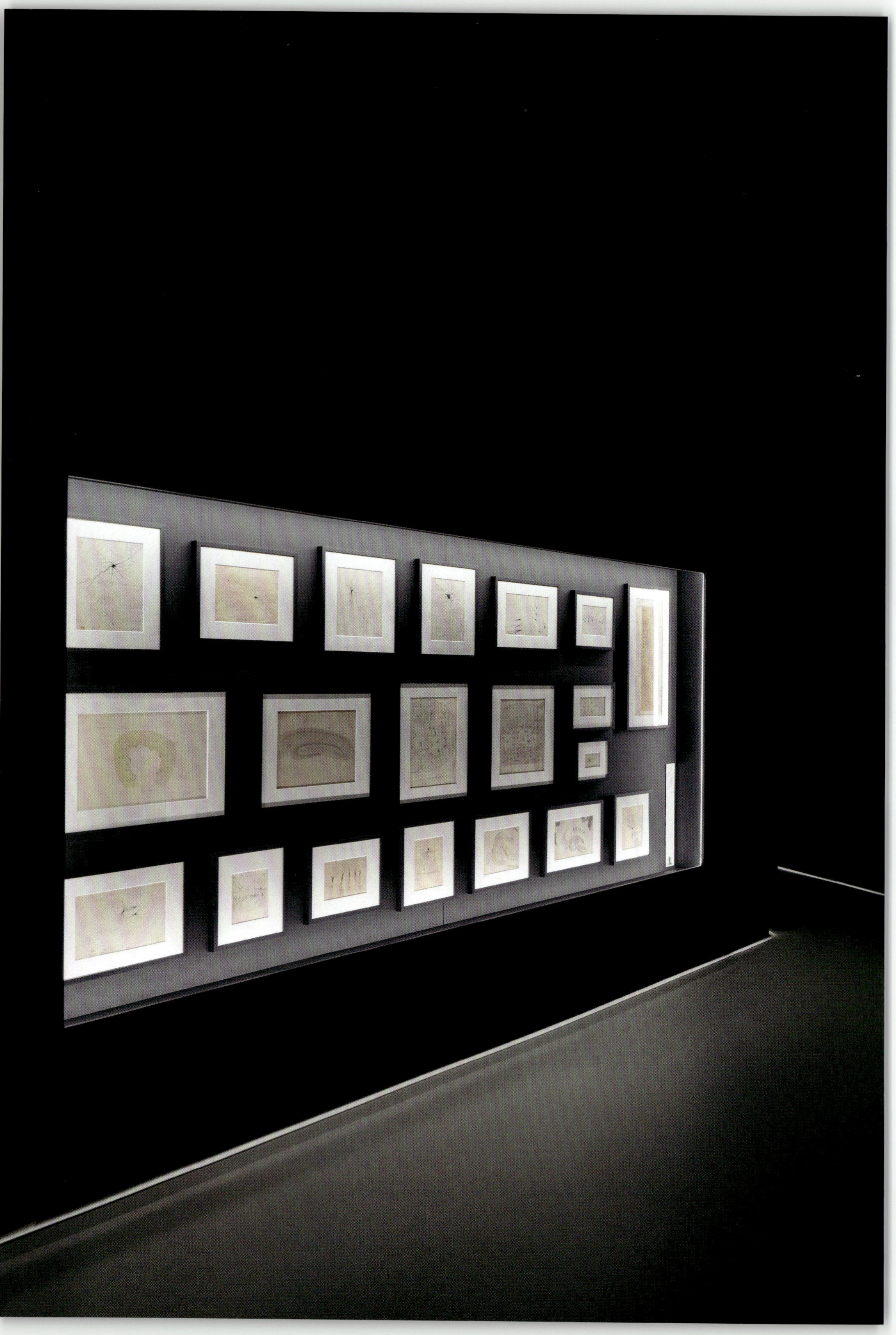

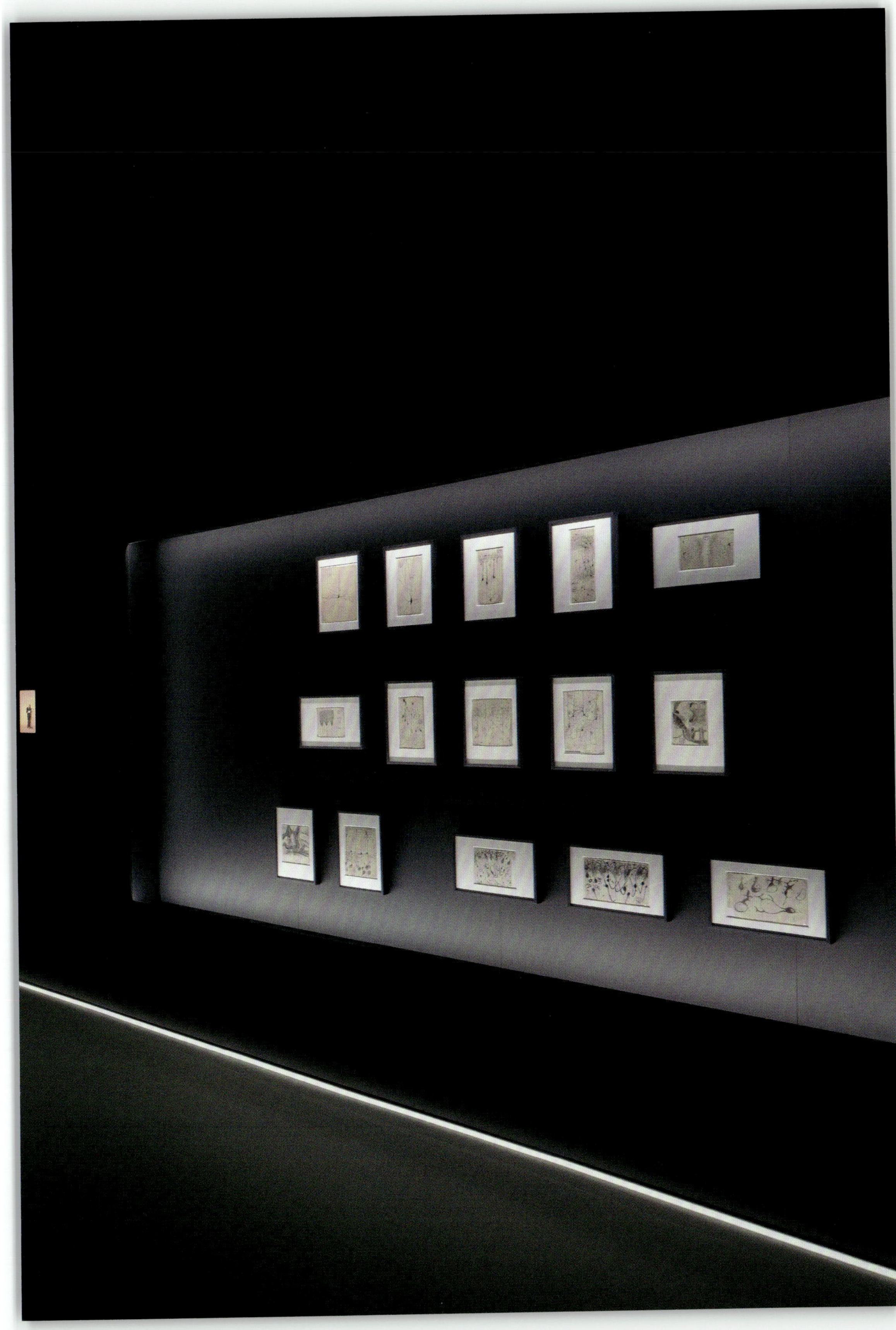

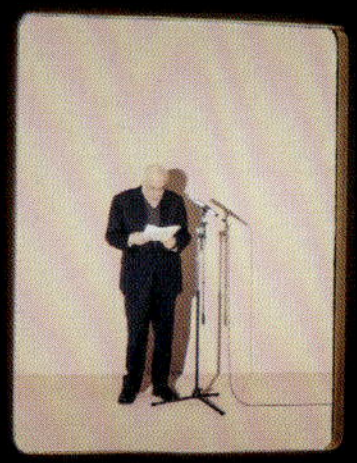

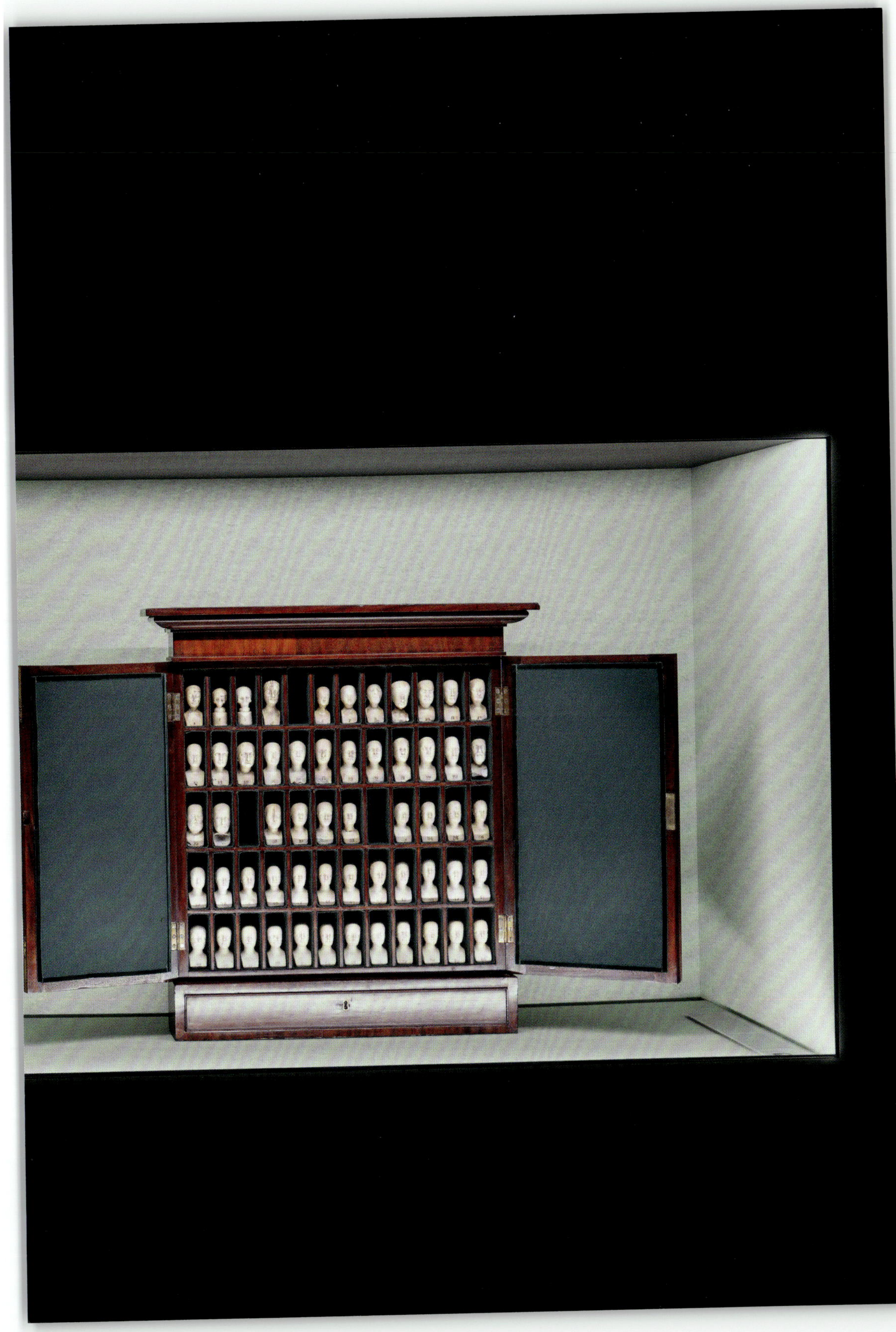

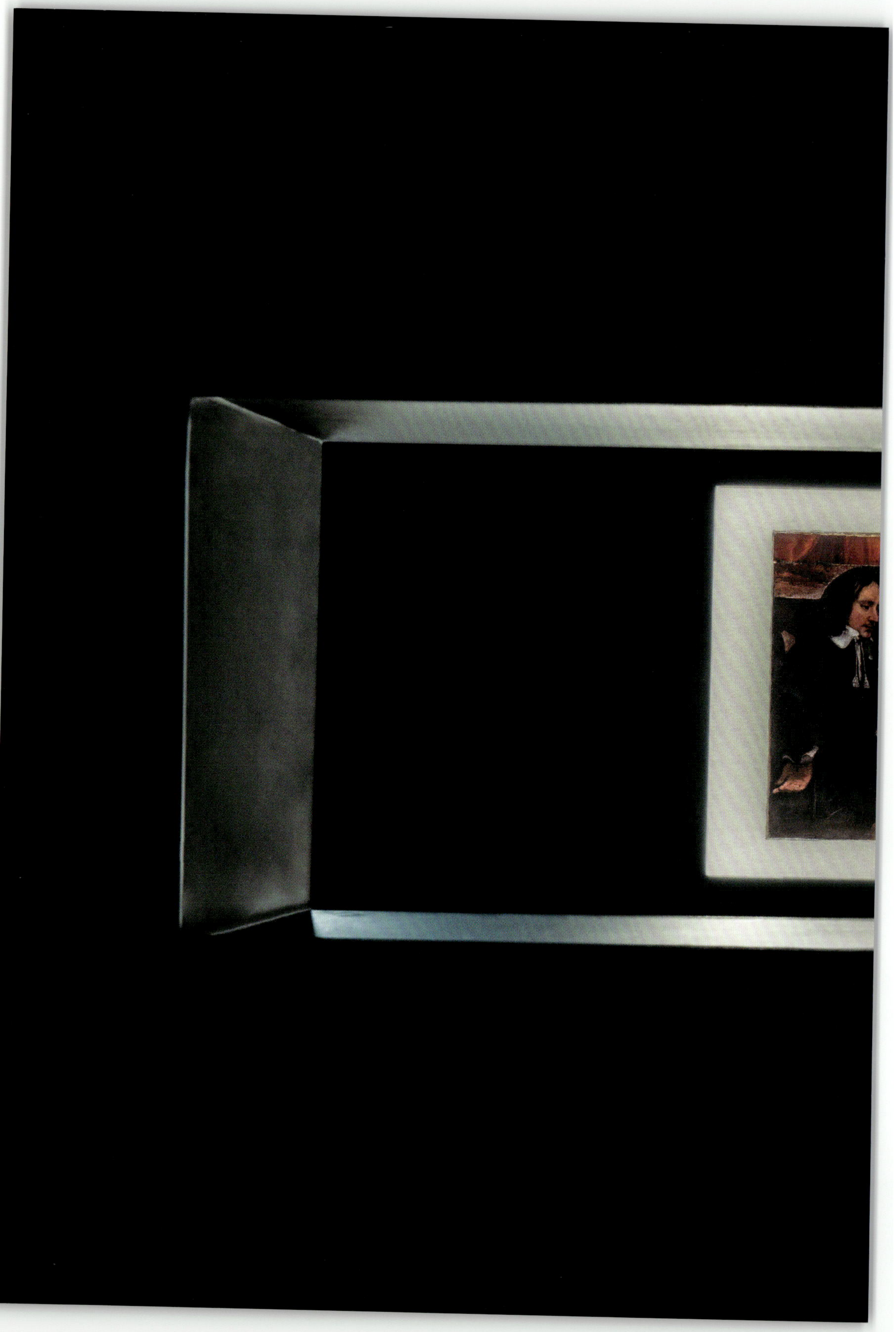

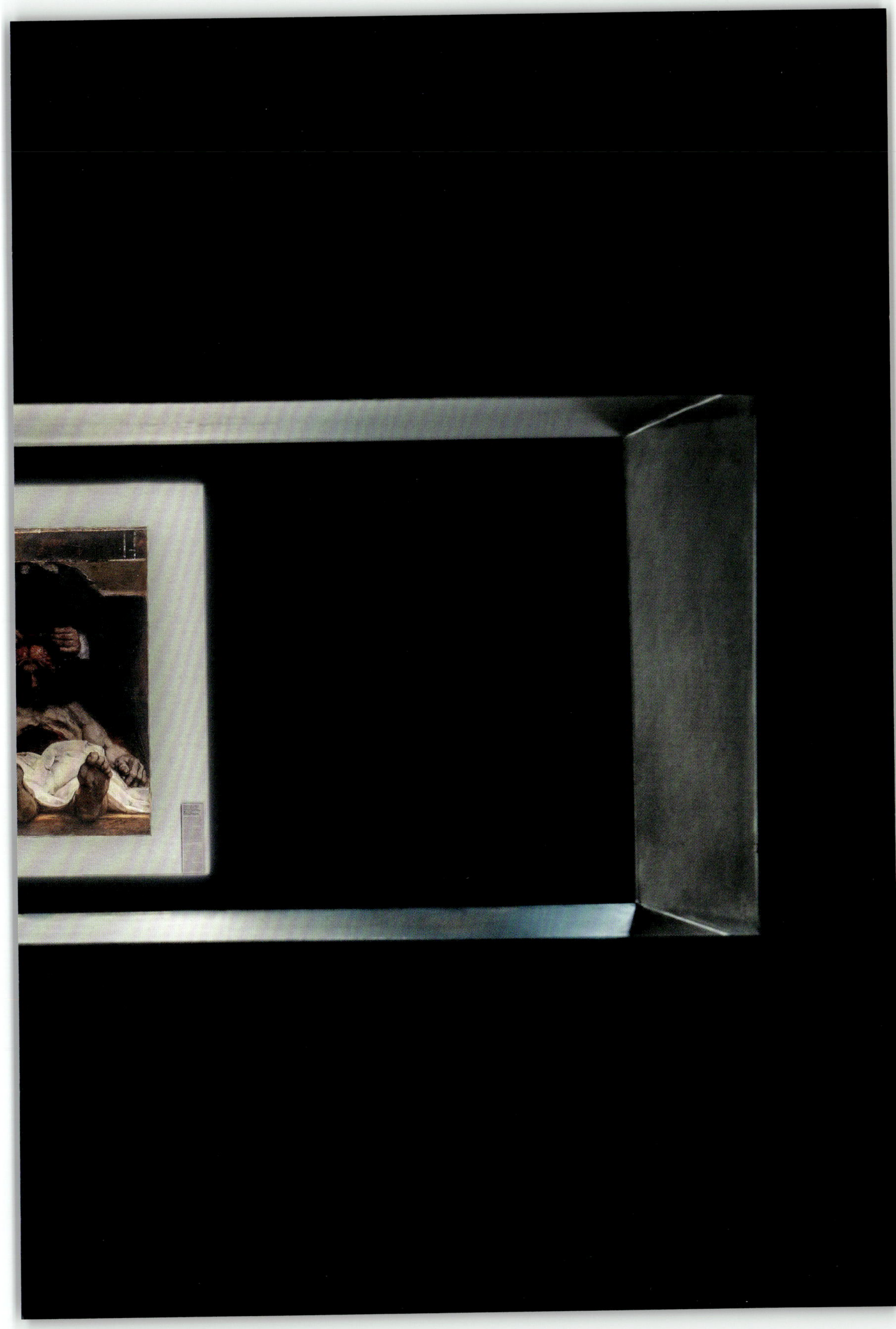

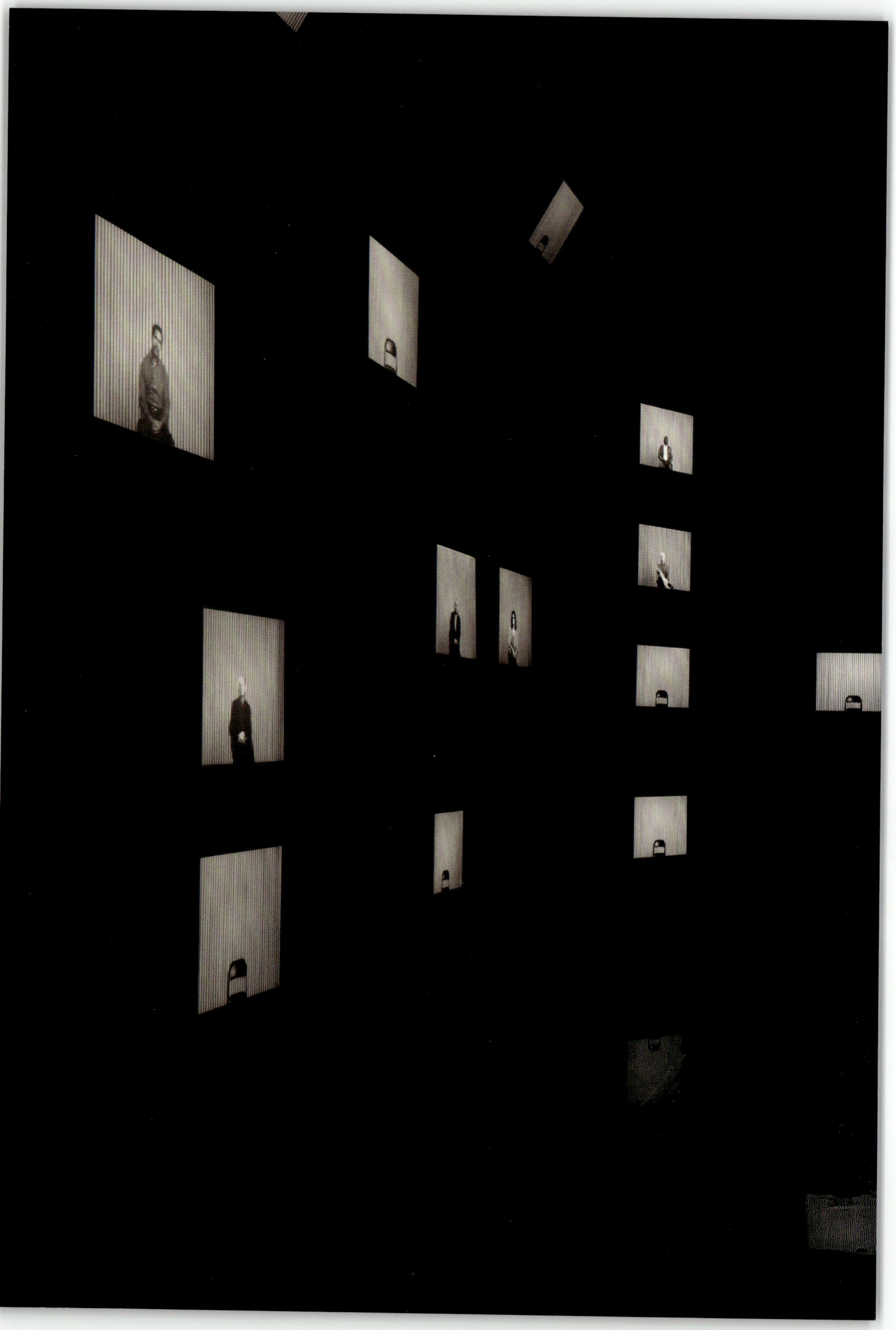

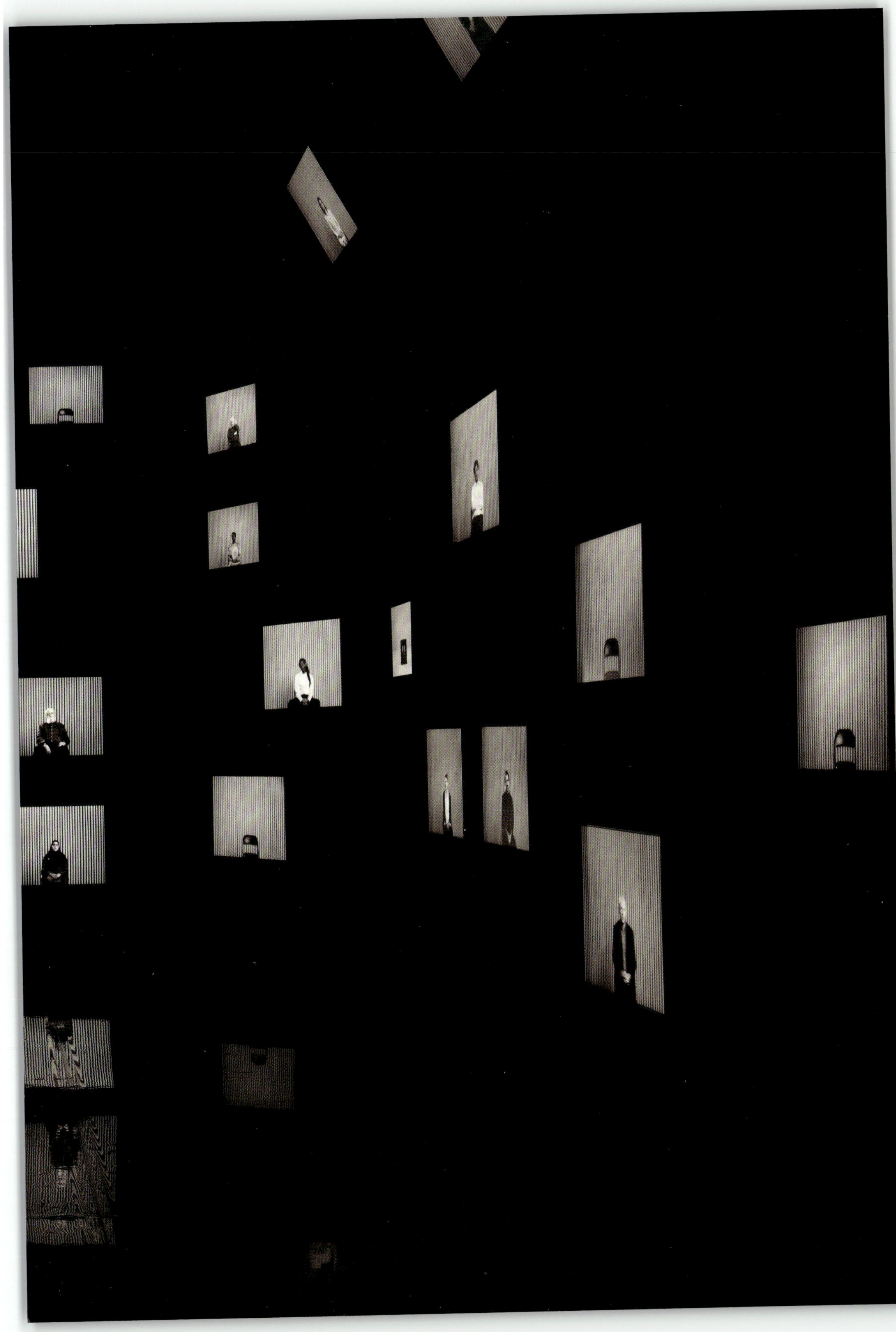

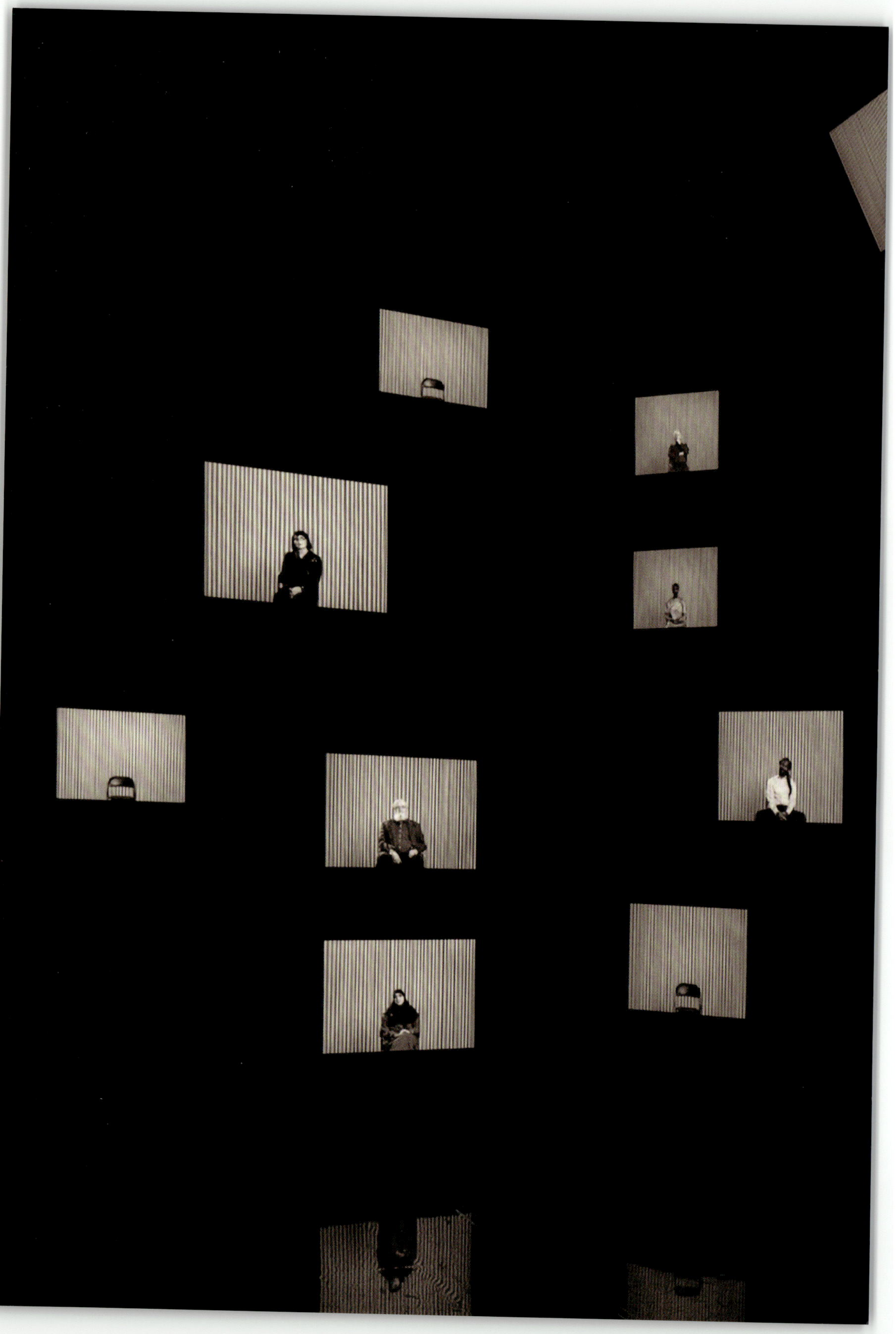

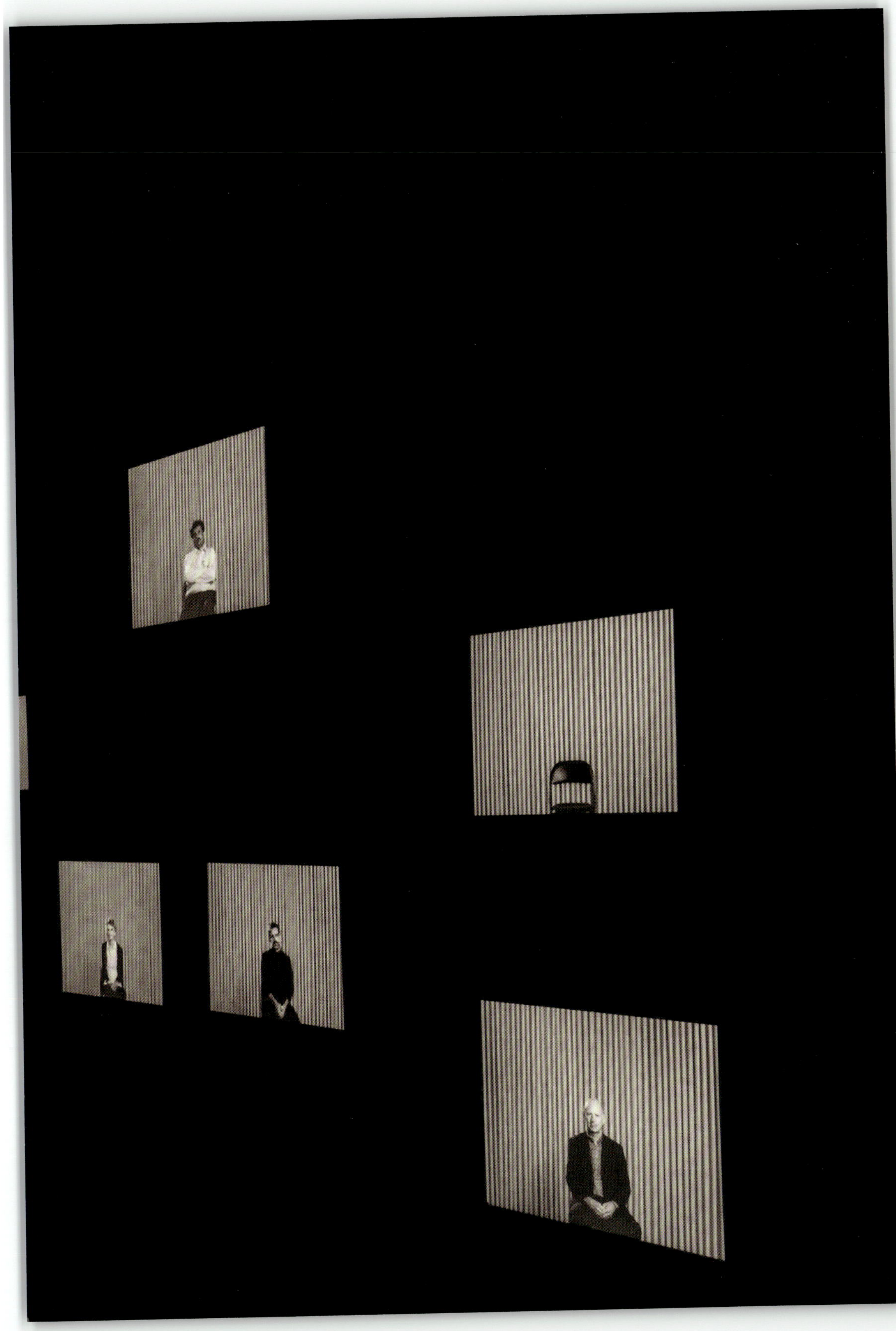

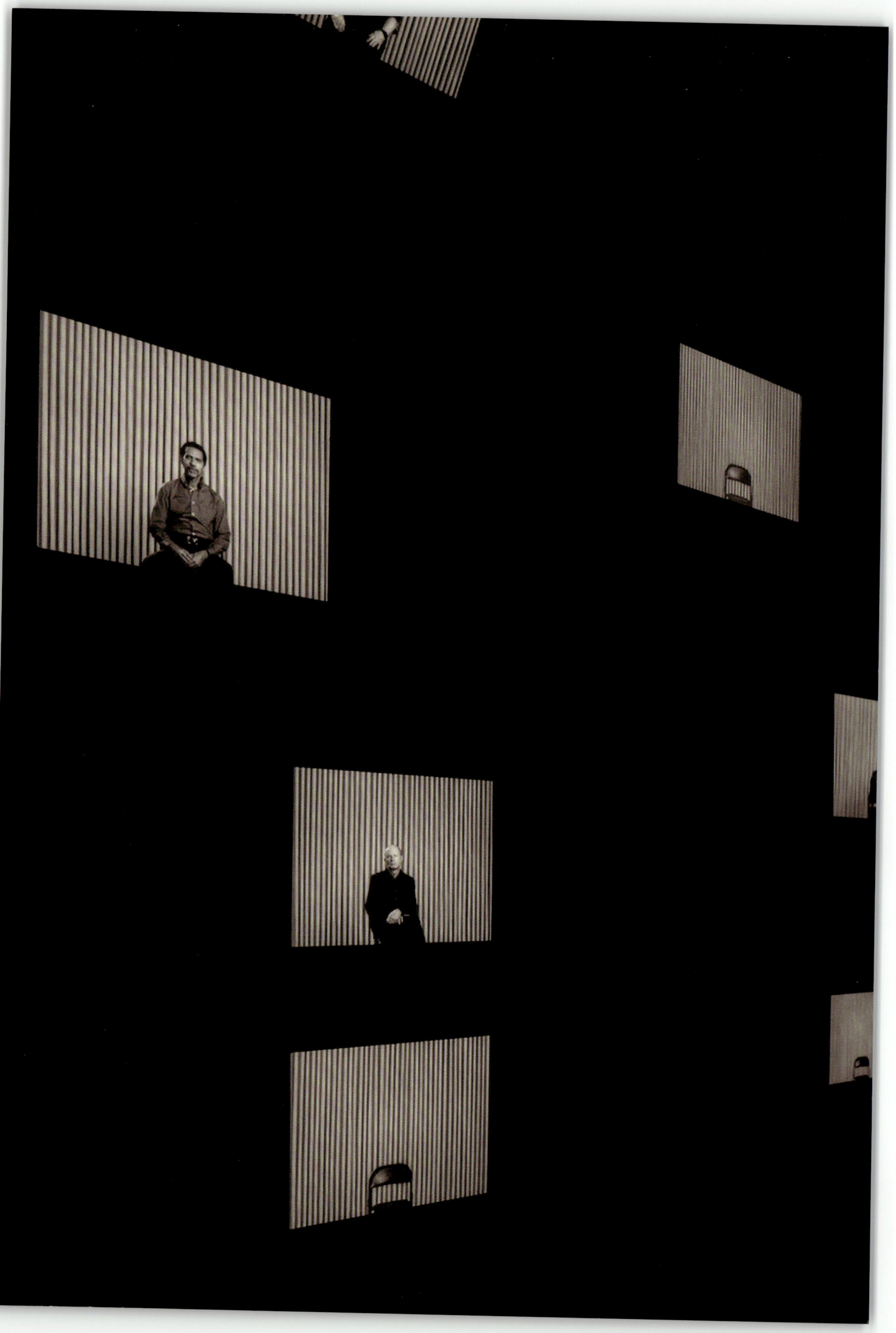

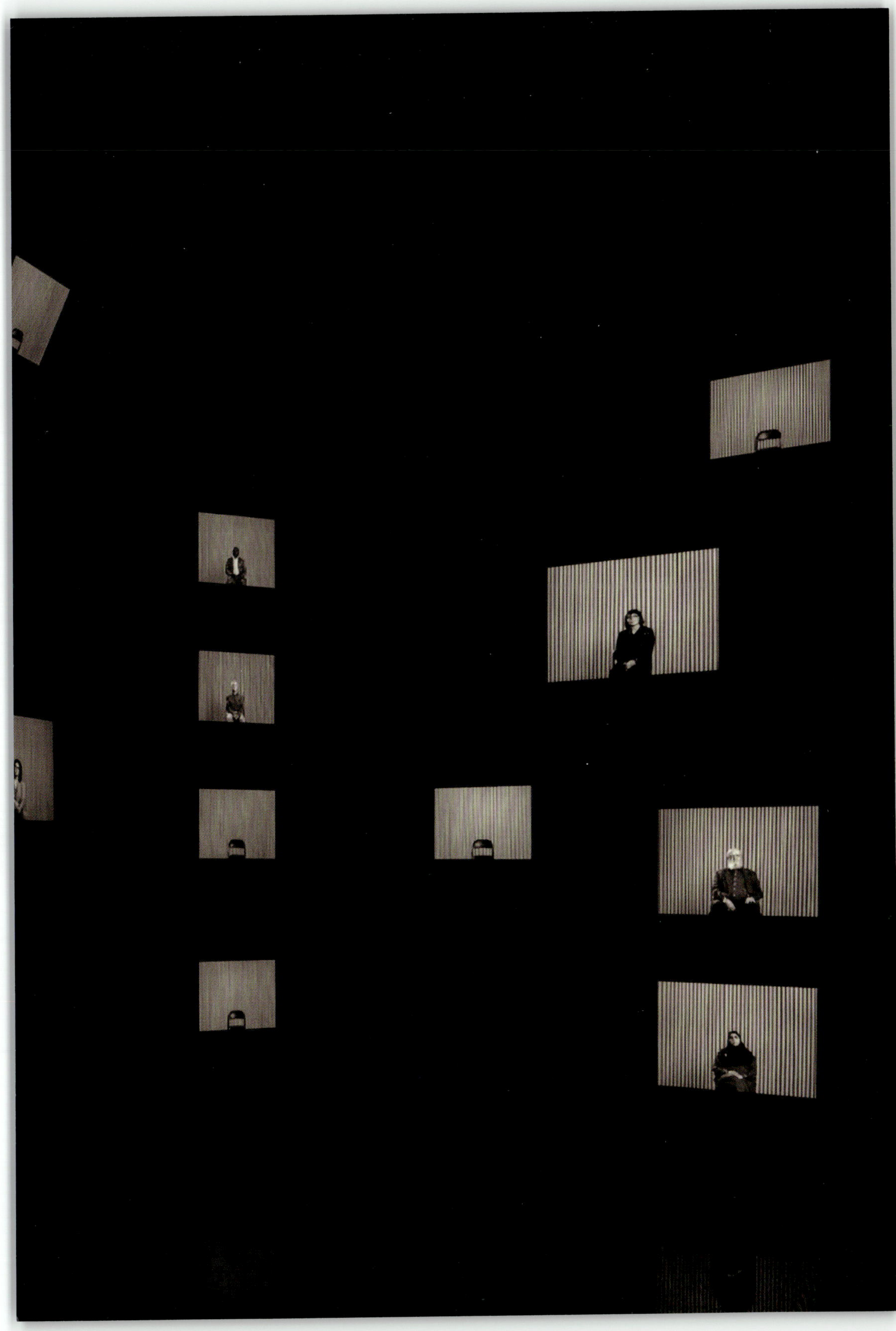

EXHIBITION VIEWS

Videos, stories, artifacts, and artworks featured in the exhibiton views

IT BEGINS WITH AN IDEA

Exhibition Checklist

LESSONS

Introduction to
the Human Brain, 2015
Suzanne S. Stensaas, Department of Neurobiology and Anatomy and Spencer S. Eccles Health Sciences Library, University of Utah, Salt Lake City
[p. 219]

David H. Hubel and Tornsten N. Wiesel's experiment on visual perception using a microelectrode implanted into the visual cortex of an anesthetized cat
Johns Hopkins University, Baltimore 1959
[p. 219]

The neurophysiology of memory explained by Wilder Penfield, neurosurgeon of McGill University Montréal; footage from the film *Gateways to the Mind*, 1958.
Produced by the Warner Brothers Pictures, released through the local offices of Bell Telephone System
[p. 219]

* * *

Andrea Moro
Impossible Languages. A Journey to the Boundaries of Babel, 2022

Stefano F. Cappa
We Are Memories, 2022

Guido Gainotti
Body and Mind in the Development of Emotions, 2022

Maria Concetta Morrone
Vision, 2022

Daniela Perani
Neuroimaging for the Exploration of the Language System, 2022

Letizia Leocani
The Sensorimotor System, 2022

* * *

Modern MRI: Details and Diffusion, 2017
Video courtesy of the USC Mark and Mary Stevens Neuroimaging and Informatics Institute (www.ini.usc.edu)
[p. 222]

Neurons and Synapses of the Medial Prefrontal Cortex, 2017
Video courtesy of the USC Mark and Mary Stevens Neuroimaging and Informatics Institute (www.ini.usc.edu)
[p. 222]

STORIES ARTIFACTS

Cylinders of Gudea, Iraq, *c.* 2120–2110 BCE
Terracotta, 61 × 32 cm (Cylinder A); 56.5 × 33 cm (Cylinder B)
Paris, Musée du Louvre, Département des Antiquités Orientales, inv. nos. MNB1512, MNB1511
Exhibition copy
[p. 228]

Inscribed with The Building of Ningîrsu's Temple, *the longest known text written in the Sumerian language, the Gudea cylinders record a dream of Gudea, who ruled the southern Mesopotamian city-state of Lagash from approximately 2150 to 2125 BCE.*

Unearthed in 1877 during excavations at Telloh, Iraq, the cylinders are among the oldest surviving records of a dream. "Profound things came suddenly to me, but the meaning of what the nocturnal vision brought to me I do not understand ... I have to tell her about this!" With these words, Gudea resolves to consult the goddess Nanshe, who explains that the dream contains the god Ningîrsu's directives concerning the construction of a new temple.

* * *

Skull of an adult Inca male with five healed trepanations, Patallacta, Cuzco, Peru, *c.* 1476–1532 (Late Horizon)
Scraped skull, 19.8 × 13.4 cm
Circular trepanations, 30–35 mm diameter
Lima, Ministerio de Cultura – Museo Nacional de Arqueología, Antropología e Historia del Perú, inv. no. 628, collection code AF-006953
Exhibition copy
[p. 232]

In trepanation, a hole is cut, scraped, or drilled into the cranium. Found throughout the world at sites dating from the late Paleolithic through the present, trepanned skulls frequently bear evidence of prior trauma, testifying to trepanation's use as an early therapeutic measure.

This skull of an adult male was discovered in a burial cave in Patallacta, an archaeological site near the Inca capital of Cuzco, Peru. Consistent in size and shape, the five healed openings are notable for the smooth, inwardly beveled edges that indicate bone regrowth and the prolonged postoperative survival of the patient.

* * *

Incan ceremonial knife (*tumi*), 13th–15th century
Bronze casting, 17 × 10.2 × 2.5 cm
Hamburg, MARKK–Museum am Rothenbaum, inv. no. 33.180:2
[p. 234]

Characterized by a semicircular blade and typically made of copper, bronze, silver, or gold, a tumi *is an Andean ceremonial knife that was used for medical operations and ritual sacrifices. This* tumi *was made by metalworkers of the Chimú civilization in northern Peru (1000–1470).*

The figure on the right presses a tumi *to the head of the central figure, who holds onto the figures on either side of him. The scene may depict a trepanation, a procedure practiced in the Andes, Europe, Africa, the Near East, and the Melanesian islands as early as the Neolithic period.*

* * *

Single-column cuneiform tablet with a drawing of the demonic creature (*bennu*), Assur (Qal'at Sherqat), Iraq, 8th century BCE, find division 14, 1914
Clay, 11.7 × 7.9 × 2.2 cm
Berlin, Vorderasiatisches Museum, Staatliche Museen zu Berlin – Preußischer Kulturbesitz, inv. no. VAT 13739 +14130
[p. 240]

Symptom descriptions, rituals, and prescriptions recorded on Mesopotamian magico-medical tablets testify to the attribution of involuntary movements and atypical behaviors to supernatural entities.

Excavated from the Haus des Beschwörungspriesters (House of the incantation priest) in ancient Assur, one side of this tablet is inscribed with a small anthropomorphic figure possessing pointy ears, wavy horns, and a forked tongue. The other side is inscribed with prescriptions for treating several neurological and psychological pathologies, including bennu, *a term used in Mesopotamian texts to refer both to demonic influence and to conditions such as epilepsy that cause convulsions or seizures.*

* * *

Shiva Nataraja, India, 18th century
Bronze, 30 × 25.6 × 9.8 cm
Stuttgart, Linden-Museum Stuttgart, inv. no. 120782
[p. 244]

Shiva Nataraja, India, 19th century
Brass casting, 23 × 20.5 × 7 cm
Stuttgart, Linden-Museum Stuttgart, inv. no. SA 04645
[p. 246]

Originating in the 9th and 10th centuries in the Tamil region of southeastern India, the sculptural typology of Shiva Nataraja represents the continuous cycle of creation and destruction.

While performing his cosmic dance, Shiva tramples the demon of ignorance, Apasmara, who is typically depicted with physical features characteristic of dwarfism. In Sanskrit, the word apasmara *refers to the negation or loss of consciousness; in Ayurveda, it refers to epilepsy. An alternate name for the figure trampled by Shiva, Muyalagan, likewise resembles the Tamil word* muyal vali *(epilepsy), so named because the rapid breathing that may follow an epileptic seizure is thought to resemble that of a* muyal *(hare) that has caught the scent of a predator.*

* * *

Edwin Smith Papyrus, plates II, III, IV, V, Thebes, Egypt, 16th–17th Dynasty, *c.* 1600 BCE
Papyrus and ink, 468 × 33 cm
New York, Courtesy of the New York Academy of Medicine Library
Exhibition copy
[p. 256]

The Edwin Smith Papyrus is one of the oldest surviving surgical texts. Its forty-eight case histories record physical examinations of patients, detailed anatomical observations, and the diagnoses, treatments, and prognoses of injuries. Cases one through nine contain the earliest known descriptions of several types of brain injuries, and cases six and seven contain the earliest known references to the brain, the meninges, and cerebrospinal fluid.

Written in Egyptian hieratic script around the 17th century BCE (and likely based on material from a thousand years earlier), the papyrus is named after an American Egyptologist and antiquities dealer who purchased it in Luxor in 1862. Originally a continuous roll, the papyrus was cut into separate sheets in the 20th century for the sake of convenience.

* * *

Ebers Papyrus, Egypt, *c.* 1500 BCE
Scroll, papyrus and ink, 30.0 × 63.5 cm
Leipzig, Papyrus- und Ostrakasammlung der Universitätsbibliothek Leipzig
Exhibition copy
[p. 257]

Named for Georg Ebers, the German Egyptologist and novelist who acquired it in 1873, the Ebers Papyrus is a compilation of medical texts comprising seven hundred natural and supernatural remedies for afflictions ranging from crocodile bites to household pests.

Dating from the first half of the 16th century BCE, the papyrus includes an account of a migraine and a description of the circulatory system that notes the function of the heart and the existence of blood vessels throughout the body.

* * *

Heart-shaped amulet engraved with chapter 30 of the *Book of the Dead*, Egypt, 25th–31st Dynasty, *c.* 712–332 BCE
Green faïence, 5 × 2.1 × 3.5 cm
Turin, Museo Egizio, inv. no. Provv. 1218
[p. 258]

Heart amulets were placed on the chests of mummies to protect the bodies of the deceased and to facilitate passage to the afterlife.

In Egyptian culture of the Late Period, the heart was believed to be the seat of an individual's feelings, consciousness, rational soul, and memories of past deeds, and the word ib *(heart) was used to indicate the mind. The heart was accordingly retained during mummification (unlike the brain, which was usually removed from the body and discarded).*

This heart-shaped amulet featuring a human head physically associates the heart and the mind. The amulet's surface is inscribed with chapter 30 of the Book of the Dead, *in which the heart of the deceased is implored not to speak against itself during the judgment ceremony in the afterlife.*

* * *

Byzantine codex reporting the text of the Hippocratic oath, 1100–1199, Urbinate Greco 64, sheets IIv–1r, 115v–116r
Papyrus manuscript, 24.5 × 18 cm
Vatican City, Biblioteca Apostolica Vaticana
Exhibition copy
[p. 262]

Attributed to Hippocrates of Cos (c. 460–370 BCE), the Hippocratic oath delineates the obligations of a physician to patients, medical students, and the physician's own medical teachers. The pledging physician vows to prescribe only beneficial treatments, to refrain from doing anyone harm or injustice, and to respect patient privacy and confidentiality.

Adaptations of the oath, which reflect the sensibilities of other cultures and eras, guide standards of medical ethics to this day. In this Byzantine codex, the oath is written in the form of a cross, reflecting the text's adaptation and use by 12th-century Christian physicians.

* * *

Hippocrates, *Hippocratis Opera Omnia* [Hippocrates' complete works], Greek and Latin edition, vol. I, Danielem, Abrahamum & Adrianum à Gaasbeeck, Leiden, 1665 (original texts *c.* 5th–4th century BCE)
Printed book, 20 × 12.8 × 5.2 cm
Venice, Biblioteca Nazionale Marciana, 67D 144
[p. 264]

Hippocrates, *Hippocratis Opera Omnia* [Hippocrates' complete works], Greek and Latin edition, vol. II, Danielem, Abrahamum & Adrianum à Gaasbeeck, Leiden, 1665 (original texts *c.* 5th–4th century BCE)
Printed book, 20 × 12.8 × 5.2 cm
Venice, Biblioteca Nazionale Marciana, 67D 145
[not illustrated]

In his treatise De morbo sacro *(On the sacred disease), the Greek physician Hippocrates of Cos (c. 460–370 BCE) proposed that diseases were caused not by supernatural forces, as was then commonly supposed, but by natural ones subject to rational understanding.*

Among the conditions whose divine etiologies Hippocrates rejected was epilepsy, known in Classical Greece as the "sacred illness." Hippocrates instead attributed epilepsy to a disorder of phlegm, one of the four bodily humors central to his understanding of human health and emotions: "It is thus with regard to the disease called sacred: it appears to me to be in no way more divine nor more sacred than other diseases ... The brain is the cause of this affliction ... When the surplus phlegm from the brain runs down through the veins, the patient loses his speech and foams at the mouth, his hands are contracted, the eyes contorted, he becomes insensible, and in some cases the bowels are emptied."

* * *

Plato, *Platonis Timaeus a Chalcidio latino sermone translatus* [Plato's *Timaeus* translated into Latin by Calcidius], Italy, 14th–15th century (original texts 4th century BCE), Vaticano Latino 2063
Illuminated manuscript on parchment, 24.4 × 18.5 × 4.4 cm
Vatican City, Biblioteca Apostolica Vaticana
Exhibited from April to August
[p. 268]

Plato, *Timaeus-interpretatio latina Chalcidii* [The *Timaeus* translated into Latin by Calcidius], 14th–17th century (original text 4th century BCE), Reginense Latino 1572
Illuminated manuscript on parchment, 38 × 26.4 × 6 cm
Vatican City, Biblioteca Apostolica Vaticana
Exhibited from August to November
[p. 269]

The dialogue Timaeus, *written around 350 BCE, presents the most complete formulation of Plato's* (c. *428–348 BCE) theory of the soul* (psyché).

The thymos, *or irascible soul, associated with emotion, is said to be located above the diaphragm in the upper torso, while the* epithymetikon, *or concupiscible soul, associated with desire for nourishment, is said to be located below the diaphragm in the lower torso. The logos, or rational soul, is the only part of the soul held to be "immortal" and "divine." It is said to be located in the head, which is described as "the most divine part, which dominates over the rest of the body in us and to which the gods gave the whole body as servant."*

* * *

Aristotle, *Liber Aristotelis de Historiis animalium* [Aristotle's *History of Animals*], Latin translation, Italy, 13rd century (original text 4th century BCE), Vaticano Latino 2095
Illuminated manuscript on parchment, 33.5 × 32.5 × 5 cm
Vatican City, Biblioteca Apostolica Vaticana
Exhibited from April to August
[p. 270]

Aristotle, *Aristotelis De animalibus libri, Theodoro Gaza intepretе* [Aristotle's *Book of Animals* translated by Theodorus Gaza], Latin translation, Italy, 15th century (original text 4th century BCE), Vaticano Latino 2094
Illuminated manuscript on parchment, 38.45 × 28.55 × 8.5 cm
Vatican City, Biblioteca Apostolica Vaticana
Exhibited from August to November
[p. 271]

The Greek philosopher Aristotle (384–322 BCE) held that every living being is a composite of matter and form, with the soul being that which gives form to a natural, or "organic," body. The soul was the cause not only of life itself, but also of any powers of perception and thought that a particular living being might possess.

In On the Soul *(c. 350 BCE), Aristotle proposed a hierarchy of souls in which the lowest position is occupied by the vegetative (or nutritive) soul, possessed by humans, animals, and plants; the middle position is occupied by the sensitive (or motor) soul, possessed only by humans and animals; and the highest position is occupied by the intellectual soul, possessed by humans alone.*

Although he provided crucial contributions to the understanding of human brain anatomy, such as the identification of the meninges and the distinction between the cerebrum and the cerebellum, Aristotle disputed his predecessors' and contemporaries' claims that the brain controls sensation and movement.

Aristotle's cardiocentric model, which had antecedents in ancient Egypt and would later find favor throughout medieval Europe and the Middle East, instead attributed sensation and movement to the heart, which was considered the central organ of the body, the source of blood, blood vessels, and body heat, and the seat of the soul itself.

* * *

Pliny the Elder, *Naturalis historia* [Natural History], Johannes de Spira, Venice, 1469 (original text *c.* 77–78)
Illuminated incunabulum with gold leaf, 41 × 28.5 cm
Venice, Biblioteca Nazionale Marciana, Inc. V. 106
Exhibited from August to November
[p. 272]

Pliny the Elder, *Naturalis historia* [Natural History], Johannes de Spira, Venice, 1469 (original text *c.* 77–78)
Pen and ink illuminated incunabulum, 44 × 30 cm
Venice, Biblioteca Nazionale Marciana, Inc. V. 45
Exhibited from April to August
[not illustrated]

The Roman author, naturalist, and military leader Pliny the Elder (23–79) defined the scope of his Naturalis Historia *as "the natural world, or life." This thirty-seven-book magnum opus, which compiles more than two thousand observations from one hundred different authors, has informed not only modern biology and physiology, but also neuroscience, psychiatry, and neurology.*

Opposing his cardiocentric contemporaries, Pliny asserted that consciousness resided in the head rather than the heart: "The brain is the most elevated of all the viscera, and the nearest to the roof of the head ... The senses hold this organ as their citadel; it is in this that are centered all the veins which spring from the heart; it is here that they terminate; this is the very culminating point of all, the regulator of the understanding."

* * *

Galen, *Prima classis humani corporis originem, formationem, dissectionem, temperaturam, facultates, facultatumque cum actiones omnes tum instrumenta & loca singula complectitur ...* [Description of the human body comprising origin, formation, dissection, temperament, capacities, and all activities as well as specific instruments and places ...], Venice, 1541–1542 (original text *c.* 1st–2nd century)
Printed book bound in hard parchment, 34.8 × 23.5 cm
Padua, Università degli Studi di Padova – Biblioteca medica centrale "Vincenzo Pinali" – Sezione antica. DUCC.I.E.1a.(PI)
[p. 278]

After completing twelve years of medical study, the physician-philosopher Galen (129–216) returned to his birthplace of Pergamon (in modern-day Turkey), where he served as surgeon to the city's gladiators, an experience that deepened his understanding of human anatomy and physiology.

Repudiating the cardiocentric model championed by Stoic philosophers, Galen contended that thought, sensation, and movement were instead functions of the brain.

Upon moving to Rome, whose laws had long prohibited the dissection of human cadavers, Galen committed himself to the systematic dissection of animal subjects, favoring pigs and other species because they did not "differ greatly in their physical nature from man."

* * *

Set of surgical instruments from the Surgeon's Domus in Piazza Ferrari, Rimini, Italy, second half of 2nd century (Roman-imperial period)
MIC–Soprintendenza ABAP di Ravenna, Forlì-Cesena e Rimini
Rimini, Museo della Città "Luigi Tonini"

Possibly a scalpel and a pterygium knife
Bronze, 13 × 1.8 × 1.9 cm
Inv. no. 151466
[p. 280 left]

Possibly two forceps with a portion belonging to a third instrument
Metal, 17.9 × 4 × 1.5 cm
Inv. no. 184318
[p. 280 center]

Double-edged scalpel blade
Iron, 5.9 × 1.3 cm
Inv. no. 184337
[p. 280 right]

Scalpel with myrtle-leaf-shaped blade used for the opening of vertebral canal
Bronze and iron, 17.7 × 1.7 × 0.9 cm

Inv. no. 184465 b
[p. 281 left]

Bow-shaped surgical instrument used for cranial trepanation
Bronze, 20 × 0.9 × 0.8 cm
Inv. no. 184360
[p. 281 center]

Lenticular chisel for cranial trepanation
Iron, 13.3 × 1.5 cm
Inv. no. 184463
[p. 281 right]

Double-edged scalpel blade
Iron, 6 × 1.1 cm
Inv. no. 184336
[not illustrated]

In 1989, the remains of a building complex constructed between the Roman Republic and the early Middle Ages were uncovered in Rimini, Italy. The complex contained the home of a famed 2nd-century military surgeon who, like his contemporary Galen, had received his medical training in Greece.

Among the items found inside the building, known today as the Surgeon's Domus, were 150 surgical instruments, most of which had been bundled together by kind. The fusion of many of the bundled instruments in the heat of a fire preserved their original arrangements, demonstrating which ones had been stored together and thus providing clues as to their functions. Displayed here are several instruments, including bow drills, scalpels, and forceps, whose uses in the treatment of cranial fractures were described by Galen.

* * *

Huangdi Neijing [The yellow Emperor's classic of internal medicine], China, 19th century (original text *c.* 3rd century BCE–3rd century CE)
Printed book, 20 × 12 cm
Private collection
[pp. 288, 290]

The Huangdi Neijing *is the foundational text of traditional Chinese medicine. The work is composed of two texts, the* Suwen *(Basic questions) and the* Lingshu *(Numinous spindle). According to the* Huangdi Neijing, *the brain is the first organ formed and the seat of mental activities, visual and auditory processing, and some bodily movements. It is not, however, among the five* zang *organs or six* fu *organs that together regulate physiological functions.*

Further, psychological states are said to be controlled not by the brain, but rather by the zang *organs, with happiness, anger, deep thought, melancholy, and fear housed in the heart, liver, spleen, lungs, and kidneys, respectively. Among the* zang-fu *organs, the heart is "the supreme monarch of all the organs" and thought to be the seat of the mind. The brain does nevertheless regulate the* zang-fu *organs, particularly in the event of their injury or long-term imbalance.*

* * *

Abū 'Alī al-Hasan ibn al-Haytham, *Kitāb al-Manāẓir* [Book of Optics], Cairo, *c.* 1083, Ms. Fatih 3212
Manuscript, 25.5 × 20 cm
Istanbul, Republic of Turkey Ministry of Culture and Tourism – Presidency of Manuscript Institution of Turkey – Süleymaniye Manuscript Library
Exhibition copy
[p. 296]

In his Kitāb al-Manāẓir, *Arab scholar Abū 'Alī al-Hasan ibn al-Haytham (965–1040), known in the West as Alhazen, contended that the eyes receive light reflected from objects, rather than emanating light themselves, as Ptolemy, Euclid, and ibn al-Haytham's contemporaries held.*

Considered the founder of modern optics, ibn al-Haytham proposed that impressions received by the retinas are then conveyed by the optic nerves to the brain, which he regarded as the true seat of all perception.

With its systematically ordered experiments, detailed anatomical descriptions, and seminal research on the psychology of visual perception, ibn al-Haytham's seven-volume treatise simultaneously transformed the understanding of vision and light and laid the foundations for the modern scientific method, influencing scientists such as Roger Bacon and Johannes Kepler.

* * *

Opticae thesaurus. Alhazeni Arabis libri septem, nunc primùm editi. Eiusdem liber De crepusculis & nubium ascensionibus. Item Vitellonis Thuringolopoli libri 10.
[Compilation of texts about optics, including the first edition of the seventh book and *De crepusculis & nubium ascensionibus* by Arab author Alhazen as well as the tenth book by Vitello Thuringopolonis], Eusebius Episcopius & haeredes Nicolai Episcopii, Basel, 1572 (original texts *c.* 10th–13th century)
Printed book, 23 × 34 cm
Cremona, Biblioteca Statale di Cremona, FA.Ingr.E.9.1, inv. no. 20473
[p. 298]

Opticae thesaurus. Alhazeni Arabis libri septem, nunc primùm editi. Eiusdem liber De crepusculis & nubium ascensionibus. Item Vitellonis Thuringolopoli libri 10.
[Collection of texts about optics, including the first edition of the seventh book and *De crepusculis & nubium ascensionibus* by Arab author Alhazen as well as the book tenth by Vitello Thuringopolonis], Eusebius Episcopius & haeredes Nicolai Episcopii Basel, 1572 (original texts *c.* 10th–13th century)
Printed book, 23 × 34 cm
Cremona, Biblioteca Statale di Cremona, FA.16.5.28, inv. no. 5867
[not illustrated]

Abū Zayd Ahmad ibn Sahl al-Balkhī, *Maṣālih al-Abdān wa al-Anfus* [Sustenance for body and soul], Baghdad, 1479 (original text *c.* 9th–10th century), Ms. Ayasofya 3741
Manuscript, 24.5 × 17 cm
Istanbul, Republic of Turkey Ministry of Culture and Tourism – Presidency of Manuscript Institution of Turkey – Süleymaniye Manuscript Library
[p. 302]

In his treatise Maṣālih al-Abdān wa al-Anfus, *Persian polymath Abū Zayd Ahmad ibn Sahl al-Balkhī (850–934) details the somatic and psychological aspects of both physical and mental illnesses. Al-Balkhī argues that because "man's construction is from both his soul and his body ... human existence cannot be healthy without the* ishtibak *(interweaving or entangling) of soul and body."*

Citing the potential for sickness of the body to cause sickness of the psyche, and vice versa, he criticizes those physicians who focus on one category of illness without adequately addressing the other. After distinguishing between depressions with and without known causes, al-Balkhī accordingly urges physicians to treat the first type by advising patients of the physical hazards of their depressed psychological states, and to treat the second type through physical means such as purification of the blood.

* * *

Abū Bakr Muhammad ibn Zakariyyā' al-Rāzī (Rhazes), *Kitāb al-Manṣūrī* [al-Mansuri's book of medicine], 1183 (original text *c.* 9th–10th century), Ms. Or. 1512
Manuscript on paper, 16 × 41 cm
Cambridge, The Syndics of Cambridge University Library
[p. 304]

Convinced that mental disorders should be treated as medical conditions, physician, philosopher, and alchemist Abū Bakr Muhammad ibn Zakariyyā' al-Rāzī (Rhazes) (c. *854–925/935, al-Rayy, modern-day Iran) pioneered an early form of psychotherapy and introduced psychiatric wards to the hospital that he directed in Baghdad.*

Al-Rāzī rejected mind-body dualism, attributing depression to disordered blood flow to the brain and incorporating diet, medication, and occupational therapy into the treatment of psychiatric conditions. His discharge plan, which stipulated that psychiatric ward patients should be given a sum of money to aid with reintegration into general society, is among the earliest recorded references to psychiatric aftercare.

Al-Rāzī's medical treatise Kitāb al-Manṣūrī *comprises ten books, the first six of which address subjects that the author considered mainly theoretical in nature (diet, hygiene, anatomy, physiology, general pathology, and surgery), the last four of which address subjects considered more practical (diagnosis, therapy, special pathology, and practical surgery). The text was widely influential throughout the Islamic world, and a Latin translation by Gerard of Cremona became one of the most widely consulted medical manuals in medieval Europe.*

Abū Bakr Muhammad ibn Zakariyyā' al-Rāzī (Rhazes), *Liber Almansoris* [al-Mansuri's book], Latin translation, 14th century (original text *c.* 9th–10th century), Ms. Plut. 73.22
Illuminated manuscript on parchment, 39.2 × 28.5 cm
Florence, Biblioteca Medicea Laurenziana
Exhibition copy from June to November
[p. 306]

Abū Bakr Muhammad ibn Zakariyyā' al-Rāzī (Rhazes), *Contenta in hoc volumine. Liber Rasis ad almansorem ... Liber de aegritudinibus puerorum eiusdem ... Libellus zoar de cura lapidis*, compilation of medical texts, Latin translation, Bonetum Locatellum Bergomensem, Venice, 1497 (original text *c.* 9th–10th century)
Incunabulum, 33 × 22 cm
Fermo, Biblioteca Civica "Romolo Spezioli," INC 4 A 2 030 838
[p. 308]

* * *

Liber Trotula [The Book of Trotula], 15th century (original texts *c.* 11th–12th century), Ms. 593
Illuminated manuscript on paper, 23.2 x 15.5 cm
Bruges Public Library
Exhibition copy during August
[p. 316]

Trocta Trotula, *De passionibus mulierum* [On female diseases], 14th century (original text *c.* 11th–12th century), Vaticano Latino 4485
Manuscript on parchment, 20.3 × 13.8 cm
Vatican City, Biblioteca Apostolica Vaticana
Exhibited from August to November
[p. 317]

Consisting of three treatises that circulated first independently and then as a single compendium, the Liber Trotula *is considered medieval Europe's most influential text on women's medicine. Its subjects include menstruation, childbirth, and women's physical constitution, with special attention paid to pathologies of the womb. Trota, who served a physician and professor at the medical school in Salerno, Italy, specialized in female reproductive health.*

Rejecting her contemporaries' belief that women should endure maximal suffering in childbirth as punishment for Eve's sin, Trota championed the use of opiates during labor. Explicit references to the brain appear in paragraph 130 of the compendium, which circulated as part of the Liber de sinthomatibus mulierum *(Book on women's diseases).*

* * *

Pieter Jansz Quast, *The Extraction of the Stone, c.* 1630
Oil on copper, 19.5 × 28.5 cm
St. Gallen, Kunstmuseum St. Gallen, donation Annette Bühler 2005
[p. 322]

Medieval European folkloric tradition attributed madness to a stone lodged in the brain. The attempt to remove this "stone of madness" through trepanation later became a popular motif in visual art, often serving as an allegory for dishonesty (represented by a quack doctor) preying upon stupidity (represented by a guileless patient).

The Extraction of the Stone *is one of several such allegorical images produced by Pieter Quast (1606–1647), a Dutch artist known for his peasant interiors, comic theatrical scenes, and representations of the five senses.*

Hieronymus Bosch, *The Extraction of the Stone of Madness, c.* 1501–1505
Oil on oak panel, 48.5 × 34.5 cm
Madrid, Museo Nacional del Prado, inv. P002056
Exhibition copy
[pp. 323, 324]

From the 15th century onwards, a popular tradition associated madness with a stone lodged in the brain. The metaphor was made literal in iconography in which an actual stone is removed from the skull via trepanation.

In The Extraction of the Stone of Madness *by the Dutch artist Hieronymus Bosch, a white-haired man in peasant's clothing is trepanned by a surgeon whose upside-down funnel hat marks him as charlatan.*

The image is inscribed with a reference to the folk belief that madness was caused by a stone lodged in the brain: Meester snijt die key ras, Myne name is lubbert Das *(Master, rid me of this stone soon, My name is Lubbert Das). The object being extracted from the patient's brain is not, however, a stone, but rather a sprouted lily bulb.*

* * *

Trilingual compendium of texts including a diagram of the brain with five cells or "ventriculi," England, first half of the 14th century, Ms. Gg.1.1
Illuminated manuscript on parchment, 13 × 34 cm
Cambridge, The Syndics of Cambridge University Library
[p. 328]

Developed in the 4th and 5th centuries by the Early Church Fathers Nemesius (350–420) and Augustine (354–430), then reworked by Persian polymath Avicenna (980–1037), the ventricular doctrine proposed that the higher faculties of the soul reside in the ventricles, or brain "cells," which were said to be spherical in shape as a result of the uniform pressure exerted on their walls by the "animal spirits."

Assembled in the early 14th century, this illuminated codex contains literary, historical, grammatical, scientific, astronomical, theological, pastoral, and devotional texts in Anglo-Norman, Middle English, and Latin.

The diagram of the human brain illustrates the five ventricles and their corresponding faculties of common sense, imagination, estimation, cogitation, and memory. It accompanies an anonymous treatise, Qualiter caput hominis situatur *(How the human head is structured), which draws upon the works of Avicenna and the Italian philosopher Thomas Aquinas.*

Alberti Magni philosophia naturalis [Albertus Magnus' natural philosophy], Michael Furter, Basel, 1506 (original text 13th century)
Printed book, 19.4 × 14.1 cm
Berlin, Staatliche Museen zu Berlin, Kupferstichkabinett, Graphische Gesellschaft zu Berlin
[p. 330]

The most influential of the ventricular doctrines, which associated the mind's faculties with the ventricles of the brain, was developed by the German bishop and philosopher Albertus Magnus (1193–1280). Albertus described a tripartite structure in which the first region was allocated to common sense and basic imagination, the second to creative imagination, fantasy, evaluation, and rational thought, and the third to memory and reminiscence.

Published as part of a sixteenth-century edition of Albertus' Philosophia naturalis, *this woodcut illustration of the ventricles testifies to the enduring influence of his theory throughout the Middle Ages and into the Renaissance.*

* * *

Leonardo da Vinci, sheet of paper with drawing of the sectioned cranium, 1489
Pen and ink, 19 × 13.7 cm
Windsor, Windsor Castle, Royal Collection Trust, inv. RCIN 919058
[not illustrated]

Leonardo da Vinci, sheet of paper with drawings of the layers of the scalp and the cerebral ventricles, *c.* 1490–1492
Red chalk, pen and ink, 20.3 × 15.3 cm
Windsor, Windsor Castle, Royal Collection Trust, inv. RCIN 912603
[not illustrated]

Leonardo da Vinci, manuscript sheet with notes and brain drawings, 1506–1508
Ink on paper, 19 × 13.5 cm
Weimar, Klassik Stiftung Weimar,

Bestand Museen
Exhibition copies
[p. 338]

The manuscript sheet presents Leonardo's exploded and section views of the human head, which show the facial nerves' direct connection to the brain through holes in the skull, as well as his notes concerning the imprensiva, *a brain ventricle that he believed assembled raw sensory data before transmitting it to the* sensus communis *for processing.*

Questioning canonical medieval categories of brain anatomy, Leonardo instead sought to gain greater understanding of the brain through observation and dissection. By adapting a technique that he used to cast bronze sculptures, Leonardo also developed a method of obtaining wax casts of the brain ventricles.

* * *

Giulio Camillo, *L'idea del theatro* [The idea of a theater], Lorenzo Torrentino Impressor Ducale, Florence, 1550
Printed book, 21.6 × 15 cm
Venice, Biblioteca Nazionale Marciana, Misc. 2087.1
[p. 342]

Giulio Camillo (c. 1480–1544) dedicated the last years of his life to the design of his Theater of Memory, which he conceived as a room-size wooden construction in the style of the Vitruvian theater. Camillo's memory device was to contain "the Ideas of everything Celestial and Inferior," thus enabling its occupant to conceptualize and memorize the entire cosmos in symbolic form.

Standing on the stage gazing out at the audience, a visitor to Camillo's theater was to visualize seven tiers, each divided into seven sectors, creating a seven-by-seven matrix within which more than 200 images were distributed. The images in Camillo's theater were conceived as mobile, thus generating a vast range of significations through their continuous recombination.

Giordano Bruno, *De umbris idearum* [The shadows of ideas], Gilles Gourbin, Paris, 1582
Printed book, 17.5 × 11.2 cm
Padua, Biblioteca Universitaria di Padova (MiC), B.16.b.60
[p. 344]

Giordano Bruno, *De umbris idearum* [The shadows of ideas], Gilles Gourbin, Paris, 1582
Printed book, 18.5 × 12 cm
Bologna, Alma Mater Studiorum – Università di Bologna, Biblioteca Universitaria di Bologna, A.V.Tav.I.F.I.414/1
[not illustrated]

Giordano Bruno, *De umbris idearum* [The shadows of ideas], Gilles Gourbin, Paris, 1582
Printed book, 17.5 × 11.2 cm
Venice, Biblioteca Nazionale Marciana, 71 C 201.1
[not illustrated]

De umbris idearum *(1582) is the first major work on memory published by Italian philosopher Giordano Bruno (1548–1600). A treatise on the fundamental principles of Brunian mnemonics, its contents included three poems, a dialogue, and instructions on the use of an alphabet of images devised to express every word of common language.*

Unlike earlier mnemonics, which conceived of their constituent images as static, Bruno's mnemonic is predicated on his belief that the soul imparts life and movement to images, thereby animating them and rendering them active subjects in dynamic scenes.

* * *

Inca khipu, Peru, 15th–16th century
Cotton and wool, twisted and knotted, 88 × 115 cm
Milan, Museo delle Culture, Collection Federico Balzarotti, inv. no. PAM 01326
Exhibition copy from October to November
[p. 352]

The khipu is a system for recording information used by the Incas and other Andean peoples as early as 7th century, and widely employed from the establishment of the Kingdom of Cuzco (12th century) to the fall of the Inca Empire (1532).

Khipus consist of a main horizontal cord with a variable number of hanging cords. In addition to the type, number, and positions of knots, the material, length, end treatment, color, and spacing of each cord carry different meanings. Combinations of these could represent numbers, dates, and accounts as well as abstract concepts. Oral records of khipus were kept by experts called khipu kamayuq *or* quipucamayos, *who passed oral knowledge of the khipu from one generation to the next.*

* * *

CNR laboratories, model in two parts of the anatomical theater of the University of Padua, 1932–1933
Polychrome wood, 50 × 21 × 44 cm each part
Milan, Museo Nazionale della Scienza e della Tecnologia Leonardo da Vinci, inv. no. IGB 16387
[p. 360]

The anatomical theater of Padua, presented here in a scale model, was the first of its kind. It was built in 1594, from the design of anatomist Fabricius ab Aquapendente, as a wooden construction incorporated into Palazzo Bo, home to the University of Padua. Its most distinctive architectural feature is a funnel-shaped 12-meter-high auditorium of six concentric rows, with space for two hundred visitors to stand.

The dissection table lies in the center, over a trapdoor where the corpse could quickly be removed from view. The professor sat at the table, with assistants and students who held candles to provide enough illumination for the dissection. The front row was reserved for the dean of the school, professors, and nobles, while the second and third rows were reserved for students. The rest were open to the general public.

M. Caimi, model of the anatomical theater of the Archiginnasio, Bologna, 18th century
Wood, 77 × 98.5 × 76.5 cm
Rome, Museo di Storia della Medicina, Dipartimento di Medicina molecolare, Sapienza Università di Roma, inv. no. I/3743
[p. 362]

Around the 16th century, public dissections became popular in Bologna, especially during the carnival season. A permanent building for anatomical demonstrations was designed by Italian architect Antonio Paolucci in 1637. Its design, presented here in a scale model, is distinct from the practical-scientific Padua type, also popular at this time, which was funnel-shaped.

The Bologna type is characterized by a representative rectangular auditorium similar to a medieval meeting place and decorated with wood paneling, statues of twelve famous physicians and twenty famous anatomists of the Bologna studio, and two spellati *("without skin"), anatomical models displaying the muscles beneath the skin, sculpted in 1734 from a design by painter Ercole Lelli.*

* * *

Charles Bell, *Illustrations of the Great Operations of Surgery: Trepan, Hernia, Amputation, Aneurism and Lithotomy*, Longman, London, 1821
Printed book with color illustrations, 27 × 39 cm
Cambridge, The Syndics of Cambridge University Library, LA.9.4
[p. 366]

Charles Bell, *Illustrations of the Great Operations of Surgery: Trepan, Hernia, Amputation, Aneurism and Lithotomy*, Longman, London, 1821
Printed book with black and white illustrations, 27 × 39 cm
Cambridge, The Syndics of Cambridge University Library, Path b.19
[p. 368]

Sir Charles Bell (1774–1842) was a Scottish surgeon and anatomist whose original observations were used to advance understandings of the nervous system, including the distinct functions of the motor and sensory nerves, seventh cranial nerve, and fifth cranial nerve. Bell also made illustrations of his patients undergoing surgeries.

Illustrations of the Great Operations of Surgery *contains detailed anatomical drawings meant to assist surgeons in mastering their technique. Of great importance is a drawing of the consequences from the rupture of the dura mater by a tumor and an illustration of a trepanation performed on a depressed skull fracture, with the removed bone fragments carefully detailed in the lower left corner of the image. The book did not identify the patients.*

* * *

Thomas Willis, *Cerebri Anatome: cui accessit nervorum descriptio et usus* [Anatomy of the brain, including description and function of nerves], Jo. Martyn & Ja. Allestry, London, 1664
Printed book with folding plates, 20.5 × 16 cm
Ministero della Cultura – Milan, Pinacoteca di Brera – Biblioteca Nazionale Braidense, BNB A.03.00433
[p. 372]

Thomas Willis, *Cerebri Anatome: cui accessit nervorum descriptio et usus* [Anatomy of the brain, including description and function of nerves], 1664
Printed book with folding plates, 17.7 × 12.1 cm
Venice, Biblioteca Nazionale Marciana, 182D 165
[p. 374]

Thomas Willis, *Cerebri Anatome: cui accessit nervorum descriptio et usus* [Anatomy of the brain, including description and function of nerves], 1664
Printed book with folding plates, 21 × 15.6 cm
Padua, Biblioteca Universitaria di Padova (MiC), A.93.a.85
[not illustrated]

Thomas Willis (1621–1675) wrote Cerebri Anatome *(1664) while a professor of natural history at Oxford. Willis believed that the rational soul, which separated humans from animals and connected them to God, could be understood through the study of anatomy. His method was devised in collaboration with physicians Richard Lower and Thomas Millington, as well as architect Christopher Wren, best known for St. Paul's Cathedral in London, who also provided the book's illustrations. The research carried out by Willis aimed to solve the "artifactual results" of the previous anatomists.*

Willis, rejecting the belief at the time that mental functions resided in the cerebral ventricles, proposed the brain as the primary seat of the rational soul and the origin of movement, memory, thought, and appetite. His observations relied on experiments carried out on slices of brain extracted from the skull as opposed to traditional methods of in situ dissection. Willis focused on the arterial circle at the base of the brain, attempting to assign a function to the anatomy he described.

In 1774–1777, the area appeared in the Bibliotheca anatomica *for the first time as the "circle of Willis."*

* * *

Francesco Calenzuoli, section of head and neck with the brain exposed, copied from Meckel, Italy, 19th century
Wax anatomical model, silk drape, wood, glass display case, 44.5 × 54.5 × 50 cm
Florence, Sistema Museale di Ateneo – Museo di Storia Naturale, Sede "La Specola,"
Inv. no vitrine. 763
[p. 380]

Wax anatomical models were used in Italy starting in the late 17th century to exhibit the recent discoveries made in anatomy.

Inspired by the models he saw as a student in Bologna, the abbot Felice Fontana worked in Florence and Pisa to produce a large number of wax models to fill the need for instructing trainee surgeons in anatomy. While the Bolognese models usually contained pieces of the skeleton, the Florentine models were made entirely of wax. Plaster casts were made from dissected cadavers, which were refined by highly specialized modelers who added blood vessels and nerves.

Fontana died in 1805, but the Florentine workshop continued its activity under the direction of modeler Clemente Susini, and of other modelers after Susini's death, in 1814. Among them were Francesco and Carlo Calenzuoli and Luigi Calamai.

* * *

Shinnin Kawaguchi, *Kaishihen* [Complete notes on the dissection of a cadaver], with illustrations by Shukuya Aoki, Japan, 1802
Manuscript, 16.5 × 22 cm
Princeton (NJ), East Asian Library and the Gest Collection, Princeton University Library, no. QM21.K38
[pp. 384, 386]

Influenced by Toyo Yamakawi's medical text Zōshi *(1759), as well as by European anatomical texts that came to Japan mainly through trade with the Dutch, the Japanese physician and surgeon Shinnin Kawaguchi (1736–1811) performed the first recorded dissection of a brain in Japan. In 1770, he was given two executed cadavers and a head to dissect. Kawaguchi performed the operations himself, on the site of the execution of the bodies.*

The Kaishihen, *complete with a large number of illustrations by painter and draughtsman Shunmei Aoki (1737–1802), was published by Kawaguchi and his colleague Gengai Ogino (1737–1806) in 1772.*

Seii Sugawara, *Fujin Zōzu* [Pictures of female internal organs], scroll with Nihonga illustrations, Japan, 1774
Manuscript, 29.5 × 790 cm
Chiba, Library of Health Sciences (Inohana Library), Chiba University
Exhibition copy
[p. 388]

Toyo Yamawaki (1705–1762) was among the first recorded physicians in Japan to perform and comment on the dissection of a human body. His dissection of an executed criminal and his observations of the body's internal anatomy, recorded in the text Zōshi *(1759), led him to question traditional Chinese medicine's descriptions of the body's functions. The work was also influenced by the anatomical atlas* Syntagma Anatomicum *(1641) by the German anatomist and botanist Johann Vesling (1598–1649).*

In 1772, Yamawaki's son Tomon Yamawaki (1736–1782) performed a full dissection of a body, whose records were illustrated by Seii Sugawara and published in 1774.

Kanzen Mikumo, Genshun Koish, Ranshū Yoshimura, *Seyakuin Kainantaizōzu* [Seyakuin's male anatomical chart], Japan, 1798
Manuscript with illustrations, ink and color on paper, 27.5 × 19 cm
Kyoto, Kyoto University, Main Library, RareBooks, no. 7-05
Exhibition copy
[pp. 390, 392, 394]

Published in the form of a kaiboemaki, *or scrolls of anatomical pictures, the* Seyakuin Kainantaizōzu *is the drawn report of an autopsy performed on a thirty-four-year-old male executed as a criminal in Japan in 1798. The dissection team included the physicians Kanzen Mikumo, Genshun Koishi, and Ranshu Yoshimura. The dissection by the physician Seyakuin is illustrated by the painter Ranshū Yoshimura (1739–1816) in thirty-six illustrated tables, with captions and explanations reporting and expanding on Dutch anatomical sources, which had come to Japan through commercial exchange.*

In 1783, Yoshimura had produced illustrations for the Heijiro Zozu, *which described an autopsy conducted by Nankei Tachibana. As a comparison of the Seyakuin with the* Heijiro Zozu *shows, the last ten years of the 18th century saw an increase in the precision of the technical, graphical, and medical aspects of anatomical practice.*

* * *

Louis Victor Leborgne's brain.
Black and white photographs, 15 × 10.5 cm
Paris, Bibliothèque de Sorbonne Université, Collections d'Anatomie Pathologique Dupuytren
[pp. 400, 402]

In 1861 the French physician Paul Broca (1824–1880) cared for the patient Louis Victor Leborgne (1809–1861), who was suffering from aphasia, an extreme difficulty in speaking. Analyzing Leborgne's brain after his death, Broca noted a lesion on the left frontal lobe of the brain that he interpreted as the main cause of the language deficit, a condition that became known as Broca aphasia.

After studying the brains of twenty-five additional patients with aphasia, Broca concluded that speech articulation was localized in the left frontal lobe. He also noted that most patients with aphasia recovered some of their abilities, suggesting that the right hemisphere was able to take over some of the functions of the left hemisphere.

* * *

Series of drawings of the nervous system by Camillo Golgi
Pavia, Università di Pavia
– Sistema Museale di Ateneo
– Museo per la Storia dell'Università di Pavia

The spinal cord, *c.* 1883–1884
India ink drawing on paper,
39.4 × 27.4 cm
[p. 406]

Purkinje neuron, *c.* 1883–1884
Ink drawing on paper, two drawings stapled together,
18.5 × 23.5 cm, 9.4 × 8.8 cm
[p. 407]

The diffuse nerve network,
c. 1883–1884
India ink drawing on light cardboard, 15.2 × 18.6 cm
[p. 408]

Human cerebral cortex section, anterior central circumvolution,
c. 1883–1884
Sketch on light cardboard,
65.5 × 25.5 cm
[not illustrated]

Neuron, *c.* 1883–1884
India ink drawing on paper,
22.5 × 30.3 cm
[not illustrated]

Neuron, *c.* 1883–1884
India ink drawing on paper,
20.2 × 27.7 cm
[not illustrated]

Neuron, *c.* 1883–1884
India ink drawing on paper,
12.3 × 7.8 cm
[not illustrated]

Neurons, *c.* 1883–1884
Pencil drawing on paper,
20.8 × 16.6 cm
[not illustrated]

Neurons of the hippocampus,
c. 1883–1884
Sketch on light cardboard,
24.2 × 21 cm
[not illustrated]

Neuron, *c.* 1883–1884
India ink drawing on paper,
22.9 × 28.5 cm
[not illustrated]

Neuron with C. Golgi's original signature, *c.* 1883–1884
Pencil drawing on paper,
21 × 21 cm
[not illustrated]

Neurons, *c.* 1883–1884
Pen drawing on paper,
29.9 × 20 cm
[not illustrated]

Neurons, *c.* 1883–1884
Pencil drawing on cardboard,
21 × 10.3 cm
[not illustrated]

Neurons, *c.* 1883–1884
Pen drawing on paper,
20.9 × 15.5 cm
[not illustrated]

Neuron, *c.* 1883–1884
India ink drawing on paper,
26.8 × 21.5 cm
[not illustrated]

The hippocampal circumvolution,
c. 1883–1884
Pen and pencil drawing on paper,
20 × 20.2 cm
[not illustrated]

The hippocampal circumvolution,
c. 1883–1884
Sketch on light cardboard,
15.2 × 18.6 cm
[not illustrated]

Vertical section fragment of a human cerebellar circumvolution,
c. 1883–1884
Sketch on light cardboard,
31.5 × 37.5 cm
[not illustrated]

Vertical section of human cerebellar circumvolution,
c. 1883–1884
Sketch on light cardboard,
46 × 32.5 cm
[not illustrated]

Series of drawings of the brain by Santiago Ramón y Cajal
Madrid, Instituto Cajal (CSIC), Legado Cajal

Section of the olfactory bulb of a cat, 1901
Pencil (graphite) and India ink on paper, 19.1 × 14.5 cm
Inv. no. CGC 3706
[p. 409]

Frontal section of the brain, 1904
Pencil (graphite) and India ink on paper,
10 × 14.5 cm
Inv. no. CGC 3116
[p. 410]

Cross section of the superior sphenoid area of the rabbit, 1901
Pencil (graphite) and India ink on paper,
23.6 × 7.8 cm
Inv. no. CGC 3857
[p. 411]

Front section of the inner ear (labyrinth) in a few-days-old bird, 1908
Pencil (graphite), India ink, gray and white watercolor on paper,
15.4 × 14 cm
Inv. no. CGC 3699
[p. 412]

Axons of the Purkinje neurons of the cerebellum, 1904
Pencil (graphite) and India ink on paper, 16.2 × 14.4 cm
Inv. no. CGC 2966
[not illustrated]

Neurons from the superficial layers of the developing cerebral cortex, 1904

Pencil (graphite)
and India ink on paper,
18.7 × 9.7 cm
Inv. no. CGC 3379
[not illustrated]

Connections between Purkinje neurons of the cerebellum, 1895
Pencil (graphite)
and India ink on paper,
11.7 × 13.2 cm
Inv. no. CGC 2994
[not illustrated]

Degeneration of Purkinje neuron axons, 1911
Pencil (graphite)
and India ink on paper,
15.2 × 12.2 cm
Inv. no. CGC 3052
[not illustrated]

Degeneration of cerebellar Purkinje neuron axons, found in a drowned man, 1911
Pencil (graphite)
and India ink on paper,
13 × 21 cm
Inv. no. CGC 3063
[not illustrated]

Front section of the mouse thalamus, 1904
Pencil (graphite)
and India ink on paper,
9.5 × 12.9 cm
Inv. no. CGC 3301
[not illustrated]

Giant pyramidal cell of the human motor region (30-year-old male), 1899
Pencil (graphite) and India ink on paper, 22.2 × 17 cm
Inv. no. CGC 3303
[not illustrated]

Granular neurons of the olfactory bulb of a 20-days-old cat, 1898
Pencil (graphite)
and India ink on paper,
10.3 × 19.2 cm
Inv. no. CGC 4535
[not illustrated]

Large neuron with short axon, from the cortical layer of medium pyramidal neurons, 1904
Pencil (graphite) and India ink on paper, 21.1 × 10.1 cm
Inv. no. CGC 3293
[not illustrated]

Various neuronal types from layers 4, 5, and 6 of the cat's visual cortex, 1921
Pencil (graphite) and India ink on paper, 20.5 × 12.5 cm
Inv. no. CGC 3749
[not illustrated]

Vacuolated Purkinje neurons taken from the cerebellum of a 25-day-old cat, 10 days after surgery, 1911
Pencil (graphite) and India ink on paper, 12.9 × 20.6 cm
Inv. no. CGC 3042
[not illustrated]

In 1873, Italian histologist Camillo Golgi (1843–1926) developed the reazione nera *or black reaction, a new method for using chemicals to stain nerve cells to make them visible in brain tissue under a microscope. With this procedure, axons (the parts of the neuron that transmit electrical pulses), dendrites (which receive communication from other cells) and glia (nonneuronal cells) turn black and show their shape.*

Golgi observed a dense network of axons and dendrites connecting brain areas through the transmission of an electrical nervous impulse. Because his black reaction stained only 3 percent of neurons, Golgi's view was limited, and he supported the popular "reticular theory" that said neuronal processes physically fuse with each other.

Fourteen years after its discovery by Golgi, Santiago Ramón y Cajal (1852–1934) used a modified version of the black reaction to challenge Golgi's interpretation of the architecture of the neural system. Cajal inferred that axons and dendrites ended freely and supported the "neuron theory" that the nervous system is composed of a large number of independent cells. Both scientists published their findings accompanied by hand-drawn illustrations of cell structures. In 1906, they shared the Nobel Prize despite their opposing theories.

* * *

Rita Levi-Montalcini, Viktor Hamburger, "Selective Growth Stimulating Effects of Mouse Sarcoma on the Sensory and Sympathetic Nervous System of the Chick Embryos," *Journal of Experimental Zoology*, vol. 116, no. 2 , The Wistar Institute of Anatomy and Biology, Philadelphia, March 1951, pp. 321–361
Printed book, 25 × 18 cm
Milan, Biblioteca di Biologia, Informatica, Chimica e Fisica dell'Università degli Studi di Milano
[p. 418]

Rita Levi-Montalcini, "Nerve Growth Factor," *Annals of the New York Academy of Sciences*, vol. 118, no. 3, New York Academy of Sciences, New York, October 1964, pp. 149–170
Printed book, 23 × 15 cm
Private collection
[p. 420]

Rectoral decree signed by dean Azzo Azzi suspending from service Rita Levi-Montalcini, volunteer assistant at the Clinic of Nervous and Mental Illnesses of the Turin University, October 18, 1938
Typescript and signed document on institutional letterhead, 32.8 × 22 cm
Turin, Archivio Storico dell'Università di Torino
Exhibition copy
[p. 422]

In 1939, after the enactment of anti-Semitic racial laws, Rita Levi-Montalcini (1909–2012) left her position at the University of Turin. By 1940, Levi-Montalcini had set up a home laboratory where, inspired by findings of the embryologist Viktor Hamburger, she studied neurogenesis in the chick embryo. In 1947, Hamburger invited her to his laboratory at Washington University in St. Louis, where she stayed for thirty years.

Levi-Montalcini's development of an in-vitro culture technique that allowed neurons to be grown outside of an embryo led to her discovery of the nerve growth factor (NGF). This discovery was driven by Levi-Montalcini's 1952 visit to Hertha Meyer's laboratories in Rio, where she observed nerve cell growth in response to the proximity of cancerous cells. The discovery was also supported by biochemist Stanley Cohen.

NGF, the first neurotrophin, promotes the growth and survival of certain categories of neurons, as well as synaptic pruning to remove nerve cells with poor connectivity. NGF has been shown to play a role in learning and memory in the adult brain, and its biological functions are still being studied. In 1986, Levi-Montalcini and Cohen shared the Nobel Prize.

* * *

Hans Berger, "Über das Elektrenkephalogramm des Menschen" [On human electroencephalography], *Archiv für Psychiatrie und Nervenkrankheiten*, vol. 87, no. 1, Springer Verlag, Vienna, 1929, pp. 527–570
Printed book, 23 × 18 cm
Pergine Valsugana (Trento), Azienda Provinciale per i Servizi Sanitari, Biblioteca dell'ex ospedale psichiatrico
[p. 428]

Hans Berger, "Über die Entstehung der Erscheinungen des großen epileptischen Anfalls" [On the origin of the phenomena of the grand epileptic seizure], *Klinische Wochenschrift*, no. 14, Springer Verlag, Vienna, February 1935, pp. 217–219
Printed book, 28.5 × 21.5 cm
Kork, Deutsches Epilepsiemuseum
[p. 430]

Hans Berger, "Das Elektrenkephalogramm des Menschen" [Human electroencephalography], *Nova Acta Leopoldina*, series 6, no. 38, Leopoldina Nationale Akademie der Wisseschaften, Halle (Saale), 1938, pp. 173–309
Printed book, 26.5 × 19.5 cm
Kork, Deutsches Epilepsiemuseum
[p. 431]

German psychiatrist Hans Berger (1873–1941) first measured electrical activity in the brain by injecting novocaine on the scalp and recording from an electrode placed in the periosteum and a galvanometer.

He recorded his first electroencephalogram, or EEG, on July 6, 1924, from a seventeen-year-old boy during a neurosurgery. Later, he was the first person to noninvasively record the electrical activity of the brain.

In 1929 he published the article Über das Elektrenkephalogramm des Menschen *describing his findings, including the results of experiments performed on his son, Klaus, and his daughter, Ilse. In this paper, Berger demonstrated the appearance and disappearance of alpha and beta waves following an epileptic seizure.*

* * *

Max Kohl, loop galvanometer for measuring brain waves, Germany, 20th century
Metal, 41.8 × 8.7 × 4 cm
Jena, Friedrich-Schiller-Universität Jena, Collection of scientific and technical devices for physics, object ID 011
[p. 432]

Luigi Galvani, *De viribus electricitatis in motu musculari commentarius* [Commentary on the effect of electricity on muscular motion], Societas Typographica, Modena, 1792
Printed book, etched plates, 29.2 × 21.4 cm
London, The Royal Society
[p. 436]

During a course of experiments on the effect of atmospheric electricity on dissected frogs, Italian physician Luigi Galvani (1737–1798) observed muscle retraction in a frog leg that was attached to a metal mount and touched with a scalpel. From this reaction and further experimentation, Galvani theorized the existence of "animal electricity" flowing from the brain to the muscles.

In 1791, he published his findings in De viribus electricitatis in motu musculari commentarius. *Galvani's results were contradicted by the Italian physicist Alessandro Volta, who did not believe in animal electricity and argued that the electrical charge was produced from the contact of the two metals conducted by a fluid. Giovanni Aldini, a nephew of and assistant to Galvani, used both theories in his work, which led to the development of various forms of electrotherapy and noninvasive brain stimulation.*

Giovanni Aldini, *De animali electricitate dissertationes duae* [Two dissertations on animal electricity], Typographia Instituti Scientiarum, Bologna, 1794
Printed book, etched plates, 31.7 × 22.5 cm
Padua, Università degli Studi di Padova, Biblioteca medica centrale "Vincenzo Pinali," Sezione antica, STM.DUCC. XVIII.G.23 (DU)
[p. 438]

Giovanni Aldini, *De animali electricitate dissertationes duae* [Two dissertations on animal electricity], Typographia Instituti Scientiarum, Bologna, 1794
Printed book, etched plates, 27 × 42 cm

Turin, Accademia delle Scienze di Torino, Misc. 363 (no.4), inv. no. PNI 12055
[not illustrated]

Giovanni Aldini, *Essai théorique et expérimental sur le galvanisme* [Theoretical and experimental essay on galvanism], volume 1, Fournier Fils, Paris, 1804
Printed book, etched plates, 21 × 14 cm
Cremona, Biblioteca Statale di Cremona, FA.20.3.80, inv. no. S.A 9724
[p. 440]

Giovanni Aldini, *Essai théorique et expérimental sur le galvanisme* [Theoretical and experimental essay on galvanism], volume 2, Fournier Fils, Paris, 1804
Printed book, etched plates, 21 × 14 cm
Cremona, Biblioteca Statale di Cremona, FA.20.3.81, inv. no. S.A 9725
[not illustrated]

Giovanni Aldini, *Essai théorique et expérimental sur le galvanisme* [Theoretical and experimental essay on galvanism], volume 1, Fournier Fils, Paris, 1804
Printed book, etched plates, 20 × 26 cm
Turin, Accademia delle Scienze di Torino, O.IX.43, inv. no. PNI 12056
[not illustrated]

Giovanni Aldini, *Essai théorique et expérimental sur le galvanisme* [Theoretical and experimental essay on galvanism], volume 1, Fournier Fils, Paris, 1804
Printed book, etched plates, 20 × 26 cm
Turin, Accademia delle Scienze di Torino, O.IX.44, inv. no. PNI-12056
[not illustrated]

Luigi Galvani's position in the dispute with Alessandro Volta on the precise nature of galvanism was defended by his nephew, Giovanni Aldini (1762–1834). Aldini conducted public experiments in which he used voltaic piles to stimulate the corpses of animals and humans, in order to demonstrate the existence of animal electricity by making body muscles spasm and facial muscles contract.

In 1804, Aldini published the accounts of his experiments in a two-volume, 680-page-long treatise called Essai théorique et expérimental sur le galvanisme. *It also contained the results of a series of studies on the use of galvanism to treat patients with mental conditions, with apparent success. This treatment can be considered an early form of electroconvulsive therapy (ECT).*

* * *

Lucio Bini – Ugo Cerletti, electroconvulsive therapy machine, Italy, 1938–1940
Metal, 96 × 47 × 38 cm
Reggio Emilia, Musei Civici di Reggio Emilia – Museo di Storia della Psichiatria, inv. no. 38
[p. 446]

Italian neuropsychiatrists Ugo Cerletti (1877–1963) and Lucio Bini (1908–1964) were the first to use an electroshock machine on a human subject to induce seizures with electricity for therapeutic effect. The use of electroshock would replace metrazol and insulin therapy and the chemical induction of seizures.

Cerletti and Bini first carried out their experiments on animals to demonstrate the harmlessness of electroconvulsive therapy (ECT). In 1938, they applied it for the first time on a patient with schizophrenia.

In the 1940s and 1950s, ECT was used widely in the treatment of drug-resistant depression as well as mania, catatonia, and schizophrenia. The application of ECT can be controlled through electrode placement, length, and intensity of the stimulus. Currently, ECT is typically delivered to anesthetized patients under scrutinized conditions.

* * *

René Descartes, *Opera Philosophica* [Philosophical works], Daniel Elzevir, Amsterdam, 1672
Printed book, 20 × 22 cm
London, Wellcome Collection, inv. no. EPB/B/20195
[p. 450]

René Descartes, *L'Homme de René Descartes et un Traité de la formation du fœtus* [René Descartes' treatise on man and the formation of the fetus], Daniel Elzevir, Amsterdam, 1677 (first edition 1664)
Printed book, 20 × 21.5 cm
London, Wellcome Collection, inv. no. EPB/B/20203/1
[p. 451]

In the first half of the 17th century, French philosopher René Descartes (1596–1650) entered the centuries-old debate on the identification of the seat of thought. Considering the soul and the body to be ontologically separate but interacting entities, he sought to specify their mode and site of interaction, identifying the latter with the pineal gland. As the only unpaired element of the brain, the pineal gland was held responsible for unifying perceptions received from the paired organs of sense. In Descartes' writings, the "little gland H"—as he termed it—is usually illustrated as a pear shape suspended in the middle of the ventricles, which he still believed to be inflated by "animal spirits" communicating the soul's spiritual autonomy to the body mechanism.

Rembrandt van Rijn, *The Anatomy Lesson of Dr. Jan Deijman* (fragment), 1656
Oil on canvas, 100 × 134 cm
Amsterdam, Amsterdam Museum, inv. no. SA 7394
Exhibition copy
[p. 452]

The only surviving fragment of a much larger group portrait, which was severely damaged by a fire in 1723, Rembrandt's painting provides unique evidence of the dissection of a human brain in the context of public anatomical demonstrations in 17th century Flanders. The scene refers to the actual lecture delivered by the praelector *and* doctor medicinae *Jan Deijman, who performed his autopsy on the corpse of Joris "Black Jack" Fonteijn, a tailor executed for robbing a textile store and threatening those present with a knife. Notably, Dr. Deijman is portrayed as he lifts the falx cerebri, showing the pineal gland. According to Descartes' authority, the latter was considered the seat of the soul, and therefore the event marked a crucial moment in the public spectacle of the dissection.*

* * *

Eugène Louis Pirodon (after André Brouillet), *A Clinical Lesson at the Salpêtrière*, 1888
Lithograph on paper, 45.5 × 59 cm
London, Freud Museum London, inv. no. LDFRD 4378
[p. 458]

In 1885, the Austrian neurologist Sigmund Freud (1856–1939) went to Paris to study under the supervision of French neurologist Jean-Martin Charcot at the Salpêtrière hospital. There, Freud took up his work on hysteria, from which his theories of psychoanalysis ultimately emerged.

The lithograph portrays Charcot as he delivers a clinical lecture on hypnosis involving a demonstration on his patient Marie "Blanche" Wittmann to a group of physicians.

In 1889, Freud returned from Paris with the print, which he hung over the couch in his consulting room, first in Vienna and then in England, where it is still preserved.

Sigmund Freud,
Die Traumdeutung, Franz Deuticke, Leipzig-Vienna, 1900
Printed book, 23.2 × 15.5 cm
London, Freud Museum London, inv. no. LDFRD 2865
[p. 459]

Sigmund Freud,
The Interpretation of Dreams, G. Allen, London, 1913
Printed book, 22.1 × 15 cm
London, Freud Museum London, inv. no. LDFRD 2887
[p. 460]

Sigmund Freud,
La interpretación de los sueños, Biblioteca Nueva, Madrid, 1931
Printed book, 23 × 15 cm
Private collection
[p. 461]

Sigmund Freud, *Yume Handan*, Shunyodo, Tokyo, 1930–1933
Printed book, 19 × 13.5 cm
London, Freud Museum London, ref. no.: LDFRD 5407
[p. 462]

Sigmund Freud,
La interpretación de los sueños, Editorial Ercilla, Santiago de Chile, 1936, volumes nos. 1 and 3
Printed books, 18 × 13 cm
Private collection
[p. 463]

Sigmund Freud, *Výklad Snů*, Julius Albert, Prague, 1937
Printed book, 23.6 × 15 cm
London, Freud Museum London, ref. no. LDFRD 5406
[p. 464]

Sigmund Freud, *L'interpretazione dei sogni*, Astrolabio, Rome, 1952
Printed book, 21 × 15.3 cm
Private collection
[p. 465]

Sigmund Freud, *Meng de jie xi*, Chih Wen Publishing Co., Taipei, 1972
Printed book, 19 × 13.5 cm
Taiwan, National Central Library, system no. 000079558, shelf mark: 175.7 8738-3
[p. 466]

In 1900, Sigmund Freud published The Interpretation of Dreams. *The book introduces the theory of the dynamic unconscious, here discussed in relation to dream interpretation. According to Freud, dreams could provide access to the unconscious processes and desires that affect a person's behavior without their conscious knowledge. These desires affect the psyche and cause neurotic symptoms with no underlying physical condition (hence the need for a means to access and interpret them).*

The early editions of The Interpretation of Dreams *from across the globe testify to the swift diffusion of Freudian theories, as well as to their variable reception in different cultural contexts.*

* * *

Korbinian Brodmann,
Vergleichende der Grosshirnrinde: in ihren Prinzipien dargestellt auf Grund des Zellenbaues
[Comparative localization studies in the brain cortex: its fundamentals represented on the basis of its cellular architecture], Johann Ambrosius Barth, Leipzig, 1909
Printed book, 24 × 16.5 cm
Turin, Biblioteca Federata di Medicina "Ferdinando Rossi" – Università degli Studi di Torino, inv. no. 2225
[p. 470]

Korbinian Brodmann,
Vergleichende der Grosshirnrinde: in ihren Prinzipien dargestellt auf Grund des Zellenbaues
[Comparative localization studies in the brain cortex: its fundamentals represented on the basis of its cellular architecture], Johann Ambrosius Barth, Leipzig, 1909
Printed book, 24 × 17 cm
Reggio Emilia, Biblioteca scientifica Carlo Livi – AUSL di Reggio Emilia
[p. 472]

The relevance of the research carried out in early 1900 by physician Korbinian Brodmann (1868–1918) is demonstrated by the reliability of his maps of the brain, still studied today.

Brodmann's name is associated with his 1909 text, Vergleichende Lokalisationslehre der Grosshirnrinde in ihren Prinzipien dargestellt auf Grund des Zellenbaues, *in which the German physician published his graphical organization of the human cortex. Brodmann created a system of cytoarchitectonics in which the cortex was subdivided into fifty-two areas, each signified by a number on the basis of its cell type and laminar structure.*

Though Brodmann's approach focused on anatomical evidence rather than on physiology or behavior, many of his brain areas have gone on to be associated with different nervous functions.

Phrenology cabinet, 19th century
Wood, felt, metal (cabinet), plaster, and glue (heads), 120 × 77.5 × 19 cm
Edinburgh, The Anatomical Museum, The University of Edinburgh,
inv. no. AM5069
[p. 474]

At the turn of the 18th century, the Viennese physicians Franz Joseph Gall (1758–1828) and Johan Gaspar Spurzheim (1776–1832) were pioneers in the field, today wholly discredited, of phrenology. Gall and Spurzheim believed that mental faculties such as instincts, sentiments, reason, and understanding were located in specific areas of the brain and that the strength or prominence of these faculties could be judged based on the size and shape of the brain areas as they pressed against the skull.

Gall thought that the more relevant an aspect was, the more prominent the volume of the corresponding area would grow, molding the skull around it and making it possible to investigate personalities and mental conditions through the palpation of the head. In 1813, Spurzheim began lecturing in England with sculptor William Bally (1796–1858).

In 1832, Bally produced a set of sixty miniature phrenological specimens to be sold as teaching tools for aspiring phrenologists, accompanied by printed descriptions written by Spurzheim. The busts were scaled down from plaster casts of actual heads that best exhibited phrenological types.

Carl D. Anderson, "The Positive Electron," *The Physical Review*, vol. 43, no. 6, The American Institute of Physics, Lancaster–New York, 1933
Printed book, 26 × 20 cm
Private collection
[p. 478]

Positrons, the first antiparticles ever to be detected, were accidentally discovered by the American physicist Carl David Anderson (1905–1991) while studying cloud-chamber photographs of cosmic rays.

In 1932, he noticed that these images contained a number of tracks apparently caused by positively charged particles too small to be protons. This fact led Anderson to the identification of positively charged particles with the same mass as electrons, which he called "positrons" or "positive electrons."

This revolutionary discovery earned Anderson the Nobel Prize in physics in 1936. It also influenced medical diagnosis, leading to the development of Positron Emission Tomography (PET), an imaging technique crucial for the study of brain molecular functions.

This is the first publication announcing the discovery of positrons, illustrated with images of Anderson's cloud chambers.

"Britain's Brains Produce First NMR Scans," *New Scientist*, vol. 80, no. 1130, London, 1978, p. 588
Printed book, 28 × 21.5 cm
Private collection
[p. 479]

J.W. Belliveau *et al.*, "Functional Mapping of the Human Visual Cortex by Magnetic Resonance Imaging," *Science*, vol. 254, no. 5032, American Association for the Advancement of Science, Washington, November 1991, pp. 716–719
Printed book, 27.5 × 21 cm
Private collection
[p. 480]

Magnetic resonance imaging (MRI) is a noninvasive imaging technique that enables the observation of anatomic structures, physiological functions, and the biochemical composition of tissues.

In the early 1970s, the methods of the chemist Paul Lauterbur and the physicist Peter Mansfield for using a gradient magnetic field to stack two-dimensional images of an object to create a three-dimensional view and for "line scan imaging" were adopted into the first MRI technology for use on human subjects. Lauterbur and Mansfield's advancements led to the first pictures of a section through a live human head. The nuclear magnetic resonance (NMR) images were recorded in 1978, by Hugh Clow and Ian Young at the EMI Central Research Laboratory (CRL), in Hayes, Middlesex.

In the early 1990s, high-resolution MRI was first applied to the imaging of human brain function by John Belliveau, leading to the invention of functional magnetic resonance imaging (fMRI).

Kenneth Kwong, J.W. Belliveau, *et al.* "Dynamic Magnetic Resonance Imaging of Human Brain Activity During Primary Sensory Stimulation," *Proceedings of the National Academy of Sciences of the United States of America*, vol. 89, no. 12, United States National Academy of Sciences, Washington, June 1992,

pp. 5675–5679
Printed book, 28 × 21 cm
Milan, Biblioteca di Biologia, Informatica, Chimica e Fisica dell'Università degli Studi di Milano
[p. 481]

Seiji Ogawa *et al.*, "Intrinsic Signal Changes Accompanying Sensory Stimulation: Functional Brain Mapping with Magnetic Resonance Imaging," *Proceedings of the National Academy of Sciences of the United States of America*, vol. 89, no. 13, United States National Academy of Sciences, Washington, 1992
Printed book, 28 × 21 cm
Milan, Biblioteca di Biologia, Informatica, Chimica e Fisica dell'Università degli Studi di Milano
[p. 482]

In the 1990s, functional magnetic resonance imaging (fMRI), which measures stimulus response and regional brain activity, was devised by John Belliveau. The technology was perfected by two independent research teams, Kenneth Kwong's at MGH NMR Center (now Martinos Center) and Seiji Ogawa's at AT&T Bell Laboratories.

Kwong and Ogawa demonstrated that cerebral activation could be imaged by exploiting endogenous contrast, a technique currently known as BOLD (Blood Oxygen Level Dependent), thereby eliminating the need for injection of an external agent.

Today, fMRI plays a crucial role in the study of brain functions ranging from primary sensory responses to cognitive and motor activities.

THE CONVERSATION MACHINE OBJECTS

Yasmin Abofoul

Image cards used in studies of perception of emotional expression.
[p. 632]

Huda Akil

Opium pipe, ivory.
[p. 691]

Opium pipe, silver.
[p. 693]

Polina Anikeeva

Macroscale models (preforms) that are drawn into microscale fibers used to modulate and record brain signals.
[p. 542]

Electromagnet used in conjunction with magnetic nanoparticles to remotely activate neural activity.
[p. 571]

A solution of magnetic nanoparticles (ferrofluid) that can be injected into the brain to mediate magnetic neural stimulation.
[p. 573]

Mouse brain in saline solution.
[not illustrated]

Anirban Bandyopadhyay

Assembly of clocks; each clock corresponds to the anatomical kinesiology—the bodily movement—of a running cheetah.
[p. 510]

Organic gel model of a human brain, containing 1,050 distinct neural networks.
[p. 565]

Graphic model of the world's largest brain rhythm database, brainrhythm.org.
[p. 622]

György Buzsáki

Macintosh SE computer used to analyze neuroscience data (1987).
[p. 530]

Sculpture featuring glucose monitor used in rats by György Buzsáki and his research team to measure the effects of hippocampal sharp wave ripples (SPW-Rs) on glucose levels during sleep.
[p. 698]

Amplifier used by György Buzsáki in studying hippocampal sharp wave ripples (SPW-R), which assist in memory consolidation during sleep.
[p. 702]

David Chalmers

Oculus Quest 2 virtual reality headset.
[p. 566]

Hair dryer that David Chalmers used as a prop at a 1996 conference to lampoon the notion of a "consciousness meter."
[p. 600]

Smartphone.
[p. 670]

Patricia Churchland

Polished section of tree root wrapped around a rock.
[p. 586]

Sculpture. Nunavit, Canada (Inuit). Whalebone vertebrae.
[p. 658]

Stanislas Dehaene

Spatial problem-solving puzzle, the object of which is to remove the tetrahedron from its enclosure.
[p. 517]

3D print of a 3 Tesla MRI scan of a human brain.
[p. 569]

Spatial problem-solving puzzle.
[p. 601]

Daniel C. Dennett

Matryoshka dolls lampooning Descartes' mind-body dualism. The large doll represents Descartes; the medium-sized doll represents what philosopher Gilbert Ryle derisively termed "the ghost in the machine"; the small doll represents the machine in the ghost.
[p. 537]

Whittled wood sculptures with movable parts, made by Daniel Dennett.
[p. 648]

Catherine Dulac

Catherine Dulac's laboratory notebook.
[p. 583]

David Erritzoe

Fern from David Erritzoe's psychopharmacology and psychedelic research clinic.
[p. 667]

Lisa Feldman Barrett

Image used in demonstration of visual perception.
[p. 661]

Photograph of Serena Williams at the 2008 US Open (Barton Silverman/*The New York Times*) used in research into the perception of emotional expression.
[p. 686]

André Fenton

Handmade microdrive (*c.* 2000) used to place and maneuver 16 electrodes in a rat's brain.
[p. 533]

Wireless, battery-operated MicroEEG machine used to record the brain's electrical activity.
[p. 637]

Karl Friston

Victorian clock with mechanical bird.
[p. 556]

Slide rule (1970s).
[p. 562]

Ali Ghazizadeh

3D-printed CT scan of a macaque monkey skull.
[p. 608]

Carl Hart

Medical device used to measure pulse, temperature, blood pressure, and SpO_2 (saturation of peripheral oxygen).
[p. 550]

Suzana Herculano-Houzel

Herculano-Houzel's Condensed Human Brain Soup and *Herculano-Houzel's Condensed Macaque Brain Soup*, colored pencil on paper.
[p. 491]

3D-printed models of minke whale, giraffe, raccoon, and ferret brains.
[p. 598]

Suzana Herculano-Houzel's laboratory notebook.
[p. 655]

Amadi O. Ihunwo

Brain knife used for dissection.
[p. 495]

Stained and plastinated (preserved) coronal section of a human brain.
[p. 582]

Stained brain sections from a Tg2576 mouse (transgenic mouse model of Alzheimer's disease).
[p. 675]

Erich Jarvis

3D model of pigeon brain.
[p. 527]

Albert Einstein action figure.
[p. 659]

Bianca Jones Marlin

Vials, gloves, hypodermic needle, and other medical implements.
[p. 506]

Glass beakers.
[p. 591]

Whole cell *in vivo* recording of an auditory neuron in a mouse exposed to mouse pup calls.
[p. 682]

Photograph of Bianca Jones Marlin as an infant with her parents.
[p. 682]

Kumi Kuroda

Video footage of mouse mother and virgin female mouse in Kumi Kuroda's laboratory experiment.
[p. 593]

Glass slides with mouse brain sections:
Left: mouse brain sections exhibiting activation patterns of a father mouse caring for unfamiliar pups.
Center: mouse brain sections in which Golgi staining has been used to highlight neuronal morphology.
Right: mouse brain sections exhibiting activation patterns of a virgin male mouse attacking unfamiliar pups.
[p. 595]

Video footage of macaque mother and infant.
[not illustrated]

Video footage of polar bear mother and cubs.
[p. 704]

Joseph E. LeDoux

Cowboy Tales: The Boo LeDoux Story. Based on interviews by Bill Jones with rodeo bull rider Boo LeDoux (Joseph LeDoux's father).
[p. 543]

Fender Stratocaster guitar lamp.
[p. 634]

Mahmoud Bukar Maina

"Fig. 1: African neuroscience publications 1996–2017." Maina, M.B. *et al.*, "Two decades of neuroscience publication trends in Africa," *Nature Communications*, vol. 12, 3429 (June 8, 2021), doi:10.1038/s41467-021-23784-8.
[p. 492]

Mahmoud Bukar Maina's laboratory notebook.
[p. 524]

Unilux-12 binocular microscope.
[p. 560]

Catherine Malabou

Mask used in funeral ceremonies in Ivory Coast (Senoufo).
[p. 497]

Eve Marder

Fine dissection tools used to prepare stomatogastric nervous system for electrophysiological recordings.
[p. 493]

Drawings by Benjamin Marder (Eve Marder's nephew) of *Cancer borealis* and other species used as laboratory animals in Eve Marder's research.
[p. 512]

Eve Marder's laboratory notebook.
October 4, 1973–January 29, 1974.
[p. 514]

Hannah Monyer

Image showing live neurons in a brain slice and the tip of a patch pipette used to analyze cellular membranes.
[p. 519]

Bust, Marie Rosalie (Mitzi) Friedmann-Otten, 1925, enamel and copper. Produced by Wiener Werkstätte.
[p. 645]

David Poeppel

Hourglass with blue sand.
[p. 603]

Supratim Ray

Set of drawing pencils.
[p. 545]

Supratim Ray, *Two Hands*, 2002. Pencil on paper.
[p. 551]

Figurine of Swami Vivekananda.
[p. 673]

Daniela Schiller

Metal sculpture by Arie Lamdan.
[p. 640]

Model of human brain.
[p. 704]

Wolf Singer

Pendulum made by Wolf Singer to demonstrate deterministic chaos.
[p. 501]

Wire model for brain sculpture made by E.R. Nele and installed at the Frankfurt Institute for Advanced Studies, Germany, which Wolf Singer co-founded in 2003.
[p. 652]

Motherboard from mainframe computer.
[p. 662]

Mitchell Joseph Valdés-Sosa

Models of the right brain hemispheres of two patients with prosopagnosia, an inability to recognize faces.
[p. 666]

Angela Vincent

Pipette used in myasthenia gravis (an autoimmune disorder that affects neuromuscular junctions) studies involving detection of acetylcholine receptor (AChR, a neurotransmitter) antibodies.
[p. 516]

Rotarod device used to measure the effects of autoimmune antibodies on mouse locomotion.
[p. 548]

Image of *Bungarus multicinctus*, the venom of which contains a neurotoxin used in myasthenia gravis studies.
[p. 581]

Video still of an individual who, when tapped on the chin with a reflex hammer, exhibited the startle reaction characteristic of myasthenia gravis.
Included in M. Hutchinson, P. Waters, J. McHugh, G. Gorman, S. O'Riordan, S. Connolly, H. Hager, P. Yu, C.-M. Becker, A. Vincent, "Progressive encephalomyelitis, rigidity, and myoclonus: a novel glycine receptor antibody." Published in *Neurology* (Oct. 14, 2008), 71(16):1291-92. Copyright © 2008, American Academy of Neurology.
[p. 585]

Charles Yang

Spatial problem-solving puzzle.
[p. 539]

Rafael Yuste

Magnifying glasses used for identifying cnidarians (aquatic invertebrates), some of which may be as small as 1 mm.
[p. 615]

Li Zhaoping

3D anaglyph glasses.
[p. 564]

IV

PRESERVING THE BRAIN

"Preserving the Brain" is the final chapter of the Fondazione Prada program "Human Brains." It has been developed in collaboration with 13 among the most outstanding neuroscience international institutions with the aims to stimulate an open and critical dialogue with scientists, experts, and scholars on the topic of neurodegenerative diseases. The involved institutions were: Harvard Medical School, Brigham and Women's Hospital, Ann Romney Center for Neurological Diseases, Boston, USA; Hôpital de la Pitié-Salpêtrière, Sorbonne University AP-HP, Neurology department and Paris Brain Institute, Paris, France; UniSR – Università Vita-Salute San Raffaele, Milan, Italy; Juntendo University Hospital, Neurology Department, Tokyo, Japan; Karolinska Institutet, Stockholm, Sweden; German Center for Neurodegenerative Diseases (DZNE) within the Helmholtz Association, Bonn, Germany; Max Planck Institute of Neurobiology, Munich, Germany; Montreal Neurological Institute-Hospital, McGill University, Montreal, Canada; Tianjin Medical University General Hospital, Neurology Department, Tianjin, China; UCSF Weill Institute for Neurosciences, University of California San Francisco, San Francisco, USA; University College London Hospitals NHS Foundation Trust, National Hospital for Neurology and Neurosurgery Faculty of Brain Sciences, Institute of Neurology, Department of Brain Repair and Rehabilitation, United Kingdom; Weizmann Institute of Science, Rehovot, Israel; Yale School of Medicine, New Haven, USA.

The Congress and a Scientific Exhibition were dedicated to ongoing research activities in the area of neurodegenerative diseases.From October 6 to October 10, 2022, more than 200 experts gathered at the Milan venue of Fondazione Prada to explore the topic of neurodegenerative diseases (NDDs) from different angles, starting with an up-to-date exchange of knowledge on epidemiological and genetic implications and molecular disease mechanisms, and ending with a debate on key concepts, possible lacunae in clinical trials, and targets for new treatments.

SCIENTIFIC DIRECTOR

Giancarlo Comi

WITH

Katerina Akassoglou, Alberto Albanese, Jack Antel, Alberto Ascherio, Sergio Baranzini, Claudio Bassetti, Mario Alberto Battaglia, Randall Bateman, Shibeshish Belachew, Hagai Bergman, Monique Breteler, Alexis Brice, Jens Bruning, Maria Chiara Carrozza, Adriano Chiò, Tim Coetzee, Giancarlo Comi, Jeffrey Cummings, Frédéric Destrebecq, Vincenzo di Lazzaro, Monica Di Luca, Donato di Monte, Alessandro di Rocco, Alexandra Durr, Lucia Faccio, Luigi Ferini Strambi, Carlo Ferrarese, Giovanni Frisoni, Thomas Gasser, Lisa Ann Gerdes, Lice Ghilardi, Brian Hafler, David Hafler, John Hardy, Hans Peter Hartung, Stephen Hauser, Michael Heneka, Nils-Göran Larsson, Hans Lassmann, Letizia Leocani, John Lehr, Chiara Liberati, Catherine Lubetzki, Xavier Montalban, Richard I. Morimoto, Elena Moro, Cath Mummery, Pierluigi Nicotera, Luigi Naldini, Jorge Oksenberg, Tomas Olsson, Daniela Perani, Fredrik Piehl, Niels Plath, Stefano Pluchino, Helen Plun Favreau, Miuccia Prada, Marco Salvetti, Jan Schadrack, Anthony Schapira, Philip Scheltens, Maximilian Schuier, Michal Schwartz, Vincenzo Silani, Fu-Dong Shi, Per Svenningsson, Fabrizio Tagliavini, Alan Thompson, Gonzalo Torres, Taiji Tsunemi, Florian Von Raison, George Vradenburg, Kristine Yaffe, Nan Zhang, Henrik Zetterberg

Forum on Neurodegenerative Diseases

Giancarlo Comi

Neurodegeneration occurs in a diverse range of diseases, including Alzheimer's Disease (AD), Parkinson's disease (PD), Multiple Sclerosis Lateral Sclerosis (ALS) and Multiple Sclerosis (MS), which are widespread and as yet incurable. Neurodegenerative diseases (NDDs) pose one of the biggest challenges to human health in the 21st century, striking the most complex organ in the human body and causing enormous damage to the most fundamental human traits: cognition and personality. Current understanding of these disorders builds upon decades of advances in different areas of neuroscience, from pathology to neurophysiology, from genetic to molecular medicine, from neuroepidemiology to clinical trials. The scientific conference saw the participation of neuroscientists, patients associations, representatives of healthcare institutions, and the pharmaceutical and biotechnology industries. The following is a summary of the highlights of this two-day conference.

Epidemiology and risk factors

Brain health is an evolving concept that has emerged in the academic literature in recent decades and has become increasingly popular within science, research, journalism, and the general public.[1] Compelling arguments for optimizing brain health are the epidemiological transition of the ageing population, which adds to the growing burden of NDDs, particularly when coupled with increasingly unhealthy lifestyles (lack of exercise, unhealthy diet, obesity, lack of sleep).

One in three people worldwide will develop a neurological disorder at some point in their life.[2] As such, optimizing brain health can not only reduce the increasing burden of NDDs, but also improve overall mental and physical health and have positive social and economic consequences. While genes do not change over time, lifestyle and environmental factors do, and these can be adjusted. The evidence and possible mechanisms of action for environmental factors in the causation of NDD were discussed during the first session.

Alzheimer's Disease (AD) is the most common cause of dementia, and its prevalence is expected to triple by 2050. Given the current lack of disease-modifying treatments and increasing awareness that symptoms develop over many years or even decades, there is growing interest in identifying effective strategies for prevention.[3] Potentially modifiable risk factors for AD include diabetes, hypertension, obesity, and smoking. Delaying the onset of symptoms by just one year could potentially reduce the prevalence of AD by more than 9 million cases over the next 40 years.[4]

It is recognized that clinical Parkinson's disease (PD), the second most common NDD, is preceded by a long prodromal period during which individuals may experience non-motor and subtle motor signs before the onset of classic motor features.[5] These prodromal features are diverse, but can for example include hyposmia, constipation, REM sleep behavior disorder and depression, and are related to physical activity. As well as healthy a dietary pattern (e.g. higher flavonoid intake) also contributes to lower mortality risk in PD.

Multiple Sclerosis (MS) is a chronic demyelinating disease of the central nervous system (CNS), and its prevalence has reportedly increased in the decades. The underlying cause of this disease is not yet fully understood, but the Epstein-Barr virus is thought to be a possible culprit. Vitamin D availability through sun exposure or diet and dietary fat intake could also play a role in the development of MS.[6] The gut-brain axis, which mediates the communication between the CNS and the gut microbiome, is a rapidly growing topic. Current advances targeting the study of the gut-brain axis provide evidence for specific gut microbiome associations that drive MS risk, course, and progression.

Genetics and epigenetics in NDD

The genetic basis for most of the common inherited NDDs has been established over the past 25 years, and subsequently a wealth of information on the biochemical, molecular, and neural pathways affected by these genetic changes has come to light.[7] Most NDDs are characterized by progressive dysfunction of specific populations of neurons, which determines the clinical presentation. Neuronal loss is usually accompanied by extra and intracellular accumulation of misfolded proteins, the hallmarks of many neurodegenerative proteinopathies, e.g. amyloid-β (Aβ) plaques and intracellular hyperphosphorylated tau protein neurofibrillary tangles in AD, alpha-synuclein-containing neuronal aggregates (Lewy bodies) in PD and transactivation response DNA binding protein 43 (TDP-43) in ALS. Identification of genetic risk variants has been facilitated by genome-wide association studies (GWAS) and meta-analyses. Polygenic risk scores developed from GWAS to predict heritability can be used for targeted screening and prevention strategies. The latest studies have also focused on the epigenetic profile characterizing these neuropathologies. Epigenetic changes include a plethora of modifications affecting gene expression without altering the DNA sequence, in turn determined by age, diet, stress, and disease state.

AD is usually categorized as either early onset or late onset, with the latter being more common. The genetics of the early form of AD are well known, with common gene variants engaged in Aβ production. However, Aβ dysregulation in late-onset "sporadic" AD is less well understood. Possible mechanisms of increased Aβ production in late AD include an increase in the amount of beta-secretase or in beta-secretase activity. On the other hand, impaired clearance of Aβ may also lead to late-onset AD through interactions with ApoE4, decreased catabolism of Aβ via reduced proteolytic enzymes or impaired transport across the blood-brain barrier (BBB).[8] What is also important to consider is the age-dependent failure of protein clearance mechanisms in these diseases (increase in damage over time).

PD is caused by the death of dopaminergic neurons containing neuromelanin in the substantia nigra in the brain, with the presence of Lewy bodies. About 5–10% of all PD cases can be attributed to monogenic forms (i.e. Mendelian PD). Most patients are sporadic cases related to a variable combination of genetic susceptibility and environmental factors. For the monogenic forms, there are several recognized genes, e.g. mutations in the alpha-synuclein gene, SNCA; mutations in leucine-rich repeat kinase 2, LRRK2; and VPS13C.[9] Furthermore, mutations in genes such as glucocerebrosidase (GBA1) fall in between a monogenic cause and a genetic susceptibility factor, and are accepted as the most commonly known genetic risk factors for PD and dementia with Lewy bodies. Knowledge of these mutations has led to clinical applications of diagnostic and genetic counseling, the definition of the prognosis and, what matters more, the exploration of many potential therapeutic targets.

The genetics underlying who develops MS have been difficult to work out. Examining more than 47,000 cases and 68,000 controls with multiple genome-wide association studies, the International Multiple Sclerosis Genetics Consortium (IMSC) identified more than 200 risk loci in MS to date.[10] Most of the genetic risk factors have a role on the immune system.

Molecular and cellular mechanisms in ageing and NDD

The understanding of the molecular and cellular mechanisms of many NDDs is still patchy, which is essential to identify potential treatment targets and strategies. Recent advances in pathological mechanisms linked to ageing were reviewed at this meeting, including loss of the ability to maintain a functional proteome (or proteostasis), mitochondrial dysfunction, disrupted axonal transport, and neuroinflammation.

Proteostasis mechanisms help proteins adopt and maintain their correct folding state during and after synthesis and, when this can no longer be achieved, they ensure that misfolded proteins are degraded. This can be particularly difficult when DNA mutations in components of proteostasis factors impair their function or alter residues that are important for a protein to fold correctly; when cells are exposed to acute stress; and, importantly, during aging. Damaged and misfolded proteins accumulate with age, impairing cell function and tissue homeostasis. The accumulation of damaged proteins contributes to several age-related diseases such as AD, PD or Huntington's disease.[11] Damaged proteins are degraded by the ubiquitin-proteasome system or through autophagy-lysosome, key components of the proteostasis network. Modulation of either proteasome activity or autophagic-lysosomal potential extends lifespan and protects organisms from symptoms associated with proteostasis disorders.

Next, the role of mitochondria, intracellular organelles that contribute to several metabolic pathways, was the focus of the discussion. Mitochondrial dysfunction constitutes one of the hallmarks of ageing and is characterized, for example, by an accumulation of mitochondrial DNA mutations, increased production of mitochondrial reactive oxygen species, and the consequent oxidative damage to nucleic acids, proteins, and lipids. Mitophagy, a mitochondrial quality control mechanism that enables the breakdown of damaged and superfluous mitochondria, prevents such detrimental effects and reinstates cellular homeostasis in response to stress. To date, there is growing evidence that mitophagy is significantly impaired in several human pathologies including ageing and age-related diseases such as NDDs. These results enrich our insights of physiological events regulating mitophagy and establish a novel pathway for drug targeting in neurodegeneration.[12]

People with type 2 diabetes have an increased risk of dementia. Beyond insulin's polyvalent effects in peripheral tissues, insulin also acts on neural circuits to control systemic metabolism via the regulation of body weight and systemic fat and glucose metabolism. Insulin activates insulin receptors in neurons and glial cells to provide a feedback signal that instructs the brain on the availability of glucose in the organism.[13] Converging data from animal and human studies suggest that impaired neuronal insulin action might be a unifying pathomechanism in the development of both type 2 diabetes mellitus and NDDs.[14] Promising evidence for metabolism-based treatment targets for NDDs was reviewed.

The final topic of this session addressed the bidirectional relationships between the CNS and the peripheral immune system and the relevance of immune components to brain ageing and NDDs. For decades, it was widely believed that the brain is excluded from peripheral immune activity and is self-sufficient for its maintenance and repair. More recently, central nervous system confine inflammation has been shown to play a major role in many NDDs. Microglia Rejuvenation of the systemic immune system, such as vaccination with CNS antigens and blocking inhibitory PD-1 immune checkpoints, is being explored in mouse models to combat cognitive impairment in AD, for example.[15] Microglia are CNS-resident immune cells belonging to the myeloid lineage of the innate arm of immunity. Upon various stimuli (misfolded proteins, aggregates, B and T cells infiltrates), microglia acquire a phagocytic function and produce proinflammatory cytokines and chemokines that contribute to the process of neurodegeneration. Novel, promising targets for therapeutic strategies for AD have been sought in the field of microglia activation.

Synapse dysfunction and plasticity loss are crucial to the pathophysiology of AD. In the cascade of AD pathogenesis, synaptic failure, triggered by Aβ and tau, has been identified as a key step and an early event in AD, preceding neurodegeneration. Available data on tackling the early stages of the disease be used as a guide to finding new therapeutic avenues.[16]

Emerging studies have also shown that fibrinogen is not only a marker of BBB disruption but also plays a causal role in a large number of animal models of neurological diseases, highlighting the potential role of fibrinogen in MS, AD, brain trauma, and nerve injury.[17] The neuropathological effects of fibrinogen in the nervous system are numerous. Fibrinogen is unique among plasma proteins owing to its molecular structure, which contains binding sites for receptors expressed by nervous system cells and for proteins that regulate important nervous system functions, making it a potential druggable target.

Considerations in design and outcomes of clinical trials

Starting from neurochemical mechanisms in NDDs on the first day of the congress, the focus during the second day was on translating this knowledge to the clinic, and ultimately to the patient. The right clinical trial is a well-conducted study with appropriate clinical and biomarker outcomes collected over an appropriate period of time, powered to reveal a statistically significant and clinically meaningful positive effect. Failures of phase III clinical trials in most NDDs have been attributed to errors in clinical trial designs. As highlighted in the previous sessions, a lack of understanding of some critical aspects of disease biology and drug action may have affected the success of development programs even when the "rights" were adhered to.[18] Lessons learned from clinical trials, even the failed ones, were discussed.

The choice of outcomes in phase III randomized clinical trials is fundamental, and a critical aspect of the protocol is that the outcome measures to evaluate treatment efficacy are appropriate.[19] Digital technologies (e-health) are not yet advanced in most NDD trials. However, adding continuous remote digital self-monitoring may take us closer to the disease course and address the invisible symptoms. As well, there is a clear unmet need for a better-defined patient-reported outcome (PRO) methodology, PROs need to climb up positions in the hierarchy of clinical trial outcomes to assess the genuine impact on patients' lives.[20]

The time course of NDDs, such as the prodromal phase, has also implications for the design of clinical trials. In the past decade, clinical trials have increasingly focused on earlier, pre-NDD stages, as intervening with NDD pathology before the accumulation of irreversible tissue damage is crucial, and moreover contrasting the disease while the individual is still highly functional is an important goal.

Early failure saves valuable resources including opportunity costs. The application of exploratory trials can therefore establish go/no-go decision points, support proof-of-concept, and dose selection, and terminate drugs that fail to show target engagement with suitable exposure and acceptable safety profiles. This is especially important for programs in academia and small biotechnology companies but can also be applied to high-risk projects in large pharmaceutical companies to achieve proof-of-concept more rapidly at lower costs than traditional approaches.[21]

There is a high rate of failure in drug development for AD with 99% of trials showing no drug-placebo difference. In many cases, BBB penetration, dose, target engagement, or rigorous interrogation of early-stage data has not been adequately pursued. Agents have been advanced to phase 3 with little or no evidence of efficacy in phase 2.[22] Today, anti-amyloid monoclonal antibodies seem to be a promising disease-modifying treatment for AD, and a large number of other interesting potential disease-altering therapies are on the way.

The most common genetic risk factors for PD are variants in the GBA gene, encoding the lysosomal enzyme glucocerebrosidase (GCase).[23] GCase deficiency leads to α-synuclein accumulation. Given the strongly implicated role of GCase in PD pathogenesis, the development of an effective treatment that can restore neural GCase levels could not only drastically improve the quality of life for patients with neuronopathic Gaucher disease, but could also potentially prevent Parkinson's onset in patients susceptible to Gaucher-associated Parkinson's disease or even idiopathic parkinsonism. Several strategies, such as enzyme replacement therapy, substrate reduction therapy, chaperones, and gene therapy are being investigated nowadays.[24]

MS is a lifelong disorder that follows a biphasic curve: ambulation is usually conserved during the relapsing-remitting phase, while impairment preferentially occurs later during the progressive phase. Although inflammation is now prevented by treatments during the early relapsing-remitting phase, uncertainty remains about their preventive effect for a late transition to secondary progression. With the advent of newer and more aggressive disease-modifying therapies, there is a real need to define endpoints that accurately determine the efficacy of therapy in MS. Composite measures, like NEDA (no evidence of progression) combining clinical symptoms with MRI and parameters for cognition and tracking brain volume loss, are increasingly implemented in clinical trials in relapsing MS, alongside clinical outcomes and biological markers, to address the heterogeneity of this disease. To face progressive MS there are some ongoing phase III clinical trials, such as those testing Bruton kinase inhibitors.

Among NDDs, Amyotrophic Lateral Sclerosis (ALS) is the most rapid to progress to fatality, and the development of new therapeutic agents for ALS has been limited by the complexity and heterogeneity of disease pathophysiology. Adaptive platform trials, as a more efficient alternative to randomized controlled trials that test one investigational product at a time, represent a novel approach and have found their way in ALS drug development, and hopefully many other NDDs will follow. In adaptive platform trials, already used in oncology, different interventions for a single disease are investigated constantly, with the aim of identifying novel treatments, biomarkers, and trial endpoints more quickly.[25]

Significant efforts are being made to identify ideal biomarkers that are sensitive to preclinical/prodromal stages, detect disease progression and predict underlying pathology, but also can speed up clinical trials. The sixth session of this congress covered the latest advances in the pursuit of highly specific but affordable and easy-to-apply biomarkers.

Magnetic resonance imaging (MRI), for example, is proven to be a very useful neuroimaging biomarker in clinical trials: changes in global brain volume, white matter and gray matter brain volume, and thalamic volume are a marker of irreversible nervous tissue damage and are used as a secondary endpoint in phase III clinical trials and as primary endpoints in phase II clinical trials in progressive MS.[26] Positron emission tomography (PET) is a non-invasive, quantitative technique that uses radiolabeled compounds (PET tracers) to characterize various pathological processes. By binding to specific targets, PET imaging can depict metabolic, biochemical, or cellular changes depending on the tracer in use. FDG-PET and amyloid-PET have proven vital in evaluating the risk of progression and drug targets for dementia.[27] In recent years Pittsburgh Compound B, an amyloid tracer, has been used also to image *in vivo* myelin and is now under evaluation as a biomarker in remyelinating interventions. Fluid biomarkers are also regularly included in clinical trials; well-known examples are cerebrospinal fluid (and more recently also plasma) levels of Aβ42, total tau, phosphorylated tau, and neurofilament light chain. Fluid biomarkers are important because they can provide information about the underlying biochemical processes occurring in the brain and about the treatment response.

The clinical and pathological heterogeneity in frontotemporal lobar degeneration (FTLD), presents a variety of challenges including accurate and timely diagnosis, prognostication, monitoring, and identification of appropriate endpoints in clinical trials. Recent advances in neuroimaging (structural MRI, PET) enhanced the understanding about the pathophysiology of FTLD, offering improved diagnosis of sporadic clinical syndromes associated with FTLD.[28]

Expanding the portfolio of therapeutic targets

Although several promising therapies are currently being investigated in clinical trials, there is an urgent need for treatments that significantly alter the course of NDDs; most therapies are currently symptomatic. Since NDDs are estimated to consume a significant part of the health budget, it is imperative that more effective, disease-modifying therapies are identified. In the last two sessions of the congress, therapeutic strategies for NDDs in the third millennium were actualized. One of the biggest challenges in treating NDDs is partly caused by difficulties in penetrating the BBB.

Very recently, amyloid antibody-based immunotherapies in AD show evidence of modification of the disease, albeit modestly and at an early stage. Other promising targets in AD include modulation of peripheral inflammation (PD-L1), or suppression of tau by active immunization or monoclonal antibodies. Novel promising approaches to slow down the progression of PD include interfering with α-synuclein via antibody immunization, inhibition of α-synuclein and misfolding, and targeting the immune system by glucagon-like peptide-1 (GLP-1) receptor agonists that inhibit activated microglia. The treatment paradigm in relapsing MS has shifted since evidence that early and potent antinflammatory treatments may consistently reduce the frequency of relapses and prevent/delay disability progression. The introduction of B-cell monoclonal antibodies resulted in a strong reduction of the accumulation of new lesions, showing that B lymphocytes play a central role in the pathogenic cascade.[29] Nevertheless, progression occurs in MS patients, sometime already in the initial phases of the disease, probably due to confined CNS inflammation trapped behind an intact BBB, so new interventions, active inside the CNS, should be combined very early with disease-modifying treatments against the peripheral immune mechanisms. There is an unmet clinical need for therapies that can arrest and reverse the persistent accumulation of disability associated with progressive forms of MS.[30]

Antisense drugs are synthetic oligonucleotides designed to bind to RNA. Antisense oligonucleotides (ASOs) typically incorporate sugar modifications to enhance binding affinity to the target RNA, increase metabolic stability, and decrease adverse effects and phosphate modifications, which enhance metabolic stability and increase protein binding to enhance tissue distribution and cellular uptake. The ASO technology has made remarkable progress over the past 10 years, with drug approval and multiple potential therapeutic agents for NDDs (such as Duchenne's, Huntington's disease, ALS) in development. The profound clinical benefit of nusinersen in spinal muscular atrophy patients validates the platform.

To bypass the BBB, corrected genetic material is directly delivered into the affected tissue using viral vectors. While a variety of viral vectors are available, the most commonly used type is the adeno-associated virus (AAV) vector. AAV owes its popularity due to its ability to reliably introduce corrected DNA into the cell via non-pathogenic infection with near-perfect specificity, reducing the need to implant excessive copies of the correctional gene into the cell.

Considerable progress has been made in the development of potential regenerative cell-based therapies for NDDs, with some initial success being neural grafting in PD. Neural stem cell transplantation has also demonstrated encouraging therapeutic potential in preclinical studies of neurological diseases such as progressive MS, and there is a growing understanding of the mechanisms of action through which they act on the injured CNS.[31,32]

Other, non-medicinal approaches, discussed included neuromodulation using non-invasive transcranial brain stimulation to counteract neurodegeneration and compensate for CNS damage caused by the disease.[33] Deep brain stimulation, a surgical procedure for PD, was also raised which can help ease motor symptoms and decrease medication needs in some patients who have reached the limit of pharmaceutical therapy.

Conclusions and future directions in NDD

The second day wrapped up with a roundtable discussion in which technology experts, pharmaceutical industry representatives, and patient advocacy associations discussed future challenges in developing new therapies, and jointly set priorities in NDD research.

Attention to multistakeholder, cross-border drug development with a shared agenda will increase the likelihood of success, reduce the risks associated with drug development, enhance the ability to attract investment and increase the likelihood of new therapies becoming available to those with or vulnerable to the emergence of NDD. Cooperation, exchange of ideas, cross-fertilization across diseases/pathologies, and a multidisciplinary approach are essential to our battle against NDD. To accomplish this, the following was proposed:

- The diversity of NDDs is underestimated and regulatory agencies require diverse populations to be included. Therefore, close collaboration with regulatory agencies is required to improve clinical trial design and outcomes.

- Since multiple disease pathologies are commonly associated with neurodegeneration, multifactorial disease therapies may prove more effective than monotherapies targeting one aspect of the disease Combination therapies or re-purposing drugs could become an attractive proposition.
- In contrast to oncology, there is truly little funding in the field of NDDs. Pooling expertise and resources through public-private partnerships, for example, could address any fragmentation and duplication of current research efforts. Raising political awareness around NDDs, which now often remain under the radar, could also help boost investment.
- Embedding digital technology (e.g. artificial intelligence) in research could better characterize disease progression patterns to inform clinical trial design. Collaborative platforms for efficient sample and data sharing (in compliance with the GDPR), leveraging and linking existing research infrastructures could be of value to accelerate the discovery of biomarkers, new diagnostics, and treatments for the benefit of people with NDDs.
- A multi-faceted approach is also needed to inspire and attract a new generation of researchers and scientists. Moreover, scientific research needs to be easily translated into medical practice, be scalable and easily integrated into healthcare systems.
- Shared investment in drug hunting is key, but apart from that, the concept of brain health should be advanced by focusing on primary prevention (physical exercise, diet, sleep, smoking) and developing a joint communication strategy to de-stigmatize dementia and other NDDs. Targeting neuroplasticity, cognitive recovery, and emotional well-being are welcome complements to therapeutics and deserve a place in specialized rehabilitation centers.
- Finally, science today requires the voice of people living with the disease, providing input and giving direction to research ideas. The value that the science of and with patient input can bring to the clinical encounter is increasingly being recognized. This approach also highlights the importance of personalized medicine, one of the key messages of the "Human Brains" project, in which the human brain is examined in the plural—as expressed by the title—to underline its intrinsic complexity and the irreducible singularity of each individual.

1 *Optimizing Brain Health Across the Life Course: WHO Position Paper* (Geneva: World Health Organization, 2022). Licence: CC BY-NC-SA 3.0 IGO, creativecommons.org/licenses/by-nc-sa/3.0/igo/.
2 Bassetti C.L.A., Endres M., Sander A., Crean M., Subramaniam S., Carvalho V. *et al.*, "The European Academy of Neurology Brain Health Strategy: One Brain, One Life, One Approach," in *European Journal of Neurology* (2022) 29, pp. 2559–2566.
3 McDade E., Llibre-Guerra J.J., Holtzman D.M., Morris J.C., Bateman R.J, "The Informed Road Map to Prevention of Alzheimer Disease: A Call to Arms," in *Molecular Neurodegeneration* (2021) 16 (1), p. 49.
4 Barnes D.E., Yaffe K., "The Projected Effect of Risk Factor Reduction on Alzheimer's Disease Prevalence," in *The Lancet Neurology* (2011), 10, pp. 819–828.
5 Ascherio A., Schwarzschild M.A., "The Epidemiology of Parkinson's Disease: Risk Factors and Prevention," in *The Lancet Neurology* (2016), 15, pp. 1257–1272.
6 Amato M.P., Derfuss T., Hemmer B., Liblau R., Montalban X., Soelberg Sørensen P. *et al.*, "Environmental Modifiable Risk Factors for Multiple Sclerosis: Report From the 2016 Ectrims Focused Workshop," in *Multiple Sclerosis Journal* (2017), 24, pp. 590–603.
7 Singleton A., Hardy J., "The Evolution of Genetics: Alzheimer's and Parkinson's Diseases," in *Neuron* (2016) 90, pp. 1154–1163.
8 Mawuenyega K.G., Sigurdson W., Ovod V., Munsell L., Kasten T., Morris J.C. *et al.*, "Decreased Clearance of Cns Beta-Amyloid in Alzheimer's Disease," in *Science* (2010), 330 (6012), p. 1774.
9 Kumar K.R., Lohmann K., Klein C., "Genetics of Parkinson's Disease and Other Movement Disorders," in *Current Opinion in Neurology* (2012), 25, pp. 466–474.
10 Patsopoulos N.A., Baranzini S.E., Santaniello A., Shoostari P., Cotsapas C., Wong G. *et al.*, "Multiple Sclerosis Genomic Map Implicates Peripheral Immune Cells and Microglia in Susceptibility," in *Science* (2019), 365 (6460), eaav7188.
11 Labbadia J., Morimoto R.I., "The Biology of Proteostasis in Aging and Disease," in *Annual Review of Biochemistry* (2015), 84, pp. 435–464.
12 Motori E., Atanassov I., Kochan S.M.V., Folz-Donahue K., Sakthivelu V., Giavalisco P. *et al.*, "Neuronal Metabolic Rewiring Promotes Resilience to Neurodegeneration Caused by Mitochondrial Dysfunction," in *Sci Adv* (2020), 6, eaba8271.
13 Porniece Kumar M., Cremer A.L., Klemm P., Steuernagel L., Sundaram S., Jais A. *et al.*, "Insulin Signalling in Tanycytes Gates Hypothalamic Insulin Uptake and Regulation of Agrp Neuron Activity," in *Nature Metabolism* (2021), 3, pp. 1662–1679.
14 Steculorum S.M., Solas M., Brüning J.C., "The Paradox of Neuronal Insulin Action and Resistance in the Development of Aging-Associated Diseases," in *Alzheimer's & Dementia* (2014), 10 (1 Suppl.), S3–11.
15 Schwartz M., Peralta Ramos J.M., Ben-Yehuda H., "A 20-Year Journey from Axonal Injury to Neurodegenerative Diseases and the Prospect of Immunotherapy for Combating Alzheimer's Disease," in *Journal of Immunology* (2020), 204, pp. 243–250.
16 Musardo S., Therin S., Pelucchi S., D'Andrea L., Stringhi R., Ribeiro A. *et al.*, "The Development of Adam10 Endocytosis Inhibitors for the Treatment of Alzheimer's Disease," in *Molecular Therapy* (2022), 30, pp. 2474–2490.
17 Petersen M.A., Ryu J.K., Akassoglou K., "Fibrinogen in Neurological Diseases: Mechanisms, Imaging and Therapeutics," in *Nature Reviews Neuroscience*, (2018), 19, pp. 283–301.
18 Cummings J., Feldman H.H., Scheltens P., "The 'Rights' of Precision Drug Development for Alzheimer's Disease," in *Alzheimer's Research & Therapy* (2019), 11, p. 76.
19 Jutten R.J., Papp K.V., Hendrix S., Ellison N., Langbaum J.B., Donohue M.C. *et al.*, "Why a Clinical Trial Is as Good as Its Outcome Measure: A Framework for the Selection and Use of Cognitive Outcome Measures for Clinical Trials of Alzheimer's Disease," in *Alzheimer's &Dementia* (2022), doi: 10.1002/alz.12773.
20 Lakin L., Davis B.E., Binns C.C., Currie K.M., Rensel M.R., " Comprehensive Approach to Management of Multiple Sclerosis: Addressing Invisible Symptoms-A Narrative Review," in *Neurology and Therapy* (2021), 10, pp. 75–98.
21 Wang T., McAuslane N., Goettsch W.G., Leufkens H.G.M., De Bruin M.L., "Challenges and Opportunities for Companies to Build HTA/Payer Perspectives Into Drug Development Through the Use of a Dynamic Target Product Profile," in *Frontiers in Pharmacology* (2022), 13, 948161, doi: 10.3389/fphar.2022.948161.
22 Friedman L.G., McKeehan N., Hara Y., Cummings J.L., Matthews D.C., Zhu J. *et al.*, "Value-Generating Exploratory Trials in Neurodegenerative Dementias," in *Neurology* (2021), 96, pp. 944–954.
23 Gegg M.E., Menozzi E., Schapira A.H.V., "Glucocerebrosidase-Associated Parkinson's Disease: Pathogenic Mechanisms and Potential Drug Treatments," in *Neurobiology of Disease* (2022), 166, 105663.
24 Elkouzi A., Vedam-Mai V., Eisinger R.S., Okun M.S. "Emerging Therapies in Parkinson's Disease-Repurposed Drugs and New Approaches," in *Nature Reviews Neurology* (2019), 15, pp. 204–223.
25 Paganoni S., Berry J.D., Quintana M., Macklin E., Saville B.R., Detry M.A. *et al.*, "Adaptive Platform Trials to Transform Amyotrophic Lateral Sclerosis Therapy Development," in *Annals of Neurology* (2022), 91, pp. 165–175.
26 Li K.R., Wu A.G., Tang Y., He X.P., Yu C.L., Wu J.M., Hu G.Q., Yu L., " The Key Role of Magnetic Resonance Imaging in the Detection of Neurodegenerative Diseases-Associated Biomarkers: A Review," in *Molecular Neurobiology* (2022), 59, pp. 5935–5954.
27 Perani D., Iaccarino L., Jacobs A.H., "Application of Advanced Brain Positron Emission Tomography-Based Molecular Imaging for a Biological Framework in Neurodegenerative Proteinopathies," in *Alzheimer's & Dementia* (Amst) (2019), 11, pp. 327–332.
28 Dev S.I., Dickerson B.C., Touroutoglou A., "Neuroimaging in Frontotemporal Lobar Degeneration: Research and Clinical Utility," in *Advances in Experimental Medicine and Biology* (2021), 1281, pp. 93–112.
29 Comi G., Bar-Or A., Lassmann H., Uccelli A., Hartung H.P., Montalban X., Sørensen P.S., Hohlfeld R., Hauser S.L., "Role of B Cells in Multiple Sclerosis and Related Disorders," in *Annals of Neurology*, 2021, 89, pp. 13–23.
30 Hauser S.L., Cree B.A.C., "Treatment of Multiple Sclerosis: A Review," in *The American Journal of Medicine* (2020), 133, pp. 1380–1390.
31 Pluchino S., Smith J.A., Peruzzotti-Jametti L., "Promises and Limitations of Neural Stem Cell Therapies for Progressive Multiple Sclerosis," in *Trends in Molecular Medicine* (2020), 26, pp. 898–912.
32 Peruzzotti-Jametti L., Pluchino S., "Targeting Mitochondrial Metabolism in Neuroinflammation: Towards a Therapy for Progressive Multiple Sclerosis," in *Trends in Molecular Medicine* (2018), 24, pp. 838–855.
33 Leocani L., Chieffo R., Gentile A., Centonze D., "Beyond Rehabilitation in Ms: Insights From Non-Invasive Brain Stimulation," in *Multiple Sclerosis Journal* (2019), 25, pp. 1363–1371.

Program

OPENING

Miuccia Prada, *President and Director of Fondazione Prada*
Giancarlo Comi, *President of the "Human Brains" Scientific Board*
Claudio Bassetti, *Neurology Department Inselspital, University Hospital, Bern, Switzerland*

SESSION 1

Neurodegenerative Diseases: Epidemiology and Risk Factors

Moderated by Fabrizio Tagliavini, *Fondazione IRCCS Istituto Neurologico Carlo Besta, Milan*

Introductory Remarks
Kristine Yaffe, *UCSF, San Francisco*

Aging and Risks For Dementias
Monique Breteler, *DZNE, Bonn*

Modifiable Risks
Alberto Ascherio, *Harvard T.H. Chan School of Public Health, Boston*

General Discussion
Microbiota as a Triggering Factor of MS? Lessons from the MS Twin Study
Lisa Ann Gerdes, *Institute of Clinical Neuroimmunology, LMU Munich*

Microbioma
Sergio Baranzini, *UCSF, San Francisco*

SESSION 2

Genetic and Epigenetic

Moderated by Sergio Baranzini, *UCSF, San Francisco*

Dementias
John Hardy, *UCL, London*

Movement Disorders
Alexis Brice, *Paris Brain Institute – ICM, Paris*

Multiple Sclerosis
Jorge Oksenberg, *UCSF, San Francisco*

General Discussion
Parkinson's Disease
Thomas Gasser, *DZNE, Bonn*

Dystonias
Taiji Tsunemi, *Juntendo University, Tokyo*

SESSION 3

Molecular and Cellular Mechanisms in Ageing and Neurodegenerative Diseases 1

Moderated by John Hardy, *UCL, London*

Proteostasis Mechanisms in Aging and Neurodegenerative Diseases
Richard I. Morimoto, *Northwestern University, Evanston*

The role of Mitochondrial Dysfunction in Neurodegeneration
Nils-Göran Larsson, *Karolinska Institutet, Stockholm*

Brain Metabolism
Jens Bruning, *Max Planck Institute, Munich*

General Discussion
Synaptic Dysfunction
Monica Di Luca, *Università degli Studi di Milano, Milan*

Autophagy – Parkinson's Disease Mitophagy
Helen Plun Favreau, *UCL, London*

SESSION 4

Molecular Mechanisms in Ageing and Neurodegenerative Diseases 2

Moderated by Fredrik Piehl, *Karolinska Institutet, Stockholm*

Immunotherapy to Defeat Alzheimer's Disease: A Transformative Understanding of Brain Immunity
Michal Schwartz, *Weizmann Institute, Rehovot*

The Immunology of Blood: Unlocking Drivers of Neurodegeneration
Katerina Akassoglou, *UCSF, San Francisco*

Modelling α-Synuclein Pathology in Vivo
Donato di Monte, *DNZE, Bonn*

General Discussion
Properties of Human Glial Cells across the Ages
Jack Antel, *McGill University, Canada*

The Lesson of Age-Related Macular Degeneration
Brian Hafler, *Yale, New Haven*

Keynote 1
Pathology of Neurodegenerative Disease: The Basis for the Understanding of Disease Mechanism
Hans Lassmann, *Center for Brain Research at the Medical University, Wien*

Working Dinner
Preserving the Brain: A Global Campaign on Modifiable Risk Factors
Moderated by Elena Moro, *CHU Grenoble*, and Giancarlo Comi, *UniSR – Università Vita-Salute San Raffaele, Milan, Italy*

Burden of Neurodegenerative Diseases
Alberto Ascherio, *Harvard T.H. Chan School of Public Health, Boston*

Environmental Factors
Carlo Ferrarese, *Università degli Studi di Milano-Bicocca, Milan*

Microbiome and Food
Sergio Baranzini, *UCSF, San Francisco*, and *Gonzalo Torres, Loyola University, Chicago*

Physical Exercise
Letizia Leocani, *UniSR – Università Vita-Salute San Raffaele, Milan, Italy*

Smoking
Tomas Olsson, *Karolinska Institutet, Stockholm*

Sleep
Luigi Ferini Strambi, *IRCCS Ospedale San Raffaele, Milan*

Keynote 2
Neurodegenerative Disorders and Sterile Inflammation
Michael Heneka, *Luxembourg Centre for Systems Biomedicine – LCSB*

SESSION 5

Clinical Outcomes in Phase II-III Clinical Trials

Moderated by Giovanni Frisoni, *Centre de la mémoire, Geneva University Hospital*

Dementia
Jeffrey Cummings, *University of Nevada, Las Vegas*

Targeting the Glucocerebrosidase Pathway: A Paradigm for Slowing Parkinson's Disease
Anthony Schapira, *UCL, London*

Multiple Sclerosis
Xavier Montalban, *Multiple Sclerosis Centre of Catalonia – Cemcat, Barcelona*

General Discussion
Amyotrophic Lateral Sclerosis (ALS)
Vincenzo Silani, *Università degli Studi di Milano, Milan*

Fronto-Temporal Dementia
Nan Zhang, *Tianjin Medical University, Tianjin*

SESSION 6

Biomarkers in Phase II-III Clinical Trials in Neurodegenerative Diseases

Moderated by Philip Scheltens, *University Medical Center, Amsterdam*

Magnetic Resonance Imaging (MRI)
Giovanni Frisoni, *Centre de la mémoire, Geneva University Hospital*

Positron Emission Tomography (PET)
Daniela Perani, *UniSR – Università Vita-Salute San Raffaele, Milan, Italy*

Body Fluids
Henrik Zetterberg, *UCL, London*

General Discussion
Bodyfluids Biomarkers in MS
Hans Peter Hartung, *Department of Neurology Heinrich-Heine-Universität Düsseldorf, Germany / Brain and Mind Center, University of Sydney*

Biomarkers in NMOSD and MOGAD
Fu-Dong Shi, *Department of Neurology, Tianjin Medical University General Hospital / Beijing Tiantan Hospital*

Keynote 3
Clinical Trials Design to Accelerate Therapy
Randall Bateman, *Washington University School of Medicine, St. Louis Missouri*

SESSION 7

Pharmacological Treatment

Moderated by David Hafler, *Yale, New Haven*

Early Phase Trials in Alzheimer's Disease: Dawn of a New Age
Cath Mummery, *UCL, London*

Parkinson's Disease
Per Svenningsson, *Karolinska Institutet, Stockholm*

Multiple Sclerosis
Stephen Hauser, *USCF, San Francisco*

General Discussion
Amyotrophic Lateral Sclerosis (ALS)
Adriano Chiò, *Università degli Studi di Torino, Turin*

Huntington Disease
Alexandra Durr, *Paris Brain Institute – ICM, Paris*

SESSION 8

Further Topics

Moderated by Alberto Albanese, *Humanitas Research Hospital, Milan*

Gene Therapy
Luigi Naldini, *San Raffaele Hospital, Milan*

Advanced Cellular and Acellular Therapeutics
Stefano Pluchino, *Cambridge Biosciences Campus, UK*

Rehabilitation – Neuromodulation
Letizia Leocani, *UniSR – Università Vita-Salute San Raffaele, Milan, Italy*

General Discussion
Brain Stimulation in Amyotrophic Lateral Sclerosis
Vincenzo di Lazzaro, *Università Campus Bio-Medico, Rome*

DBS in Parkinson
Hagai Bergman, *The Hebrew University of Jerusalem, Israel*

Round Table
Together to Accelerate Therapy for Neurodegenerative Diseases
Moderated by Giancarlo Comi, *UniSR – Università Vita-Salute San Raffaele, Milan, Italy* and
Pierluigi Nicotera, *DZNE, Bonn*

Speakers:
Mario Alberto Battaglia, *FISM*
Shibeshish Belachew, *Biogen*
Maria Chiara Carrozza, *CNR Italy*
Tim Coetzee, *National MS Society*
Frédéric Destrebecq, *European Brain Council*
Alessandro di Rocco, *Parkinson League New York*
Lucia Faccio, *Safinnova Partners*
Lice Ghilardi, *FRESCO Foundation*
John Lehr, *Parkinson's Foundation*
Chiara Liberati, *Axxam*
Catherine Lubetzki, *Paris Brain Institute, AP-HP Sorbonne University*
Niels Plath, *Biotech MUNA*
Marco Salvetti, *Università Sapienza, Rome*
Jan Schadrack, *Roche*
Maximilian Schuier, *Janssen*
Alan Thompson, *UCL, London*
Florian Von Raison, *Novartis*
George Vradenburg, *Founding Chairman of the Board, Davos Alzheimer's Collaborative*

Concluding Remarks and Future Actions

Epidemiology of Neurodegenerative Diseases

Number of People with MS, Prevalence per 100,000 People

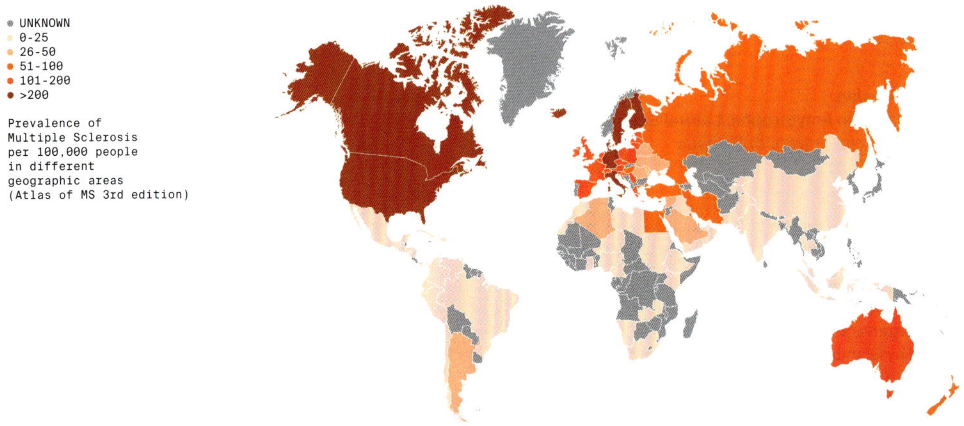

(A) Prevalence of Alzheimer disease per 1,000 men and women by age in the USA

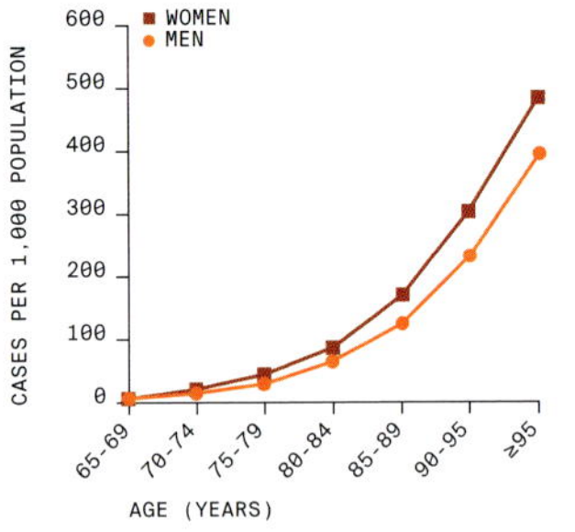

(B) Prevalence of Parkinson's disease per 100,000 men and women by age globally

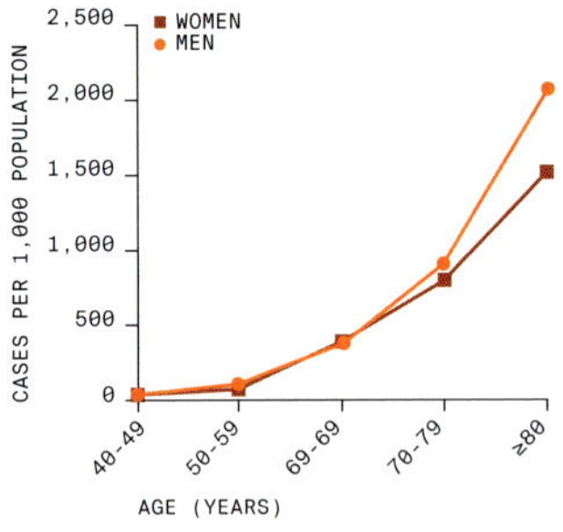

(C) Prevalence of Amyotrophic Lateral Sclerosis per 100,000 population in the USA

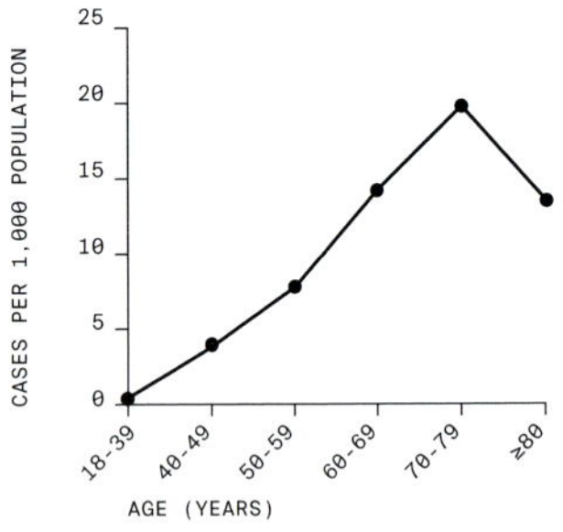

Ou et al. Nature Reviews Neurology 2019

Prevalence of Alzheimer Disease (A) and Parkinson's disease (B) per 100 by age in China.
Each point is a study; the dimension of the point reflects the sample size

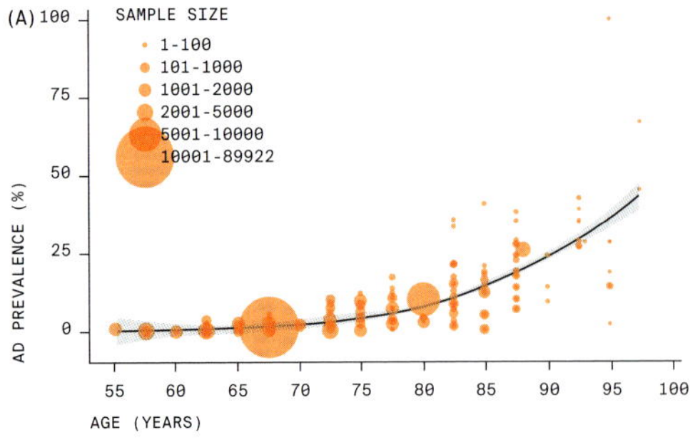

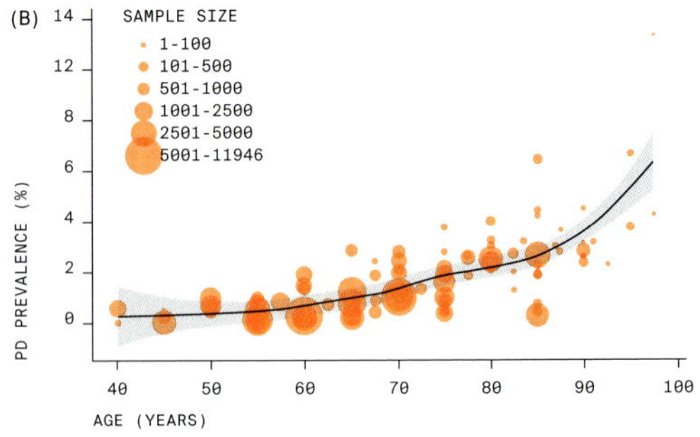

Cui et al. Frontiers in Neuroscience, 2020

"Preserving the Brain"

The exhibition was on stage at the Podium, the space at the center of Fondazione Prada's Milan venue. This exhibition browsed through a history of neuroscientific knowledge creation marked by rigor, breakthrough, and discovery, as well as flaws and uncertainties. It traces the outlines of consciousness, the gaps in scientific research and what is known and unknown in our understanding of the human brain.

The exhibition architecture involved a sequence of 14 sections supervised by the research centers. Each section explored specific research fields on NDDs through video presentations, scientific texts, visual materials, technological objects, and instruments.
The exhibition path ranged from pathophysiological mechanisms studied at the microscopic scale to the treatment of a single person. The various steps of a drug's journey were highlighted, starting with the "discovery" of therapeutic targets by basic scientific research, through the process of initial preclinical validation *in vitro* and *in vivo* models, to the various phases of clinical trials, and ending with the drug's entry in pharmacies.

Research Institutes

1 **Technology in the Study of Neurological Diseases: From a Single Cell to a Whole-Brain**
Karolinska Institutet, Stockholm, Sweden

2 **Single Cell Sequencing in Neurodegenerative Disorders**
Yale School of Medicine, New Haven, USA

3 **Role of Glial Cells in Neurodegenerative Disorders**
Montreal Neurological Institute-Hospital, McGill University, Montreal, Canada

4 **Modeling Alzheimer's Disease Using Human Brain Organoids**
Weizmann Institute of Science, Rehovot, Israel

5 **Parkinson's Disease: A Whole Body Disorder**
Juntendo University Hospital, Neurology Department, Tokyo, Japan

6 **Multiple Sclerosis: The Gut-Brain Connection**
Max Planck Institute of Neurobiology, Munich, Germany

7 **Remyelination Basic Studies and Imaging of Remyelination**
Hôpital de la Pitié-Salpêtrière, Sorbonne University AP-HP, Neurology Department and Paris Brain Institute, Paris, France

8 **Memory Driven Computing in Neurodegenerative Diseases**
German Center for Neurodegenerative Diseases (DZNE) within the Helmholtz Association, Bonn, Germany

9 **Neuroimaging of Dementia**
Tianjin Medical University General Hospital, Neurology Department, Tianjin, China

10 **Pet and Body Fluid Biomarkers in Neurodegenerative Diseases**
UniSR – Università Vita-Salute San Raffaele, Milan, Italy

11 **Neuroscape: Bridging the Gap Between Neuroscience and Technology**
UCSF Weill Institute for Neurosciences, University of California San Francisco, San Francisco, USA

12 **Neuromodulation to Treat Neurodegeneration**
UniSR – Università Vita-Salute San Raffaele, Milan, Italy

13 **Translation of Scientific Discovery to Patient Care**
Harvard Medical School, Brigham and Women's Hospital, Ann Romney Center for Neurologic Diseases, Boston, USA

14 **Translational Clinical Research**
University College London, London, UK

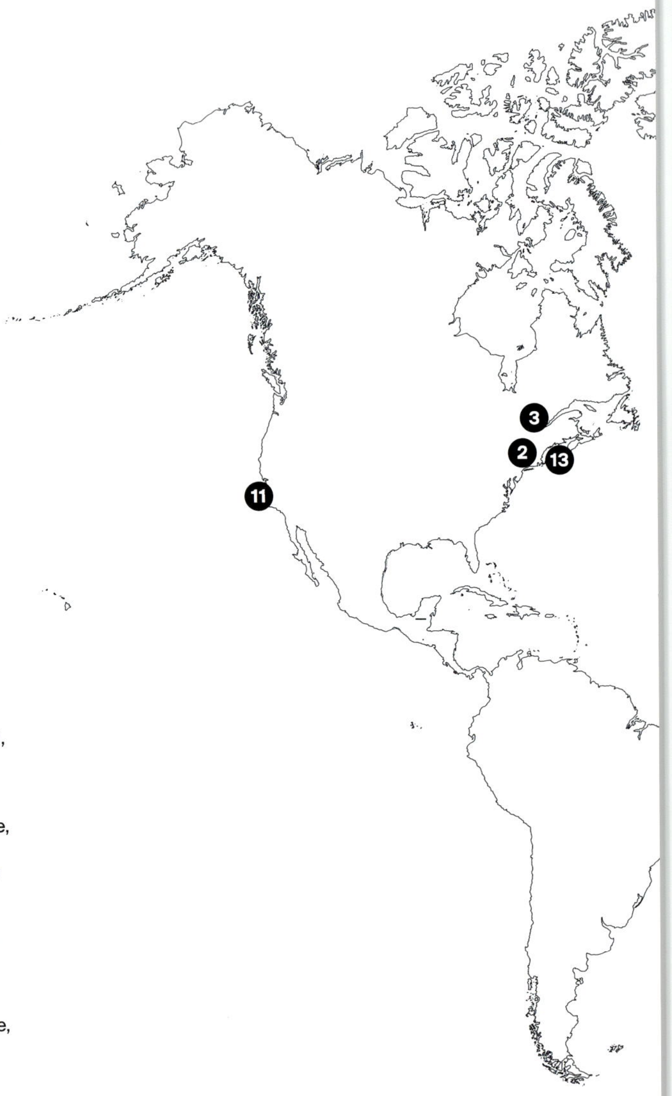

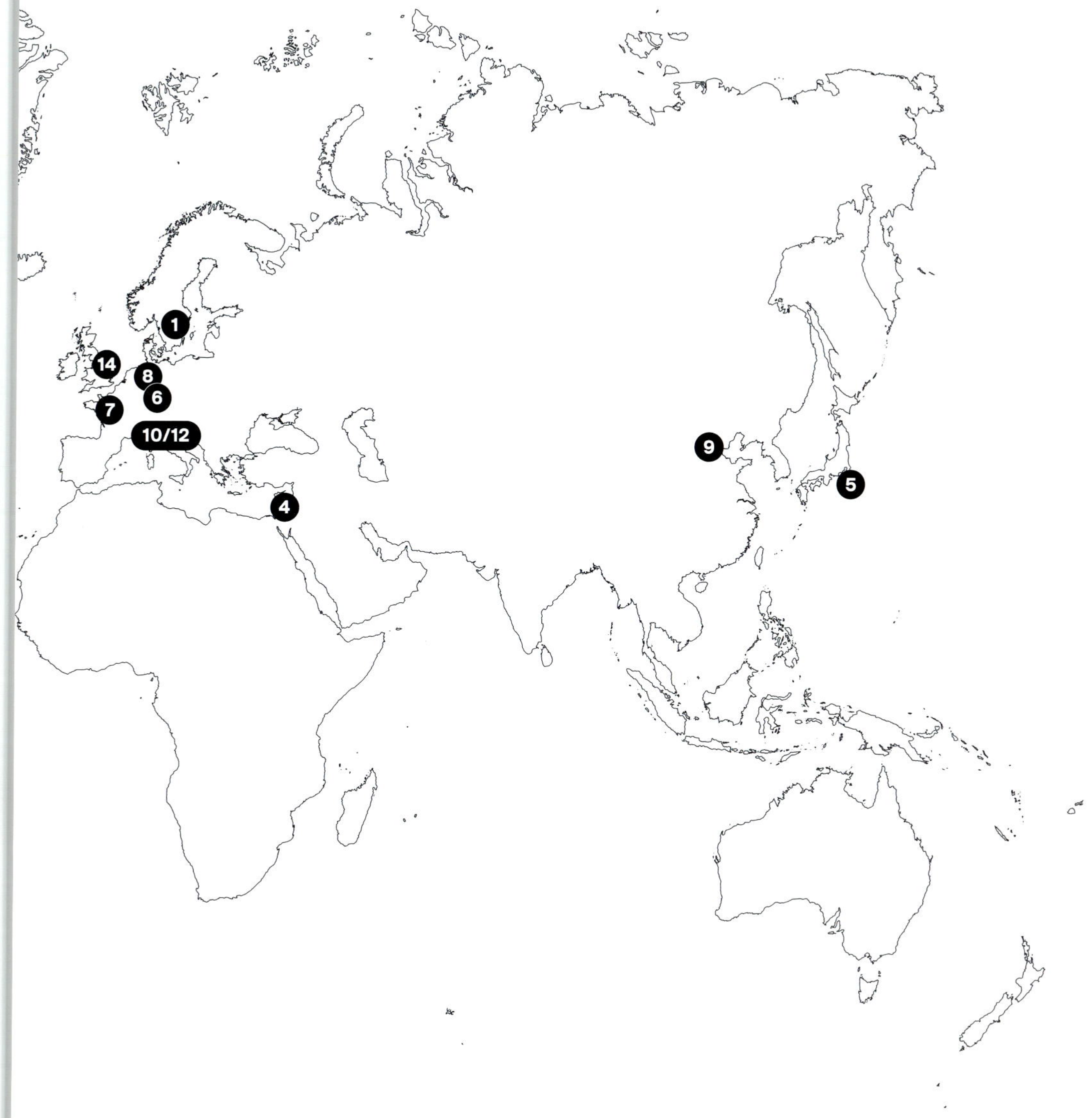
1
14
8
6
7
10/12
4
9
5

Technology in the Study of Neurological Diseases: From a Single Cell to a Whole-Brain

1

Karolinska Institutet
Stockholm, Sweden

The first observations regarding neurological pathologies date back to the mid-19th century by the father of what is today considered modern neurology, Jean-Martin Charcot (France, 1825). Back then no indication could offer the slightest hint to neurologists like Charcot about the future revolutionary development of technologies capable of unveiling the activation status of genes at single-cell level (single-cell sequencing, a technology pioneered at Karolinska Institutet), visualizing the location of a single molecule in single cells in tissues (RNA in situ hybridization) or the combination of computer-generated magnetic fields and radio waves to recreate images of organs inside the human body (magnetic resonance imaging).

The application of these technologies in neuroscience has spear-headed the understanding of the human brain in health and in disease. The images presented by Karolinska Institutet intend to guide visitors across a fascinating and eye-stimulating walk from a single cell to the complex brain through the lens of these revolutionary technologies.

1.1

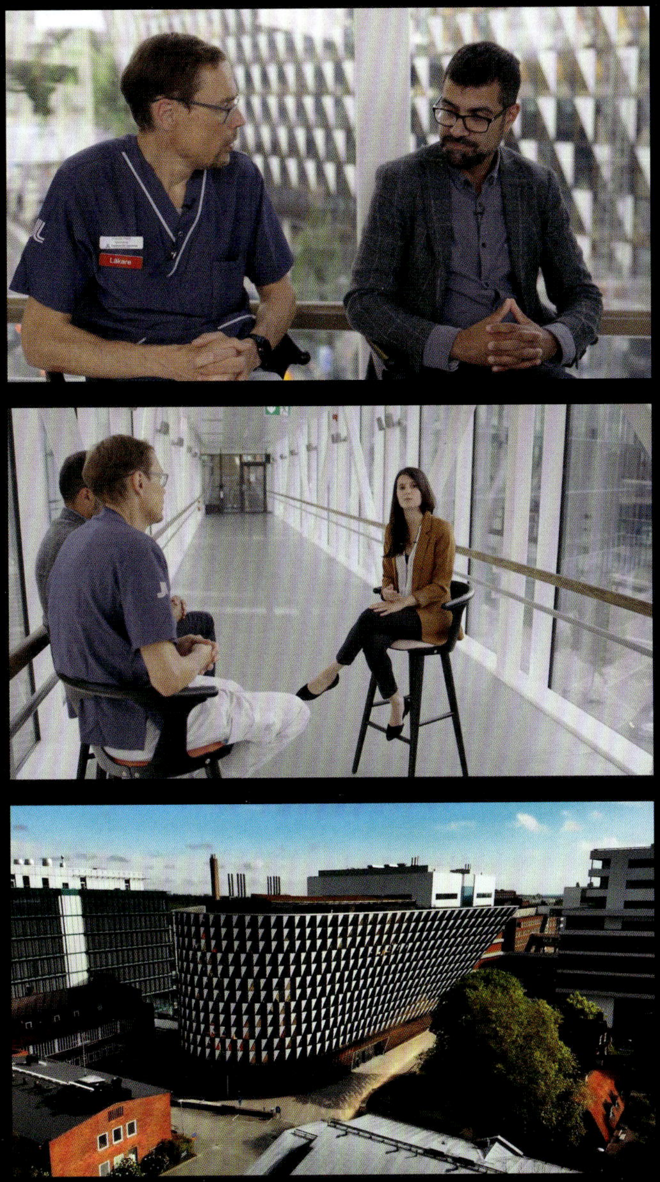

Stills from video featured in Room 1

1.2

Probing cell diversity with epigenomic information at a single-cell level in the mouse brain, 2021

Illustration used as cover for the scientific journal *Nature Biotechnology*, volume 39, issue no. 7, July 2021.

The image illustrates the decoding of genome-wide epigenomic information from thousands of individual cells from the mouse brain, using the technology called "Single-cell CUT&Tag" developed in the laboratory of Professor Gonçalo Castelo-Branco at Karolinska Institutet.

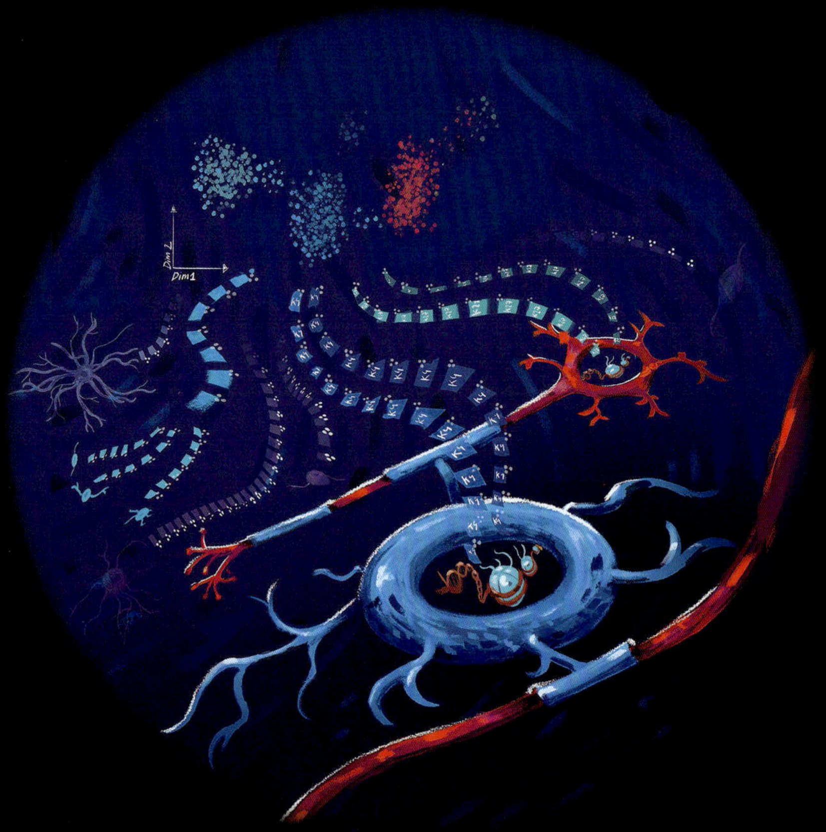

Amagoia Agirre, Marek Bartosovic,
and Gonçalo Castelo-Branco
Department of Medical Biochemistry &
Biophysics

Single Cell Sequencing in Neurodegenerative Disorders

2

Yale School of Medicine
New Haven, United States

Immune Cells in the Multiple Sclerosis Nervous System

Multiple Sclerosis (MS) is a genetically mediated autoimmune disease. Over the past half-century, we have gone from not having an understanding of the disease's underlying cause to a good working model that has been validated by clinical trials, such that early treatment can have a major effect in stopping disease flare-ups. Specifically, depleting B cells also has a dramatic effect on stopping attacks in early disease where the treatment is given to a patient every 6 months. In this regard, we are now engaged in clinical trials using B cell depletion at the very earliest stages of disease before there are any clinical manifestations. The genetic bases for MS have been elucidated and we have identified 233 common genetic variants, each with a small effect on disease risk, but that together lead to the disease. The majority of these common variants are controlling immune function and together contribute to a lower threshold for activating the immune system.

Single cell RNA sequencing is an important advance that allows us to deeply characterize the immune system without bias in order to investigate and interrogate each individual immune cell. Details of the technique are presented in the video. We extract spinal fluid from a patient using a lumbar puncture—which means inserting a needle between the vertebrae and collecting the fluid. Once extracted, the spinal fluid is rushed to the lab where it is spun down to collect the cells, and then brought to the 10x machine. We encapsulate each single T cell into a functionalized gel bead that is bar coded and mix them with enzymes and oils to create single cell microdroplets or GEMS. We then perform a chemical reaction to amplify the nucleic acid that codes for proteins that define each cell type to learn their function. The elegance of the technology is that we can identify cellular subtypes and rare cells with little bias giving a big picture of the biology underlying the disease.

This technique has elucidated new pathways underlying the cause of MS. There is still much to learn. While we have a good working model for early, relapsing remitting MS, we have little insight into the progressive phase of the disease. We also don't yet know whether very early treatment with B cell depletion prevents evolution to the progressive form of MS. Our examination of spinal fluid with these powerful new single cell technologies has revealed previously unknown pathways found in the infiltrating immune cells that are causing the disease.

2.1

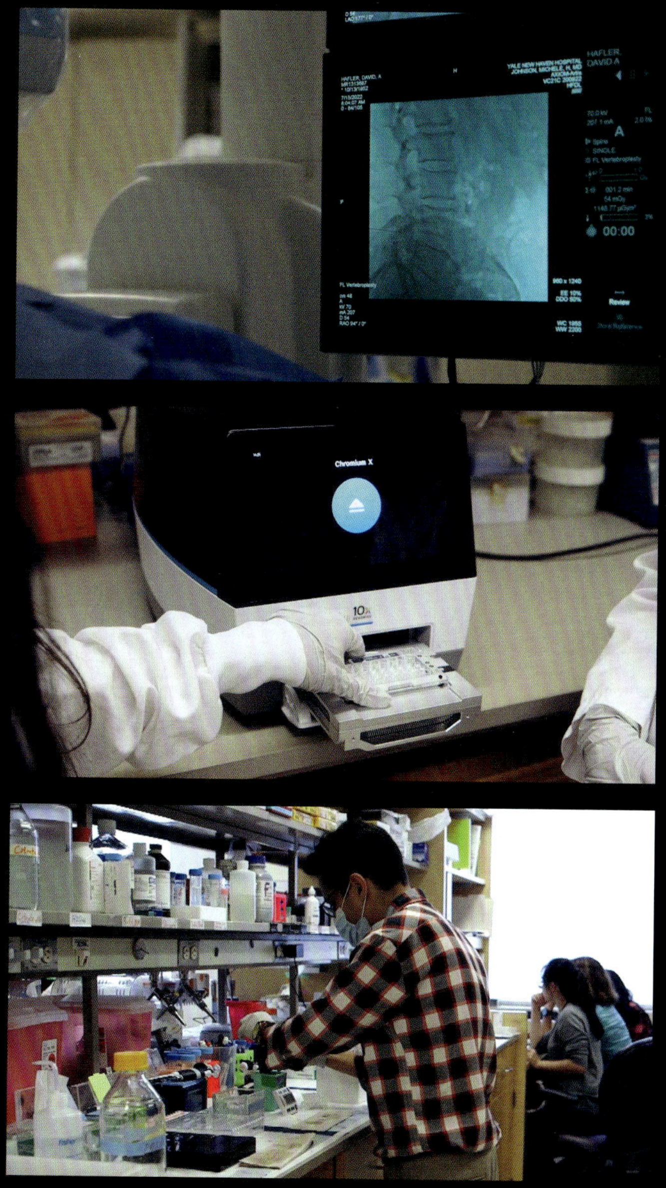

Stills from video featured in Room 2
Karyl Evans Productions LLC

2.2

Imaging captured with technology "4i to see Multiple Sclerosis brain"

Most of T cells are at the bottom right corner of this chronic active Multiple Sclerosis lesion.

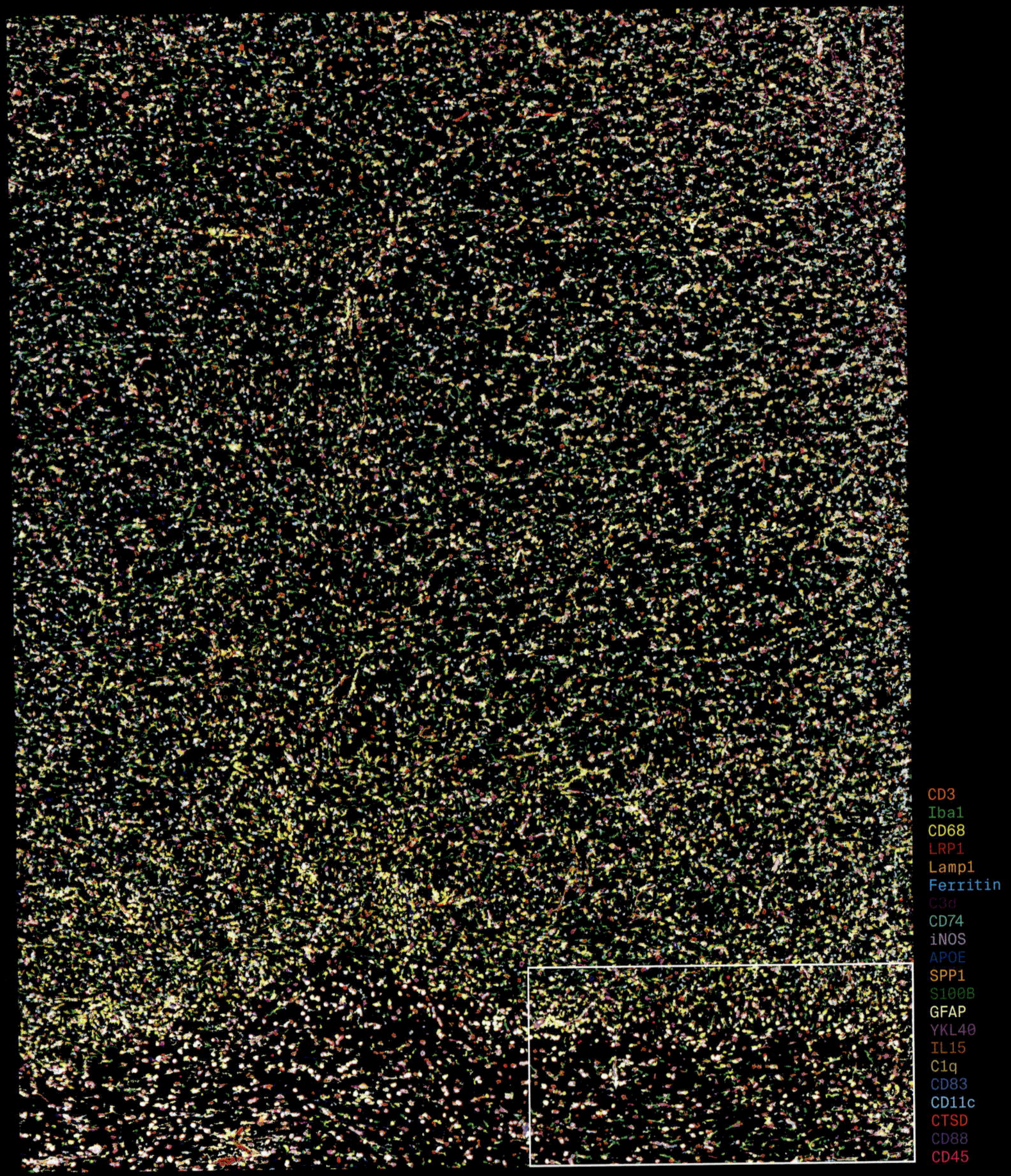

Drs. David Pitt and Tzu-Chieh Huang,
Department of Neurology,
Yale School of Medicine

2.3

A zoom-in image of a T cell. It is represented by an orange (CD3) cell structure with an aqua center (nucleus).

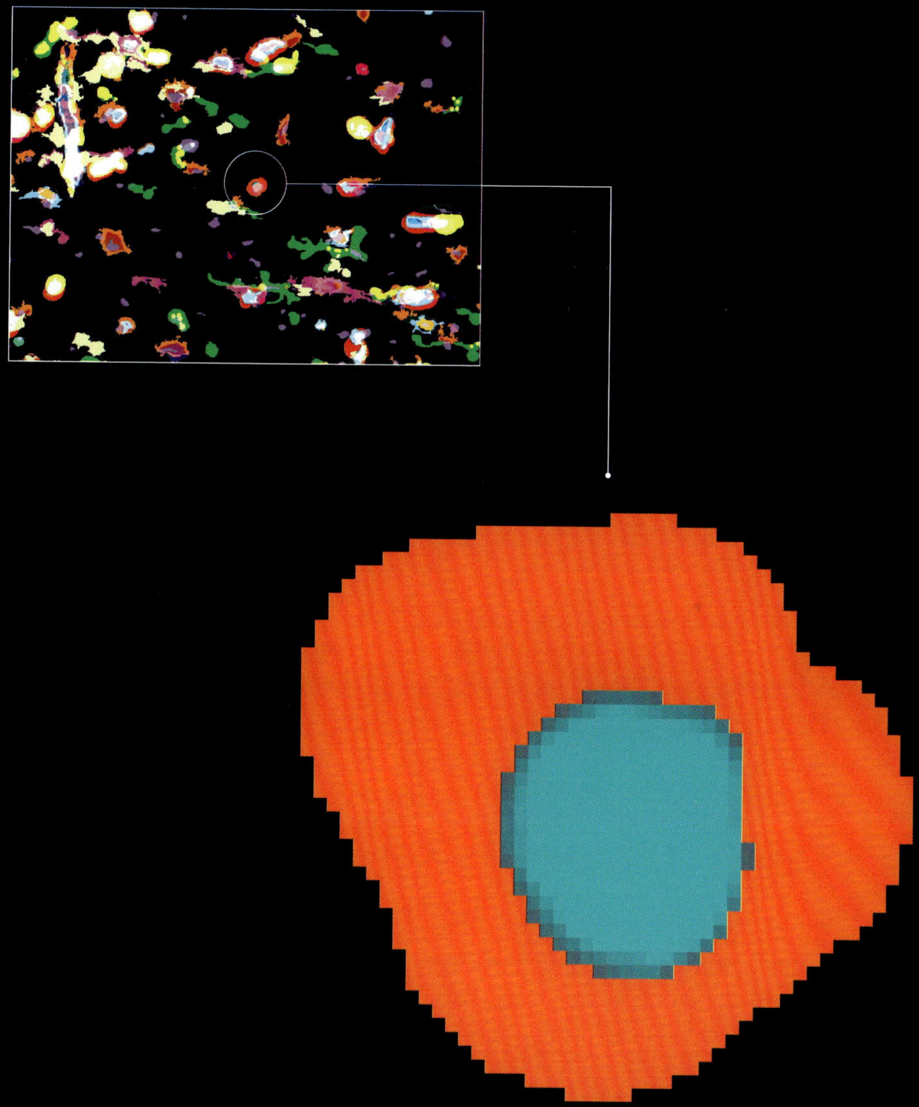

Drs. David Pitt and Tzu-Chieh Huang, Department of Neurology, Yale School of Medicine

Role of Glial Cells in Neurodegenerative Disorders

3

Montreal Neurological Institute-Hospital
McGill University
Montreal, Canada

Glial cells encompass the non-neuronal cells of the brain and spinal cord, deriving their name from the Greek word for "glue" based on the 19th-century concept of them comprising material that held the brain together. Pioneering histological work, driven by advances in techniques to stain brain tissues, showed that "glia" were comprised of distinct cells type that included astrocytes ("star cells"— Lenhossék 1893) and the "3rd element" oligodendrocytes and microglia (Del Hortega 1921; Penfield 1924). Subsequent advances in physiology–, tissue culture–, and molecular biology–based techniques have shown the functional properties of these cells under homeostatic and pathologic conditions.

Astrocytes and oligodendrocytes share a developmental lineage with neurons, all being derived from neural stem cells. Notable astrocyte functions include providing nutrients to neurons, maintaining integrity of the blood brain barrier by extending their foot processes to end on endothelial cells, regulating concentrations of neurotransmitters and electrolytes that impact on the function of neuronal synapses and axons, and providing clearance of material from the brain (glymphatics). Oligodendrocytes ensure efficient electrical conduction through the wrapping of their myelin membranes around axons and by providing trophic support to neurons.

Microglia are derived from the embryonic yolk sack, sharing features with bone marrow derived macrophages; they are the primary phagocytic cells of the brain required for clearance of dead cells and debris and for pruning of synapses. Both microglia and astrocytes are sources of molecules usually associated with the immune system, allowing them to regulate functional properties of infiltrating immune cells during the course of neuroinflammatory disorders and to directly mediate tissue injury or conversely contribute to repair. These functions are dependent on their state of activation with the latter determined by their interaction with their microenvironment and by molecules reaching the brain from the systemic compartment (e.g. from the gut microbiome). An important communication network exists between the microglia and astrocytes.

Further included amongst glial cells are NG2 progenitor cells found within the brain parenchyma and ependymal cell that line the ventricles. NG2 glial cells are recognized as progenitors for myelinating oligodendrocytes and for their interactions with neuronal synapses. Ependymal cells contribute to the integrity of the brain/cerebrospinal fluid barrier.

Understanding the dynamic properties of all these glial populations under physiologic and pathologic conditions provides the opportunity to target these for therapeutic purposes aimed at neuroprotection and repair. Genome wide screening association studies (GWAS) have identified genes linked to glial cells as underlying or contributing to a wide range of "neurodegenerative disorders." The advent of techniques to derive each of these cell types from inducible pluripotent stem cells (ipscs) generated from blood cells or fibroblasts from such individuals and "controls" can be used to assess effects of modulating expression of disease-relevant genes and response to candidate therapeutic agents.

3.1

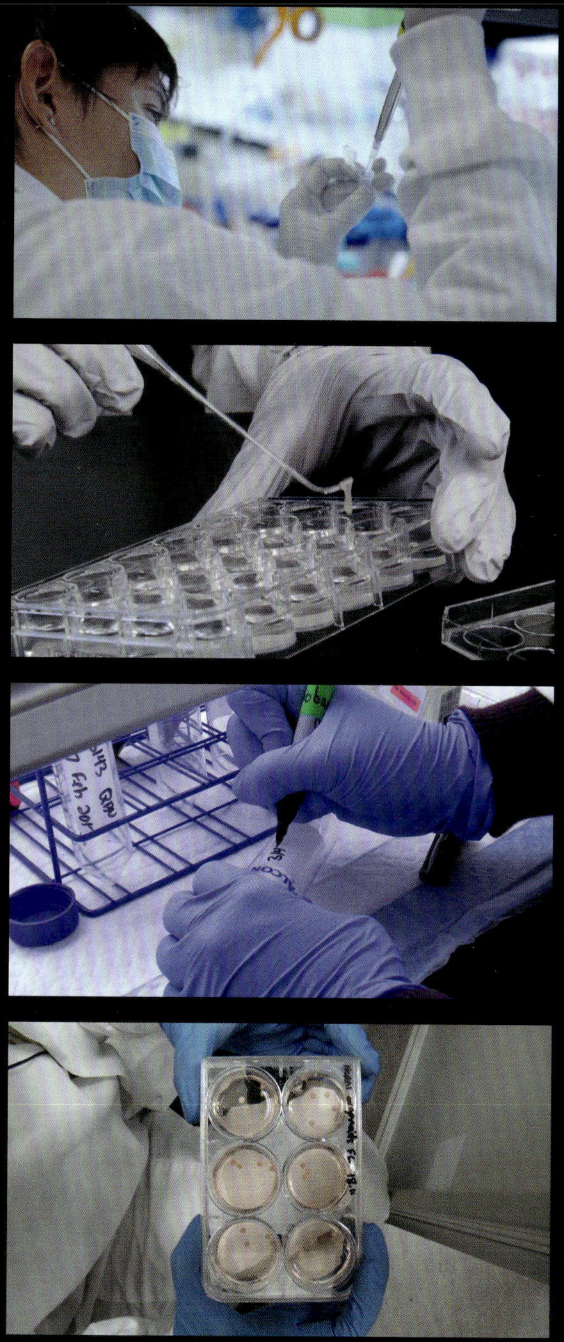

Stills from video
featured in Room 3

3.2

Illustration of a cluster of oligodendroglia in the cerebellar white matter.

The oligodendroglial vine-like, ramified prolongations run parallel to nerves fibers, which are not seen with this method of staining. The stellar-shaped cell labeled M on the right is an astrocyte. The large tubular structure at the bottom third of the illustration is a blood vessel.

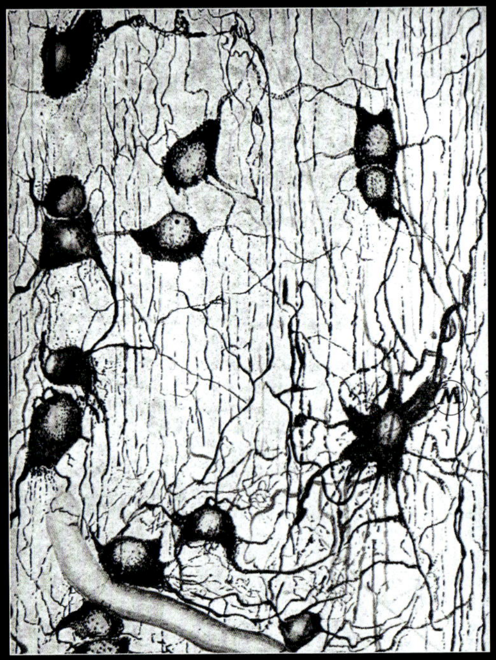

Cross section of circular-appearing nerve fibers in the cerebellar white matter demonstrating three oligo-dendroglial cells.

Their fine projections encircle the myelin sheaths, which are unstained by this method. The encircling oligodendroglial projections are best seen as the very dark string-like projections at the right lower left of the image, originating from the lowest oligodendroglial cell.

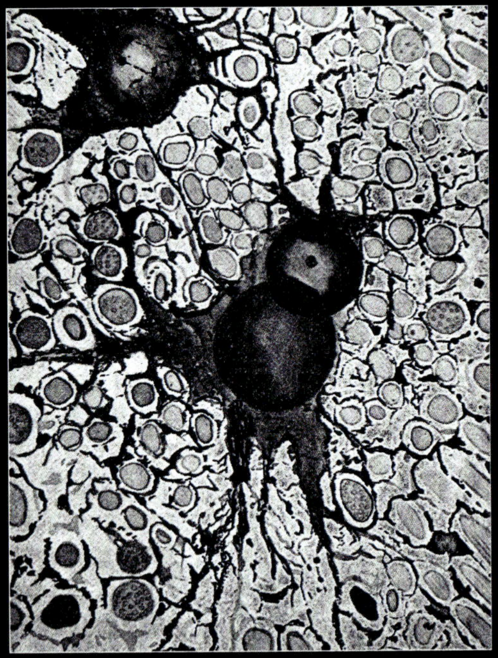

Wilder Penfield 1925,
Montreal Neurological Institute.

Microglial satellite cells around neurons of the cerebral cortex, showing their slender elongated bodies and thick projections, characteristics that Wilder Penfield relied upon to distinguish them from oligodendroglia.

Unlike oligodendroglia, whose role is to generate the myelin sheath that insulates nerve fibers, the microglia's role is to remove cerebral debris by a process of phagocytosis.

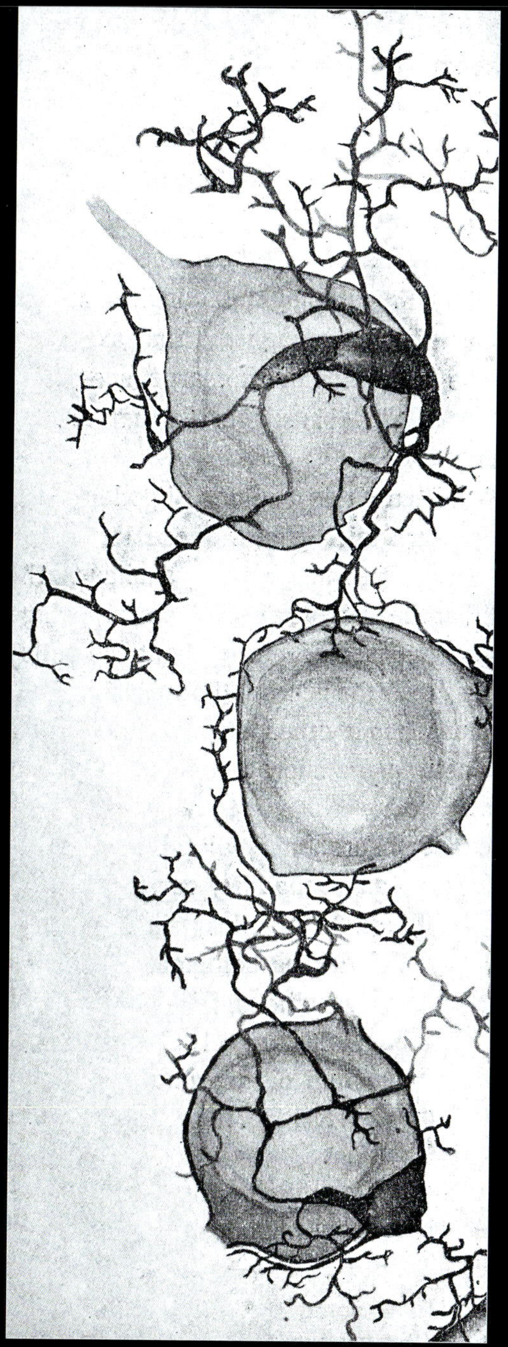

Wilder Penfield 1925.
Montreal Neurological Institute.

3.3

Human oligodendrocyte cultures

Human brain-derived oligodendrocytes isolated from surgically resected adult human brain tissue sample; grown in dissociated cell culture and immuno-stained with O4 antibody.

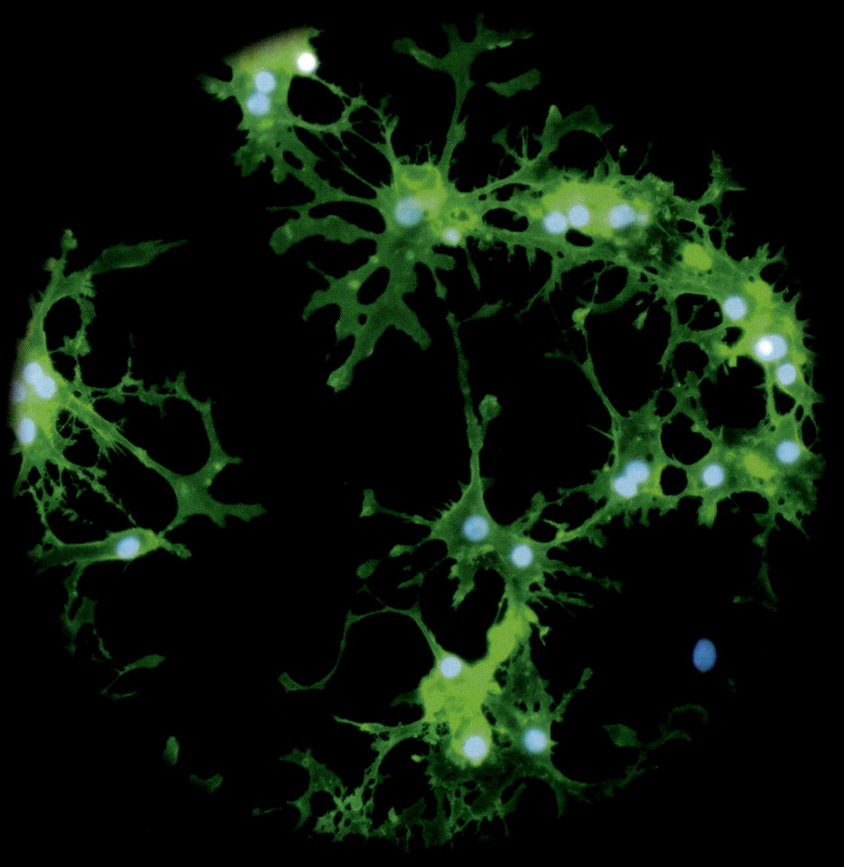

3.4

Oligodendrocytes growing on nanofibers

Human adult brain-derived oligodendrocytes (OLs) ensheathing synthetic nanofibers *in vitro*. Process extension and fiber ensheathment fulfills historical prediction of OLs as the myelin-producing cells and provides test bed to assess myelin repair promoting therapeutic agents.

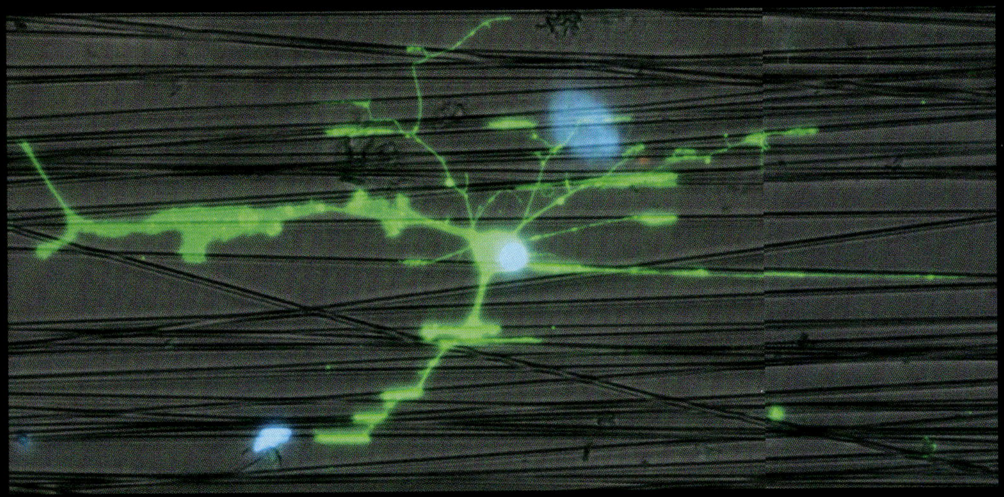

3.5

Nanofiber

Understanding the mechanisms by which oligodendrocytes produce myelin is crucial for the development of therapies to promote remyelination. Researchers at the Montreal Neurological Institute have adopted a novel method to study these processes by culturing these cells on microfabricated plastic fibers, which they wrap up with myelin just as they would wrap axons. Using high resolution microscopy and these "nanofiber cultures" we can visualize how oligodendrocytes grow, develop, and make myelin, and test candidate drugs to promote oligodendrocyte survival and myelin production. Working with experts in computer vision, streamlined processes have been developed to automatically image and analyze these cultures, expediting research into myelin biology and diseases.

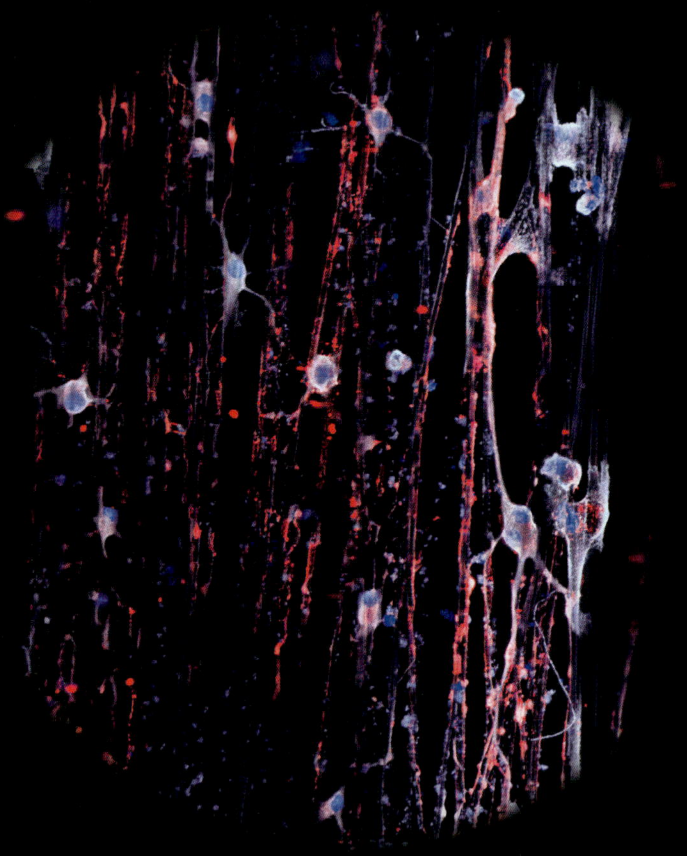

3.6

Dying oligodendrocyte

In many neurodegenerative diseases, pathological inflammation in the brain negatively affects glial cells, in some cases doing irreversible damage that prevents the cells from recovering. Understanding how cells like oligodendrocytes react to stress and then attempting to prevent or even reverse these cellular injuries is a key focus for developing therapies. Researchers at the Montreal Neurological Institute are studying these processes by taking high-resolution images and videos that reveal how glial cells react when stressed. This image shows a mature oligodendrocyte beginning to break up into "blebs," illustrating a process in which the cell takes itself apart following an injury.

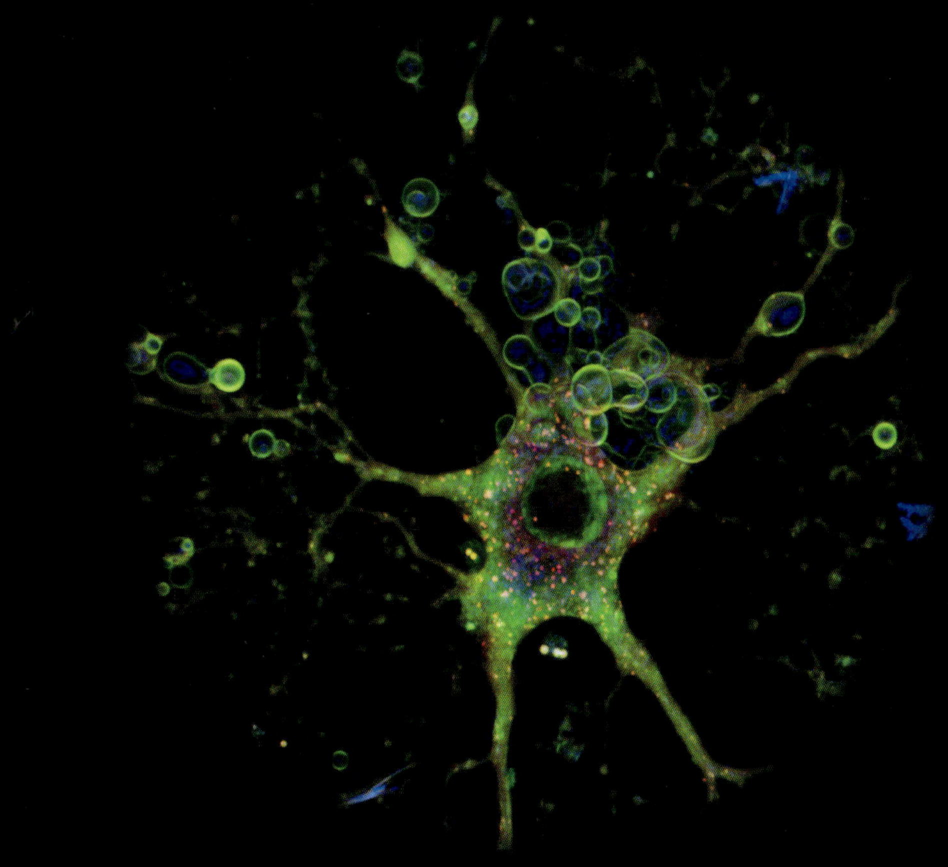

3.7

Human brain-derived astrocytes grown in dissociated cell culture and immuno-stained with GFAP antibody.

Activation state of astrocytes links with their expression of molecules that contribute to maintaining the optimal state of brain function under physiologic conditions and either take part in and/or protect from tissue injury under conditions of disease or injury.

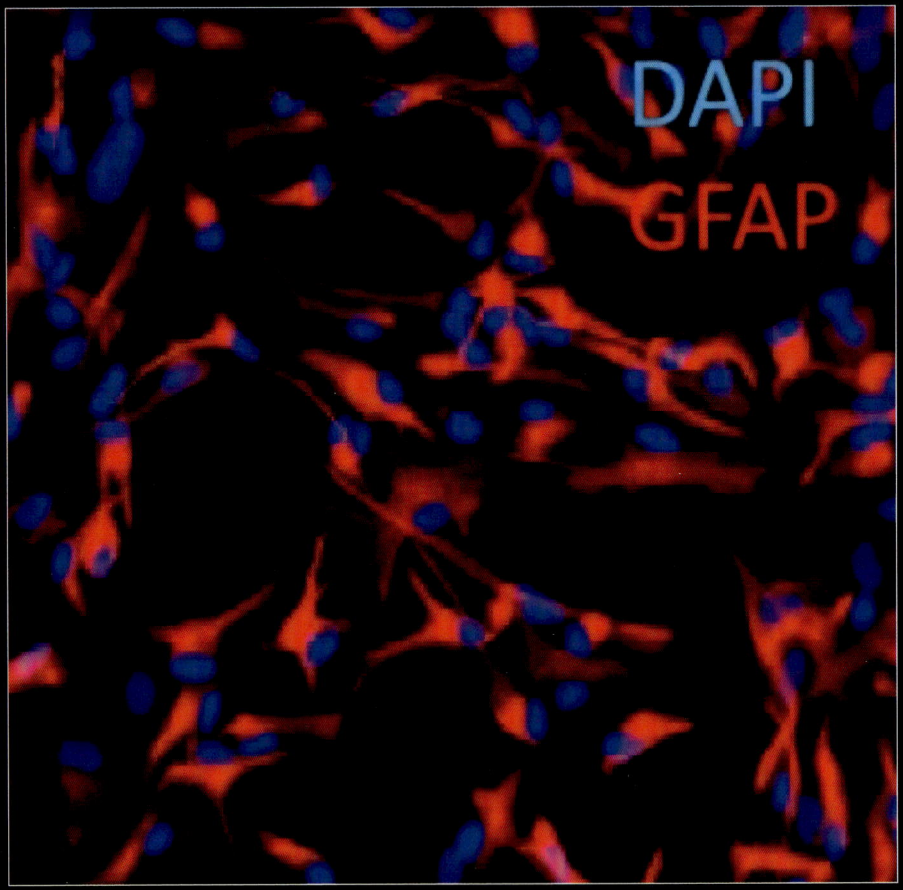

Kieran NW, J Neuroinflammation. Jan 6, 2022;19(1):10

3.8

Human adult brain-derived microglia grown in dissociated cell culture. Insert shows uptake (phagocytosis) of myelin (labeled in green) by such cells. Clearance of damaged tissue from disease or trauma creates conditions that would promote subsequent tissue repair.

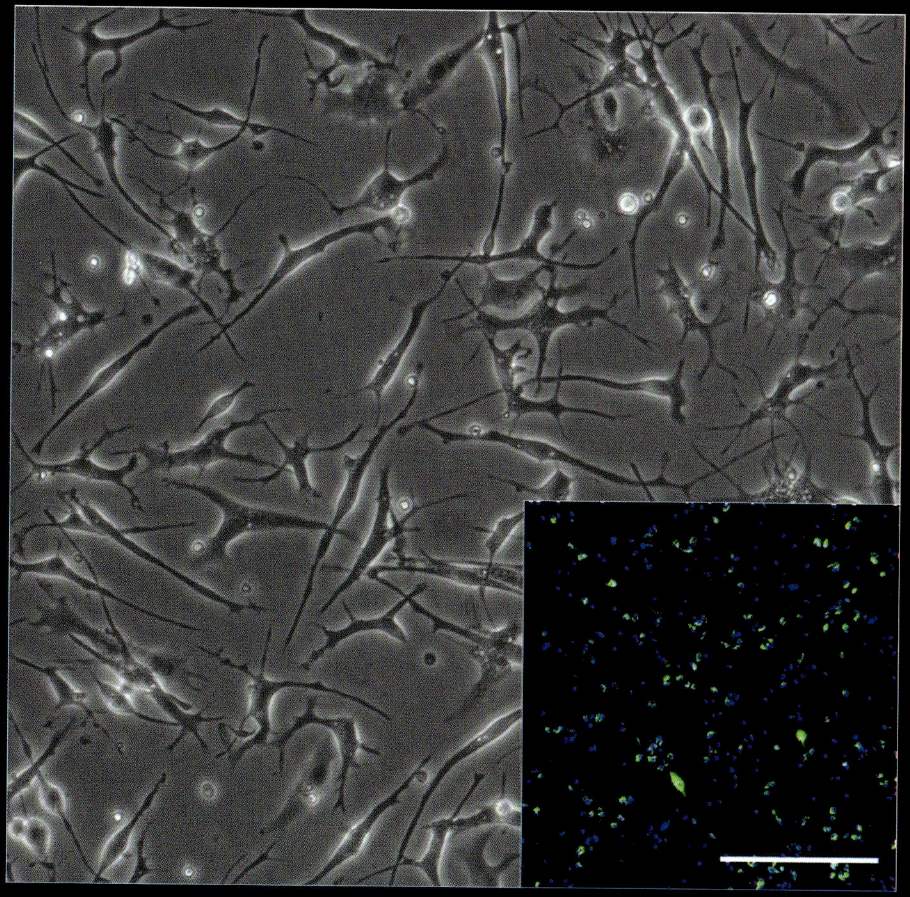

3.9

Human oligodendrocyte lineage cells derived from stem cell source (iPSC cell line). IPSC technology allows deriving specific glial cell types from patients with genetic and acquired neurologic disorders for studies of underlying disease mechanisms.

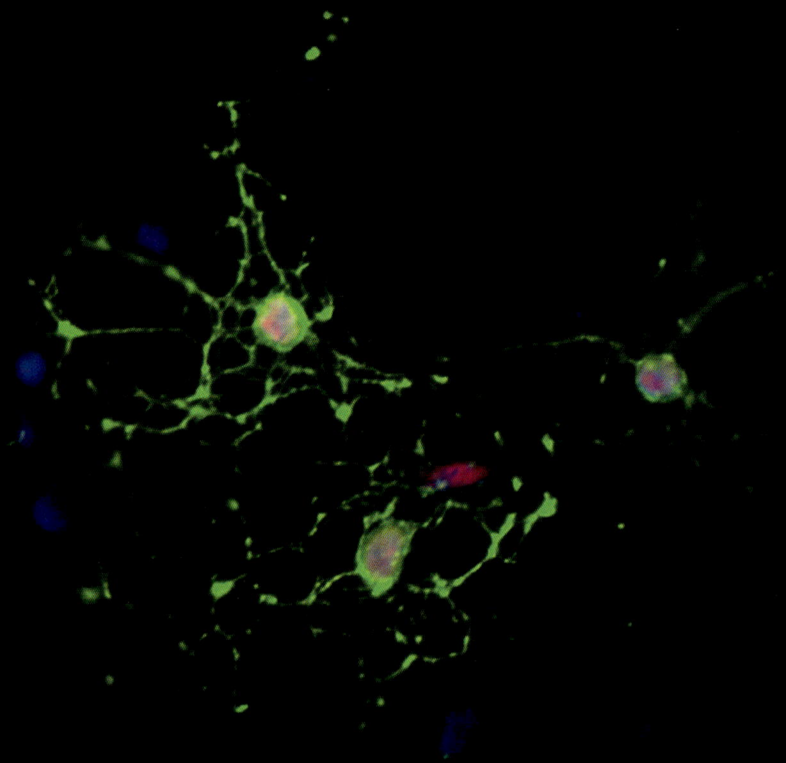

Courtesy of Experimental Drug Development Unit (EDDU) at The Montreal Neurological Institute

Modeling Alzheimer's Disease Using Human Brain Organoids

4

Weizmann Institute of Science
Rehovot, Israel

Alzheimer's Disease (AD) is a heterogeneous neurodegenerative disorder that results in the most common form of dementia. We generated mutations in presenilin 1 (PSEN1) and tau (MAPT) using genome editing. PSEN1 and MAPT are genes that are mutated in cases of familial AD (fAD) and frontal-temporal dementia. Multiple brain regions are impacted in AD, including the hippocampus. Neuropathology and dysfunction in long-term potentiation have been described in the hippocampus and are associated with cognitive deficits. Furthermore, there are accumulating data indicating a strong involvement of the immune system, and in particular the immune resident cells in the brain, the microglia. In addition, over the last few years a strong correlation between metabolic changes and neurodegeneration has been suggested; however, this has yet to be thoroughly investigated.

To better understand and identify early changes in fAD at the cell biology level we used both region specific and complex human brain organoids. Pioneering research from the lab of the late Yoshiki Sasai demonstrated that stem cells can recapitulate several features of organogenesis, including cell differentiation, spatial patterning, and morphogenesis, and has successfully generated organoids resembling different brain regions and retina. Subsequent research from Lancaster and Knoblich demonstrated that it is possible to obtain a mixed regional identity using a relatively simple media. This field has increased dramatically over the last few years with multiple protocols. Scientists can now generate connections between different brain regions by fusing structures known as "assembloids," thus mimicking a higher organization level, which may prove critical in modeling diseases. There have also been advances in regards to generating functional networks. Multiple studies have characterized the cell repertoire and diversity in organoids obtained from different protocols using single-cell (sc) analyses. Nevertheless, the understanding of brain organoid proteomics and metabolomics is still rudimentary. Our project aims to generate diverse brain organoid models that will include the hippocampus, in which microglia, the innate immune cells of the brain, will be included. The presence of immune cells is of great importance since aberrant inflammation in the CNS has been implicated as a major player in the pathogenesis of human neurodegenerative disease. We will further harness advances in genome editing and mutate PSEN1 and MAPT leading to the generation of a series of isogenic pluripotent cell lines. The mitochondria will be tagged with dual purpose tags that will enable imaging and biochemical purification in a cell-type specific manner. During recent years there have been important advances that enable biochemical purification of mitochondria for proteomics and metabolomics analyses.

Our study focuses on characterizing the pathophysiology of early stages of genetic models of fAD and on mitochondria using hippocampal brain organoids combined with microglia and analysis using live-imaging, mitochondria health parameters, proteomics, metabolomics, and transcriptomics. These studies have the potential to reveal early biomarkers and identify possible therapeutic targets.

4.1

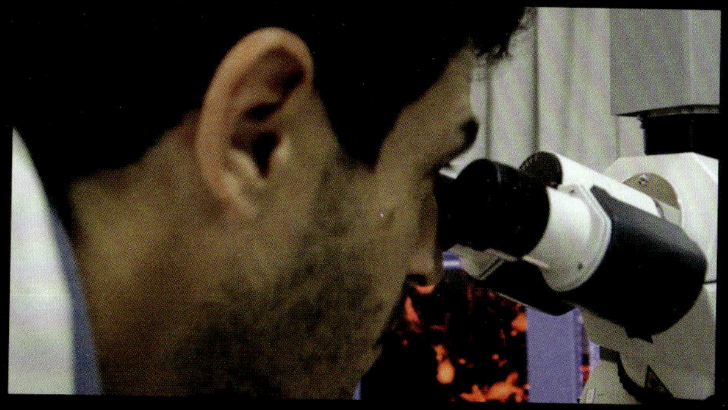

Time-lapse of an on-chip brain organoid
Human embryonic stem cells were labeled with a fluorescent red nuclear protein and a green fluorescent protein. Human on-chip brain organoids were generated by Karzburn et al. (*Nature Physics* 2018, 14: 515–522), and were subjected to time-lapse microscopy.

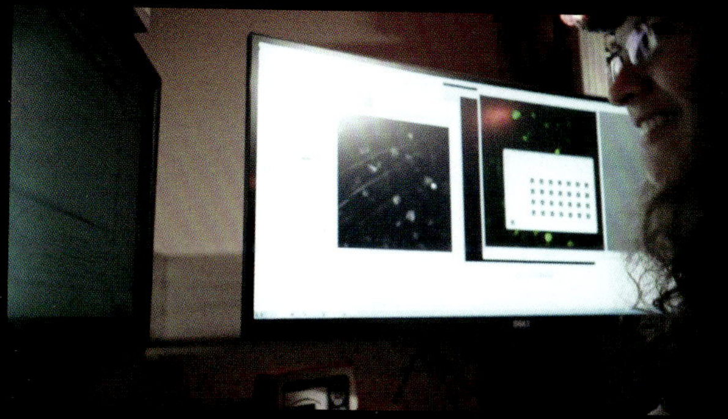

Mitochondria and lysosome dynamics in human embryonic stem cells
Time-lapse of mitochondria (red) and lysosomes (green) labeled with a fluorescent reporter (mitochondrial and lysosomal proteins tagged with red and green fluorescent proteins, respectively). The videos show a high level of motility and interaction between mitochondria and lysosomes.

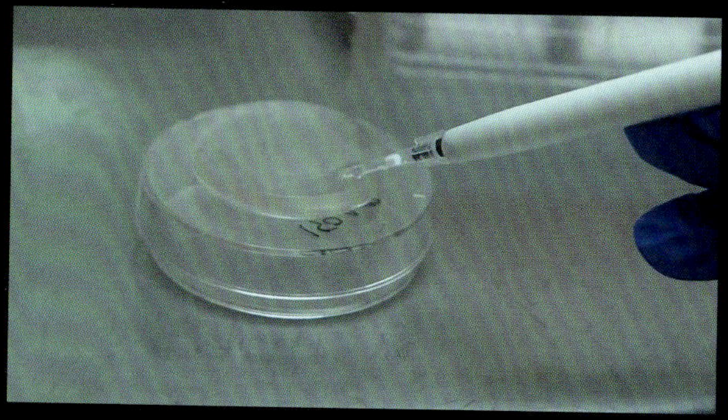

Microglia surveillance in a human hippocampal organoid model
Time-lapse of microglia cells in sliced human hippocampal organoid (MORGA). Microglia cells extend and retract their processes to constantly survey the hippocampal brain organoid.

Stills from video featured in Room 4

4.2

Schematic of sliced human hippocampal organoids integrated with microglia (MORGA)

To create this new hippocampal organoid model integrated with microglia, we applied, in parallel, two different panels of specific factors to stem cells, generating hippocampal organoids (bottom) and microglia progenitors (top).

We then transplanted microglia progenitors into the hippocampal organoid. To avoid cell death in the core of the organoids, we sliced organoids and grow them on an air-liquid interphase membrane.

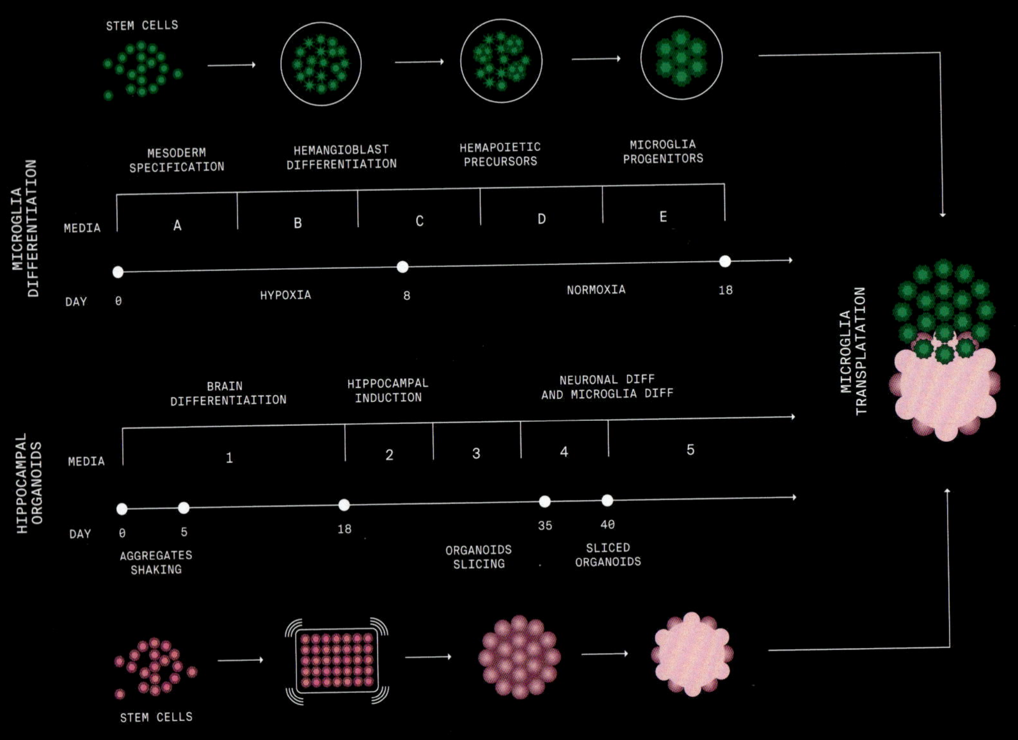

Orly Reiner's lab, Weizmann Institute of Science

4.3

From stem cells to microglia progenitors

Microglia progenitors arise as round-shaped cells from the basement of cells composed mainly of hemangioblast colonies. These floating cells show the expression of microglia and hematopoietic marker (as shown in the dot plot graph).

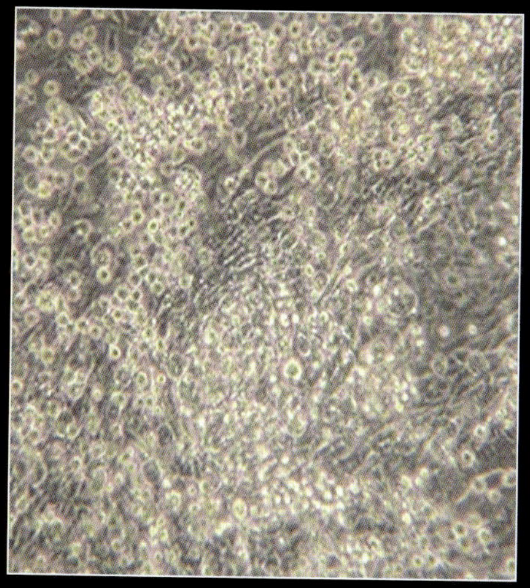

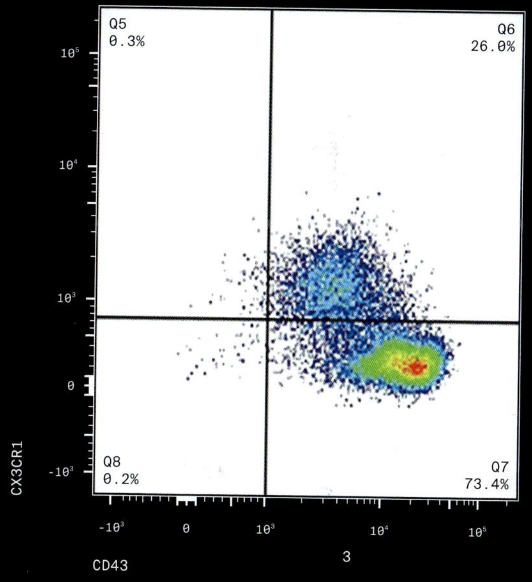

Orly Reiner's lab, Weizmann Institute of Science

4.4

Live imaging of microglia cells integrated into the human MORGA

3D reconstruction of transplanted human microglia cells (Lyn-GFP) integrated into sliced human hippocampal organoid (MORGA). Microglia cells are highly branched and ramified cells, typical of human adult microglia. The new protocol developed by Weizmann Institute of Science allows the stable integration of microglia for several months.

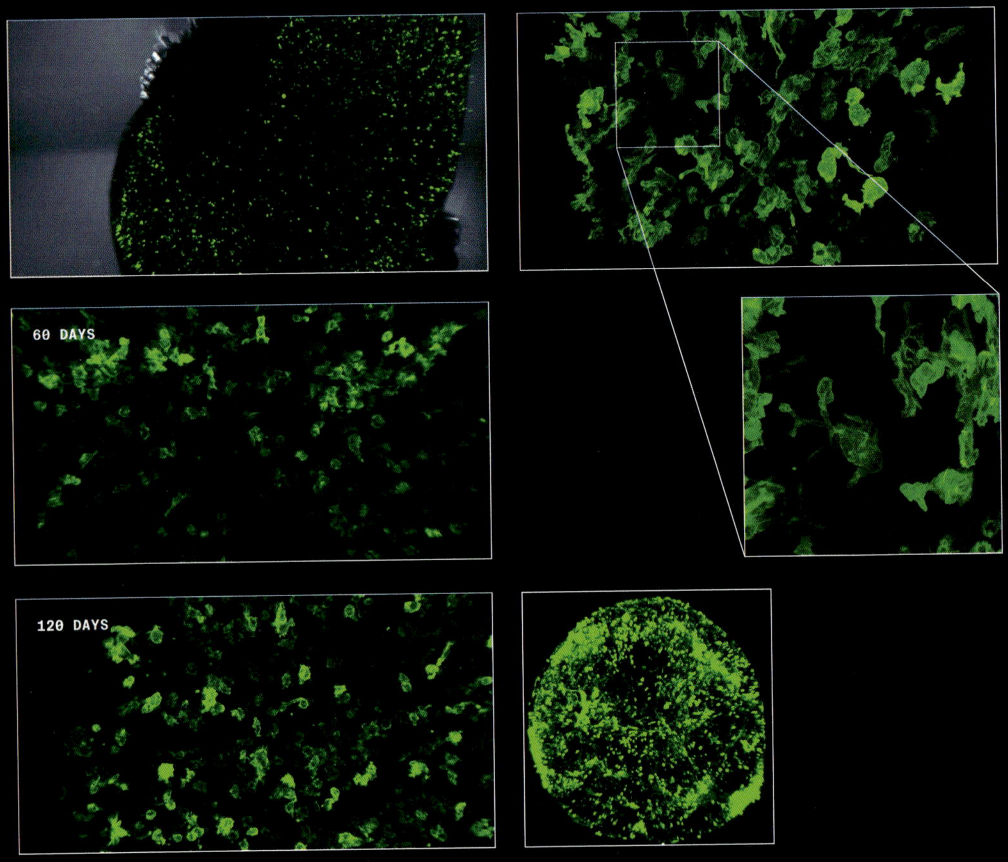

Control hippocampal organoid

Orly Reiner's lab, Weizmann Institute of Science

4.5

Advancements of the new developed hippocampal organoid model (MORGA)

Organoid slices on an air-liquid interphase membrane (A). There is no cell death in the core of sliced organoids (SO) compared to whole organoid (WO) as shown by apoptotic marker (green, cleaved Caspase 3, section B), and SOs show higher connectivity between areas as demonstrated by axonal marker (red, SMI312, section C). This model comprises several cellular populations, including neuronal progenitors (in particular, Zbtb20 positive cells are hippocampal-specific neuronal progenitors; section E), astroglial cells (GFAP, section C), post-mitotic neurons (Neun, section B), and a high degree of maturation shown by axonal (ANKG, as action initial segment marker; section E) and postsynaptic marker (PSD95, a marker of glutamatergic synapses; section F).

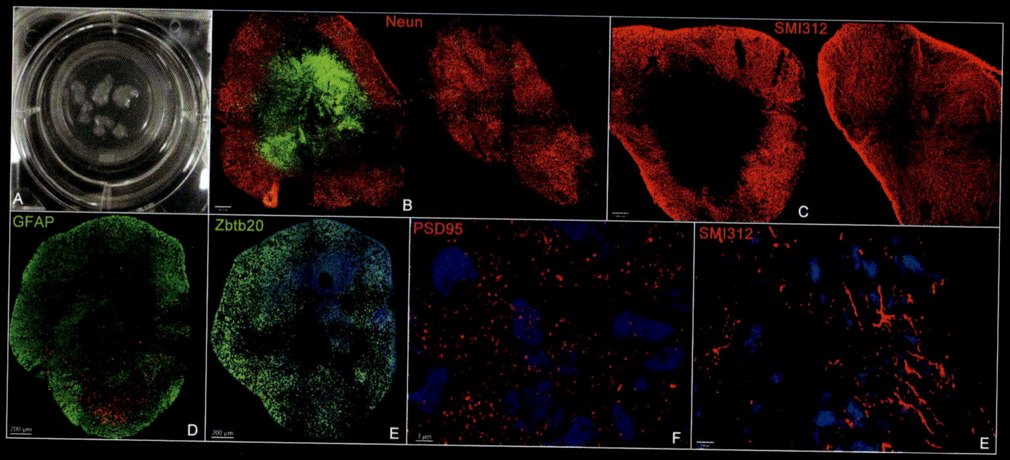

Orly Reiner's lab, Weizmann Institute of Science

4.6

Immunostaining of a hippocampal organoid

Immunohistochemistry of a hippocampal organoid at Day 160 for Neun (in red) and GFAP (in green). We can see a localized group of post-mitotic neurons within the entire organoid. A highly-folded layer of cells also suggests a choroid plexus-like structure.

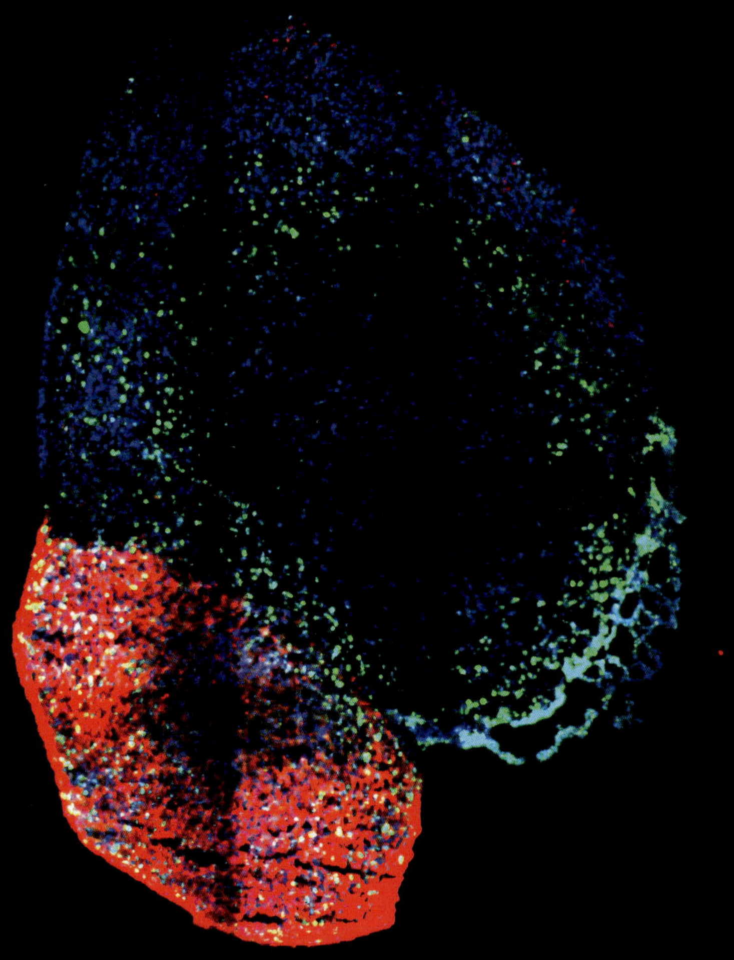

Orly Reiner's lab, Weizmann Institute of Science

Immunohistochemistry of a hippocampal organoid at Day 160 for pTAU (in red) and LEF1 (in green). We can see a localized expression of phosphorylated Tau protein. This region also has some LEF1+ cells.

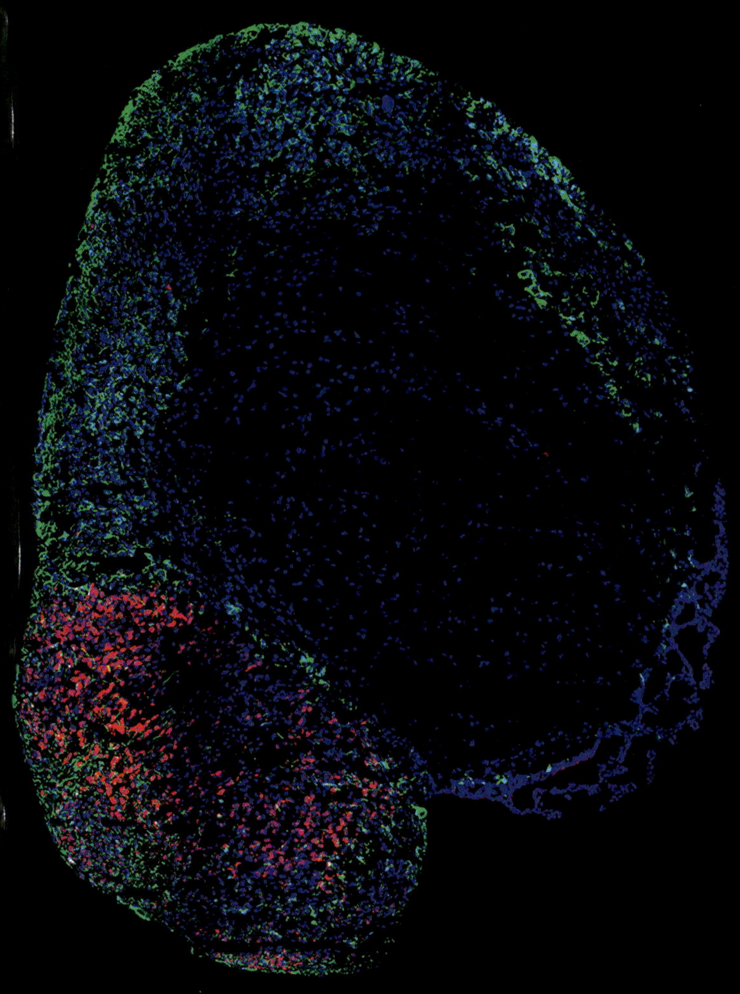

Orly Reiner's lab, Weizmann Institute of Science

Parkinson's Disease: A Whole Body Disorder

5

Juntendo University Hospital
Neurology Department
Tokyo, Japan

Parkinson's Disease (PD) is the most common movement disorder. The number of patients is around 15 million in the world. The incidence increases with age, suggesting that PD will continue to increase in the aging society.

PD has been clinically characterized by four main motor symptoms—rest tremor, akinesia, postural imbalance, and rigidity—which are mainly caused by the loss of dopaminergic neurons in substantia nigra. During the disease, patients also show a broad range of non-motor symptoms including constipation, anxiety, insomnia, impulse control disorder, and dementia that reflect pathological changes spreading in the entire body. The pathological hallmark of PD is alpha-synuclein-containing neuronal aggregates named Lewy bodies and Lewy neurites.

Currently, no treatment can achieve complete remission, however, a considerable number of medications that target dopaminergic dysfunction are available to attenuate the symptoms.

5.1

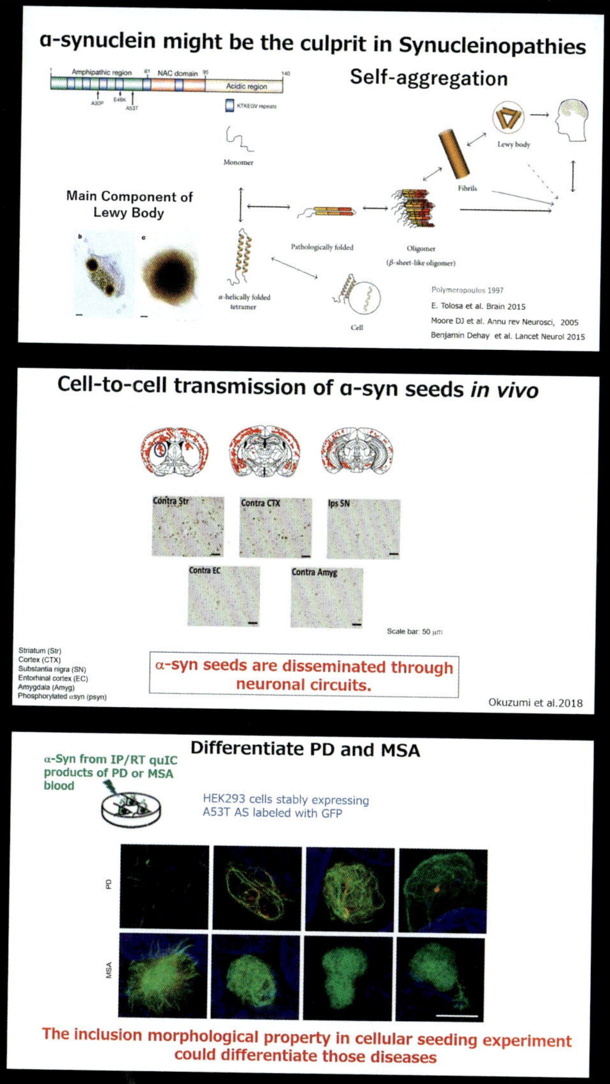

Still from the video featured in Room 5

5.2

Parkinson's Disease may be the "systemic disorder"

Skin

The study of skin in Parkinson's Disease revealed patho-physiological mechanisms.

Previous reports showed the role of alpha-synuclein in peripheral tissues, including the skin, and sebum RNA in Parkinson's Disease.

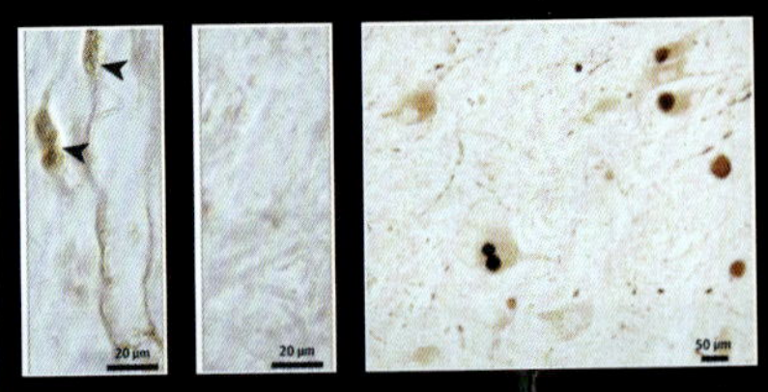

PD skin tissue CTRL skin tissue PD brain tissue

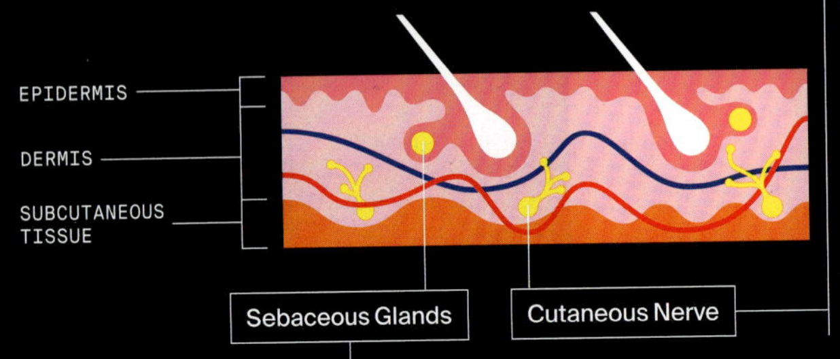

Phosphorylated alpha-synuclein depositions are detected in cutaneous autonomic nerve of Parkinson's Disease.

Sebum RNA profile in Parkinson's Disease

Changes in sebum RNA expression associated with mitochondrial function

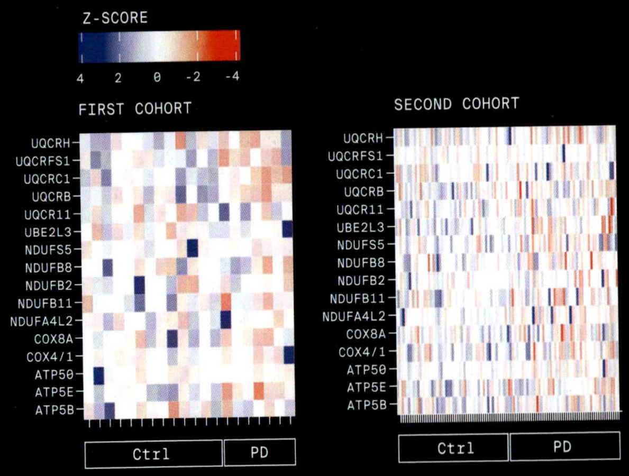

Discrimination of Parkinson's Disease using sebum RNA information and machine learning

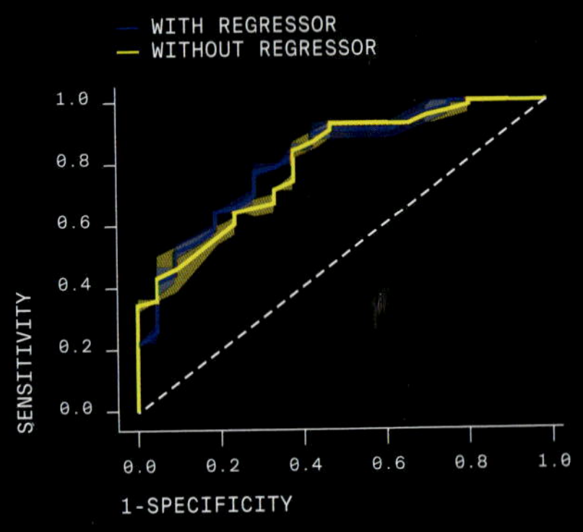

The area under the curve representing the discrimination performance was 0.793 when sebum RNA, age, and gender information were used, and by adding the severity prediction results to these, even higher performance was obtained at 0.806.

Uehara Y., Ueno S.I., Saiki S., Hattori N. et al., Sci Rep 2021; 11:18550

Gut

Gut-Brain Axis is an epoch-making concept of pathogenesis of Parkinson's Disease.

This concept may provide abundant new biomarker and therapeutic target.

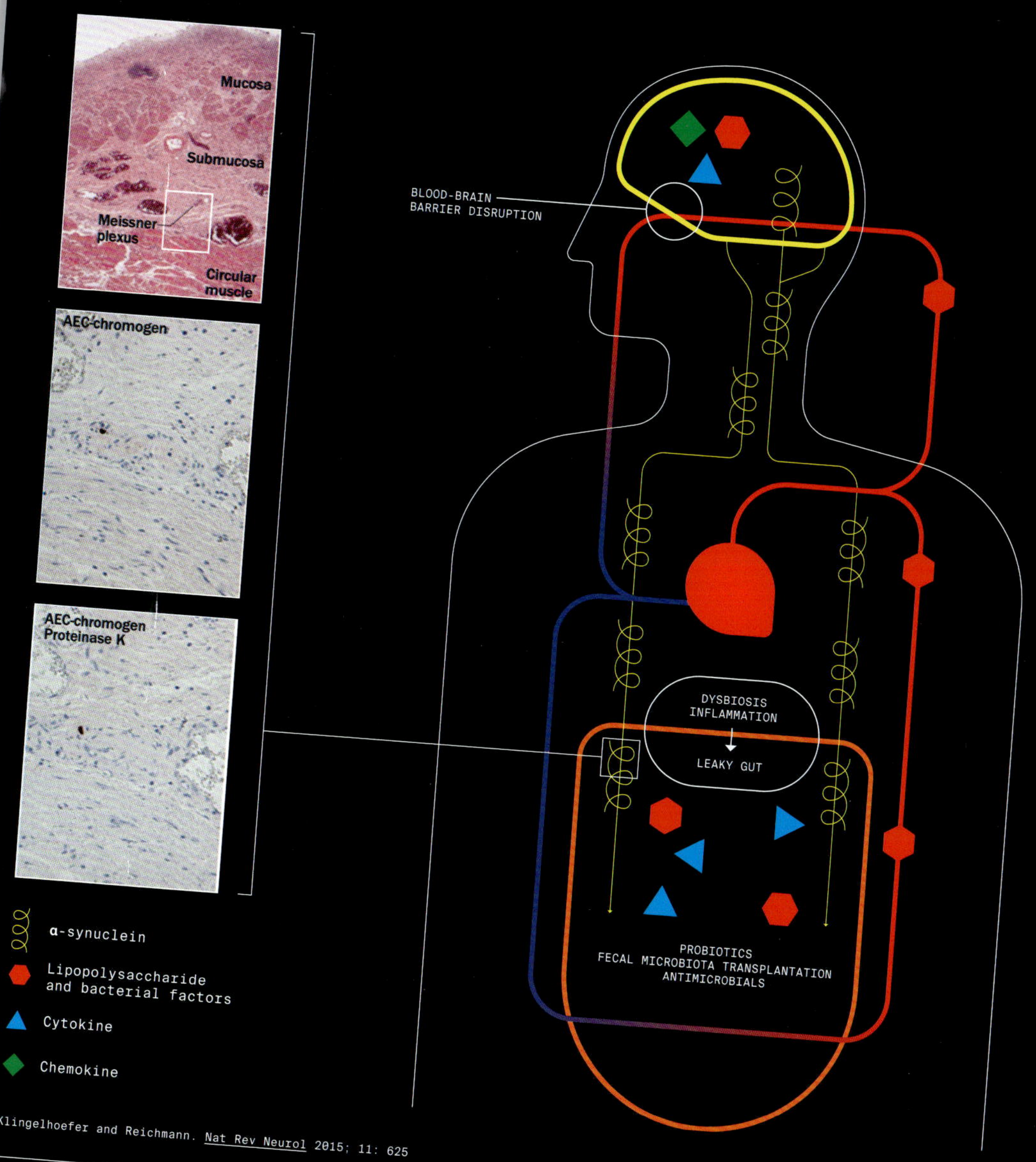

Klingelhoefer and Reichmann. Nat Rev Neurol 2015; 11: 625

5.3

Biomarkers using biological specimens in Parkinson's Disease

Summary

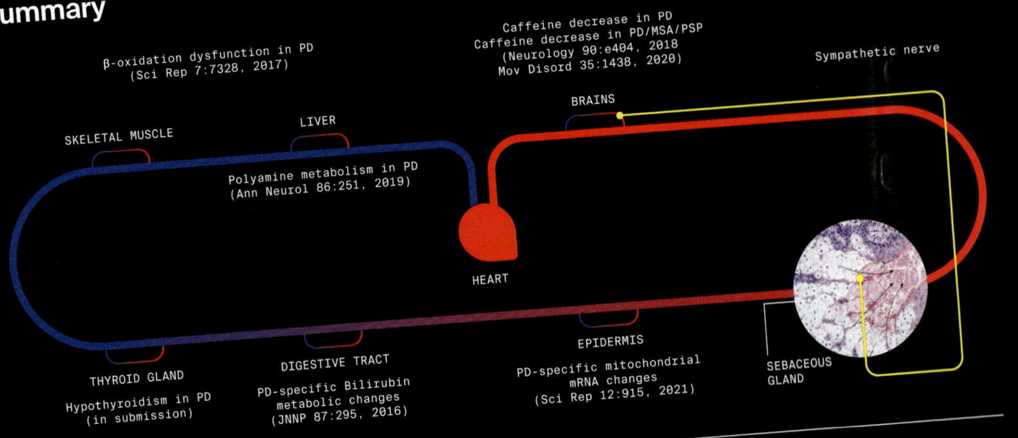

Polyamine metabolic changes in PD

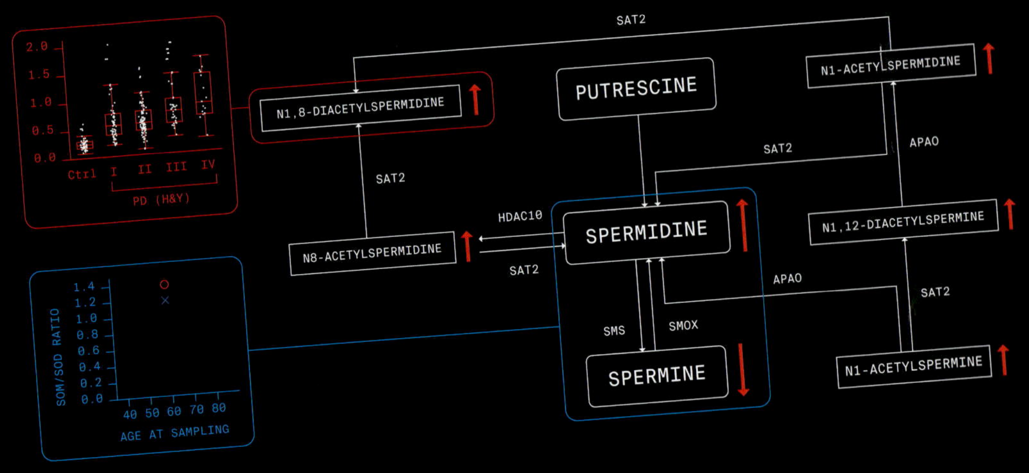

1) N1,8-diacetylspermidine is a surrogating biomarker reflecting axonal degeneration.

2) The Spm/Spd ratio is consistently suppressed in PD, compared to the control, implying of early aging of PD patients.

Ann Neurol 86:251, 2019

Expression changes of sebum mRNA in PD

A. mRNA profiling of each cohort

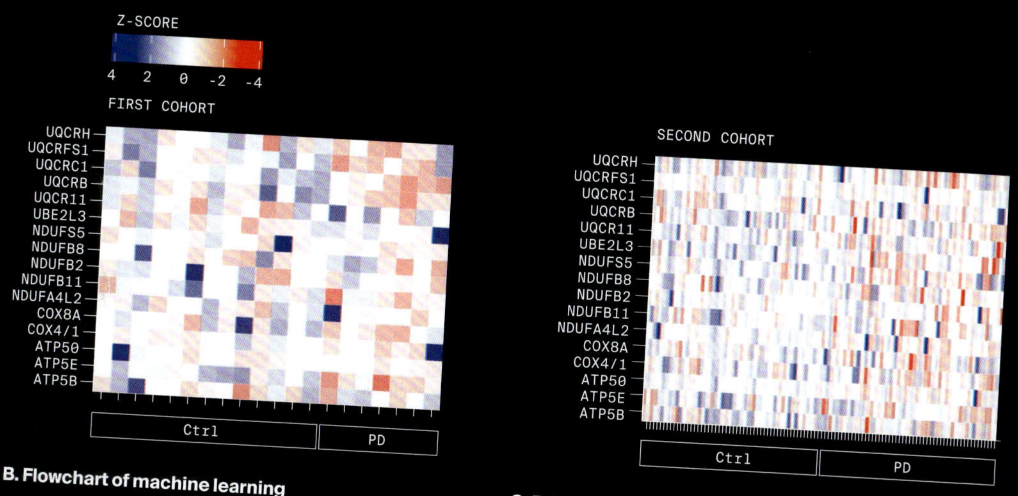

B. Flowchart of machine learning

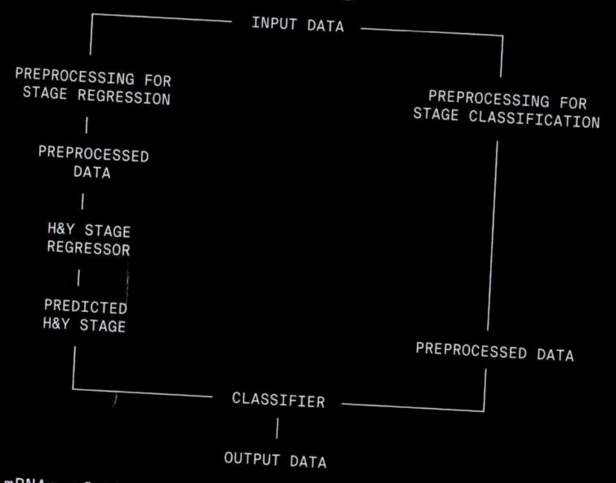

1) mRNAs of the mitochondrial respiratory chain were significantly expressed in PD sebum compared to controls.

C. ROC curve analysis

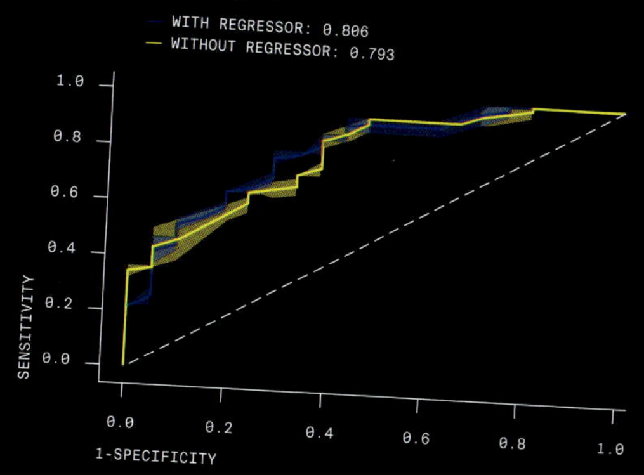

2) Relative higher diagnostic values were identified with sebum mRNA expression using machine learning.

Sci Rep 12:915, 2021

Conclusion

- We summarized PD-specific biomarkers with biological specimens.
- Serum levels of N1,8-diacetylspermidine are promising disease-surrogating biomarkers for PD.
- Non-invasive tests with sebum mRNAs with machine leaning may contribute to a PD diagnosis.

References of our published papers
- Sci Rep 12:915, 2022
- Sci Rep 11:18550, 2021
- Mov Disord 35:1438-1447, 2020
- Ann Clin Transl Neurol 7:307-317, 2020
- PLoS One 14:e0223113, 2019
- Ann Neurol 86:251-263, 2019
- Ann Clin Transl Neurol 6:525-536, 2019
- Neurology 90:e404-e411, 2018
- Sci Rep 7:7328, 2017
- J Neurol Neurosurg Psychiatry 87:295-301, 2016

5.4

Impulsive Compulsive Behaviors of Parkinson's Disease

What are Impulsive Compulsive Behaviors?

- Impulsive compulsive behaviors (ICBs) occur in around 8-40% of Parkinson's Disease (PD) patients.[1-3]
- Anti-parkinsonian therapies, especially dopamine agonists, are thought to induce ICBs.[4]
- ICBs are thought to be caused by overdose of dopaminergic therapy in the relatively spared ventral striatum, or by dopamine denervation induced hypersensitivity of D3 receptors in the ventral striatum.[5]
- Recent investigations have reported that alterations not only of the dopaminergic system, but also of brain networks might contribute to the development of ICBs in PD.[6-7]

Risk factors[4]

- Personal or familial history of alcoholism or gambling
- Personality (impulsive, novelty-seeking)
- Younger age
- Being unmarried
- Smoking
- Male – compulsive sexual behavior
- Female – compulsive shopping and binge eating

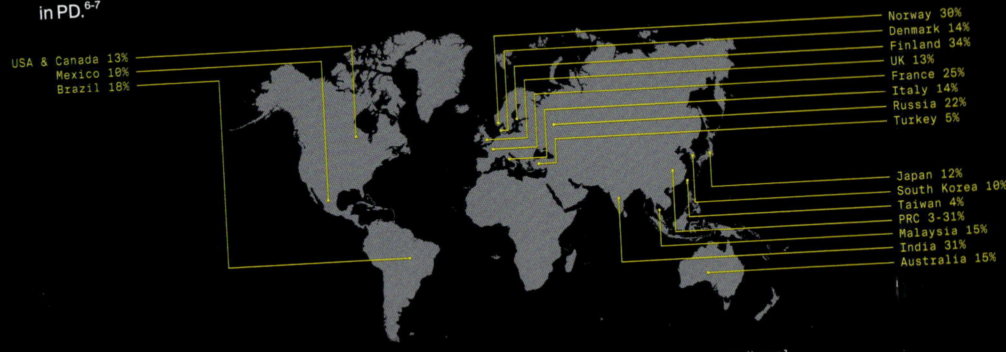

Rates of impulse control disorders (ICDs) in people with Parkinson's Disease (PD) around the world are highly variable.

Eisinger R. S. et al., Front Neurol. 2019 Feb 26;10:86

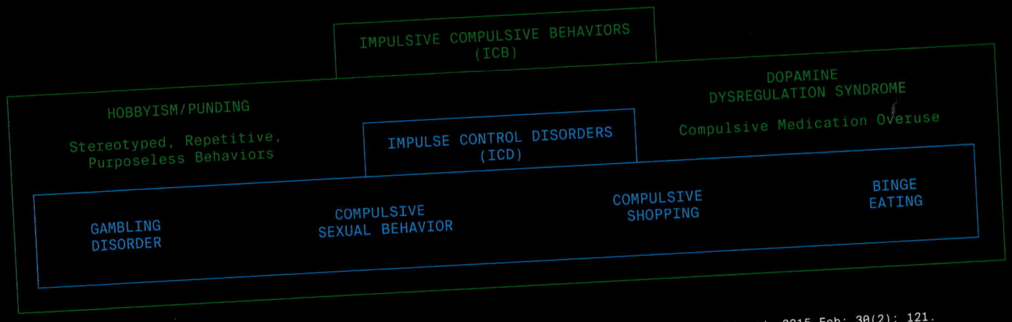

1 Weintraub D. et al., Arch Neurol. 2010;67(5): 589-595.
2 Joutsa J. et al., Parkinsonism Relat Disord. 2012; 18(10): 1079-1083.
3 Poletti M. et al., J Clin Psychopharmacol. 2013; 33(5): 691-694.
4 Weintraub D. et al., Mov Disord. 2015 Feb; 30(2): 121.
5 Vriend C., Cell Tissue Res. 2018; 373(1): 327-336.
6 Tessitore A. et al., Mov Disord. 2017; 32(12): 1710-1719.
7 Takeshige-Amano H. et al., Ann Clin Transl Neurol. 2022 Mar; 9(3): 253-263.

Parkinson's Disease patients with Impulsive Compulsive Behaviors have relatively preserved white matter microstructures

The comparison of white matter microstructures among 18 patients without ICBs (PD-nICB), 19 patients with ICBs (PD-ICB), and 20 age- and sex-matched healthy controls. Compared with HCs, PD-nICBs had significant alterations in many major white matter tracts in most parameters. In contrast, PD-ICBs had only partial changes in several parameters.

Takeshige-Amano H. et al., Ann Clin Transl Neurol. 2022 Mar;9(3):253-263

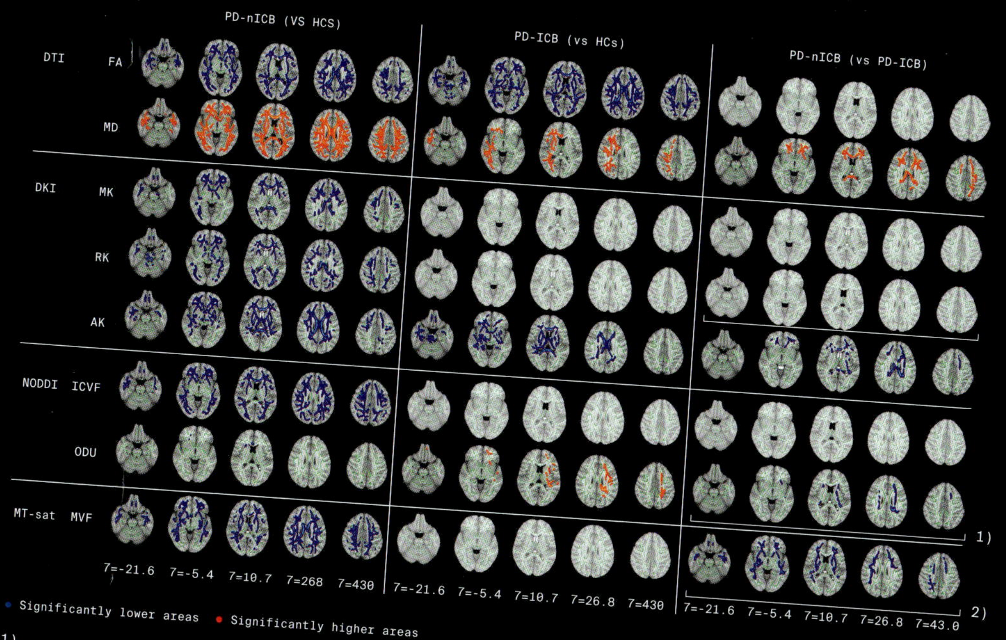

1)

Tract-Based Spatial Statistics

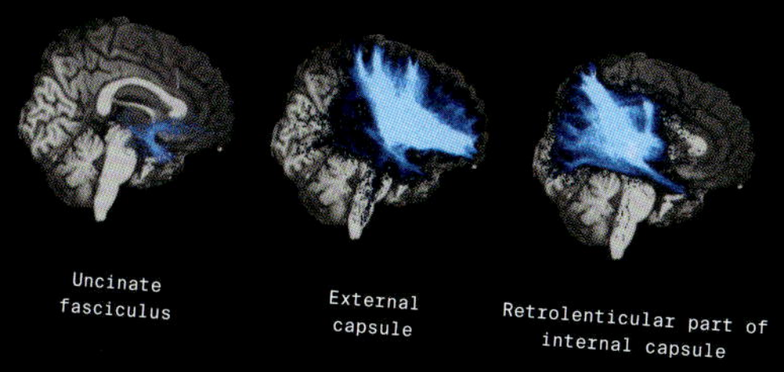

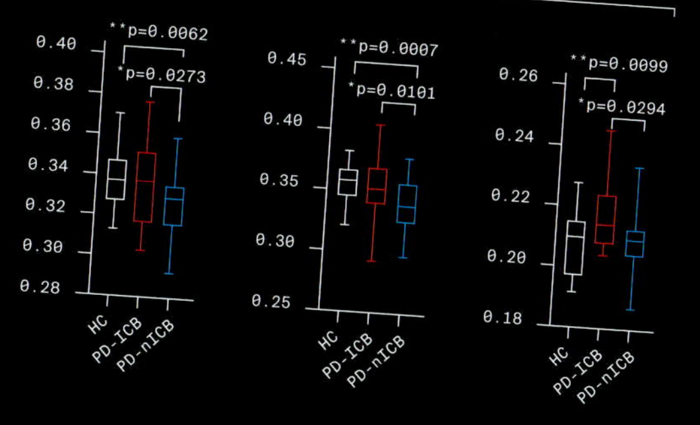

Tract-specific analysis (TSA), another way to analyze the difference of certain tracts, showed significant alteration in PD-nICB compared with PD-ICB and HCs.

2)

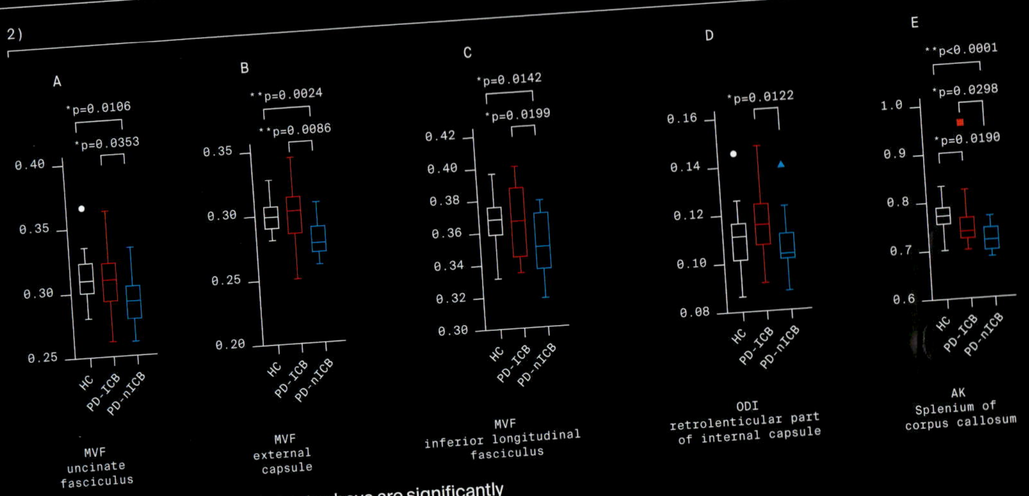

The parameters of the certain tracts above are significantly different between PD-ICB and PD-nICB. Mainly, PD-nICB had significant alteration compared with HC and PD-ICB.

Visual Emotion Recognition

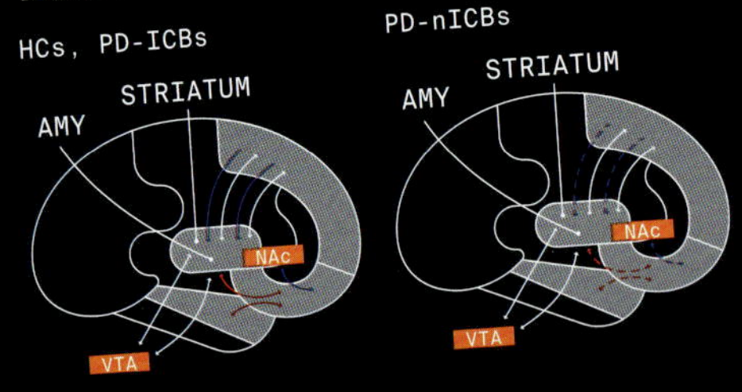

Reward system

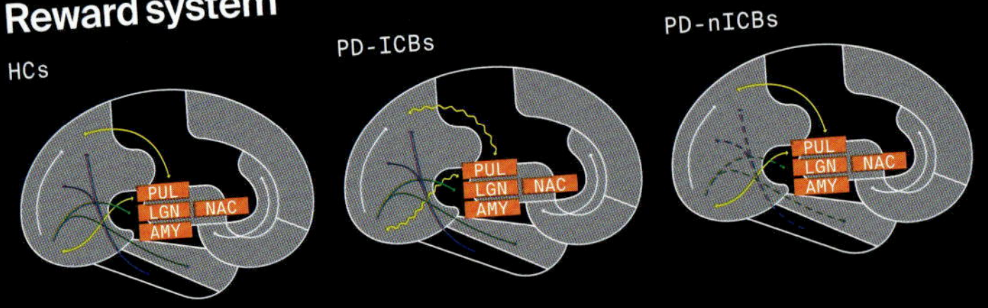

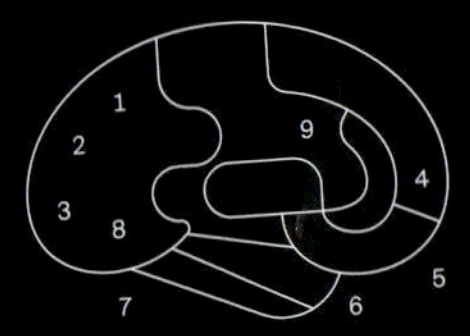

- The white matter microstructure of PD-ICB was more preserved than that of PD-nICB.
- Significant pathways associated with ICBs are roughly divided into two areas; those involved in reward system and those involved in facial emotion recognition.
- Microstructures of these pathways are spared in PD-ICB, suggesting overdosing dopaminergic therapy to these relatively spared systems might induce ICBs.

1 Parietal Association Cortex
2 Dorsal Patway
3 Occipital Association Cortex
4 Prefrontal Cortex
5 Orbifrontal Cortex
6 Fusiform gyrus
7 Mediotemporal Region
8 Visual Cortex
9 ACC

5.5

Exosomes

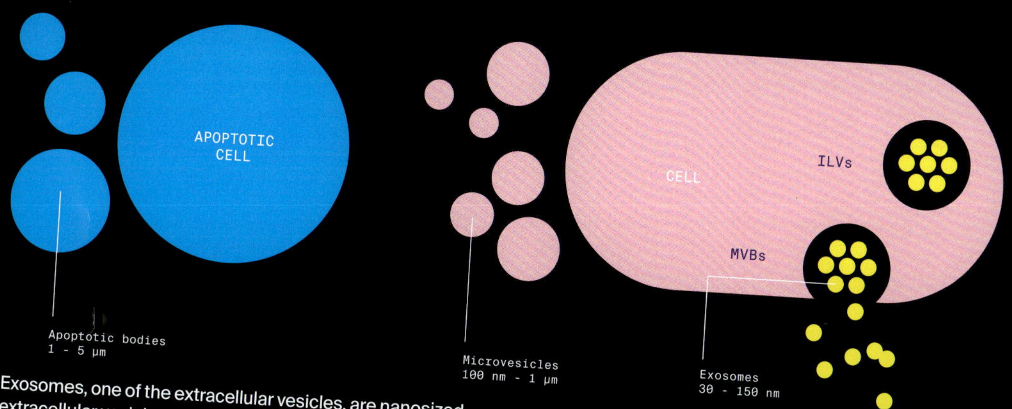

Exosomes, one of the extracellular vesicles, are nanosized extracellular vesicles of 30–150 nm in diameter.

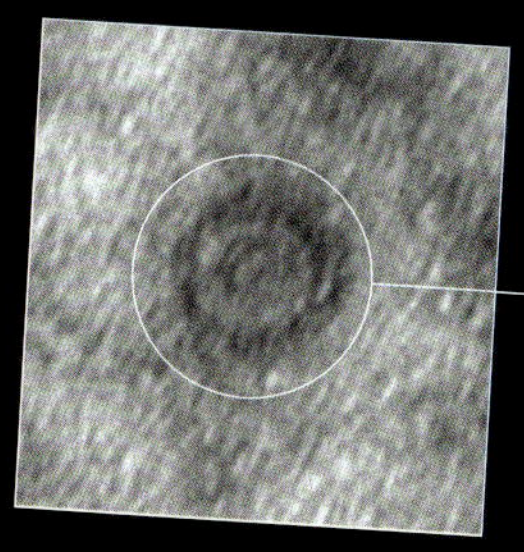

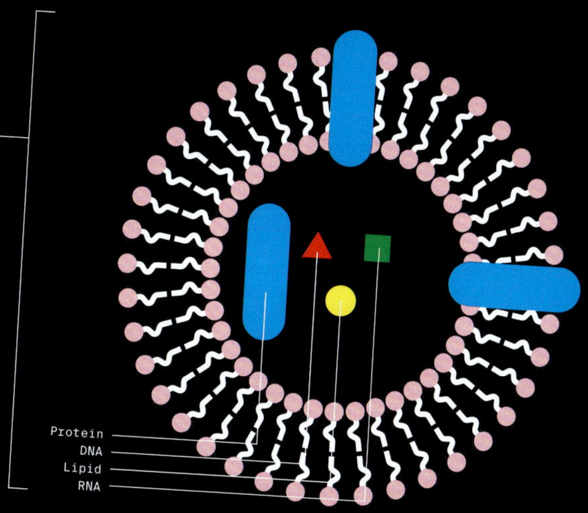

Exosomes physiologically play an important role in transmitting information to distant cells and tissues, carrying various proteins, lipids, and RNAs depending on the original cells and tissues that release them but have also been involved in the development of various diseases.

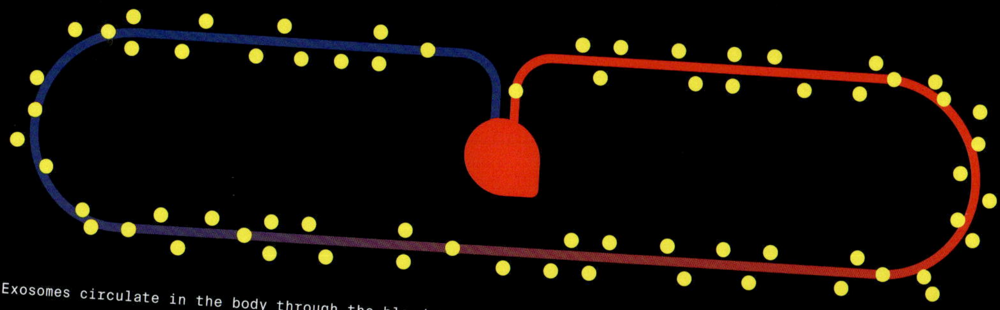

Exosomes circulate in the body through the bloodstream.

Exosomes are involved in Parkinson's pathogenesis
Kufor-Rakeb syndrome

Rare hereditary neurodegenerative disorder

Clinical symptoms:
- Onset 6 – 40 years old
- Parkinson's Disease manifestations (bradykinesia, postural instability) respond well to L-DOPA
- Dementia, supranuclear gaze palsy and spasticity caused by loss-of-function mutations in ATP13A2/PARK9

PATIENTS IN THE WORLD:	
Inuit	6
Chilean	5
Jordanian	4
Iranian	3
Chinese	2
Italian	2
Pakistani	2
Afghan	1
Brazilian	1
Japanese	1

Behrens et al. Mov Disord 2010;25:1929-1937

Diffuse brain atrophy

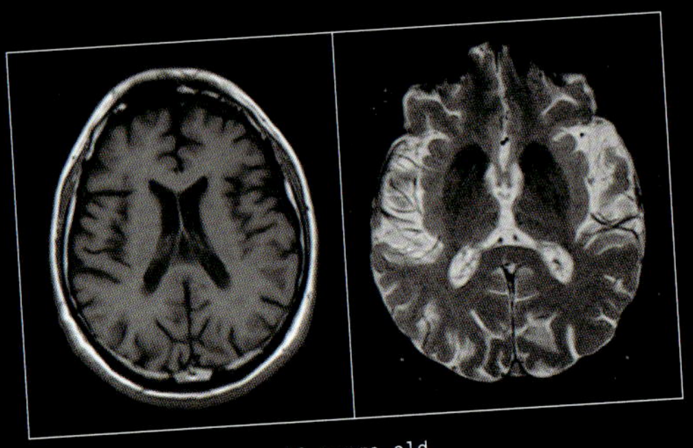
43 years old

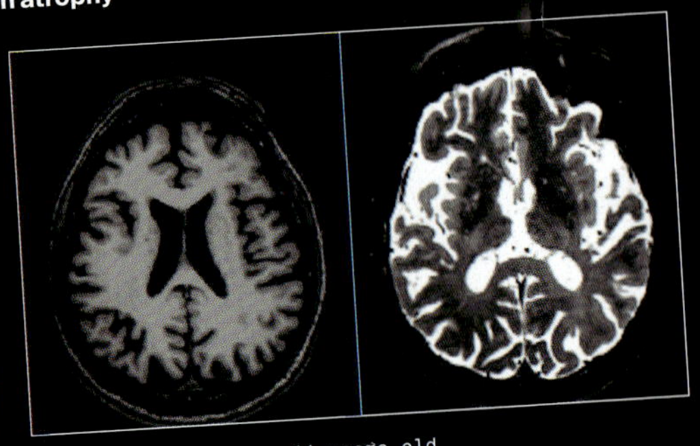
44 years old

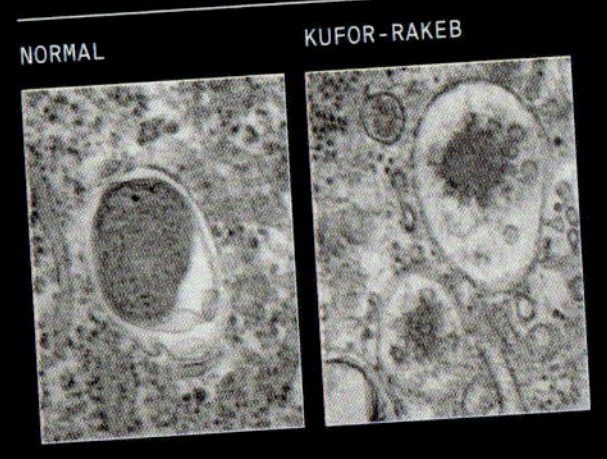

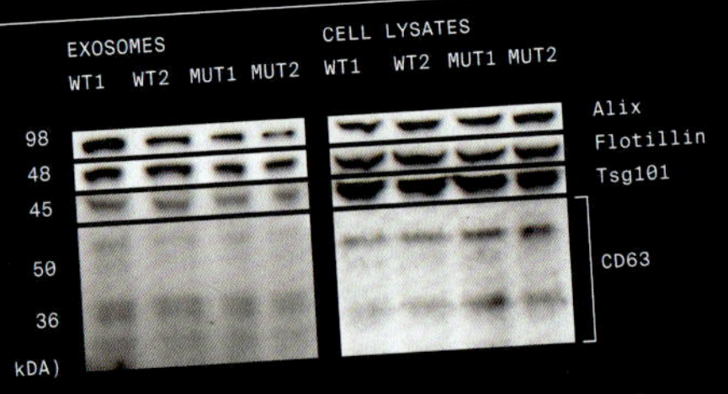

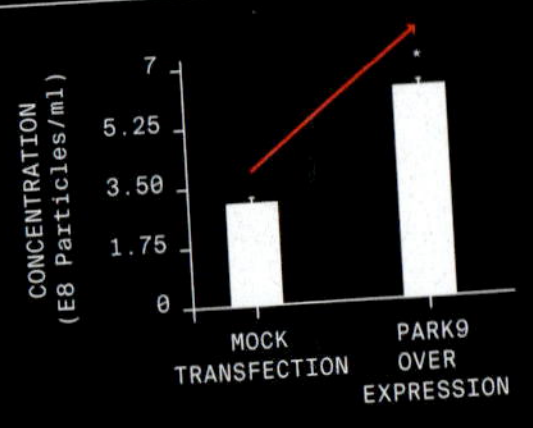

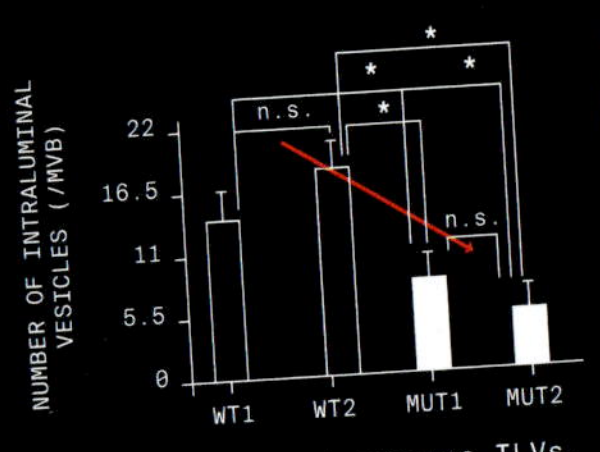

PARK9 deficiency decreases ILVs.

Tsunemi et al., J Neurosci 2014;34:15281-15287

Nanoparticle tracking analysis

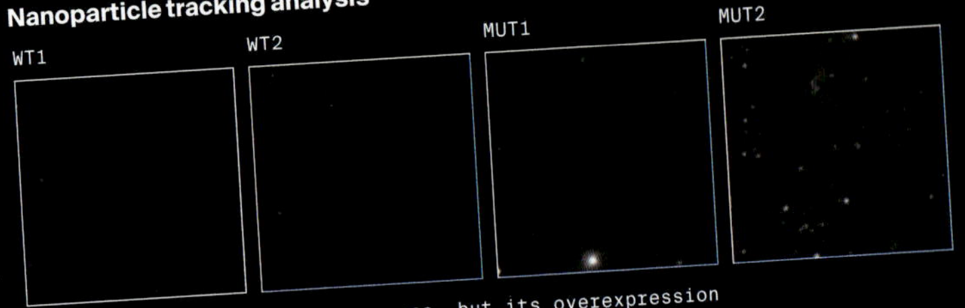

PARK9 deficiency decreases exosomes, but its overexpression increases them.

Accumulating evidence indicate that exosomes are involved in the pathogenesis of neurodegenerative disorders, which is pathologically characterized by the accumulation and propagation of misfolded proteins. One such disorder is Parkinson's Disease.

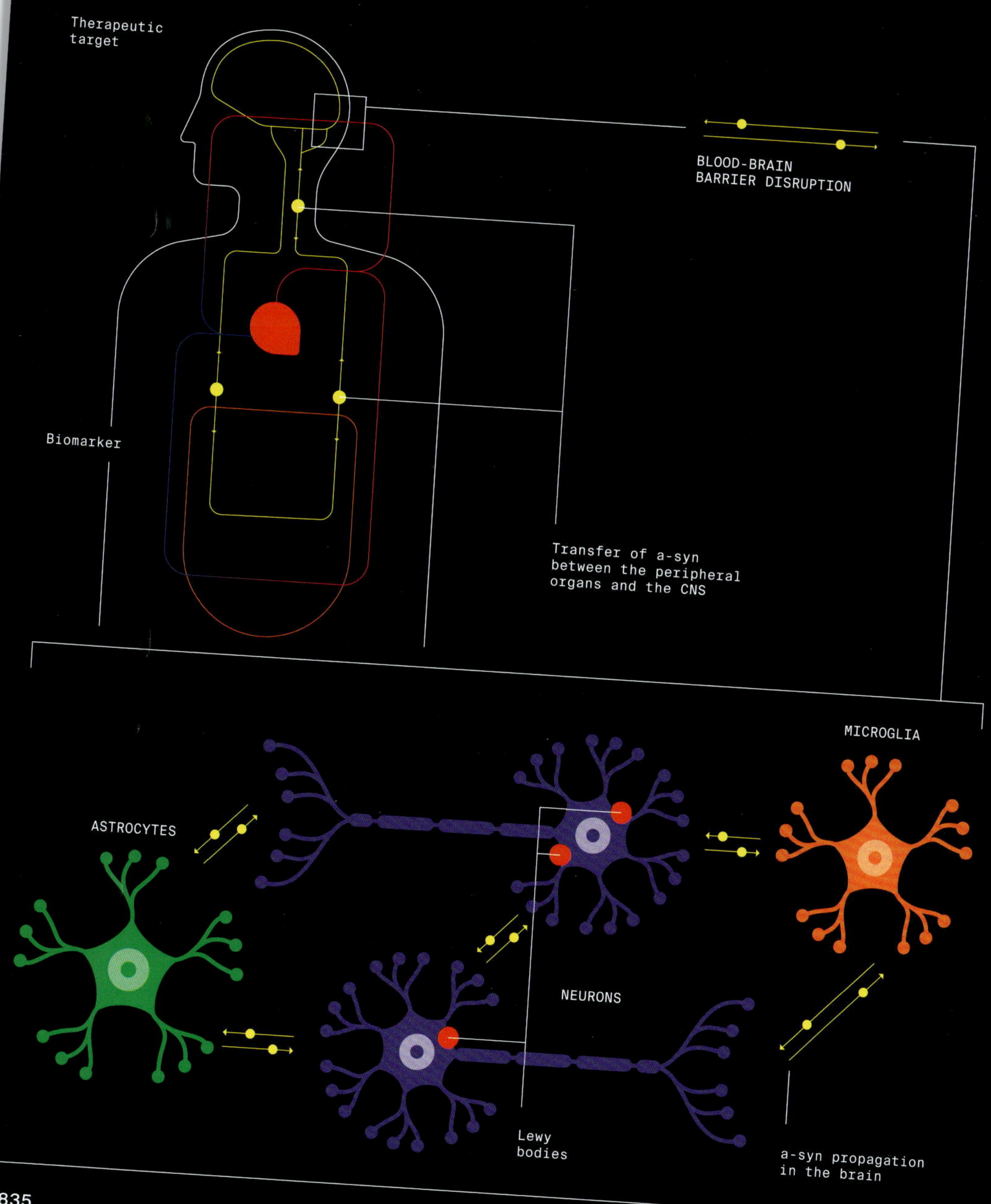

5.6

Gene

Genetics of Parkinson's Disease

The human body is made up of tens of trillions of cells (there are various theories such as 60 trillion or 37 trillion). Each cell has a nucleus, which contains DNA, the genetic material. Genes are determined by the order of the four base pairs (adenine, thymine, cytosine, and guanine) in DNA, and humans have more than 20,000 different genes.

Of course, there are also variants associated with Parkinson's Disease. If we can find the variants involved in Parkinson's Disease, we will be able to elucidate the cause of the disease and create new therapeutic agents based on the function of the genes that have the variant.

There are two ways to find variants about Parkinson's Disease: one is to study hereditary Parkinson's Disease, in which parents and children develop Parkinson's Disease. The other is to analyze a population of people with and without Parkinson's Disease.

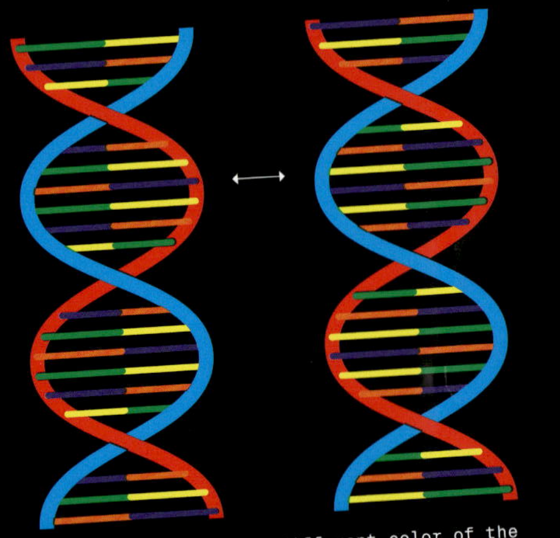

Variants (Can you see the different color of the arrows in the upper and lower DNA?)

Genome-wide association studies (GWAS) are underway to find genetic risk for sporadic Parkinson's Disease. Tens of thousands of case-control genomes have been examined to discover risk variants for Parkinson's Disease. Risk variants are present in people who do not have Parkinson's Disease, but Parkinson's Disease patients may have several risk variants or may develop the disease due to a combination of lifestyle and environmental factors.

Genes do not change in their basic design with race, gender, or age. With the exception of identical twins, however, there are small differences in genes. These differences are called VARIANTS. Most variants do not affect your life in any way, but a very few variants are known to affect your constitution and temperament.

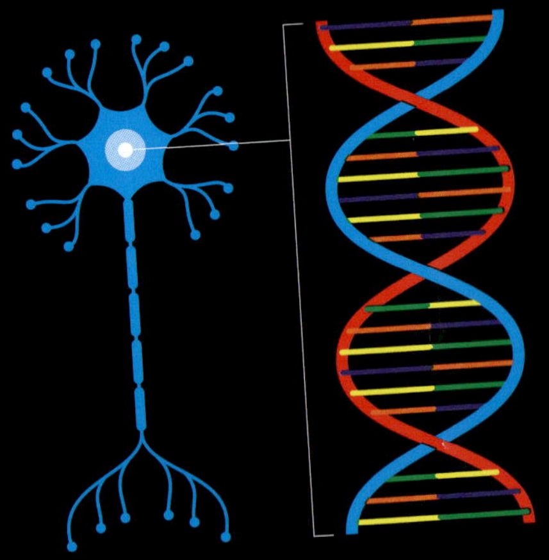

There are currently more than 200 genes known to be involved in Parkinson's Disease. In 1997, analysis of a large family in southern Italy led to the discovery of the SNCA gene as a cause of hereditary Parkinson's Disease. The SNCA gene encodes a protein called alpha-synuclein. Alpha-synuclein is a major component of the Lewy bodies found in the brain of Parkinson's Disease. Alpha-synuclein is thought to cause Parkinson's Disease when it accumulates in the brain and causes various damaging effects in the brain, leading to the death of nerve cells.

1. Family Based Study
2. Association Study

How to find variants related to Parkinson's Disease

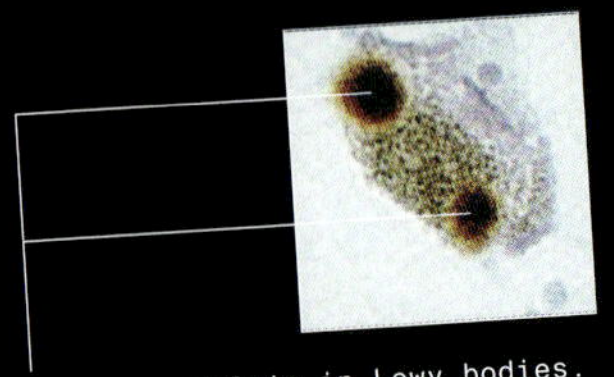

Alpha-synuclein in Lewy bodies.
(Nature 388,839)

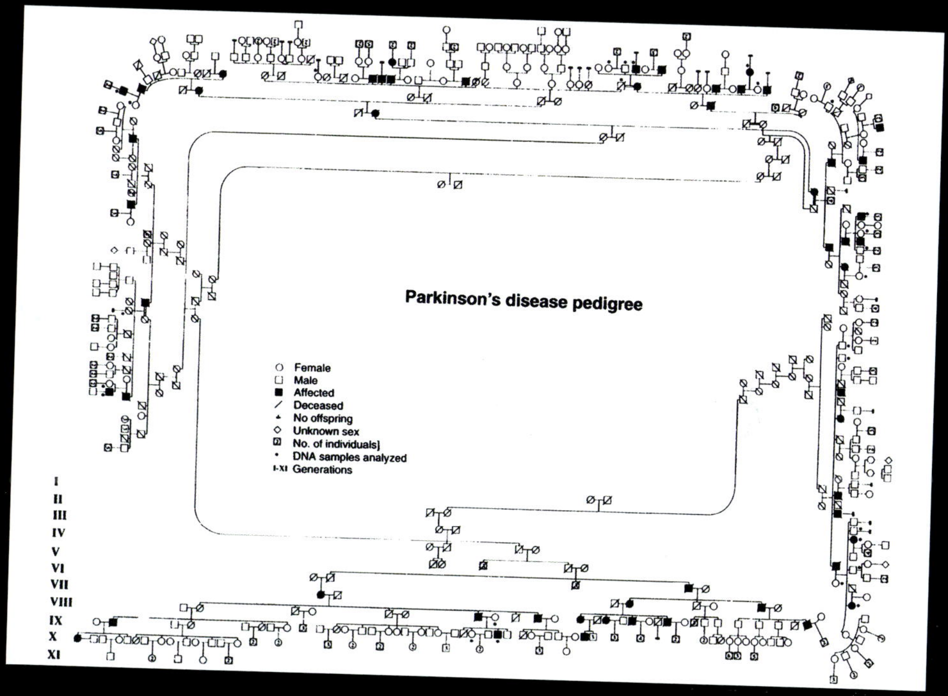

A large Italian family with SNCA.
(SCIENCE 274 1197-1199)

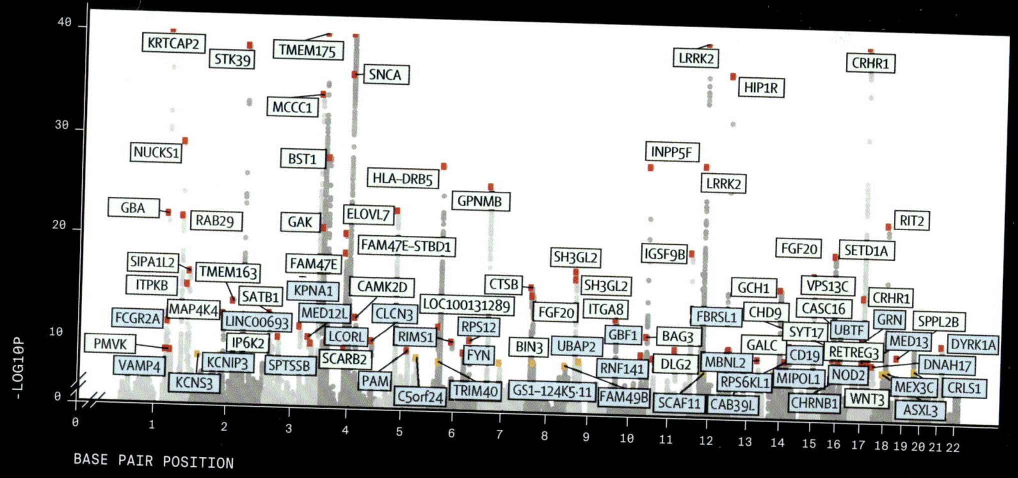

Manhattan plot (named after the towers in Manhattan) for Parkinson's Disease GWAS.
(Lancet Neurol 18: 1091-102)

5.7

Lysosome

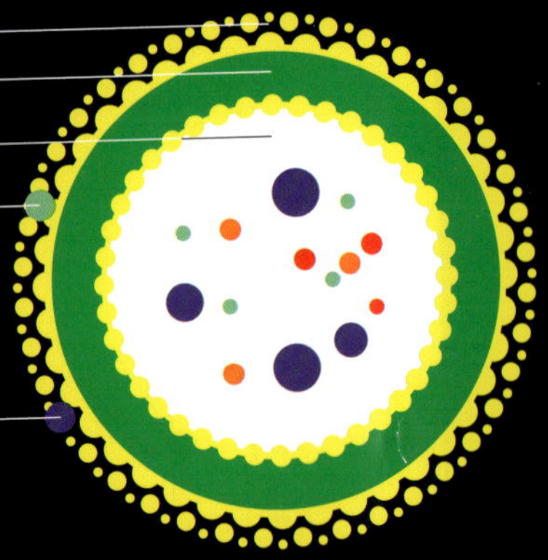

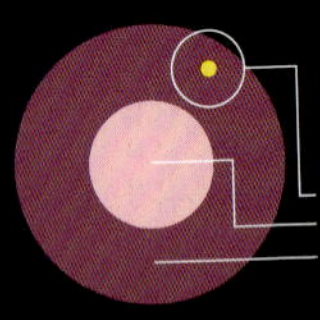

What's lysosome?

1. Lysosomes are one of the organelles of eukaryotes.
2. The lumen of lysosomes is acidified around pH 5 and contains more than 60 various hydrolases for the degradation of intracellular and extracellular components, such as proteins, lipids, nucleotides, and other organelles, for the maintenance of cellular homeostasis.
3. Degradation substrates are transported to lysosomes by pathways such as endocytosis and autophagy.

Lysosomal function and Parkinson's Disease

1. Lysosomal function is closely linked to the mechanism of Parkinson's Disease.
2. Several genes of lysosomal storage disorders, such as *glucocerebrosidase* (GBA), are genetically linked to Parkinson's Disease.
3. Alpha-synuclein is degraded in the lysosome.
4. Several cellular processes around the lysosome are also involved in Parkinson's Disease.

1. LYSOSOME
2. AUTOPHAGOSOME
3. MITOCHONDRION
4. EARLY ENDOSOME
5. LATE ENDOSOME
6. ALPHA-SYNUCLEIN
7. ALPHA-SYNUCLEIN FIBRIL
8. CLATHRINA

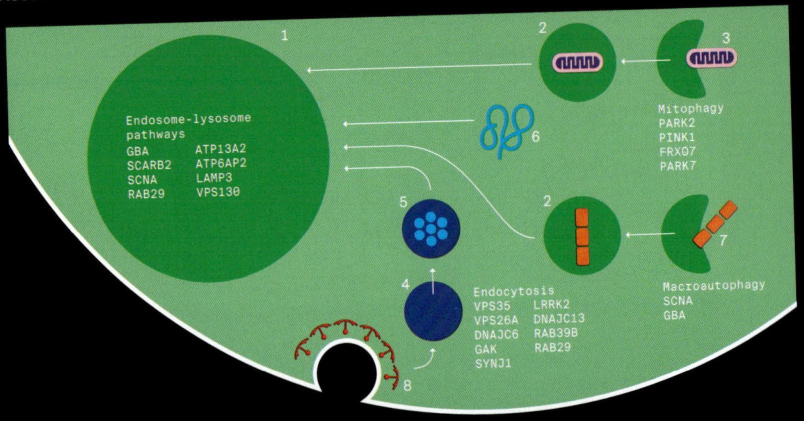

Lysosomal storage disorders

1. Lysosomal dysfunction causes hereditary lysosomal disease, which are named lysosomal storage disorders.
2. Lysosomal storage disorders are rare diseases that are caused by homozygous mutations in the genes coding lysosomal enzymes or proteins.
3. Gene mutations cause the deficiency of lysosomal enzymes or proteins, which lead to accumulation of substrates, subsequently resulting in the lysosomal dysfunction. Lysosomal storage disorders can cause multiple organs including brain.

Hair
HPS1, HPS2, HPS3, HPS4, HPSS, HPS6, HPS7, HPS8, HPS9, MYOSA, RAB27A and CTSA.

Brain
GLA, GBA, GLB1, HEXB, HEXA, GM2A, ARSA, PSAP, SMPD1, IDUA, IDS, SGSH, NAGLU, HIGSNAT, GNS,GUSB, MAN2B1, MANBA, FUCA1, AGA. NAGA, NEU1, CTSA, GNPTAB, LAMP2, SCARB2, SLC17A5,NPC1, NPC2, MCOI N1, MYOSA, IYST, PPT1, TPP1,CLN3, DNAJC5, CLN5, CLN6, MFSD8, CLN8, CLN9,CTSD, GRN, ATP13A2, CTSF, KCTD7 and GALC.

Immune deficiency
MAN2B1 and RAB27A.

Eye
GLB1, GLA, IDUA, GALNS, ARSB, GUSB, NAGA, CTSA, CTNS, NPC1, NPC2, MCOLN1, HPS1, HPS2, HPS3, HPS4, HPS5, HPS6, HPS7, HPS8, HPS9, MYO5A, LYST, CLN3, CLN5, CLN8, CLN9, CTSD, GRN, KCTD7, NEU1, TPP1, GALC, HEXA, HEXB and GMZA.

Ear
IDUA, IDS, GALNS, ARSB, GUSB, HYAL1, NAGA, CTSA, GNPTAB, SCARB2 and NEU1.

Coarse facies
IDUA, IDS, SCSH, NAGLU, HGSNAT, GNS, GUSB, MAN2B1, AGA, NAGA, CTSA, GNPTAB, SLC17A5, GUSB and GALNS.

Tongue
IDUA, IDS, ARSB and GUSB.

Thyroid
CTNS and CTSA.

Heart
GLA, IDUA, IDS, SGSH, NAGLU, GUSB, GAA, NAGA, CTSA, GNPTAB, GNPTAG, LAMP2, SLC17A5, GLB1, CLN3 and ARSB.

Lung
SMPD1, AGA, GNPTAB, NPC1, NPC2, HPS1, HPS2, and HPS4.

Hepatosplenomegaly
GBA, SMPD1, IDUA, IDS, SGSH, NAGLU, HGSNAT, GNS, ARSB, AGA, NAGA, CTSA, GNPTAB, SLC1745. NPC1 and NPC2.

Kidney
GLA, CTNS,SCARB2, NEU1 and CTSA

Gastrointestinal System
GLA, ST3GAI 5 and MCOLN1

Haematologic
GBA, SMPD1, MCOLN1, HSP1, HSP2, HSP3, HSP4, HSPS, HSP6, HSP7, HSP8, HSP9, RAB27A and LYST.

Joint
ASAH1, GLB1, IDUA,IDS, SGSH, NAGLU,HICSNAT, GNS, GALNS and ARSB

Muscle
GAA, CTNS, LAMP2 and NEU1

Hernias
IDUA, IDS, SGSH, NAGLU, HGSNAT, GNS, GALNS, GLB1, ARSB, GUSB, CTSA and GNPTAB.

Skeletal
GLB1, DBA, SMPD1, IDUA, IDS, SGSH, NAGLU, HGSNAT, GNS, GALNS, ARSB, HYAL1, MAN2B1, FUCA1, NAGA, CTSA, GNPTAB, GNPTAG, CTNS and NEU1.

Skin
GLA, ASAH1, IDS, MANBA, FUCA1, NAGA, CTSA, CTNS, HP51, HPS2, HIPS3, HPS4, HPS5, HPS6, HIPS7, HPS8, HPS9, RAB27A, LYST and GBA.

Peripheral nervous system
GLA. HEXB. SCARB2 and LYST.

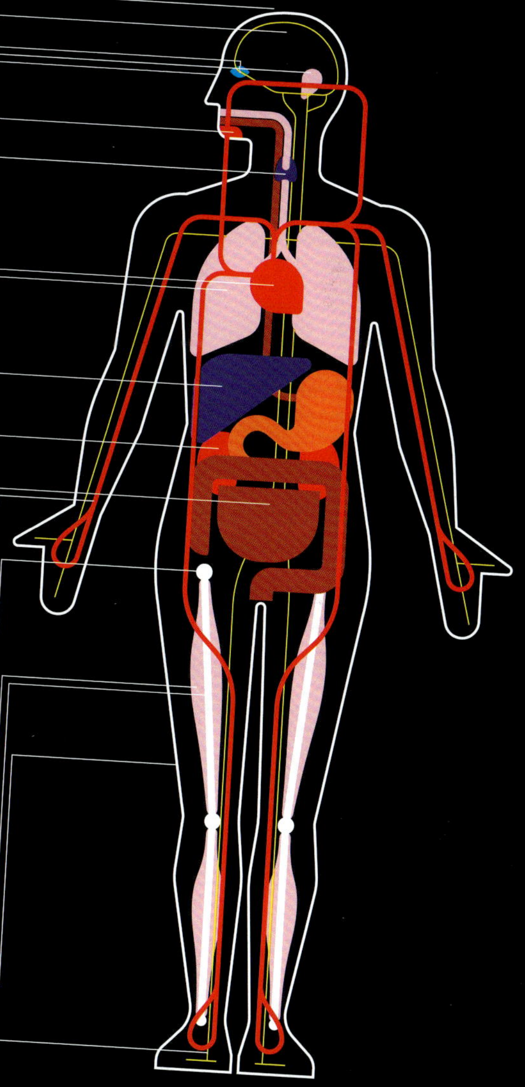

Genetic mutations that cause specific clinical manifestations
Genetic mutations that cause non-specific symptoms

5.8

Mitochondria and Parkinson's Disease

What are mitochondria?

Powerhouse of the cells that make up living organisms.

Benefits of mitochondria

Mitochondria originated from bacteria that live symbiotically with cells in ancient times. Energy was efficiently supplied by the symbiotic mitochondria, and the cell evolved into its present form.

Fatal defects of mitochondria

Mitochondria release reactive oxygen species when making energy. Reactive oxygen species are considered one of the causes of aging. When mitochondria become abnormal, excessive reactive oxygen species are produced, resulting in cell death or cancer. Mitochondria are like nuclear power plants, a necessary evil.

Parkinson's Disease

In Parkinson's Disease, the midbrain dopamine neurons gradually die many years before the onset of the disease. Mitochondrial activity of dopaminergic nerves is also impaired.

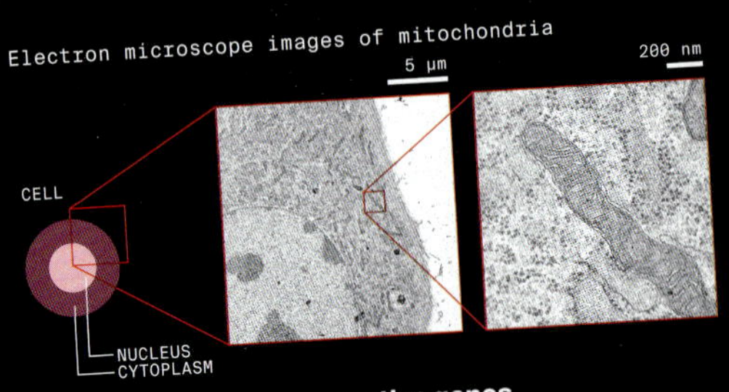

Parkinson's Disease-causative genes

Genetic analysis of Parkinson's Disease families has revealed more than 20 causative genes. Among them, PINK1, Parkin, and CHCHD2 are genes involving mitochondria.

PINK1 and Parkin manage mitochondrial quality and remove broken mitochondria.

By monitoring mitochondrial function, it is believed that the generation of reactive oxygen species is suppressed and neuronal cell death is prevented.

CHCHD2 is a component of the mitochondrial apparatus that produces energy through oxygen respiration. When CHCHD2 is destroyed, reactive oxygen species are generated, resulting in neuronal cell death.

Therapeutic trials using models of Parkinson's Disease

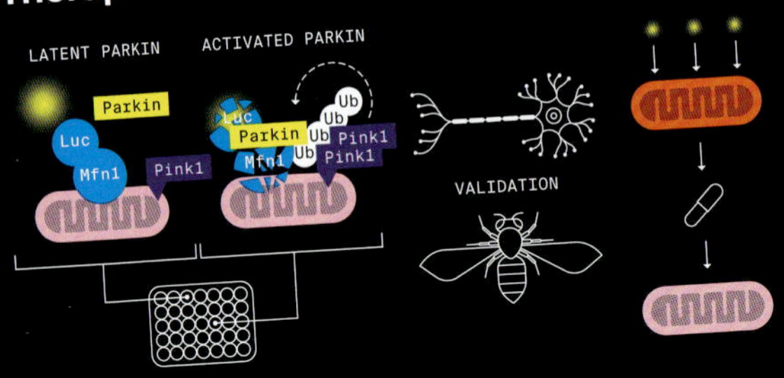

We searched for drugs that activate Parkin and were able to energize mitochondria in iPS cells and *Drosophila* (fruit fly) models.

Shiba-Fukushima, iScience 2020

By incorporating an energy producing device that is activated by light in mitochondria, we were able to suppress the degeneration of dopaminergic neurons in *Drosophila* lacking CHCHD2 gene.

Imai, Commun Biol 2019

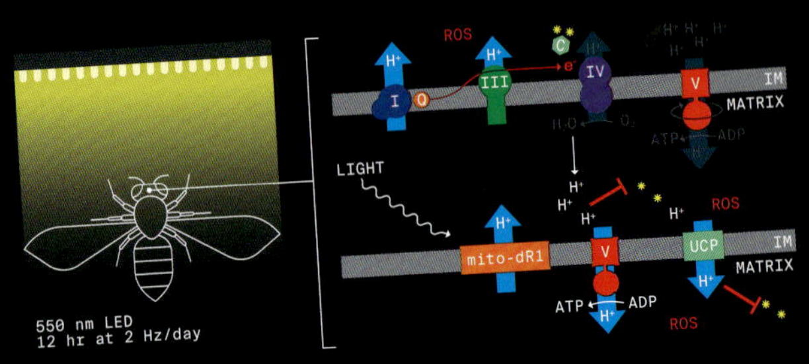

Multiple Sclerosis: The Gut-Brain Connection

6

Max Planck Institute of Neurobiology
Munich, Germany

Multiple Sclerosis (MS) is not a rare disease: it afflicts an estimated 2M persons worldwide, mostly in the *Western* world, with an increasing tendency. The underlying lesions (plaques) are distributed throughout the central nervous system, presenting inflammatory infiltrations and confluent zones of myelin destruction.

The drivers of the pathogenesis are brain-specific immune cells (CD4+ T cells), which are activated in the peripheral immune system, that break through the tight blood-brain barrier and interact with local tissue cells to ultimately destroy myelin structures that normally insulate electrically active neuronal processes, *axons*.

Peripheral activation is the critical event sparking MS pathogenesis. Very unexpectedly, at least part of these processes happen in the intestine—far away from the central nervous system. It appears that certain bacterial components of the gut *microbiota* stimulate myelin-recognizing T cells, which sit quiescently in the healthy immune system, to become fit to attack the brain tissues.

This presentation will examine the nature of the relevant gut microbes and the mechanisms they use to arm the cognate T cells to become pathogenic. Ultimately, the question of potentially new *therapeutic strategies* will be discussed.

6.1

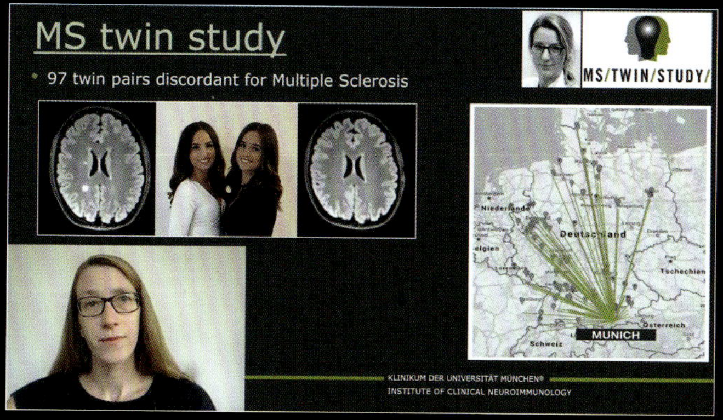

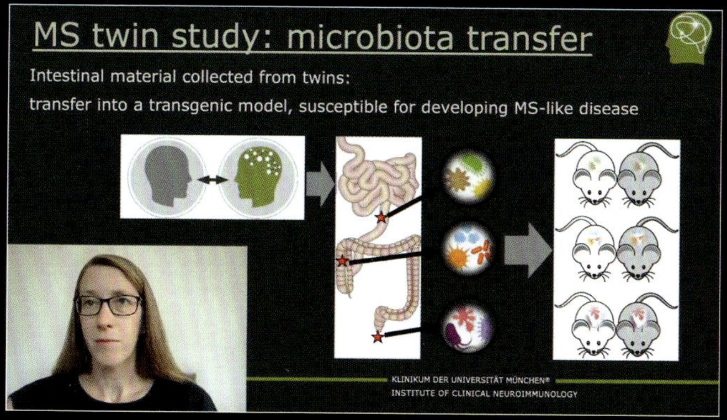

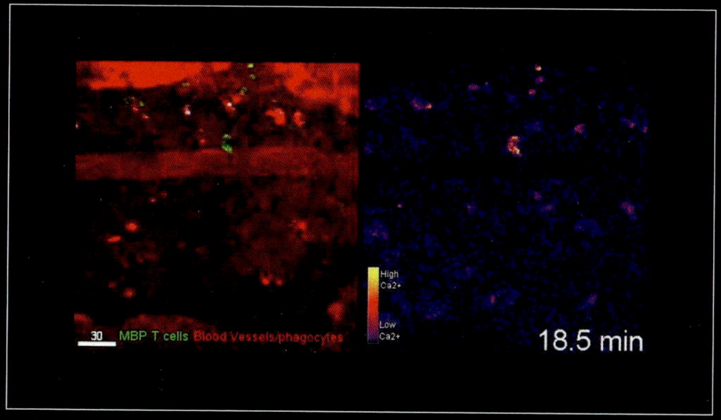

Stills from video featured in Room 6

6.2

Background

Who may contract MS?

- Incidence (Germany): >200 / 100'000
- Risk factors: genetics (HLA, "immune genes")
- Risk factors: environmental (smoking, obesity, vitamin D deficit, EBV...)
- Triggers: infection?

Microbiota signals to CNS and immune system

CNS (Conditoning target organ)

- Via bloodstream:
  - SFA (glia, GPC receptors)
  - Neurotransmitters: GABA, dopamine Endotoxin (LPS)
  - Indole-3-aldehydes (glia, AHR receptors)
- Vagus nerve
- Import via circulating immune cells

Immune system (Conditioning autoimmune effector T cells)

- Molecular mimicry
- Pro-inflammatory microenvironment
- Activation of pathogenic Th17 cells (Segmented filamentous bacteria, SFB) in ileum
- Activation of regulatory Treg cells (Bacteroides polysaccharide A, PSA) in colon

Self recognition and CNS autoimmunity

Homeostasis
CNS-reactive immune cells are present in healthy repertoire, but remain quiescent, don't access target organ (CNS)

Autoimmunity
Activation by pathogenic stimuli leads to invasion of CNS through BBB

6.3

Microbiota in guman gut segments: strictly sorted

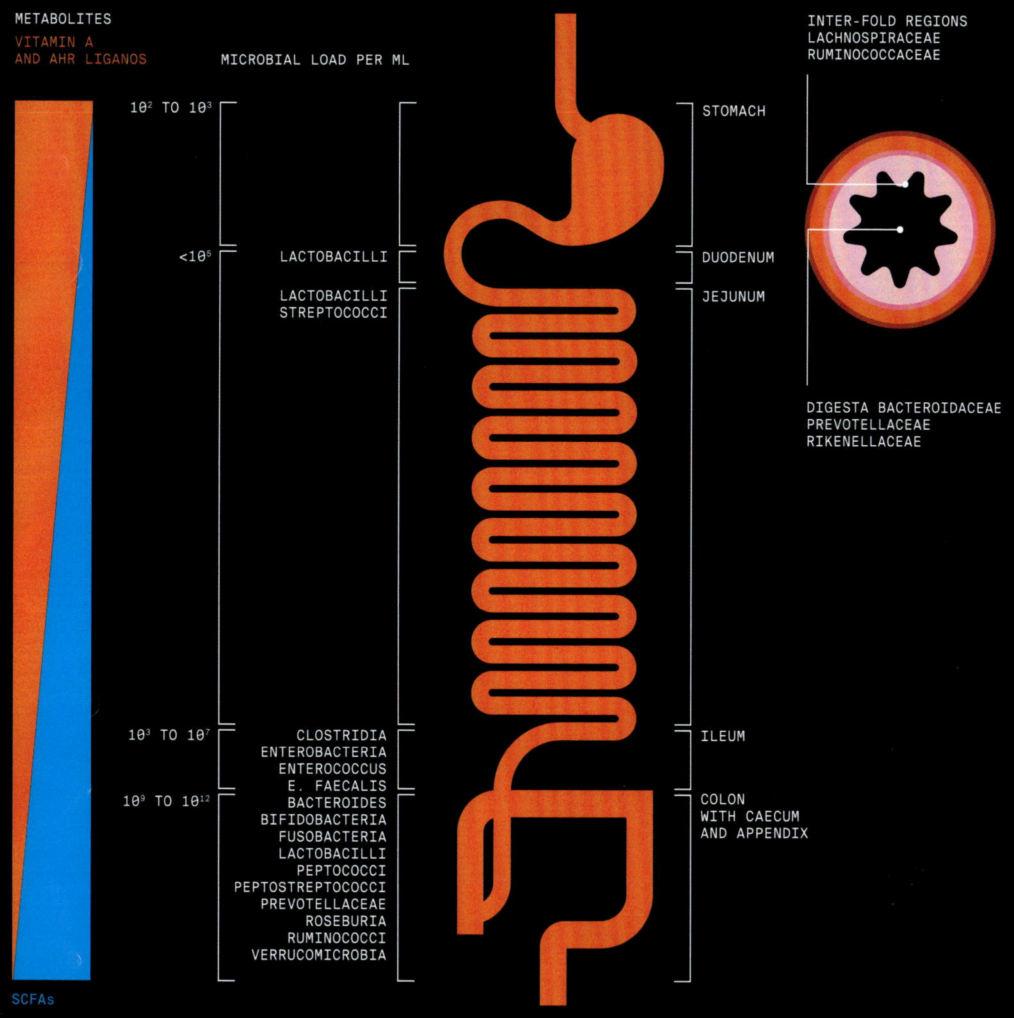

Nature Review Immunology

6.4

Mouse model of human Multiple Sclerosis

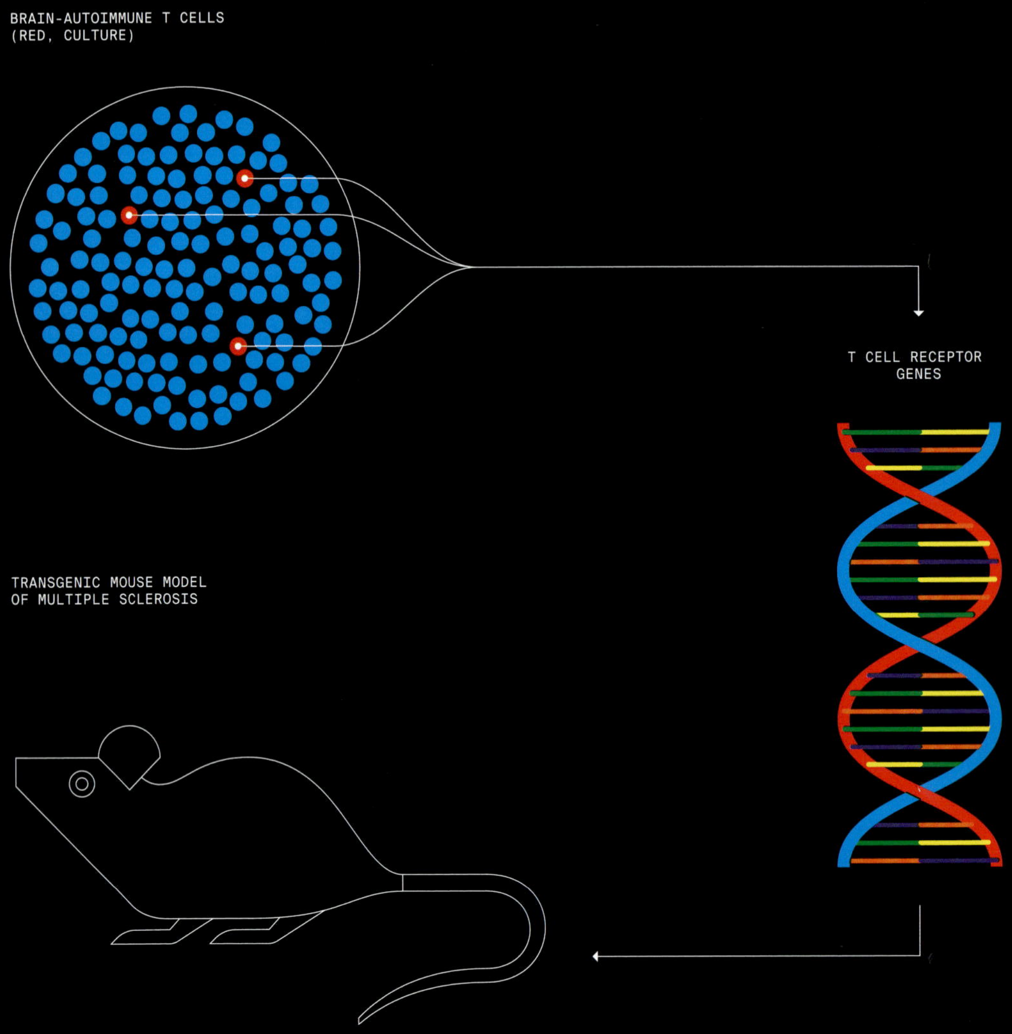

Pöllinger et al. J.Exp.Med. 2009

6.5

Gut flora triggers spontaneous MS-like brain autoimmune disease

Spontaneous "MS" in transgenic mice with healthy (left, red), and without gut flora (yellow)

Right side: "MS" development after colonization with healthy gut flora (blue)

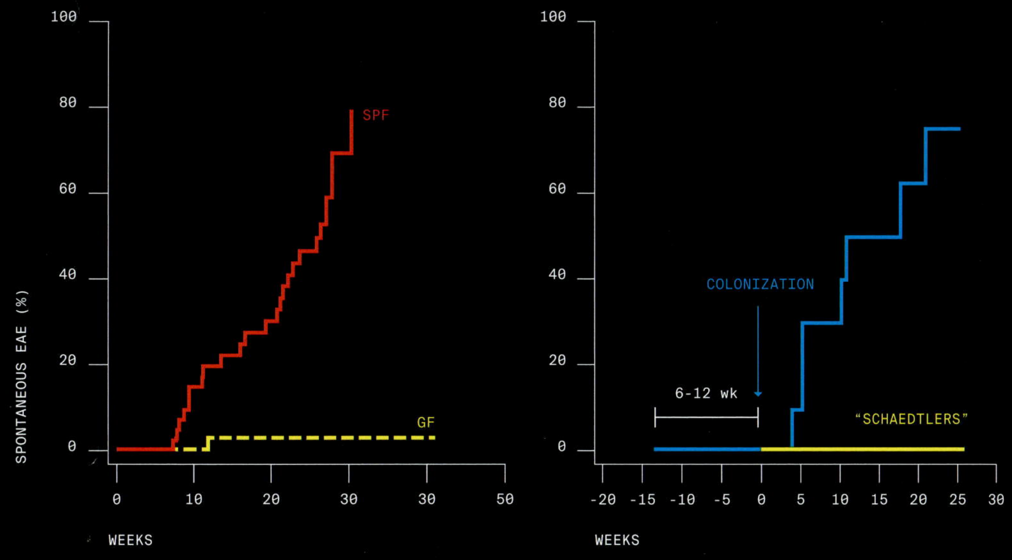

Berer et al., Nature 2009

6.6

MS-Research: from bedside to lab bench and back

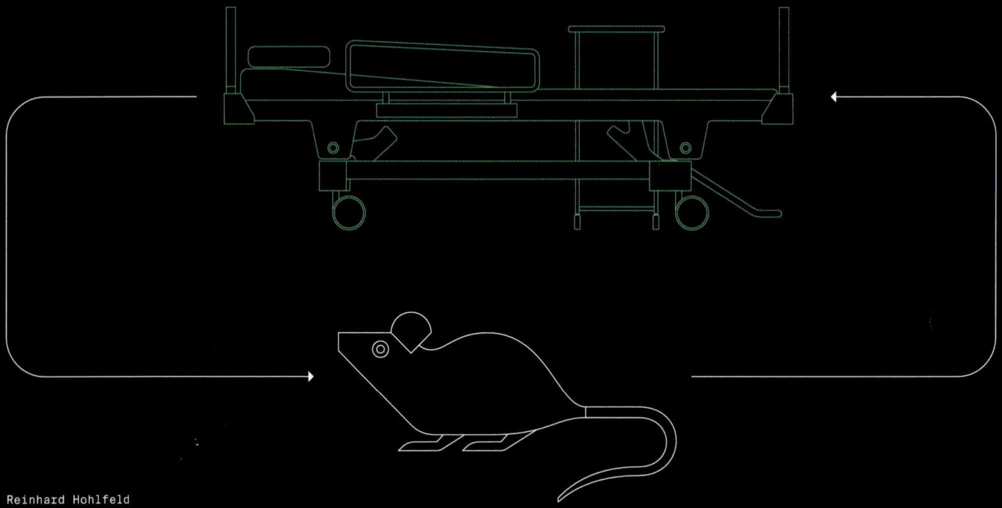

Reinhard Hohlfeld

6.7

Munich twin study

97 twin pairs discordant for Multiple Sclerosis

Prof. R. Hohlfeld
PD Dr. T. Kümpfel
Dr. L.-A. Gerdes

Supported by: DMSG
Therapieverein Multiple Sklerose
iMSMS (San Francisco)

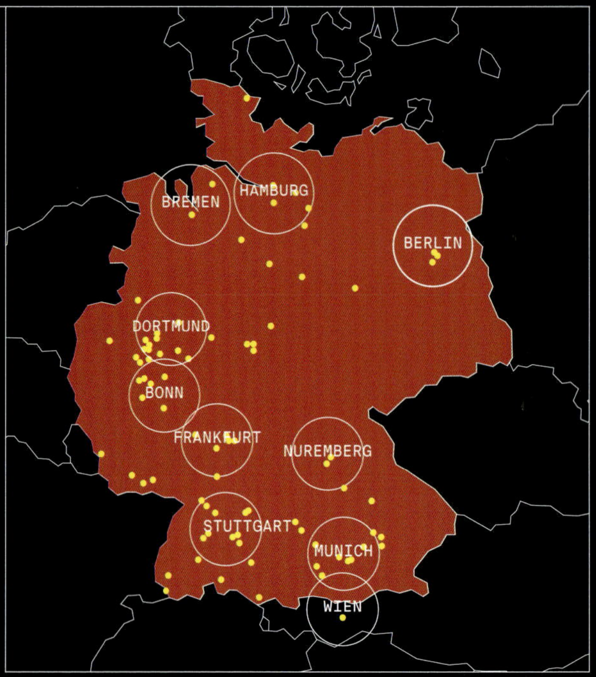

Drs. T. Kümpfel and L.-A. Gerdes

6.8

MS-Twin study: microbiota transfer human-to-RR mouse

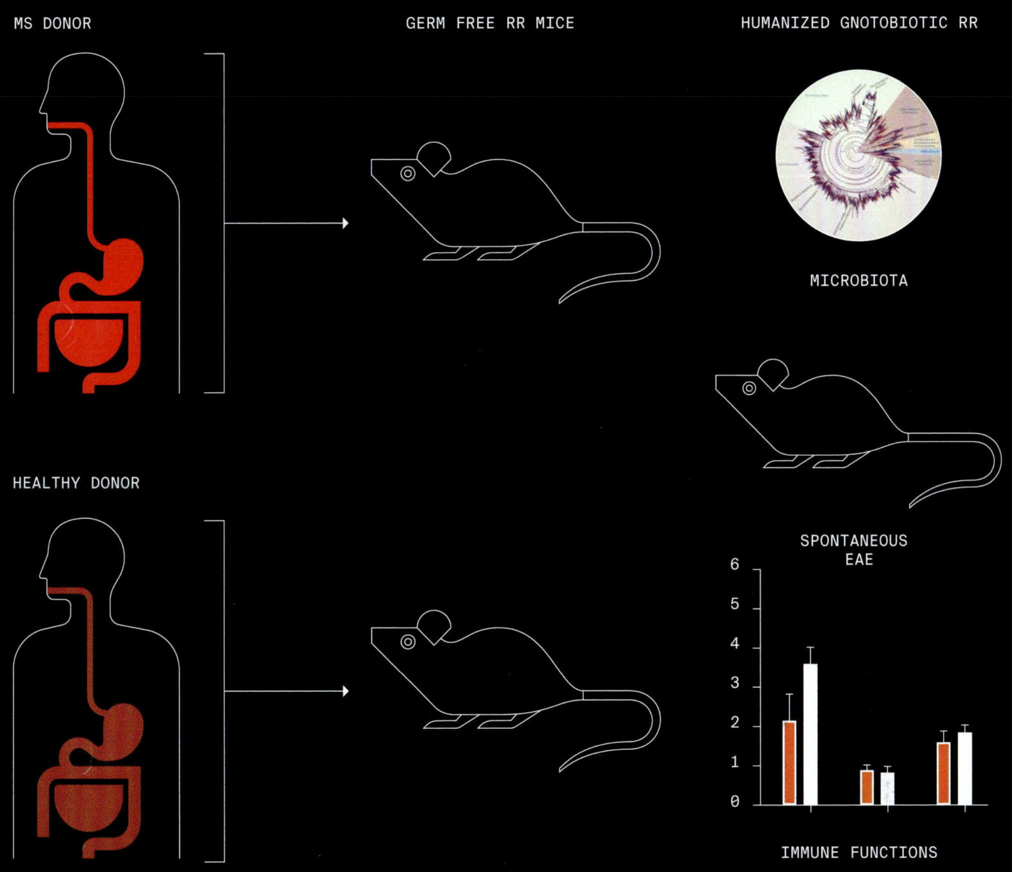

6.9

Results from microbial profiling of human intestinal microbiota

- No striking differences between groups
- Exception: Akkermansia muciniphila, a mucus eating bacterium
- Association but no causal connection

Berer et al. PNAS 2017

6.10

Human MS-derived microbiota trigger “MS” in transgenic mice

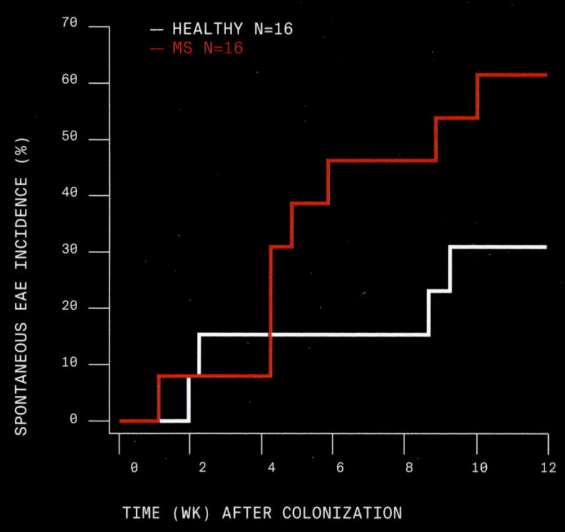

Disease triggered by intestinal Microbiota from MS (red) versus healthy twin (white)

MS-triggering microbiota from MS-Ileum: Koch’s postulates fulfilled

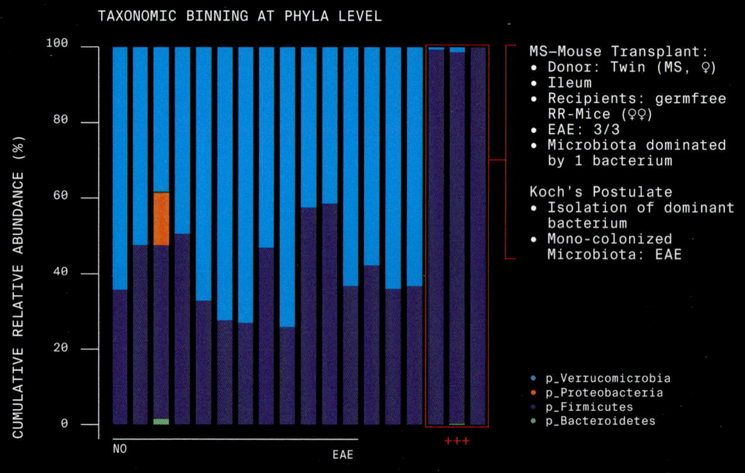

6.11

Conclusions

- MS-like mouse brain disease requires intact gut microbiota
- In human MS, gut microbial profiles may be changed
- MS-derived gut microbes trigger EAE in germfree mice
- One triggering bacterium identified

Remyelination Basic Studies and Imaging of Remyelination

7

Hôpital de la Pitié-Salpêtrière, Sorbonne University AP-HP,
Neurology Department and Paris Brain Institute
Paris, France

Promoting Repair in Multiple Sclerosis: The Next Challenge for Patients' Care

Affecting about 2,8 millions patients worldwide, and beginning mostly between the age of 25 and 35, Multiple Sclerosis is considered one of the major cause of acquired disability in young adults. It is driven by complementary components: I) inflammation of different types of immune cells infiltrating the central nervous system; II) demyelination (destruction of the myelin sheath, which surrounds the neuronal processes named axons, leading to reduced conduction velocity and increased vulnerability); III) neurodegeneration (loss or irreversible damage of neuronal cell bodies and axons). The contribution of each mechanism depends on the age of the patient, the disease duration, and its evolution: whereas focal inflammation predominates within Multiple Sclerosis plaques that are disseminated in the relapsing remitting forms of the disease, diffuse neurodegeneration is mostly the hallmark of the progressive phases. Symptoms (motor, sensory, visual symptoms, gait disorders, bladder dysfunction, cognitive deficit, among others) differ also in severity.

Major progresses were made within the last decades on the development of immunotherapies, which drastically reduce the inflammatory component of the disease and the relapse rate. But these therapies insufficiently improve the disability accrual and the progressive phase of the disease, which is related to progressive accumulation of irreversible neuronal damage, mostly independent of bouts of inflammation. A key strategy to counteract neurodegeneration is to achieve neuroprotection by enhancing myelin regeneration, hence restoring nerve conduction and metabolic support to the axon.

In this context, research has led to a better understanding of the cells in charge of remyelination in the central nervous system. These "repairing" cells are immature oligodendroglial cells (oligodendrocyte precursors or progenitors), which persist in the adult central nervous system, or mature oligodendrocytes (the myelinating cells of the central nervous system, that may survive the demyelinating insult and contribute to repair). Experimental research using complementary models of demyelination/remyelination allowed to unravel the main molecular mechanisms underlying repair, and to identify potential therapeutic targets. Further problems have emerged, including the identification of an optimal design for repair clinical trials, and of outcomes measures to quantify repair. The development of reliable imaging biomarkers assessing the efficacy of emerging promyelinating drugs is therefore an urgent need, and imaging methods applying magnetic resonance imaging (MRI) but also Positron Emission Tomography (PET, a molecular imaging technology that takes advantage of ligands that specifically bind to myelin), are actively developed.

These new methods should transform the landscape of therapeutic trials targeting central nervous system repair. Several promising studies translating efficacious pro-remyelination therapies to people with Multiple Sclerosis are currently on the way.

7.1

Repair in multiple sclerosis: from biology to clinical translation

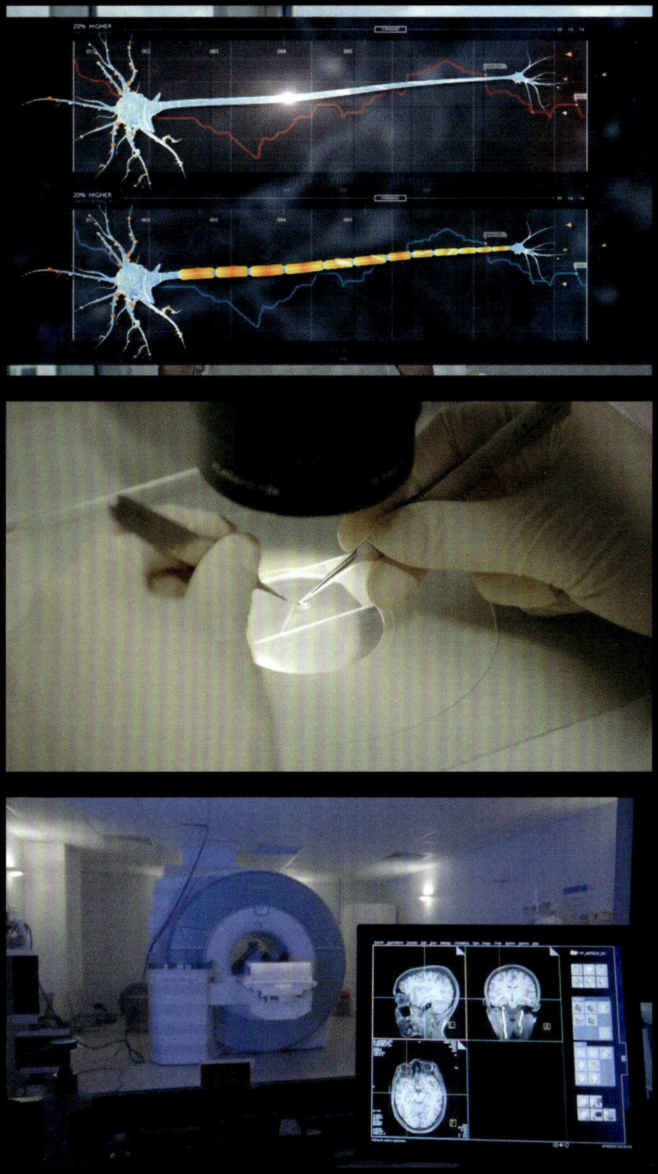

Stills from video featured in Room 7

Prof. Catherine Lubetzki, Professor of Neurology, co-team Leader at the Paris Brain Institute

7.2

The myelination of axons

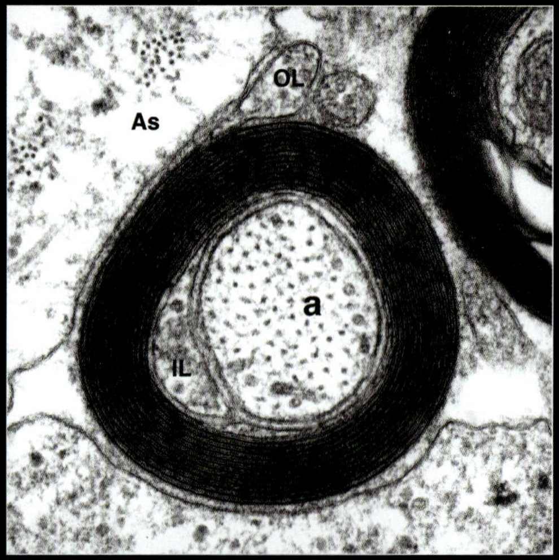

Electron microscopic view of a myelin sheath, synthetized by an oligodendrocyte (OL), wrapping around an axon (a) in the central nervous system.

Adapted from Lubetzki and Zalc, "Biology of Myelin," in Raine - McFarland - Hohlfeld, Multiple Sclerosis: a Comprehensive Text (Edinburgh/New York: Saunders/Elsevier, 2008)

Remyelination is possible after a lesion in Multiple Sclerosis

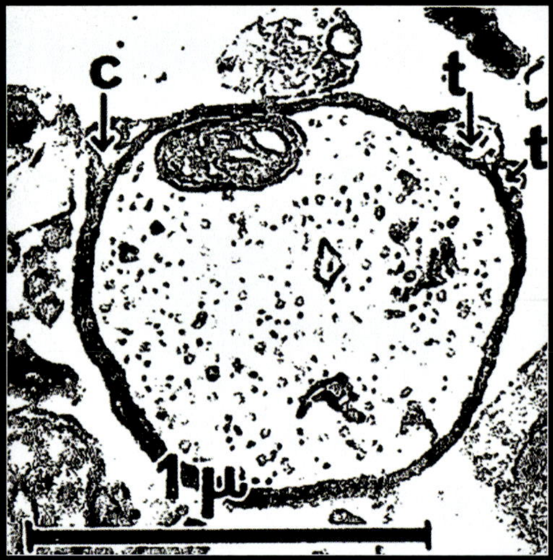

First image by electron microscopy of a remyelinated axon in a Multiple Sclerosis lesion. Note the thin meylin sheath typical of remyelination.

Périer - Grégoire, "Electron Microscopic Features of Multiple Sclerosis Lesions," in Brain, vol 88, no. 5, December 1965, pp. 937-952

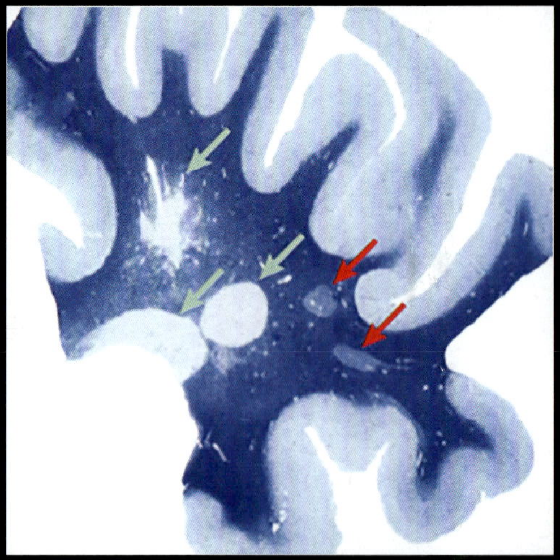

Human Multiple Sclerosis brain section. Dark blue areas correspond to normally myelinated white matter. White areas are demyelinated plaques (green arrows). Pale blue areas correspond to remyelinated plaques, named shadow plaques (red arrows).

Reproduced from Adams, A Colour Atlas of Multiple Sclerosis and Other Myelin Disorders (London: Wolfe Medical Publications, 1989)

7.3

Development and validation of complementary models

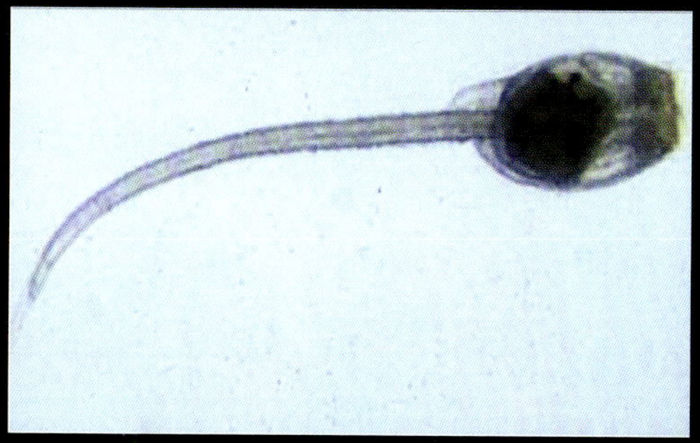

Transparent tadpole larva.

Elodie Martin, Paris Brain Institute

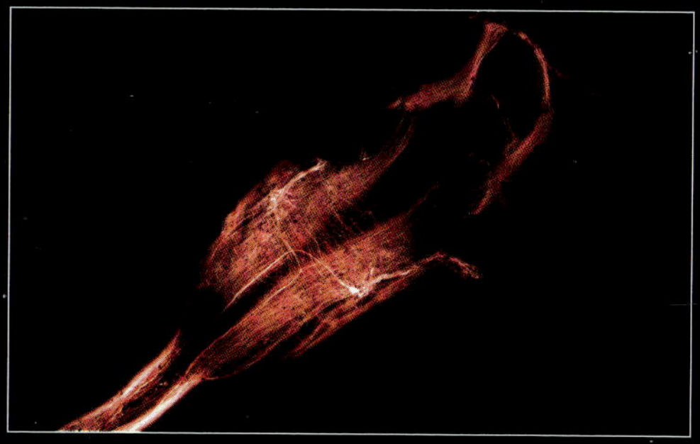

Brain of a tadpole larva. Central nervous system of Xenopus tadpole. The myelin is in red, and the axons are in white.

Elodie Martin, Paris Brain Institute

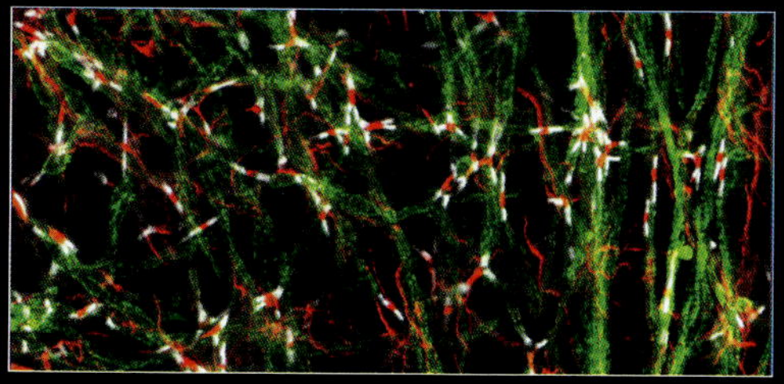

Mouse cerebellar slice in culture imaged by confocal microscopy. Myelin is in green, nodes of Ranvier in red, paranodes, flanking the nodes, in white.

Anne Desmazières, Inserm/Paris Brain Institute

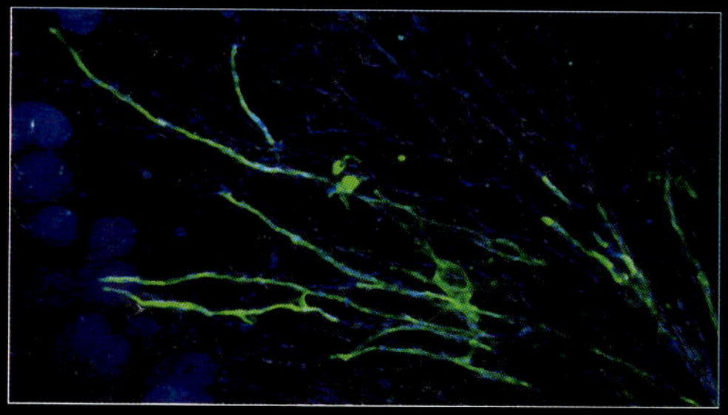

Oligodendrocyte (green) myelinating Purkinje cell axons (blue) in a cultured mouse cerebellar slice.

Anne Desmazières, Inserm/Paris Brain Institute

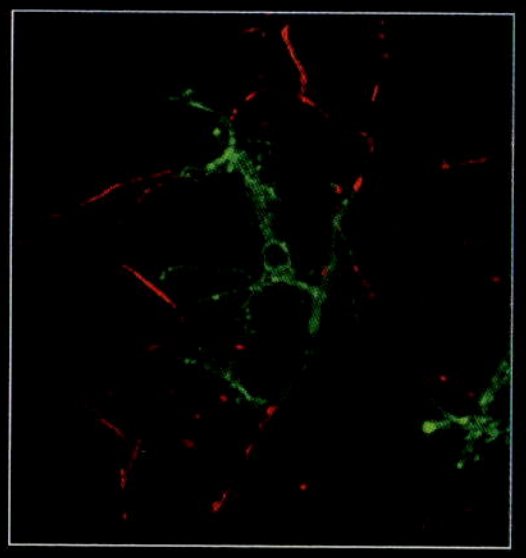

Live imaging image of an oligodendrocyte (green) myelinating an axon with nodal structures (red) in a mixed culture of hippocampal cells.

Anne Desmazières, Inserm/Paris Brain Institute

7.4

Stilbene and benzothiazole derivatives: repurposed radiotracers for myelin PET imaging

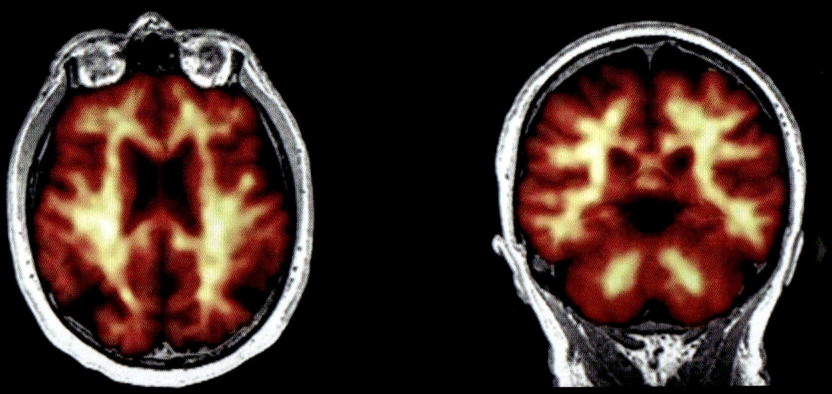

TEP-MRI of a human healthy subject using ^{18}F-florbetapir tracer binding to myelin.

Benedetta Bodini and Bruno Stankoff,
AP-HP/Sorbonne University/Paris Brain Institute

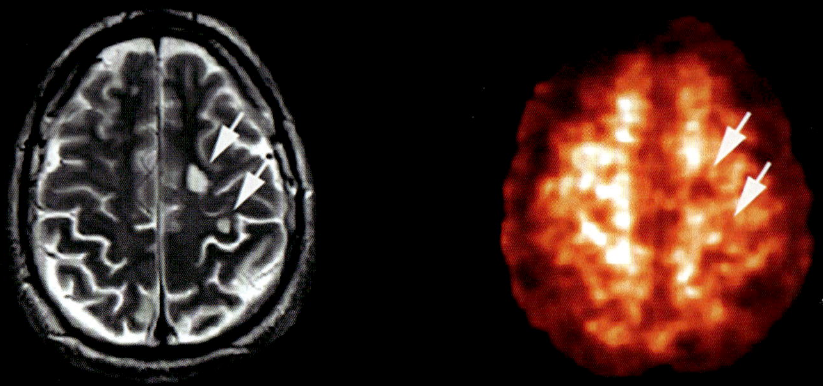

TEP-MRI using ^{11}C-PIB tracer detects demyelinated MS lesions (white arrows).

Bruno Stankoff, AP-HP/Sorbonne University/Paris Brain Institute

Four-months evolution of myelin content showing heterogeneity between patients and lesions

PET-MRI shows heterogeneous remyelination profiles in 3 subjects with Multiple Sclerosis. Demyelination in red, remyelination in blue. Note very few remyelinated areas in the patient's brain on the left, contrasting with numerous remyelinated areas in both patients in the middle and on the right.

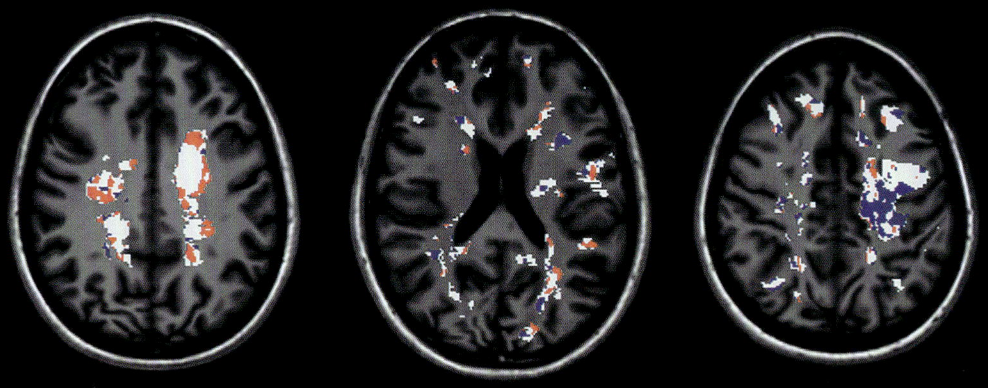

Bruno Stankoff, AP-HP/Sorbonne University/Paris Brain Institute

7.5

Charcot’s notebook

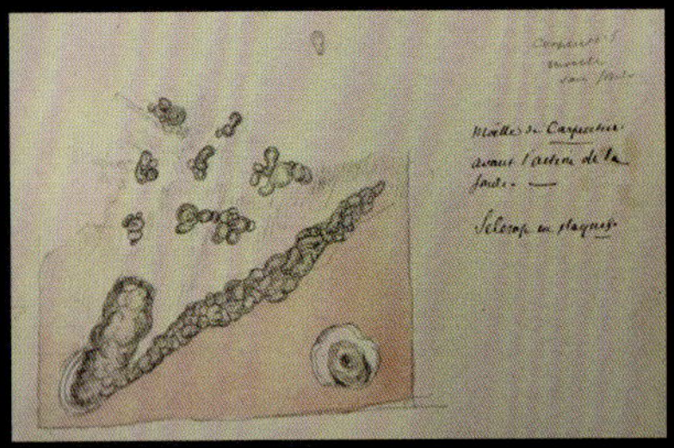

Original color drawing from Jean-Martin Charcot’s notebook showing the lymphatic sheath along a blood vessel distended by large droplets of “grease,” which correspond to myelin debris.

On the bottom right is a vessel cut transversally. On the right is Charcot’s handwritten legend: *Carpentier’s [patient’s name] spinal cord; Before soda [NaOH] treatment; Sclérose en plaques.*

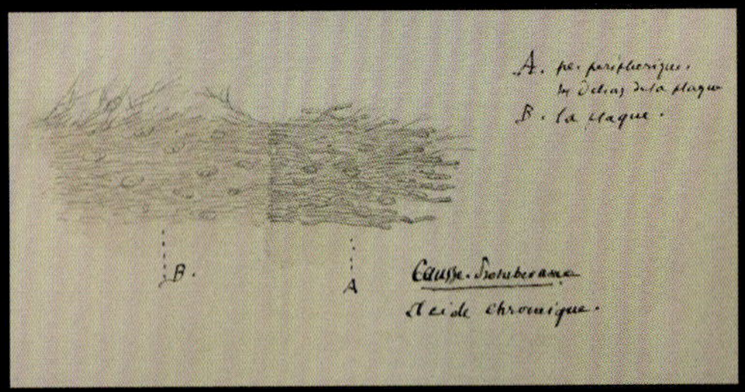

Original drawing from Jean-Martin Charcot’s notebook illustrating the limit of a lesion in the protuberance.

Protuberance is the ancient name of the pons, a part of the brain stem. The sharp limit between a lesion and the adjacent normal tissue is visible on this drawing. Chromic acid is the reagent used for the tissue’s staining.

Handwritten legend by Charcot; on the right: *A. periphery outside the plaque. B. the plaque.* Bottom: *Causse [patient’s name] Protuberance. Chromic acid.*

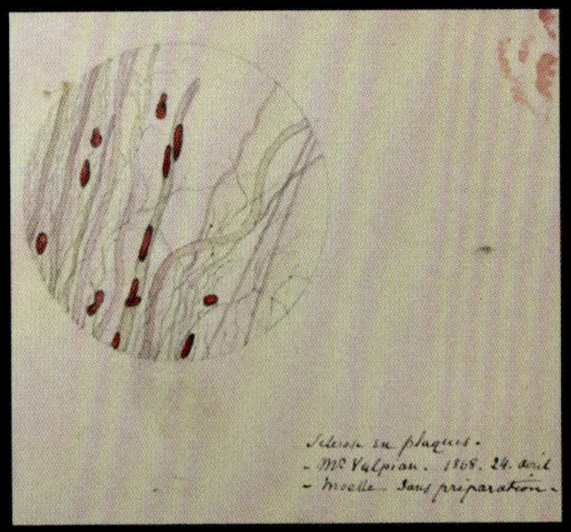

Original color drawing from Jean-Martin Charcot’s notebook illustrating the center of a sclerotic plaque.

In the center of the preparation is a blood vessel covered with many carmin positive nuclei and on each side are demyelinated axons of different diameter, lightly stained with carmin. In between the axons are thin fibrils, some associated with a nucleus. The rarefaction of axons is due to axonal degeneration.

On the right is Jean-Martin Charcot’s handwritten legend: *Sclérose en plaques; Mr Vulpian April 24, 1868* [indicating that it was Vulpian’s patient]; *Spinal cord no preparation.*

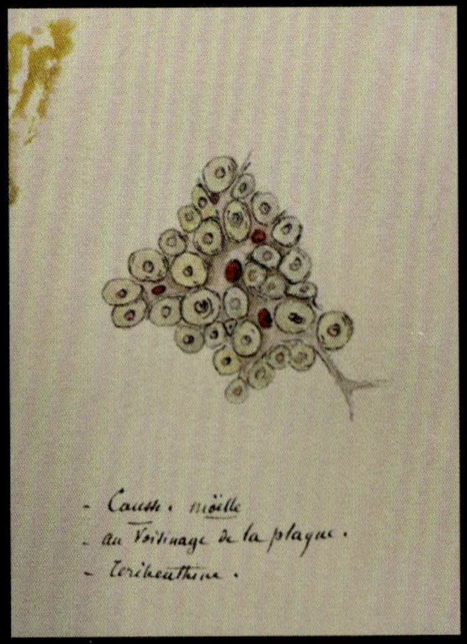

Original color drawing from Jean-Martin Charcot’s notebook showing myelinated axons on a transversal section of the spinal cord in the vicinity of a lesion. Cell nuclei are in red.

Bottom: *Causse* [patient’s name]; *Moelle (spinal cord); Around the lesion; Terébenthine (Turpentine).*

Département de Neuropathologie,
Groupe Hospitalier Pitié-Salpêtrière-Charles Foix,
Musée de l’Assistance Publique-Hôpitaux de Paris

Memory Driven Computing in Neurodegenerative Diseases

8

German Center for Neurodegenerative Diseases (DZNE)
within the Helmholtz Association
Bonn, Germany

The exhibit is part of the first prototype of "The Machine," a completely new computer architecture developed by Hewlett Packard Enterprise (HPE). Most current computers are based on the so-called "von Neumann architecture," which places the processor at the center of the computer. However, the traditional computer infrastructure fails when it comes to processing enormously large amounts of data. As a result, large amounts of data must nowadays be distributed over very large computer systems. Such distribution causes a considerable amount of computing time, which becomes prohibitive when analyzing biomedical data.

"The Machine" focuses on memory to store, compute, and process large amounts of data. It is also referred to as "memory-driven computing" or memory-centric computing. The exhibit is a so-called node of a 40-node prototype computer system for memory-centric computing. Each node has 4 terabytes (TB) of memory, which has been combined into a total of 160 TB of unified memory. The 160 TB of memory is equivalent to approximately 250,000 CDs. Currently, this is the largest pool of memory within a single computer system. "The Machine" thus allows for more efficient processing and a modular architecture. The processors are hidden under heat distributors, but the many memory bars are clearly visible. Research into neurodegenerative diseases is becoming increasingly data-driven. Microscopes and MRI machines deliver higher-resolution images, genomics data is collected and clinical and population studies provide a data avalanche that outpaces the growth of computational resources. This new type of computer is becoming a "gamechanger" as it allows to process data more easily in parallel. Working with genomics data often involves a series of tools that exchange data using files, but this I/O centric approach can be replaced using memory-driven computing. Here, tools can work more efficiently together on data in memory

The DZNE has processed genomics up to 100 times faster by changing the representation of a reference genome in memory so that it can be accessed more efficiently. Not only has this saved 60% of energy, but it also allows to study the molecular processes within a cell at higher resolution. Using high-memory computer hardware, DZNE can analyze single-cell data sets with millions of cells to understand changes in gene expressions and cellular processes during on-set and progression of dementia in patients in comparison to healthy controls.

The processing of large data sets in other domains, e.g. high resolution imagery, can often be portioned to multiple machines that need to synchronize because they do not share state. With memory-driven computing, a shared state limits overheads and reduces the needed programming, giving the researcher more time to focus on their research question.

8.1

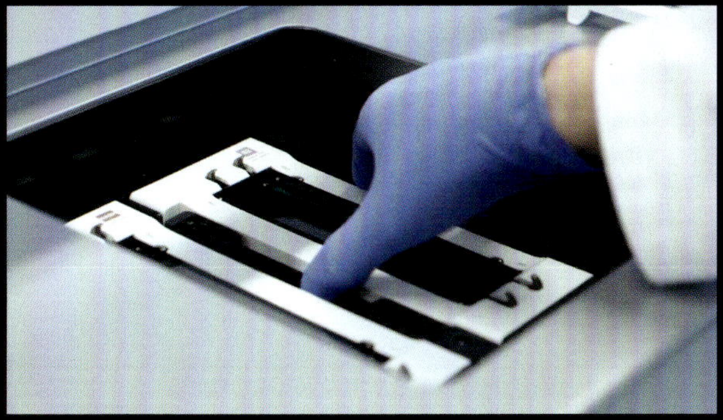

Fighting Alzheimer's Disease with Memory-Driven Computing

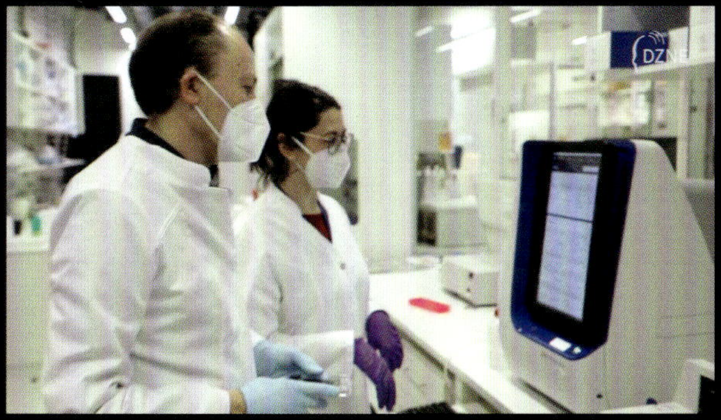

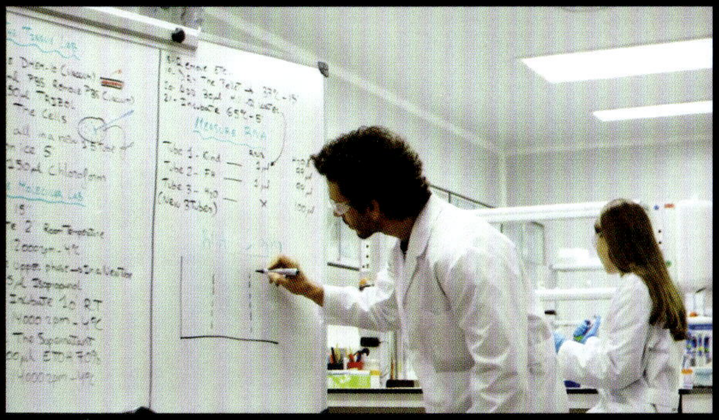

The concept of Swarm Learning

Stills from video featured in Room 8

8.2

Matthias Becker in the datacenter at DZNE in Bonn. Becker works on Modular High Performance Computing and Artificial Intelligence.

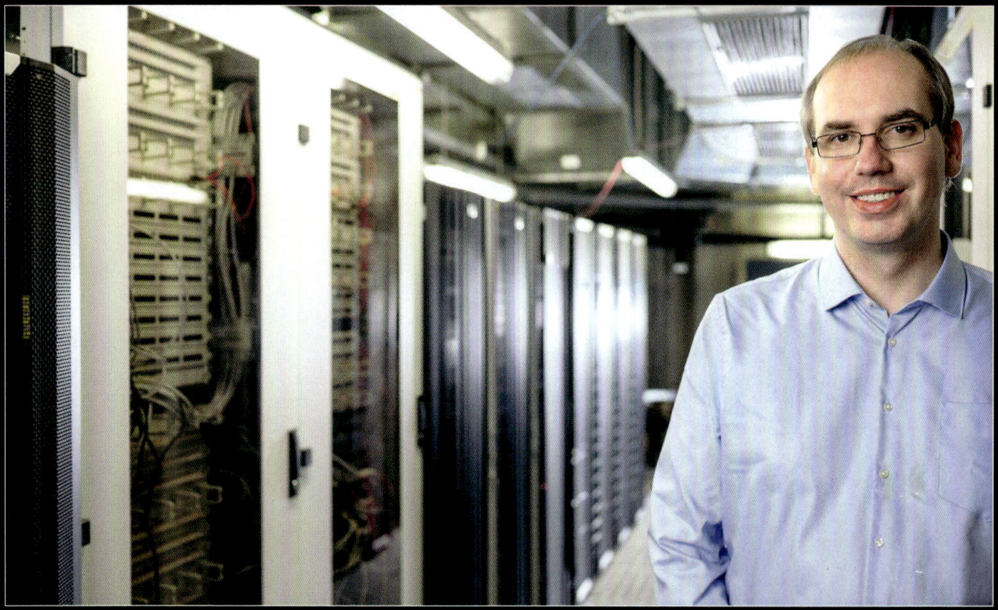

Credit: DZNE/Frommann

8.3

The main building of DZNE in Bonn.

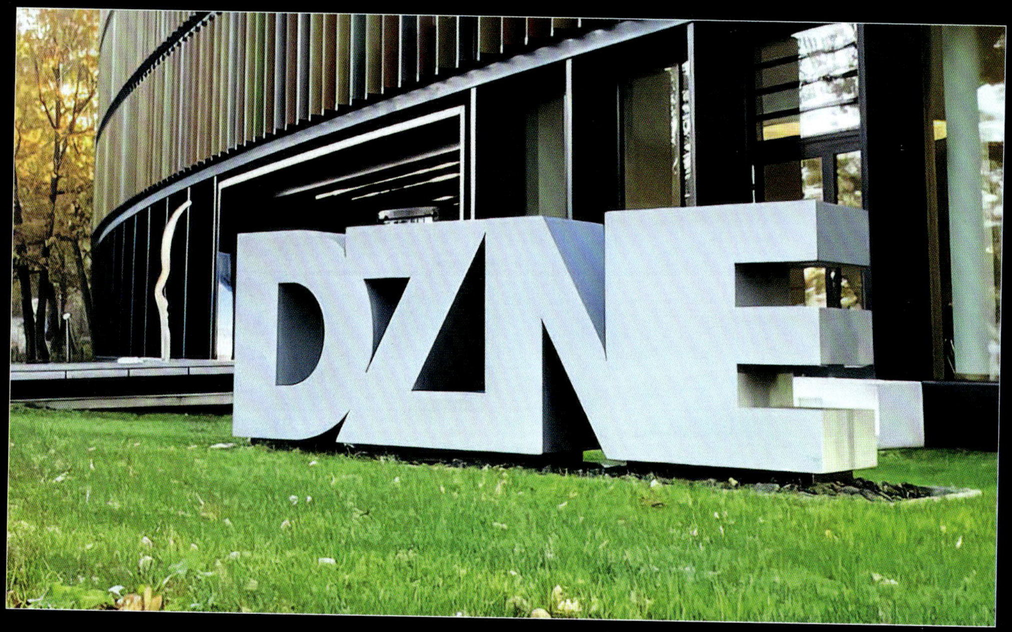

Credit: DZNE/Frommann

8.4

Various blood samples prepared to be analyzed in the lab.

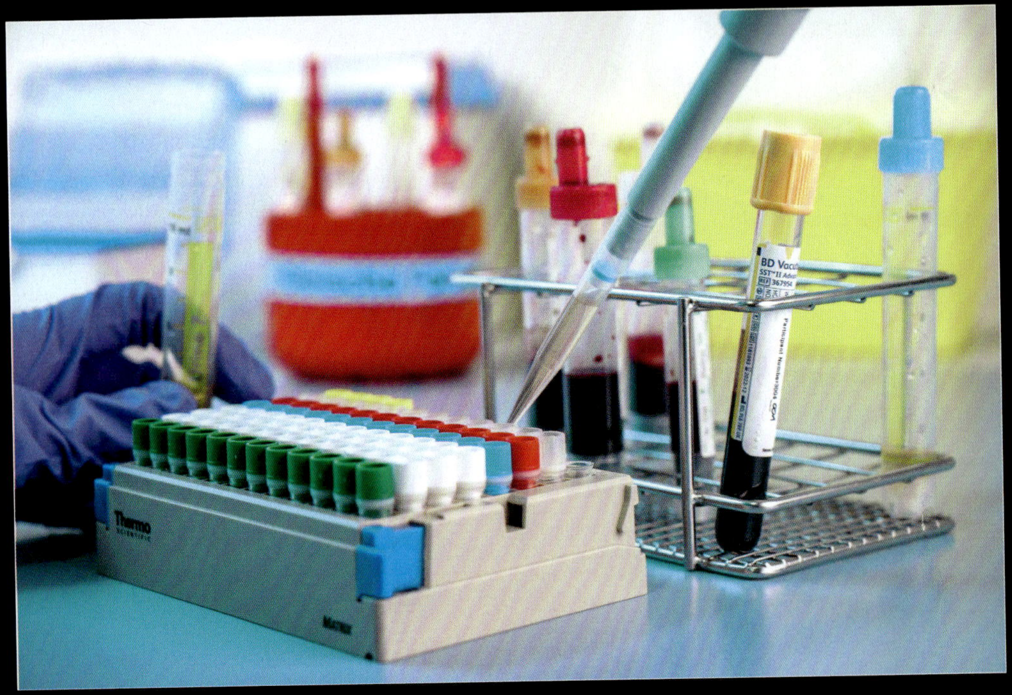

Credit: DZNE/Frommann

8.5

Daily work at DZNE lab.

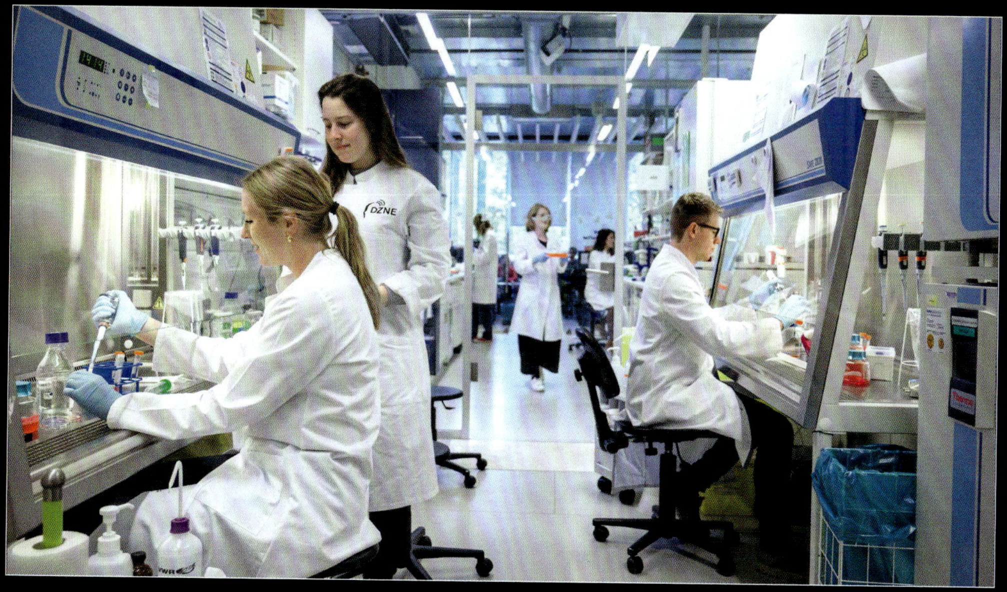

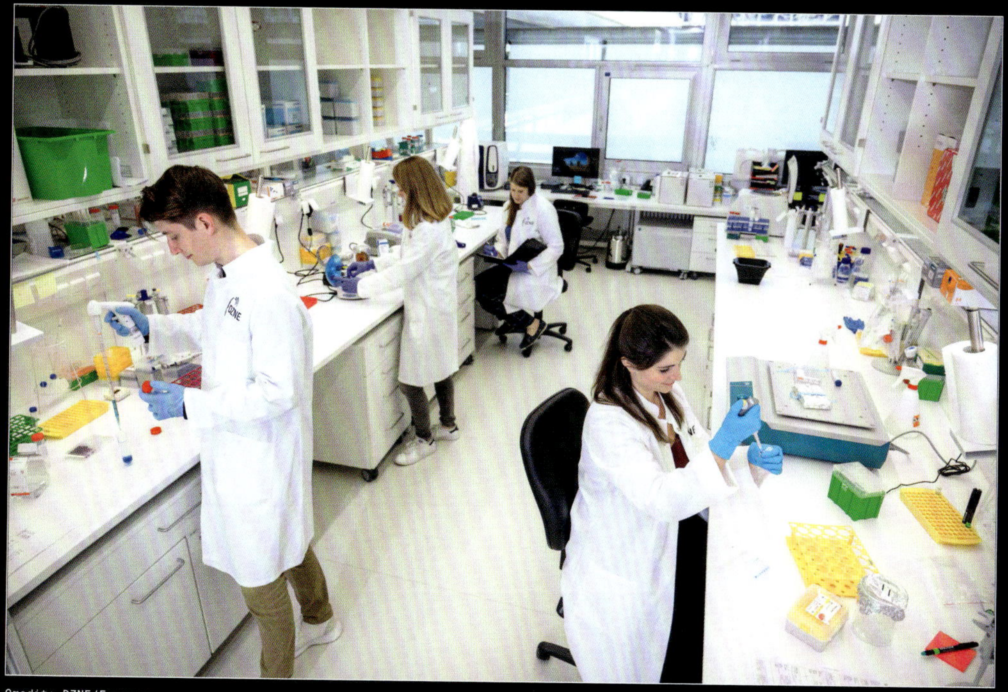

Credit: DZNE/Frommann

8.6

The main datacenter at DZNE in Bonn.

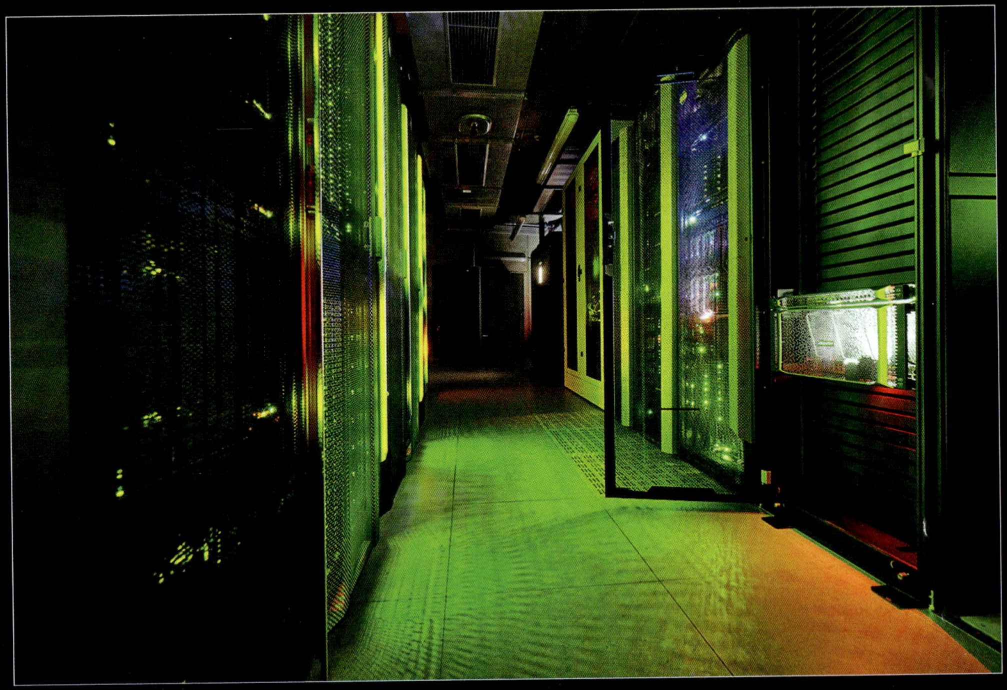

Credit: DZNE/Lannert

Neuroimaging of Dementia

9

Tianjin Medical University General Hospital,
Neurology Department
Tianjin, China

MRI and PET Scans in Patients with Alzheimer's Disease

Alzheimer's Disease (AD) is the most common cause of dementia in the elderly with a neurodegenerative continuum that includes a long period in the preclinical phase without any symptoms and an early clinical phase, which is named Mild Cognitive Impairment (MCI) or the prodromal stage, and finally a progressive phase of dementia.

Structural Magnetic Resonance Imaging (MRI) has been widely used to identify patients with AD and distinguish them from those with other causes of cognitive impairment, such as vascular dementia. Furthermore, arterial spin labeling (ASL), which is a non-invasive MRI technique to quantitatively measure Cerebral Blood Flow (CBF) using arterial blood water as an endogenous tracer, has been increasingly investigated in AD, because it has been suggested that changes in CBF plays an important role in the pathogenesis and progression of neurodegenerative diseases, even independent of their core neuropathology.

In the past decades, Positron Emission Tomography (PET) neuroimaging techniques played an important role and have been recommended to use in clinical diagnosis of dementia with appropriate tracers to assess pathophysiology *in vivo* in neurodegenerative conditions. Compared with indirect measuring of abnormal molecular expressions in the cerebrospinal fluid, PET scans are noninvasive and able to reflect topographical distribution of protein deposition or metabolic changes.

In conclusion, patients with AD had characterized patterns of brain atrophy, cerebral hypoperfusion and hypometabolism. In addition to structural scans, ASL MRI (especially processed with multivariate analysis) may provide a promising biomarker for AD diagnosis and monitoring on a prospective single-subject basis in the future. Currently, amyloid and tau depositions and metabolic patterns measured with PET scans are useful in early diagnosis and differential diagnosis in patients with cognitive impairment.

9.1

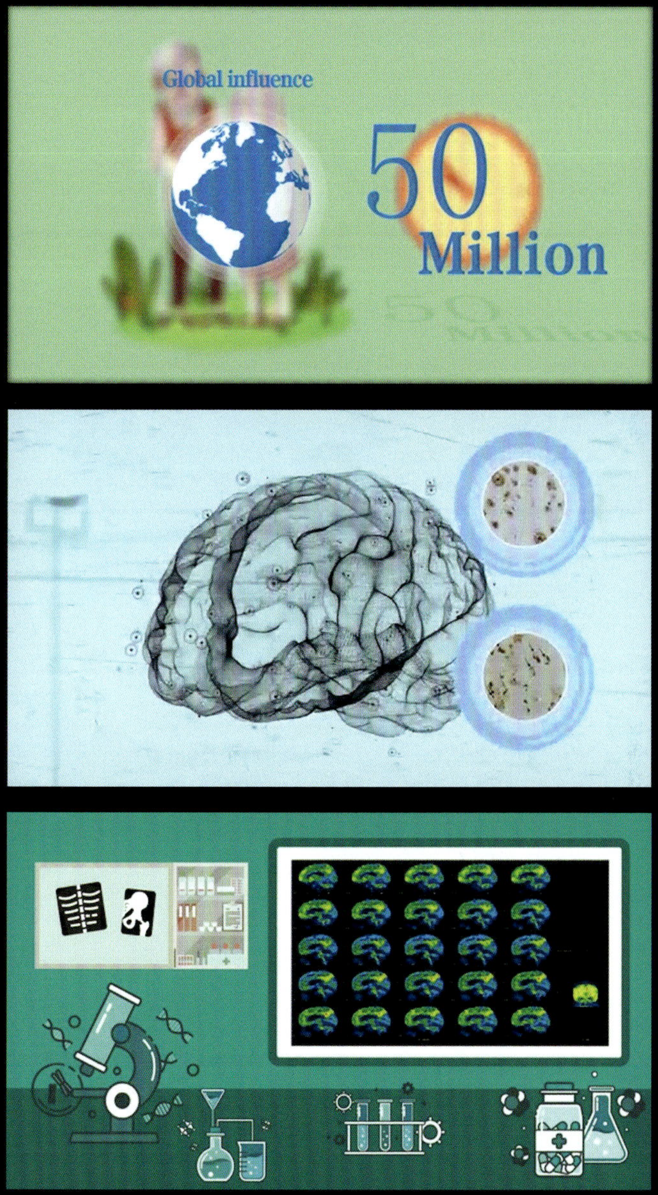

Stills from video featured in Room 9

9.2

Magnetic Resonance Imaging

Structural magnetic resonance imaging (MRI) has been wildly used to identify patients with Alzheimer's Disease (AD) and distinguish them from those with other causes of cognitive impairment, such as vascular dementia.

Brain atrophy, in particular medial temporal lobe atrophy, reflects neurodegeneration or neuronal injury and correlates with decline of cognitive and global functions in AD.

Brain atrophy on MRI in patients with unimpaired cognition and Alzheimer's Disease

Obvious atrophy in extensive brain areas, especially medial temporal lobe and hippocampus, on structural MRI (top left and right) was observed in a 67-year-old patient with AD whose cognitive testing of the Mini-Mental State Examination (MMSE) was 8/30. Mild atrophy related to aging was observed on T1WI structural MRI (bottom left and right) in a 70-year-old patient with unimpaired cognition, indicating by an MMSE score of 30/30.

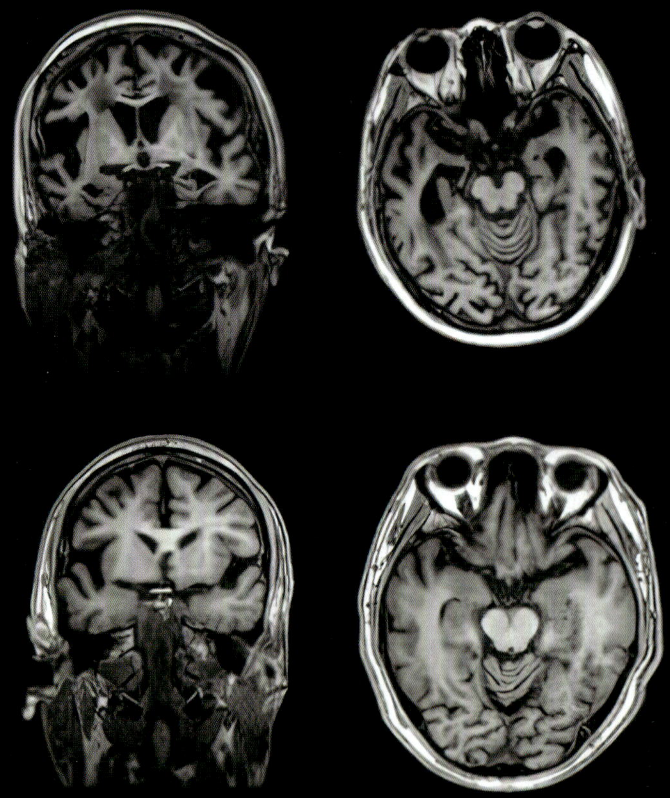

Brain atrophy on coronal MRI in a patient with Alzheimer's Disease

A 60-year-old patient who had a complaint of memory decline for two years, was diagnosed as clinically probable AD in 2011. Brain atrophy, in particular medial temporal lobe and hippocampus, was getting more and more severe during a 7-year follow up.

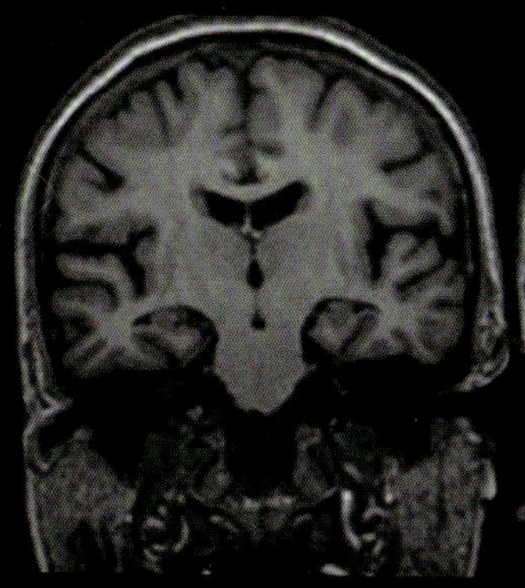

2011.09

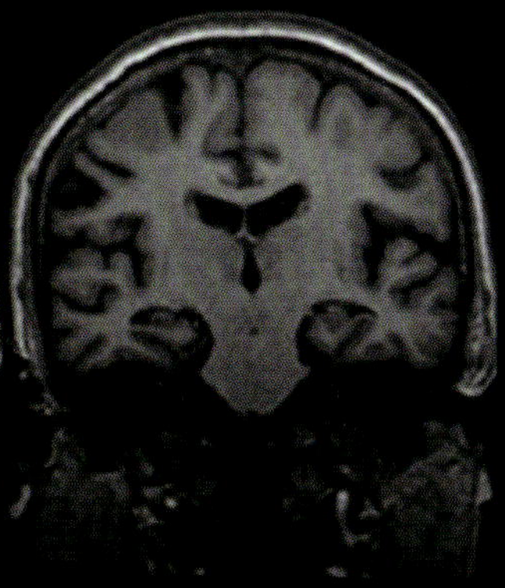

2013.03

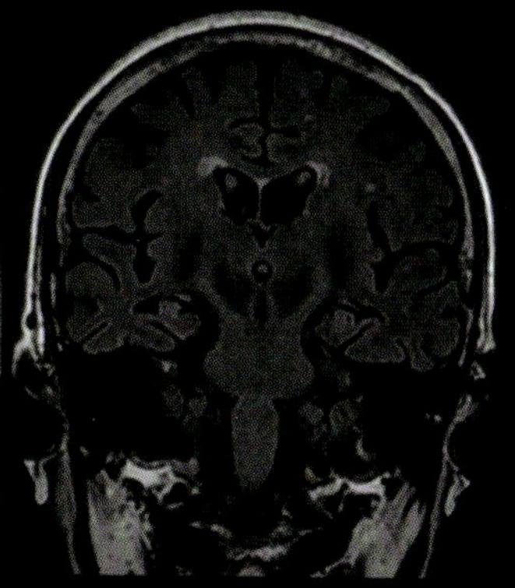

2015.07

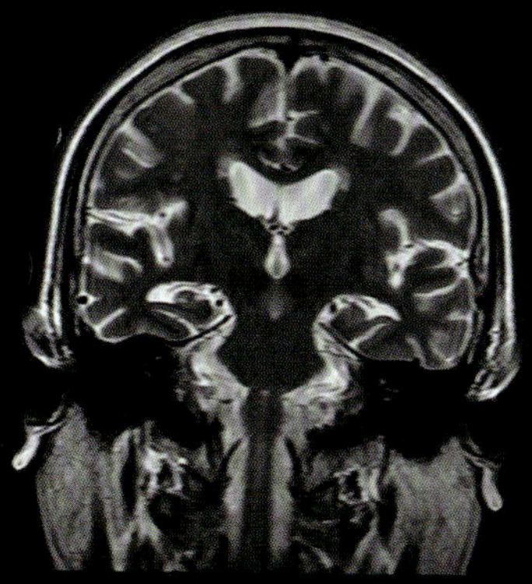

2016.05

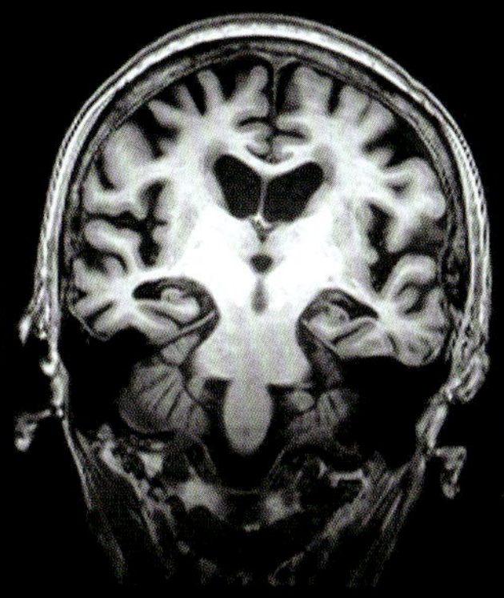

2017.11

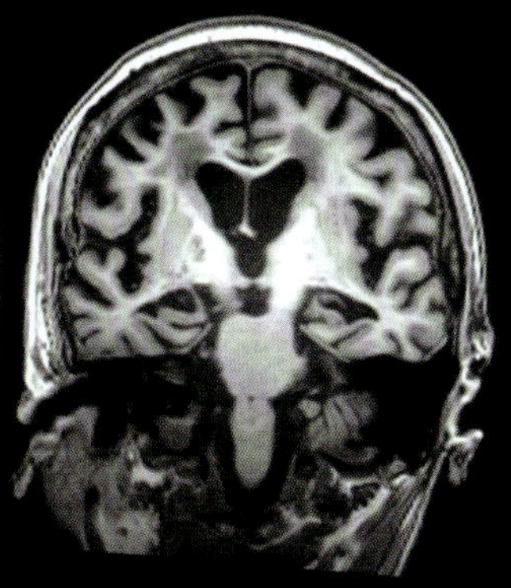

2018.07

Arterial Spin Labeling

Arterial spin labeling (ASL), which is a non-invasive MRI technique to quantitatively measure Cerebral Blood Flow (CBF) using arterial blood water as an endogenous tracer, has been increasingly investigated in AD, because it has been suggested that changes in CBF plays an important role in the pathogenesis and progression of neurodegenerative diseases, even independent of their core neuropathology. Recently, an AD-related perfusion pattern (ADRP) measured with ASL MRI was established using multivariate spatial covariance analysis. Consistent with the findings of voxel-wise univariate analysis, the ADRP featured negative loading in bilateral middle and posterior cingulate and precuneus, inferior parietal lobule, and frontal areas, and positive loading in right cerebellum and bilateral basal areas. Subject expression score of the ADRP was significantly elevated in patients with AD compared with cognitively normal individuals and showed good diagnostic accuracy for AD. Moreover, there were negative correlations between subject expression score of the ADRP and global cognitive function and performance in various cognitive domains in patients with AD.

Regional topographies of perfusion pattern measured with ASL MRI in patients with Alzheimer's Disease

This perfusion pattern was identified by scaled subprofile model (SSM) / principal component analysis (PCA).

Cool color indicates regions with decreased loading, and warm color indicates regions with increased loading in patients with AD.

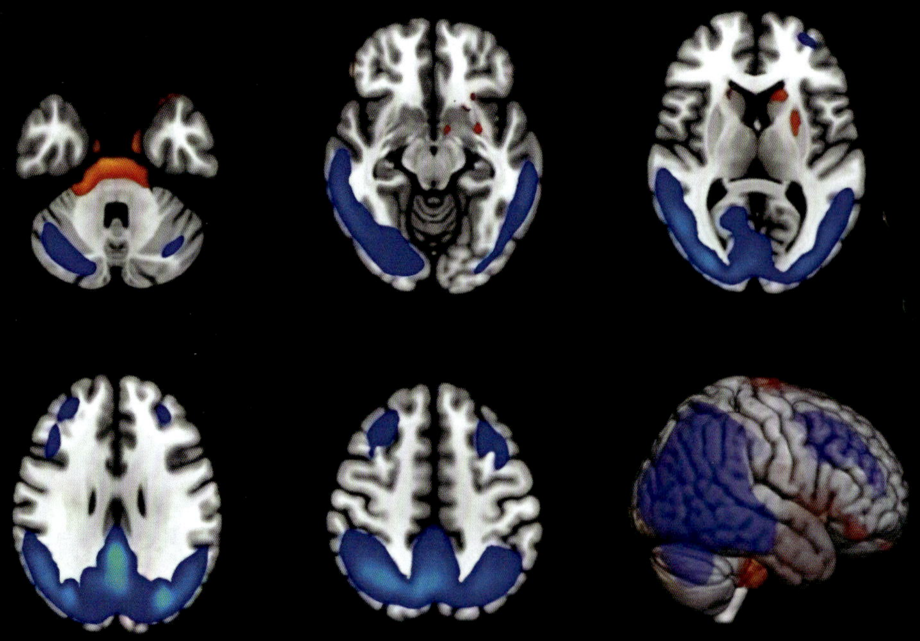

9.3

Positron Emission Tomography

In the past decades, Positron Emission Tomography (PET) neuroimaging techniques played an important role and have been recommended to use in clinical diagnosis of dementia with appropriate tracers to assess pathophysiology in vivo in neurodegenerative conditions. Similar as atrophy on anatomic MRI, cerebral glucose hypometabolism on fluorodeoxyglucose (FDG) PET is a downstream biomarker of neuronal injury and neurodegeneration.

Characteristic patterns of regional metabolic changes on FDG PET have been identified in association with common types of neurodegenerative dementia, such as hypometabolism predominantly in the temporoparietal cortex and posterior cingulate cortex in Alzheimer's Disease (AD) and hypometabolism in the frontal and anterior temporal areas in frontotemporal lobar degeneration (FTLD). There was a significant difference in the hypometabolic pattern between amyloid-positive and amyloid-negative patients with AD or mild cognitive impairment (MCI) in a retrospective analysis based on data from our memory clinic.

Metabolic features measured with Fluorodeoxyglucose-Positron Emission Tomography (FDG PET) in patients with unimpaired cognition, Mild Cognitive Impairment, and Alzheimer's Disease

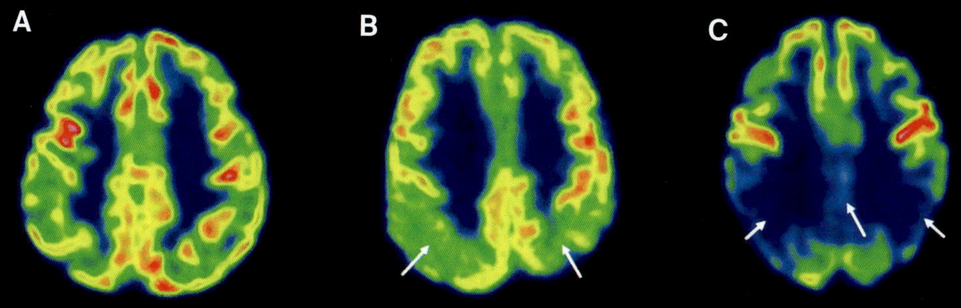

A: There is no specific metabolic change in a 51-year-old patient with unimpaired cognition, whose cognitive testing of Mini-Mental State Examination (MMSE) was 30 (total score is 30).

B: A 69-year-old patient with MCI (MMSE=24/30) showed a moderate hypometabolism in parietal lobe.

C: A characteristic metabolic pattern was observed in a 55-year-old patient with AD (MMSE=11/30), featuring predominant hypometabolism in the posterior regions, such as temporoparietal cortex and posterior cingulate, and moderate hypometabolism in frontal areas.

Both β-amyloid plaques and neurofibrillary tau deposits are the pathological hallmarks of Alzheimer's Disease

Extensive cortical plaques were observed in patients with AD at postmortem. Amyloid deposition is believed to begin long time (approximately 15-20 years) before the onset of symptoms.

The positivity of amyloid PET was 77.5% and 47.8% for individuals with clinically suspected AD or amnestic MCI, respectively, in our memory clinic. The deposition of fibrillar tau aggregates has been implicated in AD and allied neurodegenerative disorders collectively referred to as tauopathies. In AD, tau fibrillogenesis is more intimately linked to neuronal death than amyloid aggregation, and is also supposed to be initiated decades before the clinical onset of the disease. Tau accumulation was identified as involving the medial temporal lobes and limbic system at the early stages and parietal, temporal, precuneus, posterior cingulate, and frontal lobes as the disease progression on PET in patients with AD. In addition, imaging of tau pathologies is also expected to allow classification of FTLD spectrum into neuropathological subtypes.

Amyloid deposition measured with Carbon-11 (11C)-labeled Pittsburgh Compound B (PiB) PET in patients with unimpaired cognition, Mild Cognitive Impairment, and Alzheimer's Disease

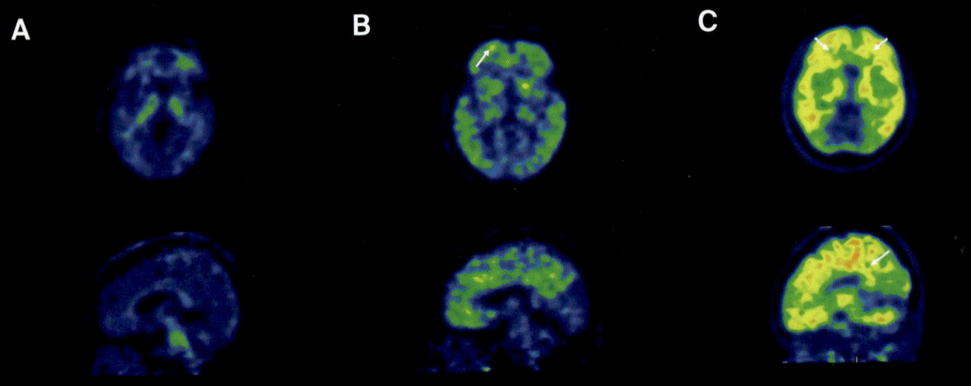

A: No obvious uptake of PiB was observed in a 64-year-old patient with unimpaired cognition, indicating by an MMSE score of 30/30.

B: There were moderate bindings of PiB in frontal and parietal lobes in a 64-year-old patient with MCI (MMSE=27/30).

C: A 61-year-old patient with AD (MMSE=19/30) showed obviously increased PiB retention in extensive cortical regions, in particular frontal lobes and temporoparietal cortex.

Tau deposition measured with ^{18}F-S16 (^{18}F-(S)-1-(4-(6-(dimethylamino) quinoxalin-2-yl)phenoxy)-3-fluoropropan-2-ol) PET in patients with unimpaired cognition, Mild Cognitive Impairment, and Alzheimer's Disease

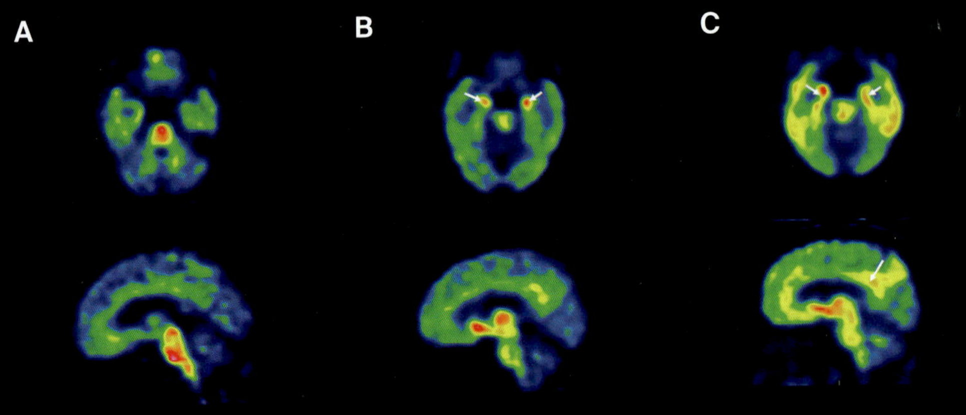

A: No obvious uptake of tau was observed in a 72-year-old patient with unimpaired cognition, indicating by an MMSE score of 30/30.

B: There were tau depositions in the medial temporal lobe and hippocampus in a 64-year-old patient with MCI (MMSE=27/30).

C: A 61-year-old patient with AD (MMSE=19/30) showed tau bindings in temporal and parietal cortex, precuneus, posterior cingulate, and a few frontal areas.

PET and Body Fluid Biomarkers in Neurodegenerative Diseases 10

UniSR – Università Vita-Salute San Raffaele
Milan, Italy

Biomarkers have a double role: to contribute to reveal potential therapeutic targets and to diagnose and to monitor diseases evolution and response to treatment. In neurodegenerative diseases, neuroimaging techniques and body fluid biomarkers provide fundamental information on the ongoing brain damage and the underlying pathophysiological mechanisms. The contribution of Positron Emission Tomography (PET), a structural and functional technique, and of serum and spinal fluid neurofilaments are here presented, highlighting the results of some studies performed at San Raffaele Hospital.

10.1

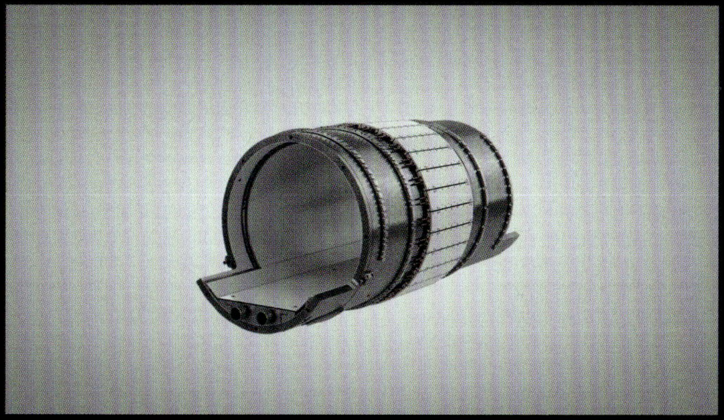

Positron emission tomography (PET)

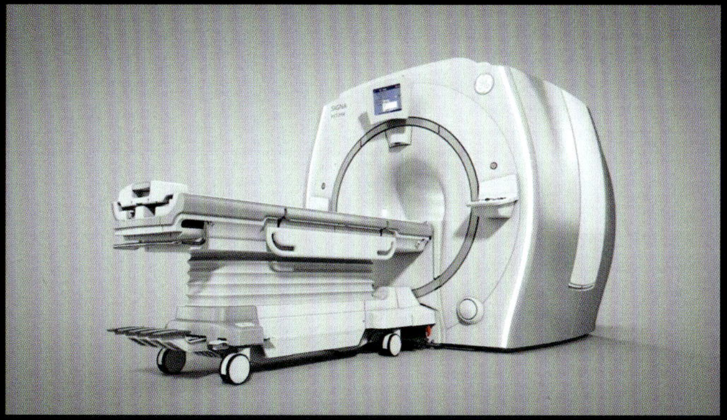

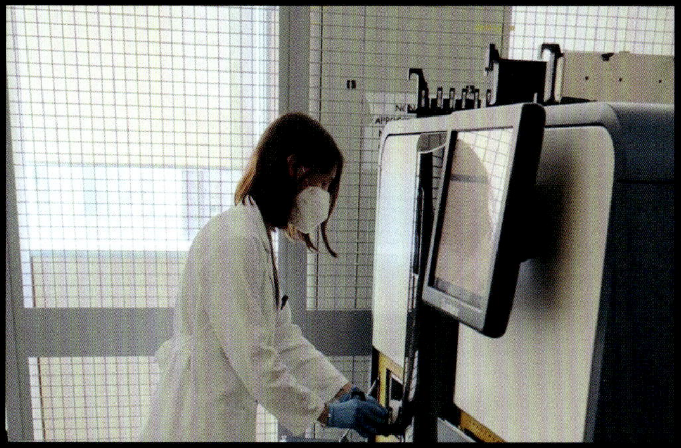

Body fluid biomarkers

Stills from video featured in Room 10

10.2

Positron Emission Tomography (PET)

PET scan and technical details

A compelling need in the field of neuro-degenerative diseases is the development and validation of biomarkers for early identification and differential diagnosis.
The availability of Positron Emission Tomography (PET) neuroimaging tools for the assessment of brain function, molecular biology and neuropathology has opened new venues in the diagnostic design and the conduction of new clinical trials. PET techniques are increasingly showing great potential in supporting clinical diagnosis also in the early and even preclinical phases of dementia.
PET techniques have indeed provided major advances in promoting novel approaches to support an early and differential dementia diagnosis.

The most recent evidence on fluorine-18 fluorodeoxyglucose-, amyloid-, tau-, and neuro-inflammation-PET tools highlight possible new perspectives in research and clinical applications in neurodegenerative conditions. 18F-FDG PET provides a well-validated key value to dementia diagnosis and prognosis and should be considered as one of the most valuable tools for monitoring neurodegenerative disease status and progression and also for selecting candidates for clinical trials.

The large amount of scientific evidence obtained by *in vivo* PET imaging of brain amyloid burden has challenged the primary role for amyloid in pathogenesis in Alzheimer dementia. One of the most recent advances for *in vivo* PET imaging is the evaluation of cerebral tau burden and the novel tau PET imaging is very promising, given its tight relationship with neurodegeneration. *In vivo* imaging of neuroinflammation has recently gained particular interest in clinical neuroscience research, and neuroinflammation has been recognized as a key player in the course of neurodegeneration. The currently available PET molecular imaging techniques allow the measurement of neuroinflammation through imaging of both astrocytes and microglia activation.

The last decades have progressively witnessed a shift from a solely clinical diagnosis to a biomarker-supported diagnosis, and molecular neuroimaging techniques such as PET have played a leading role in dementia diagnostic and prognostic work-up. An accurate quantification, together with ad hoc use of PET techniques, can nowadays detect very subtle, but significant, pathological and functional brain changes, even before clinical symptoms arise. This is crucial for early interventions, personalized care planning, and inclusion in clinical trials.

The most important advances of four of the leading PET tools (or both) in the dementia research field, namely 18F-FDG (fluorine-18 fluorodeoxyglucose), amyloid, tau, and neuroinflammation PET imaging are illustrated in this section. The use of these techniques, together with the standardization of appropriate methodologies, allow unique breakthroughs in our understanding of neurodegeneration and have remarkable implications for diagnostic algorithms and therapy monitoring. It is likely that the multi-modal integration of pathological and functional PET biomarkers will be the key proxy to the most accurate identification of both underlying pathology and phenotypic syndrome, leading to the tailoring of the most appropriate care plan and prognosis.

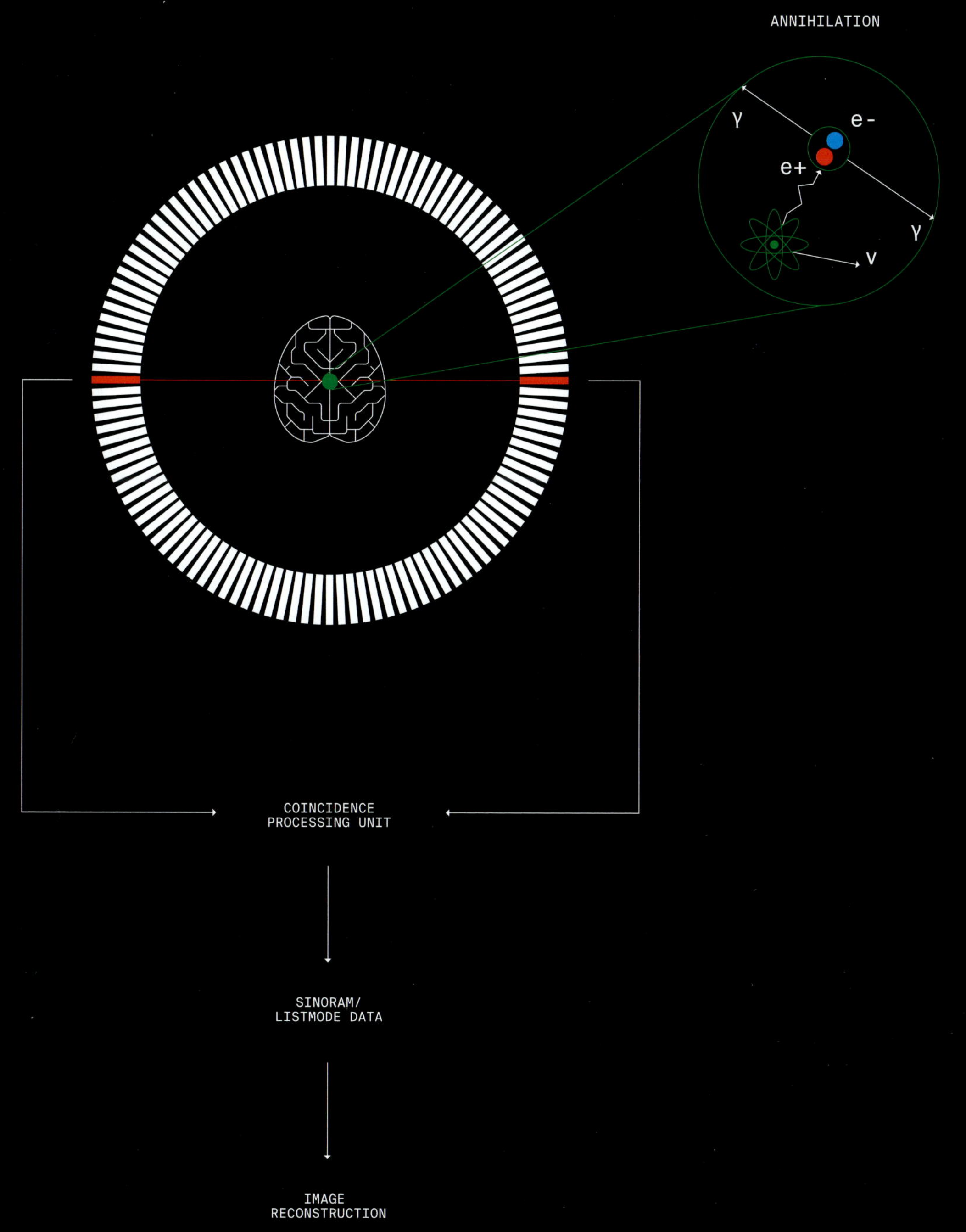
ANNIHILATION
γ
e-
e+
γ
v
COINCIDENCE
PROCESSING UNIT
SINORAM/
LISTMODE DATA
IMAGE
RECONSTRUCTION

10.3

Fluorine-18 fluorodeoxyglucose PET measure of brain glucose metabolism

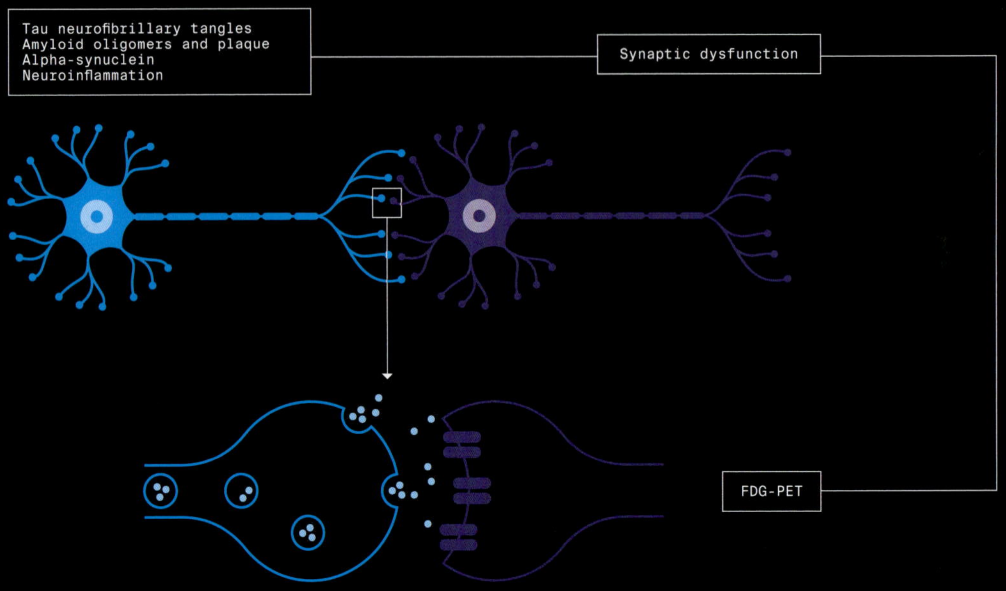

Brain Hypometabolism Patterns discriminating different neurodegenerative dementias

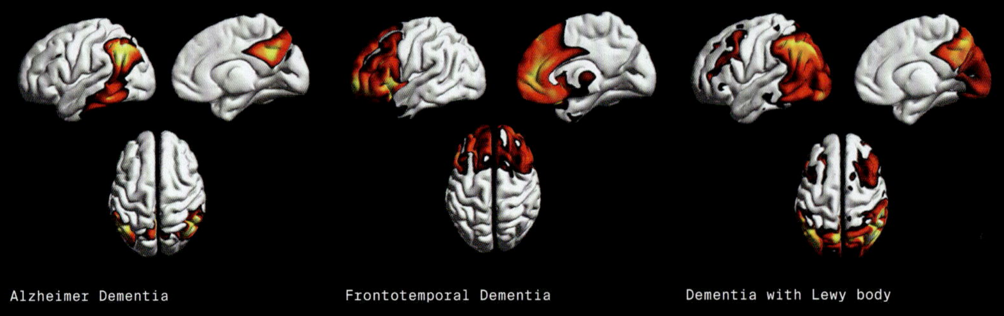

Courtesy of Prof. D. Perani, San Raffaele Scientific Institute, personal data.

10.4

PET Imaging of neuroinflammation, microglia activation

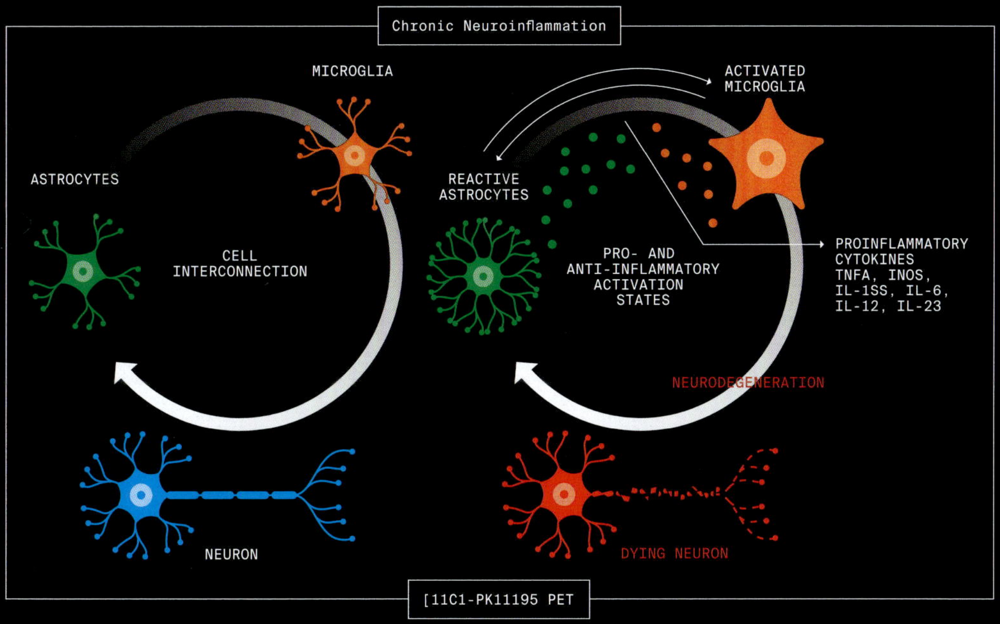

TSPO binding, a protein which is overexpressed by activated microglia and astrocytes allows the *in vivo* detection of neuroinflammatory responses

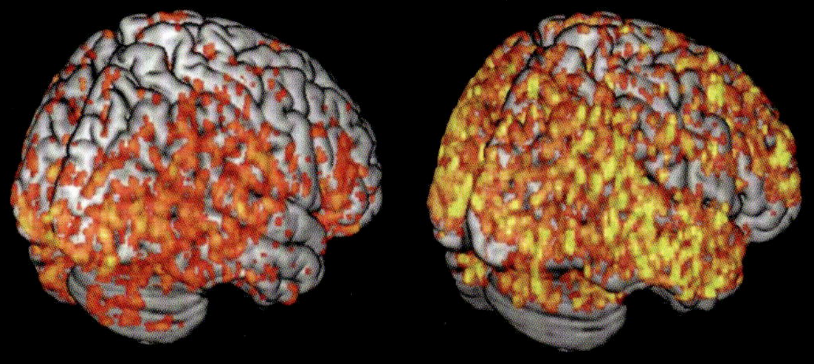

Mild Cognitive Impairment due to Alzheimer's Disease

Early onset Alzheimer's Disease dementia

Modified from Tondo et al. 2020, 2021.

10.5

PET presynaptic receptor measures of dopaminergic dysfunction in Parkinson's disease

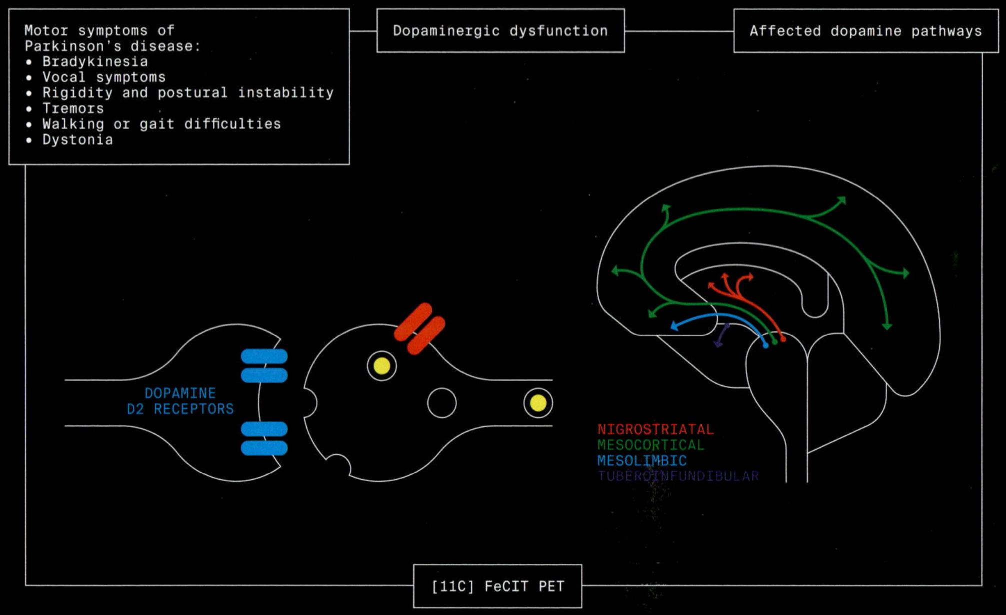

Parkinson's disease (PD) scan

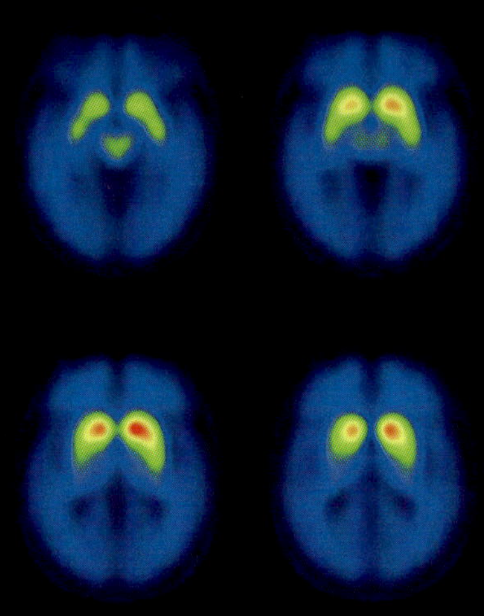

Modified from Caminiti et al. 2017

Healthy subject (HC) scan

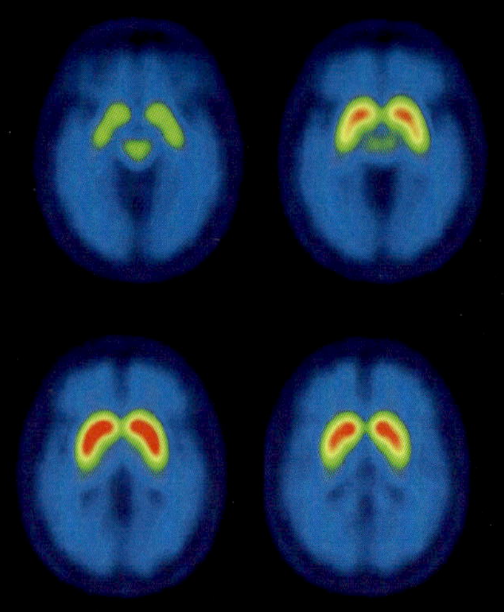

Caminiti et al. 2017 NeuroImage:Clinical

10.6

PET Biomarkers in preclinical/prodromal subjects

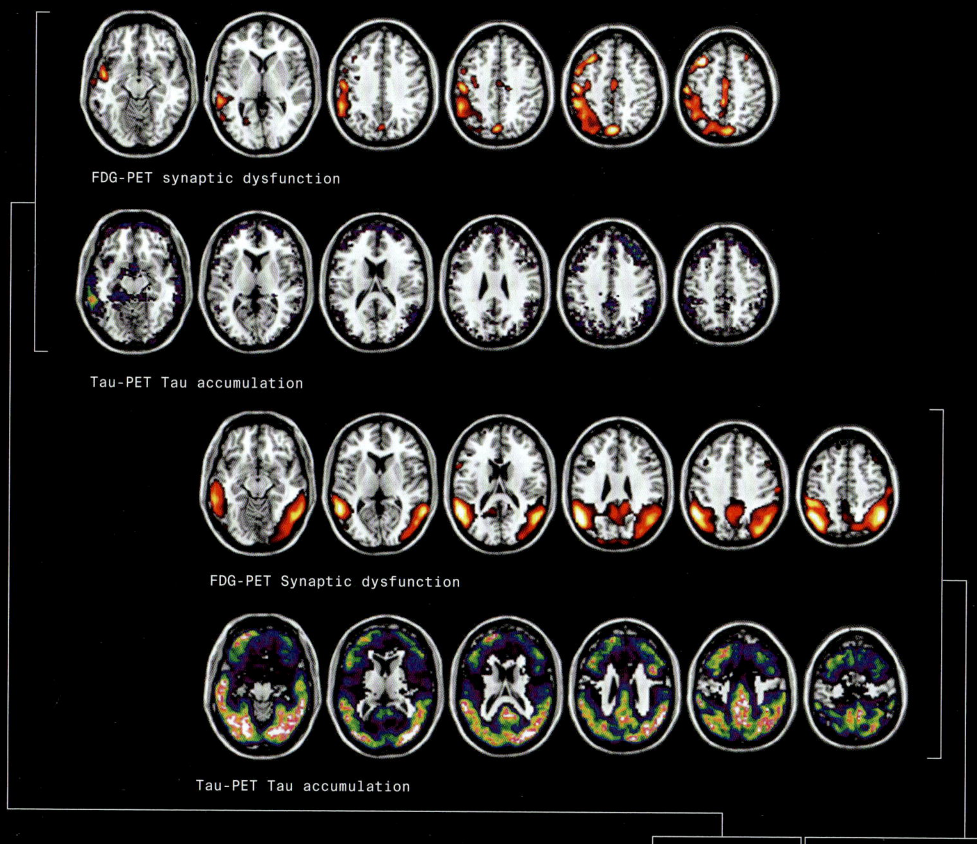

Model of dynamic biomarkers of Alzheimer's Disease
by C.R. Jack et al., 2013 (Lancet Neurol)

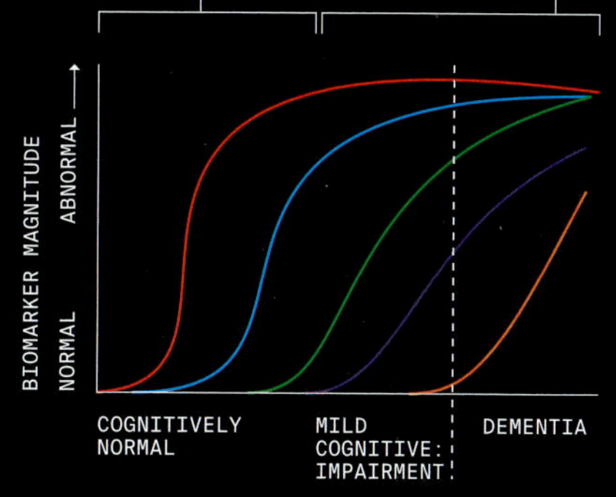

- Amyloid β (Amyloid-PET or CSF)
- Tau-mediated neural injury and dysfunction (Tau-PET and FDG-PET)
- Brain structure (MRI)
- Memory
- Clinical function

Courtesy of Profs. D. Perani, San Raffaele University and San Raffaele Scientific Institute, and V. Garibotto, Geneva University and Geneva University Hospitals

10.7

PET cerebral tau and amyloid burden in healthy aging and Alzheimer's Disease

HEALTHY BRAIN

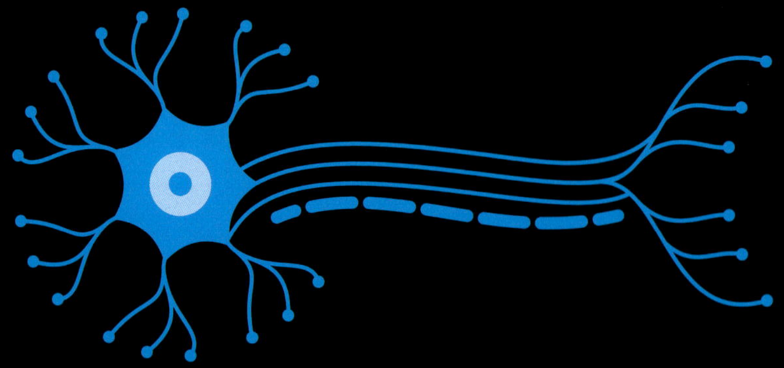

Tau-PET Negative Scan

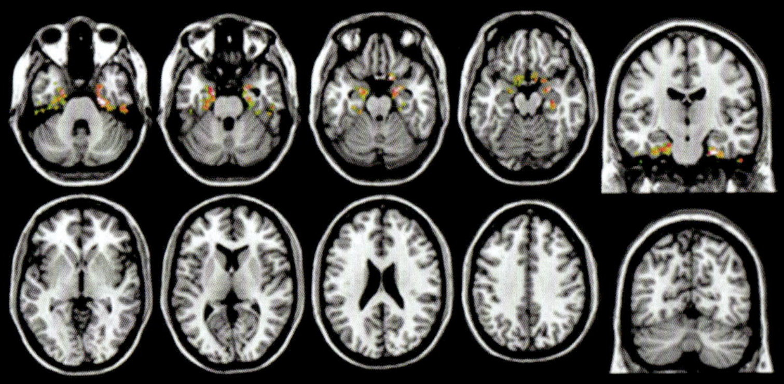

Amyloid-PET Negative Scan

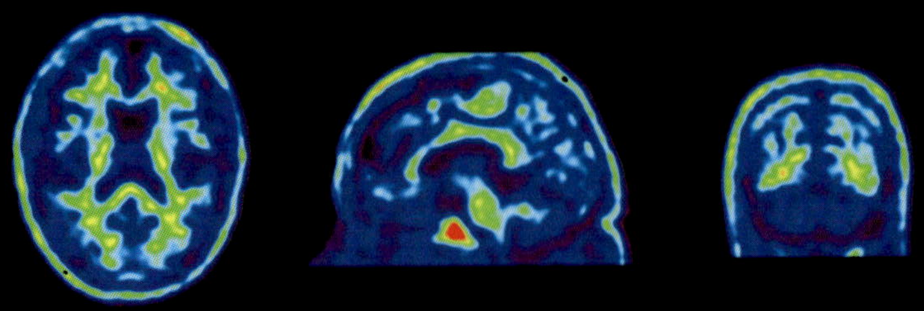

Courtesy of Profs D. Perani, San Raffaele University and San Raffaele Scientific Institute, and V. Garibotto, Geneva University and Geneva University Hospitals

ALZHEIMER'S DISEASE BRAIN

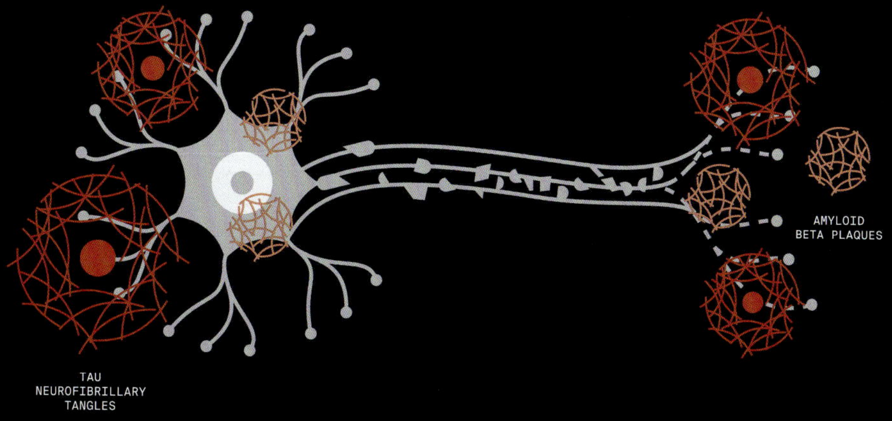

Tau-PET Positive Scan

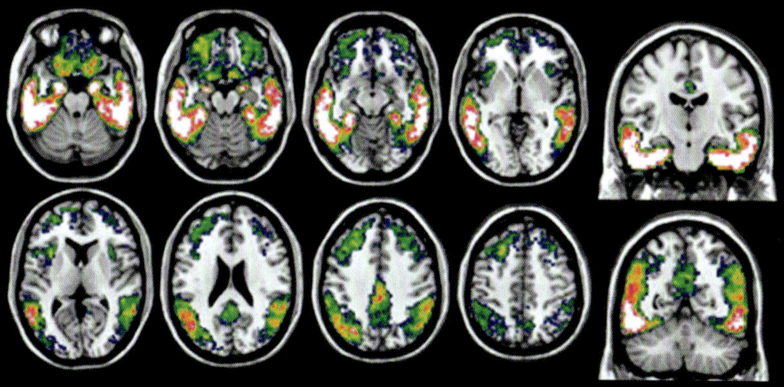

Amyloid-PET Positive Scan

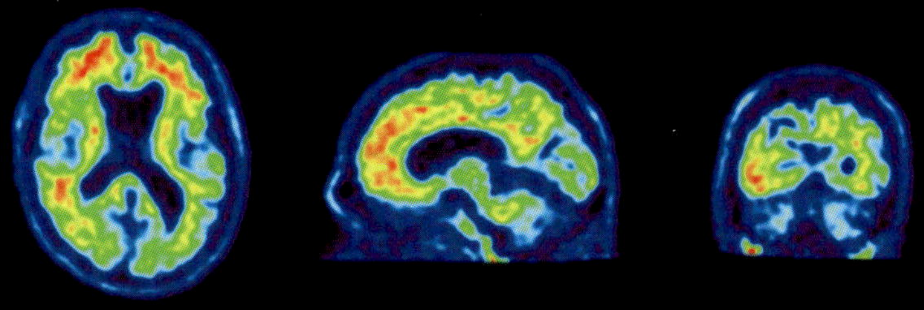

10.8

Body fluid biomarkers

Neurofilaments, a new way to measure brain degeneration

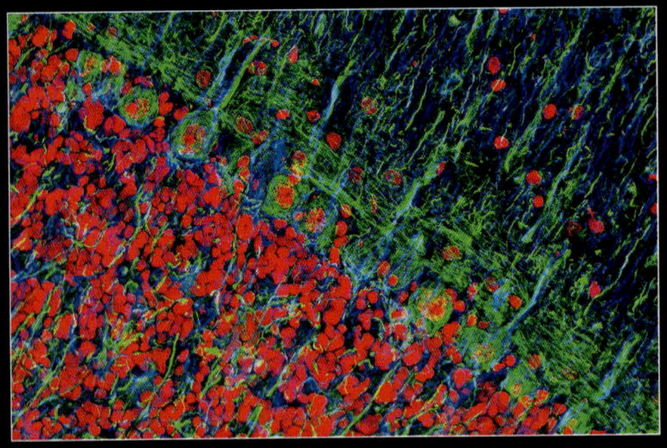

Neurodegenerative diseases are one of the major challenges of present and future medicine. These diseases currently lack efficient treatments. Research on neurodegeneration is hampered by the absence of suitable *in vitro* or *in vivo* models, and in humans, by the absence of a handy, inexpensive, and efficient way to measure brain tissue loss in real time. Neurofilaments are proteins contributing to form the cytoskeleton of neurons from the central and the peripheral nervous system. Detection of increased levels of neurofilaments in serum or plasma may represent a way to measure tissue damage in the nervous system. This would allow to diagnose the presence of a neurodegenerative disease, monitor its course, and measure the efficiency of new drugs with a simple blood test.

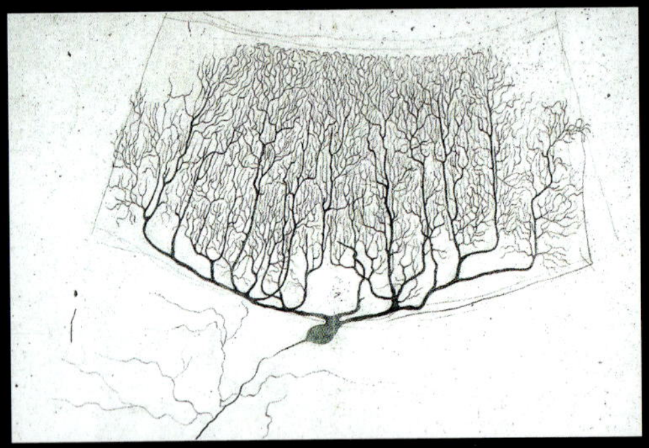

Neurofibrils in the cytoplasm of neurons were first visualized with silver stains over a century ago by the neuroanatomists Camillo Golgi, Santiago Ramón y Cajal, and others, who also noted their striking abundance in axons. For the silver staining and for the studies using the silver staining on neurons, Golgi and Cajal shared the Nobel Prize for Medicine and Physiology in 1906.

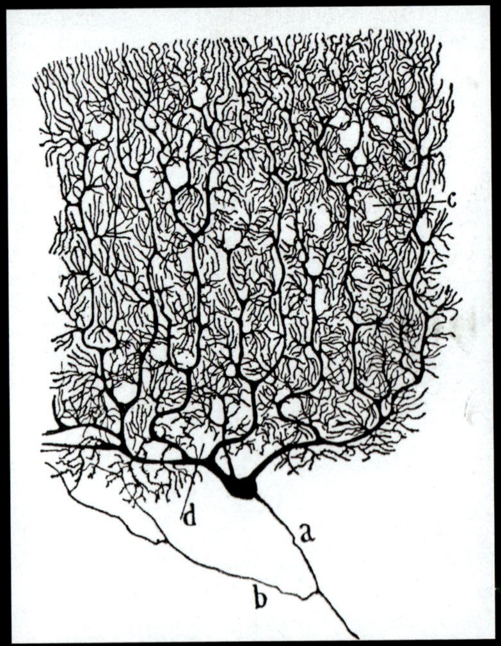

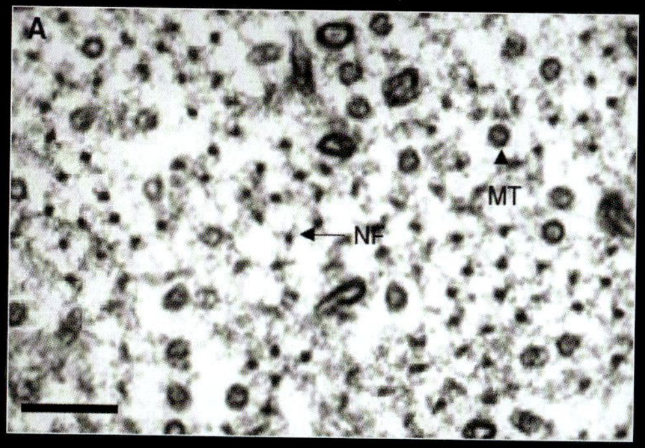

Electron microscopy analyses revealed that neurofibrils in axons are bundles of 10 nanometer-diameter filaments (Schmitt and Geren 1950), which were designated "neuro-filaments" based on their location exclusively in neurons. Neurofilaments are now known to be members of the family of intermediate filaments defined as cell-selective or cell-specific 10-nanometer filaments that are "intermediate" in diameter between actin filaments (6 nanometers) and myosin filaments (15 nanometers) found in muscle cells. In the electron micrograph above, arrows indicate neuro-filaments (NF), and microtubules (MT).

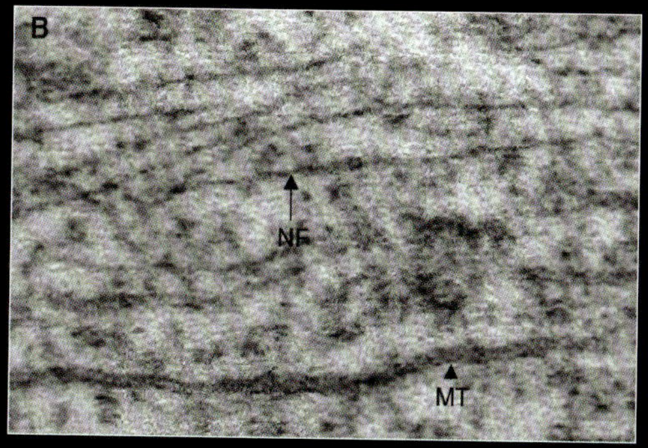

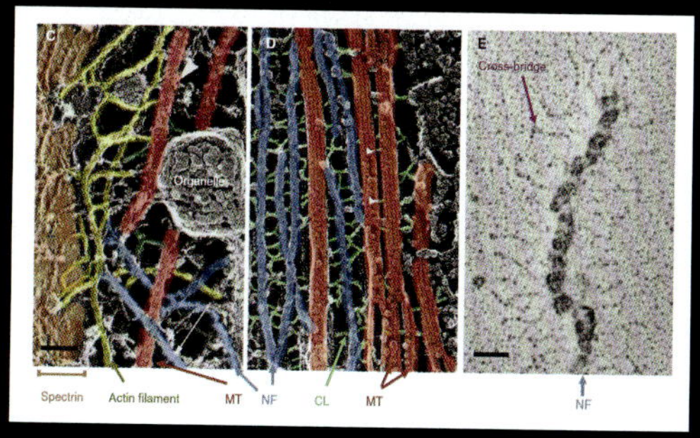

Neurofilaments can be divided in neurofilament light (NF-L), neurofilament middle (NF-M), neurofilament heavy (NF-H). Side chains forming cross-bridges are α-internexin in the central nervous system and peripherin in the peripheral nervous system. In the electron micrographs on the left, stained with pseudocolors, actin filaments are yellow, microtubules are red, neurofilaments are blue, and side chains are green.

Technologies

The most common and inexpensive technology to measure proteins in biological samples is called enzyme-linked immunosorbent assay (ELISA). It has a lower detection limit of around 15-20 picograms per milliliter. This is close to the concentration of neurofilament light measured in the blood of healthy individuals. Measuring small differences, having however clinical relevance, is therefore impossible with this technology. Only the introduction of extremely more sensitive methods, such as single molecule arrays (SIMOA) have allowed the reliable detection of neurofilament light in the blood and to measure its modulation during disease.

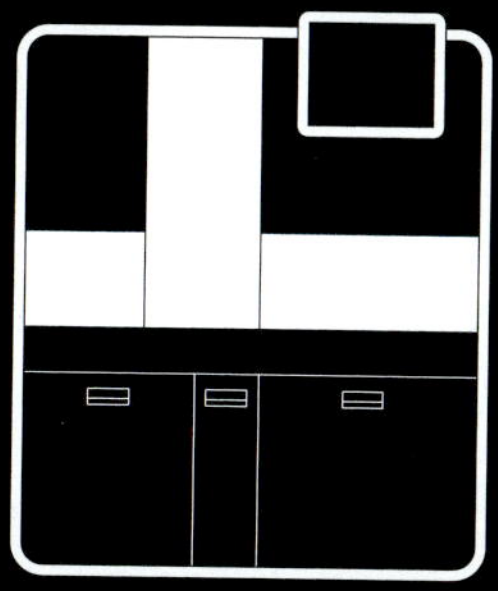

The Simoa HD-I Analyzer: A Novel Fully Automated Digital Immunoassay Analyzer with Single-Molecule Sensitivity and Multiplexing

10.9

Neurodegenerative diseases

The E280A mutation in the presenilin 1 gene is by far the most common cause of familial early onset Alzheimer's Disease. In this diagram, one can appreciate how neuro-filament light blood levels increase with age more rapidly in carriers of the mutation (red line and dots), as compared to their relatives that do not carry the mutation (blue lines and dots). Curves, however, start separating after the age of 30. In the carriers of this cohort mild cognitive impairment sets in around age 45 and dementia by age 50.

Quiroz Y.T., Zetterberg H., Reiman E.M., Chen Y., Su Y., Fox-Fuller J.T., Garcia G., Villegas A., Sepulveda-Falla D., Villada M., Arboleda-Velasquez J.F., Guzmán-Vélez E., Vila-Castelar C., Gordon B.A., Schultz S.A., Protas H.D., Ghisays V., Giraldo M., Tirado V., Baena A., Munoz C., Rios-Romenets S., Tariot P.N., Blennow K., Lopera F. "Plasma Neurofilament Light Chain in the Presenilin 1 E280A Autosomal Dominant Alzheimer's Disease Kindred: a Cross-Sectional and Longitudinal Cohort Study," Lancet Neurology, June 2020;19(6):513-521. doi: 10.1016/S1474-4422(20)30137-X. Epub May 26, 2020. PMID: 32470423; PMCID: PMC7417082

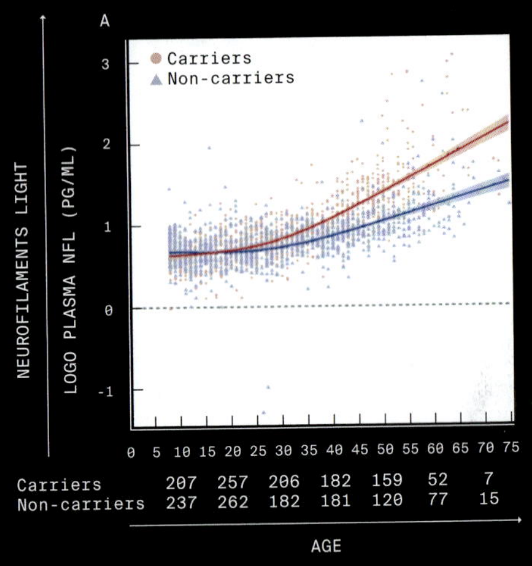

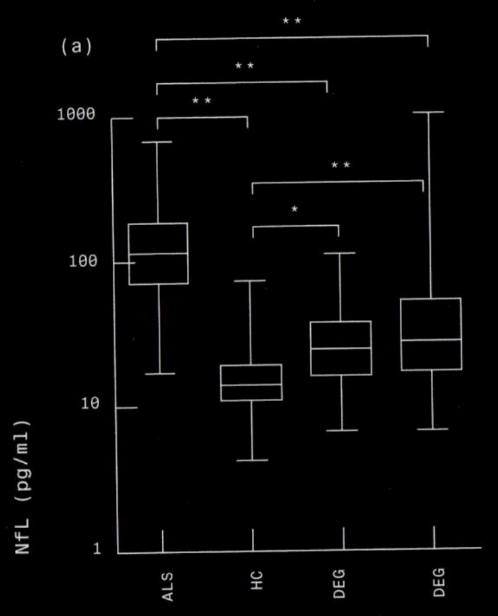

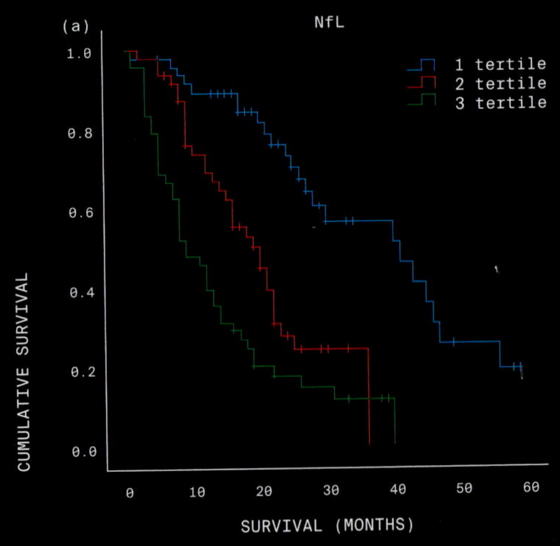

Falzone Y.M., Domi T., Mandelli A., Pozzi L., Schito P., Russo T., Barbieri A., Fazio R., Volontè M.A., Magnani G., Del Carro U., Carrera P., Malaspina A., Agosta F., Quattrini A., Furlan R., Filippi M., Riva N.,"Integrated Evaluation of a Panel of Neurochemical Biomarkers to Optimize Diagnosis and Prognosis in Amyotrophic Lateral Sclerosis," European Journal of Neurology, July 2022;29(7):1930-1939. doi: 10.1111/ene.15321. Epub March 23, 2022. PMID: 35263489; PMCID: PMC9314044

Amyotrophic Lateral Sclerosis (ALS) is a devastating neurodegenerative disease with fatal outcome. In the left diagram it is shown how neurofilament light (NF-L) blood levels have diagnostic value by differentiating persons with ALS from healthy controls (HC), other degenerative diseases (DEG), and especially from ALS mimic diseases (ALSmd). Further, neurofilament light blood levels also have prognostic value since, as shown in the right diagram, patients having values in the lower tertile (blue line) have a much longer survival compared to patients with blood values in the middle (red line), and higher (green line) tertile.

10.10

Bipolar disorder

Bipolar disorder is a psychiatric disease in which periods of depression are alternated with periods of mania, often with psychotic behavior. This condition, difficult to treat, also leads to a significantly increased risk of dementia, suggesting the presence of a deterioration of the brain associated with bipolar disorder. As shown in this study, it is possible to associate an increase in neurofilament light blood levels to the deterioration of specific white matter tracts in the brain (yellow pixels in the magnetic resonance brain images), in other words to the degeneration of specific nerve fiber bundles.

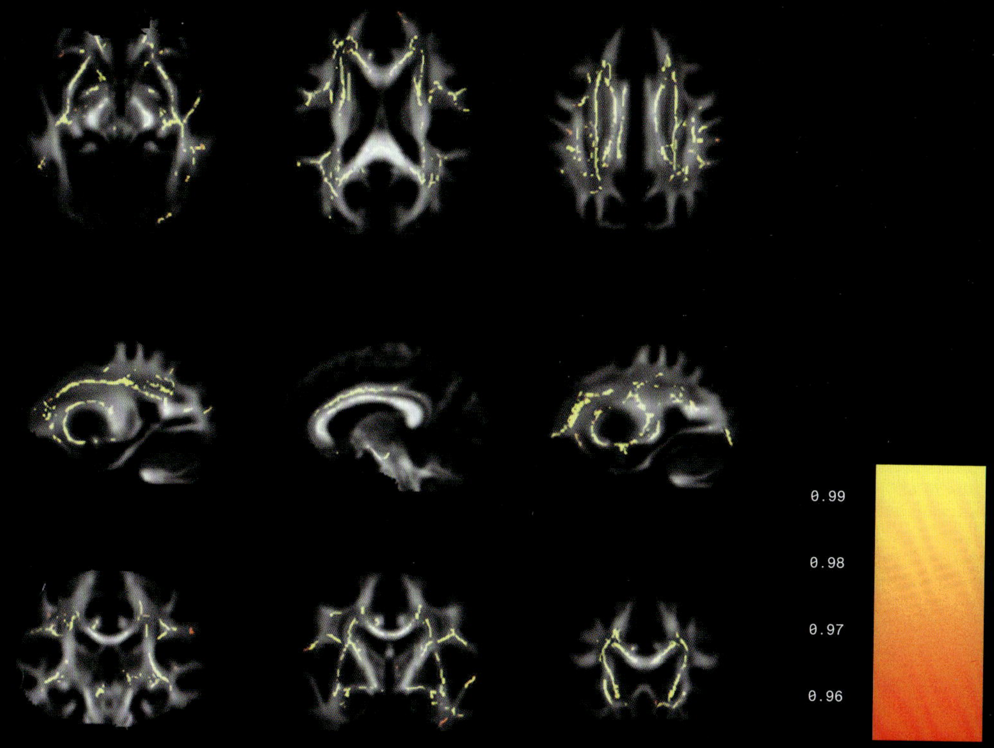

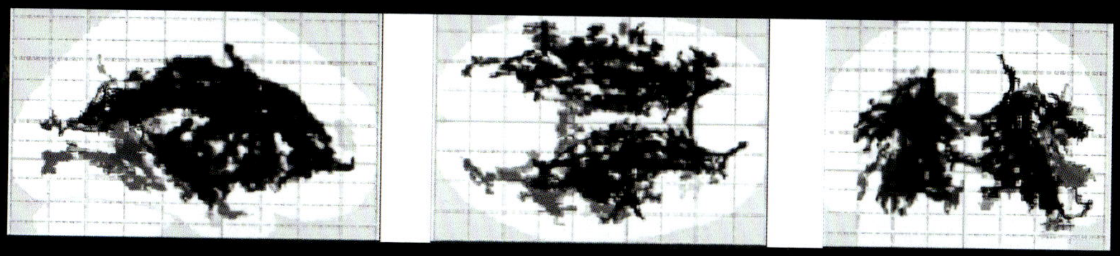

Aggio V., Fabbella L., Finardi A., Mazza E.B., Colombo C., Falini A., Benedetti F., Furlan R. "Neurofilaments Light: Possible Biomarker of Brain Modifications in Bipolar Disorder," Journal of Affective Disorders, March 1, 2022;300:243-248. doi: 10.1016/j.jad.2021.12.122. Epub December 31, 2021. PMID: 34979181

10.11

Neuroinflammation

Multiple Sclerosis (MS) is the prototypical neuroinflammatory disease of the central nervous system. Clinically isolated syndrome (CIS) is the very first clinical episode of MS, but not all CIS patients convert to defined MS. To minimize the risk to accumulate severe disability, early treatment is mandatory. It is therefore of paramount importance to identify CIS patients that will convert to defined MS to allow early treatment. In this study, the increase of neurofilament light blood levels is significantly associated with an increased hazard ratio to developing defined MS.

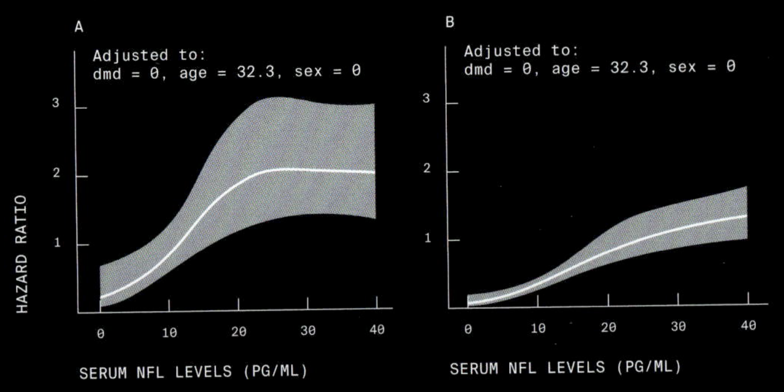

Smoothed plot of hazard ratios for clinically definite multiple sclerosis (A) and McDonald 2017 multiple sclerosis (B) according to serum neurofilament light (NfL) levels.

Once the diagnosis of Multiple Sclerosis has been established, it is of paramount importance to understand if ongoing treatments are effective or if a new relapse or worsening of the disease is going to occur soon. The detection of impending relapses would allow more effective treatments thus avoiding relapse-related disability.

This study shows that higher blood neurofilament levels are predictive of an imminent disease activity in terms of clinical relapse (diagram A), new gadolinium enhancing magnetic resonance imaging lesions (diagram B), or composite confirmed disability accumulation (cCDA; diagram C).

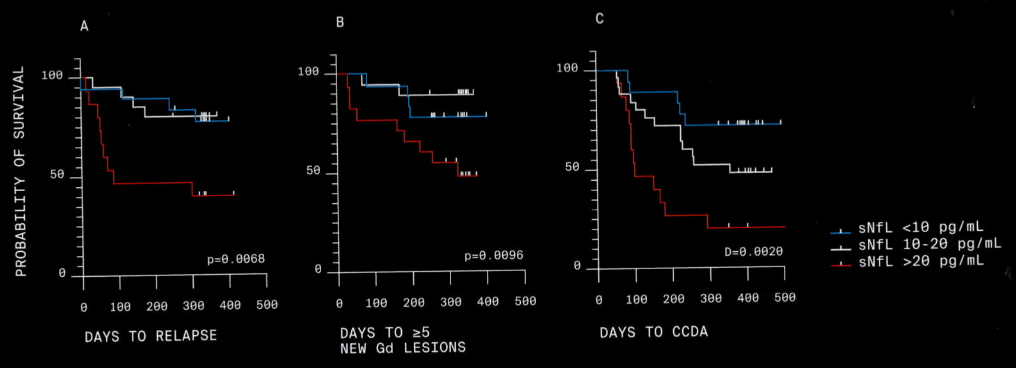

TOP: Dalla Costa G., Martinelli V., Sangalli F., Moiola L., Colombo B., Radaelli M., Leocani L., Furlan R., Comi G. "Prognostic Value of Serum Neurofilaments in Patients with Clinically Isolated Syndromes," Neurology, February 12, 2019;92(7):e733-e741. doi: 10.1212/WNL.0000000000006902. Epub January 11, 2019. PMID: 30635483; PMCID: PMC6382362

BOTTOM: Thebault S., Reaume M., Marrie R.A., Marriott J.J., Furlan R., Laroni A., Booth R.A., Uccelli A., Freedman M.S. "High or Increasing Serum NfL Is Predictive of Impending Multiple Sclerosis Relapses," Multiple Sclerosis and Related Disorders, Volume 59, 2022, 103535, ISSN 2211-0348, https://doi.org/10.1016/j.msard.2022.103535.(https://www.sciencedirect.com/science/article/pii/S2211034822000505)

Like other neurodegenerative diseases, Multiple Sclerosis also lacks efficacious treatments when it becomes progressive, and would highly benefit from a biomarker able to capture neurodegeneration. This study shows how the level of blood neurofilaments at baseline predicts the development of long-term (but not short-term) disability. A cut-off of 10.2 picograms per milliliter in the blood identifies with a sensitivity of 75% and a specificity of 67% of the persons with MS that will accumulate long-term disability.

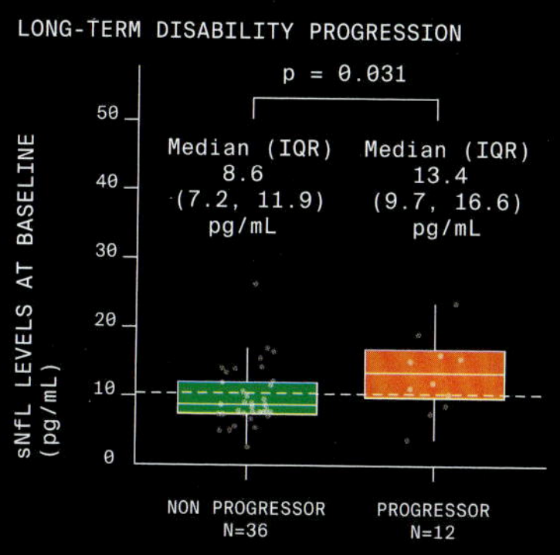

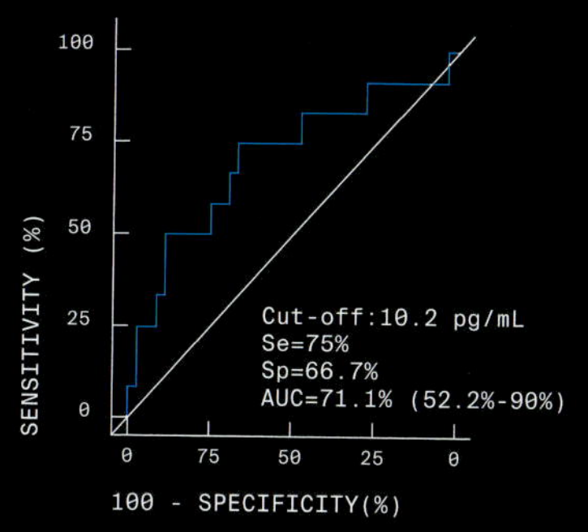

Multiple Sclerosis can be treated effectively with immunomodulators and immune suppressants. High efficacy treatments, however, are hampered by side effects. One of the most severe ones is progressive multifocal leukoencephalitis (PML), a brain infection caused by the JC virus. Early diagnosis of PML allows effective treatment and minimizes unwanted consequences and sequelae. This study shows that neurofilament blood levels are significantly increased over basal during PML and that the increase likely precedes by at least one-month clinical onset.

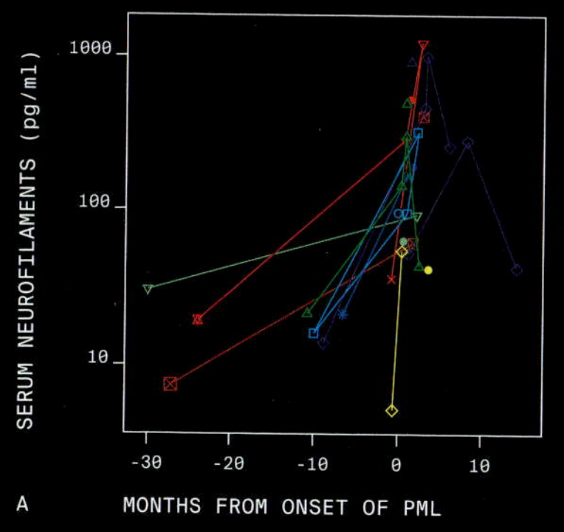

TOP: Khalil M., Pirpamer L., Hofer E. et al. "Serum Neurofilament Light Levels in Normal Aging and their Association with Morphologic Brain Changes," Nature Commununications, 11, 812 (2020). https://doi.org/10.1038/s41467-020-14612-6

BOTTOM: Dalla Costa G., Martinelli V., Moiola L., Sangalli F., Colombo B., Finardi A., Cinque P., Kolb E.M., Haghikia A., Gold R., Furlan R., Comi G. "Serum Neurofilaments Increase at Progressive Multifocal Leukoencephalopathy Onset in Natalizumab-Treated Multiple Sclerosis Patients," Annals of Neurology, April 2019;85(4):606-610. doi: 10.1002/ana.25437. Epub March 5, 2019 PMID: 30761586.

10.12

COVID-19

Especially in the first two waves of the pandemic, when vaccination was not already protecting most SARS-CoV-2 infected persons, early identification of patients that were at risk of developing severe pneumonia was a very relevant issue. Identifying patients that, despite vaccination, will possibly develop severe COVID-19, is still very relevant. This study shows that neurofilament light blood levels at admission to the hospital are highly predictive of fatal outcome or to the necessity to transfer in intensive care unit.

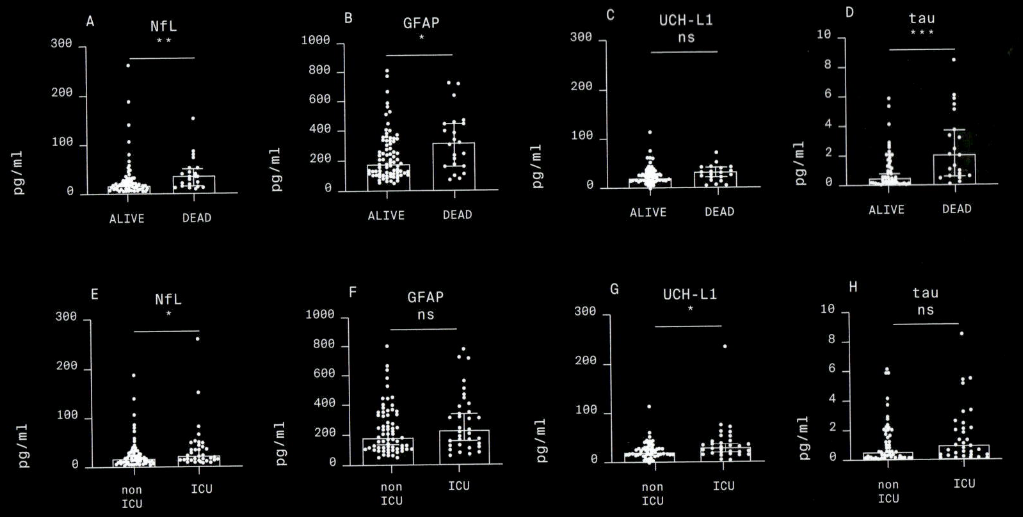

De Lorenzo R., Loré N.I., Finardi A., Mandelli A., Cirillo D.M., Tresoldi C., Benedetti F., Ciceri F., Rovere-Querini P., Comi G., Filippi M., Manfredi A.A., Furlan R. "Blood Neurofilament Light Chain and Total Tau Levels at Admission Predict Death in COVID-19 Patients," Journal of Neurology, December 2021;268(12):4436-4442. doi: 10.1007/s00415-021-10595-6. Epub May 10, 2021. PMID: 33973106; PMCID: PMC8108733

10.13

Conclusions

Take home messages:

- Blood NF-L will be soon used in the clinics to help diagnose and evaluate prognosis in several neurodegenerative and neuroinflammatory diseases.
- If blood NF-L is able to monitor neurodegeneration progression and response to treatment has still to be fully established.
- Blood NF-L measurement is blind to peripheral or central nervous system damage and to the damaging or release process.
- Thus, the real significance of very small changes in the same individual, that however may be in the range of changes caused by neurodegeneration, may have different, even physiological meaning.
- Several other confounding factors have to be taken into account.
- NF-L are already in clinical practice in many advanced centers but we still need to learn how to interpret them correctly.

Neuroscape: Bridging the Gap Between Neuroscience and Technology

11

UCSF Weill Institute for Neurosciences
University of California San Francisco
San Francisco, United States

Neuroscape is a translational neuroscience center at the University of California, San Francisco, engaged in technology creation and scientific research to better assess and optimize brain function of both healthy and impaired individuals.

Neuroscape engages in multiple areas of neuroscience research, including studies of attention, memory, perception, neuroplasticity, cognitive development, brain aging, non-invasive brain stimulation, neurofeedback, and psychedelics. Our research studies are supported by Neuroscape's technological innovations in interactive media and video games, virtual reality and multisensory integration, mobile technology and remote trial platforms, multimodal biosensing and advanced neuroimaging, machine learning and signal processing. Results from our studies are applied to advance novel approaches that better assess and improve cognition in healthy populations—children to seniors—as well as individuals with a broad range of medical conditions, such as Attention Deficit Hyperactivity Disorder, Depression, Multiple Sclerosis, and Alzheimer's Disease.

11.1

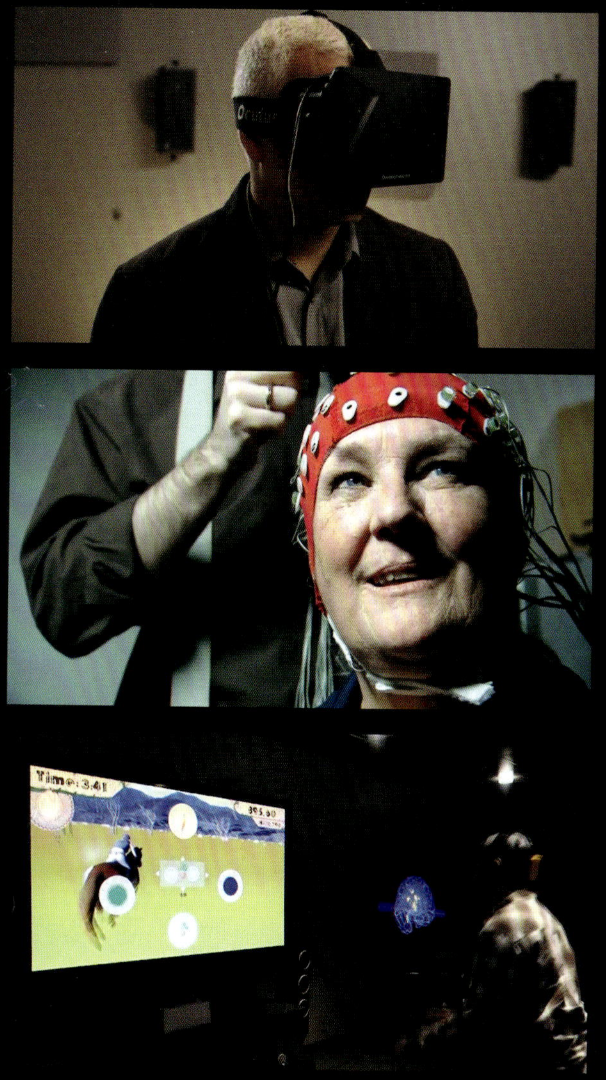

Stills from the videos featured in Room 11

11.2

VR-EEG (Virtual Reality-Electroencephalography)

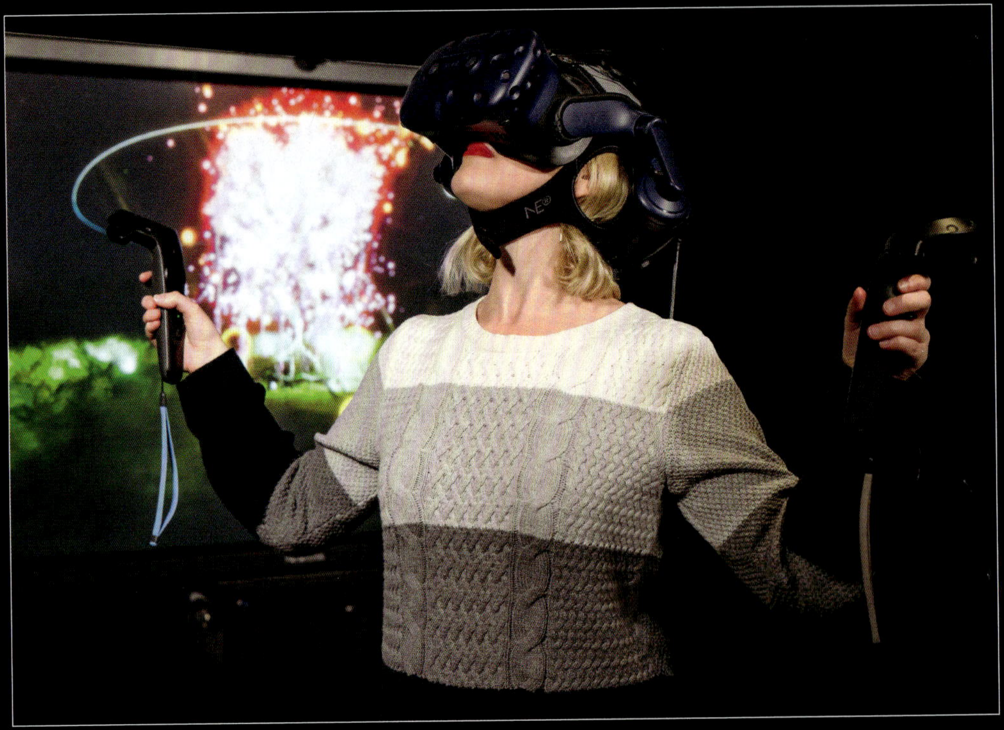

Neuroscape / UCSF Weill Institute for Neurosciences

11.3

Glass Brain Project: visualizing brain activity in 3D

Anatomically-realistic 3D brain visualization depicting real-time source-localized activity (power and "effective" connectivity) from EEG (electroencephalographic) signals. Each color represents source power and connectivity in a different frequency band (theta, alpha, beta, gamma) and the golden lines are white matter anatomical fiber tracts. Estimated information transfer between brain regions is visualized as pulses of light flowing along the fiber tracts connecting the regions.

The modeling pipeline includes MRI (Magnetic Resonance Imaging) brain scanning to generate a high-resolution 3D model of an individual's brain, skull, and scalp tissue, DTI (Diffusion Tensor Imaging) for reconstructing white matter tracts, and digital tools as BCILAB and SIFT to remove artifacts and statistically reconstruct the locations and dynamics (amplitude and multivariate Granger-causal interactions) of multiple sources of activity inside the brain from signals measured at electrodes on the scalp.

The final visualization is done in Unity (real-time 3D development platform) and allows the user to fly around and through the brain with a gamepad while seeing real-time live brain activity from a person wearing an EEG cap.

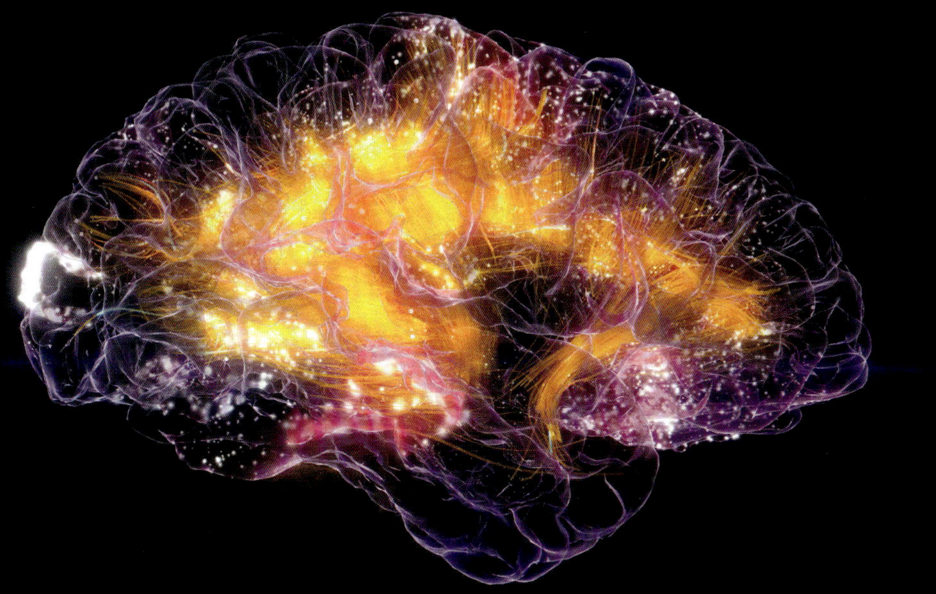

Neuroscape / UCSF Weill Institute for Neurosciences

11.4

Glass Brain Project: visualizing brain activity in 3D

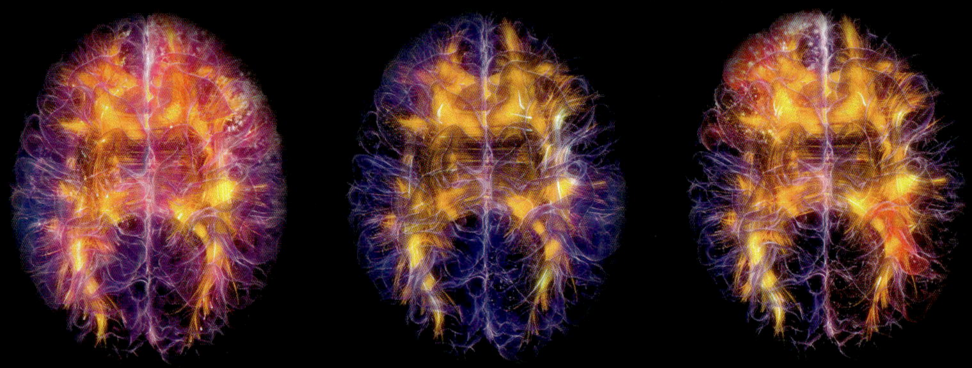

Neuroscape / UCSF Weill Institute for Neurosciences

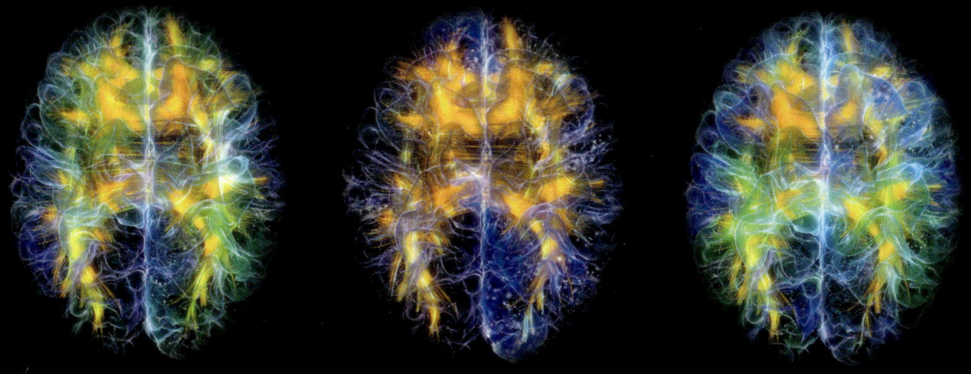

11.5

Glass Brain Project: visualizing brain activity in 3D

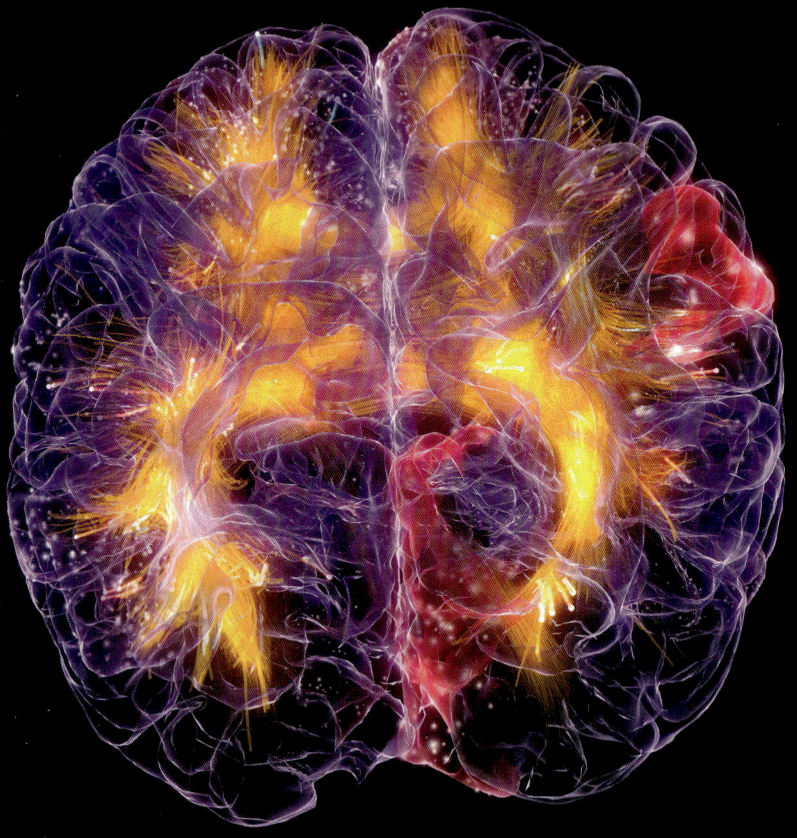

Neuroscape / UCSF Weill Institute for Neurosciences

Neuromodulation to Treat Neurodegeneration

12

UniSR – Università Vita-Salute San Raffaele
Milan, Italy

Neuronal and synaptic loss occur since our early life and continue throughout aging. In a sense, aging in itself is a neurodegenerative condition. In some individuals, owing to genetic repertoire and expression and their interaction with the environment, accelerated or specific patterns of neurodegeneration occur, leading to what is considered a pathological condition, particularly if the consequences of such neurodegenerative processes become clinically evident. However, the behavioral and cognitive consequences of neurodegenerative processes can vary remarkably among individuals and within the same person in different life circumstances and stages, owing to differences in brain reserve, cognitive reserve, and coping/adaptation strategies.

Among the different factors influencing the consequences of neurodegenerative processes, or even neurodegeneration itself, a fundamental role is played by neuroplasticity. This term is widely used to include all phenomena occurring in neuronal function and structure in response to activity, including learning and practice, or to injuries. Variability in plastic mechanisms also accounts for differences among individuals exposed to similar events, from learning opportunities to brain diseases. Therefore, one strategy to prevent neurodegenerative processes, or to counteract their consequences, involves favoring positive plasticity, as opposed to detrimental plasticity occurring when neuronal changes lead to unfavorable results. These refer to active promotion of behaviors leading to neuronal survival and connectivity, such as cognitive and motor training, but also to the use of physical stimuli that can modulate neuronal activity itself, from the delivery of natural visual, auditory or somatosensory inputs to neuromodulation interventions. Nowadays, technological advancements allow non-invasive neuromodulation approaches, such as transcranial magnetic or direct current stimulation. These methodologies, acting on plasticity mechanisms in both directions—e.g. increasing or decreasing neuronal activity—may be used to improve or restore the ideal balance of excitatory and inhibitory activity within the target brain circuits. As neuromodulation itself acts on the same plasticity mechanisms induced by natural sensory stimulation or motor and cognitive behaviors occurring for example during learning and training, it has been hypothesized—and demonstrated in several clinical and research conditions—that the ideal strategy to optimize results is to combine all these methods for synergistic effects. However, much needs to be discovered about the best combinations depending on the individual condition and preferences. Finally, as prevention in general is much more effective than any treatment, much needs to be achieved not only in the fields of medical research, but also in ethical discussions about the opportunity to potentiate apparently not (yet) impaired neuronal functions.

12.1

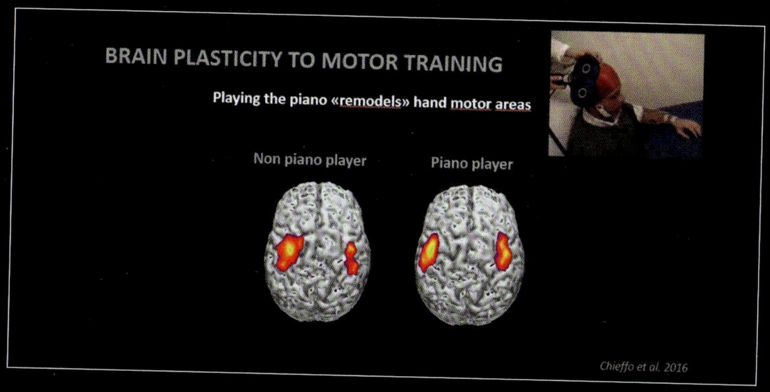

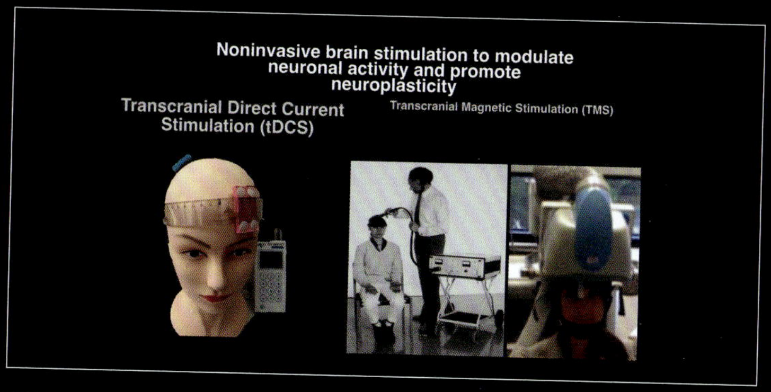

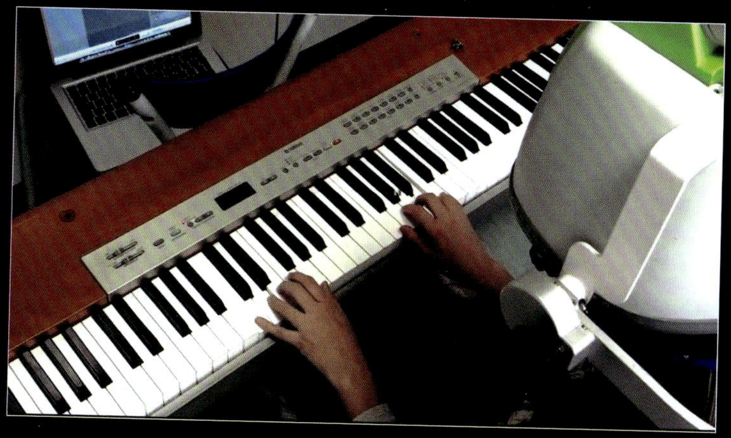

Stills from video featured in Room 12

12.2

Plasticity and learning/activity

Environmental enrichment (more stimuli/motor activity) increases neuronal arborization (left) and synapses (right), resulting in more brain connections.

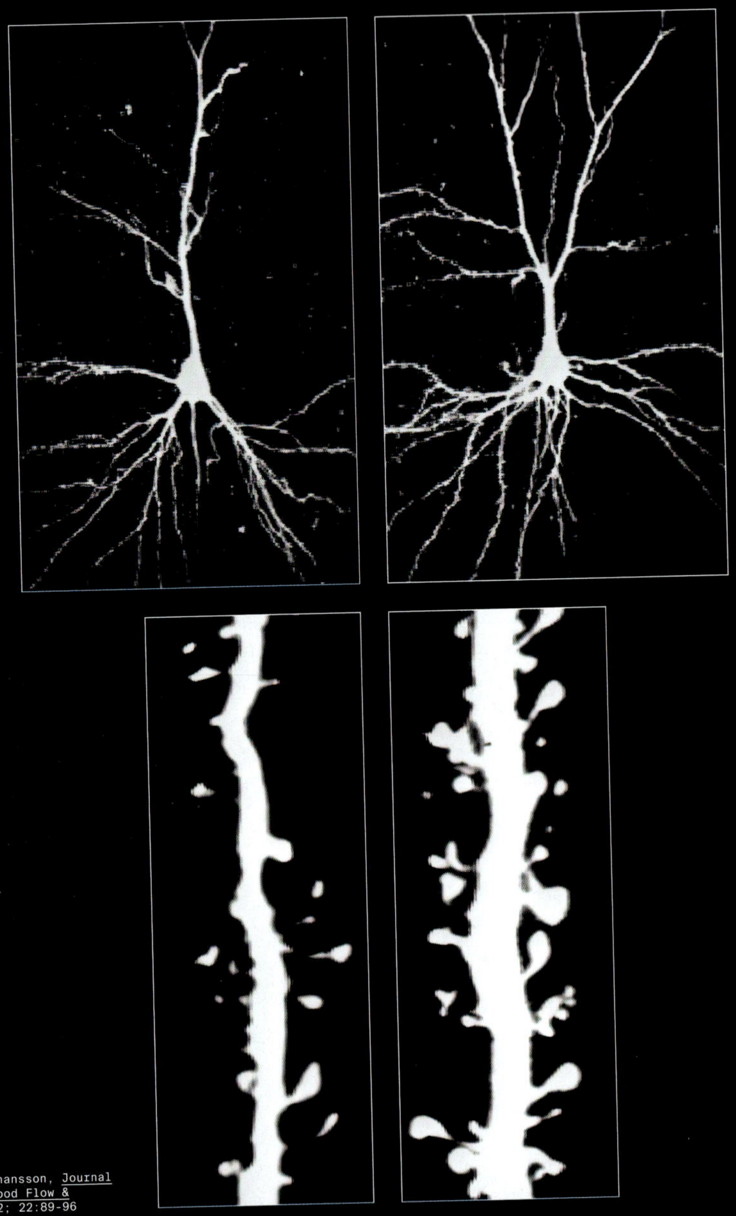

Images from:
Belicenko & Johansson, Journal of Cerebral Blood Flow & Metabolism 2002; 22:89-96

12.3

Brain plasticity to motor training

Playing the piano "remodels" hand motor areas. Muscle maps of little finger using Transcranial Magnetic Stimulation.

Non piano player

Piano player

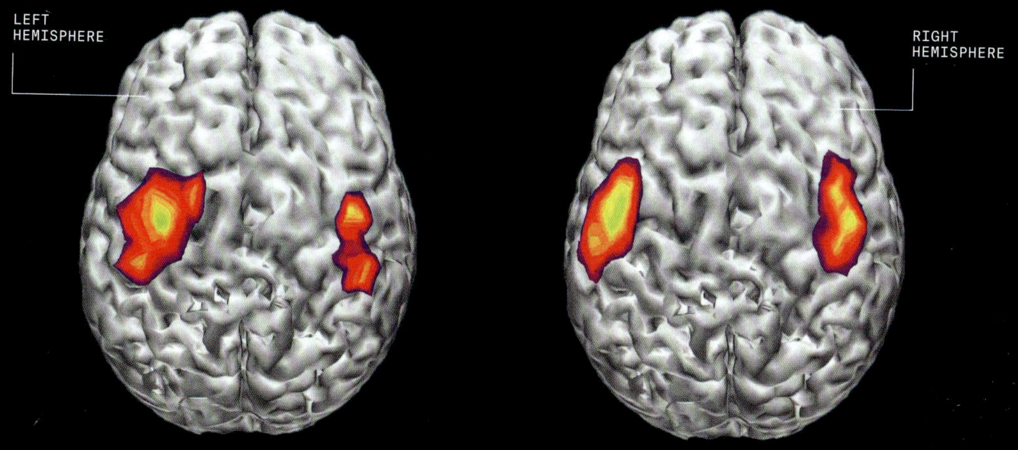

Dominant hand motor area

Non dominant hand motor area

Images from:
Motor cortical plasticity
to training started in childhood:
the example of piano players.
doi: 10.1371/journal.pone.0157952

12.4

Proprioceptive stimulation

Through mechanical modulation simultaneous with sounds, mechanoreceptors activate specific pathways entraining sensorimotor networks involved in the perception of fatigue and pain, control of muscle contraction and relaxation, as well as postural and motor activity, the latter being affected in several neurodegenerative diseases involving motor control.

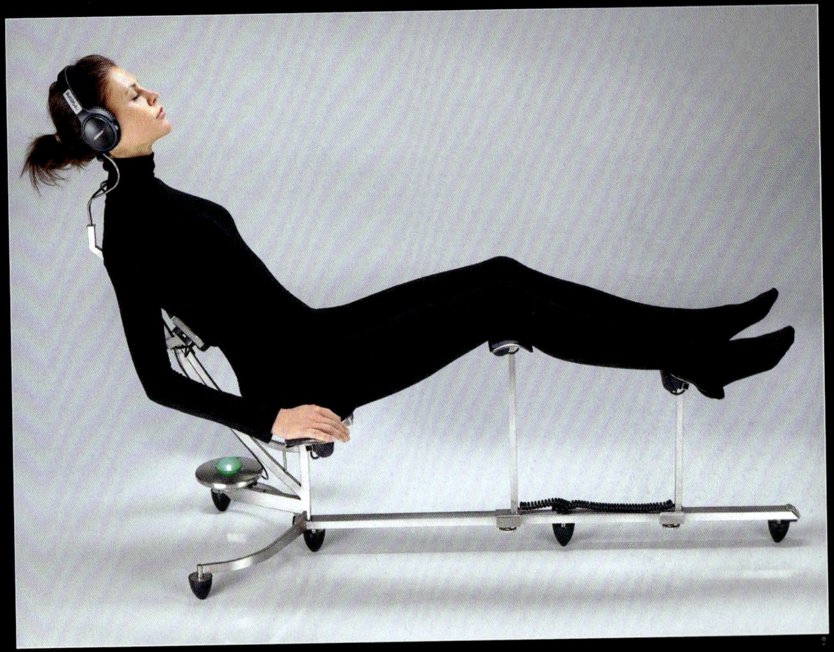

Pictured example: Keope®

12.5

Transcranial Direct Current Stimulation – tDCS

Scribonius Largus (circa 46 AD) used live torpedo fish to apply electric currents to the scalp of people suffering from headaches, reporting inducing transient torpor and local pain. Nowadays, tDCS involves the delivery of weak electric currents, generally with two electrodes (active and reference) on the scalp. In general, cathodal stimulation is considered as having inhibitory effects on neuronal activity, with opposite effects of anodal stimulation, which is considered as excitatory and thus used to increase neuronal activity and thus reinforce brain plasticity and learning/restoration of brain functions affected by neurodegenerative diseases.

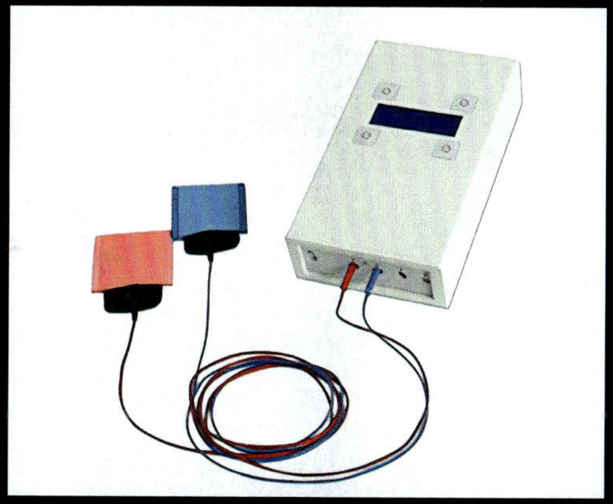

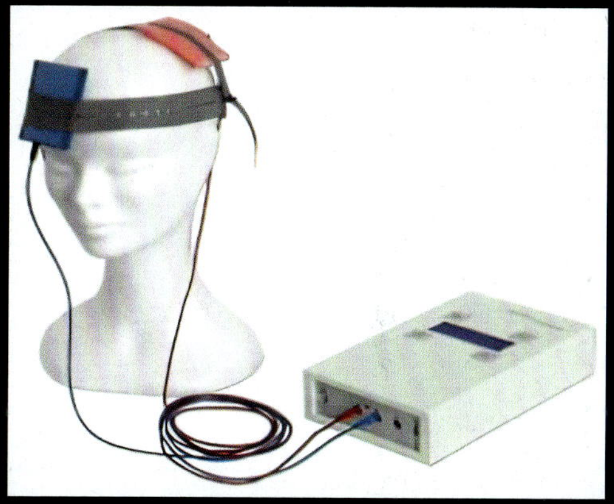

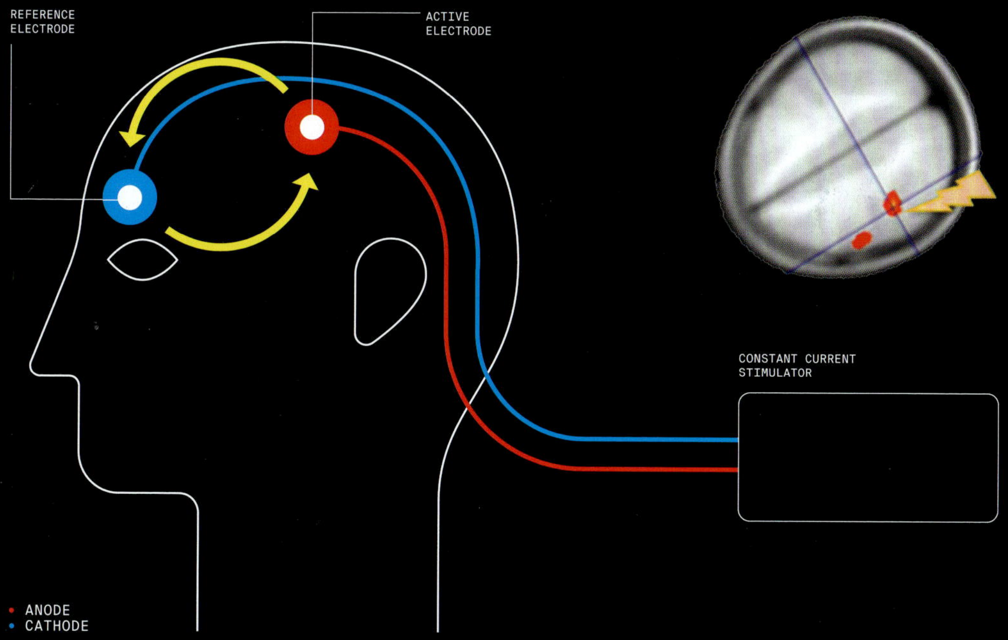

Images from:
Thair et al., Front Neurosci 2017; 11:641
Schlaug et al., Arch Neurol 2008; 65:1571-1576

12.6

Transcranial Magnetic Stimulation – TMS

A brief current (10 microseconds) at high intensity (thousands of Ampères), in a copper wire coil produces a magnetic field (up to 2 Tesla) based on electromagnetic induction (Faraday's law). The magnetic field pulse from a coil placed non-invasively, non-painfully on the scalp can generate an electric field on the brain (Lenz's law), that can reach the intensity sufficient to produce action potentials and to activate brain networks safely and painlessly.

Safety parameters for TMS in humans have been established (Rossi et al., Clin Neurophysiol 2021; 132:269-306).

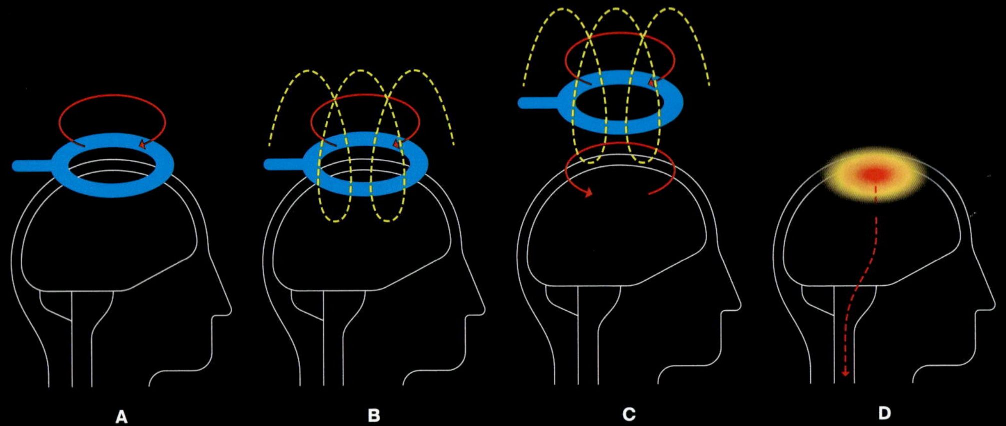

A – Electric current flowing in the coil.

B – Magnetic field generated by electric current of the coil (Faraday's law).

C – Electric current in cortical layers induced by magnetic field is opposite to the electric current flow produced by coil (Lenz's law).

D – Depolarization of cortical neurons activating the corticospinal tract during transcranial magnetic stimulation.

12.7

Transcranial Magnetic Stimulation – TMS

The first machines for clinical use were built in the mid-1980s by Barker & colleagues.

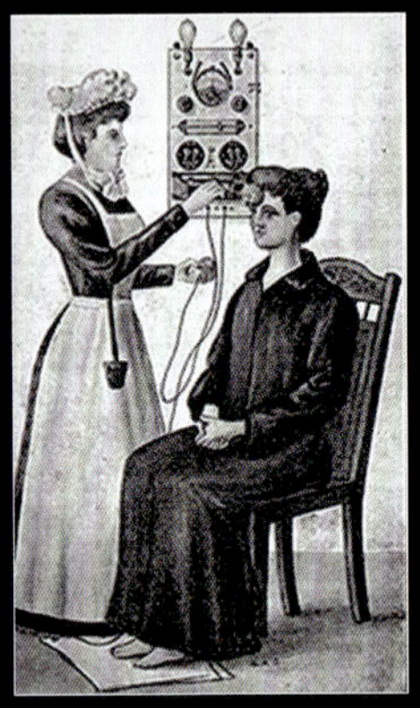

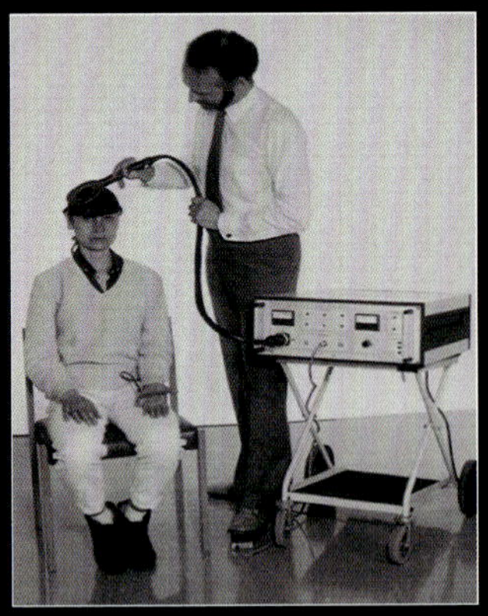

LEFT: Transcranial stimulation with the general Faradization method (Cowen 1900, p. 51, Fig. 28). Courtesy of Wellcome Library, London.

RIGHT: Anthony T. Barker in London, 1985. Courtesy of AT Barker.

Images from:
Vidal-Dourado et al., "Magnetic Fields in Noninvasive Brain Stimulation," Neuroscientist 2014; 20:112-121.

12.8

Repetitive Transcranial Magnetic Stimulation – rTMS

When applied repetitively in trains of stimuli for 10-20 minutes, repetitive TMS (rTMS) is able to modify neuronal excitability in a frequency-dependent manner (excitatory for high-frequency 10-20 stimuli per second; inhibitory for low frequency, below 1 stimulus per second). These changes outlast the time of stimulation for several minutes after a single session, with cumulative effects lasting for several weeks or months in the case of several sessions.
These properties of rTMS, referred to neuroplasticity mechanisms, have led to the use of rTMS to normalize aberrant patterns of cortical activity for treating neuropsychiatric disorders. The most established clinical use of rTMS is in the treatment of drug-resistant depression with high-frequency stimulation of the left dorsolateral prefrontal cortex.

Clinical research is currently exploring promising applications in neurodegenerative conditions, such as Multiple Sclerosis, Parkinson's Disease, Alzheimer's Disease.

Focal coil

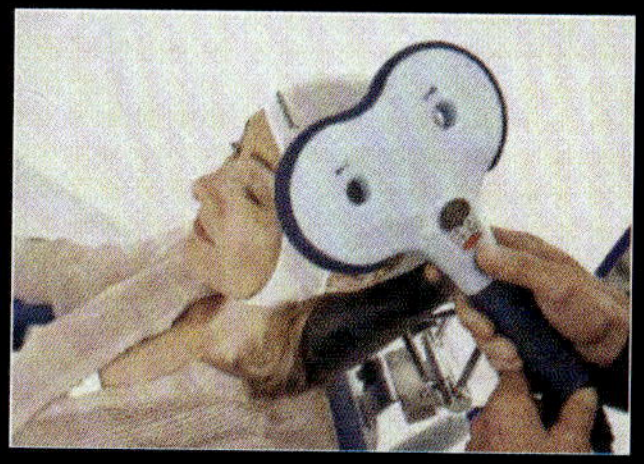
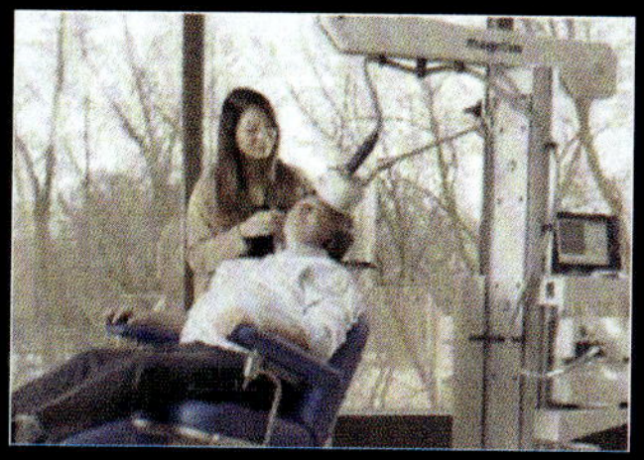
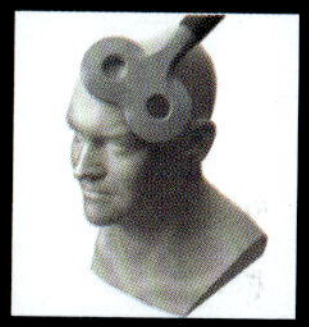
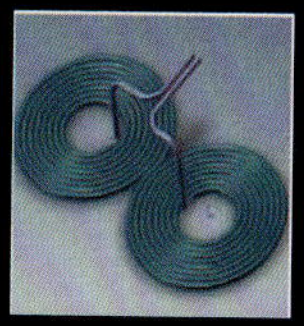

H-coil (Hesed coil)

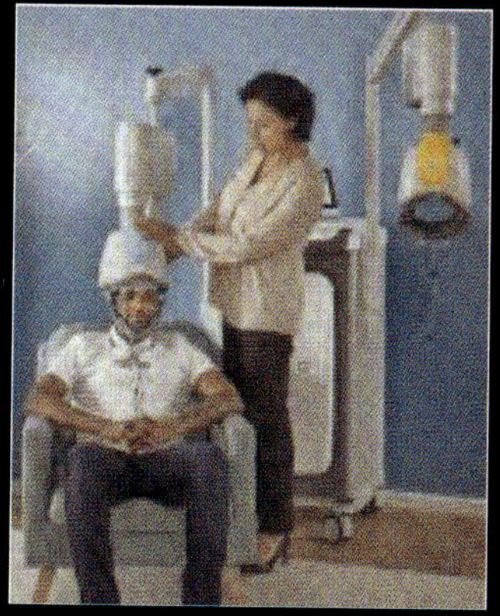
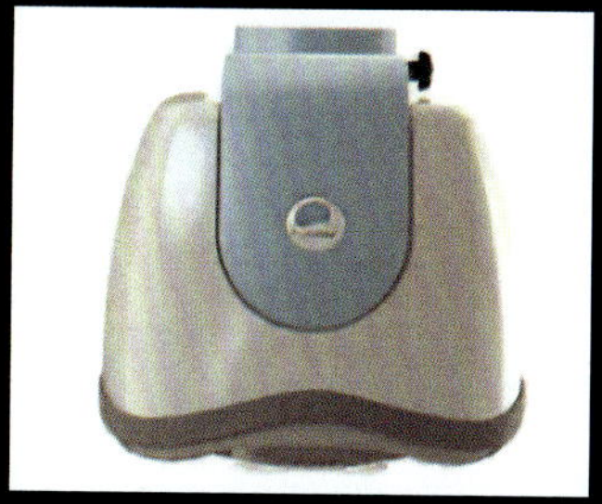
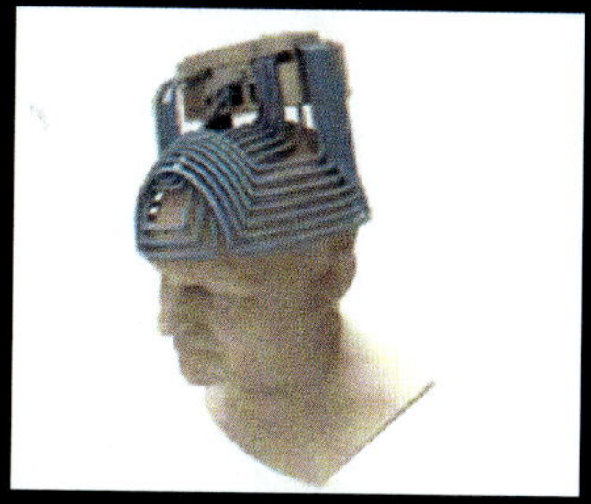

Images from:
magstim.com
brainsway.com

12.9

The coil geometry shapes the magnetic field targeting the brain

Neurodegenerative diseases may involve wide brain cognitive or motor networks, difficult to target with a focal coil in a single treatment session.

The lower limb motor regions, often dysfunctional in neurodegenerative diseases affecting long pathways like Multiple Sclerosis, are located deep under the skull (within the red circle in the figure on the right) and may be difficult to reach with a focal coil.

H-coils have been used in clinical studies on Alzheimer's Disease (Leocani et al. 2021), Parkinson's Disease (Spagnolo et al. 2021) and Multiple Sclerosis (Leocani et al. 2022).

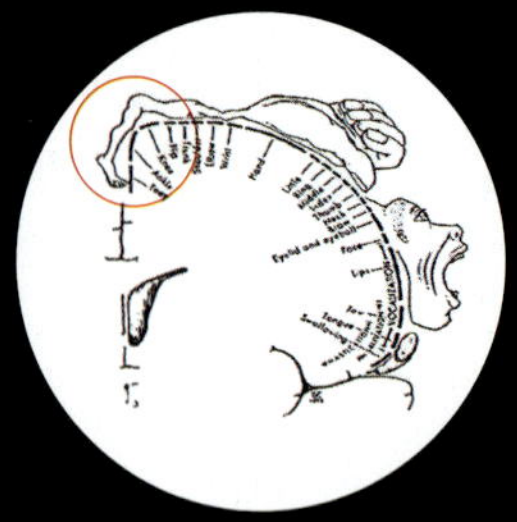

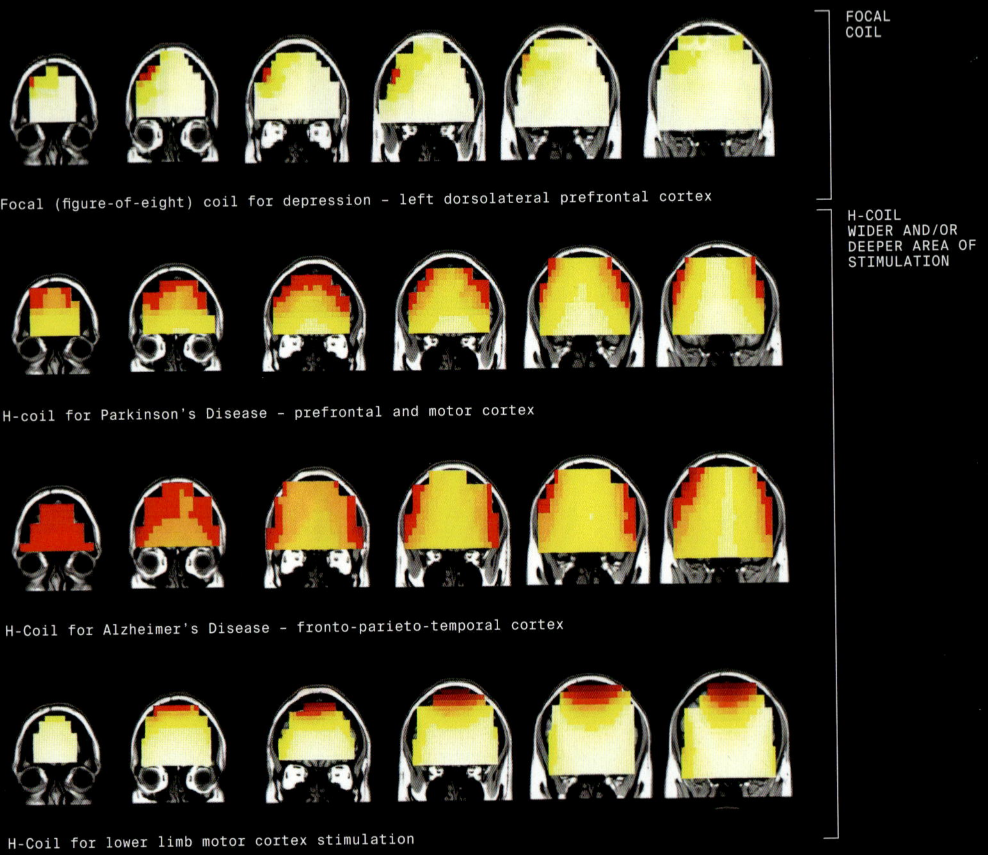

Focal (figure-of-eight) coil for depression - left dorsolateral prefrontal cortex

H-coil for Parkinson's Disease - prefrontal and motor cortex

H-Coil for Alzheimer's Disease - fronto-parieto-temporal cortex

H-Coil for lower limb motor cortex stimulation

12.10

rTMS with H-coil – examples of clinical studies comparing real and sham (apparent) stimulation

Alzheimer's Disease

Change in cognitive score (the lower, the better) after 12 sessions over 4 weeks of real or sham (apparent) rtMS.

Modified from Leocani et al., Front Neurol 2021; 11:614351

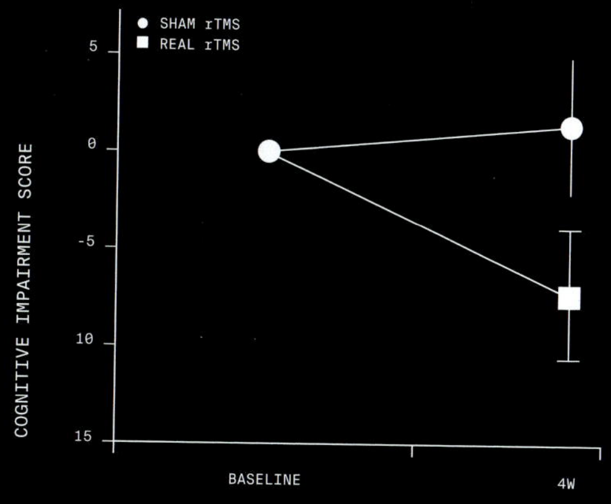

Parkinson's Disease

Percent change in tremor score (the higher, the better) after 12 treatment sessions over 4 weeks with real or sham (apparent) rtMS.

Modified from Spagnolo et al., Front Neurol 2021; 11:584713

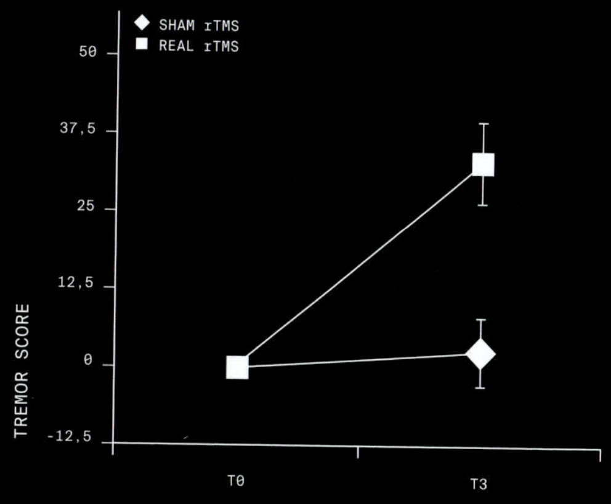

Multiple Sclerosis

Percent change in time to walk 10 meters (the lower, the better) after 11 treatment sessions over 3 weeks with real or sham (apparent) rTMS, coupled with motor rehabilitation.

Modified from Leocani et al., presented at the American Academy of Neurology, 2022 (unpublished)

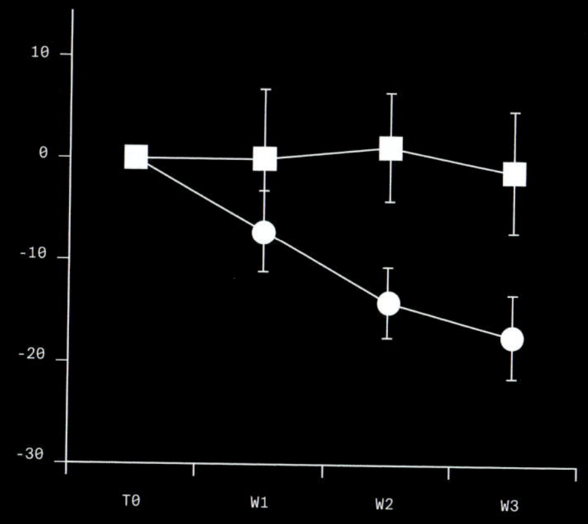

12.11

Brain stimulation and training may be combined to maximize their beneficial effects on brain plasticity

tDCS
transcranial Direct Current Stimulation

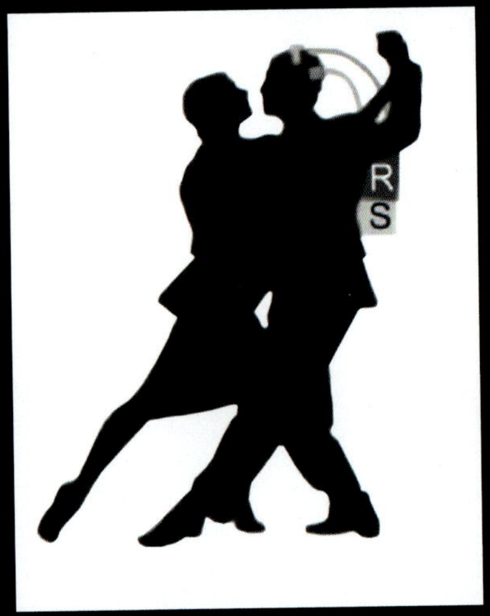

tDCS & Tango Dancing

Image from:
Kaski et al., "Applying Anodal tDCS During Tango Dancing in a Patient with Parkinson's Disease," Neurosci Lett 2014; 568:39-43

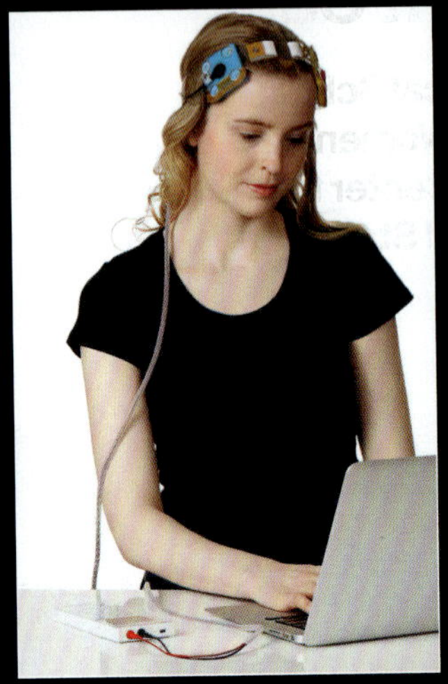

tDCS & Cognitive Training

Image from:
Soterix Medical

rTMS
repetitive Transcranial Magnetic Stimulation

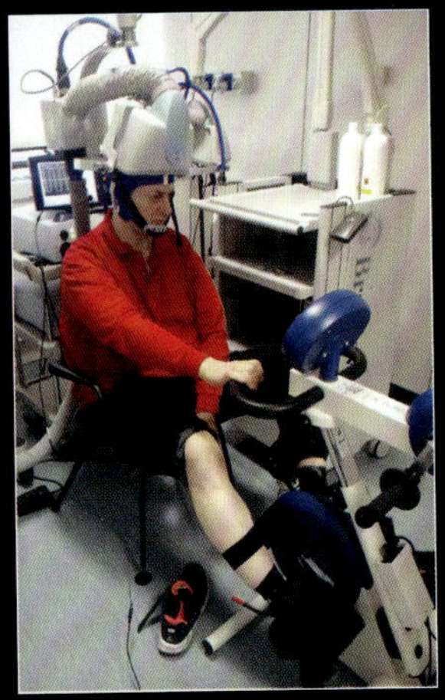

rTMS with H-coil & Cycling

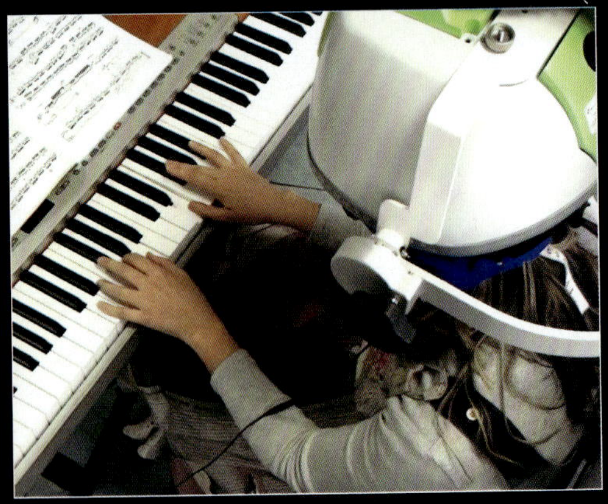

rTMS with H-coil & Piano Playing

Images with H-coil:
courtesy of Prof. Leocani (unpublished)

Translation of Scientific Discovery to Patient Care

13

Harvard Medical School
Brigham and Women's Hospital
Ann Romney Center for Neurologic Diseases
Boston, United States

The induction of tolerance is a major goal of immunotherapy. Investigations over nearly 30 years have shown that anti-CD3 monoclonal antibodies (mAbs), a fully human anti-CD3 monoclonal antibody (mAb), modulates the immune system and prevents inflammation by inducing regulatory T cells. Several studies in animals showed that Foralumab ameliorates diseases including Multiple Sclerosis (MS), diabetes, arthritis inflammatory bowel diseases, and lupus. Foralumab use was shown to be safe in humans and benefit people with mild COVID-19. Currently, two MS patients are being treated at Brigham and Women's Hospital, Boston MA, with intranasal Foralumab under Expanded Access with signs of clinical benefit. The strategy to induce oral tolerance by anti-CD3 mAb represents an exciting and novel avenue for treatment of autoimmune diseases due to the very good safety profile and the high variety of potential applications.

13.1

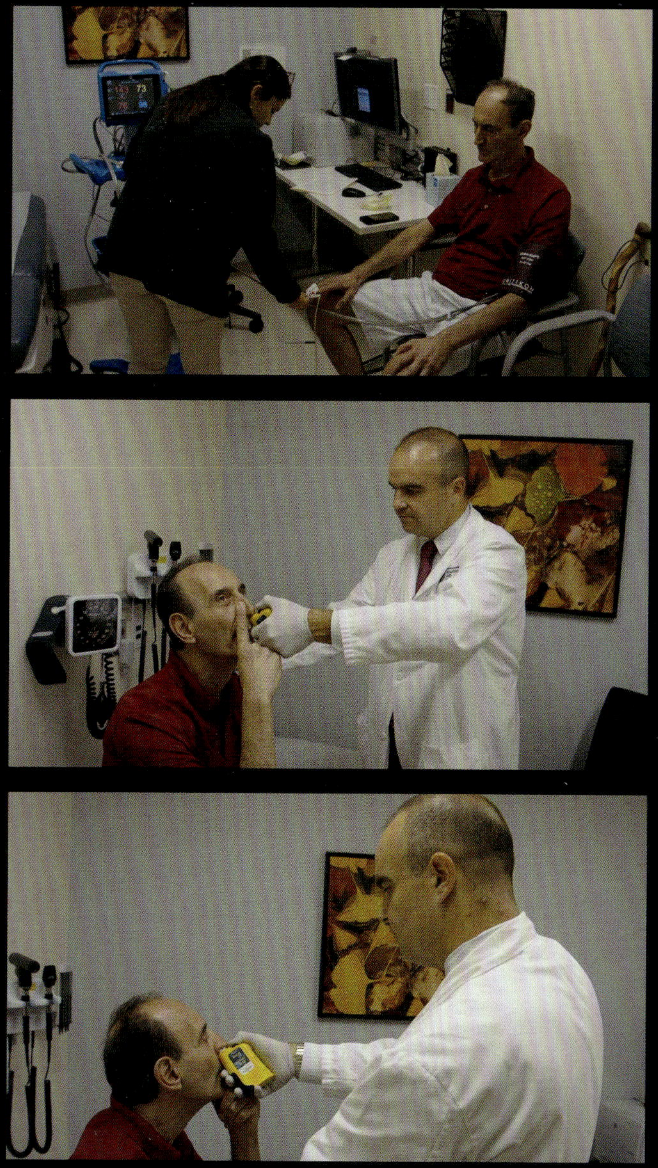

Stills from video featured in Room 13

13.2

Effect of nasal anti-CD3 (Foralumab) in Animal Model of Progressive Multiple Sclerosis

Background/Aims

The mucosal immune system is a unique tolerogenic site that provides a physiological approach for the induction of regulatory T cells. Nasal administration of anti-CD3 may modulate T cell responses and control inflammation in the central nervous system. We investigated the effect of nasally administered anti-CD3 in experimental model of Multiple Sclerosis (MS).

Methods/Design

Experimental autoimmune encephalomyelitis (EAE) was induced by immunization female mice with myelin oligodendrocyte glycoprotein peptide (MOG35-55) emulsified in complete Freund's adjuvant (at a dose of 100 mg) followed by the administration of pertussis toxin on days 0 and 2. Nasal anti-CD3 or Isotype control treatment started on day 30 and was performed daily until endpoint.

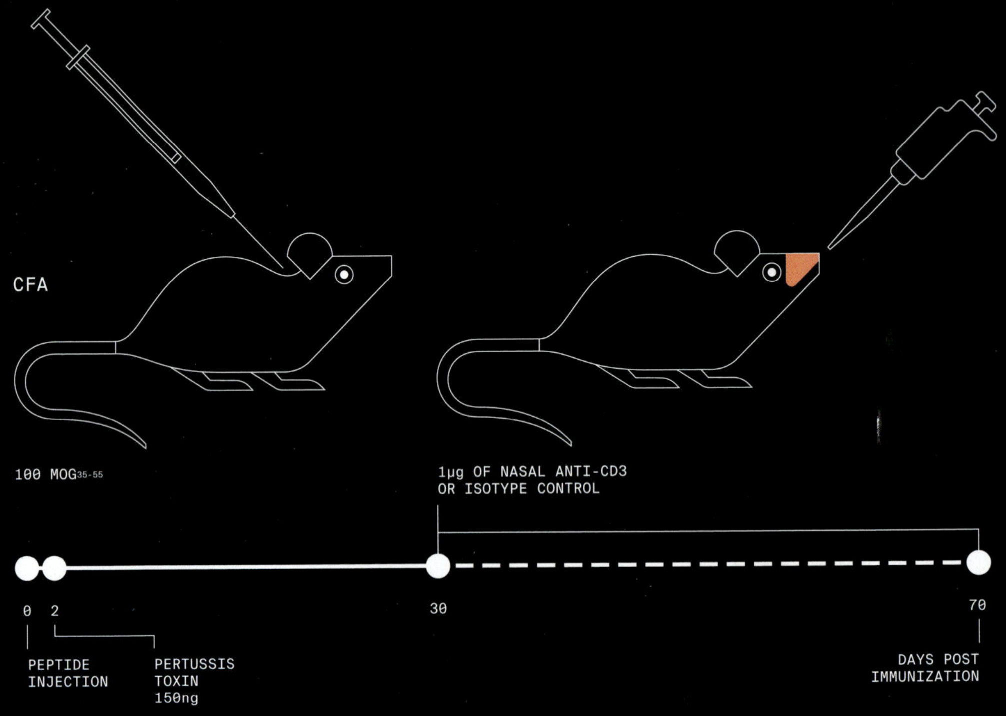

Lior Mayo[1], Andre Pires Da Cunha[1], Asaf Madi[1], Vanessa Beynon[1], Zhiping Yang[1], Jorge I. Alvarez[1], Alexandre Prat,[1] Raymond A. Sobel[1], Lester Kobzik[1], Hans Lassmann[1], Francisco J. Quintana[1], Howard L. Weiner[1]

1 Ann Romney Center for Neurologic Diseases, Brigham and Women's Hospital, Harvard Medical School, Boston, MA, United States

Results

Nasal anti-CD3 ameliorates disease in a progressive model of EAE.

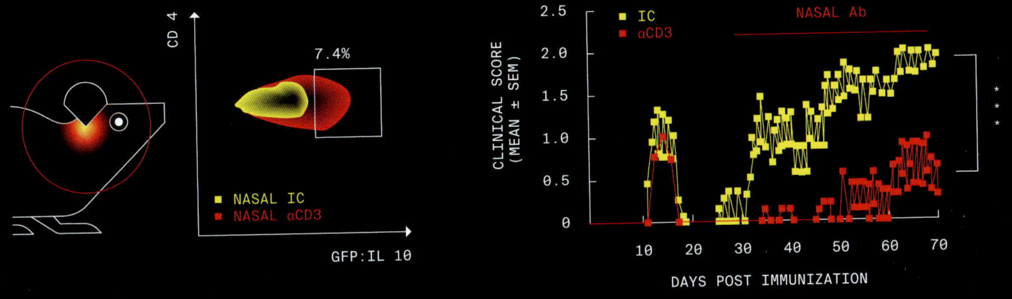

Clinical scores of EAE in NOD mice treated nasally with CD3 specific or isotype control mAbs

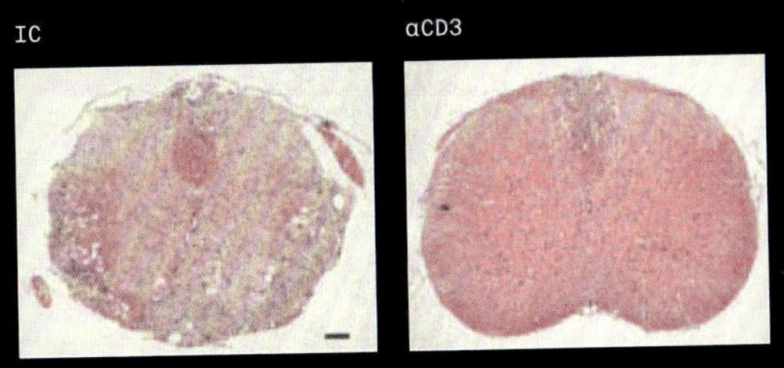

Histopathology analysis at the experimental endpoint (Day 70)

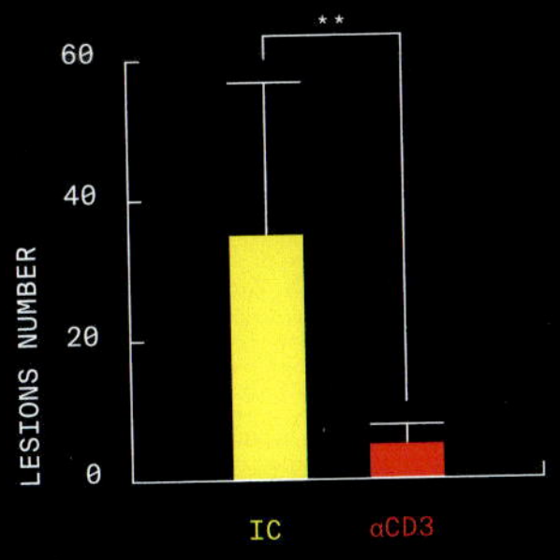

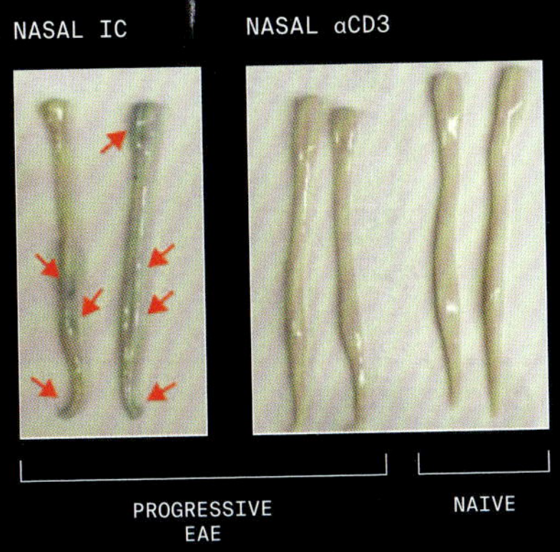

Histopathology analysis of lumbar spinal cord serial sections from EAE NOD mice treated with anti-CD3 or isotype control

Nasal anti-CD3 induces an IL-10 +LAP + T-cell that attenuates progressive EAE in an IL-10 dependent manner.

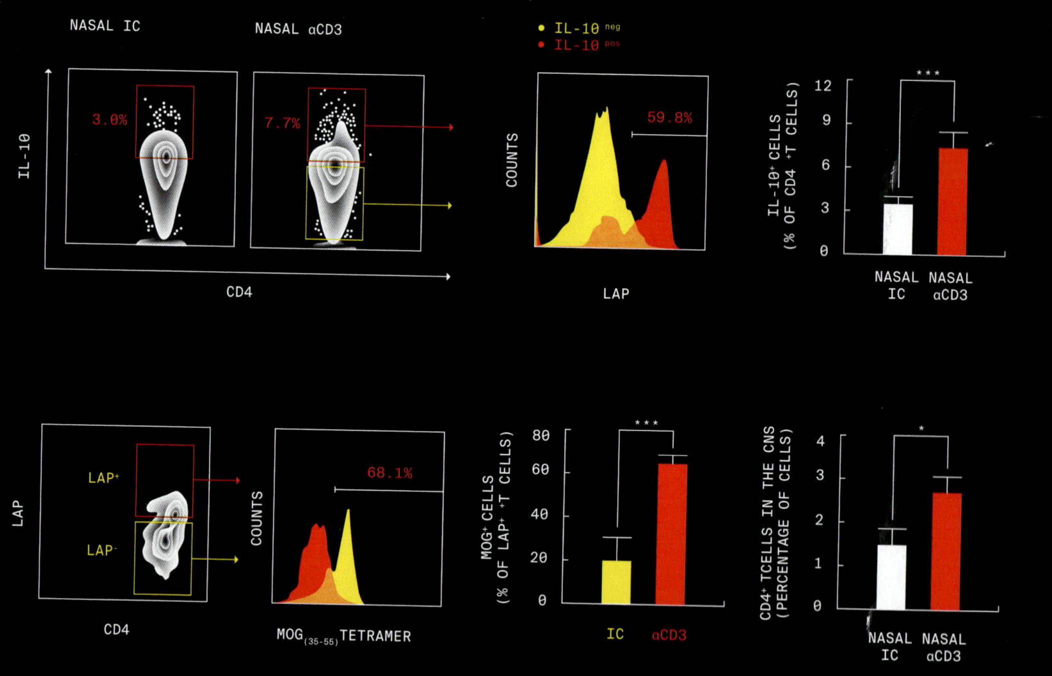

IL-10 and LAP expression by splenic CD4+ T cells was examined by fluorescence-activated cell sorting (FACS)

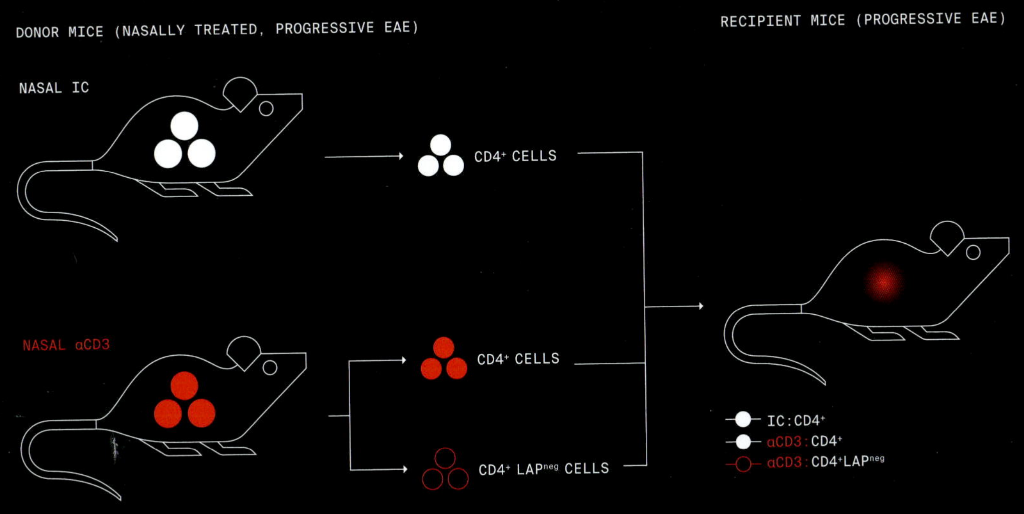

Clinical scores of EAE in NOD mice following adoptive transfer of splenic CD4+, or CD4+LAPneg T cells sorted from chronic NOD EAE mice (day 60) that were treated with anti-CD3 and Isotype controls

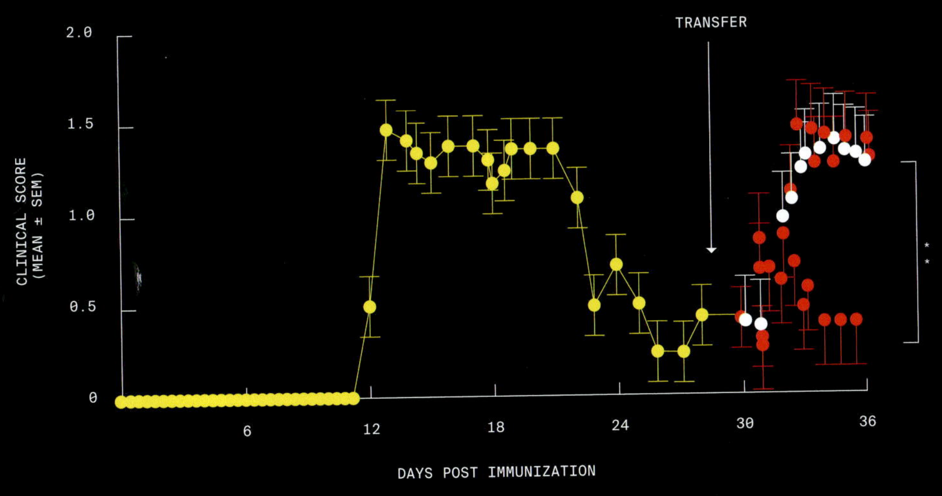

Summary of Results and Conclusion

1. Anti-CD3 treatment attenuated inflammatory milieu in the central nervous system, decreased microglia activation, reduced recruitment of peripheral monocytes, stabilization of the blood–brain barrier with less neurodegeneration.
2. Nasal administration of anti-CD3 ameliorates nervous system inflammation via il-10-dependent tr1 cells in animal models of MS. Further studies should address the effect of nasal anti-CD3 in subjects with MS.

13.3

Effect of nasal anti-CD3 (Foralumab) in healthy subjects

Background/Aims

A major challenge for the treatment of autoimmune and inflammatory diseases is the induction of an anti-inflammatory response that treats the disease condition with minimal side effects.

Healthy volunteers were recruited for a Phase I study in which safety and dosing of nasal anti-CD3 was tested.

Methods/Design

Twenty-seven healthy volunteers (nine per group) received nasal anti-CD3 (Foralumab) or placebo at a dose of 10µg, 50µg, or 250µg daily for 5 days. Safety and immune parameters were measured on days 5, 10, and 30. Blood samples at baseline (t1) and at visits scheduled for 7 (t2), 14 (t3), and 28 (t4) days after drug administration.

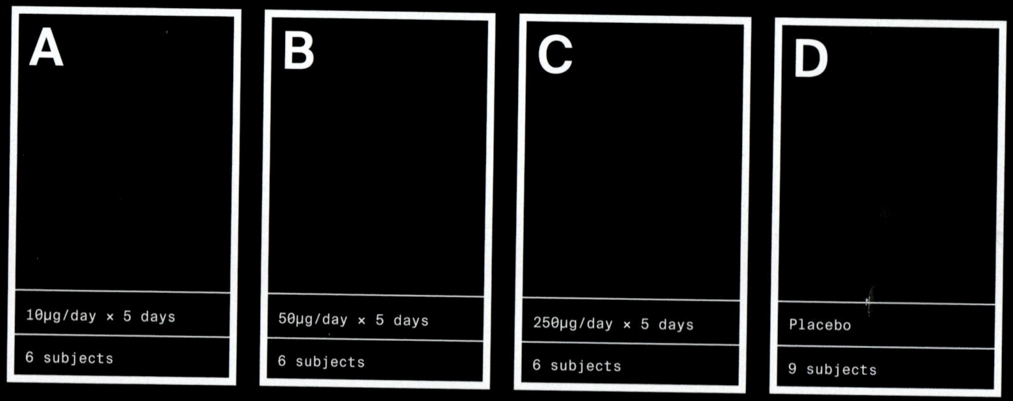

Clare Baecher-Allan[1], Junning Case[1], Katherine Hanus[1], Johnna Varghas[1], Brian C. Healy[1], Christian Gauthier[1], Taylor Saraceno[1], Kunwar Shailubhai[2], Tanuja Chitnis[1], Howard L. Weiner[1]

[1] Ann Romney Center for Neurologic Diseases, Brigham and Women's Hospital, Harvard Medical School, Boston, MA, United States
[2] Tiziana LifeScience, Doylestown, PA, United States

Results–Optimal Dosing

- Reduction in effector memory CD8+ T cells and increase in naïve CD8 + T cells in anti-CD3 treated subjects.
- Proliferation of CD8 + T cells and downmodulation of proinflammatory IFN-γ+ TNF-α + CD4 + T cells.
- Nasal anti-CD3 (Foralumab) does not modulate CD3 from the T cell surface.

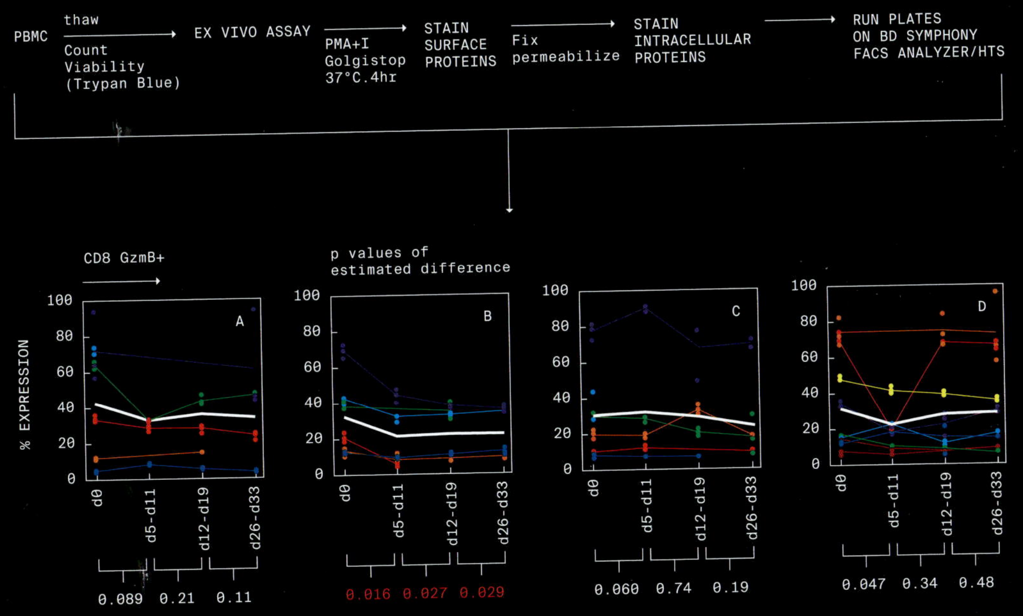

PBMC were thawed and plated for in vitro T cell stimulation and stained for immunophenotyping by fluorescence-activated cell sorting (FACS)

Summary of Results and Conclusion

1. Nasal anti-CD3 (Foralumab) affected the IgG repertoire as measured by antigen arrays. These immune effects were observed primarily at the 50μg dose. No side effects were observed and no subjects developed human anti-mouse antibodies.
2. Anti-CD3 (Foralumab) is biologically active and is safe to be administered in humans.

13.4

Effect of nasal anti-CD3 (Foralumab) in moderate COVID-19

Background/Aims

T cell responses are crucial to SARS-CoV-2 immunity. Specific CD4 and CD8 SARS-CoV-2-T cells are present in early stages of the infectious and increased over time. They showed strong reactivity to the viral proteins and play a major role in lasting immunity (Mateus et al., 2020).

The immunomodulation of T cell function by the nasal administration of anti-CD3 was shown in animal models of auto-immunity and inflammation such as MS (Mayor et al., 2016). We performed a randomized pilot study of fully-human nasal anti-CD3 (Foralumab) in patients with mild to moderate COVID-19 to determine if its immunomodulatory properties had ameliorating effects on disease (Moreira et al., 2021).

Methods/Design

Thirty-nine outpatients were randomized to three cohorts: 1) Control, no Foralumab (n=16); 2) Nasal Foralumab (100µg/day) given for 10 consecutive days with 6 mg dexamethasone given on days 1-3 (n=11); and 3) Nasal Foralumab alone (100µg/day) given for 10 consecutive days (n=12). Day 0: eligibility confirmation and start of treatment.

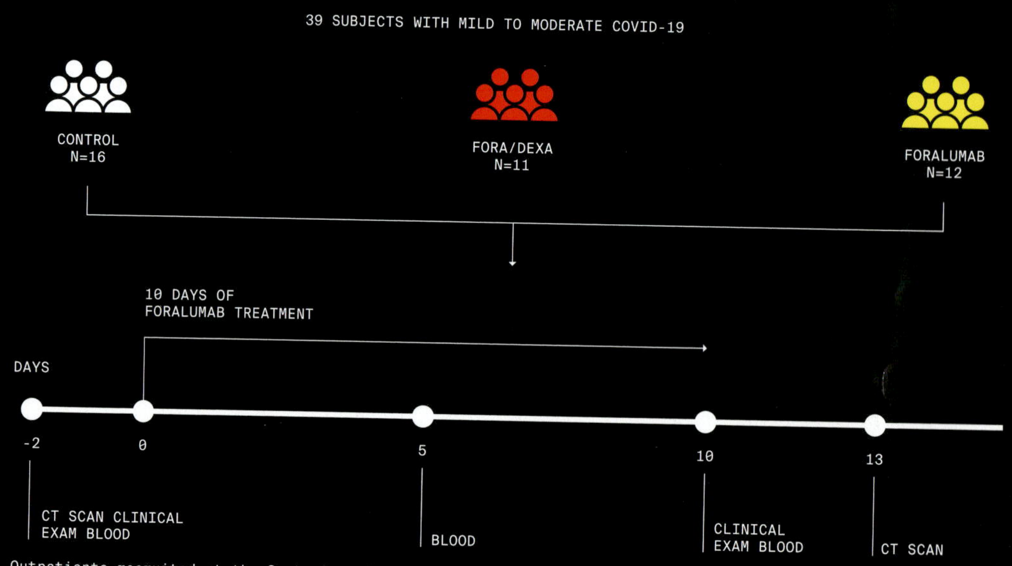

Outpatients recruited at the Santa Casa de Misericordia de Santos Hospital in Brazil.
CAAE: 38056120.1.0000.5509

Thais G. Moreira[1], Kimble T.F. Matos[2], Giovana S. De Paula[3], Thais M.M. Santana[3], Raquel G. Da Mata[3], Fernando C. Pansera[3], Andre S. Cortina[3], Marcelle G. Spinola[2], Clare M. Baecher-Allan[1], Gerson D. Keppeke[2], Jules Jacob[4], Vaseem Palejwala[4], Karen Chen[1], Saef Izzy[1], Brian C. Healey[1], Rafael M. Rezende[1], Rogerio A. Dedivitis[3], Kunwar Shailubhai[4], Howard L. Weiner[1]

1 Ann Romney Center for Neurologic Diseases, Brigham and Women's Hospital, Harvard Medical School, Boston, MA, United States
2 Escola Paulista de Medicina, Universidade Federal de São Paulo, São Paulo, Brazil
3 Santa Casa de Misericordia de Santos, Santos, Brazil
4 Tiziana LifeScience, Doylestown, PA, United States

Results

No reduction in white blood cells was observed with treatment and no serious adverse events were observed during the study.

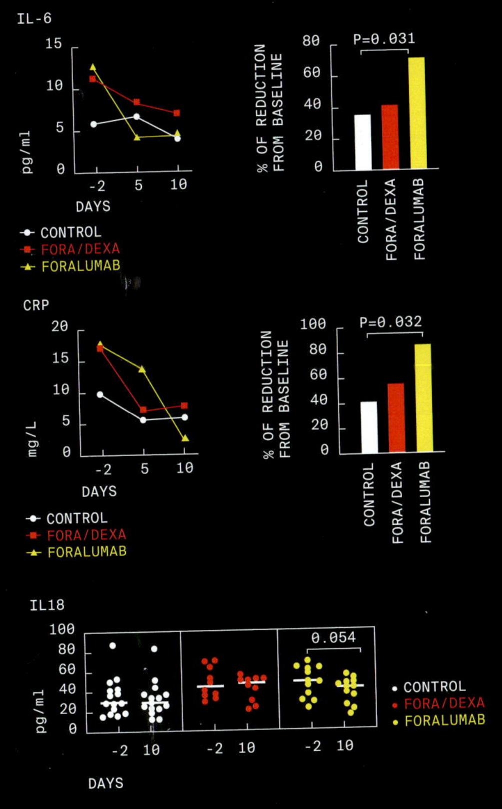

Serum quantification and percentage of reduction of blood inflammatory markers IL-6 and C-reactive protein and Il-18

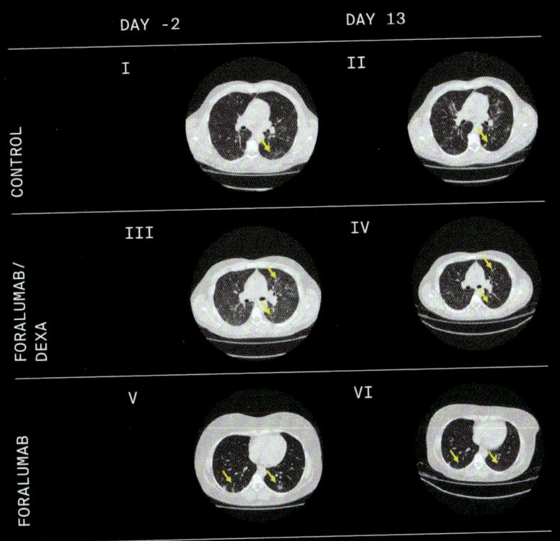

Lung CT Scan: CT chest in COVID-19 patients by treatment group

	CONTROL	FORALUMAB/ DEXA	FORALUMAB
Worsened	0/14	1/10	0/12
Stable	10/14	2/10	2/12
Improved	3/8	1/8	5/12
Marked Improvement*	1/8	6/8	5/12

Marked improvement was predominantly observed in subjects receiving Foralumab/Dexa or Foralumab alone. Control vs. Foralumab/ Dexa, p=0.01 and Control vs. Forlumab/Dexa+Foralumab, p=0.04

Summary of results and conclusion

1. We observed reduction of serum IL-6 and C-reactive protein in Foralumab alone vs. untreated or Foralumab/ Dexa treated patients. More rapid clearance of lung infiltrates as measured by chest CT was observed in Foralumab and Foralumab/Dexa treated subjects vs. those that did not receive Foralumab. Treatment was well tolerated with no severe adverse events.

2. Nasal Foralumab is well tolerated and may be of benefit in treatment of immune hyper-activity and lung involvement in COVID-19.

13.5

Effect of nasal anti-CD3 (Foralumab) in patient with Progressive Multiple Sclerosis

Background/Aims

Mucosal administration of anti-CD3 mAb by binding to the T cell receptor in the mucosal environment induces regulatory T cells and treats autoimmune and inflammatory diseases in animals including models of Multiple Sclerosis (MS).

Nasal anti-CD3 was well tolerated in healthy subjects and in moderate cases of COVID-19.

Methods/Design

50μg OF NASAL ANTI-CD3

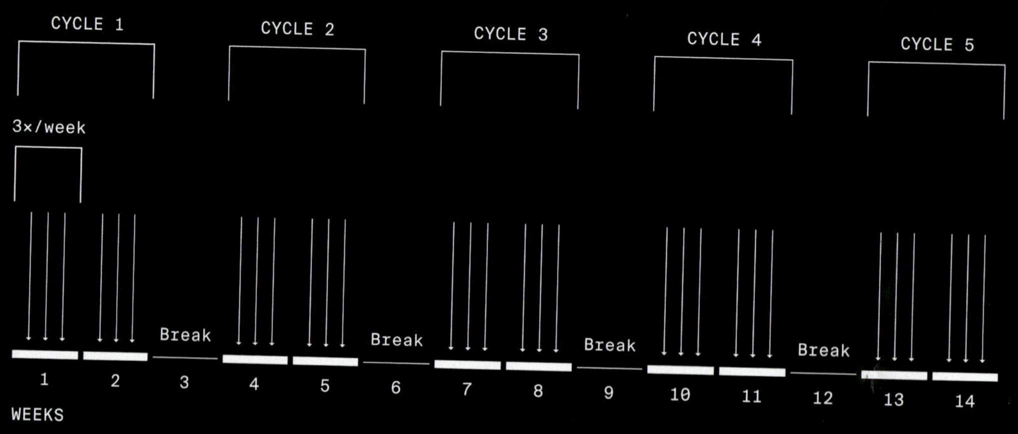

Single patient with secondary progressive MS.
61-year-old man with non-active progressive MS unresponsive to Ocrelizumb® and other MS medications.
Diagnosed with MS at age 41.
Overall continued worsening of clinical status and stable brain and spine MRI

Tanuja Chitnis[1], Tarun Singhal[1], Clare Baecher-Allan[1], Taylor Saraceno[1], Christian Gauthier[1], Anu Paul[1], Thais G. Moreira[1], Fermisk Saleh[1], Liam Murphy[1], Brian Healy[1], Kunwar Shailubhai[2], Jon Zurawski[1], Howard L. Weiner[1]

1 Ann Romney Center for Neurologic Diseases, Brigham and Women's Hospital, Harvard Medical School, Boston, MA, United States
2 Tiziana LifeScience, Doylestown, PA, United States

Results

Reduction of serum proinflammatory markers as well as reduction of microglia activation in the brain. Suggestion of increased strength on right side. No toxicity observed.

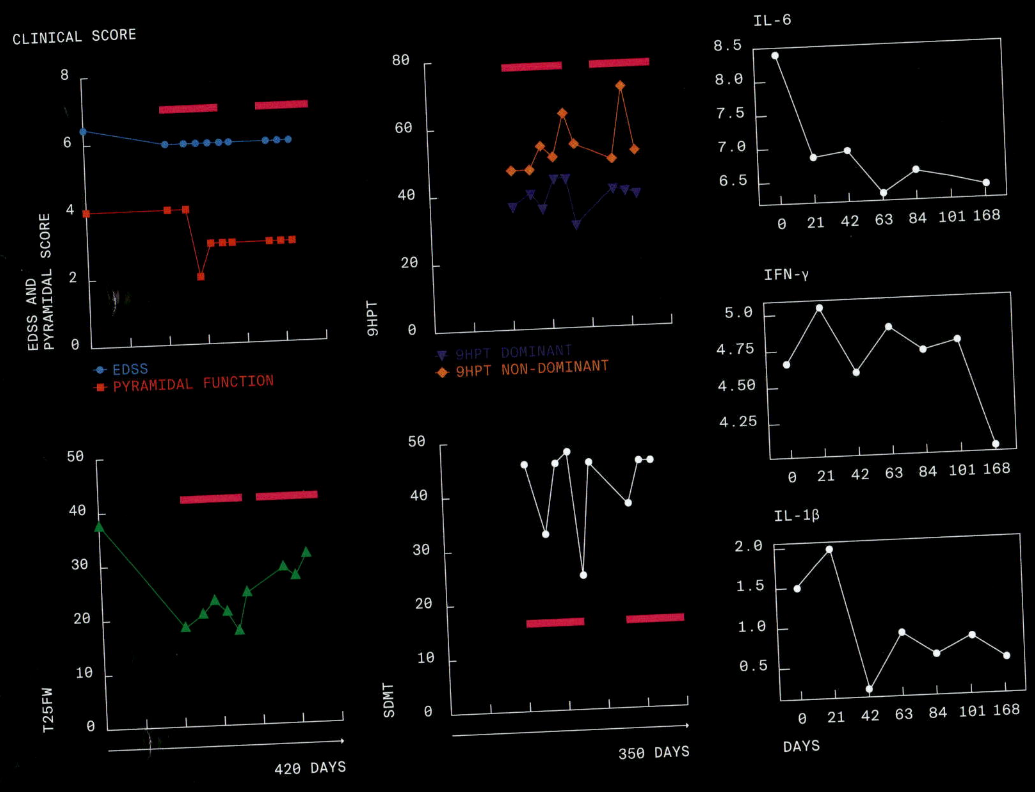

Treatment anti-CD3 (Foralumab)

Serum quantification of blood inflammatory markers

PET Scan (TSPO) showing significantly reduction of microglia activation

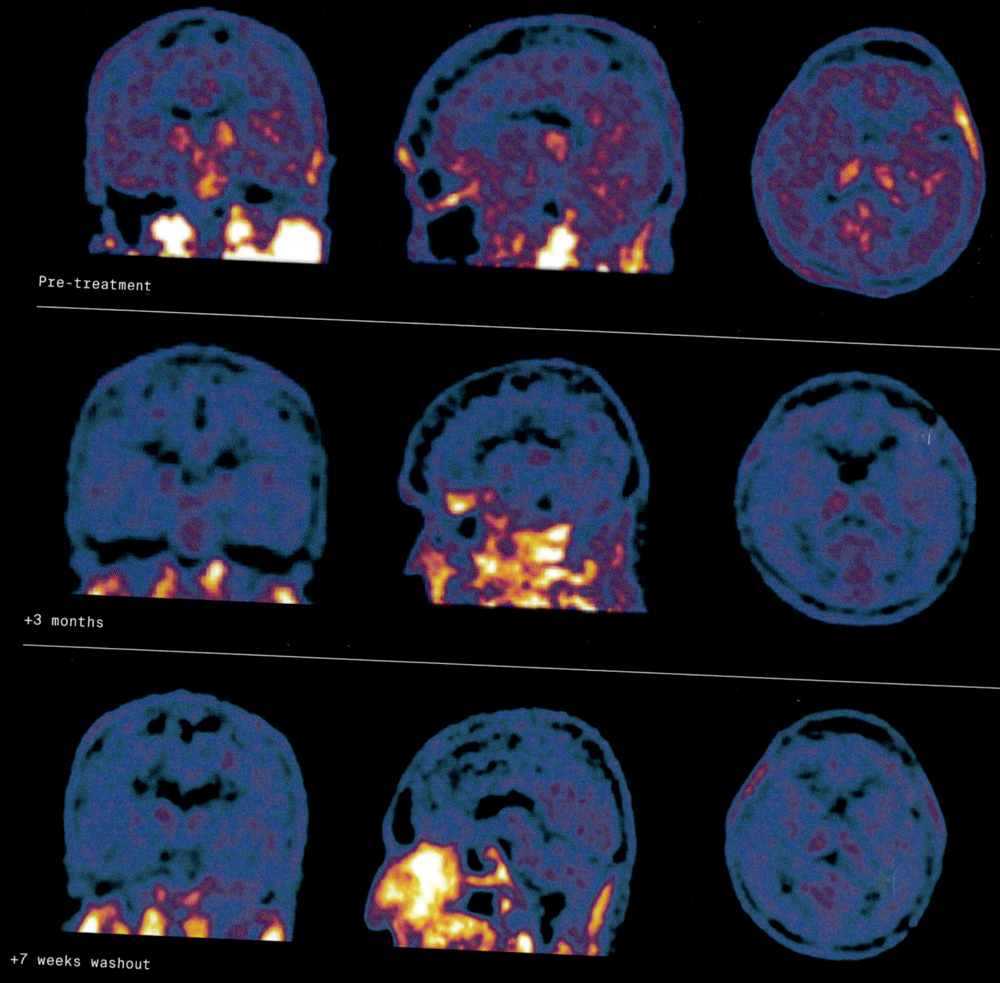

Summary of results and conclusion

1. Mucosal administration of anti-CD3 mAb by binding to the T cell receptor in the mucosal environment induces regulatory T cells and treats autoimmune and inflammatory diseases in animals including models of MS.

   Nasal anti-CD3 was well tolerated in healthy subjects and in moderate cases of COVID-19.

2. Reduction of serum proinflammatory markers as well as reduction of microglia activation in the brain. Suggestion of increased strength on right side. No toxicity observed.

Translational Clinical Research

14

University College London
London, United Kingdom

Neurodegenerative diseases such as Alzheimer's Disease cause the brain and nervous system to become progressively damaged over time. This eventually leads to dementia, disability, and death.

We are living in times where extremely promising breakthroughs are being made in scientific laboratories, both to understand what goes wrong in the brain and to develop treatments that might slow or even begin to reverse the damage that causes dementia.

However, these scientific advances have little real-life meaning unless they can be translated into benefit for people suffering with these conditions. Translational research is the process of taking a discovery or drug from the laboratory and into the patient. It is a process that requires skills and contributions from a wide range of people: vision and leadership from clinical scientists; scientific rigor, precision, and patience from those designing and executing the clinical drug trials; commitment and bravery from patients, nurses, and doctors who might be receiving or delivering a new drug that has never before been tested in humans.

At UCL this research is carried out in a highly specialist unit that bridges the worlds of clinical neurology and highly experimental neuroscience. The Center exists within and is supported by the hospital, but is run by specialist clinical academic research staff.

In this exhibition, we confront the viewers with a question. We challenge them to consider whether they would participate themselves in a clinical drug trial for a new drug that is being tested in humans for the very first time. Through a series of photos and a video installation we explain the critical importance of clinical translational research and explain who and what is involved.

14.1

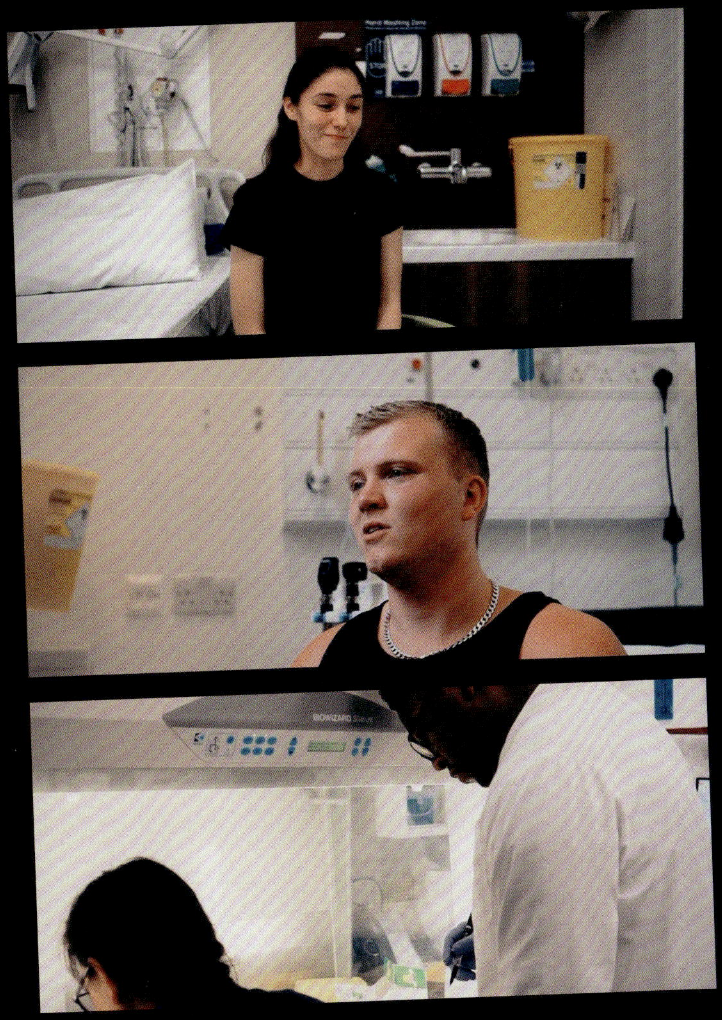

Stills from video featured in Room 14

14.2

Would you sit in this chair?

Translating potential treatments that have been developed in a scientific laboratory through to research studies involving real people requires vision and bravery. It also requires scientific rigor and attention to detail, with a particular focus on safety.

Testing new treatments usually involves some participants receiving the "active" drug and others receiving a placebo or"sham" version of the drug. Neither the research team nor the participants know which is which to avoid biasing the results.

These studies can be highly complex in design and involve collecting lots of information (data) about the safety of a treatment and how it might affect the disease it is designed to treat.

Everything that should happen in a clinical research trial is outlined in the study protocol. This is a detailed instruction manual that must be approved by scientists, doctors, medical ethicists, regulatory authorities, and interested members of the public. It is the responsibility of the principal investigator leading the study (usually a senior doctor with dual training in medicine and science) to make sure that their team understands the protocol and knows exactly what should happen, by whom, when, and how. Members of the team should be accountable and be able to demonstrate they have executed their role with honesty, accuracy, and precision. "If it isn't documented it didn't happen!"

There may be hundreds of steps in the protocol to make sure participants are treated safely and in exactly the same way every time. This rigorous consistency helps to ensure that the results of the study are due to the effect of the treatment on the human body, and not the consequence of other factors such as measurement errors or effects of the environment. It also keeps participants as safe as possible.

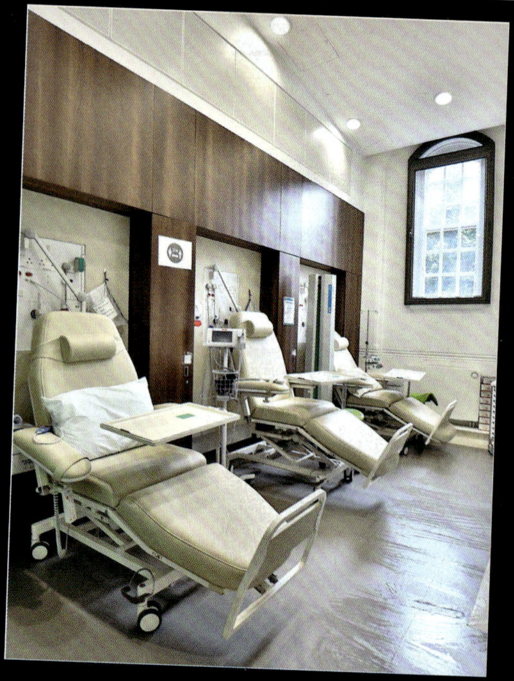

14.3

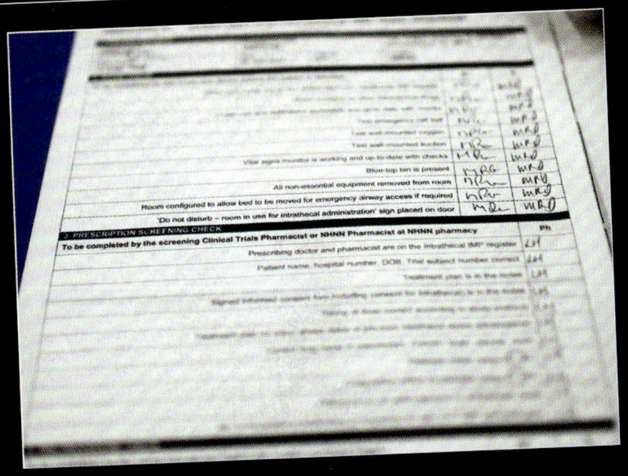

↑
Checklist.

Comprehensive checklists help to make sure all the key safety steps are followed when administering experimental drugs or executing research procedures that carry risk. Meticulous documentation greatly reduces the chance of human error.

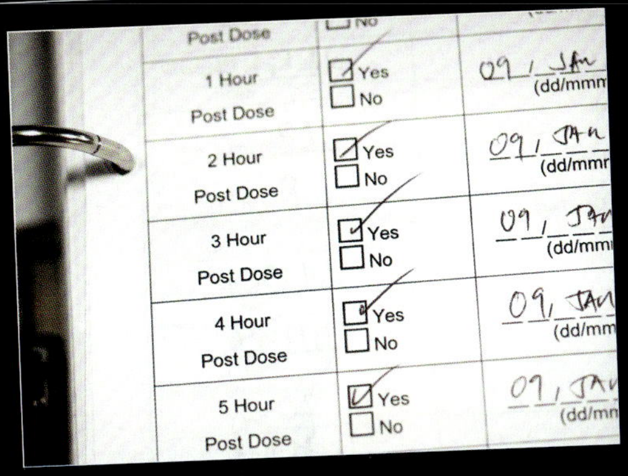

↓
Clinical examination.

Physical examination by the Clinical Trials Doctor is important for determining what effect an experimental drug might have on the brain, spinal cord, and nerves.

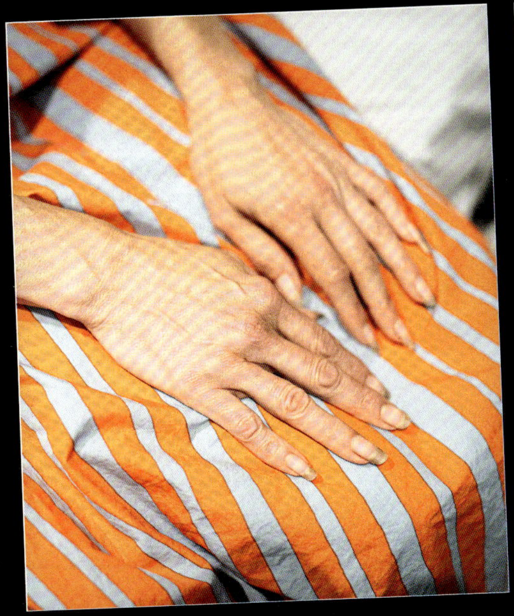

↑
"Hands"

Participants place themselves in a position of vulnerability. This may be particularly challenging in diseases that affect memory, thinking, and judgement. Effective communication and kindness can make things easier.

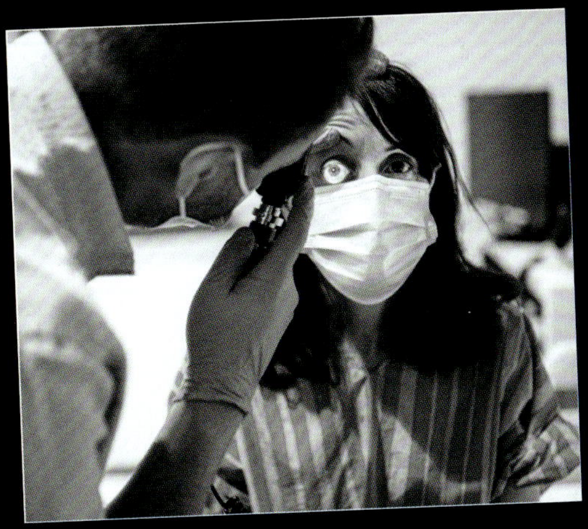

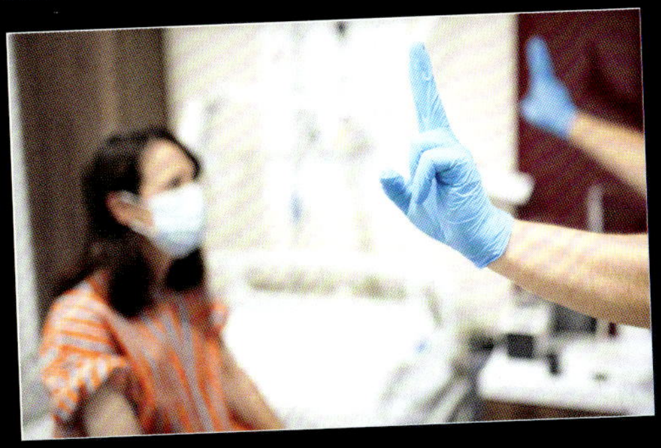

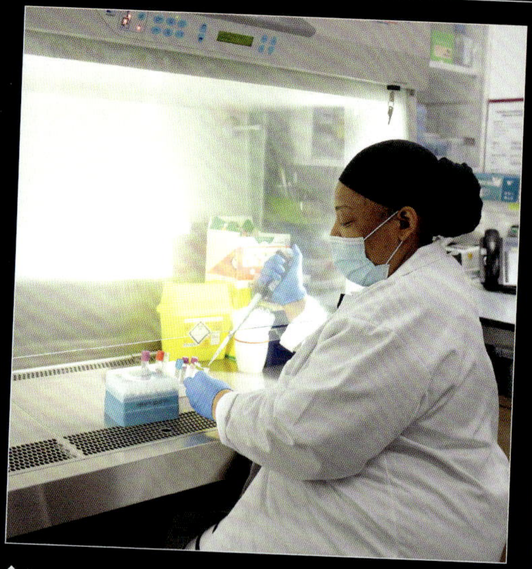

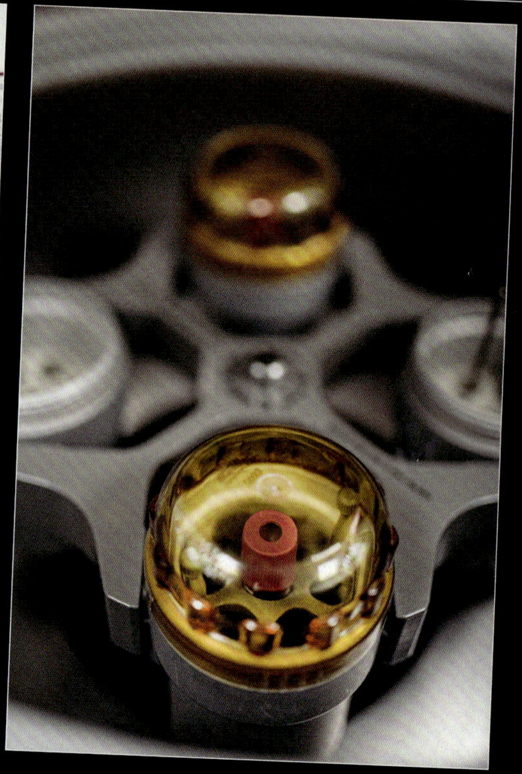

↑→
Laboratory processing

Samples are taken from participants and then processed in the laboratory. Processing includes spinning samples at high speeds to remove debris and separate different components of the sample. It is important that samples are handled in the same way so that results reflect human biology and not other influences from the environment, e.g. temperature.

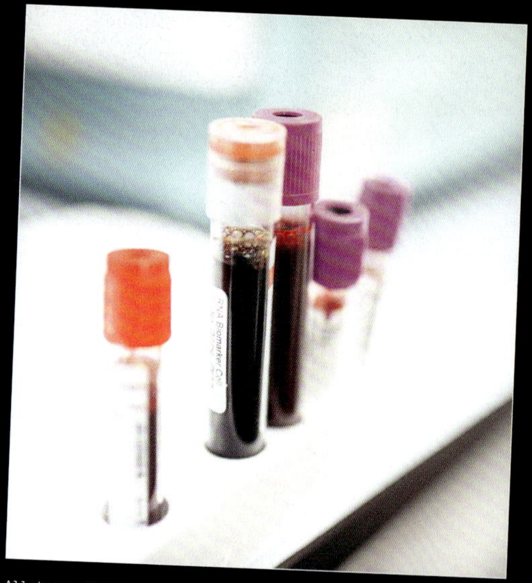

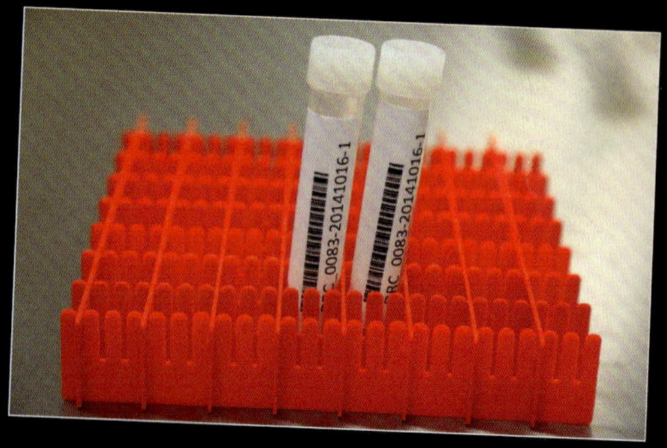

←↑
Sample tubes

Biological samples are collected including blood and cerebrospinal fluid (the fluid that bathes the brain and spinal cord). These fluids contain signatures of the disease process that can be studied and measured in the laboratory. Researchers often collect other samples such as urine, tears or semen which may be useful in the future when technology advances.

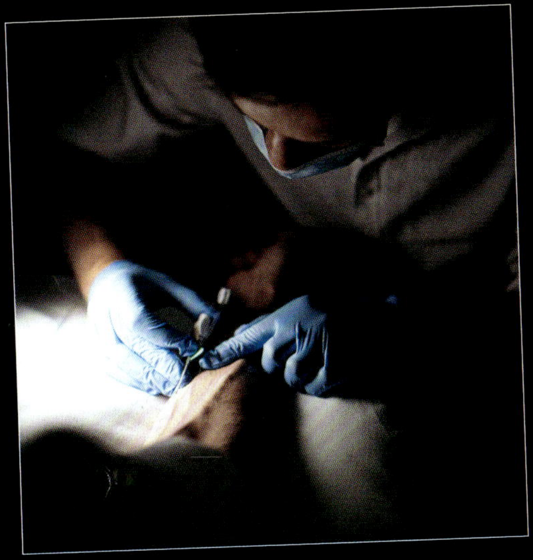

←
Venepuncture

Blood is drawn from the vein from a "healthy control" volunteer. These are individuals who do not have the disease under study, but provide an important group for comparison. This is how the research community understands what a "normal" result looks like. Blood samples are used to monitor the effects and the safety of new drugs. Technology has made it possible to detect proteins and other molecules in the blood that can be used to understand the biology of the brain as well, and to track an individual's response to a treatment.

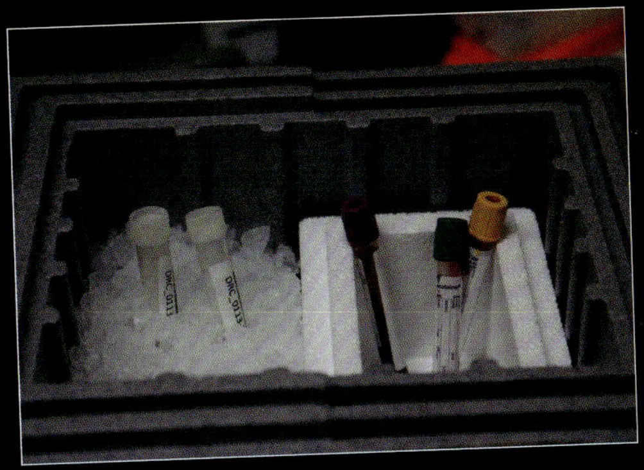

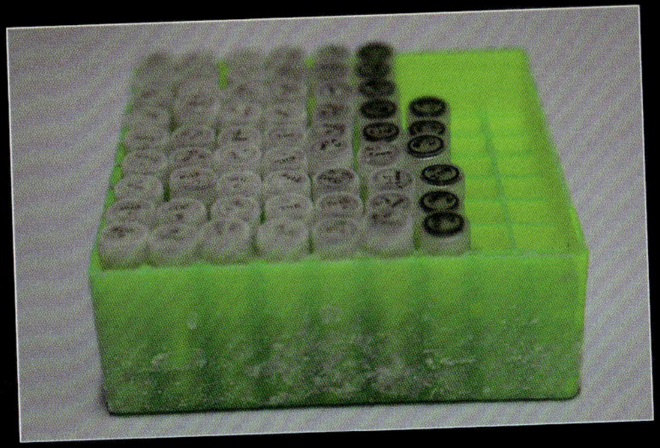

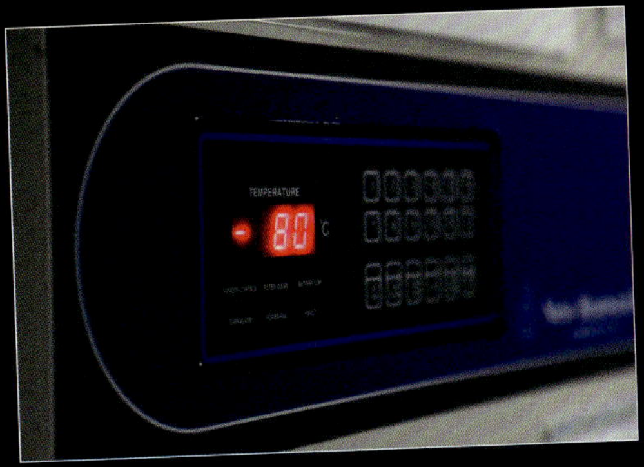

↑↗→
Dry ice

As soon as fluids are removed there is a risk they may be degraded by the environment outside of the body. To avoid this, they are often frozen at low temperatures within seconds or minutes of being taken.

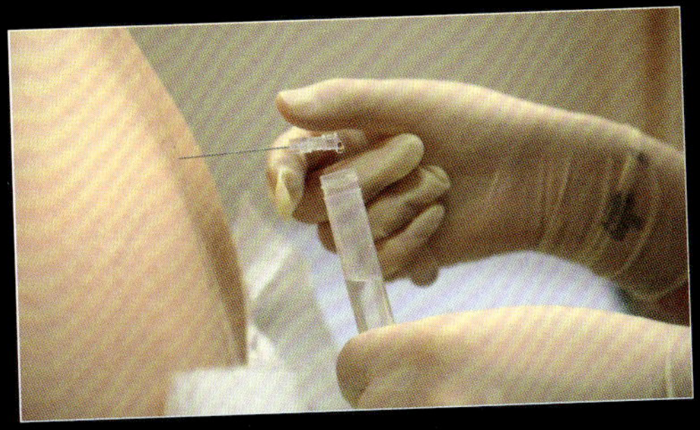

←
Lumbar puncture

Cerebrospinal fluid (CSF) is the fluid that bathes the brain and spinal cord, and can be removed easily from the lower back using a fine needle in a procedure known as a lumbar puncture. Since CSF is in direct communication with the brain it is currently the most reliable way of measuring chemical changes in the brain.

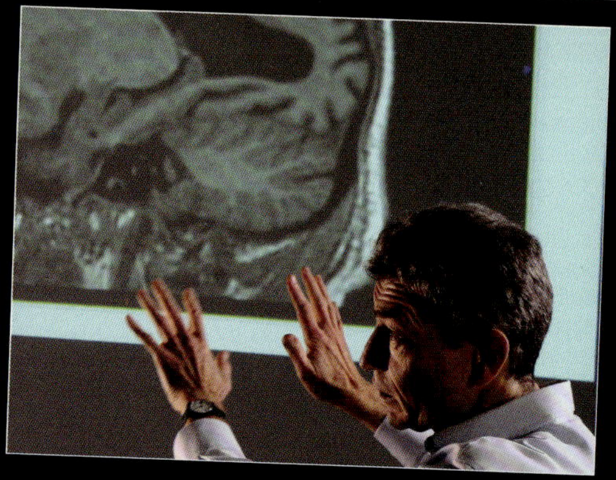

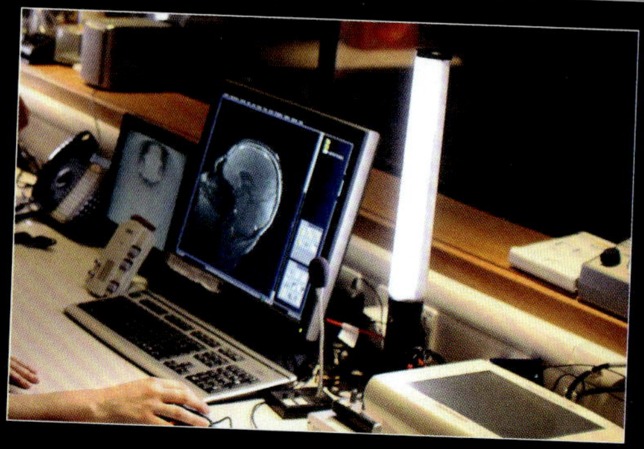

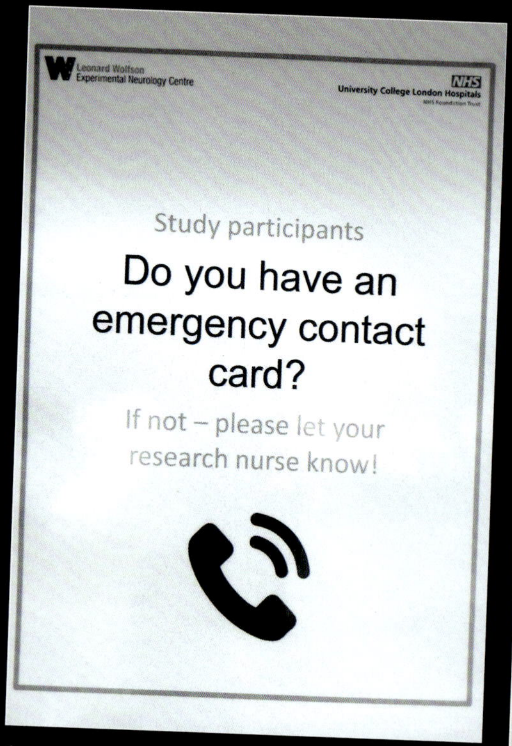

↖↑
MRI

Magnetic resonance imaging (MRI) provides exquisitely detailed images of the brain. These can be used to determine whether individuals are suitable for a research study or whether the brain displays adverse reactions to a treatment. Brain MRI is also used to measure the rate at which the brain is being affected by a neurodegenerative disease. Seeing a reduction in this rate of damage in a clinical trial may indicate that an experimental treatment is having a protective effect.

↓
Infusion pump

Drugs and fluids must be given to research participants in a consistent way. The infusion pump controls the rate of flow of a fluid as it is given to an individual. This equipment must be tested and cleaned regularly.

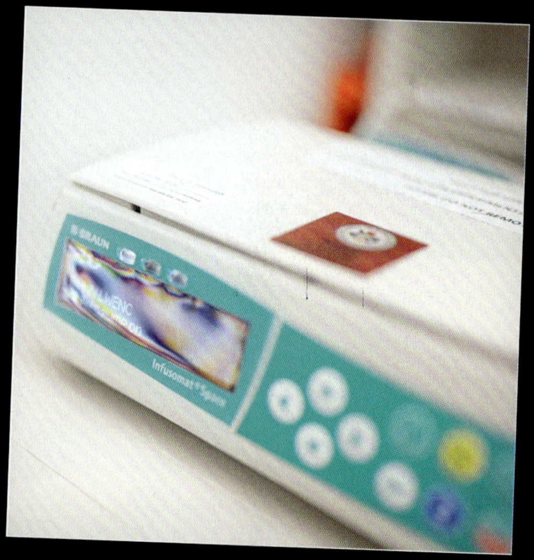

↑
Emergency poster

These are experimental treatments and it may be the first time they have been given to a human being. It is important to monitor potential adverse drug reactions, day or night. A specialist doctor is always on-call and contactable by participants.

14.4

High quality research produces high quality data. High quality data allows us to understand the effects of a new treatment and to determine whether it works or not

Breakthroughs in medical research are sometimes major and change the course of medical treatment for patients. Often, however, it is the smaller steps that incrementally improve medical care. Every research trial tells us something important, even if a treatment doesn't work. We often interrogate data for years after a trial has concluded and as new knowledge comes to light we might be able to ask new questions of a study, or re-analyze the samples collected from it. For this reason, data may be retained for a decade or more after a study has concluded.

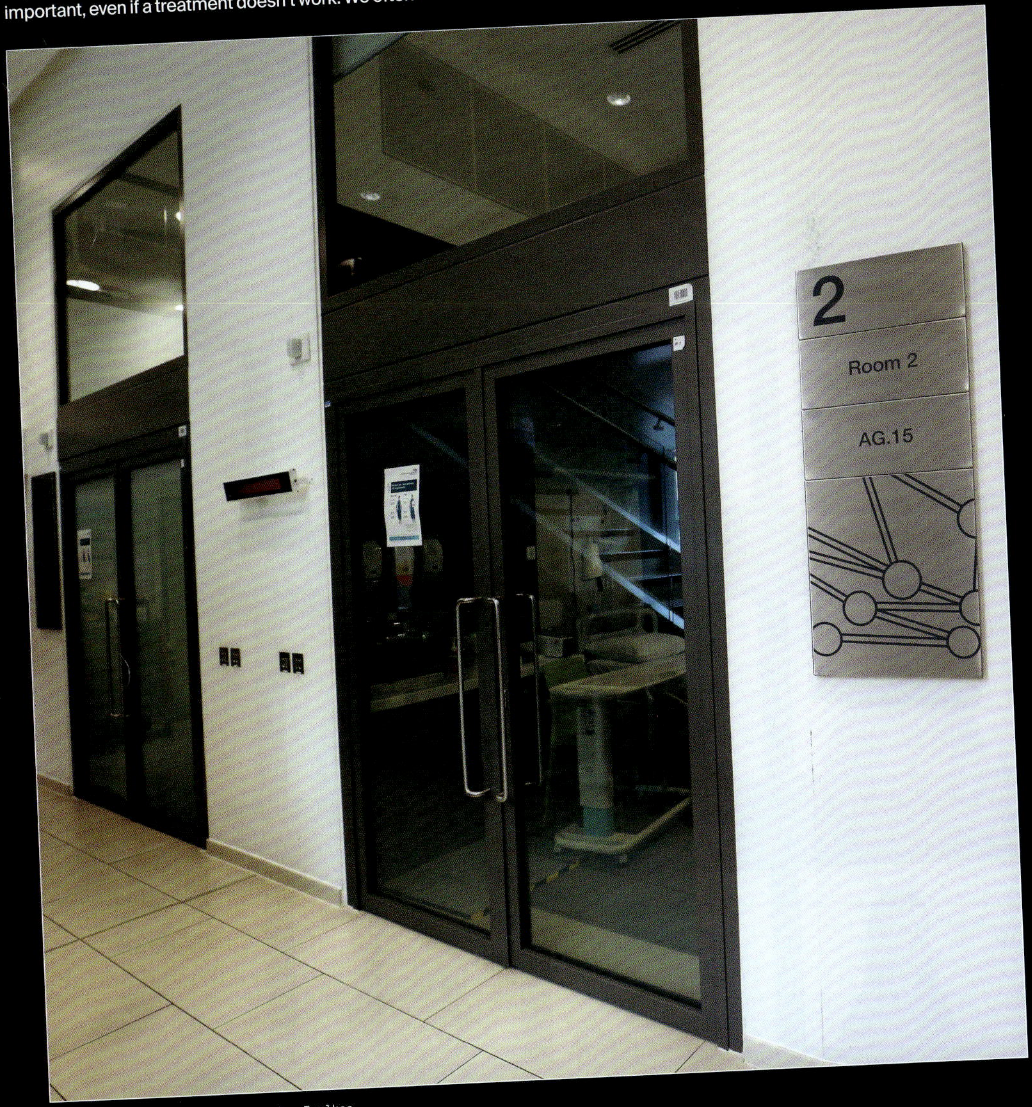

Image © Andrew Green and Alexander Foulkes

14.5

Exceptional people make translational neuroscience possible

The Clinical Research Facility is a hospital within a hospital. It is no ordinary ward, however. It is a space designed for patients and research teams to journey together into the unknown. The drugs being delivered in the Clinical Research Facility may never have been used before in mankind and a diverse range of people and skills are needed to deliver high quality translational research here. All must be specifically trained and prepared for the unknown and unexpected as they take this voyage of discovery into the unknown with the brave patients who have volunteered to receive these experimental drugs.

Each member of the team has specific tasks to carry out. For example, principal investigators design and oversee the running of the study. Pharmacists ensure that experimental medication is prepared and administered to a safe and consistent standard. The cleaner maintains a safe and clean environment for research participants and staff.

Nurses collect clinical data, take blood, administer drugs, and offer reassurance. Clinical trials doctors combine medical training and specialist research training to seek consent, examine patients, prescribe and administer experimental drugs, and monitor clinical safety. The laboratory manager runs the laboratory in which blood, cerebrospinal fluid, and other body fluids are processed into small amounts that can be stored in the freezer in preparation for later analysis. Recruitment co-ordinators select and approach individuals who might be appropriate for a research study and co-ordinate their ongoing involvement. Psychologists carry out detailed assessments of memory and thinking using a battery of specially designed paper and computer-based tests.

And the research participants step up and give their time and their bodies for the benefit of medical research. Could you sit in this chair?

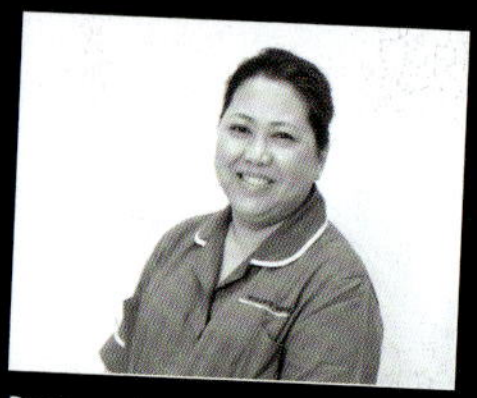
Research nurse

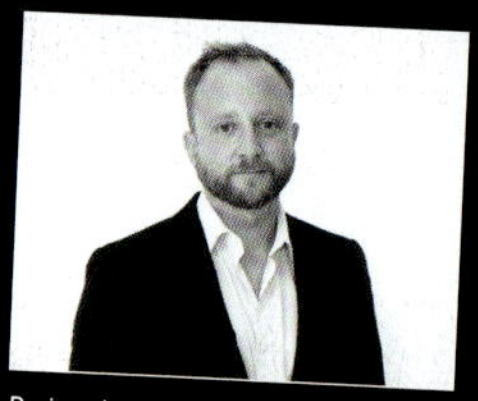
Principal investigator

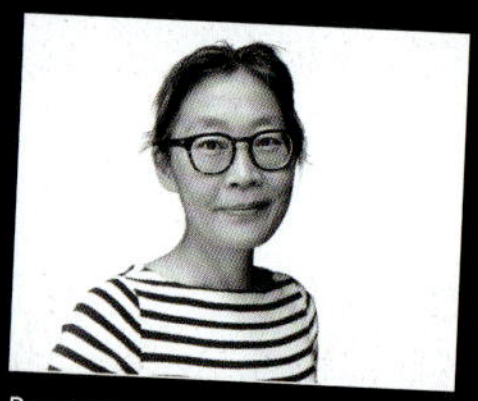
Research coordinator

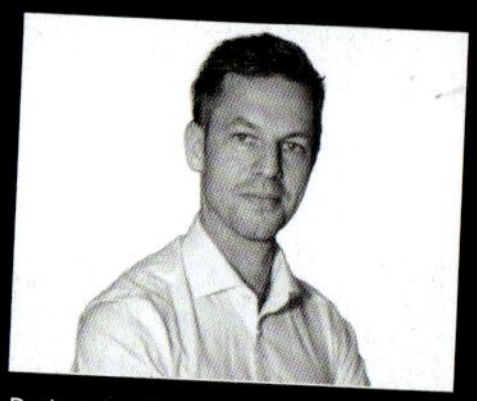
Principal investigator

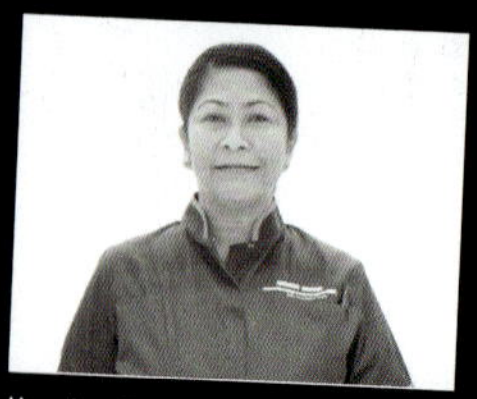
Head of research delivery

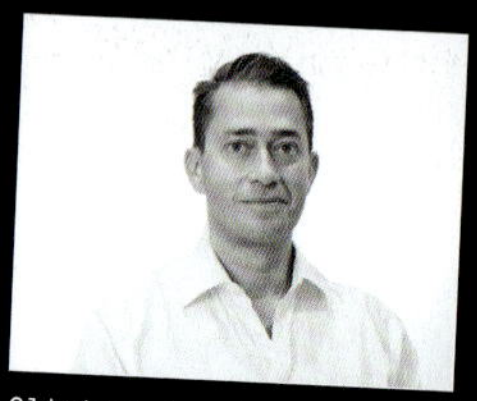
Clinical trials doctor

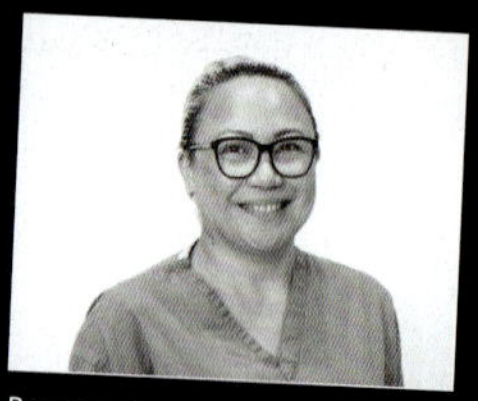
Research nurse

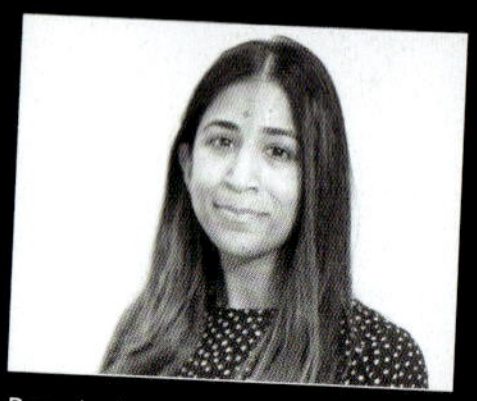
Psychologist

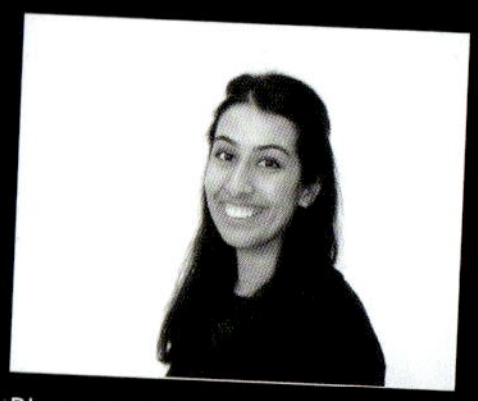
Pharmacist

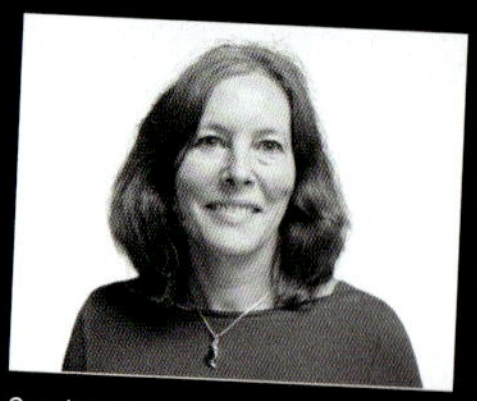
Senior research nurse

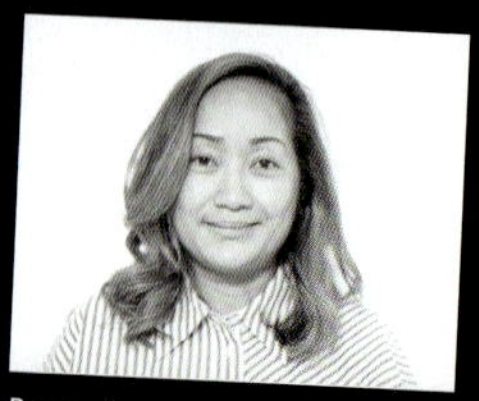
Recruitment coordinator

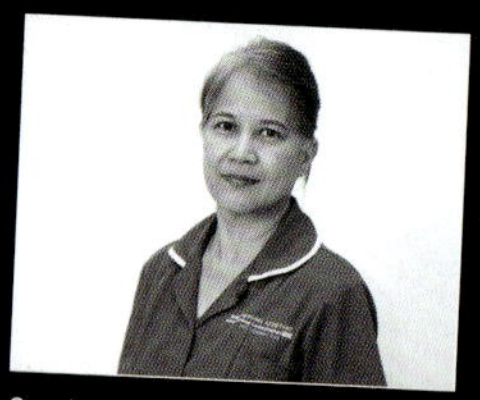
Senior health care assistant

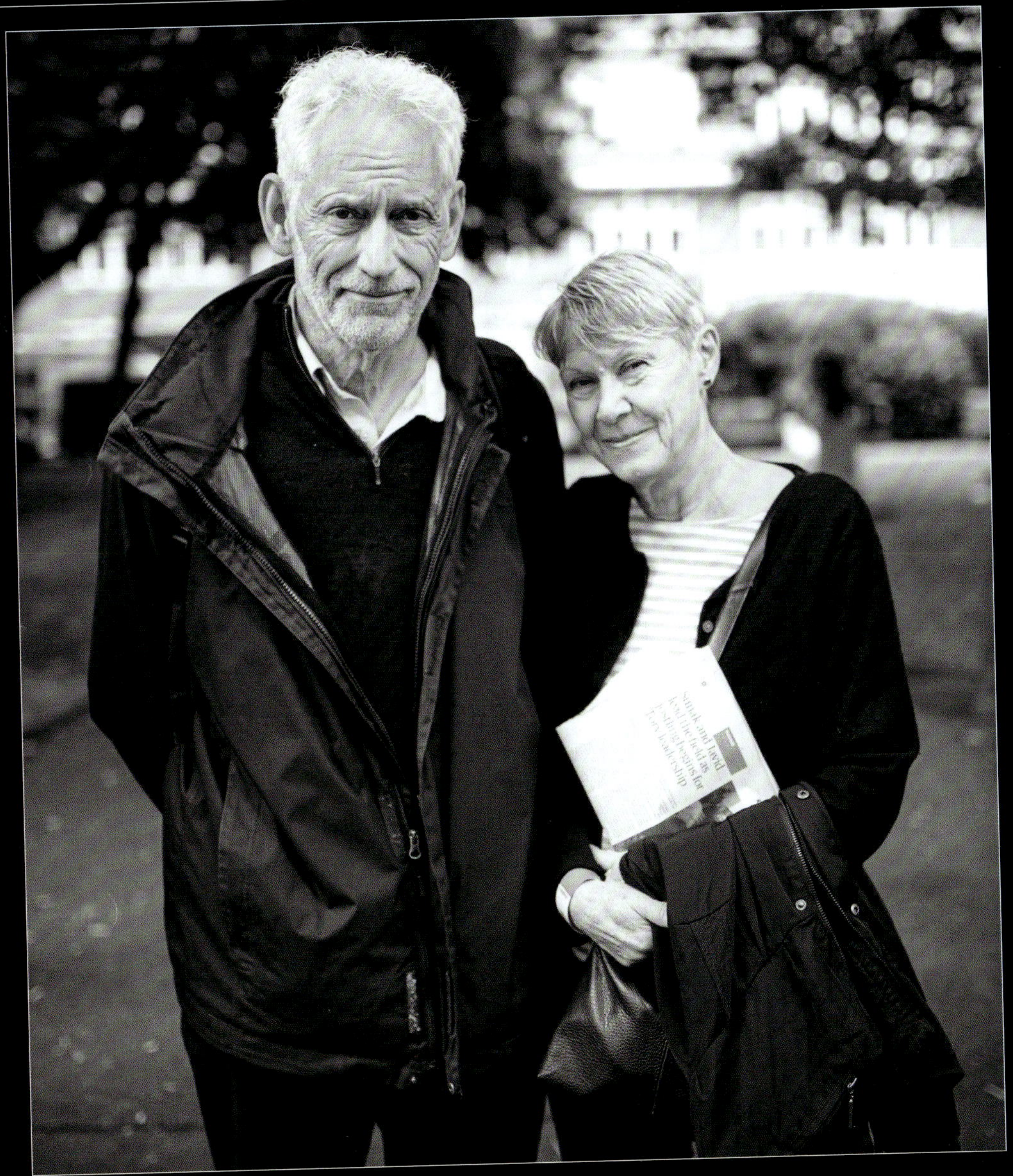

Study participants

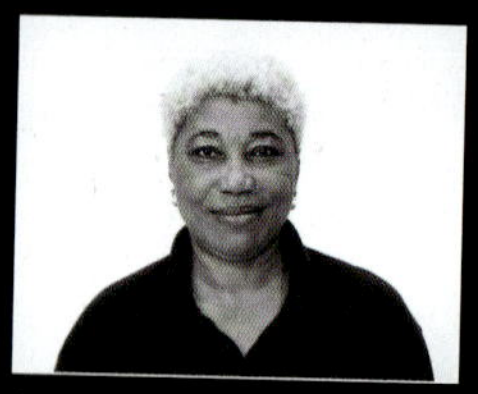
Cleaner

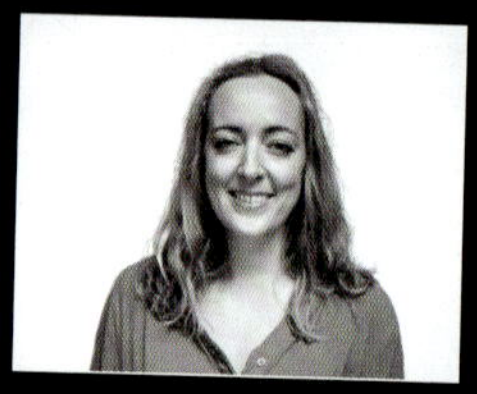
Recruitment coordinator

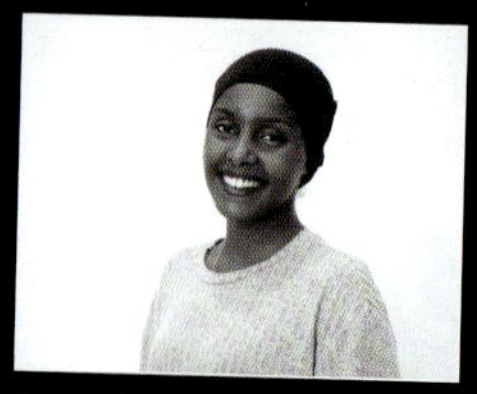
Receptionist

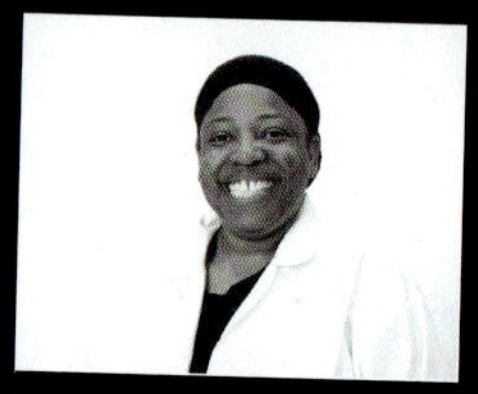
Laboratory manager

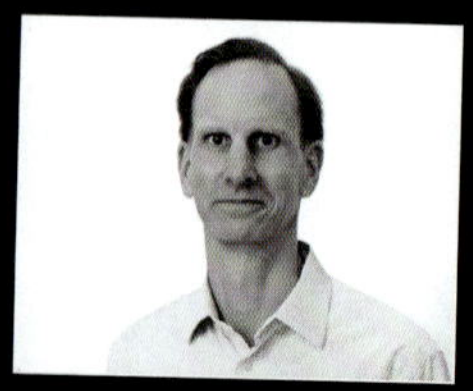
Consultant neurologist

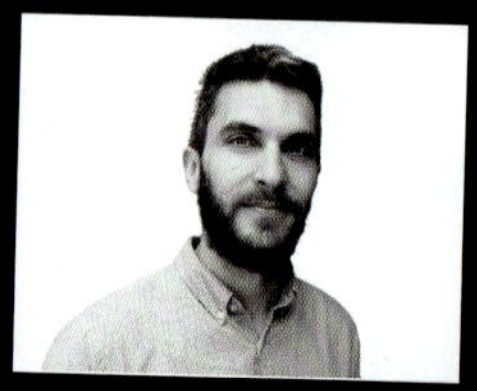
Clinical trials doctor

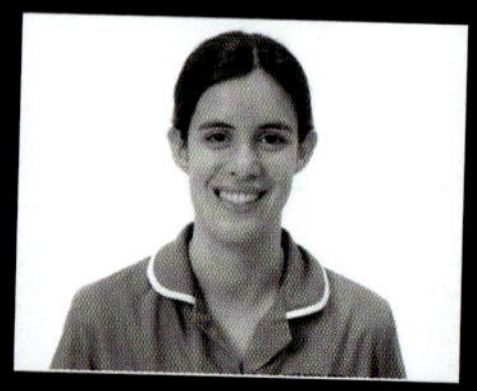
Research nurse

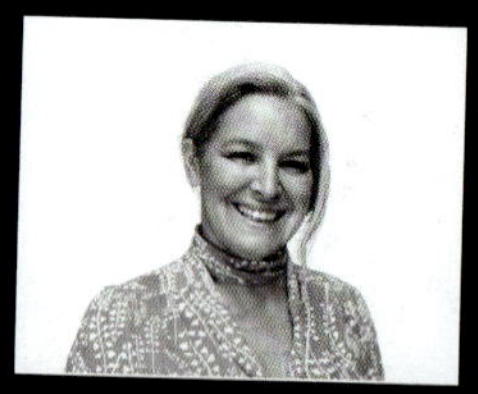
Principal investigator

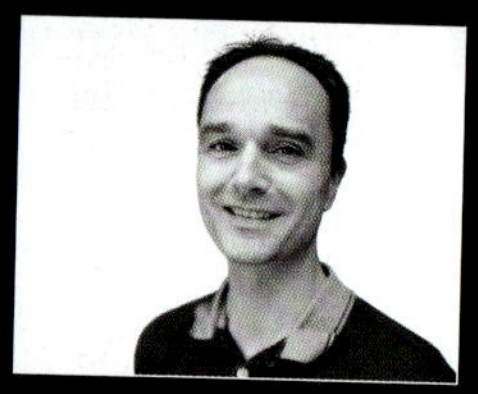
Pharmacist

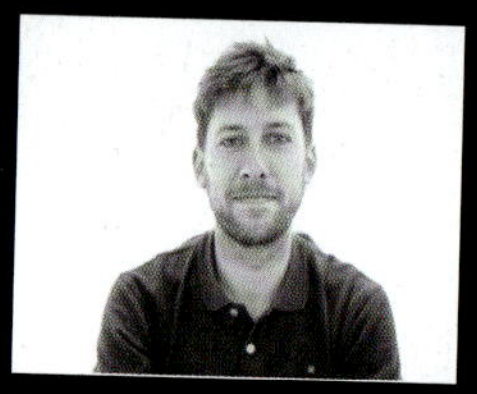
Study manager

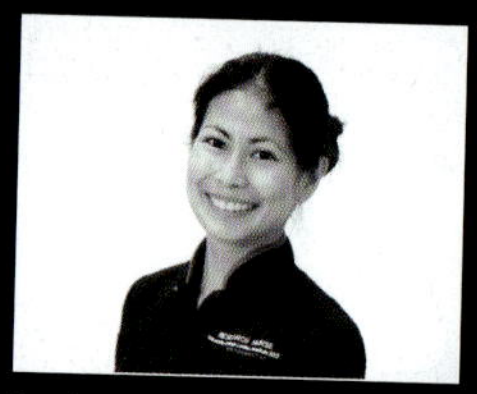
Senior research nurse

PRESERVING THE BRAIN

Fondazione Prada, Milan

September 16 – November 10, 2022

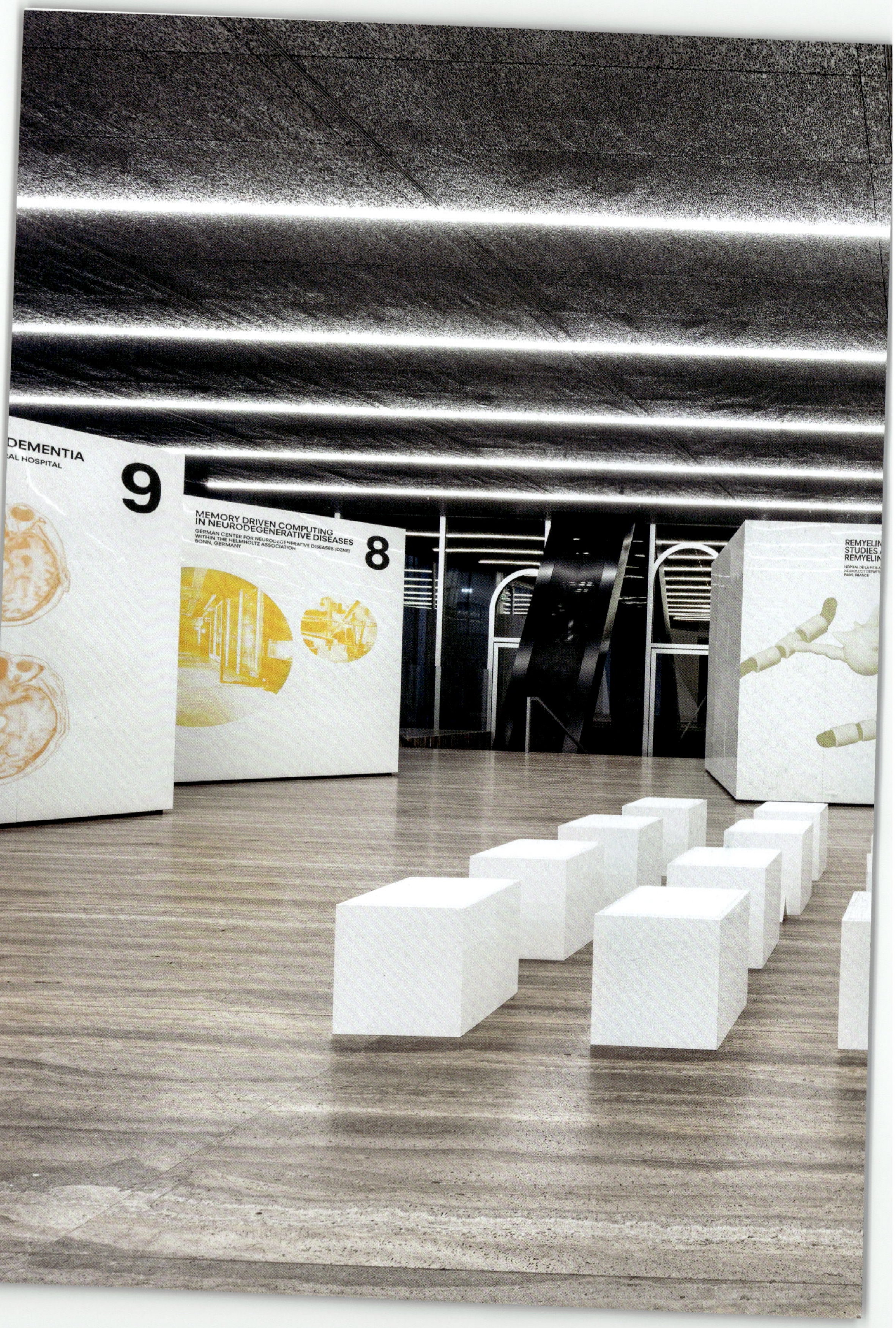
DEMENTIA
9
MEMORY DRIVEN COMPUTING
IN NEURODEGENERATIVE DISEASES
GERMAN CENTER FOR NEURODEGENERATIVE DISEASES (DZNE)
WITHIN THE HELMHOLTZ ASSOCIATION
BONN, GERMANY
8
PARIS, FRANCE

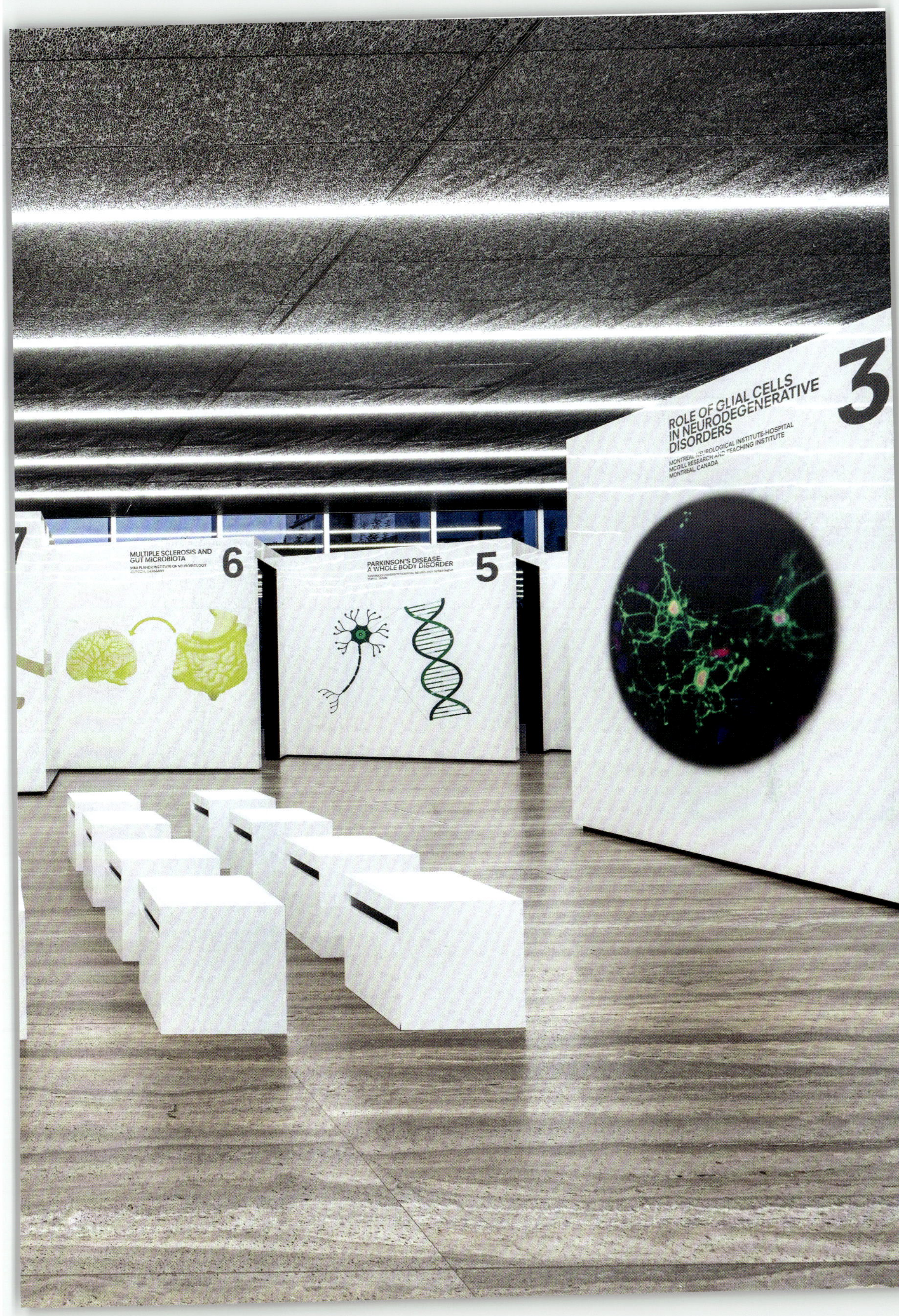
7
MULTIPLE SCLEROSIS AND GUT MICROBIOTA
MAX PLANCK INSTITUTE OF NEUROBIOLOGY
6
PARKINSON'S DISEASE: A WHOLE BODY DISORDER
5
ROLE OF GLIAL CELLS IN NEURODEGENERATIVE DISORDERS
MONTREAL, CANADA
3

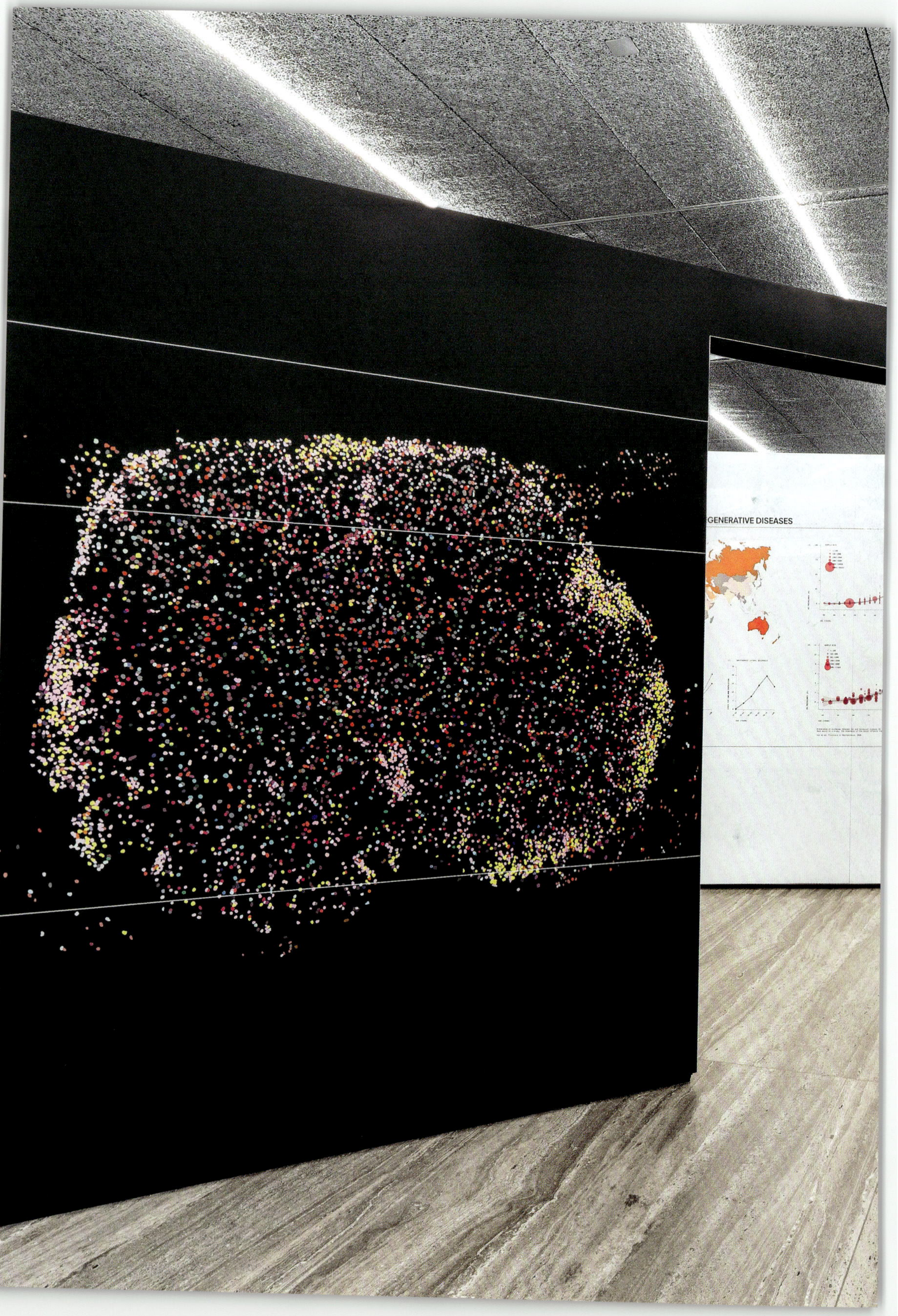
GENERATIVE DISEASES

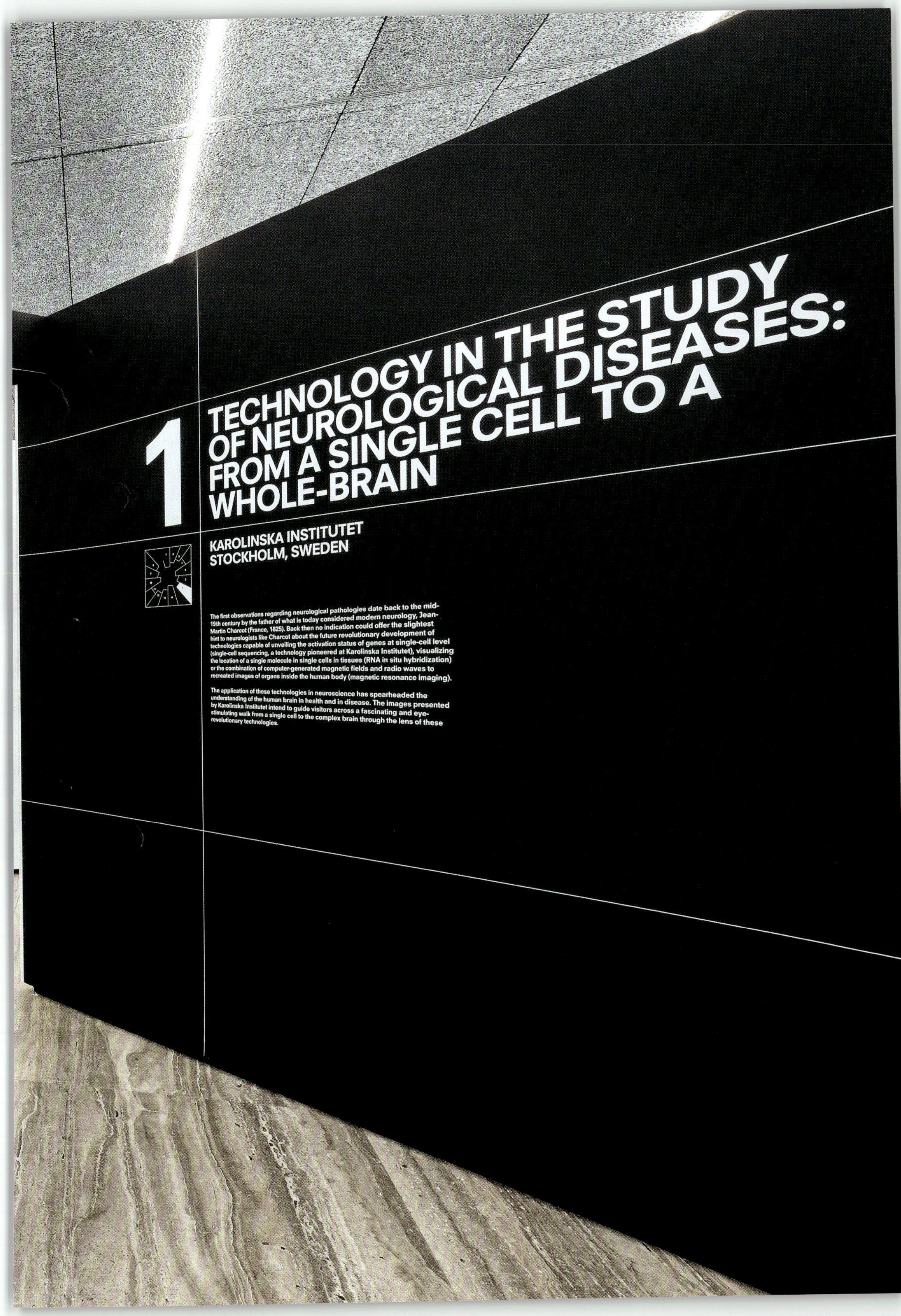
1
TECHNOLOGY IN THE STUDY
OF NEUROLOGICAL DISEASES:
FROM A SINGLE CELL TO A
WHOLE-BRAIN
KAROLINSKA INSTITUTET
STOCKHOLM, SWEDEN
The first observations regarding neurological pathologies date back to the mid-19th century by the father of what is today considered modern neurology, Jean-Martin Charcot (France, 1825). Back then no indication could offer the slightest hint to neurologists like Charcot about the future revolutionary development of technologies capable of unveiling the activation status of genes at single-cell level (single-cell sequencing, a technology pioneered at Karolinska Institutet), visualizing the location of a single molecule in single cells in tissues (RNA in situ hybridization) or the combination of computer-generated magnetic fields and radio waves to recreated images of organs inside the human body (magnetic resonance imaging).
The application of these technologies in neuroscience has spearheaded the understanding of the human brain in health and in disease. The images presented by Karolinska Institutet intend to guide visitors across a fascinating and eye-stimulating walk from a single cell to the complex brain through the lens of these revolutionary technologies.

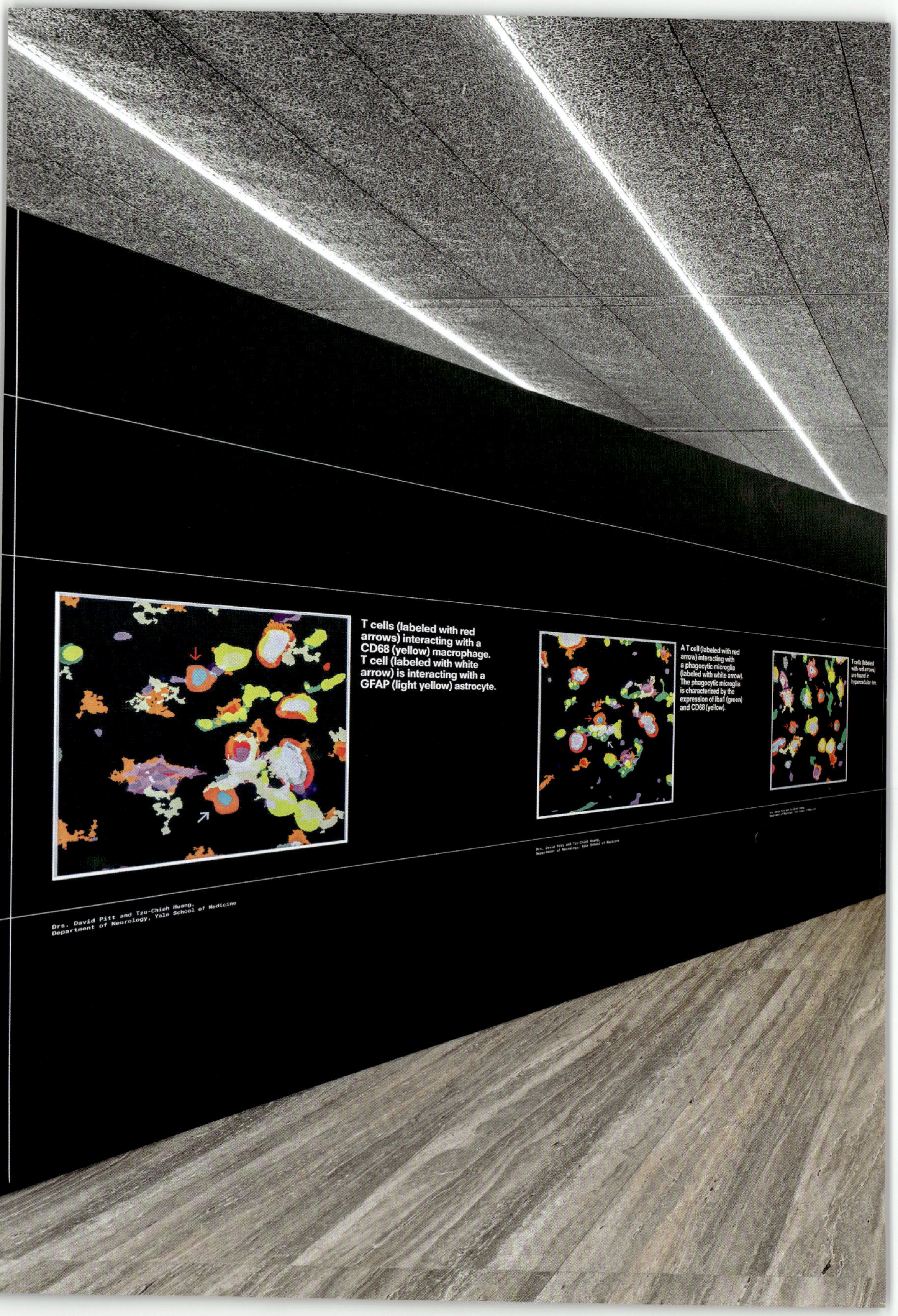
T cells (labeled with red arrows) interacting with a CD68 (yellow) macrophage. T cell (labeled with white arrow) is interacting with a GFAP (light yellow) astrocyte.
Drs. David Pitt and Tzu-Chieh Huang, Department of Neurology, Yale School of Medicine
A T cell (labeled with red arrow) interacting with a phagocytic microglia (labeled with white arrow). The phagocytic microglia is characterized by the expression of Iba1 (green) and CD68 (yellow).
Drs. David Pitt and Tzu-Chieh Huang, Department of Neurology, Yale School of Medicine
T cells (labeled with red arrows) are found in hypercellular rim.

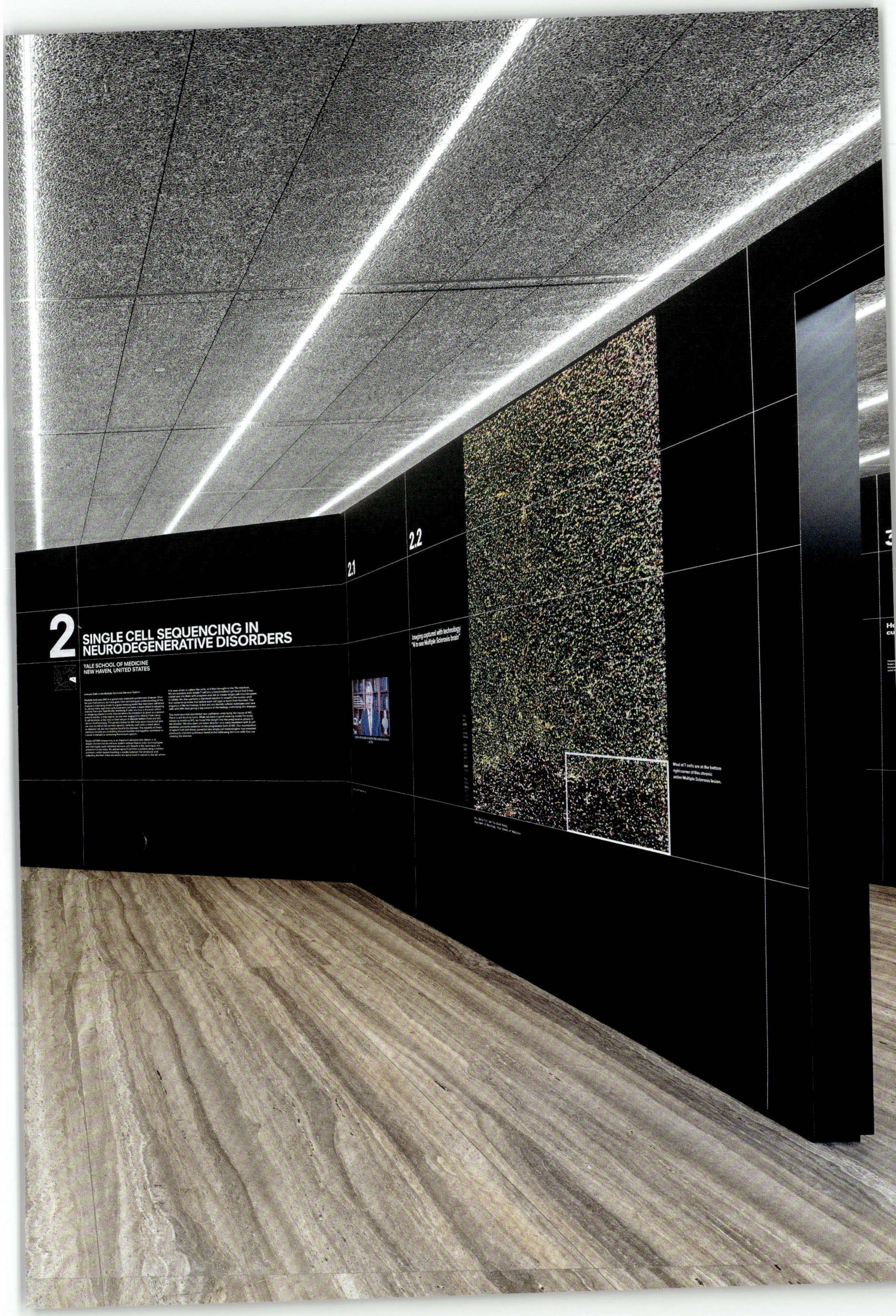
2
SINGLE CELL SEQUENCING IN NEURODEGENERATIVE DISORDERS
YALE SCHOOL OF MEDICINE
NEW HAVEN, UNITED STATES
2.1
2.2

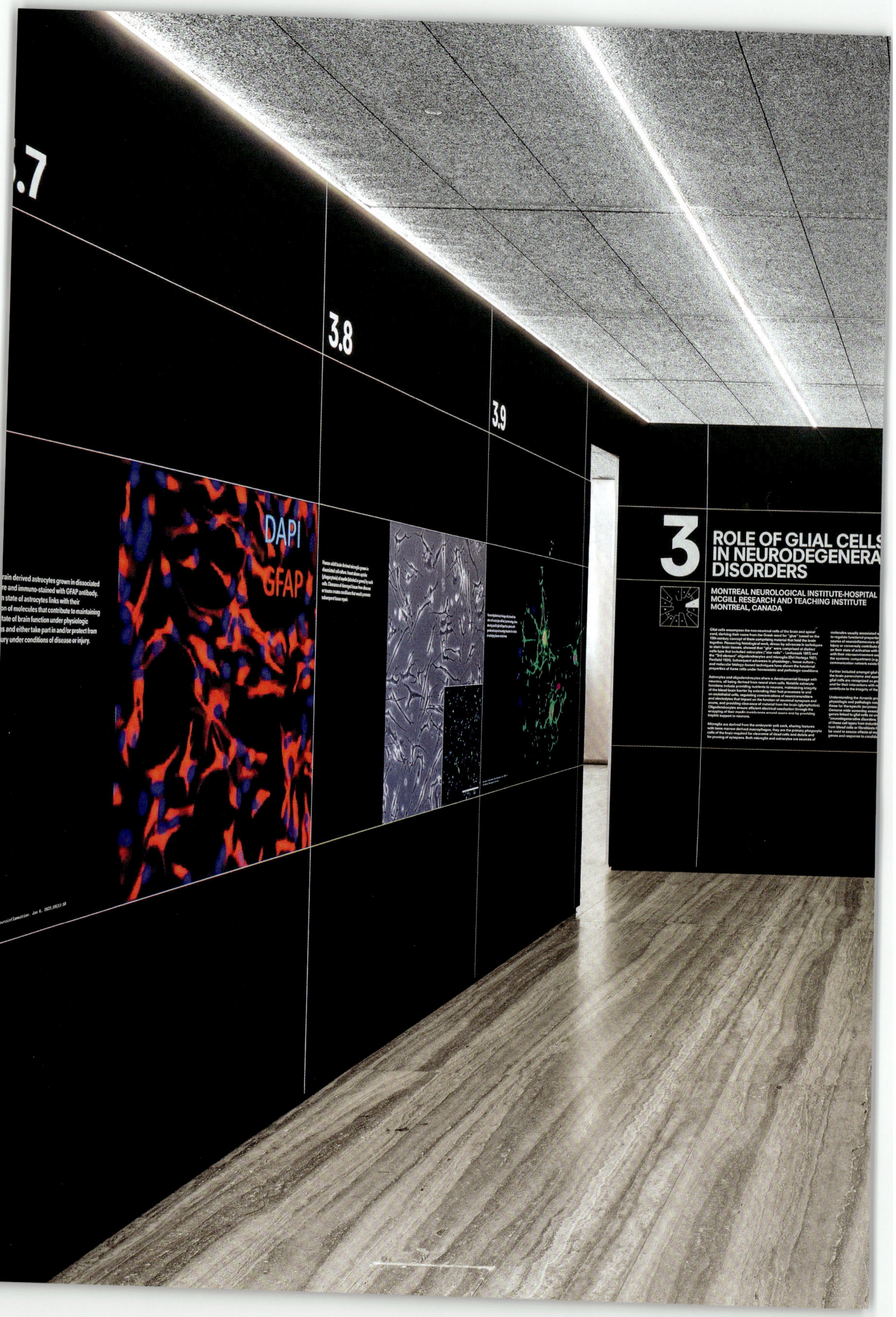
3.8
3.9
DAPI
GFAP
3
ROLE OF GLIAL CELLS
DISORDERS
MONTREAL NEUROLOGICAL INSTITUTE-HOSPITAL
MCGILL RESEARCH AND TEACHING INSTITUTE
MONTREAL, CANADA

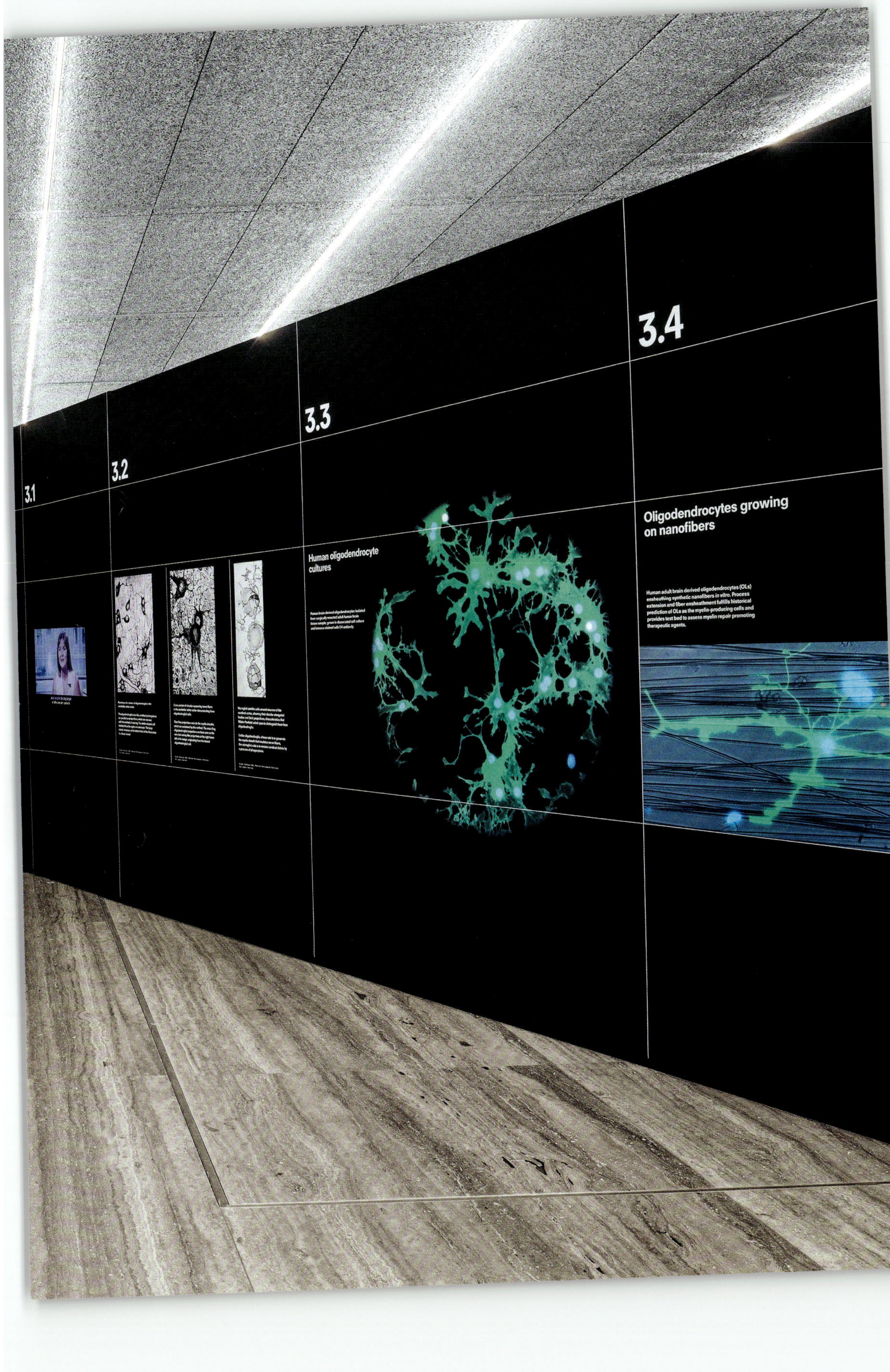
3.1
3.2
3.3
3.4
Human oligodendrocyte cultures
Oligodendrocytes growing on nanofibers
Human adult brain derived oligodendrocytes (OLs) ensheathing synthetic nanofibers in vitro. Process extension and fiber ensheathment fulfills historical prediction of OLs as the myelin-producing cells and provides test bed to assess myelin repair promoting therapeutic agents.

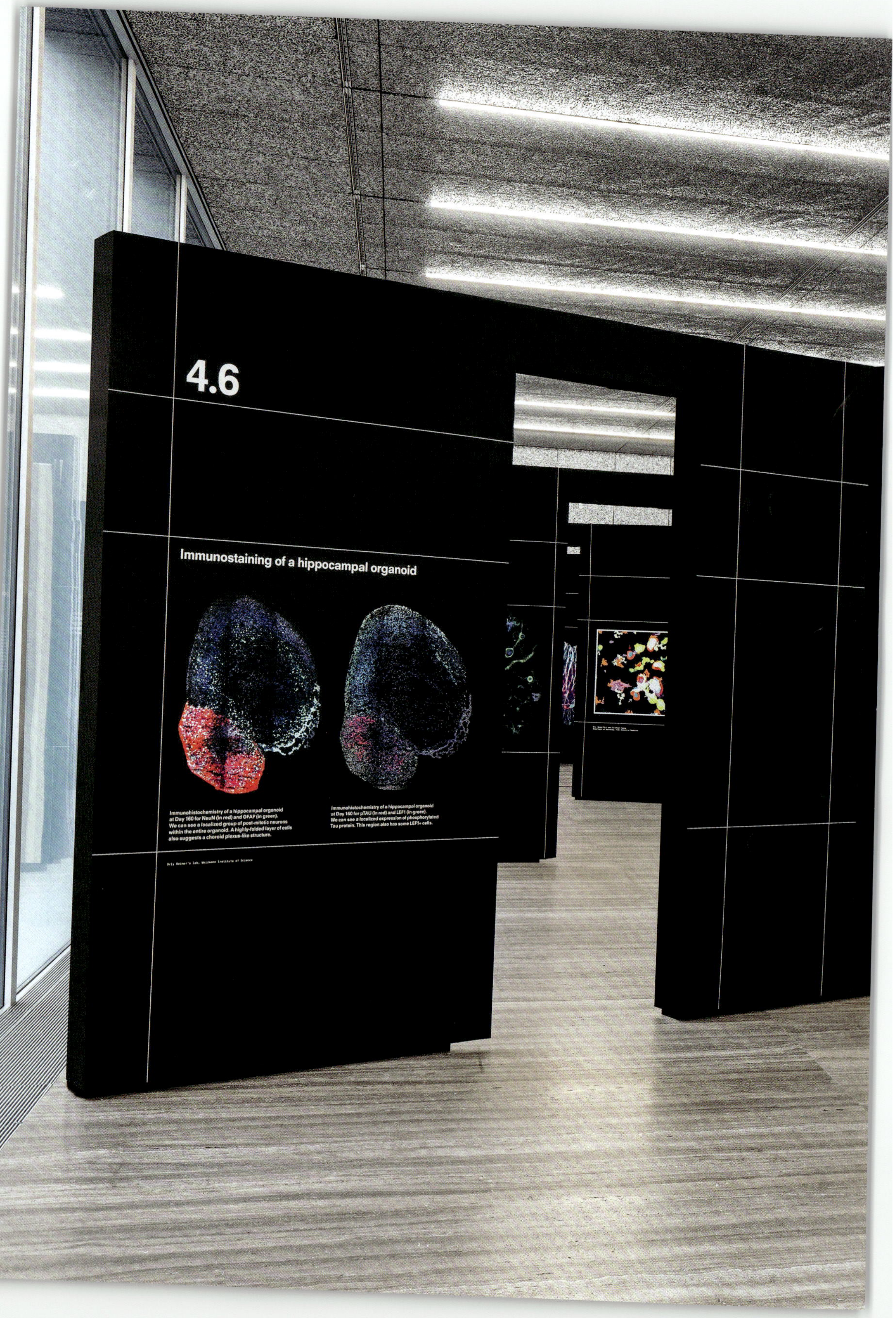
4.6
Immunostaining of a hippocampal organoid
Immunohistochemistry of a hippocampal organoid at Day 160 for NeuN (in red) and GFAP (in green). We can see a localized group of post-mitotic neurons within the entire organoid. A highly-folded layer of cells also suggests a choroid plexus-like structure.
Immunohistochemistry of a hippocampal organoid at Day 160 for pTAU (in red) and LEF1 (in green). We can see a localized expression of phosphorylated Tau protein. This region also has some LEF1+ cells.
Orly Reiner's lab, Weizmann Institute of Science

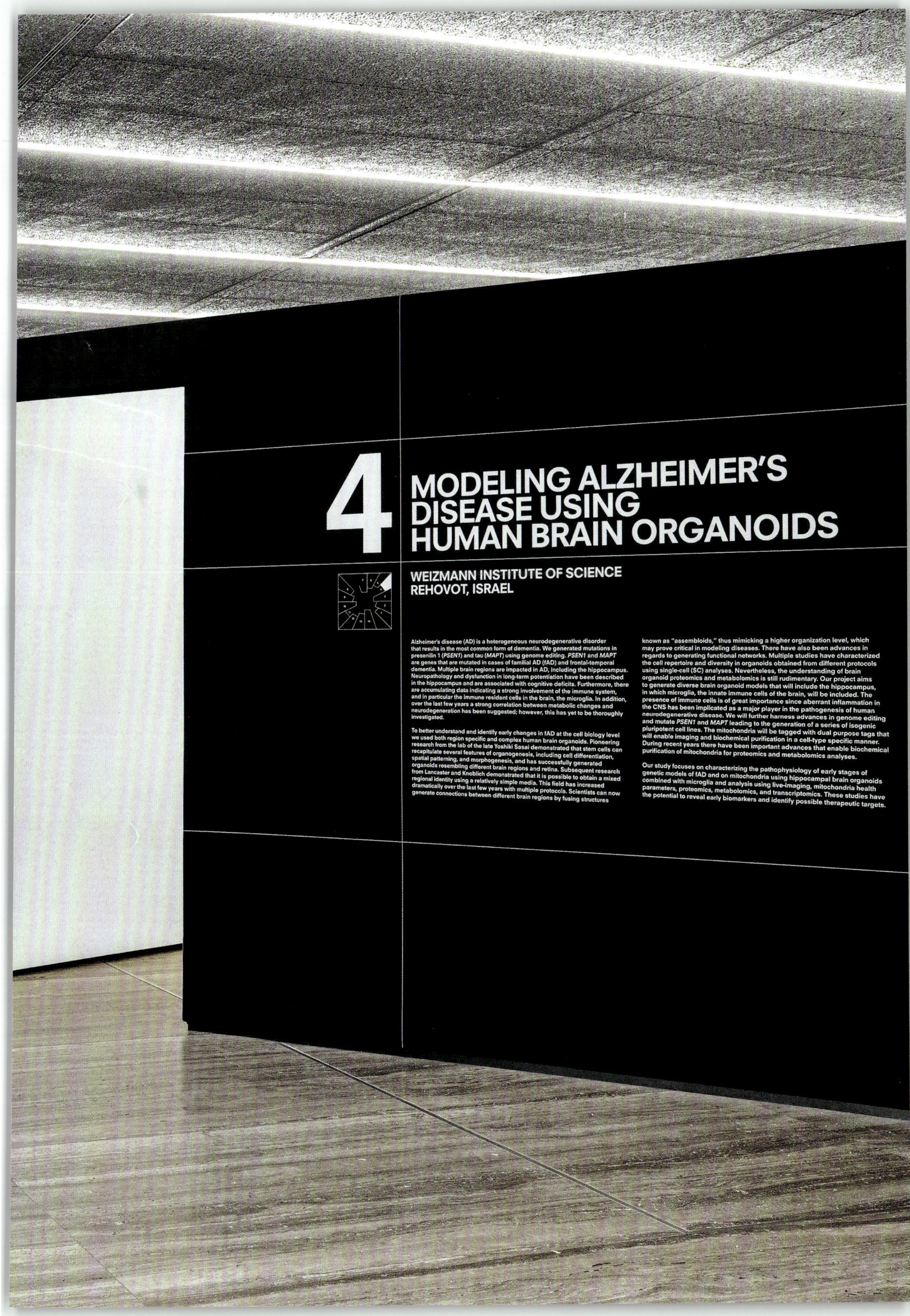
4
MODELING ALZHEIMER'S DISEASE USING HUMAN BRAIN ORGANOIDS
WEIZMANN INSTITUTE OF SCIENCE
REHOVOT, ISRAEL
Alzheimer's disease (AD) is a heterogeneous neurodegenerative disorder that results in the most common form of dementia. We generated mutations in presenilin 1 (PSEN1) and tau (MAPT) using genome editing. PSEN1 and MAPT are genes that are mutated in cases of familial AD (fAD) and frontal-temporal dementia. Multiple brain regions are impacted in AD, including the hippocampus. Neuropathology and dysfunction in long-term potentiation have been described in the hippocampus and are associated with cognitive deficits. Furthermore, there are accumulating data indicating a strong involvement of the immune system, and in particular the immune resident cells in the brain, the microglia. In addition, over the last few years a strong correlation between metabolic changes and neurodegeneration has been suggested; however, this has yet to be thoroughly investigated.
To better understand and identify early changes in fAD at the cell biology level we used both region specific and complex human brain organoids. Pioneering research from the lab of the late Yoshiki Sasai demonstrated that stem cells can recapitulate several features of organogenesis, including cell differentiation, spatial patterning, and morphogenesis, and has successfully generated organoids resembling different brain regions and retina. Subsequent research from Lancaster and Knoblich demonstrated that it is possible to obtain a mixed regional identity using a relatively simple media. This field has increased dramatically over the last few years with multiple protocols. Scientists can now generate connections between different brain regions by fusing structures known as "assembloids," thus mimicking a higher organization level, which may prove critical in modeling diseases. There have also been advances in regards to generating functional networks. Multiple studies have characterized the cell repertoire and diversity in organoids obtained from different protocols using single-cell (SC) analyses. Nevertheless, the understanding of brain organoid proteomics and metabolomics is still rudimentary. Our project aims to generate diverse brain organoid models that will include the hippocampus, in which microglia, the innate immune cells of the brain, will be included. The presence of immune cells is of great importance since aberrant inflammation in the CNS has been implicated as a major player in the pathogenesis of human neurodegenerative disease. We will further harness advances in genome editing and mutate PSEN1 and MAPT leading to the generation of a series of isogenic pluripotent cell lines. The mitochondria will be tagged with dual purpose tags that will enable imaging and biochemical purification in a cell-type specific manner. During recent years there have been important advances that enable biochemical purification of mitochondria for proteomics and metabolomics analyses.
Our study focuses on characterizing the pathophysiology of early stages of genetic models of fAD and on mitochondria using hippocampal brain organoids combined with microglia and analysis using live-imaging, mitochondria health parameters, proteomics, metabolomics, and transcriptomics. These studies have the potential to reveal early biomarkers and identify possible therapeutic targets.

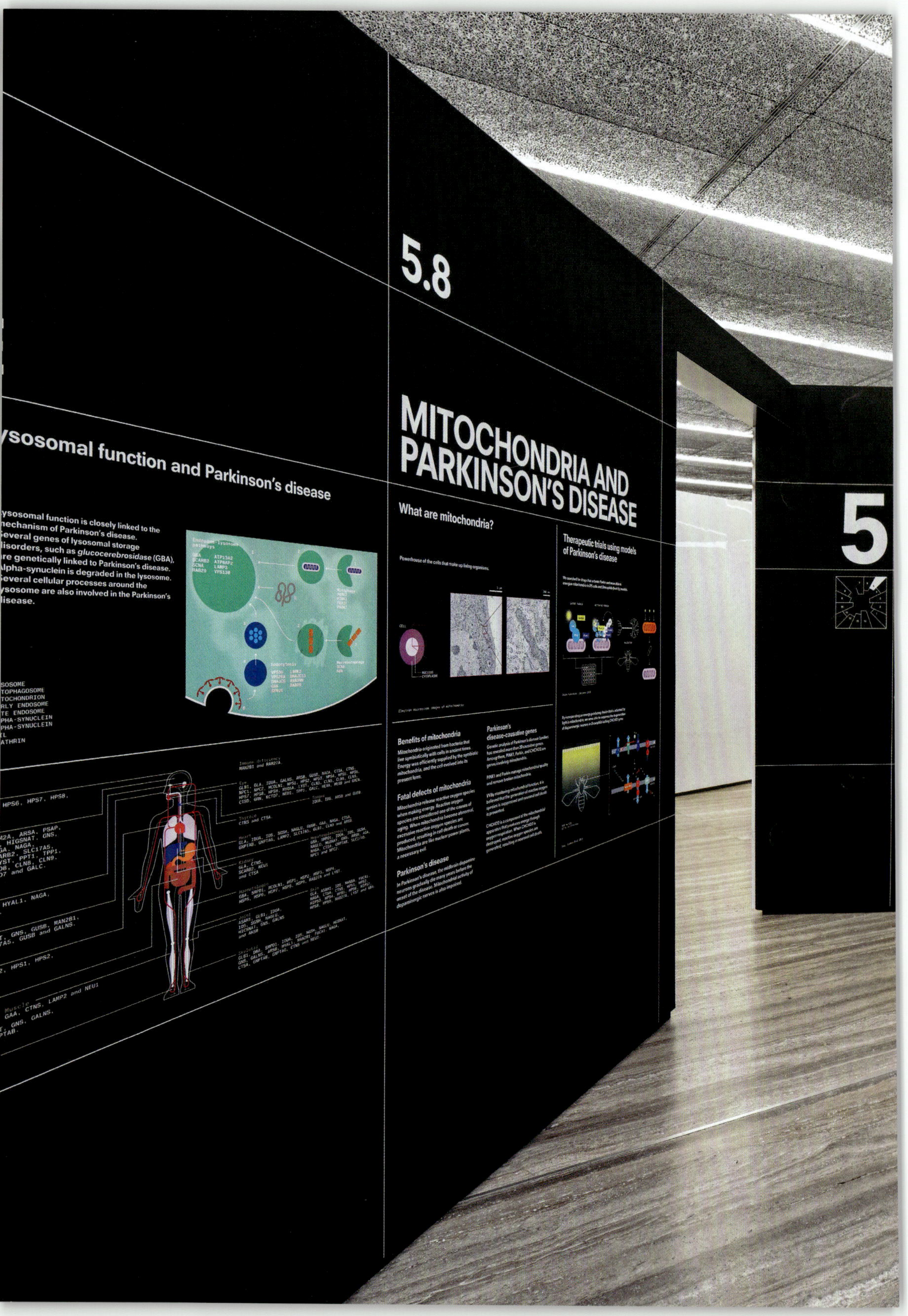
5.8
MITOCHONDRIA AND PARKINSON'S DISEASE
What are mitochondria?
ysosomal function and Parkinson's disease
Therapeutic trials using models of Parkinson's disease
Benefits of mitochondria
Fatal defects of mitochondria
Parkinson's disease
5

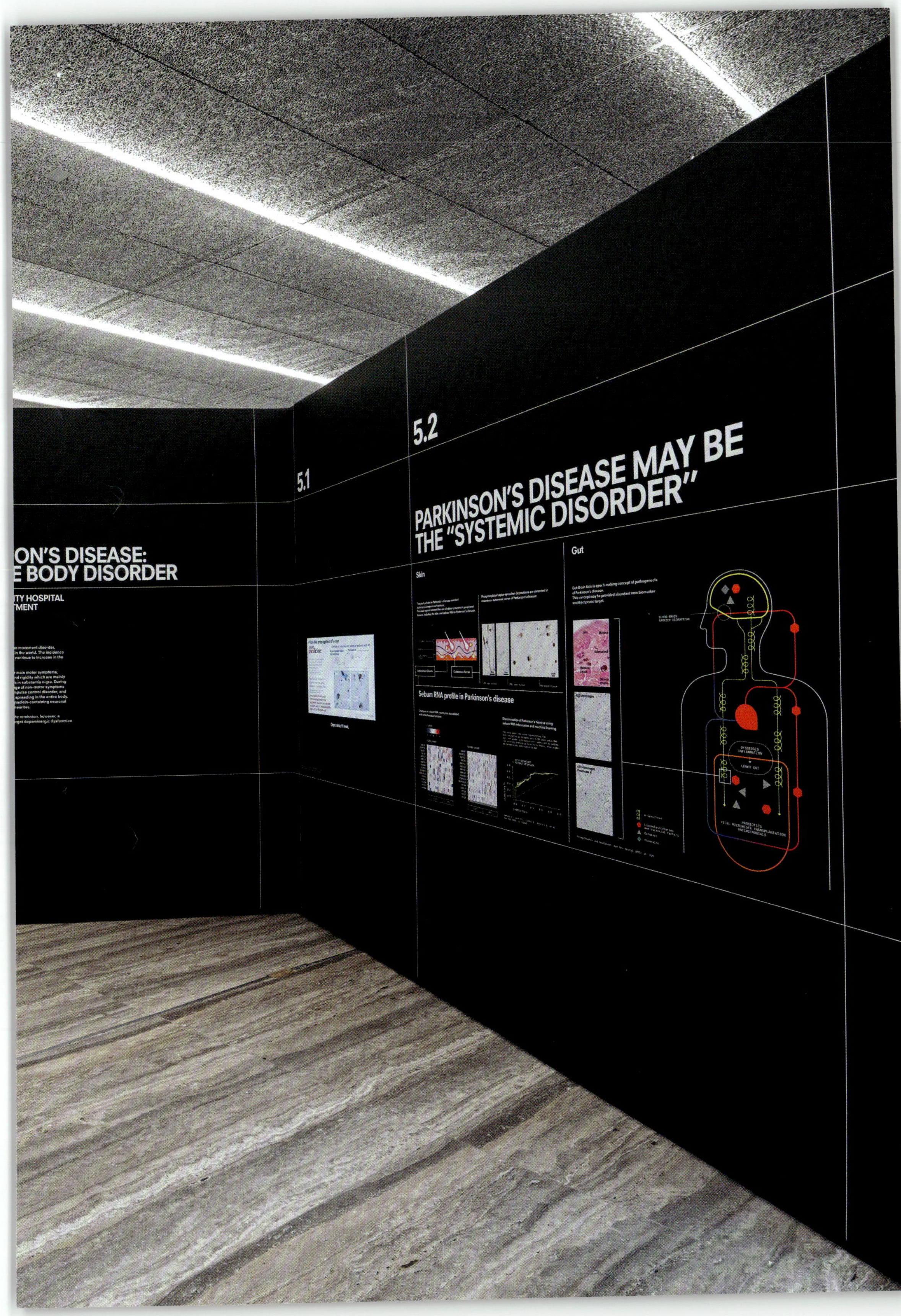

5.1
5.2
PARKINSON'S DISEASE MAY BE THE "SYSTEMIC DISORDER"
Skin
Gut
Sebum RNA profile in Parkinson's disease

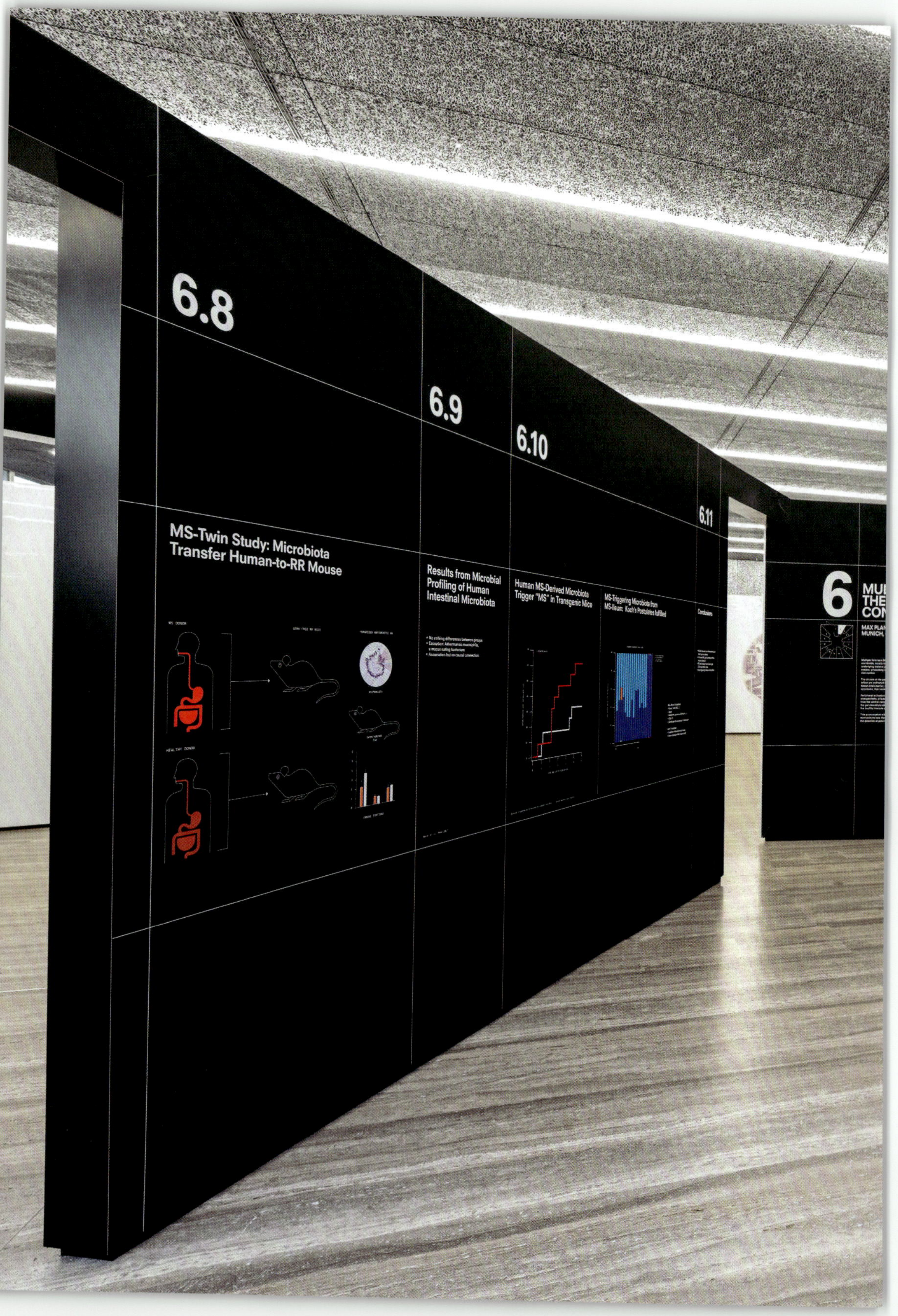
6.8
MS-Twin Study: Microbiota Transfer Human-to-RR Mouse
6.9
Results from Microbial Profiling of Human Intestinal Microbiota
6.10
Human MS-Derived Microbiota Trigger "MS" in Transgenic Mice
MS-Triggering Microbiota from MS-Ileum: Koch's Postulates fulfilled
6.11
Conclusions
6

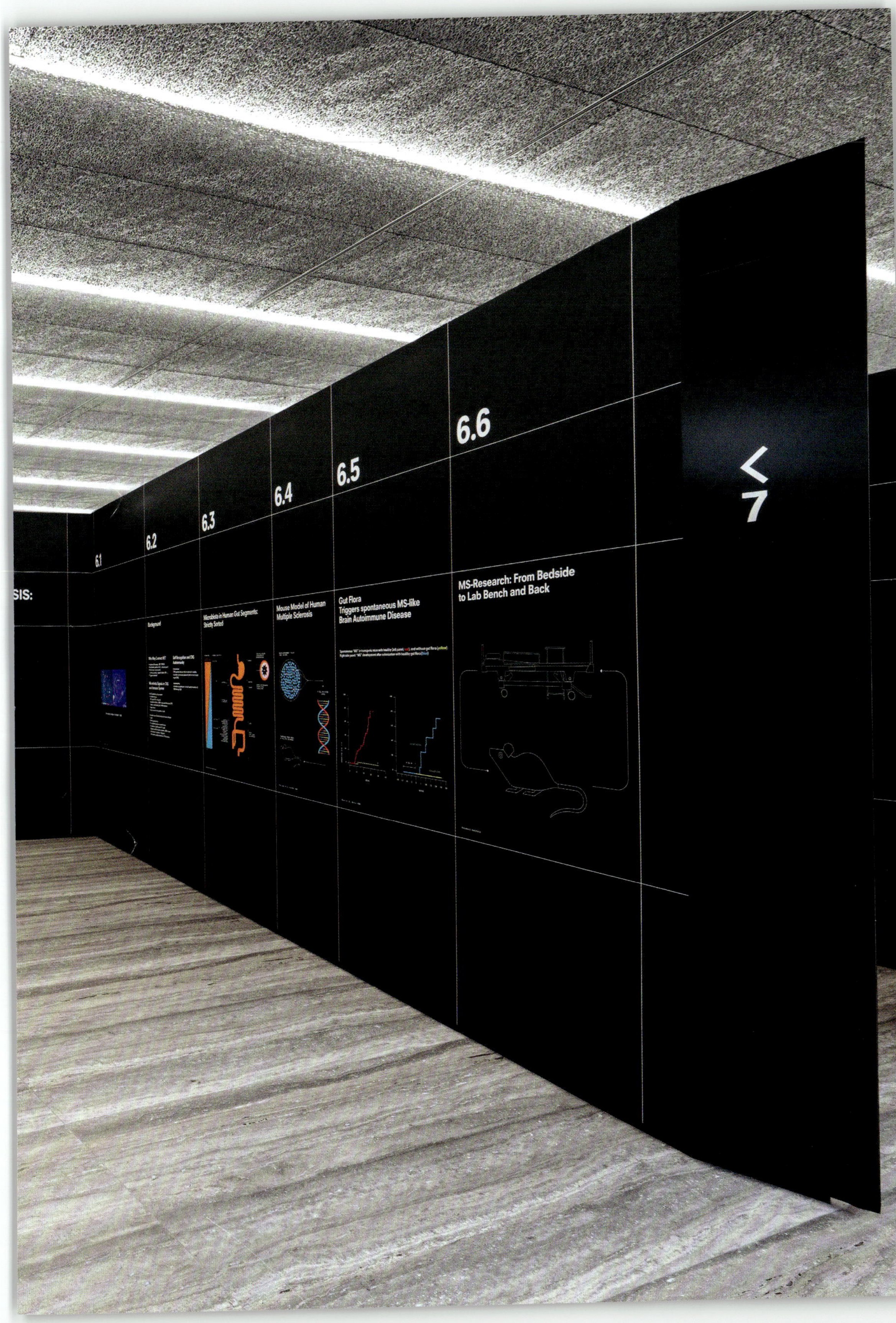
6.1
6.2
6.3
6.4
6.5
6.6
7
Mouse Model of Human Multiple Sclerosis
Gut Flora
Triggers spontaneous MS-like
Brain Autoimmune Disease
MS-Research: From Bedside
to Lab Bench and Back

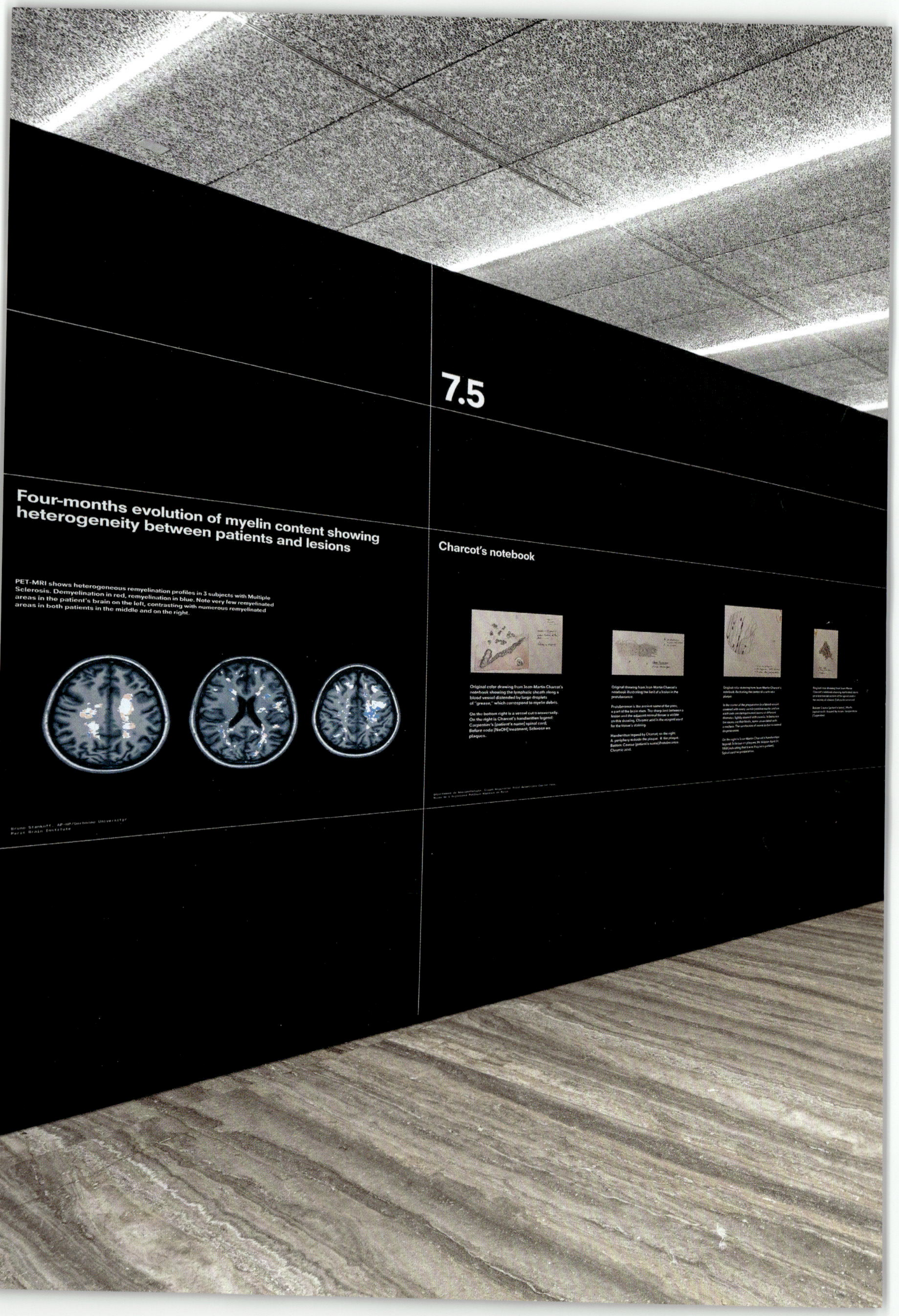
7.5
Four-months evolution of myelin content showing heterogeneity between patients and lesions
PET-MRI shows heterogeneous remyelination profiles in 3 subjects with Multiple Sclerosis. Demyelination in red, remyelination in blue. Note very few remyelinated areas in the patient's brain on the left, contrasting with numerous remyelinated areas in both patients in the middle and on the right.
Charcot's notebook
Original color drawing from Jean-Martin Charcot's notebook showing the lymphatic sheath along a blood vessel distended by large droplets of "grease," which correspond to myelin debris.
On the bottom right is a vessel cut transversally. On the right is Charcot's handwritten legend: Carpentier's [patient's name] spinal cord; Before soda [NaOH] treatment; Sclérose en plaques.
Original drawing from Jean-Martin Charcot's notebook illustrating the limit of a lesion in the protuberance.

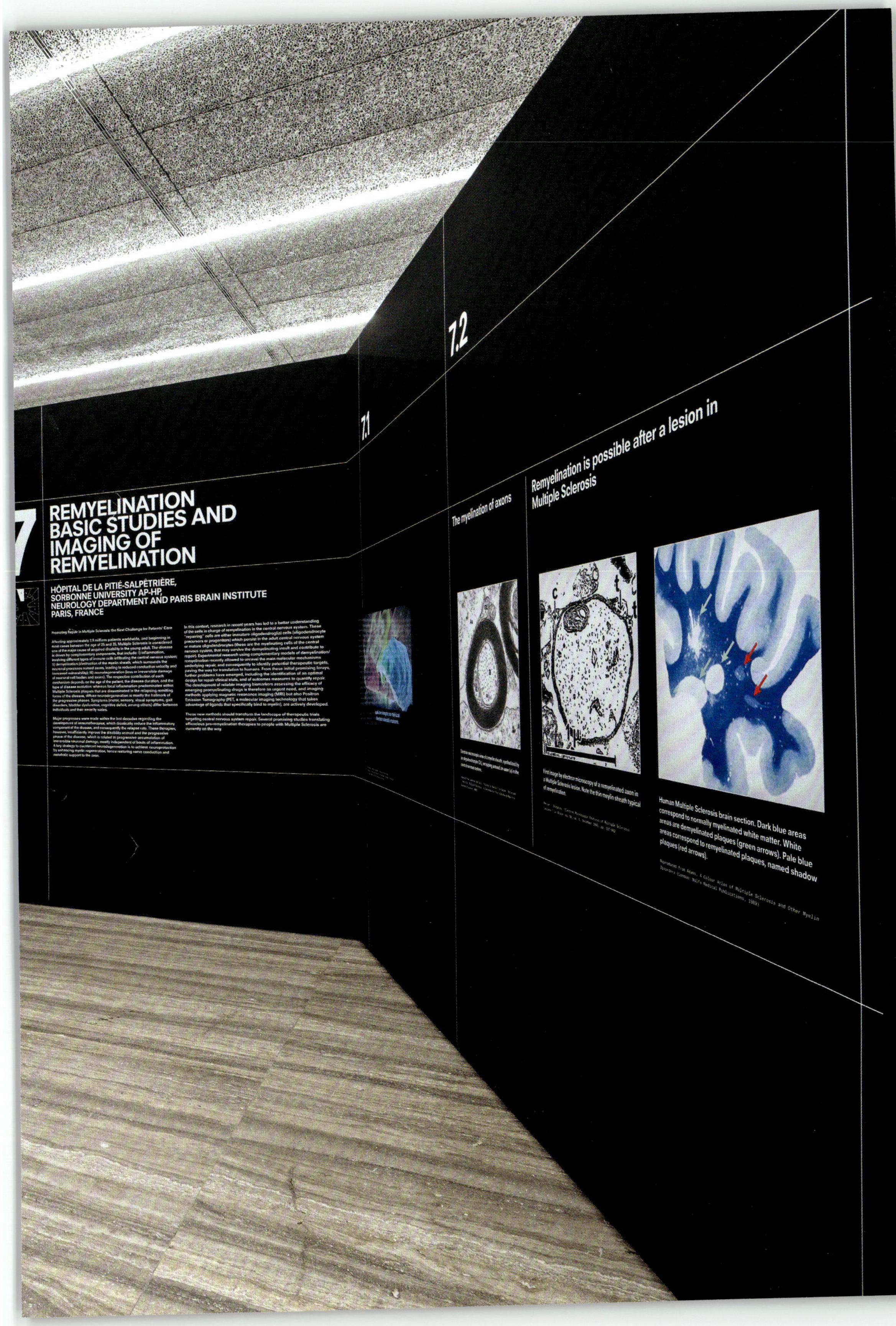
7
REMYELINATION
BASIC STUDIES AND
IMAGING OF
REMYELINATION
HÔPITAL DE LA PITIÉ-SALPÊTRIÈRE,
SORBONNE UNIVERSITY AP-HP,
NEUROLOGY DEPARTMENT AND PARIS BRAIN INSTITUTE
PARIS, FRANCE
7.1
7.2
The myelination of axons
Remyelination is possible after a lesion in Multiple Sclerosis
Human Multiple Sclerosis brain section. Dark blue areas correspond to normally myelinated white matter. White areas are demyelinated plaques (green arrows). Pale blue areas correspond to remyelinated plaques, named shadow plaques (red arrows).

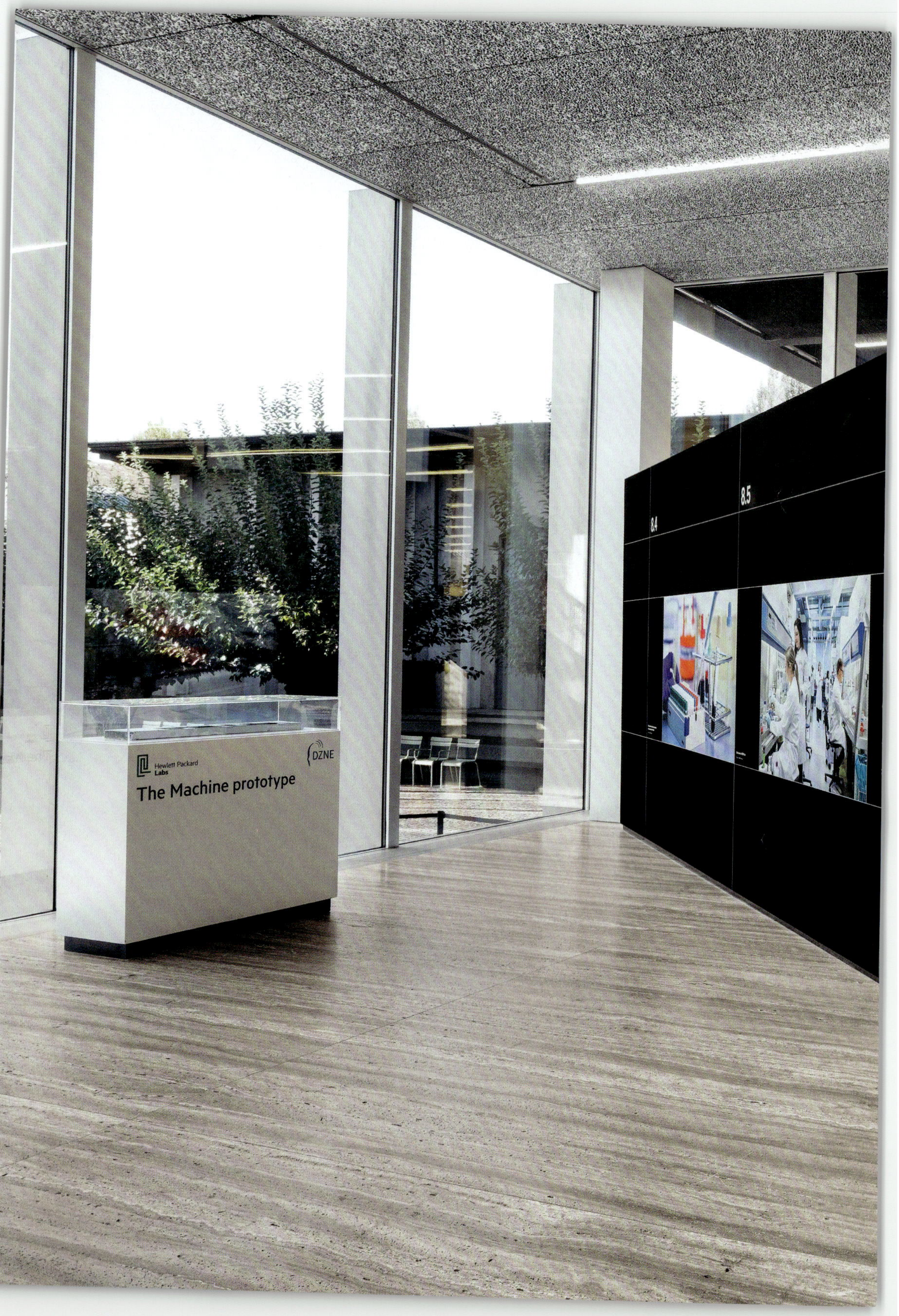
Hewlett Packard Labs
DZNE
The Machine prototype
8.4
8.5

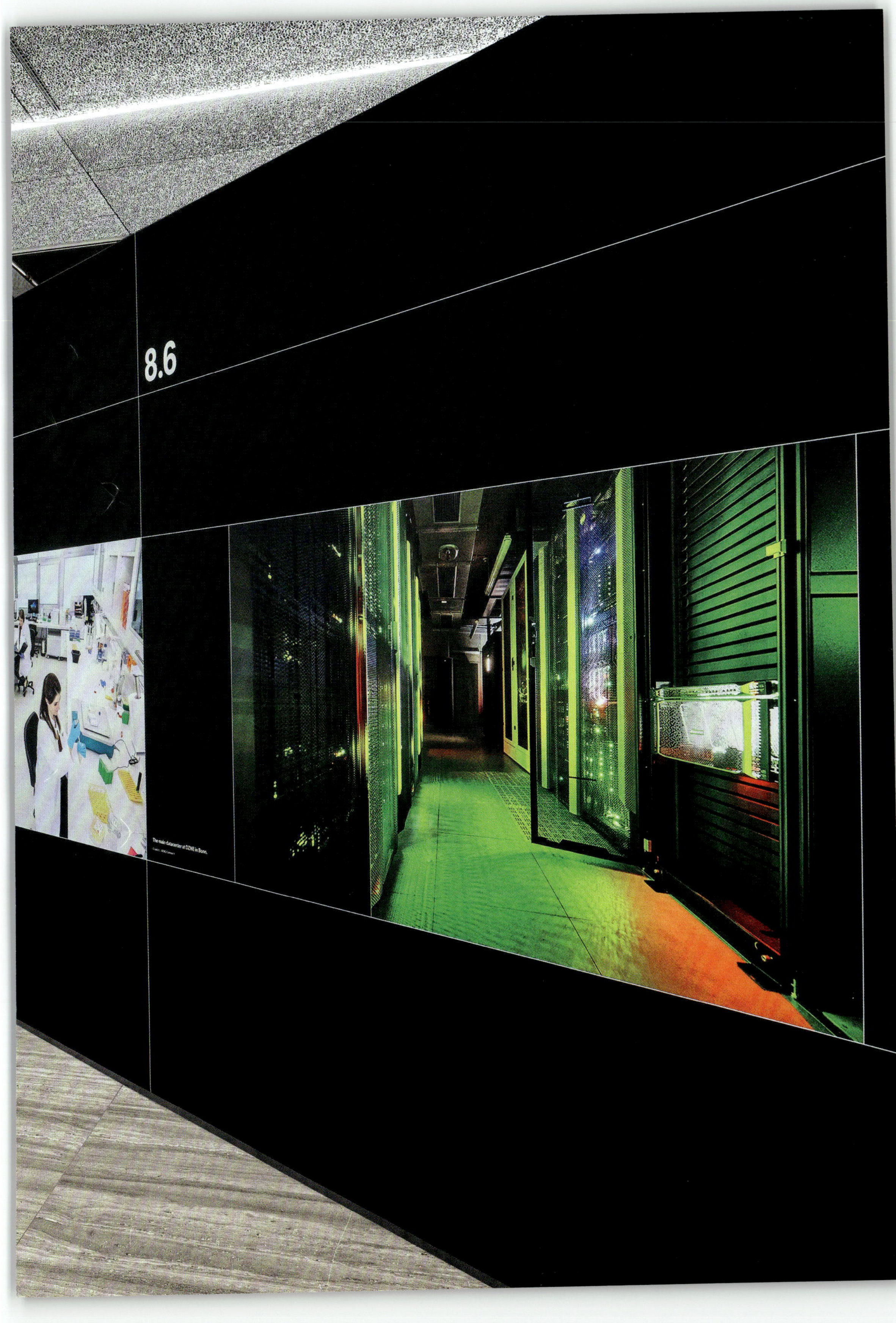
8.6
The main datacenter at DZNE in Bonn.

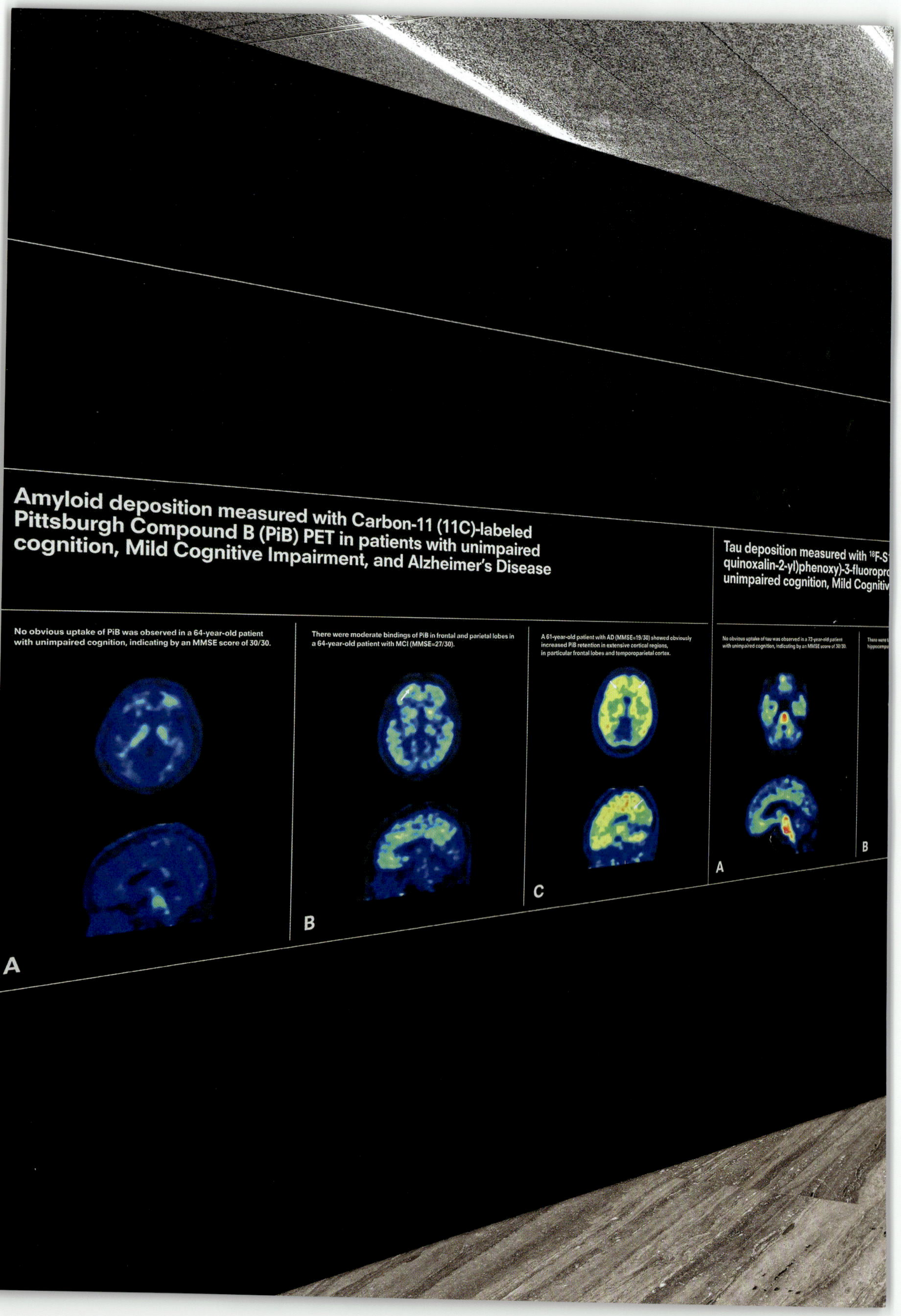
Amyloid deposition measured with Carbon-11 (11C)-labeled Pittsburgh Compound B (PiB) PET in patients with unimpaired cognition, Mild Cognitive Impairment, and Alzheimer's Disease
No obvious uptake of PiB was observed in a 64-year-old patient with unimpaired cognition, indicating by an MMSE score of 30/30.
There were moderate bindings of PiB in frontal and parietal lobes in a 64-year-old patient with MCI (MMSE=27/30).
A 61-year-old patient with AD (MMSE=19/30) showed obviously increased PiB retention in extensive cortical regions, in particular frontal lobes and temporoparietal cortex.
A
B
C
Tau deposition measured with
No obvious uptake of tau was observed in a 72-year-old patient with unimpaired cognition, indicating by an MMSE score of 30/30.
A
B

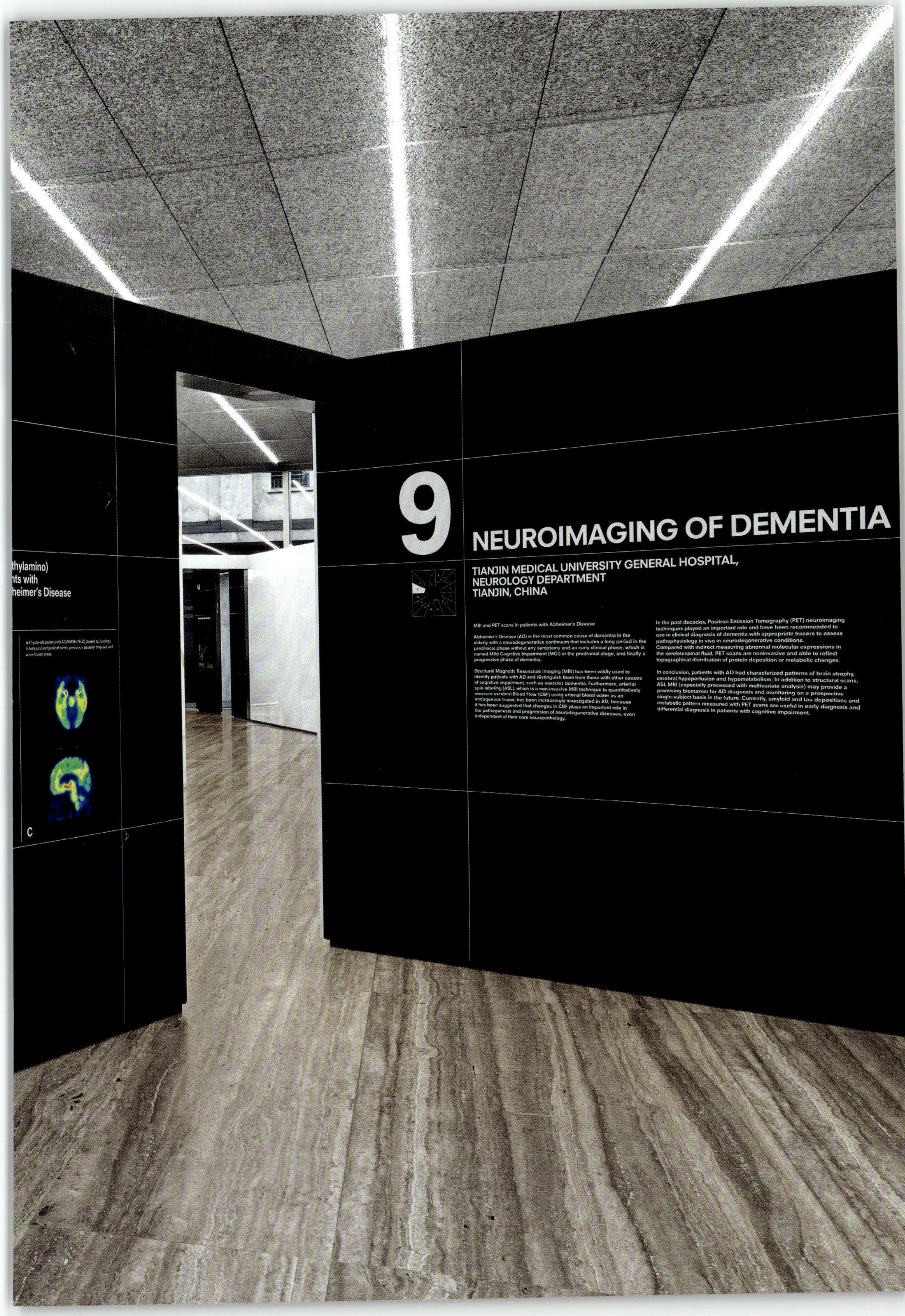
9
NEUROIMAGING OF DEMENTIA
TIANJIN MEDICAL UNIVERSITY GENERAL HOSPITAL,
NEUROLOGY DEPARTMENT
TIANJIN, CHINA
MRI and PET scans in patients with Alzheimer's Disease
Alzheimer's Disease (AD) is the most common cause of dementia in the elderly with a neurodegenerative continuum that includes a long period in the preclinical phase without any symptoms and an early clinical phase, which is named Mild Cognitive Impairment (MCI) or the prodromal stage, and finally a progressive phase of dementia.
Structural Magnetic Resonance Imaging (MRI) has been wildly used to identify patients with AD and distinguish them from those with other causes of cognitive impairment, such as vascular dementia. Furthermore, arterial spin labeling (ASL), which is a non-invasive MRI technique to quantitatively measure cerebral Blood Flow (CBF) using arterial blood water as an endogenous tracer, has been increasingly investigated in AD, because it has been suggested that changes in CBF plays an important role in the pathogenesis and progression of neurodegenerative diseases, even independent of their core neuropathology.
In the past decades, Positron Emission Tomography (PET) neuroimaging techniques played an important role and have been recommended to use in clinical diagnosis of dementia with appropriate tracers to assess pathophysiology in vivo in neurodegenerative conditions.
Compared with indirect measuring abnormal molecular expressions in the cerebrospinal fluid, PET scans are noninvasive and able to reflect topographical distribution of protein deposition or metabolic changes.
In conclusion, patients with AD had characterized patterns of brain atrophy, cerebral hypoperfusion and hypometabolism. In addition to structural scans, ASL MRI (especially processed with multivariate analysis) may provide a promising biomarker for AD diagnosis and monitoring on a prospective single-subject basis in the future. Currently, amyloid and tau depositions and metabolic pattern measured with PET scans are useful in early diagnosis and differential diagnosis in patients with cognitive impairment.
C

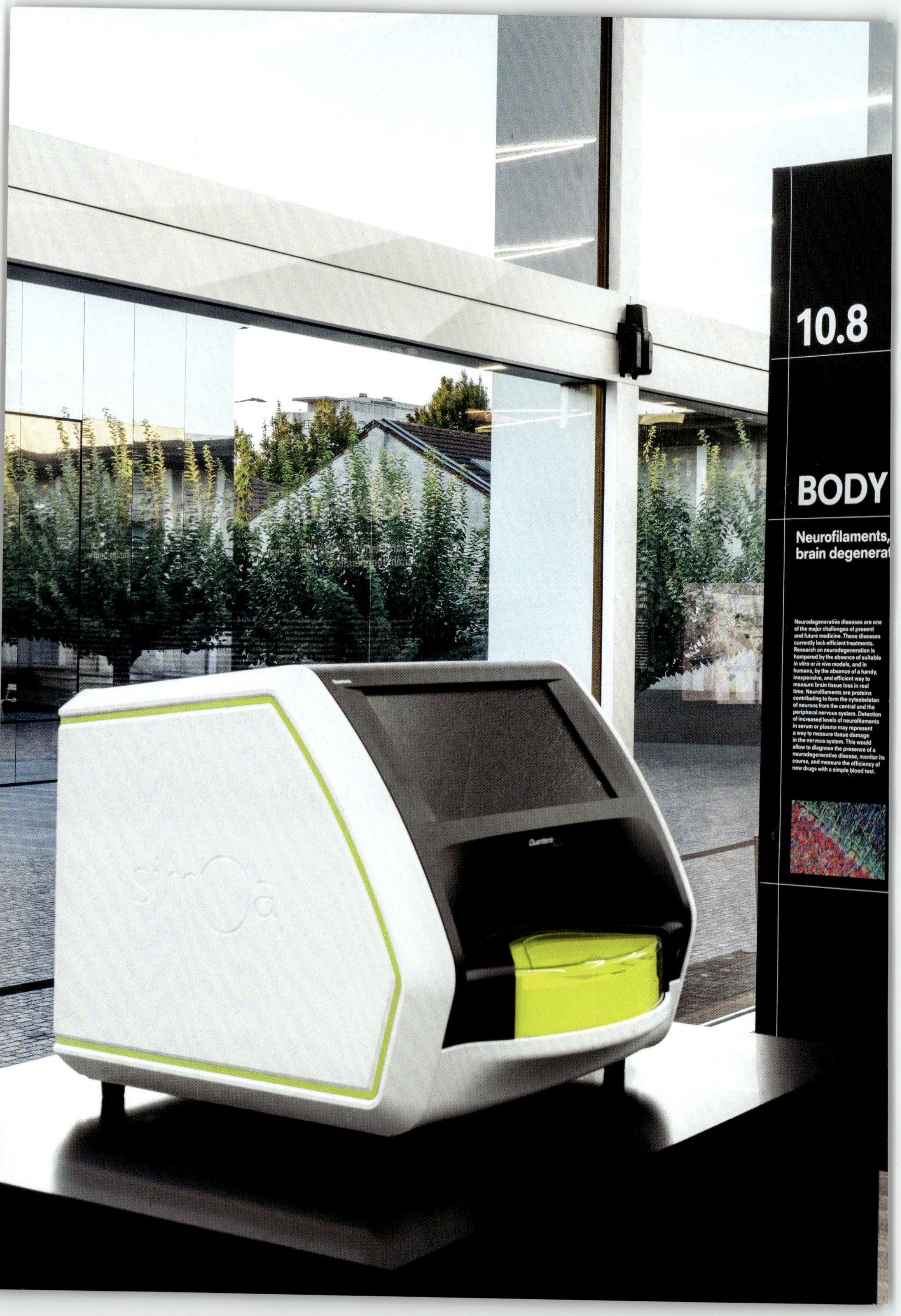
10.8
BODY
Neurofilaments,
Neurodegenerative diseases are one of the major challenges of present and future medicine. These diseases currently lack efficient treatments. Research on neurodegeneration is hampered by the absence of suitable in vitro or in vivo models, and in humans, by the absence of a handy, inexpensive, and efficient way to measure brain tissue loss in real time. Neurofilaments are proteins contributing to form the cytoskeleton of neurons from the central and the peripheral nervous system. Detection of increased levels of neurofilaments in serum or plasma may represent a way to measure tissue damage in the nervous system. This would allow to diagnose the presence of a neurodegenerative disease, monitor its course, and measure the efficiency of new drugs with a simple blood test.

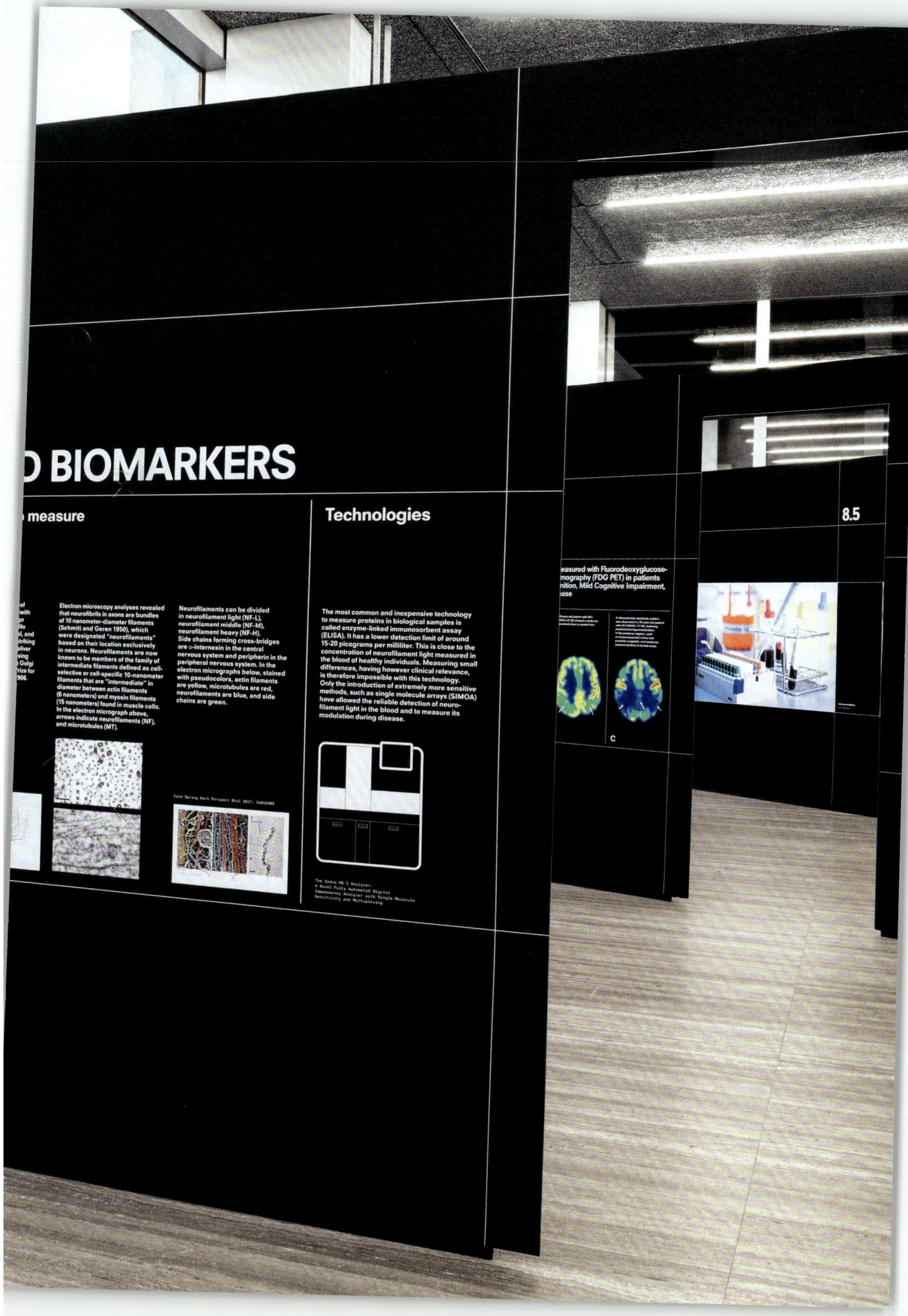
BIOMARKERS
measure
Technologies
8.5
Electron microscopy analyses revealed that neurofibrils in axons are bundles of 10 nanometer-diameter filaments (Schmitt and Geren 1950), which were designated "neurofilaments" based on their location exclusively in neurons. Neurofilaments are now known to be members of the family of intermediate filaments defined as cell-selective or cell-specific 10-nanometer filaments that are "intermediate" in diameter between actin filaments (6 nanometers) and myosin filaments (15 nanometers) found in muscle cells. In the electron micrograph above, arrows indicate neurofilaments (NF), and microtubules (MT).
Neurofilaments can be divided in neurofilament light (NF-L), neurofilament middle (NF-M), neurofilament heavy (NF-H). Side chains forming cross-bridges are α-internexin in the central nervous system and peripherin in the peripheral nervous system. In the electron micrographs below, stained with pseudocolors, actin filaments are yellow, microtubules are red, neurofilaments are blue, and side chains are green.
Cold Spring Harb Perspect Biol 2017;
The most common and inexpensive technology to measure proteins in biological samples is called enzyme-linked immunosorbent assay (ELISA). It has a lower detection limit of around 15-20 picograms per milliliter. This is close to the concentration of neurofilament light measured in the blood of healthy individuals. Measuring small differences, having however clinical relevance, is therefore impossible with this technology. Only the introduction of extremely more sensitive methods, such as single molecule arrays (SIMOA) have allowed the reliable detection of neurofilament light in the blood and to measure its modulation during disease.
The Simoa HD-1 Analyzer:
A Novel Fully Automated Digital
Immunoassay Analyzer with Single-Molecule
Sensitivity and Multiplexing
Mild Cognitive Impairment,
C

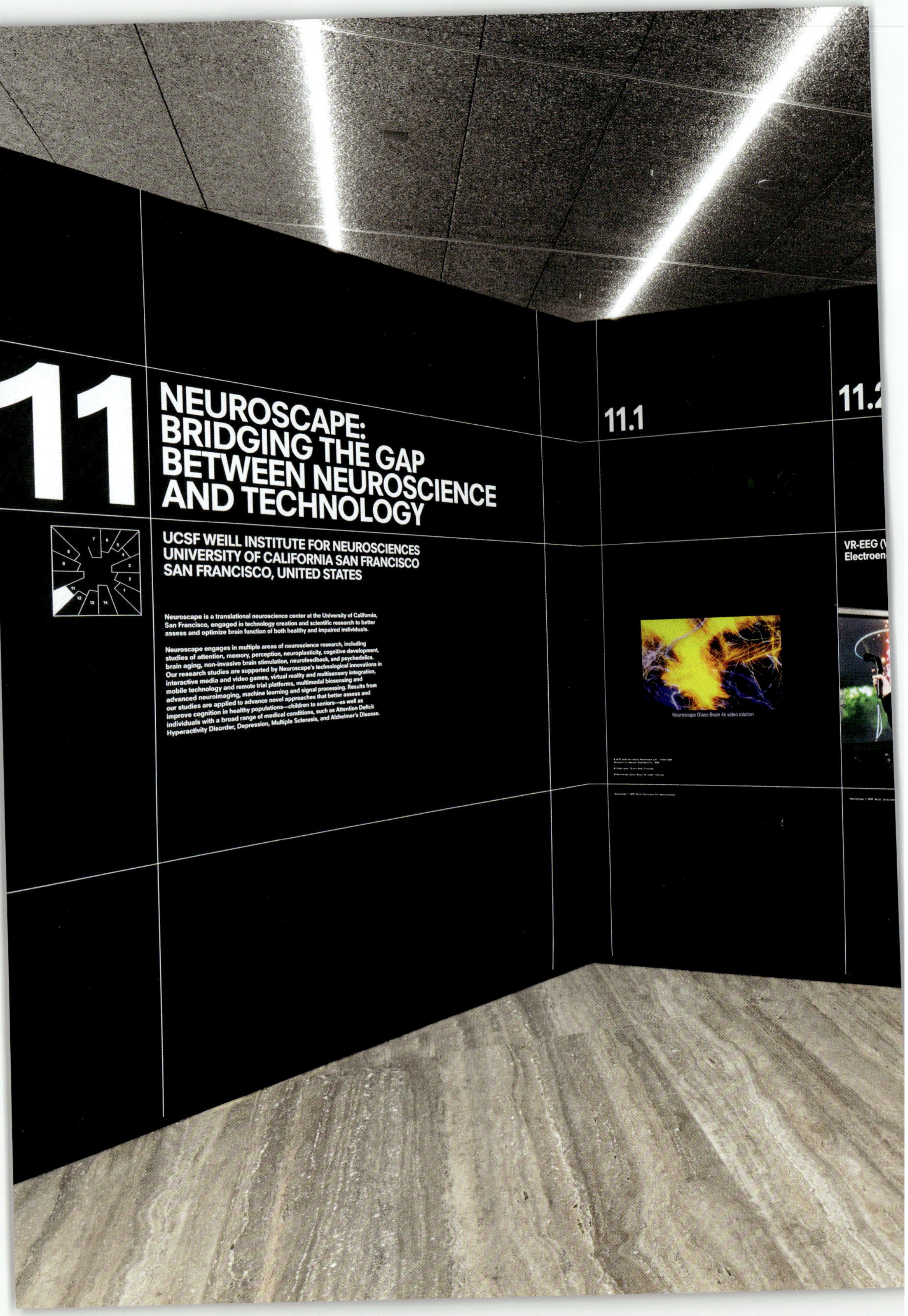

11
NEUROSCAPE: BRIDGING THE GAP BETWEEN NEUROSCIENCE AND TECHNOLOGY
UCSF WEILL INSTITUTE FOR NEUROSCIENCES
UNIVERSITY OF CALIFORNIA SAN FRANCISCO
SAN FRANCISCO, UNITED STATES
Neuroscape is a translational neuroscience center at the University of California, San Francisco, engaged in technology creation and scientific research to better assess and optimize brain function of both healthy and impaired individuals.
Neuroscape engages in multiple areas of neuroscience research, including studies of attention, memory, perception, neuroplasticity, cognitive development, brain aging, non-invasive brain stimulation, neurofeedback, and psychedelics. Our research studies are supported by Neuroscape's technological innovations in interactive media and video games, virtual reality and multisensory integration, mobile technology and remote trial platforms, multimodal biosensing and advanced neuroimaging, machine learning and signal processing. Results from our studies are applied to advance novel approaches that better assess and improve cognition in healthy populations—children to seniors—as well as individuals with a broad range of medical conditions, such as Attention Deficit Hyperactivity Disorder, Depression, Multiple Sclerosis, and Alzheimer's Disease.
11.1
Neuroscape Glass Brain 4k video rotation
11.2

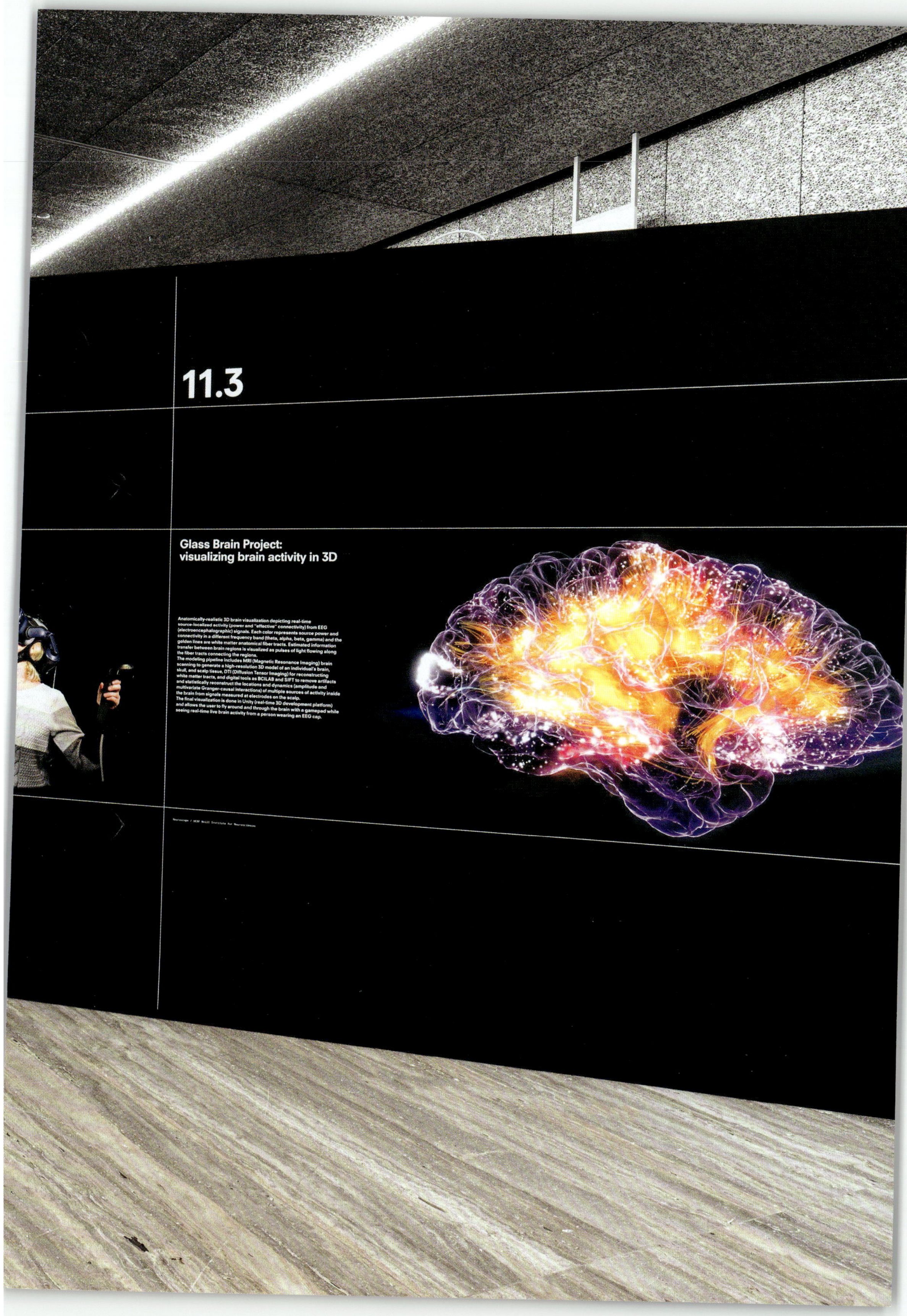
11.3
Glass Brain Project:
visualizing brain activity in 3D
Anatomically-realistic 3D brain visualization depicting real-time source-localized activity (power and "effective" connectivity) from EEG (electroencephalographic) signals. Each color represents source power and connectivity in a different frequency band (theta, alpha, beta, gamma) and the golden lines are white matter anatomical fiber tracts. Estimated information transfer between brain regions is visualized as pulses of light flowing along the fiber tracts connecting the regions.
The modeling pipeline includes MRI (Magnetic Resonance Imaging) brain scanning to generate a high-resolution 3D model of an individual's brain, skull, and scalp tissue, DTI (Diffusion Tensor Imaging) for reconstructing white matter tracts, and digital tools as BCILAB and SIFT to remove artifacts and statistically reconstruct the locations and dynamics (amplitude and multivariate Granger-causal interactions) of multiple sources of activity inside the brain from signals measured at electrodes on the scalp.
The final visualization is done in Unity (real-time 3D development platform) and allows the user to fly around and through the brain with a gamepad while seeing real-time live brain activity from a person wearing an EEG cap.

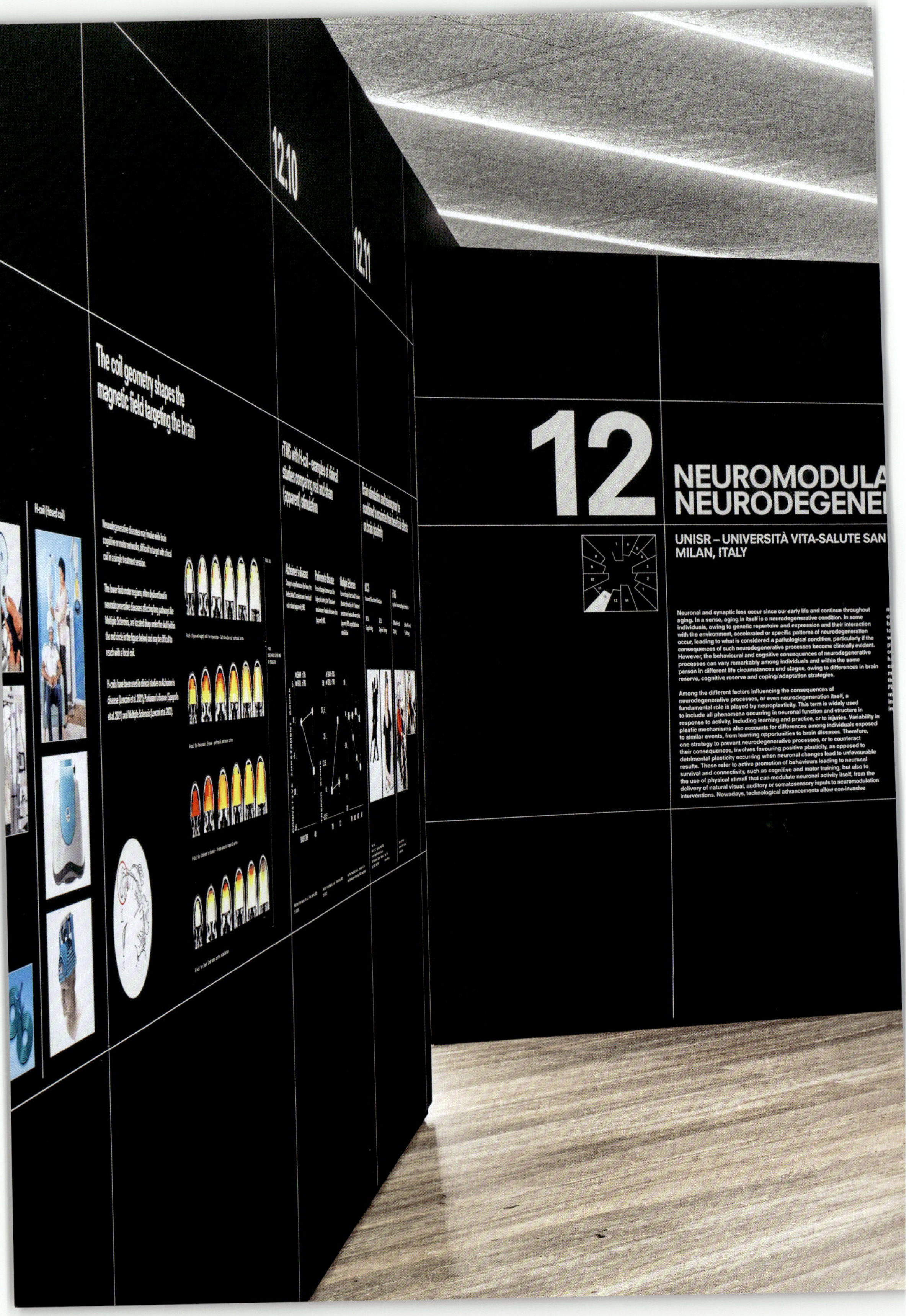

12.10
12.11
The coil geometry shapes the magnetic field targeting the brain
12
NEUROMODULA
NEURODEGENE
UNISR – UNIVERSITÀ VITA-SALUTE SAN
MILAN, ITALY
Neuronal and synaptic loss occur since our early life and continue throughout aging. In a sense, aging in itself is a neurodegenerative condition. In some individuals, owing to genetic repertoire and expression and their interaction with the environment, accelerated or specific patterns of neurodegeneration occur, leading to what is considered a pathological condition, particularly if the consequences of such neurodegenerative processes become clinically evident. However, the behavioural and cognitive consequences of neurodegenerative processes can vary remarkably among individuals and within the same person in different life circumstances and stages, owing to differences in brain reserve, cognitive reserve and coping/adaptation strategies.
Among the different factors influencing the consequences of neurodegenerative processes, or even neurodegeneration itself, a fundamental role is played by neuroplasticity. This term is widely used to include all phenomena occurring in neuronal function and structure in response to activity, including learning and practice, or to injuries. Variability in plastic mechanisms also accounts for differences among individuals exposed to similar events, from learning opportunities to brain diseases. Therefore, one strategy to prevent neurodegenerative processes, or to counteract their consequences, involves favouring positive plasticity, as opposed to detrimental plasticity occurring when neuronal changes lead to unfavourable results. These refer to active promotion of behaviours leading to neuronal survival and connectivity, such as cognitive and motor training, but also to the use of physical stimuli that can modulate neuronal activity itself, from the delivery of natural visual, auditory or somatosensory inputs to neuromodulation interventions. Nowadays, technological advancements allow non-invasive

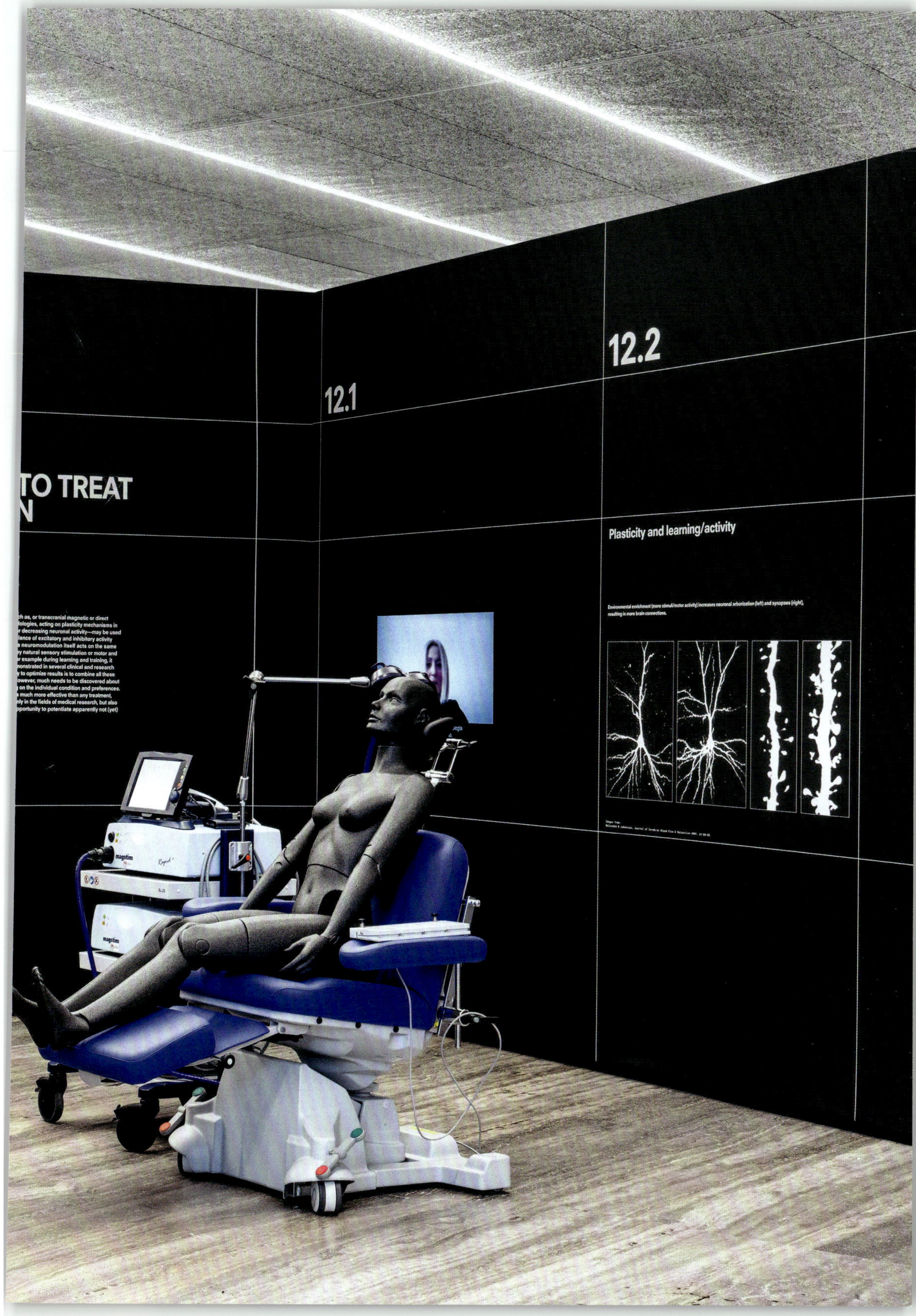
TO TREAT
N
12.1
12.2
Plasticity and learning/activity

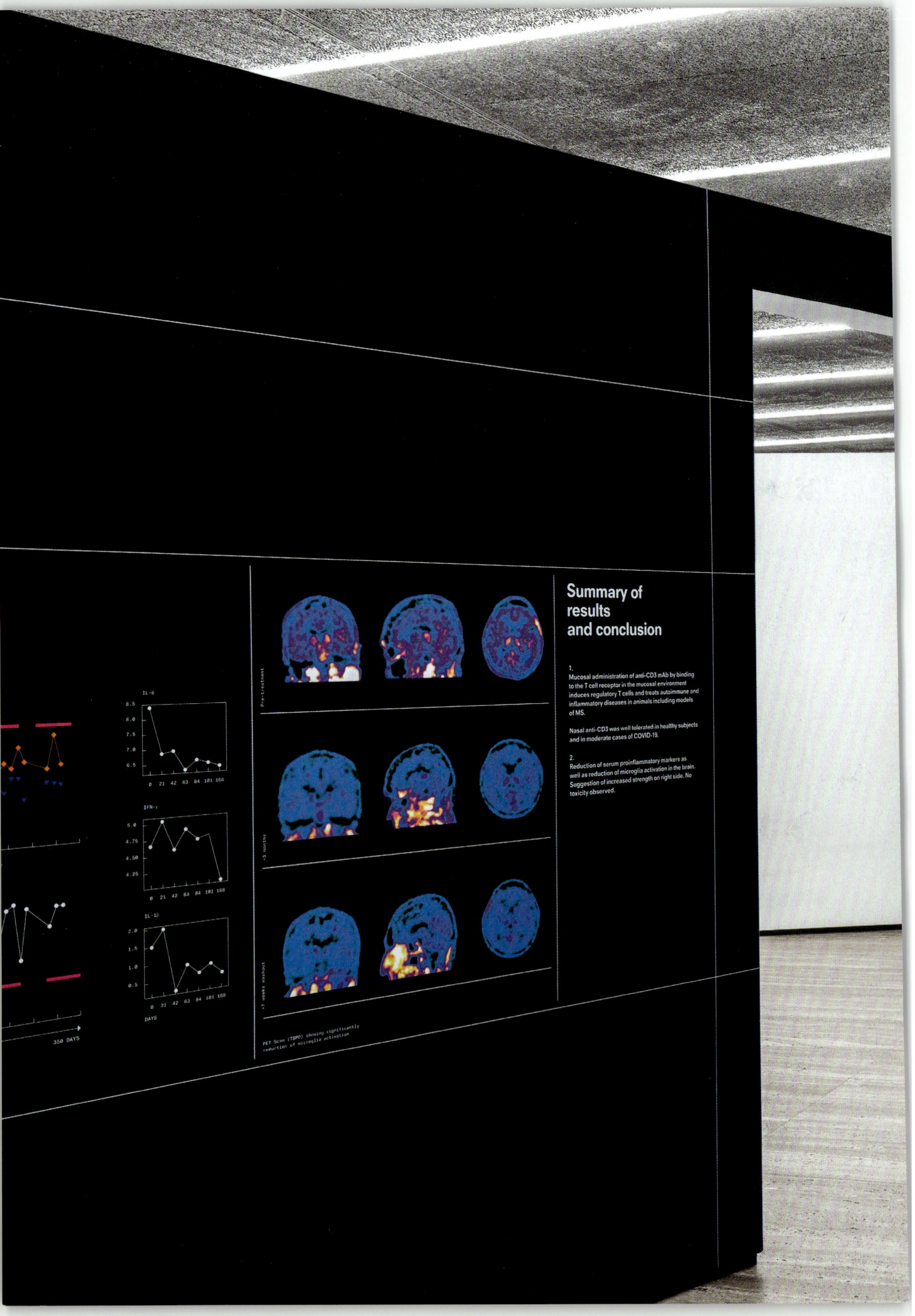
Summary of results and conclusion
1.
Mucosal administration of anti-CD3 mAb by binding to the T cell receptor in the mucosal environment induces regulatory T cells and treats autoimmune and inflammatory diseases in animals including models of MS.
Nasal anti-CD3 was well tolerated in healthy subjects and in moderate cases of COVID-19.
2.
Reduction of serum proinflammatory markers as well as reduction of microglia activation in the brain. Suggestion of increased strength on right side. No toxicity observed.

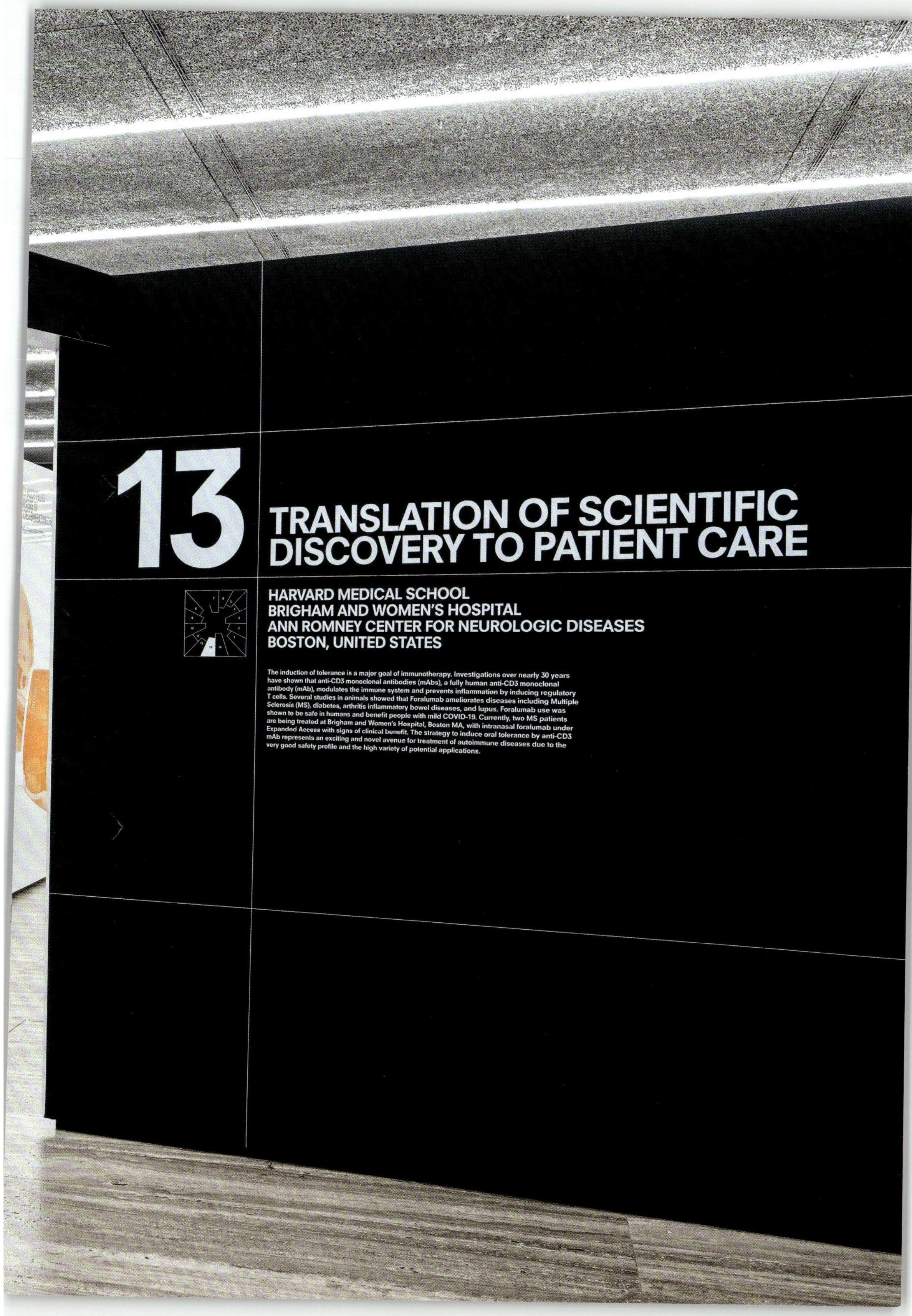
13
TRANSLATION OF SCIENTIFIC DISCOVERY TO PATIENT CARE
HARVARD MEDICAL SCHOOL
BRIGHAM AND WOMEN'S HOSPITAL
ANN ROMNEY CENTER FOR NEUROLOGIC DISEASES
BOSTON, UNITED STATES
The induction of tolerance is a major goal of immunotherapy. Investigations over nearly 30 years have shown that anti-CD3 monoclonal antibodies (mAbs), a fully human anti-CD3 monoclonal antibody (mAb), modulates the immune system and prevents inflammation by inducing regulatory T cells. Several studies in animals showed that Foralumab ameliorates diseases including Multiple Sclerosis (MS), diabetes, arthritis inflammatory bowel diseases, and lupus. Foralumab use was shown to be safe in humans and benefit people with mild COVID-19. Currently, two MS patients are being treated at Brigham and Women's Hospital, Boston MA, with intranasal foralumab under Expanded Access with signs of clinical benefit. The strategy to induce oral tolerance by anti-CD3 mAb represents an exciting and novel avenue for treatment of autoimmune diseases due to the very good safety profile and the high variety of potential applications.

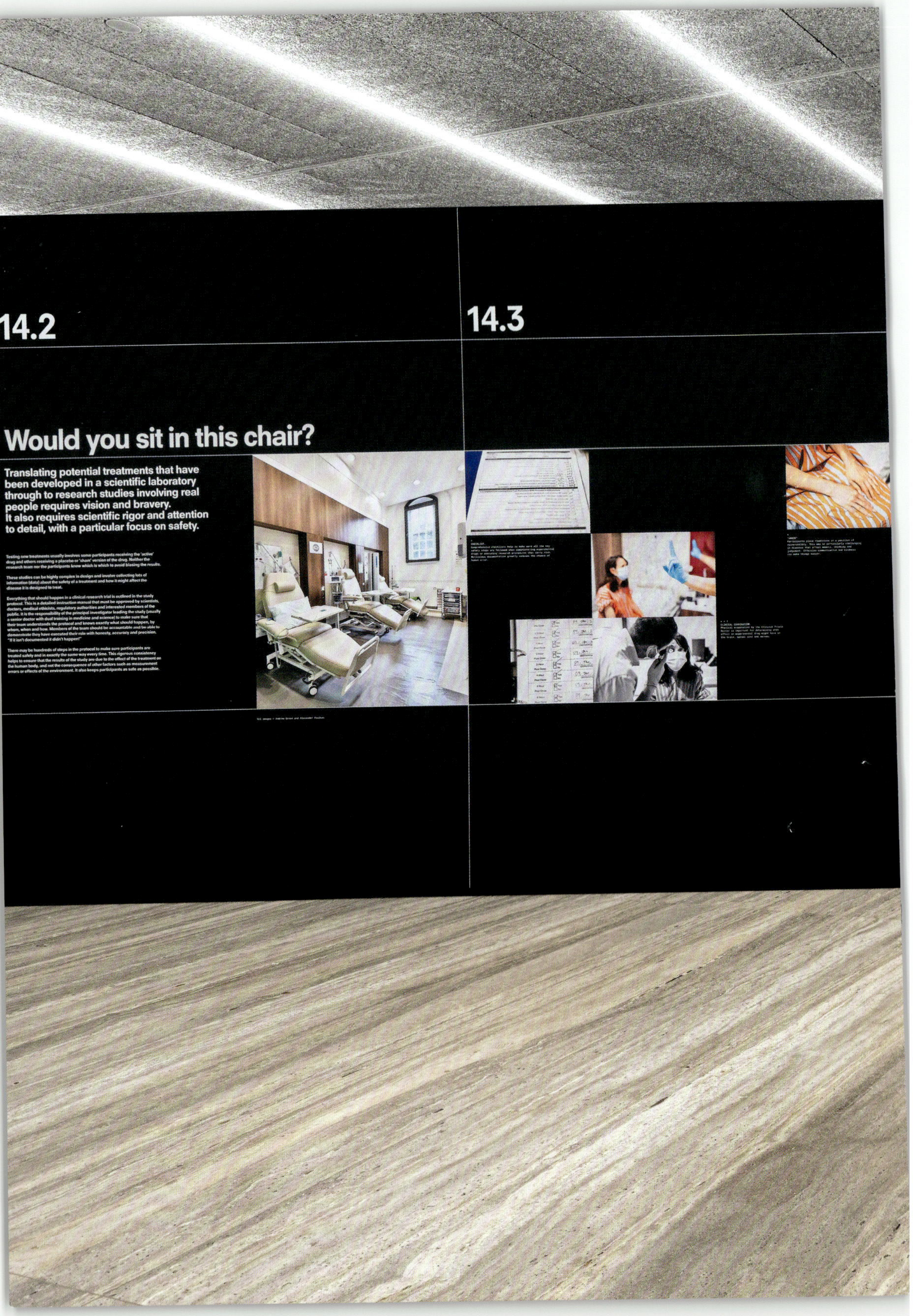
14.2
14.3
Would you sit in this chair?
Translating potential treatments that have been developed in a scientific laboratory through to research studies involving real people requires vision and bravery. It also requires scientific rigor and attention to detail, with a particular focus on safety.
Testing new treatments usually involves some participants receiving the 'active' drug and others receiving a placebo or 'sham' version of the drug. Neither the research team nor the participants know which is which to avoid biasing the results.
These studies can be highly complex in design and involve collecting lots of information (data) about the safety of a treatment and how it might affect the disease it is designed to treat.
Everything that should happen in a clinical research trial is outlined in the study protocol. This is a detailed instruction manual that must be approved by scientists, doctors, medical ethicists, regulatory authorities and interested members of the public. It is the responsibility of the principal investigator leading the study (usually a senior doctor with dual training in medicine and science) to make sure that their team understands the protocol and knows exactly what should happen, by whom, when and how. Members of the team should be accountable and be able to demonstrate they have executed their role with honesty, accuracy and precision. "If it isn't documented it didn't happen!"
There may be hundreds of steps in the protocol to make sure participants are treated safely and in exactly the same way every time. This rigorous consistency helps to ensure that the results of the study are due to the effect of the treatment on the human body, and not the consequence of other factors such as measurement errors or effects of the environment. It also keeps participants as safe as possible.

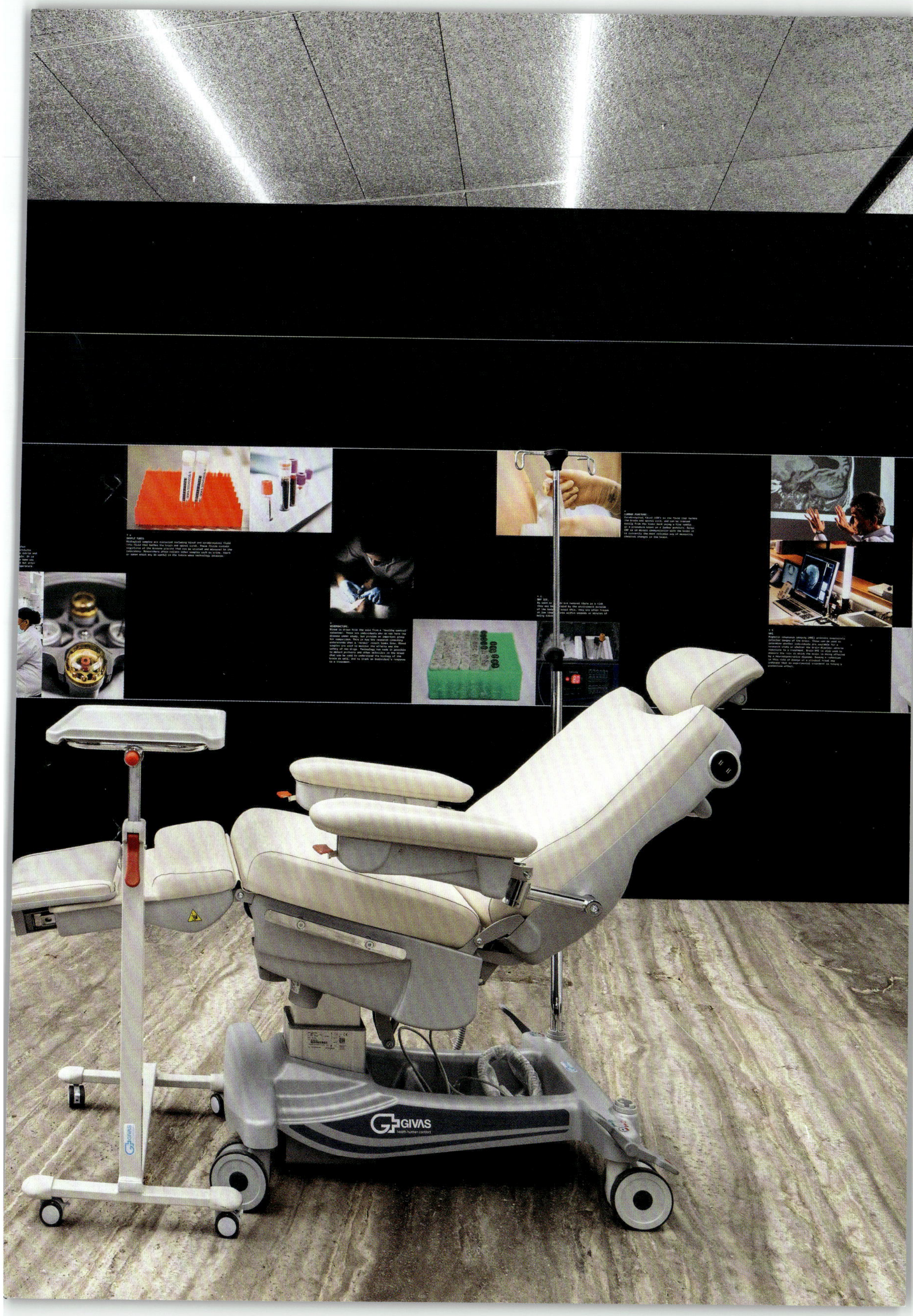
GIVAS

EXHIBITED OBJECTS

Room 1

Chromium Controller
10X Genomics

Machine that separates hundreds to tens of thousands of cells and label each of them individually in a matter of minutes. Once each cell has been labeled with a unique code, it is possible to study which genes or proteins are expressed in each individual cell

Room 5

Model of the brain
Courtesy Juntendo University Hospital, Neurology Department, Tokyo, Japan

Room 7

Three cables showing the different states of myelin throughout the disease progression. Myelin is an insulating sheath that forms around axons, conductors of electrical impulses between neurons

3D mock-up of axons (blue) and oligodendrocytes (orange) to illustrate the myelination process

Courtesy Hôpital de la Pitié-Salpêtrière, Sorbonne University AP-HP, Neurology department and Paris Brain Institute, Paris, France

Room 8

The Machine prototype
Hewlett Packard Labs
Courtesy German Center for Neurodegenerative Diseases (DZNE) within the Helmholtz Association, Bonn, Germany

Room 10

SR-X Biomarker Detection System
Quanterix

Video dedicated to the PET MR Scanner machine
Courtesy GE Healthcare, a General Electric Company

Room 11

Electroencephalogram EEG device
Cap Neurolectrics Barcelona, SLU
Oculus Quest 2

Room 12

Keope GPR chair for proprioceptive stimulation with sound and mechanic impulses
Andromeda

Soterix electrodes and device for transcranial direct current stimulation (tDCS)
Taomed

Brainsway H-coil helmet for repetitive transcranial magnetic stimulation (rTMS)
Courtesy ACeSM Onlus

Lemi chair and Magstim machine for repetitive transcranial magnetic stimulation (rTMS) and related focal coil
EMS

Room 13

Nasal Device for drug administration
Courtesy Harvard Medical School, Brigham and Women's Hospital, Ann Romney Center for Neurologic Diseases, Boston, USA

Room 14

Infusion chair
GIVAS

INDEX AND CREDITS

HUMAN BRAINS

FONDAZIONE PRADA

FOUNDERS
Miuccia Prada and Patrizio Bertelli

PRESIDENT AND DIRECTOR
Miuccia Prada

GENERAL MANAGER
Cristian Valsecchi
Carolina Agliardi, Executive Assistant

HEAD OF PROGRAMS
Chiara Costa
Susanna Cecchini, Executive Assistant

STEERING COMMITTEE
Giuliana Bruno
Giancarlo Comi
Theaster Gates
Alejandro González Iñárritu
Salvatore Settis

RESEARCH AND PUBLICATIONS
Mario Mainetti, Head of Research and Publications
Carlo Barbatti, Exhibition Curator
Niccolò Gravina, Exhibition Curator
Cornelia Mattiacci, Exhibition Curator
Cristina Tagliabue, Research Specialist
with
Ahmed Morsi

EXHIBITION DESIGN AND PRODUCTION
Alessia Salerno, Head of Exhibition Design and and Production
Teresa Bovi, Exhibition Registrar
Camilla Finzi, Exhibition Registrar
Samanta Montagna, Exhibition Registrar
Rosaria Borzacchiello, Architect
Veronica Tassone, Architect
Alessia Gennari, Performing and Live Event Producer
with
Clarenza Catullo
Stefano Pavia

COLLECTION MANAGEMENT
Giulio Sangiuliano, Collection Manager
Francesca Cattoi, Collection Registrar

PROJECT MANAGEMENT
Greta Gelmini, Project Manager
Marianna Barro, Project Management Specialist

PRESS OFFICE, COMMUNICATION, AND MARKETING
Andrea Goffo, Head of Press Office and Communication
Nicolò Scialanga, Press & Communication Officer
Roberta Palma, Digital Communication Coordinator
Katia Nicita, Events Coordinator
Marta Ronzio, Communication Specialist
Giovanni Spera, Graphic Specialist
Leonardo Gava, Press & Communication Specialist
Giulia Parolin, Digital Communication Specialist

ACCADEMIA DEI BAMBINI
Gabriele Ferraris, Consultant for Educational Program
Marta Motterlini, Coordinator

CINEMA
Paolo Moretti, Curator
Susanna Cecchini, Project Manager
Giovanni Cordini, Production Coordinator

GENERAL SECRETARY AND VISITOR SERVICES
Cristina Romanenghi, Office Manager
Cristina Fumagalli, Secretary
Katia Fraccaro, Secretary
Paolo Passaro, Back Office
Virginia Mazzucco, Visitor Services
Lucrezia Di Donfrancesco, Visitor Services
Elisabeth de Wrangel, Visitor Services

ADMINISTRATION, FINANCE AND CONTROL
Paola Magnanini, Office Manager
Dario Furgada, Accountant
Domenico Sorrentino, Accountant

LEGAL OFFICE
Chiara Bazoli, Legal Counsel

HUMAN RESOURCES
Giusy Antignani, Head of Human Resources
with
Marilena Marletta

OPERATIONS
Paolo Miano, Operation Manager
Diego Riminucci, Building and Safety Manager
Ester Viafora, Operation Specialist
with
Giuseppe Zotti

HUMAN BRAINS

SCIENTIFIC COMMITTEE

Giancarlo Comi (President)
Jubin Abutalebi
Massimo Cacciari
Viviana Kasam
Udo Kittelmann
Letizia Leocani
Andrea Moro
Daniela Perani

CULTURE AND CONSCIOUSNESS

Online conferences and activities
9.11 – 13.11.2020

SCIENTIFIC DIRECTOR

Giancarlo Comi

PROGRAM CONCEIVED BY

Giancarlo Comi
Jubin Abutalebi
Massimo Cacciari
Viviana Kasam
Andrea Moro
Daniela Perani

LECTURES BY

Katrin Amunts
Jean-Pierre Changeux
Antonio Damasio
Stanislas Dehaene
Michele Di Francesco
Eve Marder
Andrea Moro
Mavi Sanchez-Vives
Idan Segev
Ian Tattersall
Giulio Tononi

MODERATORS

Jubin Abutalebi
Massimo Cacciari
Giancarlo Comi
Viviana Kasam
Daniela Perani

CURATORIAL TEAM

Mario Mainetti
Niccolò Gravina
Cornelia Mattiacci

PRODUCTION MANAGER

Alessia Salerno

PROJECT MANAGER

Daniele Manfrè

LIVE EVENT MANAGER

Alessia Gennari

COMMUNICATION MANAGER

Andrea Goffo

PRESS OFFICE & PROMOTION

Nicolò Scialanga
Marta Ronzio

DIGITAL COMMUNICATION

Roberta Palma
Ester Pirrò

PRODUCTION

Atlantis Film&Video, Milan

TRANSLATIONS

TIP, Milan

CONVERSATIONS
Online conferences and activities
30.9.2021 – 14.4.2022

SCIENTIFIC DIRECTOR
Giancarlo Comi

PROGRAM CONCEIVED BY
Giancarlo Comi
Jubin Abutalebi
Massimo Cacciari
Viviana Kasam
Letizia Leocani
Andrea Moro
Daniela Perani

LECTURES BY
Paolo Bartolomeo
Robert C. Berwick
Ellen Bialystok
Massimo Cacciari
Michele Di Francesco
Angela D. Friederici
Guido Gainotti
Simona Ginsburg
Eva Jablonka
Jeffrey A. Kleim
Stefan Koelsch
Steven Mithen
Álvaro Pascual-Leone
Jason Rothman

MODERATOR
Alice Roberts

CURATORIAL TEAM
Mario Mainetti
Cornelia Mattiacci

PRODUCTION MANAGER
Alessia Salerno

PROJECT MANAGER
Daniele Manfrè

LIVE EVENT MANAGER
Alessia Gennari

COMMUNICATION MANAGER
Andrea Goffo

PRESS OFFICE & PROMOTION
Nicolò Scialanga
Marta Ronzio

DIGITAL COMMUNICATION
Roberta Palma
Ester Pirrò

SCIENTIFIC SECRETARIAT
Fabiana Totaro

PRODUCTION
Atlantis Film&Video, Milan

TRANSLATIONS
TIP, Milan

IT BEGINS WITH AN IDEA
Exhibition
Fondazione Prada, Venice
23.4 – 27.11.2022

SCIENTIFIC COMMITTEE
Giancarlo Comi
Jubin Abutalebi
Massimo Cacciari
Viviana Kasam
Udo Kittelmann
Letizia Leocani
Andrea Moro
Daniela Perani

CURATOR
Udo Kittelmann
in collaboration with
Taryn Simon

ASSOCIATE CURATORS
Niccolò Gravina
Mario Mainetti

SCIENTIFIC CONSULTANTS
Stefano Cappa
Michele Di Francesco
Guido Gainotti
Stephan Koelsh
Marcello Massimini
Paolo Mazzarello
Maria Concetta Morrone
Giovanni Naldi
Andrea Sereni

TARYN SIMON STUDIO
Sara Douglas
Hannah Shea

RESEARCH
Chiara Borgonovo
Niccolò Gravina
Cornelia Mattiacci
Cristina Tagliabue
and
Pia Chakraverti-Wuerthwein
Sara Douglas
Hannah Shea

ARTIFACTS STORIES VIDEOS DIRECTED BY
Taryn Simon

ARTIFACTS STORIES BY
Ayọ̀bámi Adébáyọ̀
Tash Aw
Hanan al-Shaykh
Chloe Aridjis
Mauro Javier Cárdenas

Akwaeke Emezi
Esther Freud
Rivka Galchen
Daniel Galera
Paolo Giordano
Uzodinma Iweala
Mieko Kawakami
John Keene
Daniel Kehlmann
Sheng Keyi
Katie Kitamura
Alexander Kluge
Hari Kunzru
Hervé Le Tellier
Michele Mari
Ch'aska Anka Ninawaman
Charu Nivedita
Helen Oyeyemi
Tilsa Otta
Sidarta Ribeiro
Cord Riechelmann
Salman Rushdie
Ekaterina Sedia
Leanne Shapton
Ahdaf Soueif
Maria Stepanova
McKenzie Wark

ARTIFACTS STORIES NARRATED BY
George Guidall

THE CONVERSATION MACHINE VIDEOS, INTERVIEWS AND ORCHESTRATION BY
Taryn Simon

PARTICIPANTS IN THE CONVERSATION MACHINE
Yasmin Abofoul
Huda Akil
Polina Anikeeva
Anirban Bandyopadhyay
György Buzsáki
David Chalmers
Patricia Churchland
Antonio Damasio
Stanislas Dehaene
Daniel C. Dennett
Catherine Dulac
David Erritzoe
Lisa Feldman Barrett
André Fenton
Karl Friston
Ali Ghazizadeh
Carl Hart
Suzana Herculano-Houzel
Amadi O. Ihunwo
Erich Jarvis
Bianca Jones Marlin
Kumi Kuroda
Joseph E. LeDoux
Mahmoud Bukar Maina
Catherine Malabou
Eve Marder
Hannah Monyer
David Poeppel
Supratim Ray
Daniela Schiller
Wolf Singer
Mitchell Joseph Valdés-Sosa
Angela Vincent
Charles Yang
Rafael Yuste
Li Zhaoping

ARTIFACTS STORIES AND THE CONVERSATION MACHINE PRODUCED BY
Fondazione Prada

PRODUCTION MANAGER
Alessia Salerno

PROJECT MANAGER
Daniele Manfrè

EXHIBITION REGISTRAR
Camilla Finzi

COMMUNICATION MANAGER
Andrea Goffo

PRESS OFFICE & PROMOTION
Nicolò Scialanga
Marta Ronzio

DIGITAL COMMUNICATION
Roberta Palma
Ester Pirrò

SCIENTIFIC SECRETARIAT
Fabiana Totaro

FACILITY MANAGEMENT
Romeo Scarpa
Zero4uno srl Ingegneria, Venice

DESIGN DEVELOPMENT
Sammy Zarka

GRAPHIC DESIGN
2×4, New York
Michael Rock
Donnie Luu
Gina Shin
Chiara Assanelli
Emily Green
with
Sean Dong
Oluremi Oni
and
Anna Gorichenko

EDITING & TRANSLATIONS
Malerba Editorial & Partners, Milan
Francesca Malerba
Laura Magda Barazza
Raffaella Esposito
Flavia Scotti
Timothy Stroud
Sarah Ponting
and
Evan Camfield
Sara Douglas
Hannah Shea
Aliza Watters

TRANSLATIONS (STORIES, ENGLISH AND ITALIAN)
Malerba Editorial & Partners, Milan
Madeleine Brook (stories 8, 32)
Raffaella Esposito (stories 3, 11, 26)
Ilaria Ortolina (stories 2, 4, 17, 18, 32)
and
Maria Baiocchi and Anna Tagliavini (story 31)
Ross Benjamin (story 29)
Sandra Biondo (story 14)
David Boyd (story 22)
Stefania Buonamassa (story 7)
Maria Valeria Caredda (story 6)
Catherine Cobham (story 11)
Gianluca Coci (story 22)
Anna D'Elia (story 13)
Alessandro D'Onofrio (story 20)
Hugo Doyle (story 24)
Sasha Dugdale (story 16)
Ottavio Fatica (story 19)
Marta Federici (story 23)
Nicky Harman (story 9)
Adriana Hunter (story 13)
Daniel Hahn (story 14)
Nandini Krishnan (story 4)
Alison Krögel (English adaptation, story 17)
Daniela Liberti (story 16)
Anna Nadotti (story 15)
Laura Noulian (story 27)
Gianni Pannofino (story 1)
Monica Pesetti (stories 29, 30)
Nicoletta Poo (story 5)
Silvia Pozzi (story 9)
Costanza Prinetti (story 12)
Chiara Reali (story 28)
Angela Ricci (story 8)
Anna Rusconi (story 21)
Kit Schluter (story 2)
Alex Valente (story 25)

VIDEO PRODUCTION
RadicalMedia, New York
Jon Kamen
Jaye Callahan
Courtney Ford
Naomi Gilbert
R. Hollis Smith
Hannah Shea
Josh Kanuck
J.J. Kelley
Chris Kim
Matt Mayer
Kyle McCabe
Laura O'Carroll
Lesley Young-Pavia
Jeff Tomcho
Joe Tomcho
Kelly Vigil
Jason Yue

LESSONS PRODUCTION
Alessia Gennari
with
Atlantis Film&Video, Milan

AUDIO & VIDEO COORDINATOR
Luca de Marinis

AUDIO & VIDEO INSTALLATION
3P Technologies Srl, Gessate (MI)
Akoustic Arts, Paris
Atlantis Film&Video, Milan
LG Electronics Italia s.p.a., Milan
Marco Mazzon, Musile di Piave (VE)
RCF spa, Reggio Emilia

INSTALLATION
Attitudine Forma, Turin

CONSERVATORS
Giulia Passerini
Valeria Pretto
Laura Riato
Letizia Satto
Giorgia Uraghi

PHOTOGRAPHIC DOCUMENTATION
Marco Cappelletti
Niccolò Quaresima
Patrick Toomey Neri

VIDEO DOCUMENTATION
Claudia Ferri
Variante Artistica

INSURANCE
AON, Milan
AXA, Milan
Generali Italia spa

SHIPPING
Butterfly Transport, Massa and Cozzile (PT)
Rubelli, Venice

CONSTRUCTION
CIP srl, Cerro Maggiore (MI)
Tecnolegno Allestimenti, Cormano (MI)
Tramonte Elettroimpianti Engineering, Lupia (VE)
Trevisan Impianti srl, Vicenza

FONDAZIONE PRADA WISHES TO THANK

FOR THEIR WILLING COLLABORATION AND GENEROUS LOANS:
Accademia delle Scienze di Torino, Turin; Alma Mater Studiorum, Università di Bologna, Biblioteca Universitaria, Bologna; Amsterdam Museum, Amsterdam; APSS, Azienda Provinciale per i Servizi Sanitari, Biblioteca dell'ex-ospedale psichiatrico, Pergine Valsugana; Archivio Storico dell'Università di Torino, Turin; AUSL, Biblioteca Scientifica "Carlo Livi", Reggio Emilia; Biblioteca Apostolica Vaticana, Vatican City; Biblioteca Civica "Romolo Spezioli", Fermo; Biblioteca Medicea Laurenziana, Florence; Biblioteca Nazionale Marciana, Venice; Biblioteca Statale di Cremona, Cremona; Biblioteca Universitaria di Padova (MiC), Padua; Bibliothèque de Sorbonne Université, Collections d'Anatomie Pathologique Dupuytren, Paris; Bruges Public Library, Bruges; Chiba University Library of Health Sciences (Inohana Library), Chiba; Deutsches Epilepsiemuseum, Kork; Freud Museum, London; Friedrich Schiller University, Collection of scientific and technical devices for physics, Jena; Instituto Cajal (CSIC), Legado Cajal, Madrid; Klassik Stiftung Weimar, Bestand Museen, Weimar; Kunstmuseum St. Gallen, St. Gallen; Kyoto University, Main Library, Rarebooks, Kyoto; Linden-Museum, Stuttgart; MIC – Soprintendenza ABAP di Ravenna, Forlì-Cesena and Rimini, Museo della Citta "Luigi Tonini", Rimini; Ministerio de Cultura – Museo Nacional de Arqueología, Antropología e Historia del Perú, Lima; Ministero della Cultura, Pinacoteca di Brera, Biblioteca Nazionale Braidense, Milan; Musée du Louvre, Département des Antiquités Orientales, Paris; Musei Civici di Reggio Emilia– Museo di Storia della Psichiatria, Reggio Emilia; Museo delle Culture, Milan; Museo Egizio, Turin; Museo Nacional del Prado, Madrid; Museo Nazionale della Scienza e della Tecnologia Leonardo da Vinci, Milan; Museum am Rothenbaum, MARKK, Hamburg; National Central Library, Taiwan; Papyrus- und Ostrakasammlung der Universitätsbibliothek Leipzig, Leipzig; Princeton University Library, East Asian Library and the Gest Collection, Princeton; Republic of Turkey Ministry of Culture and Tourism- Presidency of Manuscript Institution of Turkey-Süleymaniye Manuscript Library, Istanbul; Sistema Museale di Ateneo, Museo di Storia Naturale, Sede "La Specola", Florence; Staatliche Museen zu Berlin, Kupferstichkabinett, Graphische Gesellschaft zu Berlin, Berlin; Staatliche Museen zu Berlin, Preußischer Kulturbesitz, Vorderasiatisches Museum, Berlin; The New York Academy of Medicine Library, New York; The Royal Society, London; The Syndics of Cambridge University Library, Cambridge; Università degli Studi di Milano, Biblioteca di Biologia, Informatica, Chimica e Fisica, Milan; Università degli Studi di Padova, Biblioteca medica centrale "Vincenzo Pinali" – Sezione antica, Padua; Università degli Studi di Torino, Biblioteca Federata di Medicina "Ferdinando Rossi", Turin; Università di Pavia, Sistema Museale di Ateneo – Museo per la Storia dell'Università di Pavia, Pavia; Università Sapienza, Dipartimento di Medicina molecolare, Museo di Storia della medicina, Rome; University of Edinburgh, Anatomical Museum, Edinburgh; Wellcome Collection, London, and all the lenders who wish to remain anonymous

FOR THEIR SCIENTIFIC AND DOCUMENTARY CONTRIBUTIONS
Monika Green
Elif Kamisli
Zahide Özkan-Rashed
Marco Vespa

THANKS TO

Agenzia MalaTesta
Esenija Bannan
Stefania Eusebio, La Nave di Teseo
R. Hollis Smith
Courtney Ford
Josh Kanuck
Matt Mayer
Kyle McCabe
Laura O'Carroll
Giuseppe Strazzeri
Jeff Tomcho
Joe Tomcho
Kelly Vigil
Jason Yue

SPECIAL THANKS TO

Danielle S. Basset
Emilio Bizzi
Linda Buck
György Buzsáki
Robin Carhart-Harris
Noam Chomsky
Mirella Dapretto
Karl Alexander Deisseroth
Adele Diamond
Paul Ekman
Marco Iacoboni
Mitsuo Kawato
Roustem Khazipov
Nikos Logothetis
Shima Moein
Martin Monti
May-Britt Moser
Noeline Nakasujja
Miguel Nicolelis
Roger Penrose
Mu-Ming Poo
Supratim Ray
Giacomo Rizzolatti
Botond Roska
Idan Segev
Narayanan Srinivasan
Richard H. Thaler
Shuba Tole
Osvaldo Uchitel
Rafael Yuste
Milin Zhang
Huda Zoghbi

SUPPORTED BY

PRESERVING THE BRAIN
Exhibition and Forum
Fondazione Prada, Milan
16.9 – 10.11.2022

SCIENTIFIC COMMITTEE

Jack Antel
Gonçalo Castelo-Branco
Roberto Furlan
David A. Hafler
Nobutaka Hattori
Stephen L. Hauser,
Catherine Lubetzki
Pierluigi Nicotera
Fredrik Piehl
Orly Reiner
Alan Thompson
Howard Weiner
Hartmut Wekerle
Nan Zhang

A PROJECT
IN COLLABORATION WITH

German Center for Neurodegenerative Diseases (DZNE) within the Helmholtz Association, Bonn, Germany
Harvard Medical School, Brigham and Women's Hospital, Ann Romney Center for Neurologic Diseases, Boston, United States
Hôpital de la Pitié-Salpêtrière, Sorbonne University AP-HP, Neurology Department and Paris Brain Institute, Paris, France
Juntendo University Hospital, Neurology Department, Tokyo, Japan
Karolinska Institutet, Stockholm, Sweden
Max Planck Institute of Neurobiology, Munich, Germany
Montreal Neurological Institute Hospital, McGill University, Montreal, Canada
Tianjin Medical University General Hospital, Neurology Department, Tianjin, China
UCSF Weill Institute for Neurosciences, University of California San Francisco, San Francisco, United States
UniSR – Università Vita-Salute San Raffaele, Milan, Italy
University College London, London, United Kingdom
Weizmann Institute of Science, Rehovot, Israel
Yale School of Medicine, New Haven, United States

SCIENTIFIC DIRECTOR

Giancarlo Comi

CURATORIAL TEAM

Cornelia Mattiacci
Mario Mainetti
Niccolò Gravina

PRODUCTION MANAGER

Alessia Salerno

PROJECT MANAGER

Daniele Manfrè

EXHIBITION REGISTRAR

Camilla Finzi

LIVE EVENT MANAGER

Alessia Gennari

COMMUNICATION MANAGER

Andrea Goffo

PRESS OFFICE & PROMOTION

Nicolò Scialanga
Marta Ronzio
with
Leonardo Gava

DIGITAL COMMUNICATION

Roberta Palma
with
Giulia Parolin

EDITORIAL COORDINATION

Cristina Tagliabue

SCIENTIFIC SECRETARIAT

Fabiana Totaro

ORGANIZATIONAL SECRETARIAT

Creo Spa

DESIGN DEVELOPMENT

Sammy Zarka

EXHIBITION ARCHITECTURE AND GRAPHIC DESIGN

2×4, New York
Michael Rock
Christopher Kupski
Sung Joong Kim
Chiara Assanelli
Donnie Luu
Urechi Oguguo
Alessandro Costariol
Adarsh Nellore
with
Giovanni Spera

EDITING AND TRANSLATIONS
Malerba Editorial & Partners, Milan
Francesca Malerba
Laura Magda Barazza
Raffaella Esposito
Flavia Scotti
with
Timothy Stroud

VIDEO EDITING
Atlantis Film&Video, Milan

VIDEO SUBTITLES
TIP, Milan

PHOTOGRAPHIC DOCUMENTATION
Delfino Sisto Legnani – DLS Studio
Patrick Toomey Neri

VIDEO DOCUMENTATION
Jacopo Farina

INSURANCE
Aon, Milan
Axa Art, Milan

CONSTRUCTION
Cartotecnica RB, Vigliano Biellese
Italivideo Service, Madone
Masserdotti, Castel Mella

FONDAZIONE PRADA WISHES TO THANK

FOR THEIR SCIENTIFIC AND DOCUMENTARY CONTRIBUTIONS:
German Center for Neurodegenerative Diseases (DZNE) within the Helmholtz Association,
Bonn, Germany
Pierluigi Nicotera
Joachim L. Schultze
Harvard Medical School, Brigham and Women's Hospital, Ann Romney Center for Neurologic Diseases, Boston, United States
Howard Weiner
Thais Moreira
Hôpital de la Pitié-Salpêtrière,
Sorbonne University AP-HP, Neurology Department and Paris Brain Institute, Paris, France
Alexis Brice
Bassem Hassan
Catherine Lubetzki
Juntendo University Hospital, Neurology Department, Tokyo, Japan
Nobutaka Hattori
Taiji Tsunemi
Karolinska Institutet, Stockholm, Sweden
Gonçalo Castelo-Branco
Fredrik Piehl
Tony Jimenez-Beristáin
Chiara Starvaggi Cucuzza
Max Planck Institute of Neurobiology, Munich, Germany
Hartmut Wekerle
Anneli Peters
Montreal Neurological Institute-Hospital, McGill University, Montreal, Canada
Jack Antel
Stefano Stifani
Tianjin Medical University General
Hospital, Neurology Department, Tianjin, China
Nan Zhang
Jing Xu
UCSF Weill Institute for
Neurosciences, University of California San Francisco, San Francisco, United States
Stephen Hauser
Adam Gazzaley
Jon-Eric VanLeeuwen
UniSR – Università Vita-Salute San Raffaele, Milan, Italy
Roberto Furlan
Letizia Leocani
Daniela Perani
University College London, London, United Kingdom
Alan Thompson
Nick Fox
Alexander Foulkes
Ross Paterson
Weizmann Institute of Science, Rehovot, Israel
Eran Hornstein
Orly Reiner
Yale School of Medicine, New Haven, United States
David Hafler
David Pitt
Tzu-Chieh Huang

FOR THEIR GENEROUS LOANS
10XGenomics
ACeSM Onlus
Andromeda
EMS
Givas
Neuroelectrics Barcelona, SLU
Quanterix
Tao Med

THANKS TO
Nicolas Brard
Nicoletta Carbone
Marion Doucet
Karyl Evans
Allison Greco
Anita Kar
Gizel Maimon
Hannah Martens
Sandra McPherson
Kim Nijs
Maria Luisa Paleari
Isabelle Rebeix

MEDIA PARTNER
Radio24

HUMAN BRAINS. THE BOOK

GENERAL EDITOR
Fondazione Prada

ASSOCIATE EDITORS

FOR "CULTURE AND CONSCIOUSNESS"
Massimo Cacciari
in collaboration with
Giancarlo Comi

FOR "CONVERSATIONS"
Giancarlo Comi

FOR "IT BEGINS WITH AN IDEA"
Udo Kittelmann
in collaboration with
Taryn Simon

FOR "PRESERVING THE BRAIN" FORUM ON NEURODEGENERATIVE DISEASES
Giancarlo Comi

EDITORIAL TEAM
Mario Mainetti
Cornelia Mattiacci
Niccolò Gravina
Cristina Tagliabue

EDITORIAL COORDINATION
Cristina Tagliabue

GRAPHIC DESIGN
2×4, New York
Michael Rock
Donnie Luu
Chase Booker
Gina Shin

PRODUCTION EDITOR
Malerba Editorial & Partners, Milan
Francesca Malerba
Laura Magda Barazza
Raffaella Esposito
Flavia Scotti
and
Evan Camfield
Sara Douglas
Hannah Shea
Aliza Watters

TRANSLATIONS
Malerba Editorial & Partners, Milan
Timothy Stroud
Sarah Ponting
Stories:
Ross Benjamin (story 29)
David Boyd (story 22)
Madeleine Brook (story 8, 32)
Catherine Cobham (story 11)
Hugo Doyle (story 24)
Sasha Dugdale (story 16)
Nicky Harman (story 9)
Adriana Hunter (story 13)
Daniel Hahn (story 14)
Nandini Krishnan (story 4)
Alison Krögel (English adaptation, Story 17)
Kit Schluter (story 2)
Alex Valente (story 25)

PRODUCTION COORDINATION
Marco Previtali
Nensi Shapllo
with
Mara Spinelli

PUBLISHER
Fondazione Prada, Milan

REPRO
De Pedrini, Milan

PRINTER
Nava Press, Milan

COVER AND FRONTISPIECES IMAGES
2×4, New York

ISBN: 978-88-87029-88-8

This publication is typeset in the Suisse family of typefaces, and printed on Arena Natural Rough 90 g/m² (pp. 1 – 486, 971 – 986), Magnogloss 150 g/m² (pp. 487 – 762), Magnosatin 100 g/m² (pp. 763 – 970), Skinplast R 601 and Invercote g/m² (cover), Skinplast R 601 and Sirio Ultrablack 115 g/m² (endsheets), PET 100 μ (dust jacket).

PHOTO CREDITS

I. CULTURE AND CONSCIOUSNESS

©the authors for their contributions, included the materials featured in the visual excerpts reproduced in this publication

II. CONVERSATIONS

©the authors for their contributions, included the materials featured in the visual excerpts reproduced in this publication

III. IT BEGINS WITH AN IDEA

p. 219 (left column) Courtesy of Department of Neurobiology and Anatomy and Spencer S. Eccles Health Sciences Library, University of Utah, Salt Lake City; p. 219 (center column) Courtesy of Johns Hopkins University, Baltimore; p. 219 (right column) Produced by the Warner Brothers Pictures, released through the local offices of Bell Telephone System; p. 222 Video courtesy of the USC Mark and Mary Stevens Neuroimaging and Informatics Institute; p. 228 ©2023. Musée du Louvre, Paris / RMN-Grand Palais / Dist. Ph.Scala, Florence. Ph. Benjamin Soligny/ Raphael Chipault; pp. 232–233 Courtesy of Ministerio de Cultura – Museo Nacional de Arqueología, Antropología e Historia del Perú. Colección Af-006953 MNAAHP, cod. 628, ph. Enrique Quispe; p. 234 ©Museum am Rothenbaum (MARKK), Hamburg. Collection America/Brüning: inv.no. 33.180:2; p. 240 ©2023. Staatliche Museen zu Berlin – Vorderasiatisches Museum. Scala, Florence/bpk, Bildagentur fuer Kunst, Kultur und Geschichte, Berlin. Ph. Olaf M. Tessmer; pp. 244–246 ©Linden-Museum Stuttgart. Ph. Dominik Drasdow; p. 256 Courtesy of the New York Academy of Medicine Library; p. 257 Courtesy of Papyrus und Ostrakasammlung der Universitätsbibliothek Leipzig; p. 258 Courtesy of Museo Egizio, Turin. Ph. Nicola dell'Aquila/ Federico Taverni; pp. 262–263, 268–271, 317 Urb.gr.64 - ff.IIv-1r, Urb.gr.64 - ff.115v-116r, Vat.lat.2063 - ff.2v-3r; Reg.lat.1572 - ff.14v-15r, Vat.lat.2095 - ff.1v-1r, Vat.lat.2094 - ff.IIIv-1r, Vat.lat.4485 - ff.66v-67r, courtesy of Biblioteca Apostolica Vaticana, all rights reserved; pp. 264, 272, 342–343, 374–375 Courtesy of Ministero della Cultura - Biblioteca Nazionale Marciana, unauthorized reproduction prohibited; pp. 278–279, 438 Ufficio Biblioteca Digitale - Università degli Studi di Padova - Biblioteca medica centrale "Vincenzo Pinali" – Sezione antica; pp. 280–281 Courtesy of Ministero della Cultura - Soprintendenza Archeologia, Belle Arti e Paesaggio per le province di Ravenna, Forlì-Cesena, Rimini; pp. 288–291, 461, 463, 465 Courtesy of Fondazione Prada; pp. 296–297, 302–303 Ms. Fatih 3212, Ms. Ayasofya 3741, Courtesy of Republic of Turkey Ministry of Culture and Tourism - Presidency of Manuscript, Institution of Turkey - Süleymaniye Manuscript Library, Türkiye Yazma Eserler Kurumu Başkanlığı Süleymaniye Yazma Eser Kütüphanesi; pp. 298–299, 440–442 Courtesy of Biblioteca Statale di Cremona; pp. 304–305, 328–329, 366–368 Reproduced by kind permission of the Syndics of Cambridge University Library; pp. 306–307 Courtesy of Ministero della Cultura – Biblioteca Medicea Laurenziana. unauthorized reproduction prohibited; p. 308 Courtesy of Ditta Omnia Foto di Diego Pizi – San Benedetto del Tronto (AP); p. 316 Courtesy of Brugge, Openbare Bibliotheek / Bruges, Public Library; p. 322 Courtesy of Kunstmuseum St. Gallen, ph. Sebastian Stadler, 2013; pp. 323–324 ©Photographic Archive Museo Nacional del Prado; pp. 330–331 ©Kupferstichkabinett, Staatliche Museen zu Berlin, ph. Dietmar Katz; p. 338 ©Klassik Stiftung Weimar, Bestand Museen, ph. Mokansky Olaf; pp. 344, 376 Courtesy of Ministero della Cultura - Biblioteca Universitaria di Padova, unauthorized reproduction prohibited Digital imaging by Shylock e-Solutions; p. 345 ©Alma Mater Studiorum - Università di Bologna, Biblioteca Universitaria di Bologna; p. 352 ©Comune di Milano, all rights reserved; pp. 360–361 ©Museo Nazionale della Scienza e della Tecnologia Leonardo da Vinci, Milano; p. 362 Courtesy of Museo di Storia della medicina - Dipartimento di Medicina molecolare, Sapienza Università di Roma; pp. 372–373 Courtesy of Ministero della Cultura - Pinacoteca di Brera - Biblioteca Nazionale Braidense; p. 380 Courtesy of Sistema Museale di Ateneo - Museo di Storia Naturale dell'Università di Firenze, "Museo La Specola". Ph. Saulo Bambi; pp. 384–387 Courtesy of the East Asian Library and the Gest Collection, Princeton University Library; pp. 388–389 Image courtesy of the Library of Health Sciences (Inohana Library), Chiba University Library, detail; pp. 390–395 Photograph courtesy of the Main Library, Kyoto University; pp. 400–403 ©Christel Jeanne; pp. 406–408 Courtesy of Università di Pavia - Sistema Museale di Ateneo - Museo per la Storia dell'Università; pp. 409–412 Courtesy of Legado Cajal. Instituto Cajal (CSIC). Madrid; pp. 418–419 Courtesy of Biblioteca di biologia, informatica, chimica e fisica dell'Università degli studi di Milano. ©1951 Wiley-Liss, Inc., A Wiley Company; pp. 420–421 Courtesy of Fondazione Prada. Microphotographs by Mr. Cramer Lewis, Department of Illustration of the Washington University School of Medicine. ©2006, John Wiley and Sons; p. 422 Courtesy of Archivio storico dell'Università di Torino; pp. 428–429 Courtesy of APSS, Azienda Provinciale per i Servizi Sanitari, Biblioteca dell'ex-ospedale psichiatrico, Pergine Valsugana; pp. 430–431 Property of the Deutsches Epilepsiemuseum Kork; pp. 432–433 Courtesy of Collection of scientific and technical devices for physics of the Friedrich Schiller University Jena, Ph. Hannah Bayer; pp. 436–437 A. Galvani, De viribus electricitatis; p.1792, tab II-III ©The Royal Society; p. 439 Courtesy of Accademia delle Scienze di Torino; p. 446 Courtesy of Musei Civici di Reggio Emilia - Museo di Storia della Psichiatria; pp. 450–451 Courtesy of Wellcome Collection,

London; p. 452 Courtesy of Amsterdam Museum; pp. 458–460, 462, 464 ©Freud Museum London; p. 466 Courtesy of National Central Library, Taiwan; p. 470 Courtesy of Biblioteca Federata di Medicina "Ferdinando Rossi" - Università degli Studi di Torino; pp. 471–473 Courtesy of Biblioteca Carlo Livi Reggio Emilia; p. 474 ©Anatomical Museum, The University of Edinburgh; p. 478 ©2023 by the American Physical Society; p. 479 ©New Scientist; p. 480 DOI: 10.1126/science.1948051, reprinted with permission from AAAS; p. 481 DOI: doi: 10.1073/pnas.89.12.5675; p. 482 Courtesy of Biblioteca di Biologia, Informatica, Chimica e Fisica dell'Università degli Studi di Milano, DOI: 10.1073/pnas.89.13.5951.

THE CONVERSATION MACHINE

p. 225, 489–707 Videos, Interviews and Orchestration by Taryn Simon, produced by Fondazione Prada for the "Human Brains. It Begins with an Idea" project; pp. 488, 710–739 Courtesy Fondazione Prada, ph. Marco Cappelletti

IV. PRESERVING THE BRAIN

pp. 779, 781–782, 786–791, 794–798, 802–811, 814–820, 824–839, 842, 844–849, 852–857, 860–865, 868–873, 876, 878–891, 894–899, 902–911, 914–925, 928–937 Courtesy 2x4, NY; p. 939–970 Courtesy Fondazione Prada, ph. Delfino Sisto Legnani e Alessandro Saletta – DSL Studio

170,00 EURO

ISBN 978-88-87029-88-8
9 788887 029888